Extensive and Varied Exercise Sets	An abundant collection of exercises is included in an exercise set at the end of each section. Exercises are organized within categories. Your instructor will usually provide guidance on which exercises to work. The exercises in the first category, Practice Exercises, follow the same order as the section's worked examples.	The parallel order of the Practice Exercises lets you refer to the worked examples and use them as models for solving these problems.
Practice Plus Problems	This category of exercises contains more challenging problems that often require you to combine several skills or concepts.	It is important to dig in and develop your problem-solving skills. Practice Plus Exercises provide you with ample opportunity to do so.

3 Review for Quizzes and Tests

Feature	Description	Benefit
Mid-Chapter Check Points	At approximately the midway point in the chapter, an integrated set of review exercises allows you to review the skills and concepts you learned separately over several sections.	By combining exercises from the first half of the chapter, the Mid-Chapter Check Points give a comprehensive review before you move on to the material in the remainder of the chapter.
Chapter Review Grids	Each chapter contains a review chart that summarizes the definitions and concepts in every section of the chapter. Examples that illustrate these key concepts are also included in the chart.	Review this chart and you'll know the most important material in the chapter!
Chapter Review Exercises	A comprehensive collection of review exercises for each of the chapter's sections follows the review grid.	Practice makes perfect. These exercises contain the most significant problems for each of the chapter's sections.
Chapter Tests	Each chapter contains a practice test with approximately 25 problems that cover the important concepts in the chapter. Take the practice test, check your answers, and then watch the Chapter Test Prep Video CD to see worked-out solutions for any exercises you miss.	You can use the chapter test to determine whether you have mastered the material covered in the chapter.
Chapter Test Prep Video CDs	These video CDs found at the back of your text contain worked-out solutions to every exercise in each chapter test.	The videos let you review any exercises you miss on the chapter test.
Cumulative Review Exercises	Beginning with Chapter 2, each chapter concludes with a comprehensive collection of mixed cumulative review exercises. These exercises combine problems from previous chapters and the present chapter, providing an ongoing cumulative review.	Ever forget what you've learned? These exercises ensure that you are not forgetting anything as you move forward.

More Tools for Your Mathematics Success

Student Study Pack

Get math help when YOU need it! The **Student Study Pack** provides you with the ultimate set of support resources to go along with your text. Packaged at no charge with a new textbook, the **Student Study Pack** contains these invaluable tools:

☑ **Student Solutions Manual**

A printed manual containing full solutions to odd-numbered textbook exercises.

☑ **Prentice Hall Math Tutor Center**

Tutors provide one-on-one tutoring for any problem with an answer at the back of the book. You can contact the Tutor Center via a toll-free phone number, fax, or email.

☑ **CD-ROM Lecture Series**

A comprehensive set of textbook-specific CD-ROMs containing short video clips of the textbook objectives being reviewed and key examples being solved.

Tutorial and Homework Options

MYMATHLAB

MyMathLab, packaged at no charge with a new textbook, is a complete online multimedia resource to help you succeed in learning. **MyMathLab** features:

☑ The entire textbook online.

☑ Problem-solving video clips and practice exercises, correlated to the examples and exercises in the text.

☑ Online tutorial exercises with guided examples.

☑ Online homework and tests.

☑ Generates a personalized study plan based on your test results.

☑ Tracks all online homework, tests, and tutorial work you do in **MyMathLab** gradebook.

Essentials of Introductory and Intermediate Algebra for College Students

Essentials of Introductory and Intermediate Algebra for College Students

Robert Blitzer

Miami-Dade College

PEARSON

Prentice
Hall

Upper Saddle River, New Jersey 07458

Senior Acquisitions Editor: *Paul Murphy*
Editor in Chief: *Christine Hoag*
Project Manager: *Liz Covello*
Production Editor: *Prepare, Inc.*
Assistant Managing Editor: *Bayani Mendoza de Leon*
Senior Managing Editor: *Linda Mihatov Behrens*
Executive Managing Editor: *Kathleen Schiaparelli*
Media Project Manager: *Audra J. Walsh*
Media Production Editor: *Donna Crilly*
Managing Editor, Digital Supplements: *Nicole M. Jackson*
Assistant Manufacturing Manager/Buyer: *Michael Bell*
Executive Marketing Manager: *Eilish Collins Main*
Marketing Assistant: *Rebecca Alimena*
Director of Marketing: *Patrice Jones*
Development Editor: *Don Gecewicz*
Editor in Chief, Development: *Carol Trueheart*
Editorial Assistant: *Mary Burket*
Assistant Editor/Print Supplements Editor: *Christina Simoneau*
Art Director: *Maureen Eide*
Interior Design: *Running River Design*
Cover Art and Designer: *Koala Bear Design*
Art Editor: *Thomas Benfatti*
Director of Creative Services: *Paul Belfanti*
Director, Image Resource Center: *Melinda Reo*
Manager, Rights and Permissions: *Zina Arabia*
Manager, Visual Research: *Beth Brenzel*
Image Permission Coordinator: *Debbie Hewitson*
Photo Researchers: *Elaine Soares; Beaura K. Ringrose*
Art Studios: *Scientific Illustrators; Laserwords*
Compositor: *Prepare, Inc.*

© 2006 Pearson Education, Inc.
Pearson Prentice Hall
Pearson Education, Inc.
Upper Saddle River, New Jersey 07458

Printed in the United States of America
10 9 8 7 6 5 4 3 2 1

ISBN 0-13-186606-0
ISBN 0-13-186555-2 (Student Edition)

Pearson Education LTD., *London*
Pearson Education Australia PTY, Limited, *Sydney*
Pearson Education Singapore, Pte. Ltd
Pearson Education North Asia Ltd, *Hong Kong*
Pearson Education Canada, Ltd., *Toronto*
Pearson Educación de Mexico, S.A. de C.V.
Pearson Education—Japan, *Tokyo*
Pearson Education Malaysia, Pte. Ltd

CONTENTS

A BRIEF GUIDE TO GETTING THE MOST FROM THIS BOOK
Inside front cover

CHAPTER 1   THE REAL NUMBER SYSTEM   1

CHAPTER 2   LINEAR EQUATIONS AND INEQUALITIES IN ONE VARIABLE   102

CHAPTER **6** FACTORING POLYNOMIALS **382**

CHAPTER **7** RATIONAL EXPRESSIONS **441**

CHAPTER **8** BASICS OF FUNCTIONS **530**

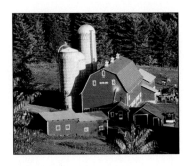

Shaded chapters are available in Blitzer,
Introductory and Intermediate Algebra for College Students, Second Edition.

PREFACE

Essentials of Introductory and Intermediate Algebra for College Students provides comprehensive, in-depth coverage of the topics required in most courses combining the study of introductory and intermediate algebra. The book is written for college students who have no previous experience in algebra and for those who need a review of basic algebraic concepts before moving on to intermediate algebra. I wrote the book to help diverse students, with different backgrounds and career plans, to succeed in a combined introductory and intermediate algebra course. *Essentials of Introductory and Intermediate Algebra for College Students* has two primary goals:

1. To help students acquire a solid foundation in the skills and concepts of introductory and intermediate algebra, without the repetition of topics in two separate texts.

2. To show students how algebra can model and solve authentic real-world problems.

One major obstacle in the way of achieving these goals is the fact that very few students actually read their textbook. This has been a regular source of frustration for me and for my colleagues in the classroom. Anecdotal evidence gathered over years highlights two basic reasons why students do not take advantage of their textbook:

- "I'll never use this information."
- "I can't follow the explanations."

I've written every page of this book with the intent of eliminating these two objections. The ideas and tools I've used to do so are described in the features that follow. These features and their benefits are highlighted for the student in "A Brief Guide to Getting the Most from This Book" that appears inside the front cover.

A Note on the Essentials Version of Introductory and Intermediate Algebra for College Students

Essentials of Introductory and Intermediate Algebra for College Students is a concise version of the Second Edition of *Introductory and Intermediate Algebra for College Students*. **The essentials version differs from the Second Edition only in terms of length**. Chapter 12 (Exponential and Logarithmic Functions), Chapter 13 (Conic Sections and Systems of Nonlinear Equations), and Chapter 14 (Sequences, Series, and Probability) have been eliminated. The omitted chapters and their sections are highlighted for your reference in the Table of Contents. The essentials version contains two appendices (Appendix C: The Circle and Appendix D: Summation Notation and the Binomial Theorem) for instructors who prefer the lighter, less expensive alternative to the Second Edition, but who still want to include these topics.

Features in the Essentials Version of Introductory and Intermediate Algebra for College Students

- **Detailed Worked-Out Examples.** Each worked example is titled, making clear the purpose of the example. Examples are clearly written and provide students with detailed step-by-step solutions. No steps are omitted and each step is thoroughly explained to the right of the mathematics.

- **Check Point Examples.** Each example is followed by a similar matched problem, called a Check Point, offering students the opportunity to test their understanding of the example by working a similar exercise. The answers to the Check Points are provided in the answer section.

- **Explanatory Voice Balloons.** Voice balloons are used in a variety of ways to demystify mathematics. They translate algebraic ideas into everyday English, help clarify problem-solving procedures, present alternative ways of understanding concepts, and connect problem solving to concepts students have already learned.

- **Applications Using Real-World Data.** Interesting and innovative applications from nearly every discipline, supported by up-to-date real-world data, are included in nearly every section.

- **Extensive and Varied Exercise Sets.** An abundant collection of exercises is included in an exercise set at the end of each section. Exercises are organized within seven category types: Practice Exercises, Practice Plus Exercises, Application Exercises, Writing in Mathematics, Technology Exercises, Critical Thinking Exercises, and Review Exercises. This format makes it easy to create well-rounded homework assignments. The order of the practice exercises is exactly the same as the order of the section's worked examples. This parallel order enables students to refer to the titled examples and their detailed explanations to achieve success working the practice exercises.

- **Practice Plus Exercises.** This category of exercises contains more challenging practice problems that often require students to combine several skills or concepts. With an average of ten practice plus problems per exercise set, instructors are provided with the option of creating assignments that take practice exercises to a more challenging level.

- **Mid-Chapter Check Points.** At approximately the midway point in each chapter, an integrated set of review exercises allows students to review and assimilate the skills and concepts they learned separately over several sections. The problems that make up the Mid-Chapter Check Points, averaging 23 exercises per check point, are of a mixed nature, requiring students to discriminate which concepts or skills to apply. The Mid-Chapter Check Points should help students bring together the different objectives covered in the first half of the chapter before they move on to the material in the remainder of the chapter.

- **Graphing.** Chapter 1 contains an introduction to graphing, a topic that is integrated throughout the book. Line, bar, circle, and rectangular coordinate graphs that use real data appear in nearly every section and exercise set. Many examples and exercises use graphs to explore relationships between data and to provide ways of visualizing a problem's solution.

- **Geometric Problem Solving.** Chapter 2 contains a section on problem solving in geometry that teaches geometric concepts that are important to a student's understanding of algebra. There is a frequent emphasis on problem solving in geometric situations, as well as on geometric models that allow students to visualize algebraic formulas.

- **Functions.** Chapter 8 (Basics of Functions), the book's transitional chapter into intermediate algebra, is devoted exclusively to functions. Because functions are the core of the intermediate algebra portion of the book, students are repeatedly shown how functions relate to equations and inequalities.

- **Chapter-Opening and Section-Opening Scenarios.** Every chapter and every section open with a scenario presenting a unique application of algebra in students' lives outside the classroom. These scenarios are revisited in the course of the chapter or section in an example, discussion, or exercise.

- **Section Objectives.** Learning objectives are clearly stated at the beginning of each section. These objectives help students recognize and focus on the section's most important ideas. The objectives are restated in the margin at their point of use.

- **Thorough, Yet Optional Technology.** Although the use of graphing utilities is optional, they are utilized in Using Technology boxes to enable students to visualize and gain numerical insight into algebraic concepts. The use of graphing utilities is also reinforced in the technology exercises appearing in the exercise sets for those who want this option. With the book's early introduction to graphing, students can look at the calculator screens in the Using Technology boxes and, with the assistance of the explanatory voice balloons, begin to understand both graphical and numerical approaches to problems even if they are not using a graphing utility in the course.

- **Study Tips.** The book's Study Tip boxes offer suggestions for problem solving, point out common errors to avoid, and provide informal hints and suggestions. These invaluable hints appear throughout the book.

- **Chapter Review Grids.** Each chapter contains a review chart that summarizes the definitions and concepts in every section of the chapter. Examples that illustrate these key concepts are also included in the chart.

- **End-of-Chapter Materials.** A comprehensive collection of review exercises for each of the chapter's sections follows the review grid. This is followed by a chapter test that enables students to test their understanding of the material covered in the chapter. Beginning with Chapter 2, each chapter concludes with a comprehensive collection of mixed cumulative review exercises.

- **Chapter Test Prep Video CDs.** Packaged at the back of the text, these video CDs provide students with step-by-step solutions for each of the exercises in the book's chapter tests.

- **Enrichment Essays.** These discussions provide historical, interdisciplinary, and otherwise interesting connections to the algebra under study, showing students that math is an interesting and dynamic discipline.

- **Discovery.** Discover for Yourself boxes, found throughout the text, encourage students to further explore algebraic concepts. These explorations are optional and their omission does not interfere with the continuity of the topic under consideration.

- **Chapter Projects.** At the end of each chapter is a collaborative activity that gives students the opportunity to work cooperatively as they think and talk about mathematics. Additional group projects can be found in the *Instructor's Resource Manual*. Many of these exercises should result in interesting group discussions.

Resources for the Instructor

Print

Annotated Instructor's Edition (ISBN: 0-13-186606-0)

- Answers appear in place on the same text page as exercises or in the Graphing Answer Section
- Answers to all exercises in the exercise sets, Mid-Chapter Check Points, Chapter Reviews, Chapter Tests, and Cumulative Reviews

Instructor's Solutions Manual (ISBN: 0-13-192178-9)

- Detailed step-by-step solutions to the even-numbered section exercises
- Solutions to every exercise (odd and even) in the Mid-Chapter Check Points, Chapter Reviews, Chapter Tests, and Cumulative Reviews
- Solution methods reflect those emphasized in the text.

Instructor's Resource Manual with Tests (ISBN: 0-13-192177-0)

- Six test forms per chapter—3 free response, 3 multiple choice
- Two *Cumulative Tests* for all even-numbered chapters
- Two *Final Exams*
- Answers to all test items
- *Mini-Lectures* for each section with brief lectures including key learning objectives, classroom examples, and teaching notes
- Additional *Activities*, two per chapter, providing short group activities in a convenient ready-to-use handout format
- *Skill Builders* providing an enhanced worksheet for each text section, including concept rules, explained examples, and extra problems for students
- Twenty *Additional Exercises* per section for added test exercises or worksheets

Media

Lab Pack CD Lecture Series (ISBN: 0-13-192180-0)

- Organized by section, *Lab Pack CD Lecture Series* contains problem-solving techniques and examples from the textbook.
- Step-by-step solutions to selected exercises from each textbook section marked with a video icon.

TestGen (ISBN: 0-13-192181-9)

- Windows and Macintosh compatible
- Algorithmically driven, text-specific testing program covering all objectives of the text
- Chapter Test file for each chapter provides algorithms specific to exercises in each *Chapter Test* from the text.
- Edit and add your own questions with the built-in question editor, which allows you to create graphs, import graphics, and insert math notation.
- Create a nearly unlimited number of tests and worksheets, as well as assorted reports and summaries.
- Networkable for administering tests and capturing grades online, or on a local area network

MyMathLab
www.mymathlab.com
An all-in-one, online tutorial, homework, assessment, and course management tool with the following features:

- Rich and flexible set of course materials, featuring free-response exercises algorithmically generated for unlimited practice and mastery
- Entire textbook online with links to multimedia resources including video clips, practice exercises, and animations that are correlated to the textbook examples and exercises
- Homework and test managers to select and assign online exercises correlated directly to the text
- A personalized Study Plan generated based on student test results. The Study Plan links directly to unlimited tutorial exercises for the areas students need to study and retest, so they can practice until they have mastered the skills and concepts.
- *MyMathLab Gradebook* allows you to track all of the online homework, tests, and tutorial work while providing grade control.
- Import *TestGen* tests

MathXL®
www.mathxl.com
A powerful online homework, tutorial, and assessment system that allows instructors to:

- Create, edit, and assign online homework and tests using algorithmically generated exercises correlated at the objective level to the textbook
- Track student work in *MathXL*'s online gradebook

Resources for the Student

Student Solutions Manual CD/Video PH Math/Tutor Center MathXL Tutorials on CD MathXL® MyMathLab Interactmath.com

Student Study Pack (ISBN: 0-13-154936-7)

Get math help when YOU need it! Available at no charge when packaged with a new text, *Student Study Pack* provides the ultimate set of support resources to go along with the text. The *Student Study Pack* includes the *Student Solutions Manual*, access to the *Prentice Hall Math Tutor Center*, and the *CD Lecture Videos*.

Print
Student Solutions Manual (ISBN: 0-13-192179-7)

- Solutions to all odd-numbered section exercises
- Solutions to every (even and odd) exercise in the Mid-Chapter Check Points, Chapter Reviews, Chapter Tests, and Cumulative Reviews
- Solution methods reflect those emphasized in the text.

Media
MyMathLab
An all-in-one, online tutorial, homework, assessment, and course management tool with the following student features:

- The entire textbook online with links to multimedia resources including video clips, practice exercises, and animations correlated to the textbook examples and exercises
- Online tutorial, homework, and tests
- A personalized Study Plan based on student test results. The Study Plan links directly to unlimited tutorial exercises for the areas students need to study and retest, so they can practice until they have mastered the skills and concepts.

MathXL®

A powerful online homework, tutorial, and assessment system that allows students to:

- Take chapter tests and receive a personalized study plan based on their test results
- See diagnosed weaknesses and link directly to tutorial exercises for the objectives they need to study and retest
- Access supplemental animations and video clips directly from selected exercises

www.InterActMath.com

- Have the power of the *MathXL®* text-specific tutorial exercises available for unlimited practice online, without an access code, and without tracking capabilities

MathXL® Tutorials on CD (ISBN: 0-13-192176-2)

An interactive tutorial that provides:

- Algorithmically generated practice exercises correlated at the objective level
- Practice exercises accompanied by an example and a guided solution
- Tutorial video clips within the exercise to help students visualize concepts
- Easy-to-use tracking of student activity and scores and printed summaries of students' progress

Chapter Test Prep Video CDs (ISBN: 0-13-192174-6)

- Provide step-by-step video solutions to each exercise in each Chapter Test in the textbook
- Packaged at no charge with the text, inside the back cover

PH Tutor Center (ISBN: 0-13-064604-0)

- Tutorial support via phone, fax, or email bundled at no charge with a new text, or purchased separately with a used book
- Staffed by developmental math faculty
- Available 5 days a week, 7 hours a day

Acknowledgments

An enormous benefit of authoring a successful series is the broad-based feedback I receive from the students, dedicated users, and reviewers. Every change to this edition is the result of their thoughtful comments and suggestions. I would like to express my appreciation to all the reviewers, whose collective insights form the backbone of this revision. In particular, I would like to thank:

Howard Anderson	*Skagit Valley College*
John Anderson	*Illinois Valley Community College*
Michael H. Andreoli	*Miami Dade College–North Campus*
Hien Bui	*Hillsborough Community College*
Warren J. Burch	*Brevard Community College*
Alice Burstein	*Middlesex Community College*
Edie Carter	*Amarillo College*
Thomas B. Clark	*Trident Technical College*
Sandra Pryor Clarkson	*Hunter College*

Bettyann Daley	*University of Delaware*
Robert A. Davies	*Cuyahoga Community College*
Paige Davis	*Lurleen B. Wallace Community College*
Ben Divers, Jr.	*Ferrum College*
Irene Doo	*Austin Community College*
Charles C. Edgar	*Onondaga Community College*
Rhoderick Fleming	*Wake Technical Community College*
Susan Forman	*Bronx Community College*
Donna Gerken	*Miami-Dade College*
Marion K. Glasby	*Anne Arundel Community College*
Sue Glascoe	*Mesa Community College*
Jay Graening	*University of Arkansas*
Robert B. Hafer	*Brevard Community College*
Mary Lou Hammond	*Spokane Community College*
Donald Herrick	*Northern Illinois University*
Beth Hooper	*Golden West College*
Tracy Hoy	*College of Lake County*
Judy Kasabian	*Lansing Community College*
Gary Kersting	*North Central Michigan College*
Gary Knippenberg	*Lansing Community College*
Mary Kochler	*Cuyahoga Community College*
Kristi Laird	*Jackson Community College*
Jennifer Lempke	*North Central Michigan College*
Sandy Lofstock	*St. Petersburg College*
Hank Martel	*Broward Community College*
Diana Martelly	*Miami-Dade College*
John Robert Martin	*Tarrant County Junior College*
Mikal McDowell	*Cedar Valley College*
Irwin Metviner	*State University of New York at Old Westbury*
Terri Moser	*Austin Community College*
Robert Musselman	*California State University, Fresno*
Kamilia Nemri	*Spokane Community College*
Allen R. Newhart	*Parkersburg Community College*
Steve O'Donnell	*Rogue Community College*
Jeff Parent	*Oakland Community College*
Kate Rozsa	*Mesa Community College*
Scott W. Satake	*Eastern Washington University*
Mike Schramm	*Indian River Community College*
Kathy Shepard	*Monroe County Community College*
Gayle Smith	*Lane Community College*
Linda Smoke	*Central Michigan University*
Dick Spangler	*Tacoma Community College*
Janette Summers	*University of Arkansas*
Robert Thornton	*Loyola University*
Lucy C. Thrower	*Francis Marion College*

Andrew Walker   *North Seattle Community College*
Margaret Williamson   *Milwaukee Area Technical College*
Roberta Yellott   *McNeese State University*
Marilyn Zopp   *McHenry County College*

Special thanks to Professors Doreen Kelly and Kate Rozsa at Mesa Community College for your work on the custom edition that resulted in many of the practice plus exercises and many of the new exercises in the Second Edition. Thank you, Brad Davis, consummate mathematician and proofreader, for preparing the answer and graphing answer sections, as well as for your invaluable edits that kept me at my mathematical and syntactical best.

Additional acknowledgments are extended to Paul Murphy, senior acquisitions editor; Liz Covello, project manager; Barbara Mack, production editor; Linda Martino, full service editor from Prepare; and Bayani Mendoza de Leon, assistant managing editor. Lastly, my thanks to the staff at Prentice Hall for their continuing support: Eilish Main, executive marketing manager; Patrice Jones, director of marketing; Chris Hoag, editor-in-chief; and to the entire Prentice Hall sales force for their confidence and enthusiasm about my books.

I hope that my love for learning, as well as my respect for the diversity of students I have taught and learned from over the years, is apparent throughout this new edition. By connecting algebra to the whole spectrum of learning, it is my intent to show students that their world is profoundly mathematical, and indeed, π is in the sky.

Robert Blitzer

TO THE STUDENT

I've written this book so that you can learn about the power of algebra and how it relates directly to your life outside the classroom. All concepts are carefully explained, important definitions and procedures are set off in boxes, and worked-out examples that present solutions in a step-by-step manner appear in every section. Each example is followed by a similar matched problem, called a Check Point, for you to try so that you can actively participate in the learning process as you read the book. (Answers to all Check Points appear in the back of the book.) Study Tips offer hints and suggestions and often point out common errors to avoid. A great deal of attention has been given to applying algebra to your life to make your learning experience both interesting and relevant.

As you begin your studies, I would like to offer some specific suggestions for using this book and for being successful in this course:

1. Read the book. Read each section with pen (or pencil) in hand. Move through the worked-out examples with great care. These examples provide a model for doing exercises in the exercise sets. As you proceed through the reading, do not give up if you do not understand every single word. Things will become clearer as you read on and see how various procedures are applied to specific worked-out examples.

2. Work problems every day and check your answers. The way to learn mathematics is by doing mathematics, which means working the Check Points and assigned exercises in the exercise sets. The more exercises you work, the better you will understand the material.

3. Review for quizzes and tests. After completing a chapter, study the chapter review chart, work the exercises in the Chapter Review, and work the exercises in the Chapter Test. Answers to all these exercises are given in the back of the book.

The methods that I've used to help you read the book, work the problems, and review for tests are described in "A Brief Guide to Getting the Most from This Book" that appears inside the front cover. Spend a few minutes reviewing the guide to familiarize yourself with the book's features and their benefits.

4. Use the resources available with this book. Additional resources to aid your study are described following the guide to getting the most from your book. These resources include a Solutions Manual, a Chapter Test Prep Video CD, MyMathLab, an online version of the book with links to multimedia resources, MathXL®, an online homework, tutorial, and assessment system of the text, and tutorial support at no charge at the PH Tutor Center.

5. Attend all lectures. No book is intended to be a substitute for valuable insights and interactions that occur in the classroom. In addition to arriving for lecture on time and being prepared, you will find it useful to read the section before it is covered in lecture. This will give you a clear idea of the new material that will be discussed.

I wrote this book in Point Reyes National Seashore, 40 miles north of San Francisco. The park consists of 75,000 acres with miles of pristine surf-washed beaches, forested ridges, and bays bordered by white cliffs. It was my hope to convey the beauty and excitement of mathematics using nature's unspoiled beauty as a source of inspiration and creativity. Enjoy the pages that follow as you empower yourself with the algebra needed to succeed in college, your career, and in your life.

Regards,
Bob
Robert Blitzer

ABOUT THE AUTHOR

Bob Blitzer is a native of Manhattan and received a Bachelor of Arts degree with dual majors in mathematics and psychology (minor: English literature) from the City College of New York. His unusual combination of academic interests led him toward a Master of Arts in mathematics from the University of Miami and a doctorate in behavioral sciences from Nova University. Bob is most energized by teaching mathematics and has taught a variety of mathematics courses at Miami-Dade College for nearly 30 years. He has received numerous teaching awards, including Innovator of the Year from the League for Innovations in the Community College, and was among the first group of recipients at Miami-Dade College for an endowed chair based on excellence in the classroom. In addition to *Essentials of Introductory and Intermediate Algebra for College Students*, Bob has written *Introductory Algebra for College Students, Intermediate Algebra for College Students, Essentials of Intermediate Algebra for College Students, Introductory and Intermediate Algebra for College Students, Algebra for College Students, Thinking Mathematically, College Algebra, Algebra and Trigonometry*, and *Precalculus*, all published by Prentice Hall.

Sitting in the biology department office, you overhear two of the professors discussing the possible adult heights of their respective children. Looking at the blackboard that they've been writing on, you see that there are formulas that can estimate the height a child will attain as an adult. If the child is x years old and h inches tall, that child's adult height, H, in inches, is approximated by one of the following formulas:

Girls:
$$H = \frac{h}{0.00028x^3 - 0.0071x^2 + 0.0926x + 0.3524}$$

Boys:
$$H = \frac{h}{0.00011x^3 - 0.0032x^2 + 0.0604x + 0.3796}.$$

You will use one of these formulas in Exercise 118 on page 91.

CHAPTER 1

The Real Number System

When you encounter formulas with symbols such as those that appear in the estimation of a child's adult height, don't panic! You are already familiar with many symbols—the smiley face, the peace symbol, the heart symbol, the dollar sign, and even symbols on your calculator or computer. In this chapter, you will become familiar with the special symbolic notation of algebra. You will see that the language of algebra describes your world and holds the power to solve many of its problems.

SECTION 1.1

Objectives

1 Convert between mixed numbers and improper fractions.

2 Write the prime factorization of a composite number.

3 Reduce or simplify fractions.

4 Multiply fractions.

5 Divide fractions.

6 Add and subtract fractions with identical denominators.

7 Add and subtract fractions with unlike denominators.

8 Solve problems involving fractions.

FRACTIONS

The Nine Justices of the U.S. Supreme Court

"If I were asked where I place the American aristocracy, I should reply without hesitation that it occupies the judicial bench and bar. Scarcely any political question arises in the United States that is not resolved, sooner or later, into a judicial question."

Alexis de Tocqueville,
Democracy in America

President Jimmy Carter had no opportunity to make an appointment to the Supreme Court. However, he selected more African Americans, Hispanics, and women for the lower federal courts than all other prior presidents combined—40 women, 37 African Americans, and 16 Hispanics. The graph in Figure 1.1 shows the number of female and minority appointments to federal judgeships for Presidents Carter, Reagan, Bush Senior, and Clinton.

We can use the numbers shown in Figure 1.1 to form various fractions. For example, the graph indicates that President Carter appointed a total of 258 federal judges. Of the 258 judges, there were 40 women. We can say that the *fraction* of women appointed to federal judgeships under Carter was $\frac{40}{258}$. Can you see how this fraction is used to refer to part (40 women) of a whole (258 total appointments)?

In a fraction, the number that is written above the fraction bar is called the **numerator**. The number below the fraction bar is called the **denominator**.

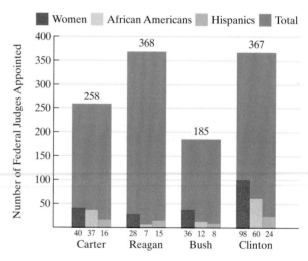

FIGURE 1.1

Source: James Burns et al., *Government by the People*, Prentice Hall, 2002

Fraction bar $\dfrac{40}{258}$ 40 is the numerator.   258 is the denominator.

The numbers 40 and 258 are examples of *natural numbers*. The **natural numbers** are the numbers that we use for counting:

$$1, 2, 3, 4, 5, \ldots.$$

The three dots after the 5 indicate that the list continues in the same manner without ending.

Fractions appear throughout algebra. We are frequently given numerical information that involves fractions. In this section, we present a brief review of operations with fractions that we will use in algebra.

1 Convert between mixed numbers and improper fractions.

Mixed Numbers and Improper Fractions A **mixed number** consists of the addition of a natural number and a fraction, expressed without the use of an addition sign. Here is an example of a mixed number:

$$3\frac{4}{5}.$$ The natural number is **3** and the fraction is $\frac{4}{5}$. $3\frac{4}{5}$ means $3 + \frac{4}{5}$.

An **improper fraction** is a fraction whose numerator is greater than its denominator. An example of an improper fraction is $\frac{19}{5}$.

The mixed number $3\frac{4}{5}$ can be converted to the improper fraction $\frac{19}{5}$ using the following procedure:

> **CONVERTING A MIXED NUMBER TO AN IMPROPER FRACTION**
>
> **1.** Multiply the denominator of the fraction by the natural number and add the numerator to this product.
>
> **2.** Place the result from step 1 over the denominator in the mixed number.

EXAMPLE 1 Converting from Mixed Number to Improper Fraction

Convert $3\frac{4}{5}$ to an improper fraction.

SOLUTION

$$3\frac{4}{5} = \frac{5 \cdot 3 + 4}{5}$$ Multiply the denominator by the natural number and add the numerator.

$$= \frac{15 + 4}{5} = \frac{19}{5}$$ Place the result over the mixed number's denominator.

STUDY TIP

After working each Check Point, check your answer in the answer section before continuing your reading.

✔ **CHECK POINT 1** Convert $2\frac{5}{8}$ to an improper fraction. $\frac{21}{8}$

An improper fraction can be converted to a mixed number using the following procedure:

> **CONVERTING AN IMPROPER FRACTION TO A MIXED NUMBER**
>
> **1.** Divide the denominator into the numerator. Record the quotient (the result of the division) and the remainder.
>
> **2.** Write the mixed number using the following form:
>
> $$\text{quotient } \frac{\text{remainder}}{\text{original denominator}}.$$
>
> natural number part   fraction part

EXAMPLE 2 Converting from an Improper Fraction to a Mixed Number

Convert $\frac{42}{5}$ to a mixed number.

SOLUTION

Step 1. **Divide the denominator into the numerator.**

$$
\begin{array}{r}
8 \quad \text{quotient} \\
5\overline{)42} \\
\underline{40} \\
2 \quad \text{remainder}
\end{array}
$$

Step 2. **Write the mixed number using quotient** $\dfrac{\text{remainder}}{\text{original denominator}}$. Thus,

$$
\frac{42}{5} = 8\frac{2}{5}.
$$

quotient ← 8   remainder → 2   original denominator → 5

CHECK POINT 2 Convert $\frac{5}{3}$ to a mixed number. $1\frac{2}{3}$

Using Variables to Express Fractions   Algebra uses letters, such as a and b, to represent numbers. Such letters are called **variables**. Throughout this section, we will use variables to express fractions. These fractions can be represented by the *variable expression*

$$
\frac{a}{b}.
$$

In this expression, the variable a represents any natural number and the variable b also represents any natural number. Can you see why $\frac{a}{b}$ is called a variable expression? Its value varies with the choice of a and the choice of b. Here are some examples:

Choose $a = 5$ and $b = 3$. $\dfrac{a}{b} = \dfrac{5}{3}$ Choose $a = 3$ and $b = 5$. $\dfrac{a}{b} = \dfrac{3}{5}$ Choose $a = 1$ and $b = 100$. $\dfrac{a}{b} = \dfrac{1}{100}.$

2   Write the prime factorization of a composite number.

Factors and Prime Factorizations   In the section opener, we saw that the fraction of women appointed to federal judgeships under President Carter was $\frac{40}{258}$. As you looked at this fraction, did you attempt to *simplify* it by *reducing it to its lowest terms*? Before we simplify fractions, let's first review how to factor the natural numbers that make up the numerator and the denominator.

To **factor** a natural number means to write it as a multiplication of two or more natural numbers. For example, 21 can be factored as $7 \cdot 3$. In the statement $7 \cdot 3 = 21$, 7 and 3 are called the **factors** and 21 is the **product**.

7 is a factor of 21. $7 \cdot 3 = 21$ The product of 7 and 3 is 21.

3 is a factor of 21.

Are 7 and 3 the only factors of 21? The answer is no because 21 can also be factored as $1 \cdot 21$. Thus, 1 and 21 are also factors of 21. The factors of 21 are 1, 3, 7, and 21.

Unlike the number 21, some natural numbers have only two factors: the number itself and 1. For example, the number 7 has only two factors: 7 (the number itself) and 1. The only way to factor 7 is $1 \cdot 7$ or, equivalently, $7 \cdot 1$. For this reason, 7 is called a *prime number*.

> **PRIME NUMBERS**   A **prime number** is a natural number greater than 1 that has only itself and 1 as factors.

The first ten prime numbers are

$$2, 3, 5, 7, 11, 13, 17, 19, 23, \text{ and } 29.$$

Can you see why the natural number 15 is not in this list? In addition to having 15 and 1 as factors ($15 = 1 \cdot 15$), it also has factors of 3 and 5 ($15 = 3 \cdot 5$). The number 15 is an example of a *composite number*.

> **COMPOSITE NUMBERS**   A **composite number** is a natural number greater than 1 that is not a prime number.

Every composite number can be expressed as the product of prime numbers. For example, the composite number 45 can be expressed as

$$45 = 3 \cdot 3 \cdot 5.$$

This product contains only prime numbers: 3 and 5.

Expressing a composite number as the product of prime numbers is called the **prime factorization** of that composite number. The prime factorization of 45 is $3 \cdot 3 \cdot 5$. The order in which we write these factors does not matter. This means that

$$45 = 3 \cdot 3 \cdot 5 \quad \text{or} \quad 45 = 5 \cdot 3 \cdot 3 \quad \text{or} \quad 45 = 3 \cdot 5 \cdot 3.$$

To find the prime factorization of a composite number, begin by selecting any two numbers whose product is the number to be factored. If one or both of the factors are not prime numbers, continue to factor each composite number. Stop when all numbers are prime.

EXAMPLE 3   Prime Factorization of a Composite Number

Find the prime factorization of 100.

SOLUTION   Begin by selecting any two numbers whose product is 100. Here is one possibility:

$$100 = 4 \cdot 25.$$

Because the factors 4 and 25 are not prime, we factor each of these composite numbers.

$$100 = 4 \cdot 25 \qquad \text{This is our first factorization.}$$
$$= 2 \cdot 2 \cdot 5 \cdot 5 \qquad \text{Factor 4 and 25.}$$

Notice that 2 and 5 are both prime. The prime factorization of 100 is $2 \cdot 2 \cdot 5 \cdot 5$. ■

 CHECK POINT 3 Find the prime factorization of 36. $2 \cdot 2 \cdot 3 \cdot 3$

3 Reduce or simplify fractions.

FIGURE 1.2

Reducing Fractions   Two fractions are **equivalent** if they represent the same value. Writing a fraction as an equivalent fraction with a smaller denominator is called **reducing a fraction**. A fraction is **reduced to its lowest terms** when the numerator and denominator have no common factors other than 1.

Look at the rectangle in Figure 1.2. Can you see that it is divided into 6 equal parts? Of these 6 parts, 4 of the parts are red. Thus, $\frac{4}{6}$ of the rectangle is red.

The rectangle in Figure 1.2 is also divided into 3 equal stacks and 2 of the stacks are red. Thus, $\frac{2}{3}$ of the rectangle is red. Because both $\frac{4}{6}$ and $\frac{2}{3}$ of the rectangle are red, we can conclude that $\frac{4}{6}$ and $\frac{2}{3}$ are equivalent fractions.

How can we show that $\frac{4}{6} = \frac{2}{3}$ without using Figure 1.2? Prime factorizations of 4 and 6 play an important role in the process. So does the **Fundamental Principle of Fractions**.

FUNDAMENTAL PRINCIPLE OF FRACTIONS   In words: The value of a fraction does not change if the numerator and denominator are divided (or multiplied) by the same nonzero number.

In algebraic language: If $\frac{a}{b}$ is a fraction and c is a nonzero number, then

$$\frac{a \cdot c}{b \cdot c} = \frac{a}{b}.$$

We use prime factorizations and the Fundamental Principle to reduce $\frac{4}{6}$ to its lowest terms as follows:

$$\frac{4}{6} = \frac{2 \cdot 2}{3 \cdot 2} = \frac{2}{3}$$

Write prime factorizations of 4 and 6.   Divide the numerator and the denominator by the common prime factor, 2.

Here is a procedure for writing a fraction in lowest terms:

REDUCING A FRACTION TO ITS LOWEST TERMS

1. Write the prime factorization of the numerator and the denominator.
2. Divide the numerator and the denominator by the greatest common factor, the product of all factors common to both.

Division lines can be used to show dividing out a fraction's numerator and denominator by common factors:

$$\frac{4}{6} = \frac{2 \cdot \cancel{2}}{3 \cdot \cancel{2}} = \frac{2}{3}.$$

STUDY TIP

When reducing a fraction to its lowest terms, only factors that are common to the numerator and the denominator can be divided out. **If you have not factored** and expressed the numerator and denominator in terms of multiplication, **do not divide out.**

Correct:

$$\frac{2 \cdot \cancel{2}}{3 \cdot \cancel{2}} = \frac{2}{3}$$

Incorrect:

$$\frac{2 + \cancel{2}}{3 + \cancel{2}} = \frac{2}{3}$$

Note that $\frac{2+2}{3+2}$ $= \frac{4}{5}$, not $\frac{2}{3}$.

EXAMPLE 4   Reducing Fractions

Reduce each fraction to its lowest terms:

   a. $\dfrac{6}{14}$ **b.** $\dfrac{15}{75}$ **c.** $\dfrac{25}{11}$ **d.** $\dfrac{11}{33}$.

SOLUTION   For each fraction, begin with the prime factorization of the numerator and the denominator.

   a. $\dfrac{6}{14} = \dfrac{3 \cdot \cancel{2}}{7 \cdot \cancel{2}} = \dfrac{3}{7}$ 2 is the greatest common factor of 6 and 14. Divide the numerator and the denominator by 2.

> Including 1 as a factor is helpful when all other factors can be divided out.

   b. $\dfrac{15}{75} = \dfrac{3 \cdot 5}{3 \cdot 25} = \dfrac{1 \cdot \cancel{3} \cdot \cancel{5}}{\cancel{3} \cdot \cancel{5} \cdot 5} = \dfrac{1}{5}$ 3 • 5, or 15, is the greatest common factor of 15 and 75. Divide the numerator and the denominator by 3 • 5.

   c. $\dfrac{25}{11} = \dfrac{5 \cdot 5}{1 \cdot 11}$

Because 11 and 25 share no common factor (other than 1), $\frac{11}{25}$ is already reduced to its lowest terms.

   d. $\dfrac{11}{33} = \dfrac{1 \cdot \cancel{11}}{3 \cdot \cancel{11}} = \dfrac{1}{3}$ 11 is the greatest common factor of 11 and 33. Divide the numerator and denominator by 11.

When reducing fractions, it may not always be necessary to write prime factorizations. In some cases, you can use inspection to find the greatest common factor of the numerator and the denominator. For example, when reducing $\frac{15}{75}$, you can use 15 rather than 3 • 5:

$$\frac{15}{75} = \frac{1 \cdot \cancel{15}}{5 \cdot \cancel{15}} = \frac{1}{5}.$$

 CHECK POINT 4 Reduce each fraction to its lowest terms:

   a. $\dfrac{10}{15}\quad \dfrac{2}{3}$ **b.** $\dfrac{42}{24}\quad \dfrac{7}{4}$ **c.** $\dfrac{13}{15}\quad \dfrac{13}{15}$ **d.** $\dfrac{9}{45}.\quad \dfrac{1}{5}$

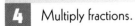 **4** Multiply fractions.

Multiplying Fractions   The result of multiplying two fractions is called their **product**.

> **MULTIPLYING FRACTIONS**   In words: The product of two or more fractions is the product of their numerators divided by the product of their denominators.
>
> In algebraic language: If $\frac{a}{b}$ and $\frac{c}{d}$ are fractions, then
>
> $$\frac{a}{b} \cdot \frac{c}{d} = \frac{a \cdot c}{b \cdot d}.$$

Here is an example that illustrates the rule in the previous box:

$$\frac{3}{8} \cdot \frac{5}{11} = \frac{3 \cdot 5}{8 \cdot 11} = \frac{15}{88}.$$

> The product of $\frac{3}{8}$ and $\frac{5}{11}$ is $\frac{15}{88}$.

> Multiply numerators and multiply denominators.

EXAMPLE 5 Multiplying Fractions

Multiply. If possible, reduce the product to its lowest terms:

a. $\frac{3}{7} \cdot \frac{2}{5}$ **b.** $5 \cdot \frac{7}{12}$ **c.** $\frac{2}{3} \cdot \frac{9}{4}$ **d.** $\left(3\frac{2}{3}\right)\left(1\frac{1}{4}\right).$

SOLUTION

a. $\frac{3}{7} \cdot \frac{2}{5} = \frac{3 \cdot 2}{7 \cdot 5} = \frac{6}{35}$

> Multiply numerators and multiply denominators.

b. $5 \cdot \frac{7}{12} = \frac{5}{1} \cdot \frac{7}{12} = \frac{5 \cdot 7}{1 \cdot 12} = \frac{35}{12}$ or $2\frac{11}{12}$

> Write 5 as $\frac{5}{1}$. Then multiply numerators and multiply denominators.

c. $\frac{2}{3} \cdot \frac{9}{4} = \frac{2 \cdot 9}{3 \cdot 4} = \frac{18}{12} = \frac{3 \cdot \cancel{6}}{2 \cdot \cancel{6}} = \frac{3}{2}$ or $1\frac{1}{2}$

> Simplify $\frac{18}{12}$; 6 is the greatest common factor of 18 and 12.

d. $\left(3\frac{2}{3}\right)\left(1\frac{1}{4}\right) = \frac{11}{3} \cdot \frac{5}{4} = \frac{11 \cdot 5}{3 \cdot 4} = \frac{55}{12}$ or $4\frac{7}{12}$

■

✔ **CHECK POINT 5** Multiply. If possible, reduce the product to its lowest terms:

a. $\frac{4}{11} \cdot \frac{2}{3}$ $\frac{8}{33}$ **b.** $6 \cdot \frac{3}{5}$ $\frac{18}{5}$ or $3\frac{3}{5}$

c. $\frac{3}{7} \cdot \frac{2}{3}$ $\frac{2}{7}$ **d.** $\left(3\frac{2}{5}\right)\left(1\frac{1}{2}\right).$ $\frac{51}{10}$ or $5\frac{1}{10}$

STUDY TIP

You can divide numerators and denominators by common factors before performing multiplication. Then multiply the remaining factors in the numerators and multiply the remaining factors in the denominators. For example,

$$\frac{7}{15} \cdot \frac{20}{21} = \frac{\cancel{7} \cdot 1}{\cancel{5} \cdot 3} \cdot \frac{\cancel{5} \cdot 4}{\cancel{7} \cdot 3} = \frac{1 \cdot 4}{3 \cdot 3} = \frac{4}{9}.$$

> 7 is the greatest common factor of 7 and 21.

> 5 is the greatest common factor of 15 and 20.

Study Tip (continued)

The divisions involving the common factors, 7 and 5, are often shown as follows:

$$\frac{7}{15} \cdot \frac{20}{21} = \frac{\overset{1}{\cancel{7}}}{\underset{3}{\cancel{15}}} \cdot \frac{\overset{4}{\cancel{20}}}{\underset{3}{\cancel{21}}} = \frac{1 \cdot 4}{3 \cdot 3} = \frac{4}{9}.$$

Divide by 7.   Divide by 5.

5 Divide fractions.

Dividing Fractions The result of dividing two fractions is called their **quotient**. A geometric figure is useful for developing a process for determining the quotient of two fractions.

Consider the division

$$\frac{4}{5} \div \frac{1}{10}.$$

We want to know how many $\frac{1}{10}$'s are in $\frac{4}{5}$. We can use Figure 1.3 to find this quotient. The rectangle is divided into fifths. The dashed lines further divide the rectangle into tenths.

Figure 1.3 shows that $\frac{4}{5}$ of the rectangle is red. How many $\frac{1}{10}$'s of the rectangle does this include? Can you see that this includes eight of the $\frac{1}{10}$ pieces? Thus, there are eight $\frac{1}{10}$'s in $\frac{4}{5}$:

$$\frac{4}{5} \div \frac{1}{10} = 8.$$

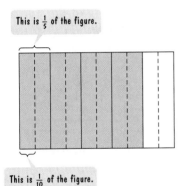

This is $\frac{1}{5}$ of the figure.

This is $\frac{1}{10}$ of the figure.

FIGURE 1.3

We can obtain the quotient 8 in the following way:

$$\frac{4}{5} \div \frac{1}{10} = \frac{4}{5} \cdot \frac{10}{1} = \frac{4 \cdot 10}{5 \cdot 1} = \frac{40}{5} = 8.$$

Change the division to multiplication.   Invert the divisor, $\frac{1}{10}$.

By inverting the divisor, $\frac{1}{10}$, and obtaining $\frac{10}{1}$, we are writing the divisor's *reciprocal*. Two fractions are **reciprocals** of each other if their product is 1. Thus, $\frac{1}{10}$ and $\frac{10}{1}$ are reciprocals because $\frac{1}{10} \cdot \frac{10}{1} = 1$.

Generalizing from the result shown above and using the word *reciprocal*, we obtain the following rule for dividing fractions:

DIVIDING FRACTIONS In words: The quotient of two fractions is the first fraction multiplied by the reciprocal of the second fraction.

In algebraic language: If $\frac{a}{b}$ and $\frac{c}{d}$ are fractions and $\frac{c}{d}$ is not 0, then

$$\frac{a}{b} \div \frac{c}{d} = \frac{a}{b} \cdot \frac{d}{c}.$$

Change division to multiplication.   Invert $\frac{c}{d}$ and write its reciprocal.

EXAMPLE 6 Dividing Fractions

Divide: **a.** $\dfrac{2}{3} \div \dfrac{7}{15}$ **b.** $\dfrac{3}{4} \div 5$ **c.** $4\dfrac{3}{4} \div 1\dfrac{1}{2}.$

SOLUTION

a. $\dfrac{2}{3} \div \dfrac{7}{15} = \dfrac{2}{3} \cdot \dfrac{15}{7} = \dfrac{2 \cdot 15}{3 \cdot 7} = \dfrac{30}{21} = \dfrac{10 \cdot \cancel{3}}{7 \cdot \cancel{3}} = \dfrac{10}{7}$ or $1\dfrac{3}{7}$

> Change division to multiplication.   Invert $\frac{7}{15}$ and write its reciprocal.   Simplify: 3 is the greatest common factor of 30 and 21.

b. $\dfrac{3}{4} \div 5 = \dfrac{3}{4} \div \dfrac{5}{1} = \dfrac{3}{4} \cdot \dfrac{1}{5} = \dfrac{3 \cdot 1}{4 \cdot 5} = \dfrac{3}{20}$

> Change division to multiplication.   Invert $\frac{5}{1}$ and write its reciprocal.

c. $4\dfrac{3}{4} \div 1\dfrac{1}{2} = \dfrac{19}{4} \div \dfrac{3}{2} = \dfrac{19}{4} \cdot \dfrac{2}{3} = \dfrac{19 \cdot 2}{4 \cdot 3} = \dfrac{38}{12} = \dfrac{19 \cdot \cancel{2}}{6 \cdot \cancel{2}} = \dfrac{19}{6}$ or $3\dfrac{1}{6}$

✔ **CHECK POINT 6** Divide:

a. $\dfrac{5}{4} \div \dfrac{3}{8}$ $\dfrac{10}{3}$ or $3\dfrac{1}{3}$ **b.** $\dfrac{2}{3} \div 3$ $\dfrac{2}{9}$ **c.** $3\dfrac{3}{8} \div 2\dfrac{1}{4}$ $\dfrac{3}{2}$ or $1\dfrac{1}{2}$

---

6 Add and subtract fractions with identical denominators.

FIGURE 1.4

Adding and Subtracting Fractions with Identical Denominators The result of adding two fractions is called their **sum**. The result of subtracting two fractions is called their **difference**. A geometric figure is useful for developing a process for determining the sum or difference of two fractions with identical denominators.

Consider the addition

$$\frac{3}{7} + \frac{2}{7}.$$

We can use Figure 1.4 to find this sum. The rectangle is divided into sevenths. On the left, $\frac{3}{7}$ of the rectangle is red. On the right, $\frac{2}{7}$ of the rectangle is red. Including both the left and the right, a total of $\frac{5}{7}$ of the rectangle is red. Thus,

$$\frac{3}{7} + \frac{2}{7} = \frac{5}{7}.$$

We can obtain the sum $\frac{5}{7}$ in the following way:

$$\frac{3}{7} + \frac{2}{7} = \frac{3 + 2}{7} = \frac{5}{7}.$$

> Add numerators and put this result over the common denominator.

Generalizing from this result gives us the following rule:

> **ADDING AND SUBTRACTING FRACTIONS WITH IDENTICAL DENOMINATORS**
> In words: The sum or difference of two fractions with identical denominators is the sum or difference of their numerators over the common denominator.
> In algebraic language: If $\frac{a}{b}$ and $\frac{c}{b}$ are fractions, then
> $$\frac{a}{b} + \frac{c}{b} = \frac{a + c}{b} \quad \text{and} \quad \frac{a}{b} - \frac{c}{b} = \frac{a - c}{b}.$$

EXAMPLE 7	Adding and Subtracting Fractions with Identical Denominators

Perform the indicated operations:

a. $\dfrac{3}{11} + \dfrac{4}{11}$ 　　　**b.** $\dfrac{11}{12} - \dfrac{5}{12}$ 　　　**c.** $5\dfrac{1}{4} - 2\dfrac{3}{4}$.

SOLUTION

a. $\dfrac{3}{11} + \dfrac{4}{11} = \dfrac{3+4}{11} = \dfrac{7}{11}$

b. $\dfrac{11}{12} - \dfrac{5}{12} = \dfrac{11-5}{12} = \dfrac{6}{12} = \dfrac{1 \cdot \cancel{6}}{2 \cdot \cancel{6}} = \dfrac{1}{2}$

c. $5\dfrac{1}{4} - 2\dfrac{3}{4} = \dfrac{21}{4} - \dfrac{11}{4} = \dfrac{21-11}{4} = \dfrac{10}{4} = \dfrac{\cancel{2} \cdot 5}{\cancel{2} \cdot 2} = \dfrac{5}{2}$ or $2\dfrac{1}{2}$ ■

✔ **CHECK POINT 7** Perform the indicated operations:

a. $\dfrac{2}{11} + \dfrac{3}{11}$ 　$\frac{5}{11}$ 　　　**b.** $\dfrac{5}{6} - \dfrac{1}{6}$ 　$\frac{2}{3}$ 　　　**c.** $3\dfrac{3}{8} - 1\dfrac{1}{8}$. 　$\frac{9}{4}$ or $2\frac{1}{4}$

---

7 Add and subtract fractions with unlike denominators.

Adding and Subtracting Fractions with Unlike Denominators How do we add or subtract fractions with different denominators? We must first rewrite them as equivalent fractions with the same denominator. We do this by using the Fundamental Principle of Fractions: The value of a fraction does not change if the numerator and the denominator are multiplied by the same nonzero number. Thus, if $\frac{a}{b}$ is a fraction and c is a nonzero number, then

$$\frac{a}{b} = \frac{a \cdot c}{b \cdot c}.$$

EXAMPLE 8	Writing an Equivalent Fraction

Write $\frac{3}{4}$ as an equivalent fraction with a denominator of 16.

SOLUTION　To obtain a denominator of 16, we must multiply the denominator of the given fraction, $\frac{3}{4}$, by 4. So that we do not change the value of the fraction, we also multiply the numerator by 4.

$$\frac{3}{4} = \frac{3 \cdot 4}{4 \cdot 4} = \frac{12}{16}$$ ■

✔ **CHECK POINT 8** Write $\frac{2}{3}$ as an equivalent fraction with a denominator of 21. 　$\frac{14}{21}$

Equivalent fractions can be used to add fractions with different denominators, such as $\frac{1}{2}$ and $\frac{1}{3}$. Figure 1.5 indicates that the sum of half the whole figure and one-third of the whole figure results in 5 parts out of 6, or $\frac{5}{6}$, of the figure. Thus,

$$\frac{1}{2} + \frac{1}{3} = \frac{5}{6}.$$

We can obtain the sum $\frac{5}{6}$ if we rewrite each fraction as an equivalent fraction with a denominator of 6.

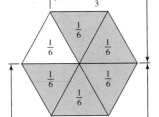

FIGURE 1.5 $\frac{1}{2} + \frac{1}{3} = \frac{5}{6}$

$$\frac{1}{2} + \frac{1}{3} = \frac{1 \cdot 3}{2 \cdot 3} + \frac{1 \cdot 2}{3 \cdot 2}$$

Rewrite each fraction as an equivalent fraction with a denominator of 6.

$$= \frac{3}{6} + \frac{2}{6}$$

Perform the multiplications. We now have a common denominator.

$$= \frac{3 + 2}{6}$$

Add the numerators and place this sum over the common denominator.

$$= \frac{5}{6}$$

Perform the addition.

When adding $\frac{1}{2}$ and $\frac{1}{3}$, there are many common denominators that we can use, such as 6, 12, 18, and so on. The given denominators, 2 and 3, divide into all of these numbers. However, the denominator 6 is the smallest number that 2 and 3 divide into. For this reason, 6 is called the *least common denominator*, abbreviated LCD.

ADDING AND SUBTRACTING FRACTIONS WITH UNLIKE DENOMINATORS

1. Rewrite the fractions as equivalent fractions with the least common denominator.

2. Add or subtract the numerators, putting this result over the common denominator.

EXAMPLE 9 Adding and Subtracting Fractions with Unlike Denominators

Perform the indicated operation: **a.** $\frac{1}{5} + \frac{3}{4}$ **b.** $\frac{3}{4} - \frac{1}{6}$ **c.** $2\frac{7}{15} - 1\frac{4}{5}$.

SOLUTION

a. Just by looking, can you can tell that the smallest number divisible by both 5 and 4 is 20? Thus, the least common denominator for the denominators 5 and 4 is 20. We rewrite both fractions as equivalent fractions with the least common denominator, 20.

$$\frac{1}{5} + \frac{3}{4} = \frac{1 \cdot 4}{5 \cdot 4} + \frac{3 \cdot 5}{4 \cdot 5}$$

To obtain denominators of 20, multiply the numerator and denominator of the first fraction by 4 and the second fraction by 5.

$$= \frac{4}{20} + \frac{15}{20}$$

Perform the multiplications.

$$= \frac{4 + 15}{20}$$

Add the numerators and put this sum over the least common denominator.

$$= \frac{19}{20}$$

Perform the addition.

b. By looking, can you tell that the smallest number divisible by both 4 and 6 is 12? Thus, the least common denominator for the denominators 4 and 6 is 12. We rewrite both fractions as equivalent fractions with the least common denominator, 12.

$$\frac{3}{4} - \frac{1}{6} = \frac{3 \cdot 3}{4 \cdot 3} - \frac{1 \cdot 2}{6 \cdot 2}$$

To obtain denominators of 12, multiply the numerator and denominator of the first fraction by 3 and the second fraction by 2.

$$= \frac{9}{12} - \frac{2}{12}$$

Perform the multiplications.

$$= \frac{9 - 2}{12}$$

Subtract the numerators and put this difference over the least common denominator.

$$= \frac{7}{12}$$

Perform the subtraction.

DISCOVER FOR YOURSELF

Try Example 9(a), $\frac{1}{5} + \frac{3}{4}$, using a common denominator of 40. Because both 5 and 4 divide into 40, 40 is a common denominator, although not the *least* common denominator. Describe what happens. What is the advantage of using the least common denominator?

c. $2\dfrac{7}{15} - 1\dfrac{4}{5} = \dfrac{37}{15} - \dfrac{9}{5}$ *Convert each mixed number to an improper fraction.*

> The smallest number divisible by both 15 and 5 is 15. Thus, the least common denominator for the denominators 15 and 5 is 15. Because the first fraction already has a denominator of 15, we only have to rewrite the second fraction.

$= \dfrac{37}{15} - \dfrac{9 \cdot 3}{5 \cdot 3}$ *To obtain denominators of 15, multiply the numerator and denominator of the second fraction by 3.*

$= \dfrac{37}{15} - \dfrac{27}{15}$ *Perform the multiplications.*

$= \dfrac{37 - 27}{15}$ *Subtract the numerators and put this difference over the common denominator.*

$= \dfrac{10}{15}$ *Perform the subtraction.*

$= \dfrac{2 \cdot \cancel{5}}{3 \cdot \cancel{5}} = \dfrac{2}{3}$ *Reduce to lowest terms.*

 CHECK POINT 9 Perform the indicated operation:

a. $\dfrac{1}{2} + \dfrac{3}{5}$ $\dfrac{11}{10}$ **b.** $\dfrac{4}{3} - \dfrac{3}{4}$ $\dfrac{7}{12}$ **c.** $3\dfrac{1}{6} - 1\dfrac{11}{12}$. $\dfrac{5}{4}$ or $1\dfrac{1}{4}$

EXAMPLE 10   **Using Prime Factorizations to Find the LCD**

Perform the indicated operation: $\dfrac{1}{15} + \dfrac{7}{24}$.

SOLUTION   We need to first find the least common denominator. Using inspection, it is difficult to determine the smallest number divisible by both 15 and 24. We will use their prime factorizations to find the least common denominator:

$$15 = 5 \cdot 3 \quad \text{and} \quad 24 = 8 \cdot 3 = 2 \cdot 2 \cdot 2 \cdot 3.$$

The different prime factors are 5, 3, and 2. The least common denominator is obtained by using the greatest number of times each factor appears in any prime factorization. Because 5 and 3 appear as prime factors and 2 is a factor of 24 three times, the least common denominator is

$$5 \cdot 3 \cdot 2 \cdot 2 \cdot 2 = 5 \cdot 3 \cdot 8 = 120.$$

Now we can rewrite both fractions as equivalent fractions with the least common denominator, 120.

$\dfrac{1}{15} + \dfrac{7}{24} = \dfrac{1 \cdot 8}{15 \cdot 8} + \dfrac{7 \cdot 5}{24 \cdot 5}$ *To obtain denominators of 120, multiply the numerator and denominator of the first fraction by 8 and the second fraction by 5.*

$= \dfrac{8}{120} + \dfrac{35}{120}$ *Perform the multiplications.*

$= \dfrac{8 + 35}{120}$ *Add the numerators and put this sum over the least common denominator.*

$= \dfrac{43}{120}$ *Perform the addition.*

 CHECK POINT 10 Perform the indicated operation: $\dfrac{3}{10} + \dfrac{7}{12}$. $\dfrac{53}{60}$

8 Solve problems involving fractions.

Applications You can hardly pick up a newspaper without being bombarded with graphs that describe how we live, what we want, and even how we are likely to die. Understanding fractions and their operations is often helpful in interpreting the information given in these graphs.

Circle graphs, also called **pie charts**, show how a quantity is divided into parts. Circle graphs are divided into pieces, called **sectors**. The sizes of the sectors show the relative sizes of the categories. Figure 1.6 is an example of a typical circle graph. The graph shows the breakdown of the number of countries in the world that are free, partly free, or not free.

World's Countries by Status of Freedom

FIGURE 1.6

Source: Larry Berman and Bruce Murphy, *Approaching Democracy*, 4th edition, Prentice Hall, 2003

EXAMPLE 11   Using a Circle Graph to Interpret Information

Use Figure 1.6 to determine the fraction of the world's countries that are not free. Reduce the answer to its lowest terms.

SOLUTION   The sector on the upper left shows that there are 48 countries that are not free. By adding the numbers in the three sectors, we see that there are a total of

$$48 + 58 + 86$$

or 192 countries.

Fraction of nonfree countries

$$= \frac{\text{number of countries that are not free}}{\text{total number of countries}}$$

$$= \frac{48}{192} = \frac{2 \cdot 24}{2 \cdot 96} = \frac{2 \cdot 2 \cdot 12}{2 \cdot 2 \cdot 48} = \frac{\cancel{2} \cdot \cancel{2} \cdot \cancel{12} \cdot 1}{\cancel{2} \cdot \cancel{2} \cdot \cancel{12} \cdot 4} = \frac{1}{4}$$

The fraction of the world's countries that are not free is $\frac{1}{4}$. ∎

✔ **CHECK POINT 11** Use Figure 1.6 to determine the fraction of the world's countries that are partly free. Reduce the answer to its lowest terms. $\frac{29}{96}$

The graph in Figure 1.6 can show the fraction of countries, rather than the actual number of countries, in each of the three sectors. If a circle graph shows a fraction within each of its sectors, the sum of the fractional parts must be 1, representing one whole circle.

EXAMPLE 12   Miss America Hair Color

The Miss America title has been awarded, with some breaks, since 1921. The graph in Figure 1.7 might disprove the often-quoted claim that "gentlemen prefer blondes." What fraction of Miss America titleholders have been blondes?

SOLUTION   We begin by finding the fraction of Miss Americas who have not been blondes. Add the fractions of brunettes, $\frac{7}{10}$, and redheads, $\frac{3}{50}$.

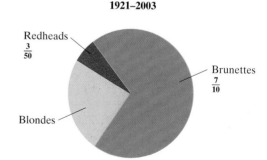

FIGURE 1.7

Source: Ben Schott, *Schott's Miscellany*, Bloomsbury, 2003

$$\frac{7}{10} + \frac{3}{50}$$ The least common denominator is 50.

$$= \frac{7 \cdot 5}{10 \cdot 5} + \frac{3}{50}$$ To obtain denominators of 50, multiply the numerator and denominator of the first fraction by 5.

$$= \frac{35}{50} + \frac{3}{50}$$ Perform the multiplications.

$$= \frac{35 + 3}{50} = \frac{38}{50}$$ Add the numerators and put this sum over the least common denominator.

$$= \frac{19 \cdot \cancel{2}}{25 \cdot \cancel{2}} = \frac{19}{25}$$ Reduce to lowest terms.

Thus, $\frac{19}{25}$ of the Miss Americas have not been blondes. The fractions representing all three hair colors in Figure 1.7 must add up to 1. Therefore, the fraction of Miss Americas who have been blondes can be found by subtracting $\frac{19}{25}$ from 1:

$$1 - \frac{19}{25} = \frac{25}{25} - \frac{19}{25} = \frac{25 - 19}{25} = \frac{6}{25}.$$

We see that $\frac{6}{25}$ of Miss America titleholders have been blondes. ∎

✔ **CHECK POINT 12** At a workshop on enhancing creativity, $\frac{1}{4}$ of the participants are musicians, $\frac{2}{5}$ are artists, $\frac{1}{10}$ are actors, and the remaining participants are writers. Find the fraction of people at the workshop who are writers. $\frac{1}{4}$

1.1 EXERCISE SET

Student Solutions Manual   CD/Video   PH Math/Tutor Center   MathXL Tutorials on CD   MathXL®   MyMathLab   Interactmath.com

Practice Exercises

In Exercises 1–6, convert each mixed number to an improper fraction.

1. $2\frac{3}{8}$ $\frac{19}{8}$ **2.** $2\frac{7}{9}$ $\frac{25}{9}$ **3.** $7\frac{3}{5}$ $\frac{38}{5}$

4. $6\frac{2}{5}$ $\frac{32}{5}$ **5.** $8\frac{7}{16}$ $\frac{135}{16}$ **6.** $9\frac{5}{16}$ $\frac{149}{16}$

In Exercises 7–12, convert each improper fraction to a mixed number.

7. $\frac{23}{5}$ $4\frac{3}{5}$ **8.** $\frac{47}{8}$ $5\frac{7}{8}$ **9.** $\frac{76}{9}$ $8\frac{4}{9}$

10. $\frac{59}{9}$ $6\frac{5}{9}$ **11.** $\frac{711}{20}$ $35\frac{11}{20}$ **12.** $\frac{788}{25}$ $31\frac{13}{25}$

In Exercises 13–28, identify each natural number as prime or composite. If the number is composite, find its prime factorization.

13. 22 $2 \cdot 11$ **14.** 15 $3 \cdot 5$ **15.** 20 $2 \cdot 2 \cdot 5$
16. 75 $3 \cdot 5 \cdot 5$ **17.** 37 prime   **18.** 23 prime
19. 36 $2 \cdot 2 \cdot 3 \cdot 3$ **20.** 100 $2 \cdot 2 \cdot 5 \cdot 5$ **21.** 140 $2 \cdot 2 \cdot 5 \cdot 7$
22. 110 $2 \cdot 5 \cdot 11$ **23.** 79 prime   **24.** 83 prime
25. 81 $3 \cdot 3 \cdot 3 \cdot 3$ **26.** 64 $2 \cdot 2 \cdot 2 \cdot 2 \cdot 2 \cdot 2$
27. 240 $2 \cdot 2 \cdot 2 \cdot 2 \cdot 3 \cdot 5$ **28.** 360 $2 \cdot 2 \cdot 2 \cdot 3 \cdot 3 \cdot 5$

In Exercises 29–40, simplify each fraction by reducing it to its lowest terms.

29. $\frac{10}{16}$ $\frac{5}{8}$ **30.** $\frac{8}{14}$ $\frac{4}{7}$ **31.** $\frac{15}{18}$ $\frac{5}{6}$ **32.** $\frac{18}{45}$ $\frac{2}{5}$

33. $\frac{35}{50}$ $\frac{7}{10}$ **34.** $\frac{45}{50}$ $\frac{9}{10}$ **35.** $\frac{32}{80}$ $\frac{2}{5}$ **36.** $\frac{75}{80}$ $\frac{15}{16}$

37. $\frac{44}{50}$ $\frac{22}{25}$ **38.** $\frac{38}{50}$ $\frac{19}{25}$ **39.** $\frac{120}{86}$ $\frac{60}{43}$ **40.** $\frac{116}{86}$ $\frac{58}{43}$

In Exercises 41–90, perform the indicated operation. Where possible, reduce the answer to its lowest terms.

41. $\frac{2}{5} \cdot \frac{1}{3}$ $\frac{2}{15}$ **42.** $\frac{3}{7} \cdot \frac{1}{4}$ $\frac{3}{28}$ **43.** $\frac{3}{8} \cdot \frac{7}{11}$ $\frac{21}{88}$

44. $\frac{5}{8} \cdot \frac{3}{11}$ $\frac{15}{88}$ **45.** $9 \cdot \frac{4}{7}$ $\frac{36}{7}$ **46.** $8 \cdot \frac{3}{7}$ $\frac{24}{7}$

47. $\frac{1}{10} \cdot \frac{5}{6}$ $\frac{1}{12}$ **48.** $\frac{1}{8} \cdot \frac{2}{3}$ $\frac{1}{12}$ **49.** $\frac{5}{4} \cdot \frac{6}{7}$ $\frac{15}{14}$

50. $\frac{7}{4} \cdot \frac{6}{11}$ $\frac{21}{22}$ **51.** $\left(3\frac{3}{4}\right)\left(1\frac{3}{5}\right)$ 6

52. $\left(2\frac{4}{5}\right)\left(1\frac{1}{4}\right)$ $3\frac{1}{2}$ or $\frac{7}{2}$ **53.** $\frac{5}{4} \div \frac{4}{3}$ $\frac{15}{16}$

54. $\frac{7}{8} \div \frac{2}{3}$ $\frac{21}{16}$ **55.** $\frac{18}{5} \div 2$ $\frac{9}{5}$ **56.** $\frac{12}{7} \div 3$ $\frac{4}{7}$

57. $2 \div \frac{18}{5}$ $\frac{5}{9}$ **58.** $3 \div \frac{12}{7}$ $\frac{7}{4}$ **59.** $\frac{3}{4} \div \frac{1}{4}$ 3

60. $\frac{3}{7} \div \frac{1}{7}$ 3 **61.** $\frac{7}{6} \div \frac{5}{3}$ $\frac{7}{10}$ **62.** $\frac{7}{4} \div \frac{3}{8}$ $\frac{14}{3}$

63. $\frac{1}{14} \div \frac{1}{7}$ $\frac{1}{2}$ **64.** $\frac{1}{8} \div \frac{1}{4}$ $\frac{1}{2}$ **65.** $6\frac{3}{5} \div 1\frac{1}{10}$ 6

66. $1\frac{3}{4} \div 2\frac{5}{8}$ $\frac{2}{3}$ **67.** $\frac{2}{11} + \frac{4}{11}$ $\frac{6}{11}$ **68.** $\frac{5}{13} + \frac{2}{13}$ $\frac{7}{13}$

69. $\dfrac{7}{12} + \dfrac{1}{12}$ $\dfrac{2}{3}$ **70.** $\dfrac{5}{16} + \dfrac{1}{16}$ $\dfrac{3}{8}$ **71.** $\dfrac{5}{8} + \dfrac{5}{8}$ $\dfrac{5}{4}$

72. $\dfrac{3}{8} + \dfrac{3}{8}$ $\dfrac{3}{4}$ **73.** $\dfrac{7}{12} - \dfrac{5}{12}$ $\dfrac{1}{6}$ **74.** $\dfrac{13}{18} - \dfrac{5}{18}$ $\dfrac{4}{9}$

75. $\dfrac{16}{7} - \dfrac{2}{7}$ 2 **76.** $\dfrac{17}{5} - \dfrac{2}{5}$ 3 **77.** $\dfrac{1}{2} + \dfrac{1}{5}$ $\dfrac{7}{10}$

78. $\dfrac{1}{3} + \dfrac{1}{5}$ $\dfrac{8}{15}$ **79.** $\dfrac{3}{4} + \dfrac{3}{20}$ $\dfrac{9}{10}$ **80.** $\dfrac{2}{5} + \dfrac{2}{15}$ $\dfrac{8}{15}$

81. $\dfrac{3}{8} + \dfrac{5}{12}$ $\dfrac{19}{24}$ **82.** $\dfrac{3}{10} + \dfrac{2}{15}$ $\dfrac{13}{30}$ **83.** $\dfrac{11}{18} - \dfrac{2}{9}$ $\dfrac{7}{18}$

84. $\dfrac{17}{18} - \dfrac{4}{9}$ $\dfrac{1}{2}$ **85.** $\dfrac{4}{3} - \dfrac{3}{4}$ $\dfrac{7}{12}$ **86.** $\dfrac{3}{2} - \dfrac{2}{3}$ $\dfrac{5}{6}$

87. $\dfrac{7}{10} - \dfrac{3}{16}$ $\dfrac{41}{80}$ **88.** $\dfrac{7}{30} - \dfrac{5}{24}$ $\dfrac{1}{40}$

89. $3\dfrac{3}{4} - 2\dfrac{1}{3}$ $1\dfrac{5}{12}$ or $\dfrac{17}{12}$ **90.** $3\dfrac{2}{3} - 2\dfrac{1}{2}$ $1\dfrac{1}{6}$ or $\dfrac{7}{6}$

Practice Plus

In Exercises 91–94, perform the indicated operation. Write the answer as a variable expression.

91. $\dfrac{3}{4} \cdot \dfrac{a}{5}$ $\dfrac{3 \cdot a}{20}$ **92.** $\dfrac{2}{3} \div \dfrac{a}{7}$ $\dfrac{14}{3 \cdot a}$

93. $\dfrac{11}{x} + \dfrac{9}{x}$ $\dfrac{20}{x}$ **94.** $\dfrac{10}{y} - \dfrac{6}{y}$ $\dfrac{4}{y}$

In Exercises 95–96, perform the indicated operations. Begin by performing operations in parentheses.

95. $\left(\dfrac{1}{2} - \dfrac{1}{3}\right) \div \dfrac{5}{8}$ $\dfrac{4}{15}$ **96.** $\left(\dfrac{1}{2} + \dfrac{1}{4}\right) \div \left(\dfrac{1}{2} + \dfrac{1}{3}\right)$ $\dfrac{9}{10}$

Application Exercises

97. The circle graph shows the results of a survey in which American adults were asked the following question:

> If you got conflicting or different reports of the same news story from television, newspapers, radio, or magazines, which would you be most inclined to believe?

The circle graph shows the fraction of American adults finding each medium the most believable.

Most Believable News Source

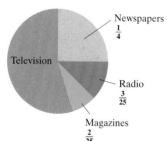

Newspapers $\dfrac{1}{4}$

Television

Radio $\dfrac{3}{25}$

Magazines $\dfrac{2}{25}$

Source: Thomas R. Dye, *Who's Running America?* *Seventh Edition,* Prentice Hall, 2002

a. What fraction of those surveyed found television the most believable news source? $\dfrac{11}{20}$

b. If 2000 people were surveyed, how many found radio the most believable news source? 240

98. The circle graph shows the results of one of the questions in a telephone poll taken for *Time*/CNN during 2002's summer of corporate scandals.

Which of the following best describes your view of the recent accounting scandals involving major American corporations?

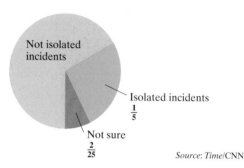

Not isolated incidents

Isolated incidents $\dfrac{1}{5}$

Not sure $\dfrac{2}{25}$

*Source: Time/*CNN

a. What fraction of those surveyed replied that the accounting scandals were not isolated incidents? $\dfrac{18}{25}$

b. If 2000 people were surveyed, how many were not sure if the accounting scandals were isolated incidents? 160

The ratio of a to b can be expressed by the fraction $\frac{a}{b}$. In Exercises 99–100, use the natural numbers shown in Figure 1.1 on page 2 to find the indicated ratio. Reduce each fraction to its lowest terms.

99. the ratio of Hispanic judges to female judges appointed by Carter $\dfrac{2}{5}$

100. the ratio of African-American judges to female judges appointed by Reagan $\dfrac{1}{4}$

A common application of fractions involves preparing food for a different number of servings than what the recipe calls for. The amount of each ingredient can be found as follows:

Amount of ingredient needed

$$= \frac{desired\ serving\ size}{recipe\ serving\ size} \times ingredient\ amount\ in\ the\ recipe.$$

Use this information to solve Exercises 101–102. Give answers as mixed numbers.

101. A chocolate-chip cookie recipe for five dozen cookies requires $\frac{3}{4}$ cup sugar. If you want to make eight dozen cookies, how much sugar is needed? $1\frac{1}{5}$ cups

102. A mix for eight servings of instant potatoes requires $2\frac{2}{3}$ cups of water. If you want to make six servings, how much water is needed? 2 cups

103. If you walk $\frac{3}{4}$ mile and then jog $\frac{2}{5}$ mile, what is the total distance covered? How much farther did you walk than jog? $\frac{23}{20}$ miles; $\frac{7}{20}$ mile farther

104. Some companies pay people extra when they work more than a regular 40-hour work week. The overtime pay is often $1\frac{1}{2}$ times the regular hourly rate. This is called *time and a half.* A summer job for students pays $12 per hour and offers time and a half for the hours worked over 40. If a student works 46 hours during one week, what is the student's total pay before taxes? $588.00

105. The legend of a map indicates that 1 inch = 16 miles. If the distance on the map between two cities is $2\frac{3}{8}$ inches, how far apart are the cities?   38 miles

Writing in Mathematics

Writing about mathematics will help you to learn mathematics. For all writing exercises in this book, use complete sentences to respond to the questions. Some writing exercises can be answered in a sentence; others require a paragraph or two. You can decide how much you need to write as long as your writing clearly and directly answers the question in the exercise. Standard references such as a dictionary and a thesaurus should be helpful.

106. Explain how to convert a mixed number to an improper fraction and give an example.   106–116. Answers will vary.

107. Explain how to convert an improper fraction to a mixed number and give an example.

108. Describe the difference between a prime number and a composite number.

109. What is meant by the prime factorization of a composite number?

110. What is the Fundamental Principle of Fractions?

111. Explain how to reduce a fraction to its lowest terms. Give an example with your explanation.

112. Explain how to multiply fractions and give an example.

113. Explain how to divide fractions and give an example.

114. Describe how to add or subtract fractions with identical denominators. Provide an example with your description.

115. Explain how to add fractions with different denominators. Use $\frac{5}{6} + \frac{1}{2}$ as an example.

116. Explain what is wrong with this statement. "If you'd like to save some money, I'll be happy to sell you my computer system for only $\frac{3}{2}$ of the price I originally paid for it."

Critical Thinking Exercises

117. Which one of the following is true?   d

a. $\frac{1}{2} + \frac{1}{5} = \frac{2}{7}$ **b.** $\frac{2+6}{2} = \frac{\cancel{2}+6}{\cancel{2}} = 6$ **c.** $\frac{1}{2} \div 4 = 2$

d. Every fraction has infinitely many equivalent fractions.

118. Shown below is a short excerpt from "The Star-Spangled Banner." The time is $\frac{3}{4}$, which means that each measure must contain notes that add up to $\frac{3}{4}$. The values of the different notes tell musicians how long to hold each note.

Use vertical lines to divide this line of "The Star-Spangled Banner" into measures.

say does that   Star-span-gled   Ban-ner   yet   wave   O'er the

See graphing answer section.

SECTION 1.2

Objectives

1 Define the sets that make up the real numbers.

2 Graph numbers on a number line.

3 Express rational numbers as decimals.

4 Classify numbers as belonging to one or more sets of the real numbers.

5 Understand and use inequality symbols.

6 Find the absolute value of a real number.

THE REAL NUMBERS

The U.N. building is designed with three golden rectangles.

The United Nations Building in New York was designed to represent its mission of promoting world harmony. Viewed from the front, the building looks like three rectangles stacked upon each other. In each rectangle, the ratio of the width to height is $\sqrt{5} + 1$ to 2, approximately 1.618 to 1. The ancient Greeks believed that such a rectangle, called a **golden rectangle**, was the most visually pleasing of all rectangles.

The ratio 1.618 to 1 is approximate because $\sqrt{5}$ is an irrational number, a special kind of real number. Irrational? Real? Let's make sense of all this by describing the kinds of numbers you will encounter in this course.

1 Define the sets that make up the real numbers.

Natural Numbers and Whole Numbers Before we describe the set of real numbers, let's be sure you are familiar with some basic ideas about sets. A **set** is a collection of objects whose contents can be clearly determined. The objects in a set are called the **elements** of the set. For example, the set of numbers used for counting can be represented by

$$\{1, 2, 3, 4, 5, \ldots\}.$$

The braces, $\{\}$, indicate that we are representing a set. This form of representing a set uses commas to separate the elements of the set. Remember that the three dots after the 5 indicate that there is no final element and that the listing goes on forever.

We have seen that the set of numbers used for counting is called the set of **natural numbers**. When we combine the number 0 with the natural numbers, we obtain the set of **whole numbers**.

NATURAL NUMBERS AND WHOLE NUMBERS

The set of **natural numbers** is $\{1, 2, 3, 4, 5, \ldots\}$.

The set of **whole numbers** is $\{0, 1, 2, 3, 4, 5, \ldots\}$.

Integers and the Number Line The whole numbers do not allow us to describe certain everyday situations. For example, if the balance in your checking account is $30 and you write a check for $35, your checking account is overdrawn by $5. We can write this as -5, read *negative* 5. The set consisting of the natural numbers, 0, and the negatives of the natural numbers is called the set of **integers**.

INTEGERS The set of **integers** is

$$\{\ldots, -4, -3, -2, -1,\ 0,\ 1, 2, 3, 4, \ldots\}.$$

$\qquad\qquad$ Negative integers $\qquad$ Positive integers

Notice that the term **positive integers** is another name for the natural numbers. The positive integers can be written in two ways:

1. Use a "+" sign. For example, $+4$ is "positive four."

2. Do not write any sign. For example, 4 is assumed to be "positive four."

EXAMPLE 1 Practical Examples of Negative Integers

Write a negative integer that describes each of the following situations:

a. A debt of $10

b. The shore surrounding the Dead Sea is 1312 feet below sea level.

SOLUTION

a. A debt of $10 can be expressed by the negative integer -10 (negative ten).

b. The shore surrounding the Dead Sea is 1312 feet below sea level, expressed as -1312. ∎

 CHECK POINT 1 Write a negative integer that describes each of the following situations:

a. A debt of $500 $\quad -500$

b. Death Valley, the lowest point in North America, is 282 feet below sea level. $\quad -282$

2 Graph numbers on a number line.

The **number line** is a graph we use to visualize the set of integers, as well as sets of other numbers. The number line is shown in Figure 1.8.

FIGURE 1.8   The number line

The number line extends indefinitely in both directions, shown by the arrows on the left and the right. Zero separates the positive numbers from the negative numbers on the number line. The positive integers are located to the right of 0, and the negative integers are located to the left of 0. Zero is neither positive nor negative. For every positive integer on a number line, there is a corresponding negative integer on the opposite side of 0.

Integers are graphed on a number line by placing a dot at the correct location for each number.

EXAMPLE 2 Graphing Integers on a Number Line

Graph: **a.** −3   **b.** 4   **c.** 0.

SOLUTION   Place a dot at the correct location for each integer.

✔ CHECK POINT 2 Graph: **a.** −4   **b.** 0   **c.** 3.   See graphing answer section.

Rational Numbers   If two integers are added, subtracted, or multiplied, the result is always another integer. This, however, is not always the case with division. For example, 10 divided by 5 is the integer 2. By contrast, 5 divided by 10 is $\frac{1}{2}$, and $\frac{1}{2}$ is not an integer. To permit divisions such as $\frac{5}{10}$, we enlarge the set of integers, calling the new collection the *rational numbers*. The set of **rational numbers** consists of all the numbers that can be expressed as a quotient of two integers, with the denominator not 0.

THE RATIONAL NUMBERS   The set of **rational numbers** is the set of all numbers that can be expressed in the form $\frac{a}{b}$, where a and b are integers and b is not equal to 0, written $b \neq 0$. The integer a is called the **numerator** and the integer b is called the **denominator**.

Here are two examples of rational numbers:

- $\frac{1}{2}$ a, the integer forming the numerator, is 1.

  b, the integer forming the denominator, is 2.

- $\frac{-3}{4}$ a, the integer forming the numerator, is −3.

  b, the integer forming the denominator, is 4.

Is the integer 5 another example of a rational number? Yes. The integer 5 can be written with a denominator of 1.

$$5 = \frac{5}{1}$$ a, the integer forming the numerator, is 5.

b, the integer forming the denominator, is 1.

All integers are also rational numbers because they can be written with a denominator of 1.

How can we express a negative mixed number, such as $-2\frac{3}{4}$, in the form $\frac{a}{b}$? Copy the negative sign and then follow the procedure discussed in the previous section:

$$-2\frac{3}{4} = -\frac{4 \cdot 2 + 3}{4} = -\frac{8 + 3}{4} = -\frac{11}{4} = \frac{-11}{4}.$$

$a = -11$

$b = 4$

Copy the negative sign from step to step and convert $2\frac{3}{4}$ to an improper fraction.

Rational numbers are graphed on a number line by placing a dot at the correct location for each number.

EXAMPLE 3 Graphing Rational Numbers on a Number Line

Graph: **a.** $\frac{7}{2}$ **b.** -4.6.

SOLUTION   Place a dot at the correct location for each rational number.

a. Because $\frac{7}{2} = 3\frac{1}{2}$, its graph is midway between 3 and 4.

b. Because $-4.6 = -4\frac{6}{10}$, its graph is $\frac{6}{10}$ of a unit to the left of -4.

CHECK POINT 3 Graph: **a.** $\frac{9}{2}$ **b.** -1.2.   See graphing answer section.

3 Express rational numbers as decimals.

Every rational number can be expressed as a fraction or a decimal. To express the fraction $\frac{a}{b}$ as a decimal, divide the denominator, b, into the numerator, a.

EXAMPLE 4 Expressing Rational Numbers as Decimals

Express each rational number as a decimal: **a.** $\frac{5}{8}$ **b.** $\frac{7}{11}$.

SOLUTION   In each case, divide the denominator into the numerator.

a.
```
   0.625
8)5.000
  4 8
   20
   16
    40
    40
     0
```

b.
```
    0.6363...
11)7.0000...
   6 6
    40
    33
    70
    66
    40
    33
    70
     ⋮
```

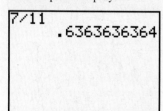
In Example 4, the decimal for $\frac{5}{8}$, namely 0.625, stops and is called a **terminating decimal**. Other examples of terminating decimals are

$$\frac{1}{4} = 0.25, \quad \frac{2}{5} = 0.4, \quad \text{and} \quad \frac{7}{8} = 0.875.$$

By contrast, the division process for $\frac{7}{11}$ results in $0.6363\ldots$, with the digits 63 repeating over and over indefinitely. To indicate this, write a bar over the digits that repeat. Thus,

$$\frac{7}{11} = 0.\overline{63}.$$

The decimal for $\frac{7}{11}$, $0.\overline{63}$, is called a **repeating decimal**. Other examples of repeating decimals are

$$\frac{1}{3} = 0.333\ldots = 0.\overline{3} \quad \text{and} \quad \frac{2}{3} = 0.666\ldots = 0.\overline{6}.$$

RATIONAL NUMBERS AND DECIMALS   Any rational number can be expressed as a decimal. The resulting decimal will either terminate (stop), or it will have a digit that repeats or a block of digits that repeat.

USING TECHNOLOGY

You can obtain decimal approximations for irrational numbers using a calculator. For example, to approximate $\sqrt{2}$, use the following keystrokes:

Many Scientific Calculators

$2 \boxed{\sqrt{}}$ or $2 \boxed{\begin{array}{c}\text{2nd}\\ \text{INV}\end{array}} \boxed{x^2}$

Many Graphing Calculators

$\boxed{\sqrt{}} \; 2 \; \boxed{\text{ENTER}}$

The display may read 1.41421356237, although your calculator may display more or fewer digits. Between which two integers would you graph $\sqrt{2}$ on a number line?

✔ **CHECK POINT 4** Express each rational number as a decimal:

a. $\frac{3}{8}$ 0.375 **b.** $\frac{5}{11}$. $0.\overline{45}$

Irrational Numbers   Can you think of a number that, when written in decimal form, neither terminates nor repeats? An example of such a number is $\sqrt{2}$ (read: "the square root of 2"). The number $\sqrt{2}$ is a number that can be multiplied by itself to obtain 2. No terminating or repeating decimal can be multiplied by itself to get 2. However, some approximations come close to 2.

• 1.4 is an approximation of $\sqrt{2}$:
$$1.4 \times 1.4 = 1.96.$$

• 1.41 is an approximation of $\sqrt{2}$:
$$1.41 \times 1.41 = 1.9881.$$

• 1.4142 is an approximation of $\sqrt{2}$:
$$1.4142 \times 1.4142 = 1.99996164.$$

Can you see how each approximation in the list is getting better? This is because the products are getting closer and closer to 2.

The number $\sqrt{2}$, whose decimal representation does not come to an end and does not have a block of repeating digits, is an example of an **irrational number**.

ENRICHMENT ESSAY

The Best and Worst of π

In 1999, two Japanese mathematicians used two different computer programs to calculate π to over 206 billion digits. The calculations took the computer 43 hours!

The most inaccurate version of π came from the 1897 General Assembly of Indiana. Bill No. 246 stated that "π was by law 4."

THE IRRATIONAL NUMBERS   Any number that can be represented on the number line that is not a rational number is called an **irrational number**. Thus, the set of irrational numbers is the set of numbers whose decimal representations are neither terminating nor repeating.

Perhaps the best known of all the irrational numbers is π (pi). This irrational number represents the distance around a circle (its circumference) divided by the diameter of the circle. In the *Star Trek* episode "Wolf in the Fold," Spock foils an evil computer by telling it to "compute the last digit in the value of π." Because π is an irrational number, there is no last digit in its decimal representation:

$$\pi = 3.1415926535897932384626433832795\ldots.$$

Because irrational numbers cannot be represented by decimals that come to an end, mathematicians use symbols such as $\sqrt{2}$, $\sqrt{3}$, and π to represent these numbers. However, **not all square roots are irrational**. For example, $\sqrt{25} = 5$ because 5 multiplied by itself is 25. Thus, $\sqrt{25}$ is a natural number, a whole number, an integer, and a rational number $\left(\sqrt{25} = \frac{5}{1}\right)$.

The Set of Real Numbers All numbers that can be represented by points on the number line are called **real numbers**. Thus, the set of real numbers is formed by combining the rational numbers and the irrational numbers. Every real number is either rational or irrational.

The sets that make up the real numbers are summarized in Table 1.1. Notice the use of the symbol $\approx$ in the examples of irrational numbers. The symbol $\approx$ means "is approximately equal to."

4 Classify numbers as belonging to one or more sets of the real numbers.

Real numbers

Rational numbers	Irrational numbers
Integers	
Whole numbers	
Natural numbers	

This diagram shows that every real number is rational or irrational.

Table 1.1	The Sets that Make Up the Real Numbers	
Name	**Description**	**Examples**
Natural numbers	$\{1, 2, 3, 4, 5, \dots\}$ These numbers are used for counting.	$2, 3, 5, 17$
Whole numbers	$\{0, 1, 2, 3, 4, 5, \dots\}$ The set of whole numbers is formed by adding 0 to the set of natural numbers.	$0, 2, 3, 5, 17$
Integers	$\{\dots, -5, -4, -3, -2, -1, 0, 1, 2, 3, 4, 5, \dots\}$ The set of integers is formed by adding negatives of the natural numbers to the set of whole numbers.	$-17, -5, -3, -2, 0,$ $2, 3, 5, 17$
Rational numbers	The set of rational numbers is the set of all numbers that can be expressed in the form $\frac{a}{b}$, where a and b are integers and b is not equal to 0, written $b \neq 0$. Rational numbers can be expressed as terminating or repeating decimals.	$-17 = \frac{-17}{1}, -5 = \frac{-5}{1}, -3, -2,$ $0, 2, 3, 5, 17,$ $\frac{2}{5} = 0.4,$ $\frac{-2}{3} = -0.6666\cdots = -0.\overline{6}$
Irrational numbers	The set of irrational numbers is the set of all numbers whose decimal representations are neither terminating nor repeating. Irrational numbers cannot be expressed as a quotient of integers.	$\sqrt{2} \approx 1.414214$ $-\sqrt{3} \approx -1.73205$ $\pi \approx 3.142$ $-\frac{\pi}{2} \approx -1.571$

EXAMPLE 5 Classifying Real Numbers

Consider the following set of numbers:

$$\left\{ -7, -\frac{3}{4}, 0, 0.\overline{6}, \sqrt{5}, \pi, 7.3, \sqrt{81} \right\}.$$

List the numbers in the set that are

   a. natural numbers.   **b.** whole numbers.   **c.** integers.
   d. rational numbers.   **e.** irrational numbers.   **f.** real numbers.

SOLUTION

   a. Natural numbers: The natural numbers are the numbers used for counting. The only natural number in the set is $\sqrt{81}$ because $\sqrt{81} = 9$. (9 multiplied by itself is 81.)

   b. Whole numbers: The whole numbers consist of the natural numbers and 0. The elements of the set that are whole numbers are 0 and $\sqrt{81}$.

c. Integers: The integers consist of the natural numbers, 0, and the negatives of the natural numbers. The elements of the set that are integers are $\sqrt{81}$, 0, and -7.

d. Rational numbers: All numbers in the set that can be expressed as the quotient of integers are rational numbers. These include $-7\left(-7 = \frac{-7}{1}\right)$, $-\frac{3}{4}$, $0\left(0 = \frac{0}{1}\right)$, and $\sqrt{81}\left(\sqrt{81} = \frac{9}{1}\right)$. Furthermore, all numbers in the set that are terminating or repeating decimals are also rational numbers. These include $0.\overline{6}$ and 7.3.

e. Irrational numbers: The irrational numbers in the set are $\sqrt{5}(\sqrt{5} \approx 2.236)$ and $\pi(\pi \approx 3.14)$. Both $\sqrt{5}$ and π are only approximately equal to 2.236 and 3.14, respectively. In decimal form, $\sqrt{5}$ and π neither terminate nor have blocks of repeating digits.

f. Real numbers: All the numbers in the given set are real numbers. ■

 CHECK POINT 5 Consider the following set of numbers:

$$\left\{-9, -1.3, 0, 0.\overline{3}, \frac{\pi}{2}, \sqrt{9}, \sqrt{10}\right\}.$$

List the numbers in the set that are

a. natural numbers. $\sqrt{9}$ b. whole numbers. $0, \sqrt{9}$

c. integers. $-9, 0, \sqrt{9}$ d. rational numbers. $-9, -1.3, 0, 0.\overline{3}, \sqrt{9}$

e. irrational numbers. $\frac{\pi}{2}, \sqrt{10}$ f. real numbers. $-9, -1.3, 0, 0.\overline{3}, \frac{\pi}{2}, \sqrt{9}, \sqrt{10}$

5 Understand and use inequality symbols.

FIGURE 1.9

Ordering the Real Numbers On the real number line, the real numbers increase from left to right. The lesser of two real numbers is the one farther to the left on a number line. The greater of two real numbers is the one farther to the right on a number line.

Look at the number line in Figure 1.9. The integers 2 and 5 are graphed. Observe that 2 is to the left of 5 on the number line. This means that 2 is less than 5:

$2 < 5$: 2 is less than 5 because 2 is to the *left* of 5 on the number line.

In Figure 1.9, we can also observe that 5 is to the right of 2 on the number line. This means that 5 is greater than 2.

$5 > 2$: 5 is greater than 2 because 5 is to the *right* of 2 on the number line.

The symbols $<$ and $>$ are called **inequality symbols**. These symbols always point to the lesser of the two real numbers when the inequality is true.

> 2 is less than 5.

> 5 is greater than 2.

$2 < 5$ *The symbol points to 2, the lesser number.*

$5 > 2$ *The symbol points to 2, the lesser number.*

EXAMPLE 6 Using Inequality Symbols

Insert either $<$ or $>$ in the shaded area between each pair of numbers to make a true statement:

a. 3   17   b. -4.5 1.2   c. -5 -83 d. $\frac{4}{5}$ $\frac{2}{3}$.

SOLUTION In each case, mentally compare the graph of the first number to the graph of the second number. If the first number is to the left of the second number, insert the symbol $<$ for "is less than." If the first number is to the right of the second number, insert the symbol $>$ for "is greater than."

a. Compare the graphs of 3 and 17 on the number line. Because 3 is to the left of 17, this means that 3 is less than 17: $3 < 17$.

b. Compare the graphs of -4.5 and 1.2. Because -4.5 is to the left of 1.2, this means that -4.5 is less than 1.2: $-4.5 < 1.2$.

c. Compare the graphs of -5 and -83. Because -5 is to the right of -83, this means that -5 is greater than -83: $-5 > -83$.

d. Compare the graphs of $\frac{4}{5}$ and $\frac{2}{3}$. To do so, convert to decimal notation or use a common denominator. Using decimal notation, $\frac{4}{5} = 0.8$ and $\frac{2}{3} = 0.\overline{6}$. Because 0.8 is to the right of $0.\overline{6}$, this means that $\frac{4}{5}$ is greater than $\frac{2}{3}$: $\frac{4}{5} > \frac{2}{3}$. ■

 CHECK POINT 6 Insert either $<$ or $>$ in the shaded area between each pair of numbers to make a true statement:

a. 14 5 $>$ **b.** -5.4 2.3 $<$ **c.** -19 -6 $<$ **d.** $\frac{1}{4}$ $\frac{1}{2}$. $<$

The symbols $<$ and $>$ may be combined with an equal sign, as shown in the table.

Symbols	Meaning	Example	Explanation
$a \leq b$	a is less than or equal to b.	$3 \leq 7$ $7 \leq 7$	Because $3 < 7$ Because $7 = 7$
$b \geq a$	b is greater than or equal to a.	$7 \geq 3$ $-5 \geq -5$	Because $7 > 3$ Because $-5 = -5$

When using the symbol $\leq$ (is less than or equal to), the inequality is a true statement if either the $<$ part or the $=$ part is true. When using the symbol $\geq$ (is greater than or equal to), the inequality is a true statement if either the $>$ part or the $=$ part is true.

EXAMPLE 7   Using Inequality Symbols

Determine whether each inequality is true or false:

a. $-7 \leq 4$ **b.** $-7 \leq -7$ **c.** $-9 \geq 6$.

SOLUTION

a. $-7 \leq 4$ is true because $-7 < 4$ is true.

b. $-7 \leq -7$ is true because $-7 = -7$ is true.

c. $-9 \geq 6$ is false because neither $-9 > 6$ nor $-9 = 6$ is true. ■

 CHECK POINT 7 Determine whether each inequality is true or false:

a. $-2 \leq 3$ true   **b.** $-2 \geq -2$ true   **c.** $-4 \geq 1$.   false

6 Find the absolute value of a real number.

Absolute Value   Suppose you are feeling a bit lazy. Instead of your usual jog, you decide to use your finger to stroll along the number line. You start at 0 and end the finger walk at -5. You have covered a distance of 5 units.

Absolute value describes distance from 0 on a number line. If a represents a real number, the symbol $|a|$ represents its absolute value, read "the absolute value of a." In terms of your walk, we can write

$$|-5| = 5.$$

The absolute value of -5 is 5 because -5 is 5 units from 0 on the number line.

ABSOLUTE VALUE   The absolute value of a real number a, denoted by $|a|$, is the distance from 0 to a on the number line. Because absolute value describes a distance, it is never negative.

EXAMPLE 8 Finding Absolute Value

Find the absolute value: **a.** $|-3|$ **b.** $|5|$ **c.** $|0|$.

SOLUTION The solution is illustrated in Figure 1.10.

a. $|-3| = 3$ *The absolute value of -3 is 3 because -3 is 3 units from 0.*

b. $|5| = 5$ *5 is 5 units from 0.*

c. $|0| = 0$ *0 is 0 units from itself.*

FIGURE 1.10 Absolute value describes distance from 0 on a number line.

Can you see that the absolute value of a real number is either positive or zero? Zero is the only real number whose absolute value is 0: $|0| = 0$. **The absolute value of any other real number is always positive**.

✔ **CHECK POINT 8** Find the absolute value:
a. $|-4|$ 4   **b.** $|6|$ 6   **c.** $|-\sqrt{2}|$. $\sqrt{2}$

1.2 EXERCISE SET

Student Solutions Manual   CD/Video   PH Math/Tutor Center   MathXL Tutorials on CD   MathXL®   MyMathLab   Interactmath.com

Practice Exercises

In Exercises 1–8, write a positive or negative integer that describes each situation.

1. Meteorology: 20° below zero -20
2. Navigation: 65 feet above sea level   65
3. Health: A gain of 8 pounds   8
4. Economics: A loss of $12,500.00 $-12,500$
5. Banking: A withdrawal of $3000.00 -3000
6. Physics: An automobile slowing down at a rate of 3 meters per second each second. -3
7. Economics: A budget deficit of 4 billion dollars -4 billion
8. Football: A 14-yard loss -14

In Exercises 9–20, start by drawing a number line that shows integers from -5 to 5. Then graph each real number on your number line. **9–20.** See graphing answer section.

9. 2   10. 5   11. -5 12. -2 13. $3\frac{1}{2}$ 14. $2\frac{1}{4}$

15. $\frac{11}{3}$ 16. $\frac{7}{3}$ 17. -1.8 18. -3.4 19. $-\frac{16}{5}$ 20. $-\frac{11}{5}$

In Exercises 21–32, express each rational number as a decimal.

21. $\frac{3}{4}$ 0.75   22. $\frac{3}{5}$ 0.6   23. $\frac{7}{20}$ 0.35

24. $\frac{3}{20}$ 0.15   25. $\frac{7}{8}$ 0.875   26. $\frac{5}{16}$ 0.3125

27. $\frac{9}{11}$ $0.\overline{81}$ 28. $\frac{3}{11}$ $0.\overline{27}$ 29. $-\frac{1}{2}$ -0.5

30. $-\frac{1}{4}$ -0.25 31. $-\frac{5}{6}$ $-0.8\overline{3}$ 32. $-\frac{7}{6}$ $-1.1\overline{6}$

In Exercises 33–36, list all numbers from the given set that are:
a. *natural numbers,* **b.** *whole numbers,* **c.** *integers,* **d.** *rational numbers,* **e.** *irrational numbers,* **f.** *real numbers.*

33. $\left\{-9, -\frac{4}{5}, 0, 0.25, \sqrt{3}, 9.2, \sqrt{100}\right\}$ **a.** $\sqrt{100}$ **b.** 0, $\sqrt{100}$
    c. $-9, 0, \sqrt{100}$ **d.** $-9, -\frac{4}{5}, 0, 0.25, 9.2, \sqrt{100}$ **e.** $\sqrt{3}$
    f. $-9, -\frac{4}{5}, 0, 0.25, \sqrt{3}, 9.2, \sqrt{100}$

34. $\left\{-7, -0.\overline{6}, 0, \sqrt{49}, \sqrt{50}\right\}$ **a.** $\sqrt{49}$ **b.** 0, $\sqrt{49}$
    c. $-7, 0, \sqrt{49}$ **d.** $-7, -0.\overline{6}, 0, \sqrt{49}$ **e.** $\sqrt{50}$
    f. $-7, -0.\overline{6}, 0, \sqrt{49}, \sqrt{50}$

35. $\left\{-11, -\frac{5}{6}, 0, 0.75, \sqrt{5}, \pi, \sqrt{64}\right\}$ **a.** $\sqrt{64}$ **b.** 0, $\sqrt{64}$
    c. $-11, 0, \sqrt{64}$ **d.** $-11, -\frac{5}{6}, 0, 0.75, \sqrt{64}$ **e.** $\sqrt{5}, \pi$
    f. $-11, -\frac{5}{6}, 0, 0.75, \sqrt{5}, \pi, \sqrt{64}$

36. $\left\{-5, -0.\overline{3}, 0, \sqrt{2}, \sqrt{4}\right\}$ **a.** $\sqrt{4}$ **b.** 0, $\sqrt{4}$ **c.** $-5, 0, \sqrt{4}$
    d. $-5, -0.\overline{3}, 0, \sqrt{4}$ **e.** $\sqrt{2}$ **f.** $-5, -0.\overline{3}, 0, \sqrt{2}, \sqrt{4}$

37. Give an example of a whole number that is not a natural number.   0
38. Give an example of an integer that is not a whole number.
    Answers will vary; example is -3.
39. Give an example of a rational number that is not an integer.
    Answers will vary; example is $\frac{1}{2}$.
40. Give an example of a rational number that is not a natural number.   Answers will vary; example is -5.
41. Give an example of a number that is an integer, a whole number, and a natural number.   Answers will vary; example is 6.

42. Give an example of a number that is a rational number, an integer, and a real number.   Answers will vary; example is 5.

43. Give an example of a number that is an irrational number and a real number.   Answers will vary; example is π.

44. Give an example of a number that is a real number, but not an irrational number.   Answers will vary; example is $\sqrt{4}$.

In Exercises 45–62, insert either $<$ or $>$ in the shaded area between each pair of numbers to make a true statement.

45. $\dfrac{1}{2}$ 2 $<$

46. 4 -3 $>$

47. 3 $-\dfrac{5}{2}$ $>$

48. 3 $\dfrac{3}{2}$ $>$

49. -4 -6 $>$

50. $-\dfrac{5}{2}$ $-\dfrac{5}{3}$ $<$

51. -2.5 1.5 $<$

52. -1.25 -0.5 $<$

53. $-\dfrac{3}{4}$ $-\dfrac{5}{4}$ $>$

54. 0 $-\dfrac{1}{2}$ $>$

55. -4.5 3 $<$

56. -5.5 2.5 $<$

57. $\sqrt{2}$ 1.5 $<$

58. $\sqrt{3}$ 2 $<$

59. $0.\overline{3}$ 0.3 $>$

60. 0.6 $0.\overline{6}$ $<$

61. $-\pi$ -3.5 $>$

62. $-\dfrac{\pi}{2}$ -2.3 $>$

In Exercises 63–70, determine whether each inequality is true or false.

63. $-5 \ge -13$ true

64. $-5 \le -8$ false

65. $-9 \ge -9$ true

66. $-14 \le -14$ true

67. $0 \ge -6$ true

68. $0 \ge -13$ true

69. $-17 \ge 6$ false

70. $-14 \ge 8$ false

In Exercises 71–78, find each absolute value.

71. $|6|$ 6

72. $|3|$ 3

73. $|-7|$ 7

74. $|-9|$ 9

75. $\left|\dfrac{5}{6}\right|$ $\dfrac{5}{6}$

76. $\left|\dfrac{4}{5}\right|$ $\dfrac{4}{5}$

77. $|-\sqrt{11}|$ $\sqrt{11}$

78. $|-\sqrt{29}|$ $\sqrt{29}$

Practice Plus

In Exercises 79–86, insert either $<$, $>$, or $=$ in the shaded area to make a true statement.

79. $|-6|$ $|-3|$ $>$

80. $|-20|$ $|-50|$ $<$

81. $\left|\dfrac{3}{5}\right|$ $|-0.6|$ $=$

82. $\left|\dfrac{5}{2}\right|$ $|-2.5|$ $=$

83. $\dfrac{30}{40} - \dfrac{3}{4}$ $\dfrac{14}{15} \cdot \dfrac{15}{14}$ $<$

84. $\dfrac{17}{18} \cdot \dfrac{18}{17}$ $\dfrac{50}{60} - \dfrac{5}{6}$ $>$

85. $\dfrac{8}{13} \div \dfrac{8}{13}$ $|-1|$ $=$

86. $|-2|$ $\dfrac{4}{17} \div \dfrac{4}{17}$ $>$

Application Exercises

Temperatures sometimes fall below zero. A combination of low temperature and wind makes it feel colder than the actual temperature. The table shows how cold it feels when low temperatures are combined with different wind speeds.

Windchill											
Wind (mph)	Temperature (°F)										
	35	30	25	20	15	10	5	0	-5	-10	-15	-20
5	31	25	19	13	7	-1	-5	-11	-16	-22	-28	-34
10	27	21	15	9	3	-4	-10	-16	-22	-28	-35	-41
15	25	19	13	6	0	-7	-13	-19	-26	-32	-39	-45
20	24	17	11	4	-2	-9	-15	-22	-29	-35	-42	-48
25	23	16	9	3	-4	-11	-17	-24	-31	-37	-44	-51

Source: National Weather Service

Use the information from the table to solve Exercises 87–88.

87. Write a negative integer that indicates how cold the temperature feels when the temperature is 15° Fahrenheit and the wind is blowing at 20 miles per hour. -2

88. Write a negative integer that indicates how cold the temperature feels when the temperature is 10° Fahrenheit and the wind is blowing at 15 miles per hour. -7

The following table shows the amount of money, in millions of dollars, collected and spent by the U.S. government from 1997 through 2002.

Year	Money Collected	Money Spent
1997	1,579,300	1,601,200
1998	1,721,800	1,652,600
1999	1,827,500	1,703,000
2000	2,025,200	1,788,800
2001	1,991,000	1,863,900
2002	1,946,100	2,052,300

Money is expressed in millions of dollars.
Source: Office of Management and Budget

Use the information from the table to solve Exercises 89–90.

89. List the years for which money collected < money spent. Was there a budget surplus or deficit in these years? *1997, 2002; budget deficit*

90. List the years for which money collected > money spent. Was there a budget surplus or deficit in these years? *1998, 1999, 2000, 2001; budget surplus*

Writing in Mathematics

Writing about mathematics will help you to learn mathematics. For all writing exercises in this book, use complete sentences to respond to the questions. Some writing exercises can be answered in a sentence; others require a paragraph or two. You can decide how much you need to write as long as your writing clearly and directly answers the question in the exercise. Standard references such as a dictionary and a thesaurus should be helpful.

91. What is a set? *91–102. Answers will vary.*

92. What are the natural numbers?

93. What are the whole numbers?

94. What are the integers?

95. How does the set of integers differ from the set of whole numbers?

96. Describe how to graph a number on the number line.

97. What is a rational number?

98. Explain how to express $\frac{3}{8}$ as a decimal.

99. Describe the difference between a rational number and an irrational number.

100. If you are given two real numbers, explain how to determine which one is the lesser.

101. Describe what is meant by the absolute value of a number. Give an example with your explanation.

102. Give an example of an everyday situation that can be described using integers but not using whole numbers.

Critical Thinking Exercises

103. Which one of the following statements is true? **c**

  a. Every rational number is an integer.

  b. Some whole numbers are not integers.

  c. Some rational numbers are not positive.

  d. Irrational numbers cannot be negative.

104. Which one of the following statements is true? **b**

  a. $\sqrt{36}$ is an irrational number.

  b. Some real numbers are not rational numbers.

  c. Some integers are not rational numbers.

  d. All whole numbers are positive.

105. Answer this question without using a calculator. Between which two consecutive integers is $-\sqrt{47}$? *−7 and −6*

106. We have used a code to describe an activity that is important to your success in school and at work. Here is the coded message:

$$(20, 18) \quad (-8, -12) \quad (0, 9) \quad (-14, -17)$$
$$(0, 11) \quad (-17, 9) \quad (-16, 14) \quad (-22, 7).$$

Read across the two rows. Each number pair represents one letter of the coded message. The first number pair represents the first letter, the second pair, the second letter, and so on. Use the following table to decode the message.
THINKING

First Decoding	Second Decoding				
Use the greater of the two numbers in each pair. If the greater number is negative, then use its absolute value.	0 = blank space	1 = A	2 = B	3 = C	4 = D
	5 = E 6 = F	7 = G	8 = H	9 = I	10 = J
	11 = K 12 = L	13 = M	14 = N	15 = O	16 = P
	17 = Q 18 = R	19 = S	20 = T	21 = U	22 = V
	23 = W 24 = X	25 = Y	26 = Z		

Technology Exercises

In Exercises 107–110, use a calculator to find a decimal approximation for each irrational number, correct to three decimal places. Between which two integers should you graph each of these numbers on the number line?

107. $\sqrt{3}$ *1.732; 1 and 2*

108. $-\sqrt{12}$ *−3.464; −4 and −3*

109. $1 - \sqrt{2}$ *−0.414; −1 and 0*

110. $2 - \sqrt{5}$ *−0.236; −1 and 0*

SECTION 1.3

ORDERED PAIRS AND GRAPHS

Objectives

1 Plot ordered pairs in the rectangular coordinate system.

2 Find coordinates of points in the rectangular coordinate system.

3 Interpret information given by line graphs.

4 Interpret information given by bar graphs.

The beginning of the seventeenth century was a time of innovative ideas and enormous intellectual progress in Europe. English theatergoers enjoyed a succession of exciting new plays by Shakespeare. William Harvey proposed the radical notion that the heart was a pump for blood rather than the center of emotion. Galileo, with his new-fangled invention called the telescope, supported the theory of Polish astronomer Copernicus that the sun, not the Earth, was the center of the solar system. Monteverdi was writing the world's first grand operas. French mathematicians Pascal and Fermat invented a new field of mathematics called probability theory.

Into this arena of intellectual electricity stepped French aristocrat René Descartes (1596–1650). Descartes, propelled by the creativity surrounding him, developed a new branch of mathematics that brought together algebra and geometry in a unified way—a way that visualized numbers as points on a graph, equations as geometric figures, and geometric figures as equations. This new branch of mathematics, called *analytic geometry*, established Descartes as one of the founders of modern thought and among the most original mathematicians and philosophers of any age. We begin this section by looking at Descartes's deceptively simple idea, called the **rectangular coordinate system** or (in his honor) the **Cartesian coordinate system**.

Points and Ordered Pairs Descartes used two number lines that intersect at right angles at their zero points, as shown in Figure 1.11. The horizontal number line is the **x-axis**. The vertical number line is the **y-axis**. The point of intersection of these axes is their zero points, called the **origin**. Positive numbers are shown to the right and above the origin. Negative numbers are shown to the left and below the origin. The axes divide the plane into four quarters, called **quadrants**. The points located on the axes are not in any quadrant.

Each point in the rectangular coordinate system corresponds to an **ordered pair** of real numbers, (x, y). Examples of such pairs are $(4, 2)$ and $(-5, -3)$. The first number in each pair, called the **x-coordinate**, denotes the distance and direction from the origin along the x-axis. The second number, called the **y-coordinate**, denotes vertical distance and direction along a line parallel to the y-axis or along the y-axis itself.

Figure 1.12 shows how we **plot**, or locate, the points corresponding to the ordered pairs $(4, 2)$ and $(-5, -3)$. We plot $(4, 2)$ by going 4 units from 0 to the right along the x-axis. Then we go 2 units up parallel to the y-axis. We plot $(-5, -3)$ by going 5 units from 0 to the left along the x-axis and 3 units down parallel to the y-axis. The phrase "the point corresponding to the ordered pair $(-5, -3)$" is often abbreviated as "the point $(-5, -3)$."

1 Plot ordered pairs in the rectangular coordinate system.

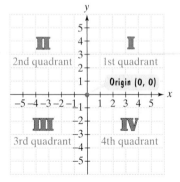

FIGURE 1.11 The rectangular coordinate system

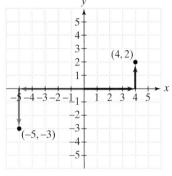

FIGURE 1.12 Plotting $(4, 2)$ and $(-5, -3)$

EXAMPLE 1 Plotting Points in the Rectangular Coordinate System

Plot the points: $A(-3, 5)$, $B(2 -4)$, $C(5, 0)$, $D(-5, -3)$, $E(0, 4)$ and $F(0, 0)$.

SOLUTION See Figure 1.13. We move from the origin and plot the points in the following way:

$A(-3, 5)$:	3 units left, 5 units up
$B(2, -4)$:	2 units right, 4 units down
$C(5, 0)$:	5 units right, 0 units up or down
$D(-5, -3)$:	5 units left, 3 units down
$E(0, 4)$:	0 units right or left, 4 units up
$F(0, 0)$:	0 units right or left, 0 units up or down

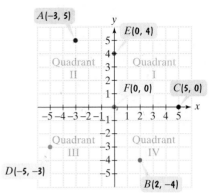

FIGURE 1.13 Plotting points

The phrase *ordered pair* is used because **order is important**. For example, the points $(2, 5)$ and $(5, 2)$ are not the same. To plot $(2, 5)$, move 2 units right and 5 units up. To plot $(5, 2)$, move 5 units right and 2 units up. The points $(2, 5)$ and $(5, 2)$ are in different locations. **The order in which coordinates appear makes a difference in a point's location**.

See graphing answer section.

✔ **CHECK POINT 1** Plot the points: $A(-2, 4)$, $B(4, -2)$, $C(-3, 0)$, and $D(0, -3)$.

2 Find coordinates of points in the rectangular coordinate system.

In the rectangular coordinate system, each ordered pair corresponds to exactly one point. Example 2 illustrates that each point in the rectangular coordinate system corresponds to exactly one ordered pair.

EXAMPLE 2 Finding Coordinates of Points

Determine the coordinates of points A, B, C, and D shown in Figure 1.14.

SOLUTION

Point	Position from the Origin	Coordinates
A	6 units left, 0 units up or down	$(-6, 0)$
B	0 units right or left, 2 units up	$(0, 2)$
C	2 units right, 0 units up or down	$(2, 0)$
D	4 units right, 2 units down	$(4, -2)$

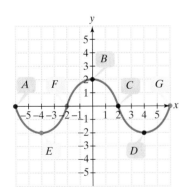

FIGURE 1.14 Finding coordinates of points

✔ **CHECK POINT 2** Determine the coordinates of points E, F, and G shown in Figure 1.14. $E(-4, -2)$ $F(-2, 0)$ $G(6, 0)$

The rectangular coordinate system lets us visualize relationships between two quantities, as shown in the next example.

FIGURE 1.15

B (8, 200); After 8 seconds, the watermelon is 200 feet above ground.

D (8.8, 0); After approximately 8.8 seconds, the watermelon is 0 feet above ground. Equivalently, the watermelon splatters on the ground after approximately 8.8 seconds.

EXAMPLE 3 An Application of the Rectangular Coordinate System

Comedian David Letterman has been known to drop watermelons out of windows and watch them fall. Suppose that he drops a watermelon from the observation deck at the top of New York's Empire State Building. (Yes, the sidewalk on the ground below the building has been cleared!) Letterman drops the melon and watches it glide past the side of the building, smiling at the instant it splatters on the ground. The points in Figure 1.15 show the height of the melon above the ground at different times. Find the coordinates of point *A*. Then interpret the coordinates in terms of the information given.

SOLUTION Let's take a few minutes to look at the rectangular coordinate system. The *x*-axis represents the time, in seconds, that the watermelon is in the air. Each mark on the *x*-axis represents one second. The *y*-axis represents the melon's height, in feet, above the ground. Each mark on the *y*-axis represents 100 feet. Because time and height are not negative, Figure 1.15 shows only the first quadrant of the rectangular coordinate system and its boundary.

Now let us find the coordinates of point *A*. Point *A* is 3 units right and 1100 units up from the origin. Thus, the coordinates of point *A* are (3, 1100). This means that after 3 seconds, the watermelon is 1100 feet above the ground. ∎

 CHECK POINT 3 Use Figure 1.15 to find the coordinates of point *B*. Then interpret the coordinates in terms of the information given.

Take another look at Figure 1.15. How can we find the coordinates of points *C* and *D*? It appears that we must estimate one or more of these coordinates and arrive at reasonable approximations.

EXAMPLE 4 Estimating Coordinates of a Point

Use Figure 1.15 to estimate the coordinates of point *C*. Then interpret the coordinates in terms of the information given.

SOLUTION Point *C* is 0 units right from the origin. Thus, its *x*-coordinate is 0. How far up from the origin is point *C*? Its distance up is approximately midway between 1200 and 1300. Thus, its *y*-coordinate is approximately 1250. A reasonable estimate is that the coordinates of point *C* are (0, 1250). This means that after 0 seconds, or at the very instant Letterman dropped the watermelon, it was approximately 1250 feet above the ground. Equivalently, the melon was dropped from a height of approximately 1250 feet. ∎

 CHECK POINT 4 Use Figure 1.15 to estimate the coordinates of point *D*. Then interpret the coordinates in terms of the information given.

STUDY TIP

Any point on the *x*-axis has a *y*-coordinate of 0. Any point on the *y*-axis has an *x*-coordinate of 0.

Throughout your study of algebra, you will see how the rectangular coordinate system is used to create graphs that show relationships between quantities. Magazines and newspapers also use graphs to display information. In the remainder of this section, we will discuss how to interpret line and bar graphs.

3 Interpret information given by line graphs.

Line Graphs Line graphs are often used to illustrate trends over time. Some measure of time, such as months or years, frequently appears on the horizontal axis. Amounts are generally listed on the vertical axis. Points are drawn to represent the given information. The graph is formed by connecting the points with line segments.

Figure 1.16 is an example of a typical line graph. The graph shows the average age at which women in the United States married for the first time from 1890 through 2002. The years are listed on the horizontal axis and the ages are listed on the vertical axis. The symbol ⭿ on the vertical axis shows that there is a break in values between 0 and 20. Thus, the first tick mark on the vertical axis represents an average age of 20.

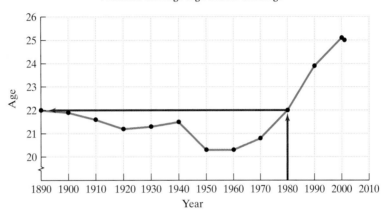

Women's Average Age of First Marriage

FIGURE 1.16
Source: U.S. Census Bureau

A line graph displays information in the first quadrant of a rectangular coordinate system. By identifying points on line graphs and their coordinates, you can interpret specific information given by the graph.

For example, the red lines in Figure 1.16 show how to find the average age at which women married for the first time in 1980.

STEP 1 Locate 1980 on the horizontal axis.

STEP 2 Locate the point above 1980.

STEP 3 Read across to the corresponding age on the vertical axis.

The age is 22. Thus, in 1980, women in the United States married for the first time at an average age of 22.

EXAMPLE 5 Using a Line Graph

The line graph in Figure 1.17 shows the fraction of federal prisoners in the United States sentenced for drug offenses from 1970 through 2001.

a. For the period shown, estimate the maximum fraction of federal prisoners sentenced for drug offenses. When did this occur?

b. Table 1.2 shows the number, in thousands, of federal prisoners in the United States for four selected years. Estimate the number of federal prisoners sentenced for drug offenses for the year in part (a).

Table 1.2

Number, in Thousands, of U.S. Federal Prisoners

Year	Federal Prisoners
1990	60
1995	90
2000	130
2001	140

Source: Bureau of Justice Statistics

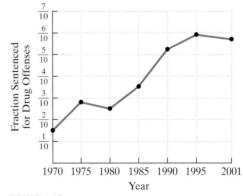

Fraction of U.S. Federal Prisoners Sentenced for Drug Offenses

FIGURE 1.17

Source: Frank Schmalleger, *Criminal Justice Today*, 7th Edition, Prentice Hall, 2003

SOLUTION

a. The maximum fraction of federal prisoners sentenced for drug offenses can be found by locating the highest point on the graph. This point lies above 1995 on the horizontal axis. Read across to the corresponding fraction on the vertical axis. The number falls slightly below $\frac{6}{10}$. It appears that $\frac{6}{10}$, or $\frac{3}{5}$, is a reasonable estimate. Thus, the maximum fraction of federal prisoners sentenced for drug offenses is approximately $\frac{3}{5}$. This occurred in 1995.

b. Table 1.2 shows that there were 90 thousand federal prisoners in 1995. The number sentenced for drug offenses can be determined as follows:

$$\text{Number sentenced for drug offenses} = \frac{3}{5} \cdot 90 = \frac{3 \cdot 90}{5} = \frac{270}{5} = 54.$$

where $\frac{3}{5}$ of the total prison population.

In 1995, approximately 54 thousand federal prisoners were sentenced for drug offenses. ■

 CHECK POINT 5 Use Figure 1.17 and Table 1.2 to estimate the number of federal prisoners sentenced for drug offenses in 1990. 30 thousand

4 Interpret information given by bar graphs.

Bar Graphs **Bar graphs** are convenient for showing comparisons among items. The bars may be vertical or horizontal, and their heights or lengths are used to show the amounts of different items. Figure 1.18 is an example of a typical bar graph. The graph shows the number of cars per 100 people in ten selected countries.

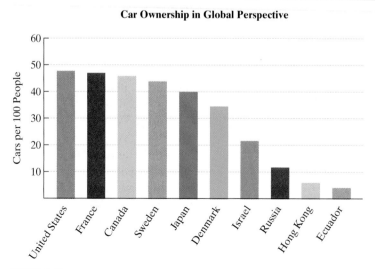

FIGURE 1.18

Source: The World Bank

EXAMPLE 6 Using a Bar Graph

Using Figure 1.18, answer the following:

a. Estimate the number of cars per 100 people in the United States.

b. Which countries have fewer than 20 cars per 100 people?

SOLUTION

a. To estimate the number of cars per 100 people in the United States, look at the top of the bar representing the United States and then read the cars-per-100-people scale on the vertical axis. The bar extends more than midway between 40 and 50, but is less than 50 by a few units. It appears that 48 is a reasonable estimate. Thus, there are approximately 48 cars per 100 people in the United States.

b. To find which countries have fewer than 20 cars per 100 people, we first locate the 20 mark on the vertical axis. Then we look for bars whose heights do not exceed 20. There are three such bars, namely, the three bars located in the right portion of the graph. The names of the countries below these bars show that Russia, Hong Kong, and Ecuador have fewer than 20 cars per 100 people. ∎

✔ **CHECK POINT 6** Using Figure 1.18, answer the following:

a. Estimate the number of cars per 100 people in Israel. 22

b. Which countries have more than 40 cars per 100 people?

United States, France, Canada, Sweden

1.3 EXERCISE SET

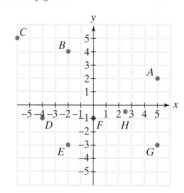

Student Solutions Manual CD/Video PH Math/Tutor Center MathXL Tutorials on CD MathXL® MyMathLab Interactmath.com

Practice Exercises

In Exercises 1–8, plot the given point in a rectangular coordinate system. Indicate in which quadrant each point lies. **1–8.** See graphing answer section.

1. $(3, 5)$ I

2. $(5, 3)$ I

3. $(-5, 1)$ II

4. $(1, -5)$ IV

5. $(-3, -1)$ III

6. $(-1, -3)$ III

7. $(6, -3.5)$ IV

8. $(-3.5, 6)$ II

In Exercises 9–24, plot the given point in a rectangular coordinate system.   **9–24.** See graphing answer section.

9. $(-3, -3)$

10. $(-5, -5)$

11. $(-2, 0)$

12. $(-5, 0)$

13. $(0, 2)$

14. $(0, 5)$

15. $(0, -3)$

16. $(0, -5)$

17. $\left(\frac{5}{2}, \frac{7}{2}\right)$

18. $\left(\frac{7}{2}, \frac{5}{2}\right)$

19. $\left(-5, \frac{3}{2}\right)$

20. $\left(-\frac{9}{2}, -4\right)$

21. $(0, 0)$

22. $\left(-\frac{5}{2}, 0\right)$

23. $\left(0, -\frac{5}{2}\right)$

24. $\left(0, \frac{7}{2}\right)$

In Exercises 25–32, give the ordered pairs that correspond to the points labeled in the figure.

25. A $(5, 2)$

26. B $(-2, 4)$

27. C $(-6, 5)$

28. D $(-4, -1)$

29. E $(-2, -3)$

30. F $(0, -1)$

31. G $(5, -3)$

32. H $(2.5, -0.5)$

33. In which quadrants are the y-coordinates positive?   I and II

34. In which quadrants are the x-coordinates negative?   II and III

35. In which quadrants do the x-coordinates and the y-coordinates have the same sign?   I and III

36. In which quadrants do the x-coordinates and the y-coordinates have opposite signs?   II and IV

Practice Plus

37. Graph four points in at least two different quadrants such that the absolute value of the x-coordinate is the y-coordinate. Answers will vary; examples are $(-2, 2), (-1, 1), (1, 1),$ $(2, 2).$

38. Graph four points in at least two different quadrants such that the absolute value of the y-coordinate is the x-coordinate. Answers will vary; examples are $(2, -2), (2, 2), (1, -1),$ $(1, 1).$

39. The point $A(-2, y)$ lies on the line that connects $(-3, 5)$ and $(1, 3)$. Estimate the y-coordinate of point A. $4\frac{1}{2}$ or $\frac{9}{2}$

40. The point $A(-1, y)$ lies on the line that connects $(-4, -1)$ and $(2, -4)$. Estimate the y-coordinate of point A. $-2\frac{1}{2}$ or $-\frac{5}{2}$

Application Exercises

A football is thrown by a quarterback to a receiver. The points in the figure show the height of the football, in feet, above the ground in terms of its distance, in yards, from the quarterback. Use this information to solve Exercises 41–46.

Distance of the Football from the Quarterback (yards)

41. Find the coordinates of point A. Then interpret the coordinates in terms of the information given. $(2, 7)$; The football is 7 feet above ground when it is 2 yards from the quarterback.

42. Find the coordinates of point B. Then interpret the coordinates in terms of the information given. $(28, 7)$; The football is 7 feet above ground when it is 28 yards from the quarterback.

43. Estimate the coordinates of point C. $(6, 9.25)$

44. Estimate the coordinates of point D. $(24, 9.5)$

45. What is the football's maximum height? What is its distance from the quarterback when it reaches its maximum height?   12 feet; 15 yards

46. What is the football's height when it is caught by the receiver? What is the receiver's distance from the quarterback when he catches the football?   5 feet; 30 yards

Afghanistan accounts for 76% of the world's illegal opium production. Opium-poppy cultivation nets big money in a country where most people earn less than $1 per day. (Source: Newsweek) The line graph shows opium-poppy cultivation, in thousands of acres, in Afghanistan from 1990 through 2002. Use the graph to solve Exercises 47–52.

Afghanistan's Opium Crop

Source: U.N. Office on Drugs and Crime

47. What are the coordinates of point A? What does this mean in terms of the information given by the graph? $(91, 125)$; In 1991, 125 thousand acres were used for cultivation.

48. What are the coordinates of point B? What does this mean in terms of the information given by the graph? $(94, 175)$; In 1994, 175 thousand acres were used for cultivation.

49. For the period shown, when did opium cultivation reach a minimum? How many thousands of acres were used to cultivate the illegal crop?   2001; 25 thousand acres

50. For the period shown, when did opium cultivation reach a maximum? How many thousands of acres were used to cultivate the illegal crop?   1999; 225 thousand acres

51. Between which two years did opium cultivation not change?   1991 and 1992

52. Between which two years did opium cultivation increase at the greatest rate? What is a reasonable estimate of the increase, in thousands of acres, used to cultivate the illegal crop during this period?   2001 and 2002; 155 thousand acres

The line graphs at the top of the next page show that oil consumption in the United States has continued to increase as domestic production of crude oil has been dropping. Use the information shown by the two line graphs to solve Exercises 53–60.

U.S. Oil Consumption and Domestic Production: 1970–2002

Source: Energy Information Administration

53. Find an estimate for oil consumption in 1985.   7.1 million barrels per day

54. Find an estimate for domestic oil production in 1985.   9.1 million barrels per day

55. In which year were oil consumption and domestic production the same? Estimate the millions of barrels of oil consumed and produced per day during that year.   1990; 7.2 million barrels per day

56. For which years did oil consumption exceed domestic production?   1991 through 2002

57. For the period shown, when did domestic oil production reach a maximum? What is a reasonable estimate for oil production during that year?   1970; 9.8 million barrels per day

58. For the period shown, when did oil consumption reach a maximum? What is a reasonable estimate for oil consumption during that year?   2002; 8.9 million barrels per day

59. Find an estimate for the difference between production and consumption in 1980.   1.7 million barrels per day

60. Find an estimate for the difference between consumption and production in 2000.   2.6 million barrels per day

You can't judge the quality of a movie by its domestic box-office gross, but you can't keep making movies that don't make money. The bar graph shows that actor Kevin Costner needs a hit. Use the information in the graph to solve Exercises 61–64.

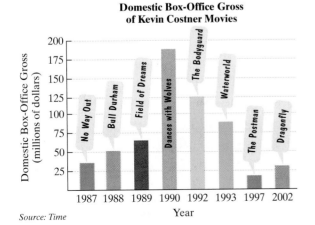

Domestic Box-Office Gross of Kevin Costner Movies

Source: Time

61. Estimate the box-office gross, in millions of dollars, of *The Bodyguard*. $120 million

62. Estimate the box-office gross, in millions of dollars, of *Waterworld*. $90 million

63. Which movies have a box-office gross of less than $50 million?   *No Way Out, The Postman, Dragonfly*

64. Which movies have a box-office gross of more than $100 million?   *Dances with Wolves, The Bodyguard*

The bar graph shows life expectancy in the United States by year of birth. Use the graph to solve Exercises 65–68.

Life Expectancy in the U.S. by Birth Year

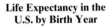

Source: U.S. Census Bureau

65. Estimate the life expectancy for men born in 1900.   48

66. Estimate the life expectancy for women born in 2050.   83

67. By approximately how many more years can women born in 1996 expect to live as compared to men born in 1950?   approximately 17 years

68. For which genders and for which birth years does life expectancy exceed 40 years but is at most 60 years?   both men and women with birth years of 1900

The bar graph shows the annual median income (the income in the middle of the list when all of the groups' incomes are arranged in order) for six groups, organized by gender and race. Use the information in the graph to solve Exercises 69–72.

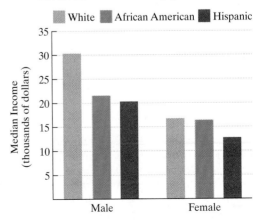

U.S. Median Income in 2001, by Gender and Race

Source: U.S. Census Bureau

(Refer to the graph on the previous page)

69. Estimate the median income, in thousands of dollars, for African-American men. $22 thousand

70. Estimate the median income, in thousands of dollars, for white women. $17 thousand

71. Estimate the difference, in thousands of dollars, between median income of Hispanic men and Hispanic women. $7 thousand

72. Estimate the difference, in thousands of dollars, between the median income of African-American men and African-American women. $6 thousand

Candidates for president of the United States keep jumping into the race earlier and earlier. The bar graph shows the average number of days before the nominating convention that candidates declared their intention to run for the office. Use the information in the graph to solve Exercises 73–74.

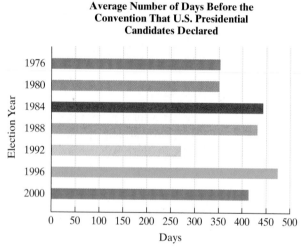

Average Number of Days Before the Convention That U.S. Presidential Candidates Declared

Source: Newsweek

73. In 1992, Clinton's 284 days, a short campaign, didn't cost him the win. Estimate the average number of days before the convention that candidates declared in 1992. By approximately how many days did Clinton's declaration exceed the 1992 average?   270; 14

74. Typical of candidates today, in 2000, Bush declared 405 days prior to the convention. Estimate the average number of days before the convention that candidates declared in 2000. By approximately how many days was Bush's declaration less than the 2000 average?   410; 5

Despite various programs over the past three decades, oil from other countries accounts for an ever-greater share of energy in the United States. The bar graph in the next column shows the fraction of oil used in the United States that was obtained from other countries for five selected years. In Exercises 75–78, use the information shown by the graph to write the year or years that fit the given description.

Oil Imports as a Fraction of U.S. Oil Consumption

Source: Energy Information Administration

75. The United States obtained more than $\frac{1}{2}$ of its oil from other countries.   2002

76. The United States obtained less than $\frac{1}{5}$ of its oil from other countries.   1962

77. The fraction of imported oil exceeded $\frac{1}{5}$ but was at most $\frac{1}{2}$.   1972, 1982, 1992

78. The fraction of imported oil exceeded $\frac{2}{5}$ but was at most $\frac{1}{2}$.   1992

79. In 2002, $\frac{1}{10}$ of U.S. imported oil was from Saudi Arabia. Use the information shown in the graph to estimate the fraction of oil used in the United States in 2002 that was imported from Saudi Arabia. $\frac{1}{20}$

80. In 2002, $\frac{2}{25}$ of U.S. imported oil was from Venezuela. Use the information shown in the graph to estimate the fraction of oil used in the United States in 2002 that was imported from Venezuela. $\frac{1}{25}$

Writing in Mathematics

81. What is the rectangular coordinate system?
81–88. Answers will vary.

82. Explain how to plot a point in the rectangular coordinate system. Give an example with your explanation.

83. Explain why $(5, -2)$ and $(-2, 5)$ do not represent the same point.

84. Explain how to find the coordinates of a point in the rectangular coordinate system.

85. Describe a line graph.

86. Describe a bar graph.

87. Describe how the information in the bar graph for Exercises 75–80 is related to the information in the line graphs for Exercises 53–60.

88. Find a graph in a newspaper, magazine, or almanac and describe what the graph illustrates.

Critical Thinking Exercises

89. a. Graph each of the following sets of points in a separate rectangular coordinate system:

$$A = \{(-3, 9), (-2, 4), (-1, 1),$$
$$(0, 0), (1, 1), (2, 4), (3, 9)\}$$
$$B = \{(9, -3), (4, -2), (1, -1),$$
$$(0, 0), (1, 1), (4, 2), (9, 3)\}.$$

See graphing answer section.

b. In which set (A or B) is each x-coordinate associated with exactly one y-coordinate? In which set are x-coordinates associated with more than one y-coordinate? Describe how these differences can be seen from your graphs in part (a). *A; B; Points line up vertically when x-coordinates are associated with more than one y-coordinate.*

90. a. Graph each of the following points:

$$\left(1, \frac{1}{2}\right), (2, 1), \left(3, \frac{3}{2}\right), (4, 2).$$

See graphing answer section.
Parts (b)–(d) can be answered by changing the sign of one or both coordinates of the points in part (a).

b. What must be done to the coordinates so that the resulting graph is a mirror-image reflection about the y-axis of your graph in part (a)? *Change the sign of each x-coordinate.*

c. What must be done to the coordinates so that the resulting graph is a mirror-image reflection about the x-axis of your graph in part (a)? *Change the sign of each y-coordinate.*

d. What must be done to the coordinates so that the resulting graph is a straight-line extension of your graph in part (a)? *Change the sign of both coordinates.*

Review Exercises

From here on, each exercise set will contain three review exercises. It is essential to review previously covered topics to improve your understanding of the topics and to help maintain your mastery of the material. If you are not certain how to solve a review exercise, turn to the section and the illustrative example given in parentheses at the end of each exercise.

91. Add: $\frac{3}{4} + \frac{2}{5}$. (Section 1.1, Example 9) $\quad \frac{23}{20}$

92. Insert $<$ or $>$ in the shaded area to make a true statement: $-\frac{1}{4} \quad 0$. (Section 1.2, Example 6) $\quad <$

93. Find the absolute value: $|-5.83|$. (Section 1.2, Example 8) 5.83

SECTION 1.4

Objectives

1 Evaluate algebraic expressions.

2 Understand and use the vocabulary of algebraic expressions.

3 Use commutative properties.

4 Use associative properties.

5 Use distributive properties.

6 Combine like terms.

7 Simplify algebraic expressions.

8 Use algebraic expressions that model reality.

BASIC RULES OF ALGEBRA

Algebraic Expressions Feeling attractive with a suntan that gives you a "healthy glow"? Think again. Direct sunlight is known to promote skin cancer. Although sunscreens protect you from burning, dermatologists are concerned with the long-term damage that results from the sun even without sunburn.

We have seen that algebra uses letters, such as x and y, to represent numbers. Recall that such letters are called **variables**. For example, we can let x represent the number of minutes that a person can stay in the sun without burning with no sunscreen. With a number 6 sunscreen, exposure time without burning is six times as long, or 6 times x. This can be written $6 \cdot x$, but it is usually expressed as $6x$. Placing a number and a letter next to one another indicates multiplication.

Notice that $6x$ combines the number 6 and the variable x using the operation of multiplication. A combination of variables and numbers using the operations of addition, subtraction, multiplication, or division, as well as powers or roots, is called an **algebraic expression**. Here are some examples of algebraic expressions:

$$x + 6, \quad x - 6, \quad 6x, \quad \frac{x}{6}, \quad 3x + 5, \quad \sqrt{x} + 7.$$

1 Evaluate algebraic expressions.

Evaluating Algebraic Expressions We can replace a variable that appears in an algebraic expression by a number. We are **substituting** the number for the variable. The process is called **evaluating the expression**. For example, we can evaluate $6x$ (from the sunscreen example) for $x = 15$. We substitute 15 for x. We obtain $6 \cdot 15$, or 90. This means if you can stay in the sun for 15 minutes without burning when you don't put on any lotion, then with a number 6 lotion, you can "cook" for 90 minutes without burning.

Many algebraic expressions involve more than one operation. The order in which we add, subtract, multiply, and divide is important. In Section 1.8, we will discuss the rules for the order in which operations should be done. For now, follow this order:

1. Perform calculations within parentheses first.
2. Perform multiplication before addition.

EXAMPLE 1 Evaluating an Algebraic Expression

The algebraic expression $2.35x + 179.5$ describes the population of the United States, in millions, x years after 1960. Evaluate the expression for $x = 40$. Describe what the answer means in practical terms.

SOLUTION We begin by substituting 40 for x. Because $x = 40$, we will be finding the U.S. population 40 years after 1960, in the year 2000.

$$2.35x + 179.5$$

Replace x with 40.

$$= 2.35(40) + 179.5$$
$$= 94 + 179.5 \qquad \text{Perform the multiplication: } 2.35(40) = 94.$$
$$= 273.5 \qquad \text{Perform the addition.}$$

According to the given algebraic expression, in 2000, the population of the United States was 273.5 million.

According to the U.S. Bureau of the Census, in 2000 the population of the United States was 281.4 million. Notice that the algebraic expression in Example 1 provides an approximate, but not an exact, description of the actual population. Many algebraic expressions approximately describe some aspect of reality, and we say that they *model* reality.

226.5; In 1980, the population of the United States was 226.5 million.

✓ **CHECK POINT 1** Evaluate: $2.35x + 179.5$ for $x = 20$.
Describe what your answer means in practical terms.

2 Understand and use the vocabulary of algebraic expressions.

The Vocabulary of Algebraic Expressions We have seen that an algebraic expression combines numbers and variables. Here is another example of an algebraic expression:

$$7x + 3.$$

The **terms** of an algebraic expression are those parts that are separated by addition. For example, the algebraic expression $7x + 3$ contains two terms, namely $7x$ and 3. Notice that a term is a number, a variable, or a number multiplied by one or more variables.

The numerical part of a term is called its **coefficient**. In the term $7x$, the 7 is the coefficient. If a term containing one or more variables is written without a coefficient, the coefficient is understood to be 1. Thus, x means $1x$ and ab means $1ab$.

A term that consists of just a number is called a **constant term**. The constant term of $7x + 3$ is 3.

The parts of each term that are multiplied are called the **factors of the term**. The factors of the term $7x$ are 7 and x.

Like terms are terms that have exactly the same variable factors. Here are two examples of like terms:

$7x$ and $3x$ These terms have the same variable factor, x.

$4y$ and $9y$.   These terms have the same variable factor, y.

By contrast, here are some examples of terms that are not like terms. These terms do not have the same variable factor.

$7x$ and 3 The variable factor of the first term is x.
     The second term has no variable factor.

$7x$ and $3y$ The variable factor of the first term is x.
     The variable factor of the second term is y.

Constant terms are like terms. Thus, the constant terms 7 and -12 are like terms.

EXAMPLE 2   Using the Vocabulary of Algebraic Expressions

Use the algebraic expression

$$4x + 7 + 5x$$

to answer the following questions:

a. How many terms are there in the algebraic expression?
b. What is the coefficient of the first term?
c. What is the constant term?
d. What are the like terms in the algebraic expression?

SOLUTION

a. Because terms are separated by addition, the algebraic expression $4x + 7 + 5x$ contains three terms.

$$4x + 7 + 5x$$

First term   Second term   Third term

b. The coefficient of the first term, $4x$, is 4.
c. The constant term in $4x + 7 + 5x$ is 7.
d. The like terms in $4x + 7 + 5x$ are $4x$ and $5x$. These terms have the same variable factor, x.   ∎

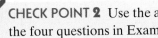 **CHECK POINT 2** Use the algebraic expression $6x + 2x + 11$ and answer each of the four questions in Example 2.   **a.** 3 terms **b.** 6 **c.** 11 **d.** $6x$ and $2x$

Equivalent Algebraic Expressions In Example 2, we considered the algebraic expression

$$4x + 7 + 5x.$$

Let's compare this expression with a second algebraic expression

$$9x + 7.$$

Evaluate each expression for some choice of x. We will select $x = 2$.

$$4x + 7 + 5x \qquad\qquad 9x + 7$$

Replace x with 2.   Replace x with 2.

$$\begin{aligned} &= 4 \cdot 2 + 7 + 5 \cdot 2 \\ &= 8 + 7 + 10 \\ &= 25 \end{aligned} \qquad\qquad \begin{aligned} &= 9 \cdot 2 + 7 \\ &= 18 + 7 \\ &= 25 \end{aligned}$$

Both algebraic expressions have the same value when $x = 2$. Regardless of what number you select for x, the algebraic expressions $4x + 7 + 5x$ and $9x + 7$ will have the same value. These expressions are called *equivalent algebraic expressions*. Two algebraic expressions that have the same value for all replacements are called **equivalent algebraic expressions**. Because $4x + 7 + 5x$ and $9x + 7$ are equivalent algebraic expressions, we write

$$4x + 7 + 5x = 9x + 7.$$

Properties of Real Numbers and Algebraic Expressions We now turn to basic properties, or rules, that you know from past experiences in working with whole numbers and fractions. These properties will be extended to include all real numbers and algebraic expressions. We will give each property a name so that we can refer to it throughout the study of algebra.

The Commutative Properties The addition or multiplication of two real numbers can be done in any order. For example, $3 + 5 = 5 + 3$ and $3 \cdot 5 = 5 \cdot 3$. Changing the order does not change the answer of a sum or a product. These facts are called **commutative properties**.

3 Use commutative properties.

THE COMMUTATIVE PROPERTIES Let a, b, and c represent real numbers, variables, or algebraic expressions.

Commutative Property of Addition

$$a + b = b + a$$

Changing order when adding does not affect the sum.

Commutative Property of Multiplication

$$ab = ba$$

Changing order when multiplying does not affect the product.

EXAMPLE 3 Using the Commutative Properties

Use the commutative properties to write an algebraic expression equivalent to each of the following: **a.** $y + 6$ **b.** $5x$.

SOLUTION

a. By the commutative property of addition, an algebraic expression equivalent to $y + 6$ is $6 + y$. Thus,

$$y + 6 = 6 + y.$$

b. By the commutative property of multiplication, an algebraic expression equivalent to $5x$ is $x5$. Thus,

$$5x = x5.$$

CHECK POINT **3** Use the commutative properties to write an algebraic expression equivalent to each of the following: **a.** $x + 14$ $14 + x$ **b.** $7y$. $y7$

EXAMPLE 4 Using the Commutative Properties

Write an algebraic expression equivalent to $13x + 8$ using

a. the commutative property of addition.

b. the commutative property of multiplication.

SOLUTION

a. By the commutative property of addition, we change the order of the terms being added. This means that an algebraic expression equivalent to $13x + 8$ is $8 + 13x$:

$$13x + 8 = 8 + 13x.$$

b. By the commutative property of multiplication, we change the order of the factors being multiplied. This means that an algebraic expression equivalent to $13x + 8$ is $x13 + 8$:

$$13x + 8 = x13 + 8.$$

 CHECK POINT **4** Write an algebraic expression equivalent to $5x + 17$ using

a. the commutative property of addition. $17 + 5x$

b. the commutative property of multiplication. $x5 + 17$

ENRICHMENT ESSAY

Commutative Words and Sentences

The commutative property states that a change in order produces no change in the answer. The words and sentences listed here suggest a characteristic of the commutative property; they read the same from left to right and from right to left!

• dad
• repaper
• never odd or even

• Go deliver a dare, vile dog!
• May a moody baby doom a yam?
• Madam, in Eden I'm Adam.

• Ma is a nun, as I am.
• A man, a plan, a canal: Panama
• Are we not drawn onward, we few, drawn onward to new era?

 4 Use associative properties.

The Associative Properties Parentheses indicate groupings. We perform operations within the parentheses first. For example,

$$(2 + 5) + 10 = 7 + 10 = 17$$

and

$$2 + (5 + 10) = 2 + 15 = 17.$$

In general, the way in which three numbers are grouped does not change their sum. It also does not change their product. These facts are called the **associative properties**.

STUDY TIP

The associative property does not hold for subtraction or division.

$$(6 - 3) - 1 \neq 6 - (3 - 1)$$
$$(8 \div 4) \div 2 \neq 8 \div (4 \div 2)$$

THE ASSOCIATIVE PROPERTIES Let a, b, and c represent real numbers, variables, or algebraic expressions.

Associative Property of Addition

$$(a + b) + c = a + (b + c)$$

Changing grouping when adding does not affect the sum.

Associative Property of Multiplication

$$(ab)c = a(bc)$$

Changing grouping when multiplying does not affect the product.

The associative properties can be used to simplify some algebraic expressions by removing the parentheses.

EXAMPLE 5 Simplifying Using the Associative Properties

Simplify: **a.** $3 + (8 + x)$ **b.** $8(4x)$.

SOLUTION

a. $3 + (8 + x)$ This is the given algebraic expression.

$= (3 + 8) + x$ Use the associative property of addition to group the first two numbers.

$= 11 + x$ Add within parentheses.

Using the commutative property of addition, this simplified algebraic expression can also be written as $x + 11$.

b. $8(4x)$ This is the given algebraic expression.

$= (8 \cdot 4)x$ Use the associative property of multiplication to group the first two numbers.

$= 32x$ Multiply within parentheses.

We can use the commutative property of multiplication and write this simplified algebraic expression as $x32$ or $x \cdot 32$. However, it is customary to express a term with its coefficient on the left. Thus, we use $32x$ as the simplified form of the algebraic expression. ■

 CHECK POINT 5 Simplify:

a. $8 + (12 + x)$ $20 + x$ or $x + 20$ **b.** $6(5x)$. $30x$

ENRICHMENT ESSAY

The Associative Property and the English Language

In the English language, sentences can take on different meanings depending on the way the words are associated with commas. Here are two examples.

• *Do not break your bread or roll in your soup.*
Do not break your bread, or roll in your soup.

• *Woman, without her man, is nothing.*
Woman, without her, man is nothing.

The next example involves the use of both basic properties to simplify an algebraic expression.

EXAMPLE 6 Using the Commutative and Associative Properties

Simplify: $7 + (x + 2)$.

SOLUTION

$7 + (x + 2)$ This is the given algebraic expression.

$= 7 + (2 + x)$ Use the commutative property to change the order of the addition.

$= (7 + 2) + x$ Use the associative property to group the first two numbers.

$= 9 + x$ Add within parentheses.

Using the commutative property of addition, an equivalent algebraic expression is $x + 9$. ■

> ✔ **CHECK POINT 6** Simplify: $8 + (x + 4)$. $12 + x$ or $x + 12$

5 Use distributive properties.

The Distributive Properties   The **distributive property** involves both multiplication and addition. The property shows how to multiply the sum of two numbers by a third number. Consider, for example, $4(7 + 3)$, which can be calculated in two ways. One way is to perform the addition within the grouping symbols and then multiply:

$$4(7 + 3) = 4(10) = 40.$$

The other way is to *distribute* the multiplication by 4 over the addition by first multiplying each number within the parentheses by 4 and then adding:

$$4\overset{\frown}{(7 + 3)} = 4 \cdot 7 + 4 \cdot 3 = 28 + 12 = 40.$$

The result in both cases is 40. Thus,

$$4(7 + 3) = 4 \cdot 7 + 4 \cdot 3.$$ Multiplication distributes over addition.

The distributive property allows us to rewrite the product of a number and a sum as the sum of two products.

> **THE DISTRIBUTIVE PROPERTY**   Let a, b, and c represent real numbers, variables, or algebraic expressions.
>
> $$a\overset{\frown}{(b + c)} = ab + ac$$
>
> Multiplication distributes over addition.

STUDY TIP

Do not confuse the distributive property with the associative property of multiplication.

Distributive:

$$4(5 + x) = 4 \cdot 5 + 4x$$
$$= 20 + 4x$$

Associative:

$$4(5 \cdot x) = (4 \cdot 5)x$$
$$= 20x$$

EXAMPLE 7   Using the Distributive Property

Multiply: $6(x + 4)$.

SOLUTION   Multiply *each term* inside the parentheses, x and 4, by the multiplier outside, 6.

$$6(x + 4) = 6x + 6 \cdot 4$$ Use the distributive property to remove parentheses.
$$= 6x + 24$$ Multiply: $6 \cdot 4 = 24$.   ■

> ✔ **CHECK POINT 7** Multiply: $5(x + 3)$. $5x + 15$

EXAMPLE 8   Using the Distributive Property

Multiply: $5(3y + 7)$.

SOLUTION   Multiply *each term* inside the parentheses, $3y$ and 7, by the multiplier outside, 5.

$$5(3y + 7) = 5 \cdot 3y + 5 \cdot 7$$ Use the distributive property to remove parentheses.
$$= 15y + 35$$ Multiply. Use the associative property of multiplication to find $5 \cdot 3y$: $5 \cdot 3y = (5 \cdot 3)y = 15y$.   ■

> ✔ **CHECK POINT 8** Multiply: $6(4y + 7)$. $24y + 42$

When using the distributive property to remove parentheses, be sure to multiply *each term* inside the parentheses by the multiplier outside.

Incorrect!

$$5(3y + 7) = 5 \cdot 3y + 7$$

| 7 must also be multiplied by 5. |

$$= 15y + 7$$

Table 1.3 shows a number of other forms of the distributive property.

Table 1.3	Other Forms of the Distributive Property	
Property	**Meaning**	**Example**
$a(b - c) = ab - ac$	Multiplication distributes over subtraction.	$5(4x - 3) = 5 \cdot 4x - 5 \cdot 3$ $= 20x - 15$
$a(b + c + d) = ab + ac + ad$	Multiplication distributes over three or more terms in parentheses.	$4(x + 10 + 3y)$ $= 4x + 4 \cdot 10 + 4 \cdot 3y$ $= 4x + 40 + 12y$
$(b + c)a = ba + ca$	Multiplication on the right distributes over addition (or subtraction).	$(x + 7)9 = x \cdot 9 + 7 \cdot 9$ $= 9x + 63$

6 Combine like terms.

Combining Like Terms   The distributive property

$$a(b + c) = ab + ac$$

lets us add and subtract like terms. To do this, we will usually apply the property in the form

$$ax + bx = (a + b)x$$

and then combine a and b. For example,

$$3x + 7x = (3 + 7)x = 10x.$$

This process is called **combining like terms**.

EXAMPLE 9   Combining Like Terms

Combine like terms: **a.** $4x + 15x$ **b.** $7a - 2a$.

SOLUTION

DISCOVER FOR YOURSELF

Can you think of a fast method that will immediately give each result in Example 9? Describe the method.

a. $4x + 15x$ These are like terms because 4x and 15x have identical variable factors.

$= (4 + 15)x$ Apply the distributive property.

$= 19x$ Add within parentheses.

b. $7a - 2a$ These are like terms because 7a and 2a have identical variable factors.

$= (7 - 2)a$ Apply the distributive property.

$= 5a$ Subtract within parentheses.

 CHECK POINT 9   Combine like terms: **a.** $7x + 3x$ $10x$ **b.** $9a - 4a$. $5a$

When combining like terms, you may find yourself leaving out the details of the distributive property. For example, you may simply write

$$7x + 3x = 10x.$$

It might be useful to think along these lines: Seven things plus three of the (same) things give ten of those things. To add like terms, add the coefficients and copy the common variable.

COMBINING LIKE TERMS MENTALLY

1. Add or subtract the coefficients of the terms.
2. Use the result of step 1 as the coefficient of the term's variable factor.

When an expression contains three or more terms, use the commutative and associative properties to group like terms. Then combine the like terms.

EXAMPLE 10 Grouping and Combining Like Terms

Simplify: **a.** $7x + 5 + 3x + 8$ **b.** $4x + 7y + 2x + 3y.$

SOLUTION

a. $7x + 5 + 3x + 8$

$= (7x + 3x) + (5 + 8)$ Rearrange terms and group the like terms using the commutative and associative properties. This step is often done mentally.

$= 10x + 13$ Combine like terms: $7x + 3x = 10x$. Combine constant terms: $5 + 8 = 13$.

b. $4x + 7y + 2x + 3y$

$= (4x + 2x) + (7y + 3y)$ Group like terms.

$= 6x + 10y$ Combine like terms by adding coefficients and keeping the variable factor. ■

 CHECK POINT 10 Simplify:

a. $8x + 7 + 10x + 3$ $18x + 10$ **b.** $9x + 6y + 5x + 2y.$ $14x + 8y$

Simplifying Algebraic Expressions An algebraic expression is **simplified** when parentheses have been removed and like terms have been combined.

7 Simplify algebraic expressions.

SIMPLIFYING ALGEBRAIC EXPRESSIONS

1. Use the distributive property to remove parentheses.
2. Rearrange terms and group like terms using commutative and associative properties. This step may be done mentally.
3. Combine like terms by combining the coefficients of the terms and keeping the same variable factor.

EXAMPLE 11 Simplifying an Algebraic Expression

Simplify: $5(3x + 7) + 6x$.

SOLUTION

$$5(3x + 7) + 6x$$
$$= 5 \cdot 3x + 5 \cdot 7 + 6x \qquad \text{Use the distributive property to remove the parentheses.}$$
$$= 15x + 35 + 6x \qquad \text{Multiply.}$$
$$= (15x + 6x) + 35 \qquad \text{Group like terms.}$$
$$= 21x + 35 \qquad \text{Combine like terms.}$$

■

✓ **CHECK POINT 11** Simplify: $7(2x + 3) + 11x$. $25x + 21$

EXAMPLE 12 Simplifying an Algebraic Expression

Simplify: $6(2x + 4y) + 10(4x + 3y)$.

SOLUTION

$$6(2x + 4y) + 10(4x + 3y)$$
$$= 6 \cdot 2x + 6 \cdot 4y + 10 \cdot 4x + 10 \cdot 3y \qquad \text{Use the distributive property to remove the parentheses.}$$
$$= 12x + 24y + 40x + 30y \qquad \text{Multiply.}$$
$$= (12x + 40x) + (24y + 30y) \qquad \text{Group like terms.}$$
$$= 52x + 54y \qquad \text{Combine like terms.}$$

■

✓ **CHECK POINT 12** Simplify: $7(4x + 3y) + 2(5x + y)$. $38x + 23y$

8 Use algebraic expressions that model reality.

Applications Have you had a good workout lately? The next example involves an algebraic expression that approximately describes, or models, your optimum heart rate during exercise.

EXAMPLE 13 Modeling Optimum Heart Rate

The optimum heart rate is the rate that a person should achieve during exercise for the exercise to be most beneficial. The algebraic expression

$$0.6(220 - a)$$

describes a person's optimum heart rate, in beats per minute, where a represents the age of the person in years.

a. Use the distributive property to rewrite the algebraic expression without parentheses.

b. Use each form of the algebraic expression to determine the optimum heart rate for a 20-year-old runner.

SOLUTION

a. $0.6(220 - a) = 0.6(220) - 0.6a \qquad \text{Use the distributive property to remove parentheses.}$
$$= 132 - 0.6a \qquad \text{Multiply: } 0.6(220) = 132.$$

b. To determine the optimum heart rate for a 20-year-old runner, substitute 20 for a in each form of the algebraic expression.

Using 0.6(220 − a):
$$0.6(220 − 20)$$
$$= 0.6(200)$$
$$= 120$$

Using 132 − 0.6a:
$$132 − 0.6(20)$$
$$= 132 − 12$$
$$= 120$$

Both forms of the algebraic expression indicate that the optimum heart rate for a 20-year-old runner is 120 beats per minute. ■

✓ **CHECK POINT 13** Use each form of the algebraic expression in Example 13 to determine the optimum heart rate for a 40-year-old runner. 108 beats per minute

---

1.4 EXERCISE SET

 Student Solutions Manual CD/Video PH Math/Tutor Center MathXL Tutorials on CD Math XL MathXL® MyMathLab Interactmath.com

Practice Exercises

In Exercises 1–10, evaluate each algebraic expression for the given value of the variable.

1. $x + 11$; $x = 4$ 15
2. $x − 7$; $x = 11$ 4
3. $8x$; $x = 10$ 80
4. $12x$; $x = 5$ 60
5. $5x + 6$; $x = 8$ 46
6. $6x + 9$; $x = 4$ 33
7. $7(x + 3)$; $x = 2$ 35
8. $8(x + 4)$; $x = 3$ 56
9. $\frac{5}{9}(F − 32)$; $F = 77$ 25
10. $\frac{5}{9}(F − 32)$; $F = 50$ 10

In Exercises 11–16, an algebraic expression is given. Use each expression to answer the following questions.

a. How many terms are there in the algebraic expression?

b. What is the numerical coefficient of the first term?

c. What is the constant term?

d. Does the algebraic expression contain like terms? If so, what are the like terms?

11. $3x + 5$ **a.** 2   **b.** 3   **c.** 5   **d.** no
12. $9x + 4$ **a.** 2   **b.** 9   **c.** 4   **d.** no
13. $x + 2 + 5x$ **a.** 3   **b.** 1   **c.** 2   **d.** yes; x and $5x$
14. $x + 6 + 7x$ **a.** 3   **b.** 1   **c.** 6   **d.** yes; x and $7x$
15. $4y + 1 + 3x$ **a.** 3   **b.** 4   **c.** 1   **d.** no
16. $8y + 1 + 10x$ **a.** 3   **b.** 8   **c.** 1   **d.** no

In Exercises 17–24, use the commutative property of addition to write an equivalent algebraic expression.

17. $y + 4$ $4 + y$
18. $x + 7$ $7 + x$
19. $5 + 3x$ $3x + 5$
20. $4 + 9x$ $9x + 4$
21. $4x + 5y$ $5y + 4x$
22. $10x + 9y$ $9y + 10x$
23. $5(x + 3)$ $5(3 + x)$
24. $6(x + 4)$ $6(4 + x)$

In Exercises 25–32, use the commutative property of multiplication to write an equivalent algebraic expression.

25. $9x$ $x9$
26. $8x$ $x8$
27. $x + y6$ $x + 6y$
28. $x + y7$ $x + 7y$
29. $7x + 23$ $x7 + 23$
30. $13x + 11$ $x13 + 11$
31. $5(x + 3)$ $(x + 3)5$
32. $6(x + 4)$ $(x + 4)6$

In Exercises 33–36, use an associative property to rewrite each algebraic expression. Once the grouping has been changed, simplify the resulting algebraic expression.

33. $7 + (5 + x)$ $(7 + 5) + x = 12 + x$
34. $9 + (3 + x)$ $(9 + 3) + x = 12 + x$
35. $7(4x)$ $(7 \cdot 4)x = 28x$
36. $8(5x)$ $(8 \cdot 5)x = 40x$

In Exercises 37–56, use a distributive property to rewrite each algebraic expression without parentheses.

37. $3(x + 5)$ $3x + 15$
38. $4(x + 6)$ $4x + 24$
39. $8(2x + 3)$ $16x + 24$
40. $9(2x + 5)$ $18x + 45$
41. $\frac{1}{3}(12 + 6r)$ $4 + 2r$
42. $\frac{1}{4}(12 + 8r)$ $3 + 2r$
43. $5(x + y)$ $5x + 5y$
44. $7(x + y)$ $7x + 7y$
45. $3(x − 2)$ $3x − 6$
46. $4(x − 5)$ $4x − 20$
47. $2(4x − 5)$ $8x − 10$
48. $6(3x − 2)$ $18x − 12$
49. $\frac{1}{2}(5x − 12)$ $\frac{5}{2}x − 6$
50. $\frac{1}{3}(7x − 21)$ $\frac{7}{3}x − 7$
51. $(2x + 7)4$ $8x + 28$
52. $(5x + 3)6$ $30x + 18$
53. $6(x + 3 + 2y)$ $6x + 18 + 12y$
54. $7(2x + 4 + y)$ $14x + 28 + 7y$
55. $5(3x − 2 + 4y)$ $15x − 10 + 20y$
56. $4(5x − 3 + 7y)$ $20x − 12 + 28y$

In Exercises 57–74, simplify each algebraic expression.

57. $7x + 10x$ $17x$ **58.** $5x + 13x$ $18x$

59. $11a - 3a$ $8a$ **60.** $14b - 5b$ $9b$

61. $3 + (x + 11)$ $14 + x$ **62.** $7 + (x + 10)$ $17 + x$

63. $5y + 3 + 6y$ $11y + 3$ **64.** $8y + 7 + 10y$ $18y + 7$

65. $2x + 5 + 7x - 4$ $9x + 1$ **66.** $7x + 8 + 2x - 3$ $9x + 5$

67. $11a + 12 + 3a + 2$ $14a + 14$

68. $13a + 15 + 2a + 11$ $15a + 26$

69. $5(3x + 2) - 4$ $15x + 6$ **70.** $2(5x + 4) - 3$ $10x + 5$

71. $12 + 5(3x - 2)$ $15x + 2$

72. $14 + 2(5x - 1)$ $10x + 12$

73. $7(3a + 2b) + 5(4a + 2b)$ $41a + 24b$

74. $11(6a + 3b) + 4(12a + 5b)$ $114a + 53b$

Practice Plus

In Exercises 75–78, name the property illustrated by each true statement.

75. $6x + (2y + 7y) = 6x + (7y + 2y)$ Commutative property of addition

76. $6x + (2y + 7y) = (2y + 7y) + 6x$ Commutative property of addition

77. $6x + (2y + 7y) = (6x + 2y) + 7y$ Associative property of addition

78. $(6x)4 = 4(6x)$ Commutative property of multiplication

In Exercises 79–82, determine if each statement is true or false. Do not use a calculator.

79. $468(787 + 289) = 787 + 289(468)$ false

80. $468(787 + 289) = 787(468) + 289(468)$ true

81. $58 \cdot 9 + 32 \cdot 9 = (58 + 32) \cdot 9$ true

82. $58 \cdot 9 \cdot 32 \cdot 9 = (58 \cdot 32) \cdot 9$ false

Application Exercises

83. Suppose you can stay in the sun for x minutes without burning when you don't put on any lotion. The algebraic expression $15x$ describes how long you can tan without burning with a number 15 lotion. Evaluate the algebraic expression for $x = 20$. Describe what the answer means in practical terms. 300; You can stay in the sun for 300 minutes without burning with a number 15 lotion.

84. Suppose that the cost of an item, excluding tax, is x dollars. The algebraic expression $0.06x$ describes the sales tax on that item. Evaluate the algebraic expression for $x = 400$. Describe what the answer means in practical terms. 24;A $400 item has sales tax of $24.

The algebraic expression

$$405x + 5565$$

gives an approximate description of credit-card debt per U.S. household x years after 1994. The actual debt per household is shown in the bar graph at the top of the next column. Use this information to solve Exercises 85–86.

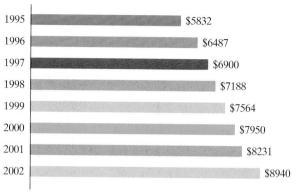

Credit-Card Debt per U.S. Household

Year	Debt
1995	$5832
1996	$6487
1997	$6900
1998	$7188
1999	$7564
2000	$7950
2001	$8231
2002	$8940

Source: Cardweb.com, Inc.

85. Evaluate the algebraic expression for $x = 8$. Describe what the answer means in practical terms. How well does the algebraic expression model the actual data for the appropriate year shown in the bar graph? 8805; In 2002, credit-card debt per U.S. household was $8805. Fairly well

86. Evaluate the algebraic expression for $x = 7$. Describe what the answer means in practical terms. How well does the algebraic expression model the actual data for the appropriate year shown in the bar graph? 8400; In 2001, credit-card debt per U.S. household was $8400. Fairly well

87. Testosterone is the hormone of choice for hundreds of thousands of middle-aged American men determined to stave off symptoms of andropause, the male version of menopause. The algebraic expression

$$2(0.18x + 0.01) + 0.02x$$

gives an approximate description of the millions of prescriptions for testosterone in the United States x years after 1997. The line graph indicates the actual number of testosterone prescriptions.

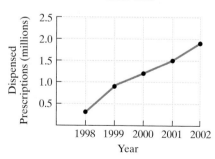

Testosterone Prescriptions in the U.S.

Source: NDC HEALTH

a. Simplify the algebraic expression. $0.38x + 0.02$

b. Use the simplified algebraic expression to find the number of testosterone prescriptions in 2002. How well does the algebraic expression model the data shown in the line graph for 2002? 1.92 million; very well

88. The equivalent algebraic expressions

$$\frac{DA + D}{24} \quad \text{and} \quad \frac{D(A + 1)}{24}$$

describe the drug dosage for children between the ages of 2 and 13. In each algebraic expression, D stands for an adult dose and A represents the child's age. If an adult dose of ibuprofen is 200 milligrams, what is the proper dose for a 12-year-old child? Use both forms of the algebraic expressions to answer the question. Which form is easier to use? The proper dose for a 12-year-old is approximately 108 mg. Answers will vary.

Writing in Mathematics

89. What is an algebraic expression? Provide an example with your description.   89–100. Answers will vary.

90. What does it mean to evaluate an algebraic expression? Provide an example with your description.

91. What is a term? Provide an example with your description.

92. What are like terms? Provide an example with your description.

93. What are equivalent algebraic expressions?

94. State a commutative property and give an example.

95. State an associative property and give an example.

96. State a distributive property and give an example.

97. Explain how to add like terms. Give an example.

98. What does it mean to simplify an algebraic expression?

99. An algebra student incorrectly used the distributive property and wrote $3(5x + 7) = 15x + 7$. If you were that student's teacher, what would you say to help the student avoid this kind of error?

100. You can transpose the letters in the word "conversation" to form the phrase "voices rant on." From "total abstainers" we can form "sit not at ale bars." What two algebraic properties do each of these transpositions (called anagrams) remind you of? Explain your answer.

Critical Thinking Exercises

101. Which one of the following statements is true?   c
  a. Subtraction is a commutative operation.
  b. $(24 \div 6) \div 2 = 24 \div (6 \div 2)$
  c. $7y + 3y = (7 + 3)y$ for any value of y
  d. $2x + 5 = 5x + 2$

102. Which one of the following statements is true?   c
  a. $a + (bc) = (a + b)(a + c)$ In words, addition can be distributed over multiplication.
  b. $4(x + 3) = 4x + 3$
  c. Not every algebraic expression can be simplified.
  d. Like terms contain the same coefficients.

103. A business that manufactures small alarm clocks has weekly fixed costs of $5000. The average cost per clock for the business to manufacture x clocks is described by

$$\frac{0.5x + 5000}{x}.$$

  a. Find the average cost when $x = 100$, 1000, and 10,000. $50.50; $5.50; $1
  b. Like all other businesses, the alarm clock manufacturer must make a profit. To do this, each clock must be sold for at least 50¢ more than what it costs to manufacture. Due to competition from a larger company, the clocks can be sold for $1.50 each and no more. Our small manufacturer can only produce 2000 clocks weekly. Does this business have much of a future? Explain. No. When producing 2000 clocks, the average cost is $3. This is more than the $1.50 maximum price at which the manufacturer can sell the clocks.

Review Exercises

104. Express $\frac{4}{9}$ as a decimal. (Section 1.2, Example 4) $0.\overline{4}$

105. Plot $(-3, -1)$ in a rectangular coordinate system. (Section 1.3, Example 1)   See graphing answer section.

106. Divide: $\frac{3}{7} \div \frac{15}{7}$. (Section 1.1, Example 6) $\frac{1}{5}$

✔ MID-CHAPTER CHECK POINT

CHAPTER 1

What You Know: We reviewed operations with fractions. We defined the real numbers and represented them as points on a number line. We used the rectangular coordinate system to represent ordered pairs of real numbers. We saw how information is displayed in line and bar graphs. Finally, we introduced some basic rules of algebra and used the commutative, associative, and distributive properties to simplify algebraic expressions.

In Exercises 1–9, perform the indicated operation or simplify the given expression. Where possible, reduce fractional answers to lowest terms.

1. $15a + 14 + 9a - 13$ $24a + 1$

2. $\frac{7}{10} - \frac{8}{15}$ $\frac{1}{6}$

3. $\frac{2}{3} \cdot \frac{3}{4}$ $\frac{1}{2}$

4. $7(9x + 3) + \frac{1}{3}(6x - 15)$ $65x + 16$

5. $\dfrac{5}{22} + \dfrac{5}{33}$ $\dfrac{25}{66}$

6. $\dfrac{3}{5} \div \dfrac{9}{10}$ $\dfrac{2}{3}$

7. $\dfrac{23}{105} - \dfrac{2}{105}$ $\dfrac{1}{5}$

8. $2\dfrac{7}{9} \div 3$ $\dfrac{25}{27}$

9. $5\dfrac{2}{9} - 3\dfrac{1}{6}$ $\dfrac{37}{18}$ or $2\dfrac{1}{18}$

10. Plot $\left(3, -5\dfrac{1}{2}\right)$ in a rectangular coordinate system. In which quadrant does the point lie? See graphing answer section, IV

In Exercises 11–13, rewrite $5(x + 3)$ as an equivalent expression using the given property.

11. the commutative property of multiplication $(x + 3)5$

12. the commutative property of addition $5(3 + x)$

13. the distributive property $5x + 15$

Use the bar graph to solve Exercises 14–15.

Percentage of Americans Ages 12 and Older Listening to Internet Radio

Source: Edison Media Research

14. Find an estimate for the difference between the percentage of Americans listening to Internet radio in 2004 and in 2001. 7%

15. For which years shown did the percentage of Americans listening to Internet radio exceed 7%? 2002, 2003, 2004

16. Insert either $<$ or $>$ in the shaded area to make a true statement:

$$-8000 \quad -8\dfrac{1}{4}. \quad <$$

17. List all the rational numbers in this set:

$$\left\{-11, -\dfrac{3}{7}, 0, 0.45, \sqrt{23}, \sqrt{25}\right\}. \quad -11, -\dfrac{3}{7}, 0, 0.45, \sqrt{25}$$

In Exercises 18–19, evaluate each algebraic expression for the given value of the variable.

18. $10x + 7; x = \dfrac{3}{5}$ 13

19. $8(x - 2); x = \dfrac{5}{2}$ 4

Use the graphs to solve Exercises 20–22.

Grades of U.S. Undergraduate College Students

Source: UCLA Higher Education Research Institute

20. What are the coordinates of point B? What does this mean in terms of the information given by the graph? (67, 20); In 1967, 20% of grades of undergraduate college students were A's.

21. For the period shown, when did the percentage of grades consisting of C or below reach a maximum? What is a reasonable estimate of the percentage of grades in this category for that year? 1970; 23%

22. Find the difference between the percentage of A's in 2003 and the percentage of C's or below in 2003. 40%

23. Find the absolute value: $|-19.3|$. 19.3

24. Express $\dfrac{1}{11}$ as a decimal. $0.\overline{09}$

25. Rewrite without parentheses:

$$8(7x - 10 + 3y). \quad 56x - 80 + 24y$$

Objectives

1 Add numbers with a number line.

2 Find sums using identity and inverse properties.

3 Add numbers without a number line.

4 Use addition rules to simplify algebraic expressions.

5 Solve applied problems using a series of additions.

ADDITION OF REAL NUMBERS

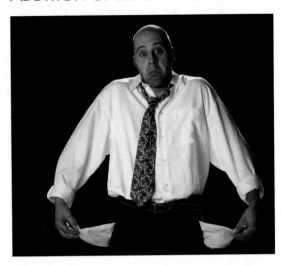

It has not been a good day! First, you lost your wallet with $30 in it. Then, to get through the day, you borrowed $10, which you somehow misplaced. Your loss of $30 followed by a loss of $10 is an overall loss of $40. This can be written

$$-30 + (-10) = -40.$$

The result of adding two or more numbers is called the **sum** of the numbers. The sum of -30 and -10 is -40. You can think of gains and losses of money to find sums. For example, to find $17 + (-13)$, think of a gain of $17 followed by a loss of $13. There is an overall gain of $4. Thus, $17 + (-13) = 4$. In the same way, to find $-17 + 13$, think of a loss of $17 followed by a gain of $13. There is an overall loss of $4, so $-17 + 13 = -4$.

1 Add numbers with a number line.

Adding with a Number Line We use the number line to help picture the addition of real numbers. Here is the procedure for finding $a + b$, the sum of a and b, using the number line:

> **USING THE NUMBER LINE TO FIND A SUM** Let a and b represent real numbers. To find $a + b$ using a number line,
> **1.** Start at a.
> **2. a.** If b is **positive**, move b units to the **right**.
>   **b.** If b is **negative**, move b units to the **left**.
>   **c.** If b is **0**, **stay** at a.
> **3.** The number where we finish on the number line represents the sum of a and b.

This procedure is illustrated in Examples 1 and 2. Think of moving to the right as a gain and moving to the left as a loss.

EXAMPLE 1 Adding Real Numbers Using a Number Line

Find the sum using a number line:

$$3 + (-5).$$

SOLUTION   We find $3 + (-5)$ using the number line in Figure 1.19.

Step 1.   We consider 3 to be the first number, represented by a in the preceding box. We start at a, or 3.

Step 2.   We consider -5 to be the second number, represented by b. Because this number is negative, we move 5 units to the left.

Step 3.   We finish at -2 on the number line. The number where we finish represents the sum of 3 and -5. Thus,

$$3 + (-5) = -2.$$

FIGURE 1.19 $3 + (-5) = -2$

Observe that if there is a gain of \$3 followed by a loss of \$5, there is an overall loss of \$2.

 CHECK POINT 1 Find the sum using a number line:

$$4 + (-7). \quad \text{−3; See graphing answer section.}$$

EXAMPLE 2   **Adding Real Numbers Using a Number Line**

Find each sum using a number line: **a.** $-3 + (-4)$ **b.** $-6 + 2$.

SOLUTION

a. To find $-3 + (-4)$, start at -3. Move 4 units to the left. We finish at -7. Thus,

$$-3 + (-4) = -7.$$

Observe that if there is a loss of \$3 followed by a loss of \$4, there is an overall loss of \$7.

b. To find $-6 + 2$, start at -6. Move 2 units to the right because 2 is positive. We finish at -4. Thus,

$$-6 + 2 = -4.$$

Observe that if there is a loss of \$6 followed by a gain of \$2, there is an overall loss of \$4.

 CHECK POINT 2 Find each sum using a number line:

a. $-1 + (-3)$ -4 **b.** $-5 + 3$. -2 See graphing answer section.

2 Find sums using identity and inverse properties.

The Number Line and Properties of Addition The number line can be used to picture some useful properties of addition. For example, let's see what happens if we add two numbers with different signs but the same absolute value. Two such numbers are 3 and -3. To find $3 + (-3)$ on a number line, we start at 3 and move 3 units to the left. We finish at 0. Thus,

$$3 + (-3) = 0.$$

Finish at 0. Move 3 units left. Start at 3.

$-5 \quad -4 \quad -3 \quad -2 \quad -1 \quad 0 \quad 1 \quad 2 \quad 3 \quad 4 \quad 5$

Numbers that are opposites, such as 3 and -3, are called *additive inverses*. **Additive inverses** are pairs of real numbers that are the same number of units from zero on the number line, but are on opposite sides of zero. Thus, -3 is the additive inverse of 3, and 5 is the additive inverse of -5. The additive inverse of 0 is 0. Other additive inverses come in pairs.

Additive inverses

$-5 \quad -4 \quad -3 \quad -2 \quad -1 \quad 0 \quad 1 \quad 2 \quad 3 \quad 4 \quad 5$

Additive inverses

In general, the sum of any real number, denoted by a, and its additive inverse, denoted by $-a$, is zero:

$$a + (-a) = 0.$$

This property is called the **inverse property of addition**. In Section 1.4, we discussed the commutative and associative properties of addition. We now add two additional properties to our previous list, shown in Table 1.4.

Table 1.4 **Identity and Inverse Properties of Addition**

Let a be a real number, a variable, or an algebraic expression.

Property	Meaning	Examples
Identity Property of Addition	Zero can be deleted from a sum. $a + 0 = a$ $0 + a = a$	• $4 + 0 = 4$ • $-3x + 0 = -3x$ • $0 + (5a + b) = 5a + b$
Inverse Property of Addition	The sum of a real number and its additive inverse gives 0, the additive identity. $a + (-a) = 0$ $(-a) + a = 0$	• $6 + (-6) = 0$ • $3x + (-3x) = 0$ • $[-(2y + 1)] + (2y + 1) = 0$

3 Add numbers without a number line.

Adding without a Number Line Now that we can picture the addition of real numbers, we look at two rules for using absolute value to add signed numbers.

ADDING TWO NUMBERS WITH THE SAME SIGN
  1. Add the absolute values.
  2. Use the common sign as the sign of the sum.

EXAMPLE 3 Adding Real Numbers

Add without using a number line:

 a. $-11 + (-15)$ **b.** $-0.2 + (-0.8)$ **c.** $-\dfrac{3}{4} + \left(-\dfrac{1}{2}\right).$

SOLUTION In each part of this example, we are adding numbers with the same sign.

 a. $-11 + (-15) = -26$ Add absolute values: $11 + 15 = 26.$

     Use the common sign.

 b. $-0.2 + (-0.8) = -1$ Add absolute values: $0.2 + 0.8 = 1.0$ or $1.$

     Use the common sign.

 c. $-\dfrac{3}{4} + \left(-\dfrac{1}{2}\right) = -\dfrac{5}{4}$ Add absolute values: $\frac{3}{4} + \frac{1}{2} = \frac{3}{4} + \frac{2}{4} = \frac{5}{4}.$

     Use the common sign.

✔ **CHECK POINT 3** Add without using a number line:

 a. $-10 + (-25)$ -35 **b.** $-0.3 + (-1.2)$ -1.5 **c.** $-\dfrac{2}{3} + \left(-\dfrac{1}{6}\right).$ $-\dfrac{5}{6}$

We also use absolute value to add two real numbers with different signs.

ADDING TWO NUMBERS WITH DIFFERENT SIGNS

 1. Subtract the smaller absolute value from the greater absolute value.
 2. Use the sign of the number with the greater absolute value as the sign of the sum.

EXAMPLE 4 Adding Real Numbers

Add without using a number line: **a.** $-13 + 4$ **b.** $-0.2 + 0.8$ **c.** $-\dfrac{3}{4} + \dfrac{1}{2}.$

SOLUTION In each part of this example, we are adding numbers with different signs.

 a. $-13 + 4 = -9$ Subtract absolute values: $13 - 4 = 9.$

     Use the sign of the number with the greater absolute value.

 b. $-0.2 + 0.8 = 0.6$ Subtract absolute values: $0.8 - 0.2 = 0.6.$

     Use the sign of the number with the greater absolute value. The sign is assumed to be positive.

 c. $-\dfrac{3}{4} + \dfrac{1}{2} = -\dfrac{1}{4}$ Subtract absolute values: $\frac{3}{4} - \frac{1}{2} = \frac{3}{4} - \frac{2}{4} = \frac{1}{4}.$

     Use the sign of the number with the greater absolute value.

✔ **CHECK POINT 4** Add without using a number line:

a. $-15 + 2$ $_{-13}$ **b.** $-0.4 + 1.6$ $_{1.2}$ **c.** $-\dfrac{2}{3} + \dfrac{1}{6}$. $_{-\frac{1}{2}}$

4 Use addition rules to simplify algebraic expressions.

Algebraic Expressions The rules for adding real numbers can be used to simplify certain algebraic expressions.

EXAMPLE 5 Simplifying Algebraic Expressions

Simplify:

   a. $-11x + 7x$

   b. $7y + (-12z) + (-9y) + 15z$

   c. $3(8 - 7x) + 5(2x - 4)$.

SOLUTION

   a. $-11x + 7x$ The given algebraic expression has two like terms. $-11x$ and $7x$ have identical variable factors.

      $= (-11 + 7)x$ Apply the distributive property.

      $= -4x$ Add within parentheses: $-11 + 7 = -4$.

   b. $7y + (-12z) + (-9y) + 15z$ The colors indicate that there are two pairs of like terms.

      $= 7y + (-9y) + (-12z) + 15z$ Arrange like terms so that they are next to one another.

      $= [7 + (-9)]y + [(-12) + 15]z$ Apply the distributive property.

      $= -2y + 3z$ Add within the grouping symbols: $7 + (-9) = -2$ and $-12 + 15 = 3$.

   c. $3(8 - 7x) + 5(2x - 4)$

      $= 3 \cdot 8 - 3 \cdot 7x + 5 \cdot 2x - 5 \cdot 4$ Use the distributive property to remove the parentheses.

      $= 24 - 21x + 10x - 20$ Multiply.

      $= (-21x + 10x) + (24 - 20)$ Group like terms.

      $= (-21 + 10)x + (24 - 20)$ Apply the distributive property.

      $= -11x + 4$ Perform operations within parentheses: $-21 + 10 = -11$ and $24 - 20 = 4$. ■

✔ **CHECK POINT 5** Simplify:

   a. $-20x + 3x$ $_{-17x}$ **b.** $3y + (-10z) + (-10y) + 16z$ $_{-7y + 6z}$

   c. $4(10 - 8x) + 4(3x - 5)$. $_{20 - 20x}$

5 Solve applied problems using a series of additions.

Applications Positive and negative numbers are used in everyday life to represent such things as gains and losses in the stock market, rising and falling temperatures, deposits and withdrawals on bank statements, and ascending and descending motion. Positive and negative numbers are used to solve applied problems involving a series of additions.

One way to add a series of positive and negative numbers is to use the commutative and associative properties.

- Add all the positive numbers.
- Add all the negative numbers.
- Add the sums obtained in the first two steps.

The next example illustrates this idea.

EXAMPLE 6 An Application of Adding Signed Numbers

A glider was towed 1000 meters into the air and then let go. It descended 70 meters into a thermal (rising bubble of warm air), which took it up 2100 meters. At this point it dropped 230 meters into a second thermal. Then it rose 1200 meters. What was its altitude at that point?

SOLUTION We use the problem's conditions to write a sum. The altitude of the glider is expressed by the following sum:

Towed to 1000 meters	then	Descended 70 meters	then	Taken up 2100 meters	then	Dropped 230 meters	then	Rose 1200 meters

$$1000 \quad + \quad (-70) \quad + \quad 2100 \quad + \quad (-230) \quad + \quad 1200.$$

$1000 + (-70) + 2100 + (-230) + 1200$ This is the sum arising from the problem's conditions.

$= (1000 + 2100 + 1200) + [(-70) + (-230)]$ Use the commutative and associative properties to group the positive and negative numbers.

$= 4300 + (-300)$ Add the positive numbers.
Add the negative numbers.

$= 4000$ Add the results.

The altitude of the glider is 4000 meters. ∎

DISCOVER FOR YOURSELF

Try working Example 6 by adding from left to right. You should still obtain 4000 for the sum. Which method do you find easier?

✔ **CHECK POINT 6** The water level of a reservoir is measured over a five-month period. During this time, the level rose 2 feet, then fell 4 feet, then rose 1 foot, then fell 5 feet, and then rose 3 feet. What was the change in the water level at the end of the five months?   The water level is down 3 feet at the end of 5 months.

1.5 EXERCISE SET

Student Solutions Manual CD/Video PH Math/Tutor Center MathXL Tutorials on CD MathXL® MyMathLab Interactmath.com

Practice Exercises

In Exercises 1–8, find each sum using a number line. **1–8.** See graphing answer section.

1. $7 + (-3)$ 4
2. $7 + (-2)$ 5
3. $-2 + (-5)$ -7
4. $-1 + (-5)$ -6
5. $-6 + 2$ -4
6. $-8 + 3$ -5
7. $3 + (-3)$ 0
8. $5 + (-5)$ 0

In Exercises 9–46, find each sum without the use of a number line.

9. $-7 + 0$ -7
10. $-5 + 0$ -5
11. $30 + (-30)$ 0
12. $15 + (-15)$ 0
13. $-30 + (-30)$ -60
14. $-15 + (-15)$ -30
15. $-8 + (-10)$ -18
16. $-4 + (-6)$ -10
17. $-0.4 + (-0.9)$ -1.3
18. $-1.5 + (-5.3)$ -6.8
19. $-\frac{7}{10} + \left(-\frac{3}{10}\right)$ -1
20. $-\frac{7}{8} + \left(-\frac{1}{8}\right)$ -1
21. $-9 + 4$ -5
22. $-7 + 3$ -4
23. $12 + (-8)$ 4
24. $13 + (-5)$ 8
25. $6 + (-9)$ -3
26. $3 + (-11)$ -8
27. $-3.6 + 2.1$ -1.5
28. $-6.3 + 5.2$ -1.1
29. $-3.6 + (-2.1)$ -5.7
30. $-6.3 + (-5.2)$ -11.5
31. $\frac{9}{10} + \left(-\frac{3}{5}\right)$ $\frac{3}{10}$
32. $\frac{7}{10} + \left(-\frac{2}{5}\right)$ $\frac{3}{10}$
33. $-\frac{5}{8} + \frac{3}{4}$ $\frac{1}{8}$
34. $-\frac{5}{6} + \frac{1}{3}$ $-\frac{1}{2}$
35. $-\frac{3}{7} + \left(-\frac{4}{5}\right)$ $-\frac{43}{35}$
36. $-\frac{3}{8} + \left(-\frac{2}{3}\right)$ $-\frac{25}{24}$
37. $4 + (-7) + (-5)$ -8
38. $10 + (-3) + (-8)$ -1
39. $85 + (-15) + (-20) + 12$ 62
40. $60 + (-50) + (-30) + 25$ 5
41. $17 + (-4) + 2 + 3 + (-10)$ 8
42. $19 + (-5) + 1 + 8 + (-13)$ 10
43. $-45 + \left(-\frac{3}{7}\right) + 25 + \left(-\frac{4}{7}\right)$ -21
44. $-50 + \left(-\frac{7}{9}\right) + 35 + \left(-\frac{11}{9}\right)$ -17
45. $3.5 + (-45) + (-8.4) + 72$ 22.1
46. $6.4 + (-35) + (-2.6) + 14$ -17.2

In Exercises 47–64, simplify each algebraic expression.

47. $-10x + 2x$ $-8x$
48. $-19x + 10x$ $-9x$
49. $25y + (-12y)$ $13y$
50. $26y + (-14y)$ $12y$
51. $-8a + (-15a)$ $-23a$
52. $-9a + (-13a)$ $-22a$
53. $-4 + 7x + 5 + (-13x)$ $1 - 6x$
54. $-5 + 8x + 3 + (-16x)$ $-2 - 8x$

55. $7b + 2 + (-b) + (-6)$ $-4 + 6b$
56. $10b + 7 + (-b) + (-15)$ $-8 + 9b$
57. $7x + (-5y) + (-9x) + 2y$ $-2x - 3y$
58. $13x + (-9y) + (-11x) + 3y$ $2x - 6y$
59. $4(5x - 3) + 6$ $20x - 6$
60. $5(2x - 3) + 4$ $10x - 11$
61. $8(3 - 4y) + 35y$ $24 + 3y$
62. $7(5 - 3y) + 25y$ $35 + 4y$
63. $6(2 - 9a) + 7(3a + 5)$ $47 - 33a$
64. $8(3 - 7a) + 4(2a + 3)$ $36 - 48a$

Practice Plus

In Exercises 65–68, find each sum.

65. $|-3 + (-5)| + |2 + (-6)|$ 12
66. $|4 + (-11)| + |-3 + (-4)|$ 14
67. $-20 + [-|15 + (-25)|]$ -30
68. $-25 + [-|18 + (-26)|]$ -33

In Exercises 69–70, insert either $<$, $>$, or $=$ in the shaded area to make a true statement.

69. $6 + [2 + (-13)]$ ⬚ $-3 + [4 + (-8)]$ $>$
70. $[(-8) + (-6)] + 10$ ⬚ $-8 + [9 + (-2)]$ $<$

Application Exercises

Solve Exercises 71–80 by writing a sum of signed numbers and adding.

71. The greatest temperature variation recorded in a day is 100 degrees in Browning, Montana, on January 23, 1916. The low temperature was $-56°$F. What was the high temperature? The high temperature was 44°.

72. In Spearfish, South Dakota, on January 22, 1943, the temperature rose 49 degrees in two minutes. If the initial temperature was $-4°$F, what was the high temperature? The high temperature was 45°.

73. The Dead Sea is the lowest elevation on earth, 1312 feet below sea level. What is the elevation of a person standing 712 feet above the Dead Sea? The elevation is 600 feet below sea level.

74. Lake Assal in Africa is 512 feet below sea level. What is the elevation of a person standing 642 feet above Lake Assal? The elevation is 130 feet above sea level.

75. The temperature at 8:00 A.M. was $-7°$F. By noon it had risen 15°F, but by 4:00 P.M. it had fallen 5°F. What was the temperature at 4:00 P.M.? The temperature at 4:00 P.M. was 3°F.

76. On three successive plays, a football team lost 15 yards, gained 13 yards, and then lost 4 yards. What was the team's total gain or loss for the three plays? The team had a total loss of 6 yards.

77. A football team started with the football at the 27-yard line, advancing toward the center of the field (the 50-yard line). Four successive plays resulted in a 4-yard gain, a 2-yard loss, an 8-yard gain, and a 12-yard loss. What was the location of the football at the end of the fourth play? The ball was at the 25-yard line at the end of the fourth play.

78. The water level of a reservoir is measured over a five-month period. At the beginning, the level is 20 feet. During this time, the level rose 3 feet, then fell 2 feet, then fell 1 foot, then fell 4 feet, and then rose 2 feet. What is the reservoir's water level at the end of the five months? The reservoir's level is 18 feet.

79. The bar graph shows the budget surplus or deficit for the United States government from 1999 through 2003.

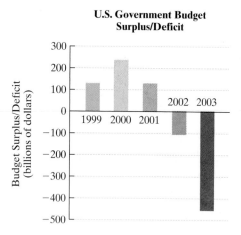

U.S. Government Budget Surplus/Deficit

Source: Budget of the U.S. Government

In 1999, the U.S. government had a budget surplus of $126 billion. This surplus then changed as follows:

    2000: increased by $110 billion

    2001: decreased by $109 billion

    2002: decreased by $233 billion

    2003: decreased by $349 billion.

What was the government's budget surplus or deficit in 2003? $455 billion deficit

80. According to *Newsweek*, the Ford Motor Company "swung from most profitable carmaker to biggest loser." The bar graph in the next column shows Ford's net profits and losses (the sum of income and expenses) from 1998 through 2002. In 1998, Ford's net profit was $22 billion. This profit then changed as follows:

    1999: decreased by $16 billion

    2000: decreased by $3 billion

    2001: decreased by $8 billion

    2002: increased by $4 billion.

What was the company's net profit or loss in 2002?
$1 billion loss

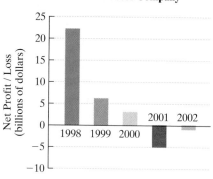

Net Profits and Losses of the Ford Motor Company

Source: Newsweek

Writing in Mathematics

81. Explain how to add two numbers with a number line. Provide an example with your explanation. **81–87.** Answers will vary.

82. What are additive inverses?

83. Describe how the inverse property of addition

$$a + (-a) = 0$$

can be shown on a number line.

84. Without using a number line, describe how to add two numbers with the same sign. Give an example.

85. Without using a number line, describe how to add two numbers with different signs. Give an example.

86. Write a problem that can be solved by finding the sum of at least three numbers, some positive and some negative. Then explain how to solve the problem.

87. Without a calculator, you can add numbers using a number line, using absolute value, or using gains and losses. Which method do you find most helpful? Why is this so?

Critical Thinking Exercises

88. Which one of the following statements is true? **c**

    a. The sum of a positive number and a negative number is a negative number.

    b. $|-9 + 2| = 9 + 2$

    c. If two numbers are both positive or both negative, then the absolute value of their sum equals the sum of their absolute values.

    d. $\dfrac{3}{4} + \left(-\dfrac{3}{5}\right) = -\dfrac{3}{20}$

89. Which one of the following statements is true? **d**

    a. The sum of a positive number and a negative number is a positive number.

    b. If one number is positive and the other negative, then the absolute value of their sum equals the sum of their absolute values.

    c. $\dfrac{3}{4} + \left(-\dfrac{2}{3}\right) = -\dfrac{1}{12}$

    d. The sum of zero and a negative number is always a negative number.

In Exercises 90–91, find the missing term.

90. $5x + \underline{} + (-11x) + (-6y) = -6x + 2y$ $8y$

91. $\underline{} + 11x + (-3y) + 3x = 7(2x - 3y)$ $-18y$

Technology Exercises

92. Use a calculator to verify any five of the sums that you found in Exercises 17–46.   **92–93.** Answers will vary.

93. Use a calculator to verify any three of the answers that you obtained in Application Exercises 71–80.

Review Exercises

94. Determine whether this inequality is true or false: $19 \geq -18$. (Section 1.2, Example 7)   true

95. Consider the set

$$\left\{-6, -\pi, 0, 0.\overline{7}, \sqrt{3}, \sqrt{4}\right\}.$$

List all numbers from the set that are **a.** natural numbers, **b.** whole numbers, **c.** integers, **d.** rational numbers, **e.** irrational numbers, **f.** real numbers. (Section 1.2, Example 5)   **a.** $\sqrt{4}$
b. $0, \sqrt{4}$ **c.** $-6, 0, \sqrt{4}$ **d.** $-6, 0, 0.\overline{7}, \sqrt{4}$ **e.** $-\pi, \sqrt{3}$
f. $-6, -\pi, 0, 0.\overline{7}, \sqrt{3}, \sqrt{4}$

96. Plot $\left(\frac{7}{2}, -\frac{5}{2}\right)$ in a rectangular coordinate system. In which quadrant does the point lie? (Section 1.3, Example 1)
IV; See graphing answer section.

SUBTRACTION OF REAL NUMBERS

People are going to live longer in the 21st century. This will put added pressure on the Social Security and Medicare systems. How insecure is Social Security's future? In this section, we use subtraction of real numbers to numerically describe one aspect of the insecurity.

The Meaning of Subtraction   Time for a new computer! Your favorite model, which normally sells for $1500, has an incredible price reduction of $600. The computer's reduced price, $900, can be expressed in two ways:

$$1500 - 600 = 900 \quad \text{or} \quad 1500 + (-600) = 900.$$

This means that

$$1500 - 600 = 1500 + (-600).$$

To subtract 600 from 1500, we add 1500 and the additive inverse of 600. Generalizing from this situation, we define subtraction as follows:

> **DEFINITION OF SUBTRACTION**   For all real numbers a and b,
>
> $$a - b = a + (-b).$$
>
> In words: To subtract b from a, add the additive inverse of b to a. The result of subtraction is called the **difference**.

1 Subtract real numbers.

A Procedure for Subtracting Real Numbers The definition of subtraction gives us a procedure for subtracting real numbers.

> **SUBTRACTING REAL NUMBERS**
> 1. Change the subtraction operation to addition.
> 2. Change the sign of the number being subtracted.
> 3. Add, using one of the rules for adding numbers with the same sign or different signs.

USING TECHNOLOGY

You can use a calculator to subtract signed numbers. Here are the keystrokes for finding $5 - (-6)$:

Scientific Calculator

$$5 \;\boxed{-}\; 6 \;\boxed{+/_-}\; \boxed{=}$$

Graphing Calculator

$$5 \;\boxed{-}\; \boxed{(-)} \; 6 \;\boxed{ENTER}$$

Here are the keystrokes for finding $-9 - (-3)$:

Scientific Calculator

$$9 \;\boxed{+/_-}\; \boxed{-}\; 3 \;\boxed{+/_-}\; \boxed{=}$$

Graphing Calculator

$$\boxed{(-)} \; 9 \;\boxed{-}\; \boxed{(-)} \; 3 \;\boxed{ENTER}$$

Don't confuse the subtraction key on a graphing calculator, $\boxed{-}$, with the sign change or additive inverse key, $\boxed{(-)}$. What happens if you do?

EXAMPLE 1 Using the Definition of Subtraction

Subtract: **a.** $7 - 10$ **b.** $5 - (-6)$ **c.** $-9 - (-3)$.

SOLUTION

a. $7 - 10 = 7 + (-10) = -3$

> Change the subtraction to addition.   Replace 10 with its additive inverse.

b. $5 - (-6) = 5 + 6 = 11$

> Change the subtraction to addition.   Replace −6 with its additive inverse.

c. $-9 - (-3) = -9 + 3 = -6$

> Change the subtraction to addition.   Replace −3 with its additive inverse.

✔ **CHECK POINT 1** Subtract:

a. $3 - 11$ -8 **b.** $4 - (-5)$ 9 **c.** $-7 - (-2)$. -5

The definition of subtraction can be applied to real numbers that are not integers.

EXAMPLE 2 Using the Definition of Subtraction

Subtract: **a.** $-5.2 - (-11.4)$ **b.** $-\dfrac{3}{4} - \dfrac{2}{3}$ **c.** $4\pi - (-9\pi)$.

SOLUTION

a. $-5.2 - (-11.4) = -5.2 + 11.4 = 6.2$

> Change the subtraction to addition.   Replace −11.4 with its additive inverse.

b. $-\dfrac{3}{4} - \dfrac{2}{3} = -\dfrac{3}{4} + \left(-\dfrac{2}{3}\right) = -\dfrac{9}{12} + \left(-\dfrac{8}{12}\right) = -\dfrac{17}{12}$

> Change the subtraction to addition.   Replace $\frac{2}{3}$ with its additive inverse.

c. $4\pi - (-9\pi) = 4\pi + 9\pi = (4 + 9)\pi = 13\pi$

> Change the subtraction to addition.   Replace −9π with its additive inverse.

Reading the symbol "−" can be a bit tricky. The way you read it depends on where it appears. For example,

$$-5.2 - (-11.4)$$

is read "negative five point two minus negative eleven point four." Read parts (b) and (c) of Example 2 aloud. When is "−" read "negative" and when is it read "minus"?

 CHECK POINT 2 Subtract:

a. $-3.4 - (-12.6)$ 9.2   **b.** $-\dfrac{3}{5} - \dfrac{1}{3}$ $-\dfrac{14}{15}$ **c.** $5\pi - (-2\pi)$. 7π

2 Simplify a series of additions and subtractions.

Problems Containing a Series of Additions and Subtractions   In some problems, several additions and subtractions occur together. We begin by converting all subtractions to additions of additive inverses, or opposites.

SIMPLIFYING A SERIES OF ADDITIONS AND SUBTRACTIONS

1. Change all subtractions to additions of additive inverses.
2. Group and then add all the positive numbers.
3. Group and then add all the negative numbers.
4. Add the results of steps 2 and 3.

EXAMPLE 3   Simplifying a Series of Additions and Subtractions

Simplify: $7 - (-5) - 11 - (-6) - 19$.

SOLUTION

$$7 - (-5) - 11 - (-6) - 19$$
$$= 7 + 5 + (-11) + 6 + (-19) \qquad \text{Write subtractions as additions of additive inverses.}$$
$$= (7 + 5 + 6) + [(-11) + (-19)] \qquad \text{Group the positive numbers. Group the negative numbers.}$$
$$= 18 + (-30) \qquad \text{Add the positive numbers. Add the negative numbers.}$$
$$= -12 \qquad \text{Add the results.} \qquad ∎$$

 CHECK POINT 3 Simplify: $10 - (-12) - 4 - (-3) - 6$.   15

---

3 Use the definition of subtraction to identify terms.

Subtraction and Algebraic Expressions   We know that the terms of an algebraic expression are separated by addition signs. Let's use this idea to identify the terms of the following algebraic expression:

$$9x - 4y - 5.$$

Because terms are separated by addition, we rewrite the algebraic expression as additions of additive inverses, or opposites. Thus,

$$9x - 4y - 5 = 9x + (-4y) + (-5).$$

The three terms of the algebraic expression are $9x$, $-4y$, and -5.

EXAMPLE 4 Using the Definition of Subtraction to Identify Terms

Identify the terms of the algebraic expression:

$$2xy - 13y - 6.$$

SOLUTION Rewrite the algebraic expression as additions of additive inverses.

$$2xy - 13y - 6 = 2xy + (-13y) + (-6)$$

First term · Second term · Third term

Because terms are separated by addition, the terms are $2xy$, $-13y$, and -6. ∎

 CHECK POINT 4 Identify the terms of the algebraic expression:

$$-6 + 4a - 7ab. \quad -6, 4a, -7ab$$

4 | Use the subtraction definition to simplify algebraic expressions.

The procedure for subtracting real numbers can be used to simplify certain algebraic expressions that involve subtraction.

EXAMPLE 5 Simplifying Algebraic Expressions

Simplify: **a.** $2 + 3x - 8x$ **b.** $-4x - 9y - 2x + 12y.$

SOLUTION

a. $2 + 3x - 8x$ This is the given algebraic expression.

$= 2 + 3x + (-8x)$ Write the subtraction as the addition of an additive inverse.

$= 2 + [3 + (-8)]x$ Apply the distributive property.

$= 2 + (-5x)$ Add within the grouping symbols

$= 2 - 5x$ Be concise and express as subtraction.

b. $-4x - 9y - 2x + 12y$ This is the given algebraic expression.

$= -4x + (-9y) + (-2x) + 12y$ Write the subtractions as the additions of additive inverses.

$= -4x + (-2x) + (-9y) + 12y$ Arrange like terms so that they are next to one another.

$= [-4 + (-2)]x + (-9 + 12)y$ Apply the distributive property.

$= -6x + 3y$ Add within the grouping symbols.

STUDY TIP

You can think of gains and losses of money to work the distributive property mentally:

- $3x - 8x = -5x$ A gain of 3 dollars followed by a loss of 8 of those dollars is a net loss of 5 of those dollars.

- $-9y + 12y = 3y$ A loss of 9 dollars followed by a gain of 12 of those dollars is a net gain of 3 of those dollars.

 CHECK POINT 5 Simplify:

a. $4 + 2x - 9x$ $4 - 7x$ **b.** $-3x - 10y - 6x + 14y.$ $-9x + 4y$

5 Solve problems involving subtraction.

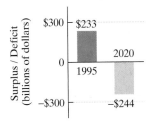

FIGURE 1.20 Social Security Annual Surplus/Deficit

Source: U.S. Office of Management and Budget

Applications Subtraction is used to solve problems in which the word "difference" appears. The difference between real numbers a and b is expressed as $a - b$.

EXAMPLE 6 An Application of Subtraction Using the Word "Difference"

The bar graph in Figure 1.20 shows that in 1995, Social Security had an annual cash surplus of $233 billion. By 2020, this amount is expected to be a negative number—a deficit of $244 billion. What is the difference between the 1995 surplus and the projected 2020 deficit?

SOLUTION

The difference | is | the 1995 surplus | minus | the 2020 deficit.

$$= 233 - (-244)$$
$$= 233 + 244 = 477$$

The difference between the 1995 surplus and the projected 2020 deficit is $477 billion. ∎

CHECK POINT 6 The peak of Mount Everest is 8848 meters above sea level. The Marianas Trench, on the floor of the Pacific Ocean, is 10,915 meters below sea level. What is the difference in elevation between the peak of Mount Everest and the Marianas Trench?   19,763 meters

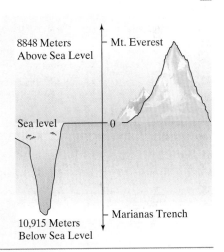

8848 Meters Above Sea Level — Mt. Everest

Sea level — 0

10,915 Meters Below Sea Level

Marianas Trench

1.6 EXERCISE SET

Student Solutions Manual   CD/Video   PH Math/Tutor Center   MathXL Tutorials on CD   MathXL®   MyMathLab   Interactmath.com

Practice Exercises

1. Consider the subtraction $5 - 12$.
  a. Find the additive inverse, or opposite, of 12. -12
  b. Rewrite the subtraction as the addition of the additive inverse of 12. $5 + (-12)$

2. Consider the subtraction $4 - 10$.
  a. Find the additive inverse, or opposite, of 10. -10
  b. Rewrite the subtraction as the addition of the additive inverse of 10. $4 + (-10)$

3. Consider the subtraction $5 - (-7)$.
  a. Find the additive inverse, or opposite, of -7. 7
  b. Rewrite the subtraction as the addition of the additive inverse of -7. $5 + 7$

4. Consider the subtraction $2 - (-8)$.
  a. Find the additive inverse, or opposite, of -8. 8
  b. Rewrite the subtraction as the addition of the additive inverse of -8. $2 + 8$

In Exercises 5–50, perform the indicated subtraction.

5. $14 - 8$ 6
6. $15 - 2$ 13

7. $8 - 14$ -6
8. $2 - 15$ -13

9. $3 - (-20)$ 23
10. $5 - (-17)$ 22

11. $-7 - (-18)$ 11
12. $-5 - (-19)$ 14

13. $-13 - (-2)$ -11
14. $-21 - (-3)$ -18

15. $-21 - 17$ -38
16. $-29 - 21$ -50

17. $-45 - (-45)$ 0   **18.** $-65 - (-65)$ 0

19. $23 - 23$ 0   **20.** $26 - 26$ 0

21. $13 - (-13)$ 26   **22.** $15 - (-15)$ 30

23. $0 - 13$ -13 **24.** $0 - 15$ -15

25. $0 - (-13)$ 13   **26.** $0 - (-15)$ 15

27. $\dfrac{3}{7} - \dfrac{5}{7}$ $-\dfrac{2}{7}$ **28.** $\dfrac{4}{9} - \dfrac{7}{9}$ $-\dfrac{1}{3}$

29. $\dfrac{1}{5} - \left(-\dfrac{3}{5}\right)$ $\dfrac{4}{5}$ **30.** $\dfrac{1}{7} - \left(-\dfrac{3}{7}\right)$ $\dfrac{4}{7}$

31. $-\dfrac{4}{5} - \dfrac{1}{5}$ -1 **32.** $-\dfrac{4}{9} - \dfrac{1}{9}$ $\dfrac{5}{9}$

33. $-\dfrac{4}{5} - \left(-\dfrac{1}{5}\right)$ $-\dfrac{3}{5}$ **34.** $-\dfrac{4}{9} - \left(-\dfrac{1}{9}\right)$ $-\dfrac{1}{3}$

35. $\dfrac{1}{2} - \left(-\dfrac{1}{4}\right)$ $\dfrac{3}{4}$ **36.** $\dfrac{2}{5} - \left(-\dfrac{1}{10}\right)$ $\dfrac{1}{2}$

37. $\dfrac{1}{2} - \dfrac{1}{4}$ $\dfrac{1}{4}$ **38.** $\dfrac{2}{5} - \dfrac{1}{10}$ $\dfrac{3}{10}$

39. $9.8 - 2.2$ 7.6   **40.** $5.7 - 3.3$ 2.4

41. $-3.1 - (-1.1)$ -2 **42.** $-4.6 - (-1.1)$ -3.5

43. $1.3 - (-1.3)$ 2.6   **44.** $1.4 - (-1.4)$ 2.8

45. $-2.06 - (-2.06)$ 0   **46.** $-3.47 - (-3.47)$ 0

47. $5\pi - 2\pi$ 3π **48.** $9\pi - 7\pi$ 2π

49. $3\pi - (-10\pi)$ 13π **50.** $4\pi - (-12\pi)$ 16π

In Exercises 51–68, simplify each series of additions and subtractions.

51. $13 - 2 - (-8)$ 19   **52.** $14 - 3 - (-7)$ 18

53. $9 - 8 + 3 - 7$ -3 **54.** $8 - 2 + 5 - 13$ -2

55. $-6 - 2 + 3 - 10$ -15 **56.** $-9 - 5 + 4 - 17$ -27

57. $-10 - (-5) + 7 - 2$ 0   **58.** $-6 - (-3) + 8 - 11$ -6

59. $-23 - 11 - (-7) + (-25)$ -52

60. $-19 - 8 - (-6) + (-21)$ -42

61. $-823 - 146 - 50 - (-832)$ -187

62. $-726 - 422 - 921 - (-816)$ -1253

63. $1 - \dfrac{2}{3} - \left(-\dfrac{5}{6}\right)$ $\dfrac{7}{6}$ **64.** $2 - \dfrac{3}{4} - \left(-\dfrac{7}{8}\right)$ $\dfrac{17}{8}$

65. $-0.16 - 5.2 - (-0.87)$ -4.49

66. $-1.9 - 3 - (-0.26)$ -4.64

67. $-\dfrac{3}{4} - \dfrac{1}{4} - \left(-\dfrac{5}{8}\right)$ $-\dfrac{3}{8}$ **68.** $-\dfrac{1}{2} - \dfrac{2}{3} - \left(-\dfrac{1}{3}\right)$ $-\dfrac{5}{6}$

In Exercises 69–72, identify the terms in each algebraic expression.

69. $-3x - 8y$ $-3x, -8y$ **70.** $-9a - 4b$ $-9a, -4b$

71. $12x - 5xy - 4$ $12x, -5xy, -4$

72. $8a - 7ab - 13$ $8a, -7ab, -13$

In Exercises 73–84, simplify each algebraic expression.

73. $3x - 9x$ $-6x$ **74.** $2x - 10x$ $-8x$

75. $4 + 7y - 17y$ $4 - 10y$ **76.** $5 + 9y - 29y$ $5 - 20y$

77. $2a + 5 - 9a$ $5 - 7a$ **78.** $3a + 7 - 11a$ $7 - 8a$

79. $4 - 6b - 8 - 3b$ $-4 - 9b$

80. $5 - 7b - 13 - 4b$ $-8 - 11b$

81. $13 - (-7x) + 4x - (-11)$ $24 + 11x$

82. $15 - (-3x) + 8x - (-10)$ $25 + 11x$

83. $-5x - 10y - 3x + 13y$ $3y - 8x$

84. $-6x - 9y - 4x + 15y$ $6y - 10x$

Practice Plus

In Exercises 85–90, find the value of each expression.

85. $-|-9 - (-6)| - (-12)$ 9   **86.** $-|-8 - (-2)| - (-6)$ 0

87. $\dfrac{5}{8} - \left(\dfrac{1}{2} - \dfrac{3}{4}\right)$ $\dfrac{7}{8}$ **88.** $\dfrac{9}{10} - \left(\dfrac{1}{4} - \dfrac{7}{10}\right)$ $\dfrac{27}{20}$ or $1\dfrac{7}{20}$

89. $|-9 - (-3 + 7)| - |-17 - (-2)|$ -2

90. $|24 - (-16)| - |-51 - (-31 + 2)|$ 18

Application Exercises

91. The peak of Mount Kilimanjaro, the highest point in Africa, is 19,321 feet above sea level. Qattara Depression, Egypt, one of the lowest points in Africa, is 436 feet below sea level. What is the difference in elevation between the peak of Mount Kilimanjaro and the Qattara Depression?   19,757 feet

92. The peak of Mount Whitney is 14,494 feet above sea level. Mount Whitney can be seen directly above Death Valley, which is 282 feet below sea level. What is the difference in elevation between these geographic locations?   14,776 feet

The bar graph shows the occupations with the greatest projected growth and the greatest projected losses from 2000 through 2010. Use the graph to solve Exercises 93–96.

Projected Changes in the Number of Jobs: 2000–2010

Source: U.S. Bureau of Labor Statistics

93. What is the difference, in thousands of jobs, between the projected increase in teachers' jobs and the projected decrease in farmers' jobs?   1039 thousand jobs

94. What is the difference, in thousands of jobs, between the projected increase in computer engineering jobs and the projected decrease in jobs for phone switchboard operators?   724 thousand jobs

95. By how many thousands of jobs does the decline for phone switchboard operators exceed the decline for farmers?   268 thousand jobs

96. By how many thousands of jobs does the decline for bank tellers exceed the decline for farmers?   269 thousand jobs

Do you enjoy cold weather? If so, try Fairbanks, Alaska. The average daily low temperature for each month in Fairbanks is shown in the bar graph. Use the graph to solve Exercises 97–100.

Each Month's Average Daily Low Temperature in Fairbanks, Alaska

Source: The Weather Channel Enterprises, Inc.

97. What is the difference between the average daily low temperatures for March and February?   21°F

98. What is the difference between the average daily low temperatures for October and November?   18°F

99. How many degrees warmer is February's average low temperature than January's average low temperature?   3°F

100. How many degrees warmer is November's average low temperature than December's average low temperature?   7°F

When a person receives a drug injected into a muscle, the concentration of the drug in the body depends on the time elapsed since the injection. The points in the rectangular system show the concentration of the drug, measured in milligrams per 100 milliliters, from the time the drug is injected until 13 hours later. Use this information to solve Exercises 101–106.

101. What is the drug's maximum concentration and when does this occur?   The maximum concentration is 0.05 milligrams per 100 milliliters and it occurs during the 3rd hour.

102. What happens by the end of 13 hours?   The concentration drops to zero milligrams per 100 milliliters.

103. What is the approximate difference between the drug's concentration 4 hours after it was injected and 1 hour after it was injected?   approximately 0.015 milligrams per 100 milliliters

104. What is the approximate difference between the drug's concentration 4 hours after it was injected and 7 hours after it was injected?   approximately 0.025 milligrams per 100 milliliters

105. When is the drug's concentration increasing?   from 0 to 3 hours

106. When is the drug's concentration decreasing?   from 3 to 13 hours

Writing in Mathematics

107. Explain how to subtract real numbers.   Answers will vary.

108. How is $4 - (-2)$ read?   four minus negative two

109. Explain how to simplify a series of additions and subtractions. Provide an example with your explanation.   **109–111.** Answers will vary.

110. Explain how to find the terms of the algebraic expression $5x - 2y - 7$.

111. Write a problem that can be solved by finding the difference between two numbers. At least one of the numbers should be negative. Then explain how to solve the problem.

Critical Thinking Exercises

112. Which one of the following statements is true?   c

    a. If a and b are negative numbers, then $a - b$ is a negative number.

    b. $7 - (-2) = 5$

    c. The difference between 0 and a negative number is always a positive number.

    d. None of the given statements is true.

113. The golden age of Athens culminated in 212 B.C. and the golden age of India culminated in A.D. 500. Determine the number of years that elapsed between these dates. (*Note:* When the calendar was reformed, the number 0 had not been invented. There was no year 0 and the year A.D. 1 followed the year 1 B.C. Calculate the difference between the years in the usual way and then use this added bit of information to modify your answer.)   711 years

114. Find the value:
$$-1 + 2 - 3 + 4 - 5 + 6 - \cdots - 99 + 100.$$ 50

Technology Exercises

115. Use a calculator to verify any five of the differences that you found in Exercises 5–46.   Answers will vary.

116. Use a calculator to verify any three of the answers that you found in Exercises 51–68.   Answers will vary.

Review Exercises

117. Graph on a number line: -4.5. (Section 1.2; Example 3)   See graphing answer section.

118. Use the commutative property of addition to write an equivalent algebraic expression: $10(a + 4)$. (Section 1.4; Example 4) $10(4 + a)$

119. Give an example of an integer that is not a natural number (Section 1.2; Example 5)   Answers will vary.

MULTIPLICATION AND DIVISION OF REAL NUMBERS

Objectives

1 Multiply real numbers.

2 Multiply more than two real numbers.

3 Find multiplicative inverses.

4 Use the definition of division.

5 Divide real numbers.

6 Simplify algebraic expressions involving multiplication.

7 Use algebraic expressions that model reality.

Technology is now promising to bring light, fast, and beautiful wheelchairs to millions of disabled people. The cost of manufacturing these radically different wheelchairs can be modeled by an algebraic expression containing division. In this section, we will see how this algebraic expression illustrates that low prices are possible with high production levels, urgently needed in this situation. There are more than half a billion people with disabilities in developing countries; an estimated 20 million need wheelchairs right now.

Multiplying Real Numbers Suppose that things go from bad to worse for Social Security, and the projected $244 billion deficit in 2020 triples by the end of the twenty-first century. The new deficit is

$$3(-244) = (-244) + (-244) + (-244) = -732$$

or $732 billion. Thus,

$$3(-244) = -732.$$

The result of the multiplication, -732, is called the **product** of 3 and -244. The numbers being multiplied, 3 and -244, are called the **factors** of the product.

　　Rules for multiplying real numbers are described in terms of absolute value. For example, $3(-244) = -732$ illustrates that the product of numbers with different signs is found by multiplying their absolute values. The product is negative.

$$3(-244) = -732$$

Multiply absolute values:
$|3| \cdot |-244| = 3 \cdot 244 = 732.$

Factors have different signs
and the product is negative.

1 Multiply real numbers.

The following rules are used to determine the sign of the product of two numbers:

THE PRODUCT OF TWO REAL NUMBERS

* The product of two real numbers with **different signs** is found by multiplying their absolute values. The product is **negative**.
* The product of two real numbers with the **same sign** is found by multiplying their absolute values. The product is **positive**.
* The product of 0 and any real number is 0. Thus, for any real number a,

$$a \cdot 0 = 0 \quad \text{and} \quad 0 \cdot a = 0.$$

EXAMPLE 1   Multiplying Real Numbers

Multiply:

a. $6(-3)$ **b.** $-\dfrac{1}{5} \cdot \dfrac{2}{3}$ **c.** $(-9)(-10)$ **d.** $(-1.4)(-2)$ **e.** $(-372)(0)$.

SOLUTION

a. $6(-3) = -18$

> Multiply absolute values: 6 · 3 = 18.

> Different signs: negative product

b. $-\dfrac{1}{5} \cdot \dfrac{2}{3} = -\dfrac{2}{15}$

> Multiply absolute values: $\frac{1}{5} \cdot \frac{2}{3} = \frac{1 \cdot 2}{5 \cdot 3} = \frac{2}{15}$.

> Different signs: negative product

c. $(-9)(-10) = 90$

> Multiply absolute values: 9 · 10 = 90.

> Same sign: positive product

d. $(-1.4)(-2) = 2.8$

> Multiply absolute values: (1.4)(2) = 2.8.

> Same sign: positive product

e. $(-372)(0) = 0$

> The product of 0 and any real number is 0: $a \cdot 0 = 0$.

 CHECK POINT 1 Multiply: **a.** $8(-5)$ $\quad -40$ **b.** $-\dfrac{1}{3} \cdot \dfrac{4}{7}$ $\quad -\dfrac{4}{21}$

c. $(-12)(-3)$ $\quad 36$ **d.** $(-1.1)(-5)$ $\quad 5.5$ **e.** $(-543)(0)$. $\quad 0$

---

2 Multiply more than two real numbers.

Multiplying More Than Two Numbers How do we perform more than one multiplication, such as

$$-4(-3)(-2)?$$

Because of the associative and commutative properties, we can order and group the numbers in any manner. Each pair of negative numbers will produce a positive product. Thus, the product of an even number of negative numbers is always positive. By contrast, the product of an odd number of negative numbers is always negative.

$$-4(-3)(-2) = -24$$

> Multiply absolute values: 4 · 3 · 2 = 24.

> Odd number of negative numbers (three): negative product

MULTIPLYING MORE THAN TWO NUMBERS

1. Assuming that no factor is zero,
 • The product of an **even** number of **negative numbers** is **positive**.
 • The product of an **odd** number of **negative numbers** is **negative**.
 The multiplication is performed by multiplying the absolute values of the given numbers.
2. If any factor is 0, the product is 0.

EXAMPLE 2 Multiplying More Than Two Numbers

Multiply: **a.** $(-3)(-1)(2)(-2)$ **b.** $(-1)(-2)(-2)(3)(-4)$.

SOLUTION

a. $(-3)(-1)(2)(-2) = -12$

> Multiply absolute values:
> $3 \cdot 1 \cdot 2 \cdot 2 = 12$.

> Odd number of negative numbers (three): negative product

b. $(-1)(-2)(-2)(3)(-4) = 48$

> Multiply absolute values:
> $1 \cdot 2 \cdot 2 \cdot 3 \cdot 4 = 48$.

> Even number of negative numbers (four): positive product

✔ **CHECK POINT 2** Multiply:

a. $(-2)(3)(-1)(4)$ 24   **b.** $(-1)(-3)(2)(-1)(5)$.   −30

Is it always necessary to count the number of negative factors when multiplying more than two numbers? No. If any factor is 0, you can immediately write 0 for the product. For example,

$$(-37)(423)(0)(-55)(-3.7) = 0.$$

> If any factor is 0, the product is 0.

③ Find multiplicative inverses.

The Meaning of Division   The result of dividing the real number a by the nonzero real number b is called the **quotient** of a and b. We can write this quotient as $a \div b$ or $\dfrac{a}{b}$.

We know that subtraction is defined in terms of addition of an additive inverse, or opposite:

$$a - b = a + (-b).$$

In a similar way, we can define division in terms of multiplication. For example, the quotient of 8 and 2 can be written as multiplication:

$$8 \div 2 = 8 \cdot \frac{1}{2}.$$

We call $\frac{1}{2}$ the *multiplicative inverse*, or *reciprocal*, of 2. Two numbers whose product is 1 are called **multiplicative inverses** or **reciprocals** of each other. Thus, the multiplicative inverse of 2 is $\frac{1}{2}$ and the multiplicative inverse of $\frac{1}{2}$ is 2 because $2 \cdot \frac{1}{2} = 1$.

EXAMPLE 3 Finding Multiplicative Inverses

Find the multiplicative inverse of each number:

a. 5   **b.** $\dfrac{1}{3}$ **c.** -4 **d.** $-\dfrac{4}{5}$.

SOLUTION

a. The multiplicative inverse of 5 is $\dfrac{1}{5}$ because $5 \cdot \dfrac{1}{5} = 1$.

b. The multiplicative inverse of $\dfrac{1}{3}$ is 3 because $\dfrac{1}{3} \cdot 3 = 1$.

c. The multiplicative inverse of -4 is $-\dfrac{1}{4}$ because $(-4)\left(-\dfrac{1}{4}\right) = 1$.

d. The multiplicative inverse of $-\dfrac{4}{5}$ is $-\dfrac{5}{4}$ because $\left(-\dfrac{4}{5}\right)\left(-\dfrac{5}{4}\right) = 1$.

 CHECK POINT 3 Find the multiplicative inverse of each number:

a. 7 $\dfrac{1}{7}$ **b.** $\dfrac{1}{8}$ 8   **c.** -6 $-\dfrac{1}{6}$ **d.** $-\dfrac{7}{13}$. $-\dfrac{13}{7}$

Can you think of a real number that has no multiplicative inverse? The number **0 has no multiplicative inverse** because 0 multiplied by any number is never 1, but always 0.

We now define division in terms of multiplication by a multiplicative inverse.

4 Use the definition of division.

> **DEFINITION OF DIVISION**   If a and b are real numbers and b is not 0, then the quotient of a and b is defined as
>
> $$a \div b = a \cdot \dfrac{1}{b}.$$
>
> In words: The quotient of two real numbers is the product of the first number and the multiplicative inverse of the second number.

EXAMPLE 4   Using the Definition of Division

Use the definition of division to find each quotient: **a.** $-15 \div 3$ **b.** $\dfrac{-20}{-4}$.

SOLUTION

a. $-15 \div 3 = -15 \cdot \dfrac{1}{3} = -5$

Change the division to multiplication.   Replace 3 with its multiplicative inverse.

b. $\dfrac{-20}{-4} = -20 \cdot \left(-\dfrac{1}{4}\right) = 5$

Change the division to multiplication.   Replace -4 with its multiplicative inverse.

 CHECK POINT 4 Use the definition of division to find each quotient:

a. $-28 \div 7$ -4 **b.** $\dfrac{-16}{-2}$.   8

5 Divide real numbers.

A Procedure for Dividing Real Numbers   Because the quotient $a \div b$ is defined as the product $a \cdot \frac{1}{b}$, the sign rules for dividing numbers are the same as the sign rules for multiplying them.

> **THE QUOTIENT OF TWO REAL NUMBERS**
> - The quotient of two real numbers with **different signs** is found by dividing their absolute values. The quotient is **negative**.
> - The quotient of two real numbers with the **same sign** is found by dividing their absolute values. The quotient is **positive**.
> - Division of a nonzero number by zero is undefined.
> - Any nonzero number divided into 0 is 0.

EXAMPLE 5   Dividing Real Numbers

Divide: **a.** $\dfrac{8}{-2}$ **b.** $-\dfrac{3}{4} \div \left(-\dfrac{5}{9}\right)$ **c.** $\dfrac{-20.8}{4}$ **d.** $\dfrac{0}{-7}.$

SOLUTION

a. $\dfrac{8}{-2} = -4$ Divide absolute values: $\frac{8}{2} = 4.$

Different signs: negative quotient

b. $-\dfrac{3}{4} \div \left(-\dfrac{5}{9}\right) = \dfrac{27}{20}$ Divide absolute values: $\frac{3}{4} \div \frac{5}{9} = \frac{3}{4} \cdot \frac{9}{5} = \frac{27}{20}.$

Same sign: positive quotient

c. $\dfrac{-20.8}{4} = -5.2$ Divide absolute values: $4\overline{)20.8}\,^{5.2}.$

Different signs: negative quotient

d. $\dfrac{0}{-7} = 0$ Any nonzero number divided into 0 is 0.

Here's another reason why division by zero is undefined. We know that

$$\frac{12}{4} = 3 \text{ because } 3 \cdot 4 = 12.$$

Now think about $\frac{-7}{0}$. If

$$\frac{-7}{0} = ? \text{ then } ? \cdot 0 = -7.$$

However, any real number multiplied by 0 is 0 and not -7. No matter what number we replace by ? in $? \cdot 0 = -7$, we can never obtain -7.

Can you see why $\frac{0}{-7}$ must be 0? The definition of division tells us that

$$\frac{0}{-7} = 0 \cdot \left(-\frac{1}{7}\right)$$

and the product of 0 and any real number is 0. By contrast, the definition of division does not allow for division by 0 because 0 does not have a multiplicative inverse. It is incorrect to write

0 does not have a multiplicative inverse.

Division by zero is not allowed or not defined. Thus, $\frac{-7}{0}$ does not represent a real number. A real number can never have a denominator of 0.

 CHECK POINT 5   Divide:

a. $\dfrac{-32}{-4}$ 8   **b.** $-\dfrac{2}{3} \div \dfrac{5}{4}$ $-\dfrac{8}{15}$ **c.** $\dfrac{21.9}{-3}$ -7.3 **d.** $\dfrac{0}{-5}.$ 0

6 Simplify algebraic expressions involving multiplication.

Multiplication and Algebraic Expressions In Section 1.4, we discussed the commutative and associative properties of multiplication. We also know that multiplication distributes over addition and subtraction. We now add some additional properties to our previous list (Table 1.5). These properties are frequently helpful in simplifying algebraic expressions.

Table 1.5 **Additional Properties of Multiplication**

Let a be a real number, a variable, or an algebraic expression.

Property	Meaning	Examples
Identity Property of Multiplication	1 can be deleted from a product. $a \cdot 1 = a$ $1 \cdot a = a$	• $\sqrt{3} \cdot 1 = \sqrt{3}$ • $1x = x$ • $1(2x + 3) = 2x + 3$
Inverse Property of Multiplication	If a is not 0: $a \cdot \dfrac{1}{a} = 1$ $\dfrac{1}{a} \cdot a = 1$ The product of a nonzero number and its multiplicative inverse, or reciprocal, gives 1, the multiplicative identity.	• $6 \cdot \dfrac{1}{6} = 1$ • $3x \cdot \dfrac{1}{3x} = 1$ (x is not 0.) • $\dfrac{1}{(y - 2)} \cdot (y - 2) = 1$ (y is not 2.)
Multiplication Property of -1	Negative 1 times a is the additive inverse, or opposite, of a. $-1 \cdot a = -a$ $a(-1) = -a$	• $-1 \cdot \sqrt{3} = -\sqrt{3}$ • $-1\left(-\frac{3}{4}\right) = \frac{3}{4}$ • $-1x = -x$ • $-(x + 4) = -1(x + 4)$ $\quad = -x - 4$
Double Negative Property	The additive inverse of $-a$ is a. $-(-a) = a$	• $-(-4) = 4$ • $-(-6y) = 6y$

In the preceding table, we used two steps to remove the parentheses from $-(x + 4)$. First, we used the multiplication property of -1.

$$-(x + 4) = -1(x + 4)$$

Then we used the distributive property, distributing -1 to each term in parentheses.

$$-1(x + 4) = (-1)x + (-1)4 = -x + (-4) = -x - 4$$

There is a fast way to obtain $-(x + 4) = -x - 4$ in just one step.

NEGATIVE SIGNS AND PARENTHESES If a negative sign precedes parentheses, remove the parentheses and change the sign of every term within the parentheses.

Here are some examples that illustrate this method.

$$-(11x + 5) = -11x - 5$$
$$-(11x - 5) = -11x + 5$$
$$-(-11x + 5) = 11x - 5$$
$$-(-11x - 5) = 11x + 5$$

EXAMPLE 6 Simplifying Algebraic Expressions

Simplify: **a.** $-2(3x)$ **b.** $6x + x$ **c.** $8a - 9a$ **d.** $-3(2x - 5)$ **e.** $-(3y - 8)$.

SOLUTION We will show all steps in the solution process. However, you probably are working many of these steps mentally.

DISCOVER FOR YOURSELF

Verify each simplification in Example 6 by substituting 5 for the variable. The value of the given expression should be the same as the value of the simplified expression. Which expression is easier to evaluate?

a. $-2(3x)$ This is the given algebraic expression.
$= (-2 \cdot 3)x$ Use the associative property and group the first two numbers.
$= -6x$ Numbers with opposite signs have a negative product.

b. $6x + x$ This is the given algebraic expression.
$= 6x + 1x$ Use the multiplication property of 1.
$= (6 + 1)x$ Apply the distributive property.
$= 7x$ Add within parentheses.

c. $8a - 9a$ This is the given algebraic expression.
$= (8 - 9)a$ Apply the distributive property.
$= -1a$ Subtract within parentheses: $8 - 9 = 8 + (-9) = -1$.
$= -a$ Apply the multiplication property of -1.

d. $-3(2x - 5)$ This is the given algebraic expression.
$= -3(2x) - (-3) \cdot (5)$ Apply the distributive property.
$= -6x - (-15)$ Multiply.
$= -6x + 15$ Subtraction is the addition of an additive inverse.

e. $-(3y - 8)$ This is the given algebraic expression.
$= -3y + 8$ Remove parentheses by changing the sign of every term inside the parentheses. ∎

✔ **CHECK POINT 6** Simplify: **a.** $-4(5x)$ $-20x$ **b.** $9x + x$ $10x$
c. $13b - 14b$ $-b$ **d.** $-7(3x - 4)$ $-21x + 28$ **e.** $-(7y - 6)$. $-7y + 6$

Before turning to applications, let's try one additional example involving simplification.

EXAMPLE 7 Simplifying an Algebraic Expression

Simplify: $5(2y - 9) - (9y - 8)$.

SOLUTION

$5(2y - 9) - (9y - 8)$ This is the given algebraic expression.
$= 5 \cdot 2y - 5 \cdot 9 - (9y - 8)$ Apply the distributive property over the first parentheses.
$= 10y - 45 - (9y - 8)$ Multiply.
$= 10y - 45 - 9y + 8$ Remove the second parentheses by changing the sign of each term within parentheses.
$= (10y - 9y) + (-45 + 8)$ Group like terms.
$= 1y + (-37)$ Combine like terms. For the variable terms, $10y - 9y = 10y + (-9y) = [10 + (-9)]y = 1y$.
$= y + (-37)$ Use the multiplication property of 1: $1y = y$.
$= y - 37$ Express addition of an additive inverse as subtraction. ∎

✔ **CHECK POINT 7** Simplify: $4(3y - 7) - (13y - 2)$ $-y - 26$

A Summary of Operations with Real Numbers Operations with real numbers are summarized in Table 1.6.

Table 1.6   Summary of Operations with Real Numbers

Signs of Numbers	Addition	Subtraction	Multiplication	Division
Both Numbers Are Positive Examples 8 and 2 2 and 8	Sum Is Always Positive $8 + 2 = 10$ $2 + 8 = 10$	Difference May Be Either Positive or Negative $8 - 2 = 6$ $2 - 8 = -6$	Product Is Always Positive $8 \cdot 2 = 16$ $2 \cdot 8 = 16$	Quotient Is Always Positive $8 \div 2 = 4$ $2 \div 8 = \frac{1}{4}$
One Number Is Positive and the Other Number Is Negative Examples 8 and -2 -8 and 2	Sum May Be Either Positive or Negative $8 + (-2) = 6$ $-8 + 2 = -6$	Difference May Be Either Positive or Negative $8 - (-2) = 10$ $-8 - 2 = -10$	Product Is Always Negative $8(-2) = -16$ $-8(2) = -16$	Quotient Is Always Negative $8 \div (-2) = -4$ $-8 \div 2 = -4$
Both Numbers Are Negative Examples -8 and -2 -2 and -8	Sum Is Always Negative $-8 + (-2) = -10$ $-2 + (-8) = -10$	Difference May Be Either Positive or Negative $-8 - (-2) = -6$ $-2 - (-8) = 6$	Product Is Always Positive $-8(-2) = 16$ $-2(-8) = 16$	Quotient Is Always Positive $-8 \div (-2) = 4$ $-2 \div (-8) = \frac{1}{4}$

7 Use algebraic expressions that model reality.

Applications Algebraic expressions that model reality frequently contain division.

EXAMPLE 8 Average Cost of Producing a Wheelchair

A company that manufactures wheelchairs has monthly fixed costs of $500,000. The average cost per wheelchair for the company to manufacture x wheelchairs per month is modeled by the algebraic expression

$$\frac{400x + 500,000}{x}.$$

Find the average cost per wheelchair for the company to manufacture

a. 10,000 wheelchairs per month.   **b.** 50,000 wheelchairs per month.

c. 100,000 wheelchairs per month.

What happens to the average cost per wheelchair as the production level increases?

SOLUTION

a. We are interested in the average cost per wheelchair for the company if 10,000 wheelchairs are manufactured per month. Because x represents the number of wheelchairs manufactured per month, we substitute $10,000$ for x in the given algebraic expression.

$$\frac{400x + 500,000}{x} = \frac{400(10,000) + 500,000}{10,000} = \frac{4,000,000 + 500,000}{10,000}$$

$$= \frac{4,500,000}{10,000} = 450$$

The average cost per wheelchair of producing 10,000 wheelchairs per month is $450.

b. Now, 50,000 wheelchairs are manufactured per month. We find the average cost per wheelchair by substituting $50,000$ for x in the given algebraic expression.

$$\frac{400x + 500,000}{x} = \frac{400(50,000) + 500,000}{50,000} = \frac{20,000,000 + 500,000}{50,000}$$

$$= \frac{20,500,000}{50,000} = 410$$

The average cost per wheelchair of producing 50,000 wheelchairs per month is $410.

c. Finally, the production level has increased to 100,000 wheelchairs per month. We find the average cost per wheelchair for the company by substituting $100,000$ for x in the given algebraic expression.

$$\frac{400x + 500,000}{x} = \frac{400(100,000) + 500,000}{100,000} = \frac{40,000,000 + 500,000}{100,000}$$

$$= \frac{40,500,000}{100,000} = 405$$

The average cost per wheelchair of producing 100,000 wheelchairs per month is $405.

As the production level increases, the average cost of producing each wheelchair decreases. This illustrates the difficulty with small businesses. It is nearly impossible to have competitively low prices when production levels are low.

The points in the rectangular coordinate system in Figure 1.21 show the relationship between production level and cost. The x-axis represents the number of wheelchairs produced per month. The y-axis represents the average cost per wheelchair for the company. The symbol $\frac{\cdot}{\cdot}$ on the y-axis indicates that there is a break in the values of y between 0 and 400. Thus, the values of y begin at 400.

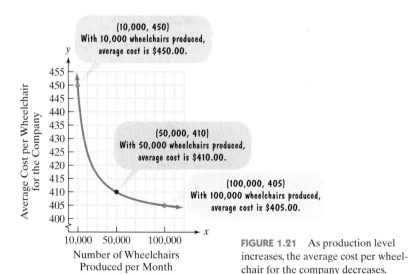

FIGURE 1.21 As production level increases, the average cost per wheelchair for the company decreases.

The three points with the voice balloons illustrate our computations in Example 8. The points in Figure 1.21 are falling from left to right. Can you see how this shows that the company's cost per wheelchair is decreasing as their production level increases?

 CHECK POINT 8 A company that manufactures running shoes has weekly fixed costs of $300,000. The average cost per pair of running shoes for the company to manufacture x pairs per week is modeled by the algebraic expression

$$\frac{30x + 300,000}{x}.$$

Find the average cost per pair of running shoes for the company to manufacture

a. 1000 pairs per week. $330   **b.** 10,000 pairs per week. $60

c. 100,000 pairs per week. $33

1.7 EXERCISE SET

 Student Solutions Manual   CD/Video   PH Math/Tutor Center   MathXL Tutorials on CD   MathXL®   MyMathLab   Interactmath.com

Practice Exercises

In Exercises 1–34, perform the indicated multiplication.

1. $5(-9)$ -45

2. $10(-7)$ -70

3. $(-8)(-3)$ 24

4. $(-9)(-5)$ 45

5. $(-3)(7)$ -21

6. $(-4)(8)$ -32

7. $(-19)(-1)$ 19

8. $(-11)(-1)$ 11

9. $0(-19)$ 0

10. $0(-11)$ 0

11. $\frac{1}{2}(-24)$ -12

12. $\frac{1}{3}(-21)$ -7

13. $\left(-\frac{3}{4}\right)(-12)$ 9

14. $\left(-\frac{4}{5}\right)(-30)$ 24

15. $-\frac{3}{5}\cdot\left(-\frac{4}{7}\right)$ $\frac{12}{35}$

16. $-\frac{5}{7}\cdot\left(-\frac{3}{8}\right)$ $\frac{15}{56}$

17. $-\frac{7}{9}\cdot\frac{2}{3}$ $-\frac{14}{27}$

18. $-\frac{5}{11}\cdot\frac{2}{7}$ $-\frac{10}{77}$

19. $3(-1.2)$ -3.6

20. $4(-1.2)$ -4.8

21. $-0.2(-0.6)$ 0.12

22. $-0.3(-0.7)$ 0.21

23. $(-5)(-2)(3)$ 30

24. $(-6)(-3)(10)$ 180

25. $(-4)(-3)(-1)(6)$ -72

26. $(-2)(-7)(-1)(3)$ -42

27. $-2(-3)(-4)(-1)$ 24

28. $-3(-2)(-5)(-1)$ 30

29. $(-3)(-3)(-3)$ -27

30. $(-4)(-4)(-4)$ -64

31. $5(-3)(-1)(2)(3)$ 90

32. $2(-5)(-2)(3)(1)$ 60

33. $(-8)(-4)(0)(-17)(-6)$ 0

34. $(-9)(-12)(-18)(0)(-3)$ 0

In Exercises 35–42, find the multiplicative inverse of each number.

35. 4 $\frac{1}{4}$

36. 3 $\frac{1}{3}$

37. $\frac{1}{5}$ 5

38. $\frac{1}{7}$ 7

39. -10 $-\frac{1}{10}$

40. -12 $-\frac{1}{12}$

41. $-\frac{2}{5}$ $-\frac{5}{2}$

42. $-\frac{4}{9}$ $-\frac{9}{4}$

In Exercises 43–46,

a. *Rewrite the division as multiplication involving a multiplicative inverse.*

b. *Use the multiplication from part (a) to find the given quotient.*

43. $-32 \div 4$ **a.** $-32\cdot\left(\frac{1}{4}\right)$

 b. -8

44. $-18 \div 6$ **a.** $-18\cdot\left(\frac{1}{6}\right)$

 b. -3

45. $\frac{-60}{-5}$ **a.** $-60\cdot\left(-\frac{1}{5}\right)$ **b.** 12

46. $\frac{-30}{-5}$ **a.** $-30\cdot\left(-\frac{1}{5}\right)$ **b.** 6

In Exercises 47–76, perform the indicated division or state that the expression is undefined.

47. $\frac{12}{-4}$ -3

48. $\frac{40}{-5}$ -8

49. $\frac{-21}{3}$ -7

50. $\frac{-60}{6}$ -10

51. $\frac{-90}{-3}$ 30

52. $\frac{-66}{-6}$ 11

53. $\frac{0}{-7}$ 0

54. $\frac{0}{-8}$ 0

55. $\frac{7}{0}$ undefined

56. $\frac{-8}{0}$ undefined

57. $-15 \div 3$ -5

58. $-80 \div 8$ -10

59. $120 \div (-10)$ -12

60. $130 \div (-10)$ -13

61. $(-180) \div (-30)$ 6

62. $(-150) \div (-25)$ 6

63. $0 \div (-4)$ 0

64. $0 \div (-10)$ 0

65. $-4 \div 0$ undefined

66. $-10 \div 0$ undefined

67. $\frac{-12.9}{3}$ -4.3

68. $\frac{-21.6}{3}$ -7.2

69. $-\dfrac{1}{2} \div \left(-\dfrac{3}{5}\right)$ $\dfrac{5}{6}$

70. $-\dfrac{1}{2} \div \left(-\dfrac{7}{9}\right)$ $\dfrac{9}{14}$

71. $-\dfrac{14}{9} \div \dfrac{7}{8}$ $-\dfrac{16}{9}$

72. $-\dfrac{5}{16} \div \dfrac{25}{8}$ $-\dfrac{1}{10}$

73. $\dfrac{1}{3} \div \left(-\dfrac{1}{3}\right)$ -1

74. $\dfrac{1}{5} \div \left(-\dfrac{1}{5}\right)$ -1

75. $6 \div \left(-\dfrac{2}{5}\right)$ -15

76. $8 \div \left(-\dfrac{2}{9}\right)$ -36

In Exercises 77–96, simplify each algebraic expression.

77. $-5(2x)$ $-10x$

78. $-9(3x)$ $-27x$

79. $-4\left(-\dfrac{3}{4}y\right)$ $3y$

80. $-5\left(-\dfrac{3}{5}y\right)$ $3y$

81. $8x + x$ $9x$

82. $12x + x$ $13x$

83. $-5x + x$ $-4x$

84. $-6x + x$ $-5x$

85. $6b - 7b$ $-b$

86. $12b - 13b$ $-b$

87. $-y + 4y$ $3y$

88. $-y + 9y$ $8y$

89. $-4(2x - 3)$ $-8x + 12$

90. $-3(4x - 5)$ $-12x + 15$

91. $-3(-2x + 4)$ $6x - 12$

92. $-4(-3x + 2)$ $12x - 8$

93. $-(2y - 5)$ $-2y + 5$

94. $-(3y - 1)$ $-3y + 1$

95. $4(2y - 3) - (7y + 2)$ $y - 14$

96. $5(3y - 1) - (14y - 2)$ $y - 3$

Practice Plus

In Exercises 97–104, write a numerical expression for each phrase.
Then simplify the numerical expression by performing the given
operations.

97. 8 added to the product of 4 and -10 $4(-10) + 8 = -32$

98. 14 added to the product of 3 and -15 $3(-15) + 14 = -31$

99. The product of -9 and -3, decreased by -2
   $(-9)(-3) - (-2) = 29$

100. The product of -6 and -4, decreased by -5
   $(-6)(-4) - (-5) = 29$

101. The quotient of -18 and the sum of -15 and 12
   $\dfrac{-18}{-15 + 12} = 6$

102. The quotient of -25 and the sum of -21 and 16
   $\dfrac{-25}{-21 + 16} = 5$

103. The difference between -6 and the quotient of 12 and -4
   $-6 - \left(\dfrac{12}{-4}\right) = -3$

104. The difference between -11 and the quotient of 20 and -5
   $-11 - \left(\dfrac{20}{-5}\right) = -7$

Application Exercises

The graph shows the millions of welfare recipients in the United
States who received cash assistance from 1994 through 2001. The
data can be modeled by the algebraic expression

$$-1.4t + 14.7,$$

which represents the number of welfare recipients, in millions, t years
after 1994. Use this algebraic expression to solve Exercises 105–106.

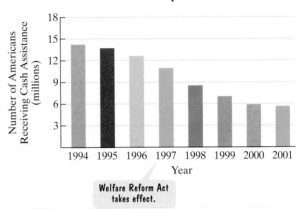

Welfare Recipients in the U.S.

Source: Department of Health and Human Services

105. According to the model, how many welfare recipients were
there in 2000? How well does the algebraic expression
model the actual data for 2000 shown in the bar graph?
6.3 million; fairly well, but slight overestimation

106. According to the model, how many welfare recipients were
there in 1997? How well does the algebraic expression
model the actual data for 1997 shown in the bar graph?
10.5 million; fairly well

In an experiment on memory, students in a language class are
asked to memorize 40 vocabulary words in Latin, a language with
which the students are not familiar. After studying the words for
one day, students are tested each day after to see how many words
they remember. The class average is taken and the results are
graphed as points in the rectangular coordinate system. Use the
points shown to solve Exercises 107–108.

Average Number of Words
Remembered over Time

107. a. Find a reasonable estimate of the number of Latin words remembered after 5 days.   11 Latin words

b. The algebraic expression

$$\frac{5x + 30}{x}$$

models the number of Latin words remembered by the students after x days. Use this expression to find the number of Latin words remembered after 5 days. How does this compare with your estimate from part (a)?   11 Latin words; it is the same.

108. a. Find a reasonable estimate of the number of Latin words remembered after 15 days.   7 Latin words

b. The algebraic expression

$$\frac{5x + 30}{x}$$

models the number of Latin words remembered by the students after x days. Use this expression to find the number of Latin words remembered after 15 days. How does this compare with your estimate from part (a)?   7 Latin words; it is the same.

In Palo Alto, California, a government agency ordered computer-related companies to contribute to a pool of money to clean up underground water supplies. (The companies had stored toxic chemicals in leaking underground containers.) The algebraic expression

$$\frac{200x}{100 - x}$$

models the cost, in tens of thousands of dollars, for removing x percent of the contaminants. Use this algebraic expression to solve Exercises 109–110.

109. a. Substitute 50 for x and find the cost, in tens of thousands of dollars, for removing 50% of the contaminants. $2,000,000 (or $200 tens of thousands)

b. Find the cost, in tens of thousands of dollars, for removing 80% of the contaminants. $8,000,000 (or $800 tens of thousands)

c. Describe what is happening to the cost of the cleanup as the percentage of contaminant removed increases.   Cost increases.

110. a. Substitute 60 for x and find the cost, in tens of thousands of dollars, for removing 60% of the contaminants. $3,000,000 (or $300 tens of thousands)

b. Find the cost, in tens of thousands of dollars, for removing 90% of the contaminants. $18,000,000 (or $1800 tens of thousands)

c. Describe what is happening to the cost of the cleanup as the percentage of contaminants removed increases.   Cost increases.

Writing in Mathematics

111. Explain how to multiply two real numbers. Provide examples with your explanation.   **111–117.** Answers will vary.

112. Explain how to determine the sign of a product that involves more than two numbers.

113. Explain how to find the multiplicative inverse of a number.

114. Why is it that 0 has no multiplicative inverse?

115. Explain how to divide real numbers.

116. Why is division by zero undefined?

117. Explain how to simplify an algebraic expression in which a negative sign precedes parentheses.

118. A politician promises to do "whatever it takes" to seize "one hundred percent" of all illegal drugs that enter the country. Suppose that the cost, in millions of dollars, for seizing x percent of the illegal drugs entering the country is modeled by the algebraic expression

$$\frac{130x}{100 - x}.$$

According to this algebraic expression, can the politician keep his or her promise? Explain your answer.   No. At 100% we have division by zero, which is undefined

Critical Thinking Exercises

119. Which one of the following statements is true?   b

a. Multiplying a negative number by a nonnegative number will always give a negative number.

b. The product of two negative numbers is always a positive number.

c. The product of -3 and 4 is 12.

d. The product of real numbers a and b is not always equal to the product of real numbers b and a.

120. Which one of the following statements is true?   c

a. The product of two negative numbers is sometimes a negative number.

b. Both the addition and the multiplication of two negative numbers result in a positive number.

c. $\left(-\frac{1}{2}\right)\left(-\frac{1}{2}\right) = \frac{1}{4}$

d. Reversing the order of the two factors in a product results in a different answer.

In Exercises 121–124, write an algebraic expression for the given English phrase.

121. The value, in cents, of x nickels $5x$

122. The distance covered by a car traveling at 50 miles per hour for x hours $50x$

123. The monthly salary, in dollars, for a person earning x dollars per year $\frac{x}{12}$

124. The fraction of people in a room who are women if there are 40 women and x men in the room $\frac{40}{x + 40}$

Technology Exercises

125. Use a calculator to verify any five of the products that you found in Exercises 1–34.   Answers will vary.

126. Use a calculator to verify any five of the quotients that you found in Exercises 47–76.   Answers will vary.

127. Simplify using a calculator:

$$0.3(4.7x - 5.9) - 0.07(3.8x - 61).   1.144x + 2.5$$

128. Use your calculator to attempt to find the quotient of -3 and 0. Describe what happens. Does the same thing occur when finding the quotient of 0 and -3? Explain the difference. Finally, what happens when you enter the quotient of 0 and itself?   Answers will vary.

Review Exercises

In Exercises 129–131, perform the indicated operation.

129. $-6 + (-3)$ (Section 1.5, Example 3) -9

130. $-6 - (-3)$ (Section 1.6, Example 1) -3

131. $-6 \div (-3)$ (Section 1.7, Example 4) 2

SECTION 1.8

EXPONENTS, ORDER OF OPERATIONS, AND MATHEMATICAL MODELS

Objectives

1 Evaluate exponential expressions.

2 Simplify algebraic expressions with exponents.

3 Use the order of operations agreement.

4 Evaluate formulas.

It's been another one of those days! Traffic is really backed up on the highway. Finally, you see the source of the traffic jam—a minor fender-bender. Still stuck in traffic, you notice that the driver appears to be quite young. This might seem like a strange observation. After all, what does a driver's age have to do with his or her chance of getting into an accident? In this section, we see how algebra describes your world, including a relationship between age and numbers of car accidents.

1 Evaluate exponential expressions.

Natural Number Exponents   Although people do a great deal of talking, the total output since the beginning of gabble to the present day, including all baby talk, love songs, and congressional debates, only amounts to about 10 million billion words. This can be expressed as 16 factors of 10, or 10^{16} words.

Exponents such as 2, 3, 4, and so on are used to indicate repeated multiplication. For example,

$$10^2 = 10 \cdot 10 = 100,$$

$$10^3 = 10 \cdot 10 \cdot 10 = 1000, \quad 10^4 = 10 \cdot 10 \cdot 10 \cdot 10 = 10,000.$$

The 10 that is repeated when multiplying is called the **base**. The small numbers above and to the right of the base are called **exponents** or **powers**. The exponent tells the number of times the base is to be used when multiplying. In 10^3, the base is 10 and the exponent is 3.

Any number with an exponent of 1 is the number itself. Thus, $10^1 = 10$.

You can use a calculator to evaluate exponential expressions. For example, to evaluate 5^3, press the following keys:

Many Scientific Calculators

5 $\boxed{y^x}$ 3 $\boxed{=}$

Many Graphing Calculators

5 $\boxed{\wedge}$ 3 $\boxed{\text{ENTER}}$

Although calculators have special keys to evaluate powers of ten and squaring bases, you can always use one of the sequences shown here.

Multiplications that are expressed in exponential notation are read as follows:

10^1: "ten to the first power"
10^2: "ten to the second power" or "ten squared"
10^3: "ten to the third power" or "ten cubed"
10^4: "ten to the fourth power"
10^5: "ten to the fifth power"
etc.

Any real number can be used as the base. Thus,

$$7^2 = 7 \cdot 7 = 49 \text{ and } (-3)^4 = (-3)(-3)(-3)(-3) = 81.$$

The bases are 7 and -3, respectively. Do not confuse $(-3)^4$ and -3^4.

$$-3^4 = -(3 \cdot 3 \cdot 3 \cdot 3) = -81$$

The negative is not taken to the power because it is not inside parentheses.

An exponent applies only to a base. A negative sign is not part of a base unless it appears in parentheses.

EXAMPLE 1 Evaluating Exponential Expressions

Evaluate: **a.** 4^2 **b.** $(-5)^3$ **c.** $(-2)^4$ **d.** -2^4.

SOLUTION

Exponent is 2.

a. $4^2 = 4 \cdot 4 = 16$ The exponent indicates that the base is used as a factor two times.

Base is 4.

We read $4^2 = 16$ as "4 to the second power is 16" or "4 squared is 16."

Exponent is 3.

b. $(-5)^3 = (-5)(-5)(-5)$ The exponent indicates that the base is used as a factor three times.

Base is –5.

    $= -125$ An odd number of negative factors yields a negative product.

We read $(-5)^3 = -125$ as "the number negative 5 to the third power is negative 125" or "negative 5 cubed is negative 125."

Exponent is 4.

c. $(-2)^4 = (-2)(-2)(-2)(-2)$ The exponent indicates the base is used as a factor four times.

Base is –2.

    $= 16$ An even number of negative factors yields a positive product.

We read $(-2)^4 = 16$ as "the number negative 2 to the fourth power is 16."

Exponent is 4.

d. $-2^4 = -(2 \cdot 2 \cdot 2 \cdot 2)$

Base is 2.

The negative is not inside parentheses and is not taken to the fourth power.

$= -16$

Multiply the twos and copy the negative.

We read $-2^4 = -16$ as "the negative of 2 raised to the fourth power is negative 16" or "the opposite, or additive inverse, of 2 raised to the fourth power is negative 16."

 CHECK POINT 1 Evaluate:

a. 6^2 36   **b.** $(-4)^3$ -64 **c.** $(-1)^4$ 1   **d.** -1^4. -1

The formal algebraic definition of a natural number exponent summarizes our discussion:

DEFINITION OF A NATURAL NUMBER EXPONENT If b is a real number and n is a natural number,

Exponent

$$b^n = \underbrace{b \cdot b \cdot b \cdot \ldots \cdot b.}_{\substack{b \text{ appears as a} \\ \text{factor } n \text{ times.}}}$$

Base

b^n is read "the nth power of b" or "b to the nth power." Thus, the nth power of b is defined as the product of n factors of b. The expression b^n is called an **exponential expression**.
    Furthermore, $b^1 = b$.

2 Simplify algebraic expressions with exponents.

Exponents and Algebraic Expressions The distributive property can be used to simplify certain algebraic expressions that contain exponents. For example, we can use the distributive property to combine like terms in the algebraic expression $4x^2 + 6x^2$:

$$4x^2 + 6x^2 = (4 + 6)x^2 = 10x^2.$$

First term with variable factor x^2

Second term with variable factor x^2

The common variable factor is x^2.

EXAMPLE 2 Simplifying Algebraic Expressions

Simplify, if possible: **a.** $7x^3 + 2x^3$ **b.** $5x^2 + x^2$ **c.** $3x^2 + 4x^3$.

SOLUTION

a. $7x^3 + 2x^3$ There are two like terms with the same variable factor, namely x^3.

$= (7 + 2)x^3$ Apply the distributive property.

$= 9x^3$ Add within parentheses.

3 Use the order of operations agreement.

b. $5x^2 + x^2$ There are two like terms with the same variable factor, namely x^2.

$= 5x^2 + 1x^2$ Use the multiplication property of 1.

$= (5 + 1)x^2$ Apply the distributive property.

$= 6x^2$ Add within parentheses.

c. $3x^2 + 4x^3$ cannot be simplified. The terms $3x^2$ and $4x^3$ are not like terms because they have different variable factors, namely x^2 and x^3. ■

 CHECK POINT **2** Simplify, if possible:

a. $16x^2 + 5x^2$ $21x^2$ **b.** $7x^3 + x^3$ $8x^3$ **c.** $10x^2 + 8x^3$.   cannot simplify

Order of Operations Suppose that you want to find the value of $3 + 7 \cdot 5$. Which procedure shown is correct?

$$3 + 7 \cdot 5 = 3 + 35 = 38 \quad \text{or} \quad 3 + 7 \cdot 5 = 10 \cdot 5 = 50$$

If you know the answer, you probably know certain rules, called the **order of operations**, to make sure that there is only one correct answer. One of these rules states that if a problem contains no parentheses or other grouping symbols, perform multiplication before addition. Thus, the procedure on the left is correct because the multiplication of 7 and 5 is done first. Then the addition is performed. The correct answer is 38.

Some problems contain grouping symbols, such as parentheses, ( ); brackets, [ ]; braces, { }; absolute value symbols, | |; or fraction bars. These grouping symbols tell us what to do first. Here are two examples:

$$\bullet \qquad (3 + 7) \cdot 5 = 10 \cdot 5 = 50$$

First, perform operations in grouping symbols.

$$\bullet \qquad 8|6 - 16| = 8|-10| = 8 \cdot 10 = 80.$$

Here are the rules for determining the order in which operations should be performed:

ORDER OF OPERATIONS

1. Perform all operations within grouping symbols.
2. Evaluate all exponential expressions.
3. Do all multiplications and divisions in the order in which they occur, working from left to right.
4. Finally, do all additions and subtractions in the order in which they occur, working from left to right.

In the third step, be sure to do all multiplications and divisions *as they occur* from left to right. For example,

$$8 \div 4 \cdot 2 = 2 \cdot 2 = 4 \qquad \text{Do the division first because it occurs first.}$$

$$8 \cdot 4 \div 2 = 32 \div 2 = 16. \qquad \text{Do the multiplication first because it occurs first.}$$

EXAMPLE 3   Using the Order of Operations

Simplify: $18 + 2 \cdot 3 - 10$.

SOLUTION   There are no grouping symbols or exponential expressions. In cases like this, we multiply and divide before adding and subtracting.

$$18 + 2 \cdot 3 - 10 = 18 + 6 - 10 \quad \text{Multiply: } 2 \cdot 3 = 6.$$
$$= 24 - 10 \quad \text{Add and subtract from left to right: } 18 + 6 = 24.$$
$$= 14 \quad \text{Subtract: } 24 - 10 = 14.$$ ■

 CHECK POINT 3 Simplify: $20 + 4 \cdot 3 - 17$.   15

EXAMPLE 4   Using the Order of Operations

Simplify: $6^2 - 24 \div 2^2 \cdot 3 - 1$.

SOLUTION   There are no grouping symbols. Thus, we begin by evaluating exponential expressions. Then we multiply or divide. Finally, we add or subtract.

$$6^2 - 24 \div 2^2 \cdot 3 - 1$$
$$= 36 - 24 \div 4 \cdot 3 - 1 \quad \text{Evaluate exponential expressions:}$$
$$\qquad\qquad\qquad\qquad 6^2 = 6 \cdot 6 = 36 \text{ and } 2^2 = 2 \cdot 2 = 4.$$
$$= 36 - 6 \cdot 3 - 1 \quad \text{Perform the multiplications and divisions}$$
$$\qquad\qquad\qquad\qquad \text{from left to right. Start with } 24 \div 4 = 6.$$
$$= 36 - 18 - 1 \quad \text{Now do the multiplication: } 6 \cdot 3 = 18.$$
$$= 18 - 1 \quad \text{Finally, perform the subtraction}$$
$$\qquad\qquad\qquad\qquad \text{from left to right: } 36 - 18 = 18.$$
$$= 17 \quad \text{Complete the subtraction: } 18 - 1 = 17.$$ ■

 CHECK POINT 4 Simplify: $7^2 - 48 \div 4^2 \cdot 5 - 2$.   32

EXAMPLE 5   Using the Order of Operations

Simplify: **a.** $(2 \cdot 5)^2$ **b.** $2 \cdot 5^2$.

SOLUTION

a. Because $(2 \cdot 5)^2$ contains grouping symbols, namely parentheses, we perform the operation within parentheses first.

$$(2 \cdot 5)^2 = 10^2 \quad \text{Multiply within parentheses:}$$
$$\qquad\qquad\qquad 2 \cdot 5 = 10.$$
$$= 100 \quad \text{Evaluate the exponential expression:}$$
$$\qquad\qquad\qquad 10^2 = 10 \cdot 10 = 100.$$

b. Because $2 \cdot 5^2$ does not contain grouping symbols, we begin by evaluating the exponential expression.

$$2 \cdot 5^2 = 2 \cdot 25 \quad \text{Evaluate the exponential expression:}$$
$$\qquad\qquad\qquad 5^2 = 5 \cdot 5 = 25.$$
$$= 50 \quad \text{Now do the multiplication:}$$
$$\qquad\qquad\qquad 2 \cdot 25 = 50.$$ ■

 CHECK POINT 5 Simplify: **a.** $(3 \cdot 2)^2$ 36   **b.** $3 \cdot 2^2$.   12

EXAMPLE 6 Using the Order of Operations

Simplify: $\left(\dfrac{1}{2}\right)^3 - \left(\dfrac{1}{2} - \dfrac{3}{4}\right)^2 (-4).$

SOLUTION Because grouping symbols appear, we perform the operation within parentheses first.

$\left(\dfrac{1}{2}\right)^3 - \left(\dfrac{1}{2} - \dfrac{3}{4}\right)^2 (-4)$

$= \left(\dfrac{1}{2}\right)^3 - \left(-\dfrac{1}{4}\right)^2 (-4)$ Work inside parentheses first:

$\qquad \dfrac{1}{2} - \dfrac{3}{4} = \dfrac{2}{4} - \dfrac{3}{4} = \dfrac{2}{4} + \left(-\dfrac{3}{4}\right) = -\dfrac{1}{4}.$

$= \dfrac{1}{8} - \dfrac{1}{16}(-4)$ Evaluate exponential expressions:

$\qquad \left(\dfrac{1}{2}\right)^3 = \dfrac{1}{2} \cdot \dfrac{1}{2} \cdot \dfrac{1}{2} = \dfrac{1}{8} \text{ and } \left(-\dfrac{1}{4}\right)^2 = \left(-\dfrac{1}{4}\right)\left(-\dfrac{1}{4}\right) = \dfrac{1}{16}.$

$= \dfrac{1}{8} - \left(-\dfrac{1}{4}\right)$ Multiply: $\dfrac{1}{16} \cdot \left(\dfrac{-4}{1}\right) = -\dfrac{4}{16} = -\dfrac{1}{4}.$

$= \dfrac{3}{8}$ Subtract: $\dfrac{1}{8} - \left(-\dfrac{1}{4}\right) = \dfrac{1}{8} + \dfrac{1}{4} = \dfrac{1}{8} + \dfrac{2}{8} = \dfrac{3}{8}.$ ∎

✔ **CHECK POINT 6** Simplify: $\left(-\dfrac{1}{2}\right)^2 - \left(\dfrac{7}{10} - \dfrac{8}{15}\right)^2 (-18).$ $\dfrac{3}{4}$

Some expressions contain many grouping symbols. An example of such an expression is $2[5(4 - 7) + 9]$. The grouping symbols are the parentheses and the brackets.

The parentheses, the innermost grouping symbols, group 4 − 7.

$$2[5(4 - 7) + 9]$$

The brackets, the outermost grouping symbols, group 5(4 − 7) + 9.

When combinations of grouping symbols appear, **perform operations within the innermost grouping symbols first**. Then work to the outside, performing operations within the outermost grouping symbols.

EXAMPLE 7 Using the Order of Operations

Simplify: $2[5(4 - 7) + 9].$

SOLUTION

$2[5(4 - 7) + 9]$

$= 2[5(-3) + 9]$ Work inside parentheses first: $4 - 7 = 4 + (-7) = -3.$

$= 2[-15 + 9]$ Work inside brackets and multiply: $5(-3) = -15.$

$= 2[-6]$ Add inside brackets: $-15 + 9 = -6.$ The resulting problem can also be expressed as $2(-6).$

$= -12$ Multiply: $2[-6] = -12.$ ∎

Parentheses can be used for both innermost and outermost grouping symbols. For example, the expression $2[5(4 - 7) + 9]$ can also be written $2(5(4 - 7) + 9)$. However, too many parentheses can be confusing. The use of both parentheses and brackets makes it easier to identify inner and outer groupings.

 CHECK POINT 7 Simplify: $4[3(6 - 11) + 5]$. -40

EXAMPLE 8   Using the Order of Operations

Simplify: $18 \div 6 + 4[5 + 2(8 - 10)^3]$.

SOLUTION

$$18 \div 6 + 4[5 + 2(8 - 10)^3]$$

$= 18 \div 6 + 4[5 + 2(-2)^3]$ Work inside parentheses first: $8 - 10 = 8 + (-10) = -2$.

$= 18 \div 6 + 4[5 + 2(-8)]$ Work inside brackets and evaluate the exponential expression: $(-2)^3 = (-2)(-2)(-2) = -8$.

$= 18 \div 6 + 4[5 + (-16)]$ Work inside brackets and multiply: $2(-8) = -16$.

$= 18 \div 6 + 4[-11]$ Work inside brackets and add: $5 + (-16) = -11$.

$= 3 + 4[-11]$ Perform the multiplications and divisions from left to right. Start with $18 \div 6 = 3$.

$= 3 + (-44)$ Now do the multiplication: $4(-11) = -44$.

$= -41$ Finally, perform the addition: $3 + (-44) = -41$.   ∎

 CHECK POINT 8 Simplify: $25 \div 5 + 3[4 + 2(7 - 9)^3]$. -31

Fraction bars are grouping symbols that separate expressions into two parts, the numerator and the denominator. Consider, for example,

The fraction bar is the grouping symbol.

The numerator is one part of the expression.

$$\frac{2(3 - 12) + 6 \cdot 4}{2^4 + 1}.$$

The denominator is the other part of the expression.

We can use brackets instead of the fraction bar. An equivalent expression is

$$[2(3 - 12) + 6 \cdot 4] \div [2^4 + 1].$$

The grouping suggests a method for simplifying expressions with fraction bars as grouping symbols:

• Simplify the numerator.
• Simplify the denominator.
• If possible, simplify the fraction.

EXAMPLE 9 Using the Order of Operations

Simplify: $\dfrac{2(3 - 12) + 6 \cdot 4}{2^4 + 1}$.

SOLUTION

$$\dfrac{2(3 - 12) + 6 \cdot 4}{2^4 + 1}$$

$$= \dfrac{2(-9) + 6 \cdot 4}{16 + 1}$$
Work inside parentheses in the numerator:
$3 - 12 = 3 + (-12) = -9$. Evaluate the exponential expression in the denominator:
$2^4 = 2 \cdot 2 \cdot 2 \cdot 2 = 16$.

$$= \dfrac{-18 + 24}{16 + 1}$$
Multiply in the numerator:
$2(-9) = -18$ and $6 \cdot 4 = 24$.

$$= \dfrac{6}{17}$$
Perform the addition in the numerator and the denominator. ∎

 CHECK POINT 9 Simplify: $\dfrac{5(4 - 9) + 10 \cdot 3}{2^3 - 1}$. $\dfrac{5}{7}$

EXAMPLE 10 Using the Order of Operations

Evaluate: $-x^2 - 7x$ for $x = -2$.

SOLUTION We begin by substituting -2 for each occurrence of x in the algebraic expression. Then we use the order of operations to evaluate the expression.

$$-x^2 - 7x$$

Replace x with -2.

$$= -(-2)^2 - 7(-2)$$

$$= -4 - 7(-2)$$
Evaluate the exponential expression:
$(-2)^2 = (-2)(-2) = 4$.

$$= -4 - (-14)$$
Multiply: $7(-2) = -14$.

$$= 10$$
Subtract:
$-4 - (-14) = -4 + 14 = 10$. ∎

 CHECK POINT 10 Evaluate: $-x^2 - 4x$ for $x = -5$. -5

Some algebraic expressions contain two sets of grouping symbols. Using the order of operations, grouping symbols are removed from innermost (parentheses) to outermost (brackets).

EXAMPLE 11 Simplifying an Algebraic Expression

Simplify: $18x^2 + 4 - [6(x^2 - 2) + 5]$.

SOLUTION

$18x^2 + 4 - [6(\overset{\frown}{x^2 - 2}) + 5]$

$= 18x^2 + 4 - [6x^2 - 12 + 5]$ Use the distributive property to remove parentheses:

$6(\overset{\frown}{x^2 - 2}) = 6x^2 - 6 \cdot 2 = 6x^2 - 12.$

$= 18x^2 + 4 - [6x^2 - 7]$ Add inside brackets: $-12 + 5 = -7$.

$= 18x^2 + 4 - 6x^2 + 7$ Remove brackets by changing the sign of each term within brackets.

$= (18x^2 - 6x^2) + 4 + 7$ Group like terms.

$= 12x^2 + 11$ Combine like terms.   ■

✔ **CHECK POINT 11** Simplify: $14x^2 + 5 - [7(x^2 - 2) + 4]$. $7x^2 + 15$

4 Evaluate formulas.

Applications: Formulas and Mathematical Models   One aim of algebra is to provide a compact, symbolic description of the world. These descriptions involve the use of *formulas*. A **formula** is a statement of equality that uses letters to express a relationship between two or more variables. For example, one variety of crickets chirps faster as the temperature rises. You can calculate the temperature by counting the number of times a cricket chirps per minute and applying the following formula:

$$T = 0.3n + 40.$$

In the formula, T is the temperature, in degrees Fahrenheit, and n is the number of cricket chirps per minute. We can use this formula to determine the temperature if you are sitting on your porch and count 80 chirps per minute. Here is how to do so:

$T = 0.3n + 40$ This is the given formula.

$T = 0.3(80) + 40$ Substitute 80 for n.

$T = 24 + 40$ Multiply: $0.3(80) = 24$.

$T = 64$.   Add.

When there are 80 cricket chirps per minute, the temperature is 64 degrees.

The process of finding formulas to describe real-world phenomena is called **mathematical modeling**. Such formulas, together with the meaning assigned to the variables, are called **mathematical models**. We often say that these formulas model, or describe, the relationship among the variables.

In creating mathematical models, we strive for both accuracy and simplicity. For example, the formula $T = 0.3n + 40$ is relatively simple to use. However, you should not get upset if you count 80 cricket chirps and the actual temperature is 62 degrees, rather than 64 degrees, as predicted by the formula. Many mathematical models give an approximate, rather than an exact, description of the relationship between variables.

Sometimes a mathematical model gives an estimate that is not a good approximation or is extended to include values of the variable that do not make sense. In these cases, we say that **model breakdown** has occurred. Here is an example:

Use the mathematical model $T = 0.3n + 40$ with $n = 1200$ (1200 cricket chirps per minute).

$$T = 0.3(1200) + 40 = 360 + 40 = 400$$

At 400° F, forget about 1200 chirps per minute! At this temperature, the cricket would "cook" and, alas, all chirping would cease.

EXAMPLE 12 Car Accidents and Age

The mathematical model

$$N = 0.4x^2 - 36x + 1000$$

approximates the number of accidents, N, per 50 million miles driven, for drivers who are x years old. The formula applies to drivers ages 16 through 74, inclusive. How many accidents, per 50 million miles driven, are there for 20-year-old drivers?

SOLUTION   In the mathematical model, x represents the age of the driver. We are interested in 20-year-old drivers. Thus, we substitute 20 for each occurrence of x. Then we use the order of operations to find N, the number of accidents per 50 million miles driven.

$N = 0.4x^2 - 36x + 1000$	This is the given mathematical model.
$N = 0.4(20)^2 - 36(20) + 1000$	Replace each occurrence of x with 20.
$N = 0.4(400) - 36(20) + 1000$	Evaluate the exponential expression: $(20)^2 = (20)(20) = 400$.
$N = 160 - 720 + 1000$	Multiply from left to right: $0.4(400) = 160$ and $36(20) = 720$.
$N = -560 + 1000$	Perform additions and subtractions from left to right. Subtract: $160 - 720 = 160 + (-720) = -560$.
$N = 440$	Add: $-560 + 1000 = 440$.

Thus, 20-year-old drivers have 440 accidents per 50 million miles driven. ■

STUDY TIP

To find

$$160 - 720 + 1000$$
$$= 160 + (-720) + 1000,$$

you can also first add the positive numbers

$$160 + 1000 = 1160$$

and then add the negative number to that result:

$$1160 + (-720) = 440.$$

How do the number of accidents for 20-year-old drivers compare to, say, the number for 45-year-old drivers? The answer is given by the points in the rectangular coordinate system in Figure 1.22. The x-coordinate of each point represents the driver's age. The y-coordinate represents the number of accidents per 50 million miles driven.

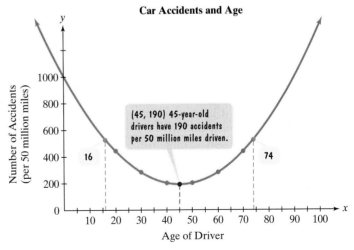

Car Accidents and Age

(45, 190) 45-year-old drivers have 190 accidents per 50 million miles driven.

FIGURE 1.22

Can you see that the point (45, 190) is lower than any of the other points? In practical terms, this indicates that 45-year-olds have the least number of car accidents, 190 per 50 million miles driven. Drivers both younger and older than 45 have more accidents per 50 million miles driven.

For 40-year-old drivers, there are 200 accidents per 50 million miles driven.

FIGURE 1.23 The Celsius scale is on the left and the Fahrenheit scale is on the right.

 CHECK POINT 12 Use the mathematical model described in Example 12, $N = 0.4x^2 - 36x + 1000$, to answer this question: How many accidents, per 50 million miles driven, are there for 40-year-old drivers?

Some formulas give an exact, rather than an approximate, relationship between variables. For example, Figure 1.23 shows temperatures on the Celsius scale and on the Fahrenheit scale. The formula

$$C = \frac{5}{9}(F - 32)$$

expresses an exact relationship between Fahrenheit temperature, F, and Celsius temperature, C.

The Formula	What the Formula Tells Us
$C = \dfrac{5}{9}(F - 32)$	If 32 is subtracted from the Fahrenheit temperature, $F - 32$, and this difference is multiplied by $\frac{5}{9}$, the resulting product, $\frac{5}{9}(F - 32)$, gives the Celsius temperature.

EXAMPLE 13 Converting from Fahrenheit to Celsius

The temperature on a warm spring day is 77°F. Use the formula $C = \frac{5}{9}(F - 32)$ to find the equivalent temperature on the Celsius scale.

SOLUTION Because the temperature is 77°F, we substitute 77 for F in the given formula. Then we use the order of operations to find the value of C.

$$C = \frac{5}{9}(F - 32)$$ This is the given formula.

$$C = \frac{5}{9}(77 - 32)$$ Replace F with 77.

$$C = \frac{5}{9}(45)$$ Work inside parentheses first: $77 - 32 = 45$.

$$C = 25$$ Multiply: $\frac{5}{9}(45) = \frac{5}{\cancel{9}} \cdot \frac{\cancel{45}}{1} = \frac{25}{1} = 25$.

Thus, 77°F is equivalent to 25°C. ■

CHECK POINT 13 The temperature on a warm summer day is 86°F. Use the formula $C = \frac{5}{9}(F - 32)$ to find the equivalent temperature on the Celsius scale.   30°C

1.8 EXERCISE SET

Student Solutions Manual   CD/Video   PH Math/Tutor Center   MathXL Tutorials on CD   MathXL®   MyMathLab   Interactmath.com

Practice Exercises

In Exercises 1–14, evaluate each exponential expression.

1. 9^2 81
2. 3^2 9
3. 4^3 64
4. 6^3 216
5. $(-4)^2$ 16
6. $(-10)^2$ 100
7. $(-4)^3$ −64
8. $(-10)^3$ −1000
9. $(-5)^4$ 625
10. $(-1)^6$ 1
11. -5^4 −625
12. -1^6 −1
13. -10^2 −100
14. -8^2 −64

In Exercises 15–28, simplify each algebraic expression, or explain why the expression cannot be simplified.

15. $7x^2 + 12x^2$ $19x^2$
16. $6x^2 + 18x^2$ $24x^2$
17. $10x^3 + 5x^3$ $15x^3$
18. $14x^3 + 8x^3$ $22x^3$
19. $8x^4 + x^4$ $9x^4$
20. $14x^4 + x^4$ $15x^4$

21. $26x^2 - 27x^2$ $-x^2$ **22.** $29x^2 - 30x^2$ $-x^2$

23. $27x^3 - 26x^3$ x^3 **24.** $30x^3 - 29x^3$ x^3

25. $5x^2 + 5x^3$ cannot be simplified

26. $8x^2 + 8x^3$ cannot be simplified

27. $16x^2 - 16x^2$ 0 **28.** $34x^2 - x^2$ $33x^2$

In Exercises 29–72, use the order of operations to simplify each expression.

29. $7 + 6 \cdot 3$ 25 **30.** $3 + 4 \cdot 5$ 23

31. $45 \div 5 \cdot 3$ 27 **32.** $40 \div 4 \cdot 2$ 20

33. $6 \cdot 8 \div 4$ 12 **34.** $8 \cdot 6 \div 2$ 24

35. $14 - 2 \cdot 6 + 3$ 5 **36.** $36 - 12 \div 4 + 2$ 35

37. $8^2 - 16 \div 2^2 \cdot 4 - 3$ 45 **38.** $10^2 - 100 \div 5^2 \cdot 2 - 1$ 91

39. $3(-2)^2 - 4(-3)^2$ -24 **40.** $5(-3)^2 - 2(-4)^2$ 13

41. $(4 \cdot 5)^2 - 4 \cdot 5^2$ 300 **42.** $(3 \cdot 5)^2 - 3 \cdot 5^2$ 150

43. $(2 - 6)^2 - (3 - 7)^2$ 0 **44.** $(4 - 6)^2 - (5 - 9)^2$ -12

45. $6(3 - 5)^3 - 2(1 - 3)^3$ -32

46. $-3(-6 + 8)^3 - 5(-3 + 5)^3$ -64

47. $[2(6 - 2)]^2$ 64 **48.** $[3(4 - 6)]^3$ -216

49. $2[5 + 2(9 - 4)]$ 30 **50.** $3[4 + 3(10 - 8)]$ 30

51. $[7 + 3(2^3 - 1)] \div 21$ $\frac{4}{3}$ **52.** $[11 - 4(2 - 3^3)] \div 37$ 3

53. $\dfrac{10 + 8}{5^2 - 4^2}$ 2 **54.** $\dfrac{6^2 - 4^2}{2 - (-8)}$ 2

55. $\dfrac{37 + 15 \div (-3)}{2^4}$ 2 **56.** $\dfrac{22 + 20 \div (-5)}{3^2}$ 2

57. $\dfrac{(-11)(-4) + 2(-7)}{7 - (-3)}$ 3 **58.** $\dfrac{-5(7 - 2) - 3(4 - 7)}{-13 - (-5)}$ 2

59. $4|10 - (8 - 20)|$ 88 **60.** $6|7 - 4 \cdot 3|$ 30

61. $8(-10) + |4(-5)|$ -60

62. $4(-15) + |3(-10)|$ -30

63. $-2^2 + 4[16 \div (3 - 5)]$ -36

64. $-3^2 + 2[20 \div (7 - 11)]$ -19

65. $24 \div \dfrac{3^2}{8 - 5} - (-6)$ 14

66. $30 \div \dfrac{5^2}{7 - 12} - (-9)$ 3

67. $\dfrac{\frac{1}{4} - \frac{1}{2}}{\frac{1}{3}}$ $-\dfrac{3}{4}$ **68.** $\dfrac{\frac{3}{5} - \frac{7}{10}}{\frac{1}{2}}$ $-\dfrac{1}{5}$

69. $-\dfrac{9}{4}\left(\dfrac{1}{2}\right) + \dfrac{3}{4} \div \dfrac{5}{6}$ $-\dfrac{9}{40}$

70. $\left[-\dfrac{4}{7} - \left(-\dfrac{2}{5}\right)\right]\left[-\dfrac{3}{8} + \left(-\dfrac{1}{9}\right)\right]$ $\dfrac{1}{12}$

71. $\dfrac{\frac{7}{9} - 3}{\frac{5}{6}} \div \dfrac{3}{2} + \dfrac{3}{4}$ $-\dfrac{37}{36}$ or $-1\dfrac{1}{36}$

72. $\dfrac{\frac{17}{25}}{\frac{3}{5} - 4} \div \dfrac{1}{5} + \dfrac{1}{2}$ $-\dfrac{1}{2}$

In Exercises 73–80, evaluate each algebraic expression for the given value of the variable.

73. $x^2 + 5x; x = 3$ 24 **74.** $x^2 - 2x; x = 6$ 24

75. $3x^2 - 8x; x = -2$ 28 **76.** $4x^2 - 2x; x = -3$ 42

77. $-x^2 - 10x; x = -1$ 9 **78.** $-x^2 - 14x; x = -1$ 13

79. $\dfrac{6y - 4y^2}{y^2 - 15}; y = 5$ -7 **80.** $\dfrac{3y - 2y^2}{y(y - 2)}; y = 5$ $-\dfrac{7}{3}$

In Exercises 81–88, simplify each algebraic expression by removing parentheses and brackets.

81. $3[5(x - 2) + 1]$ $15x - 27$

82. $4[6(x - 3) + 1]$ $24x - 68$

83. $3[6 - (y + 1)]$ $15 - 3y$

84. $5[2 - (y + 3)]$ $-5y - 5$

85. $7 - 4[3 - (4y - 5)]$ $16y - 25$

86. $6 - 5[8 - (2y - 4)]$ $10y - 54$

87. $2(3x^2 - 5) - [4(2x^2 - 1) + 3]$ $-2x^2 - 9$

88. $4(6x^2 - 3) - [2(5x^2 - 1) + 1]$ $14x^2 - 11$

Practice Plus

In Exercises 89–92, express each sentence as a single numerical expression. Then use the order of operations to simplify the expression.

89. Cube -2. Subtract this exponential expression from -10. $-10 - (-2)^3 = -2$

90. Cube -5. Subtract this exponential expression from -100. $-100 - (-5)^3 = 25$

91. Subtract 10 from 7. Multiply this difference by 2. Square this product. $[2(7 - 10)]^2 = 36$

92. Subtract 11 from 9. Multiply this difference by 2. Raise this product to the fourth power. $[2(9 - 11)]^4 = 256$

In Exercises 93–96, let x represent the number. Express each sentence as a single algebraic expression. Then simplify the expression.

93. Multiply a number by 5. Add 8 to this product. Subtract this sum from the number. $x - (5x + 8) = -4x - 8$

94. Multiply a number by 3. Add 9 to this product. Subtract this sum from the number. $x - (3x + 9) = -2x - 9$

95. Cube a number. Subtract 4 from this exponential expression. Multiply this difference by 5. $5(x^3 - 4) = 5x^3 - 20$

96. Cube a number. Subtract 6 from this exponential expression. Multiply this difference by 4. $4(x^3 - 6) = 4x^3 - 24$

Application Exercises

Medical researchers have found that the desirable heart rate, R, in beats per minute, for beneficial exercise is approximated by the mathematical models

$$R = 165 - 0.75A \qquad \text{for men}$$
$$R = 143 - 0.65A \qquad \text{for women}$$

where A is the person's age. Use these mathematical models to solve Exercises 97–98.

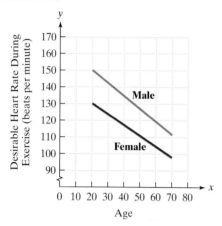

97. What is the desirable heart rate during exercise for a 40-year-old man? Identify your computation as an appropriate point in the rectangular coordinate system.   135 beats per minute; (40, 135) on blue graph

98. What is the desirable heart rate during exercise for a 40-year-old woman? Identify your computation as an appropriate point in the rectangular coordinate system.   117 beats per minute; (40, 117) on red graph

The bar graph shows the cost of Medicare, in billions of dollars, through 2005. The data can be modeled by the formula

$$N = 1.2x^2 + 15.2x + 181.4,$$

where N represents Medicare spending, in billions of dollars, x years after 1995. Use this formula to solve Exercises 99–100.

Medicare Spending

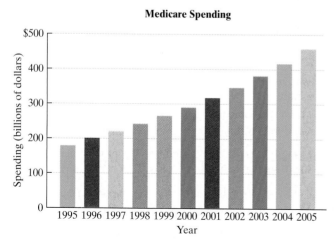

Source: Congressional Budget Office

99. According to the formula, what was the cost of Medicare, in billions of dollars, in 2000? How well does the formula describe the cost for that year shown by the bar graph? $287.4 billion; fairly well

100. According to the formula, what was the cost of Medicare, in billions of dollars, in 2005? How well does the formula describe the cost for that year shown by the bar graph? $453.4 billion; fairly well

Bariatrics is the field of medicine that deals with the overweight. Bariatric surgery closes off a large part of the stomach. As a result, patients eat less and have a diminished appetite. Celebrities like pop singer Carnie Wilson and the Today show's weatherman Al Roker have become no-longer-larger-than-life walking billboards for the operation. The bar graph shows the number of bariatric surgeries from 1992 through 2002.

Bariatric Surgery in the U.S.

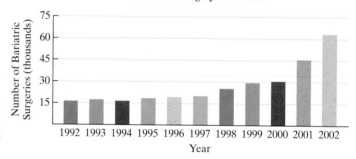

Source: American Society for Bariatric Surgery

The formula

$$N = 0.12x^3 - x^2 + 3x + 15$$

models the number of bariatric surgeries, N, in thousands, x years after 1992. Use this formula to solve Exercises 101–102.

101. According to the formula, how many bariatric surgeries were performed in 2002? How well does the formula model the data shown in the bar graph?   65 thousand; fairly well, but slight overestimation

102. According to the formula, how many bariatric surgeries were performed in 1994? How well does the formula model the data shown in the bar graph?   17.96 thousand; fairly well, but slight overestimation

The formula

$$C = \frac{5}{9}(F - 32)$$

expresses the relationship between Fahrenheit temperature, F, and Celsius temperature, C. In Exercises 103–106, use the formula to convert the given Fahrenheit temperature to its equivalent temperature on the Celsius scale.

103. 68°F   20°C   **104.** 41°F   5°C

105. −22°F   −30°C   **106.** −31°F   −35°C

Writing in Mathematics

107. Describe what it means to raise a number to a power. In your description, include a discussion of the difference between -5^2 and $(-5)^2$. **107–111.** Answers will vary.

108. Explain how to simplify $4x^2 + 6x^2$. Why is the sum not equal to $10x^4$?

109. Why is the order of operations agreement needed?

110. What is a formula?

111. The formula $F = \frac{9}{5}C + 32$ expresses the relationship between Celsius temperature, C, and Fahrenheit temperature, F. You'll be leaving the cold of winter for a vacation to Hawaii. CNN International reports a temperature in Hawaii of 30°C. Should you pack a winter coat? Use the formula to explain your answer.

Critical Thinking Exercises

112. Which one of the following is true? **d**

   a. If x is -3, then the value of $-3x - 9$ is -18.

   b. The algebraic expression $\dfrac{6x + 6}{x + 1}$ cannot have the same value when two different replacements are made for x such as $x = -3$ and $x = 2$.

   c. A miniature version of a space shuttle is an example of a mathematical model.

   d. The value of $\dfrac{|3 - 7| - 2^3}{(-2)(-3)}$ is the fraction that results when $\frac{1}{3}$ is subtracted from $-\frac{1}{3}$.

113. Simplify: $\dfrac{1}{4} - 6(2 + 8) \div \left(-\dfrac{1}{3}\right)\left(-\dfrac{1}{9}\right)$. $-\dfrac{79}{4}$

Grouping symbols can be inserted into $4 + 3 \cdot 7 - 4$ so that the resulting value is 45. By placing parentheses around the addition we obtain

$$(4 + 3) \cdot 7 - 4 = 7 \cdot 7 - 4 = 49 - 4 = 45.$$

In Exercises 114–115, insert parentheses in each expression so that the resulting value is 45.

114. $2 \cdot 3 + 3 \cdot 5$ $(2 \cdot 3 + 3) \cdot 5 = 45$

115. $2 \cdot 5 - \dfrac{1}{2} \cdot 10 \cdot 9$ $\left(2 \cdot 5 - \dfrac{1}{2} \cdot 10\right) \cdot 9 = 45$

Technology Exercises

The United States has more people in prison, as well as more people in prison per capita, than any other western industrialized nation. The bar graph shows the number of inmates in U.S. state and federal prisons in nine selected years from 1985 through 2001.

Number of Inmates in U.S. State and Federal Prisons

Source: U.S. Justice Department

The data in the graph can be modeled by each of the following formulas:

$$N = 59.5x + 505.12$$
$$N = -0.619x^2 + 69.52x + 485.28$$
$$N = -0.239x^3 + 5.288x^2 + 33.28x + 503.47.$$

In each formula, x represents the number of years after 1985 and N represents the number of inmates, in thousands. Use this information to solve Exercises 116–117.

116. Use a calculator to determine which of the three formulas serves as the best model for the inmate population in 2001. the third formula in the list

117. Use a calculator to determine which, if any, of the three formulas serves as the best model for the inmate population for the period shown by the graph. Answers will vary.

118. In this exercise, use a calculator and one of the formulas that estimates the adult height of a child given in the chapter introduction on page 1. Predict the adult height, H, in inches using the actual height of a child whose age, x, and height, h, in inches, you know. Answers will vary.

Review Exercises

119. Simplify: $-8 - 2 - (-5) + 11$ (Section 1.6, Example 3) 6

120. Multiply: $-4(-1)(-3)(2)$. (Section 1.7, Example 2) -24

121. Give an example of a real number that is not an irrational number. (Section 1.2, Example 5). Answers will vary; -3 is an example.

GROUP PROJECT

CHAPTER 1

Additional group projects can be found in the *Instructor's Resource Manual.*

One measure of physical fitness is your *resting heart rate*. Generally speaking, the more fit you are, the lower your resting heart rate. The best time to take this measurement is when you first awaken in the morning, before you get out of bed. Lie on your back with no body parts crossed and take your pulse in your neck or wrist. Use your index and second fingers and count your pulse beat for one full minute to get your resting heart rate. A resting heart rate under 48 to 57 indicates high fitness, 58 to 62, above average fitness, 63 to 70, average fitness, 71 to 82, below average fitness, and 83 or more, low fitness.

Another measure of physical fitness is your percentage of body fat. You can estimate your body fat using the following formulas:

For men: Body fat $= -98.42 + 4.15w - 0.082b$

For women: Body fat $= -76.76 + 4.15w - 0.082b$

where $w =$ waist measurement, in inches, and $b =$ total body weight, in pounds. Then divide your body fat by your total weight to get your body fat percentage. For men, less than 15% is considered athletic, 25% about average. For women, less than 22% is considered athletic, 30% about average.

Each group member should bring his or her age, resting heart rate, and body fat percentage to the group. Using the data, the group should create three graphs.

a. Create a graph that shows age and resting heart rate for group members.

b. Create a graph that shows age and body fat percentage for group members.

c. Create a graph that shows resting heart rate and body fat percentage for group members.

For each graph, select the style (line or bar) that is most appropriate.

CHAPTER 1 SUMMARY

Definitions and Concepts	Examples

Section 1.1 Fractions

Mixed Numbers and Improper Fractions

A mixed number consists of the addition of a natural number $(1, 2, 3, \ldots)$ and a fraction, expressed without the use of an addition sign. An improper fraction has a numerator that is greater than its denominator. To convert a mixed number to an improper fraction, multiply the denominator by the natural number and add the numerator. Then place this result over the original denominator.

To convert an improper fraction to a mixed number, divide the denominator into the numerator and write the mixed number using

$$\text{quotient} \frac{\text{remainder}}{\text{original denominator}}.$$

Convert $5\frac{3}{7}$ to an improper fraction.

$$5\frac{3}{7} = \frac{7 \cdot 5 + 3}{7} = \frac{35 + 3}{7} = \frac{38}{7}$$

Convert $\frac{14}{3}$ to a mixed number.

$$\frac{14}{3} = 4\frac{2}{3} \qquad 3\overline{)14} \\ \qquad\qquad \frac{12}{2}$$

A prime number is a natural number greater than 1 that has only itself and 1 as factors. A composite number is a natural number greater than 1 that is not a prime number.
The prime factorization of a composite number means to express the composite number as the product of prime numbers.

Find the prime factorization:

$$60 = 6 \cdot 10$$
$$= 2 \cdot 3 \cdot 2 \cdot 5$$

Definitions and Concepts	Examples

Section 1.1 Fractions (continued)

A fraction is reduced to its lowest terms when the numerator and denominator have no common factors other than 1. To reduce a fraction to its lowest terms, divide both the numerator and the denominator by their greatest common factor. The greatest common factor can be found by inspection or prime factorizations of the numerator and the denominator.	Reduce to lowest terms: $$\frac{8}{14} = \frac{\cancel{2} \cdot 4}{\cancel{2} \cdot 7} = \frac{4}{7}$$
Multiplying Fractions The product of two or more fractions is the product of their numerators divided by the product of their denominators.	Multiply: $$\frac{2}{7} \cdot \frac{5}{9} = \frac{2 \cdot 5}{7 \cdot 9} = \frac{10}{63}$$
Dividing Fractions The quotient of two fractions is the first multiplied by the reciprocal (or multiplicative inverse) of the second.	Divide: $$\frac{4}{9} \div \frac{3}{7} = \frac{4}{9} \cdot \frac{7}{3} = \frac{4 \cdot 7}{9 \cdot 3} = \frac{28}{27}$$
Adding and Subtracting Fractions with Identical Denominators Add or subtract numerators. Put this result over the common denominator.	Subtract: $$\frac{5}{8} - \frac{3}{8} = \frac{5 - 3}{8} = \frac{2}{8} = \frac{\cancel{2} \cdot 1}{\cancel{2} \cdot 4} = \frac{1}{4}$$
Adding and Subtracting Fractions with Unlike Denominators Rewrite the fractions as equivalent fractions with the least common denominator. Then add or subtract numerators, putting this result over the common denominator.	Add: $$\frac{3}{8} + \frac{5}{12} = \frac{3}{8} \cdot \frac{3}{3} + \frac{5}{12} \cdot \frac{2}{2}$$ The LCD is 24. $= \dfrac{9}{24} + \dfrac{10}{24} = \dfrac{19}{24}$

Section 1.2 The Real Numbers

A set is a collection of objects, called elements, whose contents can be clearly determined.	$$\{a, b, c\}$$
A line used to visualize numbers is called a number line.	 $-4\ -3\ -2\ -1\ \ 0\ \ 1\ \ 2\ \ 3\ \ 4$
Real Numbers: the set of all numbers that can be represented by points on the number line **The Sets That Make Up the Real Numbers** • Natural Numbers: $\{1, 2, 3, 4, \ldots\}$ • Whole Numbers: $\{0, 1, 2, 3, 4, \ldots\}$ • Integers: $\{\ldots, -3, -2, -1, 0, 1, 2, 3, \ldots\}$ • Rational Numbers: the set of numbers that can be expressed as the quotient of an integer and a nonzero integer; can be expressed as terminating or repeating decimals • Irrational Numbers: the set of numbers that cannot be expressed as the quotient of integers; decimal representations neither terminate nor repeat.	Given the set $$\left\{ -1.4, 0, 0.\overline{7}, \frac{9}{10}, \sqrt{2}, \sqrt{4} \right\}$$ list the • natural numbers: $\sqrt{4}$, or 2 • whole numbers: $0, \sqrt{4}$ • rational numbers: $-1.4, 0, 0.\overline{7}, \dfrac{9}{10}, \sqrt{4}$ • irrational numbers: $\sqrt{2}$ • real numbers: $$-1.4, 0, 0.\overline{7}, \frac{9}{10}, \sqrt{2}, \sqrt{4}$$

Definitions and Concepts	Examples

Section 1.2 The Real Numbers (continued)

For any two real numbers, a and b, a is less than b if a is to the left of b on the number line.

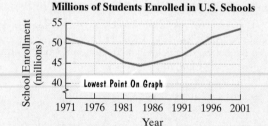

$$-4\ -3\ -2\ -1\ \ 0\ \ 1\ \ 2\ \ 3\ \ 4$$

Inequality Symbols

$<$: is less than $>$: is greater than

$\leq$: is less than or equal to $\geq$: is greater than or equal to

$-2 < 0 \qquad 0 > -2$

$0 < 2.5 \qquad 2.5 > 0$

The absolute value of a, written $|a|$, is the distance from 0 to a on the number line.

$|4| = 4 \quad |0| = 0 \quad |-6| = 6$

Section 1.3 Ordered Pairs and Graphs

The rectangular coordinate system consists of a horizontal number line, the x-axis, and a vertical number line, the y-axis, intersecting at their zero points, the origin. Each point in the system corresponds to an ordered pair of real numbers (x, y). The first number in the pair is the x-coordinate; the second number is the y-coordinate.

Plot: $(2, 3), (-5, 4), (-4, -3), (5, -2)$.

Information is often displayed using line graphs, bar graphs, and circle graphs. Line graphs are often used to illustrate trends over time. Bar graphs are convenient for showing comparisons among items.

Millions of Students Enrolled in U.S. Schools

(line graph with y-axis "School Enrollment (millions)" marked 40, 45, 50, 55 and x-axis "Year" marked 1971, 1976, 1981, 1986, 1991, 1996, 2001, labeled "Lowest Point On Graph")

Source: National Education Association

In which year did enrollment reach a minimum? 1984
Estimate enrollment for that year: 44.5 million.

Section 1.4 Basic Rules of Algebra

A letter used to represent a number is called a variable. An algebraic expression is a combination of variables, numbers, and operation symbols. Terms are separated by addition. The parts of each term that are multiplied are its factors. Like terms have the same variable factors raised to the same powers. To evaluate an algebraic expression, substitute a given number for the variable and simplify.

Evaluate: $6(x + 3) + 4x$ when $x = 5$.

Replace x with 5.

$= 6(5 + 3) + 4 \cdot 5$

$= 6(8) + 4 \cdot 5$

$= 48 + 20$

$= 68$

Definitions and Concepts	Examples

Section 1.4 Basic Rules of Algebra (continued)

Properties of Real Numbers and Algebraic Expressions

- Commutative Properties:
$$a + b = b + a$$
$$ab = ba$$
- Associative Properties:
$$(a + b) + c = a + (b + c)$$
$$(ab)c = a(bc)$$
- Distributive Properties:
$$a(b + c) = ab + ac$$
$$(b + c)a = ba + ca$$
$$a(b - c) = ab - ac$$
$$(b - c)a = ba - ca$$
$$a(b + c + d) = ab + ac + ad$$

Commutative of Addition:
$$5x + 4 = 4 + 5x$$
Commutative of Multiplication:
$$5x + 4 = x5 + 4$$
Associative of Addition:
$$6 + (4 + x) = (6 + 4) + x = 10 + x$$
Associative of Multiplication:
$$7(10x) = (7 \cdot 10)x = 70x$$
Distributive:
$$8(x + 5 + 4y) = 8x + 40 + 32y$$
Distributive to Combine Like Terms:
$$8x + 12x = (8 + 12)x = 20x$$

Simplifying Algebraic Expressions

Use the distributive property to remove grouping symbols. Then combine like terms.

$$4(5x + 7) + 13x$$
$$= 20x + 28 + 13x$$
$$= (20x + 13x) + 28$$
$$= 33x + 28$$

Section 1.5 Addition of Real Numbers

Sums on a Number Line

To find $a + b$, the sum of a and b, on a number line, start at a. If b is positive, move b units to the right. If b is negative, move b units to the left. If b is 0, stay at a. The number where we finish on the number line represents $a + b$.

$$-7 + 5 = -2$$

Start at −7. Move 5 units to the right.

Additive inverses are pairs of real numbers that are the same number of units from zero on the number line, but on opposite sides of zero.

- Identity Property of Addition:
$$a + 0 = 0 \qquad 0 + a = a$$
- Inverse Property of Addition:
$$a + (-a) = 0 \qquad (-a) + a = 0$$

The additive inverse (or opposite) of 4 is −4.
The additive inverse of −1.7 is 1.7.

Identity Property of Addition:
$$4x + 0 = 4x$$

Inverse Property of Addition:
$$4x + (-4x) = 0$$

Addition without a Number Line

To add two numbers with the same sign, add their absolute values and use their common sign. To add two numbers with different signs, subtract the smaller absolute value from the greater absolute value and use the sign of the number with the greater absolute value.

Add:
$$10 + 4 = 14$$
$$-4 + (-6) = -10$$
$$-30 + 5 = -25$$
$$12 + (-8) = 4$$

To add a series of positive and negative numbers, add all the positive numbers and add all the negative numbers. Then add the resulting positive and negative sums.

$$5 + (-3) + (-7) + 2$$
$$= (5 + 2) + [(-3) + (-7)]$$
$$= 7 + (-10)$$
$$= -3$$

Definitions and Concepts	Examples

Section 1.6 Subtraction of Real Numbers

To subtract b from a, add the additive inverse of b to a:

$$a - b = a + (-b).$$

The result is called the difference between a and b.

Subtract:

$$-7 - (-5) = -7 + 5 = -2$$

$$-\frac{3}{4} - \frac{1}{2} = -\frac{3}{4} + \left(-\frac{1}{2}\right)$$

$$= -\frac{3}{4} + \left(-\frac{2}{4}\right) = -\frac{5}{4}$$

To simplify a series of additions and subtractions, change all subtractions to additions of additive inverses. Then use the procedure for adding a series of positive and negative numbers.

Simplify:

$$-6 - 2 - (-3) + 10$$

$$= -6 + (-2) + 3 + 10$$

$$= -8 + 13$$

$$= 5$$

Section 1.7 Multiplication and Division of Real Numbers

The result of multiplying a and b, ab, is called the product of a and b. If the two numbers have different signs, the product is negative. If the two numbers have the same sign, the product is positive. If either number is 0, the product is 0.

Multiply:

$$-5(-10) = 50$$

$$\frac{3}{4}\left(-\frac{5}{7}\right) = -\frac{3}{4} \cdot \frac{5}{7} = -\frac{15}{28}$$

Assuming that no number is 0, the product of an even number of negative numbers is positive. The product of an odd number of negative numbers is negative. If any number is 0, the product is 0.

Multiply:

$$(-3)(-2)(-1)(-4) = 24$$

$$(-3)(2)(-1)(-4) = -24$$

The result of dividing the real number a by the nonzero real number b is called the quotient of a and b. If two numbers have different signs, their quotient is negative. If two numbers have the same sign, their quotient is positive. Division by zero is undefined.

Divide:

$$\frac{21}{-3} = -7$$

$$-\frac{1}{3} \div (-3) = \frac{1}{3} \cdot \frac{1}{3} = \frac{1}{9}$$

Two numbers whose product is 1 are called multiplicative inverses or reciprocals of each other. The number 0 has no multiplicative inverse.

- Identity Property of Multiplication
$$a \cdot 1 = a \qquad 1 \cdot a = a$$
- Inverse Property of Multiplication
If a is not 0:
$$a \cdot \frac{1}{a} = 1 \qquad \frac{1}{a} \cdot a = 1$$
- Multiplication Property of -1
$$-1a = -a \qquad a(-1) = -a$$
- Double Negative Property
$$-(-a) = a$$

The multiplicative inverse of 4 is $\frac{1}{4}$.
The multiplicative inverse of $-\frac{1}{3}$ is -3.

Simplify:

$$1x = x$$

$$7x \cdot \frac{1}{7x} = 1$$

$$4x - 5x = -1x = -x$$

$$-(-7y) = 7y$$

If a negative sign precedes parentheses, remove parentheses and change the sign of every term within parentheses.

Simplify:

$$-(7x - 3y + 2) = -7x + 3y - 2$$

Definitions and Concepts	Examples
Section 1.8 Exponents, Order of Operations, and Mathematical Models	
If b is a real number and n is a natural number, b^n, the nth power of b, is the product of n factors of b. Furthermore, $b^1 = b$.	Evaluate: $$8^2 = 8 \cdot 8 = 64$$ $$(-5)^3 = (-5)(-5)(-5) = -125$$
Order of Operations 1. Perform operations within grouping symbols, starting with the innermost grouping symbols. 2. Evaluate exponential expressions. 3. Multiply and divide in order from left to right. 4. Add and subtract in order from left to right.	Simplify: $$5(4-6)^2 - 2(1-3)^3$$ $$= 5(-2)^2 - 2(-2)^3$$ $$= 5(4) - 2(-8)$$ $$= 20 - (-16)$$ $$= 20 + 16 = 36$$
Some algebraic expressions contain two sets of grouping symbols: parentheses, the inner grouping symbols, and brackets, the outer grouping symbols. To simplify such expressions, use the order of operations and remove grouping symbols from innermost (parentheses) to outermost (brackets).	Simplify: $$5 - 3[2(x+1) - 7]$$ $$= 5 - 3[2x + 2 - 7]$$ $$= 5 - 3[2x - 5]$$ $$= 5 - 6x + 15$$ $$= -6x + 20$$
A formula is a statement of equality that uses letters to express a relationship between two or more variables. Formulas that describe real-world phenomena are called mathematical models. Many mathematical models give an approximate, rather than an exact, description of the relationship between variables.	The formula $N = 0.4x^2 + 0.5$ models the millions of people, N, in the United States using cable modems x years after 1996.

CHAPTER 1 REVIEW EXERCISES

1.1 *In Exercises 1–2, convert each mixed number to an improper fraction.*

1. $3\frac{2}{7}$ $\frac{23}{7}$

2. $5\frac{9}{11}$ $\frac{64}{11}$

In Exercises 3–4, convert each improper fraction to a mixed number.

3. $\frac{17}{9}$ $1\frac{8}{9}$

4. $\frac{27}{5}$ $5\frac{2}{5}$

In Exercises 5–7, identify each natural number as prime or composite. If the number is composite, find its prime factorization.

5. 60 $2 \cdot 2 \cdot 3 \cdot 5$ **6.** 63 $3 \cdot 3 \cdot 7$ **7.** 67   prime

In Exercises 8–9, simplify each fraction by reducing it to its lowest terms.

8. $\frac{15}{33}$ $\frac{5}{11}$

9. $\frac{40}{75}$ $\frac{8}{15}$

In Exercises 10–15, perform the indicated operation. Where possible, reduce the answer to its lowest terms.

10. $\frac{3}{5} \cdot \frac{7}{10}$ $\frac{21}{50}$ **11.** $\frac{4}{5} \div \frac{3}{10}$ $\frac{8}{3}$ **12.** $1\frac{2}{3} \div 6\frac{2}{3}$ $\frac{1}{4}$

13. $\frac{2}{9} + \frac{4}{9}$ $\frac{2}{3}$ **14.** $\frac{5}{6} + \frac{7}{9}$ $\frac{29}{18}$ **15.** $\frac{3}{4} - \frac{2}{15}$ $\frac{37}{60}$

16. The gas tank of a car is filled to its capacity. The first day, $\frac{1}{4}$ of the tank's gas is used for travel. The second day, $\frac{1}{3}$ of the tank's original amount of gas is used for travel. What fraction of the tank is filled with gas at the end of the second day? $\frac{5}{12}$

1.2 *In Exercises 17–18, graph each real number on a number line.*   **17–18.** See graphing answer section

17. -2.5

18. $4\frac{3}{4}$

In Exercises 19–20, express each rational number as a decimal.

19. $\frac{5}{8}$ 0.625

20. $\frac{3}{11}$ $0.\overline{27}$

21. Consider the set

$$\left\{-17, -\frac{9}{13}, 0, 0.75, \sqrt{2}, \pi, \sqrt{81}\right\}.$$

List all numbers from the set that are: **a.** natural numbers, **b.** whole numbers, **c.** integers, **d.** rational numbers, **e.** irrational numbers, **f.** real numbers.   **a.** $\sqrt{81}$ **b.** $0, \sqrt{81}$
c. $-17, 0, \sqrt{81}$ **d.** $-17, -\frac{9}{13}, 0, 0.75, \sqrt{81}$ **e.** $\sqrt{2}, \pi$
f. $-17, -\frac{9}{13}, 0, 0.75, \sqrt{2}, \pi, \sqrt{81}$

22. Give an example of an integer that is not a natural number.   Answers will vary; -2 is an example.

23. Give an example of a rational number that is not an integer.   Answers will vary; $\frac{1}{2}$ is an example.

24. Give an example of a real number that is not a rational number.   Answers will vary; π is an example.

In Exercises 25–28, insert either $<$ or $>$ in the shaded area between each pair of numbers to make a true statement.

25. -93 ▨ 17 $<$ **26.** -2 ▨ -200 $>$

27. 0 ▨ $-\frac{1}{3}$ $>$ **28.** $-\frac{1}{4}$ ▨ $-\frac{1}{5}$ $<$

In Exercises 29–30, determine whether each inequality is true or false.

29. $-13 \geq -11$ false   **30.** $-126 \leq -126$ true

In Exercises 31–32, find each absolute value.

31. $|-58|$ 58   **32.** $|2.75|$ 2.75

1.3 *In Exercises 33–36, plot the given point in a rectangular coordinate system. Indicate in which quadrant each point lies.*

33–36. See graphing answer section

33. $(1, -5)$ IV   **34.** $(4, -3)$ IV

35. $\left(\frac{7}{2}, \frac{3}{2}\right)$ I   **36.** $(-5, 2)$ II

37. Give the ordered pairs that correspond to the points labeled in the figure.

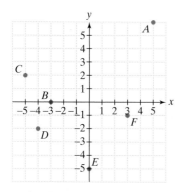

$A\,(5, 6)$ $B\,(-3, 0)$ $C\,(-5, 2)$
$D\,(-4, -2)$ $E\,(0, -5)$ $F\,(3, -1)$

We live in an era of democratic aspiration. The number of democracies worldwide is on the rise. The line graph shows the number of democracies worldwide, in four-year periods, from 1973 through 2001, including 2002. Use the graph to solve Exercises 38–42.

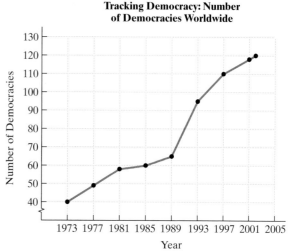

Tracking Democracy: Number of Democracies Worldwide

Source: The Freedom House

38. Find an estimate for the number of democracies in 1989.   65
39. How many more democracies were there in 2002 than in 1973?   80
40. In which four-year period did the number of democracies increase at the greatest rate?   1989–1993
41. In which four-year period did the number of democracies increase at the slowest rate?   1981–1985
42. In which year were there 49 democracies?   1977

The United States has the highest rate of television ownership in the world. The bar graph shows the seven countries in the world with the greatest number of televisions per 100 people. Use the graph to solve Exercises 43–44.

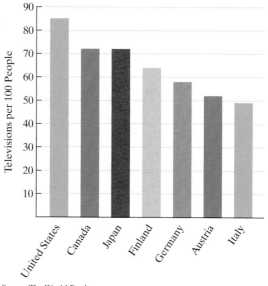

Television Ownership in Global Perspective

Source: The World Bank

43. Estimate the number of televisions per 100 people in the United States.   85

44. Which countries have more than 50 but fewer than 70 televisions per 100 people?   Finland, Germany, Austria

1.4 *In Exercises 45–46, evaluate each algebraic expression for the given value of the variable.*

45. $7x + 3; x = 10$ 73   **46.** $5(x - 4); x = 12$ 40

47. Use the commutative property of addition to write an equivalent algebraic expression: $7 + 13y$. $13y + 7$

48. Use the commutative property of multiplication to write an equivalent algebraic expression: $9(x + 7)$. $(x + 7)9$

In Exercises 49–50, use an associative property to rewrite each algebraic expression. Then simplify the resulting algebraic expression.

49. $6 + (4 + y)$ $(6 + 4) + y = 10 + y$

50. $7(10x)$ $(7 \cdot 10)x = 70x$

51. Use the distributive property to rewrite without parentheses: $6(4x - 2 + 5y)$. $24x - 12 + 30y$

In Exercises 52–53, simplify each algebraic expression.

52. $4a + 9 + 3a - 7$ $7a + 2$

53. $6(3x + 4) + 5(2x - 1)$ $28x + 19$

54. Suppose that a store is selling all computers at 25% off the regular price. If x is the regular price, the algebraic expression $x - 0.25x$ describes the sale price. Evaluate the expression when $x = 2400$. Describe what the answer means in practical terms.   1800; The sale price at 25% off is $1800 for a $2400 computer.

1.5

55. Use a number line to find the sum: $-6 + 8$.   2;   See graphing answer section

In Exercises 56–58, find each sum without the use of a number line.

56. $8 + (-11)$ -3 **57.** $-\dfrac{3}{4} + \dfrac{1}{5}$ $-\dfrac{11}{20}$

58. $7 + (-5) + (-13) + 4$ -7

In Exercises 59–60, simplify each algebraic expression.

59. $8x + (-6y) + (-12x) + 11y$ $5y - 4x$

60. $10(4 - 3y) + 28y$ $40 - 2y$

61. The Dead Sea is the lowest elevation on Earth, 1312 feet below sea level. If a person is standing 512 feet above the Dead Sea, what is that person's elevation?   800 feet below sea level

62. The water level of a reservoir is measured over a five-month period. At the beginning, the level is 25 feet. During this time, the level fell 3 feet, then rose 2 feet, then rose 1 foot, then fell 4 feet, and then rose 2 feet. What is the reservoir's water level at the end of the five months?   The reservoir's level at the end of 5 months is 23 feet.

1.6

63. Rewrite $9 - 13$ as the addition of an additive inverse. $9 + (-13)$

In Exercises 64–66, perform the indicated subtraction.

64. $-9 - (-13)$ 4   **65.** $-\dfrac{7}{10} - \dfrac{1}{2}$ $-\dfrac{6}{5}$

66. $-3.6 - (-2.1)$ -1.5

In Exercises 67–68, simplify each series of additions and subtractions.

67. $-7 - (-5) + 11 - 16$ -7

68. $-25 - 4 - (-10) + 16$ -3

69. Simplify: $3 - 6a - 8 - 2a$. $-5 - 8a$

70. What is the difference in elevation between a plane flying 26,500 feet above sea level and a submarine traveling 650 feet below sea level?   27,150 feet

1.7 *In Exercises 71–73, perform the indicated multiplication.*

71. $-7(-12)$ 84   **72.** $\dfrac{3}{5}\left(-\dfrac{5}{11}\right)$ $-\dfrac{3}{11}$

73. $5(-3)(-2)(-4)$ -120

In Exercises 74–76, perform the indicated division or state that the expression is undefined.

74. $\dfrac{45}{-5}$ -9 **75.** $-17 \div 0$ undefined

76. $-\dfrac{4}{5} \div \left(-\dfrac{2}{5}\right)$ 2

In Exercises 77–78, simplify each algebraic expression.

77. $-4\left(-\dfrac{3}{4}x\right)$ $3x$ **78.** $-3(2x - 1) - (4 - 5x)$
$-x - 1$

1.8 *In Exercises 79–81, evaluate each exponential expression.*

79. $(-6)^2$ 36   **80.** -6^2 -36 **81.** $(-2)^5$ -32

In Exercises 82–83, simplify each algebraic expression, or explain why the expression cannot be simplified.

82. $4x^3 + 2x^3$ $6x^3$ **83.** $4x^3 + 4x^2$ cannot be simplified

In Exercises 84–92, use the order of operations to simplify each expression.

84. $-40 \div 5 \cdot 2$ -16 **85.** $-6 + (-2) \cdot 5$ -16

86. $6 - 4(-3 + 2)$ 10   **87.** $28 \div (2 - 4^2)$ -2

88. $36 - 24 \div 4 \cdot 3 - 1$ 17   **89.** $-8[-4 - 5(-3)]$ -88

90. $\dfrac{6(-10 + 3)}{2(-15) - 9(-3)}$ 14   **91.** $\left(\dfrac{1}{2} + \dfrac{1}{3}\right) \div \left(\dfrac{1}{4} - \dfrac{3}{8}\right)$ $-\dfrac{20}{3}$

92. $\dfrac{1}{2} - \dfrac{2}{3} \div \dfrac{5}{9} + \dfrac{3}{10}$ $-\dfrac{2}{5}$

In Exercises 93–94, evaluate each algebraic expression for the given value of the variable.

93. $x^2 - 2x + 3; x = -1$ 6

94. $-x^2 - 7x; x = -2$ 10

In Exercises 95–96, simplify each algebraic expression.

95. $4[7(a - 1) + 2]$ $28a - 20$

96. $-6[4 - (y + 2)]$ $6y - 12$

On the average, infant girls weigh 7 pounds at birth and gain 1.5 pounds for each month for the first six months. The formula

$$W = 1.5x + 7$$

models a baby girl's weight, W, in pounds, after x months, where x is less than or equal to 6. Use the formula to solve Exercises 97–98.

Average Weight for Infant Girls

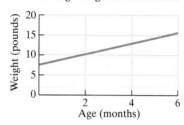

97. What does an infant girl weigh after four months? Identify your computation as an appropriate point on the line graph.
13 pounds; (4, 13)

98. What does an infant girl weigh after six months? Identify your computation as an appropriate point on the line graph.
16 pounds; (6, 16)

Among people under 40, opinions are split on downloading music: 45% think it's stealing, while 46% think it's not. (Source: Newsweek, September 22, 2003) Downloading and music CD prices that remained high through 2003 resulted in the decline of CD sales shown in the bar graph. The data can be modeled by the formula

$$N = -26x^2 + 143x + 740,$$

where N represents the number of music CD sales, in millions, x years after 1997. Use the formula to solve Exercises 99–100.

Music CD Sales in the U.S.

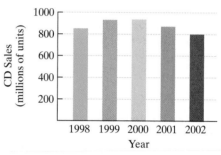

Source: RIAA

99. According to the formula, how many millions of CDs were sold in 2002? How well does the formula describe sales for that year shown by the bar graph?   805 million; fairly well, but slight overestimation

100. What do you notice about CD sales from 1998 through 2002 from the bar graph that is not obvious by looking at the formula?   Sales increase from 1998 to 2000 and decrease from 2000 to 2002.

CHAPTER 1   TEST

Remember to use your Chapter Test Prep Video CD to see the worked-out solutions to the test questions you want to review.

In Exercises 1–10, perform the indicated operation or operations.

1. $1.4 - (-2.6)$ 4

2. $-9 + 3 + (-11) + 6$ -11

3. $3(-17)$ -51

4. $\left(-\dfrac{3}{7}\right) \div \left(-\dfrac{15}{7}\right)$ $\dfrac{1}{5}$

5. $\left(3\dfrac{1}{3}\right)\left(-1\dfrac{3}{4}\right)$ $-\dfrac{35}{6}$ or $-5\dfrac{5}{6}$

6. $-50 \div 10$ -5

7. $-6 - (5 - 12)$ 1

8. $(-3)(-4) \div (7 - 10)$ -4

9. $(6 - 8)^2(5 - 7)^3$ -32

10. $\dfrac{3(-2) - 2(2)}{-2(8 - 3)}$ 1

In Exercises 11–13, simplify each algebraic expression.

11. $11x - (7x - 4)$ $4x + 4$

12. $5(3x - 4y) - (2x - y)$ $13x - 19y$

13. $6 - 2[3(x + 1) - 5]$ $10 - 6x$

14. List all the rational numbers in this set.

$$\left\{-7, -\dfrac{4}{5}, 0, 0.25, \sqrt{3}, \sqrt{4}, \dfrac{22}{7}, \pi\right\}$$

$-7, -\dfrac{4}{5}, 0, 0.25, \sqrt{4}, \dfrac{22}{7}$

15. Insert either $<$ or $>$ in the shaded area to make a true statement: -1 ___ -100. $>$

16. Find the absolute value: $|-12.8|$.   12.8

17. Plot $(-4, 3)$ in a rectangular coordinate system. In which quadrant does the point lie?   II; See graphing answer section.

18. Find the coordinates of point A in the figure.

$(-5, -2)$

In Exercises 19–20, evaluate each algebraic expression for the given value of the variable.

19. $5(x - 7); x = 4$ -15

20. $x^2 - 5x; x = -10$ 150

21. Use the commutative property of addition to write an equivalent algebraic expression: $2(x + 3)$. $2(3 + x)$

22. Use the associative property of multiplication to rewrite $-6(4x)$. Then simplify the expression.
$(-6 \cdot 4)x = -24x$

23. Use the distributive property to rewrite without parentheses: $7(5x - 1 + 2y)$. $35x - 7 + 14y$

Tule elk are introduced into a newly acquired habitat. The points in the rectangular coordinate system show the elk population every ten years. Use this information to solve Exercises 24–25.

24. Find the coordinates of point A. Then interpret the coordinates in terms of the information given. $(30, 200)$; After 30 years, there are 200 elk.

25. How many elk were introduced into the habitat?   50

26. Use the bar graph to find a reasonable estimate for the millions of DVDs sold in the United States in 2002.
725 million

Source: International Recording Media Association

27. The formula
$$T = 3(A - 20)^2 \div 50 + 10$$
describes the average running time, T, in seconds, for a person who is A years old to run the 100-yard dash. How long does it take a 30-year-old runner to run the 100-yard dash?
16 seconds

28. What is the difference in elevation between a plane flying 16,200 feet above sea level and a submarine traveling 830 feet below sea level?   17,030 feet

29. Use the graph to find a reasonable estimate, in thousands of dollars, for the average price of an existing single-family home in the United States in 1999. $135 thousand

Source: National Association of Realtors

30. The data in Exercise 29 can be modeled by the formula
$$N = 0.3x^2 + 1.2x + 92.7,$$
where N represents the average price of an existing single-family home, in thousands of dollars, x years after 1989. Use the formula to find the average price of such a home in 1999. $134.7 thousand

Note to Instructor: The Chapter 1 Test file in the TestGen program provides algorithms specifically matched to these problems so you can easily replicate this chapter test for additional practice or assessment purposes.

Reading your psychology textbook, you were intrigued by a study that found undergraduate students with an average or high sense of humor reported few increases in depression over time. By contrast, students with a low sense of humor were more likely to become depressed in response to negative life events. Is there a way to model these variables and use formulas to predict how we will respond to difficult life events?

This problem appears as Exercises 75–76 in Exercise Set 2.3.

Linear Equations and Inequalities in One Variable

The belief that humor and laughter can have positive benefits on our lives is not new. The Bible tells us, "A merry heart doeth good like a medicine, but a broken spirit drieth the bones" (Proverbs 17:22). Algebra can be used to model the influence that humor plays in our responses to negative life events. The resulting formulas predict how low- and high-humor college students will respond to these events. Formulas can be used to explain what is happening in the present and to make predictions about what might occur in the future. In this chapter, you will learn to use formulas in new ways that will help you to recognize patterns, logic, and order in a world that can appear chaotic to the untrained eye.

Objectives

1 Check whether a number is a solution to an equation.

2 Use the addition property of equality to solve equations.

3 Solve applied problems using formulas.

THE ADDITION PROPERTY OF EQUALITY

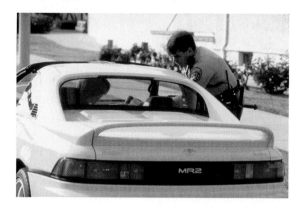

Unfortunately, many of us have been fined for driving over the speed limit. The amount of the fine depends on how fast we are speeding. Suppose that a highway has a speed limit of 60 miles per hour. The amount that speeders are fined, F, is described by the formula

$$F = 10x - 600,$$

where x is the speed in miles per hour. A friend, whom we shall call Leadfoot, borrows your car and returns a few hours later with a $400 speeding fine. Leadfoot is furious, protesting that the car was barely driven over the speed limit. Should you believe Leadfoot?

To decide if Leadfoot is telling the truth, use the formula

$$F = 10x - 600.$$

Leadfoot was fined $400, so substitute 400 for F:

$$400 = 10x - 600.$$

Now you need to find the value for x. This variable represents Leadfoot's speed, which resulted in the $400 fine.

The use of algebra in a variety of everyday applications often leads to an *equation*. An **equation** is a statement that two algebraic expressions are equal. Thus, $400 = 10x - 600$ is an example of an equation. The equal sign divides the equation into two parts, the left side and the right side:

$$\boxed{400} \quad = \quad \boxed{10x - 600}$$

Left side   Right side

The two sides of an equation can be reversed, so we can express this equation as

$$10x - 600 = 400.$$

The form of this equation is $ax + b = c$, with $a = 10$, $b = -600$, and $c = 400$. Any equation in this form is called a **linear equation in one variable**. The exponent on the variable in such an equation is 1.

In the next three sections, we will study how to solve linear equations. **Solving an equation** is the process of finding the number (or numbers) that make the equation a true statement. These numbers are called the **solutions**, or **roots**, of the equation, and we say that they **satisfy** the equation.

1 Check whether a number is a solution to an equation.

Checking Whether a Number Is a Solution to an Equation A proposed solution to an equation can be checked by substituting that number for each occurrence of the variable in the equation. If the substitution results in a true statement, the number is a solution. If the substitution results in a false statement, the number is not a solution.

EXAMPLE 1   Checking Proposed Solutions
(Is Leadfoot Telling the Truth?)

Consider the equation

$$10x - 600 = 400.$$

(Remember that x represents Leadfoot's speed that resulted in the $400 fine.) Determine whether

a. 60 is a solution.   **b.** 100 is a solution.

SOLUTION

a. To determine whether 60 is a solution to the equation, we substitute 60 for x.

$$10x - 600 = 400$$ This is the given equation.

$$10(60) - 600 \overset{?}{=} 400$$ Substitute 60 for x. The question mark over the equal sign indicates that we do not know yet if the statement is true.

$$600 - 600 \overset{?}{=} 400$$ Multiply: 10(60) = 600.

This statement is false. $$0 = 400$$ Subtract: 600 − 600 = 0.

Because the check results in a false statement, we conclude that 60 is not a solution to the given equation. (Leadfoot was not doing 60 miles per hour in your borrowed car.)

b. To determine whether 100 is a solution to the equation, we substitute 100 for x.

$$10x - 600 = 400$$ This is the given equation.

$$10(100) - 600 \overset{?}{=} 400$$ Substitute 100 for x.

$$1000 - 600 \overset{?}{=} 400$$ Multiply: 10(100) = 1000.

This statement is true. $$400 = 400$$ Subtract: 1000 − 600 = 400.

Because the check results in a true statement, we conclude that 100 is a solution to the given equation. Thus, 100 satisfies the equation. (Leadfoot was doing an outrageous 100 miles per hour, and lied with the claim that your car was barely driven over the speed limit.) ■

 CHECK POINT 1 Consider the equation $5x - 3 = 17$. Determine whether

a. 3 is a solution.   not a solution   **b.** 4 is a solution.   is a solution

---

2 Use the addition property of equality to solve equations.

Using the Addition Property of Equality to Solve Equations   Consider the equation

$$x = 11.$$

By inspection, we can see that the solution to this equation is 11. If we substitute 11 for x, we obtain the true statement $11 = 11$.

Now consider the equation

$$x - 3 = 8.$$

If we substitute 11 for x, we obtain $11 - 3 \overset{?}{=} 8$. Subtracting on the left side, we get the true statement $8 = 8$.

STUDY TIP

STUDY TIP

An equation is like a balanced scale—balanced because its two sides are equal. To maintain this balance, whatever is done to one side must also be done to the other side.

Consider $x - 3 = 8$.

The scale is balanced if the left and right sides are equal.

Add 3 to the left side.

Keep the scale balanced by adding 3 to the right side.

Thus, $x = 11$.

The equations $x - 3 = 8$ and $x = 11$ both have the same solution, namely 11, and are called *equivalent equations*. **Equivalent equations** are equations that have the same solution.

The idea in solving a linear equation is to get an equivalent equation with the variable (the letter) by itself on one side of the equal sign and a number by itself on the other side. For example, consider the equation $x - 3 = 8$. To get x by itself on the left side, add 3 to the left side, because $x - 3 + 3$ gives $x + 0$, or just x. You must then add 3 to the right side also. By doing this, we are using the **addition property of equality**.

THE ADDITION PROPERTY OF EQUALITY The same real number (or algebraic expression) may be added to both sides of an equation without changing the equation's solution. This can be expressed symbolically as follows:

$$\text{If } a = b, \text{ then } a + c = b + c.$$

EXAMPLE 2 Solving an Equation Using the Addition Property

Solve the equation: $x - 3 = 8$.

SOLUTION We can isolate the variable, x, by adding 3 to both sides of the equation.

$x - 3 = 8$	This is the given equation.
$x - 3 + 3 = 8 + 3$	Add 3 to both sides.
$x + 0 = 11$	This step is often done mentally and not listed.
$x = 11$	

By inspection, we can see that the solution to $x = 11$ is 11. To check this proposed solution, replace x with 11 in the original equation.

Check

$x - 3 = 8$	This is the original equation.
$11 - 3 \overset{?}{=} 8$	Substitute 11 for x.
$8 = 8$	Subtract: $11 - 3 = 8$.

This statement is true.

Because the check results in a true statement, we conclude that the solution to the given equation is 11. ∎

The set of an equation's solutions is called its **solution set**. Thus, the solution set of the equation in Example 2 is {11}. The solution can be expressed as 11 or, using set notation, {11}. However, do not write the solution as $x = 11$. **The solution of an equation should not be given as an equivalent equation.**

✔ **CHECK POINT 2** Solve the equation and check your proposed solution:
$$x - 5 = 12. \quad \text{17 or } \{17\}$$

When we use the addition property of equality, we add the same number to both sides of an equation. We know that subtraction is the addition of an additive inverse. Thus, the addition property also lets us subtract the same number from both sides of an equation without changing the equation's solution.

EXAMPLE 3 Subtracting the Same Number from Both Sides

Solve and check: $z + 1.4 = 2.06$.

SOLUTION

$z + 1.4 = 2.06$	This is the given equation.
$z + 1.4 - 1.4 = 2.06 - 1.4$	Subtract 1.4 from both sides. This is equivalent to adding -1.4 to both sides.
$z = 0.66$	Subtracting 1.4 from both sides eliminates 1.4 on the left.

Can you see that the solution to $z = 0.66$ is 0.66? To check this proposed solution, replace z with 0.66 in the original equation.

Check

$z + 1.4 = 2.06$	This is the original equation.
$0.66 + 1.4 \overset{?}{=} 2.06$	Substitute 0.66 for z.
$2.06 = 2.06$	This statement is true.

This true statement indicates that the solution is 0.66, or the solution set is $\{0.66\}$. ∎

 CHECK POINT 3 Solve and check: $z + 2.8 = 5.09$.   2.29 or $\{2.29\}$

When isolating the variable, we can isolate it on either the left side or the right side of an equation.

EXAMPLE 4 Isolating the Variable on the Right

Solve and check: $-\dfrac{1}{2} = x - \dfrac{2}{3}$.

SOLUTION We can isolate the variable, x, on the right side by adding $\frac{2}{3}$ to both sides of the equation.

$-\dfrac{1}{2} = x - \dfrac{2}{3}$	This is the given equation.
$-\dfrac{1}{2} + \dfrac{2}{3} = x - \dfrac{2}{3} + \dfrac{2}{3}$	Add $\frac{2}{3}$ to both sides, isolating x on the right.
$-\dfrac{3}{6} + \dfrac{4}{6} = x$	Rewrite each fraction as an equivalent fraction with a denominator of 6: $-\dfrac{1}{2} + \dfrac{2}{3} = -\dfrac{1}{2} \cdot \dfrac{3}{3} + \dfrac{2}{3} \cdot \dfrac{2}{2} = -\dfrac{3}{6} + \dfrac{4}{6}$.
$\dfrac{1}{6} = x$	Add on the left side: $-\dfrac{3}{6} + \dfrac{4}{6} = \dfrac{-3 + 4}{6} = \dfrac{1}{6}$.

STUDY TIP

The equations $a = b$ and $b = a$ have the same meaning. If you prefer, you can solve

$$-\dfrac{1}{2} = x - \dfrac{2}{3}$$

by reversing the two sides and solving

$$x - \dfrac{2}{3} = -\dfrac{1}{2}.$$

Take a moment to check the proposed solution, $\frac{1}{6}$. Substitute $\frac{1}{6}$ for x in the original equation. You should obtain $-\frac{1}{2} = -\frac{1}{2}$. This true statement indicates that the solution is $\frac{1}{6}$, or the solution set is $\left\{\frac{1}{6}\right\}$. ∎

 CHECK POINT 4 Solve and check: $-\dfrac{1}{2} = x - \dfrac{3}{4}$. $\dfrac{1}{4}$ or $\left\{\dfrac{1}{4}\right\}$

In Example 5, we combine like terms before using the addition property.

EXAMPLE 5 Combining Like Terms before Using the Addition Property

Solve and check: $5y + 3 - 4y - 8 = 6 + 9$.

SOLUTION

$$5y + 3 - 4y - 8 = 6 + 9 \qquad \text{This is the given equation.}$$

$$y - 5 = 15 \qquad \text{Combine like terms: } 5y - 4y = y, 3 - 8 = -5,$$
$$\text{and } 6 + 9 = 15.$$

$$y - 5 + 5 = 15 + 5 \qquad \text{Add 5 to both sides.}$$

$$y = 20$$

To check the proposed solution, 20, replace y with 20 in the original equation.

Check $\qquad 5y + 3 - 4y - 8 = 6 + 9 \qquad$ Be sure to use the original equation and not the simplified form in the second step above. (Why?)

$$5(20) + 3 - 4(20) - 8 \overset{?}{=} 6 + 9 \qquad \text{Substitute 20 for y.}$$

$$100 + 3 - 80 - 8 \overset{?}{=} 6 + 9 \qquad \text{Multiply on the left.}$$

$$103 - 88 \overset{?}{=} 6 + 9 \qquad \text{Combine positive numbers and combine}$$
$$\text{negative numbers on the left.}$$

$$15 = 15 \qquad \text{This statment is true.}$$

This true statement verifies that the solution is 20, or the solution set is $\{20\}$. ∎

CHECK POINT 5 Solve and check: $8y + 7 - 7y - 10 = 6 + 4$.   13 or $\{13\}$

Adding and Subtracting Variable Terms on Both Sides of an Equation   In some equations, variable terms appear on both sides. Here is an example:

$$4x = 7 + 3x.$$

A variable term, $4x$, is on the left side.   A variable term, $3x$, is on the right side.

Our goal is to isolate all the variable terms on one side of the equation. We can use the addition property of equality to do this. The property allows us to add or subtract the same variable term on both sides of an equation without changing the solution. Let's see how we can use this idea to solve $4x = 7 + 3x$.

EXAMPLE 6   Using the Addition Property to Isolate Variable Terms

Solve and check: $4x = 7 + 3x$.

SOLUTION   In the given equation, variable terms appear on both sides. We can isolate them on one side by subtracting $3x$ from both sides of the equation.

$$4x = 7 + 3x \qquad \text{This is the given equation.}$$

$$4x - 3x = 7 + 3x - 3x \qquad \text{Subtract 3x from both sides and isolate variable terms}$$
$$\text{on the left.}$$

$$x = 7 \qquad \text{Subtracting 3x from both sides eliminates 3x on the}$$
$$\text{right. On the left, } 4x - 3x = 1x = x.$$

To check the proposed solution, 7, replace x with 7 in the original equation.

Check		
	$4x = 7 + 3x$	Use the original equation.
	$4(7) \overset{?}{=} 7 + 3(7)$	Substitute 7 for x.
	$28 \overset{?}{=} 7 + 21$	Multiply: $4(7) = 28$ and $3(7) = 21$.
	$28 = 28$	This statement is true.

This true statement verifies that the solution is 7, or the solution set is $\{7\}$. ∎

 CHECK POINT 6 Solve and check: $7x = 12 + 6x$.   12 or $\{12\}$

EXAMPLE 7   Solving an Equation by Isolating the Variable

Solve and check: $3y - 9 = 2y + 6$.

SOLUTION   Our goal is to isolate variable terms on one side and constant terms on the other side. Let's begin by isolating the variable on the left.

$3y - 9 = 2y + 6$	This is the given equation.
$3y - 2y - 9 = 2y - 2y + 6$	Isolate the variable terms on the left by subtracting 2y from both sides.
$y - 9 = 6$	Subtracting 2y from both sides eliminates 2y on the right. On the left, $3y - 2y = 1y = y$.

Now we isolate the constant terms on the right by adding 9 to both sides.

$y - 9 + 9 = 6 + 9$	Add 9 to both sides.
$y = 15$	

Check		
	$3y - 9 = 2y + 6$	Use the original equation.
	$3(15) - 9 \overset{?}{=} 2(15) + 6$	Substitute 15 for y.
	$45 - 9 \overset{?}{=} 30 + 6$	Multiply: $3(15) = 45$ and $2(15) = 30$.
	$36 = 36$	This statement is true.

The solution is 15, or the solution set is $\{15\}$. ∎

 CHECK POINT 7 Solve and check: $3x - 6 = 2x + 5$.   11 or $\{11\}$

---

3 Solve applied problems using formulas.

Applications   Our next example shows how the addition property of equality can be used to find the value of a variable in a mathematical model.

EXAMPLE 8   An Application: Vocabulary and Age

There is a relationship between the number of words in a child's vocabulary, V, and the child's age, A, in months, for ages between 15 and 50 months, inclusive. This relationship can be modeled by the formula

$$V + 900 = 60A.$$

Use the formula to find the vocabulary of a child at the age of 30 months.

SOLUTION   In the formula, A represents the child's age, in months. We are interested in a 30-month-old child. Thus, we substitute 30 for A. Then we use the addition property of equality to find V, the number of words in the child's vocabulary.

$$V + 900 = 60A \qquad \text{This is the given formula.}$$
$$V + 900 = 60(30) \qquad \text{Substitute 30 for A.}$$
$$V + 900 = 1800 \qquad \text{Multiply: 60(30) = 1800.}$$
$$V + 900 - 900 = 1800 - 900 \qquad \text{Subtract 900 from both sides and solve for V.}$$
$$V = 900$$

At the age of 30 months, a child has a vocabulary of 900 words. ∎

The points in the rectangular coordinate system in Figure 2.1 allow us to "see" the formula $V + 900 = 60A$. The first coordinate of each point represents the child's age, in months. The second coordinate represents the child's vocabulary. The points are rising steadily from left to right. This shows that a typical child's vocabulary is steadily increasing with age.

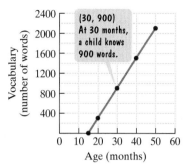

Vocabulary and Age

(30, 900) At 30 months, a child knows 900 words.

Vocabulary (number of words)

Age (months)

FIGURE 2.1

✔ **CHECK POINT 8** Use the formula $V + 900 = 60A$ to find the vocabulary of a child at the age of 50 months.   *2100 words*

2.1 EXERCISE SET

Practice Exercises

Solve each equation in Exercises 1–44 using the addition property of equality. Be sure to check your proposed solutions.

1. $x - 4 = 19$ $\{23\}$
2. $y - 5 = -18$ $\{-13\}$
3. $z + 8 = -12$ $\{-20\}$
4. $z + 13 = -15$ $\{-28\}$
5. $-2 = x + 14$ $\{-16\}$
6. $-13 = x + 11$ $\{-24\}$
7. $-17 = y - 5$ $\{-12\}$
8. $-21 = y - 4$ $\{-17\}$
9. $7 + z = 11$ $\{4\}$
10. $18 + z = 14$ $\{-4\}$
11. $-6 + y = -17$ $\{-11\}$
12. $-8 + y = -29$ $\{-21\}$
13. $x + \dfrac{1}{3} = \dfrac{7}{3}$ $\{2\}$
14. $x + \dfrac{7}{8} = \dfrac{9}{8}$ $\left\{\dfrac{1}{4}\right\}$
15. $t + \dfrac{5}{6} = -\dfrac{7}{12}$ $\left\{-\dfrac{17}{12}\right\}$
16. $t + \dfrac{2}{3} = -\dfrac{7}{6}$ $\left\{-\dfrac{11}{6}\right\}$
17. $x - \dfrac{3}{4} = \dfrac{9}{2}$ $\left\{\dfrac{21}{4}\right\}$
18. $x - \dfrac{3}{5} = \dfrac{7}{10}$ $\left\{\dfrac{13}{10}\right\}$
19. $-\dfrac{1}{5} + y = -\dfrac{3}{4}$ $\left\{-\dfrac{11}{20}\right\}$
20. $-\dfrac{1}{8} + y = -\dfrac{1}{4}$ $\left\{-\dfrac{1}{8}\right\}$
21. $3.2 + x = 7.5$ $\{4.3\}$
22. $-2.7 + w = -5.3$ $\{-2.6\}$
23. $x + \dfrac{3}{4} = -\dfrac{9}{2}$ $\left\{-\dfrac{21}{4}\right\}$
24. $r + \dfrac{3}{5} = -\dfrac{7}{10}$ $\left\{-\dfrac{13}{10}\right\}$
25. $5 = -13 + y$ $\{18\}$
26. $-11 = 8 + x$ $\{-19\}$
27. $-\dfrac{3}{5} = -\dfrac{3}{2} + s$ $\left\{\dfrac{9}{10}\right\}$
28. $\dfrac{7}{3} = -\dfrac{5}{2} + z$ $\left\{\dfrac{29}{6}\right\}$
29. $830 + y = 520$ $\{-310\}$
30. $-90 + t = -35$ $\{55\}$
31. $r + 3.7 = 8$ $\{4.3\}$
32. $x + 10.6 = -9$ $\{-19.6\}$
33. $-3.7 + m = -3.7$ $\{0\}$
34. $y + \dfrac{7}{11} = \dfrac{7}{11}$ $\{0\}$

35. $6y + 3 - 5y = 14$ $\{11\}$
36. $-3x - 5 + 4x = 9$ $\{14\}$
37. $7 - 5x + 8 + 2x + 4x - 3 = 2 + 3 \cdot 5$ $\{5\}$
38. $13 - 3r + 2 + 6r - 2r - 1 = 3 + 2 \cdot 9$ $\{7\}$
39. $7y + 4 = 6y - 9$ $\{-13\}$
40. $4r - 3 = 5 + 3r$ $\{8\}$
41. $12 - 6x = 18 - 7x$ $\{6\}$
42. $20 - 7s = 26 - 8s$ $\{6\}$
43. $4x + 2 = 3(x - 6) + 8$ $\{-12\}$
44. $7x + 3 = 6(x - 1) + 9$ $\{0\}$

Practice Plus

The equations in Exercises 45–48 contain small geometric figures that represent real numbers. Use the addition property of equality to isolate x on one side of the equation and the geometric figures on the other side.

45. $x - \square = \triangle$ $x = \triangle + \square$
46. $x + \square = \triangle$ $x = \triangle - \square$
47. $2x + \triangle = 3x + \square$ $\triangle - \square = x$
48. $6x - \triangle = 7x - \square$ $-\triangle + \square = x$

In Exercises 49–52, use the given information to write an equation. Let x represent the number described in each exercise. Then solve the equation and find the number.

49. If 12 is subtracted from a number, the result is -2. Find the number. $x - 12 = -2; 10$

50. If 23 is subtracted from a number, the result is -8. Find the number. $x - 23 = -8; 15$

51. The difference between $\dfrac{2}{5}$ of a number and 8 is $\dfrac{7}{5}$ of that number. Find the number. $\dfrac{2}{5}x - 8 = \dfrac{7}{5}x; -8$

52. The difference between 3 and $\frac{2}{7}$ of a number is $\frac{5}{7}$ of that number. Find the number. $\quad 3 - \frac{2}{7}x = \frac{5}{7}x; 3$

Application Exercises

Formulas frequently appear in the business world. For example, the cost, C, of an item (the price paid by a retailer) plus the markup, M, on that item (the retailer's profit) equals the selling price, S, of the item. The formula is

$$C + M = S.$$

Use the formula to solve Exercises 53–54.

53. The selling price of a computer is $1850. If the markup on the computer is $150, find the cost to the retailer for the computer. $1700

54. The selling price of a television is $650. If the cost to the retailer for the television is $520, find the markup. $130

The formula

$$d + 525,000 = 5000c$$

models the relationship between the annual number of deaths in the United States from heart disease, d, and the average cholesterol level, c, of blood. (Cholesterol level, c, is expressed in milligrams per deciliter of blood.) Use this formula to solve Exercises 55–56.

55. The average cholesterol level for people in the United States is 210. According to the formula, how many deaths per year from heart disease can be expected with this cholesterol level? 525,000 deaths

56. Suppose that the average cholesterol level for people in the United States could be reduced to 180. Determine the number of annual deaths from heart disease that can be expected with this reduced cholesterol level. 375,000 deaths

Statins, used to control high cholesterol, are the top-selling class of prescription drugs in the United States. The bar graph shows statin sales, in billions of dollars, from 1998 through 2002. The data can be modeled by the formula

$$S - 1.6x = 5.8,$$

where S is statin sales, in billions of dollars, x years after 1998. Use this formula to solve Exercises 57–58.

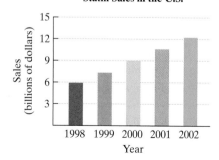

Statin Sales in the U.S.

Sales (billions of dollars) vs. Year

Source: Lawrence Berkeley
National Laboratory

57. According to the formula, what will Americans spend on statins in 2005? $17 billion

58. According to the formula, what will Americans spend on statins in 2006? $18.6 billion

Writing in Mathematics

59. What is an equation? 59–63. Answers will vary.

60. Explain how to determine whether a number is a solution to an equation.

61. State the addition property of equality and give an example.

62. Explain why $x + 2 = 9$ and $x + 2 = -6$ are not equivalent equations.

63. What is the difference between solving an equation such as

$$5y + 3 - 4y - 8 = 6 + 9$$

and simplifying an algebraic expression such as

$$5y + 3 - 4y - 8?$$

If there is a difference, which topic should be taught first? Why?

Critical Thinking Exercises

64. Which one of the following statements is true? c

a. If $y - a = -b$, then $y = a + b$.

b. If $y + 7 = 0$, then $y = 7$.

c. The solution to $4 - x = -3x$ is -2.

d. If 7 is added on one side of an equation, then it should be subtracted on the other side.

65. Write an equation with a negative solution that can be solved by adding 100 to both sides. Answers will vary. Example: $x - 100 = -101$

Technology Exercises

Use a calculator to solve each equation in Exercises 66–67.

66. $x - 7.0463 = -9.2714 \quad \{-2.2251\}$

67. $6.9825 = 4.2296 + y \quad \{2.7529\}$

Review Exercises

68. Plot $(-3, 1)$ in rectangular coordinates. In which quadrant does the point lie? (Section 1.3, Example 1)
II; See graphing answer section.

69. Simplify: $-16 - 8 \div 4 \cdot (-2)$. (Section 1.8, Example 4)
-12

70. Simplify: $3[7x - 2(5x - 1)]$. (Section 1.8, Example 11)
$6 - 9x$

SECTION 2.2

Objectives

1 Use the multiplication property of equality to solve equations.

2 Solve equations in the form $-x = c$.

3 Use the addition and multiplication properties to solve equations.

4 Solve applied problems using formulas.

THE MULTIPLICATION PROPERTY OF EQUALITY

Could you live to be 125? The number of Americans ages 100 or older could approach 850,000 by 2050. Some scientists predict that by 2100, our descendants could live to be 200 years of age. In this section, we will see how a formula can be used to make these kinds of predictions as we turn to a new property for solving linear equations.

Using the Multiplication Property of Equality to Solve Equations Can the addition property of equality be used to solve every linear equation in one variable? No. For example, consider the equation

$$\frac{x}{5} = 9.$$

We cannot isolate the variable x by adding or subtracting 5 on both sides. To get x by itself on the left side, multiply the left side by 5:

1 Use the multiplication property of equality to solve equations.

$$5 \cdot \frac{x}{5} = \left(5 \cdot \frac{1}{5}\right)x = 1x = x.$$

5 is the multiplicative inverse of $\frac{1}{5}$.

You must then multiply the right side by 5 also. By doing this, we are using the **multiplication property of equality**.

> **THE MULTIPLICATION PROPERTY OF EQUALITY** The same nonzero real number (or algebraic expression) may multiply both sides of an equation without changing the solution. This can be expressed symbolically as follows:
> If $a = b$ and $c \neq 0$, then $ac = bc$.

EXAMPLE 1 Solving an Equation Using the Multiplication Property

Solve the equation: $\frac{x}{5} = 9$.

SOLUTION We can isolate the variable, x, by multiplying both sides of the equation by 5.

$$\frac{x}{5} = 9 \qquad \text{This is the given equation.}$$

$$5 \cdot \frac{x}{5} = 5 \cdot 9 \qquad \text{Multiply both sides by 5.}$$

$$1x = 45 \qquad \text{Simplify.}$$

$$x = 45 \qquad 1x = x$$

By substituting 45 for x in the original equation, we obtain the true statement $9 = 9$. This verifies that the solution is 45, or the solution set is $\{45\}$.

✔ **CHECK POINT 1** Solve the equation: $\dfrac{x}{3} = 12$. 36 or {36}

When we use the multiplication property of equality, we multiply both sides of an equation by the same nonzero number. We know that division is multiplication by a multiplicative inverse. Thus, the multiplication property also lets us divide both sides of an equation by a nonzero number without changing the solution.

EXAMPLE 2 Dividing Both Sides by the Same Nonzero Number

Solve: **a.** $6x = 30$ **b.** $-7y = 56$ **c.** $-18.9 = 3z$.

SOLUTION In each equation, the variable is multiplied by a number. We can isolate the variable by dividing both sides of the equation by that number.

a. $6x = 30$ This is the given equation.

$\dfrac{6x}{6} = \dfrac{30}{6}$ Divide both sides by 6.

$1x = 5$ Simplify.

$x = 5$ $1x = x$

By substituting 5 for x in the original equation, we obtain the true statement $30 = 30$. The solution is 5, or the solution set is $\{5\}$.

b. $-7y = 56$ This is the given equation.

$\dfrac{-7y}{-7} = \dfrac{56}{-7}$ Divide both sides by -7.

$1y = -8$ Simplify.

$y = -8$ $1y = y$

By substituting -8 for y in the original equation, we obtain the true statement $56 = 56$. The solution is -8, or the solution set is $\{-8\}$.

c. $-18.9 = 3z$ This is the given equation.

$\dfrac{-18.9}{3} = \dfrac{3z}{3}$ Divide both sides by 3.

$-6.3 = 1z$ Simplify.

$-6.3 = z$ $1z = z$

By substituting -6.3 for z in the original equation, we obtain the true statement $-18.9 = -18.9$. The solution is -6.3, or the solution set is $\{-6.3\}$. ■

✔ **CHECK POINT 2** Solve:

a. $4x = 84$ 21 or {21}   **b.** $-11y = 44$ -4 or $\{-4\}$ **c.** $-15.5 = 5z$. -3.1 or $\{-3.1\}$

Some equations have a variable term with a fractional coefficient. Here is an example:

$$\dfrac{3}{4}y = 12.$$

The coefficient of the term $\frac{3}{4}y$ is $\frac{3}{4}$.

To isolate the variable, multiply both sides of the equation by the multiplicative inverse of the fraction. For the equation $\frac{3}{4}y = 12$, the multiplicative inverse of $\frac{3}{4}$ is $\frac{4}{3}$. Thus, we solve $\frac{3}{4}y = 12$ by multiplying both sides by $\frac{4}{3}$.

EXAMPLE 3　Using the Multiplication Property to Eliminate a Fractional Coefficient

Solve:　**a.** $\dfrac{3}{4}y = 12$　**b.** $9 = -\dfrac{3}{5}x$.

SOLUTION

a.

$$\frac{3}{4}y = 12 \qquad \text{This is the given equation.}$$

$$\frac{4}{3}\left(\frac{3}{4}y\right) = \frac{4}{3}\cdot 12 \qquad \text{Multiply both sides by } \tfrac{4}{3}\text{, the multiplicative inverse of } \tfrac{3}{4}.$$

$$1y = 16 \qquad \text{On the left, } \tfrac{4}{3}\left(\tfrac{3}{4}y\right) = \left(\tfrac{4}{3}\cdot\tfrac{3}{4}\right)y = 1y.$$
$$\text{On the right, } \tfrac{4}{3}\cdot\tfrac{12}{1} = \tfrac{48}{3} = 16.$$

$$y = 16 \qquad 1y = y$$

By substituting 16 for y in the original equation, we obtain the true statement $12 = 12$. The solution is 16, or the solution set is $\{16\}$.

b.

$$9 = -\frac{3}{5}x \qquad \text{This is the given equation.}$$

$$-\frac{5}{3}\cdot 9 = -\frac{5}{3}\left(-\frac{3}{5}x\right) \qquad \text{Multiply both sides by } -\tfrac{5}{3}\text{, the multiplicative inverse of } -\tfrac{3}{5}.$$

$$-15 = 1x \qquad \text{Simplify.}$$

$$-15 = x \qquad 1x = x$$

By substituting -15 for x in the original equation, we obtain the true statement $9 = 9$. The solution is -15, or the solution set is $\{-15\}$. ■

✔ **CHECK POINT 3** Solve:　**a.** $\dfrac{2}{3}y = 16$　24 or {24}　**b.** $28 = -\dfrac{7}{4}x$.　-16 or $\{-16\}$

STUDY TIP

The equation
$$9 = -\frac{3}{5}x$$
can be expressed as
$$9 = -\frac{3x}{5}$$
or
$$9 = \frac{-3x}{5}.$$

2 Solve equations in the form $-x = c$.

Equations and Coefficients of -1　How do we solve an equation in the form $-x = c$, such as $-x = 4$? Because the equation means $-1x = 4$, we have not yet obtained a solution. The solution of an equation is obtained from the form $x = $ some number. The equation $-x = 4$ is not yet in this form. We still need to isolate x. We can do this by multiplying or dividing both sides of the equation by -1. We will multiply by -1.

EXAMPLE 4　Solving Equations in the Form $-x = c$

Solve:　**a.** $-x = 4$　**b.** $-x = -7$.

SOLUTION　We multiply both sides of each equation by -1. This will isolate x on the left side.

a.

$$-x = 4 \qquad \text{This is the given equation.}$$

$$-1x = 4 \qquad \text{Rewrite } -x \text{ as } -1x.$$

$$(-1)(-1x) = (-1)(4) \qquad \text{Multiply both sides by } -1.$$

$$1x = -4 \qquad \text{On the left, } (-1)(-1) = 1. \text{ On the right, } (-1)(4) = -4.$$

$$x = -4 \qquad 1x = x$$

Check

$$-x = 4 \qquad \text{This is the original equation.}$$
$$-(-4) \overset{?}{=} 4 \qquad \text{Substitute } -4 \text{ for } x.$$
$$4 = 4 \qquad -(-a) = a, \text{ so } -(-4) = 4.$$

This true statement indicates that the solution is -4, or the solution set is $\{-4\}$.

b.

$$-x = -7 \qquad \text{This is the given equation.}$$
$$-1x = -7 \qquad \text{Rewrite } -x \text{ as } -1x.$$
$$(-1)(-1x) = (-1)(-7) \qquad \text{Multiply both sides by } -1.$$
$$1x = 7 \qquad (-1)(-1) = 1 \text{ and } (-1)(-7) = 7.$$
$$x = 7 \qquad 1x = x$$

By substituting 7 for x in the original equation, we obtain the true statement $-7 = -7$. The solution is 7, or the solution set is $\{7\}$. ∎

STUDY TIP

If $-x = c$, then the equation's solution is the additive inverse (the opposite) of c. For example, the solution of $-x = -7$ is the additive inverse of -7, which is 7.

✔ **CHECK POINT 4** Solve: **a.** $-x = 5$ -5 or $\{-5\}$ **b.** $-x = -3$.   3 or $\{3\}$

---

3 Use the addition and multiplication properties to solve equations.

Equations Requiring Both the Addition and Multiplication Properties When an equation does not contain fractions, we will often use the addition property of equality before the multiplication property of equality. Our overall goal is to isolate the variable with a coefficient of 1 on either the left or right side of the equation.

Here is the procedure that we will be using to solve the equations in the next three examples:

- Use the addition property of equality to isolate the variable term.
- Use the multiplication property of equality to isolate the variable.

EXAMPLE 5 Using Both the Addition and Multiplication Properties

Solve: $3x + 1 = 7$.

SOLUTION We begin by isolating the variable term, $3x$, subtracting 1 from both sides. Then we isolate the variable, x, by dividing both sides by 3.

- **Use the addition property of equality to isolate the variable term.**

$$3x + 1 = 7 \qquad \text{This is the given equation.}$$
$$3x + 1 - 1 = 7 - 1 \qquad \text{Use the addition property, subtracting 1 from both sides.}$$
$$3x = 6 \qquad \text{Simplify.}$$

- **Use the multiplication property of equality to isolate the variable.**

$$\frac{3x}{3} = \frac{6}{3} \qquad \text{Divide both sides by 3.}$$
$$x = 2 \qquad \text{Simplify.}$$

By substituting 2 for x in the original equation, we obtain the true statement $7 = 7$. The solution is 2, or the solution set is $\{2\}$. ∎

✔ **CHECK POINT 5** Solve: $4x + 3 = 27$.   6 or $\{6\}$

EXAMPLE 6 Using Both the Addition and Multiplication Properties

Solve: $-2y - 28 = 4$.

SOLUTION We begin by isolating the variable term, $-2y$, adding 28 to both sides. Then we isolate the variable, y, by dividing both sides by -2.

- **Use the addition property of equality to isolate the variable term.**

$$-2y - 28 = 4 \qquad \text{This is the given equation.}$$
$$-2y - 28 + 28 = 4 + 28 \qquad \text{Use the addition property, adding 28 to both sides.}$$
$$-2y = 32 \qquad \text{Simplify.}$$

- **Use the multiplication property of equality to isolate the variable.**

$$\frac{-2y}{-2} = \frac{32}{-2} \qquad \text{Divide both sides by } -2.$$
$$y = -16 \qquad \text{Simplify.}$$

Take a moment to substitute -16 for y in the given equation. Do you obtain the true statement $4 = 4$? The solution is -16, or the solution set is $\{-16\}$.

✔ **CHECK POINT 6** Solve: $-4y - 15 = 25$. -10 or $\{-10\}$

EXAMPLE 7 Using Both the Addition and Multiplication Properties

Solve: $3x - 14 = -2x + 6$.

SOLUTION We will use the addition property to collect all terms involving x on the left and all numerical terms on the right. Then we will isolate the variable, x, by dividing both sides by its coefficient.

- **Use the addition property of equality to isolate the variable term.**

$$3x - 14 = -2x + 6 \qquad \text{This is the given equation.}$$
$$3x + 2x - 14 = -2x + 2x + 6 \qquad \text{Add 2x to both sides.}$$
$$5x - 14 = 6 \qquad \text{Simplify.}$$
$$5x - 14 + 14 = 6 + 14 \qquad \text{Add 14 to both sides.}$$
$$5x = 20 \qquad \text{Simplify. The variable term, 5x, is isolated on the left. The numerical term, 20, is isolated on the right.}$$

- **Use the multiplication property of equality to isolate the variable.**

$$\frac{5x}{5} = \frac{20}{5} \qquad \text{Divide both sides by 5.}$$
$$x = 4 \qquad \text{Simplify.}$$

Check
$$3x - 14 = -2x + 6 \qquad \text{Use the original equation.}$$
$$3(4) - 14 \stackrel{?}{=} -2(4) + 6 \qquad \text{Substitute the proposed solution, 4, for x.}$$
$$12 - 14 \stackrel{?}{=} -8 + 6 \qquad \text{Multiply.}$$
$$-2 = -2 \qquad \text{Simplify.}$$

The true statement $-2 = -2$ verifies that the solution is 4, or the solution set is $\{4\}$. ■

✔ **CHECK POINT 7** Solve: $2x - 15 = -4x + 21$.   6 or $\{6\}$

Solve applied problems using formulas.

Applications Your life expectancy is related to the year when you were born. The bar graph in Figure 2.2 shows life expectancy in the United States by year of birth. The data for U.S. women shown by the green bars can be modeled by the formula

$$E = 0.18t + 71,$$

where E is the life expectancy for women born t years after 1950.

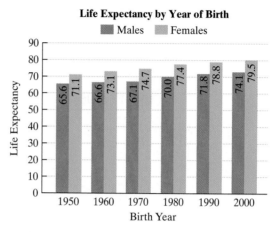

FIGURE 2.2
Source: U.S. Bureau of the Census

EXAMPLE 8 **Using the Formula For Life Expectancy**

Use the formula

$$E = 0.18t + 71$$

to determine the year of birth for which U.S. women can expect to live 87.2 years.

SOLUTION We are interested in a life expectancy of 87.2 years. We substitute 87.2 for E in the formula and solve for t. Keep in mind that t represents birth years *after* 1950.

$E = 0.18t + 71$	This is the given formula.
$87.2 = 0.18t + 71$	Replace E with 87.2.

Our goal is to isolate t.

$87.2 - 71 = 0.18t + 71 - 71$	Isolate the term containing t by subtracting 71 from both sides.
$16.2 = 0.18t$	Simplify.
$\dfrac{16.2}{0.18} = \dfrac{0.18t}{0.18}$	Divide both sides by 0.18.
$90 = t$	Simplify: $0.18\overline{)16.20}$ ... 90

The formula indicates that U.S. women born 90 years after 1950, or in 2040, can expect to live 87.2 years. ■

 CHECK POINT 8 The data for U.S. men shown by the blue bars in Figure 2.2 can be modeled by the formula $E = 0.16t + 65$, where E represents life expectancy for men born t years after 1950. Determine the year of birth for which U.S. men can expect to live 76.2 years. 2020

2.2 EXERCISE SET

Student Solutions Manual CD/Video PH Math/Tutor Center MathXL Tutorials on CD Math XL MathXL® MyMathLab MyMathLab Interactmath.com

Practice Exercises

Solve each equation in Exercises 1–28 using the multiplication property of equality. Be sure to check your proposed solutions.

1. $\frac{x}{6} = 5$ {30}

2. $\frac{x}{7} = 4$ {28}

3. $\frac{x}{-3} = 11$ {−33}

4. $\frac{x}{-5} = 8$ {−40}

5. $5y = 35$ {7}

6. $6y = 42$ {7}

7. $-7y = 63$ {−9}

8. $-4y = 32$ {−8}

9. $-28 = 8z$ {$-\frac{7}{2}$}

10. $-36 = 8z$ {$-\frac{9}{2}$}

11. $-18 = -3z$ {6}

12. $-54 = -9z$ {6}

13. $-8x = 6$ {$-\frac{3}{4}$}

14. $-8x = 4$ {$-\frac{1}{2}$}

15. $17y = 0$ {0}

16. $-16y = 0$ {0}

17. $\frac{2}{3}y = 12$ {18}

18. $\frac{3}{4}y = 15$ {20}

19. $28 = -\frac{7}{2}x$ {−8}

20. $20 = -\frac{5}{8}x$ {−32}

21. $-x = 17$ {−17}

22. $-x = 23$ {−23}

23. $-47 = -y$ {47}

24. $-51 = -y$ {51}

25. $-\frac{x}{5} = -9$ {45}

26. $-\frac{x}{5} = -1$ {5}

27. $2x - 12x = 50$ {−5}

28. $8x - 3x = -45$ {−9}

Solve each equation in Exercises 29–54 using both the addition and multiplication properties of equality. Check proposed solutions.

29. $2x + 1 = 11$ {5}

30. $2x + 5 = 13$ {4}

31. $2x - 3 = 9$ {6}

32. $3x - 2 = 9$ {$\frac{11}{3}$}

33. $-2y + 5 = 7$ {−1}

34. $-3y + 4 = 13$ {−3}

35. $-3y - 7 = -1$ {−2}

36. $-2y - 5 = 7$ {−6}

37. $12 = 4z + 3$ {$\frac{9}{4}$}

38. $14 = 5z - 21$ {7}

39. $-x - 3 = 3$ {−6}

40. $-x - 5 = 5$ {−10}

41. $6y = 2y - 12$ {−3}

42. $8y = 3y - 10$ {−2}

43. $3z = -2z - 15$ {−3}

44. $2z = -4z + 18$ {3}

45. $-5x = -2x - 12$ {4}

46. $-7x = -3x - 8$ {2}

47. $8y + 4 = 2y - 5$ {$-\frac{3}{2}$}

48. $5y + 6 = 3y - 6$ {−6}

49. $6z - 5 = z + 5$ {2}

50. $6z - 3 = z + 2$ {1}

51. $6x + 14 = 2x - 2$ {−4}

52. $9x + 2 = 6x - 4$ {−2}

53. $-3y - 1 = 5 - 2y$ {−6}

54. $-3y - 2 = -5 - 4y$ {−3}

Practice Plus

The equations in Exercises 55–58 contain small geometric figures that represent real numbers. Use the multiplication property of equality to isolate x on one side of the equation and the geometric figures on the other side.

55. $\frac{x}{\square} = \triangle$ $x = \square\triangle$

56. $\triangle = \square x$ $\frac{\triangle}{\square} = x$

57. $\triangle = -x$ $-\triangle = x$

58. $\frac{-x}{\square} = \triangle$ $x = -\square\triangle$

In Exercises 59–62, use the given information to write an equation. Let x represent the number described in each exercise. Then solve the equation and find the number.

59. If a number is multiplied by 6, the result is 10. Find the number. $6x = 10; \frac{5}{3}$

60. If a number is multiplied by −6, the result is 20. Find the number. $-6x = 20; -\frac{10}{3}$

61. If a number is divided by −9, the result is 5. Find the number. $\frac{x}{-9} = 5; -45$

62. If a number is divided by −7, the result is 8. Find the number. $\frac{x}{-7} = 8; -56$

Application Exercises

The formula

$$M = \frac{n}{5}$$

models your distance, M, in miles, from a lightning strike in a thunderstorm if it takes n seconds to hear thunder after seeing the lightning. Use this formula to solve Exercises 63–64.

63. If you are 2 miles away from the lightning flash, how long will it take the sound of thunder to reach you?   10 seconds

64. If you are 3 miles away from the lightning flash, how long will it take the sound of thunder to reach you?   15 seconds

The Mach number is a measurement of speed, named after the man who suggested it, Ernst Mach (1838–1916). The formula

$$M = \frac{A}{740}$$

indicates that the speed of an aircraft, A, in miles per hour, divided by the speed of sound, approximately 740 miles per hour, results in the Mach number, M. Use the formula to determine the speed, in miles per hour, of the aircrafts in Exercises 65–66. (Note: When an aircraft's speed increases beyond Mach 1, it is said to have broken the sound barrier.)

65.

66.

Concorde
Mach 2.03
Speed = ?

1502.2 miles per hour

SR-71 Blackbird
Mach 3.3
Speed = ?

2442 miles per hour

67. The formula $P = -0.5d + 100$ models the percentage, P, of lost hikers found in search and rescue missions when members of the search team walk parallel to one another separated by a distance of d yards. If a search and rescue team finds 70% of lost hikers, substitute 70 for P in the formula and find the parallel distance of separation between members of the search party.   60 yards

68. The formula $M = 420x + 720$ models the data for the amount of money lost to credit card fraud worldwide, M, in millions of dollars, x years after 1989. In which year did losses amount to 4080 million dollars?   1997

Writing in Mathematics

69. State the multiplication property of equality and give an example.   69–72. Answers will vary.

70. Explain how to solve the equation $-x = -50$.

71. Explain how to solve the equation $2x + 8 = 5x - 3$.

72. What might occur in the future to cause life expectancy to exceed the predictions made by the models in Example 8 and Check Point 8 on page 116?

Critical Thinking Exercises

73. Which one of the following statements is true?   d

   a. If $7x = 21$, then $x = 21 - 7$.

b. If $3x - 4 = 16$, then $3x = 12$.

c. If $3x + 7 = 0$, then $x = \dfrac{7}{3}$.

d. The solution to $6x = 0$ is not a natural number.

In Exercises 74–75, write an equation with the given characteristics.

74. The solution is a positive integer and the equation can be solved by dividing both sides by -60.   Answers will vary. Example: $-60x = -120$.

75. The solution is a negative integer and the equation can be solved by multiplying both sides by $\dfrac{4}{5}$.   Answers will vary. Example: $\dfrac{5}{4}x = -20$.

Technology Exercises

Solve each equation in Exercises 76–77. Use a calculator to help with the arithmetic. Check your solution using the calculator.

76. $3.7x - 19.46 = -9.988$ $\{2.56\}$

77. $-72.8y - 14.6 = -455.43 - 4.98y$ $\{6.5\}$

Review Exercises

78. Evaluate: $(-10)^2$. (Section 1.8, Example 1)   100

79. Evaluate: -10^2. (Section 1.8, Example 1) -100

80. Evaluate $x^3 - 4x$ for $x = -1$. (Section 1.8, Example 10)   3

SECTION 2.3

SOLVING LINEAR EQUATIONS

Objectives

1 Solve linear equations.

2 Solve linear equations containing fractions.

3 Identify equations with no solution or infinitely many solutions.

4 Solve applied problems using formulas.

Yes, we overindulged, but it was delicious. Now if we could just grow a few inches taller, we'd be back in line with our recommended weights.

In this section, we will see how algebra models the relationship between weight and height. To use this mathematical model, it would be helpful to have a systematic procedure for solving linear equations. We open the section with such a procedure.

1 Solve linear equations.

A Step-By-Step Procedure for Solving Linear Equations   Here is a step-by-step procedure for solving a linear equation in one variable. Not all of these steps are necessary to solve every equation.

SOLVING A LINEAR EQUATION

1. Simplify the algebraic expression on each side.
2. Collect all the variable terms on one side and all the constant terms on the other side.
3. Isolate the variable and solve.
4. Check the proposed solution in the original equation.

EXAMPLE 1 Solving a Linear Equation

Solve and check: $2x - 8x + 40 = 13 - 3x - 3$.

SOLUTION

Step 1. **Simplify the algebraic expression on each side.**

$$2x - 8x + 40 = 13 - 3x - 3 \qquad \text{This is the given equation.}$$
$$-6x + 40 = 10 - 3x \qquad \text{Combine like terms: } 2x - 8x = -6x$$
$$\text{and } 13 - 3 = 10.$$

Step 2. **Collect variable terms on one side and constant terms on the other side.** We will collect variable terms on the left by adding $3x$ to both sides. We will collect the numbers on the right by subtracting 40 from both sides.

$$-6x + 40 + 3x = 10 - 3x + 3x \qquad \text{Add } 3x \text{ to both sides.}$$
$$-3x + 40 = 10 \qquad \text{Simplify: } -6x + 3x = -3x.$$
$$-3x + 40 - 40 = 10 - 40 \qquad \text{Subtract 40 from both sides.}$$
$$-3x = -30 \qquad \text{Simplify.}$$

Step 3. **Isolate the variable and solve.** We isolate the variable, x, by dividing both sides by -3.

$$\frac{-3x}{-3} = \frac{-30}{-3} \qquad \text{Divide both sides by } -3.$$
$$x = 10 \qquad \text{Simplify.}$$

Step 4. **Check the proposed solution in the original equation.** Substitute 10 for x in the original equation.

$$2x - 8x + 40 = 13 - 3x - 3 \qquad \text{This is the original equation.}$$
$$2 \cdot 10 - 8 \cdot 10 + 40 \overset{2}{=} 13 - 3 \cdot 10 - 3 \qquad \text{Substitute 10 for } x.$$
$$20 - 80 + 40 \overset{2}{=} 13 - 30 - 3 \qquad \text{Perform the indicated multiplications.}$$
$$-60 + 40 \overset{2}{=} -17 - 3 \qquad \text{Subtract: } 20 - 80 = -60 \text{ and}$$
$$13 - 30 = -17.$$
$$-20 = -20 \qquad \text{Simplify.}$$

By substituting 10 for x in the original equation, we obtain the true statement $-20 = -20$. This verifies that the solution is 10, or the solution set is $\{10\}$. ■

DISCOVER FOR YOURSELF

Solve the equation in Example 1 by collecting terms with the variable on the right and numbers on the left. What do you observe?

✔ **CHECK POINT 1** Solve and check: $-7x + 25 + 3x = 16 - 2x - 3$.   6 or $\{6\}$

EXAMPLE 2 Solving a Linear Equation

Solve and check: $5x = 8(x + 3)$.

SOLUTION

Step 1. **Simplify the algebraic expression on each side.** Use the distributive property to remove parentheses on the right.

$$5x = 8(x + 3) \qquad \text{This is the given equation.}$$

$$5x = 8x + 24 \qquad \text{Use the distributive property.}$$

Step 2. **Collect variable terms on one side and constant terms on the other side.** We will work with $5x = 8x + 24$ and collect variable terms on the left by subtracting $8x$ from both sides. The only constant term, 24, is already on the right.

$$5x - 8x = 8x + 24 - 8x \qquad \text{Subtract } 8x \text{ from both sides.}$$
$$-3x = 24 \qquad \text{Simplify: } 5x - 8x = -3x.$$

Step 3. **Isolate the variable and solve.** We isolate the variable, x, by dividing both sides by -3.

$$\frac{-3x}{-3} = \frac{24}{-3} \qquad \text{Divide both sides by } -3.$$
$$x = -8 \qquad \text{Simplify.}$$

Step 4. **Check the proposed solution in the original equation.** Substitute -8 for x in the original equation.

$$5x = 8(x + 3) \qquad \text{This is the original equation.}$$
$$5(-8) \stackrel{?}{=} 8(-8 + 3) \qquad \text{Substitute } -8 \text{ for } x.$$
$$5(-8) \stackrel{?}{=} 8(-5) \qquad \text{Perform the addition in parentheses:}$$
$$\qquad\qquad\qquad -8 + 3 = -5.$$
$$-40 = -40 \qquad \text{Multiply.}$$

The true statement $-40 = -40$ verifies that -8 is the solution, or the solution set is $\{-8\}$. ■

✔ **CHECK POINT 2** Solve and check: $8x = 2(x + 6)$.   2 or {2}

EXAMPLE 3 Solving a Linear Equation

Solve and check: $2(x - 3) - 17 = 13 - 3(x + 2)$.

SOLUTION

Step 1. **Simplify the algebraic expression on each side.**

> Do not begin with 13 − 3. Multiplication (the distributive property) is applied before subtraction.

$$2(x - 3) - 17 = 13 - 3(x + 2) \qquad \text{This is the given equation.}$$
$$2x - 6 - 17 = 13 - 3x - 6 \qquad \text{Use the distributive property.}$$
$$2x - 23 = -3x + 7 \qquad \text{Combine like terms.}$$

Step 2. **Collect variable terms on one side and constant terms on the other side.** We will collect variable terms on the left by adding $3x$ to both sides. We will collect the numbers on the right by adding 23 to both sides.

$$2x - 23 + 3x = -3x + 7 + 3x \qquad \text{Add } 3x \text{ to both sides.}$$
$$5x - 23 = 7 \qquad \text{Simplify: } 2x + 3x = 5x.$$
$$5x - 23 + 23 = 7 + 23 \qquad \text{Add 23 to both sides.}$$
$$5x = 30 \qquad \text{Simplify.}$$

Step 3. **Isolate the variable and solve.** We isolate the variable, x, by dividing both sides by 5.

$$\frac{5x}{5} = \frac{30}{5} \qquad \text{Divide both sides by 5.}$$
$$x = 6 \qquad \text{Simplify.}$$

The compact, symbolic notation of algebra enables us to use a clear step-by-step method for solving equations, designed to avoid the confusion shown in the painting.

Squeak Carnwath *Equations* 1981, oil on cotton canvas 96 in. h × 72 in. w.

Step 4. **Check the proposed solution in the original equation.** Substitute 6 for x in the original equation.

$$2(x - 3) - 17 = 13 - 3(x + 2)$$ This is the original equation.

$$2(6 - 3) - 17 \stackrel{?}{=} 13 - 3(6 + 2)$$ Substitute 6 for x.

$$2(3) - 17 \stackrel{?}{=} 13 - 3(8)$$ Simplify inside parentheses.

$$6 - 17 \stackrel{?}{=} 13 - 24$$ Multiply.

$$-11 = -11$$ Subtract.

The true statement $-11 = -11$ verifies that 6 is the solution, or the solution set is $\{6\}$. ∎

 CHECK POINT 3 Solve and check: $4(2x + 1) - 29 = 3(2x - 5)$.   5 or $\{5\}$

Linear Equations with Fractions Equations are easier to solve when they do not contain fractions. How do we remove fractions from an equation? We begin by multiplying both sides of the equation by the least common denominator of any fractions in the equation. The least common denominator is the smallest number that all denominators will divide into. Multiplying every term on both sides of the equation by the least common denominator will eliminate the fractions in the equation. Example 4 shows how we "clear an equation of fractions."

EXAMPLE 4 Solving a Linear Equation Involving Fractions

Solve and check: $\dfrac{3x}{2} = \dfrac{x}{5} - \dfrac{39}{5}$.

SOLUTION The denominators are 2, 5, and 5. The smallest number that is divisible by 2, 5, and 5 is 10. We begin by multiplying both sides of the equation by 10, the least common denominator.

$$\frac{3x}{2} = \frac{x}{5} - \frac{39}{5}$$ This is the given equation.

$$10 \cdot \frac{3x}{2} = 10\left(\frac{x}{5} - \frac{39}{5}\right)$$ Multiply both sides by 10.

$$10 \cdot \frac{3x}{2} = 10 \cdot \frac{x}{5} - 10 \cdot \frac{39}{5}$$ Use the distributive property. Be sure to multiply all terms by 10.

$$\overset{5}{\cancel{10}} \cdot \frac{3x}{\underset{1}{\cancel{2}}} = \overset{2}{\cancel{10}} \cdot \frac{x}{\underset{1}{\cancel{5}}} - \overset{2}{\cancel{10}} \cdot \frac{39}{\underset{1}{\cancel{5}}}$$ Divide out common factors in the multiplications.

$$15x = 2x - 78$$ Complete the multiplications. The fractions are now cleared.

At this point, we have an equation similar to those we previously have solved. Collect the variable terms on one side and the constant terms on the other side.

$$15x - 2x = 2x - 2x - 78$$ Subtract 2x to get the variable terms on the left.

$$13x = -78$$ Simplify.

Isolate x by dividing both sides by 13.

$$\frac{13x}{13} = \frac{-78}{13}$$ Divide both sides by 13.

$$x = -6$$ Simplify.

Check the proposed solution. Substitute -6 for x in the original equation. You should obtain $-9 = -9$. This true statement verifies that the solution is -6, or the solution set is $\{-6\}$. ∎

2 Solve linear equations containing fractions.

✔ CHECK POINT 4 Solve and check: $\dfrac{x}{4} = \dfrac{2x}{3} + \dfrac{5}{6}$. −2 or {−2}

3 Identify equations with no solution or infinitely many solutions.

Linear Equations with No Solution or Infinitely Many Solutions Thus far, each equation that we have solved has had a single solution. However, some equations are not true for even one real number. Such an equation is called an **inconsistent equation**, or a **contradiction**. Here is an example of such an equation:

$$x = x + 4.$$

There is no number that is equal to itself plus 4. This equation has no solution. Its solution set is written either as

$$\{\ \} \quad \text{or} \quad \varnothing.$$

These symbols stand for the empty set, a set with no elements.

An equation that is true for all real numbers is called an **identity**. An example of an identity is

$$x + 3 = x + 2 + 1.$$

Every number plus 3 is equal to that number plus 2 plus 1. Every real number is a solution to this equation. Thus, the solution set to this equation is the set of all real numbers. This set is written as $\mathbb{R}$. In Section 2.7, you will learn a second notation for representing the set of all real numbers.

If you attempt to solve an equation with no solution, you will eliminate the variable and obtain a false statement, such as $2 = 5$. If you attempt to solve an equation that is true for every real number, you will eliminate the variable and obtain a true statement, such as $4 = 4$.

EXAMPLE 5 Solving a Linear Equation

Solve: $2x + 6 = 2(x + 4)$.

SOLUTION

$2x + 6 = 2(x + 4)$	This is the given equation.
$2x + 6 = 2x + 8$	Use the distributive property.
$2x + 6 - 2x = 2x + 8 - 2x$	Subtract 2x from both sides.
Keep reading. 6 = 8 is not the solution. $6 = 8$	Simplify.

The original equation is equivalent to the false statement $6 = 8$, which is false for every value of x. The equation is inconsistent and has no solution. You can express this by writing "no solution" or using one of the symbols for the empty set, $\{\ \}$ or $\varnothing$. ∎

✔ CHECK POINT 5 Solve: $3x + 7 = 3(x + 1)$. inconsistent, no solution; $\{\ \}$ or $\varnothing$

EXAMPLE 6 Solving a Linear Equation

Solve: $-3x + 5 + 5x = 4x - 2x + 5$.

SOLUTION

$-3x + 5 + 5x = 4x - 2x + 5$	This is the given equation.
$2x + 5 = 2x + 5$	Combine like terms: $-3x + 5x = 2x$ and $4x - 2x = 2x$.
$2x + 5 - 2x = 2x + 5 - 2x$	Subtract 2x from both sides.
Keep reading. 5 = 5 is not the solution. $5 = 5$	Simplify.

4 Solve applied problems using formulas.

The original equation is equivalent to the true statement $5 = 5$, which is true for every value of x. The equation is an identity and all real numbers are solutions. You can express this by writing "all real numbers" or using the notation $\mathbb{R}$ for the set of all real numbers. Try substituting any number of your choice for x in the original equation. You will obtain a true statement. ∎

✔ **CHECK POINT 6** Solve: $3(x - 1) + 9 = 8x + 6 - 5x$. all real numbers or $\mathbb{R}$

An equation that is not an identity, but that is true for at least one real number, is called a **conditional equation**. The equation $2x + 3 = 17$ is an example of a conditional equation. The equation is not an identity and is true only if x is 7.

Applications The next example shows how our procedure for solving equations with fractions can be used to find the value of a variable in a mathematical model.

EXAMPLE 7 Modeling Weight and Height

The formula
$$\frac{W}{2} - 3H = 53$$

models the recommended weight, W, in pounds, for a male, where H represents the man's height, in inches, over 5 feet. What is the recommended weight for a man who is 6 feet, 3 inches tall?

SOLUTION Keep in mind that H represents height in inches *above 5 feet*. A man who is 6 feet, 3 inches tall is 1 foot, 3 inches above 5 feet. Because 12 inches = 1 foot, he is 12 + 3, or 15 inches, above 5 feet tall. To find his recommended weight, we substitute 15 for H in the formula and solve for W:

$$\frac{W}{2} - 3H = 53 \qquad \text{This is the given formula.}$$

$$\frac{W}{2} - 3 \cdot 15 = 53 \qquad \text{Substitute 15 for } H.$$

$$\frac{W}{2} - 45 = 53 \qquad \text{Multiply: } 3 \cdot 15 = 45.$$

Multiply both sides of the equation by 2, the least common denominator:

$$2\left(\frac{W}{2} - 45\right) = 2 \cdot 53$$

$$2 \cdot \frac{W}{2} - 2 \cdot 45 = 2 \cdot 53 \qquad \text{Use the distributive property.}$$

$$W - 90 = 106 \qquad \text{Multiply: } 2 \cdot \frac{W}{2} = \frac{2}{1} \cdot \frac{W}{2} = \frac{2W}{2} = 1W = W.$$

$$W - 90 + 90 = 106 + 90 \qquad \text{Add 90 to both sides.}$$

$$W = 196 \qquad \text{Simplify.}$$

The recommended weight for a man whose height is 6 feet, 3 inches is 196 pounds. ∎

The points in the rectangular coordinate system in Figure 2.3 allow us to "see" the formula $\frac{W}{2} - 3H = 53$. The first coordinate of each point represents a man's height, in inches, above 5 feet. The second coordinate represents that man's recommended weight, in pounds. The points are rising steadily from left to right. This shows that the recommended weight increases as height increases.

Recommended Weights for Men

(15, 196) At 15 inches above 5 feet, a man's recommended weight is 196 pounds.

Recommended Weight (pounds) / Height above 5 feet (inches)

FIGURE 2.3

> ✔ **CHECK POINT 7** Use the formula $\frac{W}{2} - 3H = 53$ to find the recommended weight for a man who is 5 feet, 3 inches tall. 124 pounds

2.3 EXERCISE SET

Student Solutions Manual CD/Video PH Math/Tutor Center MathXL Tutorials on CD MathXL® MyMathLab Interactmath.com

Practice Exercises

In Exercises 1–30, solve each equation. Be sure to check your proposed solution by substituting it for the variable in the original equation.

1. $5x + 3x - 4x = 10 + 2$ $\{3\}$

2. $4x + 8x - 2x = 20 - 15$ $\{\frac{1}{2}\}$

3. $4x - 9x + 22 = 3x + 30$ $\{-1\}$

4. $3x + 2x + 64 = 40 - 7x$ $\{-2\}$

5. $3x + 6 - x = 8 + 3x - 6$ $\{4\}$

6. $3x + 2 - x = 6 + 3x - 8$ $\{4\}$

7. $4(x + 1) = 20$ $\{4\}$ **8.** $3(x - 2) = -6$ $\{0\}$

9. $7(2x - 1) = 42$ $\{\frac{7}{2}\}$ **10.** $4(2x - 3) = 32$ $\{\frac{11}{2}\}$

11. $38 = 30 - 2(x - 1)$ $\{-3\}$ **12.** $20 = 44 - 8(2 - x)$ $\{-1\}$

13. $2(4z + 3) - 8 = 46$ $\{6\}$ **14.** $3(3z + 5) - 7 = 89$ $\{9\}$

15. $6x - (3x + 10) = 14$ $\{8\}$ **16.** $5x - (2x + 14) = 10$ $\{8\}$

17. $5(2x + 1) = 12x - 3$ $\{4\}$ **18.** $3(x + 2) = x + 30$ $\{12\}$

19. $3(5 - x) = 4(2x + 1)$ $\{1\}$ **20.** $3(3x - 1) = 4(3 + 3x)$ $\{-5\}$

21. $8(y + 2) = 2(3y + 4)$ $\{-4\}$

22. $8(y + 3) = 3(2y + 12)$ $\{6\}$

23. $3(x + 1) = 7(x - 2) - 3$ $\{5\}$

24. $5x - 4(x + 9) = 2x - 3$ $\{-33\}$

25. $5(2x - 8) - 2 = 5(x - 3) + 3$ $\{6\}$

26. $7(3x - 2) + 5 = 6(2x - 1) + 24$ $\{3\}$

27. $6 = -4(1 - x) + 3(x + 1)$ $\{1\}$

28. $100 = -(x - 1) + 4(x - 6)$ $\{41\}$

29. $10(z + 4) - 4(z - 2) = 3(z - 1) + 2(z - 3)$ $\{-57\}$

30. $-2(z - 4) - (3z - 2) = -2 - (6z - 2)$ $\{-10\}$

Solve and check each equation in Exercises 31–46. Begin your work by rewriting each equation without fractions.

31. $\frac{x}{5} - 4 = -6$ $\{-10\}$ **32.** $\frac{x}{2} + 13 = -22$ $\{-70\}$

33. $\frac{2x}{3} - 5 = 7$ $\{18\}$ **34.** $\frac{3x}{4} - 9 = -6$ $\{4\}$

35. $\frac{2y}{3} - \frac{3}{4} = \frac{5}{12}$ $\{\frac{7}{4}\}$ **36.** $\frac{3y}{4} - \frac{2}{3} = \frac{7}{12}$ $\{\frac{5}{3}\}$

37. $\frac{x}{3} + \frac{x}{2} = \frac{5}{6}$ $\{1\}$ **38.** $\frac{x}{4} - \frac{x}{5} = 1$ $\{20\}$

39. $20 - \frac{z}{3} = \frac{z}{2}$ $\{24\}$ **40.** $\frac{z}{5} - \frac{1}{2} = \frac{z}{6}$ $\{15\}$

41. $\frac{y}{3} + \frac{2}{5} = \frac{y}{5} - \frac{2}{5}$ $\{-6\}$ **42.** $\frac{y}{12} + \frac{1}{6} = \frac{y}{2} - \frac{1}{4}$ $\{1\}$

43. $\frac{3x}{4} - 3 = \frac{x}{2} + 2$ $\{20\}$ **44.** $\frac{3x}{5} - \frac{2}{5} = \frac{x}{3} + \frac{2}{5}$ $\{3\}$

45. $\frac{x - 3}{5} - 1 = \frac{x - 5}{4}$ $\{-7\}$ **46.** $\frac{x - 2}{3} - 4 = \frac{x + 1}{4}$ $\{59\}$

In Exercises 47–64, solve each equation. Identify equations that have no solution, or equations that are true for all real numbers.

47. $3x - 7 = 3(x + 1)$ no solution; $\{\ \}$ or $\varnothing$

48. $2(x - 5) = 2x + 10$ no solution; $\{\ \}$ or $\varnothing$

49. $2(x + 4) = 4x + 5 - 2x + 3$ all real numbers; $\mathbb{R}$

50. $3(x - 1) = 8x + 6 - 5x - 9$ all real numbers; $\mathbb{R}$

51. $7 + 2(3x - 5) = 8 - 3(2x + 1)$ $\{\frac{2}{3}\}$

52. $2 + 3(2x - 7) = 9 - 4(3x + 1)$ $\{\frac{4}{3}\}$

53. $4x + 1 - 5x = 5 - (x + 4)$ all real numbers; $\mathbb{R}$

54. $5x - 5 = 3x - 7 + 2(x + 1)$ all real numbers; $\mathbb{R}$

55. $4(x + 2) + 1 = 7x - 3(x - 2)$ no solution; $\{\ \}$ or $\varnothing$

56. $5x - 3(x + 1) = 2(x + 3) - 5$ no solution; $\{\ \}$ or $\varnothing$

57. $3 - x = 2x + 3$ $\{0\}$ **58.** $5 - x = 4x + 5$ $\{0\}$

59. $\frac{x}{3} + 2 = \frac{x}{3}$ no solution; $\{\ \}$ or $\varnothing$

60. $\frac{x}{4} + 3 = \frac{x}{4}$ no solution; $\{\ \}$ or $\varnothing$

61. $\frac{x}{2} - \frac{x}{4} + 4 = x + 4$ $\{0\}$

62. $\frac{x}{2} + \frac{2x}{3} + 3 = x + 3$ $\{0\}$

63. $\frac{2}{3}x = 2 - \frac{5}{6}x$ $\{\frac{4}{3}\}$ **64.** $\frac{2}{3}x = \frac{1}{4}x - 8$ $\{-\frac{96}{5}\}$

Practice Plus

The equations in Exercises 65–66 contain small figures ($\square$, $\triangle$, and \$) that represent real numbers. Use properties of equality to isolate x on one side of the equation and the small figures on the other side.

65. $\frac{x}{\square} + \triangle = \$$ $x = \square \$ - \square \triangle$

66. $\frac{x}{\square} - \triangle = -\$$ $x = \square \triangle - \square \$$

67. If $\frac{x}{5} - 2 = \frac{x}{3}$, evaluate $x^2 - x$. 240

68. If $\frac{3x}{2} + \frac{3x}{4} = \frac{x}{4} - 4$, evaluate $x^2 - x$. 6

In Exercises 69–72, use the given information to write an equation. Let x represent the number described in each exercise. Then solve the equation and find the number.

69. When one-third of a number is added to one-fifth of the number, the sum is 16. What is the number? $\frac{x}{5} + \frac{x}{3} = 16$; 30

70. When two-fifths of a number is added to one-fourth of the number, the sum is 13. What is the number? $\frac{x}{4} + \frac{2x}{5} = 13$; 20

71. When 3 is subtracted from three-fourths of a number, the result is equal to one-half of the number. What is the number? $\frac{3x}{4} - 3 = \frac{x}{2}$; 12

72. When 30 is subtracted from seven-eighths of a number, the result is equal to one-half of the number. What is the number? $\frac{7x}{8} - 30 = \frac{x}{2}$; 80

Application Exercises

In Massachusetts, speeding fines are determined by the formula

$$F = 10(x - 65) + 50,$$

where F is the cost, in dollars, of the fine if a person is caught driving x miles per hour. Use this formula to solve Exercises 73–74.

73. If a fine comes to $250, how fast was that person speeding? 85 miles per hour

74. If a fine comes to $400, how fast was that person speeding? 100 miles per hour

The graph indicates a relationship between sense of humor and depression. Persons with a low sense of humor have higher levels of depression in response to negative life events than those with a high sense of humor.

Sense of Humor and Depression

Low-Humor Group

High-Humor Group

Group's Average Level of Depression in Response to Negative Life Events

Intensity of Negative Life Event

Low   Average   High

Source: Steven Davis and Joseph Palladino, *Psychology* *Third Edition*, Prentice Hall, 2003.

The data can be modeled by the formulas

$$D = \frac{10}{9} N + \frac{53}{9} \quad \text{low humor}$$

$$D = \frac{1}{9} N + \frac{26}{9}, \quad \text{high humor}$$

where N is the intensity of a negative life event (from 1, low, to 10, high) and D is the level of depression in response to that event. Use these formulas to solve Exercises 75–76.

75. If the low-humor group averages a level of depression of 10 in response to a negative life event, what is the intensity of that event? How is the solution shown on the line graph? 3.7; point (3.7, 10) on low-humor graph

76. If the high-humor group averages a level of depression of 3.5, or $\frac{7}{2}$, in response to a negative life event, what is the intensity of that event? How is the solution shown on the line graph? 5.5; point (5.5, 3.5) on high-humor graph

The formula

$$p = 15 + \frac{5d}{11}$$

describes the pressure of sea water, p, in pounds per square foot, at a depth of d feet below the surface. Use the formula to solve Exercises 77–78.

77. The record depth for breath-held diving, by Francisco Ferreras (Cuba) off Grand Bahama Island, on November 14, 1993, involved pressure of 201 pounds per square foot. To what depth did Ferreras descend on this ill-advised venture? (He was underwater for 2 minutes and 9 seconds!) 409.2 feet

78. At what depth is the pressure 20 pounds per square foot? 11 feet

Writing in Mathematics

79. In your own words, describe how to solve a linear equation. **79–82.** Answers will vary.

80. Explain how to solve a linear equation containing fractions.

81. Suppose that you solve $\frac{x}{5} - \frac{x}{2} = 1$ by multiplying both sides by 20, rather than the least common denominator of 5 and 2 (namely, 10). Describe what happens. If you get the correct solution, why do you think we clear the equation of fractions by multiplying by the *least* common denominator?

82. Suppose you are an algebra teacher grading the following solution on an examination:

Solve: $-3(x - 6) = 2 - x$
Solution: $-3x - 18 = 2 - x$
             $-2x - 18 = 2$
             $-2x = -16$
             $x = 8.$

You should note that 8 checks, and the solution is 8. The student who worked the problem therefore wants full credit. Can you find any errors in the solution? If full credit is 10 points, how many points should you give the student? Justify your position.

Critical Thinking Exercises

83. Which one of the following statements is true? c

a. The equation $3(x + 4) = 3(4 + x)$ has precisely one solution.

b. The equation $2y + 5 = 0$ is equivalent to $2y = 5$.

c. If $2 - 3y = 11$ and the solution to the equation is substituted into $y^2 + 2y - 3$, a number results that is neither positive nor negative.

d. The equation $x + \frac{1}{3} = \frac{1}{2}$ is equivalent to $x + 2 = 3$.

84. A woman's height, h, is related to the length of her femur, f (the bone from the knee to the hip socket), by the formula $f = 0.432h - 10.44$. Both h and f are measured in inches. A partial skeleton is found of a woman in which the femur is 16 inches long. Police find the skeleton in an area where a woman slightly over 5 feet tall has been missing for over a year. Can the partial skeleton be that of the missing woman? Explain. Yes, because the skeleton's height is about 5 feet 1 inch.

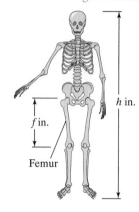

f in.

Femur

h in.

Solve each equation in Exercises 85–86.

85. $\dfrac{2x - 3}{9} + \dfrac{x - 3}{2} = \dfrac{x + 5}{6} - 1$ $\{3\}$

86. $2(3x + 4) = 3x + 2[3(x - 1) + 2]$ $\left\{\dfrac{10}{3}\right\}$

Technology Exercises

Solve each equation in Exercises 87–88. Use a calculator to help with the arithmetic. Check your solution using the calculator.

87. $2.24y - 9.28 = 5.74y + 5.42$ $\{-4.2\}$

88. $4.8y + 32.5 = 124.8 - 9.4y$ $\{6.5\}$

Review Exercises

In Exercises 89–91, insert either < or > in the box between each pair of numbers to make a true statement.

89. $-24 \ \square \ -20$ (Section 1.2, Example 6) $<$

90. $-\dfrac{1}{3} \ \square \ -\dfrac{1}{5}$ (Section 1.2, Example 6) $<$

91. Simplify: $-9 - 11 + 7 - (-3)$. (Section 1.6, Example 3) -10

SECTION

2.4

Objectives

1 Solve a formula for a variable.

2 Express a decimal as a percent.

3 Express a percent as a decimal.

4 Use the percent formula.

5 Solve applied problems involving percent change.

FORMULAS AND PERCENTS

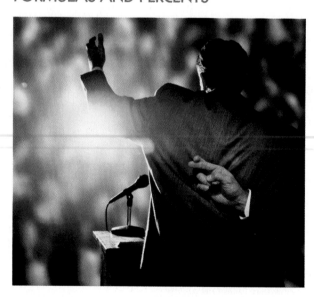

"And, if elected, it is my solemn pledge to cut your taxes by 10% for each of my first three years in office, for a total cut of 30%."

Did you know that one of the most common ways that you are given numerical information is with percents? In this section, you will learn to use a formula that will help you to understand percents, enabling you to make sense of the politician's promise.

1 Solve a formula for a variable.

Solving a Formula for One of Its Variables We know that solving an equation is the process of finding the number (or numbers) that make the equation a true statement. All of the equations we have solved contained only one letter, such as x or y.

By contrast, the formulas we have seen contain two or more letters, representing two or more variables. Here is an example:

$$C \;+\; M \;=\; S.$$

| A retailer's cost of an item | plus | the retailer's markup (profit) | equals | the items's selling price. |

We say that this formula is solved for the variable S because S is alone on one side of the equation and the other side does not contain an S.

Solving a formula for a variable means rewriting the formula so that the variable is isolated on one side of the equation. It does not mean obtaining a numerical value for that variable.

The addition and multiplication properties of equality are used to solve a formula for one of its variables. Consider the retailer's formula, $C + M = S$. How do we solve this formula for C? Use the addition property to isolate C by subtracting M from both sides:

> We need to isolate C.

$$C + M = S \qquad \text{This is the given formula.}$$
$$C + M - M = S - M \qquad \text{Subtract } M \text{ from both sides.}$$
$$C = S - M. \qquad \text{Simplify.}$$

Solved for C, the formula $C = S - M$ tells us that the cost of an item for a retailer is the item's selling price minus its markup.

To solve a formula for one of its variables, treat that variable as if it were the only variable in the equation. Think of the other variables as if they were numbers. Use the addition property of equality to isolate all terms with the specified variable on one side of the equation and all terms without the specified variable on the other side. Then use the multiplication property of equality to get the specified variable alone.

Our first example involves the formula for the area of a rectangle. The **area of a two-dimensional figure** is the number of square units it takes to fill the interior of the figure. A **square unit** is a square, each of whose sides is one unit in length, as illustrated in Figure 2.4. The figure shows that there are 12 square units contained within the rectangle. The area of the rectangle is 12 square units. Notice that the area can be determined in the following manner:

Square unit of measure

FIGURE 2.4

| Across | Down |

$$4 \text{ units} \cdot 3 \text{ units} = 4 \cdot 3 \text{ units} \cdot \text{units} = 12 \text{ square units.}$$

The area of a rectangle is the product of the distance across, its length, and the distance down, its width.

AREA OF A RECTANGLE   The area, A, of a rectangle with length l and width w is given by the formula

$$A = lw.$$

EXAMPLE 1   Solving a Formula for a Variable

Solve the formula $A = lw$ for w.

SOLUTION   Our goal is to get w by itself on one side of the formula. There is only one term with w, lw, and it is already isolated on the right side. We isolate w on the right by using the multiplication property of equality and dividing both sides by l.

> We need to isolate w.

$$A = lw \qquad \text{This is the given formula.}$$

$$\frac{A}{l} = \frac{lw}{l} \qquad \text{Isolate } w \text{ by dividing both sides by } l.$$

$$\frac{A}{l} = w \qquad \text{Simplify: } \frac{lw}{l} = 1w = w.$$

The formula solved for w is $\dfrac{A}{l} = w$ or $w = \dfrac{A}{l}$. Thus, the area of a rectangle divided by its length is equal to its width.   ∎

 CHECK POINT 1 Solve the formula $A = lw$ for l. $l = \dfrac{A}{w}$

The perimeter, P, of a two-dimensional figure is the sum of the lengths of its sides. Perimeter is measured in linear units, such as inches, feet, yards, meters, or kilometers.

Example 2 involves the perimeter of a rectangle. Because perimeter is the sum of the lengths of the sides, the perimeter of the rectangle shown in Figure 2.5 is $l + w + l + w$. This can be expressed as

$$P = 2l + 2w.$$

FIGURE 2.5   A rectangle with length l and width w

> **PERIMETER OF A RECTANGLE**   The perimeter, P, of a rectangle with length l and width w is given by the formula
>
> $$P = 2l + 2w.$$
>
> The perimeter of a rectangle is the sum of twice the length and twice the width.

EXAMPLE 2   Solving a Formula for a Variable

Solve the formula $2l + 2w = P$ for w.

SOLUTION   First, isolate $2w$ on the left by subtracting $2l$ from both sides. Then solve for w by dividing both sides by 2.

> We need to isolate w.

$$2l + 2w = P \qquad \text{This is the given formula.}$$

$$2l - 2l + 2w = P - 2l \qquad \text{Isolate } 2w \text{ by subtracting } 2l \text{ from both sides.}$$

$$2w = P - 2l \qquad \text{Simplify.}$$

$$\frac{2w}{2} = \frac{P - 2l}{2} \qquad \text{Isolate } w \text{ by dividing both sides by 2.}$$

$$w = \frac{P - 2l}{2} \qquad \text{Simplify.}$$ ∎

✔ CHECK POINT **2** Solve the formula $2l + 2w = P$ for l. $l = \dfrac{P - 2w}{2}$

EXAMPLE 3   Solving a Formula for a Variable

The total price of an article purchased on a monthly deferred payment plan is described by the following formula:

$$T = D + pm.$$

In this formula, T is the total price, D is the down payment, p is the amount of the monthly payment, and m is the number of payments. Solve the formula for p.

SOLUTION   First, isolate pm on the right by subtracting D from both sides. Then isolate p from pm by dividing both sides of the formula by m.

We need to isolate p.

$T = D + pm$	This is the given formula. We want p alone.
$T - D = D - D + pm$	Isolate pm by subtracting D from both sides.
$T - D = pm$	Simplify.
$\dfrac{T - D}{m} = \dfrac{pm}{m}$	Now isolate p by dividing both sides by m.
$\dfrac{T - D}{m} = p$	Simplify: $\dfrac{pm}{m} = \dfrac{p\overset{1}{\cancel{m}}}{\underset{1}{\cancel{m}}} = p \cdot 1 = p.$

■

✔ CHECK POINT **3** Solve the formula $T = D + pm$ for m. $m = \dfrac{T - D}{p}$

The next example has a formula that contains a fraction. To solve for a variable in a formula involving fractions, we begin by multiplying both sides by the least common denominator of all fractions in the formula. This will eliminate the fractions. Then we solve for the specified variable.

EXAMPLE 4   Solving a Formula Containing a Fraction for a Variable

Solve the formula $\dfrac{W}{2} - 3H = 53$ for W.

SOLUTION   Do you remember seeing this formula in the last section? It models the recommended weight, W, for a male, where H represents the man's height, in inches, over 5 feet. We begin by multiplying both sides of the formula by 2 to eliminate the fraction. Then we isolate the variable W.

$\dfrac{W}{2} - 3H = 53$	This is the given formula.
$2\left(\dfrac{W}{2} - 3H \right) = 2 \cdot 53$	Multiply both sides by 2.
$2 \cdot \dfrac{W}{2} - 2 \cdot 3H = 2 \cdot 53$	Use the distributive property.

We need to isolate W.

$$W - 6H = 106 \qquad \text{Simplify.}$$
$$W - 6H + 6H = 106 + 6H \qquad \text{Isolate } W \text{ by adding } 6H \text{ to both sides.}$$
$$W = 106 + 6H \qquad \text{Simplify.}$$

This form of the formula makes it easy to find a man's recommended weight, W, if we know his height, H, in inches, over 5 feet. ∎

✔ CHECK POINT 4 Solve for x: $\dfrac{x}{3} - 4y = 5$. $x = 15 + 12y$

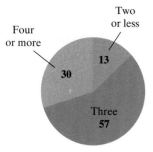

FIGURE 2.6 Number of bedrooms in privately owned single-family U.S. houses per 100 houses.

Source: U.S. Census Bureau and HUD

Before turning to a formula involving percent, let's review some of the basics of percent.

Basics of Percent **Percents** are the result of expressing numbers as a part of 100. The word *percent* means *per hundred*. For example, the circle graph in Figure 2.6 shows that 57 out of every 100 single-family homes have three bedrooms. Thus, $\frac{57}{100} = 57\%$, indicating that 57% of the houses have three bedrooms. The percent sign, %, is used to indicate the number of parts out of one hundred parts.

By definition, $57\% = \frac{57}{100}$. We can express the fraction $\frac{57}{100}$ in decimal notation as 0.57. Thus, $0.57 = 57\%$. Here is a general procedure for expressing a decimal number as a percent:

2 Express a decimal as a percent.

EXPRESSING A DECIMAL NUMBER AS A PERCENT
1. Move the decimal point two places to the right.
2. Attach a percent sign.

EXAMPLE 5 Expressing a Decimal as a Percent

Express 0.47 as a percent.

SOLUTION

Move decimal point two places right.

0.47 % ← Attach a percent sign.

Thus, $0.47 = 47\%$. ∎

✔ CHECK POINT 5 Express 0.023 as a percent. 2.3%

We reverse the procedure of Example 5 to express a percent as a decimal number.

3 Express a percent as a decimal.

EXPRESSING A PERCENT AS A DECIMAL NUMBER
1. Move the decimal point two places to the left.
2. Remove the percent sign.

STUDY TIP

Dictionaries indicate that the word *percentage* has the same meaning as the word *percent*. Use the word that sounds best in the circumstance.

EXAMPLE 6 Expressing Percents as Decimals

Express each percent as a decimal: **a.** 19%   **b.** 180%.

SOLUTION Use the two steps in the box.

a.

$$19\% = 19.\% = 0.19\%$$

The percent sign is removed.

The decimal point starts at the far right.

The decimal point is moved two places to the left.

Thus, 19% = 0.19.

b. 180% = 1.80% = 1.80 or 1.8.

CHECK POINT 6 Express each percent as a decimal:

a. 67%   0.67   **b.** 250%.   2.5

A Formula Involving Percent Percents are useful in comparing two numbers. To compare the number A to the number B using a percent P, the following formula is used:

$$A \quad \text{is} \quad P \text{ percent} \quad \text{of} \quad B.$$

$$A \quad = \quad P \quad \cdot \quad B.$$

In the formula

$$A = PB$$

B = the base number, P = the percent (in decimal form), and A = the number compared to B.

There are three basic types of percent problems that can be solved using the percent formula

$$A = PB. \qquad \text{A is P percent of B.}$$

Question	Given	Percent Formula
What is P percent of B?	P and B	Solve for A.
A is P percent of what?	A and P	Solve for B.
A is what percent of B?	A and B	Solve for P.

Let's look at an example of each type of problem.

EXAMPLE 7 Using the Percent Formula: What Is P Percent of B?

What is 8% of 20?

SOLUTION We use the formula $A = PB$: A is P percent of B. We are interested in finding the quantity A in this formula.

$$\text{What} \quad \text{is} \quad 8\% \quad \text{of} \quad 20?$$

$$A \quad = \quad 0.08 \quad \cdot \quad 20$$

Express 8% as 0.08.

$$A \quad = \quad 1.6$$

Multiply:
.08
× 20
1.60

Thus, 1.6 is 8% of 20. The answer is 1.6.

4 Use the percent formula.

CHECK POINT 7 What is 9% of 50?   4.5

EXAMPLE 8   Using the Percent Formula: A Is P Percent of What?

4 is 25% of what?

SOLUTION   We use the formula $A = PB$: A is P percent of B. We are interested in finding the quantity B in this formula.

4	is	25%	of	what?

$$4 \ = \ 0.25 \ \cdot \ B$$ Express 25% as 0.25.

$$\frac{4}{0.25} \ = \ \frac{0.25B}{0.25}$$ Divide both sides by 0.25.

$$16 \ = \ B$$ Simplify: $0.25\overline{)4.00}\,{}^{16.}$

Thus, 4 is 25% of 16. The answer is 16.   ■

 CHECK POINT 8 9 is 60% of what?   15

EXAMPLE 9   Using the Percent Formula: A Is What Percent of B?

1.3 is what percent of 26?

SOLUTION   We use the formula $A = PB$: A is P percent of B. We are interested in finding the quantity P in this formula.

1.3	is	what percent	of	26?

$$1.3 \ = \ P \ \cdot \ 26$$

$$\frac{1.3}{26} \ = \ \frac{P \cdot 26}{26}$$ Divide both sides by 26.

$$0.05 \ = \ P$$ Simplify: $26\overline{)1.30}\,{}^{0.05}$

We change 0.05 to a percent by moving the decimal point two places to the right and adding a percent sign: $0.05 = 5\%$. Thus, 1.3 is 5% of 26. The answer is 5%.   ■

CHECK POINT 9 18 is what percent of 50?   36%

---

5 Solve applied problems involving percent change.

Applications   Percents are used for comparing changes, such as increases or decreases in sales, population, prices, and production. If a quantity changes, its percent increase or percent decrease can be determined by asking the following question:

> **The change is what percent of the original amount?**

The question is answered using the percent formula as follows:

Percent Increase

$$A = P \cdot B$$

The increase   is   what percent   of   the original amount?

Percent Decrease

$$A = P \cdot B$$

The decrease   is   what percent   of   the original amount?

EXAMPLE 10   Finding Percent Decrease

A jacket regularly sells for $135.00 The sale price is $60.75. Find the percent decrease in the jacket's price.

SOLUTION   The percent decrease in price can be determined by asking the following question:

> The price decrease is what percent of the original price ($135.00)?

The price decrease is the difference between the original price and the sale price ($60.75):

$$\$135.00 - \$60.75 = \$74.25.$$

Now we use the percent formula to find the percent decrease.

The price decrease   is   what percent   of   the original price?

$$74.25 = P \cdot 135$$

$$\frac{74.25}{135} = \frac{P \cdot 135}{135} \qquad \text{Divide both sides by 135.}$$

$$0.55 = P \qquad \text{Simplify: } 135\overline{)74.25}^{\,0.55}.$$

We change 0.55 to a percent by moving the decimal point two places to the right and adding a percent sign: 0.55 = 55%. Thus, the percent decrease in the jacket's price is 55%. ∎

 CHECK POINT 10 A television regularly sells for $940. The sale price is $611. Find the percent decrease in the television's price.   35%

In our next example, we look at one of the many ways that percent can be used incorrectly.

EXAMPLE 11   Promises of a Politician

A politician states, "If you elect me to office, I promise to cut your taxes for each of my first three years in office by 10% each year, for a total reduction of 30%." Evaluate the accuracy of the politician's statement.

SOLUTION   To make things simple, let's assume that a taxpayer paid $100 in taxes in the year previous to the politician's election. A 10% reduction during year 1 is 10% of $100.

$$10\% \text{ of previous year's tax} = 10\% \text{ of } \$100 = 0.10 \cdot \$100 = \$10$$

With a 10% reduction the first year, the taxpayer will pay only $100 − $10, or $90, in taxes during the politician's first year in office.

The table on the next page shows how we calculate the new, reduced tax for each of the first three years in office.

Year	Tax paid the year before	10% reduction	Taxes paid this year
1	$100	$0.10 \cdot \$100 = \$10$	$\$100 - \$10 = \$90$
2	$ 90	$0.10 \cdot \$90 = \$9$	$\$90 - \$9 = \$81$
3	$ 81	$0.10 \cdot \$81 = \$8.10$	$\$81 - \$8.10 = \$72.90$

The tax reduction is the amount originally paid, $100.00, minus the amount paid during the politician's third year in office, $72.90:

$$\$100.00 - \$72.90 = \$27.10.$$

Now we use the percent formula to determine the percent decrease in taxes over the three years.

The tax decrease is what percent of the original tax?

$$27.1 = P \cdot 100$$

$$\frac{27.1}{100} = \frac{P \cdot 100}{100} \qquad \text{Divide both sides by 100.}$$

$$0.271 = P \qquad \text{Simplify.}$$

Change 0.271 to a percent: $0.271 = 27.1\%$. The percent decrease is 27.1%. The taxes decline by 27.1%, not by 30%. The politician is ill-informed in saying that three consecutive 10% cuts add up to a total tax cut of 30%. In our calculation, which serves as a counterexample to the promise, the total tax cut is only 27.1%. ∎

 CHECK POINT 11 Suppose you paid $1200 in taxes. During year 1, taxes decrease by 20%. During year 2, taxes increase by 20%.

a. What do you pay in taxes for year 2? $1152

Taxes during year 2 are less than originally paid. This is a 4% decrease.

b. How do your taxes for year 2 compare with what you originally paid, namely $1200? If the taxes are not the same, find the percent increase or decrease.

2.4 EXERCISE SET

Student Solutions Manual CD/Video PH Math/Tutor Center MathXL Tutorials on CD MathXL® MyMathLab Interactmath.com

Practice Exercises

In Exercises 1–26, solve each formula for the specified variable. Do you recognize the formula? If so, what does it describe?

1. $d = rt$ for r $r = \dfrac{d}{t}$

2. $d = rt$ for t $t = \dfrac{d}{r}$

3. $I = Prt$ for P $P = \dfrac{I}{rt}$

4. $I = Prt$ for r $r = \dfrac{I}{Pt}$

5. $C = 2\pi r$ for r $r = \dfrac{C}{2\pi}$

6. $C = \pi d$ for d $d = \dfrac{C}{\pi}$

7. $E = mc^2$ for m $m = \dfrac{E}{c^2}$

8. $V = \pi r^2 h$ for h $h = \dfrac{V}{\pi r^2}$

9. $y = mx + b$ for m $m = \dfrac{y - b}{x}$

10. $y = mx + b$ for x $x = \dfrac{y - b}{m}$

11. $T = D + pm$ for p $p = \dfrac{T - D}{m}$

12. $P = C + MC$ for M $M = \dfrac{P - C}{C}$

13. $A = \dfrac{1}{2}bh$ for b $b = \dfrac{2A}{h}$

14. $A = \dfrac{1}{2}bh$ for h $h = \dfrac{2A}{b}$

15. $M = \dfrac{n}{5}$ for n $n = 5M$

16. $M = \dfrac{A}{740}$ for A $A = 740M$

17. $\dfrac{c}{2} + 80 = 2F$ for c $c = 4F - 160$

18. $p = 15 + \dfrac{5d}{11}$ for d $d = \dfrac{11p - 165}{5}$

19. $A = \dfrac{1}{2}(a + b)$ for a $a = 2A - b$

20. $A = \dfrac{1}{2}(a + b)$ for b $b = 2A - a$

21. $S = P + Prt$ for r $r = \dfrac{S - P}{Pt}$

22. $S = P + Prt$ for t $t = \dfrac{S - P}{Pr}$

23. $A = \dfrac{1}{2}h(a + b)$ for b $b = \dfrac{2A}{h} - a$

24. $A = \dfrac{1}{2}h(a + b)$ for a $a = \dfrac{2A}{h} - b$

25. $Ax + By = C$ for x $x = \dfrac{C - By}{A}$

26. $Ax + By = C$ for y $y = \dfrac{C - Ax}{B}$

In Exercises 27–34, express each decimal as a percent.

27. 0.89   89%   **28.** 0.16   16%   **29.** 0.002   0.2%

30. 0.008   0.8%   **31.** 4.78   478%   **32.** 5.38   538%

33. 100   10,000%   **34.** 85   8500%

In Exercises 35–44, express each percent as a decimal.

35. 27%   0.27   **36.** 83%   0.83   **37.** 63.4%   0.634

38. 2.15%   0.0215   **39.** 170%   1.7   **40.** 360%   3.6

41. 3%   0.03   **42.** 8%   0.08   **43.** $\dfrac{1}{2}$%   0.005

44. $\dfrac{1}{4}$%   0.0025

Use the percent formula, $A = PB$: A is P percent of B, to solve Exercises 45–56.

45. What is 3% of 200?   6   **46.** What is 8% of 300?   24

47. What is 18% of 40?   7.2   **48.** What is 16% of 90?   14.4

49. 3 is 60% of what?   5   **50.** 8 is 40% of what?   20

51. 24% of what number is 40.8?   170

52. 32% of what number is 51.2?   160

53. 3 is what percent of 15?   20%

54. 18 is what percent of 90?   20%

55. What percent of 2.5 is 0.3?   12%

56. What percent of 7.5 is 0.6?   8%

57. If 5 is increased to 8, the increase is what percent of the original number?   60%

58. If 5 is increased to 9, the increase is what percent of the original number?   80%

59. If 4 is decreased to 1, the decrease is what percent of the original number?   75%

60. If 8 is decreased to 6, the decrease is what percent of the original number?   25%

Practice Plus

In Exercises 61–68, solve each equation for x.

61. $y = (a + b)x$ $x = \dfrac{y}{a + b}$

62. $y = (a - b)x$ $x = \dfrac{y}{a - b}$

63. $y = (a - b)x + 5$ $x = \dfrac{y - 5}{a - b}$

64. $y = (a + b)x - 8$ $x = \dfrac{y + 8}{a + b}$

65. $y = cx + dx$ $x = \dfrac{y}{c + d}$

66. $y = cx - dx$ $x = \dfrac{y}{c - d}$

67. $y = Ax - Bx - C$ $x = \dfrac{y + C}{A - B}$

68. $y = Ax + Bx + C$ $x = \dfrac{y - C}{A + B}$

Application Exercises

69. The average, or mean, A, of three exam grades, x, y, and z, is given by the formula

$$A = \frac{x + y + z}{3}.$$

   a. Solve the formula for z. $z = 3A - x - y$

   b. Use the formula in part (a) to solve this problem. On your first two exams, your grades are 86% and 88%: $x = 86$ and $y = 88$. What must you get on the third exam to have an average of 90%?   96%

70. The average, or mean, A, of four exam grades, x, y, z, and w, is given by the formula

$$A = \frac{x + y + z + w}{4}.$$

   a. Solve the formula for w. $w = 4A - x - y - z$

   b. Use the formula in part (a) to solve this problem. On your first three exams, your grades are 76%, 78%, and 79%: $x = 76$, $y = 78$, and $z = 79$. What must you get on the fourth exam to have an average of 80%?   87%

71. If you are traveling in your car at an average rate of r miles per hour for t hours, then the distance, d, in miles, that you travel is described by the formula $d = rt$: distance equals rate times time.

   a. Solve the formula for t. $t = \dfrac{d}{r}$

   b. Use the formula in part (a) to find the time that you travel if you cover a distance of 100 miles at an average rate of 40 miles per hour.   2.5 hours

72. The formula $F = \dfrac{9}{5}C + 32$ expresses the relationship between Celsius temperature, C, and Fahrenheit temperature, F.

   a. Solve the formula for C. $C = \dfrac{5F - 160}{9}$

   b. Use the formula from part (a) to find the equivalent Celsius temperature for a Fahrenheit temperature of 59°.   15°C

A recent Time/CNN telephone poll included never-married single women between the ages of 18 and 49 and never-married single men between the ages of 18 and 49. The circle graphs show the results for one of the questions in the poll. Use this information to solve Exercises 73–74.

If You Couldn't Find the Perfect Mate, Would You Marry Someone Else?

Single women   Single men

34%	Yes	41%
61%	No	54%
5%	Not sure	5%

Source: Time, August 28, 2000

73. There were 1200 single women who participated in the poll. How many stated they would marry someone other than the perfect mate?   408

74. There were 1200 single men who participated in the poll. How many stated they would marry someone other than the perfect mate?   492

75. In 2002, the leading cause of death in the United States was heart disease, resulting in 710,760 deaths. If 30% of all deaths were caused by heart disease, find the total number of deaths in the United States in 2002.   2,369,200

(*Source*: Department of Health and Human Services)

76. In 2001, 4% of the Hispanic population in the United States was Cuban. If the Cuban-American population at that time was approximately 1.4 million, what was the approximate U.S. Hispanic population in 2001?   35 million

In 2002, there were approximately 12,000 hate crimes reported to the FBI in the United States. The circle graph shows the breakdown of this total number. Use this information to solve Exercises 77–78.

Motivation for 12,000 U.S. Hate-Crime Incidents

Ethnicity and national orgin   1320
Sexual orientation   1680
Religion   1920
Race   7080

Source: FBI

77. What percent of hate-crime incidents were motivated by race?   59%

78. What percent of hate-crime incidents were motivated by sexual orientation?   14%

79. A charity has raised $7500, with a goal of raising $60,000. What percent of the goal has been raised?   12.5%

80. A charity has raised $225,000, with a goal of raising $500,000. What percent of the goal has been raised?   45%

81. A restaurant bill came to $60. If 15% of this bill is left as a tip, how much was the tip? $9

82. If income tax is $3502 plus 28% of taxable income over $23,000, how much is the income tax on a taxable income of $35,000? $6862

83. Suppose that the local sales tax rate is 6% and you buy a car for $16,800.
  a. How much tax is due? $1008
  b. What is the car's total cost? $17,808

84. Suppose that the local sales tax rate is 7% and you buy a graphing calculator for $96.
  a. How much tax is due? $6.72
  b. What is the calculator's total cost? $102.72

85. An exercise machine with an original price of $860 is on sale at 12% off.
  a. What is the discount amount? $103.20
  b. What is the exercise machine's sale price? $756.80

86. A dictionary that normally sells for $16.50 is on sale at 40% off.
  a. What is the discount amount? $6.60
  b. What is the dictionary's sale price? $9.90

87. A sofa regularly sells for $840. The sale price is $714. Find the percent decrease in the sofa's price.   15%

88. A fax machine regularly sells for $380. The sale price is $266. Find the percent decrease in the machine's price.   30%

89. Suppose that you put $10,000 in a rather risky investment recommended by your financial advisor. During the first year, your investment decreases by 30% of its original value. During the second year, your investment increases by 40% of its first-year value. Your advisor tells you that there must have been a 10% overall increase of your original $10,000 investment. Is your financial advisor using percentages properly? If not, what is your actual percent gain or loss of your original $10,000 investment?   no; There is a 2% loss.

90. The price of a color printer is reduced by 30% of its original price. When it still does not sell, its price is reduced by 20% of the reduced price. The salesperson informs you that there has been a total reduction of 50%. Is the salesperson using percentages properly? If not, what is the actual percent reduction from the original price?   no; There is a 44% reduction.

Writing in Mathematics

91. Explain what it means to solve a formula for a variable.
91–96. Answers will vary.

92. What is a percent?

93. Describe how to express a decimal number as a percent and give an example.

94. Describe how to express a percent as a decimal number and give an example.

95. What does the percent formula, $A = PB$, describe? Give an example of how the formula is used.

96. Describe one way in which you use percents in your daily life.

Critical Thinking Exercises

97. Which one of the following statements is true?   d

a. If $ax + b = 0$, then $x = \dfrac{b}{a}$.

b. If $A = lw$, then $w = \dfrac{l}{A}$.

c. If $A = \dfrac{1}{2}bh$, then $b = \dfrac{A}{2h}$.

d. Solving $x - y = -7$ for y gives $y = x + 7$.

98. In psychology, an intelligence quotient, Q, also called IQ, is measured by the formula

$$Q = \frac{100M}{C},$$

where M = mental age and C = chronological age. Solve the formula for C. $C = \dfrac{100M}{Q}$

99. The height, h, in feet, of water in a fountain is described by the formula

$$h = -16t^2 + 64t$$

and the velocity, v, in feet per second, of water in the fountain is described by $v = -32t + 64$. Find the time when the water's velocity is 16 feet per second, and then find the water's height at that time.   1.5 seconds; 60 feet

Review Exercises

100. Solve and check: $5x + 20 = 8x - 16$. (Section 2.2, Example 7) $\{12\}$

101. Solve and check: $5(2y - 3) - 1 = 4(6 + 2y)$. (Section 2.3, Example 3) $\{20\}$

102. Simplify: $x - 0.3x$. (Section 1.4, Example 9) $0.7x$

✔ MID-CHAPTER CHECK POINT

CHAPTER 2

What You Know: We learned a step-by-step procedure for solving linear equations, including equations with fractions. We saw that some linear equations have no solution, whereas others have all real numbers as solutions. We used the addition and multiplication properties of equality to solve formulas for variables. Finally, we worked with the percent formula $A = PB$: A is P percent of B.

1. Solve: $\dfrac{x}{2} = 12 - \dfrac{x}{4}$. $\{16\}$

2. Solve: $5x - 42 = -57$. $\{-3\}$

3. Solve for C: $H = \dfrac{EC}{825}$ $C = \dfrac{825H}{E}$

4. What is 6% of 140?   8.4

5. Solve: $\dfrac{-x}{10} = -3$. $\{30\}$

6. Solve: $1 - 3(y - 5) = 4(2 - 3y)$. $\left\{-\dfrac{8}{9}\right\}$

7. Solve for r: $S = 2\pi rh$ $r = \dfrac{S}{2\pi h}$

8. 12 is 30% of what?   40

9. Solve: $\dfrac{3y}{5} + \dfrac{y}{2} = \dfrac{5y}{4} - 3$. $\{20\}$

10. Solve: $5z + 7 = 6(z - 2) - 4(2z - 3)$ $\{-1\}$

11. Solve for x: $Ax - By = C$. $x = \dfrac{By + C}{A}$

12. Solve: $6y + 7 + 3y = 3(3y - 1)$ no solution; $\{\ \}$ or $\varnothing$

13. The formula $D = 0.12x + 5.44$ models the number of children in the United States with physical disabilities, D, in millions, x years after 2000. According to this model, in which year will there be 6.4 million children in the United States with physical disabilities?   2008

14. Solve: $10\left(\dfrac{1}{2}x + 3\right) = 10\left(\dfrac{3}{5}x - 1\right)$ $\{40\}$

15. 50 is what percent of 400?   12.5%

16. Solve: $\dfrac{3(m + 2)}{4} = 2m + 3$. $\left\{-\dfrac{6}{5}\right\}$

17. If 40 is increased to 50, the increase is what percent of the original number?   25%

18. Solve: $12w - 4 + 8w - 4 = 4(5w - 2)$.   all real numbers; $\mathbb{R}$

Objectives

1 Translate English phrases into algebraic expressions.

2 Solve algebraic word problems using linear equations.

AN INTRODUCTION TO PROBLEM SOLVING

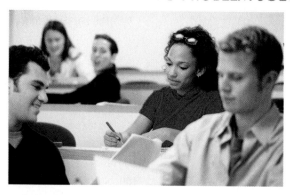

You started college with your best friend. This semester, you and your friend are taking two classes together. However, your friend often misses class and is not doing the necessary homework between classes to succeed. What can you say to your friend, who values your advice and who is in danger of flunking out of college if things continue on their present course?

Some problems have many plans for finding an answer. To solve your friend's problem, or any problem for that matter, we need to understand the problem fully, devise a plan for solving it, and then carry out the plan. However, problem solving in algebra is easier than solving the many problematic situations encountered in everyday life. Why? Algebra provides a step-by-step strategy for solving problems. As you become familiar with this strategy, you will learn to solve a wide variety of problems.

A Strategy for Solving Word Problems Using Equations Problem solving is an important part of algebra. The problems in this book are presented in English. We must translate from the ordinary language of English into the language of algebraic equations. To translate, however, we must understand the English prose and be familiar with the forms of algebraic language. Here are some general steps we will follow in solving word problems:

STRATEGY FOR SOLVING WORD PROBLEMS

Step 1 Read the problem carefully. Attempt to state the problem in your own words and state what the problem is looking for. Let x (or any variable) represent one of the unknown quantities in the problem.

Step 2 If necessary, write expressions for any other unknown quantities in the problem in terms of x.

Step 3 Write an equation in x that describes the conditions of the problem.

Step 4 Solve the equation and answer the problem's question.

Step 5 Check the solution *in the original wording* of the problem, not in the equation obtained from the words.

Take great care with step 1. Reading a word problem is not the same as reading a newspaper. Reading the problem involves slowly working your way through its parts, making notes on what is given, and perhaps rereading the problem a few times. Only at this point should you let x represent one of the quantities.

1 Translate English phrases into algebraic expressions.

The most difficult step in this process is step 3 because it involves translating verbal conditions into an algebraic equation. Translations of some commonly used English phrases are listed in Table 2.1. We choose to use x to represent the variable, but we can use any letter.

STUDY TIP

Cover the right column in Table 2.1 with a sheet of paper and attempt to formulate the algebraic expression in the column on your own. Then slide the paper down and check your answer. Work through the entire table in this manner.

Table 2.1   Algebraic Translations of English Phrases

English Phrase	Algebraic Expression
Addition	
The sum of a number and 7	$x + 7$
Five more than a number; a number plus 5	$x + 5$
A number increased by 6; 6 added to a number	$x + 6$
Subtraction	
A number minus 4	$x - 4$
A number decreased by 5	$x - 5$
A number subtracted from 8	$8 - x$
The difference between a number and 6	$x - 6$
The difference between 6 and a number	$6 - x$
Seven less than a number	$x - 7$
Seven minus a number	$7 - x$
Nine fewer than a number	$x - 9$
Multiplication	
Five times a number	$5x$
The product of 3 and a number	$3x$
Two-thirds of a number (used with fractions)	$\frac{2}{3}x$
Seventy-five percent of a number (used with decimals)	$0.75x$
Thirteen multiplied by a number	$13x$
A number multiplied by 13	$13x$
Twice a number	$2x$
Division	
A number divided by 3	$\frac{x}{3}$
The quotient of 7 and a number	$\frac{7}{x}$
The quotient of a number and 7	$\frac{x}{7}$
The reciprocal of a number	$\frac{1}{x}$
More than one operation	
The sum of twice a number and 7	$2x + 7$
Twice the sum of a number and 7	$2(x + 7)$
Three times the sum of 1 and twice a number	$3(1 + 2x)$
Nine subtracted from 8 times a number	$8x - 9$
Twenty-five percent of the sum of 3 times a number and 14	$0.25(3x + 14)$
Seven times a number, increased by 24	$7x + 24$
Seven times the sum of a number and 24	$7(x + 24)$

STUDY TIP

Here are three similar English phrases that have very different translations:

7 minus 10: $7 - 10$

7 less than 10: $10 - 7$

7 is less than 10: $7 < 10$.

Think carefully about what is expressed in English before you translate into the language of algebra.

EXAMPLE 1   Translating English Phrases into Algebraic Expressions

Write each English phrase as an algebraic expression. Let x represent the number.

a. Six subtracted from 5 times a number
b. The quotient of 9 and a number, decreased by 4 times the number

SOLUTION

a.

Six subtracted from | 5 times a number

$5x \quad - \quad 6$

The algebraic expression for "six subtracted from 5 times a number" is $5x - 6$.

b.

The quotient of 9 and a number,	decreased by	4 times the number
$\dfrac{9}{x}$	$-$	$4x$

The algebraic expression for "the quotient of 9 and a number, decreased by 4 times the number" is $\dfrac{9}{x} - 4x$. ■

 CHECK POINT 1 Write each English phrase as an algebraic expression. Let x represent the number.

a. Four times a number, increased by 6 $\quad 4x + 6$

b. The quotient of a number decreased by 4 and 9. $\quad \dfrac{x - 4}{9}$

---

2 Solve algebraic word problems using linear equations.

Applying the Strategy for Solving Word Problems Now that we've practiced writing algebraic expressions for English phrases, let's apply our five-step strategy for solving word problems.

EXAMPLE 2 Solving a Word Problem

Nine subtracted from eight times a number is 39. Find the number.

SOLUTION

Step 1. **Let x represent one of the quantities.** Because we are asked to find a number, let
$$x = \text{the number.}$$

Step 2. **Represent other quantities in terms of x.** There are no other unknown quantities to find, so we can skip this step.

Step 3. **Write an equation in x that describes the conditions.**

Nine subtracted from	eight times a number	is	39.
	$8x$ $\quad -9$	$=$	39

Step 4. **Solve the equation and answer the question.**

$$8x - 9 = 39 \qquad \text{This is the equation for the problem's conditions.}$$
$$8x - 9 + 9 = 39 + 9 \qquad \text{Add 9 to both sides.}$$
$$8x = 48 \qquad \text{Simplify.}$$
$$\frac{8x}{8} = \frac{48}{8} \qquad \text{Divide both sides by 8.}$$
$$x = 6 \qquad \text{Simplify.}$$

The number is 6.

Step 5. **Check the proposed solution in the original wording of the problem.** "Nine subtracted from eight times a number is 39." The proposed number is 6. Eight times 6 is $8 \cdot 6$, or 48. Nine subtracted from 48 is $48 - 9$, or 39. The proposed solution checks in the problem's wording, verifying that the number is 6. ■

 CHECK POINT 2 Four subtracted from six times a number is 68. Find the number. The number is 12.

EXAMPLE 3 Pet Population

Americans love their pets. The number of cats in the United States exceeds the number of dogs by 7.5 million. The number of cats and dogs combined is 114.7 million. Determine the number of dogs and cats in the United States.

SOLUTION

Step 1. **Let x represent one of the quantities.** We know something about the number of cats: the cat population exceeds the dog population by 7.5 million. This means that there are 7.5 million more cats than dogs. We will let

$$x = \text{the number (in millions) of dogs in the United States.}$$

Step 2. **Represent other quantities in terms of x.** The other unknown quantity is the number of cats. Because there are 7.5 million more cats than dogs, let

$$x + 7.5 = \text{the number (in millions) of cats in the United States.}$$

Step 3. **Write an equation in x that describes the conditions.** The number of cats and dogs combined is 114.7 million.

The number (in millions) of dogs in the U.S.	plus	the number (in millions) of cats in the U.S.	equals	114.7 million.
x	$+$	$(x + 7.5)$	$=$	114.7

Step 4. **Solve the equation and answer the question.**

$$x + (x + 7.5) = 114.7 \qquad \text{This is the equation specified by the conditions of the problem.}$$

$$2x + 7.5 = 114.7 \qquad \text{Regroup and combine like terms on the left side.}$$

$$2x + 7.5 - 7.5 = 114.7 - 7.5 \qquad \text{Subtract 7.5 from both sides.}$$

$$2x = 107.2 \qquad \text{Simplify.}$$

$$\frac{2x}{2} = \frac{107.2}{2} \qquad \text{Divide both sides by 2.}$$

$$x = 53.6 \qquad \text{Simplify.}$$

Because x represents the number (in millions) of dogs, there are 53.6 million dogs in the United States. Because $x + 7.5$ represents the number (in millions) of cats, there are $53.6 + 7.5$, or 61.1 million cats in the United States.

Step 5. **Check the proposed solution in the original wording of the problem.** The problem states that the number of cats and dogs combined is 114.7 million. By adding 53.6 million, the dog population, and 61.1 million, the cat population, we do, indeed, obtain a sum of 114.7 million. ■

 CHECK POINT 3 Americans go to great lengths to treat sick dogs. Specialized veterinary care can be costly. The average cost of a magnetic resonance imaging (MRI) scan exceeds the average cost of acupuncture by $463. Combined, the two procedures cost $687. Determine the average cost for each of these veterinary procedures. (*Source*: American Veterinary Medical Association) MRI Scan: $575; Acupuncture: $112

U.S. Pet Population

Americans spend more than $30 billion per year on their pets, from MRIs to spas to doggy diapers.

Source: American Veterinary Medical Association

EXAMPLE 4 Consecutive Integers

Two pages that face each other in a book have 145 as the sum of their page numbers. What are the page numbers?

SOLUTION

Step 1. **Let x represent one of the quantities.** We will let

$$x = \text{the page number of the page on the left.}$$

Step 2. **Represent other quantities in terms of x.** The other unknown quantity is the page number of the facing page on the right. Page numbers on facing pages are consecutive integers. Thus,

$$x + 1 = \text{the page number of the page on the right.}$$

Step 3. **Write an equation in x that describes the conditions.** The two facing pages have 145 as the sum of their page numbers.

The page number on the left	plus	the page number on the right	equals	145.
x	$+$	$(x + 1)$	$=$	145

Step 4. **Solve the equation and answer the question.**

$x + (x + 1) = 145$	This is the equation for the problem's conditions.
$2x + 1 = 145$	Regroup and combine like terms.
$2x + 1 - 1 = 145 - 1$	Subtract 1 from both sides.
$2x = 144$	Simplify.
$\dfrac{2x}{2} = \dfrac{144}{2}$	Divide both sides by 2.
$x = 72$	Simplify.

Thus,

$$\text{the page number on the left} = x = 72$$

and

$$\text{the page number on the right} = x + 1 = 72 + 1 = 73.$$

The page numbers are 72 and 73.

Step 5. **Check the proposed solution in the original wording of the problem.** The problem states that the sum of the page numbers on the facing pages is 145. By adding 72, the page number on the left, and 73, the page number on the right, we do, indeed, obtain a sum of 145. ■

 CHECK POINT 4 Two pages that face each other in a book have 193 as the sum of their page numbers. What are the page numbers? pages 96 and 97

Example 4 and Check Point 4 involved consecutive integers. By contrast, some word problems involve consecutive odd integers, such as 5, 7, and 9. Other word problems involve consecutive even integers, such as 6, 8, and 10. When working with consecutive even or consecutive odd integers, we must continually add 2 to move from one integer to the next successive integer in the list.

Table 2.2 should be helpful in solving consecutive integer problems.

Table 2.2	Consecutive Integers	
English Phrase	**Algebraic Expressions**	**Example**
Two consecutive integers	$x, x + 1$	$13, 14$
Three consecutive integers	$x, x + 1, x + 2$	$-8, -7, -6$
Two consecutive even integers	$x, x + 2$	$40, 42$
Two consecutive odd integers	$x, x + 2$	$-37, -35$
Three consecutive even integers	$x, x + 2, x + 4$	$30, 32, 34$
Three consecutive odd integers	$x, x + 2, x + 4$	$9, 11, 13$

EXAMPLE 5 Renting a Car

Rent-a-Heap Agency charges $125 per week plus $0.20 per mile to rent a small car. How many miles can you travel for $335?

SOLUTION

Step 1. **Let x represent one of the quantities.** Because we are asked to find the number of miles we can travel for $335, let

$$x = \text{the number of miles.}$$

Step 2. **Represent other quantities in terms of x.** There are no other unknown quantities to find, so we can skip this step.

Step 3. **Write an equation in x that describes the conditions.** Before writing the equation, let us consider a few specific values for the number of miles traveled. The rental charge is $125 plus $0.20 for each mile.

3 miles: The rental charge is $\$125 + \$0.20(3)$.
30 miles: The rental charge is $\$125 + \$0.20(30)$.
100 miles: The rental charge is $\$125 + \$0.20(100)$.
x miles: The rental charge is $\$125 + \$0.20x$.

The weekly charge of $125	plus	the charge of $0.20 per mile for x miles	equals	the total $335 rental charge.
125	+	0.20x	=	335

Step 4. **Solve the equation and answer the question.**

$$125 + 0.20x = 335 \qquad \text{This is the equation specified by the conditions of the problem.}$$

$$125 + 0.20x - 125 = 335 - 125 \qquad \text{Subtract 125 from both sides.}$$

$$0.20x = 210 \qquad \text{Simplify.}$$

$$\frac{0.20x}{0.20} = \frac{210}{0.20} \qquad \text{Divide both sides by 0.20.}$$

$$x = 1050 \qquad \text{Simplify.}$$

You can travel 1050 miles for $335.

Step 5. **Check the proposed solution in the original wording of the problem.** Traveling 1050 miles should result in a total rental charge of $335. The mileage charge of $0.20 per mile is

$$\$0.20(1050) = \$210.$$

Adding this to the $125 weekly charge gives a total rental charge of

$$\$125 + \$210 = \$335.$$

Because this results in the given rental charge of $335, this verifies that you can travel 1050 miles. ■

 CHECK POINT 5 A taxi charges $2.00 to turn on the meter plus $0.25 for each eighth of a mile. If you have $10.00, how many eighths of a mile can you go? How many miles is that?   32; 4 miles

We will be using the formula for the perimeter of a rectangle, $P = 2l + 2w$, in our next example. Twice the rectangle's length plus twice the rectangle's width is its perimeter.

EXAMPLE 6 Finding the Dimensions of a Soccer Field

A rectangular soccer field is twice as long as it is wide. If the perimeter of a soccer field is 300 yards, what are the field's dimensions?

SOLUTION

Step 1. **Let x represent one of the quantities.** We know something about the length; the field is twice as long as it is wide. We will let

$$x = \text{the width}.$$

Step 2. **Represent other quantities in terms of x.** Because the field is twice as long as it is wide, let

$$2x = \text{the length}.$$

Figure 2.7 illustrates the soccer field and its dimensions.

FIGURE 2.7

Step 3. **Write an equation in x that describes the conditions.** Because the perimeter of a soccer field is 300 yards,

Twice the length	plus	twice the width	is	the perimeter.
$2 \cdot 2x$	$+$	$2 \cdot x$	$=$	$300.$

Step 4. **Solve the equation and answer the question.**

$$
\begin{aligned}
2 \cdot 2x + 2 \cdot x &= 300 &&\text{This is the equation for the problem's conditions.}\\
4x + 2x &= 300 &&\text{Multiply.}\\
6x &= 300 &&\text{Combine like terms.}\\
\frac{6x}{6} &= \frac{300}{6} &&\text{Divide both sides by 6.}\\
x &= 50 &&\text{Simplify.}
\end{aligned}
$$

Thus,

$$\text{Width} = x = 50$$
$$\text{Length} = 2x = 2(50) = 100.$$

The dimensions of a soccer field are 50 yards by 100 yards.

Step 5. **Check the proposed solution in the original wording of the problem.** The perimeter of the soccer field using the dimensions that we found is 2(50 yards) + 2(100 yards) = 100 yards + 200 yards, or 300 yards. Because the problem's wording tells us that the perimeter is 300 yards, our dimensions are correct. ∎

 CHECK POINT 6 A rectangular swimming pool is three times as long as it is wide. If the perimeter of the pool is 320 feet, what are the pool's dimensions?

40 feet wide and 120 feet long

EXAMPLE 7   A Price Reduction

Your local computer store is having a sale. After a 30% price reduction, you purchase a new computer for $980. What was the computer's price before the reduction?

SOLUTION

Step 1. **Let x represent one of the quantities.** We will let x = the original price of the computer before the reduction.

Step 2. **Represent other quantities in terms of x.** There are no other unknown quantities to find, so we can skip this step.

Step 3. **Write an equation in x that describes the conditions.** The computer's original price minus the 30% reduction is the reduced price, $980.

$$x \quad - \quad 0.3x \quad = \quad 980.$$

STUDY TIP

Observe that the original price, x, reduced by 30% is $x - 0.3x$ and *not* $x - 0.3$.

Step 4. **Solve the equation and answer the question.**

$$x - 0.3x = 980 \qquad \text{This is the equation for the problem's conditions.}$$
$$0.7x = 980 \qquad \text{Combine like terms: } x - 0.3x = 1x - 0.3x = 0.7x.$$
$$\frac{0.7x}{0.7} = \frac{980}{0.7} \qquad \text{Divide both sides by 0.7.}$$
$$x = 1400 \qquad \text{Simplify: } 0.7\overline{)980.0}\,.$$

The computer's price before the reduction was $1400.

Step 5. **Check the proposed solution in the original wording of the problem.** The price before the reduction, $1400, minus the reduction in price should equal the reduced price given in the original wording, $980. The reduction in price is equal to 30% of the price before the reduction, $1400. To find the reduction, we multiply the decimal equivalent of 30%, 0.30 or 0.3, by the original price, $1400:

$$30\% \text{ of } \$1400 = (0.3)(\$1400) = \$420.$$

Now we can determine whether the calculation for the price before the reduction, $1400, minus the reduction, $420, is equal to the reduced price given in the problem, $980. We subtract:

$$\$1400 - \$420 = \$980.$$

This verifies that the price of the computer before the reduction was $1400. ∎

 CHECK POINT 7 After a 40% price reduction, an exercise machine sold for $564. What was the exercise machine's price before this reduction? $940

2.5 EXERCISE SET

Student Solutions Manual   CD/Video   PH Math/Tutor Center   MathXL Tutorials on CD   MathXL®   MyMathLab   Interactmath.com

Practice Exercises

In Exercises 1–14, let x represent the number. Write each English phrase as an algebraic expression.

1. The sum of a number and 7 $x + 7$
2. A number increased by 15 $x + 15$
3. A number subtracted from 25 $25 - x$
4. 46 less than a number $x - 46$
5. 9 decreased by 4 times a number $9 - 4x$
6. 15 less than the product of 8 and a number $8x - 15$
7. The quotient of 83 and a number $\dfrac{83}{x}$
8. The quotient of a number and 83 $\dfrac{x}{83}$
9. The sum of twice a number and 40 $2x + 40$
10. Twice the sum of a number and 40 $2(x + 40)$
11. 93 subtracted from 9 times a number $9x - 93$
12. The quotient of 13 and a number, decreased by 7 times the number $\dfrac{13}{x} - 7x$
13. Eight times the sum of a number and 14 $8(x + 14)$
14. Nine times the difference of a number and 5 $9(x - 5)$

In Exercises 15–34, let x represent the number. Use the given conditions to write an equation. Solve the equation and find the number.

15. A number increased by 60 is equal to 410. Find the number. $x + 60 = 410; 350$
16. The sum of a number and 43 is 107. Find the number. $x + 43 = 107; 64$
17. A number decreased by 23 is equal to 214. Find the number. $x - 23 = 214; 237$
18. The difference between a number and 17 is 96. Find the number. $x - 17 = 96; 113$
19. The product of 7 and a number is 126. Find the number. $7x = 126; 18$
20. The product of 8 and a number is 272. Find the number. $8x = 272; 34$
21. The quotient of a number and 19 is 5. Find the number. $\dfrac{x}{19} = 5; 95$
22. The quotient of a number and 14 is 8. Find the number. $\dfrac{x}{14} = 8; 112$
23. The sum of four and twice a number is 56. Find the number. $4 + 2x = 56; 26$
24. The sum of five and three times a number is 59. Find the number. $5 + 3x = 59; 18$
25. Seven subtracted from five times a number is 178. Find the number. $5x - 7 = 178; 37$
26. Eight subtracted from six times a number is 298. Find the number. $6x - 8 = 298; 51$
27. A number increased by 5 is two times the number. Find the number. $x + 5 = 2x; 5$
28. A number increased by 12 is four times the number. Find the number. $x + 12 = 4x; 4$
29. Twice the sum of four and a number is 36. Find the number. $2(x + 4) = 36; 14$
30. Three times the sum of five and a number is 48. Find the number. $3(5 + x) = 48; 11$
31. Nine times a number is 30 more than three times that number. Find the number. $9x = 3x + 30; 5$
32. Five more than four times a number is that number increased by 35. Find the number. $4x + 5 = x + 35; 10$
33. If the quotient of three times a number and five is increased by four, the result is 34. Find the number. $\dfrac{3x}{5} + 4 = 34; 50$
34. If the quotient of three times a number and four is decreased by three, the result is nine. Find the number. $\dfrac{3x}{4} - 3 = 9; 16$

Application Exercises

In Exercises 35–62, use the five-step strategy to solve each problem.

35. Two of the most expensive movies ever made were *Titanic* and *Waterworld*. The cost to make *Titanic* exceeded the cost to make *Waterworld* by $25 million. The combined cost to make the two movies was $375 million. Find the cost of making each of these movies.   *Titanic: $200 million; Waterworld: $175 million*

Paramount Pictures Corporation, Inc.

36. Each day, the number of births in the world exceeds the number of deaths by 229 thousand. The combined number of births and deaths is 521 thousand. Determine the number of births and the number of deaths per day.   Births: 375 thousand; Deaths: 146 thousand

Daily Growth of World Population

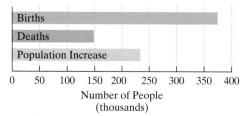

Number of People
(thousands)

Source: "Population Update" 2000

37. The circle graph shows the political ideology of U.S. college freshmen. The percentage of liberals exceeds twice that of conservatives by 4.4%. Using the displayed percents, it can be shown that liberals and conservatives combined account for 57.2% of college freshmen. Find the percentage of liberals and the percentage of conservatives. Liberals: 39.6%; Conservatives: 17.6%

**Political Ideology of
U.S. College Freshmen**

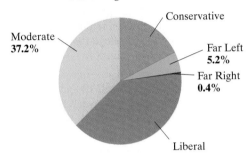

Conservative

Moderate
37.2%

Far Left
5.2%

Far Right
0.4%

Liberal

Source: The Chronicle of Higher Education

38. Commuters in one-third of the largest cities in the United States spend more than 40 hours per year, equivalent to one work week, sitting in traffic. The bar graph shows the number of hours in traffic per year for the average motorist in ten cities. The average motorist in Los Angeles spends 32 hours less than twice that of the average motorist in Miami stuck in traffic each year. In the two cities combined, 139 hours are spent by the average motorist per year in traffic. How many hours are wasted in traffic by the average motorist in Los Angeles and Miami? Los Angeles: 82 hours; Miami: 57 hours

**Hours in Traffic per Year for
the Average Motorist**

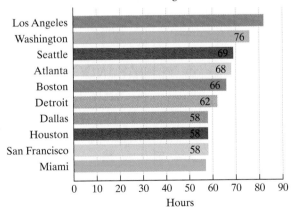

Source: Texas Transportation Institute

39. The sum of the page numbers on the facing pages of a book is 629. What are the page numbers? pages 314 and 315

40. The sum of the page numbers on the facing pages of a book is 525. What are the page numbers? pages 262 and 263

41. The highest-grossing North American concert tour was the Rolling Stones (1994), followed closely by Bruce Springsteen (2003). Combined, the two tours grossed $241 million. When expressed in millions, the earnings for the tours are consecutive integers. Find the gross, in millions, for the Rolling Stones tour and the Springsteen tour. (*Source: Rolling Stone*)
Rolling Stones: $121 million; Springsteen: $120 million

42. The first Super Bowl was played between the Green Bay Packers and the Kansas City Chiefs in 1967. Only once, in 1991, were the winning and losing scores in the Super Bowl consecutive integers. If the sum of the scores was 39, what were the scores? 19 and 20

43. Find two consecutive even integers whose sum is 66. 32 and 34

44. Find two consecutive odd integers whose sum is 72. 35 and 37

45. A car rental agency charges $200 per week plus $0.15 per mile to rent a car. How many miles can you travel in one week for $320? 800 miles

46. A car rental agency charges $180 per week plus $0.25 per mile to rent a car. How many miles can you travel in one week for $395? 860 miles

47. The average weight for female infants at birth is 7 pounds, with a monthly weight gain of 1.5 pounds. After how many months does a baby girl weigh 16 pounds? 6 months

48. The total revenue from Indian casinos in the United States has been increasing at approximately $2.2 billion per year. In 2002, Indian casinos reported a combined revenue of $12.7 billion. In which year will total revenue reach $28.1 billion? (*Source:* National Indian Gaming Commission) 2009

49. A rectangular field is four times as long as it is wide. If the perimeter of the field is 500 yards, what are the field's dimensions? 50 yards wide and 200 yards long

50. A rectangular field is five times as long as it is wide. If the perimeter of the field is 288 yards, what are the field's dimensions? 24 yards wide and 120 yards long

51. An American football field is a rectangle with a perimeter of 1040 feet. The length is 200 feet more than the width. Find the width and length of the rectangular field. 160 feet wide and 360 feet long

52. A basketball court is a rectangle with a perimeter of 86 meters. The length is 13 meters more than the width. Find the width and length of the basketball court. 15 meters wide and 28 meters long

53. A bookcase is to be constructed as shown in the figure. The length is to be 3 times the height. If 60 feet of lumber is available for the entire unit, find the length and height of the bookcase. 12 feet long and 4 feet high

54. The height of the bookcase in the figure is 3 feet longer than the length of a shelf. If 18 feet of lumber is available for the entire unit, find the length and height of the unit.   2 feet long and 5 feet high

height

length

55. After a 20% reduction, you purchase a television for $320. What was the television's price before the reduction? $400

56. After a 30% reduction, you purchase a VCR for $98. What was the VCR's price before the reduction? $140

57. The average yearly earnings of pharmacists increased by 30% from 2001 to 2002. If salaries averaged $87,100 in 2002, what was the average salary in 2001?
(*Source*: Bureau of Labor Statistics) $67,000

58. The average yearly earnings of physical education teachers increased by 40% from 2001 to 2002. If salaries averaged $63,000 in 2002, what was the average salary in 2001?
(*Source*: Bureau of Labor Statistics) $45,000

59. Including 6% sales tax, a car sold for $15,370. Find the price of the car before the tax was added. $14,500

60. Including 8% sales tax, a bed-and-breakfast inn charges $172.80 per night. Find the inn's nightly cost before the tax is added. $160

61. An automobile repair shop charged a customer $448, listing $63 for parts and the remainder for labor. If the cost of labor is $35 per hour, how many hours of labor did it take to repair the car?   11 hours

62. A repair bill on a sailboat came to $1603, including $532 for parts and the remainder for labor. If the cost of labor is $63 per hour, how many hours of labor did it take to repair the sailboat?   17 hours

Writing in Mathematics

63. In your own words, describe a step-by-step approach for solving algebraic word problems.   **63–66.** Answers will vary.

64. Many students find solving linear equations much easier than solving algebraic word problems. Discuss some of the reasons why this is the case.

65. Did you have some difficulties solving some of the problems that were assigned in this exercise set? Discuss what you did if this happened to you. Did your course of action enhance your ability to solve algebraic word problems?

66. Write an original word problem that can be solved using a linear equation. Then solve the problem.

Critical Thinking Exercises

67. Which English statement given below is correctly translated into an algebraic equation?   **d**
a. Ten pounds less than Bill's weight (x) equals 160 pounds: $10 - x = 160$.
b. Four more than five times a number (x) is one less than six times that number: $5x + 4 = 1 - 6x$.
c. Seven is three more than some number (x): $7 + 3 = x$.
d. None of the above is correctly translated.

68. Explain how to use the three percents shown on the circle graph in Exercise 37 to determine the combined percentage of liberal and conservative college freshmen.   Subtract the sum of the given percents from 100%.

69. An HMO pamphlet contains the following recommended weight for women: "Give yourself 100 pounds for the first 5 feet plus 5 pounds for every inch over 5 feet tall." Using this description, which height corresponds to an ideal weight of 135 pounds?   5 feet 7 inches

70. The rate for a particular international telephone call is $0.55 for the first minute and $0.40 for each additional minute. Determine the length of a call that costs $6.95.   17 minutes

71. In a film, the actor Charles Coburn plays an elderly "uncle" character criticized for marrying a woman when he is 3 times her age. He wittily replies, "Ah, but in 20 years time I shall only be twice her age." How old is the "uncle" and the woman?   The uncle is 60 years old and the woman is 20 years old.

72. Answer the question in the following *Peanuts* cartoon strip. (*Note*: You may not use the answer given in the cartoon!)   7 ounces

PEANUTS reprinted by permission of United Features Syndicate, Inc.

Review Exercises

73. Solve and check: $\frac{4}{5}x = -16$. (Section 2.2, Example 3) -20

74. Solve and check: $6(y - 1) + 7 = 9y - y + 1$. (Section 2.3, Example 3) 0

75. Solve for w: $V = \frac{1}{3}lwh$. (Section 2.4, Example 4) $w = \frac{3V}{lh}$

SECTION 2.6

Objectives

1 Solve problems using formulas for perimeter and area.

2 Solve problems using formulas for a circle's area and circumference.

3 Solve problems using formulas for volume.

4 Solve problems involving the angles of a triangle.

5 Solve problems involving complementary and supplementary angles.

PROBLEM SOLVING IN GEOMETRY

A portion of the human lung magnified 160 times

Geometry is about the space you live in and the shapes that surround you. You're even made of it. The human lung consists of nearly 300 spherical air sacs, geometrically designed to provide the greatest surface area within the limited volume of our bodies. Viewed in this way, geometry becomes an intimate experience.

For thousands of years, people have studied geometry in some form to obtain a better understanding of the world in which they live. A study of the shape of your world will provide you with many practical applications that will help to increase your problem-solving skills.

1 Solve problems using formulas for perimeter and area.

Geometric Formulas for Perimeter and Area Solving geometry problems often requires using basic geometric formulas. Formulas for perimeter and area are summarized in Table 2.3. Remember that perimeter is measured in linear units, such as feet or meters, and area is measured in square units, such as square feet, ft^2, or square meters, m^2.

Table 2.3	**Common Formulas for Perimeter and Area**		
Square	**Rectangle**	**Triangle**	**Trapezoid**
$A = s^2$	$A = lw$	$A = \frac{1}{2}bh$	$A = \frac{1}{2}h(a + b)$
$P = 4s$	$P = 2l + 2w$		

FIGURE 2.8 Finding the height of a triangular sail

EXAMPLE 1 Using the Formula for the Area of a Triangle

A sailboat has a triangular sail with an area of 30 square feet and a base that is 12 feet long. (See Figure 2.8.) Find the height of the sail.

SOLUTION We begin with the formula for the area of a triangle given in Table 2.3.

$$A = \frac{1}{2}bh$$ The area of a triangle is $\frac{1}{2}$ the product of its base and height.

$$30 = \frac{1}{2}(12)h$$ Substitute 30 for A and 12 for b.

$$30 = 6h$$ Simplify.

$$\frac{30}{6} = \frac{6b}{6}$$ Divide both sides by 6.

$$5 = h$$ Simplify.

The height of the sail is 5 feet.

Check

The area is $A = \frac{1}{2}bh = \frac{1}{2}(12 \text{ feet})(5 \text{ feet}) = 30$ square feet.

■

> **CHECK POINT 1** A sailboat has a triangular sail with an area of 24 square feet and a base that is 4 feet long. Find the height of the sail.   12 feet

2 Solve problems using formulas for a circle's area and circumference.

Geometric Formulas for Circumference and Area of a Circle It's a good idea to know your way around a circle. Clocks, angles, maps, and compasses are based on circles. Circles occur everywhere in nature: in ripples on water, patterns on a butterfly's wings, and cross sections of trees. Some consider the circle to be the most pleasing of all shapes.

A **circle** is a set of points in the plane equally distant from a given point, its center. Figure 2.9 shows two circles. A **radius** (plural: radii), r, is a line segment from the center to any point on the circle. For a given circle, all radii have the same length. A **diameter**, d, is a line segment through the center whose endpoints both lie on the circle. For a given circle, all diameters have the same length. In any circle, **the length of a diameter is twice the length of a radius**.

The point at which a pebble hits a flat surface of water becomes the center of a number of circular ripples.

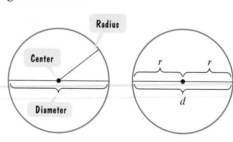

FIGURE 2.9

The words *radius* and *diameter* refer to both the line segments in Figure 2.9 as well as their linear measures. The distance around a circle (its perimeter) is called its **circumference**. Formulas for the area and circumference of a circle are given in terms of π and appear in Table 2.4. We have seen that π is an irrational number and is approximately equal to 3.14.

Table 2.4	Formulas for Circles	
Circle	**Area**	**Circumference**
	$A = \pi r^2$	$C = 2\pi r$

When computing a circle's area or circumference by hand, round π to 3.14. When using a calculator, use the $\boxed{\pi}$ key, which gives the value of π rounded to approximately 11 decimal places. In either case, calculations involving π give approximate answers. These answers can vary slightly depending on how π is rounded. The symbol $\approx$ (is approximately equal to) will be written in these calculations.

EXAMPLE 2 Finding the Area and Circumference of a Circle

Find the area and circumference of a circle with a diameter measuring 20 inches.

SOLUTION   The radius is half the diameter, so $r = \frac{20}{2} = 10$ inches.

$A = \pi r^2$	$C = 2\pi r$	*Use the formulas for area and circumference of a circle.*
$A = \pi(10)^2$	$C = 2\pi(10)$	*Substitute 10 for r.*
$A = 100\pi$	$C = 20\pi$	

The area of the circle is 100π square inches and the circumference is 20π inches. Using the fact that $\pi \approx 3.14$, the area is approximately 100(3.14), or 314 square inches. The circumference is approximately 20(3.14), or 62.8 inches. ■

 CHECK POINT 2 The diameter of a circular landing pad for helicopters is 40 feet. Find the area and circumference of the landing pad. Express answers in terms of π. Then round answers to the nearest square foot and foot, respectively.
Area = 400π feet$^2 \approx 1256$ feet^2; Circumference = 40π feet ≈ 126 feet

EXAMPLE 3 Problem Solving Using the Formula for a Circle's Area

Which one of the following is the better buy: a large pizza with a 16-inch diameter for $15.00 or a medium pizza with an 8-inch diameter for $7.50?

SOLUTION   The better buy is the pizza with the lower price per square inch. The radius of the large pizza is $\frac{1}{2} \cdot 16$ inches, or 8 inches, and the radius of the medium pizza is $\frac{1}{2} \cdot 8$ inches, or 4 inches. The area of the surface of each circular pizza is determined using the formula for the area of a circle.

$$\text{Large pizza:} \quad A = \pi r^2 = \pi(8 \text{ in.})^2 = 64\pi \text{ in.}^2 \approx 201 \text{ in.}^2$$
$$\text{Medium pizza:} \quad A = \pi r^2 = \pi(4 \text{ in.})^2 = 16\pi \text{ in.}^2 \approx 50 \text{ in.}^2$$

For each pizza, the price per square inch is found by dividing the price by the area:

$$\text{Price per square inch for large pizza} = \frac{\$15.00}{64\pi \text{ in.}^2} \approx \frac{\$15.00}{201 \text{ in.}^2} \approx \frac{\$0.07}{\text{in.}^2}$$

$$\text{Price per square inch for medium pizza} = \frac{\$7.50}{16\pi \text{ in.}^2} \approx \frac{\$7.50}{50 \text{ in.}^2} = \frac{\$0.15}{\text{in.}^2}.$$

The large pizza costs approximately $0.07 per square inch and the medium pizza costs approximately $0.15 per square inch. Thus, the large pizza is the better buy. ■

In Example 3, did you at first think that the price per square inch would be the same for the large and the medium pizzas? After all, the radius of the large pizza is twice that of the medium pizza, and the cost of the large is twice that of the medium. However, the large pizza's area, 64π square inches, is *four times the area* of the medium pizza, 16π square inches. Doubling the radius of a circle increases its area by four times the original amount.

USING TECHNOLOGY

You can use your calculator to obtain the price per square inch for each pizza in Example 3. The price per square inch for the large pizza, $\frac{15}{64\pi}$, is approximated by one of the following keystrokes:

Many Scientific Calculators

15 $\boxed{\div}$ $\boxed{(}$ 64 $\boxed{\times}$ $\boxed{\pi}$ $\boxed{)}$ $\boxed{=}$

Many Graphing Calculators

15 $\boxed{\div}$ $\boxed{(}$ 64 $\boxed{\pi}$ $\boxed{)}$ $\boxed{\text{ENTER}}$

> ✓ CHECK POINT **3** Which one of the following is the better buy: a large pizza with an 18-inch diameter for $20.00 or a medium pizza with a 14-inch diameter for $14.00?   large pizza

3 Solve problems using formulas for volume.

Geometric Formulas for Volume   A shoe box and a basketball are examples of three-dimensional figures. **Volume** refers to the amount of space occupied by such a figure. To measure this space, we begin by selecting a cubic unit. One such cubic unit, 1 cubic centimeter (cm^3), is shown in Figure 2.10.

The edges of a cube all have the same length. Other cubic units used to measure volume include 1 cubic inch (in.^{3}) and 1 cubic foot (ft^3). The volume of a solid is the number of cubic units that can be contained in the solid.

FIGURE 2.10

Formulas for volumes of three-dimensional figures are given in Table 2.5.

Table 2.5   **Common Formulas for Volume**

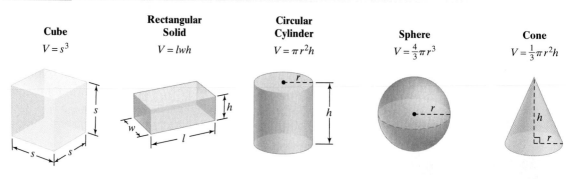

Cube	Rectangular Solid	Circular Cylinder	Sphere	Cone
$V = s^3$	$V = lwh$	$V = \pi r^2 h$	$V = \frac{4}{3}\pi r^3$	$V = \frac{1}{3}\pi r^2 h$

EXAMPLE 4   Using the Formula for the Volume of a Cylinder

A cylinder with a radius of 2 inches and a height of 6 inches has its radius doubled. (See Figure 2.11.) How many times greater is the volume of the larger cylinder than the volume of the smaller cylinder?

SOLUTION   We begin with the formula for the volume of a cylinder given in Table 2.5. Find the volume of the smaller cylinder and the volume of the larger cylinder. To compare the volumes, divide the volume of the larger cylinder by the volume of the smaller cylinder.

$$V = \pi r^2 h$$ Use the formula for the volume of a cylinder.

Radius is doubled.

$V_{\text{Smaller}} = \pi(2)^2(6)$ $V_{\text{Larger}} = \pi(4)^2(6)$ Substitute the given values.

$V_{\text{Smaller}} = \pi(4)(6)$ $V_{\text{Larger}} = \pi(16)(6)$

$V_{\text{Smaller}} = 24\pi$ $V_{\text{Larger}} = 96\pi$

The volume of the smaller cylinder is 24π cubic inches. The volume of the larger cylinder is 96π cubic inches. We use division to compare the volumes:

$$\frac{V_{\text{Larger}}}{V_{\text{Smaller}}} = \frac{96\pi}{24\pi} = \frac{4}{1}.$$

Thus, the volume of the larger cylinder is 4 times the volume of the smaller cylinder. ■

Radius: 2 inches
Height: 6 inches

Radius: 4 inches
Height: 6 inches

FIGURE 2.11   Doubling a cylinder's radius

> ✓ **CHECK POINT 4** A cylinder with a radius of 3 inches and a height of 5 inches has its height doubled. How many times greater is the volume of the larger cylinder than the volume of the smaller cylinder?   2 times

EXAMPLE 5   Applying Volume Formulas

An ice cream cone is 5 inches deep and has a radius of 1 inch. A spherical scoop of ice cream also has a radius of 1 inch. (See Figure 2.12.) If the ice cream melts into the cone, will it overflow?

SOLUTION   The ice cream will overflow if the volume of the ice cream, a sphere, is greater than the volume of the cone. Find the volume of each.

$$V_{\text{cone}} = \frac{1}{3}\pi r^2 h = \frac{1}{3}\pi(1 \text{ in.})^2 \cdot 5 \text{ in.} = \frac{5\pi}{3} \text{ in.}^3 \approx 5 \text{ in.}^3$$

$$V_{\text{sphere}} = \frac{4}{3}\pi r^3 = \frac{4}{3}\pi(1 \text{ in.})^3 = \frac{4\pi}{3} \text{ in.}^3 \approx 4 \text{ in.}^3$$

The volume of the spherical scoop of ice cream is less than the volume of the cone, so there will be no overflow.   ∎

> ✓ **CHECK POINT 5** A basketball has a radius of 4.5 inches. If the ball is filled with 350 cubic inches of air, is this enough air to fill it completely?

FIGURE 2.12

No; About 32 more cubic inches are needed.

 4   Solve problems involving the angles of a triangle.

The Angles of a Triangle   The hour hand of a clock moves from 12 to 2. The hour hand suggests a **ray**, a part of a line that has only one endpoint and extends forever in the opposite direction. An *angle* is formed as the ray in Figure 2.13 rotates from 12 to 2.

An **angle**, symbolized ∡, is made up of two rays that have a common endpoint. Figure 2.14 shows an angle. The common endpoint, *B* in the figure, is called the **vertex** of the angle. The two rays that form the angle are called its **sides**. The four ways of naming the angle are shown to the right of Figure 2.14.

FIGURE 2.14   An angle: two rays with a common endpoint

FIGURE 2.13   Clock with rays rotating to form an angle

One way to measure angles is in **degrees**, symbolized by a small, raised circle °. Think of the hour hand of a clock. From 12 noon to 12 midnight, the hour hand moves around in a complete circle. By definition, the ray has rotated through 360 degrees, or 360°. Using 360° as the amount of rotation of a ray back onto itself, a degree, 1°, is $\frac{1}{360}$ of a complete rotation.

Our next problem is based on the relationship among the three angles of any triangle.

THE ANGLES OF A TRIANGLE
The sum of the measures of the three angles of any triangle is 180°.

EXAMPLE 6   Angles of a Triangle

In a triangle, the measure of the first angle is twice the measure of the second angle. The measure of the third angle is 20° less than the second angle. What is the measure of each angle?

SOLUTION

Step 1.   **Let x represent one of the quantities.** Let

$$x = \text{the measure of the second angle.}$$

Step 2.   **Represent other quantities in terms of x.** The measure of the first angle is twice the measure of the second angle. Thus, let

$$2x = \text{the measure of the first angle.}$$

The measure of the third angle is 20° less than the second angle. Thus, let

$$x - 20 = \text{the measure of the third angle.}$$

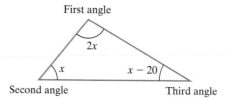

First angle

$2x$

x

$x - 20$

Second angle   Third angle

Step 3.   **Write an equation in x that describes the conditions.** Because we are working with a triangle, the sum of the measures of its three angles is 180°.

Measure of first angle	plus	measure of second angle	plus	measure of third angle	equals	180°.
$2x$	$+$	x	$+$	$(x - 20)$	$=$	180

Step 4.   **Solve the equation and answer the question.**

$2x + x + (x - 20) = 180$ This is the equation that describes the sum of the measures of the angles.

$4x - 20 = 180$ Regroup and combine like terms.

$4x - 20 + 20 = 180 + 20$ Add 20 to both sides.

$4x = 200$ Simplify.

$\dfrac{4x}{4} = \dfrac{200}{4}$ Divide both sides by 4.

$x = 50$ Simplify.

Measure of first angle $= 2x = 2 \cdot 50 = 100$
Measure of second angle $= x = 50$
Measure of third angle $= x - 20 = 50 - 20 = 30$

The angles measure 100°, 50°, and 30°.

Step 5.   **Check the proposed solution in the original wording of the problem.** The problem tells us that we are working with a triangle's angles. Thus, the sum of the measures should be 180°. Adding the three measures, we obtain 100° + 50° + 30°, giving the required sum of 180°. ∎

1st angle = 120°; 2nd angle = 40°;
3rd angle = 20°

 CHECK POINT 6 In a triangle, the measure of the first angle is three times the measure of the second angle. The measure of the third angle is 20° less than the second angle. What is the measure of each angle?

5 Solve problems involving complementary and supplementary angles.

Complementary and Supplementary Angles Two angles with measures having a sum of 90° are called **complementary angles**. For example, angles measuring 70° and 20° are complementary angles because $70° + 20° = 90°$. For angles such as those measuring 70° and 20°, each angle is a **complement** of the other: The 70° angle is the complement of the 20° angle and the 20° angle is the complement of the 70° angle. The measure of the complement can be found by subtracting the angle's measure from 90°. For example, we can find the complement of a 25° angle by subtracting 25° from 90°: $90° − 25° = 65°$. Thus, an angle measuring 65° is the complement of one measuring 25°.

Two angles with measures having a sum of 180° are called **supplementary angles**. For example, angles measuring 110° and 70° are supplementary angles because $110° + 70° = 180°$. For angles such as those measuring 110° and 70°, each angle is a **supplement** of the other: The 110° angle is the supplement of the 70° angle and the 70° angle is the supplement of the 110° angle. The measure of the supplement can be found by subtracting the angle's measure from 180°. For example, we can find the supplement of a 25° angle by subtracting 25° from 180°: $180° − 25° = 155°$. Thus, an angle measuring 155° is the supplement of one measuring 25°.

ALGEBRAIC EXPRESSIONS FOR COMPLEMENTS AND SUPPLEMENTS

Measure of an angle: x

Measure of the angle's complement: $90 − x$

Measure of the angle's supplement: $180 − x$

EXAMPLE 7 Angle Measures and Complements

The measure of an angle is 40° less than four times the measure of its complement. What is the angle's measure?

SOLUTION

Step 1. **Let x represent one of the quantities.** Let

$$x = \text{the measure of the angle.}$$

Step 2. **Represent other unknown quantities in terms of x.** Because this problem involves an angle and its complement, let

$$90 − x = \text{the measure of the complement.}$$

Step 3. **Write an equation in x that describes the conditions.**

The angle's measure   is   40° less than   four times the measure of the complement.

$$x \quad = \quad 4(90 − x) \quad − \quad 40$$

Step 4. **Solve the equation and answer the question.**

$$x = 4(90 - x) - 40 \qquad \text{This is the equation that describes the problem's conditions.}$$

$$x = 360 - 4x - 40 \qquad \text{Use the distributive property.}$$

$$x = 320 - 4x \qquad \text{Simplify: } 360 - 40 = 320.$$

$$x + 4x = 320 - 4x + 4x \qquad \text{Add 4x to both sides.}$$

$$5x = 320 \qquad \text{Simplify.}$$

$$\frac{5x}{5} = \frac{320}{5} \qquad \text{Divide both sides by 5.}$$

$$x = 64 \qquad \text{Simplify.}$$

The angle measures 64°.

Step 5. **Check the proposed solution in the original wording of the problem.** The measure of the complement is 90° − 64° = 26°. Four times the measure of the complement is 4 · 26°, or 104°. The angle's measure, 64°, is 40° less than 104°: 104° − 40° = 64°. As specified by the problem's wording, the angle's measure is 40° less than four times the measure of its complement. ∎

CHECK POINT 7 The measure of an angle is twice the measure of its complement. What is the angle's measure? 60°

2.6 EXERCISE SET

Student Solutions Manual CD/Video PH Math/Tutor Center MathXL Tutorials on CD MathXL® MyMathLab Interactmath.com

Practice Exercises

Use the formulas for perimeter and area in Table 2.3 on page 149 to solve Exercises 1–12.

In Exercises 1–2, find the perimeter and area of each rectangle.

1.

3 m

6 m

perimeter = 18 m;
area = 18 m²

2.

3 ft

4 ft

perimeter = 14 ft;
area = 12 ft²

In Exercises 3–4, find the area of each triangle.

3.

9 in. 8 in. 12 in.

14 in.

56 in²

4.

36 m 36 m

33 m

30 m

495 m²

In Exercises 5–6, find the area of each trapezoid.

5.

10 m   91 m²

9.2 m   7 m   8.5 m

16 m

6.

26 m   567 m²

18 m   18 m   21.1 m

37 m

7. A rectangular swimming pool has a width of 25 feet and an area of 1250 square feet. What is the pool's length?   50 feet

8. A rectangular swimming pool has a width of 35 feet and an area of 2450 square feet. What is the pool's length?   70 feet

9. A triangle has a base of 5 feet and an area of 20 square feet. Find the triangle's height.   8 feet

10. A triangle has a base of 6 feet and an area of 30 square feet. Find the triangle's height.   10 feet

11. A rectangle has a width of 44 centimeters and a perimeter of 188 centimeters. What is the rectangle's length?   50 centimeters

12. A rectangle has a width of 46 centimeters and a perimeter of 208 centimeters. What is the rectangle's length?   58 centimeters

Use the formulas for the area and circumference of a circle in Table 2.4 on page 150 to solve Exercises 13–18.
In Exercises 13–16, find the area and circumference of each circle. Express answers in terms of π. Then round to the nearest whole number.

13.

4 cm

area = 16π cm^2 ≈ 50 cm^2;
circumference = 8π cm ≈ 25 cm

14.

9 m

area = 81π m^2 ≈ 254 m^2;
circumference = 18π m ≈ 57 m

15.

12 yd

area = 36π yd^2 ≈ 113 yd^2
circumference = 12π yd ≈ 38 yd

16.

40 ft

area = 400π ft^2 ≈ 1257 ft^2
circumference = 40π ft ≈ 126 ft

17. The circumference of a circle is 14π inches. Find the circle's radius and diameter.   radius = 7 inches; diameter = 14 inches

18. The circumference of a circle is 16π inches. Find the circle's radius and diameter.   radius = 8 inches; diameter = 16 inches

Use the formulas for volume in Table 2.5 on page 152 to solve Exercises 19–30.
In Exercises 19–26, find the volume of each figure. Where applicable, express answers in terms of π. Then round to the nearest whole number.

19.   volume = 36 in.^{3}

4 in.

3 in.

3 in.

20.   volume = 45 cm^3

3 cm

3 cm

5 cm

21.

5 cm

6 cm

volume = 150π cm^3 ≈ 471 cm^3

22.

6 cm

8 cm

volume = 288π cm^3 ≈ 905 cm^3

23.

18 cm

volume = 972π cm^3 ≈ 3054 cm^3

24.

24 in.

volume = 2304π in.^{3} ≈ 7238 in.^{3}

25.

9 m

4 m

volume = 48π m^3 ≈ 151 m^3

26.

16 m

5 m

volume = $\frac{400}{3}\pi$ m^3 ≈ 419 m^3

27. Solve the formula for the volume of a circular cylinder for h.
$$h = \frac{V}{\pi r^2}$$

28. Solve the formula for the volume of a cone for h. $h = \dfrac{3V}{\pi r^2}$

29. A cylinder with radius 3 inches and height 4 inches has its radius tripled. How many times greater is the volume of the larger cylinder than the smaller cylinder?   9 times

30. A cylinder with radius 2 inches and height 3 inches has its radius quadrupled. How many times greater is the volume of the larger cylinder than the smaller cylinder?   16 times

Use the relationship among the three angles of any triangle to solve Exercises 31–36.

31. Two angles of a triangle have the same measure and the third angle is 30° greater than the measure of the other two. Find the measure of each angle.

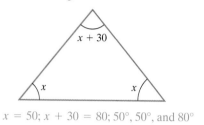

$x = 50; x + 30 = 80; 50°, 50°,$ and $80°$

32. One angle of a triangle is three times as large as another. The measure of the third angle is 40° more than that of the smallest angle. Find the measure of each angle.

$x = 28; 3x = 84; x + 40 = 68; 28°, 84°,$ and $68°$

Find the measure of each angle whose degree measure is represented in terms of x in the triangles in Exercises 33–34.

33.

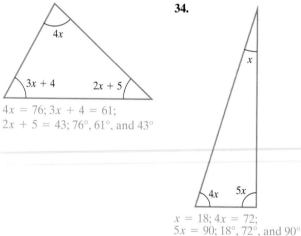

$4x = 76; 3x + 4 = 61;$
$2x + 5 = 43; 76°, 61°,$ and $43°$

34.

$x = 18; 4x = 72;$
$5x = 90; 18°, 72°,$ and $90°$

35. One angle of a triangle is twice as large as another. The measure of the third angle is 20° more than that of the smallest angle. Find the measure of each angle. $40°, 80°, 60°$

36. One angle of a triangle is three times as large as another. The measure of the third angle is 30° greater than that of the smallest angle. Find the measure of each angle. $30°, 90°, 60°$

In Exercises 37–40, find the measure of the complement of each angle.

37. 58°   32°   **38.** 41°   49°   **39.** 88°   2°   **40.** 2°   88°

In Exercises 41–44, find the measure of the supplement of each angle.

41. 132°   48°   **42.** 93°   87°   **43.** 90°   90°   **44.** 179.5°   0.5°

In Exercises 45–50, use the five-step problem-solving strategy to find the measure of the angle described.

45. The angle's measure is 60° more than that of its complement.   75°

46. The angle's measure is 78° less than that of its complement.   6°

47. The angle's measure is three times that of its supplement.   135°

48. The angle's measure is 16° more than triple that of its supplement.   139°

49. The measure of the angle's supplement is 10° more than three times that of its complement.   50°

50. The measure of the angle's supplement is 52° more than twice that of its complement.   52°

Practice Plus

In Exercises 51–53, find the area of each figure.

51.

72 m²

52.

300 m²

53.

70.5 cm²

54. Find the area of the shaded region in terms of π.

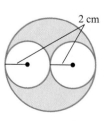

8π cm²

In Exercises 55–56, find the volume of the darkly shaded region. In Exercise 55, use the fact that the volume of a pyramid is $\frac{1}{3}$ the volume of a rectangular solid having the same base and the same height. In Exercise 56, express the answer in terms of π.

55.   168 cm³

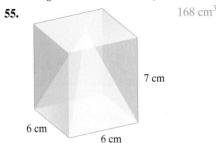

7 cm

6 cm

6 cm

56. →|2 in.|← 80π in.³

10 in.

6 in.

Application Exercises

Use the formulas for perimeter and area in Table 2.3 on page 149 to solve Exercises 57–58.

57. Taxpayers with an office in their home may deduct a percentage of their home-related expenses. This percentage is based on the ratio of the office's area to the area of the home. A taxpayer with an office in a 2200-square-foot home maintains a 20-foot by 16-foot office. If the yearly electricity bills for the home come to $4800, how much of this is deductible? $698.18

58. The lot in the figure shown, except for the house, shed, and driveway, is lawn. One bag of lawn fertilizer costs $25.00 and covers 4000 square feet.

   a. Determine the minimum number of bags of fertilizer needed for the lawn.   23 bags

   b. Find the total cost of the fertilizer. $575

60 ft

20 ft
■ 20 ft
Shed

100 ft   100 ft   200 ft

20 ft Drive

House

500 ft

Use the formulas for the area and the circumference of a circle in Table 2.4 on page 150 to solve Exercises 59–64. Round all circumference and area calculations to the nearest whole number.

59. Which one of the following is a better buy: a large pizza with a 14-inch diameter for $12.00 or a medium pizza with a 7-inch diameter for $5.00?   large pizza

60. Which one of the following is a better buy: a large pizza with a 16-inch diameter for $12.00 or two small pizzas, each with a 10-inch diameter, for $12.00?   large pizza

61. If asphalt pavement costs $0.80 per square foot, find the cost to pave the circular road in the figure shown. Round to the nearest dollar. $2262

50 ft

40 ft

62. Hardwood flooring costs $10.00 per square foot. How much will it cost (to the nearest dollar) to cover the dance floor shown in the figure with hardwood flooring? $36,566

20 ft

60 ft

63. A glass window is to be placed in a house. The window consists of a rectangle, 6 feet high by 3 feet wide, with a semicircle at the top. Approximately how many feet of stripping, to the nearest tenth of a foot, will be needed to frame the window?
approximately 19.7 feet

6 ft

3 ft

64. How many plants spaced every 6 inches are needed to surround a circular garden with a 30-foot radius?   377 plants

Use the formulas for volume in Table 2.5 on page 152 to solve Exercises 65–69. When necessary, round all volume calculations to the nearest whole number.

65. A water reservoir is shaped like a rectangular solid with a base that is 50 yards by 30 yards, and a vertical height of 20 yards. At the start of a three-month period of no rain, the reservoir was completely full. At the end of this period, the height of the water was down to 6 yards. How much water was used in the three-month period? $21,000 \text{ yd}^3$

66. A building contractor is to dig a foundation 4 yards long, 3 yards wide, and 2 yards deep for a toll booth's foundation. The contractor pays $10 per load for trucks to remove the dirt. Each truck holds 6 cubic yards. What is the cost to the contractor to have all the dirt hauled away? $40

67. Two cylindrical cans of soup sell for the same price. One can has a diameter of 6 inches and a height of 5 inches. The other has a diameter of 5 inches and a height of 6 inches. Which can contains more soup and, therefore, is the better buy?   the can with diameter of 6 inches and height of 5 inches

68. The tunnel under the English Channel that connects England and France is one of the world's longest tunnels. The Chunnel, as it is known, consists of three separate tunnels built side by side. Each is a half-cylinder that is 50,000 meters long and 4 meters high. How many cubic meters of dirt had to be removed to build the Chunnel? $1,200,000\pi \text{ m}^3 \approx 3,769,900 \text{ m}^3$

69. You are about to sue your contractor who promised to install a water tank that holds 500 gallons of water. You know that 500 gallons is the capacity of a tank that holds 67 cubic feet. The cylindrical tank has a radius of 3 feet and a height of 2 feet 4 inches. Does the evidence indicate you can win the case against the contractor if it goes to court?   Yes, the water tank is a little over 1 cubic foot too small.

Writing in Mathematics

70. Using words only, describe how to find the area of a triangle.   70–79. Answers will vary.

71. Describe the difference between the following problems: How much fencing is needed to enclose a garden? How much fertilizer is needed for the garden?

72. Describe how volume is measured. Explain why linear or square units cannot be used.

73. What is an angle?

74. If the measures of two angles of a triangle are known, explain how to find the measure of the third angle.

75. Can a triangle contain two 90° angles? Explain your answer.

76. What are complementary angles? Describe how to find the measure of an angle's complement.

77. What are supplementary angles? Describe how to find the measure of an angle's supplement?

78. Describe an application of a geometric formula involving area or volume.

79. Write and solve an original problem involving the measures of the three angles of a triangle.

Critical Thinking Exercises

80. Which one of the following is true?   b

  a. It is not possible to have a circle whose circumference is numerically equal to its area.

  b. When the measure of a given angle is added to three times the measure of its complement, the sum equals the sum of the measures of the complement and supplement of the angle.

  c. The complement of an angle that measures less than 90° is an angle that measures more than 90°.

  d. Two complementary angles cannot be equal in measure.

81. Suppose you know the cost for building a rectangular deck measuring 8 feet by 10 feet. If you decide to increase the dimensions to 12 feet by 15 feet, by how many times will the cost increase?   2.25 times

82. A rectangular swimming pool measures 14 feet by 30 feet. The pool is surrounded on all four sides by a path that is 3 feet wide. If the cost to resurface the path is $2 per square foot, what is the total cost of resurfacing the path? $600

83. What happens to the volume of a sphere if its radius is doubled?   Volume increases 8 times.

84. A scale model of a car is constructed so that its length, width, and height are each $\frac{1}{10}$ the length, width, and height of the actual car. By how many times does the volume of the car exceed its scale model?   1000 times

85. Find the measure of the angle of inclination, denoted by x in the figure, for the road leading to the bridge.   35°

Review Exercises

86. Solve for s: $P = 2s + b$. (Section 2.4, Example 3) $s = \dfrac{P-b}{2}$

87. Solve for x: $\dfrac{x}{2} + 7 = 13 - \dfrac{x}{4}$. (Section 2.3, Example 4) $\{8\}$

88. Simplify: $[3(12 \div 2^2 - 3)^2]^2$. (Section 1.8, Example 8)   0

SECTION

2.7

SOLVING LINEAR INEQUALITIES

Objectives

1 Graph the solutions of an inequality on a number line.

2 Use set-builder notation.

3 Understand properties used to solve linear inequalities.

4 Solve linear inequalities.

5 Identify inequalities with no solution or infinitely many solutions.

6 Solve problems using linear inequalities.

Do you remember Rent-a-Heap, the car rental company that charged $125 per week plus $0.20 per mile to rent a small car? In Example 5 on page 143 we asked the question: How many miles can you travel for $335? We let x represent the number of miles and set up a linear equation as follows:

$$
\underset{\text{The weekly charge of \$125}}{125} \quad \underset{\text{plus}}{+} \quad \underset{\substack{\text{the charge of} \\ \text{\$0.20 per mile} \\ \text{for } x \text{ miles}}}{0.20x} \quad \underset{\text{equals}}{=} \quad \underset{\substack{\text{the total \$335} \\ \text{rental charge.}}}{335.}
$$

Because we are limited by how much money we can spend on everything from buying clothing to renting a car, it is also possible to ask: How many miles can you travel if you can spend *at most* $335? We again let x represent the number of miles. Spending *at most* $335 means that the amount spent on the weekly rental must be *less than or equal to* $335:

$$
\underset{\text{The weekly charge of \$125}}{125} \quad \underset{\text{plus}}{+} \quad \underset{\substack{\text{the charge of} \\ \text{\$0.20 per mile} \\ \text{for } x \text{ miles}}}{0.20x} \quad \underset{\substack{\text{must be less} \\ \text{than} \\ \text{or equal to}}}{\leq} \quad \underset{\text{\$335.}}{335.}
$$

Using the commutative property of addition, we can express this inequality as

$$0.20x + 125 \leq 335.$$

The form of this inequality is $ax + b \leq c$, with $a = 0.20$, $b = 125$, and $c = 335$. Any inequality in this form is called a **linear inequality in one variable**. The symbol between $ax + b$ and c can be $\leq$ (is less than or equal to), $<$ (is less than), $\geq$ (is greater than or equal to), or $>$ (is greater than). The greatest exponent on the variable in such an inequality is 1.

In this section, we will study how to solve linear inequalities such as $0.20x + 125 \leq 335$. **Solving an inequality** is the process of finding the set of numbers that will make the inequality a true statement. These numbers are called the **solutions** of the inequality, and we say that they **satisfy** the inequality. The set of all solutions is called the **solution set** of the inequality. We begin by discussing how to graph and how to represent these solution sets.

1 Graph the solutions of an inequality on a number line.

Graphs of Inequalities There are infinitely many solutions to the inequality $x < 3$, namely, all real numbers that are less than 3. Although we cannot list all the solutions, we can make a drawing on a number line that represents these solutions. Such a drawing is called the **graph of the inequality**.

Graphs of solutions to linear inequalities are shown on a number line by shading all points representing numbers that are solutions. *Open dots* (∘) indicate endpoints that are *not solutions* and *closed dots* (·) indicate endpoints that *are solutions.*

EXAMPLE 1 Graphing Inequalities

Graph the solutions of each inequality: **a.** $x < 3$ **b.** $x \geq -1$ **c.** $-1 < x \leq 3$.

SOLUTION

a. The solutions of $x < 3$ are all real numbers that are less than 3. They are graphed on a number line by shading all points to the left of 3. The open dot at 3 indicates that 3 is not a solution, but numbers such as 2.9999 and 2.6 are. The arrow shows that the graph extends indefinitely to the left.

b. The solutions of $x \geq -1$ are all real numbers that are greater than or equal to -1. We shade all points to the right of -1 and the point for -1 itself. The closed dot at -1 shows that -1 is a solution of the given inequality. The arrow shows that the graph extends indefinitely to the right.

c. The inequality $-1 < x \leq 3$ is read "-1 is less than x *and* x is less than or equal to 3," or "x is greater than -1 *and* less than or equal to 3." The solutions of $-1 < x \leq 3$ are all real numbers between -1 and 3, not including -1 but including 3. The open dot at -1 indicates that -1 is not a solution. The closed dot at 3 shows that 3 is a solution. Shading indicates the other solutions.

CHECK POINT 1 Graph the solutions of each inequality:

a. $x < 4$ **b.** $x \geq -2$ **c.** $-4 \leq x < 1$. See graphing answer section.

2 Use set-builder notation.

Solution Sets The solutions of $x < 3$ are all real numbers that are less than 3. We can use the set concept introduced in Chapter 1 and state that the solution is the *set of all real numbers less than 3.* We use **set-builder notation** to write the solution set of $x < 3$ as

$$\{x \mid x < 3\}.$$

We read this as "the set of all x such that x is less than 3." Solutions of inequalities should be expressed in set-builder notation.

3 Understand properties used to solve linear inequalities.

Properties Used to Solve Linear Inequalities Back to our question: How many miles can you drive your Rent-a-Heap car if you can spend at most $335 per week? We answer the question by solving

$$0.20x + 125 \leq 335$$

for x. The solution procedure is nearly identical to that for solving

$$0.20x + 125 = 335.$$

Our goal is to get x by itself on the left side. We do this by subtracting 125 from both sides to isolate $0.20x$:

$$
\begin{array}{ll}
0.20x + 125 \leq 335 & \text{This is the given inequality.} \\
0.20x + 125 - 125 \leq 335 - 125 & \text{Subtract 125 from both sides.} \\
0.20x \leq 210. & \text{Simplify.}
\end{array}
$$

Finally, we isolate x from $0.20x$ by dividing both sides of the inequality by 0.20:

$$\frac{0.20x}{0.20} \leq \frac{210}{0.20}$$ *Divide both sides by 0.20.*

$$x \leq 1050.$$ *Simplify.*

With at most \$335 per week to spend, you can travel at most 1050 miles.

We started with the inequality $0.20x + 125 \leq 335$ and obtained the inequality $x \leq 1050$ in the final step. Both of these inequalities have the same solution set, namely $\{x \mid x \leq 1050\}$. Inequalities such as these, with the same solution set, are said to be **equivalent**.

We isolated x from $0.20x$ by dividing both sides of $0.20x \leq 210$ by 0.20, a positive number. Let's see what happens if we divide both sides of an inequality by a negative number. Consider the inequality $10 < 14$. Divide both 10 and 14 by -2:

$$\frac{10}{-2} = -5 \quad \text{and} \quad \frac{14}{-2} = -7.$$

Because -5 lies to the right of -7 on the number line, -5 is greater than -7:

$$-5 > -7.$$

Notice that the direction of the inequality symbol is reversed:

$$
\begin{array}{c}
10 < 14 \\
\updownarrow \\
-5 > -7
\end{array}
$$
Dividing by -2 changes the direction of the inequality symbol.

In general, **when we multiply or divide both sides of an inequality by a negative number, the direction of the inequality symbol is reversed**. When we reverse the direction of the inequality symbol, we say that we change the *sense* of the inequality.

We can isolate a variable in a linear inequality the same way we can isolate a variable in a linear equation. The following properties are used to create equivalent inequalities:

Properties of Inequalities

Property	The Property in Words	Example
The Addition Property of Inequality If $a < b$, then $a + c < b + c$. If $a < b$, then $a - c < b - c$.	If the same quantity is added to or subtracted from both sides of an inequality, the resulting inequality is equivalent to the original one.	$2x + 3 < 7$ Subtract 3: $2x + 3 - 3 < 7 - 3$ Simplify: $2x < 4$
The Positive Multiplication Property of Inequality If $a < b$ and c is positive, then $ac < bc$. If $a < b$ and c is positive, then $\dfrac{a}{c} < \dfrac{b}{c}$.	If we multiply or divide both sides of an inequality by the same positive quantity, the resulting inequality is equivalent to the original one.	$2x < 4$ Divide by 2: $\dfrac{2x}{2} < \dfrac{4}{2}$ Simplify: $x < 2$
The Negative Multiplication Property of Inequality If $a < b$ and c is negative, then $ac > bc$. If $a < b$ and c is negative, then $\dfrac{a}{c} > \dfrac{b}{c}$.	If we multiply or divide both sides of an inequality by the same negative quantity and reverse the direction of the inequality symbol, the resulting inequality is equivalent to the original one.	$-4x < 20$ Divide by -4 and reverse the sense of the inequality: $\dfrac{-4x}{-4} > \dfrac{20}{-4}$ Simplify: $x > -5$

4 Solve linear inequalities.

Solving Linear Inequalities Involving Only One Property of Inequality If you can solve a linear equation, it is likely that you can solve a linear inequality. Why? The procedure for solving linear inequalities is nearly the same as the procedure for solving linear equations, with one important exception: **When multiplying or dividing by a negative number, reverse the direction of the inequality symbol, changing the sense of the inequality.**

EXAMPLE 2 Solving a Linear Inequality

Solve and graph the solution set on a number line:
$$x + 3 < 8.$$

SOLUTION Our goal is to isolate x. We can do this by using the addition property, subtracting 3 from both sides.

$$x + 3 < 8 \qquad \text{This is the given inequality.}$$
$$x + 3 - 3 < 8 - 3 \qquad \text{Subtract 3 from both sides.}$$
$$x < 5 \qquad \text{Simplify.}$$

The solution set consists of all real numbers that are less than 5. We express this in set-builder notation as

$$\{x \mid x < 5\}. \qquad \text{This is read "the set of all } x \text{ such that } x \text{ is less than 5."}$$

The graph of the solution set is shown as follows:

<DISCOVER FOR YOURSELF>

$\{x \mid x < 3\}$

See graphing answer section.

✔ **CHECK POINT 2** Solve and graph the solution set on a number line:
$$x + 6 < 9.$$

EXAMPLE 3 Solving a Linear Inequality

Solve and graph the solution set on a number line:
$$4x - 1 \geq 3x - 6.$$

SOLUTION Our goal is to isolate all terms involving x on one side and all numerical terms on the other side, exactly as we did when solving equations. Let's begin by using the addition property to isolate variable terms on the left.

$$4x - 1 \geq 3x - 6 \qquad \text{This is the given inequality.}$$
$$4x - 3x - 1 \geq 3x - 3x - 6 \qquad \text{Subtract } 3x \text{ from both sides.}$$
$$x - 1 \geq -6 \qquad \text{Simplify.}$$

Now we isolate the numerical terms on the right. Use the addition property and add 1 to both sides.

$$x - 1 + 1 \geq -6 + 1 \qquad \text{Add 1 to both sides.}$$
$$x \geq -5 \qquad \text{Simplify.}$$

The solution set consists of all real numbers that are greater than or equal to -5. We express this in set-builder notation as

$$\{x \mid x \geq -5\}. \qquad \text{This is read "the set of all } x \text{ such that } x \text{ is greater than or equal to } -5."}$$

The graph of the solution set is shown as follows:

$\{x | x \geq -2\}$
See graphing answer section.

 CHECK POINT 3 Solve and graph the solution set on a number line:

$$8x - 2 \geq 7x - 4.$$

We solved the inequalities in Examples 2 and 3 using the addition property of inequality. Now let's practice using the multiplication property of inequality. Do not forget to reverse the direction of the inequality symbol when multiplying or dividing both sides by a negative number.

EXAMPLE 4   Solving Linear Inequalities

Solve and graph the solution set on a number line: **a.** $\frac{1}{3}x < 5$ **b.** $-3x < 21.$

SOLUTION   In each case, our goal is to isolate x. In the first inequality, this is accomplished by multiplying both sides by 3. In the second inequality, we can do this by dividing both sides by -3.

a. $\frac{1}{3}x < 5$ This is the given inequality.

$3 \cdot \frac{1}{3}x < 3 \cdot 5$ Isolate x by multiplying by 3 on both sides.
                              The symbol $<$ stays the same because we are multiplying both sides
                              by a positive number.

$x < 15$ Simplify.

The solution set is $\{x | x < 15\}$. The graph of the solution set is shown as follows:

b. $-3x < 21$ This is the given inequality.

$\dfrac{-3x}{-3} > \dfrac{21}{-3}$ Isolate x by dividing by -3 on both sides.
                              The symbol $<$ must be reversed because we are dividing both sides
                              by a negative number.

$x > -7$ Simplify.

The solution set is $\{x | x > -7\}$. The graph of the solution set is shown as follows:

 CHECK POINT 4 Solve and graph the solution set on a number line:

See graphing answer section.

a. $\frac{1}{4}x < 2$ $\{x | x < 8\}$ **b.** $-6x < 18.$ $\{x | x > -3\}$

Inequalities Requiring Both the Addition and Multiplication Properties   If an inequality does not contain fractions, it can be solved using the following procedure. Notice, again, how similar this procedure is to the procedure for solving an equation.

> ## SOLVING A LINEAR INEQUALITY
>
> **1.** Simplify the algebraic expression on each side.
>
> **2.** Use the addition property of inequality to collect all the variable terms on one side and all the constant terms on the other side.
>
> **3.** Use the multiplication property of inequality to isolate the variable and solve. Reverse the sense of the inequality when multiplying or dividing both sides by a negative number.
>
> **4.** Express the solution set in set-builder notation and graph the solution set on a number line.

EXAMPLE 5 Solving a Linear Inequality

Solve and graph the solution set on a number line:

$$4y - 7 \geq 5.$$

SOLUTION

Step 1. **Simplify each side.** Because each side is already simplified, we can skip this step.

Step 2. **Collect variable terms on one side and constant terms on the other side.** The variable term, $4y$, is already on the left. We will collect constant terms on the right by adding 7 to both sides.

$$4y - 7 \geq 5 \qquad \text{This is the given inequality.}$$
$$4y - 7 + 7 \geq 5 + 7 \qquad \text{Add 7 to both sides.}$$
$$4y \geq 12 \qquad \text{Simplify.}$$

Step 3. **Isolate the variable and solve.** We isolate the variable, y, by dividing both sides by 4. Because we are dividing by a positive number, we do not reverse the inequality symbol.

$$\frac{4y}{4} \geq \frac{12}{4} \qquad \text{Divide both sides by 4.}$$
$$y \geq 3 \qquad \text{Simplify.}$$

Step 4. **Express the solution set in set-builder notation and graph the set on a number line.** The solution set consists of all real numbers that are greater than or equal to 3, expressed in set-builder notation as $\{y | y \geq 3\}$. The graph of the solution set is shown as follows:

CHECK POINT 5 Solve and graph the solution set on a number line:

$$5y - 3 \geq 17. \quad \{y | y \geq 4\} \quad \text{See graphing answer section.}$$

> **STUDY TIP**
>
> It is possible to perform a partial check for an inequality. Select one number from the solution set. Substitute that number into the original inequality and perform the resulting computations. You should obtain a true statement.

EXAMPLE 6 Solving a Linear Inequality

Solve and graph the solution set on a number line:

$$7x + 15 \geq 13x + 51.$$

STUDY TIP

You can solve

$$7x + 15 \geq 13x + 51$$

by isolating x on the right side. Subtract $7x$ from both sides:

$$7x + 15 - 7x$$
$$\geq 13x + 51 - 7x$$
$$15 \geq 6x + 51.$$

Now subtract 51 from both sides:

$$15 - 51 \geq 6x + 51 - 51$$
$$-36 \geq 6x.$$

Finally, divide both sides by 6:

$$\frac{-36}{6} \geq \frac{6x}{6}$$
$$-6 \geq x.$$

This last inequality means the same thing as

$$x \leq -6.$$

$$\{x | x \geq 1\}$$

See graphing answer section.

SOLUTION

Step 1. **Simplify each side.** Because each side is already simplified, we can skip this step.

Step 2. **Collect variable terms on one side and constant terms on the other side.** We will collect variable terms on the left and constant terms on the right.

$7x + 15 \geq 13x + 51$	This is the given inequality.
$7x + 15 - 13x \geq 13x + 51 - 13x$	Subtract 13x from both sides.
$-6x + 15 \geq 51$	Simplify.
$-6x + 15 - 15 \geq 51 - 15$	Subtract 15 from both sides.
$-6x \geq 36$	Simplify.

Step 3. **Isolate the variable and solve.** We isolate the variable, x, by dividing both sides by -6. Because we are dividing by a negative number, we must reverse the inequality symbol.

$\dfrac{-6x}{-6} \leq \dfrac{36}{-6}$	Divide both sides by −6 and reverse the sense of the inequality.
$x \leq -6$	Simplify.

Step 4. **Express the solution set in set-builder notation and graph the set on a number line.** The solution set consists of all real numbers that are less than or equal to -6, expressed in set-builder notation as $\{x | x \leq -6\}$. The graph of the solution set is shown as follows:

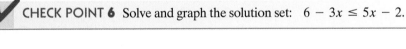

✓ **CHECK POINT 6** Solve and graph the solution set: $6 - 3x \leq 5x - 2$.

EXAMPLE 7 Solving a Linear Inequality

Solve and graph the solution set on a number line:

$$2(x - 3) + 5x \leq 8(x - 1).$$

SOLUTION

Step 1. **Simplify each side.** We use the distributive property to remove parentheses. Then we combine like terms.

$2(x - 3) + 5x \leq 8(x - 1)$	This is the given inequality.
$2x - 6 + 5x \leq 8x - 8$	Use the distributive property.
$7x - 6 \leq 8x - 8$	Add like terms on the left.

Step 2. **Collect variable terms on one side and constant terms on the other side.** We will collect variable terms on the left and constant terms on the right.

$7x - 8x - 6 \leq 8x - 8x - 8$	Subtract 8x from both sides.
$-x - 6 \leq -8$	Simplify.
$-x - 6 + 6 \leq -8 + 6$	Add 6 to both sides.
$-x \leq -2$	Simplify.

Step 3. **Isolate the variable and solve.** To isolate x in $-x \le -2$, we must eliminate the negative sign in front of the x. Because $-x$ means $-1x$, we can do this by multiplying (or dividing) both sides of the inequality by -1. We are multiplying by a negative number. Thus, we must reverse the inequality symbol.

$$(-1)(-x) \ge (-1)(-2)$$ Multiply both sides of $-x \le -2$ by -1 and reverse the sense of the inequality.

$$x \ge 2$$ Simplify.

Step 4. **Express the solution set in set-builder notation and graph the set on a number line.** The solution set consists of all real numbers that are greater than or equal to 2, expressed in set-builder notation as $\{x | x \ge 2\}$. The graph of the solution set is shown as follows:

 CHECK POINT 7 Solve and graph the solution set:

$$2(x - 3) - 1 \le 3(x + 2) - 14. \quad \{x | x \ge 1\}$$

See graphing answer section.

---

5 Identify inequalities with no solution or infinitely many solutions.

Inequalities with Unusual Solution Sets We have seen that some equations have no solution. This is also true for some inequalities. An example of such an inequality is

$$x > x + 1.$$

There is no number that is greater than itself plus 1. This inequality has no solution. Its solution set is $\varnothing$, the empty set.

By contrast, some inequalities are true for all real numbers. An example of such an inequality is

$$x < x + 1.$$

Every real number is less than itself plus 1. The solution set is $\{x | x \text{ is a real number}\}$ or $\mathbb{R}$.

If you attempt to solve an inequality that has no solution, you will eliminate the variable and obtain a false statement, such as $0 > 1$. If you attempt to solve an inequality that is true for all real numbers, you will eliminate the variable and obtain a true statement, such as $0 < 1$.

EXAMPLE 8 Solving a Linear Inequality

Solve: $3(x + 1) > 3x + 5$.

SOLUTION

$$3(x + 1) > 3x + 5$$ This is the given inequality.

$$3x + 3 > 3x + 5$$ Apply the distributive property.

$$3x + 3 - 3x > 3x + 5 - 3x$$ Subtract 3x from both sides.

Keep reading. $3 > 5$ is not the solution. $$3 > 5$$ Simplify.

The original inequality is equivalent to the false statement $3 > 5$, which is false for every value of x. The inequality has no solution. The solution set is $\emptyset$, the empty set. ∎

✔ CHECK POINT **8** Solve: $4(x + 2) > 4x + 15$.   no solution, $\emptyset$

EXAMPLE 9 Solving a Linear Inequality

Solve: $2(x + 5) \leq 5x - 3x + 14$.

SOLUTION

$$2(x + 5) \leq 5x - 3x + 14 \qquad \text{This is the given inequality.}$$

$$2x + 10 \leq 5x - 3x + 14 \qquad \text{Apply the distributive property.}$$

$$2x + 10 \leq 2x + 14 \qquad \text{Combine like terms.}$$

$$2x + 10 - 2x \leq 2x + 14 - 2x \qquad \text{Subtract 2x from both sides.}$$

 Keep reading. $10 \leq 14$ is not the solution.

$$10 \leq 14 \qquad \text{Simplify.}$$

The original inequality is equivalent to the true statement $10 \leq 14$, which is true for every value of x. The solution is the set of all real numbers, written

$$\{x \mid x \text{ is a real number}\} \text{ or } \mathbb{R}.$$ ∎

✔ CHECK POINT **9** Solve: $3(x + 1) \geq 2x + 1 + x$. $\{x \mid x$ is a real number$\}$ or $\mathbb{R}$

---

6 Solve problems using linear inequalities.

Applications As you know, different professors may use different grading systems to determine your final course grade. Some professors require a final examination; others do not. In our next example, a final exam is required *and* it also counts as two grades.

EXAMPLE 10 An Application: Final Course Grade

To earn an A in a course, you must have a final average of at least 90%. On the first four examinations, you have grades of 86%, 88%, 92%, and 84%. If the final examination counts as two grades, what must you get on the final to earn an A in the course?

SOLUTION   We will use our five-step strategy for solving algebraic word problems.

Steps 1 and 2.   **Represent unknown quantities in terms of x.** Let $x =$ your grade on the final examination.

Step 3.   **Write an inequality in x that describes the conditions.** The average of the six grades is found by adding the grades and dividing the sum by 6.

$$\text{Average} = \frac{86 + 88 + 92 + 84 + x + x}{6}$$

Because the final counts as two grades, the x (your grade on the final examination) is added twice. This is also why the sum is divided by 6.

To get an A, your average must be at least 90. This means that your average must be greater than or equal to 90.

| Your average | must be greater than or equal to | 90. |

$$\frac{86 + 88 + 92 + 84 + x + x}{6} \quad \geq \quad 90$$

Step 4. **Solve the inequality and answer the problem's question.**

$$\frac{86 + 88 + 92 + 84 + x + x}{6} \geq 90$$ This is the inequality for the given conditions.

$$\frac{350 + 2x}{6} \geq 90$$ Combine like terms in the numerator.

$$6\left(\frac{350 + 2x}{6}\right) \geq 6(90)$$ Multiply both sides by 6, clearing the fraction.

$$350 + 2x \geq 540$$ Multiply.

$$350 + 2x - 350 \geq 540 - 350$$ Subtract 350 from both sides.

$$2x \geq 190$$ Simplify.

$$\frac{2x}{2} \geq \frac{190}{2}$$ Divide both sides by 2.

$$x \geq 95$$ Simplify.

You must get at least 95% on the final examination to earn an A in the course.

Step 5. **Check.** We can perform a partial check by computing the average with any grade that is at least 95. We will use 96. If you get 96% on the final examination, your average is

$$\frac{86 + 88 + 92 + 84 + 96 + 96}{6} = \frac{542}{6} = 90\frac{1}{3}.$$

Because $90\frac{1}{3} > 90$, you earn an A in the course.

CHECK POINT 10 To earn a B in a course, you must have a final average of at least 80%. On the first three examinations, you have grades of 82%, 74%, and 78%. If the final examination counts as two grades, what must you get on the final to earn a B in the course?   You must get at least 83%.

2.7 EXERCISE SET

Student Solutions Manual · CD/Video · PH Math/Tutor Center · MathXL Tutorials on CD · Math XL MathXL® · MyMathLab MyMathLab · Interactmath.com

Practice Exercises

In Exercises 1–12, graph the solutions of each inequality on a number line.   See graphing answer section.

1. $x > 5$ **2.** $x > -3$

3. $x < -2$ **4.** $x < 0$

5. $x \geq -4$ **6.** $x \geq -6$

7. $x \leq 4.5$ **8.** $x \leq 7.5$

9. $-2 < x \leq 6$ **10.** $-3 \leq x < 6$

11. $-1 < x < 3$ **12.** $-2 \leq x \leq 0$

Describe each graph in Exercises 13–18 using set-builder notation.

13.

$\{x \mid x > -2\}$

14.

$\{x \mid x \leq -6\}$

15.

$\{x \mid x \geq 4\}$

16.

$\{x \mid x < -4\}$

17.

$\{x \mid x \geq 3\}$

18.

$\{x \mid x > 0\}$

Use the addition property of inequality to solve each inequality in Exercises 19–36. Express the solution set in set-builder notation and graph the set on a number line. See graphing answer section.

19. $x - 3 > 4$ $\{x | x > 7\}$

20. $x + 1 < 6$ $\{x | x < 5\}$

21. $x + 4 \leq 10$ $\{x | x \leq 6\}$

22. $x - 5 \geq 2$ $\{x | x \geq 7\}$

23. $y - 2 < 0$ $\{y | y < 2\}$

24. $y + 3 \geq 0$ $\{y | y \geq -3\}$

25. $3x + 4 \leq 2x + 7$ $\{x | x \leq 3\}$

26. $2x + 9 \leq x + 2$ $\{x | x \leq -7\}$

27. $5x - 9 < 4x + 7$ $\{x | x < 16\}$

28. $3x - 8 < 2x + 11$ $\{x | x < 19\}$

29. $7x - 7 > 6x - 3$ $\{x | x > 4\}$

30. $8x - 9 > 7x - 3$ $\{x | x > 6\}$

31. $x - \dfrac{2}{3} > \dfrac{1}{2}$ $\left\{x \middle| x > \dfrac{7}{6}\right\}$

32. $x - \dfrac{1}{3} \geq \dfrac{5}{6}$ $\left\{x \middle| x \geq \dfrac{7}{6}\right\}$

33. $y + \dfrac{7}{8} \leq \dfrac{1}{2}$ $\left\{y \middle| y \leq -\dfrac{3}{8}\right\}$

34. $y + \dfrac{1}{3} \leq \dfrac{3}{4}$ $\left\{y \middle| y \leq \dfrac{5}{12}\right\}$

35. $-15y + 13 > 13 - 16y$ $\{y | y > 0\}$

36. $-12y + 17 > 20 - 13y$ $\{y | y > 3\}$

Use the multiplication property of inequality to solve each inequality in Exercises 37–54. Express the solution set in set-builder notation and graph the set on a number line. See graphing answer section.

37. $\dfrac{1}{2}x < 4$ $\{x | x < 8\}$ **38.** $\dfrac{1}{2}x > 3$ $\{x | x > 6\}$

39. $\dfrac{x}{3} > -2$ $\{x | x > -6\}$ **40.** $\dfrac{x}{4} < -1$ $\{x | x < -4\}$

41. $4x < 20$ $\{x | x < 5\}$ **42.** $6x < 18$ $\{x | x < 3\}$

43. $3x \geq -21$ $\{x | x \geq -7\}$ **44.** $7x \geq -56$ $\{x | x \geq -8\}$

45. $-3x < 15$ $\{x | x > -5\}$ **46.** $-7x > 21$ $\{x | x < -3\}$

47. $-3x \geq 15$ $\{x | x \leq -5\}$ **48.** $-7x \leq 21$ $\{x | x \geq -3\}$

49. $-16x > -48$ $\{x | x < 3\}$ **50.** $-20x > -140$ $\{x | x < 7\}$

51. $-4y \leq \dfrac{1}{2}$ $\left\{y \middle| y \geq -\dfrac{1}{8}\right\}$ **52.** $-2y \leq \dfrac{1}{2}$ $\left\{y \middle| y \geq -\dfrac{1}{4}\right\}$

53. $-x < 4$ $\{x | x > -4\}$ **54.** $-x > -3$ $\{x | x < 3\}$

Use both the addition and multiplication properties of inequality to solve each inequality in Exercises 55–78. Express the solution set in set-builder notation and graph the set on a number line. See graphing answer section.

55. $2x - 3 > 7$ $\{x | x > 5\}$

56. $3x + 2 \leq 14$ $\{x | x \leq 4\}$

57. $3x + 3 < 18$ $\{x | x < 5\}$

58. $8x - 4 > 12$ $\{x | x > 2\}$

59. $3 - 7x \leq 17$ $\{x | x \geq -2\}$

60. $5 - 3x \geq 20$ $\{x | x \leq -5\}$

61. $-2x - 3 < 3$ $\{x | x > -3\}$

62. $-3x + 14 < 5$ $\{x | x > 3\}$

63. $5 - x \leq 1$ $\{x | x \geq 4\}$

64. $3 - x \geq -3$ $\{x | x \leq 6\}$

65. $2x - 5 > -x + 6$ $\left\{x \middle| x > \dfrac{11}{3}\right\}$

66. $6x - 2 \geq 4x + 6$ $\{x | x \geq 4\}$

67. $2y - 5 < 5y - 11$ $\{y | y > 2\}$

68. $4y - 7 > 9y - 2$ $\{y | y < -1\}$

69. $3(2y - 1) < 9$ $\{y | y < 2\}$

70. $4(2y - 1) > 12$ $\{y | y > 2\}$

71. $3(x + 1) - 5 < 2x + 1$ $\{x | x < 3\}$

72. $4(x + 1) + 2 \geq 3x + 6$ $\{x | x \geq 0\}$

73. $8x + 3 > 3(2x + 1) - x + 5$ $\left\{x \middle| x > \dfrac{5}{3}\right\}$

74. $7 - 2(x - 4) < 5(1 - 2x)$ $\left\{x \middle| x < -\dfrac{5}{4}\right\}$

75. $\dfrac{x}{3} - 2 \geq 1$ $\{x | x \geq 9\}$

76. $\dfrac{x}{4} - 3 \geq 1$ $\{x | x \geq 16\}$

77. $1 - \dfrac{x}{2} > 4$ $\{x | x < -6\}$

78. $1 - \dfrac{x}{2} < 5$ $\{x | x > -8\}$

In Exercises 79–88, solve each inequality. Identify inequalities that have no solution, or inequalities that are true for all real numbers.

79. $4x - 4 < 4(x - 5)$ no solution, $\varnothing$

80. $3x - 5 < 3(x - 2)$ no solution, $\varnothing$

81. $x + 3 < x + 7$ $\{x | x$ is a real number$\}$ or $\mathbb{R}$

82. $x + 4 < x + 10$ $\{x | x$ is a real number$\}$ or $\mathbb{R}$

83. $7x \leq 7(x - 2)$ no solution, $\varnothing$

84. $3x + 1 \leq 3(x - 2)$ no solution, $\varnothing$

85. $2(x + 3) > 2x + 1$ $\{x | x \text{ is a real number}\}$ or $\mathbb{R}$

86. $5(x + 4) > 5x + 10$ $\{x | x \text{ is a real number}\}$ or $\mathbb{R}$

87. $5x - 4 \le 4(x - 1)$ $\{x | x \le 0\}$

88. $6x - 3 \le 3(x - 1)$ $\{x | x \le 0\}$

Practice Plus

In Exercises 89–92, use properties of inequality to rewrite each inequality so that x is isolated on one side.

89. $3x + a > b$ $x > \dfrac{b - a}{3}$ **90.** $-2x - a \le b$ $x \ge \dfrac{a + b}{-2}$

91. $y \le mx + b$ and $m < 0$ $\dfrac{y - b}{m} \ge x$

92. $y > mx + b$ and $m > 0$ $\dfrac{y - b}{m} > x$

We know that $|x|$ represents the distance from 0 to x on a number line. In Exercises 93–96, use each sentence to describe all possible locations of x on a number line. Then rewrite the given sentence as an inequality involving $|x|$.

93. The distance from 0 to x on a number line is less than 2.   *x is between −2 and 2; $|x| < 2$*

94. The distance from 0 to x on a number line is less than 3.   *x is between −3 and 3; $|x| < 3$*

95. The distance from 0 to x on a number line is greater than 2.   *x is greater than 2 or less than −2; $|x| > 2$*

96. The distance from 0 to x on a number line is greater than 3.   *x is greater than 3 or less than −3; $|x| > 3$*

Application Exercises

The bar graph shows the percentage of wages full-time workers pay in income tax in eight selected countries. (The percents shown are averages for single earners without children.) Let x represent the percentage of wages workers pay in income tax. In Exercises 97–102, write the name of the country or countries described by the given inequality.

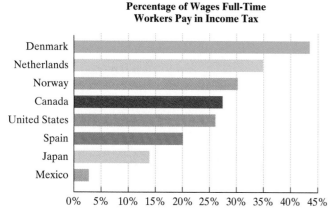

Percentage of Wages Full-Time Workers Pay in Income Tax

Denmark, Netherlands, Norway, Canada, United States, Spain, Japan, Mexico

0% 5% 10% 15% 20% 25% 30% 35% 40% 45%

Source: *The Washington Post*

97. $x \ge 30\%$ Denmark, Netherlands, Norway

98. $x > 30\%$ Denmark, Netherlands

99. $x < 20\%$ Japan, Mexico

100. $x \le 20\%$ Spain, Japan, Mexico

101. $25\% \le x < 40\%$ Nertherlands, Norway, Canada, U.S.

102. $5\% < x \le 25\%$ Spain, Japan

The line graph shows the declining consumption of cigarettes in the United States. The data shown by the graph can be modeled by

$$N = 550 - 9x,$$

where N is the number of cigarettes consumed, in billions, x years after 1988. Use this formula to solve Exercises 103–104.

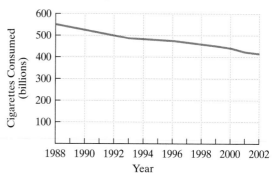

Consumption of Cigarettes in the U.S.

Source: Economic Research Service, USDA

103. Describe how many years after 1988 cigarette consumption will be less than 370 billion cigarettes each year. Which years are included in your description?   20 years; from 2009 onward

104. Describe how many years after 1988 cigarette consumption will be less than 325 billion cigarettes each year. Which years are included in your description?   25 years; from 2014 onward

105. On two examinations, you have grades of 86 and 88. There is an optional final examination, which counts as one grade. You decide to take the final in order to get a course grade of A, meaning a final average of at least 90.

   a. What must you get on the final to earn an A in the course?   at least 96

   b. By taking the final, if you do poorly, you might risk the B that you have in the course based on the first two exam grades. If your final average is less than 80, you will lose your B in the course. Describe the grades on the final that will cause this to happen.   if you get less than 66 on the final

106. On three examinations, you have grades of 88, 78, and 86. There is still a final examination, which counts as one grade.

   a. In order to get an A, your average must be at least 90. If you get 100 on the final, compute your average and determine if an A in the course is possible.   88; An A in the course is not possible.

   b. To earn a B in the course, you must have a final average of at least 80. What must you get on the final to earn a B in the course?   at least 68

107. A car can be rented from Continental Rental for $80 per week plus 25 cents for each mile driven. How many miles can you travel if you can spend at most $400 for the week? can drive up to 1280 miles

108. A car can be rented from Basic Rental for $60 per week plus 50 cents for each mile driven. How many miles can you travel if you can spend at most $600 for the week? can drive up to 1080 miles

109. An elevator at a construction site has a maximum capacity of 3000 pounds. If the elevator operator weighs 245 pounds and each cement bag weighs 95 pounds, how many bags of cement can be safely lifted on the elevator in one trip? up to 29 bags of cement

110. An elevator at a construction site has a maximum capacity of 2800 pounds. If the elevator operator weighs 265 pounds and each cement bag weighs 65 pounds, how many bags of cement can be safely lifted on the elevator in one trip? up to 39 bags of cement

Writing in Mathematics

111. When graphing the solutions of an inequality, what is the difference between an open dot and a closed dot? An open dot indicates an endpoint that is not a solution and a closed dot indicates an endpoint that is a solution.

112. When solving an inequality, when is it necessary to change the direction of the inequality symbol? Give an example. when dividing or multiplying by a negative number; Examples will vary.

113. Describe ways in which solving a linear inequality is similar to solving a linear equation. 113–115. Answers will vary.

114. Describe ways in which solving a linear inequality is different from solving a linear equation.

115. Using current trends, future costs of Medicare can be modeled by $C = 18x + 250$, where x represents the number of years after 2000 and C represents the cost of Medicare, in billions of dollars. Use the formula to write a word problem that can be solved using a linear inequality. Then solve the problem.

Critical Thinking Exercises

116. Which one of the following statements is true? d

   a. The inequality $x - 3 > 0$ is equivalent to $x < 3$.

   b. The statement "x is at most 5" is written $x < 5$.

   c. The inequality $-4x < -20$ is equivalent to $x > -5$.

   d. The statement "the sum of x and 6% of x is at least 80" is written $x + 0.06x \geq 80$.

117. A car can be rented from Basic Rental for $260 per week with no extra charge for mileage. Continental charges $80 per week plus 25 cents for each mile driven to rent the same car. How many miles should be driven in a week to make the rental cost for Basic Rental a better deal than Continental's? more than 720 miles

118. Membership in a fitness club costs $500 yearly plus $1 per hour spent working out. A competing club charges $440 yearly plus $1.75 per hour for use of their equipment. How many hours must a person work out yearly to make membership in the first club cheaper than membership in the second club? more than 80 hours

Technology Exercises

Solve each inequality in Exercises 119–120. Use a calculator to help with the arithmetic.

119. $1.45 - 7.23x > -1.442$ $\{x \mid x < 0.4\}$

120. $126.8 - 9.4y \leq 4.8y + 34.5$ $\{y \mid y \geq 6.5\}$

Review Exercises

121. 8 is 40% of what number? (Section 2.4, Example 8)   20

122. The length of a rectangle exceeds the width by 5 inches. The perimeter is 34 inches. What are the rectangle's dimensions? (Section 2.5, Example 6) length = 11 inches; width = 6 inches

123. Solve and check: $5x + 16 = 3(x + 8)$. (Section 2.3, Example 2) $\{4\}$

GROUP PROJECT

CHAPTER

2

One of the best ways to learn how to *solve* a word problem in algebra is to *design* word problems of your own. Creating a word problem makes you very aware of precisely how much information is needed to solve the problem. You must also focus on the best way to present information to a reader and on how much information to give. As you write your problem, you gain skills that will help you solve problems created by others.

   The group should design five different word problems that can be solved using an algebraic equation. All of the problems should be on different topics. For example, the group should not have more than one problem on finding a number. The group should turn in both the problems and their algebraic solutions.

Additional group projects can be found in the *Instructor's Resource Manual.*

CHAPTER 2 SUMMARY

Definitions and Concepts	Examples

Section 2.1 The Addition Property of Equality

A linear equation in one variable can be written in the form $ax + b = c$, where a is not zero.	$3x + 7 = 9$ is a linear equation.
Equivalent equations have the same solution.	$2x - 4 = 6$, $2x = 10$, and $x = 5$ are equivalent equations.

The Addition Property of Equality

Adding the same number (or algebraic expression) to or subtracting the same number (or algebraic expression) from both sides of an equation does not change its solution.

$$\bullet \quad x - 3 = 8$$
$$x - 3 + 3 = 8 + 3$$
$$x = 11$$

$$\bullet \quad x + 4 = 10$$
$$x + 4 - 4 = 10 - 4$$
$$x = 6$$

Section 2.2 The Multiplication Property of Equality

The Multiplication Property of Equality

Multiplying both sides or dividing both sides of an equation by the same nonzero real number (or algebraic expression) does not change the solution.

$$\bullet \quad \frac{x}{-5} = 6$$
$$-5\left(\frac{x}{-5}\right) = -5(6)$$
$$x = -30$$

$$\bullet \quad -50 = -5y$$
$$\frac{-50}{-5} = \frac{-5y}{-5}$$
$$10 = y$$

Equations and Coefficients of −1

If $-x = c$, multiply both sides by -1 to solve for x. The solution is the additive inverse of c.

$$-x = -12$$
$$(-1)(-x) = (-1)(-12)$$
$$x = 12$$

Using the Addition and Multiplication Properties

If an equation does not contain fractions,

• Use the addition property to isolate the variable term.
• Use the multiplication property to isolate the variable.

$$-2x - 5 = 11$$
$$-2x - 5 + 5 = 11 + 5$$
$$-2x = 16$$
$$\frac{-2x}{-2} = \frac{16}{-2}$$
$$x = -8$$

Section 2.3 Solving Linear Equations

Solving a Linear Equation

1. Simplify each side.

Solve:
$$7 - 4(x - 1) = x + 1.$$
$$7 - 4x + 4 = x + 1$$
$$-4x + 11 = x + 1$$

2. Collect all the variable terms on one side and all the constant terms on the other side.

$$-4x - x + 11 = x - x + 1$$
$$-5x + 11 = 1$$
$$-5x + 11 - 11 = 1 - 11$$
$$-5x = -10$$

Definitions and Concepts	Examples

Section 2.3 Solving Linear Equations (continued)

3. Isolate the variable and solve. (If the variable is eliminated and a false statement results, the inconsistent equation has no solution. If a true statement results, all real numbers are solutions of the identity.)

$$\frac{-5x}{-5} = \frac{-10}{-5}$$
$$x = 2$$

---

4. Check the proposed solution in the original equation.

$$7 - 4(x - 1) = x + 1$$
$$7 - 4(2 - 1) \stackrel{?}{=} 2 + 1$$
$$7 - 4(1) \stackrel{?}{=} 2 + 1$$
$$7 - 4 \stackrel{?}{=} 2 + 1$$
$$3 = 3, \text{ true}$$

The solution is 2, or the solution set is $\{2\}$.

Equations Containing Fractions

Multiply both sides (all terms) by the least common denominator. This clears the equation of fractions.

$$\frac{x}{5} + \frac{1}{2} = \frac{x}{2} - 1$$
$$10\left(\frac{x}{5} + \frac{1}{2}\right) = 10\left(\frac{x}{2} - 1\right)$$
$$10 \cdot \frac{x}{5} + 10 \cdot \frac{1}{2} = 10 \cdot \frac{x}{2} - 10 \cdot 1$$
$$2x + 5 = 5x - 10$$
$$-3x = -15$$
$$x = 5$$

The solution is 5, or the solution set is $\{5\}$.

Types of Equations

An equation that is true for all real numbers, $\mathbb{R}$, is called an identity. When solving an identity, the variable is eliminated and a true statement, such as $3 = 3$, results. An equation that is not true for even one real number is called an inconsistent equation. A false statement, such as $3 = 7$, results when solving such an equation, whose solution set is $\emptyset$, the empty set. A conditional equation is not an identity, but is true for at least one real number.

Solve:
$$4x + 5 = 4(x + 2)$$
$$4x + 5 = 4x + 8$$
$$5 = 8, \quad \text{false}$$

The inconsistent equation has no solution: $\emptyset$.

Solve:
$$5x - 4 = 5(x + 1) - 9$$
$$5x - 4 = 5x + 5 - 9$$
$$5x - 4 = 5x - 4$$
$$-4 = -4, \quad \text{true}$$

All real number satisfy the identity: $\mathbb{R}$.

Section 2.4 Formulas and Percents

To solve a formula for one of its variables, use the steps for solving a linear equation and isolate the specified variable on one side of the equation.

Solve for l:
$$w = \frac{P - 2l}{2}$$
$$2w = 2\left(\frac{P - 2l}{2}\right)$$
$$2w = P - 2l$$
$$2w - P = P - P - 2l$$
$$2w - P = -2l$$
$$\frac{2w - P}{-2} = \frac{-2l}{-2}$$
$$\frac{2w - P}{-2} = l$$

Definitions and Concepts	Examples

Section 2.4 Formulas and Percents (continued)

The word *percent* means *per hundred*. The symbol % denotes percent.	$47\% = \dfrac{47}{100} \quad 3\% = \dfrac{3}{100}$
To express a decimal as a percent, move the decimal point two places to the right and add a percent sign.	$0.37 = 37\%$ $0.006 = 0.6\%$
To express a percent as a decimal, move the decimal point two places to the left and remove the percent sign.	$250\% = 250\% = 2.5$ $4\% = 04.\% = 0.04$

A Formula Involving Percent

A is P percent of B.

$$A = P \cdot B$$

In the formula $A = PB$, P is expressed as a decimal.

• What is 5% of 20?

$$A = 0.05 \cdot 20$$
$$A = 1$$

Thus, 1 is 5% of 20.

• 6 is 30% of what?

$$6 = 0.3 \cdot B$$
$$\frac{6}{0.3} = B$$
$$20 = B$$

Thus, 6 is 30% of 20.

• 33 is what percent of 75?

$$33 = P \cdot 75$$
$$\frac{33}{75} = P$$
$$P = 0.44 = 44\%$$

Thus, 33 is 44% of 75.

Section 2.5 An Introduction to Problem Solving

Definitions and Concepts	Examples
Strategy for Solving Word Problems	The length of a rectangle exceeds the width by 3 inches. The perimeter is 26 inches. What are the rectangle's dimensions?
Step 1 Let x represent one of the quantities.	Let x = the width.
Step 2 Represent other quantities in terms of x.	$x + 3$ = the length
Step 3 Write an equation that describes the conditions.	Twice length plus twice width is perimeter. $2(x + 3) + 2x = 26$

Definitions and Concepts	Examples

Section 2.5 An Introduction to Problem Solving (continued)

Step 4 Solve the equation and answer the question.

$$2(x + 3) + 2x = 26$$
$$2x + 6 + 2x = 26$$
$$4x + 6 = 26$$
$$4x = 20$$
$$x = 5$$

The width (x) is 5 inches and the length ($x + 3$) is $5 + 3$, or 8 inches.

Step 5 Check the proposed solution in the original wording of the problem.

$$\text{Perimeter} = 2(5 \text{ in.}) + 2(8 \text{ in.})$$
$$= 10 \text{ in.} + 16 \text{ in.} = 26 \text{ in.}$$

This checks with the given perimeter.

Section 2.6 Problem Solving in Geometry

Solving geometry problems often requires using basic geometric formulas. Formulas for perimeter, area, circumference, and volume are given in Tables 2.3 (page 149), 2.4 (page 150), and 2.5 (page 152) in Section 2.6.

A sailboat's triangular sail has an area of 24 ft^2 and a base of 8 ft. Find its height.

$$A = \frac{1}{2}bh$$
$$24 = \frac{1}{2}(8)h$$
$$24 = 4h$$
$$6 = h$$

The sail's height is 6 ft.

The sum of the measures of the three angles of any triangle is 180°.

In a triangle, the first angle measures 3 times the second and the third measures 40° less than the second. Find each angle's measure.

$$\text{Second angle} = x$$
$$\text{First angle} = 3x$$
$$\text{Third angle} = x - 40$$

Sum of measures is 180°.
$$x + 3x + (x - 40) = 180$$
$$5x - 40 = 180$$
$$5x = 220$$
$$x = 44$$

The angles measure $x = 44$, $3x = 3 \cdot 44 = 132$, and $x - 40 = 44 - 40 = 4$.
The angles measure 44°, 132°, and 4°.

Two complementary angles have measures whose sum is 90°. Two supplementary angles have measures whose sum is 180°. If an angle measures x, its complement measures $90 - x$, and its supplement measures $180 - x$.

An angle measures five times its complement. Find the angle's measure.

$$x = \text{angle's measure}$$
$$90 - x = \text{measure of complement}$$
$$x = 5(90 - x)$$
$$x = 450 - 5x$$
$$6x = 450$$
$$x = 75$$

The angle measures 75°.

Definitions and Concepts	Examples

Section 2.7 Solving Linear Inequalities

A linear inequality in one variable can be written in one of these forms:

$$ax + b < c \qquad ax + b \leq c$$
$$ax + b > c \qquad ax + b \geq c$$

where a is not 0.

$3x + 6 > 12$ is a linear inequality.

Set-Builder Notation and Graphs

$\{x \mid a < x < b\}$

$\{x \mid a \leq x \leq b\}$

$\{x \mid a \leq x < b\}$

$\{x \mid x > b\}$

$\{x \mid x \leq a\}$

• Graph the solutions of $x < 4$.

$-3\ -2\ -1\ \ 0\ \ 1\ \ 2\ \ 3\ \ 4\ \ 5$

• Graph the solutions of $-2 < x \leq 1$.

$-4\ -3\ -2\ -1\ \ 0\ \ 1\ \ 2\ \ 3\ \ 4$

The Addition Property of Inequality

Adding the same number to or subtracting the same number from both sides of an inequality does not change the solutions.

$$x + 3 < 8$$
$$x + 3 - 3 < 8 - 3$$
$$x < 5$$

The Positive Multiplication Property of Inequality

Multiplying or dividing both sides of an inequality by the same positive number does not change the solutions.

$$\frac{x}{6} \geq 5$$
$$6 \cdot \frac{x}{6} \geq 6 \cdot 5$$
$$x \geq 30$$

The Negative Multiplication Property of Inequality

Multiplying or dividing both sides of an inequality by the same negative number and reversing the direction of the inequality sign does not change the solutions.

$$-3x \leq 12$$
$$\frac{-3x}{-3} \geq \frac{12}{-3}$$
$$x \geq -4$$

Solving Linear Inequalities

Use the procedure for solving linear equations. When multiplying or dividing by a negative number, reverse the direction of the inequality symbol. Express the solution set in set-builder notation and graph the set on a number line. If the variable is eliminated and a false statement results, the inequality has no solution. The solution set is ∅, the empty set. If a true statement results, the solution is the set of all real numbers: $\{x \mid x \text{ is a real number}\}$ or $\mathbb{R}$.

Solve:
$$x + 4 \geq 6x - 16$$
$$x + 4 - 6x \geq 6x - 16 - 6x$$
$$-5x + 4 \geq -16$$
$$-5x + 4 - 4 \geq -16 - 4$$
$$-5x \geq -20$$
$$\frac{-5x}{-5} \leq \frac{-20}{-5}$$
$$x \leq 4$$
$$\{x \mid x \leq 4\}$$

$-3\ -2\ -1\ \ 0\ \ 1\ \ 2\ \ 3\ \ 4\ \ 5$

CHAPTER 2 REVIEW EXERCISES

2.1 *Solve each equation in Exercises 1–5 using the addition property of equality. Be sure to check proposed solutions.*

1. $x - 10 = 22$ $\{32\}$

2. $-14 = y + 8$ $\{-22\}$

3. $7z - 3 = 6z + 9$ $\{12\}$

4. $4(x + 3) = 3x - 10$ $\{-22\}$

5. $6x - 3x - 9 + 1 = -5x + 7x - 3$ $\{5\}$

2.2 *Solve each equation in Exercises 6–13 using the multiplication property of equality. Be sure to check proposed solutions.*

6. $\dfrac{x}{8} = 10$ $\{80\}$

7. $\dfrac{y}{-8} = 7$ $\{-56\}$

8. $7z = 77$ $\{11\}$

9. $-36 = -9y$ $\{4\}$

10. $\dfrac{3}{5}x = -9$ $\{-15\}$

11. $30 = -\dfrac{5}{2}y$ $\{-12\}$

12. $-x = 25$ $\{-25\}$

13. $\dfrac{-x}{10} = -1$ $\{10\}$

Solve each equation in Exercises 14–18 using both the addition and multiplication properties of equality. Check proposed solutions.

14. $4x + 9 = 33$ $\{6\}$

15. $-3y - 2 = 13$ $\{-5\}$

16. $5z + 20 = 3z$ $\{-10\}$

17. $5x - 3 = x + 5$ $\{2\}$

18. $3 - 2x = 9 - 8x$ $\{1\}$

19. The formula $F = 1.2x + 21.6$ models the average family income, F, in thousands of dollars, for Puerto Ricans x years after 1990. How many years after 1990 is the average family income expected to reach $40.8 thousand? In which year is this expected to occur? 16 years; 2006

2.3 *Solve and check each equation in Exercises 20–28.*

20. $5x + 9 - 7x + 6 = x + 18$ $\{-1\}$

21. $3(x + 4) = 5x - 12$ $\{12\}$

22. $1 - 2(6 - y) = 3y + 2$ $\{-13\}$

23. $2(x - 4) + 3(x + 5) = 2x - 2$ $\{-3\}$

24. $-2(y - 4) - (3y - 2) = -2 - (6y - 2)$ $\{-10\}$

25. $\dfrac{2x}{3} = \dfrac{x}{6} + 1$ $\{2\}$

26. $\dfrac{x}{2} - \dfrac{1}{10} = \dfrac{x}{5} + \dfrac{1}{2}$ $\{2\}$

27. $3(8x - 1) = 6(5 + 4x)$ no solution or $\varnothing$

28. $4(2x - 3) + 4 = 8x - 8$ all real numbers or $\mathbb{R}$

29. The optimum heart rate that a person should achieve during exercise for the exercise to be most beneficial is modeled by $r = 0.6(220 - a)$, where a represents a person's age and r represents that person's optimum heart rate, in beats per minute. If the optimum heart rate is 120 beats per minute, how old is that person? 20 years old

2.4 *In Exercises 30–34, solve each formula for the specified variable.*

30. $I = Pr$ for r $r = \dfrac{I}{P}$

31. $V = \dfrac{1}{3}Bh$ for h $h = \dfrac{3V}{B}$

32. $P = 2l + 2w$ for w $w = \dfrac{P - 2l}{2}$

33. $A = \dfrac{B + C}{2}$ for B $B = 2A - C$

34. $T = D + pm$ for m $m = \dfrac{T - D}{p}$

In Exercises 35–36, express each decimal as a percent.

35. 0.72 72%

36. 0.0035 0.35%

In Exercises 37–39, express each percent as a decimal.

37. 65% 0.65

38. 150% 1.5

39. 3% 0.03

40. What is 8% of 120? 9.6

41. 90 is 45% of what? 200

42. 36 is what percent of 75? 48%

43. If 6 is increased to 12, the increase is what percent of the original number? 100%

44. If 5 is decreased to 3, the decrease is what percent of the original number? 40%

45. A college that had 40 students for each lecture course increased the number to 45 students. What is the percent increase in the number of students in a lecture course? 12.5%

46. Consider the following statement:

> My portfolio fell 10% last year, but then it rose 10% this year, so at least I recouped my losses.

Is this statement true? In particular, suppose you invested $10,000 in the stock market last year. How much money would be left in your portfolio with a 10% fall and then a 10% rise? If there is a loss, what is the percent decrease, to the nearest tenth of a percent, in your portfolio? no; There is a 1% decrease.

47. The radius is one of two bones that connect the elbow and the wrist. The formula $r = \dfrac{h}{7}$ models the length of a woman's radius, r, in inches, and her height, h, in inches.

 a. Solve the formula for h. $h = 7r$

 b. Use the formula in part (a) to find a woman's height if her radius is 9 inches long. 5 feet 3 inches

48. Every day, the average U.S. household uses 91 gallons of water flushing toilets. The circle graph on the next page shows that this represents 26% of the total number of gallons of water used per day. How many gallons of water does the average U.S. household use per day?   350 gallons

Where U.S. Households Use Water

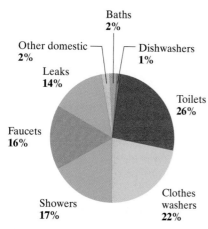

Source: American Water Works Association

2.5 *In Exercises 49–56, use the five-step strategy to solve each problem.*

49. Six times a number, decreased by 20, is four times the number. Find the number.   10

50. On average, the number of unhealthy air days per year in Los Angeles exceeds three times that of New York City by 48 days. If Los Angeles and New York combined have 268 unhealthy air days per year, determine the number of unhealthy days for the two cities.
(*Source*: Environmental Protection Agency)   New York: 55 days; Los Angeles: 213 days

51. Two pages that face each other in a book have 93 as the sum of their page numbers. What are the page numbers?   pages 46 and 47

52. The two female artists in the United States with the most platinum albums are Barbra Streisand followed by Madonna. (A platinum album represents one album sold per 266 people.) The number of platinum albums by these two singers form consecutive odd integers. Combined, they have 96 platinum albums. Determine the number of platinum albums by Streisand and the number of platinum albums by Madonna.   Streisand: 49 albums; Madonna: 47 albums

53. In 2003, the average weekly salary for workers in the United States was $612. If this amount is increasing by $15 yearly, in how many years after 2003 will the average salary reach $747? In which year will that be?   9 years; 2012

54. A bank's total monthly charge for a checking account is $6 plus $0.05 per check. If your total monthly charge is $6.90, how many checks did you write during that month?   18 checks

55. A rectangular field is three times as long as it is wide. If the perimeter of the field is 400 yards, what are the field's dimensions?   length = 150 yards; width = 50 yards

56. After a 25% reduction, you purchase a table for $180. What was the table's price before the reduction? $240

2.6 *Use a formula for area to find the area of each figure in Exercises 57–59.*

57.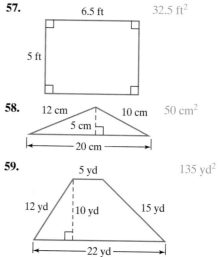
   32.5 ft²

58. 50 cm²

59. 135 yd²

60. Find the circumference and the area of a circle with a diameter of 20 meters. Round answers to the nearest whole number.
circumference = 20π m ≈ 63 m; area = 100π m² ≈ 314 m²

61. A sailboat has a triangular sail with an area of 42 square feet and a base that measures 14 feet. Find the height of the sail.   6 feet

62. A rectangular kitchen floor measures 12 feet by 15 feet. A stove on the floor has a rectangular base measuring 3 feet by 4 feet, and a refrigerator covers a rectangular area of the floor measuring 3 feet by 4 feet. How many square feet of tile will be needed to cover the kitchen floor not counting the area used by the stove and the refrigerator?   156 ft²

63. A yard that is to be covered with mats of grass is shaped like a trapezoid. The bases are 80 feet and 100 feet, and the height is 60 feet. What is the cost of putting the grass mats on the yard if the landscaper charges $0.35 per square foot? $1890

64. Which one of the following is a better buy: a medium pizza with a 14-inch diameter for $6.00 or two small pizzas, each with an 8-inch diameter, for $6.00?   medium pizza

Use a formula for volume to find the volume of each figure in Exercises 65–67. Where applicable, express answers in terms of π. Then round to the nearest whole number.

65. 60 cm³

66. 128π yd³ ≈ 402 yd³

67.

288π m$^3 \approx 905$ m^3

6 m

68. A train is being loaded with freight containers. Each box is 8 meters long, 4 meters wide, and 3 meters high. If there are 50 freight containers, how much space is needed? 4800 m^3

69. A cylindrical fish tank has a diameter of 6 feet and a height of 3 feet. How many tropical fish can be put in the tank if each fish needs 5 cubic feet of water? 16 fish

70. Find the measure of each angle of the triangle shown in the figure. $30°, 60°, 90°$

C

$3x$

$2x$ x

A B

71. In a triangle, the measure of the first angle is 15° more than twice the measure of the second angle. The measure of the third angle exceeds that of the second angle by 25°. What is the measure of each angle? $85°, 35°, 60°$

72. Find the measure of the complement of a 57° angle. $33°$

73. Find the measure of the supplement of a 75° angle. $105°$

74. How many degrees are there in an angle that measures 25° more than the measure of its complement? $57.5°$

75. The measure of the supplement of an angle is 45° less than four times the measure of the angle. Find the measure of the angle and its supplement. $45°$ and $135°$

2.7 *In Exercises 76–77, graph the solutions of each inequality on a number line.* See graphing answer section.

76. $x < -1$ **77.** $-2 < x \le 4$

Describe each graph in Exercises 78–79 using set-builder notation.

78.

$-3 \; -2 \; -1 \quad 0 \quad 1 \quad 2 \quad 3 \quad 4 \quad 5 \quad 6 \quad 7$

$\{x \mid x > 4\}$

79.

$-5 \; -4 \; -3 \; -2 \; -1 \quad 0 \quad 1 \quad 2 \quad 3 \quad 4 \quad 5$

$\{x \mid x \le -3\}$

Solve each inequality in Exercises 80–87. Express the solution set in set-builder notation and graph the set on a number line. If the inequality has no solution or is true for all real numbers, so state. It is not necessary to graph solution sets for these inequalities. See graphing answer section.

80. $2x - 5 < 3$ $\{x \mid x < 4\}$

81. $\dfrac{x}{2} > -4$ $\{x \mid x > -8\}$

82. $3 - 5x \le 18$ $\{x \mid x \ge -3\}$

83. $4x + 6 < 5x$ $\{x \mid x > 6\}$

84. $6x - 10 \ge 2(x + 3)$ $\{x \mid x \ge 4\}$

85. $4x + 3(2x - 7) \le x - 3$ $\{x \mid x \le 2\}$

86. $2(2x + 4) > 4(x + 2) - 6$ $\{x \mid x \text{ is a real number}\}$ or $\mathbb{R}$

87. $-2(x - 4) \le 3x + 1 - 5x$ no solution, $\varnothing$

88. To pass a course, a student must have an average on three examinations of at least 60. If a student scores 42 and 74 on the first two tests, what must be earned on the third test to pass the course? at least 64

89. A long distance telephone service charges 10¢ for the first minute and 5¢ for each minute thereafter. The cost, C, in cents, for a call lasting x minutes is modeled by the formula

$$C = 10 + 5(x - 1).$$

How many minutes can you talk on the phone if you do not want the cost to exceed $5, or 500¢?
no more than 99 minutes

CHAPTER 2 TEST

Remember to use your Chapter Test Prep Video CD to see the worked-out solutions to the test questions you want to review.

In Exercises 1–6, solve each equation.

1. $4x - 5 = 13$ $\left\{\frac{9}{2}\right\}$

2. $12x + 4 = 7x - 21$ $\{-5\}$

3. $8 - 5(x - 2) = x + 26$ $\left\{-\frac{4}{3}\right\}$

4. $3(2y - 4) = 9 - 3(y + 1)$ $\{2\}$

5. $\dfrac{3}{4}x = -15$ $\{-20\}$

6. $\dfrac{x}{10} + \dfrac{1}{3} = \dfrac{x}{5} + \dfrac{1}{2}$ $\left\{-\frac{5}{3}\right\}$

7. The formula $N = 2.4x + 180$ models U.S. population, N, in millions x years after 1960. How many years after 1960 is the U.S. population expected to reach 324 million? In which year is this expected to occur?   60 years; 2020

In Exercises 8–9, solve each formula for the specified variable.

8. $V = \pi r^2 h$ for h $h = \dfrac{V}{\pi r^2}$

9. $l = \dfrac{P - 2w}{2}$ for w $w = \dfrac{P - 2l}{2}$

10. What is 6% of 140? 8.4

11. 120 is 80% of what? 150

12. 12 is what percent of 240? 5%

In Exercises 13–17, solve each problem.

13. The product of 5 and a number, decreased by 9, is 306. What is the number? 63

14. In New York City, a fitness trainer earns $22,870 more per year than a preschool teacher. The yearly average salaries for fitness trainers and preschool teachers combined are $79,030. Determine the average yearly salary of a fitness trainer and a preschool teacher in New York City. (*Source: Time*, April 14, 2003) fitness trainer: $50,950; teacher: $28,080

15. A long-distance telephone plan has a monthly fee of $15.00 and a rate of $0.05 per minute. How many minutes can you chat long distance in a month for a total cost, including the $15.00, of $45.00? 600 minutes

16. A rectangular field is twice as long as it is wide. If the perimeter of the field is 450 yards, what are the field's dimensions? length = 150 yards; width = 75 yards

17. After a 20% reduction, you purchase a new Stephen King novel for $28. What was the book's price before the reduction? $35

In Exercises 18–19, find the area of each figure.

18.

517 m²

26 m 22 m 41 m

47 m

19.

30 in. 525 in.²

15 in.

40 in.

In Exercises 20–21, find the volume of each figure. Where applicable, express answers in terms of π. Then round to the nearest whole number.

20.

18 in.³

3 in.

2 in.

3 in.

21.

5 cm

7 cm

175π cm³ ≈ 550 cm³

22. What will it cost to cover a rectangular floor measuring 40 feet by 50 feet with square tiles that measure 2 feet on each side if a package of 10 tiles costs $13 per package? $650

23. A sailboat has a triangular sail with an area of 56 square feet and a base that measures 8 feet. Find the height of the sail. 14 feet

24. In a triangle, the measure of the first angle is three times that of the second angle. The measure of the third angle is 30° less than the measure of the second angle. What is the measure of each angle? 126°, 42°, 12°

25. How many degrees are there in an angle that measures 16° more than the measure of its complement? 53°

In Exercises 26–27, graph the solutions of each inequality on a number line. See graphing answer section.

26. $x > -2$

27. $-4 \leq x < 1$

28. Use set-builder notation to describe the following graph.

-5 -4 -3 -2 -1 0 1 2 3 4 5

$\{x | x \leq -1\}$

Solve each inequality in Exercises 29–31. Express the solution set in set-builder notation and graph the set on a number line. See graphing answer section.

29. $\dfrac{x}{2} < -3$ $\{x | x < -6\}$

30. $6 - 9x \geq 33$ $\{x | x \leq -3\}$

31. $4x - 2 > 2(x + 6)$ $\{x | x > 7\}$

32. A student has grades on three examinations of 76, 80, and 72. What must the student earn on a fourth examination to have an average of at least 80? at least 92

33. The length of a rectangle is 20 inches. For what widths is the perimeter greater than 56 inches? widths greater than 8 inches

Note to Instructor: The Chapter 2 Test file in the TestGen program provides algorithms specifically matched to these problems so you can easily replicate this chapter test for additional practice or assessment purposes.

CUMULATIVE REVIEW EXERCISES (CHAPTERS 1–2)

In Exercises 1–3, perform the indicated operation or operations.

1. $-8 - (12 - 16)$ -4 **2.** $(-3)(-2) + (-2)(4)$ -2

3. $(8 - 10)^3(7 - 11)^2$ -128

4. Simplify: $2 - 5[x + 3(x + 7)]$. $-103 - 20x$

5. List all the rational numbers in this set:

$$\left\{-4, -\frac{1}{3}, 0, \sqrt{2}, \sqrt{4}, \frac{\pi}{2}, 1063\right\}.$$ $-4, -\frac{1}{3}, 0, \sqrt{4}, 1063$

6. Plot $(-2, -1)$ in a rectangular coordinate system. In which quadrant does the point lie?   III; See graphing answer section.

7. Insert either $<$ or $>$ in the box to make a true statement:

$$-10,000 \,\square\, -2. \quad <$$

8. Use the distributive property to rewrite without parentheses:

$$6(4x - 1 - 5y). \quad 24x - 6 - 30y$$

The graph shows the unemployment rate in the United States from 1990 through mid-2003. Use the information in the graph to solve Exercises 9–10.

U.S. Unemployment Rate

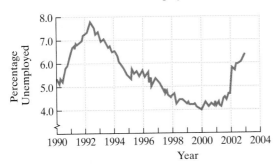

Source: Bureau of Labor Statistics

9. For the period shown, in which year was the unemployment rate at a minimum? What percentage of the work force was unemployed in that year?   2000; 4%

10. For the period shown, during which year did the unemployment rate reach a maximum? Estimate the percentage of the work force unemployed, to the nearest tenth of a percent, at that time.   1992; 7.8%

In Exercises 11–12, solve each equation.

11. $5 - 6(x + 2) = x - 14$ $\{1\}$

12. $\dfrac{x}{5} - 2 = \dfrac{x}{3}$ $\{-15\}$

13. Solve for A: $V = \dfrac{1}{3}Ah$. $A = \dfrac{3V}{h}$

14. 48 is 30% of what?   160

15. The length of a rectangular parking lot is 10 yards less than twice its width. If the perimeter of the lot is 400 yards, what are its dimensions?   length = 130 yards; width = 70 yards

16. A gas station owner makes a profit of 40 cents per gallon of gasoline sold. How many gallons of gasoline must be sold in a year to make a profit of $30,000 from gasoline sales?   75,000 gallons

17. Graph the solution set of $-2 < x \le 3$ on a number line.   See graphing answer section.

Solve each inequality in Exercises 18–19. Express the solution set in set-builder notation and graph the set on a number line.
See graphing answer section.

18. $3 - 3x > 12$ $\{x \mid x < -3\}$

19. $5 - 2(3 - x) \le 2(2x + 5) + 1$ $\{x \mid x \ge -6\}$

20. You take a summer job selling medical supplies. You are paid $600 per month plus 4% of the sales price of all the supplies you sell. If you want to earn more than $2500 per month, what value of medical supplies must you sell?   more than $47,500

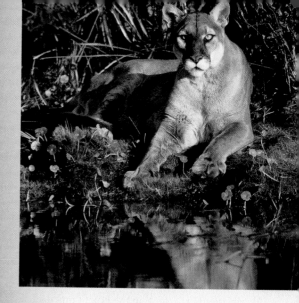

n the United States, the Department of the Interior classifies species as endangered or threatened. Endangered species are those in danger of extinction. Threatened species are those likely to become endangered species within the foreseeable future. The formula

$$E = 10x + 166$$

approximately describes the number of endangered animal species in the United States, represented by E, x years after 1980. This equation in two variables tells of ten new animal species per year classified by the federal government as endangered.

In this chapter, you will learn graphing and modeling methods for situations involving two variables. With these methods, you will be able to create formulas that model the data of your world.

SECTION 3.1

Objectives

1 Determine whether an ordered pair is a solution of an equation.

2 Find solutions of an equation in two variables.

3 Use point plotting to graph linear equations.

4 Use point plotting to graph other kinds of equations.

5 Use graphs of linear equations to solve problems.

GRAPHING EQUATIONS IN TWO VARIABLES

A picture, as they say, is worth a thousand words. Have you seen pictures of gas-guzzling cars from the 1950s, with their huge fins and overstated designs? The worst year for automobile fuel efficiency was 1958, when U.S. cars averaged a dismal 12.4 miles per gallon.

There is a formula that models fuel efficiency of U.S. cars over time. The formula is

$$y = 0.0075x^2 - 0.2672x + 14.8.$$

The variable x represents the number of years after 1940. The variable y represents the average number of miles per gallon for U.S. automobiles. Looking at the formula does not make it obvious that 1958, 18 years after 1940, was the worst year for fuel efficiency. However, if we could somehow make a picture of the formula, such as the one shown in Figure 3.1, the lowest point on the picture would reveal approximately 1958 as the year in which gas-guzzling cars averaged less than 13 miles per gallon. The shape of the graph also shows decreasing fuel efficiency from 1940 through 1958 and increasing fuel efficiency after 1958. In this chapter, we will be making pictures of equations. We can use these pictures to visualize the behavior of the variables in the equation.

FIGURE 3.1 A picture of a formula

Solutions of Equations The rectangular coordinate system allows us to visualize relationships between two variables by connecting any equation in two variables with a geometric figure. Consider, for example, the following equation in two variables:

$$x + y = 10.$$

The sum of two numbers, x and y, is 10.

Many pairs of numbers fit the description in the voice balloon, such as $x = 1$ and $y = 9$, or $x = 3$ and $y = 7$. The phrase "$x = 1$ and $y = 9$" is abbreviated using the ordered pair $(1, 9)$. Similarly, the phrase "$x = 3$ and $y = 7$" is abbreviated using the ordered pair $(3, 7)$.

A **solution of an equation in two variables**, x and y, is an ordered pair of real numbers with the following property: When the x-coordinate is substituted for x and the y-coordinate is substituted for y in the equation, we obtain a true statement. For example, $(1, 9)$ is a solution of the equation $x + y = 10$. When 1 is substituted for x and 9 is substituted for y, we obtain the true statement $1 + 9 = 10$, or $10 = 10$. Because there

1 Determine whether an ordered pair is a solution of an equation.

are infinitely many pairs of numbers that have a sum of 10, the equation $x + y = 10$ has infinitely many solutions. Each ordered-pair solution is said to **satisfy** the equation. Thus, $(1, 9)$ satisfies the equation $x + y = 10$.

EXAMPLE 1 Deciding Whether an Ordered Pair Satisfies an Equation

Determine whether each ordered pair is a solution of the equation

$$x - 4y = 14:$$

a. $(2, -3)$ **b.** $(12, 1)$.

SOLUTION

a. To determine whether $(2, -3)$ is a solution of the equation, we substitute 2 for x and -3 for y.

$x - 4y = 14$	This is the given equation.
$2 - 4(-3) \stackrel{?}{=} 14$	Substitute 2 for x and −3 for y.
$2 - (-12) \stackrel{?}{=} 14$	Multiply: 4(−3) = −12.
$14 = 14$ This statement is true.	Subtract: 2 − (−12) = 2 + 12 = 14.

Because we obtain a true statement, we conclude that $(2, -3)$ is a solution of the equation $x - 4y = 14$. Thus, $(2, -3)$ satisfies the equation.

b. To determine whether $(12, 1)$ is a solution of the equation, we substitute 12 for x and 1 for y.

$x - 4y = 14$	This is the given equation.
$12 - 4(1) \stackrel{?}{=} 14$	Substitute 12 for x and 1 for y.
$12 - 4 \stackrel{?}{=} 14$	Multiply: 4(1) = 4.
$8 = 14$ This statement is false.	Subtract: 12 − 4 = 8.

Because we obtain a false statement, we conclude that $(12, 1)$ is not a solution of $x - 4y = 14$. The ordered pair $(12, 1)$ does not satisfy the equation. ∎

 CHECK POINT 1 Determine whether each ordered pair is a solution of the equation $x - 3y = 9$:

a. $(3, -2)$ solution **b.** $(-2, 3)$. not a solution

In this chapter, we will use x and y to represent the variables of an equation in two variables. However, any two letters can be used. Solutions are still ordered pairs. The first number in an ordered pair usually replaces the variable that occurs first alphabetically. The second number in an ordered pair usually replaces the variable that occurs last alphabetically.

How do we find ordered pairs that are solutions of an equation in two variables, x and y?

2 Find solutions of an equation in two variables.

- Select a value for one of the variables.
- Substitute that value into the equation and find the corresponding value of the other variable.
- Use the values of the two variables to form an ordered pair (x, y). This pair is a solution of the equation.

EXAMPLE 2 Finding Solutions of an Equation

Find five solutions of
$$y = 2x - 1.$$

Select integers for x, starting with -2 and ending with 2.

SOLUTION We organize the process of finding solutions in the following table of values.

> Start with these values of x.

> Substitute x into $y = 2x - 1$ and compute y.

> Use values for x and y to form an ordered-pair solution.

Any numbers can be selected for x. There is nothing special about integers from -2 to 2, inclusive. We chose these values to include two negative numbers, 0, and two positive numbers. We also wanted to keep the resulting computations for y relatively simple.

x	$y = 2x - 1$	(x, y)
-2	$y = 2(-2) - 1 = -4 - 1 = -5$	$(-2, -5)$
-1	$y = 2(-1) - 1 = -2 - 1 = -3$	$(-1, -3)$
0	$y = 2 \cdot 0 - 1 = 0 - 1 = -1$	$(0, -1)$
1	$y = 2 \cdot 1 - 1 = 2 - 1 = 1$	$(1, 1)$
2	$y = 2 \cdot 2 - 1 = 4 - 1 = 3$	$(2, 3)$

Look at the ordered pairs in the last column. Five solutions of $y = 2x - 1$ are $(-2, -5)$, $(-1, -3)$, $(0, -1)$, $(1, 1)$, and $(2, 3)$. ■

 CHECK POINT 2 Find five solutions of $y = 3x + 2$. Select integers for x, starting with -2 and ending with 2. $(-2, -4), (-1, -1), (0, 2), (1, 5),$ and $(2, 8)$

3 Use point plotting to graph linear equations.

Graphing Linear Equations in the Form $y = mx + b$ In Example 2, we found five solutions of $y = 2x - 1$. We can generate as many ordered-pair solutions as desired to $y = 2x - 1$ by substituting numbers for x and then finding the corresponding values for y. The **graph of the equation** is the set of all points whose coordinates satisfy the equation.

One method for graphing an equation such as $y = 2x - 1$ is the **point-plotting method**.

THE POINT-PLOTTING METHOD FOR GRAPHING AN EQUATION IN TWO VARIABLES

1. Find several ordered pairs that are solutions of the equation.

2. Plot these ordered pairs as points in the rectangular coordinate system.

3. Connect the points with a smooth curve or line.

EXAMPLE 3 Graphing an Equation Using the Point-Plotting Method

Graph the equation: $y = 3x$.

SOLUTION

Step 1. **Find several ordered pairs that are solutions of the equation.** Because there are infinitely many solutions, we cannot list them all. To find some solutions of the equation, we select integers for x, starting with -2 and ending with 2.

Start with these values of x.	Substitute x into $y = 3x$ and compute y.	These are some solutions of $y = 3x$.
x	$y = 3x$	(x, y)
-2	$y = 3(-2) = -6$	$(-2, -6)$
-1	$y = 3(-1) = -3$	$(-1, -3)$
0	$y = 3 \cdot 0 = 0$	$(0, 0)$
1	$y = 3 \cdot 1 = 3$	$(1, 3)$
2	$y = 3 \cdot 2 = 6$	$(2, 6)$

Step 2.   **Plot these ordered pairs as points in the rectangular coordinate system.** The five ordered pairs in the table of values are plotted in Figure 3.2(a).

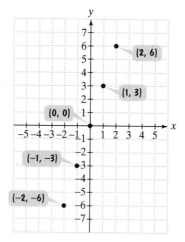

FIGURE 3.2 (a)   Some solutions of $y = 3x$ plotted as points

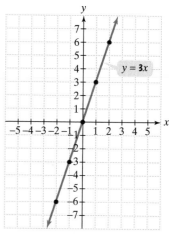

FIGURE 3.2 (b)   The graph of $y = 3x$

Step 3.   **Connect the points with a smooth curve or line.** The points lie along a straight line. The graph of $y = 3x$ is shown in Figure 3.2(b). The arrows on both ends of the line indicate that it extends indefinitely in both directions. ■

 CHECK POINT **3** Graph the equation: $y = 2x$.   See graphing answer section.

Equations like $y = 3x$ and $y = 2x$ are called **linear equations in two variables** because the graph of each equation is a line. Any equation that can be written in the form $y = mx + b$, where m and b are constants, is a linear equation in two variables. Here are examples of linear equations in two variables:

$$y = 3x \qquad\qquad y = 3x - 2$$

$$\text{or} \quad y = 3x + 0 \qquad\quad \text{or} \quad y = 3x + (-2)$$

This is in the form of $y = mx + b$ with $m = 3$ and $b = 0$.   This is in the form of $y = mx + b$ with $m = 3$ and $b = -2$.

Can you guess how the graph of the linear equation $y = 3x - 2$ compares with the graph of $y = 3x$? In Example 3, we graphed $y = 3x$. Now, let's graph the equation $y = 3x - 2$.

EXAMPLE 4 Graphing a Linear Equation in Two Variables

Graph the equation: $y = 3x - 2$.

SOLUTION

Step 1. **Find several ordered pairs that are solutions of the equation.** To find some solutions, we select integers for x, starting with -2 and ending with 2.

Start with x. Compute y. Form the ordered pair (x, y).

x	$y = 3x - 2$	(x, y)
-2	$y = 3(-2) - 2 = -6 - 2 = -8$	$(-2, -8)$
-1	$y = 3(-1) - 2 = -3 - 2 = -5$	$(-1, -5)$
0	$y = 3 \cdot 0 - 2 = 0 - 2 = -2$	$(0, -2)$
1	$y = 3 \cdot 1 - 2 = 3 - 2 = 1$	$(1, 1)$
2	$y = 3 \cdot 2 - 2 = 6 - 2 = 4$	$(2, 4)$

Step 2. **Plot these ordered pairs as points in the rectangular coordinate system.** The five ordered pairs in the table of values are plotted in Figure 3.3(a).

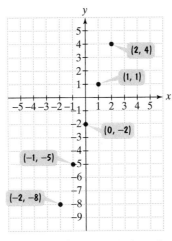

FIGURE 3.3 (a) Some solutions of $y = 3x - 2$ plotted as points

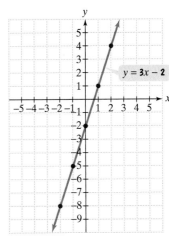

FIGURE 3.3 (b) The graph of $y = 3x - 2$

Step 3. **Connect the points with a smooth curve or line.** The points lie along a straight line. The graph of $y = 3x - 2$ is shown in Figure 3.3(b). ■

Now we are ready to compare the graphs of $y = 3x - 2$ and $y = 3x$. The graphs of both linear equations are shown in the same rectangular coordinate system in Figure 3.4. Can you see that the blue graph of $y = 3x - 2$ is parallel to the red graph of $y = 3x$ and shifted 2 units down? Instead of crossing the y-axis at $(0, 0)$, the graph now crosses the y-axis at $(0, -2)$.

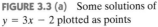

FIGURE 3.4

COMPARING GRAPHS OF LINEAR EQUATIONS   If the value of m does not change,

- The graph of $y = mx + b$ is the graph of $y = mx$ shifted b units up when b is a positive number.
- The graph of $y = mx + b$ is the graph of $y = mx$ shifted b units down when b is a negative number.

 CHECK POINT 4 Graph the equation: $y = 2x - 2$. See graphing answer section.

EXAMPLE 5 Graphing a Linear Equation in Two Variables

Graph the equation: $y = \dfrac{2}{3}x + 1$.

SOLUTION

Step 1. **Find several ordered pairs that are solutions of the equation.** Notice that m, the coefficient of x, is $\frac{2}{3}$. When m is a fraction, we will select values of x that are multiples of the denominator. In this way, we can avoid values of y that are fractions. Because the denominator of $\frac{2}{3}$ is 3, we select multiples of 3 for x. Let's use $-6, -3, 0, 3,$ and 6.

Start with multiples of 3 for x. → Compute y. → Form the ordered pair (x, y).

x	$y = \dfrac{2}{3}x + 1$	(x, y)
-6	$y = \dfrac{2}{3}(-6) + 1 = -4 + 1 = -3$	$(-6, -3)$
-3	$y = \dfrac{2}{3}(-3) + 1 = -2 + 1 = -1$	$(-3, -1)$
0	$y = \dfrac{2}{3} \cdot 0 + 1 = 0 + 1 = 1$	$(0, 1)$
3	$y = \dfrac{2}{3} \cdot 3 + 1 = 2 + 1 = 3$	$(3, 3)$
6	$y = \dfrac{2}{3} \cdot 6 + 1 = 4 + 1 = 5$	$(6, 5)$

Step 2. **Plot these ordered pairs as points in the rectangular coordinate system.** The five ordered pairs in the table of values are plotted in Figure 3.5.

Step 3. **Connect the points with a smooth curve or line.** The points lie along a straight line. The graph of $y = \frac{2}{3}x + 1$ is shown in Figure 3.5.

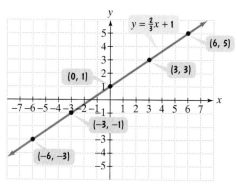

FIGURE 3.5 The graph of $y = \frac{2}{3}x + 1$

✔ **CHECK POINT 5** Graph the equation: $y = \frac{1}{2}x + 2$. See graphing answer section.

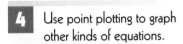

4 Use point plotting to graph other kinds of equations.

Graphing Nonlinear Equations in Two Variables Look at the picture of this gymnast. He has created a perfect balance in which the two halves of his body are mirror images of each other. Is it possible for graphs to have mirrorlike qualities? Yes. Although our next graph is not a straight line, we can obtain its cuplike U-shape using the point-plotting method for graphing an equation in two variables.

EXAMPLE 6 Graphing a Nonlinear Equation in Two Variables

Graph the equation: $y = x^2 - 4$.

SOLUTION The given equation involves two variables, x and y. However, because the variable x is squared, it is not a linear equation in two variables.

$$y = x^2 - 4$$

This is not in the form $y = mx + b$ because x is squared.

Although the graph is not a line, it is still a picture of all the ordered-pair solutions of $y = x^2 - 4$. Thus, we can use the point-plotting method to obtain the graph.

Step 1. **Find several ordered pairs that are solutions of the equation.** To find some solutions, we select integers for x, starting with -3 and ending with 3.

STUDY TIP

If the graph of an equation is not a straight line, use more solutions than when graphing lines. These extra solutions are needed to get a better general idea of the graph's shape.

Start with x.	Compute y.	Form the ordered pair (x, y).
x	$y = x^2 - 4$	(x, y)
-3	$y = (-3)^2 - 4 = 9 - 4 = 5$	$(-3, 5)$
-2	$y = (-2)^2 - 4 = 4 - 4 = 0$	$(-2, 0)$
-1	$y = (-1)^2 - 4 = 1 - 4 = -3$	$(-1, -3)$
0	$y = 0^2 - 4 = 0 - 4 = -4$	$(0, -4)$
1	$y = 1^2 - 4 = 1 - 4 = -3$	$(1, -3)$
2	$y = 2^2 - 4 = 4 - 4 = 0$	$(2, 0)$
3	$y = 3^2 - 4 = 9 - 4 = 5$	$(3, 5)$

Step 2. **Plot these ordered pairs as points in the rectangular coordinate system.** The seven ordered pairs in the table of values are plotted in Figure 3.6(a) on the next page.

Step 3. **Connect the points with a smooth curve.** The seven points are joined with a smooth curve in Figure 3.6(b). The graph of $y = x^2 - 4$ is a curve where the part of the graph to the right of the y-axis is a reflection of the part to the left of it, and vice versa. The arrows on both ends of the curve indicate that it extends indefinitely in both directions.

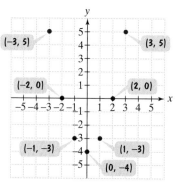

FIGURE 3.6 (a) Some solutions of $y = x^2 - 4$ plotted as points

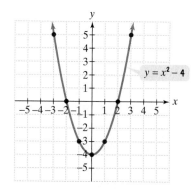

FIGURE 3.6 (b) The graph of $y = x^2 - 4$

 CHECK POINT 6 Graph the equation: $y = x^2 - 1$. Select integers for x, starting with -3 and ending with 3.   See graphing answer section.

5 Use graphs of linear equations to solve problems.

Applications   Part of the beauty of the rectangular coordinate system is that it allows us to "see" mathematical formulas and visualize the solution to a problem. This idea is demonstrated in Example 7.

EXAMPLE 7   An Application Using Graphs of Linear Equations

The toll to a bridge costs $2.50. Commuters who use the bridge frequently have the option of purchasing a monthly coupon book for $21.00. With the coupon book, the toll is reduced to $1.00. The monthly cost, y, of using the bridge x times can be described by the following formulas:

Without the coupon book:

$$y = 2.50x$$ The monthly cost, y, is $2.50 times the number of times, x, that the bridge is used.

With the coupon book:

$$y = 21 + 1 \cdot x$$ The monthly cost, y, is $21 for the book plus $1 times the number of times, x, that the bridge is used.

$$y = 21 + x$$

a. Let $x = 0, 2, 4, 10, 12, 14$, and 16. Make a table of values for each linear equation showing seven solutions for the equation.

b. Graph the equations in the same rectangular coordinate system.

c. What are the coordinates of the intersection point for the two graphs? Interpret the coordinates in practical terms.

SOLUTION

a. Tables of values showing seven solutions for each equation follow.

Without the Coupon Book				With the Coupon Book		
x	$y = 2.5x$	(x, y)		x	$y = 21 + x$	(x, y)
0	$y = 2.5(0) = 0$	$(0, 0)$		0	$y = 21 + 0 = 21$	$(0, 21)$
2	$y = 2.5(2) = 5$	$(2, 5)$		2	$y = 21 + 2 = 23$	$(2, 23)$
4	$y = 2.5(4) = 10$	$(4, 10)$		4	$y = 21 + 4 = 25$	$(4, 25)$
10	$y = 2.5(10) = 25$	$(10, 25)$		10	$y = 21 + 10 = 31$	$(10, 31)$
12	$y = 2.5(12) = 30$	$(12, 30)$		12	$y = 21 + 12 = 33$	$(12, 33)$
14	$y = 2.5(14) = 35$	$(14, 35)$		14	$y = 21 + 14 = 35$	$(14, 35)$
16	$y = 2.5(16) = 40$	$(16, 40)$		16	$y = 21 + 16 = 37$	$(16, 37)$

b. Now we are ready to graph the two equations. Because the x- and y-coordinates are nonnegative, it is only necessary to use the origin, the positive portions of the x- and y-axes, and the first quadrant of the rectangular coordinate system. The x-coordinates in our tables begin at 0 and end at 16. We will let each tick mark on the x-axis represent two units. However, the y-coordinates in our tables begin at 0 and get as large as 40 in the formula that describes the monthly cost without the coupon book. So that our y-axis does not get too long, we will let each tick mark on the y-axis represent five units. Using this setup and the two tables of values, we construct the graphs of $y = 2.5x$ and $y = 21 + x$, shown in Figure 3.7.

c. The graphs intersect at $(14, 35)$. This means that if the bridge is used 14 times in a month, the total monthly cost without the coupon book is the same as the total monthly cost with the coupon book, namely \$35. ∎

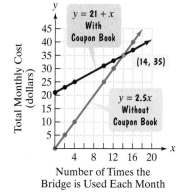

FIGURE 3.7 Options for a toll

In Figure 3.7, look at the two graphs to the right of the intersection point $(14, 35)$. The red graph of $y = 21 + x$ lies below the blue graph of $y = 2.5x$. This means that if the bridge is used more than 14 times in a month $(x > 14)$, the (red) monthly cost, y, with the coupon book is cheaper than the (blue) monthly cost, y, without the coupon book.

 CHECK POINT 7 The toll to a bridge costs \$2.00. If you use the bridge x times in a month, the monthly cost, y, is $y = 2x$. With a \$10 coupon book, the toll is reduced to \$1.00. The monthly cost, y, of using the bridge x times in a month with the coupon book is $y = 10 + x$.

a. Let $x = 0, 2, 4, 6, 8, 10,$ and 12. Make tables of values showing seven solutions of $y = 2x$ and seven solutions of $y = 10 + x$.

b. Graph the equations in the same rectangular coordinate system. *See graphing answer section.*

c. What are the coordinates of the intersection point for the two graphs? Interpret the coordinates in practical terms.

$y = 2x$			$y = 10 + x$	
x	(x, y)		x	(x, y)
0	$(0, 0)$		0	$(0, 10)$
2	$(2, 4)$		2	$(2, 12)$
4	$(4, 8)$		4	$(4, 14)$
6	$(6, 12)$		6	$(6, 16)$
8	$(8, 16)$		8	$(8, 18)$
10	$(10, 20)$		10	$(10, 20)$
12	$(12, 24)$		12	$(12, 22)$

$(10, 20)$; If the bridge is used 10 times in a month, the total monthly cost without the coupon book is the same as the monthly cost with the coupon book, namely \$20.

Mathematical Blossom

Graph of an equation in a three-dimensional Cartesian plane

This picture is a graph in a three-dimensional rectangular coordinate system. For many, the picture is more interesting than its equation:

$$z = (|x| - |y|)^2 + \frac{2|xy|}{\sqrt{x^2 + y^2}}.$$

Sometimes it's pleasant to simply get the "feel" of an equation by seeing its picture. Turning equations into visual images hints at the beauty that lies within mathematics.

USING TECHNOLOGY

Graphing calculators or graphing software packages for computers are referred to as **graphing utilities** or graphers. A graphing utility is a powerful tool that quickly generates the graph of an equation in two variables. Figure 3.8 shows two such graphs for the equations in Examples 3 and 4.

FIGURE 3.8(a) The graph of $y = 3x$

FIGURE 3.8(b) The graph of $y = 3x - 2$

What differences do you notice between these graphs and the graphs that we drew by hand? They do seem a bit "jittery." Arrows do not appear on both ends of the graphs. Furthermore, numbers are not given along the axes. For both graphs in Figure 3.8, the x-axis extends from −10 to 10 and the y-axis also extends from −10 to 10. The distance represented by each consecutive tick mark is one unit. We say that the **viewing rectangle**, or the **viewing window**, is $[-10, 10, 1]$ by $[-10, 10, 1]$.

$$[-10, \qquad 10, \qquad 1] \quad \text{by} \quad [-10, \qquad 10, \qquad 1].$$

| The minimum x-value along the x-axis is −10. | The maximum x-value along the x-axis is 10. | The scale on the x-axis is 1 unit per tick mark. | The minimum y-value along the y-axis is −10. | The maximum y-value along the y-axis is 10. | The scale on the y-axis is 1 unit per tick mark. |

To graph an equation in x and y using a graphing utility, enter the equation and specify the size of the viewing rectangle. The size of the viewing rectangle sets minimum and maximum values for both the x- and y-axes. Enter these values, as well as the values between consecutive tick marks, on the respective axes. The $[-10, 10, 1]$ by $[-10, 10, 1]$ viewing rectangle used in Figure 3.8 is called the **standard viewing rectangle**.

On most graphing utilities, the display screen is two-thirds as high as it is wide. By using a square setting, you can equally space the x and y tick marks. (This does not occur in the standard viewing rectangle.) Graphing utilities can also *zoom in* and *zoom out*. When you zoom in, you see a smaller portion of the graph, but you do so in greater detail. When you zoom out, you see a larger portion of the graph. Thus, zooming out may help you to develop a better understanding of the overall character of the graph. With practice, you will become more comfortable with graphing equations in two variables using your graphing utility. You will also develop a better sense of the size of the viewing rectangle that will reveal needed information about a particular graph.

3.1 EXERCISE SET

 Student Solutions Manual   CD/Video   PH Math/Tutor Center   MathXL Tutorials on CD Math XL MathXL® MyMathLab MyMathLab   Interactmath.com

Practice Exercises

In Exercises 1–12, determine whether each ordered pair is a solution of the given equation.

1. $y = 3x$ $(2, 3), (3, 2), (-4, -12)$ $(2, 3)$ and $(3, 2)$ are not solutions; $(-4, -12)$ is a solution.

2. $y = 4x$ $(3, 12), (12, 3), (-5, -20)$ $(12, 3)$ is not a solution; $(3, 12)$ and $(-5, -20)$ are solutions.

3. $y = -4x$ $(-5, -20), (0, 0), (9, -36)$ $(-5, -20)$ is not a solution; $(0, 0)$ and $(9, -36)$ are solutions.

4. $y = -3x$ $(-5, 15), (0, 0), (7, -21)$ $(-5, 15), (0, 0)$ and $(7, -21)$ are solutions.

5. $y = 2x + 6$ $(0, 6), (-3, 0), (2, -2)$ $(2, -2)$ is not a solution; $(0, 6)$ and $(-3, 0)$ are solutions.

6. $y = 8 - 4x$ $(8, 0), (16, -2), (3, -4)$ $(8, 0)$ and $(16, -2)$ are not solutions; $(3, -4)$ is a solution.

7. $3x + 5y = 15$ $(-5, 6), (0, 5), (10, -3)$
$(0, 5)$ is not a solution; $(-5, 6)$ and $(10, -3)$ are solutions.

8. $2x - 5y = 0$ $(-2, 0), (-10, 6), (5, 0)$
$(-2, 0), (-10, 6)$ and $(5, 0)$ are not solutions.

9. $x + 3y = 0$ $(0, 0), \left(1, \frac{1}{3}\right), \left(2, -\frac{2}{3}\right)$
$\left(1, \frac{1}{3}\right)$ is not a solution; $(0, 0)$ and $\left(2, -\frac{2}{3}\right)$ are solutions.

10. $x + 5y = 0$ $(0, 0), \left(1, \frac{1}{5}\right), \left(2, -\frac{2}{5}\right)$
$\left(1, \frac{1}{5}\right)$ is not a solution; $(0, 0)$ and $\left(2, -\frac{2}{5}\right)$ are solutions.

11. $x - 4 = 0$ $(4, 7), (3, 4), (0, -4)$
$(3, 4)$ and $(0, -4)$ are not solutions; $(4, 7)$ is a solution.

12. $y + 2 = 0$ $(0, 2), (2, 0), (0, -2)$
$(0, 2)$ and $(2, 0)$ are not solutions; $(0, -2)$ is a solution.

In Exercises 13–20, find five solutions of each equation. Select integers for x, starting with −2 and ending with 2. Organize your work in a table of values.   **13–20.** See graphing answer section.

13. $y = 12x$ **14.** $y = 14x$
15. $y = -10x$ **16.** $y = -20x$
17. $y = 8x - 5$ **18.** $y = 6x - 4$
19. $y = -3x + 7$ **20.** $y = -5x + 9$

In Exercises 21–44, graph each linear equation in two variables. Find at least five solutions in your table of values for each equation.   **21–44.** See graphing answer section.

21. $y = x$ **22.** $y = x + 1$
23. $y = x - 1$ **24.** $y = x - 2$
25. $y = 2x + 1$ **26.** $y = 2x - 1$
27. $y = -x + 2$ **28.** $y = -x + 3$
29. $y = -3x - 1$ **30.** $y = -3x - 2$
31. $y = \frac{1}{2}x$ **32.** $y = -\frac{1}{2}x$
33. $y = -\frac{1}{4}x$ **34.** $y = \frac{1}{4}x$
35. $y = \frac{1}{3}x + 1$ **36.** $y = \frac{1}{3}x - 1$
37. $y = -\frac{3}{2}x + 1$ **38.** $y = -\frac{3}{2}x + 2$
39. $y = -\frac{5}{2}x - 1$ **40.** $y = -\frac{5}{2}x + 1$
41. $y = x + \frac{1}{2}$ **42.** $y = x - \frac{1}{2}$
43. $y = 4$, or $y = 0x + 4$ **44.** $y = 3$, or $y = 0x + 3$

Graph each equation in Exercises 45–50. Find seven solutions in your table of values for each equation by using integers for x, starting with −3 and ending with 3.   **45–50.** See graphing answer section.

45. $y = x^2$ **46.** $y = x^2 - 2$
47. $y = x^2 + 1$ **48.** $y = x^2 + 2$
49. $y = 4 - x^2$ **50.** $y = 9 - x^2$

Practice Plus

In Exercises 51–54, write each sentence as a linear equation in two variables. Then graph the equation.   **51–54.** See graphing answer section.

51. The y-variable is 3 more than the x-variable. $y = x + 3$
52. The y-variable exceeds the x-variable by 4. $y = x + 4$
53. The y-variable exceeds twice the x-variable by 5. $y = 2x + 5$
54. The y-variable is 2 less than 3 times the x-variable. $y = 3x - 2$

55. At the beginning of a semester, a student purchased eight pens and six pads for a total cost of $14.50.
  a. If x represents the cost of one pen and y represents the cost of one pad, write an equation in two variables that reflects the given conditions. $8x + 6y = 14.50$
  b. If pads cost $0.75 each, find the cost of one pen. $1.25

56. A nursery offers a package of three small orange trees and four small grapefruit trees for $22.
  a. If x represents the cost of one orange tree and y represents the cost of one grapefruit tree, write an equation in two variables that reflects the given conditions. $3x + 4y = 22$
  b. If a grapefruit tree costs $2.50, find the cost of an orange tree. $4.00

Application Exercises

The graph shows the percentage of divorced Americans, by race.

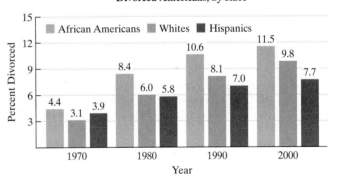

Divorced Americans, by Race

Source: U.S. Census Bureau

The data can be modeled by linear equations in two variables:

African Americans $y = 0.24x + 5.2$
Whites $y = 0.22x + 3.4$
Hispanics $y = 0.13x + 4.2$

In each model, x is the number of years after 1970 and y is the percentage of divorced Americans in the racial group. Use this information to solve Exercises 57–58.

57. a. Find four solutions of the linear equation that models the data for African Americans. Use 0, 10, 20, and 30 for x. Organize your work in a table of values. How well does the linear equation model the data shown in the bar graph?
$(0, 5.2), (10, 7.6), (20, 10), (30, 12.4)$; Answers will vary.

b. Repeat part (a) for the linear equation that models the data for whites. $(0, 3.4), (10, 5.6), (20, 7.8), (30, 10);$ Answers will vary.

c. Repeat part (a) for the linear equation that models the data for Hispanics. $(0, 4.2), (10, 5.5), (20, 6.8), (30, 8.1);$ Answers will vary.

58. a. Use the tables of values from Exercise 57 to graph any one of the three linear equations. See graphing answer section.

b. If you have not yet done so, extend your graph to include $x = 40$. Use the line to predict the percentage of divorced Americans, rounded to the nearest tenth of a percent, in this group in 2010. Answers will vary. African Americans: 14.8%; whites: 12.2%; Hispanics: 9.4%

59. A rental company charges $40.00 a day plus $0.35 per mile to rent a moving truck. The total cost, y, for a day's rental if x miles are driven is described by $y = 40 + 0.35x$. A second company charges $36.00 a day plus $0.45 per mile, so the daily cost, y, if x miles are driven is described by $y = 36 + 0.45x$. The graphs of the two equations are shown in the same rectangular coordinate system.

a. What is the x-coordinate of the intersection point of the graphs? Describe what this x-coordinate means in practical terms. 40; Both companies charge the same for 40 miles driven.

b. What is a reasonable estimate for the y-coordinate of the intersection point? Answers will vary: 55

c. Substitute the x-coordinate of the intersection point into each of the equations and find the corresponding value for y. Describe what this value represents in practical terms. How close is this value to your estimate from part (b)? 54; Both companies charge $54 for 40 miles driven.

60. The linear equation in two variables $y = 166x + 1781$ models the cost, y, in tuition and fees per year, of a four-year public college x years after 1990.

a. Find five solutions of $y = 166x + 1781$. Use 0, 5, 10, 15, and 20 for x. Organize your work in a table of values.

b. Use the solutions in part (a) to graph $y = 166x + 1781$. What does the shape of the graph indicate about the cost of a four-year public college? See graphing answer section.

61. The linear equation in two variables $y = 50x + 30,000$ models the total weekly cost, y, in dollars, for a business that manufactures x racing bicycles each week. The equation indicates that the business has weekly fixed costs of $30,000 plus a cost of $50 to manufacture each bicycle.

a. Find five solutions of $y = 50x + 30,000$. Use 0, 10, 20, 30, and 40 for x. Organize your work in a table of values.

b. Use the solutions in part (a) to graph $y = 50x + 30,000$. See graphing answer section.

Writing in Mathematics

62. How do you determine whether an ordered pair is a solution of an equation in two variables, x and y? 62–65. Answers will vary.

63. Explain how to find ordered pairs that are solutions of an equation in two variables, x and y.

64. What is the graph of an equation?

65. Explain how to graph an equation in two variables in the rectangular coordinate system.

Critical Thinking Exercises

66. Which one of the following is true? b

a. The graph of $y = 3x + 1$ is parallel to the graph of $y = 2x$, but shifted up 1 unit.

b. The graph of any equation in the form $y = mx + b$ passes through the point $(0, b)$.

c. The ordered pair $(3, 4)$ satisfies the equation

$$2y - 3x = -6.$$

d. If $(2, 5)$ satisfies an equation, then $(5, 2)$ also satisfies the equation.

Graph each equation in Exercises 67–68. Find seven solutions in your table of values by using integers for x, starting with −3 and ending with 3. 67–68. See graphing answer section.

67. $y = |x|$ **68.** $y = |x| + 1$

69. Although the level of air pollution varies from day to day and from hour to hour, during the summer the level of air pollution depends on the time of day. The equation in two variables

$$y = 0.1x^2 - 0.4x + 0.6$$

describes the level of air pollution, in parts per million (ppm), where x corresponds to the number of hours after 9 A.M.

a. Find six solutions of the equation. Select integers for x, starting with 0 and ending with 5.
$(0, 0.6), (1, 0.3), (2, 0.2), (3, 0.3), (4, 0.6), (5, 1.1)$

b. Researchers have determined that a level of 0.3 ppm or more of pollutants in the air can be hazardous to your health. Based on the six solutions in part (a), at what times between 9 A.M. and 2 P.M. should runners exercise to avoid unsafe air? between 10:00 A.M. and noon

Technology Exercises

Use a graphing utility to graph each equation in Exercises 70–73 in a standard viewing rectangle. Then use the $\boxed{\text{TRACE}}$ *feature to trace along the line and find the coordinates of two points.* 70–73. See graphing answer section.

70. $y = 2x - 1$ **71.** $y = -3x + 2$

72. $y = \frac{1}{2}x$ **73.** $y = \frac{3}{4}x - 2$

74. Use a graphing utility to verify any five of your hand-drawn graphs in Exercises 21–44. Use an appropriate viewing rectangle and the ZOOM SQUARE feature to make the graph look like the one you drew by hand.   Answers will vary.

75. The linear equation $y = 2.4x + 180$ models U.S. population, y, in millions, x years after 1960. Use a graphing utility to graph the equation in a $[0, 90, 10]$ by $[0, 500, 100]$ viewing rectangle.

What does the shape of the graph indicate about changing U.S. population over time?   See graphing answer section.

Review Exercises

76. Solve: $3x + 5 = 4(2x - 3) + 7$. (Section 2.3, Example 3) 2

77. Simplify: $3(1 - 2 \cdot 5) - (-28)$. (Section 1.8, Example 7)   1

78. Solve for h: $V = \frac{1}{3}Ah$. (Section 2.4, Example 4) $h = \frac{3V}{A}$

SECTION 3.2

GRAPHING LINEAR EQUATIONS USING INTERCEPTS

Objectives

1 Use a graph to identify intercepts.

2 Graph a linear equation in two variables using intercepts.

3 Graph horizontal or vertical lines.

Despite the nutritional disaster floating within the cans, the international sale of carbonated soft drinks continues to flourish. A can of Coca-Cola is sold every six seconds throughout the world.

The carbonated soft drinks market in the United States appears to be losing its fizz. As consumption of carbonated soft drinks declines, there is an emerging market for alternative beverages, including juices, teas, bottled waters, and sports drinks. The line graphs in Figure 3.9 show the projected U.S. beverage market share for carbonated and noncarbonated alternative beverages through 2010.

The decreasing market share for carbonated beverages can be modeled by

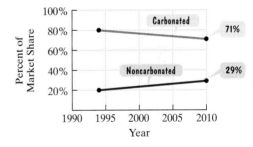

U.S. Beverage Market Share

FIGURE 3.9

Source: Beverage Digest

$$y = -0.57x + 80.$$

The variable x represents the number of years after 1994. The variable y represents the market share of carbonated beverages, measured as a percentage of all sales.

There is another way that we can write the equation $y = -0.57x + 80$. We will collect the x- and y-terms on the left side. This is done by adding $0.57x$ to both sides:

$$0.57x + y = 80.$$

The form of this equation is $Ax + By = C$.

$$0.57x \ + \ y \ = \ 80$$

| A, the coefficient of x, is 0.57. | B, the coefficient of y, is 1. | C, the constant on the right, is 80. |

All equations of the form $Ax + By = C$ are straight lines when graphed as long as A and B are not both zero. To graph linear equations of this form, we will use two important points: the *intercepts*.

1 Use a graph to identify intercepts.

Intercepts An **x-intercept** of a graph is the x-coordinate of a point where the graph intersects the x-axis. For example, look at the graph of $2x - 4y = 8$ in Figure 3.10. The graph crosses the x-axis at $(4, 0)$. Thus, the x-intercept is 4. **The y-coordinate corresponding to an x-intercept is always zero.**

A **y-intercept** of a graph is the y-coordinate of a point where the graph intersects the y-axis. The graph of $2x - 4y = 8$ in Figure 3.10 shows that the graph crosses the y-axis at $(0, -2)$. Thus, the y-intercept is -2. **The x-coordinate corresponding to a y-intercept is always zero.**

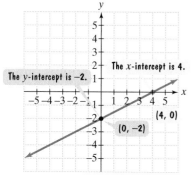

FIGURE 3.10 The graph of $2x - 4y = 8$

STUDY TIP

Mathematicians tend to use two ways to describe intercepts. Did you notice that we are using single numbers? If a graph's x-intercept is a, it passes through the point $(a, 0)$. If a graph's y-intercept is b, it passes through the point $(0, b)$.

Some books state that the x-intercept is the *point* $(a, 0)$ and the x-intercept is *at a* on the x-axis. Similarly, the y-intercept is the *point* $(0, b)$ and the y-intercept is *at b* on the y-axis. In these descriptions, the intercepts are the actual points where a graph crosses the axes.

Although we'll describe intercepts as single numbers, we'll immediately state the point on the x- or y-axis that the graph passes through. Here's the important thing to keep in mind:

> x-intercept: The corresponding y is 0.
> y-intercept: The corresponding x is 0.

EXAMPLE 1 Identifying Intercepts

Identify the x- and y-intercepts.

a.

b.

c.

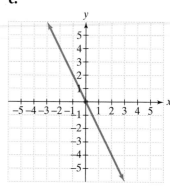

SOLUTION

a. The graph crosses the x-axis at $(-1, 0)$. Thus, the x-intercept is -1. The graph crosses the y-axis at $(0, 2)$. Thus, the y-intercept is 2.

b. The graph crosses the x-axis at $(3, 0)$, so the x-intercept is 3. This vertical line does not cross the y-axis. Thus, there is no y-intercept.

c. This graph crosses the x- and y-axes at the same point, the origin. Because the graph crosses both axes at $(0, 0)$, the x-intercept is 0 and the y-intercept is 0. ∎

✔ **CHECK POINT 1** Identify the *x*- and *y*-intercepts.

a. *x*-intercept: −3; *y*-intercept: 5   **b.** no *x*-intercept; *y*-intercept: 4   **c.** *x*-intercept: 0; *y*-intercept: 0

2 Graph a linear equation in two variables using intercepts.

Graphing Using Intercepts   An equation of the form $Ax + By = C$, where A, B, and C are integers, is called the **standard form** of the equation of a line. The equation can be graphed by finding the *x*- and *y*-intercepts, plotting the intercepts, and drawing a straight line through these points. How do we find the intercepts of a line, given its equation? Because the *y*-coordinate of the *x*-intercept is 0, to find the *x*-intercept,

- Substitute 0 for *y* in the equation.
- Solve for *x*.

EXAMPLE 2   Finding the *x*-Intercept

Find the *x*-intercept of the graph of $3x - 4y = 24$.

SOLUTION   To find the *x*-intercept, let $y = 0$ and solve for *x*.

$$3x - 4y = 24 \qquad \text{This is the given equation.}$$
$$3x - 4 \cdot 0 = 24 \qquad \text{Let } y = 0.$$
$$3x = 24 \qquad \text{Simplify: } 4 \cdot 0 = 0 \text{ and } 3x - 0 = 3x.$$
$$x = 8 \qquad \text{Divide both sides by 3.}$$

The *x*-intercept is 8. The graph of $3x - 4y = 24$ passes through the point $(8, 0)$.   ■

✔ **CHECK POINT 2**   Find the *x*-intercept of the graph of $4x - 3y = 12$.   3

Because the *x*-coordinate of the *y*-intercept is 0, to find the *y*-intercept,
- Substitute 0 for *x* in the equation.
- Solve for *y*.

EXAMPLE 3   Finding the *y*-Intercept

Find the *y*-intercept of the graph of $3x - 4y = 24$.

SOLUTION   To find the *y*-intercept, let $x = 0$ and solve for *y*.

$$3x - 4y = 24 \qquad \text{This is the given equation.}$$
$$3 \cdot 0 - 4y = 24 \qquad \text{Let } x = 0.$$
$$-4y = 24 \qquad \text{Simplify: } 3 \cdot 0 = 0 \text{ and } 0 - 4y = -4y.$$
$$y = -6 \qquad \text{Divide both sides by } -4.$$

The y-intercept is -6. The graph of $3x - 4y = 24$ passes through the point $(0, -6)$. ■

CHECK POINT 3 Find the y-intercept of the graph of $4x - 3y = 12$. -4

When graphing using intercepts, it is a good idea to use a third point, a checkpoint, before drawing the line. A checkpoint can be obtained by selecting a value for either variable, other than 0, and finding the corresponding value for the other variable. The checkpoint should lie on the same line as the x- and y-intercepts. If it does not, recheck your work and find the error.

> **USING INTERCEPTS TO GRAPH $Ax + By = C$**
>
> **1.** Find the x-intercept. Let $y = 0$ and solve for x.
> **2.** Find the y-intercept. Let $x = 0$ and solve for y.
> **3.** Find a checkpoint, a third ordered-pair solution.
> **4.** Graph the equation by drawing a line through the three points.

EXAMPLE 4   Using Intercepts to Graph a Linear Equation

Graph: $3x + 2y = 6$.

SOLUTION

Step 1.   **Find the x-intercept. Let $y = 0$ and solve for x.**

$$3x + 2 \cdot 0 = 6$$
$$3x = 6$$
$$x = 2$$

The x-intercept is 2, so the line passes through $(2, 0)$.

Step 2.   **Find the y-intercept. Let $x = 0$ and solve for y.**

$$3 \cdot 0 + 2y = 6$$
$$2y = 6$$
$$y = 3$$

The y-intercept is 3, so the line passes through $(0, 3)$.

Step 3.   **Find a checkpoint, a third ordered-pair solution.** For our checkpoint, we will let $x = 1$ and find the corresponding value for y.

$$3x + 2y = 6 \qquad \text{This is the given equation.}$$
$$3 \cdot 1 + 2y = 6 \qquad \text{Substitute 1 for x.}$$
$$3 + 2y = 6 \qquad \text{Simplify.}$$
$$2y = 3 \qquad \text{Subtract 3 from both sides.}$$
$$y = \frac{3}{2} \qquad \text{Divide both sides by 2.}$$

The checkpoint is the ordered pair $\left(1, \dfrac{3}{2}\right)$, or $(1, 1.5)$.

Step 4.   **Graph the equation by drawing a line through the three points.** The three points in Figure 3.11 lie along the same line. Drawing a line through the three points results in the graph of $3x + 2y = 6$. ■

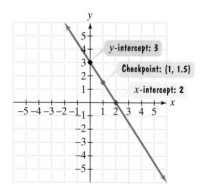

FIGURE 3.11   The graph of $3x + 2y = 6$

USING TECHNOLOGY

You can use a graphing utility to graph equations of the form $Ax + By = C$. Begin by solving the equation for y. For example, to graph $3x + 2y = 6$, solve the equation for y.

$3x + 2y = 6$	This is the equation to be graphed.
$3x - 3x + 2y = -3x + 6$	Subtract 3x from both sides.
$2y = -3x + 6$	Simplify.
$\dfrac{2y}{2} = \dfrac{-3x + 6}{2}$	Divide both sides by 2.
$y = -\dfrac{3}{2}x + 3$	Simplify.

This is the equation to enter into your graphing utility. The graph of $y = -\frac{3}{2}x + 3$ or, equivalently, $3x + 2y = 6$, is shown below in a $[-6, 6, 1]$ by $[-6, 6, 1]$ viewing rectangle.

$y = -\frac{3}{2}x + 3$

 CHECK POINT 4 Graph: $2x + 3y = 6$.   See graphing answer section.

EXAMPLE 5   Using Intercepts to Graph a Linear Equation

Graph: $2x - y = 4$.

SOLUTION

Step 1.   **Find the x-intercept. Let $y = 0$ and solve for x.**

$$2x - 0 = 4$$
$$2x = 4$$
$$x = 2$$

The x-intercept is 2, so the line passes through $(2, 0)$.

Step 2.   **Find the y-intercept. Let $x = 0$ and solve for y.**

$$2 \cdot 0 - y = 4$$
$$-y = 4$$
$$y = -4$$

The y-intercept is -4, so the line passes through $(0, -4)$.

Step 3.   **Find a checkpoint, a third ordered-pair solution.** For our checkpoint, we will let $x = 1$ and find the corresponding value for y.

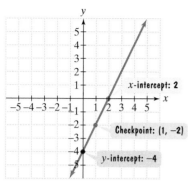

FIGURE 3.12 The graph of $2x - y = 4$

$$2x - y = 4 \qquad \text{This is the given equation.}$$
$$2 \cdot 1 - y = 4 \qquad \text{Substitute 1 for } x.$$
$$2 - y = 4 \qquad \text{Simplify.}$$
$$-y = 2 \qquad \text{Subtract 2 from both sides.}$$
$$y = -2 \qquad \text{Multiply (or divide) both sides by } -1.$$

The checkpoint is $(1, -2)$.

Step 4. **Graph the equation by drawing a line through the three points.** The three points in Figure 3.12 lie along the same line. Drawing a line through the three points results in the graph of $2x - y = 4$. ■

 CHECK POINT 5 Graph: $x - 2y = 4$. See graphing answer section.

We have seen that not all lines have two different intercepts. Some lines pass through the origin. Thus, they have an x-intercept of 0 and a y-intercept of 0. Is it possible to recognize these lines by their equations? Yes. **The graph of the linear equation $Ax + By = 0$ passes through the origin.** Notice that the constant on the right side of this equation is 0.

An equation of the form $Ax + By = 0$ can be graphed by using the origin as one point on the line. Find two other points by finding two other solutions of the equation. Select values for either variable, other than 0, and find the corresponding values for the other variable.

EXAMPLE 6   Graphing a Linear Equation of the Form $Ax + By = 0$

Graph: $x + 2y = 0$.

SOLUTION   Because the constant on the right is 0, the graph passes through the origin. The x- and y-intercepts are both 0. Remember that we are using two points and a checkpoint to determine a line. Thus, we still want to find two other points. Let $y = -1$ to find a second ordered-pair solution. Let $y = 1$ to find a third ordered-pair (checkpoint) solution.

$$x + 2y = 0 \qquad\qquad x + 2y = 0$$
$$\text{Let } y = -1. \qquad\qquad \text{Let } y = 1.$$
$$x + 2(-1) = 0 \qquad\quad x + 2 \cdot 1 = 0$$
$$x + (-2) = 0 \qquad\qquad x + 2 = 0$$
$$x = 2 \qquad\qquad\quad x = -2$$

The solutions are $(2, -1)$ and $(-2, 1)$. Plot these two points, as well as the origin—that is, $(0, 0)$. The three points in Figure 3.13 lie along the same line. Drawing a line through the three points results in the graph of $x + 2y = 0$. ■

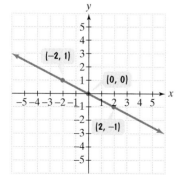

FIGURE 3.13 The graph of $x + 2y = 0$

 CHECK POINT 6 Graph: $x + 3y = 0$. See graphing answer section.

3 Graph horizontal or vertical lines.

Equations of Horizontal and Vertical Lines Some things change very little. For example, from 1985 to the present, the number of Americans participating in downhill skiing has remained relatively constant, indicated by the graph shown in Figure 3.14. Shown in the figure is a horizontal line that passes through or near most of the data points.

We can use the horizontal line in Figure 3.14 to write an equation that reasonably models the data. The y-intercept of the line is 15, so the graph passes through (0, 15). Furthermore, all points on the line have a value of y that is always 15. Thus, an equation that models the number of participants in downhill skiing for the period shown is

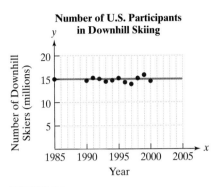

FIGURE 3.14

Source: National Ski Areas Association

$$y = 15.$$

The popularity of downhill skiing has remained relatively constant in the United States at approximately 15 million participants each year.

The equation $y = 15$ can be expressed as $0x + 1y = 15$. We know that the graph of any equation of the form $Ax + By = C$ is a line as long as A and B are not both zero. The graph of $y = 15$ suggests that when A is zero, the line is horizontal.

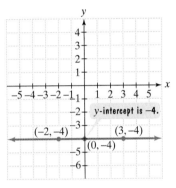

FIGURE 3.15 The graph of $y = -4$

EXAMPLE 7 Graphing a Horizontal Line

Graph the linear equation: $y = -4$.

SOLUTION All ordered pairs that are solutions of $y = -4$ have a value of y that is always -4. Any value can be used for x. Let's select three of the possible values for x: -2, 0, and 3. Using these values of x, three ordered pairs that are solutions of $y = -4$ are $(-2, -4)$, $(0, -4)$, and $(3, -4)$. Plot each of these points. Drawing a line that passes through the three points gives the horizontal line shown in Figure 3.15. ■

✔ CHECK POINT **7** Graph the linear equation: $y = 3$. See graphing answer section

Next, let's see what we can discover about the graph of an equation of the form $Ax + By = C$ when B is zero.

FIGURE 3.16 The graph $x = 5$

EXAMPLE 8 Graphing a Vertical Line

Graph the linear equation: $x = 5$.

SOLUTION All ordered pairs that are solutions of $x = 5$ have a value of x that is always 5. Any value can be used for y. Let's select three of the possible values for y: -2, 0, and 3. Using these values of y, three ordered pairs that are solutions of $x = 5$ are $(5, -2)$, $(5, 0)$, and $(5, 3)$. Drawing a line that passes through the three points gives the vertical line shown in Figure 3.16. ■

STUDY TIP

Do not confuse two-dimensional graphing and one-dimensional graphing of $x = 5$. The graph of $x = 5$ in a two-dimensional rectangular coordinate system is the vertical line in Figure 3.16 on page 203. By contrast, the graph of $x = 5$ on a one-dimensional number line representing values of x is a single point at 5:

✔ **CHECK POINT 8** Graph the linear equation: $x = -2$. See graphing answer section.

HORIZONTAL AND VERTICAL LINES The graph of a linear equation in one variable is a horizontal or vertical line.

The graph of $y = b$ is a horizontal line. The y-intercept is b.

The graph of $x = a$ is a vertical line. The x-intercept is a.

3.2 EXERCISE SET

Student Solutions Manual   CD/Video   PH Math/Tutor Center   MathXL Tutorials on CD   MathXL®   MyMathLab   Interactmath.com

Practice Exercises

In Exercises 1–8, use the graph to identify the
a. *x-intercept, or state that there is no x-intercept;*   **b.** *y-intercept, or state that there is no y-intercept.*

1.

a. 3   b. 4

2.

a. 5   b. 2

5.

a. 0   b. 0

6.

a. 0   b. 0

3.

a. −4   b. −2

4.
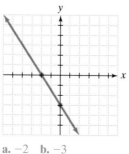
a. −2   b. −3

7.

a. no *x*-intercept   b. −2

8.
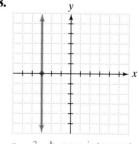
a. −3   b. no *y*-intercept

In Exercises 9–18, find the x-intercept and the y-intercept of the graph of each equation. Do not graph the equation.

9. $2x + 5y = 20$ *x*-intercept is 10; *y*-intercept is 4.

10. $2x + 6y = 30$ *x*-intercept is 15; *y*-intercept is 5.

11. $2x - 3y = 15$ *x*-intercept is $\frac{15}{2}$, or $7\frac{1}{2}$; *y*-intercept is -5.

12. $4x - 5y = 10$ *x*-intercept is $\frac{5}{2}$, or $2\frac{1}{2}$; *y*-intercept is -2.

13. $-x + 3y = -8$ *x*-intercept is 8; *y*-intercept is $-\frac{8}{3}$, or $-2\frac{2}{3}$.

14. $-x + 3y = -10$ *x*-intercept is 10; *y*-intercept is $-\frac{10}{3}$, or $-3\frac{1}{3}$.

15. $7x - 9y = 0$ *x*-intercept is 0; *y*-intercept is 0.

16. $8x - 11y = 0$ *x*-intercept is 0; *y*-intercept is 0.

17. $2x = 3y - 11$ *x*-intercept is $-\frac{11}{2}$, or $-5\frac{1}{2}$; *y*-intercept is $\frac{11}{3}$, or $3\frac{2}{3}$.

18. $2x = 4y - 13$ *x*-intercept is $-\frac{13}{2}$, or $-6\frac{1}{2}$; *y*-intercept is $\frac{13}{4}$, or $3\frac{1}{4}$.

In Exercises 19–40, use intercepts and a checkpoint to graph each equation. **19–40.** See graphing answer section.

19. $x + y = 5$

20. $x + y = 6$

21. $x + 3y = 6$

22. $2x + y = 4$

23. $6x - 9y = 18$

24. $6x - 2y = 12$

25. $-x + 4y = 6$

26. $-x + 3y = 10$

27. $2x - y = 7$

28. $2x - y = 5$

29. $3x = 5y - 15$

30. $2x = 3y + 6$

31. $25y = 100 - 50x$

32. $10y = 60 - 40x$

33. $2x - 8y = 12$

34. $3x - 6y = 15$

35. $x + 2y = 0$

36. $2x + y = 0$

37. $y - 3x = 0$

38. $y - 4x = 0$

39. $2x - 3y = -11$

40. $3x - 2y = -7$

In Exercises 41–46, write an equation for each graph.

41.

$y = 3$

42.

$y = -5$

43.

$x = -3$

44.

$x = \frac{7}{2}$ or $x = 3\frac{1}{2}$

45.

$y = 0$

46.

$x = 0$

In Exercises 47–62, graph each equation.
47–62. See graphing answer section.

47. $y = 4$

48. $y = 2$

49. $y = -2$

50. $y = -3$

51. $x = 2$

52. $x = 4$

53. $x + 1 = 0$

54. $x + 5 = 0$

55. $y - 3.5 = 0$

56. $y - 2.5 = 0$

57. $x = 0$

58. $y = 0$

59. $3y = 9$

60. $5y = 20$

61. $12 - 3x = 0$

62. $12 - 4x = 0$

Practice Plus

In Exercises 63–68, match each equation with one of the graphs shown in Exercises 1–8.

63. $3x + 2y = -6$ Exercise 4

64. $x + 2y = -4$ Exercise 3

65. $y = -2$ Exercise 7

66. $x = -3$ Exercise 8

67. $4x + 3y = 12$ Exercise 1

68. $2x + 5y = 10$ Exercise 2

In Exercises 69–70,

a. *Write a linear equation in standard form satisfying the given condition. Assume that all measurements shown in each figure are in feet.*

b. *Graph the equation in part (a). Because x and y must be non-negative (why?), limit your final graph to quadrant I and its boundaries.*

69. The perimeter of the larger rectangle is 58 feet.

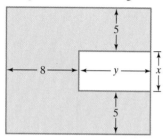

a. $x + y = 11$

b. See graphing answer section.

70. The perimeter of the shaded trapezoid is 84 feet.

a. $x - y = 17$

b. See graphing answer section.

Application Exercises

The flight of a vulture is observed for 30 seconds. The graph shows the vulture's height, in meters, during this period of time. Use the graph to solve Exercises 71–75.

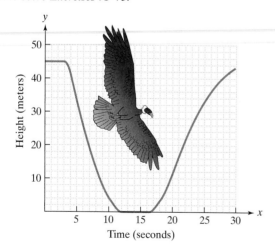

71. During which period of time is the vulture's height decreasing? from 3 to 12 seconds

72. During which period of time is the vulture's height increasing? from 17 to 30 seconds

73. What is the y-intercept? What does this mean about the vulture's height at the beginning of the observation? 45; The vulture was 45 meters above the ground when the observation started.

74. During the first three seconds of observation, the vulture's flight is graphed as a horizontal line. Write the equation of the line. What does this mean about the vulture's flight pattern during this time? $y = 45$; The vulture maintains a constant height of 45 meters during the first three seconds.

75. Use integers to write five x-intercepts of the graph. What is the vulture doing during these times? 12, 13, 14, 15, 16; The vulture is on the ground at this time.

Too late for that flu shot now! It's only 8 A.M. and you're feeling lousy. Fascinated by the way that algebra models the world (your author is projecting a bit here), you decide to construct a graph showing your body temperature from 8 A.M. through 3 P.M. You decide to let x represent the number of hours after 8 A.M. and y represent your temperature at time x. The graph is shown. Use it to solve Exercises 76–80.

76. What is the y-intercept? What does this mean about your temperature at 8 A.M.? 101; You have a temperature of 101° F at 8 A.M.

77. During which period of time is your temperature decreasing? 8 A.M. to 11 A.M.

78. Estimate your minimum temperature during the time period shown. How many hours after 8 A.M. does this occur? At what time does this occur? 98.6° F; 3; 11 A.M.

79. During which period of time is your temperature increasing? 11 A.M. to 1 P.M.

80. From five hours after 8 A.M. until seven hours after 8 A.M., your temperature is graphed as a portion of a horizontal line. Write the equation of the line. What does this mean about your temperature over this period of time? $y = 100$; Your temperature has remained constant at 100° F.

81. In the section opener, we saw that the linear equation

$$0.57x + y = 80$$

models the percentage of market share, y, for carbonated beverages of soft drinks sold in the United States x years after 1994.

a. Find the *y*-intercept of the equation's graph. Describe what this means in terms of the variables in the equation.   80; In 1994, carbonated beverages had 80% of the market share.

b. How well does your description in part (a) model the appropriate portion of the data shown in Figure 3.9 on page 197?   very well

c. Find the *x*-intercept of the equation's graph. Round to the nearest whole number. Describe what this means in terms of the variables in the equation. Does this seem reasonable or has model breakdown occurred?   140; In 2134, there will be no consumption of carbonated beverages. Model breakdown: model extended too far into the future.

82. As shown in the bar graph, the percentage of people in the United States satisfied with their lives remains relatively constant for all age groups. If *x* represents a person's age and *y* represents the percentage of people satisfied with their lives at that age, write an equation that reasonably models the data.   Answers will vary; examples: $y = 79$ or $y = 80$.

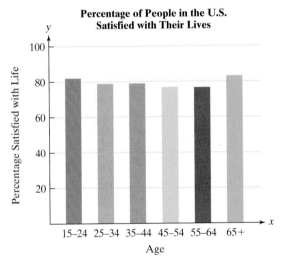

Percentage of People in the U.S. Satisfied with Their Lives

Source: Culture Shift in Advanced Industrial Society, Princeton University Press

Writing in Mathematics

83. What is an *x*-intercept of a graph?   83–92. Answers will vary.

84. What is a *y*-intercept of a graph?

85. If you are given an equation of the form $Ax + By = C$, explain how to find the *x*-intercept.

86. If you are given an equation of the form $Ax + By = C$, explain how to find the *y*-intercept.

87. Explain how to graph $Ax + By = C$ if *C* is not equal to zero.

88. Explain how to graph a linear equation of the form $Ax + By = 0$.

89. How many points are needed to graph a line? How many should actually be used? Explain.

90. Describe the graph of $y = 200$.

91. Describe the graph of $x = -100$.

92. We saw that the number of skiers in the United States has remained constant over time. Exercise 82 showed that the percentage of people satisfied with their lives remains constant for all age groups. Give another example of a real-world phenomenon that has remained relatively constant. Try writing an equation that models this phenomenon.

Critical Thinking Exercises

93. Write the equation of the line passing through the point $(5, 6)$ and parallel to the line whose equation is $y = -1$. $y = 6$

In Exercises 94–95, find the coefficients that must be placed in each shaded area so that the equation's graph will be a line with the specified intercepts.

94. $\blacksquare x + \blacksquare y = 10$; *x*-intercept = 5; *y*-intercept = 2   2; 5

95. $\blacksquare x + \blacksquare y = 12$; *x*-intercept = -2; *y*-intercept = 4 $-6; 3$

Technology Exercises

96. Use a graphing utility to verify any five of your hand-drawn graphs in Exercises 19–40. Solve the equation for *y* before entering it.   Answers will vary.

In Exercises 97–100, use a graphing utility to graph each equation. You will need to solve the equation for y before entering it. Use the equation displayed on the screen to identify the x-intercept and the y-intercept.   97–100. See graphing answer section.

97. $2x + y = 4$

98. $3x - y = 9$

99. $2x + 3y = 30$

100. $4x - 2y = -40$

Review Exercises

101. Find the absolute value: $|-13.4|$. (Section 1.2, Example 8)   13.4

102. Simplify: $7x - (3x - 5)$. (Section 1.7, Example 7) $4x + 5$

103. Graph: $-2 \leq x < 4$. (Section 2.7, Example 1)   See graphing answer section.

SECTION

3.3

Objectives

1 Compute a line's slope.

2 Use slope to show that lines are parallel.

3 Calculate rate of change in applied situations.

SLOPE

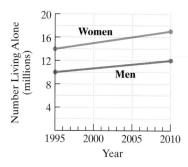

Number of People in the U.S. Living Alone

FIGURE 3.17

Source: Forrester Research

A best guess at the look of our nation in the next decades indicates that the number of men and women living alone will increase each year. Figure 3.17 shows that by 2010, approximately 12 million men and 17 million women will be living alone.

By looking at Figure 3.17, can you tell that the green graph representing women is steeper than the blue graph representing men? This indicates a greater yearly increase in the millions of women living alone than in the millions of men living alone. In this section, we will study the idea of a line's steepness and see what that has to do with how its variables are changing.

The Slope of a Line Mathematicians have developed a useful measure of the steepness of a line, called the **slope** of the line. Slope compares the vertical change (the **rise**) to the horizontal change (the **run**) when moving from one fixed point to another along the line. To calculate the slope of a line, we use a ratio that compares the change in y (the rise) to the change in x (the run).

DEFINITION OF SLOPE The **slope** of the line through the distinct points (x_1, y_1) and (x_2, y_2) is

$$\frac{\text{Change in } y}{\text{Change in } x} = \frac{\text{Rise}}{\text{Run}}$$

$$= \frac{y_2 - y_1}{x_2 - x_1},$$

where $x_2 - x_1 \neq 0$.

It is common notation to let the letter m represent the slope of a line. The letter m is used because it is the first letter of the French verb *monter*, meaning to rise, or to ascend.

ENRICHMENT ESSAY

Slope and the Streets of San Francisco

San Francisco's Filbert Street has a slope of 0.613, meaning that for every horizontal distance of 100 feet, the street ascends 61.3 feet vertically. With its 31.5° angle of inclination, the street is too steep to pave and is only accessible by wooden stairs.

EXAMPLE 1 Using the Definition of Slope

Find the slope of the line passing through each pair of points:

a. $(-3, -1)$ and $(-2, 4)$ **b.** $(-3, 4)$ and $(2, -2)$.

SOLUTION

a. Let $(x_1, y_1) = (-3, -1)$ and $(x_2, y_2) = (-2, 4)$. We obtain the slope as follows:

$$m = \frac{\text{Change in } y}{\text{Change in } x} = \frac{y_2 - y_1}{x_2 - x_1} = \frac{4 - (-1)}{-2 - (-3)} = \frac{5}{1} = 5.$$

The situation is illustrated in Figure 3.18(a). The slope of the line is 5, indicating that there is a vertical change, a rise, of 5 units for each horizontal change, a run, of 1 unit. The slope is positive and the line rises from left to right.

STUDY TIP

When computing slope, it makes no difference which point you call (x_1, y_1) and which point you call (x_2, y_2). If we let $(x_1, y_1) = (-2, 4)$ and $(x_2, y_2) = (-3, -1)$, the slope is still 5:

$$m = \frac{\text{Change in } y}{\text{Change in } x} = \frac{y_2 - y_1}{x_2 - x_1} = \frac{-1 - 4}{-3 - (-2)} = \frac{-5}{-1} = 5.$$

However, you should not subtract in one order in the numerator ($y_2 - y_1$) and then in a different order in the denominator ($x_1 - x_2$). The slope is *not* -5:

$$\frac{-1 - 4}{-2 - (-3)} = \frac{-5}{1} = -5. \quad \text{Incorrect}$$

b. We can let $(x_1, y_1) = (-3, 4)$ and $(x_2, y_2) = (2, -2)$. The slope of the line shown in Figure 3.18(b) is computed as follows:

$$m = \frac{\text{Change in } y}{\text{Change in } x} = \frac{y_2 - y_1}{x_2 - x_1} = \frac{-2 - 4}{2 - (-3)} = \frac{-6}{5} = -\frac{6}{5}.$$

The slope of the line is $-\frac{6}{5}$. For every vertical change of -6 units (6 units down), there is a corresponding horizontal change of 5 units. The slope is negative and the line falls from left to right.

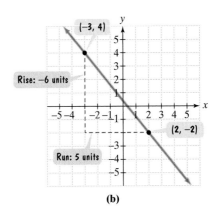

(a)   (b)

FIGURE 3.18 Visualizing slope

✔ **CHECK POINT 1** Find the slope of the line passing through each pair of points:

a. $(-3, 4)$ and $(-4, -2)$ $m = 6$ **b.** $(4, -2)$ and $(-1, 5)$. $m = -\frac{7}{5}$

EXAMPLE 2 **Using the Definition of Slope for Horizontal and Vertical Lines**

Find the slope of the line passing through each pair of points:

a. $(5, 4)$ and $(3, 4)$ **b.** $(2, 5)$ and $(2, 1)$.

SOLUTION

a. Let $(x_1, y_1) = (5, 4)$ and $(x_2, y_2) = (3, 4)$. We obtain the slope as follows:

$$m = \frac{\text{Change in } y}{\text{Change in } x} = \frac{y_2 - y_1}{x_2 - x_1} = \frac{4 - 4}{3 - 5} = \frac{0}{-2} = 0.$$

The situation is illustrated in Figure 3.19(a). Can you see that the line is horizontal? Because any two points on a horizontal line have the same y-coordinate, these lines neither rise nor fall from left to right. The change in y, $y_2 - y_1$, is always zero. Thus, **the slope of any horizontal line is zero**.

b. We can let $(x_1, y_1) = (2, 5)$ and $(x_2, y_2) = (2, 1)$. Figure 3.19(b) shows that these points are on a vertical line. We attempt to compute the slope as follows:

$$m = \frac{\text{Change in } y}{\text{Change in } x} = \frac{1 - 5}{2 - 2} = \frac{-4}{0} \quad \text{Division by zero is undefined.}$$

Because division by zero is undefined, the slope of the vertical line in Figure 3.19(b) is undefined. In general, **the slope of any vertical line is undefined**.

(a) Horizontal lines have no vertical change.

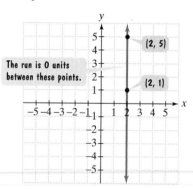

(b) Vertical lines have no horizontal change.

FIGURE 3.19 Visualizing slope

Table 3.1 summarizes four possibilities for the slope of a line.

Table 3.1	Possibilities for a Line's Slope		
Positive Slope	**Negative Slope**	**Zero Slope**	**Undefined Slope**
$m > 0$	$m < 0$	$m = 0$	m is undefined.
Line rises from left to right.	Line falls from left to right.	Line is horizontal.	Line is vertical.

✔ **CHECK POINT 2** Find the slope of the line passing through each pair of points or state that the slope is undefined:

a. $(6, 5)$ and $(2, 5)$ *m = 0*   **b.** $(1, 6)$ and $(1, 4)$.   undefined slope

2 Use slope to show that lines are parallel.

Slope and Parallel Lines Two nonintersecting lines that lie in the same plane are **parallel**. If two lines do not intersect, the ratio of the vertical change to the horizontal change is the same for each line. Because two parallel lines have the same "steepness," they must have the same slope.

SLOPE AND PARALLEL LINES

1. If two nonvertical lines are parallel, then they have the same slope.
2. If two distinct nonvertical lines have the same slope, then they are parallel.
3. Two distinct vertical lines, each with undefined slope, are parallel.

FIGURE 3.20 Using slope to show that lines are parallel

EXAMPLE 3 Using Slope to Show That Lines Are Parallel

Show that the line passing through $(1, 4)$ and $(3, 2)$ is parallel to the line passing through $(2, 8)$ and $(4, 6)$.

SOLUTION The situation is illustrated in Figure 3.20. The lines certainly look like they are parallel. Let's use equal slopes to confirm this fact. For each line, we compute the ratio of the difference in y-coordinates to the difference in x-coordinates. Be sure to subtract the coordinates in the same order.

Slope of the line through $(1, 4)$ and $(3, 2)$ is

$$\frac{4 - 2}{1 - 3} = \frac{2}{-2} = -1.$$

Slope of the line through $(2, 8)$ and $(4, 6)$ is

$$\frac{8 - 6}{2 - 4} = \frac{2}{-2} = -1.$$

Because the slopes are equal, the lines are parallel. ■

✔ **CHECK POINT 3** Show that the line passing through $(4, 2)$ and $(6, 6)$ is parallel to the line passing through $(0, -2)$ and $(1, 0)$.   Both slopes equal 2, so the lines are parallel.

3 Calculate rate of change in applied situations.

Slope as Rate of Change Slope is defined as the ratio of a change in y to a corresponding change in x. It tells how fast y is changing with respect to x. Thus, the slope of a line represents its rate of change.

Our next example shows how slope can be interpreted as a rate of change in an applied situation. When calculating slope in applied problems, keep track of the units in the numerator and the denominator.

EXAMPLE 4 Slope as a Rate of Change

The line graphs for the number of women and men living alone are shown again in Figure 3.21 at the top of the next page. Find the slope of the line segment for the women. Describe what this slope represents.

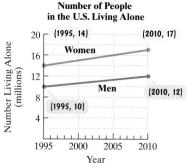

Number of People in the U.S. Living Alone

FIGURE 3.21
Source: Forrester Research

$$m = \frac{2}{15} \approx \frac{0.13 \text{ million men}}{\text{year}}$$

The slope indicates that the number of U.S. men living alone is projected to increase by 0.13 million each year. The rate of change is 0.13 million men per year.

SOLUTION We let x represent a year and y the number of women living alone in that year. The two points shown on the line segment for women have the following coordinates:

$$(1995, 14) \quad \text{and} \quad (2010, 17).$$

In 1995, 14 million U.S. women lived alone.

In 2010, 17 million U.S. women are projected to live alone.

Now we compute the slope:

The unit in the numerator is *million women.*

$$m = \frac{\text{Change in } y}{\text{Change in } x} = \frac{17 - 14}{2010 - 1995}$$

The unit in the denominator is *year.*

$$= \frac{3}{15} = \frac{1}{5} = \frac{0.2 \text{ million people}}{\text{year}}.$$

The slope indicates that the number of U.S. women living alone is projected to increase by 0.2 million each year. The rate of change is 0.2 million women per year. ∎

 CHECK POINT 4 Use the graph in Example 4 to find the slope of the line segment for the men. Express the slope correct to two decimal places and describe what it represents.

In Check Point 4, did you find that the slope of the line segment for the men is different from that of the women? The rate of change for men living alone is not equal to the rate of change for women living alone. Because of these different slopes, if you extend the line segments in Figure 3.21, the resulting lines will intersect. They are not parallel.

ENRICHMENT ESSAY

Railroads and Highways

The steepest part of Mt. Washington Cog Railway in New Hampshire has a 37% grade. This is equivalent to a slope of $\frac{37}{100}$. For every horizontal change of 100 feet, the railroad ascends 37 feet vertically. Engineers denote slope by grade, expressing slope as a percentage.

4 feet   5 feet

100 feet
4% Grade

100 feet
−5% Grade

Railroad grades are usually less than 2%, although in the mountains they may go as high as 4%. The grade of the Mt. Washington Cog Railway is phenomenal, making it necessary for locomotives to *push* single cars up its steepest part.

A Mount Washington Cog Railway locomotive pushing a single car up the steepest part of the railroad. The locomotive is about 120 years old.

3.3 EXERCISE SET

Student Solutions Manual   CD/Video   PH Math/Tutor Center   MathXL Tutorials on CD   MathXL®   MyMathLab   Interactmath.com

Practice Exercises

In Exercises 1–10, find the slope of the line passing through each pair of points or state that the slope is undefined. Then indicate whether the line through the points rises, falls, is horizontal, or is vertical.

1. $(4, 7)$ and $(8, 10)$ $m = \frac{3}{4}$; line rises

2. $(2, 1)$ and $(3, 4)$ $m = 3$; line rises

3. $(-2, 1)$ and $(2, 2)$ $m = \frac{1}{4}$; line rises

4. $(-1, 3)$ and $(2, 4)$ $m = \frac{1}{3}$; line rises

5. $(4, -2)$ and $(3, -2)$ $m = 0$; line is horizontal

6. $(4, -1)$ and $(3, -1)$ $m = 0$; line is horizontal

7. $(-2, 4)$ and $(-1, -1)$ $m = -5$; line falls

8. $(6, -4)$ and $(4, -2)$ $m = -1$; line falls

9. $(5, 3)$ and $(5, -2)$ slope is undefined; line is vertical

10. $(3, -4)$ and $(3, 5)$ slope is undefined; line is vertical

In Exercises 11–22, find the slope of each line, or state that the slope is undefined.

11.

$m = \frac{1}{2}$

12.

$m = 1$

13.

$m = -\frac{1}{3}$

14.

$m = -\frac{1}{3}$

15.

$m = -\frac{1}{2}$

16.

$m = -\frac{2}{3}$

17.

$m = -\frac{2}{3}$

18.

$m = \frac{1}{2}$

19.

$m = 0$

20.

$m = 0$

21.

undefined slope

22.
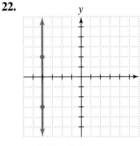
undefined slope

In Exercises 23–26, determine whether the distinct lines through each pair of points are parallel.

23. $(-2, 0)$ and $(0, 6)$; $(1, 8)$ and $(0, 5)$ $m = 3$; so lines are parallel

24. $(2, 4)$ and $(6, 1)$; $(-3, 1)$ and $(1, -2)$ $m = -\frac{3}{4}$; so lines are parallel

25. $(0, 3)$ and $(1, 5)$; $(-1, 7)$ and $(1, 10)$ $m = 2$ and $m = \frac{3}{2}$; lines are not parallel

26. $(-7, 6)$ and $(0, 4)$; $(-9, -3)$ and $(1, 5)$ $m = -\frac{2}{7}$ and $m = \frac{4}{5}$; lines are not parallel

Practice Plus

27. On the same set of axes, draw lines passing through the origin with slopes -1, $-\frac{1}{2}$, 0, $\frac{1}{3}$, and 2.   See graphing answer section.

28. On the same set of axes, draw lines with y-intercept 4 and slopes -1, $-\frac{1}{2}$, 0, $\frac{1}{3}$, and 2.   See graphing answer section.

Use slopes to solve Exercises 29–30.

29. Show that the points whose coordinates are $(-3, -3)$, $(2, -5)$, $(5, -1)$, and $(0, 1)$ are the vertices of a four-sided figure whose opposite sides are parallel. (Such a figure is called a *parallelogram*.)   Slopes of corresponding opposite sides are equal: $-\frac{2}{5}$ and $\frac{4}{3}$.

30. Show that the points whose coordinates are $(-3, 6)$, $(2, -3)$, $(11, 2)$, and $(6, 11)$ are the vertices of a four-sided figure whose opposite sides are parallel.   Slopes of corresponding opposite sides are equal: $\frac{5}{9}$ and $-\frac{9}{5}$.

31. The line passing through $(5, y)$ and $(1, 0)$ is parallel to the line joining $(2, 3)$ and $(-2, 1)$. Find y. 2

32. The line passing through $(1, y)$ and $(7, 12)$ is parallel to the line joining $(-3, 4)$ and $(-5, -2)$. Find y. -6

In Exercises 33–34, use slopes to determine whether the given points lie on a line. (Such points are said to be collinear.)

33. $(1, 2)$, $(3, 12)$, $(6, 27)$ collinear

34. $(1, 3)$, $(3, 11)$, $(6, 22)$ not collinear

Application Exercises

The graph shows online shopping per U.S. online household through 2004. Use the information provided by the graph to solve Exercises 35–36.

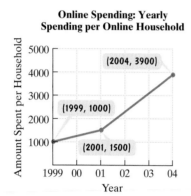

Online Spending: Yearly Spending per Online Household

(2004, 3900)
(1999, 1000)
(2001, 1500)

Source: Forrester Research

35. Find the slope of the line passing through $(1999, 1000)$ and $(2001, 1500)$. What does this represent in terms of the increase in online shopping per year? $m = 250$; The amount spent online per U.S. online household increased by $250 each year from 1999 to 2001.

36. Find the slope of the line passing through $(2001, 1500)$ and $(2004, 3900)$. What does this represent in terms of the increase in online shopping per year? $m = 800$; The amount spent online per U.S. online household increased by $800 each year from 2001 to 2004.

The graphs approximate the percentage of full-time police officers in the United States, by gender, y, for year x, where $1972 \leq x \leq 2002$. Use this information to solve Exercises 37–38.

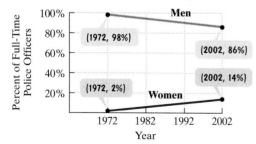

Police Officers in the U.S., by Gender

Men
(1972, 98%)
(2002, 86%)
(2002, 14%)
(1972, 2%)
Women

Source: National Center for Women and Policing

37. Find the slope of the line segment representing men. Then express the slope as a rate of change with the proper units attached. -0.4; The percentage of men is decreasing by 0.4% per year.

38. Find the slope of the line segment representing women. Then express the slope as a rate of change with the proper units attached.   0.4; The percentage of women is increasing by 0.4% per year.

The graphs approximate the mean, or average, income for well-to-do U.S. families (the highest 20%) and the lowest-income U.S. families (the lowest 20%), y, for year x, where $1980 \leq x \leq 2002$. Use this information to solve Exercises 39–40.

Mean Income of U.S. Families in 2002 Dollars

(2002, 161,800)
Highest-Income 20%
(1980, 98,000)
Lowest-Income 20%
(1980, 12,000)
(2002, 13,650)

Source: U.S. Census Bureau

39. Find the slope of the line segment representing the highest-income 20%. Then express the slope as a rate of change with the proper units attached.   2900; Mean income is increasing $2900 per year.

40. Find the slope of the line segment representing the lowest-income 20%. Then express the slope as a rate of change with the proper units attached.   75; Mean income is increasing $75 per year.

In Exercises 41–42, find the slope of each line. Then express the slope as a rate of change with the proper units attached.

41. The graph shows the cost to own and operate a full-size truck, y, in dollars, in terms of the miles the truck is driven, x.   0.40; Cost increases by $0.40 per mile driven.

Cost to Own and Operate a Full-Size Truck

(40,000, 16,000)
(20,000, 8000)

Source: Federal Highway Administration

42. The graph shows the cost to own and operate a compact car, *y*, in dollars, in terms of the miles the car is driven, *x*.
0.36; Cost increases by $0.36 per mile driven.

Cost to Own and Operate a Compact Car

(20,000, 7200)
(10,000, 3600)

Source: Federal Highway Administration

The pitch of a roof refers to the absolute value of its slope. In Exercises 43–44, find the pitch of each roof shown.

43.

6 feet   ←18 feet→

pitch $= \frac{1}{3}$

44.

6 feet
10 feet

pitch $= \frac{3}{5}$

The grade of a road or ramp refers to its slope expressed as a percent. Use this information to solve Exercises 45–46.

45. Construction laws are very specific when it comes to access ramps for the disabled. Every vertical rise of 1 foot requires a horizontal run of 12 feet. What is the grade of such a ramp? Round to the nearest tenth of a percent.   8.3%

1 foot

12 feet

46. A college campus goes beyond the standards described in Exercise 45. All wheelchair ramps on campus are designed so that every vertical rise of 1 foot is accompanied by a horizontal run of 14 feet. What is the grade of such a ramp? Round to the nearest tenth of a percent.   7.1%

Writing in Mathematics

47. What is the slope of a line?   47–52. Answers will vary.

48. Describe how to calculate the slope of a line passing through two points.

49. What does it mean if the slope of a line is zero?

50. What does it mean if the slope of a line is undefined?

51. If two lines are parallel, describe the relationship between their slopes.

52. Look back at the graph for Exercises 35–36. Do you think that the line through the points corresponding to the year 2001 and the year 2004 will model online spending per online household in the year 2040? Explain your answer.

Critical Thinking Exercises

53. Which one of the following is true?   b

   a. Slope is run divided by rise.

   b. The line through $(2, 2)$ and the origin has slope 1.

   c. A line with slope 3 can be parallel to a line with slope -3.

   d. The line through $(3, 1)$ and $(3, -5)$ has zero slope.

In Exercises 54–55, use the figure shown to make the indicated list.

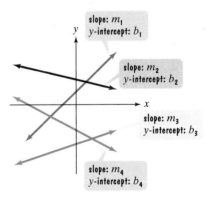

slope: m_1
y-intercept: b_1

slope: m_2
y-intercept: b_2

slope: m_3
y-intercept: b_3

slope: m_4
y-intercept: b_4

54. List the slopes m_1, m_2, m_3, and m_4 in order of decreasing size. m_1, m_3, m_2, m_4

55. List the y-intercepts b_1, b_2, b_3, and b_4 in order of decreasing size. b_2, b_1, b_4, b_3

Technology Exercises

Use a graphing utility to graph each equation in Exercises 56–59. Then use the [TRACE] *feature to trace along the line and find the coordinates of two points. Use these points to compute the line's slope.*

56. $y = 2x + 4$ $m = 2$ **57.** $y = -3x + 6$ $m = -3$

58. $y = -\frac{1}{2}x - 5$ $m = -\frac{1}{2}$

59. $y = \frac{3}{4}x - 2$ $m = \frac{3}{4}$

60. In Exercises 56–59, compare the slope that you found with the line's equation. What relationship do you observe between the line's slope and one of the constants in the equation?   The line's slope is the coefficient of x.

Review Exercises

61. A 36-inch board is cut into two pieces. One piece is twice as long as the other. How long are the pieces? (Section 2.5, Example 3)   12 inches and 24 inches

62. Simplify: $-10 + 16 \div 2(-4)$. (Section 1.8, Example 4) -42

63. Solve and graph the solution set on a number line: $2x - 3 \le 5$. (Section 2.7, Example 5) $\{x \mid x \le 4\}$; See graphing answer section.

SECTION 3.4

THE SLOPE-INTERCEPT FORM OF THE EQUATION OF A LINE

Objectives

1 Find a line's slope and y-intercept from its equation.

2 Graph lines in slope-intercept form.

3 Use slope and y-intercept to graph $Ax + By = C$.

4 Use slope and y-intercept to model data.

Children in Kampala, Uganda, discuss HIV/AIDS issues.

AIDS, famine, malaria, and civil wars have all taken their toll on Africa. In 2003, Secretary of State Colin Powell declared, "HIV is now more destructive than any army, any conflict, any weapon of mass destruction."

In this section, you will study the form of an equation that will enable you to develop a model for the growth of HIV/AIDS in Africa. Using this model, you will understand the need to fight AIDS on a continent ravaged by the virus.

1 Find a line's slope and y-intercept from its equation.

The Slope-Intercept Form of the Equation of a Line   Let's begin with an example that shows how easy it is to find a line's slope and y-intercept from its equation.

Figure 3.22 shows the graph of $y = 2x + 4$. Verify that the x-intercept is -2 by setting y equal to 0 and solving for x. Similarly, verify that the y-intercept is 4 by setting x equal to 0 and solving for y.

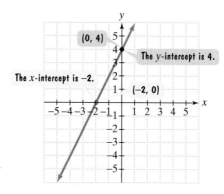

FIGURE 3.22   The graph of $y = 2x + 4$

Now that we have two points on the line, $(-2, 0)$ and $(0, 4)$, we can calculate the slope of the graph of $y = 2x + 4$.

$$\text{Slope} = \frac{\text{Change in } y}{\text{Change in } x}$$

$$= \frac{4 - 0}{0 - (-2)} = \frac{4}{2} = 2$$

We see that the slope of the line is 2, the same as the coefficient of x in the equation $y = 2x + 4$. The y-intercept is 4, the same as the constant in the equation $y = 2x + 4$.

$$y = 2x + 4$$

The slope is 2.   The y-intercept is 4.

It is not merely a coincidence that the x-coefficient is the line's slope and the constant term is the y-intercept. Let's find the equation of any nonvertical line with slope m and y-intercept b. Because the y-intercept is b, the point $(0, b)$ lies on the line. Now, let (x, y) represent any other point on the line, shown in Figure 3.23. Keep in mind that the point (x, y) is arbitrary and is not in one fixed position. By contrast, the point $(0, b)$ is fixed.

Regardless of where the point (x, y) is located, the steepness of the line in Figure 3.23 remains the same. Thus, the ratio for slope stays a constant m. This means that for all points along the line

$$m = \frac{\text{Change in } y}{\text{Change in } x} = \frac{y - b}{x - 0} = \frac{y - b}{x}.$$

We can clear the fraction by multiplying both sides by x, the least common denominator.

$$m = \frac{y - b}{x} \qquad \text{This is the slope of the line in Figure 3.23.}$$

$$mx = \frac{y - b}{x} \cdot x \qquad \text{Multiply both sides by x.}$$

$$mx = y - b \qquad \text{Simplify: } \frac{y - b}{\cancel{x}} \cdot \cancel{x} = y - b.$$

$$mx + b = y - b + b \qquad \text{Add b to both sides and solve for y.}$$

$$mx + b = y \qquad \text{Simplify.}$$

Now, if we reverse the two sides, we obtain the **slope-intercept form** of the equation of a line.

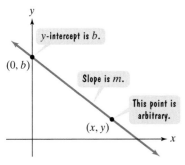

FIGURE 3.23 A line with slope m and y-intercept b

SLOPE-INTERCEPT FORM OF THE EQUATION OF A LINE   The **slope-intercept equation** of a nonvertical line with slope m and y-intercept b is

$$y = mx + b.$$

Thus, if a line's equation is written with y isolated on one side, the x-coefficient is the line's slope and the constant term is the y-intercept.

EXAMPLE 1   Finding a Line's Slope and y-Intercept from Its Equation

Find the slope and the y-intercept of the line with the given equation:

a. $y = 2x - 4$ **b.** $y = \dfrac{1}{2}x + 2$ **c.** $5x + y = 4$.

SOLUTION

a. We write $y = 2x - 4$ as $y = 2x + (-4)$. The slope is the x-coefficient and the y-intercept is the constant term.

$$y = 2x + (-4)$$

The slope is 2. The y-intercept is -4.

b. The equation $y = \frac{1}{2}x + 2$ is in the form $y = mx + b$. We can find the slope, m, by identifying the coefficient of x. We can find the y-intercept, b, by identifying the constant term.

$$y = \frac{1}{2}x + 2$$

The slope is $\frac{1}{2}$. The y-intercept is 2.

c. The equation $5x + y = 4$ is not in the form $y = mx + b$. We can obtain this form by isolating y on one side. We isolate y on the left side by subtracting $5x$ from both sides.

$$5x + y = 4 \qquad \text{This is the given equation.}$$
$$5x - 5x + y = -5x + 4 \qquad \text{Subtract 5x from both sides.}$$
$$y = -5x + 4 \qquad \text{Simplify.}$$

Now, the equation is in the form $y = mx + b$. The slope is the coefficient of x and the y-intercept is the constant term.

$$y = -5x + 4$$

The slope is -5. The y-intercept is 4.

 CHECK POINT 1 Find the slope and the y-intercept of the line with the given equation:

a. $y = 5x - 3$ $m = 5$; y-intercept is -3 **b.** $y = \frac{2}{3}x + 4$ $m = \frac{2}{3}$; y-intercept is 4

c. $7x + y = 6$. $m = -7$; y-intercept is 6

2 Graph lines in slope-intercept form.

Graphing $y = mx + b$ by Using the Slope and y-Intercept If a line's equation is written with y isolated on one side, we can use the y-intercept and the slope to obtain its graph.

GRAPHING $y = mx + b$ BY USING THE SLOPE AND y-INTERCEPT

1. Plot the point containing the y-intercept on the y-axis. This is the point $(0, b)$.
2. Obtain a second point using the slope, m. Write m as a fraction, and use rise over run, starting at the point containing the y-intercept, to plot this point.
3. Use a straightedge to draw a line through the two points. Draw arrowheads at the ends of the line to show that the line continues indefinitely in both directions.

EXAMPLE 2 Graphing by Using the Slope and y-Intercept

Graph the line whose equation is $y = 4x - 3$.

SOLUTION We write $y = 4x - 3$ in the form $y = mx + b$.

$$y = 4x + (-3)$$

The slope is 4.   The y-intercept is –3.

Now that we have identified the slope and the y-intercept, we use the three steps in the box to graph the equation.

Step 1. **Plot the point containing the y-intercept on the y-axis.** The y-intercept is -3. We plot the point $(0, -3)$, shown in Figure 3.24(a).

Step 2. **Obtain a second point using the slope, *m*. Write *m* as a fraction, and use rise over run, starting at the point containing the y-intercept, to plot this point.** The slope, 4, written as a fraction is $\frac{4}{1}$.

$$m = \frac{4}{1} = \frac{\text{Rise}}{\text{Run}}$$

We plot the second point on the line by starting at $(0, -3)$, the first point. Based on the slope, we move 4 units *up* (the rise) and 1 unit to the *right* (the run). This puts us at a second point on the line, $(1, 1)$, shown in Figure 3.24(b).

Step 3. **Use a straightedge to draw a line through the two points.** The graph of $y = 4x - 3$ is shown in Figure 3.24(c).

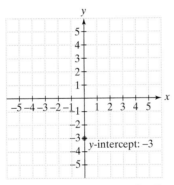

(a) The y-intercept is –3, so (0, –3) is a point on the line.

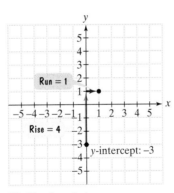

(b) The slope is 4.

(c) The graph of $y = 4x - 3$

FIGURE 3.24 Graphing $y = 4x - 3$ using the y-intercept and slope

See graphing answer section.

CHECK POINT **2** Graph the line whose equation is $y = 3x - 2$.

EXAMPLE 3 Graphing by Using the Slope and y-Intercept

Graph the line whose equation is $y = \frac{2}{3}x + 2$.

SOLUTION The equation of the line, $y = \frac{2}{3}x + 2$, is in the form $y = mx + b$. We can find the slope, m, by identifying the coefficient of x. We can find the y-intercept, b, by identifying the constant term.

$$y = \frac{2}{3}x + 2$$

The slope is $\frac{2}{3}$.

The y-intercept is 2.

Now that we have identified the slope and the y-intercept, we use the three-step procedure to graph the equation.

Step 1. **Plot the point containing the y-intercept on the y-axis.** The y-intercept is 2. We plot $(0, 2)$, shown in Figure 3.25.

Step 2. **Obtain a second point using the slope, m. Write m as a fraction, and use rise over run, starting at the point containing the y-intercept, to plot this point.** The slope, $\frac{2}{3}$, is already written as a fraction.

$$m = \frac{2}{3} = \frac{\text{Rise}}{\text{Run}}$$

We plot the second point on the line by starting at $(0, 2)$, the first point. Based on the slope, we move 2 units *up* (the rise) and 3 units to the *right* (the run). This puts us at a second point on the line, $(3, 4)$, shown in Figure 3.25.

Step 3. **Use a straightedge to draw a line through the two points.** The graph of $y = \frac{2}{3}x + 2$ is shown in Figure 3.25. ■

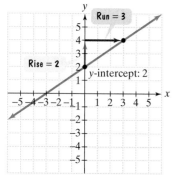

FIGURE 3.25 The graph of $y = \frac{2}{3}x + 2$

See graphing answer section.

 CHECK POINT 3 Graph the line whose equation is $y = \frac{3}{5}x + 1$.

---

3 Use slope and y-intercept to graph $Ax + By = C$.

Graphing $Ax + By = C$ by Using the Slope and y-Intercept Earlier in this chapter, we considered linear equations of the form $Ax + By = C$. We used x- and y-intercepts, as well as checkpoints, to graph these equations. It is also possible to obtain the graphs by using the slope and y-intercept. To do this, begin by solving $Ax + By = C$ for y. This will put the equation in slope-intercept form. Then use the three-step procedure to graph the equation. This is illustrated in Example 4.

EXAMPLE 4 Graphing by Using the Slope and y-Intercept

Graph the linear equation $2x + 5y = 0$ by using the slope and y-intercept.

SOLUTION We put the equation in slope-intercept form by solving for y.

$$2x + 5y = 0 \qquad \text{This is the given equation.}$$

$$2x - 2x + 5y = -2x + 0 \qquad \text{Subtract 2x from both sides.}$$

$$5y = -2x + 0 \qquad \text{Simplify.}$$

$$\frac{5y}{5} = \frac{-2x + 0}{5} \qquad \text{Divide both sides by 5.}$$

$$y = \frac{-2x}{5} + \frac{0}{5} \qquad \text{Divide each term in the numerator by 5.}$$

$$y = -\frac{2}{5}x + 0 \qquad \text{Simplify.}$$

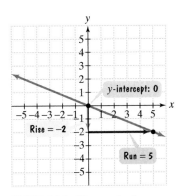

FIGURE 3.26 The graph of $2x + 5y = 0$, or $y = -\frac{2}{5}x + 0$

Now that the equation is in slope-intercept form, we can use the slope and y-intercept to obtain its graph. Examine the slope-intercept form:

$$y = -\frac{2}{5}x + 0.$$

slope: $-\frac{2}{5}$ y-intercept: 0

Note that the slope is $-\frac{2}{5}$ and the y-intercept is 0. Use the y-intercept to plot $(0, 0)$ on the y-axis. Then locate a second point by using the slope.

$$m = -\frac{2}{5} = \frac{-2}{5} = \frac{\text{Rise}}{\text{Run}}$$

Because the rise is -2 and the run is 5, move *down* 2 units and to the *right* 5 units, starting at the point $(0, 0)$. This puts us at a second point on the line, $(5, -2)$. The graph of $2x + 5y = 0$ is the line drawn through these points, shown in Figure 3.26. ∎

DISCOVER FOR YOURSELF

Obtain a second point in Example 4 by writing the slope as follows:

$$m = \frac{2}{-5} = \frac{\text{Rise}}{\text{Run}}$$

$-\frac{2}{5}$ can be expressed as $\frac{-2}{5}$ or $\frac{2}{-5}$.

Obtain a second point in Figure 3.26 by moving *up* 2 units and to the *left* 5 units, starting at $(0, 0)$. What do you observe once you graph the line?

 CHECK POINT 4 Graph the linear equation $3x + 4y = 0$ by using the slope and y-intercept.   See graphing answer section.

---

4   Use slope and y-intercept to model data.

Modeling with the Slope-Intercept Form of the Equation of a Line   If an equation in slope-intercept form models some physical situation, then the slope and y-intercept have physical interpretations. For the equation $y = mx + b$, the y-intercept, b, tells us what is happening to y when x is 0. If x represents time, the y-intercept describes the value of y at the beginning, or when time equals 0. The slope represents the rate of change in y per unit change in x.

Let's see how we can use these ideas to develop a model for data. In the previous section, we looked at line graphs for the number of U.S. men and women living alone, repeated in Figure 3.27. Let's develop a model for the data for women living alone. We let x represent the number of years after 1995. At the beginning of our data, or 0 years after 1995, 14 million women lived alone. Thus, $b = 14$. In Example 4 in the previous section, we found that $m = 0.2$ (rate of change is 0.2 million women per year). An equation of the form $y = mx + b$ that models the data is

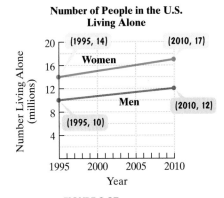

FIGURE 3.27

$$y = 0.2x + 14,$$

where y is the number, in millions, of U.S. women living alone x years after 1995.

3.4 EXERCISE SET

Student Solutions Manual CD/Video PH Math/Tutor Center MathXL Tutorials on CD MathXL® MyMathLab Interactmath.com

Practice Exercises

In Exercises 1–12, find the slope and the y-intercept of the line with the given equation.

1. $y = 3x + 2$
$m = 3$; y-intercept $= 2$

2. $y = 9x + 4$
$m = 9$; y-intercept $= 4$

3. $y = 3x - 5$
$m = 3$; y-intercept $= -5$

4. $y = 4x - 2$
$m = 4$; y-intercept $= -2$

5. $y = -\frac{1}{2}x + 5$
$m = -\frac{1}{2}$; y-intercept $= 5$

6. $y = -\frac{3}{4}x + 6$
$m = -\frac{3}{4}$; y-intercept $= 6$

7. $y = 7x$
$m = 7$; y-intercept $= 0$

8. $y = 10x$
$m = 10$; y-intercept $= 0$

9. $y = 10$
$m = 0$; y-intercept $= 10$

10. $y = 7$
$m = 0$; y-intercept $= 7$

11. $y = 4 - x$
$m = -1$; y-intercept $= 4$

12. $y = 5 - x$
$m = -1$; y-intercept $= 5$

In Exercises 13–26, begin by solving the linear equation for y. This will put the equation in slope-intercept form. Then find the slope and the y-intercept of the line with this equation.

13. $-5x + y = 7$ $y = 5x + 7$; $m = 5$; y-intercept $= 7$

14. $-9x + y = 5$ $y = 9x + 5$; $m = 9$; y-intercept $= 5$

15. $x + y = 6$ $y = -x + 6$; $m = -1$; y-intercept $= 6$

16. $x + y = 8$ $y = -x + 8$; $m = -1$; y-intercept $= 8$

17. $6x + y = 0$ $y = -6x$; $m = -6$; y-intercept $= 0$

18. $8x + y = 0$ $y = -8x$; $m = -8$; y-intercept $= 0$

19. $3y = 6x$ $y = 2x$; $m = 2$; y-intercept $= 0$

20. $3y = -9x$ $y = -3x$; $m = -3$; y-intercept $= 0$

21. $2x + 7y = 0$ $y = -\frac{2}{7}x$; $m = -\frac{2}{7}$; y-intercept $= 0$

22. $2x + 9y = 0$ $y = -\frac{2}{9}x$; $m = -\frac{2}{9}$; y-intercept $= 0$

23. $3x + 2y = 3$ $y = -\frac{3}{2}x + \frac{3}{2}$; $m = -\frac{3}{2}$; y-intercept $= \frac{3}{2}$

24. $4x + 3y = 4$ $y = -\frac{4}{3}x + \frac{4}{3}$; $m = -\frac{4}{3}$; y-intercept $= \frac{4}{3}$

25. $3x - 4y = 12$ $y = \frac{3}{4}x - 3$; $m = \frac{3}{4}$; y-intercept $= -3$

26. $5x - 2y = 10$ $y = \frac{5}{2}x - 5$; $m = \frac{5}{2}$; y-intercept $= -5$

In Exercises 27–38, graph each linear equation using the slope and y-intercept. **27–38.** See graphing answer section.

27. $y = 2x + 4$

28. $y = 3x + 1$

29. $y = -3x + 5$

30. $y = -2x + 4$

31. $y = \frac{1}{2}x + 1$

32. $y = \frac{1}{3}x + 2$

33. $y = \frac{2}{3}x - 5$

34. $y = \frac{3}{4}x - 4$

35. $y = -\frac{3}{4}x + 2$

36. $y = -\frac{2}{3}x + 4$

37. $y = -\frac{5}{3}x$

38. $y = -\frac{4}{3}x$

In Exercises 39–46, **39–46c.** See graphing answer section.

a. *Put the equation in slope-intercept form by solving for y.*

b. *Identify the slope and the y-intercept.*

c. *Use the slope and y-intercept to graph the equation.*

39. $3x + y = 0$ **a.** $y = -3x$
b. $m = -3$; y-intercept $= 0$

40. $2x + y = 0$ **a.** $y = -2x$
b. $m = -2$; y-intercept $= 0$

41. $3y = 4x$ **a.** $y = \frac{4}{3}x$
b. $m = \frac{4}{3}$; y-intercept $= 0$

42. $4y = 5x$ **a.** $y = \frac{5}{4}x$
b. $m = \frac{5}{4}$; y-intercept $= 0$

43. $2x + y = 3$ **a.** $y = -2x + 3$
b. $m = -2$; y-intercept $= 3$

44. $3x + y = 4$ **a.** $y = -3x + 4$
b. $m = -3$; y-intercept $= 4$

45. $7x + 2y = 14$ **a.** $y = -\frac{7}{2}x + 7$
b. $m = -\frac{7}{2}$; y-intercept $= 7$

46. $5x + 3y = 15$ **a.** $y = -\frac{5}{3}x + 5$
b. $m = -\frac{5}{3}$; y-intercept $= 5$

In Exercises 47–52, graph both linear equations in the same rectangular coordinate system. If the lines are parallel, explain why. **47–52.** See graphing answer section.

47. $y = 3x + 1$
$y = 3x - 3$
parallel, $m = 3$

48. $y = 2x + 4$
$y = 2x - 3$
parallel, $m = 2$

49. $y = -3x + 2$
$y = 3x + 2$
not parallel, $m = -3$ and $m = 3$

50. $y = -2x + 1$
$y = 2x + 1$
not parallel, $m = -2$ and $m = 2$

51. $x - 2y = 2$
$2x - 4y = 3$
parallel, $m = \frac{1}{2}$

52. $x - 3y = 9$
$3x - 9y = 18$
parallel, $m = \frac{1}{3}$

Practice Plus

In Exercises 53–58, write an equation in the form $y = mx + b$ of the line that is described.

53. The y-intercept is 5 and the line is parallel to the line whose equation is $3x + y = 6$. $y = -3x + 5$

54. The y-intercept is -4 and the line is parallel to the line whose equation is $2x + y = 8$. $y = -2x - 4$

55. The line has the same y-intercept as the line whose equation is $16y = 8x + 32$ and is parallel to the line whose equation is $3x + 3y = 9$. $y = -x + 2$

56. The line has the same y-intercept as the line whose equation is $2y = 6x + 8$ and is parallel to the line whose equation is $4x + 4y = 20$. $y = -x + 4$

57. The line rises from left to right. It passes through the origin and a second point with equal x- and y-coordinates. $y = x$

58. The line falls from left to right. It passes through the origin and a second point with opposite x- and y-coordinates. $y = -x$

Application Exercises

59. The formula $E = 10x + 166$ models the number of endangered animal species, E, in the United States x years after 1980.

   a. What is the slope of this model? What does it represent in this situation?   10; The number of endangered animal species in the United States has been increasing by 10 species per year.

   b. What is the y-intercept of this model? What does it represent in this situation?   166; In 1980, there were 166 endangered animal species in the United States.

   c. Use the model to find the number of endangered animal species in the United States in 1980, 1985, 1990, 1995, 2000, and 2003. How well does the model describe the actual data shown in the bar graph?   166; 216; 266; 316; 366; 396; Answers will vary.

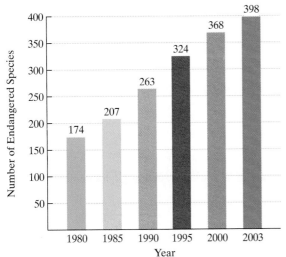

Endangered Animal Species in the U.S.

Source: U.S. Fish and Wildlife Service

60. A salesperson receives a fixed salary plus a percentage of all sales. The linear equation $y = 0.05x + 500$ describes the weekly salary, y, in dollars, in terms of weekly sales, x, also in dollars.

   a. Use the formula to find the weekly salary for sales of $0, $1, $2, $3, $4, $5, $100, and $1000. $500, $500.05, $500.10, $500.15, $500.20, $500.25, $505, and $550

   b. What is the slope of this equation? What does it represent in this situation? $m = 0.05$; The weekly salary increases $0.05 for each dollar of sales.

   c. What is the y-intercept of this equation? What does it represent in this situation? y-intercept is 500; When there are no sales, the weekly salary is $500.

61. In 2002, approximately 29 million people, including 3 million children, were living with HIV/AIDS in sub-Saharan Africa. The region accounted for 70% of the world's new HIV infections. The bar graph shows the number of people, in millions, living with the virus from 1997 through 2002. The graph in the rectangular coordinate system approximates the data.

Millions of People Living with HIV/AIDS in Sub-Saharan Africa

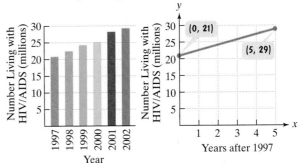

Source: U.N. AIDS

   a. According to the rectangular coordinate graph, what is the y-intercept? Describe what this represents in this situation.   21; In 1997, 21 million people were living with AIDS in sub-Saharan Africa.

   b. Use the coordinates of the two points shown to compute the slope. What does this mean in terms of rate of change?   1.6; The number of people living with AIDS in sub-Saharan Africa is increasing by 1.6 million per year.

   c. Use the y-intercept from part (a) and the slope from part (b) to write an equation that models the number of people, in millions, living with HIV in sub-Saharan Africa, y, x years after 1997. Write your model in the form $y = mx + b$. $y = 1.6x + 21$

   d. If worldwide efforts to fight AIDS in sub-Saharan Africa are not successful, use your model from part (c) to predict the number of people, in millions, who will be living with the virus in 2006.   35.4 million

62. The graph shows that the percentage of married women in the U.S. labor force who have children under 6 has been increasing steadily since 1960.

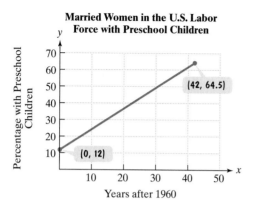

Married Women in the U.S. Labor Force with Preschool Children

(42, 64.5)

(0, 12)

Years after 1960

Source: James M. Henslin, *Essentials of Sociology*, Allyn and Bacon, 2002

a. According to the graph, what is the *y*-intercept? Describe what this represents in this situation.   12; In 1960, 12% of married women in the labor force had preschool children.

b. Use the coordinates of the two points shown to compute the slope. What does this mean in terms of rate of change? 1.25; The percentage increased by 1.25% per year.

c. Use the *y*-intercept from part (a) and the slope from part (b) to write an equation that models the percentage of married women in the labor force with preschool children, *y*, *x* years after 1960. Write your model in the form $y = mx + b$. $y = 1.25x + 12$

d. Use your model from part (c) to predict the percentage of married women in the labor force with preschool children in 2006.   69.5%

Writing in Mathematics

63. Describe how to find the slope and the *y*-intercept of a line whose equation is given.   63–65. Answers will vary.

64. Describe how to graph a line using the slope and *y*-intercept. Provide an original example with your description.

65. A formula in the form $y = mx + b$ models the cost, *y*, of a four-year college *x* years after 2003. Would you expect *m* to be positive, negative, or zero? Explain your answer.

Critical Thinking Exercises

66. Which one of the following is true?   c

a. The equation $y = mx + b$ shows that no line can have a *y*-intercept that is numerically equal to its slope.

b. Every line in the rectangular coordinate system has an equation that can be expressed in slope-intercept form.

c. The line $3x + 2y = 5$ has slope $-\frac{3}{2}$.

d. The line $2y = 3x + 7$ has a *y*-intercept of 7.

67. The relationship between Celsius temperature, *C*, and Fahrenheit temperature, *F*, can be described by a linear equation in the form $F = mC + b$. The graph of this equation contains the point $(0, 32)$: Water freezes at $0°C$ or at $32°F$. The line also contains the point $(100, 212)$: Water boils at $100°C$ or at $212°F$. Write the linear equation expressing Fahrenheit temperature in terms of Celsius temperature. $F = \frac{9}{5}C + 32$

68. The graph below indicates that lower fertility rates (the number of births per woman) are related to the percentage of the population using contraceptives. A line that best fits the data is shown. Estimate the *y*-intercept and the slope of this line. Then write the line's slope-intercept equation. Use the equation to find the number of births per woman if 90% of married women of child-bearing age used contraceptives.
y-intercept is approximately 7.4 and the slope is approximately -0.07; $y = -0.07x + 7.4$; about 1.1 births

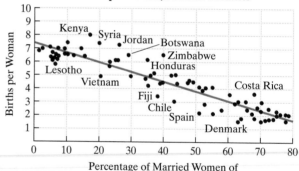

Contraceptive Prevalence and Average Number of Births per Woman, Selected Countries

Percentage of Married Women of Child-Bearing Age Using Contraceptives

Source: Population Reference Bureau

Review Exercises

69. Solve: $\dfrac{x}{2} + 7 = 13 - \dfrac{x}{4}$. (Section 2.3, Example 4)   8

70. Simplify: $3(12 \div 2^2 - 3)^2$. (Section 1.8, Example 6)   0

71. 14 is 25% of what? (Section 2.4, Example 8)   56

✔ MID-CHAPTER CHECK POINT

CHAPTER 3

What You Know: We learned to graph equations in two variables using point plotting, as well as a variety of other techniques. We used intercepts and a checkpoint to graph linear equations in the form $Ax + By = C$. We saw that the graph of a linear equation in one variable is a horizontal or a vertical line: $y = b$ graphs as a horizontal line and $x = a$ graphs as a vertical line. We determined a line's steepness, or rate of change, by computing its slope and we saw that lines with the same slope are parallel. Finally, we learned to graph linear equations in slope-intercept form, $y = mx + b$, using the slope, m, and the y-intercept, b.

In Exercises 1–3, use each graph to determine
  a. *the x-intercept, or state that there is no x-intercept.*
  b. *the y-intercept, or state that there is no y-intercept.*
  c. *the line's slope, or state that the slope is undefined.*

1.

  a. 4 **b.** 2 **c.** $-\frac{1}{2}$

2.

  a. -5 **b.** no y-intercept
  c. undefined slope

3.

  a. 0 **b.** 0 **c.** $\frac{3}{5}$

In Exercises 4–15, graph each equation in a rectangular coordinate system.   **4–15.** See graphing answer section.

4. $y = -2x$ **5.** $y = -2$ **6.** $x + y = -2$

7. $y = \frac{1}{3}x - 2$ **8.** $x = 3.5$ **9.** $4x - 2y = 8$

10. $y = 3x + 2$ **11.** $3x + y = 0$ **12.** $y = x^2 - 4$

13. $y = x - 4$ **14.** $5y = -3x$ **15.** $5y = 20$

16. Find the slope and the y-intercept of the line whose equation is $5x - 2y = 10$. $m = \frac{5}{2}$; y-intercept is -5

17. Determine whether the line through $(2, -4)$ and $(7, 0)$ is parallel to a second line through $(-4, 2)$ and $(1, 6)$.   parallel; Both have slopes of $\frac{4}{5}$.

18. The graph shows the percentage of U.S. colleges that offered distance learning by computer for selected years from 1995 through 2002.

Percentage of U.S. Colleges Offering Distance Learning by Computer

Source: International Data Corporation

The data for the years 1995 through 2002 can be modeled by the linear equation

$$y = 7.8x + 33$$

where x is the number of years after 1995 and y is the percentage of U.S. colleges offering distance learning.

  a. Find the y-intercept of this model.   33

  b. Describe what this y-intercept represents.   In 1995, 33% of U.S. colleges offered distance learning.

  c. Find the slope of this model.   7.8

  d. Describe the meaning of this slope as a rate of change.   For the years 1995 through 2002, the percentage of colleges that offered distance learning increased at a rate of 7.8% per year.

3.5

Objectives

1 Use the point-slope form to write equations of a line.

2 Find slopes and equations of parallel and perpendicular lines.

3 Model data with linear equations and make predictions.

THE POINT-SLOPE FORM OF THE EQUATION OF A LINE

Surprised by the number of people smoking cigarettes in movies and television shows made in the 1940s and 1950s? At that time, there was little awareness of the relationship between tobacco use and numerous diseases. Cigarette smoking was seen as a healthy way to relax and help digest a hearty meal. Then, in 1964, a linear equation changed everything. To understand the mathematics behind this turning point in public health, we explore another form of a line's equation.

Point-Slope Form We can use the slope of a line to obtain another useful form of the line's equation. Consider a nonvertical line that has slope m and contains the point (x_1, y_1). Now, let (x, y) represent any other point on the line, shown in Figure 3.28. Keep in mind that the point (x, y) is arbitrary and is not in one fixed position. By contrast, the point (x_1, y_1) is fixed.

Regardless of where the point (x, y) is located, the steepness of the line in Figure 3.28 remains the same. Thus, the ratio for slope stays a constant m. This means that for all points along the line,

$$m = \frac{\text{Change in } y}{\text{Change in } x} = \frac{y - y_1}{x - x_1}.$$

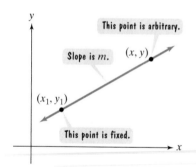

FIGURE 3.28 A line passing through (x_1, y_1) with slope m

We can clear the fraction by multiplying both sides by $x - x_1$, the least common denominator.

$$m = \frac{y - y_1}{x - x_1} \qquad \text{This is the slope of the line in Figure 3.28.}$$

$$m(x - x_1) = \frac{y - y_1}{x - x_1} \cdot (x - x_1) \qquad \text{Multiply both sides by } x - x_1.$$

$$m(x - x_1) = y - y_1 \qquad \text{Simplify: } \frac{y - y_1}{x - x_1} \cdot x - x_1 = y - y_1$$

Now, if we reverse the two sides, we obtain the **point-slope form** of the equation of a line.

> **POINT-SLOPE FORM OF THE EQUATION OF A LINE** The **point-slope equation** of a nonvertical line with slope m that passes through the point (x_1, y_1) is
>
> $$y - y_1 = m(x - x_1).$$

For example, the point-slope equation of the line passing through $(1, 4)$ with slope 2 $(m = 2)$ is

$$y - 4 = 2(x - 1).$$

1 Use the point-slope form to write equations of a line.

Using the Point-Slope Form to Write a Line's Equation If we know the slope of a line and a point not containing the y-intercept through which the line passes, the point-slope form is the equation that we should use. Once we have obtained this equation, it is customary to solve for y and write the equation in slope-intercept form. Examples 1 and 2 illustrate these ideas.

EXAMPLE 1 Writing the Point-Slope Form and the Slope-Intercept Form

Write the point-slope form and the slope-intercept form of the equation of the line with slope 4 that passes through the point $(-1, 3)$.

SOLUTION We begin with the point-slope equation of a line with $m = 4$, $x_1 = -1$, and $y_1 = 3$.

$y - y_1 = m(x - x_1)$	This is the point-slope form of the equation.
$y - 3 = 4[x - (-1)]$	Substitute the given values.
$y - 3 = 4(x + 1)$	We now have the point-slope form of the equation of the given line.

Now we solve this equation for y and write an equivalent equation in slope-intercept form $(y = mx + b)$.

$y - 3 = 4(x + 1)$	This is the point-slope equation.
$y - 3 = 4x + 4$	Use the distributive property.
$y = 4x + 7$	Add 3 to both sides.

The slope-intercept form of the line's equation is $y = 4x + 7$. ∎

✔ **CHECK POINT 1** Write the point-slope form and the slope-intercept form of the equation of the line with slope 6 that passes through the point $(2, -5)$.

$y + 5 = 6(x - 2); y = 6x - 17$

EXAMPLE 2 Writing the Point-Slope Form and the Slope-Intercept Form

A line passes through the points $(4, -3)$ and $(-2, 6)$. (See Figure 3.29.) Find the equation of the line

a. in point-slope form.   **b.** in slope-intercept form.

SOLUTION

a. To use the point-slope form, we need to find the slope. The slope is the change in the y-coordinates divided by the corresponding change in the x-coordinates.

$$m = \frac{6 - (-3)}{-2 - 4} = \frac{9}{-6} = -\frac{3}{2}$$

This is the definition of slope using $(4, -3)$ and $(-2, 6)$.

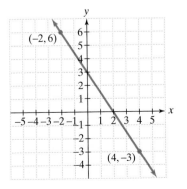

FIGURE 3.29

We can take either point on the line, $(-2, 6)$ or $(4, -3)$, to be (x_1, y_1). Let's use $(x_1, y_1) = (4, -3)$. Now, we are ready to write the point-slope equation.

$$y - y_1 = m(x - x_1) \qquad \text{This is the point-slope form of the equation.}$$

$$y - (-3) = -\frac{3}{2}(x - 4) \qquad \text{Substitute: } (x_1, y_1) = (4, -3) \text{ and } m = -\frac{3}{2}.$$

$$y + 3 = -\frac{3}{2}(x - 4) \qquad \text{Simplify.}$$

This equation is the point-slope form of the equation of the line shown in Figure 3.29 on page 227.

b. Now, we solve this equation for y and write an equivalent equation in slope-intercept form ($y = mx + b$).

$$y + 3 = -\frac{3}{2}(x - 4) \qquad \text{This is the point-slope equation.}$$

$$y + 3 = -\frac{3}{2}x + 6 \qquad \text{Use the distributive property.}$$

$$y = -\frac{3}{2}x + 3 \qquad \text{Subtract 3 from both sides.}$$

This equation is the slope-intercept form of the equation of the line shown in Figure 3.29 on page 227. ■

DISCOVER FOR YOURSELF

If you are given two points on a line, you can use either point for (x_1, y_1) when you write its point-slope equation. Rework Example 2 using $(-2, 6)$ for (x_1, y_1). Once you solve for y, you should obtain the same slope-intercept equation as the one shown in the last line of the solution to Example 2.

a. $y + 1 = -5(x + 2)$ or
$y + 6 = -5(x + 1)$
b. $y = -5x - 11$

 CHECK POINT 2 A line passes through the points $(-2, -1)$ and $(-1, -6)$. Find the equation of the line
a. in point-slope form. **b.** in slope-intercept form.

In Examples 1 and 2, we eventually write a line's equation in slope-intercept form? But where do we start our work?

Starting with $y = mx + b$	Starting with $y - y_1 = m(x - x_1)$
Begin with the slope-intercept form if you know	Begin with the point-slope form if you know
1. The slope of the line and the y-intercept.	**1.** The slope of the line and a point on the line not containing the y-intercept or
	2. Two points on the line, neither of which contains the y-intercept.

2 Find slopes and equations of parallel and perpendicular lines.

Parallel and Perpendicular Lines The next example uses the fact that parallel lines have the same slope.

EXAMPLE 3 Writing Equations of a Line Parallel to a Given Line

Write an equation of the line passing through $(-3, 1)$ and parallel to the line whose equation is $y = 2x + 1$. Express the equation in point-slope form and slope-intercept form.

SOLUTION   The situation is illustrated in Figure 3.30. We are looking for the equation of the red line shown on the left. How do we obtain this equation? Notice that the line passes through the point $(-3, 1)$. Using the point-slope form of the line's equation, we have $x_1 = -3$ and $y_1 = 1$.

$$y - y_1 = m(x - x_1)$$

$y_1 = 1$ $x_1 = -3$

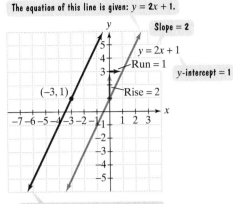

The equation of this line is given: $y = 2x + 1$.

Slope = 2

$y = 2x + 1$

Run = 1

y-intercept = 1

$(-3, 1)$

Rise = 2

We must write the equation of this line.

FIGURE 3.30

Now the only thing missing from the equation is m, the slope of the red line. Do we know anything about the slope of either line in Figure 3.30? The answer is yes; we know the slope of the dark blue line on the right, whose equation is given.

$$y = 2x + 1$$

The slope of the blue line on the right in Figure 3.30 is 2.

Parallel lines have the same slope. Because the slope of the blue line is 2, the slope of the red line, the line whose equation we must write, is also 2: $m = 2$. We now have values for x_1, y_1, and m for the red line.

$$y - y_1 = m(x - x_1)$$

$y_1 = 1$ $m = 2$ $x_1 = -3$

The point-slope form of the red line's equation is

$$y - 1 = 2[x - (-3)] \text{ or}$$

$$y - 1 = 2(x + 3).$$

Solving for y, we obtain the slope-intercept form of the equation.

$$y - 1 = 2x + 6 \qquad \text{Apply the distributive property.}$$

$$y = 2x + 7 \qquad \text{Add 1 to both sides. This is the slope-intercept form, } y = mx + b, \text{ of the equation.}$$

✔ **CHECK POINT 3** Write an equation of the line passing through $(-2, 5)$ and parallel to the line whose equation is $y = 3x + 1$. Express the equation in point-slope form and slope-intercept form. $y - 5 = 3(x + 2); y = 3x + 11$

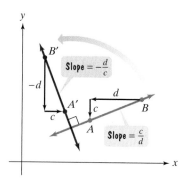

FIGURE 3.31 Slopes of perpendicular lines

Two lines that intersect at a right angle (90°) are said to be **perpendicular**, shown in Figure 3.31. The relationship between the slopes of perpendicular lines is not as obvious as the relationship between parallel lines. Figure 3.31 shows line AB, with slope $\frac{c}{d}$. Rotate line AB through 90° to the left to obtain line $A'B'$, perpendicular to line AB. The figure indicates that the rise and the run of the new line are reversed from the original line, but the rise is now negative. This means that the slope of the new line is $-\frac{d}{c}$. Notice that the product of the slopes of the two perpendicular lines is -1:

$$\left(\frac{c}{d}\right)\left(-\frac{d}{c}\right) = -1.$$

This relationship holds for all perpendicular lines and is summarized in the following box:

SLOPE AND PERPENDICULAR LINES

1. If two nonvertical lines are perpendicular, then the product of their slopes is -1.

2. If the product of the slopes of two lines is -1, then the lines are perpendicular.

3. A horizontal line having zero slope is perpendicular to a vertical line having undefined slope.

An equivalent way of stating this relationship is to say that one line is perpendicular to another line if its slope is the *negative reciprocal* of the slope of the other. For example, if a line has slope 5, any line having slope $-\frac{1}{5}$ is perpendicular to it. Similarly, if a line has slope $-\frac{3}{4}$, any line having slope $\frac{4}{3}$ is perpendicular to it.

EXAMPLE 4 Finding the Slope of a Line Perpendicular to a Given Line

Find the slope of any line that is perpendicular to the line whose equation is $x + 4y - 8 = 0$.

SOLUTION We begin by writing the equation of the given line in slope-intercept form. Solve for y.

$x + 4y - 8 = 0$	This is the given equation.
$4y = -x + 8$	To isolate the y-term, subtract x and add 8 on both sides.
$y = -\dfrac{1}{4}x + 2$	Divide both sides by 4.

Slope is $-\frac{1}{4}$.

The given line has slope $-\frac{1}{4}$. Any line perpendicular to this line has a slope that is the negative reciprocal of $-\frac{1}{4}$. Thus, the slope of any perpendicular line is 4. ∎

✓ **CHECK POINT 4** Find the slope of any line that is perpendicular to the line whose equation is $x + 3y - 12 = 0$. $m = 3$

3 Model data with linear equations and make predictions.

USING TECHNOLOGY

You can use a graphing utility to obtain a model for a scatter plot in which the data points fall on or near a straight line. The line that best fits the data is called the **regression line**. After entering the data in Figure 3.32(b), a graphing utility displays a scatter plot of the data and the regression line.

[–10, 40, 5] by [25, 39, 2]

Also displayed is the regression line's equation.

Applications Linear equations are useful for modeling data that fall on or near a line. For example, the bar graph in Figure 3.32(a) gives the median age of the U.S. population in the indicated year. (The median age is the age in the middle when all the ages of the U.S. population are arranged from youngest to oldest.) The data are displayed as a set of five points in a rectangular coordinate system in Figure 3.32(b).

The Graying of America: Median Age of the U.S. Population

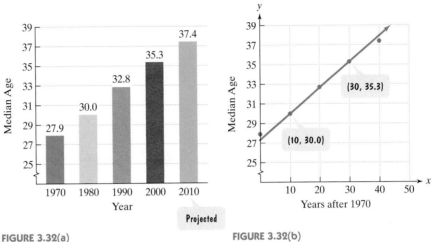

FIGURE 3.32(a)

Source: U.S. Census Bureau

FIGURE 3.32(b)

A set of points representing data is called a **scatter plot**. Also shown on the scatter plot in Figure 3.32(b) is a line that passes through or near the five points. By writing the equation of this line, we can obtain a model of the data and make predictions about the median age of the U.S. population in the future.

EXAMPLE 5   Modeling the Graying of America

Write the slope-intercept equation of the line shown in Figure 3.32(b). Use the equation to predict the median age of the U.S. population in 2020.

SOLUTION   The line in Figure 3.32(b) passes through $(10, 30.0)$ and $(30, 35.3)$. We start by finding its slope.

$$m = \frac{\text{Change in } y}{\text{Change in } x} = \frac{35.3 - 30.0}{30 - 10} = \frac{5.3}{20} = 0.265$$

The slope indicates that each year the median age of the U.S. population is increasing by 0.265 years.

Now, we write the line's slope-intercept equation.

$y - y_1 = m(x - x_1)$	Begin with the point-slope form.
$y - 30.0 = 0.265(x - 10)$	Either ordered pair can be (x_1, y_1). Let $(x_1, y_1) = (10, 30.0)$. From above, $m = 0.265$.
$y - 30.0 = 0.265x - 2.65$	Apply the distributive property.
$y = 0.265x + 27.35$	Add 30 to both sides and solve for y.

A linear equation that models the median age of the U.S. population, y, x years after 1970 is

$$y = 0.265x + 27.35.$$

Now, let's use this equation to predict the median age in 2020. Because 2020 is 50 years after 1970, substitute 50 for x and compute y.

$$y = 0.265(50) + 27.35 = 40.6$$

Our model predicts that the median age of the U.S. population in 2020 will be 40.6. ■

CHECK POINT 5 Use the data points (10, 30.0) and (20, 32.8) from Figure 3.32(b) on page 231 to write a slope-intercept equation that models the median age of the U.S. population x years after 1970. Use this model to predict the median age in 2020. $y = 0.28x + 27.2$; 41.2

ENRICHMENT ESSAY

Source: *Smoking and Health*, Washington, D.C., 1964

Cigarettes and Lung Cancer

This scatter plot shows a relationship between cigarette consumption among males and deaths due to lung cancer per million males. The data are from 11 countries and date back to a 1964 report by the U.S. Surgeon General. The scatter plot can be modeled by a line whose slope indicates an increasing death rate from lung cancer with increased cigarette consumption. At that time, the tobacco industry argued that in spite of this regression line, tobacco use is not the cause of cancer. Recent data do, indeed, show a causal effect between tobacco use and numerous diseases.

3.5 EXERCISE SET

 Student Solutions Manual   CD/Video   PH Math/Tutor Center   MathXL Tutorials on CD   MathXL®   MyMathLab   Interactmath.com

Practice Exercises

Write the point-slope form of the line satisfying each of the conditions in Exercises 1–28. Then use the point-slope form of the equation to write the slope-intercept form of the equation.

1. Slope = 3, passing through (2, 5)
   $y - 5 = 3(x - 2)$; $y = 3x - 1$
2. Slope = 6, passing through (3, 1)
   $y - 1 = 6(x - 3)$; $y = 6x - 17$
3. Slope = 5, passing through (−2, 6)
   $y - 6 = 5(x + 2)$; $y = 5x + 16$
4. Slope = 7, passing through (−4, 9)
   $y - 9 = 7(x + 4)$; $y = 7x + 37$
5. Slope = −8, passing through (−3, −2)
   $y + 2 = -8(x + 3)$; $y = -8x - 26$
6. Slope = −4, passing through (−5, −2)
   $y + 2 = -4(x + 5)$; $y = -4x - 22$
7. Slope = −12, passing through (−8, 0)
   $y - 0 = -12(x + 8)$; $y = -12x - 96$
8. Slope = −11, passing through (0, −3)
   $y + 3 = -11(x - 0)$; $y = -11x - 3$

9. Slope = −1, passing through $\left(-\frac{1}{2}, -2\right)$
   $y + 2 = -1\left(x + \frac{1}{2}\right)$; $y = -x - \frac{5}{2}$
10. Slope = −1, passing through $\left(-4, -\frac{1}{4}\right)$
    $y + \frac{1}{4} = -1(x + 4)$; $y = -x - \frac{17}{4}$
11. Slope = $\frac{1}{2}$, passing through the origin
    $y - 0 = \frac{1}{2}(x - 0)$; $y = \frac{1}{2}x$
12. Slope = $\frac{1}{3}$, passing through the origin
    $y - 0 = \frac{1}{3}(x - 0)$; $y = \frac{1}{3}x$
13. Slope = $-\frac{2}{3}$, passing through (6, −2)
    $y + 2 = -\frac{2}{3}(x - 6)$; $y = -\frac{2}{3}x + 2$
14. Slope = $-\frac{3}{5}$, passing through (10, −4)
    $y + 4 = -\frac{3}{5}(x - 10)$; $y = -\frac{3}{5}x + 2$
15. Passing through (1, 2) and (5, 10)
    $y - 2 = 2(x - 1)$ or $y - 10 = 2(x - 5)$; $y = 2x$
16. Passing through (3, 5) and (8, 15)
    $y - 5 = 2(x - 3)$ or $y - 15 = 2(x - 8)$; $y = 2x - 1$
17. Passing through (−3, 0) and (0, 3)
    $y - 0 = 1(x + 3)$ or $y - 3 = 1(x - 0)$; $y = x + 3$

18. Passing through $(-2, 0)$ and $(0, 2)$
$y - 0 = 1(x + 2)$ or $y - 2 = 1(x - 0)$; $y = x + 2$

19. Passing through $(-3, -1)$ and $(2, 4)$
$y + 1 = 1(x + 3)$ or $y - 4 = 1(x - 2)$; $y = x + 2$

20. Passing through $(-2, -4)$ and $(1, -1)$
$y + 4 = 1(x + 2)$ or $y + 1 = 1(x - 1)$; $y = x - 2$

21. Passing through $(-4, -1)$ and $(3, 4)$
$y + 1 = \frac{5}{7}(x + 4)$ or $y - 4 = \frac{5}{7}(x - 3)$; $y = \frac{5}{7}x + \frac{13}{7}$

22. Passing through $(-6, 1)$ and $(2, -5)$
$y - 1 = -\frac{3}{4}(x + 6)$ or $y + 5 = -\frac{3}{4}(x - 2)$; $y = -\frac{3}{4}x - \frac{7}{2}$

23. Passing through $(-3, -1)$ and $(4, -1)$
$y + 1 = 0(x + 3)$ or $y + 1 = 0(x - 4)$; $y = -1$

24. Passing through $(-2, -5)$ and $(6, -5)$
$y + 5 = 0(x + 2)$ or $y + 5 = 0(x - 6)$; $y = -5$

25. Passing through $(2, 4)$ with x-intercept $= -2$
$y - 4 = 1(x - 2)$ or $y - 0 = 1(x + 2)$; $y = x + 2$

26. Passing through $(1, -3)$ with x-intercept $= -1$
$y + 3 = -\frac{3}{2}(x - 1)$ or $y - 0 = -\frac{3}{2}(x + 1)$; $y = -\frac{3}{2}x - \frac{3}{2}$

27. x-intercept $= -\frac{1}{2}$ and y-intercept $= 4$
$y - 0 = 8\left(x + \frac{1}{2}\right)$ or $y - 4 = 8(x - 0)$; $y = 8x + 4$

28. x-intercept $= 4$ and y-intercept $= -2$
$y - 0 = \frac{1}{2}(x - 4)$ or $y + 2 = \frac{1}{2}(x - 0)$; $y = \frac{1}{2}x - 2$

In Exercises 29–44, the equation of a line is given. Find the slope of a line that is **a.** *parallel to the line with the given equation; and* **b.** *perpendicular to the line with the given equation.*

29. $y = 5x$ **a.** $m = 5$ **b.** $m = -\frac{1}{5}$

30. $y = 3x$ **a.** $m = 3$ **b.** $m = -\frac{1}{3}$

31. $y = -7x$ **a.** $m = -7$ **b.** $m = \frac{1}{7}$

32. $y = -9x$ **a.** $m = -9$ **b.** $m = \frac{1}{9}$

33. $y = \frac{1}{2}x + 3$ **a.** $m = \frac{1}{2}$ **b.** $m = -2$

34. $y = \frac{1}{4}x - 5$ **a.** $m = \frac{1}{4}$ **b.** $m = -4$

35. $y = -\frac{2}{5}x - 1$ **a.** $m = -\frac{2}{5}$ **b.** $m = \frac{5}{2}$

36. $y = -\frac{3}{7}x - 2$ **a.** $m = -\frac{3}{7}$ **b.** $m = \frac{7}{3}$

37. $4x + y = 7$ **a.** $m = -4$ **b.** $m = \frac{1}{4}$

38. $8x + y = 11$ **a.** $m = -8$ **b.** $m = \frac{1}{8}$

39. $2x + 4y - 8 = 0$ **a.** $m = -\frac{1}{2}$ **b.** $m = 2$

40. $3x + 2y - 6 = 0$ **a.** $m = -\frac{3}{2}$ **b.** $m = \frac{2}{3}$

41. $2x - 3y - 5 = 0$ **a.** $m = \frac{2}{3}$ **b.** $m = -\frac{3}{2}$

42. $3x - 4y + 7 = 0$ **a.** $m = \frac{3}{4}$ **b.** $m = -\frac{4}{3}$

43. $x = 6$ **a.** m is undefined.   **b.** $m = 0$

44. $y = 9$ **a.** $m = 0$ **b.** m is undefined.

In Exercises 45–48, write an equation for line L in point-slope form and slope-intercept form.

45.
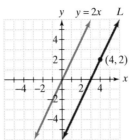
L is parallel to $y = 2x$.
$y - 2 = 2(x - 4)$;
$y = 2x - 6$

46.
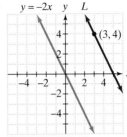
L is parallel to $y = -2x$.
$y - 4 = -2(x - 3)$;
$y = -2x + 10$

47.
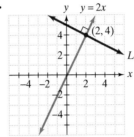
L is perpendicular to $y = 2x$.
$y - 4 = -\frac{1}{2}(x - 2)$;
$y = -\frac{1}{2}x + 5$

48.
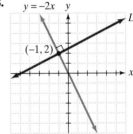
L is perpendicular to $y = -2x$.
$y - 2 = \frac{1}{2}(x + 1)$;
$y = \frac{1}{2}x + \frac{5}{2}$

In Exercises 49–56, use the given conditions to write an equation for each line in point-slope form and slope-intercept form.

49. Passing through $(-8, -10)$ and parallel to the line whose equation is $y = -4x + 3$
$y + 10 = -4(x + 8)$; $y = -4x - 42$

50. Passing through $(-2, -7)$ and parallel to the line whose equation is $y = -5x + 4$
$y + 7 = -5(x + 2)$; $y = -5x - 17$

51. Passing through $(2, -3)$ and perpendicular to the line whose equation is $y = \frac{1}{5}x + 6$
$y + 3 = -5(x - 2)$; $y = -5x + 7$

52. Passing through $(-4, 2)$ and perpendicular to the line whose equation is $y = \frac{1}{3}x + 7$
$y - 2 = -3(x + 4)$; $y = -3x - 10$

53. Passing through $(-2, 2)$ and parallel to the line whose equation is $2x - 3y - 7 = 0$
$y - 2 = \frac{2}{3}(x + 2)$; $y = \frac{2}{3}x + \frac{10}{3}$

54. Passing through $(-1, 3)$ and parallel to the line whose equation is $3x - 2y - 5 = 0$
$y - 3 = \frac{3}{2}(x + 1)$; $y = \frac{3}{2}x + \frac{9}{2}$

55. Passing through $(4, -7)$ and perpendicular to the line whose equation is $x - 2y - 3 = 0$
$y + 7 = -2(x - 4)$; $y = -2x + 1$

56. Passing through $(5, -9)$ and perpendicular to the line whose equation is $x + 7y - 12 = 0$
$y + 9 = 7(x - 5); \; y = 7x - 44$

Practice Plus

In Exercises 57–66, write an equation in slope-intercept form of the line satisfying the given conditions.

57. The line passes through $(2, 4)$ and has the same y-intercept as the line whose equation is $x - 4y = 8$. $\quad y = 3x - 2$

58. The line passes through $(2, 6)$ and has the same y-intercept as the line whose equation is $x - 3y = 18$. $\quad y = 6x - 6$

59. The line has an x-intercept at -4 and is parallel to the line containing $(3, 1)$ and $(2, 6)$. $\quad y = -5x - 20$

60. The line has an x-intercept at -6 and is parallel to the line containing $(4, -3)$ and $(2, 2)$. $\quad y = -\frac{5}{2}x - 15$

61. The line passes through $(-1, 5)$ and is perpendicular to the line whose equation is $x = 6$. $\quad y = 5$

62. The line passes through $(-2, 6)$ and is perpendicular to the line whose equation is $x = -4$. $\quad y = 6$

63. The line passes through $(-6, 4)$ and is perpendicular to the line that has an x-intercept of 2 and a y-intercept of -4.
$y = -\frac{1}{2}x + 1$

64. The line passes through $(-5, 6)$ and is perpendicular to the line that has an x-intercept of 3 and a y-intercept of -9.
$y = -\frac{1}{3}x + \frac{13}{3}$

65. The line is perpendicular to the line whose equation is $3x - 2y = 4$ and has the same y-intercept as this line.

$y = -\frac{2}{3}x - 2$

66. The line is perpendicular to the line whose equation is $4x - y = 6$ and has the same y-intercept as this line.

$y = -\frac{1}{4}x - 6$

67. What is the slope of a line that is parallel to the line whose equation is $Ax + By = C, B \neq 0$? $\quad -\frac{A}{B}$

68. What is the slope of a line that is perpendicular to the line whose equation is $Ax + By = C, A \neq 0$ and $B \neq 0$? $\quad \frac{B}{A}$

Application Exercises

69. We seem to be fed up with being lectured at about our waistlines. The points in the graph show the average weight of American adults from 1990 through 2002. Also shown is a line that passes through or near the points.

Average Weight of Americans

U.S. Average Weight (pounds)

(8, 168)

(2, 162)

Years after 1990

Source: Diabetes Care

a. Use the two points whose coordinates are shown by the voice balloons to find the point-slope equation of the line that models average weight of Americans, y, in pounds, x years after 1990. $\quad y - 162 = 1(x - 2)$ or $y - 168 = 1(x - 8)$

b. Write the equation in part (a) in slope-intercept form.
$y = x + 160$

c. Use the slope-intercept equation to predict the average weight of Americans in 2010. $\quad$ 180 pounds

70. Films may not be getting any better, but in this era of moviegoing, the number of screens available for new films and the classics has exploded. The points in the graph show the number of screens in the United States from 1995 through 2002. Also shown is a line that passes through or near the points.

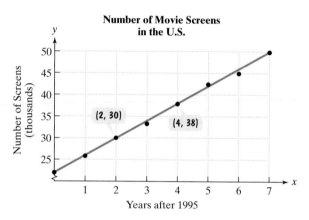

Number of Movie Screens in the U.S.

Number of Screens (thousands)

(2, 30)

(4, 38)

Years after 1995

Source: Motion Picture Association of America

a. Use the two points whose coordinates are shown by the voice balloons to find the point-slope equation of the line that models the number of screens, y, in thousands, x years after 1995. $\quad y - 30 = 4(x - 2)$ or $y - 38 = 4(x - 4)$

b. Write the equation in part (a) in slope-intercept form.
$y = 4x + 22$

c. Use the slope-intercept equation to predict the number of screens, in thousands, in 2010. $\quad$ 82 thousand

71. Is there a relationship between education and prejudice? With increased education, does a person's level of prejudice tend to decrease? The scatter plot at the top of the next column shows ten data points, each representing the number of years of school completed and the score on a test measuring prejudice for each subject. Higher scores on this 1-to-10 test indicate greater prejudice. Also shown is the regression line, the line that best fits the data. Use two points on this line to write both its point-slope and slope-intercept equations. Then use the slope-intercept equation to predict the score on the prejudice test for a person with seven years of education. $\quad$ Answers will vary. Example: $y - 3 = -\frac{2}{3}(x - 12); \; y = -\frac{2}{3}x + 11$; A person with seven years of education is predicted to have a score of 6.3.

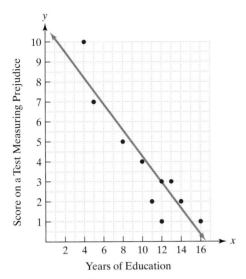

Years of Education

72. A business discovers a linear relationship between the number of shirts it can sell and the price per shirt. In particular, 20,000 shirts per week can be sold at $19 each. Raising the price to $55 causes the sales to fall to 2000 shirts per week. Write the point-slope and slope-intercept equations of the *demand line* through the ordered pairs

(20,000 shirts, $19) and (2000 shirts, $55).

Then determine the number of shirts that can be sold per week at $50 each. $y - 19 = -0.002(x - 20,000)$ or $y - 55 = -0.002(x - 2000)$; $y = -0.002x + 59$; 4500 shirts can be sold per week at $50.

73. The scatter plot shows the average number of minutes each that 16 people exercise per week and the average number of headaches per month each person experiences.
a–d. Answers will vary.

Minutes per Week Spent Exercising

a. Draw a line that fits the data so that the spread of the data points around the line is as small as possible.

b. Use the coordinates of two points along your line to write its point-slope and slope-intercept equations.

c. Use the equation in part (b) to predict the number of headaches per month for a person exercising 130 minutes per week.

d. What general observation can you make about the number of headaches per month as the number of minutes per week spent exercising increases?

Writing in Mathematics

74. Describe how to write the equation of a line if its slope and a point along the line are known.   **74–76.** Answers will vary.

75. Describe how to write the equation of a line if two points along the line are known.

76. Take a second look at the scatter plot in Exercise 71, shown on the left. Although there is a relationship between education and prejudice, we cannot necessarily conclude that increased education causes a person's level of prejudice to decrease. Offer two or more possible explanations for the data in the scatter plot.

Critical Thinking Exercises

77. Which one of the following is true?   c

a. If a line has undefined slope, then it has no equation.

b. The line whose equation is $y - 3 = 7(x + 2)$ passes through $(-3, 2)$.

c. The point-slope form cannot be applied to the line through the points $(2, -5)$ and $(2, 6)$.

d. The slope of the line whose equation is $3x + y = 7$ is 3.

78. Excited about the success of celebrity stamps, post office officials were rumored to have put forth a plan to institute two new types of thermometers. On these new scales, $°E$ represents degrees Elvis and $°M$ represents degrees Madonna. If it is known that $40°E = 25°M$, $280°E = 125°M$, and degrees Elvis is linearly related to degrees Madonna, write an equation expressing E in terms of M. $E = 2.4M - 20$

Technology Exercises

79. Use a graphing utility to graph $y = 1.75x - 2$. Select the best viewing rectangle possible by experimenting with the range settings to show that the line's slope is $\frac{7}{4}$.   Answers will vary.

80. Use a graphing utility to graph the slope-intercept equation that you wrote in Exercise 72. Then select an appropriate range setting and use the TRACE feature to graphically show the number of shirts that can be sold at $50 each.   4500 shirts; See graphing answer section.

81. a. Use the statistical menu of a graphing utility to enter the ten data points shown in the scatter plot in Exercise 71.

b. Use the DRAW menu and the scatter plot capability to draw a scatter plot of the data points like the one shown in Exercise 71.   See graphing answer section.

c. Select the linear regression option. Use your utility to obtain values for a and b for the equation of the regression line, $y = ax + b$. You may also be given a *correlation coefficient*, r. Values of r close to 1 indicate that the points can be described by a linear relationship and the regression line has a positive slope. Values of r close to -1 indicate that the points can be described by a linear relationship and the regression line has a negative slope. Values of r close to 0 indicate no linear relationship between the variables. $a = -0.6867924528$; $b = 11.01132075$; $r = -0.9214983162$.

d. Use the appropriate sequence (consult your manual) to graph the regression equation on top of the points in the scatter plot.   See graphing answer section.

Review Exercises

82. How many sheets of paper, weighing 2 grams each, can be put in an envelope weighing 4 grams if the total weight must not exceed 29 grams? (Section 2.7, Example 8, and Section 2.5, Example 5)   at most 12 sheets of paper

83. List all the natural numbers in this set:

$$\left\{ -2, 0, \frac{1}{2}, 1, \sqrt{3}, \sqrt{4} \right\}.$$

(Section 1.2, Example 5) $1, \sqrt{4}$

84. Use intercepts to graph $3x - 5y = 15$. (Section 3.2, Example 4)
See graphing answer section.

GROUP PROJECT

CHAPTER 3

Additional group projects can be found in the *Instructor's Resource Manual.*

In Example 5 on page 231, we used the data in Figure 3.32 to develop a linear equation that modeled the graying of America. For this group exercise, you might find it helpful to pattern your work after Figure 3.32 and the solution to Example 5. Group members should begin by consulting an almanac, newspaper, magazine, or the Internet to find data that lie approximately on or near a straight line. Working by hand or using a graphing utility, construct a scatter plot for the data. If working by hand, draw a line that approximately fits the data and then write its equation. If using a graphing utility, obtain the equation of the regression line. Then use the equation of the line to make a prediction about what might happen in the future. Are there circumstances that might affect the accuracy of this prediction? List some of these circumstances.

CHAPTER 3 SUMMARY

Definitions and Concepts	Examples

Section 3.1 Graphing Equations in Two Variables

An ordered pair is a solution of an equation in two variables if replacing the variables by the coordinates of the ordered pair results in a true statement.

Is $(-1, 4)$ a solution of $2x + 5y = 18$?

$$2(-1) + 5 \cdot 4 \stackrel{?}{=} 18$$
$$-2 + 20 \stackrel{?}{=} 18$$
$$18 = 18, \text{ true}$$

Thus, $(-1, 4)$ is a solution.

One method for graphing an equation in two variables is point plotting. Find several ordered-pair solutions, plot them as points, and connect the points with a smooth curve or line.

Graph: $y = 2x + 1$.

x	$y = 2x + 1$	(x, y)
-2	$y = 2(-2) + 1 = -3$	$(-2, -3)$
-1	$y = 2(-1) + 1 = -1$	$(-1, -1)$
0	$y = 2 \cdot 0 + 1 = 1$	$(0, 1)$
1	$y = 2 \cdot 1 + 1 = 3$	$(1, 3)$
2	$y = 2 \cdot 2 + 1 = 5$	$(2, 5)$

The graph of $y = 2x + 1$

Definitions and Concepts	Examples

Section 3.2 Graphing Linear Equations Using Intercepts

If a graph intersects the *x*-axis at $(a, 0)$, then *a* is an *x*-intercept.

If a graph intersects the *y*-axis at $(0, b)$, then *b* is a *y*-intercept.

An equation of the form $Ax + By = C$, where *A*, *B*, and *C* are integers, is called the standard form of the equation of a line. The graph of $Ax + By = C$ is a line that can be obtained using intercepts. To find the *x*-intercept, let $y = 0$ and solve for *x*. To find the *y*-intercept, let $x = 0$ and solve for *y*. Find a checkpoint, a third ordered-pair solution. Graph the equation by drawing a line through the three points.

Graph using intercepts: $4x + 3y = 12$.

x-intercept: $4x = 12$
                $x = 3$

y-intercept: $3y = 12$
                $y = 4$

Checkpoint: Let $x = 2$.
           $8 + 3y = 12$
              $3y = 4$
              $y = \frac{4}{3}$

The graph of $Ax + By = 0$ is a line that passes through the origin. Find two other points by finding two other solutions of the equation. Graph the equation by drawing a line through the origin and these two points.

Graph: $x + 2y = 0$.

$x = 2$: $2 + 2y = 0$
          $2y = -2$
           $y = -1$

$y = 1$: $x + 2(1) = 0$
            $x = -2$

Horizontal and Vertical Lines

The graph of $y = b$ is a horizontal line. The *y*-intercept is *b*.

The graph of $x = a$ is a vertical line. The *x*-intercept is *a*.

Definitions and Concepts	**Examples**

Section 3.3 Slope

The slope, m, of the line through the points (x_1, y_1) and (x_2, y_2) is

$$m = \frac{y_2 - y_1}{x_2 - x_1}, \quad x_2 - x_1 \neq 0.$$

If the slope is positive, the line rises from left to right. If the slope is negative, the line falls from left to right.

The slope of a horizontal line is 0. The slope of a vertical line is undefined.

If two distinct nonvertical lines have the same slope, then they are parallel.

Find the slope of the line passing through the points shown.

Let $(x_1, y_1) = (-1, 2)$ and $(x_2, y_2) = (2, -2)$.

$$m = \frac{y_2 - y_1}{x_2 - x_1} = \frac{-2 - 2}{2 - (-1)} = \frac{-4}{3} = -\frac{4}{3}.$$

Section 3.4 The Slope-Intercept Form of the Equation of a Line

The slope-intercept equation of a nonvertical line with slope m and y-intercept b is

$$y = mx + b.$$

Find the slope and the y-intercept of the line with the given equation.

• $y = -2x + 5$

  Slope –2. y-intercept is 5.

• $2x + 3y = 9$ (Solve for y)

  $3y = -2x + 9$ Subtract $2x$.

  $y = -\dfrac{2}{3}x + 3$ Divide by 3.

  Slope is $-\frac{2}{3}$. y-intercept is 3.

Graphing $y = mx + b$ Using the Slope and y-Intercept

1. Plot the point containing the y-intercept on the y-axis. This is the point $(0, b)$.
2. Use the slope, m, to obtain a second point. Write m as a fraction, and use rise over run, starting at the point containing the y-intercept, to plot this point.
3. Graph the equation by drawing a line through the two points.

Graph: $y = -\dfrac{3}{4}x + 1.$

Slope is $-\frac{3}{4}$. y-intercept is 1.

Definitions and Concepts	Examples

Section 3.5 The Point-Slope Form of the Equation of a Line

The point-slope equation of a nonvertical line with slope m that passes through the point (x_1, y_1) is $$y - y_1 = m(x - x_1).$$	Slope $= -3$, passing through $(-1, 5)$ $\boxed{m = -3}$ $\boxed{x_1 = -1}$ $\boxed{y_1 = 5}$ The line's point-slope equation is $$y - 5 = -3[x - (-1)].$$ Simplify: $$y - 5 = -3(x + 1).$$
To write the point-slope form of the line passing through two points, begin by using the points to compute the slope, m. Use either given point as (x_1, y_1) and write the point-slope equation: $$y - y_1 = m(x - x_1).$$ Solving this equation for y gives the slope-intercept form of the line's equation.	Write an equation in point-slope form and slope-intercept form of the line passing through $(-1, -3)$ and $(4, 2)$. $$m = \frac{2 - (-3)}{4 - (-1)} = \frac{2 + 3}{4 + 1} = \frac{5}{5} = 1$$ Using $(4, 2)$ as (x_1, y_1), the point-slope equation is $$y - 2 = 1(x - 4).$$ Solve for y to obtain the slope-intercept form. $$y = x - 2 \qquad \text{Add 2 to both sides.}$$
Nonvertical parallel lines have the same slope. If the product of the slopes of two lines is -1, then the lines are perpendicular. One line is perpendicular to another line if its slope is the negative reciprocal of the slope of the other.	Write point-slope and slope-intercept equations of the line passing through $(2, -1)$ $\qquad\qquad x_1 \quad y_1$ and perpendicular to $y = -\dfrac{1}{5}x + 6$. $\qquad\qquad\qquad\quad$ slope Slope, m, of perpendicular line is 5, the negative reciprocal of $-\frac{1}{5}$. $\begin{aligned} y - (-1) &= 5(x - 2) \quad \text{Point-slope equation} \\ y + 1 &= 5(x - 2) \\ y + 1 &= 5x - 10 \\ y &= 5x - 11 \end{aligned}$ Slope intercept equation

CHAPTER 3   REVIEW EXERCISES

3.1 *In Exercises 1–2, determine whether each ordered pair is a solution of the given equation.*

1. $y = 3x + 6$ $(-3, 3), (0, 6), (1, 9)$ $(-3, 3)$ is not a solution; $(0, 6)$ and $(1, 9)$ are solutions.

2. $3x - y = 12$ $(0, 4), (4, 0), (-1, 15)$ $(0, 4)$ and $(-1, 15)$ are not solutions; $(4, 0)$ is a solution.

In Exercises 3–4,   **3–6a.** See graphing answer section.

   a. *Find five solutions of each equation. Organize your work in a table of values.*

   b. *Use the five solutions in the table to graph each equation.*

3. $y = 2x - 3$ **4.** $y = \frac{1}{2}x + 1$

5. Graph the equation: $y = x^2 - 3$. Select integers for x, starting with -3 and ending with 3.

6. The linear equation in two variables $y = 5x - 41$ models the percentage of U.S. adults, y, with x years of education who are doing volunteer work.

   a. Find four solutions of the equation. Use 10, 12, 14, and 16 for x. Organize your work in a table of values.

   b. How well does the given equation model the data shown in the following table? Explain your answer.   Answers will vary.

Years of Education	10	12	14	16
Percentage Doing Volunteer Work	8.3%	18.8%	28.1%	38.4%

Source: U.S. Bureau of Labor

3.2 *In Exercises 7–9, use the graph to identify the*

   a. *x-intercept, or state that there is no x-intercept.*

   b. *y-intercept, or state that there is no y-intercept.*

7.
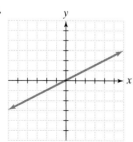
a. -2 **b.** -4

8.
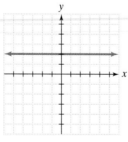
a. no *x*-intercept   **b.** 2

9.
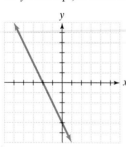
a. 0   **b.** 0

In Exercises 10–13, use intercepts to graph each equation.

10. $2x + y = 4$ **11.** $3x - 2y = 12$

12. $3x = 6 - 2y$ **13.** $3x - y = 0$

10–17. See graphing answer section.

In Exercises 14–17, graph each equation.

14. $x = 3$ **15.** $y = -5$ **16.** $y + 3 = 5$ **17.** $2x = -8$

18. The graph shows the Fahrenheit temperature, y, x hours after noon.

   a. At what time did the minimum temperature occur? What is the minimum temperature?   5 P.M., $-4°$F

   b. At what time did the maximum temperature occur? What is the maximum temperature?   8 P.M., $16°$F

   c. What are the *x*-intercepts? In terms of time and temperature, interpret the meaning of these intercepts.   4 and 6; At 4 P.M. and 6 P.M., the temperature was 0°F.

   d. What is the *y*-intercept? What does this mean in terms of time and temperature?   12; At noon, the temperature was 12°F.

   e. From 9 P.M. until midnight, the graph is shown as a horizontal line. What does this mean about the temperature over this period of time?   The temperature stayed the same, 12°F.

3.3 *In Exercises 19–22, calculate the slope of the line passing through the given points. If the slope is undefined, so state. Then indicate whether the line rises, falls, is horizontal, or is vertical.*

19. $(3, 2)$ and $(5, 1)$ $m = -\frac{1}{2}$; falls

20. $(-1, 2)$ and $(-3, -4)$ $m = 3$; rises

21. $(-3, 4)$ and $(6, 4)$ $m = 0$; horizontal

22. $(5, 3)$ and $(5, -3)$ undefined slope; vertical

In Exercises 23–26, find the slope of each line, or state that the slope is undefined.

23.

$m = \frac{3}{5}$

24.

undefined slope

25.

$m = -\frac{1}{3}$

26.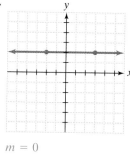

$m = 0$

In Exercises 27–28, determine whether the distinct lines through each pair of points are parallel.

27. $(-1, -3)$ and $(2, -8)$
$(8, -7)$ and $(9, 10)$
$m = -\frac{5}{3}$ and $m = 17$
so they are not parallel

28. $(5, 4)$ and $(9, 7)$
$(-6, 0)$ and $(-2, 3)$
$m = \frac{3}{4}$; parallel

29. The graph shows new AIDS diagnoses among the general U.S. population, y, for year x, where $1999 \le x \le 2003$.

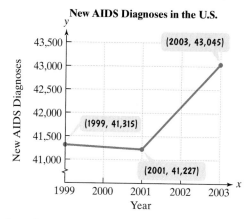

New AIDS Diagnoses in the U.S.

Source: Centers for Disease Control

a. Find the slope of the line passing through $(1999, 41{,}315)$ and $(2001, 41{,}227)$. Then express the slope as a rate of change with the proper units attached. $m = -44$; The number of new AIDS diagnoses decreased at a rate of 44 each year from 1999 to 2001.

b. Find the slope of the line passing through $(2001, 41{,}227)$ and $(2003, 43{,}045)$. Then express the slope as a rate of change. $m = 909$; The number of new AIDS diagnoses increased at a rate of 909 each year from 2001 to 2003.

c. Draw a line passing through $(1999, 41{,}315)$ and $(2003, 43{,}045)$ and find its slope. Is the slope the average of the slopes of the lines that you found in parts (a) and (b)? Explain your answer. $m = 432.5$; yes; Answers will vary.

3.4 In Exercises 30–33, find the slope and the y-intercept of the line with the given equation.

30. $y = 5x - 7$
$m = 5$; y-intercept is -7

31. $y = 6 - 4x$
$m = -4$; y-intercept is 6

32. $y = 3$
$m = 0$; y-intercept is 3

33. $2x + 3y = 6$
$m = -\frac{2}{3}$; y-intercept is 2

In Exercises 34–36, graph each linear equation using the slope and y-intercept. **34–38.** See graphing answer section.

34. $y = 2x - 4$

35. $y = \frac{1}{2}x - 1$

36. $y = -\frac{2}{3}x + 5$

In Exercises 37–38, write each equation in slope-intercept form. Then use the slope and y-intercept to graph the equation.

37. $y - 2x = 0$ $y = 2x$

38. $\frac{1}{3}x + y = 2$ $y = -\frac{1}{3}x + 2$

39. Graph $y = -\frac{1}{2}x + 4$ and $y = -\frac{1}{2}x - 1$ in the same rectangular coordinate system. Are the lines parallel? If so, explain why. Yes, they are parallel because both have slopes of $-\frac{1}{2}$ and different y-intercepts. See graphing answer section.

40. The graph shows the average age of U.S. whites, African Americans, and Americans of Hispanic origin from 1990 through 2000.

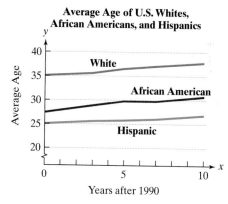

Average Age of U.S. Whites, African Americans, and Hispanics

Source: U.S. Census Bureau

a. What is the smallest y-intercept? Describe what this represents in this situation. 25; In 1990, the average age of U.S. Hispanics was 25.

b. The average age of the group with the greatest y-intercept was approximately 38 in 2000. Use the points $(0, 35)$ and $(10, 38)$ to compute the slope for this group. What does this mean about their average age for the period shown? $m = 0.3$; The average age for U.S. whites increased by a rate of about 0.3 each year from 1990 to 2000.

c. Use the slope from part (b) and the y-intercept for this group shown by the graph to write an equation that models the group's average age, y, x years after 1990. Write your model in $y = mx + b$ form. $y = 0.3x + 35$

d. Use your model from part (c) to predict the average age for the group in 2010. 41 years old

3.5 Write the point-slope form of the line satisfying the conditions in Exercises 41–44. Then use the point-slope form of the equation to write the slope-intercept form.

41. Slope $= 6$, passing through $(-4, 7)$
$y - 7 = 6(x + 4)$; $y = 6x + 31$

42. Passing through $(3, 4)$ and $(2, 1)$ $y - 4 = 3(x - 3)$ or $y - 1 = 3(x - 2)$; $y = 3x - 5$

43. Passing through $(4, -7)$ and parallel to the line whose equation is $3x + y - 9 = 0$
$y + 7 = -3(x - 4); \ y = -3x + 5$

44. Passing through $(-2, 6)$ and perpendicular to the line whose equation is $y = \frac{1}{3}x + 4$ $y - 6 = -3(x + 2); \ y = -3x$

45. You can click a mouse and bet the house. The points in the graph show the dizzying growth of online gambling. With more than 1800 sites, the industry has become the Web's biggest moneymaker.

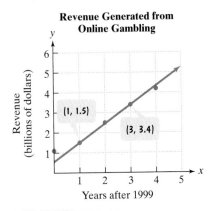

Revenue Generated from Online Gambling

(1, 1.5)

(3, 3.4)

Source: Newsweek

a. Use the two points whose coordinates are shown by the voice balloons to find the point-slope equation of the line that models the revenue from online gambling, y, in billions of dollars, x years after 1999. $y - 1.5 = 0.95(x - 1)$ or $y - 3.4 = 0.95(x - 3)$

b. Write the equation in part (a) in slope-intercept form.
$y = 0.95x + 0.55$

c. In 2003, nearly \$3.5 billion was lost on Internet bets, triggering a sharp backlash that threatened to shut down Internet wagering. If this crackdown on the industry is not successful, use your slope-intercept model to predict the billions of dollars in revenue from online gambling in 2009.   \$10.05 billion

---

CHAPTER 3 TEST

Remember to use your Chapter Test Prep Video CD to see the worked-out solutions to the test questions you want to review.

1. Determine whether each ordered pair is a solution of $4x - 2y = 10$:

$$(0, -5), \quad (-2, 1), \quad (4, 3).$$

$(-2, 1)$ is not a solution; $(0, -5)$ and $(4, 3)$ are solutions.

2. Find five solutions of $y = 3x + 1$. Organize your work in a table of values. Then use the five solutions in the table to graph the equation.   See graphing answer section.

3. Graph: $y = x^2 - 1$. Select integers for x, starting with -3 and ending with 3.   See graphing answer section.

4. Use the graph to identify the

 a. x-intercept, or state that there is no x-intercept.   2

 b. y-intercept, or state that there is no y-intercept. -3

5. Use intercepts to graph $4x - 2y = -8$.   **5–6.** See graphing answer section.

6. Graph $y = 4$ in a rectangular coordinate system.

In Exercises 7–8, calculate the slope of the line passing through the given points. If the slope is undefined, so state. Then indicate whether the line rises, falls, is horizontal, or is vertical.

7. $(-3, 4)$ and $(-5, -2)$ $m = 3$; rises

8. $(6, -1)$ and $(6, 3)$ undefined slope; vertical

9. Find the slope of the line in the figure shown or state that the slope is undefined. $m = \frac{3}{2}$

10. Determine whether the line through $(2, 4)$ and $(6, 1)$ is parallel to a second line through $(-3, 1)$ and $(1, -2)$.
parallel, $m = -\frac{3}{4}$ and different y-intercepts

In Exercises 11–12, find the slope and the y-intercept of the line with the given equation.

11. $y = -x + 10$ $m = -1$; *y*-intercept $= 10$

12. $2x + y = 6$ $m = -2$; *y*-intercept $= 6$

In Exercises 13–14, graph each linear equation using the slope and y-intercept.   **13–14.** See graphing answer section.

13. $y = \dfrac{2}{3}x - 1$ **14.** $y = -2x + 3$

In Exercises 15–16, use the given conditions to write an equation for each line in point-slope form and slope-intercept form.

15. Slope $= -2$, passing through $(-1, 4)$
$y - 4 = -2(x + 1)$; $y = -2x + 2$

16. Passing through $(2, 1)$ and $(-1, -8)$ $y - 1 = 3(x - 2)$
or $y + 8 = 3(x + 1)$; $y = 3x - 5$

17. Passing through $(-2, 3)$ and perpendicular to the line whose equation is $y = -\frac{1}{2}x - 4$
$y - 3 = 2(x + 2)$; $y = 2x + 7$

18. Passing through $(6, -4)$ and parallel to the line whose equation is $x + 2y = 5$
$y + 4 = -\frac{1}{2}(x - 6)$; $y = -\frac{1}{2}x - 1$

19. The graph shows spending per pupil in public schools. Find the slope of the line passing through $(1970, 2100)$ and $(2000, 5280)$. Describe what the slope represents in this situation. $m = 106$; Spending per pupil increased at a rate of about \$106 each year.

**Spending per Pupil in
America's Public Schools**

Source: National Education Association

20. The scatter plot shows the annual number of adult and child deaths worldwide due to AIDS from 1999 through 2003. Also shown is a line that models the data.

**Number of Deaths Worldwide
Due to AIDS**

Source: World Health Organization

a. Use the two points whose coordinates are shown by the voice balloons to find the point-slope equation of the line that models the number of worldwide deaths due to AIDS, *y*, in millions, *x* years after 1999.
$y - 2.3 = 0.3(x - 1)$ or $y - 3.2 = 0.3(x - 4)$

b. Write the equation in part (a) in slope-intercept form.
$y = 0.3x + 2$

c. Use the linear equation to predict the number of worldwide deaths due to AIDS, in millions, in 2007.
4.4 million

Note to Instructor: The Chapter 3 Test file in the TestGen program provides algorithms specifically matched to these problems so you can easily replicate this chapter test for additional practice or assessment purposes.

CUMULATIVE REVIEW EXERCISES (CHAPTERS 1–3)

1. Perform the indicated operations:

$$\frac{10 - (-6)}{3^2 - (4 - 3)}. \quad 2$$

2. Simplify: $6 - 2[3(x - 1) + 4]$. $4 - 6x$

3. List all the irrational numbers in this set:
$\left\{-3, 0, 1, \sqrt{4}, \sqrt{5}, \frac{11}{2}\right\}$. $\sqrt{5}$

In Exercises 4–5, solve each equation.

4. $6(2x - 1) - 6 = 11x + 7$ $\{19\}$

5. $x - \dfrac{3}{4} = \dfrac{1}{2}$ $\left\{\dfrac{5}{4}\right\}$

6. Solve for x: $y = mx + b$. $x = \dfrac{y - b}{m}$

7. 120 is 15% of what? 800

8. The formula $y = 4.5x - 46.7$ models the stopping distance, y, in feet, for a car traveling x miles per hour. If the stopping distance is 133.3 feet, how fast was the car traveling? 40 miles per hour

In Exercises 9–10, solve each inequality. Express the solution set in set-builder notation and graph the set on a number line.

9. $2 - 6x \geq 2(5 - x)$ $\{x | x \leq -2\}$; See graphing answer section.

10. $6(2 - x) > 12$ $\{x | x < 0\}$; See graphing answer section.

11. Evaluate $x^2 - 10x$ for $x = -3$. 39

12. Insert either $<$ or $>$ in the shaded area to make a true statement:
$$-2000 \quad \blacksquare \quad -3. \quad <$$

13. On February 8, the temperature in Manhattan at 10 P.M. was $-4°$F. By 3 A.M. the next day, the temperature had fallen $11°$, but by noon the temperature increased by $21°$. What was the temperature at noon? $6°$F

14. The amount of money owed by doctors graduating from medical school is on the rise. The formula
$$D = 4x + 30$$
models the average debt, D, in thousands of dollars, of in-debted medical-school graduates x years after 1985. Use the formula to determine in which year this debt will reach $150 thousand. 2015

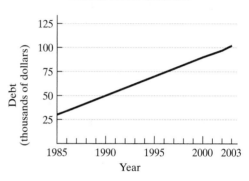

Average Debt of Indebted Medical School Graduates

Source: American Medical Association

15. The length of a rectangular football field is 14 meters more than twice the width. If the perimeter is 346 meters, find the field's dimensions. width = 53 meters; length = 120 meters

16. After a 10% weight loss, a person weighed 180 pounds. What was the weight before the loss? 200 pounds

17. A plumber charged a customer $228, listing $18 for parts and the remainder for labor. If the cost of the labor is $35 per hour, how many hours did the plumber work? 6 hours

18. In a triangle, the measure of the second angle is $20°$ greater than the measure of the first angle. The measure of the third angle is twice that of the first. What is the measure of each angle? $40°, 60°, 80°$

In Exercises 19–20, graph each equation or inequality in the rectangular coordinate system. **19–20.** See graphing answer section.

19. $2x - y = 4$ **20.** $y = -4x + 3$

Mary Katherine Campbell,
Miss America 1922

Ericka Dunlap,
Miss America 2004

You are not a great fan of beauty pageants. However, as you were channel surfing, you tuned into the Miss America festivities. Is it your imagination, or does the icon of American beauty have that lean and hungry look?

This question is addressed in the Enrichment Essay "Missing America" in Section 4.1.

C H A P T E R

4

Television, movies, and magazines place great emphasis on physical beauty. Our culture emphasizes physical appearance to such an extent that it is a central factor in the perception and judgment of others. The modern emphasis on thinness as the ideal body shape has been suggested as a major cause of eating disorders among adolescent women. In this chapter, you will learn how systems of linear equations in two variables reveal the hidden patterns of your world, including a relationship between our changing cultural values of physical attractiveness and undernutrition.

Systems of Linear Equations

SOLVING SYSTEMS OF LINEAR EQUATIONS BY GRAPHING

Key West residents Brian Goss (left), George Wallace, and Michael Mooney (right) hold on to each other as they battle 90 mph winds along Houseboat Row in Key West, FL, on Friday, Sept. 25, 1998. The three had sought shelter behind a Key West hotel as Hurricane Georges descended on the Florida Keys, but were forced to seek other shelter when the storm conditions became too rough. Hundreds of people were killed by the storm when it swept through the Caribbean.

Problems ranging from scheduling airline flights to controlling traffic flow to routing phone calls over the nation's communication network often require solutions in a matter of moments. The solution to these real-world problems can involve solving thousands of equations having thousands of variables. AT&T's domestic long-distance network involves 800,000 variables! Meteorologists describing atmospheric conditions surrounding a hurricane must solve problems involving thousands of equations rapidly and efficiently. The difference between a two-hour warning and a two-day warning is a life-and-death issue for people in the path of one of nature's most destructive forces.

Although we will not be solving 800,000 equations with 800,000 variables, we will turn our attention to two equations with two variables, such as

$$2x - 3y = -4$$
$$2x + \ y = \ \ 4.$$

The methods that we consider for solving such problems provide the foundation for solving far more complex systems with many variables.

1 Decide whether an ordered pair is a solution of a linear system.

Systems of Linear Equations and Their Solutions We have seen that all equations in the form $Ax + By = C$ are straight lines when graphed. Two such equations, such as those listed above, are called a **system of linear equations** or a **linear system**. A **solution to a system of two linear equations in two variables** is an ordered pair that satisfies both equations in the system. For example, (3, 4) satisfies the system

$$x + y = \ \ 7 \qquad \text{(3 + 4 is, indeed, 7.)}$$
$$x - y = -1. \qquad \text{(3 - 4 is, indeed, -1.)}$$

Thus, (3, 4) satisfies both equations and is a solution of the system. The solution can be described by saying that $x = 3$ and $y = 4$. The solution can also be described using set notation. The solution set of the system is $\{(3, 4)\}$—that is, the set consisting of the ordered pair (3, 4).

A system of linear equations can have exactly one solution, no solution, or infinitely many solutions. We begin with systems having exactly one solution.

EXAMPLE 1 Determining Whether Ordered Pairs Are Solutions of a System

Consider the system:

$$x + 2y = 2$$
$$x - 2y = 6.$$

Determine if each ordered pair is a solution of the system:

a. $(4, -1)$ **b.** $(-4, 3)$.

SOLUTION

a. We begin by determining whether $(4, -1)$ is a solution. Because 4 is the x-coordinate and -1 is the y-coordinate of $(4, -1)$, we replace x with 4 and y with -1.

$$x + 2y = 2 \qquad\qquad x - 2y = 6$$
$$4 + 2(-1) \stackrel{?}{=} 2 \qquad\qquad 4 - 2(-1) \stackrel{?}{=} 6$$
$$4 + (-2) \stackrel{?}{=} 2 \qquad\qquad 4 - (-2) \stackrel{?}{=} 6$$
$$2 = 2, \quad \text{true} \qquad\qquad 4 + 2 \stackrel{?}{=} 6$$
$$6 = 6, \quad \text{true}$$

The pair $(4, -1)$ satisfies both equations: It makes each equation true. Thus, the ordered pair is a solution of the system.

b. To determine whether $(-4, 3)$ is a solution, we replace x with -4 and y with 3.

$$x + 2y = 2 \qquad\qquad x - 2y = 6$$
$$-4 + 2 \cdot 3 \stackrel{?}{=} 2 \qquad\qquad -4 - 2 \cdot 3 \stackrel{?}{=} 6$$
$$-4 + 6 \stackrel{?}{=} 2 \qquad\qquad -4 - 6 \stackrel{?}{=} 6$$
$$2 = 2, \quad \text{true} \qquad\qquad -10 = 6, \quad \text{false}$$

The pair $(-4, 3)$ fails to satisfy *both* equations: It does not make both equations true. Thus, the ordered pair is not a solution of the system. ■

 CHECK POINT 1 Consider the system:

$$2x - 3y = -4$$
$$2x + y = 4.$$

Determine if each ordered pair is a solution of the system:

a. $(1, 2)$ solution **b.** $(7, 6)$.   not a solution

---

2 Solve systems of linear equations by graphing.

Solving Linear Systems by Graphing The solution of a system of linear equations can be found by graphing both of the equations in the same rectangular coordinate system. For a system with one solution, **the coordinates of the point of intersection give the system's solution.** For example, the system in Example 1,

$$x + 2y = 2$$
$$x - 2y = 6$$

is graphed in Figure 4.1. The solution of the system, $(4, -1)$, corresponds to the point of intersection of the lines.

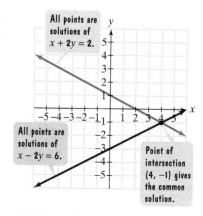

FIGURE 4.1 Visualizing a system's solution

SOLVING SYSTEMS OF TWO LINEAR EQUATIONS IN TWO VARIABLES, x AND y, BY GRAPHING

1. Graph the first equation.
2. Graph the second equation on the same axes.
3. If the lines representing the two graphs intersect at a point, determine the coordinates of this point of intersection. The ordered pair is the solution of the system.
4. Check the solution in both equations.

EXAMPLE 2   Solving a Linear System by Graphing

Solve by graphing:

$$2x + 3y = 6$$
$$2x + y = -2.$$

SOLUTION

Point of intersection: $(-3, 4)$

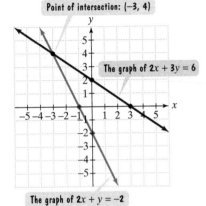

The graph of $2x + 3y = 6$

The graph of $2x + y = -2$

FIGURE 4.2

Step 1. **Graph the first equation.** We use intercepts to graph $2x + 3y = 6$.

x-intercept (Set $y = 0$.)	*y*-intercept (Set $x = 0$.)
$2x + 3 \cdot 0 = 6$	$2 \cdot 0 + 3y = 6$
$2x = 6$	$3y = 6$
$x = 3$	$y = 2$

The x-intercept is 3, so the line passes through $(3, 0)$. The y-intercept is 2, so the line passes through $(0, 2)$. The graph of $2x + 3y = 6$ is shown as the red line in Figure 4.2.

Step 2. **Graph the second equation on the same axes.** We use intercepts to graph $2x + y = -2$.

x-intercept (Set $y = 0$.)	*y*-intercept (Set $x = 0$.)
$2x + 0 = -2$	$2 \cdot 0 + y = -2$
$2x = -2$	$y = -2$
$x = -1$	

The x-intercept is -1, so the line passes through $(-1, 0)$. The y-intercept is -2, so the line passes through $(0, -2)$. The graph of $2x + y = -2$ is shown as the blue line in Figure 4.2.

Step 3. **Determine the coordinates of the intersection point. This ordered pair is the system's solution.** Using Figure 4.2, it appears that the lines intersect at $(-3, 4)$. The "apparent" solution of the system is $(-3, 4)$.

Step 4. **Check the solution in both equations.**

Check $(-3, 4)$ in $2x + 3y = 6$:	Check $(-3, 4)$ in $2x + y = -2$:
$2(-3) + 3 \cdot 4 \overset{?}{=} 6$	$2(-3) + 4 \overset{?}{=} -2$
$-6 + 12 \overset{?}{=} 6$	$-6 + 4 \overset{?}{=} -2$
$6 = 6,$ true	$-2 = -2,$ true

Because both equations are satisfied, $(-3, 4)$ is the solution of the system and $\{(-3, 4)\}$ is the solution set. ∎

✓ **CHECK POINT 2** Solve by graphing: (1, 4) or $\{(1, 4)\}$

$$2x + y = 6$$
$$2x - y = -2.$$

DISCOVER FOR YOURSELF

Must two lines intersect at exactly one point? Sketch two lines that have less than one intersection point. Now sketch two lines that have more than one intersection point. What does this say about each of these systems?

STUDY TIP

When solving linear systems by graphing, neatly drawn graphs are essential for determining points of intersection.

- Use rectangular coordinate graph paper.
- Use a ruler or straightedge.
- Use a pencil with a sharp point.

EXAMPLE 3 Solving a Linear System by Graphing

Solve by graphing:

$$y = -3x + 2$$
$$y = 5x - 6.$$

SOLUTION Each equation is in the form $y = mx + b$. Thus, we use the y-intercept, b, and the slope, m, to graph each line.

Step 1. Graph the first equation.

$$y = -3x + 2$$

The slope is –3.   The y-intercept is **2**.

The y-intercept is 2, so the line passes through $(0, 2)$. The slope is $-\frac{3}{1}$. Start at the y-intercept and move 3 units down (the rise) and 1 unit to the right (the run). The graph of $y = -3x + 2$ is shown as the red line in Figure 4.3.

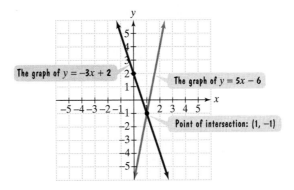

The graph of $y = -3x + 2$

The graph of $y = 5x - 6$

Point of intersection: $(1, -1)$

FIGURE 4.3

Step 2. Graph the second equation on the same axes.

$$y = 5x - 6$$

The slope is **5**.   The y-intercept is **–6**.

The y-intercept is -6, so the line passes through $(0, -6)$. The slope is $\frac{5}{1}$. Start at the y-intercept and move 5 units up (the rise) and 1 unit to the right (the run). The graph of $y = 5x - 6$ is shown as the blue line in Figure 4.3.

Step 3. Determine the coordinates of the intersection point. This ordered pair is the system's solution. Using Figure 4.3, it appears that the lines intersect at $(1, -1)$. The "apparent" solution of the system is $(1, -1)$.

Step 4. **Check the solution in both equations.**

Check $(1, -1)$ in	**Check $(1, -1)$ in**
$y = -3x + 2$:	$y = 5x - 6$:
$-1 \overset{?}{=} -3 \cdot 1 + 2$	$-1 \overset{?}{=} 5 \cdot 1 - 6$
$-1 \overset{?}{=} -3 + 2$	$-1 \overset{?}{=} 5 - 6$
$-1 = -1$, true	$-1 = -1$, true

Because both equations are satisfied, $(1, -1)$ is the solution and the system's solution set is $\{(1, -1)\}$. ∎

 CHECK POINT 3 Solve by graphing: $(3, 3)$ or $\{(3, 3)\}$

$$y = -x + 6$$
$$y = 3x - 6.$$

3 Use graphing to identify systems with no solution or infinitely many solutions.

Linear Systems Having No Solution or Infinitely Many Solutions We have seen that a system of linear equations in two variables represents a pair of lines. The lines either intersect, are parallel, or are identical. Thus, there are three possibilities for the number of solutions to a system of two linear equations.

THE NUMBER OF SOLUTIONS TO A SYSTEM OF TWO LINEAR EQUATIONS
The number of solutions to a system of two linear equations in two variables is given by one of the following. (See Figure 4.4.)

Number of Solutions	**What This Means Graphically**
Exactly one ordered-pair solution	The two lines intersect at one point.
No solution	The two lines are parallel.
Infinitely many solutions	The two lines are identical.

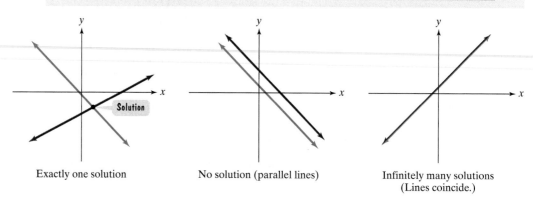

Exactly one solution   No solution (parallel lines)   Infinitely many solutions (Lines coincide.)

FIGURE 4.4 Possible graphs for a system of two linear equations in two variables

A linear system with no solution is called an **inconsistent system**. If you attempt to solve such a system by graphing, you will obtain two parallel lines. The solution set is the empty set, $\varnothing$.

EXAMPLE 4 A System with No Solution

Solve by graphing:

$$y = 2x - 1$$
$$y = 2x + 3.$$

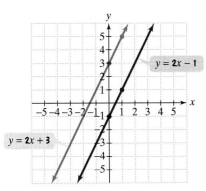

FIGURE 4.5 The graph of an inconsistent system

SOLUTION Compare the slopes and y-intercepts in the two equations.

> The lines have the same slope, **2**.

$$y = 2x - 1 \qquad y = 2x + 3$$

> The lines have different y-intercepts, **−1 and 3**.

Figure 4.5 shows the graphs of the two equations. Because both equations have the same slope, 2, but different y-intercepts, the lines are parallel. The system is inconsistent and has no solution. The solution set is the empty set, $\varnothing$. ■

 CHECK POINT 4 Solve by graphing: no solution or $\varnothing$

$$y = 3x - 2$$
$$y = 3x + 1.$$

EXAMPLE 5 A System with Infinitely Many Solutions

Solve by graphing:

$$2x + \ y = 3$$
$$4x + 2y = 6.$$

SOLUTION We use intercepts to graph each equation.

• $2x + y = 3$

x-intercept	**y-intercept**
$2x + 0 = 3$	$2 \cdot 0 + y = 3$
$2x = 3$	$y = 3$
$x = \dfrac{3}{2}$	

> Graph $\left(\frac{3}{2}, 0\right)$ and $(0, 3)$.

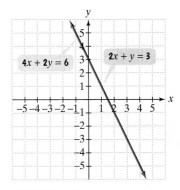

FIGURE 4.6 The graph of a system with infinitely many solutions

• $4x + 2y = 6$

x-intercept	**y-intercept**
$4x + 2 \cdot 0 = 6$	$4 \cdot 0 + 2y = 6$
$4x = 6$	$2y = 6$
$x = \dfrac{6}{4} = \dfrac{3}{2}$	$y = 3$

> Graph $\left(\frac{3}{2}, 0\right)$ and $(0, 3)$.

Both lines have the same x-intercept, $\frac{3}{2}$, or 1.5, and the same y-intercept, 3. Thus, the graphs of the two equations in the system are the same line, shown in Figure 4.6. The two equations have the same solutions. Any ordered pair that is a solution of one equation is a solution of the other, and, consequently, a solution of the system. The system has an infinite number of solutions, namely all points that are solutions of either equation. We express the solution set for the system in one of two equivalent ways:

$$\{(x, y)|2x + y = 3\}$$ The set of all ordered pairs (x, y) such that $2x + y = 3$

$$\{(x, y)|4x + 2y = 6\}.$$ The set of all ordered pairs (x, y) such that $4x + 2y = 6$ ■

Take a second look at the two equations, $2x + y = 3$ and $4x + 2y = 6$, in Example 5. If you multiply both sides of the first equation, $2x + y = 3$, by 2, you will obtain the second equation, $4x + 2y = 6$.

$$2x + y = 3$$ This is the first equation in the system.

$$2(2x + y) = 2 \cdot 3$$ Multiply both sides by 2.

$$2 \cdot 2x + 2y = 2 \cdot 3$$ Use the distributive property.

> This is the second equation in the system.

$$4x + 2y = 6$$ Simplify.

Because $2x + y = 3$ and $4x + 2y = 6$ are different forms of the same equation, these equations are called *dependent equations*. In general, the equations in a linear system with infinitely many solutions are called **dependent equations**.

✔ **CHECK POINT 5** Solve by graphing: infinitely many solutions; $\{(x, y)|x + y = 3\}$ or $\{(x, y)|2x + 2y = 6\}$

$$x + y = 3$$
$$2x + 2y = 6.$$

ENRICHMENT ESSAY

Missing America

Here she is, Miss America, the icon of American beauty. Always thin, she is becoming more so. The scatter plot in the figure shows Miss America's body-mass index, a ratio comparing weight divided by the square of height. Two lines are also shown: a line that passes near the data points and a horizontal line representing the World Health Organization's cutoff point for undernutrition. The intersection point indicates that in approximately 1978, Miss America reached this cutoff. There she goes: If the trend continues, Miss America's body-mass index could reach zero in about 320 years.

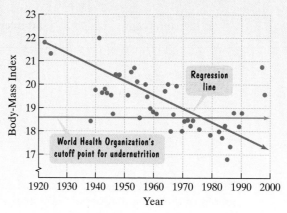

Body-Mass Index of Miss America

Source: Johns Hopkins School of Public Health

4.1 EXERCISE SET

Student Solutions Manual   CD/Video   PH Math/Tutor Center   MathXL Tutorials on CD   MathXL®   MyMathLab   Interactmath.com

Practice Exercises

In Exercises 1–10, determine whether the given ordered pair is a solution of the system.

1. $(2, -3)$ solution
$2x + 3y = -5$
$7x - 3y = 23$

2. $(-2, -5)$ solution
$6x - 2y = -2$
$3x + y = -11$

3. $\left(\dfrac{2}{3}, \dfrac{1}{9}\right)$ solution
$x + 3y = 1$
$4x + 3y = 3$

4. $\left(\dfrac{7}{25}, -\dfrac{1}{25}\right)$ solution
$4x + 3y = 1$
$3x - 4y = 1$

5. $(-5, 9)$ not a solution
$5x + 3y = 2$
$x + 4y = 14$

6. $(10, 7)$ not a solution
$6x - 5y = 25$
$4x + 15y = 13$

7. $(1400, 450)$ solution
$x - 2y = 500$
$0.03x + 0.02y = 51$

8. $(200, 700)$ solution
$-4x + y = -100$
$0.05x - 0.06y = -32$

9. $(8, 5)$ not a solution
$5x - 4y = 20$
$3y = 2x + 1$

10. $(5, -2)$ not a solution
$4x - 3y = 26$
$x = 15 - 5y$

In Exercises 11–42, solve each system by graphing. If there is no solution or an infinite number of solutions, so state, and use set notation to express solution sets.

11. $x + y = 6$ $\{(4, 2)\}$
$x - y = 2$

12. $x + y = 2$ $\{(3, -1)\}$
$x - y = 4$

13. $x + y = 1$ $\{(-1, 2)\}$
$y - x = 3$

14. $x + y = 4$ $\{(0, 4)\}$
$y - x = 4$

15. $2x - 3y = 6$ $\{(3, 0)\}$
$4x + 3y = 12$

16. $x + 2y = 2$ $\{(2, 0)\}$
$x - y = 2$

17. $4x + y = 4$ $\{(1, 0)\}$
$3x - y = 3$

18. $5x - y = 10$ $\{(2, 0)\}$
$2x + y = 4$

19. $y = x + 5$ $\{(-1, 4)\}$
$y = -x + 3$

20. $y = x + 1$ $\{(1, 2)\}$
$y = 3x - 1$

21. $y = 2x$ $\{(2, 4)\}$
$y = -x + 6$

22. $y = 2x + 1$ $\{(-1, -1)\}$
$y = -2x - 3$

23. $y = -2x + 3$ $\{(2, -1)\}$
$y = -x + 1$

24. $y = 3x - 4$ $\{(1, -1)\}$
$y = -2x + 1$

25. $y = 2x - 1$ no solution;
$y = 2x + 1$ $\varnothing$

26. $y = 3x - 1$ no solution; $\varnothing$
$y = 3x + 2$

27. $x + y = 4$ $\{(-2, 6)\}$
$x = -2$

28. $x + y = 6$ $\{(9, -3)\}$
$y = -3$

29. $x - 2y = 4$
$2x - 4y = 8$
infinitely many solutions;
$\{(x, y) | x - 2y = 4\}$ or
$\{(x, y) | 2x - 4y = 8\}$

30. $2x + 3y = 6$
$4x + 6y = 12$
infinitely many solutions;
$\{(x, y) | 2x + 3y = 6\}$ or
$\{(x, y) | 4x + 6y = 12\}$

31. $y = 2x - 1$ $\{(2, 3)\}$
$x - 2y = -4$

32. $y = -2x - 4$ $\{(0, -4)\}$
$4x - 2y = 8$

33. $x + y = 5$ no solution;
$2x + 2y = 12$ $\varnothing$

34. $x - y = 2$ no solution; $\varnothing$
$3x - 3y = -6$

35. $x - y = 0$
$y = x$
infinitely many solutions;
$\{(x, y) | x - y = 0\}$
or $\{(x, y) | y = x\}$

36. $2x - y = 0$
$y = 2x$
infinitely many solutions;
$\{(x, y) | 2x - y = 0\}$
or $\{(x, y) | y = 2x\}$

37. $x = 2$ $\{(2, 4)\}$
$y = 4$

38. $x = 3$ $\{(3, 5)\}$
$y = 5$

39. $x = 2$ no solution; $\varnothing$
$x = -1$

40. $x = 3$ no solution; $\varnothing$
$x = -2$

41. $y = 0$ no solution; $\varnothing$
$y = 4$

42. $y = 0$ no solution; $\varnothing$
$y = 5$

Practice Plus

Solve Exercises 43–50 without graphing. Find the slope and the y-intercept for the graph of each equation in the given system. Use this information (and not the equations' graphs) to determine if the system has no solution, one solution, or an infinite number of solutions.

43. $y = \dfrac{1}{2}x - 3$
$y = \dfrac{1}{2}x - 5$ $m = \dfrac{1}{2}; b = -3$ and -5; no solution

44. $y = \dfrac{3}{4}x - 2$
$y = \dfrac{3}{4}x + 1$ $m = \dfrac{3}{4}; b = -2$ and 1; no solution

45. $y = -\dfrac{1}{2}x + 4$
$3x - y = -4$ $m = -\dfrac{1}{2}$ and $3; b = 4$; one solution

46. $y = -\dfrac{1}{4}x + 3$
$4x - y = -3$ $m = -\dfrac{1}{4}$ and $4; b = 3$; one solution

47. $3x - y = 6$
$x = \dfrac{y}{3} + 2$ $m = 3; b = -6$; infinite number of solutions

48. $2x - y = 4$
$x = \dfrac{y}{2} + 2$ $m = 2; b = -4$; infinite number of solutions

49. $3x + y = 0$
$y = -3x + 1$ $m = -3; b = 0$ and 1; no solution

50. $2x + y = 0$
$y = -2x + 1$ $m = -2; b = 0$ and 1; no solution

Application Exercises

51. The graph shows the number of births in Massachusetts for women under 30 years old and women 30 years or older.

Number of Births in Massachusetts

Source: Massachusetts Department of Public Health

   a. Estimate the coordinates of the point of intersection. What does this mean in terms of the number of babies born to older mothers? Answers will vary. Example: (1996, 41): In 1996, both groups had the same number of births, 41 thousand.

   b. Describe what is happening to the number of births in Massachusetts to the right of the intersection point. There are more births to mothers 30 years and older than to mothers under 30 years old.

52. The figure shows scatter plots for the men's and women's winning times, in seconds, in the Olympic 100-meter freestyle swimming event. Also shown are lines that best fit the data, one for the men and one for the women.

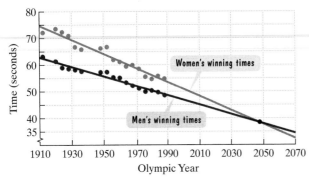

Winning Times in the Olympic 100-Meter Freestyle Swimming Event

   a. Estimate the coordinates of the point of intersection. What does this mean in terms of the women's time and the men's time? Answers may vary. Example: (2048, 39); During the 2048 Olympics, the women's and men's times will be the same at 39 seconds.

   b. Make a prediction about the swimming event in the Olympic years to the right of the intersection point. The women's winning time will be lower than the men's.

Writing in Mathematics

53. What is a system of linear equations? Provide an example with your description. 53–60. Answers will vary.

54. What is a solution of a system of linear equations?

55. Explain how to determine if an ordered pair is a solution of a system of linear equations.

56. Explain how to solve a system of linear equations by graphing.

57. What is an inconsistent system? What happens if you attempt to solve such a system by graphing?

58. Explain how a linear system can have infinitely many solutions.

59. What are dependent equations? Provide an example with your description.

60. The following system models the winning times, y, in seconds, in the Olympic 500-meter speed skating event x years after 1970:

$$y = -0.19x + 43.7 \quad \text{Women}$$
$$y = -0.16x + 39.9. \quad \text{Men}$$

Use the slope of each model to explain why the system has a solution. What does this solution represent?

Critical Thinking Exercises

61. Which one of the following statements is true? c

   a. If a linear system has graphs with equal slopes, the system must be inconsistent.

   b. If a linear system has graphs with equal y-intercepts, the system must have infinitely many solutions.

   c. If a linear system has two distinct points that are solutions, then the graphs of the system's equations have equal slopes and equal y-intercepts.

   d. It is possible for a linear system with one solution to have graphs with equal slopes.

62. Write a system of linear equations whose solution is $(5, 1)$. How many different systems are possible? Explain. Answers will vary.

63. Write a system of equations with one solution, a system of equations with no solution, and a system of equations with infinitely many solutions. Explain how you were able to think of these systems. Answers will vary.

64. Graph $y = x^2$ and $y = x + 2$ on the same axes. Find two ordered pairs that satisfy the system. Check that your answers satisfy both equations in the system. $(-1, 1)$ and $(2, 4)$

Technology Exercises

65. Verify your solutions to any five exercises from Exercises 11 through 36 by using a graphing utility to graph the two equations in the system in the same viewing rectangle. After entering the two equations, one as y_1 and the other as y_2, and graphing them, use the TRACE and ZOOM features to find the coordinates of the intersection point. (It may first be necessary to solve the equation for y before entering it.) Many graphing utilities have a special INTERSECTION feature that displays the coordinates of the intersection point once the equations are graphed. Consult your manual. Answers will vary.

Read Exercise 65. Then use a graphing utility to solve the systems in Exercises 66–73.

66. $y = 2x + 2$ $\{(1, 4)\}$
$y = -2x + 6$

67. $y = -x + 5$ $\{(6, -1)\}$
$y = x - 7$

68. $x + 2y = 2$ $\{(2, 0)\}$
$x - y = 2$

69. $2x - 3y = 6$ $\{(3, 0)\}$
$4x + 3y = 12$

70. $3x - y = 5$ $\{(0, -5)\}$
$-5x + 2y = -10$

71. $2x - 3y = 7$ $\{(2, -1)\}$
$3x + 5y = 1$

72. $y = \dfrac{1}{3}x + \dfrac{2}{3}$ $\{(7, 3)\}$
$y = \dfrac{5}{7}x - 2$

73. $y = -\dfrac{1}{2}x + 2$ $\{(-4, 4)\}$
$y = \dfrac{3}{4}x + 7$

Review Exercises

In Exercises 74–76, perform the indicated operation.

74. $-3 + (-9)$ (Section 1.7, Table 1.6) $\quad -12$

75. $-3 - (-9)$ (Section 1.7, Table 1.6) $\quad 6$

76. $-3(-9)$ (Section 1.7, Table 1.6) $\quad 27$

4.2

Objectives

1 Solve linear systems by the substitution method.

2 Use the substitution method to identify systems with no solution or infinitely many solutions.

3 Solve problems using the substitution method.

SOLVING SYSTEMS OF LINEAR EQUATIONS BY THE SUBSTITUTION METHOD

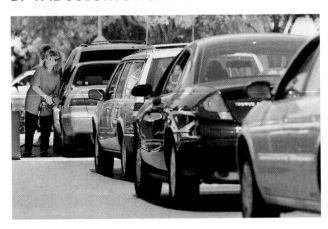

Other than outrage, what is going on at the gas pumps? Is surging demand creating the increasing oil prices? Like all things in a free market economy, the price of a commodity is based on supply and demand. In this section, we use a second method for solving linear systems, the *substitution method*, to understand this economic phenomenon.

1 Solve linear systems by the substitution method.

Eliminating a Variable Using the Substitution Method Finding the solution of a linear system by graphing equations may not be easy to do. For example, a solution of $\left(-\frac{2}{3}, \frac{157}{29}\right)$ would be difficult to "see" as an intersection point on a graph.

Let's consider a method that does not depend on finding a system's solution visually: the substitution method. This method involves converting the system to one equation in one variable by an appropriate substitution.

EXAMPLE 1 Solving a System by Substitution

Solve by the substitution method:

$$y = -x - 1$$
$$4x - 3y = 24.$$

SOLUTION

Step 1. **Solve either of the equations for one variable in terms of the other.** This step has already been done for us. The first equation, $y = -x - 1$, is solved for y in terms of x.

Step 2. **Substitute the expression from step 1 into the other equation.** We substitute the expression $-x - 1$ for y into the other equation:

$$y = \boxed{-x - 1} \quad 4x - 3\boxed{y} = 24. \qquad \text{Substitute } -x - 1 \text{ for } y.$$

This gives us an equation in one variable, namely

$$4x - 3(-x - 1) = 24.$$

The variable y has been eliminated.

Step 3. **Solve the resulting equation containing one variable.**

$$\begin{aligned}
4x - 3(-x - 1) &= 24 &\text{This is the equation containing one variable.} \\
4x + 3x + 3 &= 24 &\text{Apply the distributive property.} \\
7x + 3 &= 24 &\text{Combine like terms.} \\
7x &= 21 &\text{Subtract 3 from both sides.} \\
x &= 3 &\text{Divide both sides by 7.}
\end{aligned}$$

Step 4. **Back-substitute the obtained value into the equation from step 1.** We now know that the x-coordinate of the solution is 3. To find the y-coordinate, we back-substitute the x-value into the equation from step 1,

$$y = -x - 1. \qquad \text{This is the equation from step 1.}$$

> Substitute 3 for x.

$$y = -3 - 1$$
$$y = -4 \qquad \text{Simplify.}$$

With $x = 3$ and $y = -4$, the proposed solution is $(3, -4)$.

Step 5. **Check the proposed solution in both of the system's given equations.** Replace x with 3 and y with -4.

$$\begin{aligned}
y &= -x - 1 & 4x - 3y &= 24 \\
-4 &\overset{?}{=} -3 - 1 & 4(3) - 3(-4) &\overset{?}{=} 24 \\
-4 &= -4, \quad \text{true} & 12 + 12 &\overset{?}{=} 24 \\
& & 24 &= 24, \quad \text{true}
\end{aligned}$$

The pair $(3, -4)$ satisfies both equations. The system's solution is $(3, -4)$ and the solution set is $\{(3, -4)\}$. ∎

✔ **CHECK POINT 1** Solve by the substitution method: $\quad (3, 2) \text{ or } \{(3, 2)\}$

$$y = 5x - 13$$
$$2x + 3y = 12.$$

USING TECHNOLOGY

A graphing utility can be used to solve the system in Example 1. Graph each equation and use the intersection feature. The utility displays the solution $(3, -4)$ as $x = 3$, $y = -4$.

Intersection
X=3   Y=-4

$[-10, 10, 1]$ by $[-10, 10, 1]$

Before considering additional examples, let's summarize the steps used in the substitution method.

STUDY TIP

In step 1, you can choose which variable to isolate in which equation. If possible, solve for a variable whose coefficient is 1 or -1 to avoid working with fractions.

SOLVING LINEAR SYSTEMS BY SUBSTITUTION

1. Solve either of the equations for one variable in terms of the other. (If one of the equations is already in this form, you can skip this step.)
2. Substitute the expression found in step 1 into the *other* equation. This will result in an equation in one variable.
3. Solve the equation containing one variable.
4. Back-substitute the value found in step 3 into the equation from step 1. Simplify and find the value of the remaining variable.
5. Check the proposed solution in both of the system's given equations.

EXAMPLE 2 Solving a System by Substitution

Solve by the substitution method:

$$5x - 4y = 9$$
$$x - 2y = -3.$$

SOLUTION

Step 1. **Solve either of the equations for one variable in terms of the other.** We begin by isolating one of the variables in either of the equations. By solving for x in the second equation, which has a coefficient of 1, we can avoid fractions.

$$x - 2y = -3 \qquad \text{This is the second equation in the given system.}$$
$$x = 2y - 3 \qquad \text{Solve for x by adding 2y to both sides.}$$

Step 2. **Substitute the expression from step 1 into the other equation.** We substitute $2y - 3$ for x in the first equation.

$$x = \boxed{2y - 3} \quad 5\boxed{x} - 4y = 9$$

This gives us an equation in one variable, namely

$$5(2y - 3) - 4y = 9.$$

The variable x has been eliminated.

Step 3. **Solve the resulting equation containing one variable.**

$$5(2y - 3) - 4y = 9 \qquad \text{This is the equation containing one variable.}$$
$$10y - 15 - 4y = 9 \qquad \text{Apply the distributive property.}$$
$$6y - 15 = 9 \qquad \text{Combine like terms.}$$
$$6y = 24 \qquad \text{Add 15 to both sides.}$$
$$y = 4 \qquad \text{Divide both sides by 6.}$$

Step 4. **Back-substitute the obtained value into the equation from step 1.** Now that we have the y-coordinate of the solution, we back-substitute 4 for y in the equation $x = 2y - 3$.

$$x = 2y - 3 \qquad \text{Use the equation obtained in step 1.}$$
$$x = 2(4) - 3 \qquad \text{Substitute 4 for y.}$$
$$x = 8 - 3 \qquad \text{Multiply.}$$
$$x = 5 \qquad \text{Subtract.}$$

With $x = 5$ and $y = 4$, the proposed solution is $(5, 4)$.

Step 5. **Check.** Take a moment to show that $(5, 4)$ satisfies both given equations. The solution is $(5, 4)$ and the solution set is $\{(5, 4)\}$. ∎

 CHECK POINT 2 Solve by the substitution method: $(1, -2)$ or $\{(1, -2)\}$

$$3x + 2y = -1$$
$$x - y = 3.$$

2 Use the substitution method to identify systems with no solution or infinitely many solutions.

The Substitution Method with Linear Systems Having No Solution or Infinitely Many Solutions Recall that a linear system with no solution is called an **inconsistent system**. If you attempt to solve such a system by substitution, you will eliminate both variables. A false statement such as $0 = 17$ will be the result.

EXAMPLE 3 Using the Substitution Method on an Inconsistent System

Solve the system:
$$y + 1 = 5(x + 1)$$
$$y = 5x - 1.$$

SOLUTION The variable y is isolated in the second equation. We use the substitution method and substitute the expression for y in the first equation.

$\boxed{y} + 1 = 5(x + 1) \qquad y = \boxed{5x - 1}$ Substitute $5x - 1$ for y.

$(5x - 1) + 1 = 5(x + 1)$ This substitution gives an equation in one variable.

$5x = 5x + 5$ Simplify on the left side. Use the distributive property on the right side.

There are no values of x and y for which $0 = 5$. $0 = 5$,   false   Subtract $5x$ from both sides.

The false statement $0 = 5$ indicates that the system is inconsistent and has no solution. The solution set is the empty set, $\varnothing$. ■

✔ **CHECK POINT 3** Solve the system: no solution or $\varnothing$
$$3x + y = -5$$
$$y = -3x + 3.$$

Do you remember that the equations in a linear system with infinitely many solutions are called **dependent**? If you attempt to solve such a system by substitution, you will eliminate both variables. However, a true statement such as $5 = 5$ will be the result.

EXAMPLE 4 Using the Substitution Method on a System with Infinitely Many Solutions

Solve the system:
$$y = 3 - 2x$$
$$4x + 2y = 6.$$

SOLUTION The variable y is isolated in the first equation. We use the substitution method and substitute the expression for y in the second equation.

$y = \boxed{3 - 2x} \qquad 4x + 2\boxed{y} = 6$ Substitute $3 - 2x$ for y.

$4x + 2(3 - 2x) = 6$ This substitution gives an equation in one variable.

$4x + 6 - 4x = 6$ Apply the distributive property:

$2(3 - 2x) = 2 \cdot 3 - 2 \cdot 2x = 6 - 4x.$

$6 = 6$,   true   Simplify: $4x - 4x = 0$.

This true statement indicates that the system contains dependent equations and has infinitely many solutions. We express the solution set for the system in one of two equivalent ways:

$$\{(x, y)\,|\,y = 3 - 2x\}$$ The set of all ordered pairs (x, y) such that y = 3 − 2x

or $\{(x, y)\,|\,4x + 2y = 6\}$.   The set of all ordered pairs (x, y) such that 4x + 2y = 6 ∎

✔ **CHECK POINT 4** Solve the system: infinitely many solutions; $\{(x, y)\,|\,y = 3x - 4\}$ or $\{(x, y)\,|\,9x - 3y = 12\}$

$$y = 3x - 4$$
$$9x - 3y = 12.$$

3 Solve problems using the substitution method.

Applications An important application of systems of equations arises in connection with supply and demand. As the price of a product increases, the demand for that product decreases. However, at higher prices suppliers are willing to produce greater quantities of the product.

EXAMPLE 5   Supply and Demand Models

A chain of video stores specializes in cult films. The weekly demand and supply models for *The Rocky Horror Picture Show* are given by

$$N = -13p + 760 \qquad \text{Demand model}$$
$$N = 2p + 430 \qquad \text{Supply model}$$

in which p is the price of the video and N is the number of copies of the video sold or supplied each week to the chain of stores.

  a. How many copies of the video can be sold and supplied at $18 per copy?

  b. Find the price at which supply and demand are equal. At this price, how many copies of *Rocky Horror* can be supplied and sold each week?

SOLUTION

  a. To find how many copies of the video can be sold and supplied at $18 per copy, we substitute 18 for p in the demand and supply models.

Demand Model	**Supply Model**
$N = -13p + 760$	$N = 2p + 430$
Substitute 18 for p.	Substitute 18 for p.
$N = -13 \cdot 18 + 760 = 526$	$N = 2 \cdot 18 + 430 = 466$

At $18 per video, the chain can sell 526 copies of *Rocky Horror* in a week. The manufacturer is willing to supply 466 copies per week. This will result in a shortage of copies of the video. Under these conditions, the retail chain is likely to raise the price of the video.

b. We can find the price at which supply and demand are equal by solving the demand-supply linear system. We will use substitution, substituting $-13p + 760$ for N in the second equation.

$$N = \boxed{-13p + 760} \qquad \boxed{N} = 2p + 430 \qquad \text{Substitute } -13p + 760 \text{ for } N.$$

$$-13p + 760 = 2p + 430 \qquad \text{The resulting equation contains only one variable.}$$

$$-15p + 760 = 430 \qquad \text{Subtract } 2p \text{ from both sides.}$$

$$-15p = -330 \qquad \text{Subtract } 760 \text{ from both sides.}$$

$$p = 22 \qquad \text{Divide both sides by } -15.$$

The price at which supply and demand are equal is $22 per video. To find the value of N, the number of videos supplied and sold weekly at this price, we back-substitute 22 for p into either the demand or the supply model. We'll use both models to make sure we get the same number in each case.

Demand Model	**Supply Model**
$N = -13p + 760$	$N = 2p + 430$
Substitute **22** for p.	Substitute **22** for p.
$N = -13 \cdot 22 + 760 = 474$	$N = 2 \cdot 22 + 430 = 474$

At a price of $22 per video, 474 units of the video can be supplied and sold weekly. The intersection point, (22, 474), is shown in Figure 4.7.

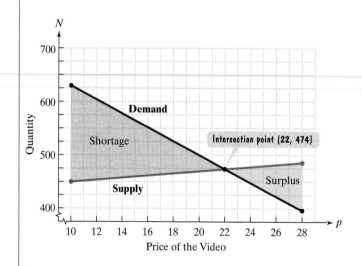

FIGURE 4.7 Priced at $22 per video, 474 copies of the video can be supplied and sold weekly. ∎

✔ CHECK POINT **5** The demand for a product is modeled by $N = -20p + 1000$ and the supply for the product by $N = 5p + 250$. In these models, p is the price of the product and N is the number of units of the product supplied or sold weekly. At what price will supply equal demand? At that price, how many units of the product will be supplied and sold each week? $30; 400 units

4.2 EXERCISE SET

Student Solutions Manual | CD/Video | PH Math/Tutor Center | MathXL Tutorials on CD | MathXL® | MyMathLab | Interactmath.com

Practice Exercises

In Exercises 1–32, solve each system by the substitution method. If there is no solution or an infinite number of solutions, so state, and use set notation to express solution sets.

1. $x + y = 4$ $\{(1, 3)\}$
$y = 3x$

2. $x + y = 6$ $\{(2, 4)\}$
$y = 2x$

3. $x + 3y = 8$ $\{(5, 1)\}$
$y = 2x - 9$

4. $2x - 3y = -13$ $\{(-2, 3)\}$
$y = 2x + 7$

5. $x + 3y = 5$ $\{(2, 1)\}$
$4x + 5y = 13$

6. $x + 2y = 5$ $\{(-5, 5)\}$
$2x - y = -15$

7. $2x - y = -5$ $\{(-1, 3)\}$
$x + 5y = 14$

8. $2x + 3y = 11$ $\{(4, 1)\}$
$x - 4y = 0$

9. $2x - y = 3$ $\{(4, 5)\}$
$5x - 2y = 10$

10. $-x + 3y = 10$ $\{(-7, 1)\}$
$2x + 8y = -6$

11. $-3x + y = -1$
$x - 2y = 4$ $\left\{\left(-\frac{2}{5}, -\frac{11}{5}\right)\right\}$

12. $-4x + y = -11$ $\left\{\left(\frac{14}{5}, \frac{1}{5}\right)\right\}$
$2x - 3y = 5$

13. $x = 9 - 2y$
$x + 2y = 13$
no solution; ∅

14. $6x + 2y = 7$
$y = 2 - 3x$
no solution; ∅

15. $y = 3x - 5$
$21x - 35 = 7y$
infinitely many solutions;
$\{(x, y)|y = 3x - 5\}$
or $\{(x, y)|21x - 35 = 7y\}$

16. $9x - 3y = 12$
$y = 3x - 4$
infinitely many solutions;
$\{(x, y)|9x - 3y = 12\}$
or $\{(x, y)|y = 3x - 4\}$

17. $5x + 2y = 0$ $\{(0, 0)\}$
$x - 3y = 0$

18. $4x + 3y = 0$ $\{(0, 0)\}$
$2x - y = 0$

19. $2x - y = 6$ $\left\{\left(\frac{17}{7}, -\frac{8}{7}\right)\right\}$
$3x + 2y = 5$

20. $2x - y = 4$ $\left\{\left(\frac{18}{7}, \frac{8}{7}\right)\right\}$
$3x - 5y = 2$

21. $2(x - 1) - y = -3$
$y = 2x + 3$
no solution; ∅

22. $x + y - 1 = 2(y - x)$
$y = 3x - 1$
infinitely many solutions;
$\{(x, y)|x + y - 1 = 2(y - x)\}$
or $\{(x, y)|y = 3x - 1\}$

23. $x = 2y + 9$ $\left\{\left(\frac{43}{5}, -\frac{1}{5}\right)\right\}$
$x = 7y + 10$

24. $x = 5y - 3$ $\left\{\left(-\frac{44}{3}, -\frac{7}{3}\right)\right\}$
$x = 8y + 4$

25. $4x - y = 100$ $\{(200, 700)\}$
$0.05x - 0.06y = -32$

26. $x + 6y = 8000$ $\{(2000, 1000)\}$
$0.3x - 0.6y = 0$

27. $y = \frac{1}{3}x + \frac{2}{3}$ $\{(7, 3)\}$
$y = \frac{5}{7}x - 2$

28. $y = -\frac{1}{2}x + 2$ $\{(-4, 4)\}$
$y = \frac{3}{4}x + 7$

29. $\frac{x}{6} - \frac{y}{2} = \frac{1}{3}$ $\{(-1, -1)\}$
$x + 2y = -3$

30. $\frac{x}{4} - \frac{y}{4} = -1$ $\{(-5, -1)\}$
$x + 4y = -9$

31. $2x - 3y = 8 - 2x$ $\{(5, 4)\}$
$3x + 4y = x + 3y + 14$

32. $3x - 4y = x - y + 4$
$2x + 6y = 5y - 4$ $\{(-1, -2)\}$

Practice Plus

In Exercises 33–38, write a system of equations describing the given conditions. Then solve the system by the substitution method and find the two numbers.

33. The sum of two numbers is 81. One number is 41 more than the other. Find the numbers. $x + y = 81, y = x + 41$; 20 and 61

34. The sum of two numbers is 62. One number is 12 more than the other. Find the numbers. $x + y = 62, y = x + 12$; 25 and 37

35. The difference between two numbers is 5. Four times the larger number is 6 times the smaller number. Find the numbers. $x - y = 5, 4x = 6y$; 10 and 15

36. The difference between two numbers is 25. Two times the larger number is 12 times the smaller number. Find the numbers. $x - y = 25, 2x = 12y$; 5 and 30

37. The difference between two numbers is 1. The sum of the larger number and twice the smaller number is 7. Find the numbers. $x - y = 1, x + 2y = 7$; 2 and 3

38. The difference between two numbers is 5. The sum of the larger number and twice the smaller number is 14. Find the numbers. $x - y = 5, x + 2y = 14$; 3 and 8

In Exercises 39–40, multiply each equation in the system by an appropriate number so that the coefficients are integers. Then solve the system by the substitution method.

39. $0.7x - 0.1y = 0.6$
$0.8x - 0.3y = -0.8$ $(2, 8)$

40. $1.25x - 0.01y = 4.5$
$0.5x - 0.02y = 1$ $(4, 50)$

Application Exercises

41. At a price of p dollars per ticket, the number of tickets to a rock concert that can be sold is given by the demand model $N = -25p + 7500$. At a price of p dollars per ticket, the number of tickets that the concert's promoters are willing to make available is given by the supply model $N = 5p + 6000$.

   a. How many tickets can be sold and supplied for $40 per ticket?   6500 sold; 6200 supplied

   b. Find the ticket price at which supply and demand are equal. At this price, how many tickets will be supplied and sold? $50; 6250 tickets

42. The weekly demand and supply models for a particular brand of scientific calculator for a chain of stores are given by the demand model $N = -53p + 1600$ and the supply model $N = 75p + 320$. In these models, p is the price of the calculator and N is the number of calculators sold or supplied each week to the stores.

   a. How many calculators can be sold and supplied at $12 per calculator?   964 sold; 1220 supplied

   b. Find the price at which supply and demand are equal. At this price, how many calculators of this type can be supplied and sold each week? $10; 1070 calculators

A business breaks even when the cost for running the business is equal to the money taken in by the business. In Exercises 43–44, determine how many units must be sold so that the business breaks even, experiencing neither loss nor profit.

43. A gasoline station has weekly costs and revenue (the money taken in by the station) that depend on the number of gallons of gasoline purchased and sold. If x gallons are purchased and sold, weekly costs are given by $y = 1.2x + 1080$ and weekly revenue by $y = 1.6x$. How many gallons of gasoline must be sold weekly for the station to break even? 2700 gallons

44. An artist has monthly costs and revenue (the money taken in by the artist) that depend on the number of ceramic pieces produced and sold. If x ceramic pieces are produced and sold, monthly costs are given by $y = 4x + 2000$ and monthly revenue by $y = 9x$. How many ceramic pieces must be sold monthly for the artist to break even? 400 ceramic pieces

45. Infant mortality for African Americans is decreasing at a faster rate than it is for whites, shown by the graphs below. Infant mortality for African Americans can be modeled by $M = -0.41x + 22$ and for whites by $M = -0.18x + 10$. In both models, x is the number of years after 1980 and M is infant mortality, measured in deaths per 1000 live births. Use these models to project when infant mortality for African Americans and whites will be the same. What will be the infant mortality for both groups at that time? 2032; about 0.6 deaths per 1000 live births or less than 1 death per 1000 live births.

Infant Mortality

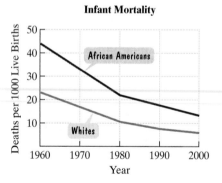

Source: National Center for Health Statistics

46. The equation $x + 10y = 2120$ models deaths from gunfire in the United States, y, in deaths per hundred thousand Americans, in year x. The equation $7x + 8y = 14,065$ models deaths from car accidents in the United States, y, in deaths per hundred thousand Americans, in year x. Solve the linear system formed by the two models. Then describe what the solution means in terms of the variables in the given models. $\{(1999, 12.5)\}$; In 1995, there were 12.5 deaths per hundred thousand Americans by gunfire and by car accidents.

Writing in Mathematics

47. Describe a problem that might arise when solving a system of equations using graphing. Assume that both equations in the system have been graphed correctly and the system has exactly one solution. 47–51. Answers will vary.

48. Explain how to solve a system of equations using the substitution method. Use $y = 3 - 3x$ and $3x + 4y = 6$ to illustrate your explanation.

49. When using the substitution method, how can you tell if a system of linear equations has no solution?

50. When using the substitution method, how can you tell if a system of linear equations has infinitely many solutions?

51. The law of supply and demand states that, in a free market economy, a commodity tends to be sold at its equilibrium price. At this price, the amount that the seller will supply is the same amount that the consumer will buy. Explain how systems of equations can be used to determine the equilibrium price.

Critical Thinking Exercises

52. Which one of the following is true? b
  a. Solving an inconsistent system by substitution results in a true statement.
  b. The line passing through the intersection of the graphs of $x + y = 4$ and $x - y = 0$ with slope $= 3$ has an equation given by $y - 2 = 3(x - 2)$.
  c. Unlike the graphing method, where solutions cannot be seen, the substitution method provides a way to visualize solutions as intersection points.
  d. To solve the system

$$2x - y = 5$$
$$3x + 4y = 7$$

  by substitution, replace y in the second equation with $5 - 2x$.

53. If $x = 3 - y - z, 2x + y - z = -6$, and $3x - y + z = 11$, find the values for x, y, and z. $x = 1, y = -3, z = 5$

54. Find the value of m that makes

$$y = mx + 3$$
$$5x - 2y = 7$$

an inconsistent system. $m = \dfrac{5}{2}$

Review Exercises

55. Graph: $4x + 6y = 12$. (Section 3.2, Example 4) See graphing answer section.

56. Solve: $4(x + 1) = 25 + 3(x - 3)$. (Section 2.3, Example 3) 12

57. List all the integers in this set: $\left\{-73, -\dfrac{2}{3}, 0, \dfrac{3}{1}, \dfrac{3}{2}, \dfrac{\pi}{1}\right\}$. (Section 1.2, Example 5) $-73, 0, \dfrac{3}{1}$

SECTION

4.3

Objectives

1 Solve linear systems by the addition method.

2 Use the addition method to identify systems with no solution or infinitely many solutions.

3 Determine the most efficient method for solving a linear system.

SOLVING SYSTEMS OF LINEAR EQUATIONS BY THE ADDITION METHOD

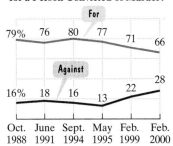

Are You in Favor of the Death Penalty for a Person Convicted of Murder?

For: 79%, 76, 80, 77, 71, 66

Against: 16%, 18, 16, 13, 22, 28

Oct. 1988, June 1991, Sept. 1994, May 1995, Feb. 1999, Feb. 2000

The graphs shown above are based on 543 adults polled nationally by *Newsweek*. If these trends continue, when will the percentage of Americans in favor of the death penalty be the same as the percentage of those who oppose it? The question can be answered by modeling the data with a system of linear equations and solving the system. However, the substitution method is not always the easiest way to solve linear systems. In this section we consider a third method for solving these systems.

1 Solve linear systems by the addition method.

Eliminating a Variable Using the Addition Method   The substitution method is most useful if one of the given equations has an isolated variable. A third, and frequently the easiest, method for solving a linear system is the addition method. Like the substitution method, the addition method involves eliminating a variable and ultimately solving an equation containing only one variable. However, this time we eliminate a variable by adding the equations.

For example, consider the following system of equations:

$$3x - 4y = 11$$
$$-3x + 2y = -7.$$

When we add these two equations, the *x*-terms are eliminated. This occurs because the coefficients of the *x*-terms, 3 and -3, are opposites (additive inverses) of each other.

$$
\begin{array}{r}
3x - 4y = 11 \\
-3x + 2y = -7 \\
\hline
\text{Add:} \quad -2y = 4
\end{array}
$$

The sum is an equation in one variable.

$$y = -2 \qquad \textit{Solve for y by dividing both sides by } -2.$$

Now we can back-substitute -2 for *y* into one of the original equations to find *x*. It does not matter which equation you use; you will obtain the same value for *x* in either case. If we use either equation, we can show that $x = 1$ and the solution $(1, -2)$ satisfies both equations in the system.

When we use the addition method, we want to obtain two equations whose sum is an equation containing only one variable. The key step is to **obtain, for one of the variables, coefficients that differ only in sign**. To do this, we may need to multiply one or both equations by some nonzero number so that the coefficients of one of the variables, *x* or *y*, become opposites. Then when the two equations are added, this variable is eliminated.

EXAMPLE 1   Solving a System by the Addition Method

Solve by the addition method:

$$x + y = 4$$
$$x - y = 6.$$

SOLUTION   The coefficients of y in the two equations, 1 and -1, differ only in sign. Therefore, by adding the two left sides and the two right sides, we can eliminate the y-terms.

$$
\begin{aligned}
x + y &= 4 \\
x - y &= 6 \\
\hline
\text{Add:}\quad 2x + 0y &= 10 \\
2x &= 10 \qquad \text{Simplify.}
\end{aligned}
$$

Now y is eliminated and we can solve $2x = 10$ for x.

$$2x = 10$$
$$x = 5 \qquad \text{Divide both sides by 2 and solve for } x.$$

We back-substitute 5 for x into one of the original equations to find y. We will use both equations to show that we obtain the same value for y in either case.

Use the first equation:

$$x + y = 4$$
$$5 + y = 4$$
$$y = -1.$$

Use the second equation:

$$x - y = 6$$
$$5 - y = 6 \qquad \text{Replace x with 5.}$$
$$-y = 1 \qquad \text{Solve for y.}$$
$$y = -1.$$

Thus, $x = 5$ and $y = -1$. The proposed solution, $(5, -1)$, can be shown to satisfy both equations in the system. Consequently, the solution is $(5, -1)$ and the solution set is $\{(5, -1)\}$. ∎

 CHECK POINT 1 Solve by the addition method:

$$x + y = 5$$
$$x - y = 9. \qquad (7, -2) \text{ or } \{(7, -2)\}$$

EXAMPLE 2   Solving a System by the Addition Method

Solve by the addition method:

$$3x - y = 11$$
$$2x + 5y = 13.$$

SOLUTION   We must rewrite one or both equations in equivalent forms so that the coefficients of the same variable (either x or y) differ only in sign. Consider the terms in y in each equation, that is, $-1y$ and $5y$. To eliminate y, we can multiply each term of the first equation by 5 and then add the equations.

$$
\begin{array}{lcl}
3x - y = 11 & \xrightarrow{\text{Multiply by 5.}} & 15x - 5y = 55 \\
2x + 5y = 13 & \xrightarrow{\text{No change}} & 2x + 5y = 13 \\
& & \hline
& \text{Add:} & 17x + 0y = 68 \qquad \text{Simplify.} \\
& & 17x = 68 \qquad \text{Divide both sides by} \\
& & x = 4 \qquad \text{17 and solve for x.}
\end{array}
$$

Thus, $x = 4$. To find y, we back-substitute 4 for x into either one of the given equations. We'll use the second equation.

$$2x + 5y = 13 \qquad \text{This is the second equation in the given system.}$$
$$2 \cdot 4 + 5y = 13 \qquad \text{Substitute 4 for x.}$$
$$8 + 5y = 13 \qquad \text{Multiply: } 2 \cdot 4 = 8.$$
$$5y = 5 \qquad \text{Subtract 8 from both sides.}$$
$$y = 1 \qquad \text{Divide both sides by 5.}$$

The solution is $(4, 1)$. Check to see that it satisfies both of the original equations in the system. The solution set is $\{(4, 1)\}$. ∎

 CHECK POINT 2 Solve by the addition method:

$$4x - y = 22$$
$$3x + 4y = 26. \qquad (6, 2) \text{ or } \{(6, 2)\}$$

Before considering additional examples, let's summarize the steps for solving linear systems by the addition method.

SOLVING LINEAR SYSTEMS BY ADDITION

1. If necessary, rewrite both equations in the form $Ax + By = C$.

2. If necessary, multiply either equation or both equations by appropriate nonzero numbers so that the sum of the x-coefficients or the sum of the y-coefficients is 0.

3. Add the equations in step 2. The sum is an equation in one variable.

4. Solve the equation in one variable.

5. Back-substitute the value obtained in step 4 into either of the given equations and solve for the other variable.

6. Check the solution in both of the original equations.

EXAMPLE 3　Solving a System by the Addition Method

Solve by the addition method:

$$3x + 2y = 48$$
$$9x - 8y = -24.$$

SOLUTION

Step 1. **Rewrite both equations in the form $Ax + By = C$.** Both equations are already in this form. Variable terms appear on the left and constants appear on the right.

Step 2. **If necessary, multiply either equation or both equations by appropriate numbers so that the sum of the x-coefficients or the sum of the y-coefficients is 0.** We can eliminate x or y. Let's eliminate x. Consider the terms in x in each equation, that is, $3x$ and $9x$. To eliminate x, we can multiply each term of the first equation by -3 and then add the equations.

$$
\begin{array}{lll}
3x + 2y = 48 & \xrightarrow{\text{Multiply by } -3.} & -9x - 6y = -144 \\
9x - 8y = -24 & \xrightarrow{\text{No change}} & \underline{9x - 8y = -24} \\
& \text{Add:} & -14y = -168
\end{array}
$$

Step 3. **Add the equations.**

Step 4. **Solve the equation in one variable.** We solve $-14y = -168$ by dividing both sides by -14.

$$\frac{-14y}{-14} = \frac{-168}{-14} \quad \text{Divide both sides by } -14.$$

$$y = 12 \quad \text{Simplify.}$$

Step 5. **Back-substitute and find the value for the other variable.** We can back-substitute 12 for y into either one of the given equations. We'll use the first one.

$$3x + 2y = 48 \qquad \text{This is the first equation in the given system.}$$
$$3x + 2(12) = 48 \qquad \text{Substitute 12 for y.}$$
$$3x + 24 = 48 \qquad \text{Multiply.}$$
$$3x = 24 \qquad \text{Subtract 24 from both sides.}$$
$$x = 8 \qquad \text{Divide both sides by 3.}$$

The solution is $(8, 12)$.

Step 6. **Check.** Take a few minutes to show that $(8, 12)$ satisfies both of the original equations in the system. The solution set is $\{(8, 12)\}$. ∎

 CHECK POINT **3** Solve by the addition method:

$$4x + 5y = 3$$
$$2x - 3y = 7. \qquad (2, -1) \text{ or } \{(2, -1)\}$$

Some linear systems have solutions that are not integers. If the value of one variable turns out to be a "messy" fraction, back-substitution might lead to cumbersome arithmetic. If this happens, you can return to the original system and use addition to find the value of the other variable.

EXAMPLE 4   Solving a System by the Addition Method

Solve by the addition method:

$$2x = 7y - 17$$
$$5y = 17 - 3x.$$

SOLUTION

Step 1. **Rewrite both equations in the form $Ax + By = C$.** We first arrange the system so that variable terms appear on the left and constants appear on the right. We obtain

$$2x - 7y = -17 \qquad \text{Subtract 7y from both sides of the first equation.}$$
$$3x + 5y = 17. \qquad \text{Add 3x to both sides of the second equation.}$$

Step 2. **If necessary, multiply either equation or both equations by appropriate numbers so that the sum of the x-coefficients or the sum of the y-coefficients is 0.** We can eliminate x or y. Let's eliminate x by multiplying the first equation by 3 and the second equation by -2.

$$
\begin{array}{l}
2x - 7y = -17 \quad \xrightarrow{\text{Multiply by 3.}} \quad 6x - 21y = -51 \\
3x + 5y = 17 \quad \xrightarrow{\text{Multiply by } -2} \quad \underline{-6x - 10y = -34} \\
\phantom{3x + 5y = 17 \quad \text{Multiply by } -2} \quad \text{Add:} \quad -31y = -85
\end{array}
$$

Step 3. **Add the equations.**

Step 4. **Solve the equation in one variable.** We solve $-31y = -85$ by dividing both sides by -31.

$$\frac{-31y}{-31} = \frac{-85}{-31} \qquad \text{Divide both sides by } -31.$$

$$y = \frac{85}{31} \qquad \text{Simplify.}$$

Step 5. **Back-substitute and find the value for the other variable.** Back-substitution of $\frac{85}{31}$ for y into either of the given equations results in cumbersome arithmetic. Instead, let's use the addition method on the given system in the form $Ax + By = C$ to find the value for x. Thus, we eliminate y by multiplying the first equation by 5 and the second equation by 7.

$$
\begin{array}{l}
2x - 7y = -17 \xrightarrow{\text{Multiply by 5.}} 10x - 35y = -85 \\
3x + 5y = 17 \xrightarrow{\text{Multiply by 7.}} \underline{21x + 35y = 119} \\
\phantom{3x + 5y = 17 \xrightarrow{\text{Multiply by 7.}}} \text{Add: } 31x = 34 \\
\phantom{3x + 5y = 17 \xrightarrow{\text{Multiply by 7.}} \text{Add: }} x = \frac{34}{31} \qquad \text{Divide both sides by 31.}
\end{array}
$$

The solution is $\left(\dfrac{34}{31}, \dfrac{85}{31}\right)$.

Step 6. **Check.** For this system, a calculator is helpful in showing that $\left(\frac{34}{31}, \frac{85}{31}\right)$ satisfies both of the original equations in the system. The solution set is $\left\{\left(\frac{34}{31}, \frac{85}{31}\right)\right\}$. ■

✔ **CHECK POINT 4** Solve by the addition method:

$$2x = 9 + 3y$$
$$4y = 8 - 3x. \qquad \left(\frac{60}{17}, -\frac{11}{17}\right) \text{ or } \left\{\left(\frac{60}{17}, -\frac{11}{17}\right)\right\}$$

2 Use the addition method to identify systems with no solution or infinitely many solutions.

The Addition Method with Linear Systems Having No Solution or Infinitely Many Solutions
As with the substitution method, if the addition method results in a false statement, the linear system is inconsistent and has no solution.

EXAMPLE 5 Using the Addition Method on an Inconsistent System

Solve the system:

$$4x + 6y = 12$$
$$6x + 9y = 12.$$

SOLUTION We can eliminate x or y. Let's eliminate x by multiplying the first equation by 3 and the second equation by -2.

$$
\begin{array}{l}
4x + 6y = 12 \xrightarrow{\text{Multiply by 3.}} 12x + 18y = 36 \\
6x + 9y = 12 \xrightarrow{\text{Multiply by -2.}} \underline{-12x - 18y = -24} \\
\phantom{6x + 9y = 12 \xrightarrow{\text{Multiply by -2.}}} \text{Add: } 0 = 12
\end{array}
$$

There are no values of x and y for which $0 = 12$.

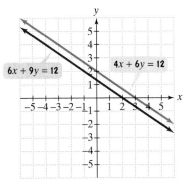

FIGURE 4.8 Visualizing the inconsistent system in Example 5

The false statement $0 = 12$ indicates that the system is inconsistent and has no solution. The solution set is the empty set, $\varnothing$. The graphs of the system's equations are shown in Figure 4.8. The lines are parallel and have no point of intersection. ■

✔ **CHECK POINT 5** Solve the system:

$$x + 2y = 4$$
$$3x + 6y = 13. \qquad \text{no solution or } \varnothing$$

If you use the addition method, how can you tell if a system has infinitely many solutions? As with the substitution method, you will eliminate both variables and obtain a true statement.

EXAMPLE 6 Using the Addition Method on a System with Infinitely Many Solutions

Solve by the addition method:

$$2x - y = 3$$
$$-4x + 2y = -6.$$

SOLUTION We can eliminate y by multiplying the first equation by 2.

$$
\begin{array}{llll}
2x - y = 3 & \xrightarrow{\text{Multiply by 2.}} & 4x - 2y = 6 \\
-4x + 2y = -6 & \xrightarrow{\text{No change}} & \underline{-4x + 2y = -6} \\
& & \text{Add:} & 0 = 0
\end{array}
$$

The true statement $0 = 0$ indicates that the system contains dependent equations and has infinitely many solutions. Any ordered pair that satisfies the first equation also satisfies the second equation. We express the solution set for the system in one of two equivalent ways:

$$\{(x, y) \mid 2x - y = 3\} \text{ or } \{(x, y) \mid -4x + 2y = -6\}. \quad \blacksquare$$

 CHECK POINT 6 Solve by the addition method:

$$x - 5y = 7$$
$$3x - 15y = 21.$$

infinitely many solutions; $\{(x, y) \mid x - 5y = 7\}$ or $\{(x, y) \mid 3x - 15y = 21\}$

3 Determine the most efficient method for solving a linear system.

Comparing the Three Solution Methods The chart that follows compares the graphing, substitution, and addition methods for solving systems of linear equations. With increased practice, it becomes easier for you to select the best method for solving a particular linear system.

Comparing Solution Methods

Method	Advantages	Disadvantages
Graphing	You can see the solutions.	If the solutions do not involve integers or are too large to be seen on the graph, it's impossible to tell exactly what the solutions are.
Substitution	Gives exact solutions. Easy to use if a variable is on one side by itself.	Solutions cannot be seen. Can introduce extensive work with fractions when no variable has a coefficient of 1 or −1.
Addition	Gives exact solutions. Easy to use even if no variable has a coefficient of 1 or −1.	Solutions cannot be seen.

4.3 EXERCISE SET

Student Solutions Manual CD/Video PH Math/Tutor Center MathXL Tutorials on CD MathXL® MyMathLab Interactmath.com

Practice Exercises

In Exercises 1–44, solve each system by the addition method. If there is no solution or an infinite number of solutions, so state, and use set notation to express solution sets.

1. $x + y = -3$
$x - y = 11$ $\{(4, -7)\}$

2. $x + y = 6$
$x - y = -2$ $\{(2, 4)\}$

3. $2x + 3y = 6$
$2x - 3y = 6$ $\{(3, 0)\}$

4. $3x + 2y = 14$
$3x - 2y = 10$ $\{(4, 1)\}$

5. $x + 2y = 7$
$-x + 3y = 18$ $\{(-3, 5)\}$

6. $2x + y = -2$
$-2x - 3y = -6$ $\{(-3, 4)\}$

7. $5x - y = 14$
$-5x + 2y = -13$ $\{(3, 1)\}$

8. $7x - 4y = 13$
$-7x + 6y = -11$ $\{\left(\frac{17}{7}, 1\right)\}$

9. $3x + y = 7$
$2x - 5y = -1$ $\{(2, 1)\}$

10. $3x - y = 11$
$2x + 5y = 13$ $\{(4, 1)\}$

11. $x + 3y = 4$
$4x + 5y = 2$ $\{(-2, 2)\}$

12. $x + 2y = -1$
$4x - 5y = 22$ $\{(3, -2)\}$

13. $-3x + 7y = 14$
$2x - y = -13$ $\{(-7, -1)\}$

14. $2x - 5y = -1$
$3x + y = 7$ $\{(2, 1)\}$

15. $3x - 14y = 6$
$5x + 7y = 10$ $\{(2, 0)\}$

16. $5x - 4y = 19$
$3x + 2y = 7$ $\{(3, -1)\}$

17. $3x - 4y = 11$
$2x + 3y = -4$ $\{(1, -2)\}$

18. $2x + 3y = -16$
$5x - 10y = 30$ $\{(-2, -4)\}$

19. $3x + 2y = -1$
$-2x + 7y = 9$ $\{(-1, 1)\}$

20. $5x + 3y = 27$
$7x - 2y = 13$ $\{(3, 4)\}$

21. $3x = 2y + 7$
$5x = 2y + 13$ $\{(3, 1)\}$

22. $9x = 25 + y$
$2y = 4 - 9x$ $\{(2, -7)\}$

23. $2x = 3y - 4$
$-6x + 12y = 6$ $\{(-5, -2)\}$

24. $5x = 4y - 8$
$3x + 7y = 14$ $\{(0, 2)\}$

25. $2x - y = 3$
$4x + 4y = -1$ $\{\left(\frac{11}{12}, -\frac{7}{6}\right)\}$

26. $3x - y = 22$
$4x + 5y = -21$ $\{\left(\frac{89}{19}, -\frac{151}{19}\right)\}$

27. $4x = 5 + 2y$
$2x + 3y = 4$ $\{\left(\frac{23}{16}, \frac{3}{8}\right)\}$

28. $3x = 4y + 1$
$4x + 3y = 1$ $\{\left(\frac{7}{25}, -\frac{1}{25}\right)\}$

29. $3x - y = 1$
$3x - y = 2$ no solution; $\varnothing$

30. $4x - 9y = -2$
$-4x + 9y = -2$ no solution; $\varnothing$

31. $x + 3y = 2$
$3x + 9y = 6$
infinitely many solutions;
$\{(x, y) | x + 3y = 2\}$ or
$\{(x, y) | 3x + 9y = 6\}$

32. $4x - 2y = 2$
$2x - y = 1$
infinitely many solutions;
$\{(x, y) | 4x - 2y = 2\}$ or
$\{(x, y) | 2x - y = 1\}$

33. $7x - 3y = 4$
$-14x + 6y = -7$
no solution; $\varnothing$

34. $2x + 4y = 5$
$3x + 6y = 6$
no solution; $\varnothing$

35. $5x + y = 2$
$3x + y = 1$ $\{\left(\frac{1}{2}, -\frac{1}{2}\right)\}$

36. $2x - 5y = -1$
$2x - y = 1$ $\{\left(\frac{3}{4}, \frac{1}{2}\right)\}$

37. $x = 5 - 3y$
$2x + 6y = 10$
infinitely many solutions;
$\{(x, y) | x = 5 - 3y\}$ or
$\{(x, y) | 2x + 6y = 10\}$

38. $4x = 36 + 8y$
$3x - 6y = 27$
infinitely many solutions;
$\{(x, y) | 4x = 36 + 8y\}$ or
$\{(x, y) | 3x - 6y = 27\}$

39. $4(3x - y) = 0$
$3(x + 3) = 10y$ $\{\left(\frac{1}{3}, 1\right)\}$

40. $2(2x + 3y) = 0$
$7x = 3(2y + 3) + 2$ $\{\left(1, -\frac{2}{3}\right)\}$

41. $x + y = 11$
$\frac{x}{5} + \frac{y}{7} = 1$ $\{(-10, 21)\}$

42. $x - y = -3$
$\frac{x}{9} - \frac{y}{7} = -1$ $\{(18, 21)\}$

43. $\frac{4}{5}x - y = -1$
$\frac{2}{5}x + y = 1$ $\{(0, 1)\}$

44. $\frac{x}{3} + y = 3$
$\frac{x}{2} - \frac{y}{4} = 1$ $\{(3, 2)\}$

In Exercises 45–56, solve each system by the method of your choice. If there is no solution or an infinite number of solutions, so state, and use set notation to express solution sets. Explain why you selected one method over the other two. **45–56.** Explanations will vary.

45. $3x - 2y = 8$
$x = -2y$ $\{(2, -1)\}$

46. $2x - y = 10$
$y = 3x$ $\{(-10, -30)\}$

47. $3x + 2y = -3$
$2x - 5y = 17$ $\{(1, -3)\}$

48. $2x - 7y = 17$
$4x - 5y = 25$ $\{(5, -1)\}$

49. $3x - 2y = 6$
$y = 3$ $\{(4, 3)\}$

50. $2x + 3y = 7$
$x = 2$ $\{(2, 1)\}$

51. $y = 2x + 1$
$y = 2x - 3$ no solution; $\varnothing$

52. $y = 2x + 4$
$y = 2x - 1$ no solution; $\varnothing$

53. $2(x + 2y) = 6$
$3(x + 2y - 3) = 0$
infinitely many solutions;
$\{(x, y) | 2(x + 2y) = 6\}$ or
$\{(x, y) | 3(x + 2y - 3) = 0\}$

54. $2(x + y) = 4x + 1$
$3(x - y) = x + y - 3$
$\{\left(\frac{1}{2}, 1\right)\}$

55. $3y = 2x$
$2x + 9y = 24$ $\{(3, 2)\}$

56. $4y = -5x$
$5x + 8y = 20$ $\{(-4, 5)\}$

Practice Plus

In Exercises 57–64, solve each system or state that the system is inconsistent or dependent.

57. $\frac{3x}{5} + \frac{4y}{5} = 1$
$\frac{x}{4} - \frac{3y}{8} = -1$ $\{(-1, 2)\}$

58. $\frac{x}{3} - \frac{y}{2} = \frac{2}{3}$
$\frac{2x}{3} + y = \frac{4}{3}$ $\{(2, 0)\}$

59. $5(x + 1) = 7(y + 1) - 7$
$6(x + 1) + 5 = 5(y + 1)$ $\{(-1, 0)\}$

60. $6x = 5(x + y + 3) - x$
$3(x - y) + 4y = 5(y + 1)$ $\{(-5, -5)\}$

61. $0.4x + y = 2.2$
$0.5x - 1.2y = 0.3$ $\{(3, 1)\}$

62. $1.25x - 1.5y = 2$
$3.5x - 1.75y = 10.5$ $\{(4, 2)\}$

63. $\dfrac{x}{2} = \dfrac{y + 8}{3}$
$\dfrac{x + 2}{2} = \dfrac{y + 11}{3}$ dependent

64. $\dfrac{x}{2} = \dfrac{y + 8}{4}$
$\dfrac{x + 3}{2} = \dfrac{y + 5}{4}$ inconsistent

Application Exercises

We opened this section with data from 1988 through 2000 showing the percentage of Americans for and against the death penalty for a person convicted of murder. The data can be modeled by the following system of equations:

$$13x + 12y = 992$$
The percent, y, in favor of the death penalty x years after 1988

$$-x + y = 16.$$
The percent, y, against the death penalty x years after 1988

Use this system to solve Exercises 65–66.

65. Use the addition method to determine in which year the percentage of Americans in favor of the death penalty will be the same as the percentage of Americans who oppose it. For that year, what percent will be for the death penalty and what percent will be against it?   2020; 48% for 48% against

66. Use the substitution method to solve Exercise 65.   2020; 48% for and 48% against

Writing in Mathematics

67. Explain how to solve a system of equations using the addition method. Use $3x + 5y = -2$ and $2x + 3y = 0$ to illustrate your explanation.   67–71. Answers will vary.

68. When using the addition method, how can you tell if a system of linear equations has no solution?

69. When using the addition method, how can you tell if a system of linear equations has infinitely many solutions?

70. Take a second look at the data about the death penalty shown on page 263. Do you think that these trends will continue? Explain your answer.

71. The formula $3239x + 96y = 134{,}014$ models the number of daily evening newspapers, y, x years after 1980. The formula $-665x + 36y = 13{,}800$ models the number of daily morning newspapers, y, x years after 1980. What is the most efficient method for solving this system? Explain why. What does the solution mean in terms of the variables in the formulas? (It is not necessary to actually solve the system.)

Critical Thinking Exercises

72. Which one of the following statements is true?   c

  a. If x can be eliminated by the addition method, y cannot be eliminated by using the original equations of the system.

  b. If $Ax + 2y = 2$ and $2x + By = 10$ have graphs that intersect at $(2, -2)$, then $A = -3$ and $B = 3$.

  c. The equations $y = x - 1$ and $x = y + 1$ are dependent.

  d. If the two equations in a linear system are $5x - 3y = 7$ and $4x + 9y = 11$, multiplying the first equation by 4, the second by 5, and then adding equations will eliminate x.

73. Solve by expressing x and y in terms of a and b:

$$x - y = a$$
$$y = 2x + b.$$ $x = -a - b; y = -2a - b$

74. The point of intersection of the graphs of the equations $Ax - 3y = 16$ and $3x + By = 7$ is $(5, -2)$. Find A and B. $A = 2, B = 4$

Technology Exercises

75. Some graphing utilities can give the solution to a system of linear equations. (Consult your manual for details.) This capability is usually accessed with the $\boxed{\text{SIMULT}}$ (simultaneous equations) feature. First, you must enter 2, for two equations in two variables. With each equation in $Ax + By = C$ form, you must then enter the coefficients for x and y and the constant term, one equation at a time. After entering all six numbers, press $\boxed{\text{SOLVE}}$. The solution will be displayed on the screen. (The x-value may be displayed as $x_1 =$ and the y-value as $x_2 =$.) Use this capability to verify the solution to any five of the exercises you solved in the practice exercises of this exercise set. Describe what happens when you use your graphing utility on a system with no solution or infinitely many solutions.   Answers will vary.

If your graphing utility has the feature described in Exercise 75, use it to solve each system in Exercises 76–78.

76. $\dfrac{1}{4}x - \dfrac{1}{4}y = -1$
$-3x + 7y = 8$ $\{(-5, -1)\}$

77. $x = 5y$
$2x - 3y = 7$ $\{(5, 1)\}$

78. $0.6x + 0.08y = 4$
$3x + 2y = 4$ $\{(8, -10)\}$

Review Exercises

79. For which number is 5 times the number equal to the number increased by 40? (Section 2.5, Example 2)   10

80. In which quadrant is $\left(-\dfrac{3}{2}, 15\right)$ located? (Section 1.3, Example 1)   II

81. Solve: $29{,}700 + 150x = 5000 + 1100x$. (Section 2.2, Example 7)   26

✔ MID-CHAPTER CHECK POINT

CHAPTER

4

What You Know: We learned how to solve systems of linear equations by graphing, by the substitution method, and by the addition method. We saw that some systems, called inconsistent systems, have no solution, whereas other systems, called dependent systems, have infinitely many solutions.

In Exercises 1–3, solve each system by graphing.

1. $3x + 2y = 6$
$2x - y = 4$ $\{(2, 0)\}$

2. $y = 2x - 1$
$y = 3x - 2$ $\{(1, 1)\}$

3. $y = 2x - 1$
$6x - 3y = 12$ no solution or $\emptyset$

In Exercises 4–15, solve each system by the method of your choice.

4. $5x - 3y = 1$
$y = 3x - 7$ $\{(5, 8)\}$

5. $6x + 5y = 7$
$3x - 7y = 13$ $\{(2, -1)\}$

6. $x = \dfrac{y}{3} - 1$
$6x + y = 21$ $\{(2, 9)\}$

7. $3x - 4y = 6$
$5x - 6y = 8$ $\{(-2, -3)\}$

8. $3x - 2y = 32$
$\dfrac{x}{5} + 3y = -1$ $\{(10, -1)\}$

9. $x - y = 3$
$2x = 4 + 2y$
no solution or $\emptyset$

10. $x = 2(y - 5)$
$4x + 40 = y - 7$
$\{(-12, -1)\}$

11. $y = 3x - 2$
$y = 2x - 9$ $\{(-7, -23)\}$

12. $2x - 3y = 4$
$3x + 4y = 0$
$\left\{\left(\dfrac{16}{17}, -\dfrac{12}{17}\right)\right\}$

13. $y - 2x = 7$
$4x = 2y - 14$
infinitely many solutions;
$\{(x, y) | y - 2x = 7\}$ or
$\{(x, y) | 4x = 2y - 14\}$

14. $4(x + 3) = 3y + 7$
$2(y - 5) = x + 5$
$\{(7, 11)\}$

15. $\dfrac{x}{2} - \dfrac{y}{5} = 1$
$y - \dfrac{x}{3} = 8$ $\{(6, 10)\}$

SECTION

4.4

PROBLEM SOLVING USING SYSTEMS OF EQUATIONS

Objectives

1 Solve problems using linear systems.

2 Solve simple interest problems.

3 Solve mixture problems.

4 Solve motion problems.

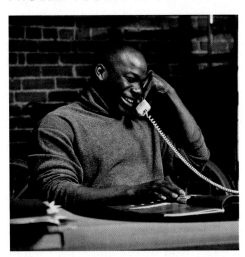

Enjoy chatting long distance on the phone? Telecommunication companies want your business. You can go online and get a list of their plans. Is there a monthly fee? What is the rate per minute? Does the plan involve a monthly minimum? In this section, you will learn to use systems of equations to select a plan that will save you the most money.

A Strategy for Solving Word Problems Using Systems of Equations   When

1   Solve problems using linear systems.

we solved problems in Chapters 2 and 3, we let x represent a quantity that was unknown. Problems in this section involve two unknown quantities. We will let x and *y* represent these quantities. We then translate from the verbal conditions of the problem to a *system* of linear equations.

EXAMPLE 1 The World's Longest Snakes

The royal python and the anaconda are the world's longest snakes. The maximum length for each of these snakes is implied by the following description:

> Three royal pythons and two anacondas measure 161 feet. The royal python's length increased by triple the anaconda's length is 119 feet. Find the maximum length for each of these snakes.

SOLUTION

Step 1. **Use variables to represent unknown quantities.** Let *x* represent the royal python's length, in feet. Let *y* represent the anaconda's length, in feet.

Step 2. **Write a system of equations describing the problem's conditions.**

Three royal pythons	plus	two anacondas	measure	161 feet.
$3x$	$+$	$2y$	$=$	161

The royal python's length	increased by	triple the anaconda's length	is	119 feet.
x	$+$	$3y$	$=$	119

Step 3. **Solve the system and answer the problem's question.** The system

$$3x + 2y = 161$$
$$x + 3y = 119$$

can be solved by substitution or addition. Substitution works well because *x* in the second equation has a coefficient of 1. We can solve for *x* by subtracting $3y$ from both sides, thereby avoiding fractions. Addition also works well; if we multiply the second equation by -3, adding equations will eliminate *x*. We will use addition.

$$
\begin{array}{ll}
3x + 2y = 161 & \xrightarrow{\text{No change}} \\
x + 3y = 119 & \xrightarrow{\text{Multiply by } -3.}
\end{array}
\qquad
\begin{array}{r}
3x + 2y = 161 \\
-3x - 9y = -357 \\
\hline
\text{Add:} \qquad -7y = -196
\end{array}
$$

$$y = \frac{-196}{-7} = 28$$

Because *y* represents the anaconda's length, we see that the anaconda is 28 feet long. Now we can find *x*, the royal python's length. We do so by back-substituting 28 for *y* in either of the system's equations.

$$
\begin{array}{ll}
x + 3y = 119 & \text{We'll use the second equation.} \\
x + 3 \cdot 28 = 119 & \text{Back-substitute 28 for } y. \\
x + 84 = 119 & \text{Multiply: } 3 \cdot 28 = 84. \\
x = 35 & \text{Subtract 84 from both sides.}
\end{array}
$$

Because $x = 35$ and $y = 28$, the royal python is 35 feet long and the anaconda is 28 feet long.

Step 4. **Check the proposed answers in the original wording of the problem.** Three royal pythons and two anacondas should measure 161 feet:

$$3(35 \text{ ft}) + 2(28 \text{ ft}) = 105 \text{ ft} + 56 \text{ ft} = 161 \text{ ft}.$$

The royal python's length increased by triple the anaconda's length should be 119 feet:

$$35 \text{ ft} + 3(28 \text{ ft}) = 35 \text{ ft} + 84 \text{ ft} = 119 \text{ ft}.$$ ■

This verifies that the royal python's and the anaconda's length are 35 feet and 28 feet, respectively.

 CHECK POINT 1 The bustard and the condor are the world's heaviest flying bird and the world's heaviest bird of prey, respectively. The maximum weight for each of these birds is implied by the following description:

Two bustards and three condors weigh 173 pounds. The bustard's weight increased by double the condor's weight is 100 pounds. What is the maximum weight for each of these birds?

A bustard weighs 46 pounds and a condor weighs 27 pounds.

EXAMPLE 2 Cholesterol and Heart Disease

The verdict is in. After years of research, the nation's health experts agree that high cholesterol in the blood is a major contributor to heart disease. Thus, cholesterol intake should be limited to 300 milligrams or less each day. Fast foods provide a cholesterol carnival. All together, two McDonald's Quarter Pounders and three Burger King Whoppers with cheese contain 520 milligrams of cholesterol. Three Quarter Pounders and one Whopper with cheese exceed the suggested daily cholesterol intake by 53 milligrams. Determine the cholesterol content in each item.

SOLUTION

Step 1. **Use variables to represent the unknown quantities.** Let x represent the cholesterol content, in milligrams, of a Quarter Pounder. Let y represent the cholesterol content, in milligrams, of a Whopper with cheese.

Step 2. **Write a system of equations describing the problem's conditions.**

About 15 million hamburgers are eaten every day in the United States.

The amount of cholesterol in 2 Quarter Pounders	plus	the amount of cholesterol in 3 Whoppers with cheese	is	520 mg.
$2x$	$+$	$3y$	$=$	520

The amount of cholesterol in 3 Quarter Pounders	plus	the amount of cholesterol in 1 Whopper with cheese	is	the suggested daily limit plus 53 mg.
$3x$	$+$	y	$=$	$300 + 53$

Step 3. **Solve the system and answer the problem's question.** The system

$$2x + 3y = 520$$
$$3x + y = 353$$

can be solved by substitution or addition. We will use addition; if we multiply the second equation by -3, adding equations will eliminate y.

$$2x + 3y = 520 \quad \xrightarrow{\text{No change}} \quad 2x + 3y = 520$$
$$3x + y = 353 \quad \xrightarrow{\text{Multiply by } -3.} \quad -9x - 3y = -1059$$
$$\text{Add:} \quad -7x = -539$$
$$x = \frac{-539}{-7} = 77$$

Because x represents the cholesterol content of a Quarter Pounder, we see that a Quarter Pounder contains 77 milligrams of cholesterol. Now we can find y, the cholesterol content of a Whopper with cheese. We do so by back-substituting 77 for x in either of the system's equations.

$$3x + y = 353 \qquad \text{We'll use the second equation.}$$
$$3(77) + y = 353 \qquad \text{Back-substitute 77 for } x.$$
$$231 + y = 353 \qquad \text{Multiply.}$$
$$y = 122 \qquad \text{Subtract 231 from both sides.}$$

Because $x = 77$ and $y = 122$, a Quarter Pounder contains 77 milligrams of cholesterol and a Whopper with cheese contains 122 milligrams of cholesterol.

Step 4. **Check the proposed answers in the original wording of the problem.** Two Quarter Pounders and three Whoppers with cheese contain

$$2(77 \text{ mg}) + 3(122 \text{ mg}) = 520 \text{ mg},$$

which checks with the given conditions. Furthermore, three Quarter Pounders and one Whopper with cheese contain

$$3(77 \text{ mg}) + 1(122 \text{ mg}) = 353 \text{ mg},$$

which does exceed the daily limit of 300 milligrams by 53 milligrams. ■

✔ **CHECK POINT 2** How do the Quarter Pounder and Whopper with cheese measure up in the calorie department? Actually, not too well. Two Quarter Pounders and three Whoppers with cheese provide 2607 calories. Even combining one of each provides enough calories to bring tears to Jenny Craig's eyes—9 calories in excess of what is allowed on a 1000 calorie-a-day diet. Find the caloric content of each item.

A Quarter Pounder has 420 calories and a Whopper with cheese has 589 calories.

EXAMPLE 3 Solar and Electric Heating Systems

The costs for two different kinds of heating systems for a three-bedroom home are given in the following table.

System	Cost to Install	Operating Cost/Year
Solar	$29,700	$150
Electric	$5000	$1100

After how many years will total costs for solar heating and electric heating be the same? What will be the cost at that time?

SOLUTION

Step 1. **Use variables to represent unknown quantities.** Let x represent the number of years the heating system is used. Let y represent the total cost for the heating system.

System	Cost to Install	Operating Cost/Year
Solar	$29,700	$150
Electric	$5000	$1100

Costs for heating systems, repeated

Step 2. **Write a system of equations describing the problem's conditions.**

Total cost for the solar system	equals	installation cost	plus	yearly operating cost	times	the number of years the system is used.
y	$=$	$29,700$	$+$	150	$\cdot$	x

Total cost for the electric system	equals	installation cost	plus	yearly operating cost	times	the number of years the system is used.
y	$=$	5000	$+$	1100	$\cdot$	x

Step 3. **Solve the system and answer the problem's question.** We want to know after how many years the total costs for the two systems will be the same. We must solve the system

$$y = 29,700 + 150x$$
$$y = 5000 + 1100x.$$

Substitution works well because y is isolated in each equation.

$$y = \boxed{29,700 + 150x} \; \boxed{y} = 5000 + 1100x \qquad \text{Substitute } 29,700 + 150x \text{ for } y.$$

$$29,700 + 150x = 5000 + 1100x \qquad \text{This substitution gives an equation in one variable.}$$

$$29,700 = 5000 + 950x \qquad \text{Subtract 150x from both sides.}$$

$$24,700 = 950x \qquad \text{Subtract 5000 from both sides.}$$

$$26 = x \qquad \text{Divide both sides by 950: } \frac{24,700}{950} = 26.$$

Because x represents the number of years the heating system is used, we see that after 26 years, the total costs for the two systems will be the same. Now we can find y, the total cost. Back-substitute 26 for x in either of the system's equations. We will use the second equation, $y = 5000 + 1100x$.

$$y = 5000 + 1100 \cdot 26 = 5000 + 28,600 = 33,600$$

Because $x = 26$ and $y = 33,600$, after 26 years, the total costs for the two systems will be the same. The cost for each system at that time will be $33,600.

Step 4. **Check the proposed answers in the original wording of the problem.** Let's verify that after 26 years, the two systems will cost the same amount. The installation cost for the solar system is $29,700 and the yearly operating cost is $150. Thus, the total cost after 26 years is

$$\$29,700 + \$150(26) = \$29,700 + \$3900 = \$33,600.$$

The installation cost for the electric system is $5000 and the yearly operating cost is $1100. Thus, the total cost after 26 years is

$$\$5000 + \$1100(26) = \$5000 + \$28,600 = \$33,600.$$

This verifies that after 26 years the two systems will cost the same amount, $33,600.

The graphs in Figure 4.9 give us a way of visualizing the solution to Example 3. The total cost of solar heating over 40 years is represented by the blue line. The total cost of electric heating over 40 years is represented by the red line. The lines intersect at $(26, 33,600)$: After 26 years, the cost for each system is the same, $33,600.

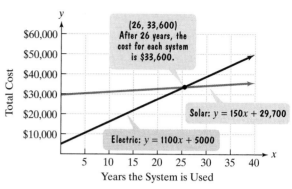

FIGURE 4.9 Visualizing total costs of solar and electric heating

Can you see that to the right of the intersection point, $(26, 33,600)$, the blue graph representing solar costs lies below the red graph representing electric costs? Thus, after 26 years, or when $x > 26$, the cost for solar heating is less than the cost for electric heating.

 CHECK POINT 3 Costs for two different kinds of heating systems for a three-bedroom house are given in the following table.

System	Cost to Install	Operating Cost/Year
Electric	$5000	$1100
Gas	$12,000	$700

After how long will total costs for electric heating and gas heating be the same? What will be the cost at that time?   17.5 years; $24,250

Next, we will solve problems involving investments, mixtures, and motion with systems of equations. We will continue using our four-step problem-solving strategy. We will also use tables to help organize the information in the problems.

2 Solve simple interest problems.

Dual Investments with Simple Interest Simple interest involves interest calculated only on the amount of money that we invest, called the **principal**. The formula $I = Pr$ is used to find the simple interest, I, earned for one year when the principal, P, is invested at an annual interest rate, r. Dual investment problems involve different amounts of money in two or more investments, each paying a different rate.

EXAMPLE 4 Solving a Dual Investment Problem

Your grandmother needs your help. She has $50,000 to invest. Part of this money is to be invested in noninsured bonds paying 15% annual interest. The rest of this money is to be invested in a government-insured certificate of deposit paying 7% annual interest. She told you that she requires $6000 per year in extra income from both of these investments. How much money should be placed in each investment?

SOLUTION

Step 1. **Use variables to represent unknown quantities.**
Let x = the amount invested in the 15% noninsured bonds.
Let y = the amount invested in the 7% certificate of deposit.

Step 2. **Write a system of equations describing the problem's conditions.** Because Grandma has $50,000 to invest,

Furthermore, Grandma requires $6000 in total interest. We can use a table to organize the information in the problem and obtain a second equation.

	Principal (amount invested)	×	Interest rate	=	Interest earned
15% Investment	x		0.15		0.15x
7% Investment	y		0.07		0.07y

The interest for the two investments combined must be $6000.

$$0.15x \quad + \quad 0.07y \quad = \quad 6000$$

Step 3. **Solve the system and answer the problem's question.** The system

$$x + y = 50{,}000$$
$$0.15x + 0.07y = 6000$$

can be solved by substitution or addition. Substitution works well because both variables in the first equation have coefficients of 1. Addition also works well; if we multiply the first equation by -0.15 or -0.07, adding equations will eliminate a variable. We will use addition.

$$
\begin{aligned}
x + y &= 50{,}000 \quad \xrightarrow{\text{Multiply by } -0.07.} && -0.07x - 0.07y = -3500 \\
0.15x + 0.07y &= 6000 \quad \xrightarrow{\text{No change}} && \underline{0.15x + 0.07y = 6000} \\
&& \text{Add:} && 0.08x = 2500 \\
&& && x = \frac{2500}{0.08} \\
&& && x = 31{,}250
\end{aligned}
$$

Because x represents the amount that should be invested at 15%, Grandma should place $31,250 in 15% noninsured bonds. Now we can find y, the amount that she should place in the 7% certificate of deposit. We do so by back-substituting 31,250 for x in either of the system's equations.

$$x + y = 50{,}000 \qquad \text{We'll use the first equation.}$$
$$31{,}250 + y = 50{,}000 \qquad \text{Back-substitute 31,250 for x.}$$
$$y = 18{,}750 \qquad \text{Subtract 31,250 from both sides.}$$

Because $x = 31{,}250$ and $y = 18{,}750$, Grandma should invest $31,250 at 15% and $18,750 at 7%.

Step 4. **Check the proposed answers in the original wording of the problem.** Has Grandma invested $50,000?

$$\$31{,}250 + \$18{,}750 = \$50{,}000$$

Yes, all her money was placed in the dual investments. Can she count on $6000 interest? The interest earned on $31,250 at 15% is ($31,250)(0.15), or $4687.50. The interest earned on $18,750 at 7% is ($18,750)(0.07), or $1312.50. The total interest is $4687.50 + $1312.50, or $6000, exactly as it should be. You've made your grandmother happy. (Now if you would just visit her more often ...)

✔ CHECK POINT 4 You inherited $5000 with the stipulation that for the first year the money had to be invested in two funds paying 9% and 11% annual interest. How much did you invest at each rate if the total interest earned for the year was $487? $3150 at 9%; $1850 at 11%

3 Solve mixture problems.

Problems Involving Mixtures   Chemists and pharmacists often have to change the concentration of solutions and other mixtures. In these situations, the amount of a particular ingredient in the solution or mixture is expressed as a percentage of the total.

EXAMPLE 5   Solving a Mixture Problem

A chemist working on a flu vaccine needs to mix a 10% sodium-iodine solution with a 60% sodium-iodine solution to obtain 50 milliliters of a 30% sodium-iodine solution. How many milliliters of the 10% solution and of the 60% solution should be mixed?

SOLUTION

Step 1.   **Use variables to represent unknown quantities.**
Let x = the number of milliliters of the 10% solution to be used in the mixture.
Let y = the number of milliliters of the 60% solution to be used in the mixture.

Step 2.   **Write a system of equations describing the problem's conditions.** The situation is illustrated in Figure 4.10.

Volume	x	y	50
Strength	10% solution	60% solution	30% solution

FIGURE 4.10

The chemist needs 50 milliliters of a 30% sodium-iodine solution. We form a table that shows the amount of sodium-iodine in each of the three solutions.

Solution	Number of milliliters	×	Percent of Sodium-Iodine	=	Amount of Sodium-Iodine
10% Solution	x		10% = 0.1		$0.1x$
60% Solution	y		60% = 0.6		$0.6y$
30% Mixture	50		30% = 0.3		$0.3(50) = 15$

The chemist needs to obtain a 50-milliliter mixture.

$$x + y = 50$$

The 50-milliliter mixture must be 30% sodium-iodine. The amount of sodium-iodine must be 30% of 50, or $(0.3)(50) = 15$ milliliters.

$$0.1x + 0.6y = 15$$

Step 3. **Solve the system and answer the problem's question.** The system

$$x + y = 50$$

$$0.1x + 0.6y = 15$$

can be solved by substitution or addition. Let's use substitution. The first equation can easily be solved for x or y. Solving for y, we obtain $y = 50 - x$.

$$y = \boxed{50 - x} \qquad 0.1x + 0.6\boxed{y} = 15$$

We substitute $50 - x$ for y in the second equation. This gives us an equation in one variable.

$0.1x + 0.6(50 - x) = 15$	This equation contains one variable, x.
$0.1x + 30 - 0.6x = 15$	Apply the distributive property.
$-0.5x + 30 = 15$	Combine like terms.
$-0.5x = -15$	Subtract 30 from both sides.
$x = \dfrac{-15}{-0.5} = 30$	Divide both sides by -0.5.

Back-substituting 30 for x in either of the system's equations ($x + y = 50$ is easier to use) gives $y = 20$. Because x represents the number of milliliters of the 10% solution and y the number of milliliters of the 60% solution, the chemist should mix 30 milliliters of the 10% solution with 20 milliliters of the 60% solution.

Step 4. **Check the proposed solution in the original wording of the problem.** The problem states that the chemist needs 50 milliliters of a 30% sodium-iodine solution. The amount of sodium-iodine in this mixture is $0.3(50)$, or 15 milliliters. The amount of sodium-iodine in 30 milliliters of the 10% solution is $0.1(30)$, or 3 milliliters. The amount of sodium-iodine in 20 milliliters of the 60% solution is $0.6(20) = 12$ milliliters. The amount of sodium-iodine in the two solutions used in the mixture is 3 milliliters + 12 milliliters, or 15 milliliters, exactly as it should be. ∎

Problems involving dual investments and problems involving mixtures are both based on the same idea: The total amount times the rate gives the amount.

| total money amount | rate of interest | amount of interest |

Dual Investment Problems: principal · rate = interest

Percents are expressed as decimals in these equations.

Mixture Problems: solution · concentration = ingredient

| total solution amount | percent of ingredient, or rate of concentration | amount of ingredient |

Our dual investment problem involved mixing two investments. Our mixture problem involved mixing two liquids. The equations in these problems are obtained from similar conditions:

Dual Investment Problems

| Interest from investment 1 | + | interest from investment 2 | = | amount of interest from mixed investments. |

Mixture Problems

| Ingredient amount in solution 1 | + | ingredient amount in solution 2 | = | amount of ingredient in mixture. |

Being aware of the similarities between dual investment and mixture problems should make you a better problem solver in a variety of situations that involve mixtures.

✔ **CHECK POINT 5** A chemist needs to mix a 12% acid solution with a 20% acid solution to obtain 160 ounces of a 15% acid solution. How many ounces of each of the acid solutions must be used? 12% solution: 100 ounces; 20% solution: 60 ounces

4 Solve motion problems.

Problems Involving Motion Suppose that you ride your bike at an average speed of 12 miles per hour. What distance do you cover in 2 hours? Your distance is the product of your speed and the time that you travel:

$$\frac{12 \text{ miles}}{\text{hour}} \times 2 \text{ hours} = 24 \text{ miles}.$$

Your distance is 24 miles. Notice how the hour units cancel. The distance is expressed in miles.

In general, the distance covered by any moving body is the product of its average speed, or rate, and its time in motion:

A FORMULA FOR MOTION

$$d = rt$$

Distance equals rate times time.

Wind and water current have the effect of increasing or decreasing a traveler's rate.

EXAMPLE 6 Solving a Motion Problem

When a small airplane flies with the wind, it can travel 450 miles in 3 hours. When the same airplane flies in the opposite direction against the wind, it takes 5 hours to fly the same distance. Find the average rate of the plane in still air and the average rate of the wind.

SOLUTION

Step 1. **Use variables to represent unknown quantities.**
Let x = the average rate of the plane in still air.
Let y = the average rate of the wind.

Step 2. **Write a system of equations describing the problem's conditions.** As it travels with the wind, the plane's rate is increased. The net rate is its rate in still air, x, plus the rate of the wind, y, given by the expression $x + y$. As it travels against the wind, the plane's rate is decreased. The net rate is its rate in still air, x, minus the rate of the wind, y, given by the expression $x - y$. Here is a chart that summarizes the problem's information and includes the increased and decreased rates.

	Rate	×	Time	=	Distance
Trip with the Wind	$x + y$		3		$3(x + y)$
Trip against the Wind	$x - y$		5		$5(x - y)$

The problem states that the distance in each direction is 450 miles. We use this information to write our system of equations.

The distance of the trip with the wind | is | 450 miles.

$$3(x + y) = 450$$

The distance of the trip against the wind | is | 450 miles.

$$5(x - y) = 450$$

Step 3. **Solve the system and answer the problem's question.** We can simplify the system by dividing both sides of the equations by 3 and 5, respectively.

$$3(x + y) = 450 \xrightarrow{\text{Divide by 3.}} x + y = 150$$
$$5(x - y) = 450 \xrightarrow{\text{Divide by 5.}} x - y = \ \ 90$$

Solve the system on the right by the addition method.

$$x + y = 150$$
$$\underline{x - y = \ \ 90}$$
Add: $$2x \ \ \ \ \ = 240$$
$$x = 120 \quad \text{Divide both sides by 2.}$$

Back-substituting 120 for x in either of the system's equations gives $y = 30$. Because $x = 120$ and $y = 30$, the average rate of the plane in still air is 120 miles per hour and the average rate of the wind is 30 miles per hour.

Step 4. **Check the proposed solution in the original wording of the problem.** Is the average rate of the plane in still air 120 miles per hour and the average rate of the wind 30 miles per hour? The problem states that the distance in each direction is 450 miles. The average rate of the plane with the wind is $120 + 30 = 150$ miles per hour. In 3 hours, it travels $150 \cdot 3$, or 450 miles, which checks with the stated condition. Furthermore, the average rate of the plane against the wind is $120 - 30 = 90$ miles per hour. In 5 hours, it travels $90 \cdot 5 = 450$ miles, which is the stated distance. ∎

 CHECK POINT 6 With the current, a motorboat can travel 84 miles in 2 hours. Against the current, the same trip takes 3 hours. Find the average rate of the boat in still water and the average rate of the current.

boat in still water: 35 miles per hour; current: 7 miles per hour

4.4 EXERCISE SET

Student Solutions Manual   CD/Video   PH Math/Tutor Center   MathXL Tutorials on CD   MathXL®   MyMathLab   Interactmath.com

Practice Exercises

In Exercises 1–4, let x represent one number and let y represent the other number. Use the given conditions to write a system of equations. Solve the system and find the numbers.

1. The sum of two numbers is 17. If one number is subtracted from the other, their difference is −3. Find the numbers.
   $x + y = 17$, $x - y = -3$; 7 and 10

2. The sum of two numbers is 5. If one number is subtracted from the other, their difference is 13. Find the numbers.
   $x + y = 5$, $x - y = 13$; 9 and −4

3. Three times a first number decreased by a second number is −1. The first number increased by twice the second number is 23. Find the numbers. $3x - y = -1$, $x + 2y = 23$; 3 and 10

4. The sum of three times a first number and twice a second number is 43. If the second number is subtracted from twice the first number, the result is −4. Find the numbers.
   $3x + 2y = 43$, $2x - y = -4$; 5 and 14

Application Exercises

5. The graph in the next column makes Super Bowl Sunday look like a day of snack food bingeing in the United States. Combined, we wolf down 10.4 million pounds of potato chips and tortilla chips. The difference between consumption of potato chips and tortilla chips is 1.2 million pounds. How many millions of pounds of potato chips and tortilla chips are consumed on Super Bowl Sunday? 5.8 million pounds of potato chips; 4.6 million pounds of tortilla chips

Millions of Pounds of Snack Food Consumed on Super Bowl Sunday

Source: Association of American Snack Foods

6. The graph at the top of the next page shows the places with the greatest number of documented fatal shark attacks from 1580 through 2002. Combined, 112 fatal attacks were recorded in the continental United States and South Africa. The difference between the number of attacks in the two countries was 26. How many documented fatal shark attacks were there in the continental United States and in South Africa from 1580 through 2002? U.S.: 69; South Africa: 43

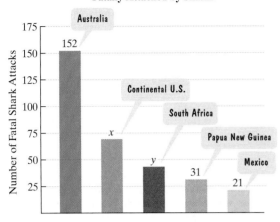

Places Where Most People Are Fatally Attacked by Sharks

Australia
152
Continental U.S.
x
South Africa
y
Papua New Guinea
31
Mexico
21

Number of Fatal Shark Attacks

Source: International Shark Attack File

The graph shows the calories in some favorite fast foods. Use the information in Exercises 7–8 to find the exact caloric content of the specified foods.

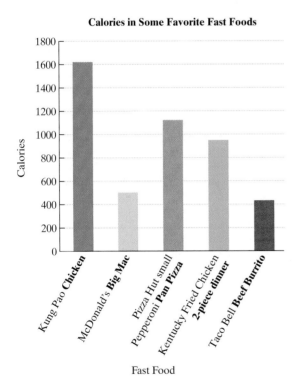

Calories in Some Favorite Fast Foods

Calories

Kung Pao Chicken
McDonald's **Big Mac**
Pizza Hut small Pepperoni **Pan Pizza**
Kentucky Fried Chicken **2-piece dinner**
Taco Bell **Beef Burrito**

Fast Food

Source: Center for Science in the Public Interest

7. One pan pizza and two beef burritos provide 1980 calories. Two pan pizzas and one beef burrito provide 2670 calories. Find the caloric content of each item.   A pan pizza has 1120 calories and a beef burrito has 430 calories.

8. One Kung Pao chicken and two Big Macs provide 2620 calories. Two Kung Pao chickens and one Big Mac provide 3740 calories. Find the caloric content of each item.   Kung Pao chicken has 1620 calories and a Big Mac has 500 calories.

9. Cholesterol intake should be limited to 300 mg or less each day. One serving of scrambled eggs from McDonalds and one Double Beef Whopper from Burger King exceed this intake by 241 mg. Two servings of scrambled eggs and three Double Beef Whoppers provide 1257 mg of cholesterol. Determine the cholesterol content in each item.   Scrambled eggs have 366 mg and a Double Beef Whopper has 175 mg.

10. Two medium eggs and three cups of ice cream contain 701 milligrams of cholesterol. One medium egg and one cup of ice cream exceed the suggested daily cholesterol intake of 300 milligrams by 25 milligrams. Determine the cholesterol content in each item.   egg: 274 mg; ice cream: 51 mg

11. In a discount clothing store, all sweaters are sold at one fixed price and all shirts are sold at another fixed price. If one sweater and three shirts cost $42, while three sweaters and two shirts cost $56, find the price of one sweater and the price of one shirt.   sweater: $12; shirt: $10

12. A restaurant purchased eight tablecloths and five napkins for $106. A week later, a tablecloth and six napkins were bought for $24. Find the cost of one tablecloth and the cost of one napkin, assuming the same prices for both purchases.   tablecloth: $12; napkin; $2

13. You are choosing between two long-distance telephone plans. Plan A has a monthly fee of $20 with a charge of $0.05 per minute for all long-distance calls. Plan B has a monthly fee of $5 with a charge of $0.10 per minute for all long-distance calls.
    a. For how many minutes of long-distance calls will the costs for the two plans be the same? What will be the cost for each plan?   300 minutes; $35
    b. If you make approximately 10 long-distance calls per month, each averaging 20 minutes, which plan should you select? Explain your answer.   Plan B; Answers will vary.

14. You are choosing between two long-distance telephone plans. Plan A has a monthly fee of $15 with a charge of $0.08 per minute for all long-distance calls. Plan B has a monthly fee of $3 with a charge of $0.12 per minute for all long-distance calls.
    a. For how many minutes of long-distance calls will the costs for the two plans be the same? What will be the cost for each plan?   300 minutes; $39
    b. If you make approximately 15 long-distance calls per month, each averaging 30 minutes, which plan should you select? Explain your answer.   Plan A; Answers will vary.

15. You are choosing between two plans at a discount warehouse. Plan A offers an annual membership fee of $100 and you pay 80% of the manufacturer's recommended list price. Plan B offers an annual membership fee of $40 and you pay 90% of the manufacturer's recommended list price. How many dollars of merchandise would you have to purchase in a year to pay the same amount under both plans? What will be the cost for each plan? $600 of merchandise; $580

16. You are choosing between two plans at a discount warehouse. Plan A offers an annual membership fee of $300 and you pay 70% of the manufacturer's recommended list price. Plan B offers an annual membership fee of $40 and you pay 90% of the manufacturer's recommended list price. How many dollars of merchandise would you have to purchase in a year to pay the same amount under both plans? What will be the cost for each plan? $1300 of merchandise; $1210

The graphs show average weekly earnings of full-time wage and salary workers 25 and older, by educational attainment. Exercises 17–18 involve the information in these graphs.

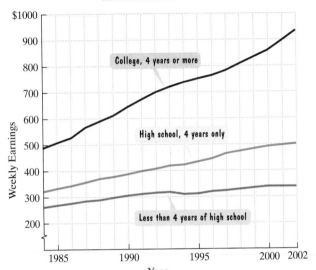

Average Weekly Earnings by Educational Attainment

Source: U.S. Bureau of Labor Statistics

17. In 1985, college graduates averaged $508 in weekly earnings. This amount has increased by approximately $25 in weekly earnings per year. By contrast, in 1985, high school graduates averaged $345 in weekly earnings. This amount has only increased by approximately $9 in weekly earnings per year. How many years after 1985 will college graduates be earning twice the amount per week that high school graduates earn? In which year will this occur? What will be the weekly earnings for each group at that time?   26 years; 2011; college grads: $1158; high school grads: $579

18. In 1985, college graduates averaged $508 in weekly earnings. This amount has increased by approximately $25 in weekly earnings per year. By contrast, in 1985, people with less than four years of high school averaged $270 in weekly earnings. This amount has only increased by approximately $4 in weekly earnings per year. How many years after 1985 will college graduates be earning three times the amount per week that people with less than four years of high school earn? (Round to the nearest whole number.) In which year will this occur? What will be the weekly earnings for each group at that time?   23 years; 2008; college grads: $1083; less than 4 years of high school: $362

19. Nutritional information for macaroni and broccoli is given in the table. How many servings of each would it take to get exactly 14 grams of protein and 48 grams of carbohydrates?   2 servings of macaroni and 4 servings of broccoli

	Macaroni	Broccoli
Protein (grams/serving)	3	2
Carbohydrates (grams/serving)	16	4

20. The calorie-nutrient information for an apple and an avocado is given in the table. How many of each should be eaten to get exactly 1000 calories and 100 grams of carbohydrates?   3 apples and 2 avocados

	One Apple	One Avocado
Calories	100	350
Carbohydrates (grams)	24	14

Exercises 21–26 involve simple interest. Use the four-step strategy to solve each problem.

21. You invested $7000 in two accounts paying 6% and 8% annual interest, respectively. If the total interest earned for the year was $520, how much was invested at each rate? $2000 at 6% and $5000 at 8%

22. You invested $11,000 in stocks and bonds, paying 5% and 8% annual interest, respectively. If the total interest earned for the year was $730, how much was invested in stocks and how much was invested in bonds?   stocks: $5000; bonds: $6000

23. You invested money in two funds. Last year, the first fund paid a dividend of 9% and the second a dividend of 3%, and you received a total of $900 in interest. This year, the first fund paid a 10% dividend and the second only 1%, and you received a total of $860 in interest. How much money did you invest in each fund?   first fund: $8000; second fund: $6000

24. You invested money in two funds. Last year, the first fund paid a dividend of 8% and the second a divident of 5%, and you received a total of $1330 in interest. This year, the first fund paid a 12% dividend and the second only 2%, and you received a total of $1500 in interest. How much money did you invest in each fund?   first fund: $11,000; second fund: $9000

25. Things did not go quite as planned. You invested $20,000, part of it in a stock that paid 12% annual interest. However, the rest of the money suffered a 5% loss. If the total annual income from both investments was $1890, how much was invested at each rate? $17,000 was invested at 12% annual interest and $3000 was invested at a 5% loss.

26. Things did not go quite as planned. You invested $30,000, part of it in a stock that paid 14% annual interest. However, the rest of the money suffered a 6% loss. If the total annual income from both investments was $200, how much was invested at each rate? $10,000 was invested at 14% annual interest and $20,000 was invested at a 6% loss.

Exercises 27–34 involve mixtures. Use the four-step strategy to solve each problem.

27. A wine company needs to blend a California wine with a 5% alcohol content and a French wine with a 9% alcohol content to obtain 200 gallons of wine with a 7% alcohol content. How many gallons of each kind of wine must be used?   California: 100 gallons; French: 100 gallons

28. A jeweler needs to mix an alloy with a 16% gold content and an alloy with a 28% gold content to obtain 32 ounces of a new alloy with a 25% gold content. How many ounces of each of the original alloys must be used?   16% gold content: 8 ounces; 28% gold content: 24 ounces

29. For thousands of years, gold has been considered one of Earth's most precious metals. One hundred percent pure gold is 24-karat gold, which is too soft to be made into jewelry. In the United States, most gold jewelry is 14-karat gold, approximately 58% gold. If 18-karat gold is 75% gold and 12-karat gold is 50% gold, how much of each should be used to make a 14-karat gold bracelet weighing 300 grams?   96 grams of 18-karat gold and 204 grams of 12-karat gold

30. In the "Peanuts" cartoon shown, solve the problem that is sending Peppermint Patty into an agitated state. How much cream and how much milk, to the nearest thousandth of a gallon, must be mixed together to obtain 50 gallons of cream that contains 12.5% butterfat?

© 1978 United Media/United Feature Syndicate, Inc.
cream: 20.93 gallons; milk: 29.07 gallons

31. The manager of a candystand at a large multiplex cinema has a popular candy that sells for $1.60 per pound. The manager notices a different candy worth $2.10 per pound that is not selling well. The manager decides to form a mixture of both types of candy to help clear the inventory of the more expensive type. How many pounds of each kind of candy should be used to create a 75-pound mixture selling for $1.90 per pound?   30 pounds of the cheaper candy and 45 pounds of the more expensive candy

32. A grocer needs to mix raisins at $2.00 per pound with granola at $3.25 per pound to obtain 10 pounds of a mixture that costs $2.50 per pound. How many pounds of raisins and how many pound of granola must be used?   6 pounds of raisins and 4 pounds of granola

33. A coin purse contains a mixture of 15 coins in nickels and dimes. The coins have a total value of $1.10. Determine the number of nickels and the number of dimes in the purse.
8 nickels and 7 dimes

34. A coin purse contains a mixture of 15 coins in dimes and quarters. The coins have a total value of $3.30. Determine the number of dimes and the number of quarters in the purse.
3 dimes and 12 quarters

Exercises 35–40 involve motion. Use the four-step strategy to solve each problem.

35. When a small plane flies with the wind, it can travel 800 miles in 5 hours. When the plane flies in the opposite direction, against the wind, it takes 8 hours to fly the same distance. Find the rate of the plane in still air and the rate of the wind.
The plane's rate in still air is 130 miles per hour and the rate of the wind is 30 miles per hour.

36. When a plane flies with the wind, it can travel 4200 miles in 6 hours. When the plane flies in the opposite direction, against the wind, it takes 7 hours to fly the same distance. Find the rate of the plane in still air and the rate of the wind.   The plane's rate in still air is 650 miles per hour and the rate of the wind is 50 miles per hour.

37. A boat's crew rowed 16 kilometers downstream, with the current, in 2 hours. The return trip upstream, against the current, covered the same distance, but took 4 hours. Find the crew's rowing rate in still water and the rate of the current.
The crew's rowing rate in still water is 6 kilometers per hour and the rate of the current is 2 kilometers per hour.

38. A motorboat traveled 36 miles downstream, with the current, in 1.5 hours. The return trip upstream, against the current, covered the same distance, but took 2 hours. Find the boat's rate in still water and the rate of the current.   The boat's rate in still water is 21 miles per hour and the rate of the current is 3 miles per hour.

39. With the current, you can canoe 24 miles in 4 hours. Against the same current, you can canoe only $\frac{3}{4}$ of this distance in 6 hours. Find your rate in still water and the rate of the current.
Rate in still water is 4.5 miles per hour and the rate of the current is 1.5 miles per hour.

40. With the current, you can row 24 miles in 3 hours. Against the same current, you can row only $\frac{2}{3}$ of this distance in 4 hours. Find your rowing rate in still water and the rate of the current.   Rate in still water is 6 miles per hour and the rate of the current is 2 miles per hour.

Writing in Mathematics

41. Describe the conditions in a problem that enable it to be solved using a system of linear equations.   41–43. Answers will vary.

42. Write a word problem that can be solved by translating to a system of linear equations. Then solve the problem.

43. Exercises 13–16 involve using systems of linear equations to compare costs of long-distance telephone plans and plans at a discount warehouse. Describe another situation that involves choosing between two options that can be modeled and solved with a linear system.

Critical Thinking Exercises

44. A set of identical twins can only be recognized by the characteristic that one always tells the truth and the other always lies. One twin tells you of a lucky number pair: "When I multiply my first lucky number by 3 and my second lucky number by 6, the addition of the resulting numbers produces a sum of 12. When I add my first lucky number and twice my second lucky number, the sum is 5." Which twin is talking?
the twin that lies

45. Tourist: "How many birds and lions do you have in your zoo?" Zookeeper: "There are 30 heads and 100 feet." Tourist: "I can't tell from that." Zookeeper: "Oh, yes, you can!" Can you? Find the number of each.   10 birds and 20 lions

46. Find the measure of each angle whose degree measure is represented with a variable. x: 55°; y: 35°; $(3y + 20)$: 125°

47. One apartment is directly above a second apartment. The resident living downstairs calls his neighbor living above him and states, "If one of you is willing to come downstairs, we'll have the same number of people in both apartments." The upstairs resident responds, "We're all too tired to move. Why don't one of you come up here? Then we'll have twice as many people up here as you've got down there." How many people are in each apartment?   There are 5 people downstairs and 7 people upstairs.

48. In Lewis Carroll's *Through the Looking Glass*, the following dialogue takes place:

> ***Tweedledum (to Tweedledee):*** The sum of your weight and twice mine is 361 pounds.

> ***Tweedledee (to Tweedledum):*** Contrawise, the sum of your weight and twice mine is 362 pounds.

Find the weight of each of the two characters.
Tweedledee: 121 pounds; Tweedledum 120 pounds

Technology Exercise

49. Select any two problems that you solved from Exercises 5–20. Use a graphing utility to graph the system of equations that you wrote for that problem. Then use the TRACE or INTERSECTION feature to show the point on the graphs that corresponds to the problem's solution.
Answers will vary.

Review Exercises

50. Find the slope of the line containing the points $(-6, 1)$ and $(2, -1)$ (Section 3.3, Example 1) $-\frac{1}{4}$

51. Add: $\frac{1}{5} + \left(-\frac{3}{4}\right)$. (Section 1.5, Example 4) $-\frac{11}{20}$

52. Graph: $y = x^2$. (Section 3.1, Example 6)
See graphing answer section.

SECTION 4.5

SYSTEMS OF LINEAR EQUATIONS IN THREE VARIABLES

Objectives

1 Verify the solution of a system of linear equations in three variables.

2 Solve systems of linear equations in three variables.

3 Identify inconsistent and dependent systems.

4 Solve problems using systems in three variables.

All animals sleep, but the length of time they sleep varies widely: Cattle sleep for only a few minutes at a time. We humans seem to need more sleep than other animals, up to eight hours a day. Without enough sleep, we have difficulty concentrating, make mistakes in routine tasks, lose energy, and feel bad-tempered. There is a relationship between hours of sleep and death rate per year per 100,000 people. How many hours of sleep will put you in the group with the minimum death rate? In this section, you will learn how to solve systems of linear equations with more than two variables to answer this question.

1 Verify the solution of a system of linear equations in three variables.

Point of intersection

P

FIGURE 4.11

Systems of Linear Equations in Three Variables and Their Solutions An equation such as $x + 2y - 3z = 9$ is called a *linear equation in three variables*. In general, any equation of the form

$$Ax + By + Cz = D,$$

where A, B, C, and D are real numbers such that A, B, and C are not all 0, is a **linear equation in three variables: *x*, *y*, and *z*.** The graph of this linear equation in three variables is a plane in three-dimensional space.

The process of solving a system of three linear equations in three variables is geometrically equivalent to finding the point of intersection (assuming that there is one) of three planes in space. (See Figure 4.11). A **solution** of a system of linear equations in three variables is an ordered triple of real numbers that satisfies all equations in the system. The **solution set** of the system is the set of all its solutions.

> **EXAMPLE 1** Determining Whether an Ordered Triple Satisfies a System
>
> Show that the ordered triple $(-1, 2, -2)$ is a solution of the system:
>
> $$x + 2y - 3z = 9$$
> $$2x - y + 2z = -8$$
> $$-x + 3y - 4z = 15.$$

SOLUTION Because -1 is the x-coordinate, 2 is the y-coordinate, and -2 is the z-coordinate of $(-1, 2, -2)$, we replace x with -1, y with 2, and z with -2 in each of the three equations.

$$x + 2y - 3z = 9$$
$$-1 + 2(2) - 3(-2) \stackrel{?}{=} 9$$
$$-1 + 4 + 6 \stackrel{?}{=} 9$$
$$9 = 9, \text{ true}$$

$$2x - y + 2z = -8$$
$$2(-1) - 2 + 2(-2) \stackrel{?}{=} -8$$
$$-2 - 2 - 4 \stackrel{?}{=} -8$$
$$-8 = -8, \text{ true}$$

$$-x + 3y - 4z = 15$$
$$-(-1) + 3(2) - 4(-2) \stackrel{?}{=} 15$$
$$1 + 6 + 8 \stackrel{?}{=} 15$$
$$15 = 15, \text{ true}$$

The ordered triple $(-1, 2, -2)$ satisfies the three equations: It makes each equation true. Thus, the ordered triple is a solution of the system. ∎

 CHECK POINT 1 Show that the ordered triple $(-1, -4, 5)$ is a solution of the system:

$$x - 2y + 3z = 22 \quad (-1) - 2(-4) + 3(5) = 22;$$
$$2x - 3y - z = 5 \quad 2(-1) - 3(-4) - 5 = 5;$$
$$3x + y - 5z = -32. \quad 3(-1) + (-4) - 5(5) = -32$$

2 Solve systems of linear equations in three variables.

Solving Systems of Linear Equations in Three Variables by Eliminating Variables The method for solving a system of linear equations in three variables is similar to that used on systems of linear equations in two variables. We use addition to eliminate any variable, reducing the system to two equations in two variables. Once we obtain a system of two equations in two variables, we use addition or substitution to eliminate another variable. The result is a single equation in one variable. We solve this equation to get the value of the remaining variable. Other variable values are found by back-substitution.

SOLVING LINEAR SYSTEMS IN THREE VARIABLES BY ELIMINATING VARIABLES

1. Reduce the system to two equations in two variables. This is usually accomplished by taking two different pairs of equations and using the addition method to eliminate the same variable from both pairs.

2. Solve the resulting system of two equations in two variables using addition or substitution. The result is an equation in one variable that gives the value of that variable.

3. Back-substitute the value of the variable found in step 2 into either of the equations in two variables to find the value of the second variable.

4. Use the values of the two variables from steps 2 and 3 to find the value of the third variable by back-substituting into one of the original equations.

5. Check the proposed solution in each of the original equations.

EXAMPLE 2   Solving a System in Three Variables

Solve the system:

$$5x - 2y - 4z = 3 \qquad \text{Equation 1}$$
$$3x + 3y + 2z = -3 \qquad \text{Equation 2}$$
$$-2x + 5y + 3z = 3. \qquad \text{Equation 3}$$

SOLUTION   There are many ways to proceed. Because our initial goal is to reduce the system to two equations in two variables, **the central idea is to take two different pairs of equations and eliminate the same variable from both pairs**.

Step 1.   **Reduce the system to two equations in two variables.** We choose any two equations and use the addition method to eliminate a variable. Let's eliminate z from Equations 1 and 2. We do so by multiplying Equation 2 by 2. Then we add equations.

(Equation 1) $5x - 2y - 4z = 3 \xrightarrow{\text{No change}} 5x - 2y - 4z = 3$

(Equation 2) $3x + 3y + 2z = -3 \xrightarrow{\text{Multiply by 2.}} \underline{6x + 6y + 4z = -6}$

Add: $11x + 4y = -3$ (Equation 4)

Now we must eliminate the *same* variable from another pair of equations. We can eliminate z from Equations 2 and 3. First, we multiply Equation 2 by -3. Next, we multiply Equation 3 by 2. Finally, we add equations.

(Equation 2) $3x + 3y + 2z = -3 \xrightarrow{\text{Multiply by } -3.} -9x - 9y - 6z = 9$

(Equation 3) $-2x + 5y + 3z = 3 \xrightarrow{\text{Multiply by 2.}} \underline{-4x + 10y + 6z = 6}$

Add: $-13x + y = 15$ (Equation 5)

Equations 4 and 5 give us a system of two equations in two variables.

Step 2.   **Solve the resulting system of two equations in two variables.** We will use the addition method to solve Equations 4 and 5 for x and y. To do so, we multiply Equation 5 by -4 and add this to Equation 4.

(Equation 4) $11x + 4y = -3 \xrightarrow{\text{No change}} 11x + 4y = -3$

(Equation 5) $-13x + y = 15 \xrightarrow{\text{Multiply by } -4.} \underline{52x - 4y = -60}$

Add: $63x = -63$

$x = -1$ Divide both sides by 63.

Step 3. **Use back-substitution in one of the equations in two variables to find the value of the second variable.** We back-substitute -1 for x in either Equation 4 or 5 to find the value of y. We will use Equation 5.

$$-13x + y = 15 \qquad \text{Equation 5}$$

$$-13(-1) + y = 15 \qquad \text{Substitute } -1 \text{ for } x.$$

$$13 + y = 15 \qquad \text{Multiply.}$$

$$y = 2 \qquad \text{Subtract 13 from both sides.}$$

Step 4. **Back-substitute the values found for two variables into one of the original equations to find the value of the third variable.** We can now use any one of the original equations and back-substitute the values of x and y to find the value for z. We will use Equation 2.

$$3x + 3y + 2z = -3 \qquad \text{Equation 2}$$

$$3(-1) + 3(2) + 2z = -3 \qquad \text{Substitute } -1 \text{ for } x \text{ and } 2 \text{ for } y.$$

$$3 + 2z = -3 \qquad \begin{array}{l}\text{Multiply and then add:}\\ 3(-1) + 3(2) = -3 + 6 = 3.\end{array}$$

$$2z = -6 \qquad \text{Subtract 3 from both sides.}$$

$$z = -3 \qquad \text{Divide both sides by 2.}$$

With $x = -1$, $y = 2$, and $z = -3$, the proposed solution is the ordered triple $(-1, 2, -3)$.

Step 5. **Check.** Check the proposed solution, $(-1, 2, -3)$, by substituting the values for x, y, and z into each of the three original equations. These substitutions yield three true statements. Thus, the solution is $(-1, 2, -3)$ and the solution set is $\{(-1, 2, -3)\}$. ■

 CHECK POINT **2** Solve the system:

$$x + 4y - z = 20$$

$$3x + 2y + z = 8$$

$$2x - 3y + 2z = -16. \qquad (1, 4, -3) \text{ or } \{(1, 4, -3)\}$$

In some examples, one of the variables is already eliminated from a given equation. In this case, the same variable should be eliminated from the other two equations, thereby making it possible to omit one of the elimination steps. We illustrate this idea in Example 3.

EXAMPLE 3 Solving a System of Equations with a Missing Term

Solve the system:

$$x + z = 8 \qquad \text{Equation 1}$$

$$x + y + 2z = 17 \qquad \text{Equation 2}$$

$$x + 2y + z = 16. \qquad \text{Equation 3}$$

$$x +   z = 8 \quad \text{Equation 1}$$
$$x + y + 2z = 17 \quad \text{Equation 2}$$
$$x + 2y + z = 16 \quad \text{Equation 3}$$

The given system (repeated)

SOLUTION

Step 1. **Reduce the system to two equations in two variables.** Because Equation 1 contains only x and z, we could eliminate y from Equations 2 and 3. This will give us two equations in x and z. To eliminate y from Equations 2 and 3, we multiply Equation 2 by -2 and add Equation 3.

$$\text{(Equation 2)} \quad x + y + 2z = 17 \xrightarrow{\text{Multiply by } -2.} -2x - 2y - 4z = -34$$
$$\text{(Equation 3)} \quad x + 2y + z = 16 \xrightarrow{\text{No change}} x + 2y + z = 16$$
$$\text{Add:} \quad -x - 3z = -18 \quad \text{(Equation 4)}$$

Equation 4 and the given Equation 1 provide us with a system of two equations in two variables.

Step 2. **Solve the resulting system of two equations in two variables.** We will solve Equations 1 and 4 for x and z.

$$x + z = 8 \quad \text{Equation 1}$$
$$-x - 3z = -18 \quad \text{Equation 4}$$
$$\text{Add:} \quad -2z = -10$$
$$z = 5 \quad \text{Divide both sides by } -2.$$

Step 3. **Use back-substitution in one of the equations in two variables to find the value of the second variable.** To find x, we back-substitute 5 for z in either Equation 1 or 4. We will use Equation 1.

$$x + z = 8 \quad \text{Equation 1}$$
$$x + 5 = 8 \quad \text{Substitute 5 for z.}$$
$$x = 3 \quad \text{Subtract 5 from both sides.}$$

Step 4. **Back-substitute the values found for two variables into one of the original equations to find the value of the third variable.** To find y, we back-substitute 3 for x and 5 for z into Equation 2 or 3. We can't use Equation 1 because y is missing in this equation. We will use Equation 2.

$$x + y + 2z = 17 \quad \text{Equation 2}$$
$$3 + y + 2(5) = 17 \quad \text{Substitute 3 for x and 5 for z.}$$
$$y + 13 = 17 \quad \text{Multiply and add.}$$
$$y = 4 \quad \text{Subtract 13 from both sides.}$$

We found that $z = 5$, $x = 3$, and $y = 4$. Thus, the proposed solution is the ordered triple $(3, 4, 5)$.

Step 5. **Check.** Substituting 3 for x, 4 for y, and 5 for z into each of the three original equations yields three true statements. Consequently, the solution is $(3, 4, 5)$ and the solution set is $\{(3, 4, 5)\}$. ∎

 CHECK POINT 3 Solve the system:

$$2y - z = 7$$
$$x + 2y + z = 17$$
$$2x - 3y + 2z = -1. \quad (4, 5, 3) \text{ or } \{(4, 5, 3)\}$$

3 | Identify inconsistent and dependent systems.

Inconsistent and Dependent Systems A system of linear equations in three variables represents three planes. The three planes need not intersect at one point. The planes may have no common point of intersection and represent an **inconsistent system** with no solution. Figure 4.12 illustrates some of the geometric possibilities for inconsistent systems.

Three planes are parallel with no common intersection point.

Two planes are parallel with no common intersection point.

Planes intersect two at a time. There is no intersection point common to all three planes.

FIGURE 4.12   Three planes may have no common point of intersection.

If you attempt to solve an inconsistent system algebraically, at some point in the solution process you will eliminate all three variables. A false statement, such as $0 = 17$, will be the result. For example, consider the system

$$\begin{aligned} 2x + 5y + z &= 12 &\text{Equation 1}\\ x - 2y + 4z &= -10 &\text{Equation 2}\\ -3x + 6y - 12z &= 20. &\text{Equation 3} \end{aligned}$$

Suppose we reduce the system to two equations in two variables by eliminating x. To eliminate x in Equations 2 and 3, we multiply Equation 2 by 3 and add Equation 3:

$$\begin{array}{l}x - 2y + 4z = -10 \quad\xrightarrow{\text{Multiply by 3.}}\quad 3x - 6y + 12z = -30\\ -3x + 6y - 12z = 20 \quad\xrightarrow{\text{No change}}\quad \underline{-3x + 6y - 12z = 20}\\ \text{Add:} 0 = -10\end{array}$$

There are no values of $x, y,$ and z for which $0 = -10$. The false statement $0 = -10$ indicates that the system is inconsistent and has no solution. The solution set is the empty set, $\varnothing$.

We have seen that a linear system that has at least one solution is called a **consistent system**. Planes that intersect at one point and planes that intersect at infinitely many points both represent consistent systems. Figure 4.13 illustrates planes that intersect at infinitely many points. The equations in these linear systems with infinitely many solutions are called **dependent**.

FIGURE 4.13   Three planes may intersect at infinitely many points.

The planes intersect along a common line.

The planes coincide.

If you attempt to solve a system with dependent equations algebraically, at some point in the solution process you will eliminate all three variables. A true statement, such as $0 = 0$, will be the result. If this occurs as you are solving a linear system, simply state that the equations are dependent.

Solve problems using systems in three variables.

Applications Systems of equations may allow us to find models for data without using a graphing utility. Three data points that do not lie on or near a line determine the graph of an equation of the form

$$y = ax^2 + bx + c, a \neq 0.$$

Such an equation has a cuplike U-shaped graph, making it ideal for modeling situations in which values of y are decreasing and then increasing.

The process of determining an equation whose graph contains given points is called **curve fitting**. In our next example, we fit the curve whose equation is $y = ax^2 + bx + c$ to three data points. Using a system of equations, we find values for $a, b,$ and c.

EXAMPLE 4 Modeling Data Relating Sleep and Death Rate

In a study relating sleep and death rate, the following data were obtained. Use the equation $y = ax^2 + bx + c$ to model the data.

x (Average Number of Hours of Sleep)	y (Death Rate per Year per 100,000 Males)
4	1682
7	626
9	967

SOLUTION We need to find values for $a, b,$ and c in $y = ax^2 + bx + c$. We can do so by solving a system of three linear equations in $a, b,$ and c. We obtain the three equations by using the values of x and y from the data as follows:

$y = ax^2 + bx + c$ *Use the quadratic function to model the data.*

When x = 4, y = 1682: $1682 = a \cdot 4^2 + b \cdot 4 + c$ or $16a + 4b + c = 1682$

When x = 7, y = 626: $626 = a \cdot 7^2 + b \cdot 7 + c$ or $49a + 7b + c = 626$

When x = 9, y = 967: $967 = a \cdot 9^2 + b \cdot 9 + c$ or $81a + 9b + c = 967.$

The easiest way to solve this system is to eliminate c from two pairs of equations, obtaining two equations in a and b. Solving this system gives $a = 104.5, b = -1501.5,$ and $c = 6016$. We now substitute the values for $a, b,$ and c into $y = ax^2 + bx + c$. The equation that models the given data is

$$y = 104.5x^2 - 1501.5x + 6016.$$ ∎

We can use the model that we obtained in Example 4 to find the death rate of males who average, say, 6 hours of sleep. Substitute 6 for x in $y = 104.5x^2 - 1501.5x + 6016$:

$$y = 104.5(6)^2 - 1501.5(6) + 6016 = 769.$$

According to the model, the death rate for males who average 6 hours of sleep is 769 deaths per 100,000 males.

USING TECHNOLOGY

The graph of

$$y = 104.5x^2 - 1501.5x + 6016$$

is displayed in a $[3, 12, 1]$ by $[500, 2000, 100]$ viewing rectangle. The utility indicates that the lowest point on the graph is approximately $(7.2, 622.5)$. Men who average 7.2 hours of sleep are in the group with the lowest death rate, approximately 622.5 deaths per 100,000 males.

Minimum
X=7.212766 Y=622.53916

CHECK POINT 4 Find the equation $y = ax^2 + bx + c$ whose graph passes through the points $(1, 4), (2, 1),$ and $(3, 4).$

$y = 3x^2 - 12x + 13$

Problems involving three unknowns can be solved using the same strategy for solving problems with two unknown quantities. You can let $x, y,$ and z represent the unknown quantities. We then translate from the verbal conditions of the problem to a system of three equations in three variables. Problems of this type are included in the exercise set that follows.

4.5 EXERCISE SET

Student Solutions Manual   CD/Video   PH Math/Tutor Center   MathXL Tutorials on CD   MathXL®   MyMathLab   Interactmath.com

Practice Exercises

In Exercises 1–4, determine if the given ordered triple is a solution of the system.

1. $(2, -1, 3)$

$\begin{aligned} x + y + z &= 4 \\ x - 2y - z &= 1 \\ 2x - y - z &= -1 \end{aligned}$ not a solution

2. $(5, -3, -2)$

$\begin{aligned} x + y + z &= 0 \\ x + 2y - 3z &= 5 \\ 3x + 4y + 2z &= -1 \end{aligned}$ solution

3. $(4, 1, 2)$

$\begin{aligned} x - 2y &= 2 \\ 2x + 3y &= 11 \\ y - 4z &= -7 \end{aligned}$ solution

4. $(-1, 3, 2)$

$\begin{aligned} x - 2z &= -5 \\ y - 3z &= -3 \\ 2x - z &= -4 \end{aligned}$ solution

Solve each system in Exercises 5–22. If there is no solution or if there are infinitely many solutions and a system's equations are dependent, so state.

5. $\begin{aligned} x + y + 2z &= 11 \\ x + y + 3z &= 14 \\ x + 2y - z &= 5 \end{aligned}$ $\{(2, 3, 3)\}$

6. $\begin{aligned} 2x + y - 2z &= -1 \\ 3x - 3y - z &= 5 \\ x - 2y + 3z &= 6 \end{aligned}$ $\{(1, -1, 1)\}$

7. $\begin{aligned} 4x - y + 2z &= 11 \\ x + 2y - z &= -1 \\ 2x + 2y - 3z &= -1 \end{aligned}$ $\{(2, -1, 1)\}$

8. $\begin{aligned} x - y + 3z &= 8 \\ 3x + y - 2z &= -2 \\ 2x + 4y + z &= 0 \end{aligned}$ $\{(1, -1, 2)\}$

9. $\begin{aligned} 3x + 2y - 3z &= -2 \\ 2x - 5y + 2z &= -2 \\ 4x - 3y + 4z &= 10 \end{aligned}$ $\{(1, 2, 3)\}$

10. $\begin{aligned} 2x + 3y + 7z &= 13 \\ 3x + 2y - 5z &= -22 \\ 5x + 7y - 3z &= -28 \end{aligned}$ $\{(-1, -2, 3)\}$

11. $\begin{aligned} 2x - 4y + 3z &= 17 \\ x + 2y - z &= 0 \\ 4x - y - z &= 6 \end{aligned}$ $\{(3, 1, 5)\}$

12. $\begin{aligned} x + z &= 3 \\ x + 2y - z &= 1 \\ 2x - y + z &= 3 \end{aligned}$ $\{(1, 1, 2)\}$

13. $\begin{aligned} 2x + y &= 2 \\ x + y - z &= 4 \\ 3x + 2y + z &= 0 \end{aligned}$ $\{(1, 0, -3)\}$

14. $\begin{aligned} x + 3y + 5z &= 20 \\ y - 4z &= -16 \\ 3x - 2y + 9z &= 36 \end{aligned}$ $\{(0, 0, 4)\}$

15. $\begin{aligned} x + y &= -4 \\ y - z &= 1 \\ 2x + y + 3z &= -21 \end{aligned}$ $\{(1, -5, -6)\}$

16. $\begin{aligned} x + y &= 4 \\ x + z &= 4 \\ y + z &= 4 \end{aligned}$ $\{(2, 2, 2)\}$

17. $\begin{aligned} 2x + y + 2z &= 1 \\ 3x - y + z &= 2 \\ x - 2y - z &= 0 \end{aligned}$ no solution or $\varnothing$

18. $\begin{aligned} 3x + 4y + 5z &= 8 \\ x - 2y + 3z &= -6 \\ 2x - 4y + 6z &= 8 \end{aligned}$ no solution or $\varnothing$

19. $5x - 2y - 5z = 1$
$10x - 4y - 10z = 2$ infinitely many solutions;
$15x - 6y - 15z = 3$ dependent equations

20. $x + 2y + z = 4$
$3x - 4y + z = 4$ infinitely many solutions;
$6x - 8y + 2z = 8$ dependent equations

21. $3(2x + y) + 5z = -1$
$2(x - 3y + 4z) = -9$ $\left\{\left(\frac{1}{2}, \frac{1}{3}, -1\right)\right\}$
$4(1 + x) = -3(z - 3y)$

22. $7z - 3 = 2(x - 3y)$
$5y + 3z - 7 = 4x$ $\left\{\left(\frac{1}{2}, 3, -2\right)\right\}$
$4 + 5z = 3(2x - y)$

In Exercises 23–26, find the equation $y = ax^2 + bx + c$ whose graph passes through the given points.

23. $(-1, 6), (1, 4) (2, 9)$ $y = 2x^2 - x + 3$

24. $(-2, 7), (1, -2), (2, 3)$ $y = 2x^2 - x - 3$

25. $(-1, -4), (1, -2), (2, 5)$ $y = 2x^2 + x - 5$

26. $(1, 3), (3, -1), (4, 0)$ $y = x^2 - 6x + 8$

In Exercises 27–28, let x represent the first number, y the second number, and z the third number. Use the given conditions to write a system of equations. Solve the system and find the numbers.

27. The sum of three numbers is 16. The sum of twice the first number, 3 times the second number, and 4 times the third number is 46. The difference between 5 times the first number and the second number is 31. Find the three numbers.
first number: 7; second number: 4; third number: 5

28. The following is known about three numbers: Three times the first number plus the second number plus twice the third number is 5. If 3 times the second number is subtracted from the sum of the first number and 3 times the third number, the result is 2. If the third number is subtracted from the sum of 2 times the first number and 3 times the second number, the result is 1. Find the numbers. first number: -1; second number: 2; third number: 3

Practice Plus

Solve each system in Exercises 29–30.

29. $\dfrac{x + 2}{6} - \dfrac{y + 4}{3} + \dfrac{z}{2} = 0$

$\dfrac{x + 1}{2} + \dfrac{y - 1}{2} - \dfrac{z}{4} = \dfrac{9}{2}$ $\{(4, 8, 6)\}$

$\dfrac{x - 5}{4} + \dfrac{y + 1}{3} + \dfrac{z - 2}{2} = \dfrac{19}{4}$

30. $\dfrac{x + 3}{2} - \dfrac{y - 1}{2} + \dfrac{z + 2}{4} = \dfrac{3}{2}$

$\dfrac{x - 5}{2} + \dfrac{y + 1}{3} - \dfrac{z}{4} = -\dfrac{25}{6}$ $\{(-3, 0, 2)\}$

$\dfrac{x - 3}{4} - \dfrac{y + 1}{2} + \dfrac{z - 3}{2} = -\dfrac{5}{2}$

In Exercises 31–32, find the equation $y = ax^2 + bx + c$ whose graph is shown. Select three points whose coordinates appear to be integers.

31.

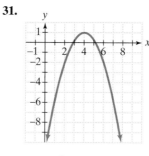

$y = -\dfrac{3}{4}x^2 + 6x - 11$

32.

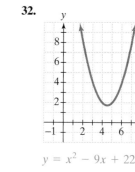

$y = x^2 - 9x + 22$

In Exercises 33–34, solve each system for (x, y, z) in terms of the nonzero constants a, b, and c.

33. $ax - by - 2cz = 21$
$ax + by + cz = 0$
$2ax - by + cz = 14$

$\left(\dfrac{8}{a}, -\dfrac{3}{b}, -\dfrac{5}{c}\right)$

34. $ax - by + 2cz = -4$
$ax + 3by - cz = 1$
$2ax + by + 3cz = 2$

$\left(-\dfrac{9}{a}, \dfrac{5}{b}, \dfrac{5}{c}\right)$

Application Exercises

35. The bar graph shows the percentage of the U.S. population that was foreign-born from 1900 through 2002.

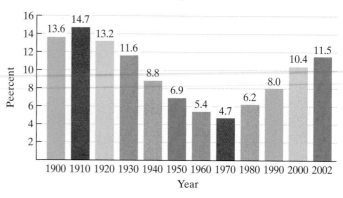

Percentage of U.S. Population That was Foreign-Born, 1900-2002

Source: U.S. Census Bureau

a. Write the data for 1900, 1970, and 2002 as ordered pairs (x, y), where x is the number of years after 1900 and y is the percentage of the U.S. population that was foreign-born in that year. $(0, 13.6), (70, 4.7), (102, 11.5)$

b. The three data points in part (a) can be modeled by the equation $y = ax^2 + bx + c$. Substitute each ordered pair into this equation, one ordered pair at a time, and write a system of linear equations in three variables that can be used to find values for a, b, and c. It is not necessary to solve the system. $0a + 0b + c = 13.6$, $4900a + 70b + c = 4.7$, $10{,}404a + 102b + c = 11.5$

36. The bar graph shows that the U.S. divorce rate increased from 1970 to 1985 and then decreased from 1985 to 2001.

U.S. Divorce Rates: Number of Divorces per 1000 People

Source: U.S. Census Bureau

a. Write the data for 1970, 1985, and 2001 as ordered pairs (x, y), where x is the number of years after 1970 and y is that year's divorce rate. $(0, 3.5)$, $(15, 5.0)$, $(31, 4.0)$

b. The three data points in part (a) can be modeled by the equation $y = ax^2 + bx + c$. Substitute each ordered pair into this equation, one ordered pair at a time, and write a system of linear equations in three variables that can be used to find values for a, b, and c. It is not necessary to solve the system. $0a + 0b + c = 3.5$, $225a + 15b + c = 5.0$, $961a + 31b + c = 4.0$

37. You throw a ball straight up from a rooftop. The ball misses the rooftop on its way down and eventually strikes the ground. A mathematical model can be used to describe the relationship for the ball's height above the ground, y, after x seconds. Consider the following data.

x, seconds after the ball is thrown	y, ball's height, in feet, above the ground
1	224
3	176
4	104

a. Find the equation $y = ax^2 + bx + c$ whose graph passes through the given points.
$y = -16x^2 + 40x + 200$

b. Use the equation in part (a) to find the value for y when $x = 5$. Describe what this means.   0; After 5 seconds, the ball hits the ground.

38. A mathematical model can be used to describe the relationship between the number of feet a car travels once the brakes are applied, y, and the number of seconds the car is in motion after the brakes are applied, x. A research firm collects the data shown at the top of the next column.

x, seconds in motion after brakes are applied	y, feet car travels once the brakes are applied
1	46
2	84
3	114

a. Find the equation $y = ax^2 + bx + c$ whose graph passes through the given points. $y = -4x^2 + 50x$

b. Use the equation in part (a) to find the value for y when $x = 6$. Describe what this means.   156; During the 6 seconds after the brakes are applied, the car has traveled 156 feet.

In Exercises 39–46, use the four-step strategy to solve each problem. Use x, y, and z to represent unknown quantities. Then translate from the verbal conditions of the problem to a system of three equations in three variables.

39. In current U.S. dollars, John D. Rockefeller's 1913 fortune of $900 million would be worth about $189 billion. The bar graph shows that Rockefeller is the wealthiest among the world's five richest people of all time. The combined estimated wealth, in current billions of U.S. dollars, of Andrew Carnegie, Cornelius Vanderbilt, and Bill Gates is $256 billion. The difference between Carnegie's estimated wealth and Vanderbilt's is $4 billion. The difference between Vanderbilt's estimated wealth and Gates's is $36 billion. Find the estimated wealth, in current billions of U.S. dollars, of Carnegie, Vanderbilt, and Gates.   Carnegie: $100 billion; Vanderbilt: $96 billion; Gates: $60 billion

The Richest People of All Time
Estimated Wealth, in Current Billions of U.S. Dollars

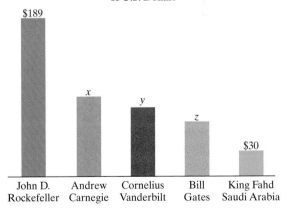

Source: Scholastic Book of World Records

40. The circle graph at the top of the next page indicates computers in use for the United States and the rest of the world. The percentage of the world's computers in Europe and Japan combined is 13% less than the percentage of the world's computers in the United States. If the percentage of the world's computers in Europe is doubled, it is only 3% more than the percentage of the world's computers in the United States. Find the percentage of the world's computers in the United States, Europe, and Japan.   United States: 43%; Europe: 23%; Japan: 7%

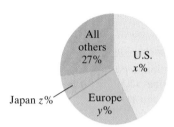

Percentage of the World's Computers:
U.S. and the World

All others 27%

U.S. $x\%$

Japan $z\%$

Europe $y\%$

Source: Jupiter Communications

41. A person invested $6700 for one year, part at 8%, part at 10%, and the remainder at 12%. The total annual income from these investments was $716. The amount of money invested at 12% was $300 more than the amount invested at 8% and 10% combined. Find the amount invested at each rate. $1200 was invested at 8%, $2000 was invested at 10%, and $3500 was invested at 12%.

42. A person invested $17,000 for one year, part at 10%, part at 12%, and the remainder at 15%. The total annual income from these investments was $2110. The amount of money invested at 12% was $1000 less than the amount invested at 10% and 15% combined. Find the amount invested at each rate. $4000 was invested at 10%, $8000 was invested at 12%, and $5000 was invested at 15%.

43. At a college production of *Streetcar Named Desire*, 400 tickets were sold. The ticket prices were $8, $10, and $12, and the total income from ticket sales was $3700. How many tickets of each type were sold if the combined number of $8 and $10 tickets sold was 7 times the number of $12 tickets sold?   200 $8 tickets, 150 $10 tickets and 50 $12 tickets were sold.

44. A certain brand of razor blades comes in packages of 6, 12, and 24 blades, costing $2, $3, and $4 per package, respectively. A store sold 12 packages containing a total of 162 razor blades and took in $35. How many packages of each type were sold?   5 $2 packages, 3 $3 packages and 4 $4 packages were sold.

45. Three foods have the following nutritional content per ounce.

	Calories	Protein (in grams)	Vitamin C (in milligrams)
Food *A*	40	5	30
Food *B*	200	2	10
Food *C*	400	4	300

If a meal consisting of the three foods allows exactly 660 calories, 25 grams of protein, and 425 milligrams of vitamin C, how many ounces of each kind of food should be used?   4 ounces of food A, 0.5 ounce of food B and 1 ounce of food C should be used.

46. A furniture company produces three types of desks: a children's model, an office model, and a deluxe model. Each desk is manufactured in three stages: cutting, construction, and finishing. The time requirements for each model and manufacturing stage are given in the following table.

	Children's model	Office model	Deluxe model
Cutting	2 hr	3 hr	2 hr
Construction	2 hr	1 hr	3 hr
Finishing	1 hr	1 hr	2 hr

Each week the company has available a maximum of 100 hours for cutting, 100 hours for construction, and 65 hours for finishing. If all available time must be used, how many of each type of desk should be produced each week?   15 children's models, 10 office models, and 20 deluxe models, should be produced each week.

Writing in Mathematics

47. What is a system of linear equations in three variables?   47–52. Answers will vary.

48. How do you determine whether a given ordered triple is a solution of a system in three variables?

49. Describe in general terms how to solve a system in three variables.

50. Describe what happens when using algebraic methods to solve an inconsistent system.

51. Describe what happens when using algebraic methods to solve a system with dependent equations.

52. AIDS is taking a deadly toll on southern Africa. Describe how to use the techniques that you learned in this section to obtain a model for African life span using projections with AIDS. Let x represent the number of years after 1985 and let y represent African life span in that year.   *Source*: United Nations

African Life Span

Projections without AIDS

Projections with AIDS

Life Span (years)

Year

Technology Exercises

53. Does your graphing utility have a feature that allows you to solve linear systems by entering coefficients and constant terms? If so, use this feature to verify the solutions to any five exercises that you worked by hand from Exercises 5–16.   53–54. Answers will vary.

54. Verify your results in Exercises 23–26 by using a graphing utility to graph each equation that you obtained. Trace along the curve and convince yourself that the three points given in the exercise lie on the equation's graph.

Critical Thinking Exercises

55. Which one of the following is true?   **c**
   a. The ordered triple $(2, 15, 14)$ is the only solution of the equation $x + y - z = 3$.
   b. The equation $x - y - z = -6$ is satisfied by $(2, -3, 5)$.
   c. If two equations in a system are $x + y - z = 5$ and $2x + 2y - 2z = 7$, then the system must be inconsistent.
   d. An equation with four variables, such as $x + 2y - 3z + 5w = 2$, cannot be satisfied by real numbers.

56. A modernistic painting consists of triangles, rectangles, and pentagons, all drawn so as to not overlap or share sides. Within each rectangle are drawn 2 red roses, and each pentagon contains 5 carnations. How many triangles, rectangles, and pentagons appear in the painting if the painting contains a total of 40 geometric figures, 153 sides of geometric figures, and 72 flowers?   13 triangles, 21 rectangles, and 6 pentagons

57. In the following triangle, the degree measures of the three interior angles and two of the exterior angles are represented with variables. Find the measure of each interior angle.

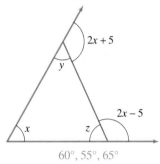

2x + 5

y

2x - 5

x   z

60°, 55°, 65°

58. Two blocks of wood having the same length and width are placed on the top and bottom of a table, as shown in (a). Length A measures 32 centimeters. The blocks are rearranged as shown in (b). Length B measures 28 centimeters. Determine the height of the table.   30 centimeters

(a)   **(b)**

Review Exercises

In Exercises 59–61, graph each equation.

59. $y = -\dfrac{3}{4}x + 3$ (Section 3.4, Example 3)   59–61. See graphing answer section.

60. $-2x + y = 6$ (Section 3.2, Example 4)

61. $y = -5$ (Section 3.2, Example 7)

GROUP PROJECT

CHAPTER 4

Additional group projects can be found in the *Instructor's Resource Manual.*

Group members should go online and obtain a list of telecommunication companies that provide residential long-distance service. The list should contain the monthly fee, the monthly minimum, and the rate per minute for each service provider.

   a. For each provider in the list, write an equation that describes the total monthly cost, y, of x minutes of long-distance phone calls.
   b. Compare two of the plans. After how many minutes of long-distance calls will the costs for the two plans be the same? Solve a linear system to obtain your answer. What will be the cost for each plan?
   c. Repeat part (b) for another two of the plans.
   d. Each person should estimate the number of minutes he or she spends talking long distance each month. Group members should assist that person in selecting the plan that will save the most amount of money. Be sure to factor in the monthly minimum, if any, when choosing a plan. Whenever possible, use the equations that you wrote in part (a).

Caution: We've left something out! Your comparisons do not take into account the in-state rates of the plan. Furthermore, if you make international calls, international rates should also be one of your criteria in choosing a plan. Problem solving in real life can get fairly complicated.

CHAPTER 4 SUMMARY

Definitions and Concepts	Examples

Section 4.1 Solving Systems of Linear Equations by Graphing

A system of linear equations in two variables, x and y, consists of two equations of the form $Ax + By = C$. A solution is an ordered pair of numbers that satisfies both equations.

Determine whether $(3, -1)$ is a solution of

$$2x + 5y = 1$$
$$4x + y = 11.$$

Replace x with 3 and y with -1 in both equations.

$$2x + 5y = 1 \qquad\qquad 4x + y = 11$$
$$2 \cdot 3 + 5(-1) \stackrel{?}{=} 1 \qquad 4 \cdot 3 + (-1) \stackrel{?}{=} 11$$
$$6 + (-5) \stackrel{?}{=} 1 \qquad 12 + (-1) \stackrel{?}{=} 11$$
$$1 = 1, \quad \text{true} \qquad 11 = 11, \quad \text{true}$$

Thus, $(3, -1)$ is a solution of the system.

Using the graphing method, a solution of a linear system is a point common to the graphs of both equations in the system. If the graphs are parallel lines, the system has no solution and is called inconsistent. If the graphs are the same line, the system has infinitely many solutions. The equations are called dependent.

Solve by graphing: $2x + y = 4$
$$x + y = 2.$$

The solution is $(2, 0)$ and the solution set is $\{(2, 0)\}$.

Section 4.2 Solving Systems of Linear Equations by the Substitution Method

To solve a linear system by the substitution method,

1. Solve one equation for one variable in terms of the other.
2. Substitute the expression for that variable into the other equation. This will result in an equation in one variable.
3. Solve the equation in one variable.
4. Back-substitute the value of the variable found in step 3 in the equation from step 1. Simply and find the value of the remaining variable.
5. Check the proposed solution in both of the system's given equations.

If both variables are eliminated and a false statement results, the system has no solution. If both variables are eliminated and a true statement results, the system has infinitely many solutions.

Solve by the substitution method:

$$y = 2x + 3$$
$$7x - 5y = -18.$$

Substitute $2x + 3$ for y in the second equation.

$$7x - 5(2x + 3) = -18$$
$$7x - 10x - 15 = -18$$
$$-3x - 15 = -18$$
$$-3x = -3$$
$$x = 1$$

Find y. Substitute 1 for x in $y = 2x + 3$.

$$y = 2 \cdot 1 + 3 = 2 + 3 = 5$$

The solution, $(1, 5)$, checks. The solution set is $\{(1, 5)\}$.

Definitions and Concepts	Examples

Section 4.3 Solving Systems of Linear Equations by the Addition Method

To solve a linear system by the addition method,

1. Write equations in $Ax + By = C$ form.

2. Multiply one or both equations by nonzero numbers so that coefficients of a variable are opposites.

3. Add equations.

4. Solve the resulting equation for a variable.

5. Back-substitute the value of the variable into either original equation and find the value of the remaining variable.

6. Check the proposed solution in both the original equations.

If both variables are eliminated and a false statement results, the system has no solution. If both variables are eliminated and a true statement results, the system has infinitely many solutions.

Solve by the addition method:

$$3x + y = -11$$
$$6x - 2y = -2.$$

Eliminate y. Multiply both sides of the first equation by 2.

$$6x + 2y = -22$$
$$\underline{6x - 2y = -2}$$
Add: $\quad 12x \quad\quad = -24$
$$x = -2$$

Find y. Back-substitute -2 for x. We'll use the first equation.

$$3(-2) + y = -11$$
$$-6 + y = -11$$
$$y = -5$$

The solution, $(-2, -5)$, checks. The solution set is $\{(-2, -5)\}$.

Section 4.4 Problem Solving Using Systems of Equations

A Problem-Solving Strategy

1. Use variables, usually x and y, to represent unknown quantities.

2. Write a system of equations describing the problem's conditions.

3. Solve the system and answer the problem's question.

4. Check proposed answers in the problem's wording.

You invested $14,000 in two stocks paying 7% and 9% interest. Total year-end interest was $1180. How much was invested at each rate?

Let x = amount invested at 7% and
y = amount invested at 9%.

amount invested at 7%		amount invested at 9%		
x	$+$	y	$=$	14,000

interest from 7% investment		interest from 9% investment		
$0.07x$	$+$	$0.09y$	$=$	1180

Solving by substitution or addition, $x = 4000$ and $y = 10,000$. Thus, $4000 was invested at 7% and $10,000 at 9%.

Section 4.5 Systems of Linear Equations in Three Variables

A system of linear equations in three variables, x, y, and z, consists of three equations of the form $Ax + By + Cz = D$. The solution set is the set consisting of the ordered triple that satisfies all three equations. The solution represents the point of intersection of three planes in space.

Is $(2, -1, 3)$ a solution of

$$3x + 5y - 2z = -5$$
$$2x + 3y - z = -2$$
$$2x + 4y + 6z = 18?$$

Replace x with 2, y with -1, and z with 3. Using the first equation, we obtain:

$$3 \cdot 2 + 5(-1) - 2(3) \stackrel{?}{=} -5$$
$$6 - 5 - 6 \stackrel{?}{=} -5$$
$$-5 = -5, \quad \textit{true}$$

The ordered triple $(2, -1, 3)$ satisfies the first equation. In a similar manner, it satisfies the other two equations and is a solution.

Definitions and Concepts	Examples

Section 4.5 Systems of Linear Equations in Three Variables (continued)

To solve a linear system in three variables by eliminating variables,

1. Reduce the system to two equations in two variables.

2. Solve the resulting system of two equations in two variables.

3. Use back-substitution in one of the equations in two variables to find the value of the second variable.

4. Back-substitute the values for two variables into one of the original equations to find the value of the third variable.

5. Check.

If all variables are eliminated and a false statement results, the system is inconsistent and has no solution. If a true statement results, the system contains dependent equations and has infinitely many solutions.

Solve:
$$2x + 3y - 2z = 0 \quad \boxed{1}$$
$$x + 2y - z = 1 \quad \boxed{2}$$
$$3x - y + z = -15. \quad \boxed{3}$$

Add equations $\boxed{2}$ and $\boxed{3}$ to eliminate z.

$$4x + y = -14 \quad \boxed{4}$$

Eliminate z again. Multiply equation $\boxed{3}$ by 2 and add to equation $\boxed{1}$.

$$8x + y = -30 \quad \boxed{5}$$

Multiply equation $\boxed{4}$ by -1 and add to equation $\boxed{5}$.

$$-4x - y = 14$$
$$\underline{8x + y = -30}$$
$$\text{Add:} \quad 4x \quad\quad = -16$$
$$x = -4$$

Substitute -4 for x in $\boxed{4}$.

$$4(-4) + y = -14$$
$$y = 2$$

Substitute -4 for x and 2 for y in $\boxed{3}$.

$$3(-4) - 2 + z = -15$$
$$-14 + z = -15$$
$$z = -1$$

Checking verifies that $(-4, 2, -1)$ is the solution and $\{(-4, 2, -1)\}$ is the solution set.

Curve Fitting

Curve fitting is determining an equation whose graph contains given points. Three points that do not lie on a line determine the graph of an equation in the form

$$y = ax^2 + bx + c.$$

Use the three given points to create a system of three equations. Solve the system to find $a, b,$ and c.

Find the equation $y = ax^2 + bx + c$ whose graph passes through the points $(-1, 2), (1, 8),$ and $(2, 14)$.
Use $y = ax^2 + bx + c$.

When $x = -1, y = 2$: $\quad 2 = a(-1)^2 + b(-1) + c$

When $x = 1, y = 8$: $\quad 8 = a \cdot 1^2 + b \cdot 1 + c$

When $x = 2, y = 14$: $\quad 14 = a \cdot 2^2 + b \cdot 2 + c$

Solving,
$$a - b + c = 2$$
$$a + b + c = 8$$
$$4a + 2b + c = 14,$$

$a = 1, b = 3,$ and $c = 4$. The equation, $y = ax^2 + bx + c$, is $y = x^2 + 3x + 4$.

CHAPTER 4 REVIEW EXERCISES

4.1 *In Exercises 1–2, determine whether the given ordered pair is a solution of the system.*

1. $(1, -5)$ solution
$4x - y = 9$
$2x + 3y = -13$

2. $(-5, 2)$ not a solution
$2x + 3y = -4$
$x - 4y = -10$

3. Does the graphing-utility screen show the solution for the following system? Explain.   No; $(-1, 3)$ does not satisfy $2x + y = -5$.

$$x + y = 2$$
$$2x + y = -5$$

Intersection
X=-1   Y=3

In Exercises 4–14, solve each system by graphing. If there is no solution or an infinite number of solutions, so state, and use set notation to express solution sets.

4. $x + y = 2$
$x - y = 6$ $\{(4, -2)\}$

5. $2x - 3y = 12$
$-2x + y = -8$ $\{(3, -2)\}$

6. $3x + 2y = 6$
$3x - 2y = 6$ $\{(2, 0)\}$

7. $y = \frac{1}{2}x$
$y = 2x - 3$ $\{(2, 1)\}$

8. $x + 2y = 2$
$y = x - 5$ $\{(4, -1)\}$

9. $x + 2y = 8$
$3x + 6y = 12$ no solution; $\varnothing$

10. $2x - 4y = 8$
$x - 2y = 4$
infinitely many solutions;
$\{(x, y) | 2x - 4y = 8\}$ or
$\{(x, y) | x - 2y = 4\}$

11. $y = 3x - 1$
$y = 3x + 2$ no solution; $\varnothing$

12. $x - y = 4$
$x = -2$ $\{(-2, -6)\}$

13. $x = 2$
$y = 5$ $\{(2, 5)\}$

14. $x = 2$
$x = 5$ no solution; $\varnothing$

4.2 *In Exercises 15–23, solve each system by the substitution method. If there is no solution or an infinite number of solutions, so state, and use set notation to express solution sets.*

15. $2x - 3y = 7$
$y = 3x - 7$ $\{(2, -1)\}$

16. $2x - y = 6$
$x = 13 - 2y$ $\{(5, 4)\}$

17. $2x - 5y = 1$
$3x + y = -7$ $\{(-2, -1)\}$

18. $3x + 4y = -13$
$5y - x = -21$ $\{(1, -4)\}$

19. $y = 39 - 3x$
$y = 2x - 61$ $\{(20, -21)\}$

20. $4x + y = 5$
$12x + 3y = 15$
infinitely many solutions;
$\{(x, y) | 4x + y = 5\}$ or
$\{(x, y) | 12x + 3y = 15\}$

21. $4x - 2y = 10$
$y = 2x + 3$ no solution; $\varnothing$

22. $x - 4 = 0$
$9x - 2y = 0$ $\{(4, 18)\}$

23. $8y = 4x$
$7x + 2y = -8$ $\left\{\left(-1, -\frac{1}{2}\right)\right\}$

24. The weekly demand and supply models for the video *Titanic* at a chain of stores that sells videos are given by the demand model $N = -60p + 1000$ and the supply model $N = 4p + 200$, in which p is the price of the video and N is the number of videos sold or supplied each week to the chain of stores. Find the price at which supply and demand are equal. At this price, how many copies of *Titanic* can be supplied and sold each week? $12.50; 250$ copies

4.3 *In Exercises 25–35, solve each system by the addition method. If there is no solution or an infinite number of solutions, so state, and use set notation to express solution sets.*

25. $x + y = 6$
$2x + y = 8$ $\{(2, 4)\}$

26. $3x - 4y = 1$
$12x - y = -11$ $\{(-1, -1)\}$

27. $3x - 7y = 13$
$6x + 5y = 7$ $\{(2, -1)\}$

28. $8x - 4y = 16$
$4x + 5y = 22$ $\{(3, 2)\}$

29. $5x - 2y = 8$
$3x - 5y = 1$ $\{(2, 1)\}$

30. $2x + 7y = 0$
$7x + 2y = 0$ $\{(0, 0)\}$

31. $x + 3y = -4$
$3x + 2y = 3$ $\left\{\left(\frac{17}{7}, -\frac{15}{7}\right)\right\}$

32. $2x + y = 5$
$2x + y = 7$ no solution; $\varnothing$

33. $3x - 4y = -1$
$-6x + 8y = 2$
infinitely many solutions;
$\{(x, y) | 3x - 4y = -1\}$ or
$\{(x, y) | -6x + 8y = 2\}$

34. $2x = 8y + 24$
$3x + 5y = 2$ $\{(4, -2)\}$

35. $5x - 7y = 2$
$3x = 4y$ $\{(-8, -6)\}$

In Exercises 36–41, solve each system by the method of your choice. If there is no solution or an infinite number of solutions, so state, and use set notation to express solution sets.

36. $3x + 4y = -8$
$2x + 3y = -5$ $\{(-4, 1)\}$

37. $6x + 8y = 39$
$y = 2x - 2$ $\left\{\left(\frac{5}{2}, 3\right)\right\}$

38. $x + 2y = 7$
$2x + y = 8$ $\{(3, 2)\}$

39. $y = 2x - 3$
$y = -2x - 1$ $\left\{\left(\frac{1}{2}, -2\right)\right\}$

40. $3x - 6y = 7$
$3x = 6y$ no solution; $\varnothing$

41. $y - 7 = 0$
$7x - 3y = 0$ $\{(3, 7)\}$

4.4

42. The bar graph shows the five countries with the longest healthy life expectancy at birth. Combined, people in Japan and Switzerland can expect to spend 146.4 years in good health. The difference between healthy life expectancy between these two countries is 0.8 years. Find the healthy life expectancy at birth in Japan and Switzerland.   Japan: 73.6 years; Switzerland: 72.8 years

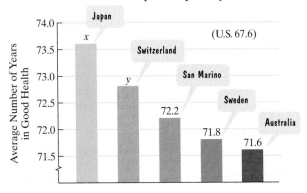

Countries with the Longest Healthy Life Expectancy

Source: World Health Organization

43. The gorilla and orangutan are the heaviest of the world's apes. Two gorillas and three orangutans weigh 1465 pounds. A gorilla's weight increased by twice an orangutan's weight is 815 pounds. Find the weight for each of these primates.   gorilla: 485 pounds; orangutan: 165 pounds

44. Health experts agree that cholesterol intake should be limited to 300 milligrams or less each day. Three ounces of shrimp and 2 ounces of scallops contain 156 milligrams of cholesterol. Five ounces of shrimp and 3 ounces of scallops contain 45 milligrams of cholesterol less than the suggested maximum daily intake. Determine the cholesterol content in an ounce of each item.   shrimp: 42 mg; scallops: 15 mg

45. The perimeter of a table tennis top is 28 feet. The difference between 4 times the length and 3 times the width is 21 feet. Find the dimensions.   9 feet by 5 feet

Length: x Width: y

46. A travel agent offers two package vacation plans. The first plan costs $360 and includes 3 days at a hotel and a rental car for 2 days. The second plan costs $500 and includes 4 days at a hotel and a rental car for 3 days. The daily charge for the room is the same under each plan, as is the daily charge for the car. Find the cost per day for the room and for the car.   room: $80; car: $60

47. You are choosing between two long-distance telephone plans. One plan has a monthly fee of $15 with a charge of $0.05 per minute for all long-distance calls. The other plan has a monthly fee of $10 with a charge of $0.075 per minute for all long-distance calls. For how many minutes of long-distance calls will the costs for the two plans be the same? What will be the cost for each plan?   200 minutes; $25

48. You invested $9000 in two funds paying 4% and 7% annual interest, respectively. At the end of the year, the total interest from these investments was $555. How much was invested at each rate? $2500 was invested at 4% and $6500 at 7%.

49. A chemist needs to mix a solution that is 34% silver nitrate with one that is 4% silver nitrate to obtain 100 milliliters of a mixture that is 7% silver nitrate. How many milliliters of each of the solutions must be used?   10 milliliters of 34% and 90 milliliters of 4% silver nitrate must be used.

50. When a plane flies with the wind, it can travel 2160 miles in 3 hours. When the plane flies in the opposite direction, against the wind, it takes 4 hours to fly the same distance. Find the rate of the plane in still air and the rate of the wind.   The rate of the plane in still air is 630 miles per hour and the rate of the wind is 90 miles per hour.

4.5

51. Is $(-3, -2, 5)$ a solution of the system

$$x + y + z = 0$$
$$2x - 3y + z = 5$$
$$4x + 2y + 4z = 3?$$ no, not a solution

Solve each system in Exercises 52–54 by eliminating variables using the addition method. If there is no solution or if there are infinitely many solutions and a system's equations are dependent, so state.

52. $2x - y + z = 1$
$3x - 3y + 4z = 5$
$4x - 2y + 3z = 4$ $\{(0, 1, 2)\}$

53. $x + 2y - z = 5$
$2x - y + 3z = 0$
$2y + z = 1$ $\{(2, 1, -1)\}$

54. $3x - 4y + 4z = 7$
$x - y - 2z = 2$
$2x - 3y + 6z = 5$ infinitely many solutions; dependent equations

55. Find the equation $y = ax^2 + bx + c$ whose graph passes through the points $(1, 4)$, $(3, 20)$, and $(-2, 25)$. $y = 3x^2 - 4x + 5$

56. The graph shows the average yearly cost for veterinary care, per animal, in the United States. The combined yearly cost for a dog, a horse, and a cat is $307. The difference between the cost for a dog and a horse is $32, and the difference between the cost for a horse and a cat is $16. Find the average yearly cost for veterinary care per dog, per horse, and per cat. per dog: $129; per horse: $97; per cat: $81

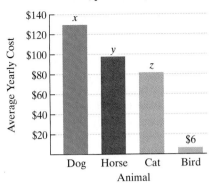

Average Yearly Cost for Veterinary Care, per Animal, in the U.S.

Source: American Veterinary Medical Association

CHAPTER 4   TEST

Remember to use your Chapter Test Prep Video CD to see the worked-out solutions to the test questions you want to review.

In Exercises 1–2, determine whether the given ordered pair is a solution of the system.

1. $(5, -5)$

$2x + y = 5$

$x + 3y = -10$ solution

2. $(-3, 2)$

$x + 5y = 7$

$3x - 4y = 1$ not a solution

In Exercises 3–4, solve each system by graphing. If there is no solution or an infinite number of solutions, so state, and use set notation to express solution sets.

3. $x + y = 6$

$4x - y = 4$ $\{(2, 4)\}$

4. $2x + y = 8$

$y = 3x - 2$ $\{(2, 4)\}$

In Exercises 5–7, solve each system by the substitution method. If there is no solution or an infinite number of solutions, so state, and use set notation to express solution sets.

5. $x = y + 4$

$3x + 7y = -18$ $\{(1, -3)\}$

6. $2x - y = 7$

$3x + 2y = 0$ $\{(2, -3)\}$

7. $2x - 4y = 3$

$x = 2y + 4$ no solution; $\varnothing$

In Exercises 8–10, solve each system by the addition method. If there is no solution or an infinite number of solutions, so state, and use set notation to express solution sets.

8. $2x + y = 2$

$4x - y = -8$ $\{(-1, 4)\}$

9. $2x + 3y = 1$

$3x + 2y = -6$ $\{(-4, 3)\}$

10. $3x - 2y = 2$

$-9x + 6y = -6$ infinitely many solutions;

$\{(x, y) | 3x - 2y = 2\}$ or $\{(x, y) | -9x + 6y = -6\}$

11. According to the U.S. Census Bureau, the two most popular female first names in the United States are Mary and Patricia.

Combined, these two names account for 3.7% of all female first names. The difference between the percentage of women named Mary and the percentage of women named Patricia is 1.5%. What percentage of all first names in the United States are Mary and what percentage are Patricia?   Mary: 2.6%; Patricia: 1.1%

12. You are choosing between two long-distance telephone plans. One plan has a monthly fee of $15 with a charge of $0.05 per minute. The other plan has a monthly fee of $5 with a charge of $0.07 per minute. For how many minutes of long-distance calls will the costs for the two plans be the same? What will be the cost for each plan?   500 minutes; $40

13. You invested $9000 in two funds paying 6% and 7% annual interest, respectively. At the end of the year, the total interest from these investments was $610. How much was invested at each rate? $2000 was invested at 6% and $7000 at 7%.

14. You need to mix a 6% peroxide solution with a 9% peroxide solution to obtain 36 ounces of an 8% peroxide solution. How many ounces of each of the solutions must be used?   12 ounces of a 6% peroxide solution and 24 ounces of a 9% peroxide solution must be used.

15. A paddleboat on the Mississippi River travels 48 miles downstream, with the current, in 3 hours. The return trip, against the current, takes the paddleboat 4 hours. Find the boat's rate in still water and the rate of the current.   rate in still water: 14 miles per hour; rate of current: 2 miles per hour

16. Solve by eliminating variables using the addition method:

$$x + y + z = 6$$
$$3x + 4y - 7z = 1$$
$$2x - y + 3z = 5.$$ $(1, 3, 2)$ or $\{(1, 3, 2)\}$

Note to Instructor: The Chapter 4 Test file in the TestGen program provides algorithms specifically matched to these problems so you can easily replicate this chapter test for additional practice or assessment purposes.

CUMULATIVE REVIEW EXERCISES (CHAPTERS 1–4)

1. Perform the indicated operations:
$$-14 - [18 - (6 - 10)]. \quad -36$$

2. Simplify: $6(3x - 2) - (x - 1)$. $\quad 17x - 11$

In Exercises 3–4, solve each equation.

3. $17(x + 3) = 13 + 4(x - 10)$ $\quad \{-6\}$

4. $\dfrac{x}{4} - 1 = \dfrac{x}{5}$ $\quad \{20\}$

5. Solve for t: $A = P + Prt$. $\quad t = \dfrac{A - P}{Pr}$

6. Solve and graph the solution set on a number line: $2x - 5 < 5x - 11$. $\quad \{x | x > 2\}$; See graphing answer section.

In Exercises 7–9, graph each equation in the rectangular coordinate system. **7–9.** See graphing answer section.

7. $x - 3y = 6$

8. $y = 4 - x^2$

9. $y = -\dfrac{3}{5}x + 2$

In Exercises 10–11, solve each linear system.

10. $3x - 4y = 8$
$4x + 5y = -10$ $\quad \{(0, -2)\}$

11. $2x - 3y = 9$
$y = 4x - 8$ $\quad \left\{ \left(\frac{3}{2}, -2 \right) \right\}$

12. Find the slope of the line passing through $(5, -6)$ and $(6, -5)$. $\quad m = 1$

13. Write the point-slope form and the slope-intercept form of the equation of the line passing through $(-1, 6)$ with slope $= -4$. $\quad y - 6 = -4(x + 1); y = -4x + 2$

14. The area of a triangle is 80 square feet. Find the height if the base is 16 feet. $\quad 10$ feet

15. If 10 pens and 15 pads cost $26, and 5 of the same pens and 10 of the same pads cost $16, find the cost of a pen and a pad. pen: $0.80; pad: $1.20

16. List all the integers in this set:
$$\left\{ -93, -\frac{7}{3}, 0, \sqrt{3}, \frac{7}{1}, \sqrt{100} \right\}. \quad -93, 0, \frac{7}{1}, \sqrt{100}$$

The graphs show the percentage of U.S. households with one computer and multiple computers. Use the information provided by the graphs to solve Exercises 17–20.

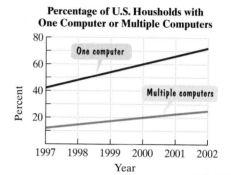

Percentage of U.S. Housholds with One Computer or Multiple Computers

Source: Forrester Research. Inc.

17. What percentage of U.S. households had multiple computers in 2000? $\quad 20\%$

18. Which graph has the greater slope? What does this mean in terms of the variables in this situation? one computer; The percentage of households with one computer is increasing faster than the percentage for multiple computers.

19. In 1997, 42% of U.S. households had one computer. This is increasing by approximately 6% per year. If this trend continues, in how many years after 1997 will 90% of U.S. households have one computer? In which year will that be? 8 years; 2005

20. The formula $y = \frac{8}{3}x + 12$ models the percentage of U.S. households, y, with multiple computers x years after 1997. Use the formula to find in which year 52% of U.S. households will have multiple computers. 2012

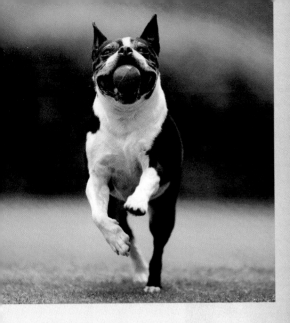

One of the joys of your life is your dog, your very special buddy. Lately, however, you've noticed that your companion is slowing down a bit. He is now 8 years old and you wonder how this translates into human years. You remember something about every year of a dog's life being equal to seven years for a human. Is there a more accurate description?

This question is addressed in Exercises 103–106 in Exercise Set 5.1.

5 CHAPTER

Exponents and Polynomials

There is a formula that models the age in human years, y, of a dog that is x years old:

$$y = -0.001618x^4 + 0.077326x^3$$
$$-1.2367x^2 + 11.460x + 2.914.$$

The algebraic expression on the right side of the formula contains variables to powers that are whole numbers and is an example of a polynomial. Much of what we do in algebra involves operations with polynomials. In this chapter, we study these operations, as well as the many applications of polynomials.

Objectives

1 Understand the vocabulary used to describe polynomials.

2 Add polynomials.

3 Subtract polynomials.

4 Use mathematical models that contain polynomials.

ADDING AND SUBTRACTING POLYNOMIALS

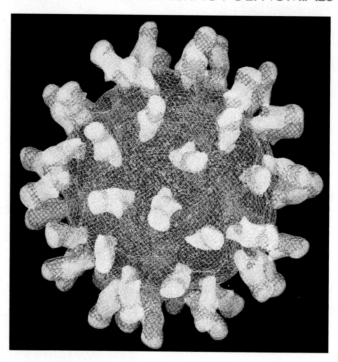

This computer-simulated model of the common cold virus was developed by researchers at Purdue University. Their discovery of how the virus infects human cells could lead to more effective treatment for the illness.

Runny nose? Sneezing? You are probably familiar with the unpleasant onset of a cold. We "catch cold" when the cold virus enters our bodies, where it multiplies. Fortunately, at a certain point the virus begins to die. The algebraic expression $-0.75x^4 + 3x^3 + 5$ describes the billions of viral particles in our bodies after x days of invasion. The expression enables mathematicians to determine the day on which there is a maximum number of viral particles and, consequently, the day we feel sickest.

The algebraic expression $-0.75x^4 + 3x^3 + 5$ is an example of a *polynomial*. A **polynomial** is a single term or the sum of two or more terms containing variables in the numerator with whole-number exponents. This particular polynomial contains three terms. Equations containing polynomials are used in such diverse areas as science, business, medicine, psychology, and sociology. In this section, we present basic ideas about polynomials. We then use our knowledge of combining like terms to find sums and differences of polynomials.

1 Understand the vocabulary used to describe polynomials.

Describing Polynomials   Consider the polynomial

$$7x^3 - 9x^2 + 13x - 6.$$

We can express this polynomial as

$$7x^3 + (-9x^2) + 13x + (-6).$$

The polynomial contains four terms. It is customary to write the terms in the order of descending powers of the variables. This is the **standard form** of a polynomial.

We begin this chapter by limiting our discussion to polynomials containing only one variable. Each term of such a polynomial in x is of the form ax^n. The **degree** of ax^n is n. For example, the degree of the term $7x^3$ is 3.

THE DEGREE OF ax^n If $a \neq 0$, the degree of ax^n is n. The degree of a nonzero constant term is 0. The constant 0 has no defined degree.

Here is an example of a polynomial and the degree of each of its four terms:

$$6x^4 - 3x^3 - 2x - 5.$$

| degree 4 | degree 3 | degree 1 | degree of nonzero constant: 0 |

Notice that the exponent on x for the term $2x$, meaning $2x^1$, is understood to be 1. For this reason, the degree of $2x$ is 1.

A polynomial which when simplified has exactly one term is called a **monomial**. A **binomial** is a polynomial that has two terms, each with a different exponent. A **trinomial** is a polynomial with three terms, each with a different exponent. Polynomials with four or more terms have no special names.

The degree of a polynomial is the highest degree of all the terms of the polynomial. For example, $4x^2 + 3x$ is a binomial of degree 2 because the degree of the first term is 2, and the degree of the other term is less than 2. Also, $7x^5 - 2x^2 + 4$ is a trinomial of degree 5 because the degree of the first term is 5, and the degrees of the other terms are less than 5.

Up to now, we have used x to represent the variable in a polynomial. However, any letter can be used. For example,

- $7x^5 - 3x^3 + 8$ is a polynomial (in x) of degree 5. Because there are three terms, the polynomial is a trinomial.

- $6y^3 + 4y^2 - y + 3$ is a polynomial (in y) of degree 3. Because there are four terms, the polynomial has no special name.

- $z^7 + \sqrt{2}$ is a polynomial (in z) of degree 7. Because there are two terms, the polynomial is a binomial.

2 Add polynomials.

Adding Polynomials Recall that *like terms* are terms containing exactly the same variables to the same powers. Polynomials are added by combining like terms. For example, we can add the monomials $-9x^3$ and $13x^3$ as follows:

$$-9x^3 + 13x^3 = (-9 + 13)x^3 = 4x^3.$$

| These like terms both contain x to the third power. | Add coefficients and keep the same variable factor, x^3. |

EXAMPLE 1 Adding Polynomials

Add: $(-9x^3 + 7x^2 - 5x + 3) + (13x^3 + 2x^2 - 8x - 6)$.

SOLUTION The like terms are $-9x^3$ and $13x^3$, containing the same variable to the same power (x^3), as well as $7x^2$ and $2x^2$ (both contain x^2), $-5x$ and $-8x$ (both contain x), and the constant terms 3 and -6. We begin by grouping these pairs of like terms.

We can use a graphing utility to check the results of polynomial addition and subtraction. For example, try graphing

$$y_1 = (-9x^3 + 7x^2 - 5x + 3)$$
$$+ (13x^3 + 2x^2 - 8x - 6)$$

and

$$y_2 = 4x^3 + 9x^2 - 13x - 3$$

on the same screen, as shown below. (You may need to experiment with the viewing rectangle so that your graph does not appear to be cut off.) Because both graphs are identical, this shows that

$$y_1 = y_2$$

or that

$$(-9x^3 + 7x^2 - 5x + 3)$$
$$+ (13x^3 + 2x^2 - 8x - 6)$$
$$= 4x^3 + 9x^2 - 13x - 3.$$

$[-5, 3, 1]$ by $[-30, 30, 3]$

$$(-9x^3 + 7x^2 - 5x + 3) + (13x^3 + 2x^2 - 8x - 6)$$
$$= (-9x^3 + 13x^3) + (7x^2 + 2x^2) + (-5x - 8x) + (3 - 6) \quad \text{Group like terms.}$$
$$= 4x^3 + 9x^2 + (-13x) + (-3) \quad\quad\quad\quad\quad\quad \text{Combine like terms.}$$
$$= 4x^3 + 9x^2 - 13x - 3 \qquad\qquad\qquad\qquad\qquad \blacksquare$$

✔ **CHECK POINT 1** Add: $(-11x^3 + 7x^2 - 11x - 5) + (16x^3 - 3x^2 + 3x - 15)$.

$5x^3 + 4x^2 - 8x - 20$

Polynomials can be added by arranging like terms in columns. Then combine like terms, column by column.

EXAMPLE 2 Adding Polynomials Vertically

Add: $(-9x^3 + 7x^2 - 5x + 3) + (13x^3 + 2x^2 - 8x - 6)$.

SOLUTION

$$
\begin{array}{rrrr}
-9x^3 & 7x^2 & -5x & 3 \\
\underline{13x^3} & \underline{2x^2} & \underline{-8x} & \underline{-6} \\
4x^3 & 9x^2 & -13x & -3
\end{array}
$$

We consider each term separately and write like terms in columns.

Add, column by column.

Now add the four sums together:

$$4x^3 + 9x^2 + (-13x) + (-3) = 4x^3 + 9x^2 - 13x - 3.$$

This is the same answer that we found in Example 1. ∎

✔ **CHECK POINT 2** Add the polynomials in Check Point 1 using a vertical format. Begin by arranging like terms in columns. $5x^3 + 4x^2 - 8x - 20$

3 Subtract polynomials.

Subtracting Polynomials We subtract real numbers by adding the opposite, or additive inverse, of the number being subtracted. For example,

$$8 - 3 = 8 + (-3) = 5.$$

Subtraction of polynomials also involves opposites. If the sum of two polynomials is 0, the polynomials are **opposites**, or **additive inverses**, of each other. Here is an example:

$$(4x^2 - 6x - 7) + (-4x^2 + 6x + 7) = 0.$$

The opposite of $4x^2 - 6x - 7$ is $-4x^2 + 6x + 7$, and vice-versa.

Observe that the opposite of $4x^2 - 6x - 7$ can be obtained by changing the sign of each of its coefficients:

Polynomial   **Opposite**

Change 4 to -4, change -6 to 6, and change -7 to 7.

$4x^2 - 6x - 7$ $-4x^2 + 6x + 7$

In general, **the opposite of a polynomial is that polynomial with the sign of every coefficient changed**. Just as we did with real numbers, we subtract one polynomial from another by adding the opposite of the polynomial being subtracted.

> **SUBTRACTING POLYNOMIALS** To subtract two polynomials, add the first polynomial and the opposite of the polynomial being subtracted.

EXAMPLE 3 Subtracting Polynomials

Subtract: $(7x^2 + 3x - 4) - (4x^2 - 6x - 7)$.

SOLUTION

$$(7x^2 + 3x - 4) - (4x^2 - 6x - 7)$$

Change the sign of each coefficient.

$$= (7x^2 + 3x - 4) + (-4x^2 + 6x + 7) \qquad \text{Add the opposite of the polynomial being subtracted.}$$

$$= (7x^2 - 4x^2) + (3x + 6x) + (-4 + 7) \qquad \text{Group like terms.}$$

$$= 3x^2 + 9x + 3 \qquad \text{Combine like terms.} \qquad \blacksquare$$

✔ **CHECK POINT 3** Subtract: $(9x^2 + 7x - 2) - (2x^2 - 4x - 6)$. $7x^2 + 11x + 4$

STUDY TIP

Be careful of the order in Example 4. For example, subtracting 2 from 5 means $5 - 2$. In general, subtracting B from A means $A - B$. The order of the resulting algebraic expression is not the same as the order in English.

EXAMPLE 4 Subtracting Polynomials

Subtract $2x^3 - 6x^2 - 3x + 9$ from $7x^3 - 8x^2 + 9x - 6$.

SOLUTION

$$(7x^3 - 8x^2 + 9x - 6) - (2x^3 - 6x^2 - 3x + 9)$$

Change the sign of each coefficient.

$$= (7x^3 - 8x^2 + 9x - 6) + (-2x^3 + 6x^2 + 3x - 9) \qquad \text{Add the opposite of the polynomial being subtracted.}$$

$$= (7x^3 - 2x^3) + (-8x^2 + 6x^2)$$

$$\quad + (9x + 3x) + (-6 - 9) \qquad \text{Group like terms.}$$

$$= 5x^3 + (-2x^2) + 12x + (-15) \qquad \text{Combine like terms.}$$

$$= 5x^3 - 2x^2 + 12x - 15 \qquad \blacksquare$$

✔ **CHECK POINT 4** Subtract $3x^3 - 8x^2 - 5x + 6$ from $10x^3 - 5x^2 + 7x - 2$.

$7x^3 + 3x^2 + 12x - 8$

Subtraction can also be performed in columns.

EXAMPLE 5 Subtracting Polynomials Vertically

Use the method of subtracting by columns to find

$$(12y^3 - 9y^2 - 11y - 3) - (4y^3 - 5y + 8).$$

SOLUTION Arrange like terms in columns.

$$
\begin{array}{r}
12y^3 - 9y^2 - 11y - 3 \\
-(4y^3 \qquad\quad - 5y + 8) \\
\hline
\end{array}
$$

Add the opposite of the polynomial being subtracted.

Leave space for the missing term.

Change the sign of each coefficient of $4y^3 - 5y + 8$.

$$
\begin{array}{r}
12y^3 - 9y^2 - 11y - 3 \\
+ \; -4y^3 \qquad\quad + 5y - 8 \\
\hline
8y^3 - 9y^2 - 6y - 11
\end{array}
$$

Combine like terms.

 CHECK POINT 5 Use the method of subtracting by columns to find

$$(8y^3 - 10y^2 - 14y - 2) - (5y^3 - 3y + 6). \quad 3y^3 - 10y^2 - 11y - 8$$

④ Use mathematical models that contain polynomials.

Applications Polynomials often appear in formulas that model real-world situations.

EXAMPLE 6 An Application: Death Rate

The formula

$$y = 0.036x^2 - 2.8x + 58.14$$

models the number of deaths per year per thousand people, y, for people who are x years old, $40 \le x \le 60$. Approximately how many people per thousand who are 50 years old die each year?

SOLUTION Because we are interested in people who are 50 years old, substitute 50 for x in the formula's polynomial.

$y = 0.036x^2 - 2.8x + 58.14$ This is the given formula.

$y = 0.036(50)^2 - 2.8(50) + 58.14$ Substitute 50 for x.

$y = 0.036(2500) - 2.8(50) + 58.14$ Evaluate the exponential expression: $50^2 = 50 \cdot 50 = 2500$.

$y = 90 - 140 + 58.14$ Perform the multiplications.

$y = 8.14 \approx 8$ Simplify.

Approximately 8 people per thousand who are 50 years old die each year.

We can use point plotting or a graphing utility to graph formulas that contain polynomials. These graphs contain only rounded curves with no sharp corners. For example, the graph of $y = 0.036x^2 - 2.8x + 58.14$, the formula from Example 6 that models the number of deaths per thousand, is shown in Figure 5.1. Our work in Example 6 can be visualized as a point on the curve.

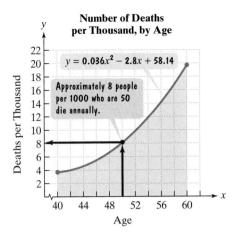

Number of Deaths per Thousand, by Age

$y = 0.036x^2 - 2.8x + 58.14$

Approximately 8 people per 1000 who are 50 die annually.

Age

FIGURE 5.1 The graph of a formula containing a polynomial

 CHECK POINT 6 Use the formula $y = 0.036x^2 - 2.8x + 58.14$ to answer this question: Approximately how many people per thousand who are 40 years old die annually? Identify your solution as a point on the curve in Figure 5.1.

approximately 4 per thousand; approximately (40, 4)

5.1 EXERCISE SET

Student Solutions Manual   CD/Video   PH Math/Tutor Center   MathXL Tutorials on CD   MathXL®   MyMathLab   Interactmath.com

Practice Exercises

In Exercises 1–16, identify each polynomial as a monomial, a binomial, or a trinomial. Give the degree of the polynomial.

1. $3x + 7$ binomial, 1
2. $5x - 2$ binomial, 1
3. $x^3 - 2x$ binomial, 3
4. $x^5 - 7x$ binomial, 5
5. $8x^2$ monomial, 2
6. $10x^2$ monomial, 2
7. 5   monomial, 0
8. 9   monomial, 0
9. $x^2 - 3x + 4$ trinomial, 2
10. $x^2 - 9x + 2$ trinomial, 2
11. $7y^2 - 9y^4 + 5$ trinomial, 4
12. $3y^2 - 14y^5 + 6$ trinomial, 5
13. $15x - 7x^3$ binomial, 3
14. $9x - 5x^3$ binomial, 3
15. $-9y^{23}$ monomial, 23
16. $-11y^{26}$ monomial, 26

In Exercises 17–38, add the polynomials.

17. $(9x + 8) + (-17x + 5)$ $-8x + 13$
18. $(8x - 5) + (-13x + 9)$ $-5x + 4$
19. $(4x^2 + 6x - 7) + (8x^2 + 9x - 2)$ $12x^2 + 15x - 9$
20. $(11x^2 + 7x - 4) + (27x^2 + 10x - 20)$ $38x^2 + 17x - 24$
21. $(7x^2 - 11x) + (3x^2 - x)$ $10x^2 - 12x$
22. $(-3x^2 + x) + (4x^2 + 8x)$ $x^2 + 9x$
23. $(4x^2 - 6x + 12) + (x^2 + 3x + 1)$ $5x^2 - 3x + 13$
24. $(-7x^2 + 8x + 3) + (2x^2 + x + 8)$ $-5x^2 + 9x + 11$
25. $(4y^3 + 7y - 5) + (10y^2 - 6y + 3)$ $4y^3 + 10y^2 + y - 2$
26. $(2y^3 + 3y + 10) + (3y^2 + 5y - 22)$ $2y^3 + 3y^2 + 8y - 12$
27. $(2x^2 - 6x + 7) + (3x^3 - 3x)$ $3x^3 + 2x^2 - 9x + 7$
28. $(4x^3 + 5x + 13) + (-4x^2 + 22)$ $4x^3 - 4x^2 + 5x + 35$

29. $(4y^2 + 8y + 11) + (-2y^3 + 5y + 2)$ $-2y^3 + 4y^2 + 13y + 13$
30. $(7y^3 + 5y - 1) + (2y^2 - 6y + 3)$ $7y^3 + 2y^2 - y + 2$
31. $(-2y^6 + 3y^4 - y^2) + (-y^6 + 5y^4 + 2y^2)$ $-3y^6 + 8y^4 + y^2$
32. $(7r^4 + 5r^2 + 2r) + (-18r^4 - 5r^2 - r)$ $-11r^4 + r$
33. $\left(9x^3 - x^2 - x - \dfrac{1}{3}\right) + \left(x^3 + x^2 + x + \dfrac{4}{3}\right)$ $10x^3 + 1$
34. $\left(12x^3 - x^2 - x + \dfrac{4}{3}\right) + \left(x^3 + x^2 + x - \dfrac{1}{3}\right)$ $13x^3 + 1$
35. $\left(\dfrac{1}{5}x^4 + \dfrac{1}{3}x^3 + \dfrac{3}{8}x^2 + 6\right) +$
$\left(-\dfrac{3}{5}x^4 + \dfrac{2}{3}x^3 - \dfrac{1}{2}x^2 - 6\right)$ $-\dfrac{2}{5}x^4 + x^3 - \dfrac{1}{8}x^2$
36. $\left(\dfrac{2}{5}x^4 + \dfrac{2}{3}x^3 + \dfrac{5}{8}x^2 + 7\right) +$
$\left(-\dfrac{4}{5}x^4 + \dfrac{1}{3}x^3 - \dfrac{1}{4}x^2 - 7\right)$ $-\dfrac{2}{5}x^4 + x^3 + \dfrac{3}{8}x^2$
37. $(0.03x^5 - 0.1x^3 + x + 0.03) +$
$(-0.02x^5 + x^4 - 0.7x + 0.3)$
$0.01x^5 + x^4 - 0.1x^3 + 0.3x + 0.33$
38. $(0.06x^5 - 0.2x^3 + x + 0.05) +$
$(-0.04x^5 + 2x^4 - 0.8x + 0.5)$
$0.02x^5 + 2x^4 - 0.2x^3 + 0.2x + 0.55$

In Exercises 39–54, use a vertical format to add the polynomials.

39. $5y^3 - 7y^2$
$\underline{6y^3 + 4y^2}$ $11y^3 - 3y^2$
40. $13x^4 - x^2$
$\underline{7x^4 + 2x^2}$ $20x^4 + x^2$

41. $3x^2 - 7x + 4$
$\underline{-5x^2 + 6x - 3}$ $\quad -2x^2 - x + 1$

42. $7x^2 - 5x - 6$
$\underline{-9x^2 + 4x + 6}$ $\quad -2x^2 - x$

43. $\frac{1}{4}x^4 - \frac{2}{3}x^3 - 5$
$\underline{-\frac{1}{2}x^4 + \frac{1}{5}x^3 + 4.7}$ $\quad -\frac{1}{4}x^4 - \frac{7}{15}x^3 - 0.3$

44. $\frac{1}{3}x^9 - \frac{1}{5}x^5 - 2.7$
$\underline{-\frac{3}{4}x^9 + \frac{2}{3}x^5 + 1}$ $\quad -\frac{5}{12}x^9 + \frac{7}{15}x^5 - 1.7$

45. $y^3 + 5y^2 - 7y - 3$
$\underline{-2y^3 + 3y^2 + 4y - 11}$ $\quad -y^3 + 8y^2 - 3y - 14$

46. $y^3 + y^2 - 7y + 9$
$\underline{-y^3 - 6y^2 - 8y + 11}$ $\quad -5y^2 - 15y + 20$

47. $4x^3 - 6x^2 + 5x - 7$
$\underline{-9x^3 \qquad\;\; - 4x + 3}$ $\quad -5x^3 - 6x^2 + x - 4$

48. $-4y^3 + 6y^2 - 8y + 11$
$\underline{2y^3 \qquad\; + 9y - \;\; 3}$ $\quad -2y^3 + 6y^2 + y + 8$

49. $7x^4 - 3x^3 + x^2$
$\underline{\qquad\quad x^3 - x^2 + 4x - 2}$ $\quad 7x^4 - 2x^3 + 4x - 2$

50. $7y^5 - 3y^3 + y^2$
$\underline{\qquad\quad 2y^3 - y^2 - 4y - 3}$ $\quad 7y^5 - y^3 - 4y - 3$

51. $7x^2 - \;\; 9x + 3$
$4x^2 + 11x - 2$ $\quad 8x^2 + 7x - 5$
$\underline{-3x^2 + \;\; 5x - 6}$

52. $7y^2 - 11y - 6$
$8y^2 + \;\; 3y + 4$ $\quad 6y^2 - 13y$
$\underline{-9y^2 - \;\; 5y + 2}$

53. $1.2x^3 - \;\; 3x^2 + 9.1$
$7.8x^3 - 3.1x^2 + \;\; 8$ $\quad 9x^3 - 4.9x^2 + 11.1$
$\underline{\qquad\quad 1.2x^2 - \;\; 6}$

54. $7.9x^3 - 6.8x^2 + 3.3$
$6.1x^3 - 2.2x^2 + 7$ $\quad 14x^3 - 4.7x^2 + 5.3$
$\underline{\qquad\quad 4.3x^2 - 5}$

In Exercises 55–74, subtract the polynomials.

55. $(x - 8) - (3x + 2)$ $\quad -2x - 10$

56. $(x - 2) - (7x + 9)$ $\quad -6x - 11$

57. $(x^2 - 5x - 3) - (6x^2 + 4x + 9)$ $\quad -5x^2 - 9x - 12$

58. $(3x^2 - 8x - 2) - (11x^2 + 5x + 4)$ $\quad -8x^2 - 13x - 6$

59. $(x^2 - 5x) - (6x^2 - 4x)$ $\quad -5x^2 - x$

60. $(3x^2 - 2x) - (5x^2 - 6x)$ $\quad -2x^2 + 4x$

61. $(x^2 - 8x - 9) - (5x^2 - 4x - 3)$ $\quad -4x^2 - 4x - 6$

62. $(x^2 - 5x + 3) - (x^2 - 6x - 8)$ $\quad x + 11$

63. $(y - 8) - (3y - 2)$ $\quad -2y - 6$

64. $(y - 2) - (7y - 9)$ $\quad -6y + 7$

65. $(6y^3 + 2y^2 - y - 11) - (y^2 - 8y + 9)$ $\quad 6y^3 + y^2 + 7y - 20$

66. $(5y^3 + y^2 - 3y - 8) - (y^2 - 8y + 11)$ $\quad 5y^3 + 5y - 19$

67. $(7n^3 - n^7 - 8) - (6n^3 - n^7 - 10)$ $\quad n^3 + 2$

68. $(2n^2 - n^7 - 6) - (2n^3 - n^7 - 8)$ $\quad -2n^3 + 2n^2 + 2$

69. $(y^6 - y^3) - (y^2 - y)$ $\quad y^6 - y^3 - y^2 + y$

70. $(y^5 - y^3) - (y^4 - y^2)$ $\quad y^5 - y^4 - y^3 + y^2$

71. $(7x^4 + 4x^2 + 5x) - (-19x^4 - 5x^2 - x)$ $\quad 26x^4 + 9x^2 + 6x$

72. $(-3x^6 + 3x^4 - x^2) - (-x^6 + 2x^4 + 2x^2)$ $\quad -2x^6 + x^4 - 3x^2$

73. $\left(\frac{3}{7}x^3 - \frac{1}{5}x - \frac{1}{3}\right) - \left(-\frac{2}{7}x^3 + \frac{1}{4}x - \frac{1}{3}\right)$ $\quad \frac{5}{7}x^3 - \frac{9}{20}x$

74. $\left(\frac{3}{8}x^2 - \frac{1}{3}x - \frac{1}{4}\right) - \left(-\frac{1}{8}x^2 + \frac{1}{2}x - \frac{1}{4}\right)$ $\quad \frac{1}{2}x^2 - \frac{5}{6}x$

In Exercises 75–88, use a vertical format to subtract the polynomials.

75. $7x + 1$
$\underline{-(3x - 5)}$ $\quad 4x + 6$

76. $4x + 2$
$\underline{-(3x - 5)}$ $\quad x + 7$

77. $7x^2 - 3$
$\underline{-(-3x^2 + 4)}$ $\quad 10x^2 - 7$

78. $9y^2 - 6$
$\underline{-(-5y^2 + 2)}$ $\quad 14y^2 - 8$

79. $7y^2 - 5y + 2$
$\underline{-(11y^2 + 2y - 3)}$ $\quad -4y^2 - 7y + 5$

80. $3x^5 - 5x^3 + 6$
$\underline{-(7x^5 + 4x^3 - 2)}$ $\quad -4x^5 - 9x^3 + 8$

81. $7x^3 + 5x^2 - 3$
$\underline{-(-2x^3 - 6x^2 + 5)}$ $\quad 9x^3 + 11x^2 - 8$

82. $3y^4 - 4y^2 + \;\; 7$
$\underline{-(-5y^4 - 6y^2 - 13)}$ $\quad 8y^4 + 2y^2 + 20$

83. $5y^3 + 6y^2 - 3y + 10$
$\underline{-(6y^3 - 2y^2 - 4y - \;\; 4)}$ $\quad -y^3 + 8y^2 + y + 14$

84. $4y^3 + 5y^2 + 7y + 11$
$\underline{-(-5y^3 + 6y^2 - 9y - \;\; 3)}$ $\quad 9y^3 - y^2 + 16y + 14$

85. $7x^4 - 3x^3 + 2x^2$
$\underline{-(\quad\; - x^3 - x^2 + x - 2)}$ $\quad 7x^4 - 2x^3 + 3x^2 - x + 2$

86. $5y^6 - 3y^3 + 2y^2$
$\underline{-(\quad\; - y^3 - y^2 - y - 1)}$ $\quad 5y^6 - 2y^3 + 3y^2 + y + 1$

87. $0.07x^3 - 0.01x^2 + 0.02x$
$\underline{-(0.02x^3 - 0.03x^2 - \qquad x)}$ $\quad 0.05x^3 + 0.02x^2 + 1.02x$

88. $0.04x^3 - 0.03x^2 + 0.05x$
$\underline{-(0.02x^3 - 0.06x^2 - \qquad x)}$ $\quad 0.02x^3 + 0.03x^2 + 1.05x$

Practice Plus

In Exercises 89–92, perform the indicated operations.

89. $[(4x^2 + 7x - 5) - (2x^2 - 10x + 3)] - (x^2 + 5x - 8)$
$x^2 + 12x$

90. $[(10x^3 - 5x^2 + 4x + 3) - (-3x^3 - 4x^2 + x)] - (7x^3 - 5x + 4)$ $\quad 6x^3 - x^2 + 8x - 1$

91. $[(4y^2 - 3y + 8) - (5y^2 + 7y - 4)] - [(8y^2 + 5y - 7) + (-10y^2 + 4y + 3)]$ $\quad y^2 - 19y + 16$

92. $[(7y^2 - 4y + 2) - (12y^2 + 3y - 5)] - [(5y^2 - 2y - 8) + (-7y^2 + 10y - 13)]$ $\quad -3y^2 - 15y + 28$

93. Subtract $x^3 - 2x^2 + 2$ from the sum of $4x^3 + x^2$ and $-x^3 + 7x - 3$. $\quad 2x^3 + 3x^2 + 7x - 5$

94. Subtract $-3x^3 - 7x + 5$ from the sum of $2x^2 + 4x - 7$ and $-5x^3 - 2x - 3$. $-2x^3 + 2x^2 + 9x - 15$

95. Subtract $-y^2 + 7y^3$ from the difference between $-5 + y^2 + 4y^3$ and $-8 - y + 7y^3$. Express the answer in standard form. $-10y^3 + 2y^2 + y + 3$

96. Subtract $-2y^2 + 8y^3$ from the difference between $-6 + y^2 + 5y^3$ and $-12 - y + 13y^3$. Express the answer in standard form. $-16y^3 + 3y^2 + y + 6$

Application Exercises

97. The common cold is caused by a rhinovirus. The polynomial

$$-0.75x^4 + 3x^3 + 5$$

describes the billions of viral particles in our bodies after x days of invasion. Find the number of viral particles, in billions, after 0 days (the time of the cold's onset when we are still feeling well), 1 day, 2 days, 3 days, and 4 days. After how many days is the number of viral particles at a maximum and, consequently, the day we feel the sickest? By when should we feel completely better? 5 billion, 7.25 billion, 17 billion, 25.25 billion, and 5 billion; 3 days; 4 days

98. The polynomial $-0.02A^2 + 2A + 22$ is used by coaches to get athletes fired up so that they can perform well. The polynomial represents the performance level related to various levels of enthusiasm, from $A = 1$ (almost no enthusiasm) to $A = 100$ (maximum level of enthusiasm). Evaluate the polynomial when $A = 20$, $A = 50$, and $A = 80$. Describe what happens to performance as we get more and more fired up. 54, 72, 54; Performance level increases, then decreases.

The graph shows cigarette consumption per U.S. adult from 1900 through 2001. The data from 1940 through 2001 can be modeled by the formula

$$y = -2.3x^2 + 135.3x + 2191,$$

where x represents years after 1940 and y represents cigarette consumption per U.S. adult. Use the formula to solve Exercises 99–100.

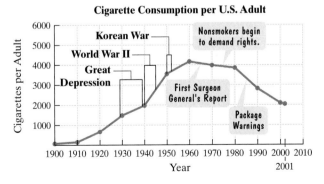

Cigarette Consumption per U.S. Adult

Source: U.S. Department of Health and Human Services

99. What was cigarette consumption per adult in 2000? How well does the formula model the actual data shown by the graph? 2029 cigarettes; Answers will vary.

100. What was cigarette consumption per adult in 1980? How well does the formula model the actual data shown by the graph? 3923 cigarettes; Answers will vary.

The wage gap is used to compare the status of women's earnings relative to men's. The wage gap is expressed as a percent and is calculated by dividing the median, or middlemost, annual earnings for women by the median annual earnings for men. The line graph shows the wage gap from 1960 through 2002. The data can be modeled by the formula

$$y = 0.012x^2 - 0.16x + 60,$$

where x represents years after 1960 and y represents median women's earnings as a percentage of median men's earnings. Use this information to solve Exercises 101–102.

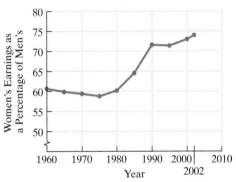

Median Women's Earnings as a Percentage of Median Men's Earnings in the U.S.

Source: U.S. Women's Bureau

101. a. Use the graph to estimate, to the nearest percent, women's earnings as a percentage of men's in 2000. 73%

　　b. Use the mathematical model to find women's earnings as a percentage of men's in 2000. 72.8%

　　c. In 2000, median annual earnings for U.S. women and men were $27,355 and $37,339, respectively. What were women's earnings as a percentage of men's? Use a calculator and round to the nearest tenth of a percent. How well do your answers in parts (a) and (b) model the actual data? 73.3%; Answers will vary.

102. a. Use the graph to estimate, to the nearest percent, women's earnings as a percentage of men's in 1970. 59% or 60%

　　b. Use the mathematical model to find women's earnings as a percentage of men's in 1970. 59.6%

　　c. In 1970, median annual earnings for U.S. women and men were $5440 and $9184, respectively. What were women's earnings as a percentage of men's? Use a calculator and round to the nearest tenth of a percent. How well do your answers in parts (a) and (b) model the actual data? 59.2%; Answers will vary.

The formula

$$y = -0.001618x^4 + 0.077326x^3 - 1.2367x^2 + 11.460x + 2.914$$

models the age in human years, y, of a dog that is x years old, where x > 1. The coefficients make it difficult to use this formula when computing by hand. However, a graph of the formula makes approximations possible. Use this information to solve Exercises 103–106.

Dog's Age in Human Years

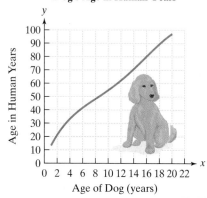

Age of Dog (years)

Source: U.C. Davis

103. a. If your dog is 6 years old, use the graph to estimate the equivalent age in human years.   42 human years

b. Use the given mathematical model to verify your estimate in part (a). Use a calculator and round to the nearest tenth of a human year.   41.8 human years

104. a. If your dog is 16 years old, use the graph to estimate the equivalent age in human years.   80 human years

b. Use the given mathematical model to verify your estimate in part (a). Use a calculator and round to the nearest tenth of a human year.   80.4 human years

105. If you are 25, use the graph to find the equivalent age for dogs.   3 dog years

106. If you are 45, use the graph to find the equivalent age for dogs.   7 dog years

Writing in Mathematics

107. What is a polynomial?   107–115. Answers will vary.

108. What is a monomial? Give an example with your explanation.

109. What is a binomial? Give an example with your explanation.

110. What is a trinomial? Give an example with your explanation.

111. What is the degree of a polynomial? Provide an example with your explanation.

112. Explain how to add polynomials.

113. Explain how to subtract polynomials.

114. A friend who is blind is having difficulty visualizing the relationship between age and deaths per thousand. Describe this relationship for your friend as age increases from 40 through 60. Use Figure 5.1 on page 311.

115. For Exercise 98, explain why performance levels do what they do as we get more and more fired up. If possible, describe an example of a time when you were too enthused and thus did poorly at something when you were hoping to do well.

Critical Thinking Exercises

116. Which one of the following is true?   d

a. In the polynomial $3x^2 - 5x + 13$, the coefficient of x is 5.

b. The degree of $3x^2 - 7x + 9x^3 + 5$ is 2.

c. $\dfrac{1}{5x^2} + \dfrac{1}{3x}$ is a binomial.

d. $(2x^2 - 8x + 6) - (x^2 - 3x + 5) = x^2 - 5x + 1$ for any value of x.

117. What polynomial must be subtracted from $5x^2 - 2x + 1$ so that the difference is $8x^2 - x + 3$? $-3x^2 - x - 2$

118. The number of people who catch a cold t weeks after January 1 is $5t - 3t^2 + t^3$. The number of people who recover t weeks after January 1 is $t - t^2 + \frac{1}{3}t^3$. Write a polynomial in standard form for the number of people who are still ill with a cold t weeks after January 1. $\frac{2}{3}t^3 - 2t^2 + 4t$

119. Explain why it is not possible to add two polynomials of degree 3 and get a polynomial of degree 4.   Answers will vary.

Review Exercises

120. Simplify: $(-10)(-7) \div (1 - 8)$. (Section 1.8, Example 8) -10

121. Subtract: $-4.6 - (-10.2)$. (Section 1.6, Example 2)   5.6

122. Solve: $3(x - 2) = 9(x + 2)$. (Section 2.3, Example 3) $\{-4\}$

MULTIPLYING POLYNOMIALS

Recent advances in our understanding of climate have changed global warming from a subject for a disaster movie (the Statue of Liberty up to its chin in water) to a serious scientific and policy issue. Global warming appears to be related to the burning of fossil fuels, which adds carbon dioxide to the atmosphere. In the next few decades, we will see whether our use of fossil fuels will add enough carbon dioxide to the atmosphere to change it (and our climate) in significant ways. In this section's essay, you will see how a polynomial models trends in global warming through 2040. In the section itself, you will learn to multiply these algebraic expressions that play a significant role in modeling your world.

Before studying how polynomials are multiplied, we must develop some rules for working with exponents.

1 Use the product rule for exponents.

The Product Rule for Exponents We have seen that exponents are used to indicate repeated multiplication. For example, 2^4, where 2 is the base and 4 is the exponent, indicates that 2 occurs as a factor four times:

$$2^4 = 2 \cdot 2 \cdot 2 \cdot 2.$$

Now consider the multiplication of two exponential expressions, such as $2^4 \cdot 2^3$. We are multiplying 4 factors of 2 and 3 factors of 2. We have a total of 7 factors of 2:

4 factors of 2   3 factors of 2

$$2^4 \cdot 2^3 = (2 \cdot 2 \cdot 2 \cdot 2) \cdot (2 \cdot 2 \cdot 2)$$

Total: 7 factors of 2

Thus, $\qquad 2^4 \cdot 2^3 = 2^7.$ Caution: $2^4 \cdot 2^3$ is not equal to $2^{4 \cdot 3}$, or 2^{12}, as might be expected.

We can quickly find the exponent, 7, of the product by adding 4 and 3, the original exponents:

$$2^4 \cdot 2^3 = 2^{4+3} = 2^7.$$

This suggests the following rule:

THE PRODUCT RULE

$$b^m \cdot b^n = b^{m+n}$$

When multiplying exponential expressions with the same base, add the exponents. Use this sum as the exponent of the common base.

EXAMPLE 1   Using the Product Rule

Multiply each expression using the product rule:

a. $2^2 \cdot 2^3$ **b.** $x^7 \cdot x^9$ **c.** $y \cdot y^5$ **d.** $y^3 \cdot y^2 \cdot y^5$.

SOLUTION

a. $2^2 \cdot 2^3 = 2^{2+3} = 2^5$ or 32

b. $x^7 \cdot x^9 = x^{7+9} = x^{16}$

c. $y \cdot y^5 = y^1 \cdot y^5 = y^{1+5} = y^6$

d. $y^3 \cdot y^2 \cdot y^5 = y^{3+2+5} = y^{10}$

■

CHECK POINT 1 Multiply each expression using the product rule:

a. $2^2 \cdot 2^4$ 2^6 or 64   **b.** $x^6 \cdot x^4$ x^{10} **c.** $y \cdot y^7$ y^8 **d.** $y^4 \cdot y^3 \cdot y^2$. y^9

2   Use the power rule for exponents.

The Power Rule for Exponents   The next property of exponents applies when an exponential expression is raised to a power. Here is an example:

$$(3^2)^4.$$

> The exponential expression 3^2 is raised to the fourth power.

There are 4 factors 3^2. Thus,

$$(3^2)^4 = 3^2 \cdot 3^2 \cdot 3^2 \cdot 3^2 = 3^{2+2+2+2} = 3^8.$$

> Add exponents when multiplying with the same base.

We can obtain the answer, 3^8, by multiplying the exponents:

$$(3^2)^4 = 3^{2 \cdot 4} = 3^8.$$

This suggests the following rule:

THE POWER RULE (POWERS TO POWERS)

$$(b^m)^n = b^{mn}$$

When an exponential expression is raised to a power, multiply the exponents. Place the product of the exponents on the base and remove the parentheses.

EXAMPLE 2   Using the Power Rule

Simplify each expression using the power rule:

a. $(2^3)^5$ **b.** $(x^6)^4$ **c.** $[(-3)^7]^5$.

3 Use the products-to-powers rule.

SOLUTION

a. $(2^3)^5 = 2^{3 \cdot 5} = 2^{15}$

b. $(x^6)^4 = x^{6 \cdot 4} = x^{24}$

c. $\left[(-3)^7\right]^5 = (-3)^{7 \cdot 5} = (-3)^{35}$ ∎

✔ **CHECK POINT 2** Simplify each expression using the power rule:

a. $(3^4)^5$ 3^{20} **b.** $(x^9)^{10}$ x^{90} **c.** $\left[(-5)^7\right]^3.$ $(-5)^{21}$

The Products-to-Powers Rule for Exponents   The next property of exponents applies when we are raising a product to a power. Here is an example:

$$(2x)^4.$$

The product $2x$ is raised to the fourth power.

There are four factors of $2x$. Thus,

$$(2x)^4 = 2x \cdot 2x \cdot 2x \cdot 2x = 2 \cdot 2 \cdot 2 \cdot 2 \cdot x \cdot x \cdot x \cdot x = 2^4 x^4.$$

We can obtain the answer, $2^4 x^4$, by raising each factor within the parentheses to the fourth power:

$$(2x)^4 = 2^4 x^4.$$

This suggests the following rule:

PRODUCTS TO POWERS

$$(ab)^n = a^n b^n$$

When a product is raised to a power, raise each factor to the power.

EXAMPLE 3   Using the Products-to-Powers Rule

Simplify each expression using the products-to-powers rule:

a. $(5x)^3$ **b.** $\left(-2y^4\right)^5.$

SOLUTION

a. $(5x)^3 = 5^3 x^3$ Raise each factor to the third power.

$= 125x^3$ $5^3 = 5 \cdot 5 \cdot 5 = 125$

b. $\left(-2y^4\right)^5 = (-2)^5(y^4)^5$ Raise each factor to the fifth power.

$= (-2)^5 y^{4 \cdot 5}$ To raise an exponential expression to a power, multiply exponents: $(b^m)^n = b^{mn}$.

$= -32y^{20}$ $(-2)^5 = (-2)(-2)(-2)(-2)(-2) = -32$ ∎

✔ **CHECK POINT 3** Simplify each expression using the products-to-powers rule:

a. $(2x)^4$ $16x^4$ **b.** $\left(-4y^2\right)^3.$ $-64y^6$

STUDY TIP

Try to avoid the following common errors that can occur when simplifying exponential expressions.

Correct	Incorrect	Description of Error
$b^3 \cdot b^4 = b^{3+4} = b^7$	$b^3 \cdot b^4 = b^{12}$	Exponents should be added, not multiplied.
$3^2 \cdot 3^4 = 3^{2+4} = 3^6$	$3^2 \cdot 3^4 = 9^{2+4} = 9^6$	The common base should be retained, not multiplied.
$(x^5)^3 = x^{5 \cdot 3} = x^{15}$	$(x^5)^3 = x^{5+3} = x^8$	Exponents should be multiplied, not added, when raising a power to a power.
$(4x)^3 = 4^3 x^3 = 64x^3$	$(4x)^3 = 4x^3$	Both factors should be cubed.

4 Multiply monomials.

Multiplying Monomials Now that we have developed three properties of exponents, we are ready to turn to polynomial multiplication. We begin with the product of two monomials, such as $-8x^6$ and $5x^3$. This product is obtained by multiplying the coefficients, -8 and 5, and then multiplying the variables by using the product rule for exponents.

$$(-8x^6)(5x^3) = -8 \cdot 5 \cdot x^6 \cdot x^3 = -8 \cdot 5x^{6+3} = -40x^9$$

Multiply coefficients and add exponents.

MULTIPLYING MONOMIALS To multiply monomials with the same variable base, multiply the coefficients and then multiply the variables. Use the product rule for exponents to multiply the variables: Keep the variable and add the exponents.

STUDY TIP

Don't confuse adding and multiplying monomials.

Addition:
$$5x^4 + 6x^4 = 11x^4$$

Multiplication:
$$\begin{aligned}(5x^4)(6x^4) &= (5 \cdot 6)(x^4 \cdot x^4) \\ &= 30x^{4+4} \\ &= 30x^8\end{aligned}$$

Only like terms can be added or subtracted, but unlike terms may be multiplied.

Addition:
$5x^4 + 3x^2$ cannot be simplified.

Multiplication:
$$\begin{aligned}(5x^4)(3x^2) &= (5 \cdot 3)(x^4 \cdot x^2) \\ &= 15x^{4+2} \\ &= 15x^6\end{aligned}$$

EXAMPLE 4 Multiplying Monomials

Multiply: **a.** $(2x)(4x^2)$ **b.** $(-10x^6)(6x^{10})$.

SOLUTION

a. $(2x)(4x^2) = (2 \cdot 4)(x \cdot x^2)$ Multiply the coefficients and multiply the variables.

$\qquad = 8x^{1+2}$ Add exponents: $b^m \cdot b^n = b^{m+n}$.

$\qquad = 8x^3$ Simplify.

b. $(-10x^6)(6x^{10}) = (-10 \cdot 6)(x^6 \cdot x^{10})$ Multiply the coefficients and multiply the variables.

$\qquad = -60x^{6+10}$ Add exponents: $b^m \cdot b^n = b^{m+n}$.

$\qquad = -60x^{16}$ Simplify. ∎

✔ **CHECK POINT 4** Multiply: **a.** $(7x^2)(10x)$ $70x^3$ **b.** $(-5x^4)(4x^5)$. $-20x^9$

5 Multiply a monomial and a polynomial.

Multiplying a Monomial and a Polynomial That Is Not a Monomial We use the distributive property to multiply a monomial and a polynomial that is not a monomial. For example,

$$3x^2(2x^3 + 5x) = 3x^2 \cdot 2x^3 + 3x^2 \cdot 5x = 3 \cdot 2x^{2+3} + 3 \cdot 5x^{2+1} = 6x^5 + 15x^3.$$

Monomial   Binomial   Multiply coefficients and add exponents.

> **MULTIPLYING A MONOMIAL AND A POLYNOMIAL THAT IS NOT A MONOMIAL** To multiply a monomial and a polynomial, use the distributive property to multiply each term of the polynomial by the monomial.

EXAMPLE 5 Multiplying a Monomial and a Polynomial

Multiply: **a.** $2x(x + 4)$ **b.** $3x^2(4x^3 - 5x + 2)$.

SOLUTION

a. $2x(x + 4) = 2x \cdot x + 2x \cdot 4$ .... Use the distributive property.

$\qquad = 2 \cdot 1x^{1+1} + 2 \cdot 4x$ .... To multiply the monomials, multiply coefficients and add exponents.

$\qquad = 2x^2 + 8x$ .... Simplify.

b. $3x^2(4x^3 - 5x + 2)$

$\qquad = 3x^2 \cdot 4x^3 - 3x^2 \cdot 5x + 3x^2 \cdot 2$ .... Use the distributive property.

$\qquad = 3 \cdot 4x^{2+3} - 3 \cdot 5x^{2+1} + 3 \cdot 2x^2$ .... To multiply the monomials, multiply coefficients and add exponents.

$\qquad = 12x^5 - 15x^3 + 6x^2$ .... Simplify.

Rectangles often make it possible to visualize polynomial multiplication. For example, Figure 5.2 shows a rectangle with length $2x$ and width $x + 4$. The area of the large rectangle is

$$2x(x + 4).$$

The sum of the areas of the two smaller rectangles is

$$2x^2 + 8x.$$

Conclusion:

$$2x(x + 4) = 2x^2 + 8x.$$

FIGURE 5.2

CHECK POINT 5 Multiply:

a. $3x(x + 5)$ $3x^2 + 15x$ **b.** $6x^2(5x^3 - 2x + 3)$. $30x^5 - 12x^3 + 18x^2$

6 Multiply polynomials when neither is a monomial.

Multiplying Polynomials When Neither Is a Monomial How do we multiply two polynomials if neither is a monomial? For example, consider

$$(2x + 3)(x^2 + 4x + 5).$$

Binomial · Trinomial

One way to perform this multiplication is to distribute $2x$ throughout the trinomial

$$2x(x^2 + 4x + 5)$$

and 3 throughout the trinomial

$$3(x^2 + 4x + 5).$$

Then combine the like terms that result. In general, the product of two polynomials is the polynomial obtained by multiplying each term of one polynomial by each term of the other polynomial and then combining like terms.

> **MULTIPLYING POLYNOMIALS WHEN NEITHER IS A MONOMIAL** Multiply each term of one polynomial by each term of the other polynomial. Then combine like terms.

EXAMPLE 6 Multiplying Binomials

Multiply: **a.** $(x + 3)(x + 2)$ **b.** $(3x + 7)(2x - 4)$.

SOLUTION We begin by multiplying each term of the second binomial by each term of the first binomial.

a. $(x + 3)(x + 2)$

$= x(x + 2) + 3(x + 2)$ Multiply the second binomial by each term of the first binomial.

$= x \cdot x + x \cdot 2 + 3 \cdot x + 3 \cdot 2$ Use the distributive property.

$= x^2 + 2x + 3x + 6$ Multiply. Note that $x \cdot x = x^1 \cdot x^1 = x^{1+1} = x^2$.

$= x^2 + 5x + 6$ Combine like terms.

b. $(3x + 7)(2x - 4)$

$= 3x(2x - 4) + 7(2x - 4)$ Multiply the second binomial by each term of the first binomial.

$= 3x \cdot 2x - 3x \cdot 4 + 7 \cdot 2x - 7 \cdot 4$ Use the distributive property.

$= 6x^2 - 12x + 14x - 28$ Multiply.

$= 6x^2 + 2x - 28$ Combine like terms. ∎

STUDY TIP

You can visualize the polynomial multiplication in Example 6(a), using the rectangle with dimensions $x + 3$ and $x + 2$.

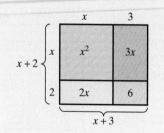

Area of large rectangle
$= (x + 3)(x + 2)$

Sum of areas of smaller rectangles
$= x^2 + 3x + 2x + 6$
$= x^2 + 5x + 6$

Conclusion:
$(x + 3)(x + 2) = x^2 + 5x + 6.$

✓ **CHECK POINT 6** Multiply:

a. $(x + 4)(x + 5)$ $x^2 + 9x + 20$ **b.** $(5x + 3)(2x - 7)$. $10x^2 - 29x - 21$

USING TECHNOLOGY

A graphing utility can be used to see if a polynomial operation has been performed correctly. For example, to check

$$(x + 3)(x + 2) = x^2 + 5x + 6,$$

graph the left and right sides on the same screen, using

$$y_1 = (x + 3)(x + 2)$$

and

$$y_2 = x^2 + 5x + 6.$$

$y_1 = (x + 3)(x + 2)$

$y_2 = x^2 + 5x + 6$

$[-6, 2, 1]$ by $[-1, 10, 1]$

As shown in the figure, both graphs are the same, verifying that the binomial multiplication was performed correctly.

EXAMPLE 7 Multiplying a Binomial and a Trinomial

Multiply: $(2x + 3)(x^2 + 4x + 5)$.

SOLUTION

$(2x + 3)(x^2 + 4x + 5)$

$= 2x(x^2 + 4x + 5) + 3(x^2 + 4x + 5)$ Multiply the trinomial by each term of the binomial.

$= 2x \cdot x^2 + 2x \cdot 4x + 2x \cdot 5 + 3x^2 + 3 \cdot 4x + 3 \cdot 5$ Use the distributive property.

$= 2x^3 + 8x^2 + 10x + 3x^2 + 12x + 15$ Multiply monomials: Multiply coefficients and add exponents.

$= 2x^3 + 11x^2 + 22x + 15$ Combine like terms:
$8x^2 + 3x^2 = 11x^2$ and
$10x + 12x = 22x$.

✔ **CHECK POINT 7** Multiply: $(5x + 2)(x^2 - 4x + 3)$. $5x^3 - 18x^2 + 7x + 6$

Another method for solving Example 7 is to use a vertical format similar to that used for multiplying whole numbers.

$$
\begin{array}{r}
x^2 + 4x + 5 \\
2x + 3 \\
\hline
3x^2 + 12x + 15 \\
2x^3 + 8x^2 + 10x \\
\hline
2x^3 + 11x^2 + 22x + 15
\end{array}
$$

Write like terms in the same column.

$3(x^2 + 4x + 5)$

$2x(x^2 + 4x + 5)$

Combine like terms.

EXAMPLE 8 Multiplying Polynomials Using a Vertical Format

Multiply: $(2x^2 - 3x)(5x^3 - 4x^2 + 7x)$.

SOLUTION To use the vertical format, it is most convenient to write the polynomial with the greatest number of terms in the top row.

$$
\begin{array}{r}
5x^3 - 4x^2 + 7x \\
2x^2 - 3x \\
\hline
\end{array}
$$

We now multiply each term in the top polynomial by the last term in the bottom polynomial.

$$
\begin{array}{r}
5x^3 - 4x^2 + 7x \\
2x^2 - 3x \\
\hline
-15x^4 + 12x^3 - 21x^2
\end{array}
$$
$\quad$ $-3x(5x^3 - 4x^2 + 7x)$

Then we multiply each term in the top polynomial by $2x^2$, the first term in the bottom polynomial. Like terms are placed in columns because the final step involves combining them.

Write like terms in the same column.

$$
\begin{array}{r}
5x^3 - 4x^2 + 7x \\
2x^2 - 3x \\
\hline
-15x^4 + 12x^3 - 21x^2 \\
10x^5 - 8x^4 + 14x^3 \\
\hline
10x^5 - 23x^4 + 26x^3 - 21x^2
\end{array}
$$

$-3x(5x^3 - 4x^2 + 7x)$

$2x^2(5x^3 - 4x^2 + 7x)$

Combine like terms, which are lined up in columns.

$6x^5 - 19x^4 + 22x^3 - 8x^2$

✔ **CHECK POINT 8** Multiply using a vertical format: $(3x^2 - 2x)(2x^3 - 5x^2 + 4x)$.

ENRICHMENT ESSAY

Is It Hot in Here Or Is It Just Me?

This is the year 2040.

Increase in Global Surface Temperature (°C)

Years after 1980

In the 1980s, a rising trend in global surface temperature was observed and the term "global warming" was coined. Scientists are more convinced than ever that burning coal, oil, and gas results in a buildup of gases and particles that trap heat and raise the planet's temperature. The average increase in global surface temperature, y, in degrees Celsius, x years after 1980 can be modeled by the polynomial formula

$$
y = \frac{21}{5,000,000}x^3 - \frac{127}{1,000,000}x^2 + \frac{1293}{50,000}x.
$$

The graph of this formula is shown above in a $[0, 60, 3]$ by $[0, 2, 0.1]$ viewing rectangle. The graph illustrates that the model predicts global warming will increase through the year 2040. Furthermore, the increasing steepness of the curve shows that global warming will increase at greater rates near the middle of the twenty-first century.

5.2 EXERCISE SET

Student Solutions Manual   CD/Video   PH Math/Tutor Center   MathXL Tutorials on CD   MathXL®   MyMathLab   Interactmath.com

Practice Exercises

In Exercises 1–8, multiply each expression using the product rule.

1. $x^{15} \cdot x^3$ x^{18} **2.** $x^{12} \cdot x^4$ x^{16}

3. $y \cdot y^{11}$ y^{12} **4.** $y \cdot y^{19}$ y^{20}

5. $x^2 \cdot x^6 \cdot x^3$ x^{11} **6.** $x^4 \cdot x^3 \cdot x^5$ x^{12}

7. $7^9 \cdot 7^{10}$ 7^{19} **8.** $8^7 \cdot 8^{10}$ 8^{17}

In Exercises 9–14, simplify each expression using the power rule.

9. $(6^9)^{10}$ 6^{90} **10.** $(6^7)^{10}$ 6^{70}

11. $(x^{15})^3$ x^{45} **12.** $(x^{12})^4$ x^{48}

13. $[(-20)^3]^3$ $(-20)^9$ **14.** $[(-50)^4]^4$ $(-50)^{16}$

In Exercises 15–24, simplify each expression using the products-to-powers rule.

15. $(2x)^3$ $8x^3$ **16.** $(4x)^3$ $64x^3$

17. $(-5x)^2$ $25x^2$ **18.** $(-6x)^2$ $36x^2$

19. $(4x^3)^2$ $16x^6$ **20.** $(6x^3)^2$ $36x^6$

21. $(-2y^6)^4$ $16y^{24}$ **22.** $(-2y^5)^4$ $16y^{20}$

23. $(-2x^7)^5$ $-32x^{35}$ **24.** $(-2x^{11})^5$ $-32x^{55}$

In Exercises 25–34, multiply the monomials.

25. $(7x)(2x)$ $14x^2$ **26.** $(8x)(3x)$ $24x^2$

27. $(6x)(4x^2)$ $24x^3$ **28.** $(10x)(3x^2)$ $30x^3$

29. $(-5y^4)(3y^3)$ $-15y^7$ **30.** $(-6y^4)(2y^3)$ $-12y^7$

31. $\left(-\dfrac{1}{2}a^3\right)\left(-\dfrac{1}{4}a^2\right)$ $\dfrac{1}{8}a^5$ **32.** $\left(-\dfrac{1}{3}a^4\right)\left(-\dfrac{1}{2}a^2\right)$ $\dfrac{1}{6}a^6$

33. $(2x^2)(-3x)(8x^4)$ $-48x^7$ **34.** $(3x^3)(-2x)(5x^6)$ $-30x^{10}$

In Exercises 35–54, find each product of the monomial and the polynomial.

35. $4x(x + 3)$ $4x^2 + 12x$ **36.** $6x(x + 5)$ $6x^2 + 30x$

37. $x(x - 3)$ $x^2 - 3x$ **38.** $x(x - 7)$ $x^2 - 7x$

39. $2x(x - 6)$ $2x^2 - 12x$ **40.** $3x(x - 5)$ $3x^2 - 15x$

41. $-4y(3y + 5)$ $-12y^2 - 20y$

42. $-5y(6y + 7)$ $-30y^2 - 35y$

43. $4x^2(x + 2)$ $4x^3 + 8x^2$ **44.** $5x^2(x + 6)$ $5x^3 + 30x^2$

45. $2y^2(y^2 + 3y)$ $2y^4 + 6y^3$ **46.** $4y^2(y^2 + 2y)$ $4y^4 + 8y^3$

47. $2y^2(3y^2 - 4y + 7)$ $6y^4 - 8y^3 + 14y^2$

48. $4y^2(5y^2 - 6y + 3)$ $20y^4 - 24y^3 + 12y^2$

49. $(3x^3 + 4x^2)(2x)$ $6x^4 + 8x^3$

50. $(4x^3 + 5x^2)(2x)$ $8x^4 + 10x^3$

51. $(x^2 + 5x - 3)(-2x)$ $-2x^3 - 10x^2 + 6x$

52. $(x^3 - 2x + 2)(-4x)$ $-4x^4 + 8x^2 - 8x$

53. $-3x^2(-4x^2 + x - 5)$ $12x^4 - 3x^3 + 15x^2$

54. $-6x^2(3x^2 - 2x - 7)$ $-18x^4 + 12x^3 + 42x^2$

In Exercises 55–78, find each product. In each case, neither factor is a monomial.

55. $(x + 3)(x + 5)$ $x^2 + 8x + 15$

56. $(x + 4)(x + 6)$ $x^2 + 10x + 24$

57. $(2x + 1)(x + 4)$ $2x^2 + 9x + 4$

58. $(2x + 5)(x + 3)$ $2x^2 + 11x + 15$

59. $(x + 3)(x - 5)$ $x^2 - 2x - 15$

60. $(x + 4)(x - 6)$ $x^2 - 2x - 24$

61. $(x - 11)(x + 9)$ $x^2 - 2x - 99$

62. $(x - 12)(x + 8)$ $x^2 - 4x - 96$

63. $(2x - 5)(x + 4)$ $2x^2 + 3x - 20$

64. $(3x - 4)(x + 5)$ $3x^2 + 11x - 20$

65. $\left(\dfrac{1}{4}x + 4\right)\left(\dfrac{3}{4}x - 1\right)$ $\dfrac{3}{16}x^2 + \dfrac{11}{4}x - 4$

66. $\left(\dfrac{1}{5}x + 5\right)\left(\dfrac{3}{5}x - 1\right)$ $\dfrac{3}{25}x^2 + \dfrac{14}{5}x - 5$

67. $(x + 1)(x^2 + 2x + 3)$ $x^3 + 3x^2 + 5x + 3$

68. $(x + 2)(x^2 + x + 5)$ $x^3 + 3x^2 + 7x + 10$

69. $(y - 3)(y^2 - 3y + 4)$ $y^3 - 6y^2 + 13y - 12$

70. $(y - 2)(y^2 - 4y + 3)$ $y^3 - 6y^2 + 11y - 6$

71. $(2a - 3)(a^2 - 3a + 5)$ $2a^3 - 9a^2 + 19a - 15$

72. $(2a - 1)(a^2 - 4a + 3)$ $2a^3 - 9a^2 + 10a - 3$

73. $(x + 1)(x^3 + 2x^2 + 3x + 4)$ $x^4 + 3x^3 + 5x^2 + 7x + 4$

74. $(x + 1)(x^3 + 4x^2 + 7x + 3)$ $x^4 + 5x^3 + 11x^2 + 10x + 3$

75. $\left(x - \dfrac{1}{2}\right)(4x^3 - 2x^2 + 5x - 6)$ $4x^4 - 4x^3 + 6x^2 - \dfrac{17}{2}x + 3$

76. $\left(x - \dfrac{1}{3}\right)(3x^3 - 6x^2 + 5x - 9)$ $3x^4 - 7x^3 + 7x^2 - \dfrac{32}{3}x + 3$

77. $(x^2 + 2x + 1)(x^2 - x + 2)$ $x^4 + x^3 + x^2 + 3x + 2$

78. $(x^2 + 3x + 1)(x^2 - 2x - 1)$ $x^4 + x^3 - 6x^2 - 5x - 1$

In Exercises 79–92, use a vertical format to find each product.

79. $x^2 - 5x + 3$
        $x + 8$ $x^3 + 3x^2 - 37x + 24$

80. $x^2 - 7x + 9$
        $x + 4$ $x^3 - 3x^2 - 19x + 36$

81. $x^2 - 3x + 9$
      $2x - 3$ $2x^3 - 9x^2 + 27x - 27$

82. $y^2 - 5y + 3$
$\underline{\qquad 4y - 5 \qquad}$ $4y^3 - 25y^2 + 37y - 15$

83. $2x^3 + x^2 + 2x + 3$
$\underline{\qquad\qquad x + 4 \qquad}$ $2x^4 + 9x^3 + 6x^2 + 11x + 12$

84. $3y^3 + 2y^2 + y + 4$
$\underline{\qquad\qquad y + 3 \qquad}$ $3y^4 + 11y^3 + 7y^2 + 7y + 12$

85. $4z^3 - 2z^2 + 5z - 4$
$\underline{\qquad\qquad 3z - 2 \qquad}$ $12z^4 - 14z^3 + 19z^2 - 22z + 8$

86. $5z^3 - 3z^2 + 4z - 3$
$\underline{\qquad\qquad 2z - 4 \qquad}$ $10z^4 - 26z^3 + 20z^2 - 22z + 12$

87. $7x^3 - 5x^2 + 6x$
$\underline{\qquad\qquad 3x^2 - 4x \qquad}$ $21x^5 - 43x^4 + 38x^3 - 24x^2$

88. $9y^3 - 7y^2 + 5y$
$\underline{\qquad\qquad 3y^2 + 5y \qquad}$ $27y^5 + 24y^4 - 20y^3 + 25y^2$

89. $2y^5 - 3y^3 + y^2 - 2y + 3$
$\underline{\qquad\qquad 2y - 1 \qquad}$
$4y^6 - 2y^5 - 6y^4 + 5y^3 - 5y^2 + 8y - 3$

90. $n^4 - n^3 + n^2 - n + 1$
$\underline{\qquad\qquad 2n + 3 \qquad}$ $2n^5 + n^4 - n^3 + n^2 - n + 3$

91. $x^2 + 7x - 3$
$\underline{\; x^2 - x - 1 \;}$ $x^4 + 6x^3 - 11x^2 - 4x + 3$

92. $x^2 + 6x - 4$
$\underline{\; x^2 - x - 2 \;}$ $x^4 + 5x^3 - 12x^2 - 8x + 8$

Practice Plus

In Exercises 93–100, perform the indicated operations.

93. $(x + 4)(x - 5) - (x + 3)(x - 6)$ $2x - 2$

94. $(x + 5)(x - 6) - (x + 2)(x - 9)$ $6x - 12$

95. $4x^2(5x^3 + 3x - 2) - 5x^3(x^2 - 6)$ $15x^5 + 42x^3 - 8x^2$

96. $3x^2(6x^3 + 2x - 3) - 4x^3(x^2 - 5)$ $14x^5 + 26x^3 - 9x^2$

97. $(y + 1)(y^2 - y + 1) + (y - 1)(y^2 + y + 1)$ $2y^3$

98. $(y + 1)(y^2 - y + 1) - (y - 1)(y^2 + y + 1)$ 2

99. $(y + 6)^2 - (y - 2)^2$ $16y + 32$

100. $(y + 5)^2 - (y - 4)^2$ $18y + 9$

Application Exercises

101. Find a trinomial for the area of the rectangular rug shown here whose sides are $x + 5$ feet and $2x - 3$ feet:

$2x^2 + 7x - 15$ square feet

$x + 5$

$2x - 3$

102. The base of a triangular sail is $4x$ feet and its height is $3x + 10$ feet. Write a binomial in terms of x for the area of the sail. $6x^2 + 20x$ square feet

In Exercises 103–104,

a. *Express the area of the large rectangle as the product of two binomials.*

b. *Find the sum of the areas of the four smaller rectangles.*

c. *Use polynomial multiplication to show that your expressions for area in parts (a) and (b) are equal.*

103.

a. $(2x + 1)(x + 2)$
b. $2x^2 + 5x + 2$
c. $(2x + 1)(x + 2) = 2x^2 + 5x + 2$

104.

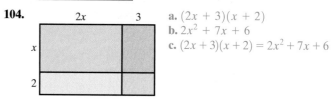

a. $(2x + 3)(x + 2)$
b. $2x^2 + 7x + 6$
c. $(2x + 3)(x + 2) = 2x^2 + 7x + 6$

Writing in Mathematics

105. Explain the product rule for exponents. Use $2^3 \cdot 2^5$ in your explanation.   **105–113.** Answers will vary.

106. Explain the power rule for exponents. Use $(3^2)^4$ in your explanation.

107. Explain how to simplify an expression that involves a product raised to a power. Provide an example with your explanation.

108. Explain how to multiply monomials. Give an example.

109. Explain how to multiply a monomial and a polynomial that is not a monomial. Give an example.

110. Explain how to multiply polynomials when neither is a monomial. Give an example.

111. Explain the difference between performing these two operations:

$$2x^2 + 3x^2 \quad \text{and} \quad (2x^2)(3x^2).$$

112. Discuss situations in which a vertical format, rather than a horizontal format, is useful for multiplying polynomials.

113. Describe one change that might alter the prediction about global warming given by the model and graph on page 322.

Critical Thinking Exercises

114. Which one of the following is true?　c

    a. $4x^3 \cdot 3x^4 = 12x^{12}$

    b. $5x^2 \cdot 4x^6 = 9x^8$

    c. $(y - 1)(y^2 + y + 1) = y^3 - 1$

    d. Some polynomial multiplications can only be performed by using a vertical format.

115. Find a polynomial in descending powers of x representing the area of the shaded region.　$8x + 16$

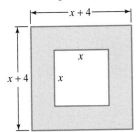

116. Find each of the products in parts (a)–(c).

    a. $(x - 1)(x + 1)$　$x^2 - 1$

    b. $(x - 1)(x^2 + x + 1)$　$x^3 - 1$

    c. $(x - 1)(x^3 + x^2 + x + 1)$　$x^4 - 1$

    d. Using the pattern found in parts (a)–(c), find $(x - 1)(x^4 + x^3 + x^2 + x + 1)$ without actually multiplying.　$x^5 - 1$

117. Find the missing factor.

$$(\underline{\quad})\left(-\frac{1}{4}xy^3\right) = 2x^5y^3 \qquad -8x^4$$

Review Exercises

118. Solve: $4x - 7 > 9x - 2$. (Section 2.7, Example 6)
$\{x \mid x < -1\}$

119. Graph $3x - 2y = 6$ using intercepts. (Section 3.2, Example 4)　See graphing answer section.

120. Find the slope of the line passing through the points $(-2, 8)$ and $(1, 6)$. (Section 3.3, Example 1)　$m = -\frac{2}{3}$

SPECIAL PRODUCTS

Objectives

1 Use FOIL in polynomial multiplication.

2 Multiply the sum and difference of two terms.

3 Find the square of a binomial sum.

4 Find the square of a binomial difference.

Let's cut to the chase. Are there fast methods for finding products of polynomials? Yes. In this section, we use the distributive property to develop patterns that will let you multiply certain binomials quite rapidly.

1 Use FOIL in polynomial multiplication.

The Product of Two Binomials: FOIL Frequently, we need to find the product of two binomials. One way to perform this multiplication is to distribute each term in the first binomial through the second binomial. For example, we can find the product of the binomials $3x + 2$ and $4x + 5$ as follows:

$$(3x + 2)(4x + 5) = 3x(4x + 5) + 2(4x + 5)$$

    Distribute $3x$ over $4x + 5$.   Distribute 2 over $4x + 5$.

$$= 3x(4x) + 3x(5) + 2(4x) + 2(5)$$

$$= 12x^2 + 15x + 8x + 10.$$

We can also find the product of $3x + 2$ and $4x + 5$ using a method called FOIL, which is based on our work shown above. Any two binomials can be quickly multiplied by using the FOIL method, in which **F** represents the product of the **first** terms in each binomial, **O** represents the product of the **outside** terms, **I** represents the product of the **inside** terms, and **L** represents the product of the **last**, or second, terms in each binomial. For example, we can use the FOIL method to find the product of the binomials $3x + 2$ and $4x + 5$ as follows:

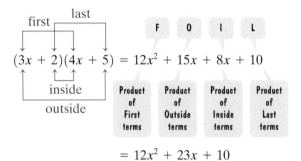

$$= 12x^2 + 23x + 10 \qquad \text{Combine like terms.}$$

In general, here's how to use the FOIL method to find the product of $ax + b$ and $cx + d$:

USING THE FOIL METHOD TO MULTIPLY BINOMIALS

EXAMPLE 1 Using the FOIL Method

Multiply: $(x + 3)(x + 4)$.

SOLUTION

$$\text{F: First terms} \quad = x \cdot x = x^2 \quad (x + 3)(x + 4)$$

$$\text{O: Outside terms} = x \cdot 4 = 4x \quad (x + 3)(x + 4)$$

$$\text{I: Inside terms} \quad = 3 \cdot x = 3x \quad (x + 3)(x + 4)$$

$$\text{L: Last terms} \quad = 3 \cdot 4 = 12 \quad (x + 3)(x + 4)$$

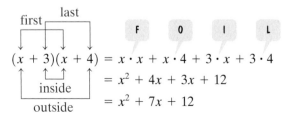

$$= x^2 + 4x + 3x + 12$$
$$= x^2 + 7x + 12 \qquad \text{Combine like terms.}$$

CHECK POINT **1** Multiply: $(x + 5)(x + 6)$. $x^2 + 11x + 30$

EXAMPLE 2 Using the FOIL Method

Multiply: $(3x + 4)(5x - 3)$.

SOLUTION

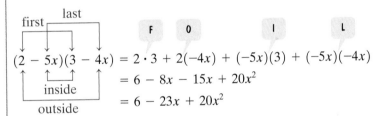

$$(3x + 4)(5x - 3) = 3x \cdot 5x + 3x(-3) + 4 \cdot 5x + 4(-3)$$
$$= 15x^2 - 9x + 20x - 12$$
$$= 15x^2 + 11x - 12 \qquad \text{Combine like terms.} \ \blacksquare$$

 CHECK POINT **2** Multiply: $(7x + 5)(4x - 3)$. $28x^2 - x - 15$

EXAMPLE 3 Using the FOIL Method

Multiply: $(2 - 5x)(3 - 4x)$.

SOLUTION

$$(2 - 5x)(3 - 4x) = 2 \cdot 3 + 2(-4x) + (-5x)(3) + (-5x)(-4x)$$
$$= 6 - 8x - 15x + 20x^2$$
$$= 6 - 23x + 20x^2 \qquad \text{Combine like terms.}$$

The product can also be expressed in standard form as $20x^2 - 23x + 6$. $\blacksquare$

 CHECK POINT **3** Multiply: $(4 - 2x)(5 - 3x)$. $6x^2 - 22x + 20$

2 Multiply the sum and difference of two terms.

Multiplying the Sum and Difference of Two Terms We can use the FOIL method to multiply $A + B$ and $A - B$ as follows:

$$(A + B)(A - B) = A^2 - AB + AB - B^2 = A^2 - B^2.$$

Notice that the outside and inside products have a sum of 0 and the terms cancel. The FOIL multiplication provides us with a quick rule for multiplying the sum and difference of two terms, referred to as a special-product formula.

THE PRODUCT OF THE SUM AND DIFFERENCE OF TWO TERMS

$$(A + B)(A - B) = A^2 - B^2$$

| The product of the sum and the difference of the same two terms | is | the square of the first term minus the square of the second term. |

EXAMPLE 4 Finding the Product of the Sum and Difference of Two Terms

Find each product by using the preceding rule:

a. $(4y + 3)(4y - 3)$ b. $(3x - 7)(3x + 7)$ c. $(5a^4 + 6)(5a^4 - 6)$.

SOLUTION Use the special-product formula shown.

$$(A + B)(A - B) \quad = \quad A^2 \quad - \quad B^2$$

First term squared $-$ Second term squared $=$ Product

a. $(4y + 3)(4y - 3) \quad = \quad (4y)^2 \quad - \quad 3^2 \quad = \quad 16y^2 - 9$
b. $(3x - 7)(3x + 7) \quad = \quad (3x)^2 \quad - \quad 7^2 \quad = \quad 9x^2 - 49$
c. $(5a^4 + 6)(5a^4 - 6) \quad = \quad (5a^4)^2 \quad - \quad 6^2 \quad = \quad 25a^8 - 36$

■

✔ **CHECK POINT 4** Find each product:

a. $(7y + 8)(7y - 8)$ $49y^2 - 64$ b. $(4x - 5)(4x + 5)$ $16x^2 - 25$
c. $(2a^3 + 3)(2a^3 - 3)$. $4a^6 - 9$

The Square of a Binomial Let's now find $(A + B)^2$, the square of a binomial sum. To do so, we begin with the FOIL method and look for a general rule.

③ Find the square of a binomial sum.

F   O   I   L

$$(A + B)^2 = (A + B)(A + B) = A \cdot A + A \cdot B + A \cdot B + B \cdot B$$
$$= A^2 + 2AB + B^2$$

This result implies the following rule, which is another example of a special-product formula:

STUDY TIP

Caution! The square of a sum is *not* the sum of the squares.

$(A + B)^2 \neq A^2 + B^2$

The middle term $2AB$ is missing.

$(x + 3)^2 \neq x^2 + 9$

Incorrect!

Show that $(x + 3)^2$ and $x^2 + 9$ are not equal by substituting 5 for x in each expression and simplifying.

THE SQUARE OF A BINOMIAL SUM

$$(A + B)^2 \quad = \quad A^2 \quad + \quad 2AB \quad + \quad B^2$$

The square of a binomial sum   is   first term squared   plus   2 times the product of the terms   plus   last term squared.

EXAMPLE 5 Finding the Square of a Binomial Sum

Square each binomial using the preceding rule:

a. $(x + 3)^2$ b. $(3x + 7)^2$.

SOLUTION Use the special-product formula shown.

$$(A + B)^2 = A^2 + 2AB + B^2$$

	(First Term)^2	+	2 · Product of the Terms	+	(Last Term)^2	= Product
a. $(x + 3)^2 =$	x^2	+	$2 \cdot x \cdot 3$	+	3^2	$= x^2 + 6x + 9$
b. $(3x + 7)^2 =$	$(3x)^2$	+	$2(3x)(7)$	+	7^2	$= 9x^2 + 42x + 49$

■

 CHECK POINT 5 Square each binomial:

a. $(x + 10)^2$ $x^2 + 20x + 100$ **b.** $(5x + 4)^2$. $25x^2 + 40x + 16$

4 Find the square of a binomial difference.

Using the FOIL method on $(A - B)^2$, the square of a binomial difference, we obtain the following rule:

THE SQUARE OF A BINOMIAL DIFFERENCE

$$(A - B)^2 = A^2 - 2AB + B^2$$

| The square of a binomial difference | is | first term squared | minus | 2 times the product of the terms | plus | last term squared. |

EXAMPLE 6 Finding the Square of a Binomial Difference

Square each binomial using the preceding rule:

a. $(x - 4)^2$ **b.** $(5y - 6)^2$.

SOLUTION Use the special-product formula shown.

$$(A - B)^2 = A^2 - 2AB + B^2$$

	(First Term)^2	−	2 · Product of the Terms	+	(Last Term)^2	= Product
a. $(x - 4)^2 =$	x^2	−	$2 \cdot x \cdot 4$	+	4^2	$= x^2 - 8x + 16$
b. $(5y - 6)^2 =$	$(5y)^2$	−	$2(5y)(6)$	+	6^2	$= 25y^2 - 60y + 36$

■

 CHECK POINT 6 Square each binomial:

a. $(x - 9)^2$ $x^2 - 18x + 81$ **b.** $(7x - 3)^2$. $49x^2 - 42x + 9$

Figure 5.3 makes it possible to visualize the square of a binomial sum. The area of the large square is

$$(A + B)(A + B) \quad \text{or} \quad (A + B)^2.$$

The sum of the areas of the four smaller rectangles that make up the large square is

$$A^2 + AB + AB + B^2$$

or

$$A^2 + 2AB + B^2.$$

Conclusion:

$$(A + B)^2 = A^2 + 2AB + B^2.$$

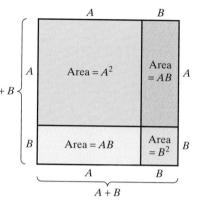

FIGURE 5.3

The following box summarizes the FOIL method and the three special products. The special products occur so frequently in algebra that it is convenient to memorize the form or pattern of these formulas.

FOIL and Special Products
Let A, B, C, and D be real numbers, variables, or algebraic expressions.

FOIL	*Example*
F O I L	**F O I L**
$(A + B)(C + D) = AC + AD + BC + BD$	$(2x + 3)(4x + 5) = (2x)(4x) + (2x)(5) + (3)(4x) + (3)(5)$
	$= 8x^2 + 10x + 12x + 15$
	$= 8x^2 + 22x + 15$
Sum and Difference of Two Terms	*Example*
$(A + B)(A - B) = A^2 - B^2$	$(2x + 3)(2x - 3) = (2x)^2 - 3^2$
	$= 4x^2 - 9$
Square of a Binomial	*Example*
$(A + B)^2 = A^2 + 2AB + B^2$	$(2x + 3)^2 = (2x)^2 + 2(2x)(3) + 3^2$
	$= 4x^2 + 12x + 9$
$(A - B)^2 = A^2 - 2AB + B^2$	$(2x - 3)^2 = (2x)^2 - 2(2x)(3) + 3^2$
	$= 4x^2 - 12x + 9$

5.3 EXERCISE SET

Student Solutions Manual   CD/Video   PH Math/Tutor Center   MathXL Tutorials on CD   MathXL®   MyMathLab   Interactmath.com

Practice Exercises

In Exercises 1–24, use the FOIL method to find each product. Express the product in descending powers of the variable.

1. $(x + 4)(x + 6)$ $x^2 + 10x + 24$

2. $(x + 8)(x + 2)$ $x^2 + 10x + 16$

3. $(y - 7)(y + 3)$ $y^2 - 4y - 21$

4. $(y - 3)(y + 4)$ $y^2 + y - 12$

5. $(2x - 3)(x + 5)$ $2x^2 + 7x - 15$

6. $(3x - 5)(x + 7)$ $3x^2 + 16x - 35$

7. $(4y + 3)(y - 1)$ $4y^2 - y - 3$

8. $(5y + 4)(y - 2)$ $5y^2 - 6y - 8$

9. $(2x - 3)(5x + 3)$ $10x^2 - 9x - 9$

10. $(2x - 5)(7x + 2)$ $14x^2 - 31x - 10$

11. $(3y - 7)(4y - 5)$ $12y^2 - 43y + 35$

12. $(4y - 5)(7y - 4)$ $28y^2 - 51y + 20$

13. $(7 + 3x)(1 - 5x)$ $-15x^2 - 32x + 7$

14. $(2 + 5x)(1 - 4x)$ $-20x^2 - 3x + 2$

15. $(5 - 3y)(6 - 2y)$ $6y^2 - 28y + 30$

16. $(7 - 2y)(10 - 3y)$ $6y^2 - 41y + 70$

17. $(5x^2 - 4)(3x^2 - 7)$ $15x^4 - 47x^2 + 28$

18. $(7x^2 - 2)(3x^2 - 5)$ $21x^4 - 41x^2 + 10$

19. $(6x - 5)(2 - x)$ $-6x^2 + 17x - 10$

20. $(4x - 3)(2 - x)$ $-4x^2 + 11x - 6$

21. $(x + 5)(x^2 + 3)$ $x^3 + 5x^2 + 3x + 15$

22. $(x + 4)(x^2 + 5)$ $x^3 + 4x^2 + 5x + 20$

23. $(8x^3 + 3)(x^2 + 5)$ $8x^5 + 40x^3 + 3x^2 + 15$

24. $(7x^3 + 5)(x^2 + 2)$ $7x^5 + 14x^3 + 5x^2 + 10$

In Exercises 25–44, multiply using the rule for finding the product of the sum and difference of two terms.

25. $(x + 3)(x - 3)$ $x^2 - 9$

26. $(y + 5)(y - 5)$ $y^2 - 25$

27. $(3x + 2)(3x - 2)$ $9x^2 - 4$

28. $(2x + 5)(2x - 5)$ $4x^2 - 25$

29. $(3r - 4)(3r + 4)$ $9r^2 - 16$

30. $(5z - 2)(5z + 2)$ $25z^2 - 4$

31. $(3 + r)(3 - r)$ $9 - r^2$

32. $(4 + s)(4 - s)$ $16 - s^2$

33. $(5 - 7x)(5 + 7x)$ $25 - 49x^2$

34. $(4 - 3y)(4 + 3y)$ $16 - 9y^2$

35. $\left(2x + \dfrac{1}{2}\right)\left(2x - \dfrac{1}{2}\right)$ $4x^2 - \dfrac{1}{4}$

36. $\left(3y + \dfrac{1}{3}\right)\left(3y - \dfrac{1}{3}\right)$ $9y^2 - \dfrac{1}{9}$

37. $(y^2 + 1)(y^2 - 1)$ $y^4 - 1$

38. $(y^2 + 2)(y^2 - 2)$ $y^4 - 4$

39. $(r^3 + 2)(r^3 - 2)$ $r^6 - 4$

40. $(m^3 + 4)(m^3 - 4)$ $m^6 - 16$

41. $(1 - y^4)(1 + y^4)$ $1 - y^8$

42. $(2 - s^5)(2 + s^5)$ $4 - s^{10}$

43. $(x^{10} + 5)(x^{10} - 5)$ $x^{20} - 25$

44. $(x^{12} + 3)(x^{12} - 3)$ $x^{24} - 9$

In Exercises 45–62, multiply using the rule for the square of a binomial.

45. $(x + 2)^2$ $x^2 + 4x + 4$

46. $(x + 5)^2$ $x^2 + 10x + 25$

47. $(2x + 5)^2$ $4x^2 + 20x + 25$

48. $(5x + 2)^2$ $25x^2 + 20x + 4$

49. $(x - 3)^2$ $x^2 - 6x + 9$

50. $(x - 6)^2$ $x^2 - 12x + 36$

51. $(3y - 4)^2$ $9y^2 - 24y + 16$

52. $(4y - 3)^2$ $16y^2 - 24y + 9$

53. $(4x^2 - 1)^2$ $16x^4 - 8x^2 + 1$

54. $(5x^2 - 3)^2$ $25x^4 - 30x^2 + 9$

55. $(7 - 2x)^2$ $49 - 28x + 4x^2$

56. $(9 - 5x)^2$ $81 - 90x + 25x^2$

57. $\left(2x + \dfrac{1}{2}\right)^2$ $4x^2 + 2x + \dfrac{1}{4}$

58. $\left(3x + \dfrac{1}{3}\right)^2$ $9x^2 + 2x + \dfrac{1}{9}$

59. $\left(4y - \dfrac{1}{4}\right)^2$ $16y^2 - 2y + \dfrac{1}{16}$

60. $\left(2y - \dfrac{1}{2}\right)^2$ $4y^2 - 2y + \dfrac{1}{4}$

61. $(x^8 + 3)^2$ $x^{16} + 6x^8 + 9$

62. $(x^8 + 5)^2$ $x^{16} + 10x^8 + 25$

In Exercises 63–82, multiply using the method of your choice.

63. $(x - 1)(x^2 + x + 1)$ $x^3 - 1$

64. $(x + 1)(x^2 - x + 1)$ $x^3 + 1$

65. $(x - 1)^2$ $x^2 - 2x + 1$

66. $(x + 1)^2$ $x^2 + 2x + 1$

67. $(3y + 7)(3y - 7)$ $9y^2 - 49$

68. $(4y + 9)(4y - 9)$ $16y^2 - 81$

69. $3x^2(4x^2 + x + 9)$ $12x^4 + 3x^3 + 27x^2$

70. $5x^2(7x^2 + x + 6)$ $35x^4 + 5x^3 + 30x^2$

71. $(7y + 3)(10y - 4)$ $70y^2 + 2y - 12$

72. $(8y + 3)(10y - 5)$ $80y^2 - 10y - 15$

73. $(x^2 + 1)^2$ $x^4 + 2x^2 + 1$

74. $(x^2 + 2)^2$ $x^4 + 4x^2 + 4$

75. $(x^2 + 1)(x^2 + 2)$ $x^4 + 3x^2 + 2$

76. $(x^2 + 2)(x^2 + 3)$ $x^4 + 5x^2 + 6$

77. $(x^2 + 4)(x^2 - 4)$ $x^4 - 16$

78. $(x^2 + 5)(x^2 - 5)$ $x^4 - 25$

79. $(2 - 3x^5)^2$ $4 - 12x^5 + 9x^{10}$

80. $(2 - 3x^6)^2$ $4 - 12x^6 + 9x^{12}$

81. $\left(\frac{1}{4}x^2 + 12\right)\left(\frac{3}{4}x^2 - 8\right)$ $\frac{3}{16}x^4 + 7x^2 - 96$

82. $\left(\frac{1}{4}x^2 + 16\right)\left(\frac{3}{4}x^2 - 4\right)$ $\frac{3}{16}x^4 + 11x^2 - 64$

In Exercises 83–88, find the area of each shaded region. Write the answer as a polynomial in descending powers of x.

83.

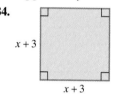

$x + 1$

$x + 1$

$x^2 + 2x + 1$

84.

$x + 3$

$x + 3$

$x^2 + 6x + 9$

85. $4x^2 - 9$

$2x + 3$

$2x - 3$

86. $16x^2 - 9$

$4x + 3$

$4x - 3$

87. $6x + 22$

$x + 9$

$x + 5$

$x + 3$ $x + 1$

88. $4x + 10$

$x + 4$

$x + 2$

$x + 3$ $x + 1$

Practice Plus

In Exercises 89–96, multiply by the method of your choice.

89. $[(2x + 3)(2x - 3)]^2$ $16x^4 - 72x^2 + 81$

90. $[(3x + 2)(3x - 2)]^2$ $81x^4 - 72x^2 + 16$

91. $(4x^2 + 1)[(2x + 1)(2x - 1)]$ $16x^4 - 1$

92. $(9x^2 + 1)[(3x + 1)(3x - 1)]$ $81x^4 - 1$

93. $(x + 2)^3$ $x^3 + 6x^2 + 12x + 8$

94. $(x + 4)^3$ $x^3 + 12x^2 + 48x + 64$

95. $[(x + 3) - y][(x + 3) + y]$ $x^2 + 6x + 9 - y^2$

96. $[(x + 5) - y][(x + 5) + y]$ $x^2 + 10x + 25 - y^2$

Application Exercises

The square garden shown in the figure measures x yards on each side. The garden is to be expanded so that one side is increased by 2 yards and an adjacent side is increased by 1 yard. The graph shows the area of the expanded garden, y, in terms of the length of one of its original square sides, x. Use this information to solve Exercises 97–100.

1 yard

$x + 1$

x yards

x yards

$x + 2$

2 yards

y

100

75

50

25

1 2 3 4 5 6 7 8 9 10

x

Area of Expanded Garden (square yards)

Measure of Original Garden's Sides (yards)

97. Write a product of two binomials that expresses the area of the larger garden. $(x + 1)(x + 2)$ square yards

98. Write a polynomial in descending powers of x that expresses the area of the larger garden. $(x^2 + 3x + 2)$ square yards

99. If the original garden measures 6 yards on a side, use your expression from Exercise 97 to find the area of the larger garden. Then identify your solution as a point on the graph shown.   56 square yards; (6, 56)

100. If the original garden measures 8 yards on a side, use your polynomial from Exercise 98 to find the area of the larger garden. Then identify your solution as a point on the graph shown.   90 square yards; (8, 90)

The square painting in the figure measures x inches on each side. The painting is uniformly surrounded by a frame that measures 1 inch wide. Use this information to solve Exercises 101–102.

1 inch

1 inch   1 inch

x

1 inch

x

101. Write a polynomial in descending powers of x that expresses the area of the square that includes the painting and the frame. $(x^2 + 4x + 4)$ square inches

102. Write an algebraic expression that describes the area of the frame. (*Hint:* The area of the frame is the area of the square that includes the painting and the frame minus the area of the painting.) $(4x + 4)$ square inches

Writing in Mathematics

103. Explain how to multiply two binomials using the FOIL method. Give an example with your explanation.   103–107. Answers will vary.

104. Explain how to find the product of the sum and difference of two terms. Give an example with your explanation.

105. Explain how to square a binomial sum. Give an example with your explanation.

106. Explain how to square a binomial difference. Give an example with your explanation.

107. Explain why the graph for Exercises 97–100 is shown only in quadrant I.

Critical Thinking Exercises

108. Which one of the following is true?   b

a. $(3 + 4)^2 = 3^2 + 4^2$

b. $(2y + 7)^2 = 4y^2 + 28y + 49$

c. $(3x^2 + 2)(3x^2 - 2) = 9x^2 - 4$

d. $(x - 5)^2 = x^2 - 5x + 25$

109. What two binomials must be multiplied using the FOIL method to give a product of $x^2 - 8x - 20$? $(x - 10)$ and $(x + 2)$

110. Express the volume of the box as a polynomial in standard form. $4x^3 - 36x^2 + 80x$

x

8 − 2x   10 − 2x

111. Express the area of the plane figure shown as a polynomial in standard form. $x^2 + 2x$

x

x

x − 1

x + 3

Technology Exercises

In Exercises 112–115, use a graphing utility to graph each side of the equation in the same viewing rectangle. (Call the left side y_1 and the right side y_2.) If the graphs coincide, verify that the multiplication has been performed correctly. If the graphs do not appear to coincide, this indicates that the multiplication is incorrect. In these exercises, correct the right side of the equation. Then graph the left side and the corrected right side to verify that the graphs coincide.

112. $(x + 1)^2 = x^2 + 1$; Use a $[-5, 5, 1]$ by $[0, 20, 1]$ viewing rectangle.   Change $x^2 + 1$ to $x^2 + 2x + 1$.

113. $(x + 2)^2 = x^2 + 2x + 4$; Use a $[-6, 5, 1]$ by $[0, 20, 1]$ viewing rectangle.   Change $x^2 + 2x + 4$ to $x^2 + 4x + 4$.

114. $(x + 1)(x - 1) = x^2 - 1$; Use a $[-6, 5, 1]$ by $[-2, 18, 1]$ viewing rectangle.   114–115. Graphs coincide.

115. $(x - 2)(x + 2) + 4 = x^2$; Use a $[-6, 5, 1]$ by $[-2, 18, 1]$ viewing rectangle.

Review Exercises

In Exercises 116–118, solve each system by the method of your choice.

116. $2x + 3y = 1$

$y = 3x - 7$

(Section 4.2, Example 1) $\{(2, -1)\}$

117. $3x + 4y = 7$

$2x + 7y = 9$

(Section 4.3, Example 3) $\{(1, 1)\}$

118. $4x + y + 2z = 6$

$x + y + z = 1$

$x - 3y + z = 5$

(Section 4.5, Example 2) $\left\{\left(\frac{3}{2}, -1, \frac{1}{2}\right)\right\}$

Objectives

1 Evaluate polynomials in several variables.

2 Understand the vocabulary of polynomials in two variables.

3 Add and subtract polynomials in several variables.

4 Multiply polynomials in several variables.

POLYNOMIALS IN SEVERAL VARIABLES

The next time you visit a lumberyard and go rummaging through piles of wood, think *polynomials*, although polynomials a bit different from those we have encountered so far. The construction industry uses a polynomial in two variables to determine the number of board feet that can be manufactured from a tree with a diameter of x inches and a length of y feet. This polynomial is

$$\frac{1}{4}x^2y - 2xy + 4y.$$

We call a polynomial containing two or more variables a **polynomial in several variables**. These polynomials can be evaluated, added, subtracted, and multiplied just like polynomials that contain only one variable.

1 Evaluate polynomials in several variables.

Evaluating a Polynomial in Several Variables Two steps can be used to evaluate a polynomial in several variables.

EVALUATING A POLYNOMIAL IN SEVERAL VARIABLES

1. Substitute the given value for each variable.

2. Perform the resulting computation using the order of operations.

EXAMPLE 1 Evaluating a Polynomial in Two Variables

Evaluate $2x^3y + xy^2 + 7x - 3$ for $x = -2$ and $y = 3$.

SOLUTION We begin by substituting -2 for x and 3 for y in the polynomial.

$$2x^3y + xy^2 + 7x - 3$$ — This is the given polynomial.

$$= 2(-2)^3 \cdot 3 + (-2) \cdot 3^2 + 7(-2) - 3$$ — Replace x with -2 and y with 3.

$$= 2(-8) \cdot 3 + (-2) \cdot 9 + 7(-2) - 3$$ — Evaluate exponential expressions: $(-2)^3 = (-2)(-2)(-2) = -8$ and $3^2 = 3 \cdot 3 = 9$.

$$= -48 + (-18) + (-14) - 3$$ — Perform the indicated multiplications.

$$= -83$$ — Add from left to right. ∎

CHECK POINT 1 Evaluate $3x^3y + xy^2 + 5y + 6$ for $x = -1$ and $y = 5$. -9

2 Understand the vocabulary of polynomials in two variables.

Describing Polynomials in Two Variables

In this section, we will limit our discussion of polynomials in several variables to two variables.

In general, a **polynomial in two variables**, x and y, contains the sum of one or more monomials in the form ax^ny^m. The constant, a, is the **coefficient**. The exponents, n and m, represent whole numbers. The **degree** of the monomial ax^ny^m is $n + m$. We'll use the polynomial from the construction industry to illustrate these ideas.

The coefficients are $\frac{1}{4}$, -2, and **4**.

$$\frac{1}{4}x^2y \;-\; 2xy \;+\; 4y$$

| Degree of monomial: $2+1=3$ | Degree of monomial: $1+1=2$ | Degree of monomial: 1 |

The **degree of a polynomial in two variables** is the highest degree of all its terms. For the preceding polynomial, the degree is 3.

EXAMPLE 2 Using the Vocabulary of Polynomials

Determine the coefficient of each term, the degree of each term, and the degree of the polynomial:

$$7x^2y^3 - 17x^4y^2 + xy - 6y^2 + 9.$$

SOLUTION

Think of xy as $1x^1y^1$.

Term	Coefficient	Degree (Sum of Exponents on the Variables)
$7x^2y^3$	7	$2 + 3 = 5$
$-17x^4y^2$	-17	$4 + 2 = 6$
xy	1	$1 + 1 = 2$
$-6y^2$	-6	2
9	9	0

The degree of the polynomial is the highest degree of all its terms, which is 6. ■

 CHECK POINT 2 Determine the coefficient of each term, the degree of each term, and the degree of the polynomial:

$$8x^4y^5 - 7x^3y^2 - x^2y - 5x + 11.$$

polynomial degree: 9

Term	Coefficient	Degree
$8x^4y^5$	8	9
$-7x^3y^2$	-7	5
$-x^2y$	-1	3
$-5x$	-5	1
11	11	0

3 Add and subtract polynomials in several variables.

Adding and Subtracting Polynomials in Several Variables

Polynomials in several variables are added by combining like terms. For example, we can add the monomials $-7xy^2$ and $13xy^2$ as follows:

$$-7xy^2 + 13xy^2 = (-7 + 13)xy^2 = 6xy^2.$$

These like terms both contain the variable factors x and y^2.   Add coefficients and keep the same variable factors, xy^2.

EXAMPLE 3   Adding Polynomials in Two Variables

Add: $(6xy^2 - 5xy + 7) + (9xy^2 + 2xy - 6)$.

SOLUTION

$$(6xy^2 - 5xy + 7) + (9xy^2 + 2xy - 6)$$
$$= (6xy^2 + 9xy^2) + (-5xy + 2xy) + (7 - 6) \qquad \text{Group like terms.}$$
$$= 15xy^2 - 3xy + 1$$

Combine like terms by adding coefficients and keeping the same variable factors. ∎

$2x^2y + 2xy - 4$

✔ **CHECK POINT 3** Add: $(-8x^2y - 3xy + 6) + (10x^2y + 5xy - 10)$.

We subtract polynomials in two variables just as we did when subtracting polynomials in one variable. Add the first polynomial and the opposite of the polynomial being subtracted.

EXAMPLE 4   Subtracting Polynomials in Two Variables

Subtract:

$$(5x^3 - 9x^2y + 3xy^2 - 4) - (3x^3 - 6x^2y - 2xy^2 + 3).$$

SOLUTION

$$(5x^3 - 9x^2y + 3xy^2 - 4) - (3x^3 - 6x^2y - 2xy^2 + 3)$$

Change the sign of each coefficient.

$$= (5x^3 - 9x^2y + 3xy^2 - 4) + (-3x^3 + 6x^2y + 2xy^2 - 3) \qquad \begin{array}{l}\text{Add the opposite of}\\\text{the polynomial being}\\\text{subtracted.}\end{array}$$

$$= (5x^3 - 3x^3) + (-9x^2y + 6x^2y) + (3xy^2 + 2xy^2) + (-4 - 3) \qquad \text{Group like terms.}$$
$$= 2x^3 - 3x^2y + 5xy^2 - 7$$

Combine like terms by adding coefficients and keeping the same variable factors. ∎

$3x^3 + 2x^2y + 5xy^2 - 10$

✔ **CHECK POINT 4** Subtract: $(7x^3 - 10x^2y + 2xy^2 - 5) - (4x^3 - 12x^2y - 3xy^2 + 5)$.

---

4 Multiply polynomials in several variables.

Multiplying Polynomials in Several Variables   The product of monomials forms the basis of polynomial multiplication. As with monomials in one variable, multiplication can be done mentally by multiplying coefficients and adding exponents on variables with the same base.

EXAMPLE 5   Multiplying Monomials

Multiply: $(7x^2y)(5x^3y^2)$.

SOLUTION

$$(7x^2y)(5x^3y^2)$$
$$= (7 \cdot 5)(x^2 \cdot x^3)(y \cdot y^2) \quad \text{This regrouping can be worked mentally.}$$
$$= 35x^{2+3}y^{1+2} \quad \text{Multiply coefficients and add exponents on variables with same base.}$$
$$= 35x^5y^3 \quad \text{Simplify.} \quad \blacksquare$$

 CHECK POINT **5** Multiply: $(6xy^3)(10x^4y^2)$. $60x^5y^5$

How do we multiply a monomial and a polynomial that is not a monomial? As we did with polynomials in one variable, multiply each term of the polynomial by the monomial.

EXAMPLE 6   Multiplying a Monomial and a Polynomial

Multiply: $3x^2y(4x^3y^2 - 6x^2y + 2)$.

SOLUTION

$$3x^2y(4x^3y^2 - 6x^2y + 2)$$
$$= 3x^2y \cdot 4x^3y^2 - 3x^2y \cdot 6x^2y + 3x^2y \cdot 2 \quad \text{Use the distributive property.}$$
$$= 12x^{2+3}y^{1+2} - 18x^{2+2}y^{1+1} + 6x^2y \quad \text{Multiply coefficients and add exponents on variables with the same base.}$$
$$= 12x^5y^3 - 18x^4y^2 + 6x^2y \quad \text{Simplify.} \quad \blacksquare$$

 CHECK POINT **6** Multiply: $6xy^2(10x^4y^5 - 2x^2y + 3)$. $60x^5y^7 - 12x^3y^3 + 18xy^2$

FOIL and the special-products formulas can be used to multiply polynomials in several variables.

EXAMPLE 7   Multiplying Polynomials in Two Variables

Multiply: **a.** $(x + 4y)(3x - 5y)$ **b.** $(5x + 3y)^2$.

SOLUTION   We will perform the multiplication in part (a) using the FOIL method. We will multiply in part (b) using the formula for the square of a binomial, $(A + B)^2$.

a. $(x + 4y)(3x - 5y)$ Multiply these binomials using the FOIL method.

      F   O   I   L

$$= (x)(3x) + (x)(-5y) + (4y)(3x) + (4y)(-5y)$$
$$= 3x^2 - 5xy + 12xy - 20y^2$$
$$= 3x^2 + 7xy - 20y^2 \quad \text{Combine like terms.}$$

$$(A + B)^2 = A^2 + 2 \cdot A \cdot B + B^2$$

b. $(5x + 3y)^2 = (5x)^2 + 2(5x)(3y) + (3y)^2$
$$= 25x^2 + 30xy + 9y^2 \quad \blacksquare$$

 CHECK POINT **7** Multiply:

     a. $(7x - 6y)(3x - y)$ $21x^2 - 25xy + 6y^2$ **b.** $(2x + 4y)^2$. $4x^2 + 16xy + 16y^2$

EXAMPLE 8 Multiplying Polynomials in Two Variables

Multiply: **a.** $(4x^2y + 3y)(4x^2y - 3y)$ **b.** $(x + y)(x^2 - xy + y^2)$.

SOLUTION We perform the multiplication in part (a) using the formula for the product of the sum and difference of two terms. We perform the multiplication in part (b) by multiplying each term of the trinomial, $x^2 - xy + y^2$, by x and y, respectively, and then adding like terms.

$$(A + B) \cdot (A - B) \quad = \quad A^2 \quad - \quad B^2$$

a. $(4x^2y + 3y)(4x^2y - 3y) = (4x^2y)^2 - (3y)^2$
$$= 16x^4y^2 - 9y^2$$

b. $(x + y)(x^2 - xy + y^2)$
$$= x(x^2 - xy + y^2) + y(x^2 - xy + y^2)$$

Multiply the trinomial by each term of the binomial.

$$= x \cdot x^2 - x \cdot xy + x \cdot y^2 + y \cdot x^2 - y \cdot xy + y \cdot y^2$$

Use the distributive property.

$$= x^3 - x^2y + xy^2 + x^2y - xy^2 + y^3$$

Add exponents on variables with the same base.

$$= x^3 + y^3$$

Combine like terms: $-x^2y + x^2y = 0$ and $xy^2 - xy^2 = 0$. ∎

✔ **CHECK POINT 8** Multiply:

a. $(6xy^2 + 5x)(6xy^2 - 5x)$ $36x^2y^4 - 25x^2$
b. $(x - y)(x^2 + xy + y^2)$. $x^3 - y^3$

---

7. polynomial degree: 9

Term	Coefficient	Degree
x^3y^2	1	5
$-5x^2y^7$	-5	9
$6y^2$	6	2
-3	-3	0

8. polynomial degree: 10

Term	Coefficient	Degree
$12x^4y$	12	5
$-5x^3y^7$	-5	10
$-x^2$	-1	2
4	4	0

5.4 EXERCISE SET

Student Solutions Manual   CD/Video   PH Math/Tutor Center   MathXL Tutorials on CD   MathXL®   MyMathLab   Interactmath.com

Practice Exercises

In Exercises 1–6, evaluate each polynomial for $x = 2$ and $y = -3$.

1. $x^2 + 2xy + y^2$ 1
2. $x^2 + 3xy + y^2$ -5
3. $xy^3 - xy + 1$ -47
4. $x^3y - xy + 2$ -16
5. $2x^2y - 5y + 3$ -6
6. $3x^2y - 4y + 5$ -19

In Exercises 7–8, determine the coefficient of each term, the degree of each term, and the degree of the polynomial.

7. $x^3y^2 - 5x^2y^7 + 6y^2 - 3$ **8.** $12x^4y - 5x^3y^7 - x^2 + 4$

In Exercises 9–20, add or subtract as indicated.

9. $(5x^2y - 3xy) + (2x^2y - xy)$ $7x^2y - 4xy$

10. $(-2x^2y + xy) + (4x^2y + 7xy)$ $2x^2y + 8xy$

11. $(4x^2y + 8xy + 11) + (-2x^2y + 5xy + 2)$
$2x^2y + 13xy + 13$

12. $(7x^2y + 5xy + 13) + (-3x^2y + 6xy + 4)$
$4x^2y + 11xy + 17$

13. $(7x^4y^2 - 5x^2y^2 + 3xy) + (-18x^4y^2 - 6x^2y^2 - xy)$
$-11x^4y^2 - 11x^2y^2 + 2xy$

14. $(6x^4y^2 - 10x^2y^2 + 7xy) + (-12x^4y^2 - 3x^2y^2 - xy)$
$-6x^4y^2 - 13x^2y^2 + 6xy$

15. $(x^3 + 7xy - 5y^2) - (6x^3 - xy + 4y^2)$
$-5x^3 + 8xy - 9y^2$

16. $(x^4 - 7xy - 5y^3) - (6x^4 - 3xy + 4y^3)$
$-5x^4 - 4xy - 9y^3$

17. $(3x^4y^2 + 5x^3y - 3y) - (2x^4y^2 - 3x^3y - 4y + 6x)$
$x^4y^2 + 8x^3y + y - 6x$

18. $(5x^4y^2 + 6x^3y - 7y) - (3x^4y^2 - 5x^3y - 6y + 8x)$
$2x^4y^2 + 11x^3y - y - 8x$

19. $(x^3 - y^3) - (-4x^3 - x^2y + xy^2 + 3y^3)$
$5x^3 + x^2y - xy^2 - 4y^3$

20. $(x^3 - y^3) - (-6x^3 + x^2y - xy^2 + 2y^3)$
$7x^3 - x^2y + xy^2 - 3y^3$

21. Add: $\quad 5x^2y^2 - 4xy^2 + 6y^2$
$\quad\quad\quad \underline{-8x^2y^2 + 5xy^2 - \quad y^2} \quad -3x^2y^2 + xy^2 + 5y^2$

22. Add: $\quad 7a^2b^2 - 5ab^2 + 6b^2$
$\quad\quad\quad \underline{-10a^2b^2 + 6ab^2 + 6b^2} \quad -3a^2b^2 + ab^2 + 12b^2$

23. Subtract: $\quad 3a^2b^4 - 5ab^2 + 7ab$
$\quad\quad\quad \underline{-(-5a^2b^4 - 8ab^2 - \quad ab)} \quad 8a^2b^4 + 3ab^2 + 8ab$

24. Subtract: $\quad 13x^2y^4 - 17xy^2 + xy$
$\quad\quad\quad \underline{-(-7x^2y^4 - \quad 8xy^2 - xy)} \quad 20x^2y^4 - 9xy^2 + 2xy$

25. Subtract $11x - 5y$ from the sum of $7x + 13y$ and $-26x + 19y$. $\quad -30x + 37y$

26. Subtract $23x - 5y$ from the sum of $6x + 15y$ and $x - 19y$. $\quad -16x + y$

In Exercises 27–76, find each product.

27. $(5x^2y)(8xy)$ $\quad 40x^3y^2$

28. $(10x^2y)(5xy)$ $\quad 50x^3y^2$

29. $(-8x^3y^4)(3x^2y^5)$ $\quad -24x^5y^9$

30. $(7x^4y^5)(-10x^7y^{11})$ $\quad -70x^{11}y^{16}$

31. $9xy(5x + 2y)$ $\quad 45x^2y + 18xy^2$

32. $7xy(8x + 3y)$ $\quad 56x^2y + 21xy^2$

33. $5xy^2(10x^2 - 3y)$ $\quad 50x^3y^2 - 15xy^3$

34. $6x^2y(5x^2 - 9y)$ $\quad 30x^4y - 54x^2y^2$

35. $4ab^2(7a^2b^3 + 2ab)$ $\quad 28a^3b^5 + 8a^2b^3$

36. $2ab^2(20a^2b^3 + 11ab)$ $\quad 40a^3b^5 + 22a^2b^3$

37. $-b(a^2 - ab + b^2)$ $\quad -a^2b + ab^2 - b^3$

38. $-b(a^3 - ab + b^3)$ $\quad -a^3b + ab^2 - b^4$

39. $(x + 5y)(7x + 3y)$ $\quad 7x^2 + 38xy + 15y^2$

40. $(x + 9y)(6x + 7y)$ $\quad 6x^2 + 61xy + 63y^2$

41. $(x - 3y)(2x + 7y)$ $\quad 2x^2 + xy - 21y^2$

42. $(3x - y)(2x + 5y)$ $\quad 6x^2 + 13xy - 5y^2$

43. $(3xy - 1)(5xy + 2)$ $\quad 15x^2y^2 + xy - 2$

44. $(7xy + 1)(2xy - 3)$ $\quad 14x^2y^2 - 19xy - 3$

45. $(2x + 3y)^2$ $\quad 4x^2 + 12xy + 9y^2$

46. $(2x + 5y)^2$ $\quad 4x^2 + 20xy + 25y^2$

47. $(xy - 3)^2$ $\quad x^2y^2 - 6xy + 9$

48. $(xy - 5)^2$ $\quad x^2y^2 - 10xy + 25$

49. $(x^2 + y^2)^2$ $\quad x^4 + 2x^2y^2 + y^4$

50. $(2x^2 + y^2)^2$ $\quad 4x^4 + 4x^2y^2 + y^4$

51. $(x^2 - 2y^2)^2$ $\quad x^4 - 4x^2y^2 + 4y^4$

52. $(x^2 - y^2)^2$ $\quad x^4 - 2x^2y^2 + y^4$

53. $(3x + y)(3x - y)$ $\quad 9x^2 - y^2$

54. $(x + 5y)(x - 5y)$ $\quad x^2 - 25y^2$

55. $(ab + 1)(ab - 1)$ $\quad a^2b^2 - 1$

56. $(ab + 2)(ab - 2)$ $\quad a^2b^2 - 4$

57. $(x + y^2)(x - y^2)$ $\quad x^2 - y^4$

58. $(x^2 + y)(x^2 - y)$ $\quad x^4 - y^2$

59. $(3a^2b + a)(3a^2b - a)$ $\quad 9a^4b^2 - a^2$

60. $(5a^2b + a)(5a^2b - a)$ $\quad 25a^4b^2 - a^2$

61. $(3xy^2 - 4y)(3xy^2 + 4y)$ $\quad 9x^2y^4 - 16y^2$

62. $(7xy^2 - 10y)(7xy^2 + 10y)$ $\quad 49x^2y^4 - 100y^2$

63. $(a + b)(a^2 - b^2)$ $\quad a^3 - ab^2 + a^2b - b^3$

64. $(a - b)(a^2 + b^2)$ $\quad a^3 + ab^2 - a^2b - b^3$

65. $(x + y)(x^2 + 3xy + y^2)$ $\quad x^3 + 4x^2y + 4xy^2 + y^3$

66. $(x + y)(x^2 + 5xy + y^2)$ $\quad x^3 + 6x^2y + 6xy^2 + y^3$

67. $(x - y)(x^2 - 3xy + y^2)$ $\quad x^3 - 4x^2y + 4xy^2 - y^3$

68. $(x - y)(x^2 - 4xy + y^2)$ $\quad x^3 - 5x^2y + 5xy^2 - y^3$

69. $(xy + ab)(xy - ab)$ $\quad x^2y^2 - a^2b^2$

70. $(xy + ab^2)(xy - ab^2)$ $\quad x^2y^2 - a^2b^4$

71. $(x^2 + 1)(x^4y + x^2 + 1)$ $\quad x^6y + x^4y + x^4 + 2x^2 + 1$

72. $(x^2 + 1)(xy^4 + y^2 + 1)$ $\quad x^3y^4 + x^2y^2 + x^2 + xy^4 + y^2 + 1$

73. $(x^2y^2 - 3)^2$ $\quad x^4y^4 - 6x^2y^2 + 9$

74. $(x^2y^2 - 5)^2$ $\quad x^4y^4 - 10x^2y^2 + 25$

75. $(x + y + 1)(x + y - 1)$ $\quad x^2 + 2xy + y^2 - 1$

76. $(x + y + 1)(x - y + 1)$ $\quad x^2 + 2x - y^2 + 1$

In Exercises 77–80 write a polynomial in two variables that describes the total area of each shaded region. Express each polynomial as the sum or difference of terms.

77.

$3x + 5y$

$x + y$

$3x^2 + 8xy + 5y^2$

78.

$x + 3y$

$x + 3y$

$x^2 + 6xy + 9y^2$

79.

$x \quad y$

$2xy + y^2$

x

y

80.

$x^2y^2 - 16$

$xy - 4$

xy 4

Practice Plus

In Exercises 81–86, find each product. As we said in the Section 5.3 opener, cut to the chase in each part of the polynomial multiplication: Use only the special-product formula for the sum and difference of two terms or the formulas for the square of a binomial.

81. $[(x^3y^3 + 1)(x^3y^3 - 1)]^2$ $x^{12}y^{12} - 2x^6y^6 + 1$

82. $[(1 - a^3b^3)(1 + a^3b^3)]^2$ $1 - 2a^6b^6 + a^{12}b^{12}$

83. $(xy - 3)^2(xy + 3)^2$ (Do not begin by squaring a binomial.) $x^4y^4 - 18x^2y^2 + 81$

84. $(ab - 4)^2(ab + 4)^2$ (Do not begin by squaring a binomial.) $a^4b^4 - 32a^2b^2 + 256$

85. $[x + y + z][x - (y + z)]$ $x^2 - y^2 - 2yz - z^2$

86. $(a - b - c)(a + b + c)$ $a^2 - b^2 - 2bc - c^2$

Application Exercises

87. The number of board feet, N, that can be manufactured from a tree with a diameter of x inches and a length of y feet is modeled by the formula

$$N = \frac{1}{4}x^2y - 2xy + 4y.$$

A building contractor estimates that 3000 board feet of lumber is needed for a job. The lumber company has just milled a fresh load of timber from 20 trees that averaged 10 inches in diameter and 16 feet in length. Is this enough to complete the job? If not, how many additional board feet of lumber is needed?   no; need 120 more board feet

88. The storage shed shown in the figure has a volume given by the polynomial

$$2x^2y + \frac{1}{2}\pi x^2y.$$

a. A small business is considering having a shed installed like the one shown in the figure. The shed's height, $2x$, is 26 feet and its length, y, is 27 feet. Using $x = 13$ and $y = 27$, find the volume of the storage shed.   about 16,294 cubic feet

b. The business requires at least 18,000 cubic feet of storage space. Should they construct the storage shed described in part (a)?   No, they need more space.

An object that is falling or vertically projected into the air has its height, in feet, above the ground given by

$$s = -16t^2 + v_0t + s_0,$$

where s is the height, in feet, v_0 is the original velocity of the object, in feet per second, t is the time the object is in motion, in seconds, and s_0 is the height, in feet, from which the object is dropped or projected. The figure shows that a ball is thrown straight up from a rooftop at an original velocity of 80 feet per second from a height of 96 feet. The ball misses the rooftop on its way down and eventually strikes the ground. Use the formula and this information to solve Exercises 89–91.

$v_0 = 80$

$s_0 = 96$ ft

89. How high above the ground will the ball be 2 seconds after being thrown?   192 feet

90. How high above the ground will the ball be 4 seconds after being thrown?   160 feet

91. How high above the ground will the ball be 6 seconds after being thrown? Describe what this means in practical terms.   0 feet; The ball hits the ground.

The graph visually displays the information about the thrown ball described in Exercises 89–91. The horizontal axis represents the ball's time in motion, in seconds. The vertical axis represents the ball's height above the ground, in feet. Use the graph to solve Exercises 92–97.

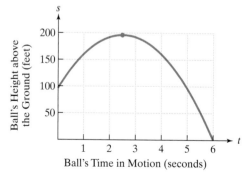

92. During which time period is the ball rising?   0 to 2.5 seconds

93. During which time period is the ball falling?   2.5 to 6 seconds

94. Identify your answer from Exercise 90 as a point on the graph.   (4, 160)

95. Identify your answer from Exercise 89 as a point on the graph.   (2, 192)

96. After how many seconds does the ball strike the ground?
6 seconds

97. After how many seconds does the ball reach its maximum height above the ground? What is a reasonable estimate of this maximum height?   2.5 seconds; 196 feet

Writing in Mathematics

98. What is a polynomial in two variables? Provide an example with your description.   98–100. Answers will vary.

99. Explain how to find the degree of a polynomial in two variables.

100. Suppose that you take up sky diving. Explain how to use the formula for Exercises 89–91 to determine your height above the ground at every instant of your fall.

Critical Thinking Exercises

101. Which one of the following is true?   c

   a. The degree of $5x^{24} - 3x^{16}y^9 - 7xy^2 + 6$ is 24.

   b. In the polynomial $4x^2y + x^3y^2 + 3x^2y^3 + 7y$, the term x^3y^2 has degree 5 and no numerical coefficient.

   c. $(2x + 3 - 5y)(2x + 3 + 5y) = 4x^2 + 12x + 9 - 25y^2$

   d. $(6x^2y - 7xy - 4) - (6x^2y + 7xy - 4) = 0$

In Exercises 102–103, find a polynomial in two variables that describes the area of the shaded region of each figure. Write the polynomial as the sum or difference of terms.

102.
$-x^2 + 18xy + 80y^2$

103.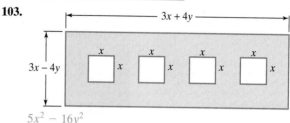
$5x^2 - 16y^2$

104. Use the formulas for the volume of a rectangular solid and a cylinder to derive the polynomial in Exercise 88 that describes the volume of the storage building.   Answers will vary.

Review Exercises

105. Solve for W: $R = \dfrac{L + 3W}{2}$. (Section 2.4, Example 4)
$$W = \frac{2R - L}{3}$$

106. Subtract: $-6.4 - (-10.2)$. (Section 1.6, Example 2)   3.8

107. Write the point-slope and slope-intercept equations of a line passing through the point $(-2, 5)$ and parallel to the line whose equation is $3x - y = 9$. (Section 3.5, Example 3)
$y - 5 = 3(x + 2)$; $y = 3x + 11$

 MID-CHAPTER CHECK POINT

CHAPTER 5

What You Know: We learned to add, subtract, and multiply polynomials. We used a number of fast methods for finding products of polynomials, including the FOIL method for multiplying binomials, a special-product formula for the product of the sum and difference of two terms $[(A + B)(A - B) = A^2 - B^2]$, and special-product formulas for squaring binomials $[(A + B)^2 = A^2 + 2AB + B^2; (A - B)^2 = A^2 - 2AB + B^2]$. Finally, we applied all of these operations to polynomials in several variables.

In Exercises 1–21, perform the indicated operations.

1. $(11x^2y^3)(-5x^2y^3)$ $-55x^4y^6$

2. $11x^2y^3 - 5x^2y^3$ $6x^2y^3$

3. $(3x + 5)(4x - 7)$ $12x^2 - x - 35$

4. $(3x + 5) - (4x - 7)$ $-x + 12$

5. $(2x - 5)(x^2 - 3x + 1)$ $2x^3 - 11x^2 + 17x - 5$

6. $(2x - 5) + (x^2 - 3x + 1)$ $x^2 - x - 4$

7. $(8x - 3)^2$ $64x^2 - 48x + 9$

8. $(-10x^4)(-7x^5)$ $70x^9$

9. $(x^2 + 2)(x^2 - 2)$ $x^4 - 4$

10. $(x^2 + 2)^2$ $x^4 + 4x^2 + 4$

11. $(9a - 10b)(2a + b)$ $18a^2 - 11ab - 10b^2$

12. $7x^2(10x^3 - 2x + 3)$ $70x^5 - 14x^3 + 21x^2$

13. $(3a^2b^3 - ab + 4b^2) - (-2a^2b^3 - 3ab + 5b^2)$
$5a^2b^3 + 2ab - b^2$

14. $2(3y - 5)(3y + 5)$ $18y^2 - 50$

15. $(-9x^3 + 5x^2 - 2x + 7) + (11x^3 - 6x^2 + 3x - 7)$
$2x^3 - x^2 + x$

16. $10x^2 - 8xy - 3(y^2 - xy)$ $10x^2 - 5xy - 3y^2$

17. $(-2x^5 + x^4 - 3x + 10) - (2x^5 - 6x^4 + 7x - 13)$
$-4x^5 + 7x^4 - 10x + 23$

18. $(x + 3y)(x^2 - 3xy + 9y^2)$ $x^3 + 27y^3$

19. $(5x^4 + 4)(2x^3 - 1)$ $10x^7 - 5x^4 + 8x^3 - 4$

20. $(y - 6z)^2$ $y^2 - 12yz + 36z^2$

21. $(2x + 3)(2x - 3) - (5x + 4)(5x - 4)$ $-21x^2 + 7$

DIVIDING POLYNOMIALS

To play the part of Charlie Chaplin, actor Robert Downey Jr. (1965–) learned to pantomime, speak two British dialects, and play left-handed tennis. His problems with substance abuse have fueled the debate over whether the illness should be handled by our criminal justice system or by health care professionals.

As you learn more mathematics, you will discover new ways to describe your world. Almost anything that you can think of involving variables can be modeled by a formula. For example, a polynomial models the annual number of drug convictions in the United States and another polynomial models drug arrests. By dividing the respective polynomials, we obtain an algebraic expression that describes the conviction rate for drug arrests.

In the next two sections, you will learn how to divide polynomials. Before turning to polynomial division, we must develop some additional rules for working with exponents.

1   Use the quotient rule for exponents.

The Quotient Rule for Exponents   Consider the quotient of two exponential expressions, such as the quotient of 2^7 and 2^3. We are dividing 7 factors of 2 by 3 factors of 2. We are left with 4 factors of 2:

7 factors of 2

$$\frac{2^7}{2^3} = \frac{2 \cdot 2 \cdot 2 \cdot 2 \cdot 2 \cdot 2 \cdot 2}{2 \cdot 2 \cdot 2} = \frac{\cancel{2} \cdot \cancel{2} \cdot \cancel{2} \cdot 2 \cdot 2 \cdot 2 \cdot 2}{\cancel{2} \cdot \cancel{2} \cdot \cancel{2}} = 2 \cdot 2 \cdot 2 \cdot 2$$

3 factors of 2   Divide out pairs of factors: $\frac{2}{2} = 1$.   4 factors of 2

Thus,

$$\frac{2^7}{2^3} = 2^4.$$

We can quickly find the exponent, 4, on the quotient by subtracting the original exponents:

$$\frac{2^7}{2^3} = 2^{7-3}.$$

This suggests the following rule:

THE QUOTIENT RULE

$$\frac{b^m}{b^n} = b^{m-n}, \quad b \neq 0$$

When dividing exponential expressions with the same nonzero base, subtract the exponent in the denominator from the exponent in the numerator. Use this difference as the exponent of the common base.

EXAMPLE 1 Using the Quotient Rule

Divide each expression using the quotient rule:

  a. $\dfrac{2^8}{2^4}$ **b.** $\dfrac{x^{13}}{x^3}$ **c.** $\dfrac{y^{15}}{y}$.

SOLUTION

  a. $\dfrac{2^8}{2^4} = 2^{8-4} = 2^4$ or 16

  b. $\dfrac{x^{13}}{x^3} = x^{13-3} = x^{10}$

  c. $\dfrac{y^{15}}{y} = \dfrac{y^{15}}{y^1} = y^{15-1} = y^{14}$

■

✔ **CHECK POINT 1** Divide each expression using the quotient rule:

  a. $\dfrac{5^{12}}{5^4}$ 5^8 **b.** $\dfrac{x^9}{x^2}$ x^7 **c.** $\dfrac{y^{20}}{y}$. y^{19}

---

2 Use the zero-exponent rule.

Zero as an Exponent A nonzero base can be raised to the 0 power. The quotient rule can be used to help determine what zero as an exponent should mean. Consider the quotient of b^4 and b^4, where b is not zero. We can determine this quotient in two ways.

$$\frac{b^4}{b^4} = 1 \qquad\qquad \frac{b^4}{b^4} = b^{4-4} = b^0$$

Any nonzero expression divided by itself is 1.   Use the quotient rule and subtract exponents.

This means that b^0 must equal 1.

> **THE ZERO-EXPONENT RULE**   If b is any real number other than 0,
> $$b^0 = 1.$$

EXAMPLE 2 Using the Zero-Exponent Rule

Use the zero-exponent rule to simplify each expression:

  a. 7^0 **b.** $(-5)^0$ **c.** -5^0 **d.** $10x^0$ **e.** $(10x)^0$.

SOLUTION

  a. $7^0 = 1$ Any nonzero number raised to the 0 power is 1.

  b. $(-5)^0 = 1$ Any nonzero number raised to the 0 power is 1.

c. $-5^0 = -1$ $\qquad$ $-5^0 = -(5^0) = -1$

Only 5 is raised to the 0 power.

d. $10x^0 = 10 \cdot 1 = 10$

Only x is raised to the 0 power.

e. $(10x)^0 = 1$

The entire expression, $10x$, is raised to the 0 power.

 CHECK POINT 2 Use the zero-exponent rule to simplify each expression:

a. 14^0 $\quad$ 1 $\qquad$ **b.** $(-10)^0$ $\quad$ 1 $\qquad$ **c.** -10^0 $\quad$ -1 $\qquad$ **d.** $20x^0$ $\quad$ 20 $\qquad$ **e.** $(20x)^0$. $\quad$ 1

3 Use the quotients-to-powers rule.

The Quotients-to-Powers Rule for Exponents We have seen that when a product is raised to a power, we raise every factor in the product to the power:

$$(ab)^n = a^n b^n.$$

There is a similar property for raising a quotient to a power.

> **QUOTIENTS TO POWERS** If a and b are real numbers and b is nonzero, then
> $$\left(\frac{a}{b}\right)^n = \frac{a^n}{b^n}.$$

When a quotient is raised to a power, raise the numerator to the power and divide by the denominator raised to the power.

EXAMPLE 3 $\quad$ Using the Quotients-to-Powers Rule

Simplify each expression using the quotients-to-powers rule:

a. $\left(\dfrac{x}{4}\right)^2$ $\qquad$ **b.** $\left(\dfrac{x^2}{5}\right)^3$ $\qquad$ **c.** $\left(\dfrac{2a^3}{b^4}\right)^5$.

SOLUTION

a. $\left(\dfrac{x}{4}\right)^2 = \dfrac{x^2}{4^2} = \dfrac{x^2}{16}$ $\qquad$ Square the numerator and the denominator.

b. $\left(\dfrac{x^2}{5}\right)^3 = \dfrac{(x^2)^3}{5^3} = \dfrac{x^{2\cdot3}}{5\cdot5\cdot5} = \dfrac{x^6}{125}$ $\qquad$ Cube the numerator and the denominator.

c. $\left(\dfrac{2a^3}{b^4}\right)^5 = \dfrac{(2a^3)^5}{(b^4)^5}$ $\qquad$ Raise the numerator and the denominator to the fifth power.

$\qquad\qquad = \dfrac{2^5(a^3)^5}{(b^4)^5}$ $\qquad$ Raise each factor in the numerator to the fifth power.

$\qquad\qquad = \dfrac{2^5 a^{3\cdot5}}{b^{4\cdot5}}$ $\qquad$ To raise exponential expressions to powers, multiply exponents: $(b^m)^n = b^{mn}$.

$\qquad\qquad = \dfrac{32a^{15}}{b^{20}}$ $\qquad$ Simplify.

CHECK POINT 3 Simplify each expression using the quotients-to-powers rule:

a. $\left(\dfrac{x}{5}\right)^2$ $\quad \dfrac{x^2}{25}$ **b.** $\left(\dfrac{x^4}{2}\right)^3$ $\quad \dfrac{x^{12}}{8}$ **c.** $\left(\dfrac{2a^{10}}{b^3}\right)^4$. $\quad \dfrac{16a^{40}}{b^{12}}$

STUDY TIP

Try to avoid the following common errors that can occur when simplifying exponential expressions.

Correct	Incorrect	Description of Error
$\dfrac{2^{20}}{2^4} = 2^{20-4} = 2^{16}$		Exponents should be subtracted, not divided.
$-8^0 = -1$		Only 8 is raised to the 0 power.
$\left(\dfrac{x}{5}\right)^2 = \dfrac{x^2}{5^2} = \dfrac{x^2}{25}$		The numerator and denominator must both be squared.

---

4 Divide monomials.

Dividing Monomials Now that we have developed three additional properties of exponents, we are ready to turn to polynomial division. We begin with the quotient of two monomials, such as $16x^{14}$ and $8x^2$. This quotient is obtained by dividing the coefficients, 16 and 8, and then dividing the variables using the quotient rule for exponents.

$$\frac{16x^{14}}{8x^2} = \frac{16}{8}x^{14-2} = 2x^{12}$$

> Divide coefficients and subtract exponents.

DIVIDING MONOMIALS To divide monomials, divide the coefficients and then divide the variables. Use the quotient rule for exponents to divide the variables: Keep the variable and subtract the exponents.

EXAMPLE 4 Dividing Monomials

Divide: **a.** $\dfrac{-12x^8}{4x^2}$ **b.** $\dfrac{2x^3}{8x^3}$ **c.** $\dfrac{15x^5y^4}{3x^2y}$.

SOLUTION

a. $\dfrac{-12x^8}{4x^2} = \dfrac{-12}{4}x^{8-2} = -3x^6$

b. $\dfrac{2x^3}{8x^3} = \dfrac{2}{8}x^{3-3} = \dfrac{1}{4}x^0 = \dfrac{1}{4} \cdot 1 = \dfrac{1}{4}$

c. $\dfrac{15x^5y^4}{3x^2y} = \dfrac{15}{3}x^{5-2}y^{4-1} = 5x^3y^3$

STUDY TIP

Look at the solution to Example 4(b). Rather than subtracting exponents for division that results in a 0 exponent, you might prefer to divide out x^3.

CHECK POINT 4 Divide:

a. $\dfrac{-20x^{12}}{10x^4}$ $\quad -2x^8$ **b.** $\dfrac{3x^4}{15x^4}$ $\quad \dfrac{1}{5}$ **c.** $\dfrac{9x^6y^5}{3xy^2}$. $\quad 3x^5y^3$

5 Check polynomial division.

Checking Division of Polynomial Problems The answer to a division problem can be checked. For example, consider the following problem:

Dividend: the polynomial you are dividing into

$$\frac{15x^5y^4}{3x^2y} = 5x^3y^3.$$

Quotient: the answer to your division problem

Divisor: the polynomial you are dividing by

The quotient is correct if the product of the divisor and the quotient is the dividend. Is the quotient shown in the preceding equation correct?

$$(3x^2y)(5x^3y^3) = 3 \cdot 5x^{2+3}y^{1+3} = 15x^5y^4.$$

Divisor   Quotient   This is the dividend.

Because the product of the divisor and the quotient is the dividend, the answer to the division problem is correct.

CHECKING DIVISION OF POLYNOMIALS To check a quotient in a division problem, multiply the divisor and the quotient. If this product is the dividend, the quotient is correct.

6 Divide a polynomial by a monomial.

Dividing a Polynomial That Is Not a Monomial by a Monomial To divide a polynomial by a monomial, we divide each term of the polynomial by the monomial. For example,

polynomial dividend

$$\frac{10x^8 + 15x^6}{5x^3} = \frac{10x^8}{5x^3} + \frac{15x^6}{5x^3} = \frac{10}{5}x^{8-3} + \frac{15}{5}x^{6-3} = 2x^5 + 3x^3.$$

monomial divisor   Divide the first term by $5x^3$.   Divide the second term by $5x^3$.

STUDY TIP

Try to avoid this common error:

Incorrect:

$$\frac{x^4 - \overset{1}{\cancel{x}}}{\cancel{x}} = \frac{x^4 - 1}{1} = x^4 - 1$$

Correct:

$$\frac{x^4 - x}{x} = \frac{x^4}{x} - \frac{x}{x}$$

$$= x^{4-1} - x^{1-1}$$

Don't leave out the 1.

$$= x^3 - x^0$$

$$= x^3 - 1$$

Is the quotient correct? Multiply the divisor and the quotient.

$$5x^3(2x^5 + 3x^3) = 5x^3 \cdot 2x^5 + 5x^3 \cdot 3x^3$$

$$= 5 \cdot 2x^{3+5} + 5 \cdot 3x^{3+3} = 10x^8 + 15x^6$$

Because this product gives the dividend, the quotient is correct.

DIVIDING A POLYNOMIAL THAT IS NOT A MONOMIAL BY A MONOMIAL To divide a polynomial by a monomial, divide each term of the polynomial by the monomial.

EXAMPLE 5 Dividing a Polynomial by a Monomial

Find the quotient: $(-12x^8 + 4x^6 - 8x^3) \div 4x^2$.

SOLUTION

$$\frac{-12x^8 + 4x^6 - 8x^3}{4x^2}$$ Rewrite the division in a vertical format.

$$= \frac{-12x^8}{4x^2} + \frac{4x^6}{4x^2} - \frac{8x^3}{4x^2}$$ Divide each term of the polynomial by the monomial.

$$= \frac{-12}{4}x^{8-2} + \frac{4}{4}x^{6-2} - \frac{8}{4}x^{3-2}$$ Divide coefficients and subtract exponents.

$$= -3x^6 + x^4 - 2x$$ Simplify.

To check the answer, multiply the divisor and the quotient.

$$4x^2(-3x^6 + x^4 - 2x) = 4x^2(-3x^6) + 4x^2 \cdot x^4 - 4x^2(2x)$$

Divisor   Quotient

$$= 4(-3)x^{2+6} + 4x^{2+4} - 4 \cdot 2x^{2+1}$$

$$= -12x^8 + 4x^6 - 8x^3$$

This is the dividend.

Because the product of the divisor and the quotient is the dividend, the answer—that is, the quotient—is correct. ∎

$-5x^7 + 2x^3 - 3x$

 CHECK POINT 5 Find the quotient: $(-15x^9 + 6x^5 - 9x^3) \div 3x^2$.

EXAMPLE 6 Dividing a Polynomial by a Monomial

Divide: $\dfrac{16x^5 - 9x^4 + 8x^3}{2x^3}$.

SOLUTION

$$\frac{16x^5 - 9x^4 + 8x^3}{2x^3}$$ This is the given polynomial division.

$$= \frac{16x^5}{2x^3} - \frac{9x^4}{2x^3} + \frac{8x^3}{2x^3}$$ Divide each term by $2x^3$.

$$= \frac{16}{2}x^{5-3} - \frac{9}{2}x^{4-3} + \frac{8}{2}x^{3-3}$$ Divide coefficients and subtract exponents. Did you immediately write the last term as 4?

$$= 8x^2 - \frac{9}{2}x + 4x^0$$ Simplify.

$$= 8x^2 - \frac{9}{2}x + 4$$ $x^0 = 1$, so $4x^0 = 4 \cdot 1 = 4$.

Check the answer by showing that the product of the divisor and the quotient is the dividend. ∎

 CHECK POINT 6 Divide: $\dfrac{25x^9 - 7x^4 + 10x^3}{5x^3}$. $5x^6 - \frac{7}{5}x + 2$

$\boxed{\text{EXAMPLE 7}}$ **Dividing Polynomials in Two Variables**

Divide: $(15x^5y^4 - 3x^3y^2 + 9x^2y) \div 3x^2y$.

SOLUTION

$$\frac{15x^5y^4 - 3x^3y^2 + 9x^2y}{3x^2y}$$

Rewrite the division in a vertical format.

$$= \frac{15x^5y^4}{3x^2y} - \frac{3x^3y^2}{3x^2y} + \frac{9x^2y}{3x^2y}$$

Divide each term of the polynomial by the monomial.

$$= \frac{15}{3}x^{5-2}y^{4-1} - \frac{3}{3}x^{3-2}y^{2-1} + \frac{9}{3}x^{2-2}y^{1-1}$$

Divide coefficients and subtract exponents.

$$= 5x^3y^3 - xy + 3$$

Simplify.

Check the answer by showing that the product of the divisor and the quotient is the dividend. ∎

✔ **CHECK POINT 7** Divide: $(18x^7y^6 - 6x^2y^3 + 60xy^2) \div 6xy^2$. $\quad 3x^6y^4 - xy + 10$

5.5 EXERCISE SET

 Student Solutions Manual   CD/Video   PH Math/Tutor Center   MathXL Tutorials on CD   MathXL®   MyMathLab   Interactmath.com

Practice Exercises

In Exercises 1–10, divide each expression using the quotient rule. Express any numerical answers in exponential form.

1. $\dfrac{3^{20}}{3^5}$ $\quad 3^{15}$

2. $\dfrac{3^{30}}{3^{10}}$ $\quad 3^{20}$

3. $\dfrac{x^6}{x^2}$ $\quad x^4$

4. $\dfrac{x^8}{x^4}$ $\quad x^4$

5. $\dfrac{y^{13}}{y^5}$ $\quad y^8$

6. $\dfrac{y^{19}}{y^6}$ $\quad y^{13}$

7. $\dfrac{5^6 \cdot 2^8}{5^3 \cdot 2^4}$ $\quad 5^3 \cdot 2^4$

8. $\dfrac{3^6 \cdot 2^8}{3^3 \cdot 2^4}$ $\quad 3^3 \cdot 2^4$

9. $\dfrac{x^{100}y^{50}}{x^{25}y^{10}}$ $\quad x^{75}y^{40}$

10. $\dfrac{x^{200}y^{40}}{x^{25}y^{10}}$ $\quad x^{175}y^{30}$

In Exercises 11–24, use the zero-exponent rule to simplify each expression.

11. 2^0 $\quad 1$

12. 4^0 $\quad 1$

13. $(-2)^0$ $\quad 1$

14. $(-4)^0$ $\quad 1$

15. -2^0 $\quad -1$

16. -4^0 $\quad -1$

17. $100y^0$ $\quad 100$

18. $200y^0$ $\quad 200$

19. $(100y)^0$ $\quad 1$

20. $(200y)^0$ $\quad 1$

21. $-5^0 + (-5)^0$ $\quad 0$

22. $-6^0 + (-6)^0$ $\quad 0$

23. $-\pi^0 - (-\pi)^0$ $\quad -2$

24. $-\sqrt{3^0} - (-\sqrt{3})^0$ $\quad -2$

In Exercises 25–36, simplify each expression using the quotients-to-powers rule. If possible, evaluate exponential expressions.

25. $\left(\dfrac{x}{3}\right)^2$ $\quad \dfrac{x^2}{9}$

26. $\left(\dfrac{x}{5}\right)^2$ $\quad \dfrac{x^2}{25}$

27. $\left(\dfrac{x^2}{4}\right)^3$ $\quad \dfrac{x^6}{64}$

28. $\left(\dfrac{x^2}{3}\right)^3$ $\quad \dfrac{x^6}{27}$

29. $\left(\dfrac{2x^3}{5}\right)^2$ $\quad \dfrac{4x^6}{25}$

30. $\left(\dfrac{3x^4}{7}\right)^2$ $\quad \dfrac{9x^8}{49}$

31. $\left(\dfrac{-4}{3a^3}\right)^3$ $\quad -\dfrac{64}{27a^9}$

32. $\left(\dfrac{-5}{2a^3}\right)^3$ $\quad -\dfrac{125}{8a^9}$

33. $\left(\dfrac{-2a^7}{b^4}\right)^5$ $\quad -\dfrac{32a^{35}}{b^{20}}$

34. $\left(\dfrac{-2a^8}{b^3}\right)^5$ $\quad -\dfrac{32a^{40}}{b^{15}}$

35. $\left(\dfrac{x^2y^3}{2z}\right)^4$ $\quad \dfrac{x^8y^{12}}{16z^4}$

36. $\left(\dfrac{x^3y^2}{2z}\right)^4$ $\quad \dfrac{x^{12}y^8}{16z^4}$

In Exercises 37–52, divide the monomials. Check each answer by showing that the product of the divisor and the quotient is the dividend.

37. $\dfrac{30x^{10}}{10x^5}$ $\quad 3x^5$

38. $\dfrac{45x^{12}}{15x^4}$ $\quad 3x^8$

39. $\dfrac{-8x^{22}}{4x^2}$ $\quad -2x^{20}$

40. $\dfrac{-15x^{40}}{3x^4}$ $\quad -5x^{36}$

41. $\dfrac{-9y^8}{18y^5}$ $\quad -\dfrac{1}{2}y^3$

42. $\dfrac{-15y^{13}}{45y^9}$ $\quad -\dfrac{1}{3}y^4$

43. $\dfrac{7y^{17}}{5y^5}$ $\quad \dfrac{7}{5}y^{12}$

44. $\dfrac{9y^{19}}{7y^{11}}$ $\quad \dfrac{9}{7}y^8$

45. $\dfrac{30x^7y^5}{5x^2y}$ $\quad 6x^5y^4$

46. $\dfrac{40x^9y^5}{2x^2y}$ $\quad 20x^7y^4$

47. $\dfrac{-18x^{14}y^2}{36x^2y^2}$ $\quad -\dfrac{1}{2}x^{12}$

48. $\dfrac{-15x^{16}y^2}{45x^2y^2}$ $\quad -\dfrac{1}{3}x^{14}$

49. $\dfrac{9x^{20}y^{20}}{7x^{20}y^{20}}$ $\quad \dfrac{9}{7}$

50. $\dfrac{7x^{30}y^{30}}{15x^{30}y^{30}}$ $\quad \dfrac{7}{15}$

51. $\dfrac{-5x^{10}y^{12}z^6}{50x^2y^3z^2}$ $-\dfrac{1}{10}x^8y^9z^4$ **52.** $\dfrac{-8x^{12}y^{10}z^4}{40x^2y^3z^2}$ $-\dfrac{1}{5}x^{10}y^7z^2$

In Exercises 53–78, divide the polynomial by the monomial. Check each answer by showing that the product of the divisor and the quotient is the dividend.

53. $\dfrac{10x^4 + 2x^3}{2}$ $5x^4 + x^3$ **54.** $\dfrac{20x^4 + 5x^3}{5}$ $4x^4 + x^3$

55. $\dfrac{14x^4 - 7x^3}{7x}$ $2x^3 - x^2$ **56.** $\dfrac{24x^4 - 8x^3}{8x}$ $3x^3 - x^2$

57. $\dfrac{y^7 - 9y^2 + y}{y}$ $y^6 - 9y + 1$ **58.** $\dfrac{y^8 - 11y^3 + y}{y}$ $y^7 - 11y^2 + 1$

59. $\dfrac{24x^3 - 15x^2}{-3x}$ $-8x^2 + 5x$ **60.** $\dfrac{10x^3 - 20x^2}{-5x}$ $-2x^2 + 4x$

61. $\dfrac{18x^5 + 6x^4 + 9x^3}{3x^2}$ $6x^3 + 2x^2 + 3x$

62. $\dfrac{18x^5 + 24x^4 + 12x^3}{6x^2}$ $3x^3 + 4x^2 + 2x$

63. $\dfrac{12x^4 - 8x^3 + 40x^2}{4x}$ $3x^3 - 2x^2 + 10x$

64. $\dfrac{49x^4 - 14x^3 + 70x^2}{-7x}$ $-7x^3 + 2x^2 - 10x$

65. $(4x^2 - 6x) \div x$ $4x - 6$

66. $(16y^2 - 8y) \div y$ $16y - 8$

67. $\dfrac{30z^3 + 10z^2}{-5z}$ $-6z^2 - 2z$

68. $\dfrac{12y^4 - 42y^2}{-4y}$ $-3y^3 + \dfrac{21}{2}y$

69. $\dfrac{8x^3 + 6x^2 - 2x}{2x}$ $4x^2 + 3x - 1$

70. $\dfrac{9x^3 + 12x^2 - 3x}{3x}$ $3x^2 + 4x - 1$

71. $\dfrac{25x^7 - 15x^5 - 5x^4}{5x^3}$ $5x^4 - 3x^2 - x$

72. $\dfrac{49x^7 - 28x^5 - 7x^4}{7x^3}$ $7x^4 - 4x^2 - x$

73. $\dfrac{18x^7 - 9x^6 + 20x^5 - 10x^4}{-2x^4}$ $-9x^3 + \dfrac{9}{2}x^2 - 10x + 5$

74. $\dfrac{25x^8 - 50x^7 + 3x^6 - 40x^5}{-5x^5}$ $-5x^3 + 10x^2 - \dfrac{3}{5}x + 8$

75. $\dfrac{12x^2y^2 + 6x^2y - 15xy^2}{3xy}$ $4xy + 2x - 5y$

76. $\dfrac{18a^3b^2 - 9a^2b - 27ab^2}{9ab}$ $2a^2b - a - 3b$

77. $\dfrac{20x^7y^4 - 15x^3y^2 - 10x^2y}{-5x^2y}$ $-4x^5y^3 + 3xy + 2$

78. $\dfrac{8x^6y^3 - 12x^8y^2 - 4x^{14}y^6}{-4x^6y^2}$ $-2y + 3x^2 + x^8y^4$

Practice Plus

In Exercises 79–82, simplify each expression.

79. $\dfrac{2x^3(4x + 2) - 3x^2(2x - 4)}{2x^2}$ $4x^2 - x + 6$

80. $\dfrac{6x^3(3x - 1) + 5x^2(6x - 3)}{3x^2}$ $6x^2 + 8x - 5$

81. $\left(\dfrac{18x^2y^4}{9xy^2}\right) - \left(\dfrac{15x^5y^6}{5x^4y^4}\right)$ $-xy^2$

82. $\left(\dfrac{9x^3 + 6x^2}{3x}\right) - \left(\dfrac{12x^2y^2 - 4xy^2}{2xy^2}\right)$ $3x^2 - 4x + 2$

83. Divide the sum of $(y + 5)^2$ and $(y + 5)(y - 5)$ by $2y$. $y + 5$

84. Divide the sum of $(y + 4)^2$ and $(y + 4)(y - 4)$ by $2y$. $y + 4$

In Exercises 85–86, the variable n in each exponent represents a natural number. Divide the polynomial by the monomial. Then use polynomial multiplication to check the quotient.

85. $\dfrac{12x^{15n} - 24x^{12n} + 8x^{3n}}{4x^{3n}}$ $3x^{12n} - 6x^{9n} + 2$

86. $\dfrac{35x^{10n} - 15x^{8n} + 25x^{2n}}{5x^{2n}}$ $7x^{8n} - 3x^{6n} + 5$

Application Exercises

87. The polynomial
$$28t^4 - 711t^3 + 5963t^2 - 1695t + 27{,}424$$
models the annual number of drug arrests in the United States t years after 1984. The polynomial
$$6t^4 - 207t^3 + 2128t^2 - 6622t + 15{,}220$$
models the annual number of drug convictions in the United States t years after 1984.

  a. Write an algebraic expression that describes the conviction rate for drug arrests in the United States t years after 1984. $\dfrac{6t^4 - 207t^3 + 2128t^2 - 6622t + 15{,}220}{28t^4 - 711t^3 + 5963t^2 - 1695t + 27{,}424}$

  b. Can the polynomial division for the model in part (a) be performed using the methods that you learned in this section? Explain your answer. No; the divisor is not a monomial.

88. The polynomial
$$0.0067t^2 + 2.56t + 250.39$$
models the U.S. population, in millions, t years after 1990. The polynomial
$$-35t^2 + 1160t + 10{,}890$$
models revenue, in millions of dollars, from video sales and rentals in the United States t years after 1990.

  a. Write an algebraic expression that describes the average amount, in dollars, that each person in the United States spent on video sales and rentals t years after 1990. $\dfrac{-35t^2 + 1160t + 10{,}890}{0.0067t^2 + 2.56t + 250.39}$

  b. Can the polynomial division for the model in part (a) be performed using the methods that you learned in this section? Explain your answer. No; the divisor is not a monomial.

Writing in Mathematics

89. Explain the quotient rule for exponents. Use $\dfrac{3^6}{3^2}$ in your explanation. **89–95.** Answers will vary.

90. Explain how to find any nonzero number to the 0 power.

91. Explain the difference between $(-7)^0$ and -7^0.

92. Explain how to simplify an expression that involves a quotient raised to a power. Provide an example with your explanation.

93. Explain how to divide monomials. Give an example.

94. Explain how to divide a polynomial that is not a monomial by a monomial. Give an example.

95. Are the expressions

$$\frac{12x^2 + 6x}{3x} \quad \text{and} \quad 4x + 2$$

equal for every value of x? Explain.

Critical Thinking Exercises

96. Which one of the following is true? **d**

   a. $x^{10} \div x^2 = x^5$ for all nonzero real numbers x.

   b. $\dfrac{12x^3 - 6x}{2x} = 6x^2 - 6x$

   c. $\dfrac{x^2 + x}{x} = x$

   d. If a polynomial in x of degree 6 is divided by a monomial in x of degree 2, the degree of the quotient is 4.

97. What polynomial, when divided by $3x^2$, yields the trinomial $6x^6 - 9x^4 + 12x^2$ as a quotient? $18x^8 - 27x^6 + 36x^4$

In Exercises 98–99, find the missing coefficients and exponents designated by question marks.

98. $\dfrac{?x^8 - ?x^6}{3x^?} = 3x^5 - 4x^3$ $\dfrac{9x^8 - 12x^6}{3x^3}$

99. $\dfrac{3x^{14} - 6x^{12} - ?x^7}{?x^?} = -x^7 + 2x^5 + 3$ $\dfrac{3x^{14} - 6x^{12} - 9x^7}{-3x^7}$

Review Exercises

100. Find the absolute value: $|-20.3|$. (Section 1.2, Example 8)
20.3

101. Express $\frac{7}{8}$ as a decimal. (Section 1.2, Example 4)   0.875

102. Graph: $y = \dfrac{1}{3}x + 2$. (Section 3.4, Example 3)

See graphing answer section.

SECTION 5.6

LONG DIVISION OF POLYNOMIALS; SYNTHETIC DIVISION

Objectives

1 Use long division to divide by a polynomial containing more than one term.

2 Divide polynomials using synthetic division.

For those of you who are dog lovers, you might still be thinking of the polynomial formula that models the age in human years, y, of a dog that is x years old, namely

$$y = -0.001618x^4 + 0.077326x^3 - 1.2367x^2 + 11.460x + 2.914.$$

Suppose that you are in your twenties, say 25. What is Fido's equivalent age? To answer this question, we must substitute 25 for y and solve the resulting polynomial equation for x:

$$25 = -0.001618x^4 + 0.077326x^3 - 1.2367x^2 + 11.460x + 2.914.$$

Don't panic! We won't be solving an equation as complicated as this—yet. You will learn to solve polynomial equations in which the highest power of the variable is 4 in more advanced algebra courses. Part of the method for solving such an equation involves the division of a polynomial by a binomial, such as

$$x + 3\overline{)x^2 + 10x + 21}.$$

Divisor has two terms and is a binomial.   The polynomial dividend has three terms and is a trinomial.

In this section, you will learn how to divide by a polynomial containing more than one term.

1 Use long division to divide by a polynomial containing more than one term.

The Steps in Dividing a Polynomial by a Binomial Dividing a polynomial by a binomial may remind you of long division.

DISCOVER FOR YOURSELF

Divide 3983 by 26 without the use of a calculator. Describe the process of the division using the four steps—*divide, multiply, subtract,* and *bring down.* What do you observe about this process? When does it come to an end?

When a divisor has more than one term, the four steps used to divide whole numbers—**divide, multiply, subtract, bring down the next term**—form the repetitive procedure for polynomial long division.

EXAMPLE 1 Dividing a Polynomial by a Binomial

Divide $x^2 + 10x + 21$ by $x + 3$.

SOLUTION The following steps illustrate how polynomial division is very similar to numerical division.

$$x + 3 \overline{)x^2 + 10x + 21}$$

Arrange the terms of the dividend ($x^2 + 10x + 21$) and the divisor ($x + 3$) in descending powers of x.

$$\begin{array}{r} x \\ x + 3 \overline{)x^2 + 10x + 21} \end{array}$$

Divide x^2 (the first term in the dividend) by x (the first term in the divisor): $\dfrac{x^2}{x} = x$. Align like terms.

$x(x + 3) = x^2 + 3x.$

$$\begin{array}{r} x \\ x + 3 \overline{)x^2 + 10x + 21} \\ x^2 + 3x \end{array}$$

Multiply each term in the divisor ($x + 3$) by x, aligning terms of the product under like terms in the dividend.

$$\begin{array}{r} x \\ x + 3 \overline{)x^2 + 10x + 21} \\ \ominus x^2 \ominus 3x \\ \hline 7x \end{array}$$

Subtract $x^2 + 3x$ from $x^2 + 10x$ by changing the sign of each term in the lower expression and adding.

Change signs of the polynomial being subtracted.

$$\begin{array}{r} x \\ x + 3 \overline{)x^2 + 10x + 21} \\ x^2 + 3x \downarrow \\ \hline 7x + 21 \end{array}$$

Bring down 21 from the original dividend and add algebraically to form a new dividend.

$$\begin{array}{r} x + 7 \\ x + 3 \overline{)x^2 + 10x + 21} \\ x^2 + 3x \\ \hline 7x + 21 \end{array}$$

Find the second term of the quotient. Divide the first term of $7x + 21$ by x, the first term of the divisor: $\dfrac{7x}{x} = 7$.

$$x + 3 \overline{)\,x^2 + 10x + 21\,}$$

$7(x + 3) = 7x + 21$

$$\frac{x + 7}{x + 3 \,\overline{)\, x^2 + 10x + 21\,}}$$
$$\underline{x^2 + 3x}$$
$$7x + 21$$
$$\underline{\ominus 7x \ominus 21}$$
$$0 \quad \text{Remainder}$$

Multiply the divisor $(x + 3)$ by 7, aligning under like terms in the new dividend. Then subtract to obtain the remainder of 0.

The quotient is $x + 7$ and the remainder is 0. We will not list a remainder of 0 in the answer. Thus,

$$\frac{x^2 + 10x + 21}{x + 3} = x + 7.$$

After performing polynomial long division, the answer can be checked. Find the product of the divisor and the quotient, and add the remainder. If the result is the dividend, the answer to the division problem is correct. For example, let's check our work in Example 1.

$$(x^2 + 10x + 21) \div (x + 3) = x + 7$$

Dividend   Divisor   Quotient we wish to check

Multiply the divisor and the quotient and add the remainder, 0:

$$(x + 3)(x + 7) + 0 = x^2 + 7x + 3x + 21 + 0 = x^2 + 10x + 21$$

Divisor   Quotient   Remainder   This is the dividend.

Because we obtained the dividend, the quotient is correct.

CHECK POINT 1 Divide $x^2 + 14x + 45$ by $x + 9$. $x + 5$

Before considering additional examples, let's summarize the general procedure for dividing by a polynomial that contains more than one term.

LONG DIVISION OF POLYNOMIALS

1. **Arrange the terms** of both the dividend and the divisor in descending powers of any variable.

2. **Divide** the first term in the dividend by the first term in the divisor. The result is the first term of the quotient.

3. **Multiply** every term in the divisor by the first term in the quotient. Write the resulting product beneath the dividend with like terms lined up.

4. **Subtract** the product from the dividend.

5. **Bring down** the next term in the original dividend and write it next to the remainder to form a new dividend.

6. Use this new expression as the dividend and repeat this process until the remainder can no longer be divided. This will occur when the degree of the remainder (the highest exponent on a variable in the remainder) is less than the degree of the divisor.

In our next division, we will obtain a nonzero remainder.

EXAMPLE 2 Long Division of Polynomials

Divide: $\dfrac{7x - 9 - 4x^2 + 4x^3}{2x - 1}$.

SOLUTION We begin by writing the dividend in descending powers of x.

$$7x - 9 - 4x^2 + 4x^3 = 4x^3 - 4x^2 + 7x - 9$$

> Think of 9 as $9x^0$. The powers descend from 3 to 0.

$$2x - 1 \overline{)4x^3 - 4x^2 + 7x - 9}$$

This is the problem with the dividend in descending powers of x.

$$\begin{array}{r} 2x^2 \\ 2x - 1 \overline{)4x^3 - 4x^2 + 7x - 9} \end{array}$$

Divide: $\dfrac{4x^3}{2x} = 2x^2$.

$$2x^2(2x - 1) = 4x^3 - 2x^2$$

$$\begin{array}{r} 2x^2 \\ 2x - 1 \overline{)4x^3 - 4x^2 + 7x - 9} \\ 4x^3 - 2x^2 \end{array}$$

Multiply: $2x^2(2x - 1) = 4x^3 - 2x^2$.

$$\begin{array}{r} 2x^2 \\ 2x - 1 \overline{)4x^3 - 4x^2 + 7x - 9} \\ \ominus 4x^3 \oplus 2x^2 \\ \hline -2x^2 \end{array}$$

Subtract: $4x^3 - 4x^2 - (4x^3 - 2x^2)$
$= 4x^3 - 4x^2 - 4x^3 + 2x^2$
$= -2x^2$.

> Change signs of the polynomial being subtracted.

$$\begin{array}{r} 2x^2 \\ 2x - 1 \overline{)4x^3 - 4x^2 + 7x - 9} \\ 4x^3 - 2x^2 \\ \hline -2x^2 + 7x \end{array}$$

Bring down 7x. The new dividend is $-2x^2 + 7x$.

$$\begin{array}{r} 2x^2 - x \\ 2x - 1 \overline{)4x^3 - 4x^2 + 7x - 9} \\ 4x^3 - 2x^2 \\ \hline -2x^2 + 7x \end{array}$$

Divide: $\dfrac{-2x^2}{2x} = -x$.

$$-x(2x - 1) = -2x^2 + x$$

$$\begin{array}{r} 2x^2 - x \\ 2x - 1 \overline{)4x^3 - 4x^2 + 7x - 9} \\ 4x^3 - 2x^2 \\ \hline -2x^2 + 7x \\ -2x^2 + x \end{array}$$

Multiply: $-x(2x - 1) = -2x^2 + x$.

$$\begin{array}{r} 2x^2 - x \\ 2x - 1 \overline{)4x^3 - 4x^2 + 7x - 9} \\ 4x^3 - 2x^2 \\ \hline -2x^2 + 7x \\ \oplus -2x^2 \ominus x \\ \hline 6x \end{array}$$

Subtract:
$-2x^2 + 7x - (-2x^2 + x)$
$= -2x^2 + 7x + 2x^2 - x$
$= 6x$.

$$
\begin{array}{r}
2x^2 - x \\
2x - 1 \overline{)\, 4x^3 - 4x^2 + 7x - 9} \\
\underline{4x^3 - 2x^2} \\
-2x^2 + 7x \\
\underline{-2x^2 + x} \\
6x - 9
\end{array}
$$

Bring down -9. *The new dividend is* $6x - 9$.

$$
\begin{array}{r}
2x^2 - x + 3 \\
2x - 1 \overline{)\, 4x^3 - 4x^2 + 7x - 9} \\
\underline{4x^3 - 2x^2} \\
-2x^2 + 7x \\
\underline{-2x^2 + x} \\
6x - 9
\end{array}
$$

Divide: $\dfrac{6x}{2x} = 3$.

$3(2x - 1) = 6x - 3$

$$
\begin{array}{r}
2x^2 - x + 3 \\
2x - 1 \overline{)\, 4x^3 - 4x^2 + 7x - 9} \\
\underline{4x^3 - 2x^2} \\
-2x^2 + 7x \\
\underline{-2x^2 + x} \\
6x - 9 \\
6x - 3
\end{array}
$$

Multiply: $3(2x - 1) = 6x - 3$.

$$
\begin{array}{r}
2x^2 - x + 3 \\
2x - 1 \overline{)\, 4x^3 - 4x^2 + 7x - 9} \\
\underline{4x^3 - 2x^2} \\
-2x^2 + 7x \\
\underline{-2x^2 + x} \\
6x - 9 \\
\underline{\ominus 6x \oplus 3} \\
-6
\end{array}
$$

Subtract:
$6x - 9 - (6x - 3)$
$= 6x - 9 - 6x + 3$
$= -6$

Remainder

The quotient is $2x^2 - x + 3$ and the remainder is -6. When there is a nonzero remainder, as in this example, list the quotient, plus the remainder above the divisor. Thus,

$$
\frac{7x - 9 - 4x^2 + 4x^3}{2x - 1} = 2x^2 - x + 3 + \frac{-6}{2x - 1}
$$

Remainder above divisor

Quotient

or

$$
\frac{7x - 9 - 4x^2 + 4x^3}{2x - 1} = 2x^2 - x + 3 - \frac{6}{2x - 1}.
$$

Check this result by showing that the product of the divisor and the quotient,

$$
(2x - 1)(2x^2 - x + 3),
$$

plus the remainder, -6, is the dividend, $7x - 9 - 4x^2 + 4x^3$. ∎

CHECK POINT 2 Divide: $\dfrac{6x + 8x^2 - 12}{2x + 3}$. $\quad 4x - 3 - \dfrac{3}{2x + 3}$

If a power of a variable is missing in a dividend, add that power of the variable with a coefficient of 0 and then divide. In this way, like terms will be aligned as you carry out the division.

EXAMPLE 3   Dividing a Polynomial with Missing Terms

Divide: $\dfrac{8x^3 - 1}{2x - 1}$.

SOLUTION   We write the dividend, $8x^3 - 1$, as

$$8x^3 + 0x^2 + 0x - 1.$$

> Use a coefficient of 0 with missing terms.

By doing this, we will keep all like terms aligned.

$$4x^2(2x - 1) = 8x^3 - 4x^2 \qquad 2x - 1 \overline{) 8x^3 + 0x^2 + 0x - 1 }$$
$$\ominus \, 8x^3 \oplus 4x^2$$
$$4x^2 + 0x$$

Divide $\left(\dfrac{8x^3}{2x} = 4x^2\right)$, multiply, subtract, and bring down the next term.
The new dividend is $4x^2 + 0x$.

$$2x(2x - 1) = 4x^2 - 2x \qquad 2x - 1 \overline{) 8x^3 + 0x^2 + 0x - 1 }$$
$$8x^3 - 4x^2$$
$$4x^2 + 0x$$
$$\ominus \, 4x^2 \oplus 2x$$
$$2x - 1$$

Divide $\left(\dfrac{4x^2}{2x} = 2x\right)$, multiply $[2x(2x - 1) = 4x^2 - 2x]$, subtract, and bring down the next term.
The new dividend is $2x - 1$.

$$1(2x - 1) = 2x - 1 \qquad 2x - 1 \overline{) 8x^3 + 0x^2 + 0x - 1 }$$
$$8x^3 - 4x^2$$
$$4x^2 + 0x$$
$$4x^2 - 2x$$
$$2x - 1$$
$$\ominus \, 2x \oplus 1$$
$$0$$

Divide $\left(\dfrac{2x}{2x} = 1\right)$, multiply $[1(2x - 1) = 2x - 1]$, and subtract. The remainder is 0.

Thus,

$$\frac{8x^3 - 1}{2x - 1} = 4x^2 + 2x + 1.$$

Check this result by showing that the product of the divisor and the quotient,

$$(2x - 1)(4x^2 + 2x + 1)$$

plus the remainder, 0, is the dividend, $8x^3 - 1$. ∎

USING TECHNOLOGY

The graphs of $y_1 = \dfrac{8x^3 - 1}{2x - 1}$ and $y_2 = 4x^2 + 2x + 1$ are shown below.

$y_1 = \dfrac{8x^3 - 1}{2x - 1}$

$y_2 = 4x^2 + 2x + 1$

$[-3, 3, 1]$ by $[-1, 15, 1]$

The graphs coincide. Thus,

$$\frac{8x^3 - 1}{2x - 1} = 4x^2 + 2x + 1.$$

✔ **CHECK POINT 3** Divide: $\dfrac{x^3 - 1}{x - 1}$. $x^2 + x + 1$

EXAMPLE 4   Long Division of Polynomials

Divide: $6x^4 + 5x^3 + 3x - 5$ by $3x^2 - 2x$.

SOLUTION   We write the dividend, $6x^4 + 5x^3 + 3x - 5$, as $6x^4 + 5x^3 + 0x^2 + 3x - 5$ to keep all like terms aligned.

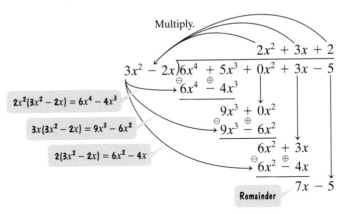

The division process is finished because the degree of $7x - 5$, which is 1, is less than the degree of the divisor $3x^2 - 2x$, which is 2. The answer is

$$\frac{6x^4 + 5x^3 + 3x - 5}{3x^2 - 2x} = 2x^2 + 3x + 2 + \frac{7x - 5}{3x^2 - 2x}.$$ ∎

✔ **CHECK POINT 4** Divide: $2x^4 + 3x^3 - 7x - 10$ by $x^2 - 2x$. $2x^2 + 7x + 14 + \dfrac{21x - 10}{x^2 - 2x}$

Dividing Polynomials Using Synthetic Division   We can use **synthetic division** to divide polynomials if the divisor is of the form $x - c$. This method provides a quotient more quickly than long division. Let's compare the two methods showing $x^3 + 4x^2 - 5x + 5$ divided by $x - 3$.

② Divide polynomials using synthetic division.

Long Division

$$\begin{array}{r}
x^2 + 7x + 16 \\
x - 3 \overline{)\, x^3 + 4x^2 - 5x + 5} \\
\underline{x^3 - 3x^2} \\
7x^2 - 5x \\
\underline{7x^2 - 21x} \\
16x + 5 \\
\underline{16x - 48} \\
53
\end{array}$$

Quotient

Dividend

Divisor
$x - c$;
$c = 3$

Remainder

Synthetic Division

$$\begin{array}{r|rrrr}
3 & 1 & 4 & -5 & 5 \\
 &   & 3 & 21 & 48 \\
\hline
 & 1 & 7 & 16 & 53
\end{array}$$

Notice the relationship between the polynomials in the long division process and the numbers that appear in synthetic division.

These are the coefficients of the dividend $x^3 + 4x^2 - 5x + 5$.

The divisor is $x - 3$.
This is 3, or c, in $x - c$.

$$3 \mid \begin{array}{rrrr} 1 & 4 & -5 & 5 \\ & 3 & 21 & 48 \\ \hline 1 & 7 & 16 & 53 \end{array}$$

These are the coefficients of the quotient $x^2 + 7x + 16$.

This is the remainder.

Now let's look at the steps involved in synthetic division.

SYNTHETIC DIVISION   To divide a polynomial by $x - c$:

Example

1. Arrange polynomials in descending powers, with a 0 coefficient for any missing term.

$$x - 3 \overline{)x^3 + 4x^2 - 5x + 5}$$

2. Write c for the divisor, $x - c$. To the right, write the coefficients of the dividend.

$$3 \mid \begin{array}{rrrr} 1 & 4 & -5 & 5 \end{array}$$

3. Write the leading coefficient of the dividend on the bottom row.

$$3 \mid \begin{array}{rrrr} 1 & 4 & -5 & 5 \end{array}$$
↓ Bring down 1 .
$$1$$

4. Multiply c (in this case, 3) times the value just written on the bottom row. Write the product in the next column in the second row.

$$3 \mid \begin{array}{rrrr} 1 & 4 & -5 & 5 \\ & 3 & & \\ \hline 1 & & & \end{array}$$

Multiply by 3: **3 · 1 = 3.**

5. Add the values in this new column, writing the sum in the bottom row.

$$3 \mid \begin{array}{rrrr} 1 & 4 & -5 & 5 \\ & 3 & & \\ \hline 1 & 7 & & \end{array}$$ Add.

6. Repeat this series of multiplications and additions until all columns are filled in.

$$3 \mid \begin{array}{rrrr} 1 & 4 & -5 & 5 \\ & 3 & 21 & \\ \hline 1 & 7 & 16 & \end{array}$$ Add.

Multiply by 3: **3 · 7 = 21.**

$$3 \mid \begin{array}{rrrr} 1 & 4 & -5 & 5 \\ & 3 & 21 & 48 \\ \hline 1 & 7 & 16 & 53 \end{array}$$ Add.

Multiply by 3: **3 · 16 = 48.**

7. Use the numbers in the last row to write the quotient, plus the remainder above the divisor. **The degree of the first term of the quotient is one less than the degree of the first term of the dividend.** The final value in this row is the remainder.

Written from
1 7 16 **53**
the last row of the synthetic division

$$1x^2 + 7x + 16 + \frac{53}{x - 3}$$

$$x - 3 \overline{)x^3 + 4x^2 - 5x + 5}$$

EXAMPLE 5 | Using Synthetic Division

Use synthetic division to divide $5x^3 + 6x + 8$ by $x + 2$.

SOLUTION The divisor must be in the form $x - c$. Thus, we write $x + 2$ as $x - (-2)$. This means that $c = -2$. Writing a 0 coefficient for the missing x^2-term in the dividend, we can express the division as follows:

$$x - (-2)\overline{)5x^3 + 0x^2 + 6x + 8}.$$

Now we are ready to set up the problem so that we can use synthetic division.

Use the coefficients of the dividend
$5x^3 + 0x^2 + 6x + 8$ in descending powers of x.

This is c in
$x - (-2)$. $-2\underline{|}\ \ 5\ \ 0\ \ 6\ \ 8$

We begin the synthetic division process by bringing down 5. This is followed by a series of multiplications and additions.

1. Bring down 5.

$$-2\underline{|}\ \ 5\ \ 0\ \ 6\ \ 8$$
$$\ \ \ \ \ \ \ 5$$

2. Multiply: $-2(5) = -10$.

$$-2\underline{|}\ \ 5\ \ \ \ 0\ \ 6\ \ \ 8$$
$$\ \ \ \ \ \ \ \ \ -10$$
$$\ \ \ \ \ \ 5$$

Multiply by -2.

3. Add: $0 + (-10) = -10$.

$$-2\underline{|}\ \ 5\ \ \ \ 0\ |\ 6\ \ 8$$
$$\ \ \ \ \ \ \ -10\ |\ \text{Add.}$$
$$\ \ \ \ \ 5\ \ -10$$

4. Multiply: $-2(-10) = 20$.

$$-2\underline{|}\ \ 5\ \ \ \ 0\ \ 6\ \ \ \ 8$$
$$\ \ \ \ \ \ \ -10\ \ 20$$
$$\ \ \ \ \ 5\ \ -10$$

Multiply by -2.

5. Add: $6 + 20 = 26$.

$$-2\underline{|}\ \ 5\ \ \ \ 0\ \ 6\ |\ 8$$
$$\ \ \ \ \ \ \ -10\ \ 20\ |\ \text{Add.}$$
$$\ \ \ \ \ 5\ \ -10\ \ 26$$

6. Multiply: $-2(26) = -52$.

$$-2\underline{|}\ \ 5\ \ \ \ 0\ \ 6\ \ \ \ 8$$
$$\ \ \ \ \ \ \ -10\ \ 20\ \ -52$$
$$\ \ \ \ \ 5\ \ -10\ \ 26$$

Multiply by -2.

7. Add: $8 + (-52) = -44$.

$$-2\underline{|}\ \ 5\ \ \ \ 0\ \ 6\ \ \ \ 8\ |$$
$$\ \ \ \ \ \ \ -10\ \ 20\ \ -52\ |\ \text{Add.}$$
$$\ \ \ \ \ 5\ \ -10\ \ 26\ \ -44$$

The numbers in the last row represent the coefficients of the quotient and the remainder. The degree of the first term of the quotient is one less than that of the dividend. Because the degree of the dividend, $5x^3 + 6x + 8$, is 3, the degree of the quotient is 2. This means that the 5 in the last row represents $5x^2$.

$$-2\underline{|}\ \ 5\ \ \ \ 0\ \ 6\ \ \ \ 8$$
$$\ \ \ \ \ \ \ -10\ \ 20\ \ -52$$
$$\ \ \ \ \ 5\ \ -10\ \ 26\ \ -44$$

The quotient is   The remainder
$5x^2 - 10x + 26$.   is -44.

Thus,

$$x + 2 \overline{)5x^3 + 6x + 8} \quad 5x^2 - 10x + 26 - \frac{44}{x + 2}$$

 CHECK POINT 5 Use synthetic division to divide

$$x^3 - 7x - 6 \text{ by } x + 2. \quad x^2 - 2x - 3$$

5.6 EXERCISE SET

Student Solutions Manual   CD/Video   PH Math/Tutor Center   MathXL Tutorials on CD   Math XL   MyMathLab   Interactmath.com
MathXL®   MyMathLab

Practice Exercises

In Exercises 1–40, divide as indicated. Check each answer by showing that the product of the divisor and the quotient, plus the remainder, is the dividend.

1. $\dfrac{x^2 + 6x + 8}{x + 2}$ $x + 4$

2. $\dfrac{x^2 + 7x + 10}{x + 5}$ $x + 2$

3. $\dfrac{2x^2 + x - 10}{x - 2}$ $2x + 5$

4. $\dfrac{2x^2 + 13x + 15}{x + 5}$ $2x + 3$

5. $\dfrac{x^2 - 5x + 6}{x - 3}$ $x - 2$

6. $\dfrac{x^2 - 2x - 24}{x + 4}$ $x - 6$

7. $\dfrac{2y^2 + 5y + 2}{y + 2}$ $2y + 1$

8. $\dfrac{2y^2 - 13y + 21}{y - 3}$ $2y - 7$

9. $\dfrac{x^2 - 3x + 4}{x + 2}$ $x - 5 + \dfrac{14}{x + 2}$

10. $\dfrac{x^2 - 7x + 5}{x + 3}$ $x - 10 + \dfrac{35}{x + 3}$

11. $\dfrac{5y + 10 + y^2}{y + 2}$ $y + 3 + \dfrac{4}{y + 2}$

12. $\dfrac{-8y + y^2 - 9}{y - 3}$ $y - 5 - \dfrac{24}{y - 3}$

13. $\dfrac{x^3 - 6x^2 + 7x - 2}{x - 1}$ $x^2 - 5x + 2$

14. $\dfrac{x^3 + 3x^2 + 5x + 3}{x + 1}$ $x^2 + 2x + 3$

15. $\dfrac{12y^2 - 20y + 3}{2y - 3}$ $6y - 1$ **16.** $\dfrac{4y^2 - 8y - 5}{2y + 1}$ $2y - 5$

17. $\dfrac{4a^2 + 4a - 3}{2a - 1}$ $2a + 3$ **18.** $\dfrac{2b^2 - 9b - 5}{2b + 1}$ $b - 5$

19. $\dfrac{3y - y^2 + 2y^3 + 2}{2y + 1}$ $y^2 - y + 2$

20. $\dfrac{9y + 18 - 11y^2 + 12y^3}{4y + 3}$ $3y^2 - 5y + 6$

21. $\dfrac{6x^2 - 5x - 30}{2x - 5}$ $3x + 5 - \dfrac{5}{2x - 5}$

22. $\dfrac{4y^2 + 8y + 3}{2y - 1}$ $2y + 5 + \dfrac{8}{2y - 1}$

23. $\dfrac{x^3 + 4x - 3}{x - 2}$ $x^2 + 2x + 8 + \dfrac{13}{x - 2}$

24. $\dfrac{x^3 + 2x^2 - 3}{x - 2}$ $x^2 + 4x + 8 + \dfrac{13}{x - 2}$

25. $\dfrac{4y^3 + 8y^2 + 5y + 9}{2y + 3}$ $2y^2 + y + 1 + \dfrac{6}{2y + 3}$

26. $\dfrac{2y^3 - y^2 + 3y + 2}{2y + 1}$ $y^2 - y + 2$

27. $\dfrac{6y^3 - 5y^2 + 5}{3y + 2}$ $2y^2 - 3y + 2 + \dfrac{1}{3y + 2}$

28. $\dfrac{4y^3 + 3y + 5}{2y - 3}$ $2y^2 + 3y + 6 + \dfrac{23}{2y - 3}$

29. $\dfrac{27x^3 - 1}{3x - 1}$ $9x^2 + 3x + 1$

30. $\dfrac{8x^3 + 27}{2x + 3}$ $4x^2 - 6x + 9$

31. $\dfrac{81 - 12y^3 + 54y^2 + y^4 - 108y}{y - 3}$ $y^3 - 9y^2 + 27y - 27$

32. $\dfrac{8y^3 + y^4 + 16 + 32y + 24y^2}{y + 2}$ $y^3 + 6y^2 + 12y + 8$

33. $\dfrac{4y^2 + 6y}{2y - 1}$ $2y + 4 + \dfrac{4}{2y - 1}$

34. $\dfrac{10x^2 - 3x}{x + 3}$ $10x - 33 + \dfrac{99}{x + 3}$

35. $\dfrac{y^4 - 2y^2 + 5}{y - 1}$ $y^3 + y^2 - y - 1 + \dfrac{4}{y - 1}$

36. $\dfrac{y^4 - 6y^2 + 3}{y - 1}$ $y^3 + y^2 - 5y - 5 - \dfrac{2}{y - 1}$

37. $(4x^4 + 3x^3 + 4x^2 + 9x - 6) \div (x^2 + 3)$ $4x^2 + 3x - 8 + \dfrac{18}{x^2 + 3}$

38. $(3x^5 - x^3 + 4x^2 - 12x - 8) \div (x^2 - 2)$

$3x^3 + 5x + 4 - \dfrac{2x}{x^2 - 2}$

39. $(15x^4 + 3x^3 + 4x^2 + 4) \div (3x^2 - 1)$ $5x^2 + x + 3 + \dfrac{x + 7}{3x^2 - 1}$

40. $(18x^4 + 9x^3 + 3x^2) \div (3x^2 + 1)$ $6x^2 + 3x - 1 + \dfrac{1 - 3x}{3x^2 + 1}$

In Exercises 41–58, divide using synthetic division. In the first two exercises, begin the process as shown.

41. $(2x^2 + x - 10) \div (x - 2)$ $\underline{2|}$ $2\ \ 1\ \ -10$ $2x + 5$

42. $(x^2 + x - 2) \div (x - 1)$ $\underline{1|}$ $1\ \ 1\ \ -2$ $x + 2$

43. $(3x^2 + 7x - 20) \div (x + 5)$ $3x - 8 + \dfrac{20}{x + 5}$

44. $(5x^2 - 12x - 8) \div (x + 3)$ $5x - 27 + \dfrac{73}{x + 3}$

45. $(4x^3 - 3x^2 + 3x - 1) \div (x - 1)$ $4x^2 + x + 4 + \dfrac{3}{x - 1}$

46. $(5x^3 - 6x^2 + 3x + 11) \div (x - 2)$ $5x^2 + 4x + 11 + \dfrac{33}{x - 2}$

47. $(6x^5 - 2x^3 + 4x^2 - 3x + 1) \div (x - 2)$

$6x^4 + 12x^3 + 22x^2 + 48x + 93 + \dfrac{187}{x - 2}$

48. $(x^5 + 4x^4 - 3x^2 + 2x + 3) \div (x - 3)$

$x^4 + 7x^3 + 21x^2 + 60x + 182 + \dfrac{549}{x - 3}$

49. $(x^2 - 5x - 5x^3 + x^4) \div (5 + x)$

$x^3 - 10x^2 + 51x - 260 + \dfrac{1300}{5 + x}$

50. $(x^2 - 6x - 6x^3 + x^4) \div (6 + x)$

$x^3 - 12x^2 + 73x - 444 + \dfrac{2664}{6 + x}$

51. $(3x^3 + 2x^2 - 4x + 1) \div \left(x - \dfrac{1}{3}\right)$ $3x^2 + 3x - 3$

52. $(2x^4 - x^3 + 2x^2 - 3x + 1) \div \left(x - \dfrac{1}{2}\right)$ $2x^3 + 2x - 2$

53. $\dfrac{x^5 + x^3 - 2}{x - 1}$ $x^4 + x^3 + 2x^2 + 2x + 2$

54. $\dfrac{x^7 + x^5 - 10x^3 + 12}{x + 2}$

$x^6 - 2x^5 + 5x^4 - 10x^3 + 10x^2 - 20x + 40 - \dfrac{68}{x + 2}$

55. $\dfrac{x^4 - 256}{x - 4}$ $x^3 + 4x^2 + 16x + 64$

56. $\dfrac{x^7 - 128}{x - 2}$ $x^6 + 2x^5 + 4x^4 + 8x^3 + 16x^2 + 32x + 64$

57. $\dfrac{2x^5 - 3x^4 + x^3 - x^2 + 2x - 1}{x + 2}$

$2x^4 - 7x^3 + 15x^2 - 31x + 64 - \dfrac{129}{x + 2}$

58. $\dfrac{x^5 - 2x^4 - x^3 + 3x^2 - x + 1}{x - 2}$ $x^4 - x^2 + x + 1 + \dfrac{3}{x - 2}$

Practice Plus

In Exercises 59–68, divide as indicated.

59. $\dfrac{x^4 + y^4}{x + y}$ $x^3 - x^2y + xy^2 - y^3 + \dfrac{2y^4}{x + y}$

60. $\dfrac{x^5 + y^5}{x + y}$ $x^4 - x^3y + x^2y^2 - xy^3 + y^4$

61. $\dfrac{3x^4 + 5x^3 + 7x^2 + 3x - 2}{x^2 + x + 2}$ $3x^2 + 2x - 1$

62. $\dfrac{x^4 - x^3 - 7x^2 - 7x - 2}{x^2 - 3x - 2}$ $x^2 + 2x + 1$

63. $\dfrac{4x^3 - 3x^2 + x + 1}{x^2 + x + 1}$ $4x - 7 + \dfrac{4x + 8}{x^2 + x + 1}$

64. $\dfrac{x^4 - x^2 + 1}{x^2 + x + 1}$ $x^2 - x - 1 + \dfrac{2x + 2}{x^2 + x + 1}$

65. $\dfrac{x^5 - 1}{x^2 - x + 2}$ $x^3 + x^2 - x - 3 + \dfrac{-x + 5}{x^2 - x + 2}$

66. $\dfrac{5x^5 - 7x^4 + 3x^3 - 20x^2 + 28x - 12}{x^3 - 4}$ $5x^2 - 7x + 3$

67. $\dfrac{4x^3 - 7x^2y - 16xy^2 + 3y^3}{x - 3y}$ $4x^2 + 5xy - y^2$

68. $\dfrac{12x^3 - 19x^2y + 13xy^2 - 10y^3}{4x - 5y}$ $3x^2 - xy + 2y^2$

69. Divide the difference between $4x^3 + x^2 - 2x + 7$ and $3x^3 - 2x^2 - 7x + 4$ by $x + 1$. $x^2 + 2x + 3$

70. Divide the difference between $4x^3 + 2x^2 - x - 1$ and $2x^3 - x^2 + 2x - 5$ by $x + 2$. $2x^2 - x - 1 + \dfrac{6}{x + 2}$

Application Exercises

71. Write a simplified polynomial that represents the length of the rectangle. $x^2 + 2x + 3$ units

The width is $x + 1$ units.

The area is $x^3 + 3x^2 + 5x + 3$ square units.

72. Write a simplified polynomial that represents the measure of the base of the parallelogram. $2x^2 + 3x - 4$ units

The height is $2x + 3$ units.

The area is $4x^3 + 12x^2 + x - 12$ square units.

You just signed a contract for a new job. The salary for the first year is $30,000 and there is to be a percent increase in your salary each year. The algebraic expression

$$\frac{30{,}000x^n - 30{,}000}{x - 1}$$

describes your total salary over n years, where x is the sum of 1 and the yearly percent increase, expressed as a decimal. Use this information to solve Exercises 73–74.

73. a. Use the given expression and write a quotient of polynomials that describes your total salary over three years.

$$\frac{30{,}000x^3 - 30{,}000}{x - 1}$$

   b. Simplify the expression in part (a) by performing the division. $30{,}000x^2 + 30{,}000x + 30{,}000$

   c. Suppose you are to receive an increase of 5% per year. Thus, x is the sum of 1 and 0.05, or 1.05. Substitute 1.05 for x in the expression in part (a) as well as in the simplified form of the expression in part (b). Evaluate each expression. What is your total salary over the three-year period? $94,575

74. a. Use the given expression and write a quotient of polynomials that describes your total salary over four years.

$$\frac{30{,}000x^4 - 30{,}000}{x - 1}$$

   b. Simplify the expression in part (a) by performing the division. $30{,}000x^3 + 30{,}000x^2 + 30{,}000x + 30{,}000$

   c. Suppose you are to receive an increase of 8% per year. Thus, x is the sum of 1 and 0.08, or 1.08. Substitute 1.08 for x in the expression in part (a) as well as in the simplified form of the expression in part (b). Evaluate each expression. What is your total salary over the four-year period? $135,183.36

Writing in Mathematics

75. In your own words, explain how to divide a polynomial by a binomial. Use $\dfrac{x^2 + 4}{x + 2}$ in your explanation.   75–78. Answers will vary.

76. When dividing a polynomial by a binomial, explain when to stop dividing.

77. After dividing a polynomial by a binomial, explain how to check the answer.

78. When dividing a binomial into a polynomial with missing terms, explain the advantage of writing the missing terms with zero coefficients.

Critical Thinking Exercises

79. Which one of the following is true?   b

   a. If $4x^2 + 25x - 3$ is divided by $4x + 1$, the remainder is 9.
   b. If polynomial division results in a remainder of zero, then the product of the divisor and the quotient is the dividend.
   c. The degree of a polynomial is the power of the term that appears in the first position.
   d. When a polynomial is divided by a binomial, the division process stops when the last term of the dividend is brought down.

80. When a certain polynomial is divided by $2x + 4$, the quotient is

$$x - 3 + \frac{17}{2x + 4}.$$

   What is the polynomial? $2x^2 - 2x + 5$

81. Find the number k such that when $16x^2 - 2x + k$ is divided by $2x - 1$, the remainder is 0. -3

82. Describe the pattern that you observe in the following quotients and remainders.

$$\frac{x^3 - 1}{x + 1} = x^2 - x + 1 - \frac{2}{x + 1}$$

$$\frac{x^5 - 1}{x + 1} = x^4 - x^3 + x^2 - x + 1 - \frac{2}{x + 1}$$

   Use this pattern to find $\dfrac{x^7 - 1}{x + 1}$. Verify your result by dividing.

   Answers will vary; $x^6 - x^5 + x^4 - x^3 + x^2 - x + 1 - \dfrac{2}{x + 1}$

Technology Exercises

In Exercises 83–87, use a graphing utility to determine whether the divisions have been performed correctly. Graph each side of the given equation in the same viewing rectangle. The graphs should coincide. If they do not, correct the expression on the right side by using polynomial division. Then use your graphing utility to show that the division has been performed correctly.

83. $\dfrac{x^2 - 4}{x - 2} = x + 2$ Graphs coincide.

84. $\dfrac{x^2 - 25}{x - 5} = x - 5$ $x - 5$ should be $x + 5$.

85. $\dfrac{2x^2 + 13x + 15}{x - 5} = 2x + 3$ $2x + 3$ should be $2x + 23 + \frac{130}{x - 5}$.

86. $\dfrac{6x^2 + 16x + 8}{3x + 2} = 2x - 4$ $2x - 4$ should be $2x + 4$.

87. $\dfrac{x^3 + 3x^2 + 5x + 3}{x + 1} = x^2 - 2x + 3$ $x^2 - 2x + 3$ should
be $x^2 + 2x + 3$.

Review Exercises

88. Solve the system:
$$7x - 6y = 17$$
$$3x + y = 18$$
(Section 4.3, Example 2) $\{(5, 3)\}$

89. What is 6% of 20? (Section 2.4, Example 7)   1.2

90. Solve: $\dfrac{x}{3} + \dfrac{2}{5} = \dfrac{x}{5} - \dfrac{2}{5}$ (Section 2.3, Example 4) $\{-6\}$

SECTION 5.7

NEGATIVE EXPONENTS AND SCIENTIFIC NOTATION

Objectives

1 Use the negative exponent rule.

2 Simplify exponential expressions.

3 Convert from scientific notation to decimal notation.

4 Convert from decimal notation to scientific notation.

5 Compute with scientific notation.

6 Solve applied problems with scientific notation.

You are listening to a discussion of the country's 6.8 trillion dollar deficit. It seems that this is a real problem, but then you realize that you don't really know what this number means. How can you look at this deficit in the proper perspective? If the national debt were evenly divided among all citizens of the country, how much would each citizen have to pay?

In the new millennium, literacy with numbers, called **numeracy**, will be a necessary skill for functioning in a meaningful way personally, professionally, and as a citizen. In this section, you will learn to use exponents to provide a way of putting large and small numbers into perspective.

1   Use the negative exponent rule.

Negative Integers as Exponents   A nonzero base can be raised to a negative power. The quotient rule can be used to help determine what a negative integer as an exponent should mean. Consider the quotient of b^3 and b^5, where b is not zero. We can determine this quotient in two ways.

$$\frac{b^3}{b^5} = \frac{1 \cdot \cancel{b} \cdot \cancel{b} \cdot \cancel{b}}{\cancel{b} \cdot \cancel{b} \cdot \cancel{b} \cdot b \cdot b} = \frac{1}{b^2} \qquad \frac{b^3}{b^5} = b^{3-5} = b^{-2}$$

After dividing out pairs of factors, we have two factors of b in the denominator.

Use the quotient rule and subtract exponents.

Notice that $\dfrac{b^3}{b^5}$ equals both b^{-2} and $\dfrac{1}{b^2}$. This means that b^{-2} must equal $\dfrac{1}{b^2}$. This example is a special case of the **negative exponent rule**.

> **THE NEGATIVE EXPONENT RULE** If b is any real number other than 0 and n is a natural number, then
>
> $$b^{-n} = \frac{1}{b^n}.$$

STUDY TIP

A negative exponent does not make the value of an expression negative. For example,

$$7^{-2} = \frac{1}{7^2} = \frac{1}{49}$$

is positive. Avoid these common errors:

Incorrect!

EXAMPLE 1 Using the Negative Exponent Rule

Use the negative exponent rule to write each expression with a positive exponent. Then simplify the expression.

a. 7^{-2} **b.** 4^{-3} **c.** $(-2)^{-4}$ **d.** -2^{-4} **e.** 5^{-1}

SOLUTION

a. $7^{-2} = \dfrac{1}{7^2} = \dfrac{1}{7\cdot 7} = \dfrac{1}{49}$

b. $4^{-3} = \dfrac{1}{4^3} = \dfrac{1}{4\cdot 4\cdot 4} = \dfrac{1}{64}$

c. $(-2)^{-4} = \dfrac{1}{(-2)^4} = \dfrac{1}{(-2)(-2)(-2)(-2)} = \dfrac{1}{16}$

d. $-2^{-4} = -\dfrac{1}{2^4} = -\dfrac{1}{2\cdot 2\cdot 2\cdot 2} = -\dfrac{1}{16}$

> The negative is not inside parentheses and is not taken to the −4 power.

e. $5^{-1} = \dfrac{1}{5^1} = \dfrac{1}{5}$

■

✔ **CHECK POINT 1** Use the negative exponent rule to write each expression with a positive exponent. Then simplify the expression.

a. 6^{-2} $\dfrac{1}{6^2} = \dfrac{1}{36}$ **b.** 5^{-3} $\dfrac{1}{5^3} = \dfrac{1}{125}$

c. $(-3)^{-4}$ $\dfrac{1}{(-3)^4} = \dfrac{1}{81}$ **d.** -3^{-4} $-\dfrac{1}{3^4} = -\dfrac{1}{81}$ **e.** 8^{-1} $\dfrac{1}{8^1} = \dfrac{1}{8}$

Negative exponents can also appear in denominators. For example,

$$\frac{1}{2^{-10}} = \frac{1}{\dfrac{1}{2^{10}}} = 1 \div \frac{1}{2^{10}} = 1 \cdot \frac{2^{10}}{1} = 2^{10}.$$

In general, if a negative exponent appears in a denominator, an expression can be written with a positive exponent using

$$\frac{1}{b^{-n}} = b^n.$$

For example,

$$\frac{1}{2^{-3}} = 2^3 = 8 \qquad \text{and} \qquad \frac{1}{(-6)^{-2}} = (-6)^2 = 36.$$

> Change only the sign of the exponent and not the sign of the base, −6.

NEGATIVE EXPONENTS IN NUMERATORS AND DENOMINATORS If b is any real number other than 0 and n is a natural number, then

$$b^{-n} = \frac{1}{b^n} \qquad \text{and} \qquad \frac{1}{b^{-n}} = b^n.$$

When a negative number appears as an exponent, switch the position of the base (from numerator to denominator or from denominator to numerator) and make the exponent positive. The sign of the base does not change.

EXAMPLE 2 Using Negative Exponents

Write each expression with positive exponents only. Then simplify, if possible.

a. $\dfrac{4^{-3}}{5^{-2}}$ b. $\left(\dfrac{3}{4}\right)^{-2}$ c. $\dfrac{1}{4x^{-3}}$ d. $\dfrac{x^{-5}}{y^{-1}}$

SOLUTION

a. $\dfrac{4^{-3}}{5^{-2}} = \dfrac{5^2}{4^3} = \dfrac{5 \cdot 5}{4 \cdot 4 \cdot 4} = \dfrac{25}{64}$

> Switch the position of the bases and make the exponents positive.

b. $\left(\dfrac{3}{4}\right)^{-2} = \dfrac{3^{-2}}{4^{-2}} = \dfrac{4^2}{3^2} = \dfrac{4 \cdot 4}{3 \cdot 3} = \dfrac{16}{9}$

> Switch the position of the bases and make the exponents positive.

c. $\dfrac{1}{4x^{-3}} = \dfrac{x^3}{4}$

> Switch the position of the base and make the exponent positive. Note that only x is raised to the −3 power.

d. $\dfrac{x^{-5}}{y^{-1}} = \dfrac{y^1}{x^5} = \dfrac{y}{x^5}$

■

✔ **CHECK POINT 2** Write each expression with positive exponents only. Then simplify, if possible.

a. $\dfrac{2^{-3}}{7^{-2}} \quad \dfrac{7^2}{2^3} = \dfrac{49}{8}$ b. $\left(\dfrac{4}{5}\right)^{-2} \quad \dfrac{5^2}{4^2} = \dfrac{25}{16}$ c. $\dfrac{1}{7y^{-2}} \quad \dfrac{y^2}{7}$ d. $\dfrac{x^{-1}}{y^{-8}} \quad \dfrac{y^8}{x^1} = \dfrac{y^8}{x}$

2 Simplify exponential expressions.

Simplifying Exponential Expressions Properties of exponents are used to simplify exponential expressions. An exponential expression is **simplified** when

- No parentheses appear.
- No powers are raised to powers.
- Each base occurs only once.
- No negative or zero exponents appear.

SIMPLIFYING EXPONENTIAL EXPRESSIONS

Example

1. If necessary, remove parentheses using
$$(ab)^n = a^n b^n \quad \text{or} \quad \left(\frac{a}{b}\right)^n = \frac{a^n}{b^n}.$$

$(xy)^3 = x^3 y^3$

2. If necessary, simplify powers to powers using
$$(b^m)^n = b^{mn}.$$

$(x^4)^3 = x^{4 \cdot 3} = x^{12}$

3. If necessary, be sure that each base appears only once, using
$$b^m \cdot b^n = b^{m+n} \quad \text{or} \quad \frac{b^m}{b^n} = b^{m-n}.$$

$x^4 \cdot x^3 = x^{4+3} = x^7$

4. If necessary, rewrite exponential expressions with zero powers as 1 ($b^0 = 1$). Furthermore, write the answer with positive exponents using
$$b^{-n} = \frac{1}{b^n} \quad \text{or} \quad \frac{1}{b^{-n}} = b^n.$$

$\dfrac{x^5}{x^8} = x^{5-8} = x^{-3} = \dfrac{1}{x^3}$

The following examples show how to simplify exponential expressions. In each example, assume that any variable in a denominator is not equal to zero.

EXAMPLE 3 Simplifying an Exponential Expression

Simplify: $x^{-9} \cdot x^4$.

SOLUTION

$$x^{-9} \cdot x^4 = x^{-9+4} \qquad b^m \cdot b^n = b^{m+n}$$
$$= x^{-5} \qquad \text{The base, x, now appears only once.}$$
$$= \frac{1}{x^5} \qquad b^{-n} = \frac{1}{b^n}$$

✔ **CHECK POINT 3** Simplify: $x^{-12} \cdot x^2$. $\quad \dfrac{1}{x^{10}}$

EXAMPLE 4 Simplifying Exponential Expressions

Simplify:

a. $\dfrac{x^4}{x^9}$ **b.** $\dfrac{25x^6}{5x^8}$ **c.** $\dfrac{10y^7}{-2y^{10}}$.

SOLUTION

a. $\dfrac{x^4}{x^9} = x^{4-9} = x^{-5} = \dfrac{1}{x^5}$

b. $\dfrac{25x^6}{5x^8} = \dfrac{25}{5} \cdot \dfrac{x^6}{x^8} = 5x^{6-8} = 5x^{-2} = \dfrac{5}{x^2}$

c. $\dfrac{10y^7}{-2y^{10}} = \dfrac{10}{-2} \cdot \dfrac{y^7}{y^{10}} = -5y^{7-10} = -5y^{-3} = -\dfrac{5}{y^3}$

✔ **CHECK POINT 4** Simplify:

a. $\dfrac{x^2}{x^{10}}$ $\dfrac{1}{x^8}$ **b.** $\dfrac{75x^3}{5x^9}$ $\dfrac{15}{x^6}$ **c.** $\dfrac{50y^8}{-25y^{14}}$ $-\dfrac{2}{y^6}$

EXAMPLE 5   Simplifying an Exponential Expression

Simplify: $\dfrac{(5x^3)^2}{x^{10}}$.

SOLUTION

$\dfrac{(5x^3)^2}{x^{10}} = \dfrac{5^2(x^3)^2}{x^{10}}$ Raise each factor in the product to the second power. Parentheses are removed using $(ab)^n = a^n b^n$.

$= \dfrac{5^2 x^{3 \cdot 2}}{x^{10}}$ Multiply powers to powers using $(b^m)^n = b^{mn}$.

$= \dfrac{25x^6}{x^{10}}$ Simplify.

$= 25x^{6-10}$ When dividing with the same base, subtract exponents: $\dfrac{b^m}{b^n} = b^{m-n}$.

$= 25x^{-4}$ Simplify. The base, x, now appears only once.

$= \dfrac{25}{x^4}$ Rewrite with a positive exponent using $b^{-n} = \dfrac{1}{b^n}$.

✔ **CHECK POINT 5** Simplify: $\dfrac{(6x^4)^2}{x^{11}}$. $\dfrac{36}{x^3}$

EXAMPLE 6   Simplifying an Exponential Expression

Simplify: $\left(\dfrac{x^5}{x^2}\right)^{-3}$.

SOLUTION
Method 1. Remove parentheses first by raising the numerator and denominator to the -3 power.

$$\left(\frac{x^5}{x^2}\right)^{-3} = \frac{(x^5)^{-3}}{(x^2)^{-3}}$$

Use $\left(\dfrac{a}{b}\right)^n = \dfrac{a^n}{b^n}$ and raise the numerator and denominator to the -3 power.

$$= \frac{x^{5(-3)}}{x^{2(-3)}}$$

Multiply powers to powers using $(b^m)^n = b^{mn}$.

$$= \frac{x^{-15}}{x^{-6}}$$

Simplify.

$$= x^{-15-(-6)}$$

When dividing with the same base, subtract the exponent in the denominator from the exponent in the numerator: $\dfrac{b^m}{b^n} = b^{m-n}$.

$$= x^{-9}$$

Subtract: $-15 - (-6) = -15 + 6 = -9$. The base, x, now appears only once.

$$= \frac{1}{x^9}$$

Rewrite with a positive exponent using $b^{-n} = \dfrac{1}{b^n}$.

Method 2. First perform the division within the parentheses.

$$\left(\frac{x^5}{x^2}\right)^{-3} = (x^{5-2})^{-3}$$

Within parentheses, divide by subtracting exponents: $\dfrac{b^m}{b^n} = b^{m-n}$.

$$= (x^3)^{-3}$$

Simplify. The base, x, now appears only once.

$$= x^{3(-3)}$$

Multiply powers to powers: $(b^m)^n = b^{mn}$.

$$= x^{-9}$$

Simplify.

$$= \frac{1}{x^9}$$

Rewrite with a positive exponent using $b^{-n} = \dfrac{1}{b^n}$.

Which method do you prefer? ■

CHECK POINT 6 Simplify: $\left(\dfrac{x^8}{x^4}\right)^{-5} \cdot \dfrac{1}{x^{20}}$

Scientific Notation   By the end of 2003, the national debt of the United States was about $6.8 trillion. This is the amount of money the government has had to borrow over the years, mostly by selling bonds, because it has spent more than it has collected in taxes. A stack of $1 bills equaling the national debt would rise to twice the distance from Earth to the moon. Because a trillion is 10^{12}, the national debt can be expressed as

$$6.8 \times 10^{12}.$$

The number 6.8×10^{12} is written in a form called *scientific notation*.

> **SCIENTIFIC NOTATION**   A positive number is written in **scientific notation** when it is expressed in the form
>
> $$a \times 10^n,$$
>
> where *a* is a number greater than or equal to 1 and less than 10 ($1 \le a < 10$) and *n* is an integer.

It is customary to use the multiplication symbol, ×, rather than a dot, when writing a number in scientific notation.

Here are two examples of numbers in scientific notation:

- Each day, 2.6×10^7 pounds of dust from the atmosphere settle on Earth.
- The length of the AIDS virus is 1.1×10^{-4} millimeter.

3 Convert from scientific notation to decimal notation.

We can use n, the exponent on the 10 in $a \times 10^n$, to change a number in scientific notation to decimal notation. If n is **positive**, move the decimal point in a to the **right** n places. If n is **negative**, move the decimal point in a to the **left** $|n|$ places.

EXAMPLE 7 Converting from Scientific to Decimal Notation

Write each number in decimal notation:

a. 2.6×10^7 **b.** 1.1×10^{-4}

SOLUTION In each case, we use the exponent on the 10 to move the decimal point. In part (a), the exponent is positive, so we move the decimal point to the right. In part (b), the exponent is negative, so we move the decimal point to the left.

a. $2.6 \times 10^7 = 26,000,000$

$n = 7$

Move the decimal point 7 places to the right.

b. $1.1 \times 10^{-4} = 0.00011$

$n = -4$

Move the decimal point $|-4|$ places, or 4 places, to the left.

 CHECK POINT 7 Write each number in decimal notation:

a. 7.4×10^9 7,400,000,000   **b.** 3.017×10^{-6}.   0.000003017

---

4 Convert from decimal notation to scientific notation.

To convert a positive number from decimal notation to scientific notation, we reverse the procedure of Example 7.

CONVERTING FROM DECIMAL TO SCIENTIFIC NOTATION Write the number in the form $a \times 10^n$.

- Determine a, the numerical factor. Move the decimal point in the given number to obtain a number greater than or equal to 1 and less than 10.
- Determine n, the exponent on 10^n. The absolute value of n is the number of places the decimal point was moved. The exponent n is positive if the given number is greater than 10 and negative if the given number is between 0 and 1.

You can use your calculator's $\boxed{\text{EE}}$ (enter exponent) or $\boxed{\text{EXP}}$ key to convert from decimal to scientific notation. Here is how it's done for 0.000023:

Many Scientific Calculators

Keystrokes	Display
.000023 $\boxed{\text{EE}}$ $\boxed{=}$	2.3 − 05

Many Graphing Calculators

Use the mode setting for scientific notation.

Keystrokes	Display
.000023 $\boxed{\text{ENTER}}$	2.3E−5

EXAMPLE 8 Converting from Decimal Notation to Scientific Notation

Write each number in scientific notation:

a. 4,600,000   **b.** 0.000023.

SOLUTION

a. $4{,}600{,}000 = 4.6 \times 10^6$

This number is greater than 10, so n is positive in $a \times 10^n$.	Move the decimal point in 4,600,000 to get $1 \le a < 10$.	The decimal point moved 6 places from 4,600,000 to 4.6.

b. $0.000023 = 2.3 \times 10^{-5}$

This number is less than 1, so n is negative in $a \times 10^n$.	Move the decimal point in 0.000023 to get $1 \le a < 10$.	The decimal point moved 5 places from 0.000023 to 2.3.

 CHECK POINT 8 Write each number in scientific notation:

a. 7,410,000,000 7.41×10^9 **b.** 0.000000092. 9.2×10^{-8}

5 Compute with scientific notation.

Computations with Scientific Notation   Properties of exponents are used to perform computations with numbers that are expressed in scientific notation.

COMPUTATIONS WITH NUMBERS IN SCIENTIFIC NOTATION

Multiplication

$$(a \times 10^n) \times (b \times 10^m) = (a \times b) \times 10^{n+m}$$

Add the exponents on 10 and multiply the other parts of the numbers separately.

Division

$$\frac{a \times 10^n}{b \times 10^m} = \left(\frac{a}{b}\right) \times 10^{n-m}$$

Subtract the exponents on 10 and divide the other parts of the numbers separately.

Exponentiation

$$(a \times 10^n)^m = a^n \times 10^{nm}$$

Multiply exponents on 10 and raise the other part of the number to the power.

After the computation is completed, the answer may require an adjustment before it is back in scientific notation.

EXAMPLE 9 Computations with Scientific Notation

Perform the indicated computations, writing the answers in scientific notation:

a. $(4 \times 10^5)(2 \times 10^9)$ **b.** $\dfrac{1.2 \times 10^6}{4.8 \times 10^{-3}}$ **c.** $(5 \times 10^{-4})^3$.

SOLUTION

a. $(4 \times 10^5)(2 \times 10^9) = (4 \times 2) \times (10^5 \times 10^9)$ Regroup.

$= 8 \times 10^{5+9}$ Add the exponents on 10 and multiply the other parts.

$= 8 \times 10^{14}$ Simplify.

b. $\dfrac{1.2 \times 10^6}{4.8 \times 10^{-3}} = \left(\dfrac{1.2}{4.8}\right) \times \left(\dfrac{10^6}{10^{-3}}\right)$ Regroup.

$= 0.25 \times 10^{6-(-3)}$ Subtract the exponents on 10 and divide the other parts.

$= 0.25 \times 10^9$ Simplify. Because 0.25 is not between 1 and 10, it must be written in scientific notation.

$= 2.5 \times 10^{-1} \times 10^9$ $0.25 = 2.5 \times 10^{-1}$

$= 2.5 \times 10^{-1+9}$ Add the exponents on 10.

$= 2.5 \times 10^8$ Simplify.

c. $(5 \times 10^{-4})^3 = 5^3 \times (10^{-4})^3$ $(ab)^n = a^n b^n$. Cube each factor in parentheses.

$= 125 \times 10^{-4 \cdot 3}$ Multiply the exponents and cube the other part of the number.

$= 125 \times 10^{-12}$ Simplify. 125 must be written in scientific notation.

$= 1.25 \times 10^2 \times 10^{-12}$ $125 = 1.25 \times 10^2$

$= 1.25 \times 10^{2+(-12)}$ Add the exponents on 10.

$= 1.25 \times 10^{-10}$ Simplify. ∎

✔ **CHECK POINT 9** Perform the indicated computation, writing the answers in scientific notation:

a. $(3 \times 10^8)(2 \times 10^2)$ 6×10^{10} **b.** $\dfrac{8.4 \times 10^7}{4 \times 10^{-4}}$ 2.1×10^{11}

c. $(4 \times 10^{-2})^3$. 6.4×10^{-5}

6 Solve applied problems with scientific notation.

Applications: Putting Numbers in Perspective   Due to tax cuts and spending increases, the United States began accumulating large deficits in the 1980s. To finance the deficit, the government had borrowed $6.8 trillion as of the end of 2003. The graph in Figure 5.4 shows the national debt increasing over time.

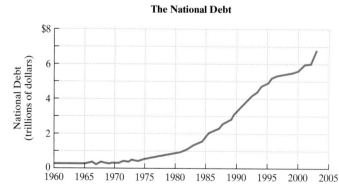

The National Debt

FIGURE 5.4
Source: Office of Management and Budget

Example 10 shows how we can use scientific notation to comprehend the meaning of a number such as 6.8 trillion.

EXAMPLE 10   The National Debt

As of November 2003, the national debt was \$6.8 trillion, or 6.8×10^{12} dollars. At that time, the U.S. population was approximately 290,000,000 (290 million), or 2.9×10^8. If the national debt was evenly divided among every individual in the United States, how much would each citizen have to pay?

SOLUTION   The amount each citizen must pay is the total debt, 6.8×10^{12} dollars, divided by the number of citizens, 2.9×10^8.

$$\frac{6.8 \times 10^{12}}{2.9 \times 10^8} = \left(\frac{6.8}{2.9}\right) \times \left(\frac{10^{12}}{10^8}\right)$$

$$\approx 2.3 \times 10^{12-8}$$

$$= 2.3 \times 10^4$$

$$= 23{,}000$$

Every U.S. citizen would have to pay about \$23,000 to the federal government to pay off the national debt.   ∎

 CHECK POINT 10   Approximately 2×10^4 people run in the New York City Marathon each year. Each runner runs a distance of 26 miles. Write the total distance covered by all the runners (assuming that each person completes the marathon) in scientific notation. 5.2×10^5 miles

ENRICHMENT ESSAY

Earthquakes and Exponents

The earthquake that ripped through northern California on October 17, 1989, measured 7.1 on the Richter scale, killed more than 60 people, and injured more than 2400. Shown here is San Francisco's Marina district, where shock waves tossed houses off their foundations and into the street.

The Richter scale is misleading because it is not actually a 1 to 8, but rather a 1 to 10 million scale. Each level indicates a tenfold increase in magnitude from the previous level, making a 7.0 earthquake a million times greater than a 1.0 quake.

The following is a translation of the Richter scale:

Richter number (R)	Magnitude (10^{R-1})
1	$10^{1-1} = 10^0 = 1$
2	$10^{2-1} = 10^1 = 10$
3	$10^{3-1} = 10^2 = 100$
4	$10^{4-1} = 10^3 = 1000$
5	$10^{5-1} = 10^4 = 10{,}000$
6	$10^{6-1} = 10^5 = 100{,}000$
7	$10^{7-1} = 10^6 = 1{,}000{,}000$
8	$10^{8-1} = 10^7 = 10{,}000{,}000$

5.7 EXERCISE SET

Student Solutions Manual CD/Video PH Math/Tutor Center MathXL Tutorials on CD MathXL® MyMathLab Interactmath.com

Practice Exercises

In Exercises 1–28, write each expression with positive exponents only. Then simplify, if possible.

1. 8^{-2} $\dfrac{1}{8^2} = \dfrac{1}{64}$

2. 9^{-2} $\dfrac{1}{9^2} = \dfrac{1}{81}$

3. 5^{-3} $\dfrac{1}{5^3} = \dfrac{1}{125}$

4. 4^{-3} $\dfrac{1}{4^3} = \dfrac{1}{64}$

5. $(-6)^{-2}$ $\dfrac{1}{(-6)^2} = \dfrac{1}{36}$

6. $(-7)^{-2}$ $\dfrac{1}{(-7)^2} = \dfrac{1}{49}$

7. -6^{-2} $-\dfrac{1}{6^2} = -\dfrac{1}{36}$

8. -7^{-2} $-\dfrac{1}{7^2} = -\dfrac{1}{49}$

9. 4^{-1} $\dfrac{1}{4^1} = \dfrac{1}{4}$

10. 6^{-1} $\dfrac{1}{6^1} = \dfrac{1}{6}$

11. $2^{-1} + 3^{-1}$ $\dfrac{1}{2^1} + \dfrac{1}{3^1} = \dfrac{1}{2} + \dfrac{1}{3} = \dfrac{5}{6}$

12. $3^{-1} - 6^{-1}$ $\dfrac{1}{3^1} - \dfrac{1}{6^1} = \dfrac{1}{3} - \dfrac{1}{6} = \dfrac{1}{6}$

13. $\dfrac{1}{3^{-2}}$ $3^2 = 9$

14. $\dfrac{1}{4^{-3}}$ $4^3 = 64$

15. $\dfrac{1}{(-3)^{-2}}$ $(-3)^2 = 9$

16. $\dfrac{1}{(-2)^{-2}}$ $(-2)^2 = 4$

17. $\dfrac{2^{-3}}{8^{-2}}$ $\dfrac{8^2}{2^3} = 8$

18. $\dfrac{4^{-3}}{2^{-2}}$ $\dfrac{2^2}{4^3} = \dfrac{1}{16}$

19. $\left(\dfrac{1}{4}\right)^{-2}$ $\dfrac{4^2}{1^2} = 16$

20. $\left(\dfrac{1}{5}\right)^{-2}$ $\dfrac{5^2}{1^2} = 25$

21. $\left(\dfrac{3}{5}\right)^{-3}$ $\dfrac{5^3}{3^3} = \dfrac{125}{27}$

22. $\left(\dfrac{3}{4}\right)^{-3}$ $\dfrac{4^3}{3^3} = \dfrac{64}{27}$

23. $\dfrac{1}{6x^{-5}}$ $\dfrac{x^5}{6}$

24. $\dfrac{1}{8x^{-6}}$ $\dfrac{x^6}{8}$

25. $\dfrac{x^{-8}}{y^{-1}}$ $\dfrac{y^1}{x^8} = \dfrac{y}{x^8}$

26. $\dfrac{x^{-12}}{y^{-1}}$ $\dfrac{y^1}{x^{12}} = \dfrac{y}{x^{12}}$

27. $\dfrac{3}{(-5)^{-3}}$ $3 \cdot (-5)^3 = -375$

28. $\dfrac{4}{(-3)^{-3}}$ $4 \cdot (-3)^3 = -108$

In Exercises 29–78, simplify each exponential expression. Assume that variables represent nonzero real numbers.

29. $x^{-8} \cdot x^3$ $\dfrac{1}{x^5}$

30. $x^{-11} \cdot x^5$ $\dfrac{1}{x^6}$

31. $(4x^{-5})(2x^2)$ $\dfrac{8}{x^3}$

32. $(5x^{-7})(3x^3)$ $\dfrac{15}{x^4}$

33. $\dfrac{x^3}{x^9}$ $\dfrac{1}{x^6}$

34. $\dfrac{x^5}{x^{12}}$ $\dfrac{1}{x^7}$

35. $\dfrac{y}{y^{100}}$ $\dfrac{1}{y^{99}}$

36. $\dfrac{y}{y^{50}}$ $\dfrac{1}{y^{49}}$

37. $\dfrac{30z^5}{10z^{10}}$ $\dfrac{3}{z^5}$

38. $\dfrac{45z^4}{15z^{12}}$ $\dfrac{3}{z^8}$

39. $\dfrac{-8x^3}{2x^7}$ $-\dfrac{4}{x^4}$

40. $\dfrac{-15x^4}{3x^9}$ $-\dfrac{5}{x^5}$

41. $\dfrac{-9a^5}{27a^8}$ $-\dfrac{1}{3a^3}$

42. $\dfrac{-15a^8}{45a^{13}}$ $-\dfrac{1}{3a^5}$

43. $\dfrac{7w^5}{5w^{13}}$ $\dfrac{7}{5w^8}$

44. $\dfrac{7w^8}{9w^{14}}$ $\dfrac{7}{9w^6}$

45. $\dfrac{x^3}{(x^4)^2}$ $\dfrac{1}{x^5}$

46. $\dfrac{x^5}{(x^3)^2}$ $\dfrac{1}{x}$

47. $\dfrac{y^{-3}}{(y^4)^2}$ $\dfrac{1}{y^{11}}$

48. $\dfrac{y^{-5}}{(y^3)^2}$ $\dfrac{1}{y^{11}}$

49. $\dfrac{(4x^3)^2}{x^8}$ $\dfrac{16}{x^2}$

50. $\dfrac{(5x^3)^2}{x^7}$ $\dfrac{25}{x}$

51. $\dfrac{(6y^4)^3}{y^{-5}}$ $216y^{17}$

52. $\dfrac{(4y^5)^3}{y^{-4}}$ $64y^{19}$

53. $\left(\dfrac{x^4}{x^2}\right)^{-3}$ $\dfrac{1}{x^6}$

54. $\left(\dfrac{x^6}{x^2}\right)^{-3}$ $\dfrac{1}{x^{12}}$

55. $\left(\dfrac{4x^5}{2x^2}\right)^{-4}$ $\dfrac{1}{16x^{12}}$

56. $\left(\dfrac{6x^7}{2x^2}\right)^{-4}$ $\dfrac{1}{81x^{20}}$

57. $(3x^{-1})^{-2}$ $\dfrac{x^2}{9}$

58. $(4x^{-1})^{-2}$ $\dfrac{x^2}{16}$

59. $(-2y^{-1})^{-3}$ $-\dfrac{y^3}{8}$

60. $(-3y^{-1})^{-3}$ $-\dfrac{y^3}{27}$

61. $\dfrac{2x^5 \cdot 3x^7}{15x^6}$ $\dfrac{2x^6}{5}$

62. $\dfrac{3x^3 \cdot 5x^{14}}{20x^{14}}$ $\dfrac{3x^3}{4}$

63. $(x^3)^5 \cdot x^{-7}$ x^8

64. $(x^4)^3 \cdot x^{-5}$ x^7

65. $(2y^3)^4 y^{-6}$ $16y^6$

66. $(3y^4)^3 y^{-7}$ $27y^5$

67. $\dfrac{(y^3)^4}{(y^2)^7}$ $\dfrac{1}{y^2}$

68. $\dfrac{(y^2)^5}{(y^3)^4}$ $\dfrac{1}{y^2}$

69. $(y^{10})^{-5}$ $\dfrac{1}{y^{50}}$

70. $(y^{20})^{-5}$ $\dfrac{1}{y^{100}}$

71. $(a^4 b^5)^{-3}$ $\dfrac{1}{a^{12}b^{15}}$

72. $(a^5 b^3)^{-4}$ $\dfrac{1}{a^{20}b^{12}}$

73. $(a^{-2} b^6)^{-4}$ $\dfrac{a^8}{b^{24}}$

74. $(a^{-7} b^2)^{-5}$ $\dfrac{a^{35}}{b^{10}}$

75. $\left(\dfrac{x^2}{2}\right)^{-2}$ $\dfrac{4}{x^4}$

76. $\left(\dfrac{x^2}{2}\right)^{-3}$ $\dfrac{8}{x^6}$

77. $\left(\dfrac{x^2}{y^3}\right)^{-3}$ $\dfrac{y^9}{x^6}$

78. $\left(\dfrac{x^3}{y^2}\right)^{-4}$ $\dfrac{y^8}{x^{12}}$

In Exercises 79–90, write each number in decimal notation without the use of exponents.

79. 8.7×10^2 870

80. 2.75×10^3 2750

81. 9.23×10^5 923,000

82. 7.24×10^4 72,400

83. 3.4×10^0 3.4

84. 9.115×10^0 9.115

85. 7.9×10^{-1} 0.79

86. 8.6×10^{-1} 0.86

87. 2.15×10^{-2} 0.0215

88. 3.14×10^{-2} 0.0314

89. 7.86×10^{-4} 0.000786

90. 4.63×10^{-5} 0.0000463

In Exercises 91–106, write each number in scientific notation.

91. 32,400 3.24×10^4

92. 327,000 3.27×10^5

93. 220,000,000 2.2×10^8

94. 370,000,000,000 3.7×10^{11}

95. 713 7.13×10^2

96. 623 6.23×10^2

97. 6751 6.751×10^3

98. 9832 9.832×10^3

99. 0.0027 2.7×10^{-3}

100. 0.00083 8.3×10^{-4}

101. 0.0000202 2.02×10^{-5}

102. 0.00000103 1.03×10^{-6}

103. 0.005 5×10^{-3}

104. 0.006 6×10^{-3}

105. 3.14159 3.14159×10^0

106. 2.71828 2.71828×10^0

In Exercises 107–126, perform the indicated computations. Write the answers in scientific notation.

107. $(2 \times 10^3)(3 \times 10^2)$ 6×10^5

108. $(3 \times 10^4)(3 \times 10^2)$ 9×10^6

109. $(2 \times 10^5)(8 \times 10^3)$ 1.6×10^9

110. $(4 \times 10^3)(5 \times 10^4)$ 2×10^8

111. $\dfrac{12 \times 10^6}{4 \times 10^2}$ 3×10^4

112. $\dfrac{20 \times 10^{20}}{10 \times 10^{10}}$ 2×10^{10}

113. $\dfrac{15 \times 10^4}{5 \times 10^{-2}}$ 3×10^6

114. $\dfrac{18 \times 10^2}{9 \times 10^{-3}}$ 2×10^5

115. $\dfrac{15 \times 10^{-4}}{5 \times 10^2}$ 3×10^{-6}

116. $\dfrac{18 \times 10^{-2}}{9 \times 10^3}$ 2×10^{-5}

117. $\dfrac{180 \times 10^6}{2 \times 10^3}$ 9×10^4

118. $\dfrac{180 \times 10^8}{2 \times 10^4}$ 9×10^5

119. $\dfrac{3 \times 10^4}{12 \times 10^{-3}}$ 2.5×10^6

120. $\dfrac{5 \times 10^2}{20 \times 10^{-3}}$ 2.5×10^4

121. $(5 \times 10^2)^3$ 1.25×10^8

122. $(4 \times 10^3)^2$ 1.6×10^7

123. $(3 \times 10^{-2})^4$ 8.1×10^{-7}

124. $(2 \times 10^{-3})^5$ 3.2×10^{-14}

125. $(4 \times 10^6)^{-1}$ 2.5×10^{-7}

126. $(5 \times 10^4)^{-1}$ 2×10^{-5}

Practice Plus

In Exercises 127–134, simplify each exponential expression. Assume that variables represent nonzero real numbers.

127. $\dfrac{(x^{-2}y)^{-3}}{(x^2y^{-1})^3}$ 1

128. $\dfrac{(xy^{-2})^{-2}}{(x^{-2}y)^{-3}}$ $\dfrac{y^7}{x^8}$

129. $(2x^{-3}yz^{-6})(2x)^{-5}$ $\dfrac{y}{16x^8z^6}$

130. $(3x^{-4}yz^{-7})(3x)^{-3}$ $\dfrac{y}{9x^7z^7}$

131. $\left(\dfrac{x^3y^4z^5}{x^{-3}y^{-4}z^{-5}}\right)^{-2}$ $\dfrac{1}{x^{12}y^{16}z^{20}}$

132. $\left(\dfrac{x^4y^5z^6}{x^{-4}y^{-5}z^{-6}}\right)^{-4}$ $\dfrac{1}{x^{32}y^{40}z^{48}}$

133. $\dfrac{(2^{-1}x^{-2}y^{-1})^{-2}(2x^{-4}y^3)^{-2}(16x^{-3}y^3)^0}{(2x^{-3}y^{-5})^2}$ $\dfrac{x^{18}y^6}{4}$

134. $\dfrac{(2^{-1}x^{-3}y^{-1})^{-2}(2x^{-6}y^4)^{-2}(9x^3y^{-3})^0}{(2x^{-4}y^{-6})^2}$ $\dfrac{x^{26}y^6}{4}$

In Exercises 135–138, perform the indicated computations. Express answers in scientific notation.

135. $(5 \times 10^3)(1.2 \times 10^{-4}) \div (2.4 \times 10^2)$ 2.5×10^{-3}

136. $(2 \times 10^2)(2.6 \times 10^{-3}) \div (4 \times 10^3)$ 1.3×10^{-4}

137. $\dfrac{(1.6 \times 10^4)(7.2 \times 10^{-3})}{(3.6 \times 10^8)(4 \times 10^{-3})}$ 8×10^{-5}

138. $\dfrac{(1.2 \times 10^6)(8.7 \times 10^{-2})}{(2.9 \times 10^6)(3 \times 10^{-3})}$ 1.2×10^1

Application Exercises

In Exercises 139–142, rewrite the number in each statement in scientific notation.

139. King Mongkut of Siam (the king in the musical *The King and I*) had 9200 wives. 9.2×10^3

140. The top-selling music album of all time is "Their Greatest Hits" by the Eagles, selling 28,000,000 copies. 2.8×10^7

141. The volume of a bacterium is 0.00000000000000025 cubic meter. 2.5×10^{-16}

142. Home computers can perform a multiplication in 0.00000000036 second. 3.6×10^{-10}

In Exercises 143–146, use 10^6 for one million and 10^9 for one billion to rewrite the number in each statement in scientific notation.

143. In 2003, Americans consumed 600 million Big Macs at McDonald's. 6×10^8

144. In 2002, the United States imported 550 million barrels of crude oil from Saudi Arabia. 5.5×10^8

In Exercises 145–146, which appear on the next page, use the numbers displayed in the following graph.

Defense Spending: U.S. and 20 Next Top-Spending Nations

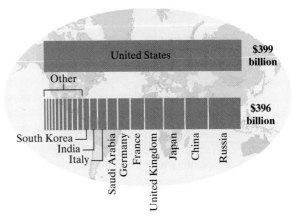

Source: U.S. Defense Department; Center for Defense Information

145. In 2003, the United States spent $399 billion on defense. 3.99×10^{11}

146. In 2003, the United States spent more on defense than the 20 next top-spending nations combined. These nations spent $396 billion on defense. 3.96×10^{11}

147. In 2002, the United States government spent approximately $2 trillion, or 2×10^{12} dollars. Use the graph to write the amount that it spent on Social Security in scientific notation. $\$4 \times 10^{11}$

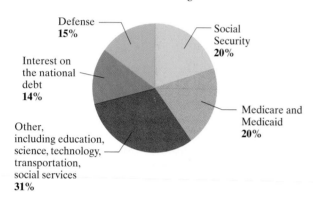

2002 Federal Budget

Defense 15%
Interest on the national debt 14%
Other, including education, science, technology, transportation, social services 31%
Social Security 20%
Medicare and Medicaid 20%

Source: U.S. Office of Management and Budget

148. Americans say they lead active lives, but for the 205 million of us ages 18 and older, walking is often as strenuous as it gets. Use the graph to write the number of Americans whose lifestyle is not very active. Express the answer in scientific notation. 2.05×10^{7} Americans

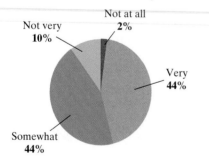

How Active Is Your Lifestyle?

Not very 10%
Not at all 2%
Very 44%
Somewhat 44%

Source: Discovery Health Media

149. If the population of the United States is 2.9×10^{8} and each person spends about $120 per year on ice cream, express the total annual spending on ice cream in scientific notation. $\$3.48 \times 10^{10}$

150. A human brain contains 3×10^{10} neurons and a gorilla brain contains 7.5×10^{9} neurons. How many times as many neurons are in the brain of a human as in the brain of a gorilla? 4 times

Use the motion formula $d = rt$, distance equals rate times time, and the fact that light travels at the rate of 1.86×10^{5} miles per second, to solve Exercises 151–152.

151. If the moon is approximately 2.325×10^{5} miles from Earth, how many seconds does it take moonlight to reach Earth? 1.25 seconds

152. If the sun is approximately 9.14×10^{7} miles from Earth, how many seconds, to the nearest tenth of a second, does it take sunlight to reach Earth? approximately 4.9 seconds

Writing in Mathematics

153. Explain the negative exponent rule and give an example. 153–158. Answers will vary.

154. How do you know if an exponential expression is simplified?

155. How do you know if a number is written in scientific notation?

156. Explain how to convert from scientific to decimal notation and give an example.

157. Explain how to convert from decimal to scientific notation and give an example.

158. Describe one advantage of expressing a number in scientific notation over decimal notation.

Critical Thinking Exercises

159. Which one of the following is true? b

   a. $4^{-2} < 4^{-3}$

   b. $5^{-2} > 2^{-5}$

   c. $(-2)^{4} = 2^{-4}$

   d. $5^{2} \cdot 5^{-2} > 2^{5} \cdot 2^{-5}$

160. Which one of the following is true? d

   a. $534.7 = 5.347 \times 10^{3}$

   b. $\dfrac{8 \times 10^{30}}{4 \times 10^{-5}} = 2 \times 10^{25}$

   c. $(7 \times 10^{5}) + (2 \times 10^{-3}) = 9 \times 10^{2}$

   d. $(4 \times 10^{3}) + (3 \times 10^{2}) = 4.3 \times 10^{3}$

161. Give an example of a number where there is no advantage to using scientific notation instead of decimal notation. Explain why this is the case. Answers will vary.

162. The mad Dr. Frankenstein has gathered enough bits and pieces (so to speak) for $2^{-1} + 2^{-2}$ of his creature-to-be. Write a fraction that represents the amount of his creature that must still be obtained. $\dfrac{1}{4}$

Technology Exercises

163. Use a calculator in a fraction mode to check any five of your answers in Exercises 1–22. 163–166. Answers will vary.

164. Use a calculator to check any three of your answers in Exercises 79–90.

165. Use a calculator to check any three of your answers in Exercises 91–106.

166. Use a calculator with an $\boxed{\text{EE}}$ or $\boxed{\text{EXP}}$ key to check any four of your computations in Exercises 107–126. Display the result of the computation in scientific notation.

Review Exercises

167. Solve: $8 - 6x > 4x - 12$. (Section 2.7, Example 6)
$\{x | x < 2\}$

168. Simplify: $24 \div 8 \cdot 3 + 28 \div (-7)$. (Section 1.8, Example 8) 5

169. List the whole numbers in this set:

$$\left\{-4, -\frac{1}{5}, 0, \pi, \sqrt{16}, \sqrt{17}\right\}.$$

(Section 1.2, Example 5) $0, \sqrt{16}$

GROUP PROJECT

CHAPTER 5

Additional group projects can be found in the *Instructor's Resource Manual.*

Putting Numbers into Perspective. A large number can be put into perspective by comparing it with another number. For example, we put the $6.8 trillion national debt into perspective by comparing it to the number of U.S. citizens. The total distance covered by all the runners in the New York City Marathon (Check Point 10 on page 371) can be put into perspective by comparing this distance with, say, the distance from New York to San Francisco.

For this project, each group member should consult an almanac, a newspaper, or the World Wide Web to find a number greater than one million. Explain to other members of the group the context in which the large number is used. Express the number in scientific notation. Then put the number into perspective by comparing it with another number.

CHAPTER 5 SUMMARY

Definitions and Concepts	Examples

Section 5.1 Adding and Subtracting Polynomials

A polynomial is a single term or the sum of two or more terms containing variables in the numerator with whole number exponents. A monomial is a simplified polynomial with exactly one term; a binomial has exactly two terms; a trinomial has exactly three terms. The degree of a polynomial is the highest power of all the terms. The standard form of a polynomial is written in descending powers of the variable.	Polynomials Monomial: $2x^5$ Degree is 5. Binomial: $6x^3 + 5x$ Degree is 3. Trinomial: $7x + 4x^2 - 5$ Degree is 2.
To add polynomials, add like terms.	$(6x^3 + 5x^2 - 7x) + (-9x^3 + x^2 + 6x)$ $= (6x^3 - 9x^3) + (5x^2 + x^2) + (-7x + 6x)$ $= -3x^3 + 6x^2 - x$
The opposite, or additive inverse, of a polynomial is that polynomial with the sign of every coefficient changed. To subtract two polynomials, add the first polynomial and the opposite of the polynomial being subtracted.	$(5y^3 - 9y^2 - 4) - (3y^3 - 12y^2 - 5)$ $= (5y^3 - 9y^2 - 4) + (-3y^3 + 12y^2 + 5)$ $= (5y^3 - 3y^3) + (-9y^2 + 12y^2) + (-4 + 5)$ $= 2y^3 + 3y^2 + 1$

Definitions and Concepts	Examples

Section 5.2 Multiplying Polynomials

Properties of Exponents
Product Rule: $b^m \cdot b^n = b^{m+n}$
Power Rule: $(b^m)^n = b^{mn}$
Products to Powers: $(ab)^n = a^n b^n$

$$x^3 \cdot x^8 = x^{3+8} = x^{11}$$
$$(x^3)^8 = x^{3 \cdot 8} = x^{24}$$
$$(-5x^2)^3 = (-5)^3(x^2)^3 = -125x^6$$

To multiply monomials, multiply coefficients and add exponents.

$$(-6x^4)(3x^{10}) = -6 \cdot 3x^{4+10} = -18x^{14}$$

To multiply a monomial and a polynomial, multiply each term of the polynomial by the monomial.

$$2x^4(3x^2 - 6x + 5)$$
$$= 2x^4 \cdot 3x^2 - 2x^4 \cdot 6x + 2x^4 \cdot 5$$
$$= 6x^6 - 12x^5 + 10x^4$$

To multiply polynomials when neither is a monomial, multiply each term of one polynomial by each term of the other polynomial. Then combine like terms.

$$(2x + 3)(5x^2 - 4x + 2)$$
$$= 2x(5x^2 - 4x + 2) + 3(5x^2 - 4x + 2)$$
$$= 10x^3 - 8x^2 + 4x + 15x^2 - 12x + 6$$
$$= 10x^3 + 7x^2 - 8x + 6$$

Section 5.3 Special Products

The FOIL method may be used when multiplying two binomials: First terms multiplied. Outside terms multiplied. Inside terms multiplied. Last terms multiplied.

$$\begin{array}{cccc} F & O & I & L \end{array}$$
$$(3x + 7)(2x - 5) = 3x \cdot 2x + 3x(-5) + 7 \cdot 2x + 7(-5)$$
$$= 6x^2 - 15x + 14x - 35$$
$$= 6x^2 - x - 35$$

The Product of the Sum and Difference of Two Terms

$$(A + B)(A - B) = A^2 - B^2$$

$$(4x + 7)(4x - 7) = (4x)^2 - 7^2$$
$$= 16x^2 - 49$$

The Square of a Binomial Sum

$$(A + B)^2 = A^2 + 2AB + B^2$$

$$(x^2 + 6)^2 = (x^2)^2 + 2 \cdot x^2 \cdot 6 + 6^2$$
$$= x^4 + 12x^2 + 36$$

The Square of a Binomial Difference

$$(A - B)^2 = A^2 - 2AB + B^2$$

$$(9x - 3)^2 = (9x)^2 - 2 \cdot 9x \cdot 3 + 3^2$$
$$= 81x^2 - 54x + 9$$

Section 5.4 Polynomials in Several Variables

To evaluate a polynomial in several variables, substitute the given value for each variable and perform the resulting computation.

Evaluate $4x^2y + 3xy - 2x$ for $x = -1$ and $y = -3$.

$$4x^2y + 3xy - 2x$$
$$= 4(-1)^2(-3) + 3(-1)(-3) - 2(-1)$$
$$= 4(1)(-3) + 3(-1)(-3) - 2(-1)$$
$$= -12 + 9 + 2 = -1$$

Definitions and Concepts	Examples

Section 5.4 Polynomials in Several Variables (continued)

For a polynomial in two variables, the degree of a term is the sum of the exponents on its variables. The degree of the polynomial is the highest degree of all its terms.

$$7x^2y + 12x^4y^3 - 17x^5 + 6$$

degree: $2 + 1 = 3$ degree: $4 + 3 = 7$ degree: 5 degree: 0

Degree of polynomial $= 7$

Polynomials in several variables are added, subtracted, and multiplied using the same rules for polynomials in one variable.

$$(5x^2y^3 - xy + 4y^2) - (8x^2y^3 - 6xy - 2y^2)$$
$$= (5x^2y^3 - xy + 4y^2) + (-8x^2y^3 + 6xy + 2y^2)$$
$$= (5x^2y^3 - 8x^2y^3) + (-xy + 6xy) + (4y^2 + 2y^2)$$
$$= -3x^2y^3 + 5xy + 6y^2$$

F   O   I   L

$$(3x - 2y)(x - y) = 3x \cdot x + 3x(-y) + (-2y)x + (-2y)(-y)$$
$$= 3x^2 - 3xy - 2xy + 2y^2$$
$$= 3x^2 - 5xy + 2y^2$$

Section 5.5 Dividing Polynomials

Additional Properties of Exponents

Quotient Rule: $\dfrac{b^m}{b^n} = b^{m-n}, \quad b \neq 0$

Zero-Exponent Rule: $b^0 = 1, \quad b \neq 0$

Quotients to Powers: $\left(\dfrac{a}{b}\right)^n = \dfrac{a^n}{b^n}, \quad b \neq 0$

$$\frac{x^{12}}{x^4} = x^{12-4} = x^8$$

$$(-3)^0 = 1 \qquad -3^0 = -(3^0) = -1$$

$$\left(\frac{y^2}{4}\right)^3 = \frac{(y^2)^3}{4^3} = \frac{y^{2\cdot3}}{4\cdot4\cdot4} = \frac{y^6}{64}$$

To divide monomials, divide coefficients and subtract exponents.

$$\frac{-40x^{40}}{20x^{20}} = \frac{-40}{20}x^{40-20} = -2x^{20}$$

To divide a polynomial by a monomial, divide each term of the polynomial by the monomial.

$$\frac{8x^6 - 4x^3 + 10x}{2x}$$
$$= \frac{8x^6}{2x} - \frac{4x^3}{2x} + \frac{10x}{2x}$$
$$= 4x^{6-1} - 2x^{3-1} + 5x^{1-1} = 4x^5 - 2x^2 + 5$$

Section 5.6 Long Division of Polynomials; Synthetic Division

To divide by a polynomial containing more than one term, begin by arranging all polynomials in descending powers of the variable. If a power of a variable is missing, add that power with a coefficient of 0. Repeat the four steps—divide, multiply, subtract, bring down the next term—until the degree of the remainder is less than the degree of the divisor.

Divide: $(2x^3 - x^2 - 7) \div (x - 2)$.

$$
\begin{array}{r}
2x^2 + 3x + 6 \\
x - 2 \overline{)\,2x^3 - x^2 + 0x - 7} \\
\underline{2x^3 - 4x^2} \\
3x^2 + 0x \\
\underline{3x^2 - 6x} \\
6x - 7 \\
\underline{6x - 12} \\
5
\end{array}
$$

The answer is $2x^2 + 3x + 6 + \dfrac{5}{x - 2}$.

Definitions and Concepts	**Examples**

Section 5.6 Long Division of Polynomials; Synthetic Division (continued)

A shortcut to long division, called synthetic division, can be used to divide a polynomial by a binomial of the form $x - c$.

Here is the division problem shown at the bottom of page 377 using synthetic division.

$$(2x^3 - x^2 - 7) \div (x - 2)$$

Coefficients of the dividend,
$2x^3 - x^2 + 0x - 7$

This is c in $x - c$.
For $x - 2$, c is 2.

$$\begin{array}{c|cccc} 2 & 2 & -1 & 0 & -7 \\ & & 4 & 6 & 12 \\ \hline & 2 & 3 & 6 & 5 \end{array}$$

Coefficients of quotient   Remainder

The answer is $2x^2 + 3x + 6 + \dfrac{5}{x - 2}$.

Section 5.7 Negative Exponents and Scientific Notation

Negative Exponents in Numerators and Denominators

If $b \neq 0$, $b^{-n} = \dfrac{1}{b^n}$ and $\dfrac{1}{b^{-n}} = b^n$.

$$6^{-2} = \frac{1}{6^2} = \frac{1}{36}$$

$$\frac{1}{(-2)^{-4}} = (-2)^4 = 16$$

$$\left(\frac{2}{3}\right)^{-3} = \frac{2^{-3}}{3^{-3}} = \frac{3^3}{2^3} = \frac{27}{8}$$

An exponential expression is simplified when
- No parentheses appear.
- No powers are raised to powers.
- Each base occurs only once.
- No negative or zero exponents appear.

Simplify: $\dfrac{(2x^4)^3}{x^{18}}$.

$$\frac{(2x^4)^3}{x^{18}} = \frac{2^3(x^4)^3}{x^{18}} = \frac{8x^{4 \cdot 3}}{x^{18}} = \frac{8x^{12}}{x^{18}} = 8x^{12-18} = 8x^{-6} = \frac{8}{x^6}$$

A positive number in scientific notation is expressed as $a \times 10^n$, where $1 \leq a < 10$ and n is an integer.

Write 2.9×10^{-3} in decimal notation.

$$2.9 \times 10^{-3} = .0029 = 0.0029$$

Write $16{,}000$ in scientific notation.

$$16{,}000 = 1.6 \times 10^4$$

Use properties of exponents with base 10

$$10^m \cdot 10^n = 10^{m+n}, \quad \frac{10^m}{10^n} = 10^{m-n}, \quad \text{and} \quad (10^m)^n = 10^{mn}$$

to perform computations with scientific notation.

$$(5 \times 10^3)(4 \times 10^{-8})$$
$$= 5 \cdot 4 \times 10^{3-8}$$
$$= 20 \times 10^{-5}$$
$$= 2 \times 10^1 \times 10^{-5} = 2 \times 10^{-4}$$

CHAPTER 5   REVIEW EXERCISES

5.1 In Exercises 1–3, identify each polynomial as a monomial, binomial, or trinomial. Give the degree of the polynomial.

1. $7x^4 + 9x$ binomial, 4   **2.** $3x + 5x^2 - 2$ trinomial, 2

3. $16x$ monomial, 1

In Exercises 4–8, add or subtract as indicated.

4. $(-6x^3 + 7x^2 - 9x + 3) + (14x^3 + 3x^2 - 11x - 7)$
  $8x^3 + 10x^2 - 20x - 4$

5. $(9y^3 - 7y^2 + 5) + (4y^3 - y^2 + 7y - 10)$
  $13y^3 - 8y^2 + 7y - 5$

6. $(5y^2 - y - 8) - (-6y^2 + 3y - 4)$ $11y^2 - 4y - 4$

7. $(13x^4 - 8x^3 + 2x^2) - (5x^4 - 3x^3 + 2x^2 - 6)$
$8x^4 - 5x^3 + 6$

8. Subtract $x^4 + 7x^2 - 11x$ from $-13x^4 - 6x^2 + 5x$.
$-14x^4 - 13x^2 + 16x$

In Exercises 9–11, add or subtract as indicated.

9. Add. $\dfrac{7y^4 - 6y^3 + 4y^2 - 4y}{y^3 - y^2 + 3y - 4}$ $7y^4 - 5y^3 + 3y^2 - y - 4$

10. Subtract. $\dfrac{7x^2 - 9x + 2}{-(4x^2 - 2x - 7)}$ $3x^2 - 7x + 9$

11. Subtract. $\dfrac{5x^3 - 6x^2 - 9x + 14}{-(-5x^3 + 3x^2 - 11x + 3)}$ $10x^3 - 9x^2 + 2x + 11$

12. The polynomial $104.5x^2 - 1501.5x + 6016$ models the death rate per year per 100,000 men for men averaging x hours of sleep each night. Evaluate the polynomial when $x = 10$. Describe what the answer means in practical terms.   1451; 1451 per 100,000 is the death rate for men averaging 10 hours of sleep each night.

5.2 *In Exercises 13–17, simplify each expression.*

13. $x^{20} \cdot x^3$ x^{23} **14.** $y \cdot y^5 \cdot y^8$ y^{14} **15.** $(x^{20})^5$ x^{100}

16. $(10y)^2$ $100y^2$ **17.** $(-4x^{10})^3$ $-64x^{30}$

In Exercises 18–26, find each product.

18. $(5x)(10x^3)$ $50x^4$ **19.** $(-12y^7)(3y^4)$ $-36y^{11}$

20. $(-2x^5)(-3x^4)(5x^3)$ $30x^{12}$

21. $7x(3x^2 + 9)$ $21x^3 + 63x$

22. $5x^3(4x^2 - 11x)$ $20x^5 - 55x^4$

23. $3y^2(-7y^2 + 3y - 6)$ $-21y^4 + 9y^3 - 18y^2$

24. $2y^5(8y^3 - 10y^2 + 1)$ $16y^8 - 20y^7 + 2y^5$

25. $(x + 3)(x^2 - 5x + 2)$ $x^3 - 2x^2 - 13x + 6$

26. $(3y - 2)(4y^2 + 3y - 5)$ $12y^3 + y^2 - 21y + 10$

In Exercises 27–28, use a vertical format to find each product.

27. $\dfrac{y^2 - 4y + 7}{3y - 5}$ $3y^3 - 17y^2 + 41y - 35$

28. $\dfrac{4x^3 - 2x^2 - 6x - 1}{2x + 3}$ $8x^4 + 8x^3 - 18x^2 - 20x - 3$

5.3 *In Exercises 29–41, find each product.*

29. $(x + 6)(x + 2)$ $x^2 + 8x + 12$

30. $(3y - 5)(2y + 1)$ $6y^2 - 7y - 5$

31. $(4x^2 - 2)(x^2 - 3)$ $4x^4 - 14x^2 + 6$

32. $(5x + 4)(5x - 4)$ $25x^2 - 16$

33. $(7 - 2y)(7 + 2y)$ $49 - 4y^2$

34. $(y^2 + 1)(y^2 - 1)$ $y^4 - 1$

35. $(x + 3)^2$ $x^2 + 6x + 9$

36. $(3y + 4)^2$ $9y^2 + 24y + 16$

37. $(y - 1)^2$ $y^2 - 2y + 1$

38. $(5y - 2)^2$ $25y^2 - 20y + 4$

39. $(x^2 + 4)^2$ $x^4 + 8x^2 + 16$

40. $(x^2 + 4)(x^2 - 4)$ $x^4 - 16$

41. $(x^2 + 4)(x^2 - 5)$ $x^4 - x^2 - 20$

42. Write a polynomial in descending powers of x that represents the area of the shaded region. $x^2 + 7x + 12$

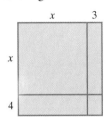

43. The parking garage shown in the figure below measures 30 yards by 20 yards. The length and the width are each increased by a fixed amount, x yards. Write a trinomial that describes the area of the expanded garage.
$x^2 + 50x + 600$ yards^2

5.4

44. Evaluate $2x^3y - 4xy^2 + 5y + 6$ for $x = -1$ and $y = 2$.   28

45. Determine the coefficient of each term, the degree of each term, and the degree of the polynomial:
$4x^2y + 9x^3y^2 - 17x^4 - 12$. $4x^2y$: 4, 3; $9x^3y^2$: 9, 5; $-17x^4$: -17, 4; -12: -12, 0; polynomial degree $= 5$

In Exercises 46–55, perform the indicated operations.

46. $(7x^2 - 8xy + y^2) + (-8x^2 - 9xy + 4y^2)$
$-x^2 - 17xy + 5y^2$

47. $(13x^3y^2 - 5x^2y - 9x^2) - (11x^3y^2 - 6x^2y - 3x^2 + 4)$
$2x^3y^2 + x^2y - 6x^2 - 4$

48. $(-7x^2y^3)(5x^4y^6)$ $-35x^6y^9$

49. $5ab^2(3a^2b^3 - 4ab)$ $15a^3b^5 - 20a^2b^3$

50. $(x + 7y)(3x - 5y)$ $3x^2 + 16xy - 35y^2$

51. $(4xy - 3)(9xy - 1)$ $36x^2y^2 - 31xy + 3$

52. $(3x + 5y)^2$ $9x^2 + 30xy + 25y^2$

53. $(xy - 7)^2$ $x^2y^2 - 14xy + 49$

54. $(7x + 4y)(7x - 4y)$ $49x^2 - 16y^2$

55. $(a - b)(a^2 + ab + b^2)$ $a^3 - b^3$

5.5 *In Exercises 56–62, simplify each expression.*

56. $\dfrac{6^{40}}{6^{10}}$ 6^{30} **57.** $\dfrac{x^{18}}{x^3}$ x^{15} **58.** $(-10)^0$ 1

59. -10^0 -1 **60.** $400x^0$ 400 **61.** $\left(\dfrac{x^4}{2}\right)^3$ $\dfrac{x^{12}}{8}$

62. $\left(\dfrac{-3}{2y^6}\right)^4$ $\dfrac{81}{16y^{24}}$

In Exercises 63–67, divide and check each answer.

63. $\dfrac{-15y^8}{3y^2}$ $-5y^6$ **64.** $\dfrac{40x^8y^6}{5xy^3}$ $8x^7y^3$

65. $\dfrac{18x^4 - 12x^2 + 36x}{6x}$ $3x^3 - 2x + 6$

66. $\dfrac{30x^8 - 25x^7 - 40x^5}{-5x^3}$ $-6x^5 + 5x^4 + 8x^2$

67. $\dfrac{27x^3y^2 - 9x^2y - 18xy^2}{3xy}$ $9x^2y - 3x - 6y$

5.6 *In Exercises 68–71, divide and check each answer.*

68. $\dfrac{2x^2 + 3x - 14}{x - 2}$ $2x + 7$

69. $\dfrac{2x^3 - 5x^2 + 7x + 5}{2x + 1}$ $x^2 - 3x + 5$

70. $\dfrac{x^3 - 2x^2 - 33x - 7}{x - 7}$ $x^2 + 5x + 2 + \dfrac{7}{x - 7}$

71. $\dfrac{y^3 - 27}{y - 3}$ $y^2 + 3y + 9$

72. $(4x^4 + 6x^3 + 3x - 1) \div (2x^2 + 1)$ $2x^2 + 3x - 1$

In Exercises 73–75, divide using synthetic division.

73. $(4x^3 - 3x^2 - 2x + 1) \div (x + 1)$ $4x^2 - 7x + 5 - \dfrac{4}{x + 1}$

74. $(3x^4 - 2x^2 - 10x - 20) \div (x - 2)$ $3x^3 + 6x^2 + 10x + 10$

75. $(x^4 + 16) \div (x + 4)$ $x^3 - 4x^2 + 16x - 64 + \dfrac{272}{x + 4}$

5.7 *In Exercises 76–80, write each expression with positive exponents only and then simplify.*

76. 7^{-2} $\dfrac{1}{7^2} = \dfrac{1}{49}$ **77.** $(-4)^{-3}$ $\dfrac{1}{(-4)^3} = -\dfrac{1}{64}$

78. $2^{-1} + 4^{-1}$ $\dfrac{1}{2^1} + \dfrac{1}{4^1} = \dfrac{1}{2} + \dfrac{1}{4} = \dfrac{3}{4}$

79. $\dfrac{1}{5^{-2}}$ $5^2 = 25$ **80.** $\left(\dfrac{2}{5}\right)^{-3}$ $\dfrac{5^3}{2^3} = \dfrac{125}{8}$

In Exercises 81–89, simplify each exponential expression. Assume that variables in denominators do not equal zero.

81. $\dfrac{x^3}{x^9}$ $\dfrac{1}{x^6}$ **82.** $\dfrac{30y^6}{5y^8}$ $\dfrac{6}{y^2}$ **83.** $(5x^{-7})(6x^2)$ $\dfrac{30}{x^5}$

84. $\dfrac{x^4 \cdot x^{-2}}{x^{-6}}$ x^8 **85.** $\dfrac{(3y^3)^4}{y^{10}}$ $81y^2$ **86.** $\dfrac{y^{-7}}{(y^4)^3}$ $\dfrac{1}{y^{19}}$

87. $(2x^{-1})^{-3}$ $\dfrac{x^3}{8}$ **88.** $\left(\dfrac{x^7}{x^4}\right)^{-2}$ $\dfrac{1}{x^6}$ **89.** $\dfrac{(y^3)^4}{(y^{-2})^4}$ y^{20}

In Exercises 90–92, write each number in decimal notation without the use of exponents.

90. 2.3×10^4 $23,000$ **91.** 1.76×10^{-3} 0.00176

92. 9×10^{-1} 0.9

In Exercises 93–96, write each number in scientific notation.

93. $73,900,000$ 7.39×10^7 **94.** 0.00062 6.2×10^{-4}

95. 0.38 3.8×10^{-1} **96.** 3.8 3.8×10^0

In Exercises 97–99, perform the indicated computation. Write the answers in scientific notation.

97. $(6 \times 10^{-3})(1.5 \times 10^6)$ 9×10^3

98. $\dfrac{2 \times 10^2}{4 \times 10^{-3}}$ 5×10^4 **99.** $(4 \times 10^{-2})^2$ 1.6×10^{-3}

100. A microsecond is 10^{-6} second and a nanosecond is 10^{-9} second. How many nanoseconds make a microsecond? 1000 nanoseconds

101. The world's population is approximately 6.3×10^9 people. Current projections double this population in 40 years. Write the population 40 years from now in scientific notation. 1.26×10^{10} people

---

CHAPTER 5 TEST

Remember to use your Chapter Test Prep Video CD to see the worked-out solutions to the test questions you want to review.

1. Identify $9x + 6x^2 - 4$ as a monomial, binomial, or trinomial. Give the degree of the polynomial.   trinomial, 2

In Exercises 2–3, add or subtract as indicated.

2. $(7x^3 + 3x^2 - 5x - 11) + (6x^3 - 2x^2 + 4x - 13)$ $13x^3 + x^2 - x - 24$

3. $(9x^3 - 6x^2 - 11x - 4) - (4x^3 - 8x^2 - 13x + 5)$ $5x^3 + 2x^2 + 2x - 9$

In Exercises 4–10, find each product.

4. $(-7x^3)(5x^8)$ $-35x^{11}$

5. $6x^2(8x^3 - 5x - 2)$ $48x^5 - 30x^3 - 12x^2$

6. $(3x + 2)(x^2 - 4x - 3)$ $3x^3 - 10x^2 - 17x - 6$

7. $(3y + 7)(2y - 9)$ $6y^2 - 13y - 63$

8. $(7x + 5)(7x - 5)$ $49x^2 - 25$

9. $(x^2 + 3)^2$ $x^4 + 6x^2 + 9$

10. $(5x - 3)^2$ $25x^2 - 30x + 9$

11. Evaluate $4x^2y + 5xy - 6x$ for $x = -2$ and $y = 3$. 30

In Exercises 12–14, perform the indicated operations.

12. $(8x^2y^3 - xy + 2y^2) - (6x^2y^3 - 4xy - 10y^2)$ $2x^2y^3 + 3xy + 12y^2$

13. $(3a - 7b)(4a + 5b)$ $12a^2 - 13ab - 35b^2$

14. $(2x + 3y)^2$ $4x^2 + 12xy + 9y^2$

In Exercises 15–17, divide and check each answer.

15. $\dfrac{-25x^{16}}{5x^4}$ $-5x^{12}$

16. $\dfrac{15x^4 - 10x^3 + 25x^2}{5x}$ $3x^3 - 2x^2 + 5x$

17. $\dfrac{2x^3 - 3x^2 + 4x + 4}{2x + 1}$ $x^2 - 2x + 3 + \dfrac{1}{2x + 1}$

18. Divide using synthetic division: $3x^3 - 4x^2 + 7$

$(3x^4 + 11x^3 - 20x^2 + 7x + 35) \div (x + 5)$.

In Exercises 19–20, write each expression with positive exponents only and then simplify.

19. 10^{-2} $\dfrac{1}{10^2} = \dfrac{1}{100}$

20. $\dfrac{1}{4^{-3}}$ $4^3 = 64$

In Exercises 21–26, simplify each expression.

21. $(-3x^2)^3$ $-27x^6$

22. $\dfrac{20x^3}{5x^8}$ $\dfrac{4}{x^5}$

23. $(-7x^{-8})(3x^2)$ $-\dfrac{21}{x^6}$

24. $\dfrac{(2y^3)^4}{y^8}$ $16y^4$

25. $(5x^{-4})^{-2}$ $\dfrac{x^8}{25}$

26. $\left(\dfrac{x^{10}}{x^5}\right)^{-3}$ $\dfrac{1}{x^{15}}$

27. Write 3.7×10^{-4} in decimal notation.   0.00037

28. Write 7,600,000 in scientific notation. 7.6×10^6

In Exercises 29–30, perform the indicated computation. Write the answers in scientific notation.

29. $(4.1 \times 10^2)(3 \times 10^{-5})$ 1.23×10^{-2}

30. $\dfrac{8.4 \times 10^6}{4 \times 10^{-2}}$ 2.1×10^8

31. Write a polynomial in descending powers of x that represents the area of the figure. $x^2 + 10x + 16$

Note to Instructor: The Chapter 5 Test file in the TestGen program provides algorithms specifically matched to these problems so you can easily replicate this chapter test for additional practice or assessment purposes.

CUMULATIVE REVIEW EXERCISES (CHAPTERS 1–5)

In Exercises 1–2, perform the indicated operation or operations.

1. $(-7)(-5) \div (12 - 3)$ $\dfrac{35}{9}$

2. $(3 - 7)^2(9 - 11)^3$ -128

3. What is the difference in elevation between a plane flying 14,300 feet above sea level and a submarine traveling 750 feet below sea level?   15,050 feet

In Exercises 4–5, solve each equation.

4. $2(x + 3) + 2x = x + 4$ $\left\{-\dfrac{2}{3}\right\}$

5. $\dfrac{x}{5} - \dfrac{1}{3} = \dfrac{x}{10} - \dfrac{1}{2}$ $\left\{-\dfrac{5}{3}\right\}$

6. The length of a rectangular sign is 2 feet less than three times its width. If the perimeter of the sign is 28 feet, what are its dimensions?   10 feet by 4 feet

7. Solve: $7 - 8x \le -6x - 5$. Express the solution set in set-builder notation and graph the solution set on a number line. $\{x \mid x \ge 6\}$; See graphing answer section.

8. You invested $6000 in two stocks paying 12% and 14% annual interest, respectively. At the end of the year, the total interest from these investments was $772. How much was invested at each rate? $3400 was invested at 12% and $2600 at 14%.

9. You need to mix a solution that is 70% antifreeze with one that is 30% antifreeze to obtain 20 liters of a mixture that is 60% antifreeze. How many liters of each of the solutions must be used?   15 liters of 70% antifreeze solution and 5 liters of 30% antifreeze solution must be used.

10. Graph $y = -\dfrac{2}{5}x + 2$ using the slope and y-intercept.
10–11. See graphing answer section.

11. Graph $x - 2y = 4$ using intercepts.

12. Find the slope of the line passing through the points $(-3, 2)$ and $(2, -4)$. Is the line rising, falling, horizontal, or vertical? $m = -\dfrac{6}{5}$; falling

13. The slope of a line is -2 and the line passes through the point $(3, -1)$. Write the line's equation in point-slope form and slope-intercept form. $y + 1 = -2(x - 3); y = -2x + 5$

In Exercises 14–15, solve each system by the method of your choice.

14. $3x + 2y = 10$
$4x - 3y = -15$ $\{(0, 5)\}$

15. $2x + 3y = -6$
$y = 3x - 13$ $\{(3, -4)\}$

16. You are choosing between two long-distance telephone plans. One has a monthly fee of $15 with a charge of $0.05 per minute for all long-distance calls. The other plan has a monthly fee of $5 with a charge of $0.07 per minute for all long-distance calls. For how many minutes of long-distance calls will the costs for the two plans be the same? What will be the cost for each plan?   500 minutes; $40

17. Write in scientific notation: 0.0024. 2.4×10^{-3}

18. Subtract: $(9x^5 - 3x^3 + 2x - 7) - (6x^5 + 3x^3 - 7x - 9)$.
$3x^5 - 6x^3 + 9x + 2$

19. Divide: $\dfrac{x^3 + 3x^2 + 5x + 3}{x + 1}$. $x^2 + 2x + 3$

20. Simplify: $\dfrac{(3x^2)^4}{x^{10}}$. $\dfrac{81}{x^2}$

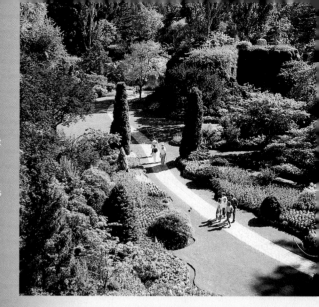

CHAPTER

6

Landscaped parks often form an oasis of greenery and calm in which to escape from the bustle of city life. Flowers, trees, ponds, and fountains provide a natural setting that mirrors the interest that many city dwellers have in their environment.

The role of polynomials in landscape design is explored in Exercises 84–85 in Exercise Set 6.6.

Factoring Polynomials

Have you ever thought about creating attractive and inviting home landscaping? Algebra and geometry play an important role in landscape design. In this chapter, you will see how rewriting a polynomial sum or difference in terms of multiplication can be used in the creation of horticultural masterpieces.

SECTION
6.1

Objectives

1 Factor monomials.

2 Find the greatest common factor.

3 Factor out the greatest common factor of a polynomial.

4 Factor by grouping.

THE GREATEST COMMON FACTOR AND FACTORING BY GROUPING

A two-year-old boy is asked, "Do you have a brother?" He answers, "Yes." "What is your brother's name?" "Tom." Asked if Tom has a brother, the two-year-old replies, "No." The child can go in the direction from self to brother, but he cannot reverse this direction and move from brother back to self.

As our intellects develop, we learn to reverse the direction of our thinking. Reversibility of thought is found throughout algebra. For example, we can multiply polynomials and show that

$$5x(2x + 3) = 10x^2 + 15x.$$

We can also reverse this process and express the resulting polynomial as

$$10x^2 + 15x = 5x(2x + 3).$$

Factoring a polynomial containing the sum of monomials means finding an equivalent expression that is a product.

Factoring $10x^2 + 15x$

Sum of monomials

Equivalent expression that is a product

$$10x^2 + 15x = 5x(2x + 3)$$

The factors of $10x^2 + 15x$ are $5x$ and $2x + 3$.

In this chapter, we will be factoring over the set of integers, meaning that the coefficients in the factors are integers. Polynomials that cannot be factored using integer coefficients are called **prime polynomials** over the set of integers.

Factoring Monomials   Factoring a monomial means finding two monomials whose product gives the original monomial. For example, $30x^2$ can be factored in a number of different ways, such as

$30x^2 = (5x)(6x)$	The factors are $5x$ and $6x$.
$30x^2 = (15x)(2x)$	The factors are $15x$ and $2x$.
$30x^2 = (10x^2)(3)$	The factors are $10x^2$ and 3.
$30x^2 = (-6x)(-5x)$.	The factors are $-6x$ and $-5x$.

Observe that each part of the factorization is called a *factor* of the given monomial.

1 Factor monomials.

2 Find the greatest common factor.

Factoring Out the Greatest Common Factor We use the distributive property to multiply a monomial and a polynomial of two or more terms. When we factor, we reverse this process, expressing the polynomial as a product.

Multiplication	Factoring
$a(b + c) = ab + ac$	$ab + ac = a(b + c)$

Here is a specific example:

Multiplication	Factoring
$5x(2x + 3)$	$10x^2 + 15x$
$= 5x \cdot 2x + 5x \cdot 3$	$= 5x \cdot 2x + 5x \cdot 3$
$= 10x^2 + 15x$	$= 5x(2x + 3)$.

In the process of finding an equivalent expression for $10x^2 + 15x$ that is a product, we used the fact that $5x$ is a factor of both $10x^2$ and $15x$. The factoring on the right shows that $5x$ is a *common factor* for all the terms of the binomial $10x^2 + 15x$.

In any factoring problem, the first step is to look for the *greatest common factor*. The **greatest common factor**, abbreviated GCF, is an expression of the highest degree that divides each term of the polynomial. Can you see that $5x$ is the greatest common factor of $10x^2 + 15x$? 5 is the greatest integer that divides 10 and 15. Furthermore, x is the greatest expression that divides x^2 and x.

The variable part of the greatest common factor always contains the smallest power of a variable that appears in all terms of the polynomial. For example, consider the polynomial

$$10x^2 + 15x.$$

> x^1, or x, is the variable raised to the smallest exponent.

We see that x is the variable part of the greatest common factor, $5x$.

EXAMPLE 1 Finding the Greatest Common Factor

Find the greatest common factor of each list of terms:

a. $6x^3$ and $10x^2$ **b.** $15y^5$, $-9y^4$, and $27y^3$ **c.** x^5y^3, x^4y^4, and x^3y^2.

SOLUTION Use numerical coefficients to determine the coefficient of the GCF. Use variable factors to determine the variable factor of the GCF.

> 2 is the greatest integer that divides 6 and 10.

a. $6x^3$ and $10x^2$

> x^2 is the variable raised to the smallest exponent.

We see that 2 is the coefficient of the GCF and x^2 is the variable factor of the GCF. Thus, the GCF of $6x^3$ and $10x^2$ is $2x^2$.

> 3 is the greatest integer that divides 15, −9, and 27.

b. $15y^5$, $-9y^4$,   and $27y^3$

> y^3 is the variable raised to the smallest exponent.

We see that 3 is the coefficient of the GCF and y^3 is the variable factor of the GCF. Thus, the GCF of $15y^5$, $-9y^4$, and $27y^3$ is $3y^3$.

> x^3 is the variable, x, raised to the smallest exponent.

c. $x^5y^3,$ $x^4y^4,$ and x^3y^2

> y^2 is the variable, y, raised to the smallest exponent.

Because all terms have coefficients of 1, 1 is the greatest integer that divides these coefficients. Thus, 1 is the coefficient of the GCF. The voice balloons show that x^3 and y^2 are the variable factors of the GCF. Thus, the GCF of x^5y^3, x^4y^4, and x^3y^2 is x^3y^2. ∎

 CHECK POINT 1 Find the greatest common factor of each list of terms:

a. $18x^3$ and $15x^2$ $3x^2$ **b.** $-20x^2, 12x^4,$ and $40x^3$ $4x^2$

c. $x^4y, x^3y^2,$ and $x^2y.$ x^2y

---

3 | Factor out the greatest common factor of a polynomial.

When we factor a monomial from a polynomial, we determine the greatest common factor of all terms in the polynomial. Sometimes there may not be a GCF other than 1. When a GCF other than 1 exists, we use the following procedure:

FACTORING A MONOMIAL FROM A POLYNOMIAL

1. Determine the greatest common factor of all terms in the polynomial.
2. Express each term as the product of the GCF and its other factor.
3. Use the distributive property to factor out the GCF.

EXAMPLE 2 Factoring Out the Greatest Common Factor

Factor: $5x^2 + 30$.

SOLUTION   The GCF of $5x^2$ and 30 is 5.

$$5x^2 + 30$$
$$= 5 \cdot x^2 + 5 \cdot 6 \qquad \text{Express each term as the product of the GCF and its other factor.}$$
$$= 5(x^2 + 6) \qquad \text{Factor out the GCF.}$$

Because factoring reverses the process of multiplication, all factoring results can be checked by multiplying.

$$5(x^2 + 6) = 5 \cdot x^2 + 5 \cdot 6 = 5x^2 + 30$$

The factoring is correct because multiplication gives us the original polynomial. ∎

STUDY TIP

When we express $5x^2 + 30$ as $5 \cdot x^2 + 5 \cdot 6$, we have factored the *terms* of the binomial, but not the binomial itself. The factorization of the binomial is not complete until we write

$$5(x^2 + 6).$$

Now we have expressed the binomial *as a product*.

 CHECK POINT 2 Factor: $6x^2 + 18$. $6(x^2 + 3)$

EXAMPLE 3 Factoring Out the Greatest Common Factor

Factor: $18x^3 + 27x^2$.

SOLUTION We begin by determining the greatest common factor.

9 is the greatest integer that divides 18 and 27.

$$18x^3 \quad \text{and} \quad 27x^2$$

x^2 is the variable raised to the smallest exponent.

The GCF of the two terms in the polynomial is $9x^2$.

$$18x^3 + 27x^2$$
$$= 9x^2(2x) + 9x^2(3) \qquad \text{Express each term as the product of the GCF and its other factor.}$$
$$= 9x^2(2x + 3) \qquad \text{Factor out the GCF.}$$

We can check this factorization by multiplying $9x^2$ and $2x + 3$, obtaining the original polynomial as the answer. ■

 CHECK POINT 3 Factor: $25x^2 + 35x^3$. $5x^2(5 + 7x)$

DISCOVER FOR YOURSELF

What happens if you factor out $3x^2$ rather than $9x^2$ from $18x^3 + 27x^2$? Although $3x^2$ is a common factor of the two terms, it is not the *greatest* common factor. Remove $3x^2$ from $18x^3 + 27x^2$ and describe what happens with the second factor. Now factor again. Make the final result look like the factorization in Example 3. What is the advantage of factoring out the greatest common factor rather than just a common factor?

EXAMPLE 4 Factoring Out the Greatest Common Factor

Factor: $16x^5 - 12x^4 + 4x^3$.

SOLUTION First, determine the greatest common factor.

4 is the greatest integer that divides 16, −12, and 4.

$$16x^5, \quad -12x^4, \quad \text{and} \quad 4x^3$$

x^3 is the variable raised to the smallest exponent.

The GCF of the three terms of the polynomial is $4x^3$.

$$16x^5 - 12x^4 + 4x^3$$
$$= 4x^3 \cdot 4x^2 - 4x^3 \cdot 3x + 4x^3 \cdot 1 \qquad \text{Express each term as the product of the GCF and its other factor.}$$
$$= 4x^3(4x^2 - 3x + 1) \qquad \text{Factor out the GCF.}$$

Don't leave out the 1.

■

 CHECK POINT 4 Factor: $15x^5 + 12x^4 - 27x^3$. $3x^3(5x^2 + 4x - 9)$

EXAMPLE 5 Factoring Out the Greatest Common Factor

Factor: $27x^2y^3 - 9xy^2 + 81xy$.

SOLUTION First, determine the greatest common factor.

> 9 is the greatest integer that divides 27, −9, and 81.

$$27x^2y^3, \quad -9xy^2, \quad \text{and} \quad 81xy$$

> The variables raised to the smallest exponents are x and y.

The GCF of the three terms of the polynomial is $9xy$.

$$27x^2y^3 - 9xy^2 + 81xy$$
$$= 9xy \cdot 3xy^2 - 9xy \cdot y + 9xy \cdot 9 \qquad \text{Express each term as the product of the GCF and its other factor.}$$
$$= 9xy(3xy^2 - y + 9) \qquad \text{Factor out the GCF.} \qquad \blacksquare$$

✔ **CHECK POINT 5** Factor: $8x^3y^2 - 14x^2y + 2xy$. $2xy(4x^2y - 7x + 1)$

---

 4 Factor by grouping.

Factoring by Grouping Up to now, we have factored a monomial from a polynomial. By contrast, in our next example, the greatest common factor of the polynomial is a binomial.

EXAMPLE 6   Factoring Out the Greatest Common Binomial Factor

Factor:

  a. $x^2(x + 3) + 5(x + 3)$ **b.** $x(y + 1) - 2(y + 1)$.

SOLUTION   Let's identify the common binomial factor in each part of the problem.

$$x^2(x + 3) \quad \text{and} \quad 5(x + 3) \qquad\qquad x(y + 1) \quad \text{and} \quad -2(y + 1)$$

> The GCF, a binomial, is $x + 3$.   The GCF, a binomial, is $y + 1$.

We factor out these common binomial factors as follows.

  a. $x^2(x + 3) + 5(x + 3)$
  $= (x + 3)x^2 + (x + 3)5$ Express each term as the product of the GCF and its other factor, in that order. Hereafter, we omit this step.
  $= (x + 3)(x^2 + 5)$ Factor out the GCF, $x + 3$.

  b. $x(y + 1) - 2(y + 1)$ The GCF is $y + 1$.
  $= (y + 1)(x - 2)$ Factor out the GCF. $\blacksquare$

✔ **CHECK POINT 6** Factor:
  a. $x^2(x + 1) + 7(x + 1)$ $(x+1)(x^2+7)$ **b.** $x(y + 4) - 7(y + 4)$. $(y+4)(x-7)$

Some polynomials have only a greatest common factor of 1. However, by a suitable grouping of the terms, it still may be possible to factor. This process, called **factoring by grouping**, is illustrated in Example 7.

EXAMPLE 7 Factoring by Grouping

Factor: $x^3 + 4x^2 + 3x + 12$.

SOLUTION   There is no factor other than 1 common to all terms. However, we can group terms that have a common factor:

$$\boxed{x^3 + 4x^2} \quad + \quad \boxed{3x + 12}.$$

Common factor is x^2.

Common factor is 3.

We now factor the given polynomial as follows:

$$x^3 + 4x^2 + 3x + 12$$

$$= (x^3 + 4x^2) + (3x + 12) \qquad \text{Group terms with common factors.}$$

$$= x^2(x + 4) + 3(x + 4) \qquad \text{Factor out the greatest common factor from the grouped terms. The remaining two terms have } x + 4 \text{ as a common binomial factor.}$$

$$= (x + 4)(x^2 + 3). \qquad \text{Factor out the GCF, } x + 4.$$

Thus, $x^3 + 4x^2 + 3x + 12 = (x + 4)(x^2 + 3)$. Check the factorization by multiplying the right side of the equation using the FOIL method. Because the factorization is correct, you should obtain the original polynomial. ∎

✔ CHECK POINT 7 Factor: $x^3 + 5x^2 + 2x + 10$. $(x + 5)(x^2 + 2)$

DISCOVER FOR YOURSELF

In Example 7, group the terms as follows:

$$(x^3 + 3x) + (4x^2 + 12).$$

Factor out the greatest common factor from each group and complete the factoring process. Describe what happens. What can you conclude?

FACTORING BY GROUPING

1. Group terms that have a common monomial factor. There will usually be two groups. Sometimes the terms must be rearranged.
2. Factor out the common monomial factor from each group.
3. Factor out the remaining common binomial factor (if one exists).

EXAMPLE 8 Factoring by Grouping

Factor: $xy + 5x - 4y - 20$.

SOLUTION   There is no factor other than 1 common to all terms. However, we can group terms that have a common factor:

$$\boxed{xy + 5x} \quad + \quad \boxed{-4y - 20}.$$

Common factor is x:
$xy + 5x = x(y + 5)$.

Use −4, rather than 4, as the common factor: $-4y - 20 = -4(y + 5)$. In this way, the common binomial factor, $y + 5$, appears.

The voice balloons illustrate that it is sometimes necessary to factor out a negative number from a grouping to obtain a common binomial factor for the two groupings. We now factor the given polynomial as follows:

$$xy + 5x - 4y - 20$$

$$= x(y + 5) - 4(y + 5) \qquad \text{Factor } x \text{ and } -4, \text{ respectively, from each grouping.}$$

$$= (y + 5)(x - 4). \qquad \text{Factor out the GCF, } y + 5.$$

Thus, $xy + 5x - 4y - 20 = (y + 5)(x - 4)$. Using the commutative property of multiplication, the factorization can also be expressed as $(x - 4)(y + 5)$. Multiply these factors using the FOIL method to verify that, regardless of the order, these are the correct factors. ∎

✔ **CHECK POINT 8** Factor: $xy + 3x - 5y - 15.$ $(y + 3)(x - 5)$

6.1 EXERCISE SET

Student Solutions Manual CD/Video PH Math/Tutor Center MathXL Tutorials on CD MathXL® MyMathLab Interactmath.com

Practice Exercises

In Exercises 1–6, find three factorizations for each monomial.
1–6. Answers will vary.

1. $8x^3$ Examples are $(2x)(4x^2), (4x)(2x^2),$ and $(8x)(x^2).$

2. $20x^4$ Examples are $(4x^2)(5x^2), (5x)(4x^3),$ and $(20x)(x^3).$

3. $-12x^5$ Examples are $(-4x^3)(3x^2), (2x^2)(-6x^3),$ and $(-3)(4x^5).$

4. $-15x^6$ Examples are $(-3x^2)(5x^4), (5x^3)(-3x^3),$ and $(-15)(x^6).$

5. $36x^4$ Examples are $(6x^2)(6x^2), (-2x)(-18x^3),$ and $(4x^3)(9x).$

6. $27x^5$ Examples are $(3x^2)(9x^3), (-9x)(-3x^4),$ and $(27)(x^5).$

In Exercises 7–18, find the greatest common factor of each list of terms.

7. 4 and $8x$ 4
8. 5 and $15x$ 5
9. $12x^2$ and $8x$ $4x$
10. $20x^2$ and $15x$ $5x$
11. $-2x^4$ and $6x^3$ $2x^3$
12. $-3x^4$ and $6x^3$ $3x^3$
13. $9y^5, 18y^2,$ and $-3y$ $3y$
14. $10y^5, 20y^2,$ and $-5y$ $5y$
15. $xy, xy^2,$ and xy^3 xy
16. $x^2y, 3x^3y,$ and $6x^2$ x^2
17. $16x^5y^4, 8x^6y^3,$ and $20x^4y^5$ $4x^4y^3$
18. $18x^5y^4, 6x^6y^3,$ and $12x^4y^5$ $6x^4y^3$

In Exercises 19–54, factor each polynomial using the greatest common factor. If there is no common factor other than 1 and the polynomial cannot be factored, so state.

19. $8x + 8$ $8(x + 1)$
20. $9x + 9$ $9(x + 1)$
21. $4y - 4$ $4(y - 1)$
22. $5y - 5$ $5(y - 1)$

23. $5x + 30$ $5(x + 6)$
24. $10x + 30$ $10(x + 3)$
25. $30x - 12$ $6(5x - 2)$
26. $32x - 24$ $8(4x - 3)$
27. $x^2 + 5x$ $x(x + 5)$
28. $x^2 + 6x$ $x(x + 6)$
29. $18y^2 + 12$ $6(3y^2 + 2)$
30. $20y^2 + 15$ $5(4y^2 + 3)$
31. $14x^3 + 21x^2$ $7x^2(2x + 3)$
32. $6x^3 + 15x^2$ $3x^2(2x + 5)$
33. $13y^2 - 25y$ $y(13y - 25)$
34. $11y^2 - 30y$ $y(11y - 30)$
35. $9y^4 + 27y^6$ $9y^4(1 + 3y^2)$
36. $10y^4 + 15y^6$ $5y^4(2 + 3y^2)$
37. $8x^2 - 4x^4$ $4x^2(2 - x^2)$
38. $12x^2 - 4x^4$ $4x^2(3 - x^2)$
39. $12y^2 + 16y - 8$ $4(3y^2 + 4y - 2)$
40. $15y^2 - 3y + 9$ $3(5y^2 - y + 3)$
41. $9x^4 + 18x^3 + 6x^2$ $3x^2(3x^2 + 6x + 2)$
42. $32x^4 + 2x^3 + 8x^2$ $2x^2(16x^2 + x + 4)$
43. $100y^5 - 50y^3 + 100y^2$ $50y^2(2y^3 - y + 2)$
44. $26y^5 - 13y^3 + 39y^2$ $13y^2(2y^3 - y + 3)$
45. $10x - 20x^2 + 5x^3$ $5x(2 - 4x + x^2)$
46. $6x - 4x^2 + 2x^3$ $2x(3 - 2x + x^2)$
47. $11x^2 - 23$ cannot be factored
48. $12x^2 - 25$ cannot be factored
49. $6x^3y^2 + 9xy$ $3xy(2x^2y + 3)$
50. $4x^2y^3 + 6xy$ $2xy(2xy^2 + 3)$
51. $30x^2y^3 - 10xy^2 + 20xy$ $10xy(3xy^2 - y + 2)$
52. $27x^2y^3 - 18xy^2 + 45x^2y$ $9xy(3xy^2 - 2y + 5x)$
53. $32x^3y^2 - 24x^3y - 16x^2y$ $8x^2y(4xy - 3x - 2)$
54. $18x^3y^2 - 12x^3y - 24x^2y$ $6x^2y(3xy - 2x - 4)$

In Exercises 55–66, factor each polynomial using the greatest common binomial factor.

55. $x(x + 5) + 3(x + 5)$ $(x + 5)(x + 3)$

56. $x(x + 7) + 10(x + 7)$ $(x + 7)(x + 10)$

57. $x(x + 2) - 4(x + 2)$ $(x + 2)(x - 4)$

58. $x(x + 3) - 8(x + 3)$ $(x + 3)(x - 8)$

59. $x(y + 6) - 7(y + 6)$ $(y + 6)(x - 7)$

60. $x(y + 9) - 11(y + 9)$ $(y + 9)(x - 11)$

61. $3x(x + y) - (x + y)$ $(x + y)(3x - 1)$

62. $7x(x + y) - (x + y)$ $(x + y)(7x - 1)$

63. $4x(3x + 1) + 3x + 1$ $(3x + 1)(4x + 1)$

64. $5x(2x + 1) + 2x + 1$ $(2x + 1)(5x + 1)$

65. $7x^2(5x + 4) + 5x + 4$ $(5x + 4)(7x^2 + 1)$

66. $9x^2(7x + 2) + 7x + 2$ $(7x + 2)(9x^2 + 1)$

In Exercises 67–84, factor by grouping.

67. $x^2 + 2x + 4x + 8$ $(x + 2)(x + 4)$

68. $x^2 + 3x + 5x + 15$ $(x + 3)(x + 5)$

69. $x^2 + 3x - 5x - 15$ $(x - 5)(x + 3)$

70. $x^2 + 7x - 4x - 28$ $(x + 7)(x - 4)$

71. $x^3 - 2x^2 + 5x - 10$ $(x^2 + 5)(x - 2)$

72. $x^3 - 3x^2 + 4x - 12$ $(x^2 + 4)(x - 3)$

73. $x^3 - x^2 + 2x - 2$ $(x^2 + 2)(x - 1)$

74. $x^3 + 6x^2 - 2x - 12$ $(x^2 - 2)(x + 6)$

75. $xy + 5x + 9y + 45$ $(y + 5)(x + 9)$

76. $xy + 6x + 2y + 12$ $(y + 6)(x + 2)$

77. $xy - x + 5y - 5$ $(y - 1)(x + 5)$

78. $xy - x + 7y - 7$ $(y - 1)(x + 7)$

79. $3x^2 - 6xy + 5xy - 10y^2$ $(x - 2y)(3x + 5y)$

80. $10x^2 - 12xy + 35xy - 42y^2$ $(5x - 6y)(2x + 7y)$

81. $3x^3 - 2x^2 - 6x + 4$ $(3x - 2)(x^2 - 2)$

82. $4x^3 - x^2 - 12x + 3$ $(4x - 1)(x^2 - 3)$

83. $x^2 - ax - bx + ab$ $(x - a)(x - b)$

84. $x^2 + ax + bx + ab$ $(x + a)(x + b)$

Practice Plus

In Exercises 85–92, factor each polynomial.

85. $24x^3y^3z^3 + 30x^2y^2z + 18x^2yz^2$ $6x^2yz(4xy^2z^2 + 5y + 3z)$

86. $16x^2y^2z^2 + 32x^2yz^2 + 24x^2yz$ $8x^2yz(2yz + 4z + 3)$

87. $x^3 - 4 + 3x^3y - 12y$ $(x^3 - 4)(1 + 3y)$

88. $x^3 - 5 + 2x^3y - 10y$ $(x^3 - 5)(1 + 2y)$

89. $4x^5(x + 1) - 6x^3(x + 1) - 8x^2(x + 1)$
$2x^2(x + 1)(2x^3 - 3x - 4)$

90. $8x^5(x + 2) - 10x^3(x + 2) - 2x^2(x + 2)$
$2x^2(x + 2)(4x^3 - 5x - 1)$

91. $3x^5 - 3x^4 + x^3 - x^2 + 5x - 5$ $(x - 1)(3x^4 + x^2 + 5)$

92. $7x^5 - 7x^4 + x^3 - x^2 + 3x - 3$ $(x - 1)(7x^4 + x^2 + 3)$

The figures for Exercises 93–94 show one or more circles drawn inside a square. Write a polynomial that represents the shaded area in each figure. Then factor the polynomial.

93.

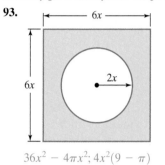

$36x^2 - 4\pi x^2; \, 4x^2(9 - \pi)$

94.

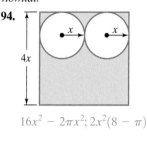

$16x^2 - 2\pi x^2; \, 2x^2(8 - \pi)$

Application Exercises

95. An explosion causes debris to rise vertically with an initial velocity of 64 feet per second. The polynomial $64x - 16x^2$ describes the height of the debris above the ground, in feet, after x seconds.

   a. Find the height of the debris after 3 seconds.   48 feet

   b. Factor the polynomial. $16x(4 - x)$

   c. Use the factored form of the polynomial in part (b) to find the height of the debris after 3 seconds. Do you get the same answer as you did in part (a)? If so, does this prove that your factorization is correct? Explain.   48 feet; yes; no; Answers will vary.

96. An explosion causes debris to rise vertically with an initial velocity of 72 feet per second. The polynomial $72x - 16x^2$ describes the height of the debris above the ground, in feet, after x seconds.

   a. Find the height of the debris after 4 seconds.   32 feet

   b. Factor the polynomial. $8x(9 - 2x)$

   c. Use the factored form of the polynomial in part (b) to find the height of the debris after 4 seconds. Do you get the same answer as you did in part (a)? If so, does this prove that your factorization is correct? Explain.   32 feet; yes; no; Answers will vary.

In Exercises 97–98, write a polynomial for the length of each rectangle.

97.

Length is $x^3 - 2$.

Area is $5x^4 - 10x$ square units. $5x$

98.

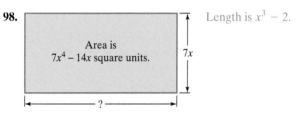

Length is $x^3 - 2$.

Area is $7x^4 - 14x$ square units. $7x$

Writing in Mathematics

99. What is factoring?   99–104. Answers will vary.

100. What is a prime polynomial?

101. Explain how to find the greatest common factor of a list of terms. Give an example with your explanation.

102. Use an example and explain how to factor out the greatest common factor of a polynomial.

103. Suppose that a polynomial contains four terms and can be factored by grouping. Explain how to obtain the factorization.

104. Write a sentence that uses the word "factor" as a noun. Then write a sentence that uses the word "factor" as a verb.

Critical Thinking Exercises

105. Which one of the following is true?   d

   a. Because a monomial contains one term, it follows that a monomial can be factored in precisely one way.

   b. The GCF for $8x^3 - 16x^2$ is $8x$.

   c. The integers 10 and 31 have no GCF.

   d. $-4x^2 + 12x$ can be factored as $-4x(x - 3)$ or $4x(-x + 3)$.

106. Suppose you receive x dollars in January. Each month thereafter, you receive $100 more than you received the month before. Write a factored polynomial that describes the total dollar amount you receive from January through April. $4(x + 150)$

In Exercises 107–108, write a polynomial that fits the given description. Do not use a polynomial that appears in this section or in the exercise set.

107. The polynomial has four terms and can be factored using a greatest common factor that has both a coefficient and a variable.   107–108. Answers will vary.

108. The polynomial has four terms and can be factored by grouping.

Technology Exercises

In Exercises 109–111, use a graphing utility to graph each side of the equation in the same viewing rectangle. Do the graphs coincide? If so, this means that the polynomial on the left side has been factored correctly. If not, factor the polynomial correctly and then use your graphing utility to verify the factorization.

109. $-3x - 6 = -3(x - 2)$ $-3(x - 2)$ should be $-3(x + 2)$.

110. $x^2 - 2x + 5x - 10 = (x - 2)(x - 5)$ $(x - 2)(x - 5)$ should be $(x - 2)(x + 5)$.

111. $x^2 + 2x + x + 2 = x(x + 2) + 1$ $x(x + 2) + 1$ should be $(x + 2)(x + 1)$.

Review Exercises

112. Multiply: $(x + 7)(x + 10)$. (Section 5.3, Example 1)
$x^2 + 17x + 70$

113. Solve the system by graphing:
$$2x - \ y = -4$$
$$x - 3y = \ 3.$$

(Section 4.1, Example 2) $\{(-3, -2)\}$

114. Write the point-slope form of a line passing through $(-7, 2)$ and $(-4, 5)$. Then use the point-slope equation to write the slope-intercept equation. (Section 3.5, Example 2)
$y - 2 = 1(x + 7)$ or $y - 5 = 1(x + 4)$; $y = x + 9$

SECTION 6.2

FACTORING TRINOMIALS WHOSE LEADING COEFFICIENT IS ONE

Objective

1 Factor trinomials of the form $x^2 + bx + c$.

Not afraid of heights and cutting-edge excitement? How about skydiving? Behind your exhilarating experience is the world of algebra. After you jump from the airplane, your height above the ground at every instant of your fall can be described by a formula involving a variable that is squared. At a height of approximately 2000 feet, you'll need to open your parachute. How can you determine when you must do so?

The answer to this critical question involves using the factoring technique presented in this section. In Section 6.6, in which applications are discussed, this technique is applied to models involving the height of any free-falling object—in this case, you.

1 Factor trinomials of the form $x^2 + bx + c$.

A Strategy for Factoring $x^2 + bx + c$ In Section 5.3, we used the FOIL method to multiply two binomials. The product was often a trinomial. The following are some examples:

| **Factored Form** | **F** | **O** | **I** | **L** | **Trinomial Form** |

$$(x + 3)(x + 4) = x^2 + 4x + 3x + 12 = x^2 + 7x + 12$$
$$(x - 3)(x - 4) = x^2 - 4x - 3x + 12 = x^2 - 7x + 12$$
$$(x + 3)(x - 5) = x^2 - 5x + 3x - 15 = x^2 - 2x - 15.$$

Observe that each trinomial is of the form $x^2 + bx + c$, where the coefficient of the squared term is 1. Our goal in this section is to start with the trinomial form and, assuming that it is factorable, return to the factored form.

The first FOIL multiplication shown above indicates that $(x + 3)(x + 4) = x^2 + 7x + 12$. Let's reverse the sides of this equation:

$$x^2 + 7x + 12 = (x + 3)(x + 4).$$

We can make several important observations about the factors on the right side.

The first term of each factor is x. The product of the First terms is $x \cdot x = x^2$.

3 and 4 are factors of 12. The product of the Last terms is $3 \cdot 4 = 12$.

The sum of the Outside and Inside products is $4x + 3x = 7x$.

These observations provide us with a procedure for factoring $x^2 + bx + c$.

A STRATEGY FOR FACTORING $x^2 + bx + c$

1. Enter x as the first term of each factor.

$$(x \quad)(x \quad) = x^2 + bx + c$$

2. List pairs of factors of the constant c.

3. Try various combinations of these factors as the second term in each set of parentheses. Select the combination in which the sum of the Outside and Inside products is equal to bx.

$$(x + \square)(x + \square) = x^2 + bx + c$$

I

O

Sum of O + I

4. Check your work by multiplying the factors using the FOIL method. You should obtain the original trinomial.

If none of the possible combinations yield an Outside product and an Inside product whose sum is equal to bx, the trinomial cannot be factored using integers and is called **prime** over the set of integers.

> ### EXAMPLE 1 Factoring a Trinomial in $x^2 + bx + c$ Form
>
> Factor: $x^2 + 6x + 8$.
>
> **SOLUTION**
>
> **Step 1.** **Enter x as the first term of each factor.**
>
> $$x^2 + 6x + 8 = (x \qquad)(x \qquad)$$
>
> To find the second term of each factor, we must find two integers whose product is 8 and whose sum is 6.
>
> **Step 2.** **List pairs of factors of the constant, 8.**
>
> | Factors of 8 | 8, 1 | 4, 2 | −8, −1 | −4, −2 |
> |---|---|---|---|---|
>
> **Step 3.** **Try various combinations of these factors.** The correct factorization of $x^2 + 6x + 8$ is the one in which the sum of the Outside and Inside products is equal to $6x$. Here is a list of the possible factorizations:
>
> | Possible Factorizations of $x^2 + 6x + 8$ | Sum of Outside and Inside Products (Should Equal $6x$) |
> |---|---|
> | $(x + 8)(x + 1)$ | $x + 8x = 9x$ |
> | $(x + 4)(x + 2)$ | $2x + 4x = 6x$ |
> | $(x − 8)(x − 1)$ | $−x − 8x = −9x$ |
> | $(x − 4)(x − 2)$ | $−2x − 4x = −6x$ |
>
> This is the required middle term.
>
> Thus, $x^2 + 6x + 8 = (x + 4)(x + 2)$.
>
> **Step 4.** **Check this result by multiplying the right side using the FOIL method.** You should obtain the original trinomial. Because of the commutative property, we can also say that
>
> $$x^2 + 6x + 8 = (x + 2)(x + 4).$$ ∎

USING TECHNOLOGY

If a polynomial contains one variable, a graphing utility can be used to check its factorization. For example, the factorization in Example 1

$$x^2 + 6x + 8 = (x + 4)(x + 2)$$

can be checked graphically or numerically.

Graphic Check

Use the GRAPH feature. Graph $y_1 = x^2 + 6x + 8$ and $y_2 = (x + 4)(x + 2)$ on the same screen. Because the graphs are identical, the factorization appears to be correct.

$y_1 = x^2 + 6x + 8$

$y_2 = (x + 4)(x + 2)$

$[−7, 1, 1]$ by $[−2, 12, 1]$

Numeric Check

Use the TABLE feature. Enter $y_1 = x^2 + 6x + 8$ and $y_2 = (x + 4)(x + 2)$ and press TABLE. Two columns of values are shown, one for y_1 and one for y_2. Because the corresponding values are equal regardless of how far up or down we scroll, the factorization is correct.

X	Y₁	Y₂
-3	-1	-1
-2	0	0
-1	3	3
0	8	8
1	15	15
2	24	24
3	35	35

X= -3

✔ CHECK POINT **1** Factor: $x^2 + 5x + 6$. $(x + 2)(x + 3)$

EXAMPLE 2 Factoring a Trinomial in $x^2 + bx + c$ Form

Factor: $x^2 - 5x + 6$.

SOLUTION

Step 1. **Enter x as the first term of each factor.**

$$x^2 - 5x + 6 = (x \quad)(x \quad)$$

To find the second term of each factor, we must find two integers whose product is 6 and whose sum is -5.

Step 2. **List pairs of factors of the constant, 6.**

Factors of 6	6, 1	3, 2	−6, −1	−3, −2

Step 3. **Try various combinations of these factors.** The correct factorization of $x^2 - 5x + 6$ is the one in which the sum of the Outside and Inside products is equal to $-5x$. Here is a list of the possible factorizations:

Possible Factorizations of $x^2 - 5x + 6$	Sum of Outside and Inside Products (Should Equal $-5x$)
$(x + 6)(x + 1)$	$x + 6x = 7x$
$(x + 3)(x + 2)$	$2x + 3x = 5x$
$(x - 6)(x - 1)$	$-x - 6x = -7x$
$(x - 3)(x - 2)$	$-2x - 3x = -5x$

This is the required middle term.

STUDY TIP

To factor $x^2 + bx + c$ when c is positive, find two numbers with the same sign as the middle term.

$x^2 + 6x + 8 = (x + 2)(x + 4)$

Same signs

$x^2 - 5x + 6 = (x - 3)(x - 2)$

Same signs

Thus, $x^2 - 5x + 6 = (x - 3)(x - 2)$. Verify this result using the FOIL method. ∎

In factoring a trinomial of the form $x^2 + bx + c$, you can speed things up by listing the factors of c and then finding their sums. We are interested in a sum of b. For example, in factoring $x^2 - 5x + 6$, we are interested in the factors of 6 whose sum is -5.

Factors of 6	6, 1	3, 2	−6, −1	−3, −2
Sum of Factors	7	5	−7	−5

This is the desired sum.

Thus, $x^2 - 5x + 6 = (x - 3)(x - 2)$.

✔ CHECK POINT **2** Factor: $x^2 - 6x + 8$. $(x - 4)(x - 2)$

EXAMPLE 3 Factoring a Trinomial in $x^2 + bx + c$ Form

Factor: $x^2 + 2x - 35$.

SOLUTION

Step 1. **Enter x as the first term of each factor.**

$$x^2 + 2x - 35 = (x \quad)(x \quad)$$

To find the second term of each factor, we must find two integers whose product is -35 and whose sum is 2.

Step 2. **List pairs of factors of the constant, -35.**

Factors of -35	$-35, 1$	$-7, 5$	$35, -1$	$7, -5$

Step 3. **Try various combinations of these factors.** We are looking for the factors whose sum is 2.

Factors of -35	$-35, 1$	$-7, 5$	$35, -1$	$7, -5$
Sum of Factors	-34	-2	34	2

This is the desired sum.

Thus, $x^2 + 2x - 35 = (x + 7)(x - 5)$.

Step 4. **Verify the factorization using the FOIL method.**

$$(x + 7)(x - 5) = x^2 - 5x + 7x - 35 = x^2 + 2x - 35$$

Because the product of the factors is the original polynomial, the factorization is correct. ∎

✔ **CHECK POINT 3** Factor: $x^2 + 3x - 10$. $(x + 5)(x - 2)$

EXAMPLE 4 Factoring a Trinomial Whose Leading Coefficient Is One

Factor: $y^2 - 2y - 99$.

SOLUTION

Step 1. **Enter y as the first term of each factor.**

$$y^2 - 2y - 99 = (y \quad)(y \quad)$$

To find the second term of each factor, we must find two integers whose product is -99 and whose sum is -2.

Step 2. **List pairs of factors of the constant, -99.**

Factors of -99	$-99, 1$	$-11, 9$	$-33, 3$	$99, -1$	$11, -9$	$33, -3$

Step 3. **Try various combinations of these factors.** We are interested in factors whose sum is -2.

Factors of -99	$-99, 1$	$-11, 9$	$-33, 3$	$99, -1$	$11, -9$	$33, -3$
Sum of Factors	-98	-2	-30	98	2	30

This is the desired sum.

Thus, $y^2 - 2y - 99 = (y - 11)(y + 9)$. Verify this result using the FOIL method. ∎

✔ **CHECK POINT 4** Factor: $y^2 - 6y - 27$. $(y - 9)(y + 3)$

STUDY TIP

To factor $x^2 + bx + c$ when c is negative, find two numbers with opposite signs whose sum is the coefficient of the middle term.

$$x^2 + 2x - 35 = (x + 7)(x - 5)$$

Negative   Opposite signs

$$y^2 - 2y - 99 = (y - 11)(y + 9)$$

Negative   Opposite signs

EXAMPLE 5 Trying to Factor a Trinomial in $x^2 + bx + c$ Form

Factor: $x^2 + x - 5$.

SOLUTION

Step 1. **Enter x as the first term of each factor.**

$$x^2 + x - 5 = (x \quad)(x \quad)$$

To find the second term of each factor, we must find two integers whose product is -5 and whose sum is 1.

Steps 2 and 3. **List pairs of factors of the constant, -5, and try various combinations of these factors.** We are interested in factors whose sum is 1.

Factors of -5	$-5, 1$	$5, -1$
Sum of Factors	-4	4

No pair gives the desired sum, 1.

Because neither pair has a sum of 1, $x^2 + x - 5$ cannot be factored using integers. This trinomial is prime. ■

 CHECK POINT 5 Factor: $x^2 + x - 7$. *cannot factor over the integers; prime*

EXAMPLE 6 Factoring a Trinomial in Two Variables

Factor: $x^2 - 5xy + 6y^2$.

SOLUTION

Step 1. **Enter x as the first term of each factor.** Because the last term of the trinomial contains y^2, the second term of each factor must contain y.

$$x^2 - 5xy + 6y^2 = (x \quad ?y)(x \quad ?y)$$

The question marks indicate that we are looking for the coefficients of y in each factor. To find these coefficients, we must find two integers whose product is 6 and whose sum is -5.

Steps 2 and 3. **List pairs of factors of the coefficient of the last term, 6, and try various combinations of these factors.** We are interested in factors whose sum is -5.

Factors of 6	6, 1	3, 2	$-6, -1$	$-3, -2$
Sum of Factors	7	5	-7	-5

This is the desired sum.

Thus, $x^2 - 5xy + 6y^2 = (x - 3y)(x - 2y)$.

Step 4. **Verify the factorization using the FOIL method.**

$$(x - 3y)(x - 2y) = x^2 - 2xy - 3xy + 6y^2 = x^2 - 5xy + 6y^2$$

Because the product of the factors is the original polynomial, the factorization is correct. ■

 CHECK POINT 6 Factor: $x^2 - 4xy + 3y^2$. *$(x - 3y)(x - y)$*

Some polynomials can be factored using more than one technique. **Always begin by looking for a greatest common factor** and, if there is one, factor it out. A polynomial is **factored completely** when it is written as the product of prime polynomials.

EXAMPLE 7 Factoring Completely

Factor: $3x^3 - 15x^2 - 42x$.

SOLUTION   The GCF of the three terms of the polynomial is $3x$. We begin by factoring out $3x$. Then we factor the remaining trinomial by the methods of this section.

$$3x^3 - 15x^2 - 42x$$
$$= 3x(x^2 - 5x - 14) \quad \text{Factor out the GCF.}$$
$$= 3x(x \quad)(x \quad) \quad \text{Begin factoring } x^2 - 5x - 14. \text{ Find two integers whose product is } -14 \text{ and whose sum is } -5.$$
$$= 3x(x - 7)(x + 2) \quad \text{The integers are } -7 \text{ and } 2.$$

Thus,

$$3x^3 - 15x^2 - 42x = 3x(x - 7)(x + 2).$$

Be sure to include the GCF in the factorization.

How can we check this factorization? We will multiply the binomials using the FOIL method. Then use the distributive property and multiply each term of this product by $3x$. If the factorization is correct, we should obtain the original polynomial.

$$3x(x - 7)(x + 2) = 3x(x^2 + 2x - 7x - 14) = 3x(x^2 - 5x - 14) = 3x^3 - 15x^2 - 42x$$

Use the FOIL method on $(x - 7)(x + 2)$.

This is the original polynomial.

The factorization is correct.

 CHECK POINT 7 Factor: $2x^3 + 6x^2 - 56x$. $2x(x - 4)(x + 7)$

6.2 EXERCISE SET

Student Solutions Manual   CD/Video   PH Math/Tutor Center   MathXL Tutorials on CD   MathXL®   MyMathLab   Interactmath.com

Practice Exercises

In Exercises 1–42, factor each trinomial, or state that the trinomial is prime. Check each factorization using FOIL multiplication.

1. $x^2 + 7x + 6$
  $(x + 6)(x + 1)$

2. $x^2 + 9x + 1$
  $(x + 8)(x + 1)$

3. $x^2 + 7x + 10$
  $(x + 2)(x + 5)$

4. $x^2 + 9x + 14$
  $(x + 2)(x + 7)$

5. $x^2 + 11x + 10$
  $(x + 1)(x + 10)$

6. $x^2 + 13x + 12$
  $(x + 1)(x + 12)$

7. $x^2 - 7x + 12$
  $(x - 4)(x - 3)$

8. $x^2 - 13x + 40$
  $(x - 5)(x - 8)$

9. $x^2 - 12x + 36$
  $(x - 6)(x - 6)$

10. $x^2 - 8x + 16$
  $(x - 4)(x - 4)$

11. $y^2 - 8y + 15$
  $(y - 3)(y - 5)$

12. $y^2 - 8y + 7$
  $(y - 7)(y - 1)$

13. $x^2 + 3x - 10$
  $(x + 5)(x - 2)$

14. $x^2 + 3x - 28$
  $(x + 7)(x - 4)$

15. $y^2 + 10y - 39$
  $(y + 13)(y - 3)$

16. $y^2 + 5y - 24$
  $(y + 8)(y - 3)$

17. $x^2 - 2x - 15$
  $(x - 5)(x + 3)$

18. $x^2 - 4x - 5$
  $(x - 5)(x + 1)$

19. $x^2 - 2x - 8$
  $(x - 4)(x + 2)$

20. $x^2 - 5x - 6$
  $(x - 6)(x + 1)$

21. $x^2 + 4x + 12$
prime

22. $x^2 + 4x + 5$
prime

23. $y^2 - 16y + 48$
$(y - 4)(y - 12)$

24. $y^2 - 10y + 21$
$(y - 7)(y - 3)$

25. $x^2 - 3x + 6$
prime

26. $x^2 + 4x - 10$
prime

27. $w^2 - 30w - 64$
$(w - 32)(w + 2)$

28. $w^2 + 12w - 64$
$(w + 16)(w - 4)$

29. $y^2 - 18y + 65$
$(y - 5)(y - 13)$

30. $y^2 - 22y + 72$
$(y - 18)(y - 4)$

31. $r^2 + 12r + 27$
$(r + 3)(r + 9)$

32. $r^2 - 15r - 16$
$(r - 16)(r + 1)$

33. $y^2 - 7y + 5$
prime

34. $y^2 - 15y + 5$
prime

35. $x^2 + 7xy + 6y^2$
$(x + 6y)(x + y)$

36. $x^2 + 6xy + 8y^2$
$(x + 2y)(x + 4y)$

37. $x^2 - 8xy + 15y^2$
$(x - 3y)(x - 5y)$

38. $x^2 - 9xy + 14y^2$
$(x - 2y)(x - 7y)$

39. $x^2 - 3xy - 18y^2$
$(x - 6y)(x + 3y)$

40. $x^2 - xy - 30y^2$
$(x - 6y)(x + 5y)$

41. $a^2 - 18ab + 45b^2$
$(a - 15b)(a - 3b)$

42. $a^2 - 18ab + 80b^2$
$(a - 10b)(a - 8b)$

In Exercises 43–66, factor completely.

43. $3x^2 + 15x + 18$
$3(x + 2)(x + 3)$

44. $3x^2 + 21x + 36$
$3(x + 3)(x + 4)$

45. $4y^2 - 4y - 8$
$4(y - 2)(y + 1)$

46. $3y^2 + 3y - 18$
$3(y + 3)(y - 2)$

47. $10x^2 - 40x - 600$
$10(x - 10)(x + 6)$

48. $2x^2 + 10x - 48$
$2(x + 8)(x - 3)$

49. $3x^2 - 33x + 54$
$3(x - 2)(x - 9)$

50. $2x^2 - 14x + 24$
$2(x - 3)(x - 4)$

51. $2r^3 + 6r^2 + 4r$
$2r(r + 2)(r + 1)$

52. $2r^3 + 8r^2 + 6r$
$2r(r + 3)(r + 1)$

53. $4x^3 + 12x^2 - 72x$
$4x(x + 6)(x - 3)$

54. $3x^3 - 15x^2 + 18x$
$3x(x - 3)(x - 2)$

55. $2r^3 + 8r^2 - 64r$
$2r(r + 8)(r - 4)$

56. $3r^3 - 9r^2 - 54r$
$3r(r - 6)(r + 3)$

57. $y^4 + 2y^3 - 80y^2$
$y^2(y + 10)(y - 8)$

58. $y^4 - 12y^3 + 35y^2$
$y^2(y - 5)(y - 7)$

59. $x^4 - 3x^3 - 10x^2$
$x^2(x - 5)(x + 2)$

60. $x^4 - 22x^3 + 120x^2$
$x^2(x - 12)(x - 10)$

61. $2w^4 - 26w^3 - 96w^2$
$2w^2(w - 16)(w + 3)$

62. $3w^4 + 54w^3 + 135w^2$
$3w^2(w + 3)(w + 15)$

63. $15xy^2 + 45xy - 60x$
$15x(y + 4)(y - 1)$

64. $20x^2y - 100xy + 120y$
$20y(x - 3)(x - 2)$

65. $x^5 + 3x^4y - 4x^3y^2$
$x^3(x + 4y)(x - y)$

66. $x^3y - 2x^2y^2 - 3xy^3$
$xy(x - 3y)(x + y)$

Practice Plus

In Exercises 67–76, factor completely.

67. $2x^2y^2 - 32x^2yz + 30x^2z^2$ $2x^2(y - 15z)(y - z)$

68. $2x^2y^2 - 30x^2yz + 28x^2z^2$ $2x^2(y - 14z)(y - z)$

69. $(a + b)x^2 + (a + b)x - 20(a + b)$
$(a + b)(x + 5)(x - 4)$

70. $(a + b)x^2 - 13(a + b)x + 36(a + b)$
$(a + b)(x - 4)(x - 9)$

71. $x^2 + 0.5x + 0.06$ $(x + 0.3)(x + 0.2)$

72. $x^2 - 0.5x - 0.06$ $(x - 0.6)(x + 0.1)$

73. $x^2 - \frac{2}{5}x + \frac{1}{25}$ $\left(x - \frac{1}{5}\right)\left(x - \frac{1}{5}\right)$

74. $x^2 + \frac{2}{3}x + \frac{1}{9}$ $\left(x + \frac{1}{3}\right)\left(x + \frac{1}{3}\right)$

75. $-x^2 - 3x + 40$ $-(x + 8)(x - 5)$

76. $-x^2 - 4x + 45$ $-(x - 5)(x + 9)$

Application Exercises

77. You dive directly upward from a board that is 32 feet high. After t seconds, your height above the water is described by the polynomial

$$-16t^2 + 16t + 32.$$

   a. Factor the polynomial completely. Begin by factoring -16 from each term. $-16(t - 2)(t + 1)$

   b. Evaluate both the original polynomial and its factored form for $t = 2$. Do you get the same answer for each evaluation? Describe what this answer means.   0; yes;
   After 2 seconds, you hit the water.

78. You dive directly upward from a board that is 48 feet high. After t seconds, your height above the water is described by the polynomial

$$-16t^2 + 32t + 48.$$

   a. Factor the polynomial completely. Begin by factoring -16 from each term. $-16(t - 3)(t + 1)$

   b. Evaluate both the original polynomial and its factored form for $t = 3$. Do you get the same answer for each evaluation? Describe what this answer means.   0; yes;
   After 3 seconds, you hit the water.

Writing in Mathematics

79. Explain how to factor $x^2 + 8x + 15$. **79–82.** Answers will vary.

80. Give two helpful suggestions for factoring $x^2 - 5x + 6$.

81. In factoring $x^2 + bx + c$, describe how the last terms in each factor are related to b and c.

82. Without actually factoring and without multiplying the given factors, explain why the following factorization is not correct:

$$x^2 + 46x + 513 = (x - 27)(x - 19).$$

Critical Thinking Exercises

83. Which one of the following is true?   c

   a. A factor of $x^2 + x + 20$ is $x + 5$.

   b. A trinomial can never have two identical factors.

   c. A factor of $y^2 + 5y - 24$ is $y - 3$.

   d. $x^2 + 4 = (x + 2)(x + 2)$

In Exercises 84–85, find all positive integers b so that the trinomial can be factored.

84. $x^2 + bx + 15$ 8, 16

85. $x^2 + 4x + b$ 3, 4

86. Factor: $x^{2n} + 20x^n + 99$. $(x^n + 9)(x^n + 11)$

87. Factor $x^3 + 3x^2 + 2x$. If x represents an integer, use the factorization to describe what the trinomial represents. $x(x + 1)(x + 2)$; the product of three consecutive integers

88. A box with no top is to be made from an 8-inch by 6-inch piece of metal by cutting identical squares from each corner and turning up the sides. (See the figure). The volume of the box is modeled by the polynomial $4x^3 - 28x^2 + 48x$. Factor the polynomial completely. Then use the dimensions given on the box and show that its volume is equivalent to the factorization that you obtain. $4x(x - 3)(x - 4)$

Technology Exercises

In Exercises 89–92, use the GRAPH *or* TABLE *feature of a graphing utility to determine if the polynomial on the left side of each equation has been correctly factored. If the graphs of y_1 and y_2 coincide, or if their corresponding table values are equal, this means that the polynomial on the left side has been correctly factored. If not, factor the trinomial correctly and then use your graphing utility to verify the factorization.*

89. $x^2 - 5x + 6 = (x - 2)(x - 3)$ correctly factored

90. $2x^2 + 2x - 12 = 2(x - 3)(x + 2)$ $2(x - 3)(x + 2)$ should be $2(x + 3)(x - 2)$.

91. $x^2 - 2x + 1 = (x + 1)(x - 1)$ $(x + 1)(x - 1)$ should be $(x - 1)(x - 1)$.

92. $2x^2 + 8x + 6 = (x + 3)(x + 1)$ $(x + 3)(x + 1)$ should be $2(x + 3)(x + 1)$.

Review Exercises

93. Multiply: $(2x + 3)(x - 2)$. (Section 5.3, Example 2) $2x^2 - x - 6$

94. Multiply: $(3x + 4)(3x + 1)$. (Section 5.3, Example 2) $9x^2 + 15x + 4$

95. Solve: $4(x - 2) = 3x + 5$. (Section 2.3, Example 2) $\{13\}$

FACTORING TRINOMIALS WHOSE LEADING COEFFICIENT IS NOT ONE

George Tooker, American, born 1920. "Farewell" 1966, egg tempera on gessoed masonite, 61×60.1 cm. (24.034×23.679 in.) P.967.76. Hood Museum of Art. Dartmouth College, Hanover, New Hampshire; gift of Pennington Haile, Class of 1924. ©George Tooker/DC Moore Gallery.

The special significance of the number 1 is reflected in our language. "One," "an," and "a" mean the same thing. The words "unit," "unity," "union," "unique," and "universal" are derived from the Latin word for "one." For the ancient Greeks, 1 was the indivisible unit from which all other numbers arose.

The Greeks' philosophy of 1 applies to our work in this section. Factoring trinomials whose leading coefficient is 1 is the basic technique from which other methods of factoring $ax^2 + bx + c$, where a is not equal to 1, follow.

1 Factor trinomials by trial and error.

Factoring by the Trial-and-Error Method   How do we factor a trinomial such as $3x^2 - 20x + 28$? Notice that the leading coefficient is 3. We must find two binomials whose product is $3x^2 - 20x + 28$. The product of the First terms must be $3x^2$:

$$(3x \quad)(x \quad).$$

From this point on, the factoring strategy is exactly the same as the one we use to factor trinomials whose leading coefficient is 1.

A STRATEGY FOR FACTORING $ax^2 + bx + c$ Assume, for the moment, that there is no greatest common factor.

1. Find two First terms whose product is ax^2:

$$(\square x + \quad)(\square x + \quad) = ax^2 + bx + c.$$

2. Find two Last terms whose product is c:

$$(\square x + \square)(\square x + \square) = ax^2 + bx + c.$$

3. By trial and error, perform steps 1 and 2 until the sum of the Outside product and the Inside product is bx:

$$(\square x + \square)(\square x + \square) = ax^2 + bx + c.$$

$$\text{Sum of O} + \text{I}$$

If no such combinations exist, the polynomial is prime.

STUDY TIP

The *error* part of the factoring strategy plays an important role in the process. If you do not get the correct factorization the first time, this is not a bad thing. This error is often helpful in leading you to the correct factorization.

EXAMPLE 1   Factoring a Trinomial Whose Leading Coefficient Is Not One

Factor: $3x^2 - 20x + 28$.

SOLUTION

Step 1.   **Find two First terms whose product is $3x^2$.**

$$3x^2 - 20x + 28 = (3x \quad)(x \quad)$$

Step 2.   **Find two Last terms whose product is 28.** The number 28 has pairs of factors that are either both positive or both negative. Because the middle term, $-20x$, is negative, both factors must be negative. The negative factorizations of 28 are $-1(-28)$, $-2(-14)$, and $-4(-7)$.

Step 3.   **Try various combinations of these factors.** The correct factorization of $3x^2 - 20x + 28$ is the one in which the sum of the Outside and Inside products is equal to $-20x$. Here is a list of the possible factorizations:

STUDY TIP

With practice, you will find that it is not necessary to list every possible factorization of the trinomial. As you practice factoring, you will be able to narrow down the list of possible factors to just a few. When it comes to factoring, practice makes perfect. (Sorry about the cliché.)

Possible Factorizations of $3x^2 - 20x + 28$	Sum of Outside and Inside Products (Should Equal $-20x$)
$(3x - 1)(x - 28)$	$-84x - x = -85x$
$(3x - 28)(x - 1)$	$-3x - 28x = -31x$
$(3x - 2)(x - 14)$	$-42x - 2x = -44x$
$(3x - 14)(x - 2)$	$-6x - 14x = -20x$
$(3x - 4)(x - 7)$	$-21x - 4x = -25x$
$(3x - 7)(x - 4)$	$-12x - 7x = -19x$

This is the required middle term.

Thus,

$$3x^2 - 20x + 28 = (3x - 14)(x - 2) \quad \text{or} \quad (x - 2)(3x - 14).$$

Show that this factorization is correct by multiplying the factors with the FOIL method. You should obtain the original trinomial. ∎

 CHECK POINT 1 Factor: $5x^2 - 14x + 8$. $(5x - 4)(x - 2)$

EXAMPLE 2 Factoring a Trinomial Whose Leading Coefficient Is Not One

Factor: $8x^2 - 10x - 3$.

SOLUTION

Step 1. **Find two First terms whose product is $8x^2$.**

$$8x^2 - 10x - 3 \stackrel{?}{=} (8x \quad)(x \quad)$$
$$8x^2 - 10x - 3 \stackrel{?}{=} (4x \quad)(2x \quad)$$

Step 2. **Find two Last terms whose product is -3.** The possible factorizations are $1(-3)$ and $-1(3)$.

Step 3. **Try various combinations of these factors.** The correct factorization of $8x^2 - 10x - 3$ is the one in which the sum of the Outside and Inside products is equal to $-10x$. Here is a list of the possible factorizations:

Possible Factorizations of $8x^2 - 10x - 3$	Sum of Outside and Inside Products (Should Equal $-10x$)
$(8x + 1)(x - 3)$	$-24x + x = -23x$
$(8x - 3)(x + 1)$	$8x - 3x = 5x$
$(8x - 1)(x + 3)$	$24x - x = 23x$
$(8x + 3)(x - 1)$	$-8x + 3x = -5x$
$(4x + 1)(2x - 3)$	$-12x + 2x = -10x$
$(4x - 3)(2x + 1)$	$4x - 6x = -2x$
$(4x - 1)(2x + 3)$	$12x - 2x = 10x$
$(4x + 3)(2x - 1)$	$-4x + 6x = 2x$

This is the required middle term.

Thus,

$$8x^2 - 10x - 3 = (4x + 1)(2x - 3) \quad \text{or} \quad (2x - 3)(4x + 1).$$

Use FOIL multiplication to check either of these factorizations. ∎

 CHECK POINT 2 Factor: $6x^2 + 19x - 7$. $(3x - 1)(2x + 7)$

Here are some suggestions for reducing the list of possible factorizations for $ax^2 + bx + c$:

1. If b is relatively small, avoid the larger factors of a.

2. If c is positive, the signs in both binomial factors must match the sign of b.

3. If the trinomial has no common factor, no binomial factor can have a common factor.

4. Reversing the signs in the binomial factors reverses the sign of bx, the middle term.

EXAMPLE 3 Factoring a Trinomial in Two Variables

Factor: $2x^2 - 7xy + 3y^2$.

SOLUTION

Step 1. **Find two First terms whose product is $2x^2$.**

$$2x^2 - 7xy + 3y^2 = (2x \quad)(x \quad)$$

Step 2. **Find two Last terms whose product is $3y^2$.** The possible factorizations are $(y)(3y)$ and $(-y)(-3y)$.

Step 3. **Try various combinations of these factors.** The correct factorization of $2x^2 - 7xy + 3y^2$ is the one in which the sum of the Outside and Inside products is equal to $-7xy$. Here is a list of possible factorizations:

Possible Factorizations of $2x^2 - 7xy + 3y^2$	Sum of Outside and Inside Products (Should Equal $-7xy$)
$(2x + 3y)(x + y)$	$2xy + 3xy = 5xy$
$(2x + y)(x + 3y)$	$6xy + xy = 7xy$
$(2x - 3y)(x - y)$	$-2xy - 3xy = -5xy$
$(2x - y)(x - 3y)$	$-6xy - xy = -7xy$

This is the required middle term.

Thus,

$$2x^2 - 7xy + 3y^2 = (2x - y)(x - 3y) \quad \text{or} \quad (x - 3y)(2x - y).$$

Use FOIL multiplication to check either of these factorizations. ∎

 CHECK POINT 3 Factor: $3x^2 - 13xy + 4y^2$. $(3x - y)(x - 4y)$

---

2 Factor trinomials by grouping.

Factoring by the Grouping Method A second method for factoring $ax^2 + bx + c$, $a \neq 0$, is called the **grouping method**. The method involves both trial and error, as well as grouping. The trial and error in factoring $ax^2 + bx + c$ depends on finding two numbers, p and q, for which $p + q = b$. Then we factor $ax^2 + px + qx + c$ by using grouping.

Let's see how this works by looking at our factorization in Example 2:

$$8x^2 - 10x - 3 = (2x - 3)(4x + 1).$$

If we multiply by using FOIL on the right, we obtain:

$$(2x - 3)(4x + 1) = 8x^2 + 2x - 12x - 3.$$

In this case, the desired numbers, p and q, are $p = 2$ and $q = -12$. Compare these numbers with ac and b in the given polynomial:

$$8x^2 - 10x - 3.$$

$ac = 8(-3) = -24$

$a = 8$ $b = -10$ $c = -3$

Can you see that p and q, 2 and -12, are factors of ac, or -24? Furthermore, p and q have a sum of b, namely -10. By expressing the middle term, $-10x$, in terms of p and q, we can factor by grouping as follows:

$$8x^2 - 10x - 3$$
$$= 8x^2 + (2x - 12x) - 3 \qquad \text{Rewrite } -10x \text{ as } 2x - 12x.$$
$$= (8x^2 + 2x) + (-12x - 3) \qquad \text{Group terms.}$$
$$= 2x(4x + 1) - 3(4x + 1) \qquad \text{Factor from each group.}$$
$$= (4x + 1)(2x - 3) \qquad \text{Factor out the common binomial factor.}$$

As we obtained in Example 2,

$$8x^2 - 10x - 3 = (4x + 1)(2x - 3).$$

Generalizing from this example, here's how to factor a trinomial by grouping:

FACTORING $ax^2 + bx + c$ USING GROUPING ($a \neq 1$)

1. Multiply the leading coefficient, a, and the constant, c.
2. Find the factors of ac whose sum is b.
3. Rewrite the middle term, bx, as a sum or difference using the factors from step 2.
4. Factor by grouping.

EXAMPLE 4 Factoring by Grouping

Factor by grouping: $2x^2 - x - 6$.

SOLUTION The trinomial is of the form $ax^2 + bx + c$.

$$2x^2 - x - 6$$
$$a = 2 \quad b = -1 \quad c = -6$$

Step 1. **Multiply the leading coefficient, a, and the constant, c.** Using $a = 2$ and $c = -6$,

$$ac = 2(-6) = -12.$$

Step 2. **Find the factors of ac whose sum is b.** We want the factors of -12 whose sum is b, or -1. The factors of -12 whose sum is -1 are -4 and 3.

Step 3. **Rewrite the middle term, $-x$, as a sum or difference using the factors from step 2, -4 and 3.**

$$2x^2 - x - 6 = 2x^2 - 4x + 3x - 6$$

Step 4. **Factor by grouping.**

$$= (2x^2 - 4x) + (3x - 6) \qquad \text{Group terms.}$$
$$= 2x(x - 2) + 3(x - 2) \qquad \text{Factor from each group.}$$
$$= (x - 2)(2x + 3) \qquad \text{Factor out the common binomial factor.}$$

Thus,

$$2x^2 - x - 6 = (x - 2)(2x + 3) \quad \text{or} \quad (2x + 3)(x - 2). \qquad \blacksquare$$

In step 2, we discovered that the desired numbers were -4 and 3, and we wrote $-x$ as $-4x + 3x$. What happens if we write $-x$ as $3x - 4x$? Use factoring by grouping on

$$2x^2 - x - 6$$
$$= 2x^2 + 3x - 4x - 6.$$

Describe what happens.

✓ **CHECK POINT 4** Factor by grouping: $3x^2 - x - 10$. $\quad (3x + 5)(x - 2)$

EXAMPLE 5 Factoring by Grouping

Factor by grouping: $8x^2 - 22x + 5$.

SOLUTION   The trinomial is of the form $ax^2 + bx + c$.

$$8x^2 - 22x + 5$$

$a = 8$ $b = -22$ $c = 5$

Step 1. **Multiply the leading coefficient, *a*, and the constant, *c*.** Using $a = 8$ and $c = 5$, $ac = 8 \cdot 5 = 40$.

Step 2. **Find the factors of *ac* whose sum is *b*.** We want the factors of 40 whose sum is b, or -22. The factors of 40 whose sum is -22 are -2 and -20.

Step 3. **Rewrite the middle term, $-22x$, as a sum or difference using the factors from step 2, -2 and -20.**

$$8x^2 - 22x + 5 = 8x^2 - 2x - 20x + 5$$

Step 4. **Factor by grouping.**

$$= (8x^2 - 2x) + (-20x + 5) \quad \text{Group terms.}$$
$$= 2x(4x - 1) - 5(4x - 1) \quad \text{Factor from each group.}$$
$$= (4x - 1)(2x - 5) \quad \text{Factor out the common binomial factor.}$$

Thus,

$$8x^2 - 22x + 5 = (4x - 1)(2x - 5) \quad \text{or} \quad (2x - 5)(4x - 1).$$ ■

✔ **CHECK POINT 5** Factor by grouping: $8x^2 - 10x + 3$. $(2x - 1)(4x - 3)$

Factoring Completely   Always begin the process of factoring a polynomial by looking for a greatest common factor. If there is one, **factor out the GCF first**. After doing this, you should attempt to factor the remaining trinomial by one of the methods presented in this section.

EXAMPLE 6 Factoring Completely

Factor completely: $15y^4 + 26y^3 + 7y^2$.

SOLUTION   We will first factor out a common monomial factor from the polynomial and then factor the resulting trinomial by the methods of this section. The GCF of the three terms is y^2.

$$15y^4 + 26y^3 + 7y^2 = y^2(15y^2 + 26y + 7) \quad \text{Factor out the GCF.}$$
$$= y^2(5y + 7)(3y + 1) \quad \text{Factor } 15y^2 + 26y + 7 \text{ using trial and error or grouping.}$$

Thus,

$$15y^4 + 26y^3 + 7y^2 = y^2(5y + 7)(3y + 1) \quad \text{or} \quad y^2(3y + 1)(5y + 7).$$

Be sure to include the GCF, y^2, in the factorization. ■

✔ **CHECK POINT 6** Factor completely: $5y^4 + 13y^3 + 6y^2$. $y^2(5y + 3)(y + 2)$

6.3 EXERCISE SET

Student Solutions Manual   CD/Video   PH Math/Tutor Center   MathXL Tutorials on CD   MathXL®   MyMathLab   Interactmath.com

Practice Exercises

In Exercises 1–58, use the method of your choice to factor each trinomial, or state that the trinomial is prime. Check each factorization using FOIL multiplication.

1. $2x^2 + 5x + 3$
$(2x + 3)(x + 1)$

2. $3x^2 + 5x + 2$
$(3x + 2)(x + 1)$

3. $3x^2 + 13x + 4$
$(3x + 1)(x + 4)$

4. $2x^2 + 7x + 3$
$(2x + 1)(x + 3)$

5. $2x^2 + 11x + 12$
$(2x + 3)(x + 4)$

6. $2x^2 + 19x + 35$
$(2x + 5)(x + 7)$

7. $5y^2 - 16y + 3$
$(5y - 1)(y - 3)$

8. $5y^2 - 17y + 6$
$(5y - 2)(y - 3)$

9. $3y^2 + y - 4$
$(3y + 4)(y - 1)$

10. $3y^2 - y - 4$
$(3y - 4)(y + 1)$

11. $3x^2 + 13x - 10$
$(3x - 2)(x + 5)$

12. $3x^2 + 14x - 5$
$(3x - 1)(x + 5)$

13. $3x^2 - 22x + 7$
$(3x - 1)(x - 7)$

14. $3x^2 - 10x + 7$
$(3x - 7)(x - 1)$

15. $5y^2 - 16y + 3$
$(5y - 1)(y - 3)$

16. $5y^2 - 8y + 3$
$(5y - 3)(y - 1)$

17. $3x^2 - 17x + 10$
$(3x - 2)(x - 5)$

18. $3x^2 - 25x - 28$
$(3x - 28)(x + 1)$

19. $6w^2 - 11w + 4$
$(3w - 4)(2w - 1)$

20. $6w^2 - 17w + 12$
$(3w - 4)(2w - 3)$

21. $8x^2 + 33x + 4$
$(8x + 1)(x + 4)$

22. $7x^2 + 43x + 6$
$(7x + 1)(x + 6)$

23. $5x^2 + 33x - 14$
$(5x - 2)(x + 7)$

24. $3x^2 + 22x - 16$
$(3x - 2)(x + 8)$

25. $14y^2 + 15y - 9$
$(7y - 3)(2y + 3)$

26. $6y^2 + 7y - 24$
$(3y + 8)(2y - 3)$

27. $6x^2 - 7x + 3$
prime

28. $9x^2 + 3x + 2$
prime

29. $25z^2 - 30z + 9$
$(5z - 3)(5z - 3)$

30. $9z^2 + 12z + 4$
$(3z + 2)(3z + 2)$

31. $15y^2 - y - 2$
$(3y + 1)(5y - 2)$

32. $15y^2 + 13y - 2$
$(15y - 2)(y + 1)$

33. $5x^2 + 2x + 9$
prime

34. $3x^2 - 5x + 1$
prime

35. $10y^2 + 43y - 9$
$(5y - 1)(2y + 9)$

36. $16y^2 - 46y + 15$
$(8y - 3)(2y - 5)$

37. $8x^2 - 2x - 1$
$(4x + 1)(2x - 1)$

38. $8x^2 - 22x + 5$
$(4x - 1)(2x - 5)$

39. $9y^2 - 9y + 2$
$(3y - 1)(3y - 2)$

40. $9y^2 + 5y - 4$
$(9y - 4)(y + 1)$

41. $20x^2 + 27x - 8$
$(5x + 8)(4x - 1)$

42. $15x^2 - 19x + 6$
$(5x - 3)(3x - 2)$

43. $2x^2 + 3xy + y^2$
$(2x + y)(x + y)$

44. $3x^2 + 4xy + y^2$
$(3x + y)(x + y)$

45. $3x^2 + 5xy + 2y^2$
$(3x + 2y)(x + y)$

46. $3x^2 + 11xy + 6y^2$
$(3x + 2y)(x + 3y)$

47. $2x^2 - 9xy + 9y^2$
$(2x - 3y)(x - 3y)$

48. $3x^2 + 5xy - 2y^2$
$(3x - y)(x + 2y)$

49. $6x^2 - 5xy - 6y^2$
$(2x - 3y)(3x + 2y)$

50. $6x^2 - 7xy - 5y^2$
$(3x - 5y)(2x + y)$

51. $15x^2 + 11xy - 14y^2$
$(3x - 2y)(5x + 7y)$

52. $15x^2 - 31xy + 10y^2$
$(5x - 2y)(3x - 5y)$

53. $2a^2 + 7ab + 5b^2$
$(2a + 5b)(a + b)$

54. $2a^2 + 5ab + 2b^2$
$(2a + b)(a + 2b)$

55. $15a^2 - ab - 6b^2$
$(3a - 2b)(5a + 3b)$

56. $3a^2 - ab - 14b^2$
$(3a - 7b)(a + 2b)$

57. $12x^2 - 25xy + 12y^2$
$(3x - 4y)(4x - 3y)$

58. $12x^2 + 7xy - 12y^2$
$(4x - 3y)(3x + 4y)$

In Exercises 59–86, factor completely.

59. $4x^2 + 26x + 30$
$2(2x + 3)(x + 5)$

60. $4x^2 - 18x - 10$
$2(2x + 1)(x - 5)$

61. $9x^2 - 6x - 24$
$3(3x + 4)(x - 2)$

62. $12x^2 - 33x + 21$
$3(4x - 7)(x - 1)$

63. $4y^2 + 2y - 30$
$2(2y - 5)(y + 3)$

64. $36y^2 + 6y - 12$
$6(2y - 1)(3y + 2)$

65. $9y^2 + 33y - 60$
$3(3y - 4)(y + 5)$

66. $16y^2 - 16y - 12$
$4(2y - 3)(2y + 1)$

67. $3x^3 + 4x^2 + x$
$x(3x + 1)(x + 1)$

68. $3x^3 + 14x^2 + 8x$
$x(3x + 2)(x + 4)$

69. $2x^3 - 3x^2 - 5x$
$x(2x - 5)(x + 1)$

70. $6x^3 + 4x^2 - 10x$
$2x(3x + 5)(x - 1)$

71. $9y^3 - 39y^2 + 12y$
$3y(3y - 1)(y - 4)$

72. $10y^3 + 12y^2 + 2y$
$2y(5y + 1)(y + 1)$

73. $60z^3 + 40z^2 + 5z$
$5z(6z + 1)(2z + 1)$

74. $80z^3 + 80z^2 - 60z$
$20z(2z - 1)(2z + 3)$

75. $15x^4 - 39x^3 + 18x^2$
$3x^2(5x - 3)(x - 2)$

76. $24x^4 + 10x^3 - 4x^2$
$2x^2(4x - 1)(3x + 2)$

77. $10x^5 - 17x^4 + 3x^3$
$x^3(2x - 3)(5x - 1)$

78. $15x^5 - 2x^4 - x^3$
$x^3(5x + 1)(3x - 1)$

79. $6x^2 - 3xy - 18y^2$
$3(2x + 3y)(x - 2y)$

80. $4x^2 + 14xy + 10y^2$
$2(2x + 5y)(x + y)$

81. $12x^2 + 10xy - 8y^2$
$2(2x - y)(3x + 4y)$

82. $24x^2 + 3xy - 27y^2$
$3(8x + 9y)(x - y)$

83. $8x^2y + 34xy - 84y$
$2y(4x - 7)(x + 6)$

84. $6x^2y - 2xy - 60y$
$2y(3x - 10)(x + 3)$

85. $12a^2b - 46ab^2 + 14b^3$
$2b(2a - 7b)(3a - b)$

86. $12a^2b - 34ab^2 + 14b^3$
$2b(3a - 7b)(2a - b)$

Practice Plus

In Exercises 87–90, factor completely.

87. $30(y + 1)x^2 + 10(y + 1)x - 20(y + 1)$
$10(y + 1)(x + 1)(3x - 2)$

88. $6(y + 1)x^2 + 33(y + 1)x + 15(y + 1)$
$3(y + 1)(2x + 1)(x + 5)$

89. $-32x^2y^4 + 20xy^4 + 12y^4$ $-4y^4(8x + 3)(x - 1)$

90. $-10x^2y^4 + 14xy^4 + 12y^4$ $-2y^4(5x + 3)(x - 2)$

91. a. Factor $2x^2 - 5x - 3$. $(2x + 1)(x - 3)$

   b. Use the factorization in part (a) to factor
$$2(y + 1)^2 - 5(y + 1) - 3.$$
   Then simplify each factor. $(2y + 3)(y - 2)$

92. a. Factor $3x^2 + 5x - 2$. $(3x - 1)(x + 2)$

   b. Use the factorization in part (a) to factor
$$3(y + 1)^2 + 5(y + 1) - 2.$$
   Then simplify each factor. $(3y + 2)(y + 3)$

93. Divide $3x^3 - 11x^2 + 12x - 4$ by $x - 2$. Use the quotient to factor $3x^3 - 11x^2 + 12x - 4$ completely.
$(x - 2)(3x - 2)(x - 1)$

94. Divide $2x^3 + x^2 - 13x + 6$ by $x - 2$. Use the quotient to factor $2x^3 + x^2 - 13x + 6$ completely.
$(x - 2)(2x - 1)(x + 3)$

Application Exercises

It is possible to construct geometric models for factorizations so that you can see the factoring. This idea is developed in Exercises 95–96.

95. Consider the following figure.

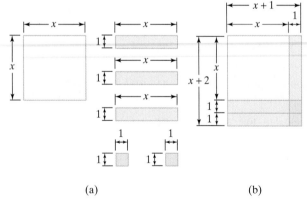

(a)   (b)

   a. Write a trinomial that expresses the sum of the areas of the six rectangular pieces shown in figure (a).
   $x^2 + 3x + 2$

   b. Express the area of the large rectangle in figure (b) as the product of two binomials. $(x + 2)(x + 1)$

   c. Are the pieces in figures (a) and (b) the same? Set the expressions that you wrote in parts (a) and (b) equal to each other. What factorization is illustrated?
   $x^2 + 3x + 2 = (x + 2)(x + 1)$

96. Copy the figure and cut out the six pieces. Use the pieces to create a geometric model for the factorization
$$2x^2 + 3x + 1 = (2x + 1)(x + 1)$$
by forming a large rectangle using all the pieces.   See graphing answer section.

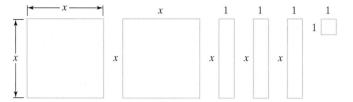

Writing in Mathematics

97. Explain how to factor $2x^2 - x - 1$.   **97–100.** Answers will vary.

98. Why is it a good idea to factor out the GCF first and then use other methods of factoring? Use $3x^2 - 18x + 15$ as an example. Discuss what happens if one first uses trial and error to factor as two binomials rather than first factoring out the GCF.

99. In factoring $3x^2 - 10x - 8$, a student lists $(3x - 2)(x + 4)$ as a possible factorization. Use FOIL multiplication to determine if this factorization is correct. If it is not correct, describe how the correct factorization can quickly be obtained using these factors.

100. Explain why $2x - 10$ cannot be one of the factors in the correct factorization of $6x^2 - 19x + 10$.

Critical Thinking Exercises

101. Which one of the following is true?   **a**

   a. Once a GCF is factored from $18y^2 - 6y + 6$, the remaining trinomial factor is prime.

   b. One factor of $12x^2 - 13x + 3$ is $4x + 3$.

   c. One factor of $4y^2 - 11y - 3$ is $y + 3$.

   d. The trinomial $3x^2 + 2x + 1$ has relatively small coefficients and therefore can be factored.

In Exercises 102–103, find all integers b so that the trinomial can be factored.

102. $3x^2 + bx + 2$ $5, 7, -5, -7$

103. $2x^2 + bx + 3$ $5, 7, -5, -7$

104. Factor: $3x^{10} - 4x^5 - 15$. $(3x^5 + 5)(x^5 - 3)$

105. Factor: $2x^{2n} - 7x^n - 4$. $(2x^n + 1)(x^n - 4)$

Review Exercises

In Exercises 106–108, perform the indicated operations.

106. $(9x + 10)(9x - 10)$ (Section 5.3, Example 4) $81x^2 - 100$

107. $(4x + 5y)^2$ (Section 5.3, Example 5) $16x^2 + 40xy + 25y^2$

108. $(x + 2)(x^2 - 2x + 4)$ (Section 5.2, Example 7) $x^3 + 8$

✔ MID-CHAPTER CHECK POINT

CHAPTER

6

What You Know: We learned to factor out a polynomial's greatest common factor and to use grouping to factor polynomials with four terms. We factored polynomials with three terms, beginning with trinomials with leading coefficient 1 and moving on to $ax^2 + bx + c$, with $a \neq 1$. We saw that the factoring process should begin by looking for a GCF and, if there is one, factoring it out first.

In Exercises 1–12, factor completely, or state that the polynomial is prime.

1. $x^5 + x^4$ $x^4(x + 1)$

2. $x^2 + 7x - 18$ $(x + 9)(x - 2)$

3. $x^2y^3 - x^2y^2 + x^2y$ $x^2y(y^2 - y + 1)$

4. $x^2 - 2x + 4$ prime

5. $7x^2 - 22x + 3$ $(7x - 1)(x - 3)$

6. $x^3 + 5x^2 + 3x + 15$ $(x^2 + 3)(x + 5)$

7. $2x^3 - 11x^2 + 5x$ $x(2x - 1)(x - 5)$

8. $xy - 7x - 4y + 28$ $(x - 4)(y - 7)$

9. $x^2 - 17xy + 30y^2$ $(x - 15y)(x - 2y)$

10. $25x^2 - 25x - 14$ $(5x + 2)(5x - 7)$

11. $16x^2 - 70x + 24$ $2(8x - 3)(x - 4)$

12. $3x^2 + 10xy + 7y^2$ $(3x + 7y)(x + y)$

SECTION

6.4

FACTORING SPECIAL FORMS

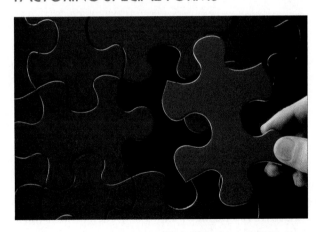

Objectives

1 Factor the difference of two squares.

2 Factor perfect square trinomials.

3 Factor the sum and difference of two cubes.

Do you enjoy solving puzzles? The process is a natural way to develop problem-solving skills that are important to every area of our lives. Engaging in problem solving for sheer pleasure releases chemicals in the brain that enhance our feeling of well-being. Perhaps this is why puzzles date back 12,000 years.

In this section, we develop factoring techniques by reversing the formulas for special products discussed in Chapter 5. These factorizations can be visualized by fitting pieces of a puzzle together to form rectangles.

1 Factor the difference of two squares.

Factoring the Difference of Two Squares   A method for factoring the difference of two squares is obtained by reversing the special product for the sum and difference of two terms.

Visualizing the Factoring for the Difference of Two Squares

Yellow Shaded Area:

$$A^2 - B^2$$

Yellow Shaded Area:

$$(A + B)(A - B)$$

The yellow shaded areas are the same. Conclusion:

$$A^2 - B^2 = (A + B)(A - B).$$

THE DIFFERENCE OF TWO SQUARES If A and B are real numbers, variables, or algebraic expressions, then

$$A^2 - B^2 = (A + B)(A - B).$$

In words: The difference of the squares of two terms factors as the product of a sum and a difference of those terms.

EXAMPLE 1 Factoring the Difference of Two Squares

Factor:

   a. $x^2 - 4$ **b.** $81x^2 - 49$.

SOLUTION We must express each term as the square of some monomial. Then we use the formula for factoring $A^2 - B^2$.

   a. $x^2 - 4 = x^2 - 2^2 = (x + 2)(x - 2)$

$$A^2 - B^2 = (A + B)(A - B)$$

   b. $81x^2 - 49 = (9x)^2 - 7^2 = (9x + 7)(9x - 7)$ ■

✔ **CHECK POINT 1** Factor:

   a. $x^2 - 81$ $(x + 9)(x - 9)$ **b.** $36x^2 - 25$. $(6x + 5)(6x - 5)$

Can $x^2 - 5$ be factored using integers and the formula for factoring $A^2 - B^2$? No. The number 5 is not the square of an integer. Thus, $x^2 - 5$ is prime over the set of integers.

EXAMPLE 2 Factoring the Difference of Two Squares

Factor:

   a. $9 - 16x^{10}$ **b.** $25x^2 - 4y^2$.

SOLUTION Begin by expressing each term as the square of some monomial. Then use the formula for factoring $A^2 - B^2$.

   a. $9 - 16x^{10} = 3^2 - (4x^5)^2 = (3 + 4x^5)(3 - 4x^5)$

$$A^2 - B^2 = (A + B)(A - B)$$

   b. $25x^2 - 4y^2 = (5x)^2 - (2y)^2 = (5x + 2y)(5x - 2y)$ ■

✔ **CHECK POINT 2** Factor:

   a. $25 - 4x^{10}$ $(5 + 2x^5)(5 - 2x^5)$ **b.** $100x^2 - 9y^2$. $(10x + 3y)(10x - 3y)$

When factoring, always check first for common factors. If there are common factors, factor out the GCF and then factor the resulting polynomial.

EXAMPLE 3 Factoring Out the GCF and Then Factoring the Difference of Two Squares

Factor:

  a. $12x^3 - 3x$ **b.** $80 - 125x^2$.

SOLUTION

  a. $12x^3 - 3x = 3x(4x^2 - 1) = 3x[(2x)^2 - 1^2] = 3x(2x + 1)(2x - 1)$

Factor out the GCF. $A^2 - B^2 = (A + B)(A - B)$

  b. $80 - 125x^2 = 5(16 - 25x^2) = 5[4^2 - (5x)^2] = 5(4 + 5x)(4 - 5x)$

 CHECK POINT 3 Factor:
  a. $18x^3 - 2x$ $2x(3x + 1)(3x - 1)$ **b.** $72 - 18x^2$. $18(2 + x)(2 - x)$

We have seen that a polynomial is factored completely when it is written as the product of prime polynomials. To be sure that you have factored completely, check to see whether any of the factors can be factored.

EXAMPLE 4 A Repeated Factorization

Factor completely: $x^4 - 81$.

SOLUTION

$x^4 - 81 = (x^2)^2 - 9^2$ Express as the difference of two squares.

$\qquad = (x^2 + 9)(x^2 - 9)$ The factors are the sum and the difference of the expressions being squared.

$\qquad = (x^2 + 9)(x^2 - 3^2)$ The factor $x^2 - 9$ is the difference of two squares and can be factored.

$\qquad = (x^2 + 9)(x + 3)(x - 3)$ The factors of $x^2 - 9$ are the sum and the difference of the expressions being squared. ∎

> **STUDY TIP**
>
> Factoring $x^4 - 81$ as
>
> $(x^2 + 9)(x^2 - 9)$
>
> is not a complete factorization. The second factor, $x^2 - 9$, is itself a difference of two squares and can be factored.

Are you tempted to further factor $x^2 + 9$, the sum of two squares, in Example 4? Resist the temptation! **The sum of two squares, $A^2 + B^2$, with no common factor other than 1 is a prime polynomial over the integers.**

 CHECK POINT 4 Factor completely: $81x^4 - 16$. $(9x^2 + 4)(3x + 2)(3x - 2)$

Factoring Perfect Square Trinomials   Our next factoring technique is obtained by reversing the special products for squaring binomials. The trinomials that are factored using this technique are called **perfect square trinomials**.

2 Factor perfect square trinomials.

Visualizing the Factoring for a Perfect Square Trinomial

Area:

$$(A + B)^2$$

Sum of Areas:

$$A^2 + 2AB + B^2$$

Conclusion:

$$A^2 + 2AB + B^2 = (A + B)^2$$

FACTORING PERFECT SQUARE TRINOMIALS Let A and B be real numbers, variables, or algebraic expressions.

1. $A^2 + 2AB + B^2 = (A + B)^2$

Same sign

2. $A^2 - 2AB + B^2 = (A - B)^2$

Same sign

The two items in the box show that perfect square trinomials come in two forms: one in which the middle term is positive and one in which the middle term is negative. Here's how to recognize a perfect square trinomial:

1. The first and last terms are squares of monomials or integers.

2. The middle term is twice the product of the expressions being squared in the first and last terms.

EXAMPLE 5 Factoring Perfect Square Trinomials

Factor:

a. $x^2 + 6x + 9$ **b.** $x^2 - 16x + 64$ **c.** $25x^2 - 60x + 36$.

SOLUTION

a. $x^2 + 6x + 9 = x^2 + 2 \cdot x \cdot 3 + 3^2 = (x + 3)^2$ The middle term has a positive sign.

$$A^2 + 2AB + B^2 = (A + B)^2$$

b. $x^2 - 16x + 64 = x^2 - 2 \cdot x \cdot 8 + 8^2 = (x - 8)^2$ The middle term has a negative sign.

$$A^2 - 2AB + B^2 = (A - B)^2$$

c. We suspect that $25x^2 - 60x + 36$ is a perfect square trinomial because $25x^2 = (5x)^2$ and $36 = 6^2$. The middle term can be expressed as twice the product of $5x$ and 6.

$$25x^2 - 60x + 36 = (5x)^2 - 2 \cdot 5x \cdot 6 + 6^2 = (5x - 6)^2$$

$$A^2 - 2AB + B^2 = (A - B)^2$$

■

✔ **CHECK POINT 5** Factor:

a. $x^2 + 14x + 49$ $(x + 7)^2$ **b.** $x^2 - 6x + 9$ $(x - 3)^2$

c. $16x^2 - 56x + 49$. $(4x - 7)^2$

EXAMPLE 6 Factoring a Perfect Square Trinomial in Two Variables

Factor: $16x^2 + 40xy + 25y^2$.

SOLUTION Observe that $16x^2 = (4x)^2$, $25y^2 = (5y)^2$, and $40xy$ is twice the product of $4x$ and $5y$. Thus, we have a perfect square trinomial.

$$16x^2 + 40xy + 25y^2 = (4x)^2 + 2 \cdot 4x \cdot 5y + (5y)^2 = (4x + 5y)^2$$

$$A^2 \; + \; 2AB \; + \; B^2 \; = \; (A \; + \; B)^2$$

✔ **CHECK POINT 6** Factor: $4x^2 + 12xy + 9y^2$. $\quad (2x + 3y)^2$

Factoring the Sum and Difference of Two Cubes We can use the following formulas to factor the sum or the difference of two cubes:

---

3 Factor the sum and difference of two cubes.

FACTORING THE SUM AND DIFFERENCE OF TWO CUBES

1. Factoring the Sum of Two Cubes

$$A^3 + B^3 = (A + B)(A^2 - AB + B^2)$$

Same sign   Opposite signs

2. Factoring the Difference of Two Cubes

$$A^3 - B^3 = (A - B)(A^2 + AB + B^2)$$

Same sign   Opposite signs

You can use a graphing utility to verify the factorization in Example 7. Enter

$$y_1 = x^3 + 8$$

and

$$y_2 = (x + 2)(x^2 - 2x + 4).$$

Graphic Check

The graphs are identical.

$y_2 = (x + 2)(x^2 - 2x + 4)$

$y_1 = x^3 + 8$

$[-3, 3, 1]$ by $[-30, 30, 3]$

Numeric Check

No matter how far up or down we scroll, $y_1 = y_2$.

X	Y₁	Y₂
-3	-19	-19
-2	0	0
-1	7	7
0	8	8
1	9	9
2	16	16
3	35	35

X = -3

EXAMPLE 7   Factoring the Sum of Two Cubes

Factor: $x^3 + 8$.

SOLUTION We must express each term as the cube of some monomial. Then we use the formula for factoring $A^3 + B^3$.

$$x^3 + 8 = x^3 + 2^3 = (x + 2)(x^2 - x \cdot 2 + 2^2) = (x + 2)(x^2 - 2x + 4)$$

$$A^3 \; + \; B^3 \; = \; (A \; + \; B) \; (A^2 \; - \; AB \; + \; B^2)$$

✔ **CHECK POINT 7** Factor: $x^3 + 27$. $\quad (x + 3)(x^2 - 3x + 9)$

EXAMPLE 8   Factoring the Difference of Two Cubes

Factor: $27 - y^3$.

SOLUTION Express each term as the cube of some monomial. Then use the formula for factoring $A^3 - B^3$.

$$27 - y^3 = 3^3 - y^3 = (3 - y)(3^2 + 3y + y^2) = (3 - y)(9 + 3y + y^2)$$

$$A^3 - B^3 \; = \; (A \; - \; B) \; (A^2 \; + \; AB \; + \; B^2)$$

✔ **CHECK POINT 8** Factor: $1 - y^3$. $\quad (1 - y)(1 + y + y^2)$

EXAMPLE 9 Factoring the Sum of Two Cubes

Factor: $64x^3 + 125$.

SOLUTION Express each term as the cube of some monomial. Then use the formula for factoring $A^3 + B^3$.

$$64x^3 + 125 = (4x)^3 + 5^3 = (4x + 5)[(4x)^2 - (4x)(5) + 5^2]$$

$$A^3 + B^3 = (A + B)(A^2 - AB + B^2)$$

$$= (4x + 5)(16x^2 - 20x + 25)$$

 CHECK POINT 9 Factor: $125x^3 + 8$. $(5x + 2)(25x^2 - 10x + 4)$

6.4 EXERCISE SET

Student Solutions Manual   CD/Video   PH Math/Tutor Center   MathXL Tutorials on CD   MathXL®   MyMathLab   Interactmath.com

Practice Exercises

In Exercises 1–26, factor each difference of two squares.

1. $x^2 - 25$
$(x + 5)(x - 5)$

2. $x^2 - 16$
$(x + 4)(x - 4)$

3. $y^2 - 1$
$(y + 1)(y - 1)$

4. $y^2 - 9$
$(y + 3)(y - 3)$

5. $4x^2 - 9$
$(2x + 3)(2x - 3)$

6. $9x^2 - 25$
$(3x + 5)(3x - 5)$

7. $25 - x^2$
$(5 + x)(5 - x)$

8. $16 - x^2$
$(4 + x)(4 - x)$

9. $1 - 49x^2$
$(1 + 7x)(1 - 7x)$

10. $1 - 64x^2$
$(1 + 8x)(1 - 8x)$

11. $9 - 25y^2$
$(3 + 5y)(3 - 5y)$

12. $16 - 49y^2$
$(4 + 7y)(4 - 7y)$

13. $x^4 - 9$
$(x^2 + 3)(x^2 - 3)$

14. $x^4 - 25$
$(x^2 + 5)(x^2 - 5)$

15. $49y^4 - 16$
$(7y^2 + 4)(7y^2 - 4)$

16. $49y^4 - 25$
$(7y^2 + 5)(7y^2 - 5)$

17. $x^{10} - 9$
$(x^5 + 3)(x^5 - 3)$

18. $x^{10} - 1$
$(x^5 + 1)(x^5 - 1)$

19. $25x^2 - 16y^2$
$(5x + 4y)(5x - 4y)$

20. $9x^2 - 25y^2$
$(3x + 5y)(3x - 5y)$

21. $x^4 - y^{10}$
$(x^2 + y^5)(x^2 - y^5)$

22. $x^{14} - y^4$
$(x^7 + y^2)(x^7 - y^2)$

23. $x^4 - 16$
$(x^2 + 4)(x + 2)(x - 2)$

24. $x^4 - 1$
$(x^2 + 1)(x + 1)(x - 1)$

25. $16x^4 - 81$
$(4x^2 + 9)(2x + 3)(2x - 3)$

26. $81x^4 - 1$
$(9x^2 + 1)(3x + 1)(3x - 1)$

In Exercises 27–40, factor completely, or state that the polynomial is prime.

27. $2x^2 - 18$
$2(x + 3)(x - 3)$

28. $5x^2 - 45$
$5(x + 3)(x - 3)$

29. $2x^3 - 72x$
$2x(x + 6)(x - 6)$

30. $2x^3 - 8x$
$2x(x + 2)(x - 2)$

31. $x^2 + 36$
prime

32. $x^2 + 4$
prime

33. $3x^3 + 27x$
$3x(x^2 + 9)$

34. $3x^3 + 15x$
$3x(x^2 + 5)$

35. $18 - 2y^2$
$2(3 + y)(3 - y)$

36. $32 - 2y^2$
$2(4 + y)(4 - y)$

37. $3y^3 - 48y$
$3y(y + 4)(y - 4)$

38. $3y^3 - 75y$
$3y(y + 5)(y - 5)$

39. $18x^3 - 2x$
$2x(3x + 1)(3x - 1)$

40. $20x^3 - 5x$
$5x(2x + 1)(2x - 1)$

In Exercises 41–62, factor any perfect square trinomials, or state that the polynomial is prime.

41. $x^2 + 2x + 1$
$(x + 1)^2$

42. $x^2 + 4x + 4$
$(x + 2)^2$

43. $x^2 - 14x + 49$
$(x - 7)^2$

44. $x^2 - 10x + 25$
$(x - 5)^2$

45. $x^2 - 2x + 1$
$(x - 1)^2$

46. $x^2 - 4x + 4$
$(x - 2)^2$

47. $x^2 + 22x + 121$
$(x + 11)^2$

48. $x^2 + 24x + 144$
$(x + 12)^2$

49. $4x^2 + 4x + 1$
$(2x + 1)^2$

50. $9x^2 + 6x + 1$
$(3x + 1)^2$

51. $25y^2 - 10y + 1$
$(5y - 1)^2$

52. $64y^2 - 16y + 1$
$(8y - 1)^2$

53. $x^2 - 10x + 100$
prime

54. $x^2 - 7x + 49$
prime

55. $x^2 + 14xy + 49y^2$
$(x + 7y)^2$

56. $x^2 + 16xy + 64y^2$
$(x + 8y)^2$

57. $x^2 - 12xy + 36y^2$
$(x - 6y)^2$

58. $x^2 - 18xy + 81y^2$
$(x - 9y)^2$

59. $x^2 - 8xy + 64y^2$
prime

60. $x^2 + 9xy + 16y^2$
prime

61. $16x^2 - 40xy + 25y^2$
$(4x - 5y)^2$

62. $9x^2 + 48xy + 64y^2$
$(3x + 8y)^2$

In Exercises 63–70, factor completely.

63. $12x^2 - 12x + 3$
$3(2x - 1)^2$

64. $18x^2 + 24x + 8$
$2(3x + 2)^2$

65. $9x^3 + 6x^2 + x$
$x(3x + 1)^2$

66. $25x^3 - 10x^2 + x$
$x(5x - 1)^2$

67. $2y^2 - 4y + 2$
$2(y - 1)^2$

68. $2y^2 - 40y + 200$
$2(y - 10)^2$

69. $2y^3 + 28y^2 + 98y$
$2y(y + 7)^2$

70. $50y^3 + 20y^2 + 2y$
$2y(5y + 1)^2$

In Exercises 71–88, factor using the formula for the sum or difference of two cubes.

71. $x^3 + 1$
$(x + 1)(x^2 - x + 1)$

72. $x^3 + 64$
$(x + 4)(x^2 - 4x + 16)$

73. $x^3 - 27$
$(x - 3)(x^2 + 3x + 9)$

74. $x^3 - 64$
$(x - 4)(x^2 + 4x + 16)$

75. $8y^3 - 1$
$(2y - 1)(4y^2 + 2y + 1)$

76. $27y^3 - 1$
$(3y - 1)(9y^2 + 3y + 1)$

77. $27x^3 + 8$
$(3x + 2)(9x^2 - 6x + 4)$

78. $125x^3 + 8$
$(5x + 2)(25x^2 - 10x + 4)$

79. $x^3y^3 - 64$
$(xy - 4)(x^2y^2 + 4xy + 16)$

80. $x^3y^3 - 27$
$(xy - 3)(x^2y^2 + 3xy + 9)$

81. $27y^4 + 8y$
$y(3y + 2)(9y^2 - 6y + 4)$

82. $64y - y^4$
$y(4 - y)(16 + 4y + y^2)$

83. $54 - 16y^3$
$2(3 - 2y)(9 + 6y + 4y^2)$

84. $128 - 250y^3$
$2(4 - 5y)(16 + 20y + 25y^2)$

85. $64x^3 + 27y^3$
$(4x + 3y)(16x^2 - 12xy + 9y^2)$

86. $8x^3 + 27y^3$
$(2x + 3y)(4x^2 - 6xy + 9y^2)$

87. $125x^3 - 64y^3$
$(5x - 4y)(25x^2 + 20xy + 16y^2)$

88. $125x^3 - y^3$
$(5x - y)(25x^2 + 5xy + y^2)$

Practice Plus

In Exercises 89–96, factor completely.

89. $25x^2 - \dfrac{4}{49}$
$\left(5x + \dfrac{2}{7}\right)\left(5x - \dfrac{2}{7}\right)$

90. $16x^2 - \dfrac{9}{25}$
$\left(4x + \dfrac{3}{5}\right)\left(4x - \dfrac{3}{5}\right)$

91. $y^4 - \dfrac{y}{1000}$
$y\left(y - \dfrac{1}{10}\right)\left(y^2 + \dfrac{y}{10} + \dfrac{1}{100}\right)$

92. $y^4 - \dfrac{y}{8}$
$y\left(y - \dfrac{1}{2}\right)\left(y^2 + \dfrac{y}{2} + \dfrac{1}{4}\right)$

93. $0.25x - x^3$
$x(0.5 + x)(0.5 - x)$

94. $0.64x - x^3$
$x(0.8 + x)(0.8 - x)$

95. $(x + 1)^2 - 25$
$(x + 6)(x - 4)$

96. $(x + 2)^2 - 49$
$(x + 9)(x - 5)$

97. Divide $x^3 - x^2 - 5x - 3$ by $x - 3$. Use the quotient to factor $x^3 - x^2 - 5x - 3$ completely. $(x - 3)(x + 1)^2$

98. Divide $x^3 + 4x^2 - 3x - 18$ by $x - 2$. Use the quotient to factor $x^3 + 4x^2 - 3x - 18$ completely. $(x - 2)(x + 3)^2$

Application Exercises

In Exercises 99–102, find the formula for the area of the shaded region and express it in factored form.

99.

$x^2 - 25 = (x + 5)(x - 5)$

100.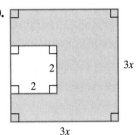

$9x^2 - 4 = (3x + 2)(3x - 2)$

101.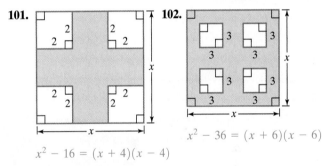

$x^2 - 16 = (x + 4)(x - 4)$

102.

$x^2 - 36 = (x + 6)(x - 6)$

Writing in Mathematics

103. Explain how to factor the difference of two squares. Provide an example with your explanation. **103–106.** Answers will vary.

104. What is a perfect square trinomial and how is it factored?

105. Explain why $x^2 - 1$ is factorable, but $x^2 + 1$ is not.

106. Explain how to factor $x^3 + 1$.

Critical Thinking Exercises

107. Which one of the following is true? **b**

a. Because $x^2 - 25 = (x + 5)(x - 5)$, then $x^2 + 25 = (x - 5)(x + 5)$.

b. All perfect square trinomials are squares of binomials.

c. Any polynomial that is the sum of two squares is prime.

d. The polynomial $16x^2 + 20x + 25$ is a perfect square trinomial.

108. Where is the error in this "proof" that $2 = 0$?

$a = b$	Suppose that a and b are any equal real numbers.
$a^2 = b^2$	Square both sides of the equation.
$a^2 - b^2 = 0$	Subtract b^2 from both sides.
$2(a^2 - b^2) = 2 \cdot 0$	Multiply both sides by 2.
$2(a^2 - b^2) = 0$	On the right side, $2 \cdot 0 = 0$.
$2(a + b)(a - b) = 0$	Factor $a^2 - b^2$.
$2(a + b) = 0$	Divide both sides by $a - b$.
$2 = 0$	Divide both sides by $a + b$.

$a - b = 0$ and division by 0 is not permitted.

In Exercises 109–112, factor each polynomial.

109. $x^2 - y^2 + 3x + 3y$ $(x + y)(x - y + 3)$

110. $x^{2n} - 25y^{2n}$ $(x^n + 5y^n)(x^n - 5y^n)$

111. $4x^{2n} + 12x^n + 9$ $(2x^n + 3)^2$

112. $(x + 3)^2 - 2(x + 3) + 1$ $[(x + 3) - 1]^2$ or $(x + 2)^2$

In Exercises 113–114, find all integers k so that the trinomial is a perfect square trinomial.

113. $9x^2 + kx + 1$ $6, -6$ **114.** $64x^2 - 16x + k$ 1

Technology Exercises

In Exercises 115–118, use the $\boxed{\text{GRAPH}}$ *or* $\boxed{\text{TABLE}}$ *feature of a graphing utility to determine if the polynomial on the left side of each equation has been correctly factored. If the graphs of* y_1 *and* y_2 *coincide, or if their corresponding table values are equal, this means that the polynomial on the left side has been correctly factored. If not, factor the polynomial correctly and then use your graphing utility to verify the factorization.*

115. $4x^2 - 9 = (4x + 3)(4x - 3)$ $(4x + 3)(4x - 3)$ should be $(2x + 3)(2x - 3)$.

116. $x^2 - 6x + 9 = (x - 3)^2$ correctly factored

117. $4x^2 - 4x + 1 = (4x - 1)^2$ $(4x - 1)^2$ should be $(2x - 1)^2$.

118. $x^3 - 1 = (x - 1)(x^2 - x + 1)$ $(x - 1)(x^2 - x + 1)$ should be $(x - 1)(x^2 + x + 1)$.

Review Exercises

119. Simplify: $(2x^2y^3)^4(5xy^2)$. (Section 5.7, Example 5) $80x^9y^{14}$

120. Subtract: $(10x^2 - 5x + 2) - (14x^2 - 5x - 1)$. (Section 5.1, Example 3) $-4x^2 + 3$

121. Divide: $\dfrac{6x^2 + 11x - 10}{3x - 2}$.   (Section 5.6, Example 1) $2x + 5$

6.5

Objectives

1 Recognize the appropriate method for factoring a polynomial.

2 Use a general strategy for factoring polynomials.

A GENERAL FACTORING STRATEGY

Yogi Berra, catcher and renowned hitter for the New York Yankees (1946–1963), said it best: "If you don't know where you're going, you'll probably end up someplace else." When it comes to factoring, it's easy to know where you're going. Why? In this section, you will learn a step-by-step strategy that provides a plan and direction for solving factoring problems.

A Strategy for Factoring Polynomials   It is important to practice factoring a wide variety of polynomials so that you can quickly select the appropriate technique. The polynomial is factored completely when all its polynomial factors, except possibly the monomial factors, are prime. Because of the commutative property, the order of the factors does not matter.

Here is a general strategy for factoring polynomials:

1 Recognize the appropriate method for factoring a polynomial.

A STRATEGY FOR FACTORING A POLYNOMIAL

1. If there is a common factor other than 1, factor out the GCF.

2. Determine the number of terms in the polynomial and try factoring as follows:

   a. If there are two terms, can the binomial be factored by one of the following special forms?

      Difference of two squares: $A^2 - B^2 = (A + B)(A - B)$

      Sum of two cubes: $A^3 + B^3 = (A + B)(A^2 - AB + B^2)$

      Difference of two cubes: $A^3 - B^3 = (A - B)(A^2 + AB + B^2)$

   b. If there are three terms, is the trinomial a perfect square trinomial? If so, factor by one of the following special forms:

$$A^2 + 2AB + B^2 = (A + B)^2$$
$$A^2 - 2AB + B^2 = (A - B)^2.$$

    If the trinomial is not a perfect square trinomial, try factoring by trial and error or grouping.

   c. If there are four or more terms, try factoring by grouping.

3. Check to see if any factors with more than one term in the factored polynomial can be factored further. If so, factor completely.

4. Check by multiplying.

2 Use a general strategy for factoring polynomials.

The following examples and those in the exercise set are similar to the previous factoring problems. One difference is that although these polynomials may be factored using the techniques we have studied in this chapter, each must be factored using at least two techniques. Also different is that these factorizations are not all of the same type. They are intentionally mixed to promote the development of a general factoring strategy.

EXAMPLE 1   Factoring a Polynomial

Factor: $4x^4 - 16x^2$.

SOLUTION

Step 1.   **If there is a common factor, factor out the GCF.** Because $4x^2$ is common to both terms, we factor it out.

$$4x^4 - 16x^2 = 4x^2(x^2 - 4) \qquad \text{Factor out the GCF.}$$

Step 2.   **Determine the number of terms and factor accordingly.** The factor $x^2 - 4$ has two terms. It is the difference of two squares: $x^2 - 2^2$. We factor using the special form for the difference of two squares and rewrite the GCF.

$$4x^4 - 16x^2 = 4x^2(x + 2)(x - 2) \qquad \text{Use } A^2 - B^2 = (A + B)(A - B)$$
$$\text{on } x^2 - 4\text{: } A = x \text{ and } B = 2.$$

Step 3.   **Check to see if any factors with more than one term can be factored further.** No factor with more than one term can be factored further, so we have factored completely.

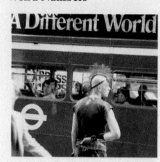
Step 4. **Check by multiplying.**

$$4x^2(x + 2)(x - 2) = 4x^2(x^2 - 4) = 4x^4 - 16x^2$$

This is the original polynomial, so the factorization is correct.

✔ **CHECK POINT 1** Factor: $5x^4 - 45x^2$. $5x^2(x + 3)(x - 3)$

EXAMPLE 2 Factoring a Polynomial

Factor: $3x^2 - 6x - 45$.

SOLUTION

Step 1. **If there is a common factor, factor out the GCF.** Because 3 is common to all terms, we factor it out.

$$3x^2 - 6x - 45 = 3(x^2 - 2x - 15) \quad \text{Factor out the GCF.}$$

Step 2. **Determine the number of terms and factor accordingly.** The factor $x^2 - 2x - 15$ has three terms, but it is not a perfect square trinomial. We factor it using trial and error.

$$3x^2 - 6x - 45 = 3(x^2 - 2x - 15) = 3(x - 5)(x + 3)$$

Step 3. **Check to see if factors can be factored further.** In this case, they cannot, so we have factored completely.

Step 4. **Check by multiplying.**

$$3(x - 5)(x + 3) = 3(x^2 - 2x - 15) = 3x^2 - 6x - 45$$

FOIL

This is the original polynomial, so the factorization is correct.

✔ **CHECK POINT 2** Factor: $4x^2 - 16x - 48$. $4(x - 6)(x + 2)$

EXAMPLE 3 Factoring a Polynomial

Factor: $7x^5 - 7x$.

SOLUTION

Step 1. **If there is a common factor, factor out the GCF.** Because $7x$ is common to both terms, we factor it out.

$$7x^5 - 7x = 7x(x^4 - 1) \quad \text{Factor out the GCF.}$$

Step 2. **Determine the number of terms and factor accordingly.** The factor $x^4 - 1$ has two terms. This binomial can be expressed as $(x^2)^2 - 1^2$, so it can be factored as the difference of two squares.

$$7x^5 - 7x = 7x(x^4 - 1) = 7x(x^2 + 1)(x^2 - 1) \quad \begin{array}{l}\text{Use } A^2 - B^2 = (A + B)(A - B) \\ \text{on } x^4 - 1: A = x^2 \text{ and } B = 1.\end{array}$$

Step 3. **Check to see if factors can be factored further.** We note that $(x^2 - 1)$ is also the difference of two squares, $x^2 - 1^2$, so we continue factoring.

$$7x^5 - 7x = 7x(x^2 + 1)(x + 1)(x - 1)$$ Factor $x^2 - 1$ as the difference of two squares.

Step 4. **Check by multiplying.**

$$7x(x^2 + 1)(x + 1)(x - 1) = 7x(x^2 + 1)(x^2 - 1) = 7x(x^4 - 1) = 7x^5 - 7x$$

We obtain the original polynomial, so the factorization is correct. ■

✔ **CHECK POINT 3** Factor: $4x^5 - 64x$. $4x(x^2 + 4)(x + 2)(x - 2)$

EXAMPLE 4 Factoring a Polynomial

Factor: $x^3 - 5x^2 - 4x + 20$.

SOLUTION

Step 1. **If there is a common factor, factor out the GCF.** Other than 1, there is no common factor.

Step 2. **Determine the number of terms and factor accordingly.** There are four terms. We try factoring by grouping.

$$x^3 - 5x^2 - 4x + 20$$

$$= (x^3 - 5x^2) + (-4x + 20)$$ Group terms with common factors.

$$= x^2(x - 5) - 4(x - 5)$$ Factor from each group.

$$= (x - 5)(x^2 - 4)$$ Factor out the common binomial factor, $x - 5$.

Step 3. **Check to see if factors can be factored further.** We note that $(x^2 - 4)$ is the difference of two squares, $x^2 - 2^2$, so we continue factoring.

$$x^3 - 5x^2 - 4x + 20 = (x - 5)(x + 2)(x - 2)$$ Factor $x^2 - 4$ as the difference of two squares.

We have factored completely because no factor with more than one term can be factored further.

Step 4. **Check by multiplying.**

F O I L

$$(x - 5)(x + 2)(x - 2) = (x - 5)(x^2 - 4) = x^3 - 4x - 5x^2 + 20$$

$$= x^3 - 5x^2 - 4x + 20$$

We obtain the original polynomial, so the factorization is correct. ■

USING TECHNOLOGY

You can use a graphing utility to check the factorization in Example 4. Enter the given polynomial and its complete factorization:

$$y_1 = x^3 - 5x^2 - 4x + 20 \quad \text{and} \quad y_2 = (x - 5)(x + 2)(x - 2).$$

Graphic Check

The graphs are identical.

Numeric Check

No matter how far up or down we scroll, $y_1 = y_2$.

$[-6, 6, 1]$ by $[-30, 30, 3]$

Caution: Keep in mind that a graphing utility cannot verify that a polynomial is factored completely. If you enter $(x - 5)(x^2 - 4)$ for y_2, which is not a complete factorization, you will still obtain identical graphs or tables with equal corresponding values.

✔ **CHECK POINT 4** Factor: $x^3 - 4x^2 - 9x + 36$. $(x - 4)(x + 3)(x - 3)$

EXAMPLE 5 Factoring a Polynomial

Factor: $2x^3 - 24x^2 + 72x$.

SOLUTION

Step 1. **If there is a common factor, factor out the GCF.** Because $2x$ is common to all terms, we factor it out.

$$2x^3 - 24x^2 + 72x = 2x(x^2 - 12x + 36) \quad \text{Factor out the GCF.}$$

Step 2. **Determine the number of terms and factor accordingly.** The factor $x^2 - 12x + 36$ has three terms. Is it a perfect square trinomial? Yes. The first term, x^2, is the square of a monomial. The last term, 36, or 6^2, is the square of an integer. The middle term involves twice the product of x and 6. We factor using $A^2 - 2AB + B^2 = (A - B)^2$.

$$2x^3 - 24x^2 + 72x = 2x(x^2 - 12x + 36)$$

$$= 2x(x^2 - 2 \cdot x \cdot 6 + 6^2) \quad \text{The second factor is a perfect square trinomial.}$$

$$A^2 - 2\ A\ B + B^2$$

$$= 2x(x - 6)^2 \quad A^2 - 2AB + B^2 = (A - B)^2$$

Step 3. **Check to see if factors can be factored further.** In this problem, they cannot, so we have factored completely.

Step 4.   **Check by multiplying.**

$$2x(x - 6)^2 = 2x(x^2 - 12x + 36) = 2x^3 - 24x^2 + 72x$$

We obtain the original polynomial, so the factorization is correct. ∎

✔ **CHECK POINT 5** Factor: $3x^3 - 30x^2 + 75x$. $3x(x - 5)^2$

EXAMPLE 6   Factoring a Polynomial

Factor: $3x^5 + 24x^2$.

SOLUTION

Step 1.   **If there is a common factor, factor out the GCF.** Because $3x^2$ is common to both terms, we factor it out.

$$3x^5 + 24x^2 = 3x^2(x^3 + 8)$$ Factor out the GCF.

Step 2.   **Determine the number of terms and factor accordingly.** The factor $x^3 + 8$ has two terms. This binomial can be expressed as $x^3 + 2^3$, so it can be factored as the sum of two cubes.

$$3x^5 + 24x^2 = 3x^2(\underbrace{x^3 + 2^3}_{A^3 + B^3})$$ Express $x^3 + 8$ as the sum of two cubes.

$$= 3x^2\underbrace{(x + 2)(x^2 - 2x + 4)}_{(A + B)(A^2 - AB + B^2)}$$ Factor the sum of two cubes.

Step 3.   **Check to see if factors can be factored further.** In this problem, they cannot, so we have factored completely.

Step 4.   **Check by multiplying.**

$$3x^2(x + 2)(x^2 - 2x + 4) = 3x^2[x(x^2 - 2x + 4) + 2(x^2 - 2x + 4)]$$

$$= 3x^2(x^3 - 2x^2 + 4x + 2x^2 - 4x + 8)$$

$$= 3x^2(x^3 + 8) = 3x^5 + 24x^2$$

We obtain the original polynomial, so the factorization is correct. ∎

✔ **CHECK POINT 6** Factor: $2x^5 + 54x^2$. $2x^2(x + 3)(x^2 - 3x + 9)$

DISCOVER FOR YOURSELF

In Examples 1–6, substitute 1 for the variable in both the given polynomial and in its factored form. Evaluate each expression. What do you observe? Do this for a second value of the variable. Is this a complete check or only a partial check of the factorization? Explain.

EXAMPLE 7   Factoring a Polynomial in Two Variables

Factor: $32x^4y - 2y^5$.

SOLUTION

Step 1.   **If there is a common factor, factor out the GCF.** Because $2y$ is common to both terms, we factor it out.

$$32x^4y - 2y^5 = 2y(16x^4 - y^4)$$ Factor out the GCF.

Step 2. **Determine the number of terms and factor accordingly.** The factor $16x^4 - y^4$ has two terms. It is the difference of two squares: $(4x^2)^2 - (y^2)^2$. We factor using the special form for the difference of two squares.

$$32x^4y - 2y^5 = 2y[\underbrace{(4x^2)^2 - (y^2)^2}_{A^2 - B^2}]$$
Express $16x^4 - y^4$ as the difference of two squares.

$$= 2y\underbrace{(4x^2 + y^2)}_{(A + B)}\underbrace{(4x^2 - y^2)}_{(A - B)}$$
$A^2 - B^2 = (A + B)(A - B)$

Step 3. **Check to see if factors can be factored further.** We note that the last factor, $4x^2 - y^2$, is also the difference of two squares, $(2x)^2 - y^2$, so we continue factoring.

$$32x^4y - 2y^5 = 2y(4x^2 + y^2)(2x + y)(2x - y)$$

Step 4. **Check by multiplying.** Multiply the factors in the factorization and verify that you obtain the original polynomial. ■

✔ CHECK POINT **7** Factor: $3x^4y - 48y^5$. $3y(x^2 + 4y^2)(x + 2y)(x - 2y)$

EXAMPLE 8 Factoring a Polynomial in Two Variables

Factor: $18x^3 + 48x^2y + 32xy^2$.

SOLUTION

Step 1. **If there is a common factor, factor out the GCF.** Because $2x$ is common to all terms, we factor it out.

$$18x^3 + 48x^2y + 32xy^2 = 2x(9x^2 + 24xy + 16y^2)$$

Step 2. **Determine the number of terms and factor accordingly.** The factor $9x^2 + 24xy + 16y^2$ has three terms. Is it a perfect square trinomial? Yes. The first term, $9x^2$, or $(3x)^2$, and the last term, $16y^2$, or $(4y)^2$, are squares of monomials. The middle term, $24xy$, is twice the product of $3x$ and $4y$. We factor using $A^2 + 2AB + B^2 = (A + B)^2$.

$$18x^3 + 48x^2y + 32xy^2 = 2x(9x^2 + 24xy + 16y^2)$$

$$= 2x[\underbrace{(3x)^2 + 2 \cdot 3x \cdot 4y + (4y)^2}_{A^2 + 2 \cdot A \cdot B + B^2}]$$
The second factor is a perfect square trinomial.

$$= 2x(3x + 4y)^2$$
$A^2 + 2AB + B^2 = (A + B)^2$

Step 3. **Check to see if factors can be factored further.** In this problem, they cannot, so we have factored completely.

Step 4. **Check by multiplication.** Multiply the factors in the factorization and verify that you obtain the original polynomial.

✔ CHECK POINT **8** Factor: $12x^3 + 36x^2y + 27xy^2$. $3x(2x + 3y)^2$

6.5 EXERCISE SET

Student Solutions Manual   CD/Video   PH Math/Tutor Center   MathXL Tutorials on CD   MathXL®   MyMathLab   Interactmath.com

Practice Exercises

In Exercises 1–62, factor completely, or state that the polynomial is prime. Check factorizations using multiplication or a graphing utility.

1. $5x^3 - 20x$
$5x(x + 2)(x - 2)$

2. $4x^3 - 100x$
$4x(x + 5)(x - 5)$

3. $7x^3 + 7x$
$7x(x^2 + 1)$

4. $6x^3 + 24x$
$6x(x^2 + 4)$

5. $5x^2 - 5x - 30$
$5(x - 3)(x + 2)$

6. $5x^2 - 15x - 50$
$5(x - 5)(x + 2)$

7. $2x^4 - 162$
$2(x^2 + 9)(x + 3)(x - 3)$

8. $7x^4 - 7$
$7(x^2 + 1)(x + 1)(x - 1)$

9. $x^3 + 2x^2 - 9x - 18$
$(x + 2)(x + 3)(x - 3)$

10. $x^3 + 3x^2 - 25x - 75$
$(x + 3)(x + 5)(x - 5)$

11. $3x^3 - 24x^2 + 48x$
$3x(x - 4)^2$

12. $5x^3 - 20x^2 + 20x$
$5x(x - 2)^2$

13. $2x^5 + 2x^2$
$2x^2(x + 1)(x^2 - x + 1)$

14. $2x^5 + 128x^2$
$2x^2(x + 4)(x^2 - 4x + 16)$

15. $6x^2 + 8x$
$2x(3x + 4)$

16. $21x^2 - 35x$
$7x(3x - 5)$

17. $2y^2 - 2y - 112$
$2(y - 8)(y + 7)$

18. $6x^2 - 6x - 12$
$6(x - 2)(x + 1)$

19. $7y^4 + 14y^3 + 7y^2$
$7y^2(y + 1)^2$

20. $2y^4 + 28y^3 + 98y^2$
$2y^2(y + 7)^2$

21. $y^2 + 8y - 16$
prime

22. $y^2 - 18y - 81$
prime

23. $16y^2 - 4y - 2$
$2(4y + 1)(2y - 1)$

24. $32y^2 + 4y - 6$
$2(8y - 3)(2y + 1)$

25. $r^2 - 25r$
$r(r - 25)$

26. $3r^2 - 27r$
$3r(r - 9)$

27. $4w^2 + 8w - 5$
$(2w + 5)(2w - 1)$

28. $35w^2 - 2w - 1$
$(7w + 1)(5w - 1)$

29. $x^3 - 4x$
$x(x + 2)(x - 2)$

30. $9x^3 - 9x$
$9x(x + 1)(x - 1)$

31. $x^2 + 64$
prime

32. $y^2 + 36$
prime

33. $9y^2 + 13y + 4$
$(9y + 4)(y + 1)$

34. $20y^2 + 12y + 1$
$(2y + 1)(10y + 1)$

35. $y^3 + 2y^2 - 4y - 8$
$(y + 2)(y + 2)(y - 2)$

36. $y^3 + 2y^2 - y - 2$
$(y + 2)(y + 1)(y - 1)$

37. $16y^2 + 24y + 9$
$(4y + 3)^2$

38. $25y^2 + 20y + 4$
$(5y + 2)^2$

39. $4y^3 - 28y^2 + 40y$
$4y(y - 5)(y - 2)$

40. $7y^3 - 21y^2 + 14y$
$7y(y - 2)(y - 1)$

41. $y^5 - 81y$
$y(y^2 + 9)(y + 3)(y - 3)$

42. $y^5 - 16y$
$y(y^2 + 4)(y + 2)(y - 2)$

43. $20a^4 - 45a^2$
$5a^2(2a + 3)(2a - 3)$

44. $48a^4 - 3a^2$
$3a^2(4a + 1)(4a - 1)$

45. $12y^2 - 11y + 2$
$(4y - 1)(3y - 2)$

46. $21x^2 - 25x - 4$
$(7x + 1)(3x - 4)$

47. $9y^2 - 64$
$(3y + 8)(3y - 8)$

48. $100y^2 - 49$
$(10y + 7)(10y - 7)$

49. $9y^2 + 64$
prime

50. $100y^2 + 49$
prime

51. $2y^3 + 3y^2 - 50y - 75$
$(2y + 3)(y + 5)(y - 5)$

52. $12y^3 + 16y^2 - 3y - 4$
$(3y + 4)(2y + 1)(2y - 1)$

53. $2r^3 + 30r^2 - 68r$
$2r(r + 17)(r - 2)$

54. $3r^3 - 27r^2 - 210r$
$3r(r + 5)(r - 14)$

55. $8x^5 - 2x^3$
$2x^3(2x + 1)(2x - 1)$

56. $y^9 - y^5$
$y^5(y^2 + 1)(y + 1)(y - 1)$

57. $3x^2 + 243$
$3(x^2 + 81)$

58. $27x^2 + 75$
$3(9x^2 + 25)$

59. $x^4 + 8x$
$x(x + 2)(x^2 - 2x + 4)$

60. $x^4 + 27x$
$x(x + 3)(x^2 - 3x + 9)$

61. $2y^5 - 2y^2$
$2y^2(y - 1)(y^2 + y + 1)$

62. $2y^5 - 128y^2$
$2y^2(y - 4)(y^2 + 4y + 16)$

Exercises 63–92 contain polynomials in several variables. Factor each polynomial completely and check using multiplication.

63. $6x^2 + 8xy$
$2x(3x + 4y)$

64. $21x^2 - 35xy$
$7x(3x - 5y)$

65. $xy - 7x + 3y - 21$
$(y - 7)(x + 3)$

66. $xy - 5x + 2y - 10$
$(y - 5)(x + 2)$

67. $x^2 - 3xy - 4y^2$
$(x - 4y)(x + y)$

68. $x^2 - 4xy - 12y^2$
$(x - 6y)(x + 2y)$

69. $72a^3b^2 + 12a^2 - 24a^4b^2$
$12a^2(6ab^2 + 1 - 2a^2b^2)$

70. $24a^4b + 60a^3b^2 + 150a^2b^3$
$6a^2b(4a^2 + 10ab + 25b^2)$

71. $3a^2 + 27ab + 54b^2$
$3(a + 6b)(a + 3b)$

72. $3a^2 + 15ab + 18b^2$
$3(a + 3b)(a + 2b)$

73. $48x^4y - 3x^2y$
$3x^2y(4x + 1)(4x - 1)$

74. $16a^3b^2 - 4ab^2$
$4ab^2(2a + 1)(2a - 1)$

75. $6a^2b + ab - 2b$
$b(3a + 2)(2a - 1)$

76. $16a^2 - 32ab + 12b^2$
$4(2a - b)(2a - 3b)$

77. $7x^5y - 7xy^5$
$7xy(x^2 + y^2)(x + y)(x - y)$

78. $3x^4y^2 - 3x^2y^2$
$3x^2y^2(x + 1)(x - 1)$

79. $10x^3y - 14x^2y^2 + 4xy^3$
$2xy(5x - 2y)(x - y)$

80. $18x^3y + 57x^2y^2 + 30xy^3$
$3xy(2x + 5y)(3x + 2y)$

81. $2bx^2 + 44bx + 242b$
$2b(x + 11)^2$

82. $3xz^2 - 72xz + 432x$
$3x(z - 12)^2$

83. $15a^2 + 11ab - 14b^2$
$(5a + 7b)(3a - 2b)$

84. $25a^2 + 25ab + 6b^2$
$(5a + 3b)(5a + 2b)$

85. $36x^3y - 62x^2y^2 + 12xy^3$
$2xy(9x - 2y)(2x - 3y)$

86. $10a^4b^2 - 15a^3b^3 - 25a^2b^4$
$5a^2b^2(2a - 5b)(a + b)$

87. $a^2y - b^2y - a^2x + b^2x$
$(y - x)(a + b)(a - b)$

88. $bx^2 - 4b + ax^2 - 4a$
$(b + a)(x + 2)(x - 2)$

89. $9ax^3 + 15ax^2 - 14ax$
$ax(3x + 7)(3x - 2)$

90. $4ay^3 - 12ay^2 + 9ay$
$ay(2y - 3)^2$

91. $81x^4y - y^5$
$y(9x^2 + y^2)(3x + y)(3x - y)$

92. $16x^4y - y^5$
$y(4x^2 + y^2)(2x + y)(2x - y)$

Practice Plus

In Exercises 93–102, factor completely.

93. $10x^2(x + 1) - 7x(x + 1) - 6(x + 1)$
$(x + 1)(5x - 6)(2x + 1)$

94. $12x^2(x - 1) - 4x(x - 1) - 5(x - 1)$
$(x - 1)(6x - 5)(2x + 1)$

95. $6x^4 + 35x^2 - 6$
$(x^2 + 6)(6x^2 - 1)$

96. $7x^4 + 34x^2 - 5$
$(x^2 + 5)(7x^2 - 1)$

97. $(x - 7)^2 - 4a^2$
$(x - 7 + 2a)(x - 7 - 2a)$

98. $(x - 6)^2 - 9a^2$
$(x - 6 + 3a)(x - 6 - 3a)$

99. $x^2 + 8x + 16 - 25a^2$
$(x + 4 + 5a)(x + 4 - 5a)$

100. $x^2 + 14x + 49 - 16a^2$
$(x + 7 + 4a)(x + 7 - 4a)$

101. $y^7 + y$
$y(y^2 + 1)(y^4 - y^2 + 1)$

102. $(y + 1)^3 + 1$
$(y + 2)(y^2 + y + 1)$

Application Exercises

103. A rock is dropped from the top of a 256-foot cliff. The height, in feet, of the rock above the water *after t seconds* is modeled by the polynomial $256 - 16t^2$. Factor this expression completely. $16(4 + t)(4 - t)$

256 feet

104. The building shown in the figure has a height represented by *x* feet. The building's base is a square and its volume is $x^3 - 60x^2 + 900x$ cubic feet. Express the building's dimensions in terms of *x*. $(x - 30)$ feet by $(x - 30)$ feet by *x* feet

x

105. Express the area of the shaded ring shown in the figure in terms of π. Then factor this expression completely.
$\pi b^2 - \pi a^2$; $\pi(b + a)(b - a)$

a

b

Writing in Mathematics

106. Describe a strategy that can be used to factor polynomials.
106–108. Answers will vary.

107. Describe some of the difficulties in factoring polynomials. What suggestions can you offer to overcome these difficulties?

108. You are about to take a great picture of fog rolling into San Francisco from the middle of the Golden Gate Bridge, 400 feet above the water. Whoops! You accidently lean too far over the safety rail and drop your camera. The height, in feet, of the camera after *t* seconds is modeled by the polynomial $400 - 16t^2$. The factored form of the polynomial is $16(5 + t)(5 - t)$. Describe something about your falling camera that is easier to see from the factored form, $16(5 + t)(5 - t)$, than from the form $400 - 16t^2$.

Critical Thinking Exercises

109. Which one of the following is true? **d**

a. $x^2 - 9 = (x - 3)^2$ for any real number *x*.

b. The polynomial $4x^2 + 100$ is the sum of two squares and therefore cannot be factored.

c. If the general factoring strategy is used to factor a polynomial, at least two factorizations are necessary before the given polynomial is factored completely.

d. Once a common monomial factor is removed from $3xy^3 + 9xy^2 + 21xy$, the remaining trinomial factor cannot be factored further.

In Exercises 110–114, factor completely.

110. $3x^5 - 21x^3 - 54x$ $\quad 3x(x^2 + 2)(x + 3)(x - 3)$

111. $5y^5 - 5y^4 - 20y^3 + 20y^2$ $\quad 5y^2(y - 1)(y + 2)(y - 2)$

112. $4x^4 - 9x^2 + 5$ $\quad (4x^2 - 5)(x + 1)(x - 1)$

113. $(x + 5)^2 - 20(x + 5) + 100$ $\quad (x - 5)^2$

114. $3x^{2n} - 27y^{2n}$ $\quad 3(x^n + 3y^n)(x^n - 3y^n)$

Technology Exercises

In Exercises 115–119, use the GRAPH *or* TABLE *feature of a graphing utility to determine if the polynomial on the left side of each equation has been correctly factored. If not, factor the polynomial correctly and then use your graphing utility to verify the factorization.*

115. $4x^2 - 12x + 9 = (4x - 3)^2$; $[-5, 5, 1]$ by $[0, 20, 1]$
$(4x - 3)^2$ should be $(2x - 3)^2$.

116. $3x^3 - 12x^2 - 15x = 3x(x + 5)(x - 1)$; $[-5, 7, 1]$ by $[-80, 80, 10]$
$3x(x + 5)(x - 1)$ should be $3x(x - 5)(x + 1)$.

117. $6x^2 + 10x - 4 = 2(3x - 1)(x + 2)$; $[-5, 5, 1]$ by $[-20, 20, 2]$ correctly factored

118. $x^4 - 16 = (x^2 + 4)(x + 2)(x - 2)$; $[-5, 5, 1]$ by $[-20, 20, 2]$ correctly factored

119. $2x^3 + 10x^2 - 2x - 10 = 2(x + 5)(x^2 + 1)$; $[-8, 4, 1]$ by $[-100, 100, 10]$
$2(x + 5)(x^2 + 1)$ should be $2(x + 5)(x + 1)(x - 1)$.

Review Exercises

120. Factor: $9x^2 - 16$. (Section 6.4, Example 1)
$(3x + 4)(3x - 4)$

121. Graph using intercepts: $5x - 2y = 10$.
(Section 3.2, Example 4)   See graphing answer section.

122. The second angle of a triangle measures three times that of the first angle's measure. The third angle measures 80° more than the first. Find the measure of each angle. (Section 2.6, Example 6) $20°, 60°, 100°$

SECTION

6.6

Objectives

1 Use the zero-product principle.

2 Solve quadratic equations by factoring.

3 Solve problems with quadratic equations.

SOLVING QUADRATIC EQUATIONS BY FACTORING

The alligator, an endangered species, was the subject of a protection program at Florida's Everglades National Park. Park rangers used the formula

$$P = -10x^2 + 475x + 3500$$

to estimate the alligator population, P, after x years of the protection program. Their goal was to bring the population up to 7250. To find out how long the program had to be continued for this to happen, we substitute 7250 for P in the formula and solve for x:

$$7250 = -10x^2 + 475x + 3500.$$

Do you see how this equation differs from a linear equation? The highest exponent on x is 2. Solving such an equation involves finding the numbers that will make the equation a true statement. In this section, we use factoring to solve equations in the form $ax^2 + bx + c = 0$. We also look at applications of these equations.

The Standard Form of a Quadratic Equation   We begin by defining a quadratic equation.

> **DEFINITION OF A QUADRATIC EQUATION**   A **quadratic equation** in x is an equation that can be written in the **standard form**
>
> $$ax^2 + bx + c = 0,$$
>
> where a, b, and c are real numbers, with $a \neq 0$. A quadratic equation in x is also called a **second-degree polynomial equation** in x.

Here is an example of a quadratic equation in standard form:

$$x^2 - 7x + 10 = 0.$$

$a = 1$ $b = -7$ $c = 10$

1 Use the zero-product principle.

Solving Quadratic Equations by Factoring We can factor the left side of the quadratic equation $x^2 - 7x + 10 = 0$. We obtain $(x - 5)(x - 2) = 0$. If a quadratic equation has zero on one side and a factored expression on the other side, it can be solved using the **zero-product principle**.

THE ZERO-PRODUCT PRINCIPLE If the product of two algebraic expressions is zero, then at least one of the factors is equal to zero.

If $AB = 0$, then $A = 0$ or $B = 0$.

For example, consider the equation $(x - 5)(x - 2) = 0$. According to the zero-product principle, this product can be zero only if at least one of the factors is zero. We set each individual factor equal to zero and solve each resulting equation for x.

$$(x - 5)(x - 2) = 0$$
$$x - 5 = 0 \quad \text{or} \quad x - 2 = 0$$
$$x = 5 \qquad\qquad x = 2$$

We can check each of the proposed solutions, 5 and 2, in the original quadratic equation, $x^2 - 7x + 10 = 0$. Substitute each one separately for x in the equation.

Check 5:

$$x^2 - 7x + 10 = 0$$
$$5^2 - 7 \cdot 5 + 10 \stackrel{?}{=} 0$$
$$25 - 35 + 10 \stackrel{?}{=} 0$$
$$0 = 0, \quad \text{true}$$

Check 2:

$$x^2 - 7x + 10 = 0$$
$$2^2 - 7 \cdot 2 + 10 \stackrel{?}{=} 0$$
$$4 - 14 + 10 \stackrel{?}{=} 0$$
$$0 = 0, \quad \text{true}$$

The resulting true statements indicate that the solutions are 5 and 2. Note that with a quadratic equation, we can have two solutions, compared to the linear equation that usually had one.

EXAMPLE 1 Using the Zero-Product Principle

Solve the equation: $(3x - 1)(x + 2) = 0$.

SOLUTION The product $(3x - 1)(x + 2)$ is equal to zero. By the zero-product principle, the only way that this product can be zero is if at least one of the factors is zero. Thus,

$$3x - 1 = 0 \quad \text{or} \quad x + 2 = 0.$$
$$3x = 1 \qquad\qquad x = -2 \qquad \textit{Solve each equation for x.}$$
$$x = \frac{1}{3}$$

Because each linear equation has a solution, the original equation, $(3x - 1)(x + 2) = 0$, has two solutions, $\frac{1}{3}$ and -2. Check these solutions by substituting each one separately into the given equation. The equation's solution set is $\left\{-2, \frac{1}{3}\right\}$. ∎

✓ CHECK POINT 1 Solve the equation: $(2x + 1)(x - 4) = 0$. $-\dfrac{1}{2}$ and 4, or $\left\{-\dfrac{1}{2}, 4\right\}$

2 Solve quadratic equations by factoring.

In Example 1 and Check Point 1, the given equations were in factored form. Here is a procedure for solving a quadratic equation when we must first do the factoring.

SOLVING A QUADRATIC EQUATION BY FACTORING

1. If necessary, rewrite the equation in the standard form $ax^2 + bx + c = 0$, moving all terms to one side, thereby obtaining zero on the other side.
2. Factor.
3. Apply the zero-product principle, setting each factor equal to zero.
4. Solve the equations in step 3.
5. Check the solutions in the original equation.

EXAMPLE 2 Solving a Quadratic Equation by Factoring

Solve: $2x^2 + 7x - 4 = 0$.

SOLUTION

Step 1. **Move all terms to one side and obtain zero on the other side.** All terms are already on the left and zero is on the other side, so we can skip this step.

Step 2. **Factor.**

$$2x^2 + 7x - 4 = 0$$
$$(2x - 1)(x + 4) = 0$$

Steps 3 and 4. **Set each factor equal to zero and solve each resulting equation.**

$$2x - 1 = 0 \quad \text{or} \quad x + 4 = 0$$
$$2x = 1 \qquad\qquad x = -4$$
$$x = \frac{1}{2}$$

Step 5. **Check the solutions in the original equation.**

STUDY TIP

Do not confuse factoring a polynomial with solving a quadratic equation by factoring.

INCORRECT!

Factor: $2x^2 + 7x - 4.$

$(2x - 1)(x + 4)$

$2x - 1 = 0$ or $x + 4 = 0$

$x = \frac{1}{2} \qquad x = -4$

Check $\dfrac{1}{2}$:

$$2x^2 + 7x - 4 = 0$$
$$2\left(\frac{1}{2}\right)^2 + 7\left(\frac{1}{2}\right) - 4 \overset{?}{=} 0$$
$$2\left(\frac{1}{4}\right) + 7\left(\frac{1}{2}\right) - 4 \overset{?}{=} 0$$
$$\frac{1}{2} + \frac{7}{2} - 4 \overset{?}{=} 0$$
$$4 - 4 \overset{?}{=} 0$$
$$0 = 0, \quad \text{true}$$

Check -4:

$$2x^2 + 7x - 4 = 0$$
$$2(-4)^2 + 7(-4) - 4 \overset{?}{=} 0$$
$$2(16) + 7(-4) - 4 \overset{?}{=} 0$$
$$32 + (-28) - 4 \overset{?}{=} 0$$
$$4 - 4 \overset{?}{=} 0$$
$$0 = 0, \quad \text{true}$$

The solutions are -4 and $\frac{1}{2}$, and the solution set is $\left\{-4, \frac{1}{2}\right\}$. ∎

✓ **CHECK POINT 2** Solve: $x^2 - 6x + 5 = 0$. 1 and 5, or $\{1, 5\}$

USING TECHNOLOGY

[−5, 2, 1] by [−11, 2, 1]

x-intercept is −4.

x-intercept is $\frac{1}{2}$.

You can use a graphing utility to check the real number solutions of a quadratic equation. **The solutions of $ax^2 + bx + c = 0$ correspond to the x-intercepts for the graph of $y = ax^2 + bx + c$.** For example, to check the solutions of $2x^2 + 7x - 4 = 0$, graph $y = 2x^2 + 7x - 4$. The U-shaped, cuplike, graph is shown on the left. The x-intercepts are -4 and $\frac{1}{2}$, verifying -4 and $\frac{1}{2}$ as the solutions.

EXAMPLE 3   Solving a Quadratic Equation by Factoring

Solve: $3x^2 = 2x$.

SOLUTION

Step 1. **Move all terms to one side and obtain zero on the other side.** Subtract $2x$ from both sides and write the equation in standard form.

$$3x^2 - 2x = 2x - 2x$$
$$3x^2 - 2x = 0$$

Step 2. **Factor.** We factor out x from the two terms on the left side.

$$3x^2 - 2x = 0$$
$$x(3x - 2) = 0$$

Steps 3 and 4. **Set each factor equal to zero and solve the resulting equations.**

$$x = 0 \quad \text{or} \quad 3x - 2 = 0$$
$$3x = 2$$
$$x = \frac{2}{3}$$

Step 5. **Check the solutions in the original equation.**

STUDY TIP

Avoid dividing both sides of $3x^2 = 2x$ by x. You will obtain $3x = 2$ and, consequently, $x = \frac{2}{3}$. The other solution, 0, is lost. We can divide both sides of an equation by any *nonzero* real number. If x is zero, we lose the second solution.

Check 0:

$$3x^2 = 2x$$
$$3 \cdot 0^2 \stackrel{?}{=} 2 \cdot 0$$
$$0 = 0, \quad \text{true}$$

Check $\frac{2}{3}$:

$$3x^2 = 2x$$
$$3\left(\frac{2}{3}\right)^2 \stackrel{?}{=} 2\left(\frac{2}{3}\right)$$
$$3\left(\frac{4}{9}\right) \stackrel{?}{=} 2\left(\frac{2}{3}\right)$$
$$\frac{4}{3} = \frac{4}{3}, \quad \text{true}$$

The solutions are 0 and $\frac{2}{3}$, and the solution set is $\left\{0, \frac{2}{3}\right\}$.

✓ **CHECK POINT 3** Solve: $4x^2 = 2x$.   0 and $\frac{1}{2}$, or $\left\{0, \frac{1}{2}\right\}$

EXAMPLE 4 Solving a Quadratic Equation by Factoring

Solve: $x^2 = 6x - 9$.

SOLUTION

Step 1. **Move all terms to one side and obtain zero on the other side.** To obtain zero on the right, we subtract $6x$ and add 9 on both sides.

$$x^2 - 6x + 9 = 6x - 6x - 9 + 9$$
$$x^2 - 6x + 9 = 0$$

Step 2. **Factor.** The trinomial on the left side is a perfect square trinomial: $x^2 - 6x + 9 = x^2 - 2 \cdot x \cdot 3 + 3^2$. We factor using $A^2 - 2AB + B^2 = (A - B)^2$: $A = x$ and $B = 3$.

$$x^2 - 6x + 9 = 0$$
$$(x - 3)^2 = 0$$

Steps 3 and 4. **Set each factor equal to zero and solve the resulting equations.** Because both factors are the same, it is only necessary to set one of them equal to zero.

$$x - 3 = 0$$
$$x = 3$$

Step 5. **Check the solution in the original equation.**

Check 3:

$$x^2 = 6x - 9$$
$$3^2 \stackrel{?}{=} 6 \cdot 3 - 9$$
$$9 \stackrel{?}{=} 18 - 9$$
$$9 = 9, \quad \text{true}$$

The solution is 3 and the solution set is $\{3\}$. ∎

✔ **CHECK POINT 4** Solve: $x^2 = 10x - 25$. 5 or $\{5\}$

EXAMPLE 5 Solving a Quadratic Equation by Factoring

Solve: $9x^2 = 16$.

SOLUTION

Step 1. **Move all terms to one side and obtain zero on the other side.** Subtract 16 from both sides and write the equation in standard form.

$$9x^2 - 16 = 16 - 16$$
$$9x^2 - 16 = 0$$

Step 2. **Factor.** The binomial on the left side is the difference of two squares: $9x^2 - 16 = (3x)^2 - 4^2$. We factor using $A^2 - B^2 = (A + B)(A - B)$: $A = 3x$ and $B = 4$.

$$9x^2 - 16 = 0$$
$$(3x + 4)(3x - 4) = 0$$

Steps 3 and 4. Set each factor equal to zero and solve the resulting equations. We use the zero-product principle to solve $(3x + 4)(3x - 4) = 0$.

$$3x + 4 = 0 \quad \text{or} \quad 3x - 4 = 0$$
$$3x = -4 \qquad\qquad 3x = 4$$
$$x = -\frac{4}{3} \qquad\qquad x = \frac{4}{3}$$

Step 5. Check the solutions in the original equation. Do this now and verify that the solutions of $9x^2 = 16$ are $-\frac{4}{3}$ and $\frac{4}{3}$. The equation's solution set is $\left\{-\frac{4}{3}, \frac{4}{3}\right\}$. ∎

✔ **CHECK POINT 5** Solve: $16x^2 = 25$. $-\frac{5}{4}$ and $\frac{5}{4}$, or $\left\{-\frac{5}{4}, \frac{5}{4}\right\}$

EXAMPLE 6   Solving a Quadratic Equation by Factoring

Solve: $(x - 2)(x + 3) = 6$.

SOLUTION

Step 1. Move all terms to one side and obtain zero on the other side. We write the equation in standard form by multiplying out the product on the left side and then subtracting 6 from both sides.

$(x - 2)(x + 3) = 6$	This is the given equation.
$x^2 + 3x - 2x - 6 = 6$	Use the FOIL method.
$x^2 + x - 6 = 6$	Simplify.
$x^2 + x - 6 - 6 = 6 - 6$	Subtract 6 from both sides.
$x^2 + x - 12 = 0$	Simplify.

Step 2. Factor.

$$x^2 + x - 12 = 0$$
$$(x + 4)(x - 3) = 0$$

Steps 3 and 4. Set each factor equal to zero and solve the resulting equations.

$$x + 4 = 0 \quad \text{or} \quad x - 3 = 0$$
$$x = -4 \qquad\qquad x = 3$$

Step 5. Check the solutions in the original equation. Do this now and verify that the solutions are -4 and 3. The equation's solution set is $\{-4, 3\}$. ∎

✔ **CHECK POINT 6** Solve: $(x - 5)(x - 2) = 28$. -2 and 9, or $\{-2, 9\}$

---

3 Solve problems with quadratic equations.

Applications of Quadratic Equations Solving quadratic equations by factoring can be used to answer questions about variables contained in mathematical models.

EXAMPLE 7   Modeling Motion

You throw a ball straight up from a rooftop 160 feet high with an initial speed of 48 feet per second. The formula

$$h = -16t^2 + 48t + 160$$

describes the ball's height above the ground, h, in feet, t seconds after you throw it. The ball misses the rooftop on its way down and eventually strikes the ground. The situation is illustrated in Figure 6.1. How long will it take for the ball to hit the ground?

SOLUTION The ball hits the ground when h, its height above the ground, is 0 feet. Thus, we substitute 0 for h in the given formula and solve for t.

FIGURE 6.1

$h = -16t^2 + 48t + 160$	This is the formula that models the ball's height.
$0 = -16t^2 + 48t + 160$	Substitute 0 for h.

It is easier to factor a trinomial with a positive leading coefficient. Thus, if a negative squared term appears in a quadratic equation, we make it positive by multiplying both sides of the equation by -1.

$$-1 \cdot 0 = -1(-16t^2 + 48t + 160)$$
$$0 = 16t^2 - 48t - 160$$

Do you see that each term on the right side of the equation changed sign? The left side of the equation remained zero. Now we continue to solve the equation.

$$16t^2 - 48t - 160 = 0 \qquad \text{Reverse the two sides of the equation. This step is optional.}$$

$$16(t^2 - 3t - 10) = 0 \qquad \text{Factor out the GCF, 16.}$$

Do not set the constant, 16, equal to zero: $16 \neq 0$.
$$16(t - 5)(t + 2) = 0 \qquad \text{Factor the trinomial.}$$

$$t - 5 = 0 \quad \text{or} \quad t + 2 = 0 \qquad \text{Set each variable factor equal to 0.}$$

$$t = 5 \qquad\qquad t = -2 \qquad \text{Solve for } t.$$

Because we begin describing the ball's height at $t = 0$, we discard the solution $t = -2$. The ball hits the ground after 5 seconds. ∎

Figure 6.2 shows the graph of the formula $h = -16t^2 + 48t + 160$. The horizontal axis is labeled t, for the ball's time in motion. The vertical axis is labeled h, for the ball's height above the ground. Because time and height are both positive, the model is graphed in quadrant I only.

FIGURE 6.2

The graph visually shows what we discovered algebraically: The ball hits the ground after 5 seconds. The graph also reveals that the ball reaches its maximum height, nearly 200 feet, after 1.5 seconds. Then the ball begins to fall.

 CHECK POINT 7 Use the formula $h = -16t^2 + 48t + 160$ to determine when the ball's height is 192 feet. Identify your solutions as points on the graph in Figure 6.2.   1 second and 2 seconds; (1, 192) and (2, 192)

In our next example, we use our five-step strategy for solving word problems.

EXAMPLE 8 Solving a Problem About a Rectangle's Area

An architect is allowed no more than 15 square meters to add a small bedroom to a house. Because of the room's design in relationship to the existing structure, the width of its rectangular floor must be 7 meters less than two times the length. Find the precise length and width of the rectangular floor of maximum area that the architect is permitted.

SOLUTION

Step 1. **Let x represent one of the quantities.** We know something about the width: It must be 7 meters less than two times the length. We will let

$$x = \text{the length of the floor.}$$

Step 2. **Represent other quantities in terms of x.** Because the width must be 7 meters less than two times the length, let

$$2x - 7 = \text{the width of the floor.}$$

The problem is illustrated in Figure 6.3.

Current house

x
$2x - 7$

Bedroom addition

FIGURE 6.3

Step 3. **Write an equation that describes the conditions.** Because the architect is allowed no more than 15 square meters, an area of 15 square meters is the maximum area permitted. The area of a rectangle is the product of its length and its width.

Length of the floor	times	Width of the floor	is	the area.
x	$\cdot$	$(2x - 7)$	$=$	15

Step 4. **Solve the equation and answer the question.**

$$x(2x - 7) = 15 \qquad \text{This is the equation for the problem's conditions.}$$

$$2x^2 - 7x = 15 \qquad \text{Use the distributive property.}$$

$$2x^2 - 7x - 15 = 0 \qquad \text{Subtract 15 from both sides.}$$

$$(2x + 3)(x - 5) = 0 \qquad \text{Factor.}$$

$$2x + 3 = 0 \quad \text{or} \quad x - 5 = 0 \qquad \text{Set each factor equal to zero.}$$

$$2x = -3 \qquad \qquad x = 5 \qquad \text{Solve the resulting equations.}$$

$$x = -\frac{3}{2}$$

A rectangle cannot have a negative length. Thus,

$$\text{Length} = x = 5$$
$$\text{Width} = 2x - 7 = 2 \cdot 5 - 7 = 10 - 7 = 3.$$

The architect is permitted a room of maximum area whose length is 5 meters and whose width is 3 meters.

Step 5. **Check the proposed solution in the original wording of the problem.** The area of the floor using the dimensions that we found is

$$A = lw = (5 \text{ meters})(3 \text{ meters}) = 15 \text{ square meters.}$$

Because the problem's wording tells us that the maximum area permitted is 15 square meters, our dimensions are correct. ∎

 CHECK POINT 8 The length of a rectangular sign is 3 feet longer than the width. If the sign's area is 54 square feet, find its length and width.   length is 9 feet; width is 6 feet

---

6.6 EXERCISE SET

Student Solutions Manual   CD/Video   PH Math/Tutor Center   MathXL Tutorials on CD   MathXL®   MyMathLab   Interactmath.com

Practice Exercises

In Exercises 1–8, solve each equation using the zero-product principle.

1. $x(x + 7) = 0$
$\{-7, 0\}$

2. $x(x - 3) = 0$
$\{0, 3\}$

3. $(x - 6)(x + 4) = 0$
$\{-4, 6\}$

4. $(x - 3)(x + 8) = 0$
$\{-8, 3\}$

5. $(x - 9)(5x + 4) = 0$
$\left\{-\frac{4}{5}, 9\right\}$

6. $(x + 7)(3x - 2) = 0$
$\left\{-7, \frac{2}{3}\right\}$

7. $10(x - 4)(2x + 9) = 0$
$\left\{-\frac{9}{2}, 4\right\}$

8. $8(x - 5)(3x + 11) = 0$
$\left\{-\frac{11}{3}, 5\right\}$

In Exercises 9–56, use factoring to solve each quadratic equation. Check by substitution or by using a graphing utility and identifying x-intercepts.

9. $x^2 + 8x + 15 = 0$
$\{-5, -3\}$

10. $x^2 + 5x + 6 = 0$
$\{-3, -2\}$

11. $x^2 - 2x - 15 = 0$
$\{-3, 5\}$

12. $x^2 + x - 42 = 0$
$\{-7, 6\}$

13. $x^2 - 4x = 21$ $\{-3, 7\}$

14. $x^2 + 7x = 18$ $\{-9, 2\}$

15. $x^2 + 9x = -8$
$\{-8, -1\}$

16. $x^2 - 11x = -10$
$\{1, 10\}$

17. $x^2 + 4x = 0$ $\{-4, 0\}$

18. $x^2 - 6x = 0$ $\{0, 6\}$

19. $x^2 - 5x = 0$ $\{0, 5\}$

20. $x^2 + 3x = 0$ $\{-3, 0\}$

21. $x^2 = 4x$ $\{0, 4\}$

22. $x^2 = 8x$ $\{0, 8\}$

23. $2x^2 = 5x$ $\left\{0, \frac{5}{2}\right\}$

24. $3x^2 = 5x$ $\left\{0, \frac{5}{3}\right\}$

25. $3x^2 = -5x$ $\left\{-\frac{5}{3}, 0\right\}$

26. $2x^2 = -3x$ $\left\{-\frac{3}{2}, 0\right\}$

27. $x^2 + 4x + 4 = 0$ $\{-2\}$

28. $x^2 + 6x + 9 = 0$ $\{-3\}$

29. $x^2 = 12x - 36$ $\{6\}$

30. $x^2 = 14x - 49$ $\{7\}$

31. $4x^2 = 12x - 9$ $\left\{\frac{3}{2}\right\}$

32. $9x^2 = 30x - 25$ $\left\{\frac{5}{3}\right\}$

33. $2x^2 = 7x + 4$ $\left\{-\frac{1}{2}, 4\right\}$

34. $3x^2 = x + 4$ $\left\{-1, \frac{4}{3}\right\}$

35. $5x^2 = 18 - x$ $\left\{-2, \frac{9}{5}\right\}$

36. $3x^2 = 15 + 4x$ $\left\{-\frac{5}{3}, 3\right\}$

37. $x^2 - 49 = 0$ $\{-7, 7\}$

38. $x^2 - 25 = 0$ $\{-5, 5\}$

39. $4x^2 - 25 = 0$ $\left\{-\frac{5}{2}, \frac{5}{2}\right\}$

40. $9x^2 - 100 = 0$ $\left\{-\frac{10}{3}, \frac{10}{3}\right\}$

41. $81x^2 = 25$ $\left\{-\frac{5}{9}, \frac{5}{9}\right\}$

42. $25x^2 = 49$ $\left\{-\frac{7}{5}, \frac{7}{5}\right\}$

43. $x(x - 4) = 21$ $\{-3, 7\}$

44. $x(x - 3) = 18$ $\{-3, 6\}$

45. $4x(x + 1) = 15$ $\left\{-\frac{5}{2}, \frac{3}{2}\right\}$

46. $x(3x + 8) = -5$ $\left\{-\frac{5}{3}, -1\right\}$

47. $(x - 1)(x + 4) = 14$ $\{-6, 3\}$

48. $(x - 3)(x + 8) = -30$ $\{-3, -2\}$

49. $(x + 1)(2x + 5) = -1$ $\left\{-2, -\frac{3}{2}\right\}$

50. $(x + 3)(3x + 5) = 7$ $\left\{-4, -\frac{2}{3}\right\}$

51. $y(y + 8) = 16(y - 1)$ $\{4\}$

52. $y(y + 9) = 4(2y + 5)$ $\{-5, 4\}$

53. $4y^2 + 20y + 25 = 0$ $\left\{-\frac{5}{2}\right\}$

54. $4y^2 + 44y + 121 = 0$ $\left\{-\frac{11}{2}\right\}$

55. $64w^2 = 48w - 9$ $\left\{\frac{3}{8}\right\}$ **56.** $25w^2 = 80w - 64$ $\left\{\frac{8}{5}\right\}$

Practice Plus

In Exercises 57–66, solve each equation and check your solutions.

57. $(x - 4)(x^2 + 5x + 6) = 0$ $\{-3, -2, 4\}$

58. $(x - 5)(x^2 - 3x + 2) = 0$ $\{1, 2, 5\}$

59. $x^3 - 36x = 0$ $\{0, -6, 6\}$

60. $x^3 - 4x = 0$ $\{0, -2, 2\}$

61. $y^3 + 3y^2 + 2y = 0$ $\{-2, -1, 0\}$

62. $y^3 + 2y^2 - 3y = 0$ $\{-3, 0, 1\}$

63. $2(x - 4)^2 + x^2 = x(x + 50) - 46x$ $\{2, 8\}$

64. $(x - 4)(x - 5) + (2x + 3)(x - 1) = x(2x - 25) - 13$
$\{-15, -2\}$

65. $(x - 2)^2 - 5(x - 2) + 6 = 0$ $\{4, 5\}$

66. $(x - 3)^2 + 2(x - 3) - 8 = 0$ $\{-1, 5\}$

Application Exercises

A ball is thrown straight up from a rooftop 300 feet high. The formula

$$h = -16t^2 + 20t + 300$$

describes the ball's height above the ground, h, in feet, t seconds after it was thrown. The ball misses the rooftop on its way down and eventually strikes the ground. The graph of the formula is shown, with tick marks omitted along the horizontal axis. Use the formula to solve Exercises 67–69.

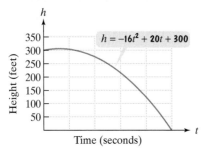

67. How long will it take for the ball to hit the ground? Use this information to provide tick marks with appropriate numbers along the horizontal axis in the figure shown.   5 seconds; Each tick represents 1 second.

68. When will the ball's height be 304 feet? Identify the solution as a point on the graph.   0.25 second and 1 second; (0.25, 304) and (1, 304)

69. When will the ball's height be 276 feet? Identify the solution as a point on the graph.   2 seconds; (2, 276)

An explosion causes debris to rise vertically with an initial speed of 72 feet per second. The formula

$$h = -16t^2 + 72t$$

describes the height of the debris above the ground, h, in feet, t seconds after the explosion. Use this information to solve Exercises 70–71.

70. How long will it take for the debris to hit the ground? $\frac{9}{2}$ seconds or 4.5 seconds

71. When will the debris be 32 feet above the ground? $\frac{1}{2}$ second and 4 seconds

The formula

$$N = 2x^2 + 22x + 320$$

models the number of inmates, N, in thousands, in U.S. state and federal prisons x years after 1980. The graph of the formula is shown in a $[0, 20, 1]$ by $[0, 1600, 100]$ viewing rectangle. Use the formula to solve Exercises 72–73.

$[0, 20, 1]$ by $[0, 1600, 100]$

72. In which year were there 740 thousand inmates in U.S. state and federal prisons? Identify the solution as a point on the graph shown.   1990; (10, 740)

73. In which year were there 1100 thousand inmates in U.S. state and federal prisons? Identify the solution as a point on the graph shown.   1995; (15, 1100)

The alligator, an endangered species, is the subject of a protection program. The formula

$$P = -10x^2 + 475x + 3500$$

models the alligator population, P, after x years of the protection program, where $0 \le x \le 12$. Use the formula to solve Exercises 74–75.

74. After how long is the population up to 5990?   6 years

75. After how long is the population up to 7250?   10 years

The graph of the alligator population is shown over time. Use the graph to solve Exercises 76–77.

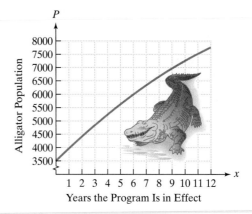

76. Identify your solution in Exercise 74 as a point on the graph.   (6, 5990)

77. Identify your solution in Exercise 75 as a point on the graph.   (10, 7250)

The formula

$$N = \frac{t^2 - t}{2}$$

describes the number of football games, N, that must be played in a league with t teams if each team is to play every other team once. Use this information to solve Exercises 78–79.

78. If a league has 36 games scheduled, how many teams belong to the league, assuming that each team plays every other team once?   9 teams

79. If a league has 45 games scheduled, how many teams belong to the league, assuming that each team plays every other team once?   10 teams

80. The length of a rectangular garden is 5 feet greater than the width. The area of the rectangle is 300 square feet. Find the length and the width.   Length is 20 feet; width is 15 feet.

81. A rectangular parking lot has a length that is 3 yards greater than the width. The area of the parking lot is 180 square yards. Find the length and the width.   Length is 15 yards; width is 12 yards.

82. Each end of a glass prism is a triangle with a height that is 1 inch shorter than twice the base. If the area of the triangle is 60 square inches, how long are the base and height?   Base is 8 inches; height is 15 inches.

83. Great white sharks have triangular teeth with a height that is 1 centimeter longer than the base. If the area of one tooth is 15 square centimeters, find its base and height.   Base is 5 centimeters; height is 6 centimeters.

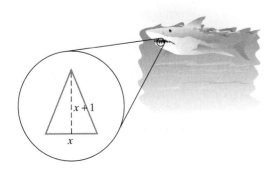

84. A vacant rectangular lot is being turned into a community vegetable garden measuring 15 meters by 12 meters. A path of uniform width is to surround the garden. If the area of the lot is 378 square meters, find the width of the path surrounding the garden.   3 meters

85. As part of a landscaping project, you put in a flower bed measuring 10 feet by 12 feet. You plan to surround the bed with a uniform border of low-growing plants.

a. Write a polynomial that describes the area of the uniform border that surrounds your flower bed. (*Hint:* The area of the border is the area of the large rectangle shown in the figure minus the area of the flower bed.) $4x^2 + 44x$

b. The low-growing plants surrounding the flower bed require 1 square foot each when mature. If you have 168 of these plants, how wide a strip around the flower bed should you prepare for the border?   3 feet

Writing in Mathematics

86. What is a quadratic equation?   86–88. Answers will vary.

87. Explain how to solve $x^2 + 6x + 8 = 0$ using factoring and the zero-product principle.

88. If $(x + 2)(x - 4) = 0$ indicates that $x + 2 = 0$ or $x - 4 = 0$, explain why $(x + 2)(x - 4) = 6$ does not mean $x + 2 = 6$ or $x - 4 = 6$. Could we solve the equation using $x + 2 = 3$ and $x - 4 = 2$ because $3 \cdot 2 = 6$?

Critical Thinking Exercises

89. Which one of the following is true?   **d**

    a. If $(x + 3)(x - 4) = 2$, then $x + 3 = 0$ or $x - 4 = 0$.

    b. The solutions of the equation $4(x - 5)(x + 3) = 0$ are $4, 5$, and -3.

    c. Equations solved by factoring always have two different solutions.

    d. Both 0 and $-\pi$ are solutions of the equation $x(x + \pi) = 0$.

90. Write a quadratic equation in standard form whose solutions are -3 and 5. $x^2 - 2x - 15 = 0$

In Exercises 91–93, solve each equation.

91. $x^3 - x^2 - 16x + 16 = 0$ $-4, 1,$ and 4

92. $3^{x^2 - 9x + 20} = 1$ 4 and 5 **93.** $(x^2 - 5x + 5)^3 = 1$ 4 and 1

In Exercises 94–97, match each equation with its graph. The graphs are labeled (a) through (d).

94. $y = x^2 - x - 2$ **c**   **95.** $y = x^2 + x - 2$ **a**

96. $y = x^2 - 4$ **d**   **97.** $y = x^2 - 4x$ **b**

a.

b.

c.

d.

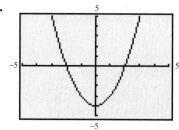

Technology Exercises

In Exercises 98–102, use the x-intercepts for the graph in a $[-10, 10, 1]$ by $[-13, 10, 1]$ viewing rectangle to solve the quadratic equation. Check by substitution.

98. Use the graph of $y = x^2 + 3x - 4$ to solve

$$x^2 + 3x - 4 = 0.$$ -4 and 1

99. Use the graph of $y = x^2 + x - 6$ to solve

$$x^2 + x - 6 = 0.$$ -3 and 2

100. Use the graph of $y = (x - 2)(x + 3) - 6$ to solve

$$(x - 2)(x + 3) - 6 = 0.$$ -4 and 3

101. Use the graph of $y = x^2 - 2x + 1$ to solve

$$x^2 - 2x + 1 = 0.$$ 1

102. Use the technique of identifying x-intercepts on a graph generated by a graphing utility to check any five equations that you solved in Exercises 9–56.   Answers will vary.

103. If you have access to a calculator that solves quadratic equations, consult the owner's manual to determine how to use this feature. Then use your calculator to solve any five of the equations in Exercises 9–56.   Answers will vary.

Review Exercises

104. Solve:

$$\begin{aligned} 4x + 2y - z &= 12 \\ 3x - y + z &= 6 \\ x + 3z &= 14 \end{aligned}$$

(Section 4.5, Example 3) $\{(2, 4, 4)\}$

105. Simplify: $\left(\dfrac{8x^4}{4x^7}\right)^2$. (Section 5.7, Example 6) $\dfrac{4}{x^6}$

106. Solve: $5x + 28 = 6 - 6x$. (Section 2.2, Example 7) $\{-2\}$

GROUP PROJECT

6

Additional group projects can be found in the *Instructor's Resource Manual*.

Group members are on the board of a condominium association. The condominium has just installed a small 35-foot-by-30-foot pool. Your job is to choose a material to surround the pool to create a border of uniform width.

a. Begin by writing an algebraic expression for the area, in square feet, of the border around the pool. (*Hint*: The border's area is the combined area of the pool and border minus the area of the pool.)

b. You must select one of the following options for the border.

Options for the Border	Price
Cement	$6 per square foot
Outdoor carpeting	$5 per square foot plus $10 per foot to install edging around the rectangular border
Brick	$8 per square foot plus a $60 charge for delivering the bricks

Write an algebraic expression for the cost of installing the border for each of these options.

c. You would like the border to be 5 feet wide. Use the algebraic expressions in part (b) to find the cost of the border for each of the three options.

d. You would prefer not to use cement. However, the condominium association is limited by a $5000 budget. Given this limitation, approximately how wide can the border be using outdoor carpeting or brick? Which option should you select and why?

CHAPTER 6 SUMMARY

Definitions and Concepts	Examples

Section 6.1 The Greatest Common Factor and Factoring by Grouping

Factoring a polynomial containing the sum of monomials means finding an equivalent expression that is a product. The greatest common factor, GCF, is an expression that divides every term of the polynomial. The variable part of the GCF contains the smallest power of a variable that appears in all terms of the polynomial.	Find the GCF of $16x^2y$, $20x^3y^2$, and $8x^2y^3$. The GCF of 16, 20, and 8 is 4. The GCF of x^2, x^3, and x^2 is x^2. The GCF of y, y^2, and y^3 is y. $$\text{GCF} = 4 \cdot x^2 \cdot y = 4x^2y$$
To factor a monomial from a polynomial, express each term as the product of the GCF and its other factor. Then use the distributive property to factor out the GCF.	$$16x^2y + 20x^3y^2 + 8x^2y^3$$ $$= 4x^2y \cdot 4 + 4x^2y \cdot 5xy + 4x^2y \cdot 2y^2$$ $$= 4x^2y(4 + 5xy + 2y^2)$$
To factor by grouping, factor out the GCF from each group. Then factor out the remaining common factor.	$$xy + 5x - 3y - 15$$ $$= x(y + 5) - 3(y + 5)$$ $$= (y + 5)(x - 3)$$

Definitions and Concepts	Examples

Section 6.2 Factoring Trinomials Whose Leading Coefficient Is One

To factor a trinomial of the form $x^2 + bx + c$, find two numbers whose product is c and whose sum is b. The factorization is $$(x + \text{one number})(x + \text{other number}).$$	Factor: $x^2 + 9x + 20$. Find two numbers whose product is 20 and whose sum is 9. The numbers are 4 and 5. $$x^2 + 9x + 20 = (x + 4)(x + 5)$$

Section 6.3 Factoring Trinomials Whose Leading Coefficient Is Not One

To factor $ax^2 + bx + c$ by trial and error, try various combinations of factors of ax^2 and c until a middle term of bx is obtained for the sum of outside and inside products.	Factor: $3x^2 + 7x - 6$. Factors of $3x^2$: $3x, x$ Factors of -6: 1 and -6, -1 and 6, $\qquad\qquad\qquad$ 2 and -3, -2 and 3. A possible combination of these factors is $$(3x - 2)(x + 3).$$ Sum of outside and inside products should equal $7x$. $$9x - 2x = 7x$$ Thus, $3x^2 + 7x - 6 = (3x - 2)(x + 3)$.
To factor $ax^2 + bx + c$ by grouping, find the factors of ac whose sum is b. Write bx using these factors. Then factor by grouping.	Factor: $3x^2 + 7x - 6$. Find the factors of $3(-6)$, or -18, whose sum is 7. They are 9 and -2. $$3x^2 + 7x - 6$$ $$= 3x^2 + 9x - 2x - 6$$ $$= 3x(x + 3) - 2(x + 3) = (x + 3)(3x - 2)$$

Section 6.4 Factoring Special Forms

The Difference of Two Squares $$A^2 - B^2 = (A + B)(A - B)$$	$9x^2 - 25y^2$ $$= (3x)^2 - (5y)^2 = (3x + 5y)(3x - 5y)$$
Perfect Square Trinomials $$A^2 + 2AB + B^2 = (A + B)^2$$ $$A^2 - 2AB + B^2 = (A - B)^2$$	$x^2 + 16x + 64 = x^2 + 2 \cdot x \cdot 8 + 8^2 = (x + 8)^2$ $25x^2 - 30x + 9 = (5x)^2 - 2 \cdot 5x \cdot 3 + 3^2 = (5x - 3)^2$
Sum and Difference of Cubes $$A^3 + B^3 = (A + B)(A^2 - AB + B^2)$$ $$A^3 - B^3 = (A - B)(A^2 + AB + B^2)$$	$8x^3 - 125 = (2x)^3 - 5^3$ $$= (2x - 5)[(2x)^2 + 2x \cdot 5 + 5^2]$$ $$= (2x - 5)(4x^2 + 10x + 25)$$

Definitions and Concepts	Examples

Section 6.5 A General Factoring Strategy

A Factoring Strategy

1. Factor out the GCF.

2. a. If two terms, try

$$A^2 - B^2 = (A + B)(A - B)$$
$$A^3 + B^3 = (A + B)(A^2 - AB + B^2)$$
$$A^3 - B^3 = (A - B)(A^2 + AB + B^2).$$

  b. If three terms, try

$$A^2 + 2AB + B^2 = (A + B)^2$$
$$A^2 - 2AB + B^2 = (A - B)^2.$$

  If not a perfect square trinomial, try trial and error or grouping.

  c. If four terms, try factoring by grouping.

3. See if any factors can be factored further.

4. Check by multiplying.

Factor: $2x^4 + 10x^3 - 8x^2 - 40x$.
The GCF is $2x$.

$$2x^4 + 10x^3 - 8x^2 - 40x$$
$$= 2x(x^3 + 5x^2 - 4x - 20)$$

> Four terms: Try grouping.

$$= 2x[x^2(x + 5) - 4(x + 5)]$$
$$= 2x(x + 5)(x^2 - 4)$$

> This can be factored further.

$$= 2x(x + 5)(x + 2)(x - 2)$$

Section 6.6 Solving Quadratic Equations by Factoring

The Zero-Product Principle
If $AB = 0$, then $A = 0$ or $B = 0$.

Solve: $(x - 6)(x + 10) = 0$
$$x - 6 = 0 \quad \text{or} \quad x + 10 = 0$$
$$x = 6 \qquad\qquad x = -10$$
The solutions are -10 and 6, and the solution set is $\{-10, 6\}$.

A quadratic equation in x is an equation that can be written in the standard form

$$ax^2 + bx + c = 0, \quad a \neq 0.$$

To solve by factoring, write the equation in standard form, factor, set each factor equal to zero, and solve each resulting equation. Check proposed solutions in the original equation.

Solve: $4x^2 + 9x = 9$.
$$4x^2 + 9x - 9 = 0$$
$$(4x - 3)(x + 3) = 0$$
$$4x - 3 = 0 \quad \text{or} \quad x + 3 = 0$$
$$x = \frac{3}{4} \qquad\qquad x = -3$$
The solutions are -3 and $\frac{3}{4}$, and the solution set is $\left\{-3, \frac{3}{4}\right\}$.

CHAPTER 6 REVIEW EXERCISES

6.1 *In Exercises 1–5, factor each polynomial by using the greatest common factor. If there is no common factor other than 1 and the polynomial cannot be factored, so state.*

1. $30x - 45$ $15(2x - 3)$

2. $12x^3 + 16x^2 - 400x$ $4x(3x^2 + 4x - 100)$

3. $30x^4y + 15x^3y + 5x^2y$ $5x^2y(6x^2 + 3x + 1)$

4. $7(x + 3) - 2(x + 3)$ $5(x + 3)$

5. $7x^2(x + y) - (x + y)$ $(7x^2 - 1)(x + y)$

In Exercises 6–9, factor by grouping.

6. $x^3 + 3x^2 + 2x + 6$ $(x^2 + 2)(x + 3)$

7. $xy + y + 4x + 4$ $(x + 1)(y + 4)$

8. $x^3 + 5x + x^2 + 5$ $(x^2 + 5)(x + 1)$

9. $xy + 4x - 2y - 8$ $(x - 2)(y + 4)$

6.2 *In Exercises 10–17, factor completely, or state that the trinomial is prime.*

10. $x^2 - 3x + 2$ $(x - 2)(x - 1)$

11. $x^2 - x - 20$ $(x - 5)(x + 4)$

12. $x^2 + 19x + 48$ $(x + 3)(x + 16)$

13. $x^2 - 6xy + 8y^2$ $(x - 4y)(x - 2y)$

14. $x^2 + 5x - 9$ prime

15. $x^2 + 16xy - 17y^2$ $(x + 17y)(x - y)$

16. $3x^2 + 6x - 24$ $3(x + 4)(x - 2)$

17. $3x^3 - 36x^2 + 33x$ $3x(x - 11)(x - 1)$

6.3 *In Exercises 18–26, factor completely, or state that the trinomial is prime.*

18. $3x^2 + 17x + 10$ $(x + 5)(3x + 2)$

19. $5y^2 - 17y + 6$ $(y - 3)(5y - 2)$

20. $4x^2 + 4x - 15$ $(2x + 5)(2x - 3)$

21. $5y^2 + 11y + 4$ prime

22. $8x^2 + 8x - 6$ $2(2x + 3)(2x - 1)$

23. $2x^3 + 7x^2 - 72x$ $x(2x - 9)(x + 8)$

24. $12y^3 + 28y^2 + 8y$ $4y(3y + 1)(y + 2)$

25. $2x^2 - 7xy + 3y^2$ $(2x - y)(x - 3y)$

26. $5x^2 - 6xy - 8y^2$ $(5x + 4y)(x - 2y)$

6.4 *In Exercises 27–30, factor each difference of two squares completely.*

27. $4x^2 - 1$ $(2x + 1)(2x - 1)$

28. $81 - 100y^2$ $(9 + 10y)(9 - 10y)$

29. $25a^2 - 49b^2$ $(5a + 7b)(5a - 7b)$

30. $z^4 - 16$ $(z^2 + 4)(z + 2)(z - 2)$

In Exercises 31–34, factor completely, or state that the polynomial is prime.

31. $2x^2 - 18$ $2(x + 3)(x - 3)$

32. $x^2 + 1$ prime

33. $9x^3 - x$ $x(3x + 1)(3x - 1)$

34. $18xy^2 - 8x$ $2x(3y + 2)(3y - 2)$

In Exercises 35–41, factor any perfect square trinomials, or state that the polynomial is prime.

35. $x^2 + 22x + 121$ $(x + 11)^2$

36. $x^2 - 16x + 64$ $(x - 8)^2$

37. $9y^2 + 48y + 64$ $(3y + 8)^2$

38. $16x^2 - 40x + 25$ $(4x - 5)^2$

39. $25x^2 + 15x + 9$ prime

40. $36x^2 + 60xy + 25y^2$ $(6x + 5y)^2$

41. $25x^2 - 40xy + 16y^2$ $(5x - 4y)^2$

In Exercises 42–45, factor using the formula for the sum or difference of two cubes.

42. $x^3 - 27$ $(x - 3)(x^2 + 3x + 9)$

43. $64x^3 + 1$ $(4x + 1)(16x^2 - 4x + 1)$

44. $54x^3 - 16y^3$ $2(3x - 2y)(9x^2 + 6xy + 4y^2)$

45. $27x^3y + 8y$ $y(3x + 2)(9x^2 - 6x + 4)$

In Exercises 46–47, find the formula for the area of the shaded region and express it in factored form.

46.

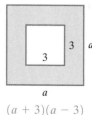

$(a + 3)(a - 3)$

47.

$(a + 2b)(a - 2b)$

48. The figure shows a geometric interpretation of a factorization. Use the sum of the areas of the four pieces on the left and the area of the square on the right to write the factorization that is illustrated. $A^2 + 2A + 1 = (A + 1)^2$

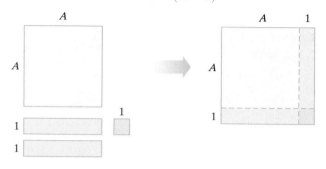

6.5 *In Exercises 49–81, factor completely, or state that the polynomial is prime.*

49. $x^3 - 8x^2 + 7x$ $x(x - 7)(x - 1)$

50. $10y^2 + 9y + 2$ $(5y + 2)(2y + 1)$

51. $128 - 2y^2$ $2(8 + y)(8 - y)$

52. $9x^2 + 6x + 1$ $(3x + 1)^2$

53. $20x^7 - 36x^3$ $4x^3(5x^4 - 9)$

54. $x^3 - 3x^2 - 9x + 27$ $(x - 3)^2(x + 3)$

55. $y^2 + 16$ prime

56. $2x^3 + 19x^2 + 35x$ $x(2x + 5)(x + 7)$

57. $3x^3 - 30x^2 + 75x$ $3x(x - 5)^2$

58. $3x^5 - 24x^2$ $3x^2(x - 2)(x^2 + 2x + 4)$

59. $4y^4 - 36y^2$ $4y^2(y + 3)(y - 3)$

60. $5x^2 + 20x - 105$ $5(x + 7)(x - 3)$

61. $9x^2 + 8x - 3$ prime

62. $10x^5 - 44x^4 + 16x^3$ $2x^3(5x - 2)(x - 4)$

63. $100y^2 - 49$ $(10y + 7)(10y - 7)$

64. $9x^5 - 18x^4$ $9x^4(x - 2)$

65. $x^4 - 1$ $(x^2 + 1)(x + 1)(x - 1)$

66. $2y^3 - 16$ $2(y - 2)(y^2 + 2y + 4)$

67. $x^3 + 64$ $(x + 4)(x^2 - 4x + 16)$

68. $6x^2 + 11x - 10$ $(3x - 2)(2x + 5)$

69. $3x^4 - 12x^2$ $3x^2(x + 2)(x - 2)$

70. $x^2 - x - 90$ $(x - 10)(x + 9)$

71. $25x^2 + 25xy + 6y^2$ $(5x + 2y)(5x + 3y)$

72. $x^4 + 125x$ $x(x + 5)(x^2 - 5x + 25)$

73. $32y^3 + 32y^2 + 6y$ $2y(4y + 3)(4y + 1)$

74. $2y^2 - 16y + 32$ $2(y - 4)^2$

75. $x^2 - 2xy - 35y^2$ $(x + 5y)(x - 7y)$

76. $x^2 + 7x + xy + 7y$ $(x + y)(x + 7)$

77. $9x^2 + 24xy + 16y^2$ $(3x + 4y)^2$

78. $2x^4y - 2x^2y$ $2x^2y(x + 1)(x - 1)$

79. $100y^2 - 49z^2$ $(10y + 7z)(10y - 7z)$

80. $x^2 + xy + y^2$ prime

81. $3x^4y^2 - 12x^2y^4$ $3x^2y^2(x + 2y)(x - 2y)$

6.6 *In Exercises 82–83, solve each equation using the zero-product principle.*

82. $x(x - 12) = 0$ $\{0, 12\}$

83. $3(x - 7)(4x + 9) = 0$ $\left\{-\dfrac{9}{4}, 7\right\}$

In Exercises 84–92, use factoring to solve each quadratic equation.

84. $x^2 + 5x - 14 = 0$ $\{-7, 2\}$ **85.** $5x^2 + 20x = 0$ $\{-4, 0\}$

86. $2x^2 + 15x = 8$ $\left\{-8, \dfrac{1}{2}\right\}$ **87.** $x(x - 4) = 32$ $\{-4, 8\}$

88. $(x + 3)(x - 2) = 50$ $\{-8, 7\}$

89. $x^2 = 14x - 49$ $\{7\}$ **90.** $9x^2 = 100$ $\left\{-\dfrac{10}{3}, \dfrac{10}{3}\right\}$

91. $3x^2 + 21x + 30 = 0$ $\{-5, -2\}$

92. $3x^2 = 22x - 7$ $\left\{\dfrac{1}{3}, 7\right\}$

93. You dive from a board that is 32 feet above the water. The formula
$$h = -16t^2 + 16t + 32$$

describes your height above the water, h, in feet, t seconds after you dive. How long will it take you to hit the water?   2 seconds

94. The length of a rectangular sign is 3 feet longer than the width. If the sign has space for 40 square feet of advertising, find its length and its width.   length: 8 feet; width: 5 feet

95. The square lot shown here is being turned into a garden with a 3-meter path at one end. If the area of the garden is 88 square meters, find the dimensions of the square lot.   11 meters by 11 meters

---

![CHAPTER 6 TEST] Remember to use your Chapter Test Prep Video CD to see the worked-out solutions to the test questions you want to review.

In Exercises 1–21, factor completely, or state that the polynomial is prime.

1. $x^2 - 9x + 18$ $(x - 3)(x - 6)$

2. $x^2 - 14x + 49$ $(x - 7)^2$

3. $15y^4 - 35y^3 + 10y^2$ $5y^2(3y - 1)(y - 2)$

4. $x^3 + 2x^2 + 3x + 6$ $(x^2 + 3)(x + 2)$

5. $x^2 - 9x$ $x(x - 9)$

6. $x^3 + 6x^2 - 7x$ $x(x + 7)(x - 1)$

7. $14x^2 + 64x - 30$ $2(7x - 3)(x + 5)$

8. $25x^2 - 9$ $(5x + 3)(5x - 3)$

9. $x^3 + 8$ $(x + 2)(x^2 - 2x + 4)$

10. $x^2 - 4x - 21$ $(x + 3)(x - 7)$

11. $x^2 + 4$ prime

12. $6y^3 + 9y^2 + 3y$ $3y(2y + 1)(y + 1)$

13. $4y^2 - 36$ $4(y + 3)(y - 3)$

14. $16x^2 + 48x + 36$ $4(2x + 3)^2$

15. $2x^4 - 32$ $2(x^2 + 4)(x + 2)(x - 2)$

16. $36x^2 - 84x + 49$ $(6x - 7)^2$

17. $7x^2 - 50x + 7$ $(7x - 1)(x - 7)$

18. $x^3 + 2x^2 - 5x - 10$ $(x^2 - 5)(x + 2)$

19. $12y^3 - 12y^2 - 45y$ $3y(2y + 3)(2y - 5)$

20. $y^3 - 125$ $(y - 5)(y^2 + 5y + 25)$

21. $5x^2 - 5xy - 30y^2$ $5(x - 3y)(x + 2y)$

In Exercises 22–27, solve each quadratic equation.

22. $x^2 + 2x - 24 = 0$ -6 and 4, or $\{-6, 4\}$

23. $3x^2 - 5x = 2$ $-\dfrac{1}{3}$ and 2, or $\left\{-\dfrac{1}{3}, 2\right\}$

24. $x(x - 6) = 16$ -2 and 8, or $\{-2, 8\}$

25. $6x^2 = 21x$ 0 and $\dfrac{7}{2}$, or $\left\{0, \dfrac{7}{2}\right\}$

26. $16x^2 = 81$ $-\dfrac{9}{4}$ and $\dfrac{9}{4}$, or $\left\{-\dfrac{9}{4}, \dfrac{9}{4}\right\}$

27. $(5x + 4)(x - 1) = 2$ -1 and $\dfrac{6}{5}$, or $\left\{-1, \dfrac{6}{5}\right\}$

28. Find a formula for the area of the shaded region and express it in factored form. $x^2 - 4 = (x + 2)(x - 2)$

29. A model rocket is launched from a height of 96 feet. The formula
$$h = -16t^2 + 80t + 96$$

describes the rocket's height, h, in feet, t seconds after it was launched. How long will it take the rocket to reach the ground?   6 seconds

30. The length of a rectangular garden is 6 feet longer than its width. If the area of the garden is 55 square feet, find its length and its width.   length: 11 feet; width: 5 feet

Note to Instructor: The Chapter 6 Test file in the TestGen program provides algorithms specifically matched to these problems so you can easily replicate this chapter test for additional practice or assessment purposes.

CUMULATIVE REVIEW EXERCISES (CHAPTERS 1–6)

1. Simplify: $6[5 + 2(3 - 8) - 3]$. -48

2. Solve: $4(x - 2) = 2(x - 4) + 3x$. $\{0\}$

3. Solve: $\dfrac{x}{2} - 1 = \dfrac{x}{3} + 1$. $\{12\}$

4. Solve and express the solution set in set-builder notation. Graph the solution set on a number line. $\{x \mid x < -2\}$; See graphing answer section.
$$5 - 5x > 2(5 - x) + 1$$

5. Find the measures of the angles of a triangle whose two base angles have equal measure and whose third angle is 10° less than three times the measure of a base angle. $38°, 38°, 104°$

6. A dinner for six people cost $159, including a 6% tax. What was the dinner's cost before tax? $150

7. Graph using the slope and y-intercept: $y = -\frac{3}{5}x + 3$.   See graphing answer section.

8. Write the point-slope form of the line passing through $(2, -4)$ and $(3, 1)$. Then use the point-slope form of the equation to write the slope-intercept equation.
$y + 4 = 5(x - 2)$ or $y - 1 = 5(x - 3)$; $y = 5x - 14$

9. Solve the system:
$$\begin{aligned} x - 2y + 2z &= 4 \\ 3x - y + 4z &= 4 \\ 2x + y - 3z &= 5. \end{aligned}$$ $\{(2, -2, -1)\}$

10. Solve the system:
$$\begin{aligned} 5x + 2y &= 13 \\ y &= 2x - 7. \end{aligned}$$ $\{(3, -1)\}$

11. Solve the system:
$$\begin{aligned} 2x + 3y &= 5 \\ 3x - 2y &= -4. \end{aligned}$$ $\left\{\left(-\dfrac{2}{13}, \dfrac{23}{13}\right)\right\}$

12. Subtract: $\dfrac{4}{5} - \dfrac{9}{8}$. $-\dfrac{13}{40}$

In Exercises 13–15, perform the indicated operations.

13. $\dfrac{6x^5 - 3x^4 + 9x^2 + 27x}{3x}$ $2x^4 - x^3 + 3x + 9$

14. $(3x - 5y)(2x + 9y)$ $6x^2 + 17xy - 45y^2$

15. $\dfrac{6x^3 + 5x^2 - 34x + 13}{3x - 5}$ $2x^2 + 5x - 3 - \dfrac{2}{3x - 5}$

16. Write 0.0071 in scientific notation. 7.1×10^{-3}

In Exercises 17–19, factor completely.

17. $3x^2 + 11x + 6$ $(3x + 2)(x + 3)$

18. $y^5 - 16y$ $y(y^2 + 4)(y + 2)(y - 2)$

19. $4x^2 + 12x + 9$ $(2x + 3)^2$

20. The length of a rectangle is 2 feet greater than its width. If the rectangle's area is 24 square feet, find its dimensions.
length: 6 feet; width: 4 feet

At the start of the twenty-first century, we are plagued by questions about the environment. Will we run out of gas? How hot will it get? Will there be neighborhoods where the air is pristine? Can we make garbage disappear? Will there be any wilderness left? Which wild animals will become extinct? How much will it cost to clean up toxic wastes from our rivers so that they can safely provide food, recreation, and enjoyment of wildlife for the millions who live along and visit their shores?

The role of algebraic fractions in modeling environmental issues is introduced in Section 7.1 and developed in Example 6 in Section 7.6.

7 CHAPTER

Rational Expressions

W hen making decisions on public policies dealing with the environment, two important questions are

- What are the costs?
- What are the benefits?

Algebraic fractions play an important role in modeling the costs. By learning to work with these fractional expressions, you will gain new insights into phenomena as diverse as the dosage of drugs prescribed for children, inventory costs for a business, the cost of environmental cleanup, and even the shape of our heads.

7.1

Objectives

1 Find numbers for which a rational expression is undefined.

2 Simplify rational expressions.

3 Solve applied problems involving rational expressions.

RATIONAL EXPRESSIONS AND THEIR SIMPLIFICATION

How do we describe the costs of reducing environmental pollution? We often use algebraic expressions involving quotients of polynomials. For example, the algebraic expression

$$\frac{250x}{100 - x}$$

describes the cost, in millions of dollars, to remove x percent of the pollutants that are discharged into a river. Removing a modest percentage of pollutants, say 40%, is far less costly than removing a substantially greater percentage, such as 95%. We see this by evaluating the algebraic expression for $x = 40$ and $x = 95$.

Evaluating $\dfrac{250x}{100 - x}$ for

$x = 40$:	$x = 95$:
Cost is $\dfrac{250(40)}{100 - 40} \approx 167.$	Cost is $\dfrac{250(95)}{100 - 95} = 4750.$

The cost increases from approximately $167 million to a possibly prohibitive $4750 million, or $4.75 billion. Costs spiral upward as the percentage of removed pollutants increases.

Many algebraic expressions that describe costs of environmental projects are examples of *rational expressions*. In this section, we introduce rational expressions and their simplification.

DISCOVER FOR YOURSELF

What happens if you try substituting 100 for x in

$$\frac{250x}{100 - x}?$$

What does this tell you about the cost of cleaning up all of the river's pollutants?

1 Find numbers for which a rational expression is undefined.

Excluding Numbers from Rational Expressions A **rational expression** is the quotient of two polynomials. Some examples are

$$\frac{x - 2}{4}, \quad \frac{4}{x - 2}, \quad \frac{x}{x^2 - 1}, \quad \text{and} \quad \frac{x^2 + 1}{x^2 + 2x - 3}.$$

Rational expressions indicate division, and division by zero is undefined. This means that **we must exclude any value or values of the variable that make a denominator zero**. For example, consider the rational expression

$$\frac{4}{x - 2}.$$

When x is replaced with 2, the denominator is 0 and the expression is undefined.

$$\text{If } x = 2: \quad \frac{4}{x - 2} = \frac{4}{2 - 2} = \frac{4}{0} \quad \boxed{\begin{array}{c}\text{Division by zero}\\\text{is undefined.}\end{array}}$$

Notice that if x is replaced by a number other than 2, such as 1, the expression is defined because the denominator is nonzero.

$$\text{If } x = 1: \quad \frac{4}{x - 2} = \frac{4}{1 - 2} = \frac{4}{-1} = -4.$$

Thus, only 2 must be excluded as a replacement for x in the rational expression $\dfrac{4}{x - 2}$.

USING TECHNOLOGY

We can use the $\boxed{\text{TABLE}}$ feature of a graphing utility to verify our work with $\dfrac{4}{x - 2}$. Enter

$$y_1 = 4 \;\boxed{\div}\; \boxed{(}\;\boxed{(}\;\boxed{x}\;\boxed{-}\;\boxed{2}\;\boxed{)}$$

and press $\boxed{\text{TABLE}}$.

$$y_1 = \frac{4}{x - 2}$$

This verifies that if $x = 1$, the value of $\dfrac{4}{x - 2}$ is -4.

This verifies that 2 must be excluded as a replacment for x.

EXCLUDING VALUES FROM RATIONAL EXPRESSIONS   If a variable in a rational expression is replaced by a number that causes the denominator to be 0, that number must be excluded as a replacement for the variable. The rational expression is undefined at any value that produces a denominator of 0.

How do we determine the value or values of the variable for which a rational expression is undefined? Set the denominator equal to 0 and then solve the resulting equation for the variable.

EXAMPLE 1   Determining Numbers for Which Rational Expressions Are Undefined

Find all the numbers for which the rational expression is undefined:

a. $\dfrac{6x + 12}{7x - 28}$ **b.** $\dfrac{2x + 6}{x^2 + 3x - 10}.$

SOLUTION   In each case, we set the denominator equal to 0 and solve.

$$\frac{6x + 12}{7x - 28} \qquad \boxed{\begin{array}{c}\text{Exclude values of } x\\\text{that make these}\\\text{denominators 0.}\end{array}} \qquad \frac{2x + 6}{x^2 + 3x - 10}$$

a. $7x - 28 = 0$ Set the denominator of $\dfrac{6x + 12}{7x - 28}$ equal to 0.

$\qquad 7x = 28$ Add 28 to both sides.

$\qquad x = 4$ Divide both sides by 7.

Thus, $\dfrac{6x + 12}{7x - 28}$ is undefined for $x = 4$.

b. $x^2 + 3x - 10 = 0$ Set the denominator of $\dfrac{2x + 6}{x^2 + 3x - 10}$ equal to 0.

$\qquad (x + 5)(x - 2) = 0$ Factor.

$\qquad x + 5 = 0$ or $x - 2 = 0$ Set each factor equal to 0.

$\qquad\qquad x = -5 \qquad\qquad x = 2$ Solve the resulting equations.

Thus, $\dfrac{2x + 6}{x^2 + 3x - 10}$ is undefined for $x = -5$ and $x = 2$.

USING TECHNOLOGY

When using a graphing utility to graph an equation containing a rational expression, you might not be pleased with the quality of the display. Compare these two graphs of
$y = \dfrac{6x + 12}{7x - 28}$.

The graph on the left was obtained using the $\boxed{\text{DOT}}$ mode in a $[-3, 10, 1]$ by $[-10, 10, 1]$ viewing rectangle. Examine the behavior of the graph near $x = 4$, the number for which the rational expression is undefined. The values of the rational expression are decreasing as the values of x get closer to 4 on the left and increasing as the values of x get closer to 4 on the right. However, there is no point on the graph corresponding to $x = 4$. Would you agree that this behavior is better illustrated in the hand-drawn graph on the right?

 CHECK POINT 1 Find all the numbers for which the rational expression is undefined:

a. $\dfrac{7x - 28}{8x - 40}$ $x = 5$ **b.** $\dfrac{8x - 40}{x^2 + 3x - 28}$. $x = -7$ and $x = 4$

Is every rational expression undefined for at least one number? No. Consider

$$\dfrac{x - 2}{4}.$$

Because the denominator is not zero for any value of x, the rational expression is defined for all real numbers. Thus, it is not necessary to exclude any values for x.

2 Simplify rational expressions.

Simplifying Rational Expressions A rational expression is **simplified** if its numerator and denominator have no common factors other than 1 or −1. The following principle is used to simplify a rational expression:

FUNDAMENTAL PRINCIPLE OF RATIONAL EXPRESSIONS If P, Q, and R are polynomials, and Q and R are not 0,

$$\frac{PR}{QR} = \frac{P}{Q}.$$

As you read the Fundamental Principle, can you see why $\dfrac{PR}{QR}$ is not simplified? The numerator and denominator have a common factor, the polynomial R. By dividing the numerator and the denominator by the common factor, R, we obtain the simplified form $\dfrac{P}{Q}$. This is often shown as follows:

$$\frac{P\overset{1}{R}}{Q\underset{1}{R}} = \frac{P}{Q}.$$

Observe that

$$\frac{PR}{QR} = \frac{P}{Q} \cdot \frac{R}{R} = \frac{P}{Q} \cdot 1 = \frac{P}{Q}.$$

The following procedure can be used to simplify rational expressions:

SIMPLIFYING RATIONAL EXPRESSIONS

1. Factor the numerator and the denominator completely.

2. Divide both the numerator and the denominator by any common factors.

EXAMPLE 2 Simplifying a Rational Expression

Simplify: $\dfrac{5x + 35}{20x}$.

SOLUTION

$$\frac{5x + 35}{20x} = \frac{5(x + 7)}{5 \cdot 4x}$$

Factor the numerator and denominator. Because the denominator is 20x, x ≠ 0.

$$= \frac{\overset{1}{5}(x + 7)}{\underset{1}{5} \cdot 4x}$$

Divide out the common factor of 5.

$$= \frac{x + 7}{4x}$$

∎

✔ **CHECK POINT 2** Simplify: $\dfrac{7x + 28}{21x}$. $\dfrac{x + 4}{3x}$

EXAMPLE 3 Simplifying a Rational Expression

Simplify: $\dfrac{x^3 + x^2}{x + 1}$.

SOLUTION

$$\frac{x^3 + x^2}{x + 1} = \frac{x^2(x + 1)}{x + 1}$$ Factor the numerator. Because the denominator is $x + 1$, $x \ne -1$.

$$= \frac{x^2\cancel{(x + 1)}^{\,1}}{\cancel{x + 1}_{\,1}}$$ Divide out the common factor of $x + 1$.

$$= x^2$$

Simplifying a rational expression can change the numbers that make it undefined. For example, we just showed that

$$\frac{x^3 + x^2}{x + 1} = x^2.$$

This is undefined for $x = -1$.

This simplified form is defined for all real numbers.

Thus, to equate the two expressions, we must restrict the values for x in the simplified expression to exclude -1. We can write

$$\frac{x^3 + x^2}{x + 1} = x^2, \quad x \ne -1.$$

Hereafter, we will assume that the simplified rational expression is equal to the original rational expression for all real numbers except those for which either denominator is 0.

USING TECHNOLOGY

A graphing utility can be used to verify that

$$\frac{x^3 + x^2}{x + 1} = x^2, \quad x \ne -1.$$

Enter $y_1 = \dfrac{x^3 + x^2}{x + 1}$ and $y_2 = x^2$.

Graphic Check

The graphs of y_1 and y_2 appear to be identical. You can use the $\boxed{\text{TRACE}}$ feature to trace y_1 and show that it is undefined for $x = -1$.

$y_2 = x^2$

$y_1 = \dfrac{x^3 + x^2}{x + 1}$

$[-10, 10, 1]$ by $[-10, 10, 1]$

Numeric Check

No matter how far up or down we scroll, if $x \ne -1$, $y_1 = y_2$. If $x = -1$, y_1 is undefined, although the value of y_2 is 1.

X	Y1	Y2
-3	9	9
-2	4	4
-1	ERROR	1
0	0	0
1	1	1
2	4	4
3	9	9

X= -3

CHECK POINT 3 Simplify: $\dfrac{x^3 - x^2}{7x - 7}$. $\dfrac{x^2}{7}$

EXAMPLE 4 Simplifying a Rational Expression

Simplify: $\dfrac{x^2 + 6x + 5}{x^2 - 25}$.

SOLUTION $\dfrac{x^2 + 6x + 5}{x^2 - 25} = \dfrac{(x + 5)(x + 1)}{(x + 5)(x - 5)}$

Factor the numerator and denominator.
Because the denominator is
$(x + 5)(x - 5),\ x \neq -5$ and $x \neq 5$.

$= \dfrac{\overset{1}{\cancel{(x + 5)}}(x + 1)}{\underset{1}{\cancel{(x + 5)}}(x - 5)}$

Divide out the common factor of x + 5.

$= \dfrac{x + 1}{x - 5}$ ∎

✔ **CHECK POINT 4** Simplify: $\dfrac{x^2 - 1}{x^2 + 2x + 1}$. $\dfrac{x - 1}{x + 1}$

STUDY TIP

When simplifying rational expressions, only *factors* that are common to the *entire numerator* and the *entire denominator* can be divided out. **It is incorrect to divide out common terms from the numerator and denominator.**

Incorrect!

The first two expressions, $\dfrac{x + 5}{x + 7}$ and $\dfrac{x^2 - 4}{4}$, have no common factors in their numerators and denominators. Thus, these rational expressions are in simplified form. The rational expression $\dfrac{x^2 - 9}{x - 3}$ can be simplified as follows:

Correct

$$\frac{x^2 - 9}{x - 3} = \frac{(x + 3)\overset{1}{\cancel{(x - 3)}}}{\underset{1}{\cancel{x - 3}}} = x + 3.$$

Divide out the common factor, x − 3.

Factors That Are Opposites How do we simplify rational expressions that contain factors in the numerator and denominator that are opposites, or additive inverses? Here is an example of such an expression:

$$\frac{x - 3}{3 - x}.$$

The numerator and denominator are opposites. They differ only in their signs.

Factor out −1 from either the numerator or the denominator. Then divide out the common factor.

$$\frac{x - 3}{3 - x} = \frac{-1(-x + 3)}{3 - x}$$

Factor −1 from the numerator. Notice how the sign of each term in the polynomial x − 3 changes.

$$= \frac{-1(3 - x)}{3 - x}$$

In the numerator, use the commutative property to rewrite −x + 3 as 3 − x.

$$= \frac{-1\overset{1}{\cancel{(3 - x)}}}{\underset{1}{\cancel{3 - x}}}$$

Divide out the common factor of 3 − x.

$$= -1$$

Our result, −1, suggests a useful property that is stated at the top of the next page.

SIMPLIFYING RATIONAL EXPRESSIONS WITH OPPOSITE FACTORS IN THE NUMERATOR AND DENOMINATOR The quotient of two polynomials that have opposite signs and are additive inverses is -1.

EXAMPLE 5   Simplifying a Rational Expression

Simplify: $\dfrac{4x^2 - 25}{15 - 6x}$.

SOLUTION

$$\frac{4x^2 - 25}{15 - 6x} = \frac{(2x + 5)(2x - 5)}{3(5 - 2x)}$$

Factor the numerator and denominator.

$$= \frac{(2x + 5)\overset{-1}{\cancel{(2x - 5)}}}{3\cancel{(5 - 2x)}}$$

The quotient of polynomials with opposite signs is -1.

$$= \frac{-(2x + 5)}{3} \quad \text{or} \quad -\frac{2x + 5}{3} \quad \text{or} \quad \frac{-2x - 5}{3}$$

Each of these forms is an acceptable answer.

✔ **CHECK POINT 5** Simplify: $\dfrac{9x^2 - 49}{28 - 12x}$. $-\dfrac{3x + 7}{4}$

---

3 Solve applied problems involving rational expressions.

Applications The equation

$$y = \frac{250x}{100 - x}$$

models the cost, in millions of dollars, to remove x percent of the pollutants that are discharged into a river. This equation contains the rational expression that we looked at in the opening to this section. Do you remember how costs were spiraling upward as the percentage of removed pollutants increased?

Is it possible to clean up the river completely? To do this, we must remove 100% of the pollutants. The problem is that the rational expression is undefined for $x = 100$.

$$y = \frac{250x}{100 - x}$$

If $x = 100$, the value of the denominator is 0.

Notice how the graph of $y = \dfrac{250x}{100 - x}$, shown in Figure 7.1, approaches but never touches the dashed vertical line $x = 100$, our undefined value. The graph continues to rise more and more steeply, visually showing the escalating costs. By never touching the dashed vertical line, the graph illustrates that no amount of money will be enough to remove all pollutants from the river.

No amount of money can remove 100% of the pollutants.

$y = \dfrac{250x}{100 - x}$

Cost (millions of dollars)

Percentage of Pollutants Removed

FIGURE 7.1

7.1 EXERCISE SET

Student Solutions Manual

CD/Video

PH Math/Tutor Center

MathXL Tutorials on CD

Math XL®

MyMathLab

Interactmath.com

Practice Exercises

In Exercises 1–20, find all numbers for which each rational expression is undefined. If the rational expression is defined for all real numbers, so state.

1. $\dfrac{5}{2x}$ $x = 0$

2. $\dfrac{11}{3x}$ $x = 0$

3. $\dfrac{x}{x - 8}$ $x = 8$

4. $\dfrac{x}{x - 6}$ $x = 6$

5. $\dfrac{13}{5x - 20}$ $x = 4$

6. $\dfrac{17}{6x - 30}$ $x = 5$

7. $\dfrac{x + 3}{(x + 9)(x - 2)}$
$x = -9$ and $x = 2$

8. $\dfrac{x + 5}{(x + 7)(x - 9)}$
$x = -7$ and $x = 9$

9. $\dfrac{4x}{(3x - 17)(x + 3)}$
$x = \dfrac{17}{3}$ and $x = -3$

10. $\dfrac{8x}{(4x - 19)(x + 2)}$
$x = \dfrac{19}{4}$ and $x = -2$

11. $\dfrac{x + 5}{x^2 + x - 12}$
$x = -4$ and $x = 3$

12. $\dfrac{7x - 14}{x^2 - 9x + 20}$
$x = 4$ and $x = 5$

13. $\dfrac{x + 5}{5}$
defined for all real numbers

14. $\dfrac{x + 7}{7}$
defined for all real numbers

15. $\dfrac{y + 3}{4y^2 + y - 3}$
$y = -1$ and $y = \dfrac{3}{4}$

16. $\dfrac{y + 8}{6y^2 - y - 2}$
$y = -\dfrac{1}{2}$ and $y = \dfrac{2}{3}$

17. $\dfrac{y + 5}{y^2 - 25}$
$y = -5$ and $y = 5$

18. $\dfrac{y + 7}{y^2 - 49}$
$y = -7$ and $y = 7$

19. $\dfrac{5}{x^2 + 1}$
defined for all real numbers

20. $\dfrac{8}{x^2 + 4}$
defined for all real numbers

In Exercises 21–76, simplify each rational expression. If the rational expression cannot be simplified, so state.

21. $\dfrac{14x^2}{7x}$ $2x$

22. $\dfrac{9x^2}{6x}$ $\dfrac{3x}{2}$

23. $\dfrac{5x - 15}{25}$ $\dfrac{x - 3}{5}$

24. $\dfrac{7x + 21}{49}$ $\dfrac{x + 3}{7}$

25. $\dfrac{2x - 8}{4x}$ $\dfrac{x - 4}{2x}$

26. $\dfrac{3x - 9}{6x}$ $\dfrac{x - 3}{2x}$

27. $\dfrac{3}{3x - 9}$ $\dfrac{1}{x - 3}$

28. $\dfrac{12}{6x - 18}$ $\dfrac{2}{x - 3}$

29. $\dfrac{-15}{3x - 9}$ $\dfrac{5}{x - 3}$ (with minus sign) $-\dfrac{5}{x - 3}$

30. $\dfrac{-21}{7x - 14}$ $-\dfrac{3}{x - 2}$

31. $\dfrac{3x + 9}{x + 3}$ 3

32. $\dfrac{5x - 10}{x - 2}$ 5

33. $\dfrac{x + 5}{x^2 - 25}$ $\dfrac{1}{x - 5}$

34. $\dfrac{x + 4}{x^2 - 16}$ $\dfrac{1}{x - 4}$

35. $\dfrac{2y - 10}{3y - 15}$ $\dfrac{2}{3}$

36. $\dfrac{6y + 18}{11y + 33}$ $\dfrac{6}{11}$

37. $\dfrac{x + 1}{x^2 - 2x - 3}$ $\dfrac{1}{x - 3}$

38. $\dfrac{x + 2}{x^2 - x - 6}$ $\dfrac{1}{x - 3}$

39. $\dfrac{4x - 8}{x^2 - 4x + 4}$ $\dfrac{4}{x - 2}$

40. $\dfrac{x^2 - 12x + 36}{4x - 24}$ $\dfrac{x - 6}{4}$

41. $\dfrac{y^2 - 3y + 2}{y^2 + 7y - 18}$ $\dfrac{y - 1}{y + 9}$

42. $\dfrac{y^2 + 5y + 4}{y^2 - 4y - 5}$ $\dfrac{y + 4}{y - 5}$

43. $\dfrac{2y^2 - 7y + 3}{2y^2 - 5y + 2}$ $\dfrac{y - 3}{y - 2}$

44. $\dfrac{3y^2 + 4y - 4}{6y^2 - y - 2}$ $\dfrac{y + 2}{2y + 1}$

45. $\dfrac{2x + 3}{2x + 5}$
cannot be simplified

46. $\dfrac{3x + 7}{3x + 10}$
cannot be simplified

47. $\dfrac{x^2 + 12x + 36}{x^2 - 36}$ $\dfrac{x + 6}{x - 6}$

48. $\dfrac{x^2 - 14x + 49}{x^2 - 49}$ $\dfrac{x - 7}{x + 7}$

49. $\dfrac{x^3 - 2x^2 + x - 2}{x - 2}$
$x^2 + 1$

50. $\dfrac{x^3 + 4x^2 - 3x - 12}{x + 4}$
$x^2 - 3$

51. $\dfrac{x^3 - 8}{x - 2}$ $x^2 + 2x + 4$

52. $\dfrac{x^3 - 125}{x^2 - 25}$ $\dfrac{x^2 + 5x + 25}{x + 5}$

53. $\dfrac{(x - 4)^2}{x^2 - 16}$ $\dfrac{x - 4}{x + 4}$

54. $\dfrac{(x + 5)^2}{x^2 - 25}$ $\dfrac{x + 5}{x - 5}$

55. $\dfrac{x}{x + 1}$
cannot be simplified

56. $\dfrac{x}{x + 7}$
cannot be simplified

57. $\dfrac{x + 4}{x^2 + 16}$
cannot be simplified

58. $\dfrac{x + 5}{x^2 + 25}$
cannot be simplified

59. $\dfrac{x - 5}{5 - x}$ -1

60. $\dfrac{x - 7}{7 - x}$ -1

61. $\dfrac{2x - 3}{3 - 2x}$ -1

62. $\dfrac{5x - 4}{4 - 5x}$ -1

63. $\dfrac{x - 5}{x + 5}$
cannot be simplified

64. $\dfrac{x - 7}{x + 7}$
cannot be simplified

65. $\dfrac{4x - 6}{3 - 2x}$ -2

66. $\dfrac{9x - 15}{5 - 3x}$ -3

67. $\dfrac{4 - 6x}{3x^2 - 2x}$ $-\dfrac{2}{x}$

68. $\dfrac{9 - 15x}{5x^2 - 3x}$ $-\dfrac{3}{x}$

69. $\dfrac{x^2 - 1}{1 - x}$ $-x - 1$

70. $\dfrac{x^2 - 4}{2 - x}$ $-x - 2$

71. $\dfrac{y^2 - y - 12}{4 - y}$ $-y - 3$ **72.** $\dfrac{y^2 - 7y + 12}{3 - y}$ $-y + 4$

73. $\dfrac{x^2 y - x^2}{x^3 - x^3 y}$ $-\dfrac{1}{x}$ **74.** $\dfrac{xy - 2x}{3y - 6}$ $\dfrac{x}{3}$

75. $\dfrac{x^2 + 2xy - 3y^2}{2x^2 + 5xy - 3y^2}$ $\dfrac{x - y}{2x - y}$ **76.** $\dfrac{x^2 + 3xy - 10y^2}{3x^2 - 7xy + 2y^2}$ $\dfrac{x + 5y}{3x - y}$

Practice Plus

In Exercises 77–84, simplify each rational expression.

77. $\dfrac{x^2 - 9x + 18}{x^3 - 27}$ $\dfrac{x - 6}{x^2 + 3x + 9}$ **78.** $\dfrac{x^3 - 8}{x^2 + 2x - 8}$ $\dfrac{x^2 + 2x + 4}{x + 4}$

79. $\dfrac{9 - y^2}{y^2 - 3(2y - 3)}$ $\dfrac{3 + y}{3 - y}$ **80.** $\dfrac{16 - y^2}{y(y - 8) + 16}$ $\dfrac{4 + y}{4 - y}$

81. $\dfrac{xy + 2y + 3x + 6}{x^2 + 5x + 6}$ $\dfrac{y + 3}{x + 3}$ **82.** $\dfrac{xy + 4y - 7x - 28}{x^2 + 11x + 28}$ $\dfrac{y - 7}{x + 7}$

83. $\dfrac{8x^2 + 4x + 2}{1 - 8x^3}$ $\dfrac{2}{1 - 2x}$ **84.** $\dfrac{x^3 - 3x^2 + 9x}{x^3 + 27}$ $\dfrac{x}{x + 3}$

Application Exercises

85. The rational expression

$$\frac{130x}{100 - x}$$

describes the cost, in millions of dollars, to inoculate x percent of the population against a particular strain of flu.

a. Evaluate the expression for $x = 40$, $x = 80$, and $x = 90$. Describe the meaning of each evaluation in terms of percentage inoculated and cost.   It costs $86.67 million to inoculate 40% of the population. It costs $520 million to inoculate 80% of the population. It costs $1170 million to inoculate 90% of the population.

b. For what value of x is the expression undefined? $x = 100$

c. What happens to the cost as x approaches 100%? How can you interpret this observation?   The cost keeps rising; No amount of money will be enough to inoculate 100% of the population.

86. The rational expression

$$\frac{60,000x}{100 - x}$$

describes the cost, in dollars, to remove x percent of the air pollutants in the smokestack emission of a utility company that burns coal to generate electricity.

a. Evaluate the expression for $x = 20$, $x = 50$, and $x = 80$. Describe the meaning of each evaluation in terms of percentage of pollutants removed and cost.   It costs $15,000 to remove 20% of the air pollutants. It costs $60,000 to remove 50% of the air pollutants. It costs $240,000 to remove 80% of the air pollutants.

b. For what value of x is the expression undefined? $x = 100$

c. What happens to the cost as x approaches 100%? How can you interpret this observation?   The cost keeps rising; No amount of money will be enough to remove 100% of the air pollutants.

Doctors use the rational expression

$$\frac{DA}{A + 12}$$

to determine the dosage of a drug prescribed for children. In this expression, A = the child's age and D = the adult dosage. Use the expression to solve Exercises 87–88.

87. If the normal adult dosage of medication is 1000 milligrams, what dosage should an 8-year-old child receive?   400 milligrams

88. If the normal adult dosage of medication is 1000 milligrams, what dosage should a 4-year-old child receive?   250 milligrams

89. A company that manufactures bicycles has costs given by the equation

$$C = \frac{100x + 100,000}{x}$$

in which x is the number of bicycles manufactured and C is the cost to manufacture each bicycle.

a. Find the cost per bicycle when manufacturing 500 bicycles. $300

b. Find the cost per bicycle when manufacturing 4000 bicycles. $125

c. Does the cost per bicycle increase or decrease as more bicycles are manufactured? Explain why this happens.   decrease; Answers will vary.

90. A company that manufactures small canoes has costs given by the equation

$$C = \frac{20x + 20,000}{x}$$

in which x is the number of canoes manufactured and C is the cost to manufacture each canoe.

a. Find the cost per canoe when manufacturing 100 canoes. $220

b. Find the cost per canoe when manufacturing 10,000 canoes. $22

c. Does the cost per canoe increase or decrease as more canoes are manufactured? Explain why this happens.   decrease; Answers will vary.

A drug is injected into a patient and the concentration of the drug in the bloodstream is monitored. The drug's concentration, y, in milligrams per liter, after x hours is modeled by

$$y = \frac{5x}{x^2 + 1}.$$

The graph of this equation, obtained with a graphing utility, is shown in the figure in a $[0, 10, 1]$ *by* $[0, 3, 1]$ *viewing rectangle. Use this information to solve Exercises 91–92.*

$$y = \frac{5x}{x^2 + 1}$$

Drug's Concentration (milligrams per liter)

Hours after Injection

$[0, 10, 1]$ by $[0, 3, 1]$

91. Use the equation to find the drug's concentration after 3 hours. Then identify the point on the equation's graph that conveys this information.   1.5 milligrams per liter; (3, 1.5)

92. Use the graph of the equation to find after how many hours the drug reaches its maximum concentration. Then use the equation to find the drug's concentration at this time.   after 1 hour; 2.5 milligrams per liter

Body-mass index takes both weight and height into account when assessing whether an individual is underweight or overweight. The formula for body-mass index, BMI, *is*

$$\text{BMI} = \frac{703w}{h^2},$$

where w is weight, in pounds, and h is height, in inches. In adults, normal values for the BMI *are between 20 and 25, inclusive. Values below 20 indicate that an individual is underweight and values above 30 indicate that an individual is obese. Use this information to solve Exercises 93–94.*

93. Calculate the BMI, to the nearest tenth, for a 145-pound person who is 5 feet 10 inches tall. Is this person underweight?   20.8; no

94. Calculate the BMI, to the nearest tenth, for a 150-pound person who is 5 feet 6 inches tall. Is this person overweight?   24.2; no

95. The bar graph shows the total number of crimes in the United States, in millions, from 1995 through 2001.

Crime in the U.S.

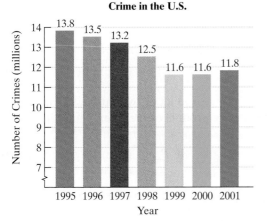

Number of Crimes (millions)

13.8 13.5 13.2 12.5 11.6 11.6 11.8

1995 1996 1997 1998 1999 2000 2001

Year

Source: FBI

The polynomial $3.7t + 257.4$ describes the U.S. population, in millions, t years after 1994. The polynomial $-0.4t + 14.2$ describes the number of crimes in the United States, in millions, t years after 1994.

a. Write a rational expression that describes the crime rate in the United States t years after 1994. $\dfrac{-0.4t + 14.2}{3.7t + 257.4}$

b. According to the rational expression in part (a), what was the crime rate in 2001? Round to two decimal places. How many crimes does this indicate per 100,000 inhabitants?   0.04; 4000

c. According to the FBI, there were 4161 crimes per 100,000 U.S. inhabitants in 2001. How well does the rational expression that you evaluated in part (b) model this number?   fairly well; Answers will vary.

Writing in Mathematics

96. What is a rational expression? Give an example with your explanation.   96–101. Answers will vary.

97. Explain how to find the number or numbers, if any, for which a rational expression is undefined.

98. Explain how to simplify a rational expression.

99. Explain how to simplify a rational expression with opposite factors in the numerator and denominator.

100. A politician claims that each year the crime rate in the United States is decreasing. Explain how to use the polynomials in Exercise 95 to verify this claim.

101. Use the graph shown for Exercises 91–92 to write a description of the drug's concentration over time. In your description, try to convey as much information as possible that is displayed visually by the graph.

Critical Thinking Exercises

102. Which one of the following is true?   d

a. $\dfrac{x + 5}{x} = 5$ **b.** $\dfrac{x^2 + 3}{3} = x^2 + 1$

c. $\dfrac{3x + 9}{3x + 13} = \dfrac{9}{13}$

d. The expression $\dfrac{-3y - 6}{y + 2}$ reduces to the consecutive integer that follows -4.

103. Write a rational expression that cannot be simplified.   103–104. Answers will vary.

104. Write a rational expression that is undefined for $x = -4$.

105. Write a rational expression with $x^2 - x - 6$ in the numerator that can be simplified to $x - 3$. $\dfrac{x^2 - x - 6}{x + 2}$

Technology Exercises

In Exercises 106–109, use the GRAPH *or* TABLE *feature of a graphing utility to determine if the rational expression has been correctly simplified. If the simplification is wrong, correct it and then verify your answer using the graphing utility.*

106. $\dfrac{3x + 15}{x + 5} = 3, \quad x \neq -5$ correctly simplified

107. $\dfrac{2x^2 - x - 1}{x - 1} = 2x^2 - 1, \; x \neq 1$ $2x^2 - 1$ should be $2x + 1$.

108. $\dfrac{x^2 - x}{x} = x^2 - 1, \; x \neq 0$ $x^2 - 1$ should be $x - 1$.

109. Use a graphing utility to verify the graph in Figure 7.1 on page 448. $\boxed{\text{TRACE}}$ along the graph as x approaches 100. What do you observe?   Answers will vary.

Review Exercises

110. Multiply: $\dfrac{5}{6} \cdot \dfrac{9}{25}$. (Section 1.1, Example 5) $\dfrac{3}{10}$

111. Divide: $\dfrac{2}{3} \div 4$. (Section 1.1, Example 6) $\dfrac{1}{6}$

112. Solve by the addition method:
$$2x - 5y = -2$$
$$3x + 4y = 20. \text{ (Section 4.3, Example 3)} \quad \{(4, 2)\}$$

7.2

Objectives

1 Multiply rational expressions.

2 Divide rational expressions.

MULTIPLYING AND DIVIDING RATIONAL EXPRESSIONS

Highbrow wit in conjunction with lowbrow comedy characterize Stephen Sondheim's *A Funny Thing Happened on the Way to the Forum*. The musical is based on the plays of Plautus, comic dramatist of ancient Rome.

Your psychology class is learning various techniques to double what we remember over time. At the beginning of the course, students memorize 40 words in Latin, a language with which they are not familiar. The rational expression

$$\frac{5t + 30}{t}$$

models the class average for the number of words remembered after t days, where $t \geq 1$. If the techniques are successful, what will be the new memory model?

The new model can be found by multiplying the given rational expression by 2. In this section, you will see that we multiply rational expressions in the same way that we multiply rational numbers. Thus, we multiply numerators and multiply denominators. The rational expression for doubling what the class remembers over time is

$$\frac{2}{1} \cdot \frac{5t + 30}{t} = \frac{2(5t + 30)}{1 \cdot t} = \frac{2 \cdot 5t + 2 \cdot 30}{t} = \frac{10t + 60}{t}.$$

1 Multiply rational expressions.

Multiplying Rational Expressions   The product of two rational expressions is the product of their numerators divided by the product of their denominators.

> **MULTIPLYING RATIONAL EXPRESSIONS**   If P, Q, R, and S are polynomials, where $Q \neq 0$ and $S \neq 0$, then
> $$\frac{P}{Q} \cdot \frac{R}{S} = \frac{PR}{QS}.$$

EXAMPLE 1 Multiplying Rational Expressions

Multiply: $\dfrac{7}{x+3} \cdot \dfrac{x-2}{5}$.

SOLUTION

$$\frac{7}{x+3} \cdot \frac{x-2}{5} = \frac{7(x-2)}{(x+3)5}$$ Multiply numerators. Multiply denominators. $(x \neq -3)$

$$= \frac{7x-14}{5x+15}$$ ■

✔ **CHECK POINT 1** Multiply: $\dfrac{9}{x+4} \cdot \dfrac{x-5}{2}$. $\dfrac{9x-45}{2x+8}$

Here is a step-by-step procedure for multiplying rational expressions. Before multiplying, divide out any factors common to both a numerator and a denominator.

MULTIPLYING RATIONAL EXPRESSIONS

1. Factor all numerators and denominators completely.
2. Divide numerators and denominators by common factors.
3. Multiply the remaining factors in the numerators and multiply the remaining factors in the denominators.

EXAMPLE 2 Multiplying Rational Expressions

Multiply: $\dfrac{x-3}{x+5} \cdot \dfrac{10x+50}{7x-21}$.

SOLUTION

$$\frac{x-3}{x+5} \cdot \frac{10x+50}{7x-21}$$

$$= \frac{x-3}{x+5} \cdot \frac{10(x+5)}{7(x-3)}$$ Factor as many numerators and denominators as possible.

$$= \frac{\overset{1}{\cancel{x-3}}}{\underset{1}{\cancel{x+5}}} \cdot \frac{10\overset{1}{\cancel{(x+5)}}}{7\underset{1}{\cancel{(x-3)}}}$$ Divide numerators and denominators by common factors.

$$= \frac{10}{7}$$ Multiply the remaining factors in the numerators and denominators. ■

✔ **CHECK POINT 2** Multiply: $\dfrac{x+4}{x-7} \cdot \dfrac{3x-21}{8x+32}$. $\dfrac{3}{8}$

EXAMPLE 3 Multiplying Rational Expressions

Multiply: $\dfrac{x-7}{x-1} \cdot \dfrac{x^2-1}{3x-21}$.

SOLUTION

$$\dfrac{x-7}{x-1} \cdot \dfrac{x^2-1}{3x-21}$$

$$= \dfrac{x-7}{x-1} \cdot \dfrac{(x+1)(x-1)}{3(x-7)} \qquad \text{Factor as many numerators and denominators as possible.}$$

$$= \dfrac{\overset{1}{\cancel{x-7}}}{\underset{1}{\cancel{x-1}}} \cdot \dfrac{(x+1)\overset{1}{\cancel{(x-1)}}}{3\underset{1}{\cancel{(x-7)}}} \qquad \text{Divide numerators and denominators by common factors.}$$

$$= \dfrac{x+1}{3} \qquad \text{Multiply the remaining factors in the numerators and denominators.} \qquad ■$$

✔ **CHECK POINT 3** Multiply: $\dfrac{x-5}{x-2} \cdot \dfrac{x^2-4}{9x-45}$.

EXAMPLE 4 Multiplying Rational Expressions

Multiply: $\dfrac{4x+8}{6x-3x^2} \cdot \dfrac{3x^2-4x-4}{9x^2-4}$.

SOLUTION

$$\dfrac{4x+8}{6x-3x^2} \cdot \dfrac{3x^2-4x-4}{9x^2-4}$$

$$= \dfrac{4(x+2)}{3x(2-x)} \cdot \dfrac{(3x+2)(x-2)}{(3x+2)(3x-2)} \qquad \text{Factor as many numerators and denominators as possible.}$$

$$= \dfrac{4(x+2)}{3x\underset{1}{\cancel{(2-x)}}} \cdot \dfrac{\overset{1}{\cancel{(3x+2)}}\overset{-1}{\cancel{(x-2)}}}{\underset{1}{\cancel{(3x+2)}}(3x-2)} \qquad \text{Divide numerators and denominators by common factors. Because } 2-x \text{ and } x-2 \text{ have opposite signs, their quotient is } -1.$$

$$= \dfrac{-4(x+2)}{3x(3x-2)} \ \text{ or } \ -\dfrac{4(x+2)}{3x(3x-2)} \qquad \text{Multiply the remaining factors in the numerators and denominators.}$$

> It is not necessary to carry out these multiplications.

■

✔ **CHECK POINT 4** Multiply: $\dfrac{5x+5}{7x-7x^2} \cdot \dfrac{2x^2+x-3}{4x^2-9}$.

 Divide rational expressions.

Dividing Rational Expressions   The quotient of two rational expressions is the product of the first expression and the multiplicative inverse, or reciprocal, of the second. The reciprocal is found by interchanging the numerator and the denominator.

DIVIDING RATIONAL EXPRESSIONS If P, Q, R, and S are polynomials, where $Q \neq 0$, $R \neq 0$, and $S \neq 0$, then

$$\frac{P}{Q} \div \frac{R}{S} = \frac{P}{Q} \cdot \frac{S}{R} = \frac{PS}{QR}.$$

Change division to multiplication.   Replace $\frac{R}{S}$ with its reciprocal by interchanging numerator and denominator.

Thus, **we find the quotient of two rational expressions by inverting the divisor and multiplying**. For example,

$$\frac{x}{7} \div \frac{6}{y} = \frac{x}{7} \cdot \frac{y}{6} = \frac{xy}{42}.$$

Change the division to multiplication.   Replace $\frac{6}{y}$ with its reciprocal by interchanging numerator and denominator.

EXAMPLE 5   Dividing Rational Expressions

Divide: $(x + 5) \div \dfrac{x - 2}{x + 9}$.

SOLUTION

$$(x + 5) \div \frac{x - 2}{x + 9} = \frac{x + 5}{1} \cdot \frac{x + 9}{x - 2}$$ Invert the divisor and multiply.

$$= \frac{(x + 5)(x + 9)}{x - 2}$$ Multiply the factors in the numerators and denominators. We need not carry out the multiplication in the numerator. ∎

 CHECK POINT 5 Divide: $(x + 3) \div \dfrac{x - 4}{x + 7}.$ $\dfrac{(x + 3)(x + 7)}{x - 4}$

EXAMPLE 6   Dividing Rational Expressions

Divide: $\dfrac{x^2 - 2x - 8}{x^2 - 9} \div \dfrac{x - 4}{x + 3}.$

SOLUTION

$$\frac{x^2 - 2x - 8}{x^2 - 9} \div \frac{x - 4}{x + 3}$$

$$= \frac{x^2 - 2x - 8}{x^2 - 9} \cdot \frac{x + 3}{x - 4}$$ Invert the divisor and multiply.

$$= \frac{(x - 4)(x + 2)}{(x + 3)(x - 3)} \cdot \frac{x + 3}{x - 4}$$ Factor as many numerators and denominators as possible.

$$= \frac{\overset{1}{\cancel{(x - 4)}}(x + 2)}{\cancel{(x + 3)}(x - 3)} \cdot \frac{\overset{1}{\cancel{(x + 3)}}}{\cancel{(x - 4)}}$$ Divide numerators and denominators by common factors.

$$= \frac{x + 2}{x - 3}$$ Multiply the remaining factors in the numerators and the denominators. ∎

✔ **CHECK POINT 6** Divide: $\dfrac{x^2 + 5x + 6}{x^2 - 25} \div \dfrac{x + 2}{x + 5}$. $\dfrac{x + 3}{x - 5}$

EXAMPLE 7   Dividing Rational Expressions

Divide: $\dfrac{y^2 + 7y + 12}{y^2 + 9} \div (7y^2 + 21y)$.

SOLUTION

$$\dfrac{y^2 + 7y + 12}{y^2 + 9} \div \dfrac{7y^2 + 21y}{1}$$
   It is helpful to write the divisor with a denominator of 1.

$$= \dfrac{y^2 + 7y + 12}{y^2 + 9} \cdot \dfrac{1}{7y^2 + 21y}$$
   Invert the divisor and multiply.

$$= \dfrac{(y + 4)(y + 3)}{y^2 + 9} \cdot \dfrac{1}{7y(y + 3)}$$
   Factor as many numerators and denominators as possible.

$$= \dfrac{(y + 4)\cancel{(y + 3)}^{1}}{y^2 + 9} \cdot \dfrac{1}{7y\cancel{(y + 3)}_{1}}$$
   Divide numerators and denominators by common factors.

$$= \dfrac{y + 4}{7y(y^2 + 9)}$$
   Multiply the remaining factors in the numerators and the denominators.

■

✔ **CHECK POINT 7** Divide: $\dfrac{y^2 + 3y + 2}{y^2 + 1} \div (5y^2 + 10y)$. $\dfrac{y + 1}{5y(y^2 + 1)}$

7.2 EXERCISE SET

 Student Solutions Manual   CD/Video   PH Math/Tutor Center   MathXL Tutorials on CD   Math**XL**   MathXL®   **MyMathLab**   MyMathLab   Interactmath.com

Practice Exercises

In Exercises 1–32, multiply as indicated.

1. $\dfrac{4}{x + 3} \cdot \dfrac{x - 5}{9}$ $\dfrac{4x - 20}{9x + 27}$ **2.** $\dfrac{8}{x - 2} \cdot \dfrac{x + 5}{3}$ $\dfrac{8x + 40}{3x - 6}$

3. $\dfrac{x}{3} \cdot \dfrac{12}{x + 5}$ $\dfrac{4x}{x + 5}$ **4.** $\dfrac{x}{5} \cdot \dfrac{30}{x - 4}$ $\dfrac{6x}{x - 4}$

5. $\dfrac{3}{x} \cdot \dfrac{4x}{15}$ $\dfrac{4}{5}$ **6.** $\dfrac{7}{x} \cdot \dfrac{5x}{35}$ 1

7. $\dfrac{x - 3}{x + 5} \cdot \dfrac{4x + 20}{9x - 27}$ $\dfrac{4}{9}$ **8.** $\dfrac{x - 2}{x + 9} \cdot \dfrac{5x + 45}{2x - 4}$ $\dfrac{5}{2}$

9. $\dfrac{x^2 + 9x + 14}{x + 7} \cdot \dfrac{1}{x + 2}$ 1   **10.** $\dfrac{x^2 + 9x + 18}{x + 6} \cdot \dfrac{1}{x + 3}$ 1

11. $\dfrac{x^2 - 25}{x^2 - 3x - 10} \cdot \dfrac{x + 2}{x}$ $\dfrac{x + 5}{x}$

12. $\dfrac{x^2 - 49}{x^2 - 4x - 21} \cdot \dfrac{x + 3}{x}$ $\dfrac{x + 7}{x}$

13. $\dfrac{4y + 30}{y^2 - 3y} \cdot \dfrac{y - 3}{2y + 15}$ $\dfrac{2}{y}$ **14.** $\dfrac{9y + 21}{y^2 - 2y} \cdot \dfrac{y - 2}{3y + 7}$ $\dfrac{3}{y}$

15. $\dfrac{y^2 - 7y - 30}{y^2 - 6y - 40} \cdot \dfrac{2y^2 + 5y + 2}{2y^2 + 7y + 3}$ $\dfrac{y + 2}{y + 4}$

16. $\dfrac{3y^2 + 17y + 10}{3y^2 - 22y - 16} \cdot \dfrac{y^2 - 4y - 32}{y^2 - 8y - 48}$ $\dfrac{y + 5}{y - 12}$

17. $(y^2 - 9) \cdot \dfrac{4}{y - 3}$ $4(y + 3)$ **18.** $(y^2 - 16) \cdot \dfrac{3}{y - 4}$ $3(y + 4)$

19. $\dfrac{x^2 - 5x + 6}{x^2 - 2x - 3} \cdot \dfrac{x^2 - 1}{x^2 - 4}$ $\dfrac{x - 1}{x + 2}$

20. $\dfrac{x^2 + 5x + 6}{x^2 + x - 6} \cdot \dfrac{x^2 - 9}{x^2 - x - 6}$ $\dfrac{x + 3}{x - 2}$

21. $\dfrac{x^3 - 8}{x^2 - 4} \cdot \dfrac{x + 2}{3x}$ $\dfrac{x^2 + 2x + 4}{3x}$

22. $\dfrac{x^2 + 6x + 9}{x^3 + 27} \cdot \dfrac{1}{x + 3}$ $\dfrac{1}{x^2 - 3x + 9}$

23. $\dfrac{(x - 2)^3}{(x - 1)^3} \cdot \dfrac{x^2 - 2x + 1}{x^2 - 4x + 4}$ $\dfrac{x - 2}{x - 1}$

24. $\dfrac{(x + 4)^3}{(x + 2)^3} \cdot \dfrac{x^2 + 4x + 4}{x^2 + 8x + 16}$ $\dfrac{x + 4}{x + 2}$

25. $\dfrac{6x + 2}{x^2 - 1} \cdot \dfrac{1 - x}{3x^2 + x}$ $-\dfrac{2}{x(x + 1)}$

26. $\dfrac{8x + 2}{x^2 - 9} \cdot \dfrac{3 - x}{4x^2 + x}$ $-\dfrac{2}{x(x + 3)}$

27. $\dfrac{25 - y^2}{y^2 - 2y - 35} \cdot \dfrac{y^2 - 8y - 20}{y^2 - 3y - 10}$ $-\dfrac{y - 10}{y - 7}$

28. $\dfrac{2y}{3y - y^2} \cdot \dfrac{2y^2 - 9y + 9}{8y - 12}$ $-\dfrac{1}{2}$

29. $\dfrac{x^2 - y^2}{x} \cdot \dfrac{x^2 + xy}{x + y}$ $(x - y)(x + y)$

30. $\dfrac{4x - 4y}{x} \cdot \dfrac{x^2 + xy}{x^2 - y^2}$ 4

31. $\dfrac{x^2 + 2xy + y^2}{x^2 - 2xy + y^2} \cdot \dfrac{4x - 4y}{3x + 3y}$ $\dfrac{4(x + y)}{3(x - y)}$

32. $\dfrac{x^2 - y^2}{x + y} \cdot \dfrac{x + 2y}{2x^2 - xy - y^2}$ $\dfrac{x + 2y}{2x + y}$

In Exercises 33–64, divide as indicated.

33. $\dfrac{x}{7} \div \dfrac{5}{3}$ $\dfrac{3x}{35}$

34. $\dfrac{x}{3} \div \dfrac{3}{8}$ $\dfrac{8x}{9}$

35. $\dfrac{3}{x} \div \dfrac{12}{x}$ $\dfrac{1}{4}$

36. $\dfrac{x}{5} \div \dfrac{20}{x}$ $\dfrac{x^2}{100}$

37. $\dfrac{15}{x} \div \dfrac{3}{2x}$ 10

38. $\dfrac{9}{x} \div \dfrac{3}{4x}$ 12

39. $\dfrac{x + 1}{3} \div \dfrac{3x + 3}{7}$ $\dfrac{7}{9}$

40. $\dfrac{x + 5}{7} \div \dfrac{4x + 20}{9}$ $\dfrac{9}{28}$

41. $\dfrac{7}{x - 5} \div \dfrac{28}{3x - 15}$ $\dfrac{3}{4}$

42. $\dfrac{4}{x - 6} \div \dfrac{40}{7x - 42}$ $\dfrac{7}{10}$

43. $\dfrac{x^2 - 4}{x} \div \dfrac{x + 2}{x - 2}$ $\dfrac{(x - 2)^2}{x}$

44. $\dfrac{x^2 - 4}{x - 2} \div \dfrac{x + 2}{4x - 8}$ $4(x - 2)$

45. $(y^2 - 16) \div \dfrac{y^2 + 3y - 4}{y^2 + 4}$ $\dfrac{(y - 4)(y^2 + 4)}{y - 1}$

46. $(y^2 + 4y - 5) \div \dfrac{y^2 - 25}{y + 7}$ $\dfrac{(y - 1)(y + 7)}{y - 5}$

47. $\dfrac{y^2 - y}{15} \div \dfrac{y - 1}{5}$ $\dfrac{y}{3}$

48. $\dfrac{y^2 - 2y}{15} \div \dfrac{y - 2}{5}$ $\dfrac{y}{3}$

49. $\dfrac{4x^2 + 10}{x - 3} \div \dfrac{6x^2 + 15}{x^2 - 9}$ $\dfrac{2(x + 3)}{3}$

50. $\dfrac{x^2 + x}{x^2 - 4} \div \dfrac{x^2 - 1}{x^2 + 5x + 6}$ $\dfrac{x(x + 3)}{(x - 2)(x - 1)}$

51. $\dfrac{x^2 - 25}{2x - 2} \div \dfrac{x^2 + 10x + 25}{x^2 + 4x - 5}$ $\dfrac{x - 5}{2}$

52. $\dfrac{x^2 - 4}{x^2 + 3x - 10} \div \dfrac{x^2 + 5x + 6}{x^2 + 8x + 15}$ 1

53. $\dfrac{y^3 + y}{y^2 - y} \div \dfrac{y^3 - y^2}{y^2 - 2y + 1}$ $\dfrac{y^2 + 1}{y^2}$

54. $\dfrac{3y^2 - 12}{y^2 + 4y + 4} \div \dfrac{y^3 - 2y^2}{y^2 + 2y}$ $\dfrac{3}{y}$

55. $\dfrac{y^2 + 5y + 4}{y^2 + 12y + 32} \div \dfrac{y^2 - 12y + 35}{y^2 + 3y - 40}$ $\dfrac{y + 1}{y - 7}$

56. $\dfrac{y^2 + 4y - 21}{y^2 + 3y - 28} \div \dfrac{y^2 + 14y + 48}{y^2 + 4y - 32}$ $\dfrac{y - 3}{y + 6}$

57. $\dfrac{2y^2 - 128}{y^2 + 16y + 64} \div \dfrac{y^2 - 6y - 16}{3y^2 + 30y + 48}$ 6

58. $\dfrac{3y + 12}{y^2 + 3y} \div \dfrac{y^2 + y - 12}{9y - y^3}$ -3

59. $\dfrac{2x + 2y}{3} \div \dfrac{x^2 - y^2}{x - y}$ $\dfrac{2}{3}$

60. $\dfrac{5x + 5y}{7} \div \dfrac{x^2 - y^2}{x - y}$ $\dfrac{5}{7}$

61. $\dfrac{x^2 - y^2}{8x^2 - 16xy + 8y^2} \div \dfrac{4x - 4y}{x + y}$ $\dfrac{(x + y)^2}{32(x - y)^2}$

62. $\dfrac{4x^2 - y^2}{x^2 + 4xy + 4y^2} \div \dfrac{4x - 2y}{3x + 6y}$ $\dfrac{3(2x + y)}{2(x + 2y)}$

63. $\dfrac{xy - y^2}{x^2 + 2x + 1} \div \dfrac{2x^2 + xy - 3y^2}{2x^2 + 5xy + 3y^2}$ $\dfrac{y(x + y)}{(x + 1)^2}$

64. $\dfrac{x^2 - 4y^2}{x^2 + 3xy + 2y^2} \div \dfrac{x^2 - 4xy + 4y^2}{x + y}$ $\dfrac{1}{x - 2y}$

Practice Plus

In Exercises 65–72, perform the indicated operation or operations.

65. $\left(\dfrac{y - 2}{y^2 - 9y + 18} \cdot \dfrac{y^2 - 4y - 12}{y + 2}\right) \div \dfrac{y^2 - 4}{y^2 + 5y + 6}$ $\dfrac{y + 3}{y - 3}$

66. $\left(\dfrac{6y^2 + 31y + 18}{3y^2 - 20y + 12} \cdot \dfrac{2y^2 - 15y + 18}{6y^2 + 35y + 36}\right) \div \dfrac{2y^2 - 13y + 15}{9y^2 + 15y + 4}$ $\dfrac{(3y + 2)(3y + 1)}{(3y - 2)(y - 5)}$

67. $\dfrac{3x^2 + 3x - 60}{2x - 8} \div \left(\dfrac{30x^2}{x^2 - 7x + 10} \cdot \dfrac{x^3 + 3x^2 - 10x}{25x^3}\right)$ $\dfrac{5(x - 5)}{4}$

68. $\dfrac{5x^2 - x}{3x + 2} \div \left(\dfrac{6x^2 + x - 2}{10x^2 + 3x - 1} \cdot \dfrac{2x^2 - x - 1}{2x^2 - x}\right)$ $\dfrac{x^2(5x - 1)^2}{(3x + 2)^2(x - 1)}$

69. $\dfrac{x^2 + xz + xy + yz}{x - y} \div \dfrac{x + z}{x + y}$ $\dfrac{(x + y)^2}{x - y}$

70. $\dfrac{x^2 - xz + xy - yz}{x - y} \div \dfrac{x - z}{y - x}$ $-(x + y)$ or $-x - y$

71. $\dfrac{3xy + ay + 3xb + ab}{9x^2 - a^2} \div \dfrac{y^3 + b^3}{6x - 2a}$ $\dfrac{2}{y^2 - by + b^2}$

72. $\dfrac{5xy - ay - 5xb + ab}{25x^2 - a^2} \div \dfrac{y^3 - b^3}{15x + 3a}$ $\dfrac{3}{y^2 + by + b^2}$

Application Exercises

73. In the Section 7.1 opener, we used

$$\dfrac{250x}{100 - x}$$

to describe the cost, in millions of dollars, to remove x percent of the pollutants that are discharged into the river. We were wrong. The cost will be half of what we originally anticipated. Write a rational expression that represents the reduced cost. $\dfrac{125x}{100 - x}$

74. We originally thought that the cost, in dollars, to manufacture each of x bicycles was

$$\frac{100x + 100,000}{x}.$$

We were wrong. We can manufacture each bicycle at half of what we originally anticipated. Write a rational expression that represents the reduced cost. $\dfrac{50(x + 1000)}{x}$

Writing in Mathematics

75. Explain how to multiply rational expressions. *75–77. Answers will vary.*

76. Explain how to divide rational expressions.

77. In dividing polynomials

$$\frac{P}{Q} \div \frac{R}{S},$$

why is it necessary to state that polynomial R is not equal to 0?

Critical Thinking Exercises

78. Which one of the following is true? **b**

a. $5 \div x = \dfrac{1}{5} \cdot x$ for any nonzero number x.

b. $\dfrac{4}{x} \div \dfrac{x - 2}{x} = \dfrac{4}{x - 2}$ if $x \neq 0$ and $x \neq 2$.

c. $\dfrac{x - 5}{6} \cdot \dfrac{3}{5 - x} = \dfrac{1}{2}$ for any value of x except 5.

d. The quotient of two rational expressions can be found by dividing their numerators and dividing their denominators.

79. Find the missing polynomials: $\dfrac{\quad}{\quad} \cdot \dfrac{3x - 12}{2x} = \dfrac{3}{2}.$

numerator: x; denominator: $x - 4$

80. Find the missing polynomials: $-\dfrac{1}{2x - 3} \div \dfrac{\quad}{\quad} = \dfrac{1}{3}.$

numerator: -3; denominator: $2x - 3$

81. Divide:

$$\frac{9x^2 - y^2 + 15x - 5y}{3x^2 + xy + 5x} \div \frac{3x + y}{9x^3 + 6x^2y + xy^2}.$$

$(3x - y)(3x + y)$

Technology Exercises

In Exercises 82–85, use the $\boxed{\text{GRAPH}}$ *or* $\boxed{\text{TABLE}}$ *feature of a graphing utility to determine if the multiplication or division has been performed correctly. If the answer is wrong, correct it and then verify your correction using the graphing utility.*

82. $\dfrac{x^2 + x}{3x} \cdot \dfrac{6x}{x + 1} = 2x$ correct answer

83. $\dfrac{x^3 - 25x}{x^2 - 3x - 10} \cdot \dfrac{x + 2}{x} = x + 5$ correct answer

84. $\dfrac{x^2 - 9}{x + 4} \div \dfrac{x - 3}{x + 4} = x - 3$ $x - 3$ should be $x + 3$.

85. $(x - 5) \div \dfrac{2x^2 - 11x + 5}{4x^2 - 1} = 2x - 1$ $2x - 1$ should be $2x + 1$.

Review Exercises

86. Solve: $2x + 3 < 3(x - 5)$. (Section 2.7, Example 6) $\{x | x > 18\}$

87. Factor completely: $3x^2 - 15x - 42$. (Section 6.5, Example 2) $3(x - 7)(x + 2)$

88. Solve: $x(2x + 9) = 5$. (Section 6.6, Example 6) $\left\{-5, \dfrac{1}{2}\right\}$

SECTION **7.3**

ADDING AND SUBTRACTING RATIONAL EXPRESSIONS WITH THE SAME DENOMINATOR

Objectives

1 Add rational expressions with the same denominator.

2 Subtract rational expressions with the same denominator.

3 Add and subtract rational expressions with opposite denominators.

Are you long, medium, or round? Your skull, that is. The varying shapes of the human skull create glorious diversity in the human species. By learning to add and subtract rational expressions with the same denominator, you will obtain an expression that models this diversity.

1 Add rational expressions with the same denominator.

Addition when Denominators Are the Same To add rational numbers having the same denominators, such as $\frac{2}{9}$ and $\frac{5}{9}$, we add the numerators and place the sum over the common denominator:

$$\frac{2}{9} + \frac{5}{9} = \frac{2 + 5}{9} = \frac{7}{9}.$$

We add rational expressions with the same denominator in an identical manner.

ADDING RATIONAL EXPRESSIONS WITH COMMON DENOMINATORS

If $\dfrac{P}{R}$ and $\dfrac{Q}{R}$ are rational expressions, then

$$\frac{P}{R} + \frac{Q}{R} = \frac{P + Q}{R}.$$

To add rational expressions with the same denominator, add numerators and place the sum over the common denominator. If possible, simplify the result.

EXAMPLE 1 Adding Rational Expressions when Denominators Are the Same

Add: $\dfrac{2x - 1}{3} + \dfrac{x + 4}{3}$.

SOLUTION

$$\frac{2x - 1}{3} + \frac{x + 4}{3} = \frac{2x - 1 + x + 4}{3} \qquad \text{Add numerators. Place this sum over the common denominator.}$$

$$= \frac{3x + 3}{3} \qquad \text{Combine like terms.}$$

$$= \frac{\overset{1}{\cancel{3}}(x + 1)}{\underset{1}{\cancel{3}}} \qquad \text{Factor and simplify.}$$

$$= x + 1$$

✔ **CHECK POINT 1** Add: $\dfrac{3x - 2}{5} + \dfrac{2x + 12}{5}$.   *x + 2*

EXAMPLE 2 Adding Rational Expressions when Denominators Are the Same

Add: $\dfrac{x^2}{x^2 - 9} + \dfrac{9 - 6x}{x^2 - 9}$.

SOLUTION

$$\frac{x^2}{x^2 - 9} + \frac{9 - 6x}{x^2 - 9} = \frac{x^2 + 9 - 6x}{x^2 - 9} \qquad \text{Add numerators. Place this sum over the common denominator.}$$

$$= \frac{x^2 - 6x + 9}{x^2 - 9} \qquad \text{Write the numerator in descending powers of x.}$$

$$= \frac{(x - 3)\overset{1}{\cancel{(x - 3)}}}{(x + 3)\underset{1}{\cancel{(x - 3)}}} \qquad \text{Factor and simplify. What values of x are not permitted?}$$

$$= \frac{x - 3}{x + 3}$$

CHECK POINT 2 Add: $\dfrac{x^2}{x^2 - 25} + \dfrac{25 - 10x}{x^2 - 25}$. $\dfrac{x - 5}{x + 5}$

2 Subtract rational expressions with the same denominator.

Subtraction when Denominators Are the Same The following box shows how to subtract rational expressions with the same denominator:

SUBTRACTING RATIONAL EXPRESSIONS WITH COMMON DENOMINATORS

If $\dfrac{P}{R}$ and $\dfrac{Q}{R}$ are rational expressions, then

$$\frac{P}{R} - \frac{Q}{R} = \frac{P - Q}{R}.$$

To subtract rational expressions with the same denominator, subtract numerators and place the difference over the common denominator. If possible, simplify the result.

EXAMPLE 3 Subtracting Rational Expressions when Denominators Are the Same

Subtract:

a. $\dfrac{2x + 3}{x + 1} - \dfrac{x}{x + 1}$ **b.** $\dfrac{5x + 1}{x^2 - 9} - \dfrac{4x - 2}{x^2 - 9}$.

SOLUTION

a. $\dfrac{2x + 3}{x + 1} - \dfrac{x}{x + 1} = \dfrac{2x + 3 - x}{x + 1}$ Subtract numerators. Place this difference over the common denominator.

$= \dfrac{x + 3}{x + 1}$ Combine like terms.

b. $\dfrac{5x + 1}{x^2 - 9} - \dfrac{4x - 2}{x^2 - 9} = \dfrac{5x + 1 - (4x - 2)}{x^2 - 9}$ Subtract numerators and include parentheses to indicate that both terms are subtracted. Place this difference over the common denominator.

$= \dfrac{5x + 1 - 4x + 2}{x^2 - 9}$ Remove parentheses and then change the sign of each term.

$= \dfrac{x + 3}{x^2 - 9}$ Combine like terms.

$= \dfrac{\overset{1}{x + 3}}{\underset{1}{(x + 3)}(x - 3)}$ Factor and simplify ($x \neq -3$ and $x \neq 3$).

$= \dfrac{1}{x - 3}$

USING TECHNOLOGY

To check Example 3(b) numerically, enter

$$y_1 = \frac{5x + 1}{x^2 - 9} - \frac{4x - 2}{x^2 - 9}$$

$$y_2 = \frac{1}{x - 3}$$

and use the TABLE feature. If $x \neq -3$ and $x \neq 3$, no matter how far up or down we scroll, $y_1 = y_2$.

X	Y1	Y2
-3	ERROR	-.1667
-2	-.2	-.2
-1	-.25	-.25
0	-.3333	-.3333
1	-.5	-.5
2	-1	-1
3	ERROR	ERROR

X= -3

 CHECK POINT 3 Subtract:

a. $\dfrac{4x + 5}{x + 7} - \dfrac{x}{x + 7}$ $\dfrac{3x + 5}{x + 7}$ **b.** $\dfrac{3x^2 + 4x}{x - 1} - \dfrac{11x - 4}{x - 1}$. $3x - 4$

STUDY TIP

When a numerator is being subtracted, be sure to **subtract every term in that expression.**

The − sign applies to the entire numerator, $4x − 2$.

Insert parentheses to indicate this.

The sign of every term of $4x − 2$ changes.

$$\frac{5x + 1}{x^2 - 9} - \frac{4x - 2}{x^2 - 9} = \frac{5x + 1 - (4x - 2)}{x^2 - 9} = \frac{5x + 1 - 4x + 2}{x^2 - 9}$$

The entire numerator of the second rational expression must be subtracted. Avoid the common error of subtracting only the first term.

Incorrect!

−2 must also be subtracted.

$$\frac{5x + 1}{x^2 - 9} - \frac{4x - 2}{x^2 - 9} = \frac{5x + 1 - 4x - 2}{x^2 - 9}$$

EXAMPLE 4 **Subtracting Rational Expressions when Denominators Are the Same**

Subtract: $\dfrac{20y^2 + 5y + 1}{6y^2 + y - 2} - \dfrac{8y^2 - 12y - 5}{6y^2 + y - 2}$.

SOLUTION

$$\frac{20y^2 + 5y + 1}{6y^2 + y - 2} - \frac{8y^2 - 12y - 5}{6y^2 + y - 2}$$

Don't forget the parentheses.

$$= \frac{20y^2 + 5y + 1 - (8y^2 - 12y - 5)}{6y^2 + y - 2}$$

Subtract numerators. Place this difference over the common denominator.

$$= \frac{20y^2 + 5y + 1 - 8y^2 + 12y + 5}{6y^2 + y - 2}$$

Remove parentheses and then change the sign of each term.

$$= \frac{(20y^2 - 8y^2) + (5y + 12y) + (1 + 5)}{6y^2 + y - 2}$$

Group like terms. This step is usually performed mentally.

$$= \frac{12y^2 + 17y + 6}{6y^2 + y - 2}$$

Combine like terms.

$$= \frac{\overset{1}{\cancel{(3y + 2)}}(4y + 3)}{\underset{1}{\cancel{(3y + 2)}}(2y - 1)}$$

Factor and simplify.

$$= \frac{4y + 3}{2y - 1}$$

∎

✓ **CHECK POINT 4** Subtract: $\dfrac{y^2 + 3y - 6}{y^2 - 5y + 4} - \dfrac{4y - 4 - 2y^2}{y^2 - 5y + 4}$. $\dfrac{3y + 2}{y - 4}$

3 Add and subtract rational expressions with opposite denominators.

Addition and Subtraction when Denominators Are Opposites How do we add or subtract rational expressions when denominators are opposites, or additive inverses? Here is an example of this type of addition problem:

$$\frac{x^2}{x - 5} + \frac{4x + 5}{5 - x}.$$

> These denominators are opposites.
> The differ only in their signs.

Multiply the numerator and the denominator of either of the rational expressions by -1. Then they will both have the same denominator.

EXAMPLE 5 Adding Rational Expressions when Denominators Are Opposites

Add: $\dfrac{x^2}{x - 5} + \dfrac{4x + 5}{5 - x}.$

SOLUTION

$$\frac{x^2}{x - 5} + \frac{4x + 5}{5 - x}$$

$$= \frac{x^2}{x - 5} + \frac{(-1)}{(-1)} \cdot \frac{4x + 5}{5 - x} \qquad \text{Multiply the numerator and denominator of the second rational expression by } -1.$$

$$= \frac{x^2}{x - 5} + \frac{-4x - 5}{-5 + x} \qquad \text{Perform the multiplications by } -1 \text{ by changing every term's sign.}$$

$$= \frac{x^2}{x - 5} + \frac{-4x - 5}{x - 5} \qquad \text{Rewrite } -5 + x \text{ as } x - 5. \text{ Both rational expressions have the same denominator.}$$

$$= \frac{x^2 + (-4x - 5)}{x - 5} \qquad \text{Add numerators. Place this sum over the common denominator.}$$

$$= \frac{x^2 - 4x - 5}{x - 5} \qquad \text{Remove parentheses.}$$

$$= \frac{\overset{1}{\cancel{(x - 5)}}(x + 1)}{\underset{1}{\cancel{x - 5}}} \qquad \text{Factor and simplify.}$$

$$= x + 1$$

■

✔ **CHECK POINT 5** Add: $\dfrac{x^2}{x - 7} + \dfrac{4x + 21}{7 - x}.$ $x + 3$

ADDING AND SUBTRACTING RATIONAL EXPRESSIONS WITH OPPOSITE DENOMINATORS When one denominator is the additive inverse of the other, first multiply either rational expression by $\frac{-1}{-1}$ to obtain a common denominator.

EXAMPLE 6 Subtracting Rational Expressions when Denominators Are Opposites

Subtract: $\dfrac{5x - x^2}{x^2 - 4x - 3} - \dfrac{3x - x^2}{3 + 4x - x^2}$.

SOLUTION We note that $x^2 - 4x - 3$ and $3 + 4x - x^2$ are opposites. We multiply the second rational expression by $\dfrac{-1}{-1}$.

$$\frac{(-1)}{(-1)} \cdot \frac{3x - x^2}{3 + 4x - x^2} = \frac{-3x + x^2}{-3 - 4x + x^2}$$
Multiply the numerator and denominator by -1 by changing every term's sign.

$$= \frac{x^2 - 3x}{x^2 - 4x - 3}$$
Write the numerator and the denominator in descending powers of x.

We now return to the original subtraction problem.

$$\frac{5x - x^2}{x^2 - 4x - 3} - \frac{3x - x^2}{3 + 4x - x^2}$$
This is the given problem.

$$= \frac{5x - x^2}{x^2 - 4x - 3} - \frac{x^2 - 3x}{x^2 - 4x - 3}$$
Replace the second rational expression by the form obtained through multiplication by $\dfrac{-1}{-1}$.

$$= \frac{5x - x^2 - (x^2 - 3x)}{x^2 - 4x - 3}$$
Subtract numerators. Place this difference over the common denominator. Don't forget parentheses!

$$= \frac{5x - x^2 - x^2 + 3x}{x^2 - 4x - 3}$$
Remove parentheses and then change the sign of each term.

$$= \frac{-2x^2 + 8x}{x^2 - 4x - 3}$$
Combine like terms in the numerator. Although the numerator can be factored, further simplification is not possible. ∎

✔ **CHECK POINT 6** Subtract: $\dfrac{7x - x^2}{x^2 - 2x - 9} - \dfrac{5x - 3x^2}{9 + 2x - x^2}$. $\dfrac{-4x^2 + 12x}{x^2 - 2x - 9}$

7.3 EXERCISE SET

Student Solutions Manual   CD/Video   PH Math/Tutor Center   MathXL Tutorials on CD   Math XL   MathXL®   MyMathLab   MyMathLab   Interactmath.com

Practice Exercises

In Exercises 1–38, add or subtract as indicated. Simplify the result, if possible.

1. $\dfrac{7x}{13} + \dfrac{2x}{13}$

2. $\dfrac{3x}{17} + \dfrac{8x}{17}$ $\dfrac{11x}{17}$

3. $\dfrac{8x}{15} + \dfrac{x}{15}$ $\dfrac{3x}{5}$

4. $\dfrac{9x}{24} + \dfrac{x}{24}$ $\dfrac{5x}{12}$

5. $\dfrac{x - 3}{12} + \dfrac{5x + 21}{12}$

6. $\dfrac{x + 4}{9} + \dfrac{2x - 25}{9}$ $\dfrac{x - 7}{3}$

7. $\dfrac{4}{x} + \dfrac{2}{x}$ $\dfrac{6}{x}$

8. $\dfrac{5}{x} + \dfrac{13}{x}$ $\dfrac{18}{x}$

9. $\dfrac{8}{9x} + \dfrac{13}{9x}$ $\dfrac{7}{3x}$

10. $\dfrac{4}{9x} + \dfrac{11}{9x}$ $\dfrac{5}{3x}$

11. $\dfrac{5}{x + 3} + \dfrac{4}{x + 3}$ $\dfrac{9}{x + 3}$

12. $\dfrac{8}{x + 6} + \dfrac{10}{x + 6}$ $\dfrac{18}{x + 6}$

13. $\dfrac{x}{x - 3} + \dfrac{4x + 5}{x - 3}$ $\dfrac{5x + 5}{x - 3}$

14. $\dfrac{x}{x - 4} + \dfrac{9x + 7}{x - 4}$ $\dfrac{10x + 7}{x - 4}$

15. $\dfrac{4x + 1}{6x + 5} + \dfrac{8x + 9}{6x + 5}$ 2

16. $\dfrac{3x + 2}{3x + 4} + \dfrac{3x + 6}{3x + 4}$ 2

17. $\dfrac{y^2+7y}{y^2-5y}+\dfrac{y^2-4y}{y^2-5y}$ $\dfrac{2y+3}{y-5}$

18. $\dfrac{y^2-2y}{y^2+3y}+\dfrac{y^2+y}{y^2+3y}$ $\dfrac{2y-1}{y+3}$

19. $\dfrac{4y-1}{5y^2}+\dfrac{3y+1}{5y^2}$ $\dfrac{7}{5y}$

20. $\dfrac{y+2}{6y^3}+\dfrac{3y-2}{6y^3}$ $\dfrac{2}{3y^2}$

21. $\dfrac{x^2-2}{x^2+x-2}+\dfrac{2x-x^2}{x^2+x-2}$ $\dfrac{2}{x+2}$

22. $\dfrac{x^2+9x}{4x^2-11x-3}+\dfrac{3x-5x^2}{4x^2-11x-3}$ $-\dfrac{4x}{4x+1}$

23. $\dfrac{x^2-4x}{x^2-x-6}+\dfrac{4x-4}{x^2-x-6}$ $\dfrac{x-2}{x-3}$

24. $\dfrac{x}{2x+7}-\dfrac{2}{2x+7}$ $\dfrac{x-2}{2x+7}$ 25. $\dfrac{3x}{5x-4}-\dfrac{4}{5x-4}$ $\dfrac{3x-4}{5x-4}$

26. $\dfrac{x}{x-1}-\dfrac{1}{x-1}$ 1 27. $\dfrac{4x}{4x-3}-\dfrac{3}{4x-3}$ 1

28. $\dfrac{2y+1}{3y-7}-\dfrac{y+8}{3y-7}$ $\dfrac{y-7}{3y-7}$ 29. $\dfrac{14y}{7y+2}-\dfrac{7y-2}{7y+2}$ 1

30. $\dfrac{2x+3}{3x-6}-\dfrac{3-x}{3x-6}$ $\dfrac{x}{x-2}$ 31. $\dfrac{3x+1}{4x-2}-\dfrac{x+1}{4x-2}$ $\dfrac{x}{2x-1}$

32. $\dfrac{x^3-3}{2x^4}-\dfrac{7x^3-3}{2x^4}$ $-\dfrac{3}{x}$ 33. $\dfrac{3y^2-1}{3y^3}-\dfrac{6y^2-1}{3y^3}$ $-\dfrac{1}{y}$

34. $\dfrac{y^2+3y}{y^2+y-12}-\dfrac{y^2-12}{y^2+y-12}$ $\dfrac{3}{y-3}$

35. $\dfrac{4y^2+5}{9y^2-64}-\dfrac{y^2-y+29}{9y^2-64}$ $\dfrac{y+3}{3y+8}$

36. $\dfrac{2y^2+6y+8}{y^2-16}-\dfrac{y^2-3y-12}{y^2-16}$ $\dfrac{y+5}{y-4}$

37. $\dfrac{6y^2+y}{2y^2-9y+9}-\dfrac{2y+9}{2y^2-9y+9}-\dfrac{4y-3}{2y^2-9y+9}$ $\dfrac{3y+2}{y-3}$

38. $\dfrac{3y^2-2}{3y^2+10y-8}-\dfrac{y+10}{3y^2+10y-8}-\dfrac{y^2-6y}{3y^2+10y-8}$ $\dfrac{2y-3}{3y-2}$

In Exercises 39–64, denominators are additive inverses. Add or subtract as indicated. Simplify the result, if possible.

39. $\dfrac{4}{x-3}+\dfrac{2}{3-x}$ $\dfrac{2}{x-3}$ 40. $\dfrac{6}{x-5}+\dfrac{2}{5-x}$ $\dfrac{4}{x-5}$

41. $\dfrac{6x+7}{x-6}+\dfrac{3x}{6-x}$ $\dfrac{3x+7}{x-6}$

42. $\dfrac{6x+5}{x-2}+\dfrac{4x}{2-x}$ $\dfrac{2x+5}{x-2}$

43. $\dfrac{5x-2}{3x-4}+\dfrac{2x-3}{4-3x}$ $\dfrac{3x+1}{3x-4}$

44. $\dfrac{9x-1}{7x-3}+\dfrac{6x-2}{3-7x}$ $\dfrac{3x+1}{7x-3}$

45. $\dfrac{x^2}{x-2}+\dfrac{4}{2-x}$ $x+2$ 46. $\dfrac{x^2}{x-3}+\dfrac{9}{3-x}$ $x+3$

47. $\dfrac{y-3}{y^2-25}+\dfrac{y-3}{25-y^2}$ 0 48. $\dfrac{y-7}{y^2-16}+\dfrac{7-y}{16-y^2}$ $\dfrac{2y-14}{y^2-16}$

49. $\dfrac{6}{x-1}-\dfrac{5}{1-x}$ $\dfrac{11}{x-1}$ 50. $\dfrac{10}{x-2}-\dfrac{6}{2-x}$ $\dfrac{16}{x-2}$

51. $\dfrac{10}{x+3}-\dfrac{2}{-x-3}$ $\dfrac{12}{x+3}$ 52. $\dfrac{11}{x+7}-\dfrac{5}{-x-7}$ $\dfrac{16}{x+7}$

53. $\dfrac{y}{y-1}-\dfrac{1}{1-y}$ $\dfrac{y+1}{y-1}$ 54. $\dfrac{y}{y-4}-\dfrac{4}{4-y}$ $\dfrac{y+4}{y-4}$

55. $\dfrac{3-x}{x-7}-\dfrac{2x-5}{7-x}$ $\dfrac{x-2}{x-7}$ 56. $\dfrac{4-x}{x-9}-\dfrac{3x-8}{9-x}$ $\dfrac{2x-4}{x-9}$

57. $\dfrac{x-2}{x^2-25}-\dfrac{x-2}{25-x^2}$ $\dfrac{2x-4}{x^2-25}$

58. $\dfrac{x-8}{x^2-16}-\dfrac{x-8}{16-x^2}$ $\dfrac{2x-16}{x^2-16}$

59. $\dfrac{x}{x-y}+\dfrac{y}{y-x}$ 1 60. $\dfrac{2x-y}{x-y}+\dfrac{x-2y}{y-x}$ $\dfrac{x+y}{x-y}$

61. $\dfrac{2x}{x^2-y^2}+\dfrac{2y}{y^2-x^2}$ $\dfrac{2}{x+y}$

62. $\dfrac{2y}{x^2-y^2}+\dfrac{2x}{y^2-x^2}$ $-\dfrac{2}{x+y}$

63. $\dfrac{x^2-2}{x^2+6x-7}+\dfrac{19-4x}{7-6x-x^2}$ $\dfrac{x-3}{x-1}$

64. $\dfrac{2x+3}{x^2-x-30}+\dfrac{x-2}{30+x-x^2}$ $\dfrac{1}{x-6}$

Practice Plus

In Exercises 65–72, perform the indicated operation or operations. Simplify the result, if possible.

65. $\dfrac{6b^2-10b}{16b^2-48b+27}+\dfrac{7b^2-20b}{16b^2-48b+27}-\dfrac{6b-3b^2}{16b^2-48b+27}$ $\dfrac{4b}{4b-3}$

66. $\dfrac{22b+15}{12b^2+52b-9}+\dfrac{30b-20}{12b^2+52b-9}-\dfrac{4-2b}{12b^2+52b-9}$ $\dfrac{9}{2b+9}$

67. $\dfrac{2y}{y-5}-\left(\dfrac{2}{y-5}+\dfrac{y-2}{y-5}\right)$ $\dfrac{y}{y-5}$

68. $\dfrac{3x}{(x+1)^2}-\left[\dfrac{5x+1}{(x+1)^2}-\dfrac{3x+2}{(x+1)^2}\right]$ $\dfrac{1}{x+1}$

69. $\dfrac{b}{ac+ad-bc-bd}-\dfrac{a}{ac+ad-bc-bd}$ $-\dfrac{1}{c+d}$

70. $\dfrac{y}{ax+bx-ay-by}-\dfrac{x}{ax+bx-ay-by}$ $-\dfrac{1}{a+b}$

71. $\dfrac{(y-3)(y+2)}{(y+1)(y-4)} - \dfrac{(y+2)(y+3)}{(y+1)(4-y)} - \dfrac{(y+5)(y-1)}{(y+1)(4-y)}$

$\dfrac{3y^2 + 8y - 5}{(y+1)(y-4)}$

72. $\dfrac{(y+1)(2y-1)}{(y-2)(y-3)} + \dfrac{(y+2)(y-1)}{(y-2)(y-3)} - \dfrac{(y+5)(2y+1)}{(3-y)(2-y)}$

$\dfrac{y^2 - 9y - 8}{(y-2)(y-3)}$

Application Exercises

73. Anthropologists and forensic scientists classify skulls using

$$\dfrac{L + 60W}{L} - \dfrac{L - 40W}{L},$$

where L is the skull's length and W is its width.

a. Express the classification as a single rational expression.
$\dfrac{100W}{L}$

b. If the value of the rational expression in part (a) is less than 75, a skull is classified as long. A medium skull has a value between 75 and 80, and a round skull has a value over 80. Use your rational expression from part (a) to classify a skull that is 5 inches wide and 6 inches long.
round

74. The temperature, in degrees Fahrenheit, of a dessert placed in a freezer for t hours is modeled by

$$\dfrac{t + 30}{t^2 + 4t + 1} - \dfrac{t - 50}{t^2 + 4t + 1}.$$

a. Express the temperature as a single rational expression.
$\dfrac{80}{t^2 + 4t + 1}$

b. Use your rational expression from part (a) to find the temperature of the dessert, to the nearest hundredth of a degree, after 1 hour and after 2 hours. 13.33°F; 6.15°F

In Exercises 75–76, find the perimeter of each rectangle.

75.
$\dfrac{5}{x+3}$ meters
10 meters
$\dfrac{5x + 10}{x + 3}$ meters

76. $\dfrac{7}{x+4}$ inches
8 inches
$\dfrac{4x + 9}{x + 4}$ inches

Writing in Mathematics

77. Explain how to add rational expressions when denominators are the same. Give an example with your explanation.
77–80. Answers will vary.

78. Explain how to subtract rational expressions when denominators are the same. Give an example with your explanation.

79. Describe two similarities between the following problems:
$$\dfrac{3}{8} + \dfrac{1}{8} \quad \text{and} \quad \dfrac{x}{x^2 - 1} + \dfrac{1}{x^2 - 1}.$$

80. Explain how to add rational expressions when denominators are opposites. Use an example to support your explanation.

Critical Thinking Exercises

81. Which one of the following is true? **d**

a. The sum of two rational expressions with the same denominator can be found by adding numerators, adding denominators, and then simplifying.

b. $\dfrac{4}{b} - \dfrac{2}{-b} = -\dfrac{2}{b}$

c. The difference between two rational expressions with the same denominator can always be simplified.

d. $\dfrac{2x + 1}{x - 7} + \dfrac{3x + 1}{x - 7} - \dfrac{5x + 2}{x - 7} = 0$

In Exercises 82–83, perform the indicated operations. Simplify the result if possible.

82. $\left(\dfrac{3x - 1}{x^2 + 5x - 6} - \dfrac{2x - 7}{x^2 + 5x - 6} \right) \div \dfrac{x + 2}{x^2 - 1} \quad \dfrac{x + 1}{x + 2}$

83. $\left(\dfrac{3x^2 - 4x + 4}{3x^2 + 7x + 2} - \dfrac{10x + 9}{3x^2 + 7x + 2} \right) \div \dfrac{x - 5}{x^2 - 4} \quad x - 2$

In Exercises 84–88, find the missing expression.

84. $\dfrac{2x}{x + 3} + \dfrac{\rule{1cm}{0.3mm}}{x + 3} = \dfrac{4x + 1}{x + 3} \quad 2x + 1$

85. $\dfrac{3x}{x + 2} - \dfrac{\rule{1cm}{0.3mm}}{x + 2} = \dfrac{6 - 17x}{x + 2} \quad 20x - 6$

86. $\dfrac{6}{x - 2} + \dfrac{\rule{1cm}{0.3mm}}{2 - x} = \dfrac{13}{x - 2} \quad -7$

87. $\dfrac{a^2}{a - 4} - \dfrac{\rule{1cm}{0.3mm}}{a - 4} = a + 3 \quad a + 12$

88. $\dfrac{3x}{x - 5} + \dfrac{\rule{1cm}{0.3mm}}{5 - x} = \dfrac{7x + 1}{x - 5} \quad -4x - 1$

Technology Exercises

In Exercises 89–91, use the $\boxed{\text{GRAPH}}$ *or* $\boxed{\text{TABLE}}$ *feature of a graphing utility to determine if the subtraction has been performed correctly. If the answer is wrong, correct it and then verify your correction using the graphing utility.*

89. $\dfrac{3x + 6}{2} - \dfrac{x}{2} = x + 3$ correct answer

90. $\dfrac{x^2 + 4x + 3}{x + 2} - \dfrac{5x + 9}{x + 2} = x - 2, x \neq -2$

   $x - 2$ should be $x - 3$.

91. $\dfrac{x^2 - 13}{x + 4} - \dfrac{3}{x + 4} = x + 4, x \neq -4$ $x + 4$ should be $x - 4$.

Review Exercises

92. Subtract: $\dfrac{13}{15} - \dfrac{8}{45}$. (Section 1.1, Example 9) $\dfrac{31}{45}$

93. Factor completely: $81x^4 - 1$. (Section 6.4, Example 4)

   $(9x^2 + 1)(3x + 1)(3x - 1)$

94. Divide: $\dfrac{3x^3 + 2x^2 - 26x - 15}{x + 3}$. (Section 5.6, Example 2)

   $3x^2 - 7x - 5$

ADDING AND SUBTRACTING RATIONAL EXPRESSIONS WITH DIFFERENT DENOMINATORS

Objectives

1 Find the least common denominator.

2 Add and subtract rational expressions with different denominators.

When my aunt asked how I liked my five-year-old nephew, I replied "medium rare." Unfortunately, my little joke did not get me out of baby sitting for the Dennis the Menace of our family. Now the little squirt doesn't want to go to bed because his head hurts. Does my aunt have any aspirin? What is the proper dosage for a child his age?

In this section's exercise set, you will use two formulas that model drug dosage for children. Before working with these models, we continue drawing on your experience from arithmetic to add and subtract rational expressions that have different denominators.

1 Find the least common denominator.

Finding the Least Common Denominator We can gain insight into adding rational expressions with different denominators by looking closely at what we do when adding fractions with different denominators. For example, suppose that we want to add $\frac{1}{2}$ and $\frac{2}{3}$. We must first write the fractions with the same denominator. We look for the smallest number that contains both 2 and 3 as factors. This number, 6, is then used as the *least common denominator*, or LCD.

The **least common denominator** of several rational expressions is a polynomial consisting of the product of all prime factors in the denominators, with each factor raised to the greatest power of its occurrence in any denominator.

FINDING THE LEAST COMMON DENOMINATOR

1. Factor each denominator completely.
2. List the factors of the first denominator.
3. Add to the list in step 2 any factors of the second denominator that do not appear in the list.
4. Form the product of each different factor from the list in step 3. This product is the least common denominator.

EXAMPLE 1 Finding the Least Common Denominator

Find the LCD of $\dfrac{7}{6x^2}$ and $\dfrac{2}{9x}$.

SOLUTION

Step 1.   **Factor each denominator completely.**

$$6x^2 = 3 \cdot 2x^2 \quad (\text{or } 3 \cdot 2 \cdot x \cdot x)$$
$$9x = 3 \cdot 3x$$

Step 2.   **List the factors of the first denominator.**

$$3, 2, x^2 \quad (\text{or } 3, 2, x, x)$$

Step 3.   **Add any unlisted factors from the second denominator.** Two factors from $3 \cdot 3x$ are already in our list. These factors include x and one factor of 3. We add the other factor of 3 to our list. We have

$$3, 3, 2, x^2.$$

Step 4.   **The least common denominator is the product of all factors in the final list.** Thus,

$$3 \cdot 3 \cdot 2x^2$$

or $18x^2$ is the least common denominator. ∎

✔ **CHECK POINT 1** Find the LCD of $\dfrac{3}{10x^2}$ and $\dfrac{7}{15x}$. $\quad 30x^2$

EXAMPLE 2 Finding the Least Common Denominator

Find the LCD of $\dfrac{3}{x+1}$ and $\dfrac{5}{x-1}$.

SOLUTION

Step 1.   **Factor each denominator completely.**

$$x + 1 = 1(x + 1)$$
$$x - 1 = 1(x - 1)$$

Step 2.   **List the factors of the first denominator.**

$$1, x + 1$$

Step 3. **Add any unlisted factors from the second denominator.** We listed 1 and $x + 1$ as factors of the first denominator, $1(x + 1)$. The factors of the second denominator, $1(x - 1)$, include 1 and $x - 1$. One factor, 1, is already in our list, but the other factor, $x - 1$, is not. We add $x - 1$ to the list. We have

$$1, x + 1, x - 1.$$

Step 4. **The least common denominator is the product of all factors in the final list.** Thus,

$$1(x + 1)(x - 1)$$

or $(x + 1)(x - 1)$ is the least common denominator of $\dfrac{3}{x + 1}$ and $\dfrac{5}{x - 1}$. ■

 CHECK POINT **2** Find the LCD of $\dfrac{2}{x + 3}$ and $\dfrac{4}{x - 3}$. $(x + 3)(x - 3)$

EXAMPLE 3 Finding the Least Common Denominator

Find the LCD of

$$\frac{7}{5x^2 + 15x} \quad \text{and} \quad \frac{9}{x^2 + 6x + 9}.$$

SOLUTION

Step 1. **Factor each denominator completely.**

$$5x^2 + 15x = 5x(x + 3)$$
$$x^2 + 6x + 9 = (x + 3)^2$$

Step 2. **List the factors of the first denominator.**

$$5, x, (x + 3)$$

Step 3. **Add any unlisted factors from the second denominator.** The second denominator is $(x + 3)^2$ or $(x + 3)(x + 3)$. One factor of $x + 3$ is already in our list, but the other factor is not. We add $x + 3$ to the list. We have

$$5, x, (x + 3), (x + 3).$$

Step 4. **The least common denominator is the product of all factors in the final list.** Thus,

$$5x(x + 3)(x + 3) \quad \text{or} \quad 5x(x + 3)^2$$

is the least common denominator. ■

 CHECK POINT **3** Find the LCD of $\dfrac{9}{7x^2 + 28x}$ and $\dfrac{11}{x^2 + 8x + 16}$.

$7x(x + 4)(x + 4)$ or $7x(x + 4)^2$

---

2 Add and subtract rational expressions with different denominators.

Adding and Subtracting Rational Expressions with Different Denominators
Finding the least common denominator for two (or more) rational expressions is the first step needed to add or subtract the expressions. For example, to add $\frac{1}{2}$ and $\frac{2}{3}$, we first determine that the LCD is 6. Then we write each fraction in terms of the LCD.

$$\frac{1}{2} + \frac{2}{3} = \frac{1}{2} \cdot \frac{3}{3} + \frac{2}{3} \cdot \frac{2}{2}$$

Multiply the numerator and denominator of each fraction by whatever extra factors are required to form 6, the LCD.

$\frac{3}{3} = 1$ and $\frac{2}{2} = 1$. Multiplying by 1 does not change a fraction's value.

$$= \frac{3}{6} + \frac{4}{6}$$

$$= \frac{3 + 4}{6}$$

Add numerators. Place this sum over the LCD.

$$= \frac{7}{6}$$

We follow the same steps in adding or subtracting rational expressions with different denominators.

ADDING AND SUBTRACTING RATIONAL EXPRESSIONS THAT HAVE DIFFERENT DENOMINATORS

1. Find the LCD of the rational expressions.

2. Rewrite each rational expression as an equivalent expression whose denominator is the LCD. To do so, multiply the numerator and the denominator of each rational expression by any factor(s) needed to convert the denominator into the LCD.

3. Add or subtract numerators, placing the resulting expression over the LCD.

4. If possible, simplify the resulting rational expression.

EXAMPLE 4 Adding Rational Expressions with Different Denominators

Add: $\dfrac{7}{6x^2} + \dfrac{2}{9x}$.

SOLUTION

Step 1. **Find the least common denominator.** In Example 1, we found that the LCD for these rational expressions is $18x^2$.

Step 2. **Write equivalent expressions with the LCD as denominators.** We must rewrite each rational expression with a denominator of $18x^2$.

$$\frac{7}{6x^2} \cdot \frac{3}{3} = \frac{21}{18x^2} \qquad \frac{2}{9x} \cdot \frac{2x}{2x} = \frac{4x}{18x^2}$$

Multiply the numerator and denominator by 3 to get $18x^2$, the LCD.

Multiply the numerator and denominator by $2x$ to get $18x^2$, the LCD.

Because $\dfrac{3}{3} = 1$ and $\dfrac{2x}{2x} = 1$, we are not changing the value of either rational expression, only its appearance.

Now we are ready to perform the indicated addition.

$$\frac{7}{6x^2} + \frac{2}{9x}$$ *This is the given problem. The LCD is $18x^2$.*

$$= \frac{7}{6x^2} \cdot \frac{3}{3} + \frac{2}{9x} \cdot \frac{2x}{2x}$$ *Write equivalent expressions with the LCD.*

$$= \frac{21}{18x^2} + \frac{4x}{18x^2}$$

Steps 3 and 4. Add numerators, putting this sum over the LCD. Simplify if possible.

$$= \frac{21 + 4x}{18x^2} \quad \text{or} \quad \frac{4x + 21}{18x^2}$$

The numerator is prime and further simplification is not possible. ∎

✔ **CHECK POINT 4** Add: $\dfrac{3}{10x^2} + \dfrac{7}{15x}$. $\dfrac{9 + 14x}{30x^2}$

EXAMPLE 5 Adding Rational Expressions with Different Denominators

Add: $\dfrac{3}{x + 1} + \dfrac{5}{x - 1}$.

SOLUTION

Step 1. Find the least common denominator. The factors of the denominators are $x + 1$ and $x - 1$. In Example 2, we found that the LCD is $(x + 1)(x - 1)$.

Step 2. Write equivalent expressions with the LCD as denominators.

$$\frac{3}{x + 1} + \frac{5}{x - 1}$$

$$= \frac{3(x - 1)}{(x + 1)(x - 1)} + \frac{5(x + 1)}{(x + 1)(x - 1)}$$ *Multiply each numerator and denominator by the extra factor required to form $(x + 1)(x - 1)$, the LCD.*

Steps 3 and 4. Add numerators, putting this sum over the LCD. Simplify if possible.

$$= \frac{3(x - 1) + 5(x + 1)}{(x + 1)(x - 1)}$$

$$= \frac{3x - 3 + 5x + 5}{(x + 1)(x - 1)}$$ *Use the distributive property to multiply and remove grouping symbols.*

$$= \frac{8x + 2}{(x + 1)(x - 1)}$$ *Combine like terms: $3x + 5x = 8x$ and $-3 + 5 = 2$.*

We can factor 2 from the numerator of the answer in Example 5 to obtain

$$\frac{2(4x + 1)}{(x + 1)(x - 1)}.$$

Because the numerator and denominator do not have any common factors, further simplification is not possible. In this section, unless there is a common factor in the numerator and denominator, we will leave an answer's numerator in unfactored form and the denominator in factored form.

 CHECK POINT **5** Add: $\dfrac{2}{x + 3} + \dfrac{4}{x - 3}.$ $\quad \dfrac{6x + 6}{(x + 3)(x - 3)}$

EXAMPLE 6 | Subtracting Rational Expressions with Different Denominators

Subtract: $\dfrac{x}{x + 3} - 1.$

SOLUTION

Step 1. **Find the least common denominator.** We know that 1 means $\frac{1}{1}$. The factor of the first denominator is $x + 3$. Adding the factor of the second denominator, 1, the LCD is $1(x + 3)$ or $x + 3$.

Step 2. **Write equivalent expressions with the LCD as denominators.**

$$\frac{x}{x + 3} - 1$$

$$= \frac{x}{x + 3} - \frac{1}{1} \qquad \text{Write 1 as } \tfrac{1}{1}.$$

$$= \frac{x}{x + 3} - \frac{1(x + 3)}{1(x + 3)} \qquad \begin{array}{l}\text{Multiply the numerator and denominator} \\ \text{of } \tfrac{1}{1} \text{ by the extra factor required to form} \\ x + 3, \text{ the LCD.}\end{array}$$

Steps 3 and 4. **Subtract numerators, putting this difference over the LCD. Simplify if possible.**

$$= \frac{x - (x + 3)}{x + 3}$$

$$= \frac{x - x - 3}{x + 3} \qquad \begin{array}{l}\text{Remove parentheses and then} \\ \text{change the sign of each term.}\end{array}$$

$$= \frac{-3}{x + 3} \quad \text{or} \quad -\frac{3}{x + 3} \qquad \text{Simplify.} \qquad \blacksquare$$

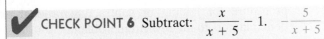 CHECK POINT **6** Subtract: $\dfrac{x}{x + 5} - 1.$ $\quad -\dfrac{5}{x + 5}$

EXAMPLE 7 Subtracting Rational Expressions with Different Denominators

Subtract: $\dfrac{y+2}{4y+16} - \dfrac{2}{y^2+4y}$.

SOLUTION

Step 1. **Find the least common denominator.** Start by factoring the denominators.

$$4y + 16 = 4(y + 4)$$

$$y^2 + 4y = y(y + 4)$$

The factors of the first denominator are 4 and $y + 4$. The only factor from the second denominator that is unlisted is y. Thus, the least common denominator is $4y(y + 4)$.

Step 2. **Write equivalent expressions with the LCD as denominators.**

$$\frac{y+2}{4y+16} - \frac{2}{y^2+4y}$$

$$= \frac{y+2}{4(y+4)} - \frac{2}{y(y+4)} \qquad \text{Factor denominators.}$$
$$\text{The LCD is } 4y(y + 4).$$

$$= \frac{(y+2)y}{4y(y+4)} - \frac{2 \cdot 4}{4y(y+4)} \qquad \text{Multiply each numerator and denominator by the extra factor required to form } 4y(y + 4), \text{ the LCD.}$$

Steps 3 and 4. **Subtract numerators, putting this difference over the LCD. Simplify if possible.**

$$= \frac{(y+2)y - 2 \cdot 4}{4y(y+4)}$$

$$= \frac{y^2 + 2y - 8}{4y(y+4)} \qquad \text{Use the distributive property:}$$
$$(y + 2)y = y^2 + 2y. \text{ Multiply: } 2 \cdot 4 = 8.$$

$$= \frac{\overset{1}{\cancel{(y+4)}}(y-2)}{4y\underset{1}{\cancel{(y+4)}}} \qquad \text{Factor and simplify.}$$

$$= \frac{y-2}{4y}$$

∎

✔ **CHECK POINT 7** Subtract: $\dfrac{5}{y^2-5y} - \dfrac{y}{5y-25}$. $-\dfrac{5+y}{5y}$

In some situations, after factoring denominators, a factor in one denominator is the opposite of a factor in the other denominator. When this happens, we can use the following procedure:

ADDING AND SUBTRACTING RATIONAL EXPRESSIONS WHEN DENOMINATORS CONTAIN OPPOSITE FACTORS When one denominator contains the opposite factor of the other, first multiply either rational expression by $\frac{-1}{-1}$. Then apply the procedure for adding or subtracting rational expressions that have different denominators to the rewritten problem.

EXAMPLE 8 Adding Rational Expressions with Opposite Factors in the Denominators

Add: $\dfrac{x^2 - 2}{2x^2 - x - 3} + \dfrac{x - 2}{3 - 2x}$.

SOLUTION

Step 1. **Find the least common denominator.** Start by factoring the denominators.

$$2x^2 - x - 3 = (2x - 3)(x + 1)$$

$$3 - 2x = 1(3 - 2x)$$

Do you see that $2x - 3$ and $3 - 2x$ are opposite factors? Thus, we multiply either rational expression by $\frac{-1}{-1}$. We will use the second rational expression, resulting in $2x - 3$ in the denominator.

$$\frac{x^2 - 2}{2x^2 - x - 3} + \frac{x - 2}{3 - 2x}$$

$$= \frac{x^2 - 2}{(2x - 3)(x + 1)} + \frac{(-1)}{(-1)} \cdot \frac{x - 2}{3 - 2x}$$

Factor the first denominator. Multiply the second rational expression by $\frac{-1}{-1}$.

$$= \frac{x^2 - 2}{(2x - 3)(x + 1)} + \frac{-x + 2}{-3 + 2x}$$

Perform the multiplications by -1 by changing every term's sign.

$$= \frac{x^2 - 2}{(2x - 3)(x + 1)} + \frac{2 - x}{2x - 3}$$

The LCD of our rewritten addition problem is $(2x - 3)(x + 1)$.

Step 2. **Write equivalent expressions with the LCD as denominators.**

$$= \frac{x^2 - 2}{(2x - 3)(x + 1)} + \frac{(2 - x)(x + 1)}{(2x - 3)(x + 1)}$$

Multiply the numerator and denominator of the second rational expression by the extra factor required to form $(2x - 3)(x + 1)$, the LCD.

Steps 3 and 4. **Add numerators, putting this sum over the LCD. Simplify if possible.**

$$= \frac{x^2 - 2 + (2 - x)(x + 1)}{(2x - 3)(x + 1)}$$

$$= \frac{x^2 - 2 + 2x + 2 - x^2 - x}{(2x - 3)(x + 1)}$$

Use the FOIL method to multiply $(2 - x)(x + 1)$.

$$= \frac{(x^2 - x^2) + (2x - x) + (-2 + 2)}{(2x - 3)(x + 1)}$$

Group like terms.

$$= \frac{x}{(2x - 3)(x + 1)}$$

Combine like terms. ∎

DISCOVER FOR YOURSELF

In Example 8, the denominators can be factored as follows:

$$2x^2 - x - 3 = (2x - 3)(x + 1)$$
$$3 - 2x = -1(2x - 3).$$

Using these factorizations, what is the LCD? Solve Example 8 by obtaining this LCD in each rational expression. Then combine the expressions. How does your solution compare with the one shown on the right?

✔ CHECK POINT 8 Add: $\dfrac{4x}{x^2 - 25} + \dfrac{3}{5 - x}$. $\dfrac{x - 15}{(x + 5)(x - 5)}$

7.4 EXERCISE SET

Student Solutions Manual   CD/Video   PH Math/Tutor Center   MathXL Tutorials on CD   MathXL®   MyMathLab   Interactmath.com

Practice Exercises

In Exercises 1–16, find the least common denominator of the rational expressions.

1. $\dfrac{7}{15x^2}$ and $\dfrac{13}{24x}$ $\quad 120x^2$

2. $\dfrac{11}{25x^2}$ and $\dfrac{17}{35x}$ $\quad 175x^2$

3. $\dfrac{8}{15x^2}$ and $\dfrac{5}{6x^5}$ $\quad 30x^5$

4. $\dfrac{7}{15x^2}$ and $\dfrac{11}{24x^5}$ $\quad 120x^5$

5. $\dfrac{4}{x-3}$ and $\dfrac{7}{x+1}$ $\quad (x-3)(x+1)$

6. $\dfrac{2}{x-5}$ and $\dfrac{3}{x+7}$ $\quad (x-5)(x+7)$

7. $\dfrac{5}{7(y+2)}$ and $\dfrac{10}{y}$ $\quad 7y(y+2)$

8. $\dfrac{8}{11(y+5)}$ and $\dfrac{12}{y}$ $\quad 11y(y+5)$

9. $\dfrac{17}{x+4}$ and $\dfrac{18}{x^2-16}$ $\quad (x+4)(x-4)$

10. $\dfrac{3}{x-6}$ and $\dfrac{4}{x^2-36}$ $\quad (x-6)(x+6)$

11. $\dfrac{8}{y^2-9}$ and $\dfrac{14}{y(y+3)}$ $\quad y(y+3)(y-3)$

12. $\dfrac{14}{y^2-49}$ and $\dfrac{12}{y(y-7)}$ $\quad y(y+7)(y-7)$

13. $\dfrac{7}{y^2-1}$ and $\dfrac{y}{y^2-2y+1}$ $\quad (y+1)(y-1)(y-1)$

14. $\dfrac{9}{y^2-25}$ and $\dfrac{y}{y^2-10y+25}$ $\quad (y+5)(y-5)(y-5)$

15. $\dfrac{3}{x^2-x-20}$ and $\dfrac{x}{2x^2+7x-4}$ $\quad (x-5)(x+4)(2x-1)$

16. $\dfrac{7}{x^2-5x-6}$ and $\dfrac{x}{x^2-4x-5}$ $\quad (x-6)(x+1)(x-5)$

In Exercises 17–82, add or subtract as indicated. Simplify the result, if possible.

17. $\dfrac{3}{x}+\dfrac{5}{x^2}$ $\quad \dfrac{3x+5}{x^2}$

18. $\dfrac{4}{x}+\dfrac{8}{x^2}$ $\quad \dfrac{4x+8}{x^2}$

19. $\dfrac{2}{9x}+\dfrac{11}{6x}$ $\quad \dfrac{37}{18x}$

20. $\dfrac{5}{6x}+\dfrac{7}{8x}$ $\quad \dfrac{41}{24x}$

21. $\dfrac{4}{x}+\dfrac{7}{2x^2}$ $\quad \dfrac{8x+7}{2x^2}$

22. $\dfrac{10}{x}+\dfrac{3}{5x^2}$ $\quad \dfrac{50x+3}{5x^2}$

23. $6+\dfrac{1}{x}$ $\quad \dfrac{6x+1}{x}$

24. $3+\dfrac{1}{x}$ $\quad \dfrac{3x+1}{x}$

25. $\dfrac{2}{x}+9$ $\quad \dfrac{2+9x}{x}$

26. $\dfrac{7}{x}+4$ $\quad \dfrac{7+4x}{x}$

27. $\dfrac{x-1}{6}+\dfrac{x+2}{3}$ $\quad \dfrac{x+1}{2}$

28. $\dfrac{x+3}{2}+\dfrac{x+5}{4}$ $\quad \dfrac{3x+11}{4}$

29. $\dfrac{4}{x}+\dfrac{3}{x-5}$ $\quad \dfrac{7x-20}{x(x-5)}$

30. $\dfrac{3}{x}+\dfrac{4}{x-6}$ $\quad \dfrac{7x-18}{x(x-6)}$

31. $\dfrac{2}{x-1}+\dfrac{3}{x+2}$ $\quad \dfrac{5x+1}{(x-1)(x+2)}$

32. $\dfrac{3}{x-2}+\dfrac{4}{x+3}$ $\quad \dfrac{7x+1}{(x-2)(x+3)}$

33. $\dfrac{2}{y+5}+\dfrac{3}{4y}$ $\quad \dfrac{11y+15}{4y(y+5)}$

34. $\dfrac{3}{y+1}+\dfrac{2}{3y}$ $\quad \dfrac{11y+2}{3y(y+1)}$

35. $\dfrac{x}{x+7}-1$ $\quad -\dfrac{7}{x+7}$

36. $\dfrac{x}{x+6}-1$ $\quad -\dfrac{6}{x+6}$

37. $\dfrac{7}{x+5}-\dfrac{4}{x-5}$ $\quad \dfrac{3x-55}{(x+5)(x-5)}$

38. $\dfrac{8}{x+6}-\dfrac{2}{x-6}$ $\quad \dfrac{6x-60}{(x+6)(x-6)}$

39. $\dfrac{2x}{x^2-16}+\dfrac{x}{x-4}$ $\quad \dfrac{x^2+6x}{(x-4)(x+4)}$

40. $\dfrac{4x}{x^2-25}+\dfrac{x}{x+5}$ $\quad \dfrac{x^2-x}{(x+5)(x-5)}$

41. $\dfrac{5y}{y^2-9}-\dfrac{4}{y+3}$ $\quad \dfrac{y+12}{(y+3)(y-3)}$

42. $\dfrac{8y}{y^2-16}-\dfrac{5}{y+4}$ $\quad \dfrac{3y+20}{(y+4)(y-4)}$

43. $\dfrac{7}{x-1}-\dfrac{3}{(x-1)^2}$ $\quad \dfrac{7x-10}{(x-1)(x-1)}$

44. $\dfrac{5}{x+3}-\dfrac{2}{(x+3)^2}$ $\quad \dfrac{5x+13}{(x+3)(x+3)}$

45. $\dfrac{3y}{4y-20}+\dfrac{9y}{6y-30}$ $\quad \dfrac{9y}{4(y-5)}$

46. $\dfrac{4y}{5y-10}+\dfrac{3y}{10y-20}$ $\quad \dfrac{11y}{10(y-2)}$

47. $\dfrac{y+4}{y}-\dfrac{y}{y+4}$ $\quad \dfrac{8y+16}{y(y+4)}$

48. $\dfrac{y}{y-5}-\dfrac{y-5}{y}$ $\quad \dfrac{10y-25}{y(y-5)}$

49. $\dfrac{2x+9}{x^2-7x+12}-\dfrac{2}{x-3}$ $\quad \dfrac{17}{(x-3)(x-4)}$

50. $\dfrac{3x+7}{x^2-5x+6}-\dfrac{3}{x-3}$ $\quad \dfrac{13}{(x-3)(x-2)}$

51. $\dfrac{3}{x^2-1}+\dfrac{4}{(x+1)^2}$ $\quad \dfrac{7x-1}{(x+1)(x+1)(x-1)}$

52. $\dfrac{6}{x^2 - 4} + \dfrac{2}{(x + 2)^2}$ $\dfrac{8x + 8}{(x + 2)(x + 2)(x - 2)}$

53. $\dfrac{3x}{x^2 + 3x - 10} - \dfrac{2x}{x^2 + x - 6}$ $\dfrac{x^2 - x}{(x + 3)(x - 2)(x + 5)}$

54. $\dfrac{x}{x^2 - 2x - 24} - \dfrac{x}{x^2 - 7x + 6}$ $-\dfrac{5x}{(x - 6)(x - 1)(x + 4)}$

55. $\dfrac{y}{y^2 + 2y + 1} + \dfrac{4}{y^2 + 5y + 4}$ $\dfrac{y^2 + 8y + 4}{(y + 4)(y + 1)(y + 1)}$

56. $\dfrac{y}{y^2 + 5y + 6} + \dfrac{4}{y^2 - y - 6}$ $\dfrac{y^2 + y + 12}{(y - 3)(y + 2)(y + 3)}$

57. $\dfrac{x - 5}{x + 3} + \dfrac{x + 3}{x - 5}$ $\dfrac{2x^2 - 4x + 34}{(x + 3)(x - 5)}$

58. $\dfrac{x - 7}{x + 4} + \dfrac{x + 4}{x - 7}$ $\dfrac{2x^2 - 6x + 65}{(x + 4)(x - 7)}$

59. $\dfrac{5}{2y^2 - 2y} - \dfrac{3}{2y - 2}$ $\dfrac{5 - 3y}{2y(y - 1)}$

60. $\dfrac{7}{5y^2 - 5y} - \dfrac{2}{5y - 5}$ $\dfrac{7 - 2y}{5y(y - 1)}$

61. $\dfrac{4x + 3}{x^2 - 9} - \dfrac{x + 1}{x - 3}$ $-\dfrac{x^2}{(x + 3)(x - 3)}$

62. $\dfrac{2x - 1}{x + 6} - \dfrac{6 - 5x}{x^2 - 36}$ $\dfrac{2x^2 - 8x}{(x + 6)(x - 6)}$

63. $\dfrac{y^2 - 39}{y^2 + 3y - 10} - \dfrac{y - 7}{y - 2}$ $\dfrac{2}{y + 5}$

64. $\dfrac{y^2 - 6}{y^2 + 9y + 18} - \dfrac{y - 4}{y + 6}$ $\dfrac{1}{y + 3}$

65. $4 + \dfrac{1}{x - 3}$ $\dfrac{4x - 11}{x - 3}$ **66.** $7 + \dfrac{1}{x - 5}$ $\dfrac{7x - 34}{x - 5}$

67. $3 - \dfrac{3y}{y + 1}$ $\dfrac{3}{y + 1}$ **68.** $7 - \dfrac{4y}{y + 5}$ $\dfrac{3y + 35}{y + 5}$

69. $\dfrac{9x + 3}{x^2 - x - 6} + \dfrac{x}{3 - x}$ $\dfrac{-x^2 + 7x + 3}{(x - 3)(x + 2)}$

70. $\dfrac{x^2 + 9x}{x^2 - 2x - 3} + \dfrac{5}{3 - x}$ $\dfrac{x^2 + 4x - 5}{(x - 3)(x + 1)}$

71. $\dfrac{x + 3}{x^2 + x - 2} - \dfrac{2}{x^2 - 1}$ $\dfrac{x^2 + 2x - 1}{(x + 1)(x - 1)(x + 2)}$

72. $\dfrac{x}{x^2 - 10x + 25} - \dfrac{x - 4}{2x - 10}$ $\dfrac{-x^2 + 11x - 20}{2(x - 5)(x - 5)}$

73. $\dfrac{y + 3}{5y^2} - \dfrac{y - 5}{15y}$ $\dfrac{-y^2 + 8y + 9}{15y^2}$

74. $\dfrac{y - 7}{3y^2} - \dfrac{y - 2}{12y}$ $\dfrac{-y^2 + 6y - 28}{12y^2}$

75. $\dfrac{x + 3}{3x + 6} + \dfrac{x}{4 - x^2}$ $\dfrac{x^2 - 2x - 6}{3(x + 2)(x - 2)}$

76. $\dfrac{x + 7}{4x + 12} + \dfrac{x}{9 - x^2}$ $\dfrac{x^2 - 21}{4(x + 3)(x - 3)}$

77. $\dfrac{y}{y^2 - 1} + \dfrac{2y}{y - y^2}$ $\dfrac{-y - 2}{(y + 1)(y - 1)}$

78. $\dfrac{y}{y^2 - 1} + \dfrac{5y}{y - y^2}$ $\dfrac{-4y - 5}{(y + 1)(y - 1)}$

79. $\dfrac{x - 1}{x} + \dfrac{y + 1}{y}$ $\dfrac{x + 2xy - y}{xy}$

80. $\dfrac{x + 2}{y} + \dfrac{y - 2}{x}$ $\dfrac{x^2 + 2x - 2y + y^2}{xy}$

81. $\dfrac{3x}{x^2 - y^2} - \dfrac{2}{y - x}$ $\dfrac{5x + 2y}{(x + y)(x - y)}$

82. $\dfrac{7x}{x^2 - y^2} - \dfrac{3}{y - x}$ $\dfrac{10x + 3y}{(x + y)(x - y)}$

Practice Plus

In Exercises 83–92, perform the indicated operation or operations. Simplify the result, if possible.

83. $\dfrac{x + 6}{x^2 - 4} - \dfrac{x + 3}{x + 2} + \dfrac{x - 3}{x - 2}$ $\dfrac{-x + 6}{(x + 2)(x - 2)}$

84. $\dfrac{x + 8}{x^2 - 9} - \dfrac{x + 2}{x + 3} + \dfrac{x - 2}{x - 3}$ $\dfrac{3x + 8}{(x + 3)(x - 3)}$

85. $\dfrac{5}{x^2 - 25} + \dfrac{4}{x^2 - 11x + 30} - \dfrac{3}{x^2 - x - 30}$

   $\dfrac{6x + 5}{(x + 5)(x - 5)(x - 6)}$

86. $\dfrac{3}{x^2 - 49} + \dfrac{2}{x^2 - 15x + 56} - \dfrac{5}{x^2 - x - 56}$

   $\dfrac{25}{(x + 7)(x - 7)(x - 8)}$

87. $\dfrac{x + 6}{x^3 - 27} - \dfrac{x}{x^3 + 3x^2 + 9x}$ $\dfrac{9}{(x - 3)(x^2 + 3x + 9)}$

88. $\dfrac{x + 8}{x^3 - 8} - \dfrac{x}{x^3 + 2x^2 + 4x}$ $\dfrac{10}{(x - 2)(x^2 + 2x + 4)}$

89. $\dfrac{9y + 3}{y^2 - y - 6} + \dfrac{y}{3 - y} + \dfrac{y - 1}{y + 2}$ $\dfrac{3}{y - 3}$

90. $\dfrac{7y - 2}{y^2 - y - 12} + \dfrac{2y}{4 - y} + \dfrac{y + 1}{y + 3}$

   $\dfrac{-y^2 - 2y - 6}{(y - 4)(y + 3)}$ or $-\dfrac{y^2 + 2y + 6}{(y - 4)(y + 3)}$

91. $\dfrac{3}{x^2 + 4xy + 3y^2} - \dfrac{5}{x^2 - 2xy - 3y^2} + \dfrac{2}{x^2 - 9y^2}$

   $\dfrac{-22y}{(x + 3y)(x - 3y)(x + y)}$

92. $\dfrac{5}{x^2 + 3xy + 2y^2} - \dfrac{7}{x^2 - xy - 2y^2} + \dfrac{4}{x^2 - 4y^2}$

   $\dfrac{2x - 20y}{(x + 2y)(x - 2y)(x + y)}$

Application Exercises

Two formulas that approximate the dosage of a drug prescribed for children are

$$\text{Young's Rule: } C = \frac{DA}{A + 12}$$

$$\text{and Cowling's Rule: } C = \frac{D(A + 1)}{24}.$$

In each formula, A = the child's age, in years, D = an adult dosage, and C = the proper child's dosage. The formulas apply for ages 2 through 13, inclusive. Use the formulas to solve Exercises 93–96.

93. Use Young's rule to find the difference in a child's dosage for an 8-year-old child and a 3-year-old child. Express the answer as a single rational expression in terms of D. Then describe what your answer means in terms of the variables in the model. $\frac{D}{5}$; Answers will vary.

94. Use Young's rule to find the difference in a child's dosage for a 10-year-old child and a 3-year-old child. Express the answer as a single rational expression in terms of D. Then describe what your answer means in terms of the variables in the model. $\frac{14D}{55}$; Answers will vary.

95. For a 12-year-old child, what is the difference in the dosage given by Cowling's rule and Young's rule? Express the answer as a single rational expression in terms of D. Then describe what your answer means in terms of the variables in the models. $\frac{D}{24}$; Answers will vary.

96. Use Cowling's rule to find the difference in a child's dosage for a 12-year-old child and a 10-year-old child. Express the answer as a single rational expression in terms of D. Then describe what your answer means in terms of the variables in the model. $\frac{D}{12}$; Answers will vary.

The graphs illustrate Young's rule and Cowling's rule when the dosage of a drug prescribed for an adult is 1000 milligrams. Use the graphs to solve Exercises 97–100.

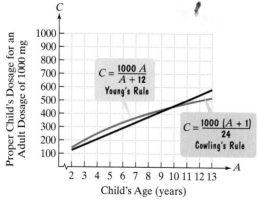

97. Does either formula consistently give a smaller dosage than the other? If so, which one? no

98. Is there an age at which the dosage given by one formula becomes greater than the dosage given by the other? If so, what is a reasonable estimate of that age? yes; 10 years

99. For what age under 11 is the difference in dosage given by the two formulas the greatest? 5 years

100. For what age over 11 is the difference in dosage given by the two formulas the greatest? 13 years

In Exercises 101–102, express the perimeter of each rectangle as a single rational expression.

101. $\frac{x}{x+3}$; $\frac{x}{x+4}$; $\dfrac{4x^2 + 14x}{(x + 3)(x + 4)}$

102. $\frac{x}{x+5}$; $\frac{x}{x+6}$; $\dfrac{4x^2 + 22x}{(x + 5)(x + 6)}$

Writing in Mathematics

103. Explain how to find the least common denominator for denominators of $x^2 - 100$ and $x^2 - 20x + 100$. 103–104. Answers will vary.

104. Explain how to add rational expressions that have different denominators. Use $\dfrac{3}{x + 5} + \dfrac{7}{x + 2}$ in your explanation.

Explain the error in Exercises 105–106. Then rewrite the right side of the equation to correct the error that now exists.

105. $\dfrac{1}{x} + \dfrac{2}{5} = \dfrac{3}{x + 5}$ Answers will vary; $\dfrac{3}{x + 5}$ should be $\dfrac{5 + 2x}{5x}$.

106. $\dfrac{1}{x} + 7 = \dfrac{1}{x + 7}$ Answers will vary; $\dfrac{1}{x + 7}$ should be $\dfrac{1 + 7x}{x}$.

107. The formulas in Exercises 93–96 relate the dosage of a drug prescribed for children to the child's age. Describe another factor that might be used when determining a child's dosage. Is this factor more or less important than age? Explain why. Answers will vary.

Critical Thinking Exercises

108. Which one of the following is true? d

a. $x - \dfrac{1}{5} = \dfrac{4}{5}x$

b. The LCD of $\dfrac{1}{x}$ and $\dfrac{2x}{x - 1}$ is $x^2 - 1$.

c. $\dfrac{1}{x} + \dfrac{x}{1} = \dfrac{1}{\overset{1}{\cancel{x}}} + \dfrac{\overset{1}{\cancel{x}}}{1} = 1 + 1 = 2$

d. $\dfrac{2}{x} + 1 = \dfrac{2 + x}{x}, x \neq 0$

In Exercises 109–110, perform the indicated operations. Simplify the result, if possible.

109. $\dfrac{y^2 + 5y + 4}{y^2 + 2y - 3} \cdot \dfrac{y^2 + y - 6}{y^2 + 2y - 3} - \dfrac{2}{y - 1}$ $\dfrac{y^3 + y^2 - 10y - 2}{(y - 1)(y - 1)(y + 3)}$

110. $\left(\dfrac{1}{x + h} - \dfrac{1}{x} \right) \div h$ $-\dfrac{1}{x(x + h)}$

In Exercises 111–112, find the missing rational expression.

111. $\dfrac{2}{x - 1} + \underline{} = \dfrac{2x^2 + 3x - 1}{x^2(x - 1)}$ $\dfrac{3x - 1}{x^2(x - 1)}$

112. $\dfrac{4}{x - 2} - \underline{} = \dfrac{2x + 8}{(x - 2)(x + 1)}$ $\dfrac{2}{x + 1}$

Review Exercises

113. Multiply: $(3x + 5)(2x - 7)$. (Section 5.3, Example 2)
$6x^2 - 11x - 35$

114. Graph: $3x - y = 3$. (Section 3.2, Example 5)
See graphing answer section.

115. Write the slope-intercept form of the equation of the line passing through $(-3, -4)$ and $(1, 0)$. (Section 3.5, Example 2) $y = x - 1$

✔ **MID-CHAPTER CHECK POINT**

CHAPTER 7

What You Know: We learned that it is necessary to exclude any value or values of a variable that make the denominator of a rational expression zero. We learned to simplify rational expressions by dividing the numerator and the denominator by common factors. We performed a variety of operations with rational expressions, including multiplication, division, addition, and subtraction.

1. Find all numbers for which $\dfrac{x^2 - 4}{x^2 - 2x - 8}$ is undefined. $x = -2$ and $x = 4$

In Exercises 2–4, simplify each rational expression.

2. $\dfrac{3x^2 - 7x + 2}{6x^2 + x - 1}$ $\dfrac{x - 2}{2x + 1}$

3. $\dfrac{9 - 3y}{y^2 - 5y + 6}$ $-\dfrac{3}{y - 2}$ or $\dfrac{3}{2 - y}$

4. $\dfrac{16w^3 - 24w^2}{8w^4 - 12w^3}$ $\dfrac{2}{w}$

In Exercises 5–20, perform the indicated operations. Simplify the result, if possible.

5. $\dfrac{7x - 3}{x^2 + 3x - 4} - \dfrac{3x + 1}{x^2 + 3x - 4}$ $\dfrac{4}{x + 4}$

6. $\dfrac{x + 2}{2x - 4} \cdot \dfrac{8}{x^2 - 4}$ $\dfrac{4}{(x - 2)^2}$

7. $1 + \dfrac{7}{x - 2}$ $\dfrac{x + 5}{x - 2}$

8. $\dfrac{2x^2 + x - 1}{2x^2 - 7x + 3} \div \dfrac{x^2 - 3x - 4}{x^2 - x - 6}$ $\dfrac{x + 2}{x - 4}$

9. $\dfrac{1}{x^2 + 2x - 3} + \dfrac{1}{x^2 + 5x + 6}$ $\dfrac{2x + 1}{(x + 2)(x + 3)(x - 1)}$

10. $\dfrac{17}{x - 5} + \dfrac{x + 8}{5 - x}$ $\dfrac{9 - x}{x - 5}$

11. $\dfrac{4y^2 - 1}{9y - 3y^2} \cdot \dfrac{y^2 - 7y + 12}{2y^2 - 7y - 4}$ $-\dfrac{2y - 1}{3y}$ or $\dfrac{1 - 2y}{3y}$

12. $\dfrac{y}{y + 1} - \dfrac{2y}{y + 2}$ $\dfrac{-y^2}{(y + 1)(y + 2)}$

13. $\dfrac{w^2 + 6w + 5}{7w^2 - 63} \div \dfrac{w^2 + 10w + 25}{7w + 21}$ $\dfrac{w + 1}{(w - 3)(w + 5)}$

14. $\dfrac{2z}{z^2 - 9} - \dfrac{5}{z^2 + 4z + 3}$ $\dfrac{2z^2 - 3z + 15}{(z + 3)(z - 3)(z + 1)}$

15. $\dfrac{z + 2}{3z - 1} + \dfrac{5}{(3z - 1)^2}$ $\dfrac{3z^2 + 5z + 3}{(3z - 1)^2}$

16. $\dfrac{8}{x^2 + 4x - 21} + \dfrac{3}{x + 7}$ $\dfrac{3x - 1}{(x + 7)(x - 3)}$

17. $\dfrac{x^4 - 27x}{x^2 - 9} \cdot \dfrac{x + 3}{x^2 + 3x + 9}$ x

18. $\dfrac{x - 1}{x^2 - x - 2} - \dfrac{x + 2}{x^2 + 4x + 3}$ $\dfrac{2x + 1}{(x + 1)(x - 2)(x + 3)}$

19. $\dfrac{x^2 - 2xy + y^2}{x + y} \div \dfrac{x^2 - xy}{5x + 5y}$ $\dfrac{5x - 5y}{x}$

20. $\dfrac{5}{x + 5} + \dfrac{x}{x - 4} - \dfrac{11x - 8}{x^2 + x - 20}$ $\dfrac{x + 3}{x + 5}$

SECTION

7.5

Objectives

1 Simplify complex rational expressions by dividing.

2 Simplify complex rational expressions by multiplying by the LCD.

COMPLEX RATIONAL EXPRESSIONS

Do you drive to and from campus each day? If the one-way distance of your round-trip commute is d, then your average rate, or speed, is given by the expression

$$\frac{2d}{\dfrac{d}{r_1} + \dfrac{d}{r_2}}$$

in which r_1 and r_2 are your average rates on the outgoing and return trips, respectively. Do you notice anything unusual about this expression? It has two separate rational expressions in its denominator.

Complex rational expressions, also called **complex fractions**, have numerators or denominators containing one or more rational expressions. Here is another example of such an expression:

$$\cfrac{1 + \dfrac{1}{x}}{1 - \dfrac{1}{x}}.$$

Numerator / Main fraction bar / Denominator / Separate rational expressions occur in the numerator and denominator.

In this section, we study two methods for simplifying complex rational expressions.

1 Simplify complex rational expressions by dividing.

Simplifying by Rewriting Complex Rational Expressions as a Quotient of Two Rational Expressions One method for simplifying a complex rational expression is to combine its numerator into a single expression and combine its denominator into a single expression. Then perform the division by inverting the denominator and multiplying.

> ### SIMPLIFYING A COMPLEX RATIONAL EXPRESSION BY DIVIDING
> **1.** If necessary, add or subtract to get a single rational expression in the numerator.
> **2.** If necessary, add or subtract to get a single rational expression in the denominator.
> **3.** Perform the division indicated by the main fraction bar: Invert the denominator of the complex rational expression and multiply.
> **4.** If possible, simplify.

The following examples illustrate the use of this first method:

EXAMPLE 1 Simplifying a Complex Rational Expression

Simplify:

$$\frac{\dfrac{1}{3} + \dfrac{2}{5}}{\dfrac{2}{5} - \dfrac{1}{3}}.$$

SOLUTION Let's first identify the parts of this complex rational expression.

Numerator

Main fraction bar
$$\frac{\dfrac{1}{3} + \dfrac{2}{5}}{\dfrac{2}{5} - \dfrac{1}{3}}$$

Denominator

Step 1. **Add to get a single rational expression in the numerator.**

$$\frac{1}{3} + \frac{2}{5} = \frac{1 \cdot 5}{3 \cdot 5} + \frac{2 \cdot 3}{5 \cdot 3} = \frac{5}{15} + \frac{6}{15} = \frac{11}{15}$$

The LCD is 3 · 5, or 15.

Step 2. **Subtract to get a single rational expression in the denominator.**

$$\frac{2}{5} - \frac{1}{3} = \frac{2 \cdot 3}{5 \cdot 3} - \frac{1 \cdot 5}{3 \cdot 5} = \frac{6}{15} - \frac{5}{15} = \frac{1}{15}$$

The LCD is 15.

Steps 3 and 4. **Perform the division indicated by the main fraction bar: Invert and multiply. If possible, simplify.**

$$\frac{\dfrac{1}{3} + \dfrac{2}{5}}{\dfrac{2}{5} - \dfrac{1}{3}} = \frac{\dfrac{11}{15}}{\dfrac{1}{15}} = \frac{11}{15} \cdot \frac{15}{1} = \frac{11}{\cancel{15}} \cdot \frac{\cancel{15}^{1}}{1} = 11$$

Invert and multiply.

■

✔ **CHECK POINT 1** Simplify: $\dfrac{\dfrac{1}{4} + \dfrac{2}{3}}{\dfrac{2}{3} - \dfrac{1}{4}}.$ $\dfrac{11}{5}$

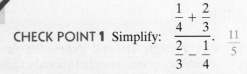

EXAMPLE 2 Simplifying a Complex Rational Expression

Simplify:

$$\frac{1 + \dfrac{1}{x}}{1 - \dfrac{1}{x}}.$$

SOLUTION

Step 1. **Add to get a single rational expression in the numerator.**

$$1 + \frac{1}{x} = \frac{1}{1} + \frac{1}{x} = \frac{1 \cdot x}{1 \cdot x} + \frac{1}{x} = \frac{x}{x} + \frac{1}{x} = \frac{x + 1}{x}$$

The LCD is 1 · x, or x.

Step 2. **Subtract to get a single rational expression in the denominator.**

$$1 - \frac{1}{x} = \frac{1}{1} - \frac{1}{x} = \frac{1 \cdot x}{1 \cdot x} - \frac{1}{x} = \frac{x}{x} - \frac{1}{x} = \frac{x - 1}{x}$$

The LCD is 1 · x, or x.

Steps 3 and 4. **Perform the division indicated by the main fraction bar: Invert and multiply. If possible, simplify.**

$$\frac{1 + \dfrac{1}{x}}{1 - \dfrac{1}{x}} = \frac{\dfrac{x + 1}{x}}{\dfrac{x - 1}{x}} = \frac{x + 1}{x} \cdot \frac{x}{x - 1} = \frac{x + 1}{\overset{}{\underset{1}{\cancel{x}}}} \cdot \frac{\overset{1}{\cancel{x}}}{x - 1} = \frac{x + 1}{x - 1}$$

Invert and multiply.

 CHECK POINT 2 Simplify: $\dfrac{2 - \dfrac{1}{x}}{2 + \dfrac{1}{x}}.$ $\dfrac{2x - 1}{2x + 1}$

EXAMPLE 3 Simplifying a Complex Rational Expression

Simplify:

$$\frac{\dfrac{1}{xy}}{\dfrac{1}{x} + \dfrac{1}{y}}.$$

SOLUTION

Step 1. **Get a single rational expression in the numerator.** The numerator, $\dfrac{1}{xy}$, already contains a single rational expression, so we can skip this step.

Step 2. **Add to get a single rational expression in the denominator.**

$$\frac{1}{x} + \frac{1}{y} = \frac{1 \cdot y}{x \cdot y} + \frac{1 \cdot x}{y \cdot x} = \frac{y}{xy} + \frac{x}{xy} = \frac{y + x}{xy}$$

The LCD is xy.

Steps 3 and 4. **Perform the division indicated by the main fraction bar: Invert and multiply. If possible, simplify.**

$$\frac{\dfrac{1}{xy}}{\dfrac{1}{x} + \dfrac{1}{y}} = \frac{\dfrac{1}{xy}}{\dfrac{y+x}{xy}} = \frac{1}{xy} \cdot \frac{xy}{y+x} = \frac{1}{\cancel{xy}} \cdot \frac{\cancel{xy}^{\,1}}{y+x} = \frac{1}{y+x}$$

Invert and multiply.

■

✔ **CHECK POINT 3** Simplify: $\dfrac{\dfrac{1}{x} - \dfrac{1}{y}}{\dfrac{1}{xy}}$. $y - x$

Simplifying Complex Rational Expressions by Multiplying by the LCD A second method for simplifying a complex rational expression is to find the least common denominator of all the rational expressions in its numerator and denominator. Then multiply each term in its numerator and denominator by this least common denominator. Because we are multiplying by a form of 1, we will obtain an equivalent expression that does not contain fractions in its numerator or denominator.

2 Simplify complex rational expressions by multiplying by the LCD.

> ### SIMPLIFYING A COMPLEX RATIONAL EXPRESSION BY MULTIPLYING BY THE LCD
>
> **1.** Find the LCD of all rational expressions within the complex rational expression.
> **2.** Multiply both the numerator and the denominator of the complex rational expression by this LCD.
> **3.** Use the distributive property and multiply each term in the numerator and denominator by this LCD. Simplify. No fractional expressions should remain in the numerator and denominator.
> **4.** If possible, factor and simplify.

We now rework Examples 1, 2, and 3 using the method of multiplying by the LCD. Compare the two simplification methods to see if there is one method that you prefer.

EXAMPLE 4 Simplifying a Complex Rational Expression by the LCD Method

Simplify:

$$\frac{\dfrac{1}{3} + \dfrac{2}{5}}{\dfrac{2}{5} - \dfrac{1}{3}}.$$

SOLUTION The denominators in the complex rational expression are 3, 5, 5, and 3. The LCD is $3 \cdot 5$, or 15. Multiply both the numerator and denominator of the complex rational expression by 15.

$$\frac{\frac{1}{3} + \frac{2}{5}}{\frac{2}{5} - \frac{1}{3}} = \frac{15}{15} \cdot \frac{\left(\frac{1}{3} + \frac{2}{5}\right)}{\left(\frac{2}{5} - \frac{1}{3}\right)} = \frac{15 \cdot \frac{1}{3} + 15 \cdot \frac{2}{5}}{15 \cdot \frac{2}{5} - 15 \cdot \frac{1}{3}} = \frac{5 + 6}{6 - 5} = \frac{11}{1} = 11$$

$\frac{15}{15} = 1$, so we are not changing the complex fraction's value.

✔ **CHECK POINT 4** Simplify by the LCD method: $\dfrac{\frac{1}{4} + \frac{2}{3}}{\frac{2}{3} - \frac{1}{4}}$. $\frac{11}{5}$

EXAMPLE 5 Simplifying a Complex Rational Expression by the LCD Method

Simplify:

$$\frac{1 + \frac{1}{x}}{1 - \frac{1}{x}}.$$

SOLUTION The denominators in the complex rational expression are $1, x, 1$, and x.

$$\frac{1 + \frac{1}{x}}{1 - \frac{1}{x}} = \frac{\frac{1}{1} + \frac{1}{x}}{\frac{1}{1} - \frac{1}{x}} \qquad \text{Denominators}$$

Denominators

The LCD is $1 \cdot x$, or x. Multiply both the numerator and denominator of the complex rational expression by x.

$$\frac{1 + \frac{1}{x}}{1 - \frac{1}{x}} = \frac{x}{x} \cdot \frac{\left(1 + \frac{1}{x}\right)}{\left(1 - \frac{1}{x}\right)} = \frac{x \cdot 1 + x \cdot \frac{1}{x}}{x \cdot 1 - x \cdot \frac{1}{x}} = \frac{x + 1}{x - 1}$$

✔ **CHECK POINT 5** Simplify by the LCD method: $\dfrac{2 - \frac{1}{x}}{2 + \frac{1}{x}}$. $\frac{2x - 1}{2x + 1}$

| EXAMPLE 6 | Simplifying a Complex Rational Expression by the LCD Method |

Simplify:

$$\frac{\dfrac{1}{xy}}{\dfrac{1}{x}+\dfrac{1}{y}}.$$

SOLUTION   The denominators in the complex rational expression are xy, x, and y. The LCD is xy. Multiply both the numerator and denominator of the complex rational expression by xy.

$$\frac{\dfrac{1}{xy}}{\dfrac{1}{x}+\dfrac{1}{y}} = \frac{xy}{xy}\cdot\frac{\left(\dfrac{1}{xy}\right)}{\left(\dfrac{1}{x}+\dfrac{1}{y}\right)} = \frac{xy\cdot\dfrac{1}{xy}}{xy\cdot\dfrac{1}{x}+xy\cdot\dfrac{1}{y}} = \frac{1}{y+x}.$$

✔

CHECK POINT 6   Simplify by the LCD method: $\dfrac{\dfrac{1}{x}-\dfrac{1}{y}}{\dfrac{1}{xy}}.$ $y-x$

7.5 EXERCISE SET

Student Solutions Manual   CD/Video   PH Math/Tutor Center   MathXL Tutorials on CD   MathXL®   MyMathLab   Interactmath.com

Practice Exercises

In Exercises 1–40, simplify each complex rational expression by the method of your choice.

1. $\dfrac{\dfrac{1}{2}+\dfrac{1}{4}}{\dfrac{1}{2}+\dfrac{1}{3}}$ $\dfrac{9}{10}$

2. $\dfrac{\dfrac{1}{3}+\dfrac{1}{4}}{\dfrac{1}{3}+\dfrac{1}{6}}$ $\dfrac{7}{6}$

3. $\dfrac{5+\dfrac{2}{5}}{7-\dfrac{1}{10}}$ $\dfrac{18}{23}$

4. $\dfrac{1+\dfrac{3}{5}}{2-\dfrac{1}{4}}$ $\dfrac{32}{35}$

5. $\dfrac{\dfrac{2}{5}-\dfrac{1}{3}}{\dfrac{2}{3}-\dfrac{3}{4}}$ $-\dfrac{4}{5}$

6. $\dfrac{\dfrac{1}{2}-\dfrac{1}{4}}{\dfrac{3}{8}+\dfrac{1}{16}}$ $\dfrac{4}{7}$

7. $\dfrac{\dfrac{3}{4}-x}{\dfrac{3}{4}+x}$ $\dfrac{3-4x}{3+4x}$

8. $\dfrac{\dfrac{2}{3}-x}{\dfrac{2}{3}+x}$ $\dfrac{2-3x}{2+3x}$

9. $\dfrac{7-\dfrac{2}{x}}{5+\dfrac{1}{x}}$ $\dfrac{7x-2}{5x+1}$

10. $\dfrac{8+\dfrac{3}{x}}{1-\dfrac{7}{x}}$ $\dfrac{8x+3}{x-7}$

11. $\dfrac{2+\dfrac{3}{y}}{1-\dfrac{7}{y}}$ $\dfrac{2y+3}{y-7}$

12. $\dfrac{4-\dfrac{7}{y}}{3-\dfrac{2}{y}}$ $\dfrac{4y-7}{3y-2}$

13. $\dfrac{\dfrac{1}{y}-\dfrac{3}{2}}{\dfrac{1}{y}+\dfrac{3}{4}}$ $\dfrac{4-6y}{4+3y}$

14. $\dfrac{\dfrac{1}{y}-\dfrac{3}{4}}{\dfrac{1}{y}+\dfrac{2}{3}}$ $\dfrac{12-9y}{12+8y}$

15. $\dfrac{\dfrac{x}{5}-\dfrac{5}{x}}{\dfrac{1}{5}+\dfrac{1}{x}}$ $x-5$

16. $\dfrac{\dfrac{3}{x}+\dfrac{x}{3}}{\dfrac{x}{3}-\dfrac{3}{x}}$ $\dfrac{9+x^2}{x^2-9}$

17. $\dfrac{1+\dfrac{1}{x}}{1-\dfrac{1}{x^2}}$ $\dfrac{x}{x-1}$

18. $\dfrac{1+\dfrac{2}{x}}{1-\dfrac{4}{x^2}}$ $\dfrac{x}{x-2}$

19. $\dfrac{\dfrac{1}{7}-\dfrac{1}{y}}{\dfrac{7-y}{7}}$ $-\dfrac{1}{y}$

20. $\dfrac{\dfrac{1}{9}-\dfrac{1}{y}}{\dfrac{9-y}{9}}$ $-\dfrac{1}{y}$

21. $\dfrac{x+\dfrac{2}{y}}{\dfrac{x}{y}}$ $\dfrac{xy+2}{x}$

22. $\dfrac{x-\dfrac{2}{y}}{\dfrac{x}{y}}$ $\dfrac{xy-2}{x}$

23. $\dfrac{\dfrac{1}{x}+\dfrac{1}{y}}{xy}\qquad \dfrac{y+x}{x^2y^2}$

24. $\dfrac{\dfrac{1}{x}+\dfrac{1}{y}}{x+y}\qquad \dfrac{1}{xy}$

25. $\dfrac{\dfrac{x}{y}+\dfrac{1}{x}}{\dfrac{y}{x}+\dfrac{1}{x}}\qquad \dfrac{x^2+y}{y(y+1)}$

26. $\dfrac{\dfrac{1}{x}+\dfrac{1}{y}}{\dfrac{1}{x}-\dfrac{1}{y}}\qquad \dfrac{y+x}{y-x}$

27. $\dfrac{\dfrac{1}{y}+\dfrac{2}{y^2}}{\dfrac{2}{y}+1}\qquad \dfrac{1}{y}$

28. $\dfrac{\dfrac{1}{y}+\dfrac{3}{y^2}}{\dfrac{3}{y}+1}\qquad \dfrac{1}{y}$

29. $\dfrac{\dfrac{12}{x^2}-\dfrac{3}{x}}{\dfrac{15}{x}-\dfrac{9}{x^2}}\qquad \dfrac{4-x}{5x-3}$

30. $\dfrac{\dfrac{8}{x^2}-\dfrac{2}{x}}{\dfrac{10}{x}-\dfrac{6}{x^2}}\qquad \dfrac{4-x}{5x-3}$

31. $\dfrac{2+\dfrac{6}{y}}{1-\dfrac{9}{y^2}}\qquad \dfrac{2y}{y-3}$

32. $\dfrac{3+\dfrac{12}{y}}{1-\dfrac{16}{y^2}}\qquad \dfrac{3y}{y-4}$

33. $\dfrac{\dfrac{1}{x+2}}{1+\dfrac{1}{x+2}}\qquad \dfrac{1}{x+3}$

34. $\dfrac{\dfrac{1}{x-2}}{1-\dfrac{1}{x-2}}\qquad \dfrac{1}{x-3}$

35. $\dfrac{x-5+\dfrac{3}{x}}{x-7+\dfrac{2}{x}}\qquad \dfrac{x^2-5x+3}{x^2-7x+2}$

36. $\dfrac{x+9-\dfrac{7}{x}}{x-6+\dfrac{4}{x}}\qquad \dfrac{x^2+9x-7}{x^2-6x+4}$

37. $\dfrac{\dfrac{3}{xy^2}+\dfrac{2}{x^2y}}{\dfrac{1}{x^2y}+\dfrac{2}{xy^3}}\qquad \dfrac{2y^2+3xy}{y^2+2x}$

38. $\dfrac{\dfrac{2}{x^3y}+\dfrac{5}{xy^4}}{\dfrac{5}{x^3y}-\dfrac{3}{xy}}\qquad \dfrac{2y^3+5x^2}{5y^3-3x^2y^3}$

39. $\dfrac{\dfrac{3}{x+1}-\dfrac{3}{x-1}}{\dfrac{5}{x^2-1}}\qquad -\dfrac{6}{5}$

40. $\dfrac{\dfrac{3}{x+2}-\dfrac{3}{x-2}}{\dfrac{5}{x^2-4}}\qquad -\dfrac{12}{5}$

Practice Plus

In Exercises 41–48, simplify each complex rational expression.

41. $\dfrac{\dfrac{6}{x^2+2x-15}-\dfrac{1}{x-3}}{\dfrac{1}{x+5}+1}\qquad \dfrac{1-x}{(x-3)(x+6)}$

42. $\dfrac{\dfrac{1}{x-2}-\dfrac{6}{x^2+3x-10}}{1+\dfrac{1}{x-2}}\qquad \dfrac{1}{x+5}$

43. $\dfrac{y^{-1}-(y+5)^{-1}}{5}\qquad \dfrac{1}{y(y+5)}$

44. $\dfrac{y^{-1}-(y+2)^{-1}}{2}\qquad \dfrac{1}{y(y+2)}$

45. $\dfrac{\dfrac{1}{1-\dfrac{1}{x}}-1}{}\qquad \dfrac{1}{x-1}$

46. $\dfrac{\dfrac{1}{1-\dfrac{1}{x+1}}-1}{}\qquad \dfrac{1}{x}$

47. $\dfrac{1}{1+\dfrac{1}{1+\dfrac{1}{x}}}\qquad \dfrac{x+1}{2x+1}$

48. $\dfrac{1}{1+\dfrac{1}{1+\dfrac{1}{2}}}\qquad \dfrac{3}{5}$

Application Exercises

49. The average rate on a round-trip commute having a one-way distance d is given by the complex rational expression

$$\dfrac{2d}{\dfrac{d}{r_1}+\dfrac{d}{r_2}}\qquad \dfrac{2r_1r_2}{r_1+r_2};\ 34\tfrac{2}{7}\ \text{mph}$$

in which r_1 and r_2 are the average rates on the outgoing and return trips, respectively. Simplify the expression. Then find your average rate if you drive to campus averaging 40 miles per hour and return home on the same route averaging 30 miles per hour.

50. If two electrical resistors with resistances R_1 and R_2 are connected in parallel (see the figure), then the total resistance in the circuit is given by the complex rational expression

$$\dfrac{1}{\dfrac{1}{R_1}+\dfrac{1}{R_2}}.$$

Simplify the expression. Then find the total resistance if $R_1 = 10$ ohms and $R_2 = 20$ ohms.

$\dfrac{R_1R_2}{R_1+R_2};\ \dfrac{20}{3}$ ohms

Writing in Mathematics

51. What is a complex rational expression? Give an example with your explanation. **51–53.** Answers will vary.

52. Describe two ways to simplify $\dfrac{\dfrac{3}{x}+\dfrac{2}{x^2}}{\dfrac{1}{x^2}+\dfrac{2}{x}}$.

53. Which method do you prefer for simplifying complex rational expressions? Why?

Critical Thinking Exercises

54. Which one of the following is true? **b**

   a. The fraction $\dfrac{31{,}729{,}546}{72{,}578{,}112}$ is a complex rational expression.

b. $\dfrac{y - \dfrac{1}{2}}{y + \dfrac{3}{4}} = \dfrac{4y - 2}{4y + 3}$ for any value of y except $-\dfrac{3}{4}$.

c. $\dfrac{\dfrac{1}{4} - \dfrac{1}{3}}{\dfrac{1}{3} + \dfrac{1}{6}} = \dfrac{1}{12} \div \dfrac{3}{6} = \dfrac{1}{6}$

d. Some complex rational expressions cannot be simplified by both methods discussed in this section.

55. In one short sentence, five words or less, explain what

$$\dfrac{\dfrac{1}{x} + \dfrac{1}{x^2} + \dfrac{1}{x^3}}{\dfrac{1}{x^4} + \dfrac{1}{x^5} + \dfrac{1}{x^6}}$$

does to each number x.   It cubes x.

In Exercises 56–57, simplify completely.

56. $\dfrac{\dfrac{2y}{2 + \dfrac{2}{y}} + \dfrac{y}{1 + \dfrac{1}{y}}}{}$ $\dfrac{2y^2}{y + 1}$

57. $\dfrac{1 + \dfrac{1}{y} - \dfrac{6}{y^2}}{1 - \dfrac{5}{y} + \dfrac{6}{y^2}} - \dfrac{1 - \dfrac{1}{y}}{1 - \dfrac{2}{y} - \dfrac{3}{y^2}}$ $\dfrac{5y + 3}{(y - 3)(y + 1)}$

Technology Exercises

In Exercises 58–60, use the GRAPH *or* TABLE *feature of a graphing utility to determine if the simplification is correct. If the answer is wrong, correct it and then verify your corrected simplification using the graphing utility.*

58. $\dfrac{x - \dfrac{1}{2x + 1}}{1 - \dfrac{x}{2x + 1}} = 2x - 1$ correct answer

59. $\dfrac{\dfrac{1}{x} + 1}{\dfrac{1}{x}} = 2$ 2 should be $1 + x$.

60. $\dfrac{\dfrac{1}{x} + \dfrac{1}{3}}{\dfrac{1}{3x}} = x + \dfrac{1}{3}$ $x + \dfrac{1}{3}$ should be $3 + x$.

Review Exercises

61. Factor completely: $2x^3 - 20x^2 + 50x$. (Section 6.5, Example 2) $2x(x - 5)^2$

62. Solve: $2 - 3(x - 2) = 5(x + 5) - 1$. (Section 2.3, Example 3) $\{-2\}$

63. Multiply: $(x + y)(x^2 - xy + y^2)$. (Section 5.2, Example 7) $x^3 + y^3$

SECTION

7.6

Objectives

1 Solve rational equations.

2 Solve problems involving formulas with rational expressions.

3 Solve a formula with a rational expression for a variable.

SOLVING RATIONAL EQUATIONS

The time has come to clean up the river. Suppose that the government has committed $375 million for this project. We know that

$$y = \dfrac{250x}{100 - x}$$

models the cost, in millions of dollars, to remove x percent of the river's pollutants. What percentage of pollutants can be removed for $375 million?

In order to determine the percentage, we use the given model. The government has committed $375 million, so substitute 375 for y:

$$375 = \frac{250x}{100 - x} \quad \text{or} \quad \frac{250x}{100 - x} = 375.$$

> The equation contains a rational expression.

Now we need to solve the equation and find the value for x. This variable represents the percentage of pollutants that can be removed for $375 million.

A **rational**, or **fractional**, **equation** is an equation containing one or more rational expressions. The preceding equation is an example of a rational equation. Do you see that there is a variable in a denominator? This is a characteristic of many rational equations. In this section, you will learn a procedure for solving such equations.

Solving Rational Equations We have seen that the LCD is used to add and subtract rational expressions. By contrast, when solving rational equations, **the LCD is used as a multiplier that clears an equation of fractions**.

1 Solve rational equations.

USING TECHNOLOGY

We can use a graphing utility to verify the solution to Example 1. Graph each side of the equation:

$$y_1 = \frac{x}{4}$$

$$y_2 = \frac{1}{4} + \frac{x}{6}.$$

Trace along the lines or use the utility's intersection feature. The solution, as shown below, is the first coordinate of the point of intersection. Thus, the solution is 3.

$[-5, 5, 1]$ by $[-2, 2, 1]$

EXAMPLE 1 Solving a Rational Equation

Solve: $\dfrac{x}{4} = \dfrac{1}{4} + \dfrac{x}{6}$.

SOLUTION The LCD of 4, 4, and 6 is 12. To clear the equation of fractions, we multiply both sides by 12.

$$\frac{x}{4} = \frac{1}{4} + \frac{x}{6} \qquad \text{This is the given equation.}$$

$$12\left(\frac{x}{4}\right) = 12\left(\frac{1}{4} + \frac{x}{6}\right) \qquad \begin{array}{l}\text{Multiply both sides by 12, the LCD of all}\\ \text{the fractions in the equation.}\end{array}$$

$$12 \cdot \frac{x}{4} = 12 \cdot \frac{1}{4} + 12 \cdot \frac{x}{6} \qquad \text{Use the distributive property on the right side.}$$

$$3x = 3 + 2x \qquad \text{Simplify: } \frac{\overset{3}{\cancel{12}}}{1} \cdot \frac{x}{\cancel{4}} = 3x; \ \frac{\overset{3}{\cancel{12}}}{1} \cdot \frac{1}{\cancel{4}} = 3; \ \frac{\overset{2}{\cancel{12}}}{1} \cdot \frac{x}{\cancel{6}} = 2x.$$

$$x = 3 \qquad \text{Subtract } 2x \text{ from both sides.}$$

Substitute 3 for x in the original equation. You should obtain the true statement $\dfrac{3}{4} = \dfrac{3}{4}$. This verifies that the solution is 3 and the solution set is $\{3\}$. ∎

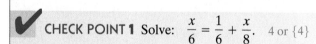 **CHECK POINT 1** Solve: $\dfrac{x}{6} = \dfrac{1}{6} + \dfrac{x}{8}$.   4 or $\{4\}$

In Example 1, we solved a rational equation with constants in the denominators. Now, let's consider an equation such as

$$\frac{1}{x} = \frac{1}{5} + \frac{3}{2x}.$$

Can you see how this equation differs from the rational equation that we solved earlier? The variable, x, appears in two of the denominators. The procedure for solving this equation still involves multiplying each side by the least common denominator. However, we must avoid any values of the variable that make a denominator zero. For example, examine the denominators in the equation:

$$\frac{1}{x} = \frac{1}{5} + \frac{3}{2x}.$$

This denominator would equal zero if $x = 0$.   This denominator would equal zero if $x = 0$.

We see that x cannot equal zero. With this in mind, let's solve the equation.

EXAMPLE 2   Solving a Rational Equation

Solve: $\frac{1}{x} = \frac{1}{5} + \frac{3}{2x}$.

SOLUTION   The denominators are x, 5, and $2x$. The least common denominator is $10x$. We begin by multiplying both sides of the equation by $10x$. We will also write the restriction that x cannot equal zero to the right of the equation.

$$\frac{1}{x} = \frac{1}{5} + \frac{3}{2x}, \quad x \neq 0 \qquad \text{This is the given equation.}$$

$$10x \cdot \frac{1}{x} = 10x\left(\frac{1}{5} + \frac{3}{2x}\right) \qquad \text{Multiply both sides by 10x.}$$

$$10x \cdot \frac{1}{x} = 10x \cdot \frac{1}{5} + 10x \cdot \frac{3}{2x} \qquad \text{Use the distributive property. Be sure to multiply all terms by 10x.}$$

$$10\cancel{x} \cdot \frac{1}{\cancel{x}} = \overset{2}{\cancel{10}}x \cdot \frac{1}{\cancel{5}}_{1} + \overset{5}{\cancel{10}}\cancel{x} \cdot \frac{3}{\cancel{2x}}_{1} \qquad \text{Divide out common factors in the multiplications.}$$

$$10 = 2x + 15 \qquad \text{Simplify.}$$

Observe that the resulting equation,

$$10 = 2x + 15$$

is now cleared of fractions. With the variable term, $2x$, already on the right, we will collect constant terms on the left by subtracting 15 from both sides.

$$-5 = 2x \qquad \text{Subtract 15 from both sides.}$$

$$-\frac{5}{2} = x \qquad \text{Divide both sides by 2.}$$

We check our solution by substituting $-\frac{5}{2}$ into the original equation or by using a calculator. With a calculator, evaluate each side of the equation for $x = -\frac{5}{2}$, or for $x = -2.5$. Note that the original restriction that $x \neq 0$ is met. The solution is $-\frac{5}{2}$ and the solution set is $\left\{-\frac{5}{2}\right\}$.   ∎

 CHECK POINT 2 Solve: $\frac{5}{2x} = \frac{17}{18} - \frac{1}{3x}$.   3 or {3}

The following steps may be used to solve a rational equation:

SOLVING RATIONAL EQUATIONS

1. List restrictions on the variable. Avoid any values of the variable that make a denominator zero.

2. Clear the equation of fractions by multiplying both sides by the LCD of all rational expressions in the equation.

3. Solve the resulting equation.

4. Reject any proposed solution that is in the list of restrictions on the variable. Check other proposed solutions in the original equation.

EXAMPLE 3 Solving a Rational Equation

Solve: $x + \dfrac{1}{x} = \dfrac{5}{2}$.

SOLUTION

Step 1. List restrictions on the variable.

$$x + \frac{1}{x} = \frac{5}{2}$$

> This denominator would equal 0 if $x = 0$.

The restriction is $x \neq 0$.

Step 2. Multiply both sides by the LCD. The denominators are x and 2. Thus, the LCD is $2x$. We multiply both sides by $2x$.

$$x + \frac{1}{x} = \frac{5}{2}, \quad x \neq 0 \qquad \text{This is the given equation.}$$

$$2x\left(x + \frac{1}{x}\right) = 2x\left(\frac{5}{2}\right) \qquad \text{Multiply both sides by the LCD.}$$

$$2x \cdot x + 2x \cdot \frac{1}{x} = 2x \cdot \frac{5}{2} \qquad \begin{array}{l}\text{Use the distributive property} \\ \text{on the left side.}\end{array}$$

$$2x^2 + 2 = 5x \qquad \text{Simplify.}$$

Step 3. Solve the resulting equation. Can you see that we have a quadratic equation? Write the equation in standard form and solve for x.

$$2x^2 - 5x + 2 = 0 \qquad \text{Subtract 5x from both sides.}$$

$$(2x - 1)(x - 2) = 0 \qquad \text{Factor.}$$

$$2x - 1 = 0 \quad \text{or} \quad x - 2 = 0 \qquad \text{Set each factor equal to 0.}$$

$$2x = 1 \qquad\qquad x = 2 \qquad \text{Solve the resulting equations.}$$

$$x = \frac{1}{2}$$

Step 4. Check proposed solutions in the original equation. The proposed solutions, $\frac{1}{2}$ and 2, are not part of the restriction that $x \neq 0$. Neither makes a denominator in the original equation equal to zero.

Check $\dfrac{1}{2}$:   Check 2:

$$x + \frac{1}{x} = \frac{5}{2}$$ $$x + \frac{1}{x} = \frac{5}{2}$$

$$\frac{1}{2} + \frac{1}{\frac{1}{2}} \stackrel{?}{=} \frac{5}{2}$$ $$2 + \frac{1}{2} \stackrel{?}{=} \frac{5}{2}$$

$$\frac{1}{2} + 2 \stackrel{?}{=} \frac{5}{2}$$ $$\frac{4}{2} + \frac{1}{2} \stackrel{?}{=} \frac{5}{2}$$

$$\frac{1}{2} + \frac{4}{2} \stackrel{?}{=} \frac{5}{2}$$ $$\frac{5}{2} = \frac{5}{2}, \text{ true}$$

$$\frac{5}{2} = \frac{5}{2}, \text{ true}$$

The solutions are $\dfrac{1}{2}$ and 2, and the solution set is $\left\{\dfrac{1}{2}, 2\right\}$. ∎

✔ **CHECK POINT 3** Solve: $x + \dfrac{6}{x} = -5$. -3 and -2, or $\{-3, -2\}$

EXAMPLE 4 Solving a Rational Equation

Solve: $\dfrac{3x}{x^2 - 9} + \dfrac{1}{x - 3} = \dfrac{3}{x + 3}$.

SOLUTION

Step 1. List restrictions on the variable. By factoring denominators, it makes it easier to see values that make denominators zero.

$$\frac{3x}{(x + 3)(x - 3)} + \frac{1}{x - 3} = \frac{3}{x + 3}$$

| This denominator is zero | This denominator | This denominator |
| if $x = -3$ or $x = 3$. | is zero if $x = 3$. | is zero if $x = -3$. |

The restrictions are $x \neq -3$ and $x \neq 3$.

Step 2. Multiply both sides by the LCD. The LCD is $(x + 3)(x - 3)$.

$$\frac{3x}{(x + 3)(x - 3)} + \frac{1}{x - 3} = \frac{3}{x + 3}, \quad x \neq -3, x \neq 3$$ This is the given equation with a denominator factored.

$$(x + 3)(x - 3)\left[\frac{3x}{(x + 3)(x - 3)} + \frac{1}{x - 3}\right] = (x + 3)(x - 3) \cdot \frac{3}{x + 3}$$ Multiply both sides by the LCD.

$$\cancel{(x + 3)}\,\cancel{(x - 3)} \cdot \frac{3x}{\cancel{(x + 3)}\,\cancel{(x - 3)}} + (x + 3)\cancel{(x - 3)} \cdot \frac{1}{\cancel{x - 3}}$$

$$= \cancel{(x + 3)}(x - 3) \cdot \frac{3}{\cancel{x + 3}}$$ Use the distributive property on the left side.

$$3x + (x + 3) = 3(x - 3)$$ Simplify.

Step 3. **Solve the resulting equation.**

$$3x + (x + 3) = 3(x - 3)$$ This is the equation cleared of fractions.

                     Combine like terms on the left side.

$$4x + 3 = 3x - 9$$ Use the distributive property on the right side.

$$x + 3 = -9$$ Subtract 3x from both sides.

$$x = -12$$ Subtract 3 from both sides.

Step 4. **Check proposed solutions in the original equation.** The proposed solution, -12, is not part of the restriction that $x \neq -3$ and $x \neq 3$. Substitute -12 for x in the given equation and show that -12 is the solution. The equation's solution set is $\{-12\}$. ∎

✔ **CHECK POINT 4** Solve: $\dfrac{11}{x^2 - 25} + \dfrac{4}{x + 5} = \dfrac{3}{x - 5}$.   24 or {24}

EXAMPLE 5   Solving a Rational Equation

Solve: $\dfrac{8x}{x + 1} = 4 - \dfrac{8}{x + 1}$.

SOLUTION

Step 1. **List restrictions on the variable.**

$$\frac{8x}{x + 1} = 4 - \frac{8}{x + 1}$$

These denominators are zero if $x = -1$.

The restriction is $x \neq -1$.

Step 2. **Multiply both sides by the LCD.** The LCD is $x + 1$.

$$\frac{8x}{x + 1} = 4 - \frac{8}{x + 1}, \quad x \neq -1$$ This is the given equation.

$$(x + 1) \cdot \frac{8x}{x + 1} = (x + 1)\left[4 - \frac{8}{x + 1}\right]$$ Multiply both sides by the LCD.

$$\cancel{(x + 1)} \cdot \frac{8x}{\cancel{x + 1}} = (x + 1) \cdot 4 - \cancel{(x + 1)} \cdot \frac{8}{\cancel{x + 1}}$$ Use the distributive property on the right side.

$$8x = 4(x + 1) - 8$$ Simplify.

Step 3. **Solve the resulting equation.**

$$8x = 4(x + 1) - 8$$ This is the equation cleared of fractions.

$$8x = 4x + 4 - 8$$ Use the distributive property on the right side.

$$8x = 4x - 4$$ Simplify.

$$4x = -4$$ Subtract 4x from both sides.

$$x = -1$$ Divide both sides by 4.

STUDY TIP

Reject any proposed solution that causes any denominator in a rational equation to equal 0.

Step 4. **Check proposed solutions.** The proposed solution, -1, is *not* a solution because of the restriction that $x \neq -1$. Notice that -1 makes both of the denominators zero in the original equation. There is *no solution to this equation.* The solution set is ∅, the empty set. ∎

✔ CHECK POINT **5** Solve: $\dfrac{x}{x-3} = \dfrac{3}{x-3} + 9.$ no solution or ∅

STUDY TIP

It is important to distinguish between adding and subtracting rational expressions and solving rational equations. We *simplify* sums and differences of terms. On the other hand, we *solve* equations. This is shown in the following two problems, both with an LCD of $3x$.

Adding Rational Expressions

Simplify:

$$\frac{5}{3x} + \frac{3}{x}.$$

$$= \frac{5}{3x} + \frac{3}{x} \cdot \frac{3}{3}$$

$$= \frac{5}{3x} + \frac{9}{3x}$$

$$= \frac{5+9}{3x}$$

$$= \frac{14}{3x}$$

Solving Rational Equations

Solve:

$$\frac{5}{3x} + \frac{3}{x} = 1.$$

$$3x\left(\frac{5}{3x} + \frac{3}{x}\right) = 3x \cdot 1$$

$$3x \cdot \frac{5}{3x} + 3x \cdot \frac{3}{x} = 3x$$

$$5 + 9 = 3x$$

$$14 = 3x$$

$$\frac{14}{3} = x$$

---

2 Solve problems involving formulas with rational expressions.

Applications of Rational Equations   Rational equations can be solved to answer questions about variables contained in mathematical models.

EXAMPLE 6 A Government-Funded Cleanup

The formula

$$y = \frac{250x}{100 - x}$$

models the cost, y, in millions of dollars, to remove x percent of a river's pollutants. If the government commits $375 million for this project, what percentage of pollutants can be removed?

SOLUTION   Substitute 375 for y and solve the resulting rational equation for x.

$$375 = \frac{250x}{100 - x}$$ The LCD is $100 - x$.

$$(100 - x)375 = \cancel{(100 - x)} \cdot \frac{250x}{\cancel{100 - x}}$$ Multiply both sides by the LCD.

$$375(100 - x) = 250x$$ Simplify.

$$37{,}500 - 375x = 250x$$ Use the distributive property on the left side.

$$37{,}500 = 625x$$ Add 375x to both sides.

$$\frac{37{,}500}{625} = \frac{625x}{625}$$ Divide both sides by 625.

$$60 = x$$ Simplify.

If the government spends $375 million, 60% of the river's pollutants can be removed. ∎

CHECK POINT 6 Use the model in Example 6 to answer this question: If govern-ment funding is increased to $750 million, what percentage of pollutants can be removed? 75%

3 Solve a formula with a rational expression for a variable.

Solving a Formula for a Variable Formulas and mathematical models frequently contain rational expressions. We solve for a specified variable using the procedure for solving rational equations. The goal is to get the specified variable alone on one side of the equation. To do so, collect all terms with this variable on one side and all other terms on the other side. It is sometimes necessary to factor out the variable you are solving for.

EXAMPLE 7 Solving for a Variable in a Formula

If you wear glasses, did you know that each lens has a measurement called its focal length, f? When an object is in focus, its distance from the lens, p, and the distance from the lens to your retina, q, satisfy the formula

$$\frac{1}{p} + \frac{1}{q} = \frac{1}{f}.$$

FIGURE 7.2

(See Figure 7.2.) Solve this formula for p.

SOLUTION Our goal is to isolate the variable p. We begin by multiplying both sides by the least common denominator, pqf, to clear the equation of fractions.

$$\frac{1}{p} + \frac{1}{q} = \frac{1}{f} \qquad \text{This is the given formula.}$$

$$pqf\left(\frac{1}{p} + \frac{1}{q}\right) = pqf\left(\frac{1}{f}\right) \qquad \text{Multiply both sides by } pqf, \text{ the LCD.}$$

$$pqf\left(\frac{1}{p}\right) + pqf\left(\frac{1}{q}\right) = pqf\left(\frac{1}{f}\right) \qquad \begin{array}{l}\text{Use the distributive property on the left} \\ \text{side.}\end{array}$$

$$qf + pf = pq \qquad \text{Simplify.}$$

Observe that the formula is now cleared of fractions. Collect terms with p, the specified variable, on one side of the equation. To do so, subtract pf from both sides.

$$qf + pf = pq \qquad \text{This is the equation cleared of fractions.}$$
$$qf = pq - pf \qquad \text{Subtract } pf \text{ from both sides.}$$
$$qf = p(q - f) \qquad \text{Factor out } p, \text{ the specified variable.}$$
$$\frac{qf}{q - f} = \frac{p(q - f)}{q - f} \qquad \text{Divide both sides by } q - f \text{ and solve for } p.$$
$$\frac{qf}{q - f} = p \qquad \text{Simplify.}$$

∎

CHECK POINT 7 Solve $\dfrac{1}{x} + \dfrac{1}{y} = \dfrac{1}{z}$ for x. $x = \dfrac{yz}{y - z}$

7.6 EXERCISE SET

Student Solutions Manual   CD/Video   PH Math/Tutor Center   MathXL Tutorials on CD   MathXL®   MyMathLab   Interactmath.com

Practice Exercises

In Exercises 1–44, solve each rational equation. If an equation has no solution, so state.

1. $\dfrac{x}{3} = \dfrac{x}{2} - 2$ $\{12\}$

2. $\dfrac{x}{5} = \dfrac{x}{6} + 1$ $\{30\}$

3. $\dfrac{4x}{3} = \dfrac{x}{18} - \dfrac{x}{6}$ $\{0\}$

4. $\dfrac{5x}{4} = \dfrac{x}{12} - \dfrac{x}{2}$ $\{0\}$

5. $2 - \dfrac{8}{x} = 6$ $\{-2\}$

6. $1 - \dfrac{9}{x} = 4$ $\{-3\}$

7. $\dfrac{2}{x} + \dfrac{1}{3} = \dfrac{4}{x}$ $\{6\}$

8. $\dfrac{5}{x} + \dfrac{1}{3} = \dfrac{6}{x}$ $\{3\}$

9. $\dfrac{2}{x} + 3 = \dfrac{5}{2x} + \dfrac{13}{4}$ $\{-2\}$

10. $\dfrac{7}{2x} = \dfrac{5}{3x} + \dfrac{22}{3}$ $\left\{\dfrac{1}{4}\right\}$

11. $\dfrac{2}{3x} + \dfrac{1}{4} = \dfrac{11}{6x} - \dfrac{1}{3}$ $\{2\}$

12. $\dfrac{5}{2x} - \dfrac{8}{9} = \dfrac{1}{18} - \dfrac{1}{3x}$ $\{3\}$

13. $\dfrac{6}{x + 3} = \dfrac{4}{x - 3}$ $\{15\}$

14. $\dfrac{7}{x + 1} = \dfrac{5}{x - 3}$ $\{13\}$

15. $\dfrac{x - 2}{2x} + 1 = \dfrac{x + 1}{x}$ $\{4\}$

16. $\dfrac{7x - 4}{5x} = \dfrac{9}{5} - \dfrac{4}{x}$ $\{8\}$

17. $x + \dfrac{6}{x} = -7$ $\{-6, -1\}$

18. $x + \dfrac{7}{x} = -8$ $\{-7, -1\}$

19. $\dfrac{x}{5} - \dfrac{5}{x} = 0$ $\{-5, 5\}$

20. $\dfrac{x}{4} - \dfrac{4}{x} = 0$ $\{-4, 4\}$

21. $x + \dfrac{3}{x} = \dfrac{12}{x}$ $\{-3, 3\}$

22. $x + \dfrac{3}{x} = \dfrac{19}{x}$ $\{-4, 4\}$

23. $\dfrac{4}{y} - \dfrac{y}{2} = \dfrac{7}{2}$ $\{-8, 1\}$

24. $\dfrac{4}{3y} - \dfrac{1}{3} = y$ $\left\{-\dfrac{4}{3}, 1\right\}$

25. $\dfrac{x - 4}{x} = \dfrac{15}{x + 4}$ $\{-1, 16\}$

26. $\dfrac{x - 1}{2x + 3} = \dfrac{6}{x - 2}$ $\{-1, 16\}$

27. $\dfrac{1}{x - 1} + 5 = \dfrac{11}{x - 1}$ $\{3\}$

28. $\dfrac{3}{x + 4} - 7 = \dfrac{-4}{x + 4}$ $\{-3\}$

29. $\dfrac{8y}{y + 1} = 4 - \dfrac{8}{y + 1}$ no solution or $\varnothing$

30. $\dfrac{2}{y - 2} = \dfrac{y}{y - 2} - 2$ no solution or $\varnothing$

31. $\dfrac{3}{x - 1} + \dfrac{8}{x} = 3$ $\left\{\dfrac{2}{3}, 4\right\}$

32. $\dfrac{2}{x - 2} + \dfrac{4}{x} = 2$ $\{1, 4\}$

33. $\dfrac{3y}{y - 4} - 5 = \dfrac{12}{y - 4}$ no solution or $\varnothing$

34. $\dfrac{10}{y + 2} = 3 - \dfrac{5y}{y + 2}$ no solution or $\varnothing$

35. $\dfrac{1}{x} + \dfrac{1}{x - 3} = \dfrac{x - 2}{x - 3}$ $\{1\}$

36. $\dfrac{1}{x - 1} + \dfrac{2}{x} = \dfrac{x}{x - 1}$ $\{2\}$

37. $\dfrac{x + 1}{3x + 9} + \dfrac{x}{2x + 6} = \dfrac{2}{4x + 12}$ $\left\{\dfrac{1}{5}\right\}$

38. $\dfrac{3}{2y - 2} + \dfrac{1}{2} = \dfrac{2}{y - 1}$ $\{2\}$

39. $\dfrac{4y}{y^2 - 25} + \dfrac{2}{y - 5} = \dfrac{1}{y + 5}$ $\{-3\}$

40. $\dfrac{1}{x + 4} + \dfrac{1}{x - 4} = \dfrac{22}{x^2 - 16}$ $\{11\}$

41. $\dfrac{1}{x - 4} - \dfrac{5}{x + 2} = \dfrac{6}{x^2 - 2x - 8}$ no solution or $\varnothing$

42. $\dfrac{6}{x + 3} - \dfrac{5}{x - 2} = \dfrac{-20}{x^2 + x - 6}$ $\{7\}$

43. $\dfrac{2}{x + 3} - \dfrac{2x + 3}{x - 1} = \dfrac{6x - 5}{x^2 + 2x - 3}$ $\left\{-6, -\dfrac{1}{2}\right\}$

44. $\dfrac{x - 3}{x - 2} + \dfrac{x + 1}{x + 3} = \dfrac{2x^2 - 15}{x^2 + x - 6}$ $\{4\}$

In Exercises 45–58, solve each formula for the specified variable.

45. $\dfrac{V_1}{V_2} = \dfrac{P_2}{P_1}$ for P_1 (chemistry) $P_1 = \dfrac{P_2 V_2}{V_1}$

46. $\dfrac{V_1}{V_2} = \dfrac{P_2}{P_1}$ for V_2 (chemistry) $V_2 = \dfrac{V_1 P_1}{P_2}$

47. $\dfrac{1}{p} + \dfrac{1}{q} = \dfrac{1}{f}$ for f (optics) $f = \dfrac{pq}{q + p}$

48. $\dfrac{1}{p} + \dfrac{1}{q} = \dfrac{1}{f}$ for q (optics) $q = \dfrac{pf}{p - f}$

49. $P = \dfrac{A}{1 + r}$ for r (investment) $r = \dfrac{A - P}{P}$

50. $S = \dfrac{a}{1 - r}$ for r (mathematics) $r = \dfrac{S - a}{S}$ or $r = -\dfrac{a - S}{S}$

51. $F = \dfrac{Gm_1 m_2}{d^2}$ for m_1 (physics) $m_1 = \dfrac{Fd^2}{Gm_2}$

52. $F = \dfrac{Gm_1 m_2}{d^2}$ for m_2 (physics) $m_2 = \dfrac{Fd^2}{Gm_1}$

53. $z = \dfrac{x - \bar{x}}{s}$ for x (statistics) $x = \bar{x} + zs$

54. $z = \dfrac{x - \bar{x}}{s}$ for s (statistics) $s = \dfrac{x - \bar{x}}{z}$

55. $I = \dfrac{E}{R + r}$ for R (electronics) $R = \dfrac{E - Ir}{I}$

56. $I = \dfrac{E}{R + r}$ for r (electronics) $r = \dfrac{E - IR}{I}$

57. $f = \dfrac{f_1 f_2}{f_1 + f_2}$ for f_1 (optics) $f_1 = \dfrac{f f_2}{f_2 - f}$

58. $f = \dfrac{f_1 f_2}{f_1 + f_2}$ for f_2 (optics) $f_2 = \dfrac{f f_1}{f_1 - f}$

Practice Plus

In Exercises 59–66, solve or simplify, whichever is appropriate.

59. $\dfrac{x^2 - 10}{x^2 - x - 20} = 1 + \dfrac{7}{x - 5}$ $\{-3\}$

60. $\dfrac{x^2 + 4x - 2}{x^2 - 2x - 8} = 1 + \dfrac{4}{x - 4}$ $\{1\}$

61. $\dfrac{x^2 - 10}{x^2 - x - 20} - 1 - \dfrac{7}{x - 5}$ $\dfrac{-6x - 18}{(x - 5)(x + 4)}$

62. $\dfrac{x^2 + 4x - 2}{x^2 - 2x - 8} - 1 - \dfrac{4}{x - 4}$ $\dfrac{2x - 2}{(x - 4)(x + 2)}$

63. $5y^{-2} + 1 = 6y^{-1}$ $\{1, 5\}$ **64.** $3y^{-2} + 1 = 4y^{-1}$ $\{1, 3\}$

65. $\dfrac{3}{y + 1} - \dfrac{1}{1 - y} = \dfrac{10}{y^2 - 1}$ $\{3\}$

66. $\dfrac{4}{y - 2} - \dfrac{1}{2 - y} = \dfrac{25}{y + 6}$ $\{4\}$

Application Exercises

A company that manufactures wheelchairs has fixed costs of $500,000. The average cost per wheelchair, C, for the company to manufacture x wheelchairs per month is modeled by the formula

$$C = \dfrac{400x + 500,000}{x}.$$

Use this mathematical model to solve Exercises 67–68.

67. How many wheelchairs per month can be produced at an average cost of $450 per wheelchair?   10,000 wheelchairs

68. How many wheelchairs per month can be produced at an average cost of $405 per wheelchair?   100,000 wheelchairs

In Palo Alto, California, a government agency ordered computer-related companies to contribute to a pool of money to clean up underground water supplies. (The companies had stored toxic chemicals in leaking underground containers.) The formula

$$C = \dfrac{2x}{100 - x}$$

models the cost, C, in millions of dollars, for removing x percent of the contaminants. Use this mathematical model to solve Exercises 69–70.

69. What percentage of the contaminants can be removed for $2 million?   50%

70. What percentage of the contaminants can be removed for $8 million?   80%

We have seen that Young's rule

$$C = \dfrac{DA}{A + 12}$$

can be used to approximate the dosage of a drug prescribed for children. In this formula, A = the child's age, in years, D = an adult dosage, and C = the proper child's dosage. Use this formula to solve Exercises 71–72.

71. When the adult dosage is 1000 milligrams, a child is given 300 milligrams. What is that child's age? Round to the nearest year.   5 years old

72. When the adult dosage is 1000 milligrams, a child is given 500 milligrams. What is that child's age?   12 years old

A grocery store sells 4000 cases of canned soup per year. By averaging costs to purchase soup and to pay storage costs, the owner has determined that if x cases are ordered at a time, the yearly inventory cost, C, can be modeled by

$$C = \dfrac{10,000}{x} + 3x.$$

The graph of this model is shown below. Use this information to solve Exercises 73–74.

73. How many cases should be ordered at a time for yearly inventory costs to be $350? Identify your solutions as points on the graph.   either 50 or approximately 67 cases; (50, 350) or $\left(66\frac{2}{3}, 350\right)$

74. How many cases should be ordered at a time for yearly inventory costs to be $790? Identify your solutions as points on the graph.   either approximately 13 or 250 cases; $\left(13\frac{1}{3}, 790\right)$ or (250, 790)

In baseball, a player's batting average is the total number of hits divided by the total number of times at bat. Use this information to solve Exercises 75–76.

75. A player has 12 hits after 40 times at bat. How many additional consecutive times must the player hit the ball to achieve a batting average of 0.440?   10 more hits

76. A player has eight hits after 50 times at bat. How many additional consecutive times must the player hit the ball to achieve a batting average of 0.250?   6 more hits

Writing in Mathematics

77. What is a rational equation?   77–82. Answers will vary.

78. Explain how to solve a rational equation.

79. Explain how to find restrictions on the variable in a rational equation.

80. Why should restrictions on the variable in a rational equation be listed before you begin solving the equation?

81. Describe similarities and differences between the procedures needed to solve the following problems:

$$\text{Add: } \dfrac{2}{x} + \dfrac{3}{4} \qquad \text{Solve: } \dfrac{2}{x} + \dfrac{3}{4} = 1.$$

82. The equation

$$P = \frac{72,900}{100x^2 + 729}$$

models the percentage of people in the United States, P, who have x years of education and are unemployed. Use this model to write a problem that can be solved with a rational equation. It is not necessary to solve the problem.

Critical Thinking Exercises

83. Which one of the following is true?   **b**

a. $\frac{1}{x} + \frac{1}{6} = 6x\left(\frac{1}{x} + \frac{1}{6}\right) = 6 + x$

b. If a is any real number, the equation $\frac{a}{x} + 1 = \frac{a}{x}$ has no solution.

c. All real numbers satisfy the equation $\frac{3}{x} - \frac{1}{x} = \frac{2}{x}$.

d. To solve $\frac{5}{3x} + \frac{3}{x} = 1$, we must first add the rational expressions on the left side.

In Exercises 84–85, solve each rational equation.

84. $\frac{x + 1}{2x^2 - 11x + 5} = \frac{x - 7}{2x^2 + 9x - 5} - \frac{2x - 6}{x^2 - 25}$ $\{-3, 2\}$

85. $\left(\frac{x + 1}{x + 7}\right)^2 \div \left(\frac{x + 1}{x + 7}\right)^4 = 0.$ no solution or ∅

86. Find b so that the solution of

$$\frac{7x + 4}{b} + 13 = x$$

is -6. $b = 2$

Technology Exercises

In Exercises 87–89, use a graphing utility to solve each rational equation. Graph each side of the equation in the given viewing rectangle. The solution is the first coordinate of the point(s) of intersection. Check by direct substitution.

87. $\frac{x}{2} + \frac{x}{4} = 6$

$[-5, 10, 1]$ by $[-5, 10, 1]$ $\{8\}$

88. $\frac{50}{x} = 2x$

$[-10, 10, 1]$ by $[-20, 20, 2]$ $\{-5, 5\}$

89. $x + \frac{6}{x} = -5$

$[-10, 10, 1]$ by $[-10, 10, 1]$ $\{-3, -2\}$

Review Exercises

90. Factor completely: $x^4 + 2x^3 - 3x - 6$. (Section 6.1, Example 7) $(x^3 - 3)(x + 2)$

91. Simplify: $(3x^2)(-4x^{-10})$. (Section 5.7, Example 3) $-\frac{12}{x^8}$

92. Simplify: $-5[4(x - 2) - 3]$. (Section 1.8, Example 11) $-20x + 55$

7.7

APPLICATIONS USING RATIONAL EQUATIONS AND PROPORTIONS

Objectives

1 Solve problems involving motion.

2 Solve problems involving work.

3 Solve problems involving proportions.

4 Solve problems involving similar triangles.

The possibility of seeing a blue whale, the largest mammal ever to grace the earth, increases the excitement of gazing out over the ocean's swell of waves. Blue whales were hunted to near extinction in the last half of the nineteenth and the first half of the twentieth centuries. Using a method for estimating wildlife populations that we discuss in this section, by the mid-1960s it was determined that the world population of blue whales was less than 1000. This led the International Whaling Commission to prevent their extinction. A dramatic increase in blue whale sightings indicates an ongoing increase in their population and the success of the killing ban.

1 Solve problems involving motion.

Problems Involving Motion   We have seen that the distance, d, covered by any moving body is the product of its average rate, r, and its time in motion, t: $d = rt$. Rational expressions appear in motion problems when the conditions of the problem involve the time traveled. We can obtain an expression for t, the time traveled, by dividing both sides of $d = rt$ by r.

$$d = rt \qquad \text{Distance equals rate times time.}$$

$$\frac{d}{r} = \frac{rt}{r} \qquad \text{Divide both sides by } r.$$

$$\frac{d}{r} = t \qquad \text{Simplify.}$$

TIME IN MOTION

$$t = \frac{d}{r}$$

$$\text{Time traveled} = \frac{\text{Distance traveled}}{\text{Rate of travel}}$$

Downstream (with the Current)

←——— 15 miles ———→

Upstream (against the Current)

←——— 9 miles ———→

EXAMPLE 1   A Motion Problem Involving Time

In still water, your small boat averages 8 miles per hour. It takes you the same amount of time to travel 15 miles downstream, with the current, as 9 miles upstream, against the current. What is the rate of the water's current?

SOLUTION

Step 1.   **Let x represent one of the quantities.** Let

$$x = \text{the rate of the current.}$$

Step 2.   **Represent other quantities in terms of x.** We still need expressions for the rate of your boat with the current and the rate against the current. Traveling with the current, the boat's rate in still water, 8 miles per hour, is increased by the current's rate, x miles per hour. Thus,

$$8 + x = \text{the boat's rate with the current.}$$

Traveling against the current, the boat's rate in still water, 8 miles per hour, is decreased by the current's rate, x miles per hour. Thus,

$$8 - x = \text{the boat's rate against the current.}$$

Step 3.   **Write an equation that describes the conditions.** By reading the problem again, we discover that the crucial idea is that the time spent going 15 miles with the current equals the time spent going 9 miles against the current. This information is summarized in the following table.

	Distance	Rate	Time = $\dfrac{\text{Distance}}{\text{Rate}}$
With the current	15	$8 + x$	$\dfrac{15}{8 + x}$
Against the current	9	$8 - x$	$\dfrac{9}{8 - x}$

These two times are equal.

We are now ready to write an equation that describes the problem's conditions.

The time spent going 15 miles with the current	equals	the time spent going 9 miles against the current.

$$\frac{15}{8 + x} = \frac{9}{8 - x}$$

Step 4. **Solve the equation and answer the question.**

$$\frac{15}{8 + x} = \frac{9}{8 - x}$$ This is the equation for the problem's conditions.

$$\cancel{(8 + x)}(8 - x) \cdot \frac{15}{\cancel{8 + x}} = (8 + x)\cancel{(8 - x)} \cdot \frac{9}{\cancel{8 - x}}$$ Multiply both sides by the LCD, $(8 + x)(8 - x)$.

$$15(8 - x) = 9(8 + x)$$ Simplify.

$$120 - 15x = 72 + 9x$$ Use the distributive property.

$$120 = 72 + 24x$$ Add 15x to both sides.

$$48 = 24x$$ Subtract 72 from both sides.

$$2 = x$$ Divide both sides by 24.

The rate of the water's current is 2 miles per hour.

Step 5. **Check the proposed solution in the original wording of the problem.** Does it take you the same amount of time to travel 15 miles downstream as 9 miles upstream if the current is 2 miles per hour? Keep in mind that your rate in still water is 8 miles per hour.

Time required to travel 15 miles with the current $= \dfrac{\text{Distance}}{\text{Rate}} = \dfrac{15}{8 + 2} = \dfrac{15}{10} = 1\dfrac{1}{2}$ hours

Time required to travel 9 miles against the current $= \dfrac{\text{Distance}}{\text{Rate}} = \dfrac{9}{8 - 2} = \dfrac{9}{6} = 1\dfrac{1}{2}$ hours

These times are the same, which checks with the original conditions of the problem. ■

✔ **CHECK POINT 1** Forget the small boat! This time we have you canoeing on the Colorado River. In still water, your average canoeing rate is 3 miles per hour. It takes you the same amount of time to travel 10 miles downstream, with the current, as 2 miles upstream, against the current. What is the rate of the water's current? 2 miles per hour

2 Solve problems involving work.

Problems Involving Work You are thinking of designing your own Web site. You estimate that it will take 30 hours to do the job. In 1 hour, $\dfrac{1}{30}$ of the job is completed. In 2 hours, $\dfrac{2}{30}$, or $\dfrac{1}{15}$, of the job is completed. In 3 hours, the fractional part of the job done is $\dfrac{3}{30}$, or $\dfrac{1}{10}$. In x hours, the fractional part of the job that you can complete is $\dfrac{x}{30}$.

Your friend, who has experience developing Web sites, took 20 hours working on his own to design an impressive site. You wonder about the possibility of working together. How long would it take both of you to design your Web site?

Problems involving work usually have two people working together to complete a job. The amount of time it takes each person to do the job working alone is frequently known. The question deals with how long it will take both people working together to complete the job.

In work problems, **the number 1 represents one whole job completed**. For example, the completion of your Web site is represented by 1. Equations in work problems are based on the following condition:

| Fractional part of the job done by the first person | + | fractional part of the job done by the second person | = | 1 (one whole job completed). |

EXAMPLE 2 Solving a Problem Involving Work

You can design a Web site in 30 hours. Your friend can design the same site in 20 hours. How long will it take to design the Web site if you both work together?

SOLUTION

Step 1. **Let x represent one of the quantities.** Let x = the time, in hours, for you and your friend to design the Web site together.

Step 2. **Represent other quantities in terms of x.** Because there are no other unknown quantities, we can skip this step.

Step 3. **Write an equation that describes the conditions.** We construct a table to help find the fractional part of the task completed by you and your friend in x hours.

	Fractional part of job completed in 1 hour	Time working together	Fractional part of job completed in x hours
You	$\dfrac{1}{30}$	x	$\dfrac{x}{30}$
Your friend	$\dfrac{1}{20}$	x	$\dfrac{x}{20}$

You can design the site in 30 hours.

Your friend can design the site in 20 hours.

| Fractional part of the job done by you | + | fractional part of the job done by your friend | = | one whole job. |

$$\frac{x}{30} \quad + \quad \frac{x}{20} \quad = \quad 1$$

Step 4. **Solve the equation and answer the question.**

$$\frac{x}{30} + \frac{x}{20} = 1 \qquad \text{This is the equation for the problem's conditions.}$$

$$60\left(\frac{x}{30} + \frac{x}{20}\right) = 60 \cdot 1 \qquad \text{Multiply both sides by 60, the LCD.}$$

$$60 \cdot \frac{x}{30} + 60 \cdot \frac{x}{20} = 60 \qquad \text{Use the distributive property on the left side.}$$

$$2x + 3x = 60 \qquad \text{Simplify: } \frac{\overset{2}{\cancel{60}}}{1} \cdot \frac{x}{\underset{1}{\cancel{30}}} = 2x \text{ and } \frac{\overset{3}{\cancel{60}}}{1} \cdot \frac{x}{\underset{1}{\cancel{20}}} = 3x.$$

$$5x = 60 \qquad \text{Combine like terms.}$$

$$x = 12 \qquad \text{Divide both sides by 5.}$$

If you both work together, you can design your Web site in 12 hours.

Step 5. **Check the proposed solution in the original wording of the problem.** Will you both complete the job in 12 hours? Because you can design the site in 30 hours, in 12 hours, you can complete $\frac{12}{30}$, or $\frac{2}{5}$, of the job. Because your friend can design the site in 20 hours, in 12 hours, he can complete $\frac{12}{20}$, or $\frac{3}{5}$, of the job. Notice that $\frac{2}{5} + \frac{3}{5} = 1$, which represents the completion of the entire job, or one whole job. ∎

STUDY TIP

Let

$a = $ the time it takes person A to do a job working alone

$b = $ the time it takes person B to do the same job working alone.

If x represents the time it takes for A and B to complete the entire job working together, then the situation can be modeled by the rational equation

$$\frac{x}{a} + \frac{x}{b} = 1.$$

 CHECK POINT 2 One person can paint the outside of a house in 8 hours. A second person can do it in 4 hours. How long will it take them to do the job if they work together? $2\frac{2}{3}$ hours or 2 hours 40 minutes

---

③ Solve problems involving proportions.

Problems Involving Proportions A **ratio** compares quantities by division. For example, this year's entering class at a medical school contains 60 women and 30 men. The ratio of women to men is $\frac{60}{30}$. We can express this ratio as a fraction reduced to lowest terms:

$$\frac{60}{30} = \frac{\cancel{30} \cdot 2}{\cancel{30} \cdot 1} = \frac{2}{1}.$$

This ratio can be expressed as 2:1, or 2 to 1.

A **proportion** is a statement that says that two ratios are equal. If the ratios are $\frac{a}{b}$ and $\frac{c}{d}$, then the proportion is

$$\frac{a}{b} = \frac{c}{d}.$$

We can clear this rational equation of fractions by multiplying both sides by bd, the least common denominator:

$$\frac{a}{b} = \frac{c}{d} \qquad \text{This is the given proportion.}$$

$$bd \cdot \frac{a}{b} = bd \cdot \frac{c}{d} \qquad \text{Multiply both sides by } bd (b \neq 0 \text{ and } d \neq 0). \text{ Then simplify. On the}$$

$$\text{left, } \frac{\cancel{b}d}{1} \cdot \frac{a}{\cancel{b}} = da = ad. \text{ On the right, } \frac{b\cancel{d}}{1} \cdot \frac{c}{\cancel{d}} = bc.$$

$$ad = bc$$

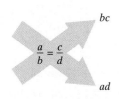

The cross-products principle: $ad = bc$

We see that the following principle is true for any proportion:

THE CROSS-PRODUCTS PRINCIPLE FOR PROPORTIONS
If $\dfrac{a}{b} = \dfrac{c}{d}$, then $ad = bc$. ($b \neq 0$ and $d \neq 0$)
The cross products ad and bc are equal.

For example, if $\frac{2}{3} = \frac{6}{9}$, we see that $2 \cdot 9 = 3 \cdot 6$, or $18 = 18$.
Here is a procedure for solving problems involving proportions:

SOLVING APPLIED PROBLEMS USING PROPORTIONS
1. Read the problem and represent the unknown quantity by x (or any letter).
2. Set up a proportion by listing the given ratio on one side and the ratio with the unknown quantity on the other side. Each respective quantity should occupy the same corresponding position on each side of the proportion.
3. Drop units and apply the cross-products principle.
4. Solve for x and answer the question.

EXAMPLE 3 Applying Proportions: Calculating Taxes

The property tax on a house with an assessed value of $65,000 is $825. Determine the property tax on a house with an assessed value of $180,000, assuming the same tax rate.

SOLUTION

Step 1. **Represent the unknown by x.** Let $x =$ the tax on the $180,000 house.

Step 2. **Set up a proportion.** We will set up a proportion comparing taxes to assessed value.

Given ratio $\left\{\dfrac{\$825}{\$65,000}\right. = \dfrac{\$x}{\$180,000}$ ← Unknown ← Given quantity

Step 3. **Drop the units and apply the cross-products principle.** We drop the dollar signs and begin to solve for x.

$$\frac{825}{65,000} = \frac{x}{180,000}$$ This is the proportion for the problem's conditions.

$$65,000x = (825)(180,000)$$ Apply the cross-products principle.

$$65,000x = 148,500,000$$ Multiply.

Step 4. **Solve for x and answer the question.**

$$\frac{65,000x}{65,000} = \frac{148,500,000}{65,000}$$ Divide both sides by 65,000.

$$x \approx 2284.62$$ Round the value of x to the nearest cent.

The property tax on the $180,000 house is approximately $2284.62. ∎

 CHECK POINT 3 The property tax on a house with an assessed value of $45,000 is $600. Determine the property tax on a house with an assessed value of $112,500, assuming the same tax rate. $1500

STUDY TIP

Here are three other correct proportions you can use in step 2:

- $\dfrac{\$65,000 \text{ value}}{\$825 \text{ tax}} = \dfrac{\$180,000 \text{ value}}{\$x \text{ tax}}$
- $\dfrac{\$65,000 \text{ value}}{\$180,000 \text{ value}} = \dfrac{\$825 \text{ tax}}{\$x \text{ tax}}$
- $\dfrac{\$180,000 \text{ value}}{\$65,000 \text{ value}} = \dfrac{\$x \text{ tax}}{\$825 \text{ tax}}$.

Each proportion gives the same cross product obtained in step 3.

Sampling in Nature The method that was used to estimate the blue whale population described in the section opener is called the **capture-recapture method**. Because it is impossible to count each individual animal within a population, wildlife biologists randomly catch and tag a given number of animals. Sometime later they recapture a second sample of animals and count the number of recaptured tagged animals. The total size of the wildlife population is then estimated using the following proportion:

$$\underset{\substack{\text{Initially}\\\text{unknown}\\(x)\ \longrightarrow}}{}\ \frac{\text{Original number of tagged animals}}{\text{Total number of animals in the population}} = \left.\frac{\text{Number of recaptured tagged animals}}{\text{Number of animals in second sample}}\right\}\ \begin{array}{l}\text{Known}\\\text{ratio}\end{array}.$$

Although this is called the capture-recapture method, it is not necessary to recapture animals to observe whether or not they are tagged. This could be done from a distance, with binoculars for instance.

EXAMPLE 4   Applying Proportions: Estimating Wildlife Population

Wildlife biologists catch, tag, and then release 135 deer back into a wildlife refuge. Two weeks later they observe a sample of 140 deer, 30 of which are tagged. Assuming the ratio of tagged deer in the sample holds for all deer in the refuge, approximately how many deer are in the refuge?

SOLUTION

Step 1.   **Represent the unknown by x.** Let x = the total number of deer in the refuge.

Step 2.   **Set up a proportion.**

$$\underset{\text{Unknown} \longrightarrow}{}\ \frac{\substack{\text{Original number}\\\text{of tagged deer}}}{\substack{\text{Total number}\\\text{of deer}}}\quad \underset{\text{equals}}{=}\quad \left.\frac{\substack{\text{Number of tagged deer}\\\text{in the observed sample}}}{\substack{\text{Total number of deer}\\\text{in the observed sample}}}\right\}\ \begin{array}{l}\text{Known}\\\text{ratio}\end{array}$$

$$\frac{135}{x} = \frac{30}{140}$$

Steps 3 and 4.   **Apply the cross-products principle, solve, and answer the question.**

$$\frac{135}{x} = \frac{30}{140} \qquad \text{This is the proportion for the problem's conditions.}$$

$$(135)(140) = 30x \qquad \text{Apply the cross-products principle.}$$

$$18,900 = 30x \qquad \text{Multiply.}$$

$$\frac{18,900}{30} = \frac{30x}{30} \qquad \text{Divide both sides by 30.}$$

$$630 = x \qquad \text{Simplify.}$$

There are approximately 630 deer in the refuge. ■

 CHECK POINT 4 Wildlife biologists catch, tag, and then release 120 deer back into a wildlife refuge. Two weeks later they observe a sample of 150 deer, 25 of which are tagged. Assuming the ratio of tagged deer in the sample holds for all deer in the refuge, approximately how many deer are in the refuge?   720 deer

4 Solve problems involving similar triangles.

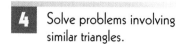

Pedestrian Crossing

Similar Triangles Shown in the margin is an international road sign. This sign is shaped just like the actual sign, although its size is smaller. Figures that have the same shape, but not the same size, are used in **scale drawings**. A scale drawing always pictures the exact shape of the object that the drawing represents. Architects, engineers, landscape gardeners, and interior decorators use scale drawings in planning their work.

Figures that have the same shape, but not necessarily the same size, are called **similar figures**. In Figure 7.3, triangles *ABC* and *DEF* are similar. Angles *A* and *D* measure the same number of degrees and are called **corresponding angles**. Angles *C* and *F* are corresponding angles, as are angles *B* and *E*. Angles with the same number of tick marks in Figure 7.3 are the corresponding angles.

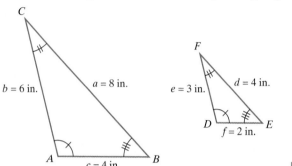

FIGURE 7.3

The sides opposite the corresponding angles are called **corresponding sides**. Although the measures of corresponding angles are equal, corresponding sides may or may not be the same length. For the triangles in Figure 7.3, each side in the smaller triangle is half the length of the corresponding side in the larger triangle.

The triangles in Figure 7.3 illustrate what it means to be **similar triangles**. **Corresponding angles have the same measure and the ratios of the lengths of the corresponding sides are equal.** For the triangles in Figure 7.3, each of these ratios is equal to 2:

$$\frac{a}{d} = \frac{8}{4} = 2 \qquad \frac{b}{e} = \frac{6}{3} = 2 \qquad \frac{c}{f} = \frac{4}{2} = 2.$$

In similar triangles, the lengths of the corresponding sides are proportional. Thus,

$$\frac{a}{d} = \frac{b}{e} = \frac{c}{f}.$$

If we know that two triangles are similar, we can set up a proportion to solve for the length of an unknown side.

EXAMPLE 5 Using Similar Triangles

The triangles in Figure 7.4 are similar. Find the missing length, *x*.

FIGURE 7.4

SOLUTION Because the triangles are similar, their corresponding sides are proportional.

Left side of △ I.
$$\frac{9}{x} = \frac{16}{24}$$
Bottom side of △ I.

Corresponding side on left of △ II.   Corresponding side on bottom of △ II.

We solve this rational equation by multiplying both sides by the LCD, 24x. (You can also apply the cross-products principle for solving proportions.)

$$24x \cdot \frac{9}{x} = 24x \cdot \frac{16}{24} \qquad \text{Multiply both sides by the LCD, 24x.}$$

$$24 \cdot 9 = 16x \qquad \text{Simplify.}$$

$$216 = 16x \qquad \text{Multiply: } 24 \cdot 9 = 216.$$

$$13.5 = x \qquad \text{Divide both sides by 16.}$$

The missing length, x, is 13.5 meters. ■

> **CHECK POINT 5** The similar triangles in the figure are positioned so that they have the same orientation. Find the missing length, x.   32 inches

How can we quickly determine if two triangles are similar? **If the measures of two angles of one triangle are equal to those of two angles of a second triangle, then the two triangles are similar.** If the triangles are similar, then their corresponding sides are proportional.

EXAMPLE 6   Problem Solving Using Similar Triangles

A man who is 6 feet tall is standing 10 feet from the base of a lamppost. (See Figure 7.5.) The man's shadow has a length of 4 feet. How tall is the lamppost?

SOLUTION   The drawing in Figure 7.6 makes the similarity of the triangles easier to see. The large triangle with the lamppost on the left and the small triangle with the man on the left both contain 90° angles. They also share an angle. Thus, two angles of the large triangle are equal in measure to two angles of the small triangle. This means that the triangles are similar and their corresponding sides are proportional. We begin by letting x represent the height of the lamppost, in feet. Because corresponding sides of similar triangles are proportional,

We solve for x by multiplying both sides by the LCD, 12.

$$12 \cdot \frac{x}{6} = 12 \cdot \frac{14}{4} \qquad \text{Multiply both sides by the LCD, 12.}$$

$$2x = 42 \qquad \text{Simplify: } \frac{\overset{2}{\cancel{12}}}{1} \cdot \frac{x}{\cancel{6}} = 2x \text{ and } \frac{\overset{3}{\cancel{12}}}{1} \cdot \frac{14}{\cancel{4}} = 42.$$

$$x = 21 \qquad \text{Divide both sides by 2.}$$

The lamppost is 21 feet tall. ■

FIGURE 7.5

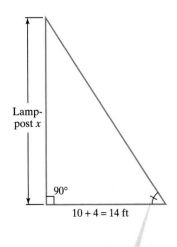

FIGURE 7.6

CHECK POINT 6 Find the height of the lookout tower using the figure that lines up the top of the tower with the top of a stick that is 2 yards long.   32 yards

h

2 yd

3.5 yd

56 yd

7.7 EXERCISE SET

 Student Solutions Manual   CD/Video   PH Math/Tutor Center   MathXL Tutorials on CD   Math XL   MathXL®   MyMathLab   MyMathLab   Interactmath.com

Practice and Application Exercises

Use rational equations to solve Exercises 1–10. Each exercise is a problem involving motion.

1. How bad is the heavy traffic? You can walk 10 miles in the same time that it takes to travel 15 miles by car. If the car's rate is 3 miles per hour faster than your walking rate, find the average rate of each.   Walking rate is 6 miles per hour; car rate is 9 miles per hour.

	Distance	Rate	Time = $\dfrac{\text{Distance}}{\text{Rate}}$
Walking	10	x	$\dfrac{10}{x}$
Car in Heavy Traffic	15	$x + 3$	$\dfrac{15}{x + 3}$

2. You can travel 40 miles on motorcycle in the same time that it takes to travel 15 miles on bicycle. If your motorcycle's rate is 20 miles per hour faster than your bicycle's, find the average rate for each.   Motorcycle rate is 32 miles per hour; bicycle rate is 12 miles per hour.

	Distance	Rate	Time = $\dfrac{\text{Distance}}{\text{Rate}}$
Motorcycle	40	$x + 20$	$\dfrac{40}{x + 20}$
Bicycle	15	x	$\dfrac{15}{x}$

3. A jogger runs 4 miles per hour faster downhill than uphill. If the jogger can run 5 miles downhill in the same time that it takes to run 3 miles uphill, find the jogging rate in each direction.   Downhill rate is 10 miles per hour; uphill rate is 6 miles per hour.

4. A truck can travel 120 miles in the same time that it takes a car to travel 180 miles. If the truck's rate is 20 miles per hour slower than the car's, find the average rate for each.   Truck rate is 40 miles per hour; car rate is 60 miles per hour.

5. In still water, a boat averages 15 miles per hour. It takes the same amount of time to travel 20 miles downstream, with the current, as 10 miles upstream, against the current. What is the rate of the water's current?   Water's current is 5 miles per hour.

6. In still water, a boat averages 18 miles per hour. It takes the same amount of time to travel 33 miles downstream, with the current, as 21 miles upstream, against the current. What is the rate of the water's current?   Water's current is 4 miles per hour.

7. As part of an exercise regimen, you walk 2 miles on an indoor track. Then you jog at twice your walking speed for another 2 miles. If the total time spent walking and jogging is 1 hour, find the walking and jogging rates.   Walking rate is 3 miles per hour; jogging rate is 6 miles per hour.

8. The joys of the Pacific Coast! You drive 90 miles along the Pacific Coast Highway and then take a 5-mile run along a hiking trail in Point Reyes National Seashore. Your driving rate is nine times that of your running rate. If the total time for driving and running is 3 hours, find the average rate driving and the average rate running.   Driving rate is 45 miles per hour; running rate is 5 miles per hour.

9. The water's current is 2 miles per hour. A boat can travel 6 miles downstream, with the current, in the same amount of time it travels 4 miles upstream, against the current. What is the boat's average rate in still water?   Still water rate is 10 miles per hour.

10. The water's current is 2 miles per hour. A canoe can travel 6 miles downstream, with the current, in the same amount of time it travels 2 miles upstream, against the current. What is the canoe's average rate in still water?   Still water rate is 4 miles per hour.

Use a rational equation to solve Exercises 11–16. Each exercise is a problem involving work.

11. You must leave for campus in 10 minutes or you will be late for class. Unfortunately, you are snowed in. You can shovel the driveway in 20 minutes and your brother claims he can do it in 15 minutes. If you shovel together, how long will it take to clear the driveway? Will this give you enough time before you have to leave? It will take about 8.6 minutes, which will give you enough time.

12. You promised your parents that you would wash the family car. You have not started the job and they are due home in 16 minutes. You can wash the car in 40 minutes and your sister claims she can do it in 30 minutes. If you work together, how long will it take to do the job? Will this give you enough time before your parents return? It will take about 17.1 minutes, which is not enough time.

13. The MTV crew will arrive in one week and begin filming the city for *The Real World Kalamazoo*. The mayor is desperate to clean the city streets before filming begins. Two teams are available, one that requires 400 hours and one that requires 300 hours. If the teams work together, how long will it take to clean all of Kalamazoo's streets? Is this enough time before the cameras begin rolling? It will take about 171.4 hours, which is not enough time.

14. A hurricane strikes and a rural area is without food or water. Three crews arrive. One can dispense needed supplies in 10 hours, a second in 15 hours, and a third in 20 hours. How long will it take all three crews working together to dispense food and water? It will take about 4.6 hours, or 4 hours 37 minutes.

15. A pool can be filled by one pipe in 4 hours and by a second pipe in 6 hours. How long will it take using both pipes to fill the pool? It will take 2.4 hours, or 2 hours 24 minutes.

16. A pool can be filled by one pipe in 3 hours and by a second pipe in 6 hours. How long will it take using both pipes to fill the pool? It will take 2 hours.

Use a proportion to solve each problem in Exercises 17–24.

17. The tax on a property with an assessed value of $65,000 is $720. Find the tax on a property with an assessed value of $162,500. $1800

18. The maintenance bill for a shopping center containing 180,000 square feet is $45,000. What is the bill for a store in the center that is 4800 square feet? $1200

19. St. Paul Island in Alaska has 12 fur seal rookeries (breeding places). In 1961, to estimate the fur seal pup population in the Gorbath rookery, 4963 fur seal pups were tagged in early August. In late August, a sample of 900 pups was observed and 218 of these were found to have been previously tagged. Estimate the total number of fur seal pups in this rookery. 20,489 fur seal pups

20. To estimate the number of bass in a lake, wildlife biologists tagged 50 bass and released them in the lake. Later they netted 108 bass and found that 27 of them were tagged. Approximately how many bass are in the lake? 200 bass

21. The ratio of monthly child support to a father's yearly income is 1:40. How much should a father earning $38,000 annually pay in monthly child support? $950

22. The amount of garbage is proportional to the population. Dallas, Texas, has a population of 1.2 million and creates 38.4 million pounds of garbage each week. Find the amount of weekly garbage produced by New York City with a population of 8 million. 256 million pounds

23. Height is proportional to foot length. A person whose foot length is 10 inches is 67 inches tall. In 1951, photos of large footprints were published. Some believed that these footprints were made by the "Abominable Snowman." Each footprint was 23 inches long. If indeed they belonged to the Abominable Snowman, how tall is the critter? 154.1 inches

Roger Patterson comparing his foot with a plaster cast of a footprint of the purported "Bigfoot" that Mr. Patterson said he sighted in a California forest in 1967.

24. A person's hair length is proportional to the number of years it has been growing. After 2 years, a person's hair grows 8 inches. The longest moustache on record was grown by Kalyan Sain of India. Sain grew his moustache for 17 years. How long was it? 68 inches

In Exercises 25–30, use similar triangles and the fact that corresponding sides are proportional to find the length of the side marked with an x.

25.

5 inches

26.

8 inches

27.

6 meters

28.

1.25 inches

29.

16 inches

30.

6 feet

Use similar triangles to solve Exercises 31–32.

31. A tree casts a shadow 12 feet long. At the same time, a vertical rod 8 feet high casts a shadow 6 feet long. How tall is the tree?   16 feet

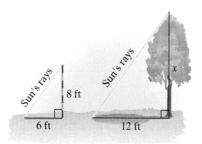

32. A person who is 5 feet tall is standing 80 feet from the base of a tree. The tree casts an 86-foot shadow. The person's shadow is 6 feet in length. What is the tree's height?   about 71.7 feet

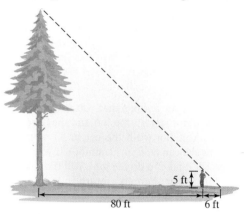

Writing in Mathematics

33. What is the relationship among time traveled, distance traveled, and rate of travel?   **33–41.** Answers will vary.

34. If you know how many hours it takes for you to do a job, explain how to find the fractional part of the job you can complete in x hours.

35. If you can do a job in 6 hours and your friend can do the same job in 3 hours, explain how to find how long it takes to complete the job working together. It is not necessary to solve the problem.

36. When two people work together to complete a job, describe one factor that can result in more or less time than the time given by the rational equations we have been using.

37. What is a proportion? Give an example with your description.

38. What are similar triangles?

39. If the ratio of the corresponding sides of two similar triangles is 1 to 1 $\left(\frac{1}{1}\right)$, what must be true about the triangles?

40. What are corresponding angles in similar triangles?

41. Describe how to identify the corresponding sides in similar triangles.

Critical Thinking Exercises

42. Two skiers begin skiing along a trail at the same time. The faster skier averages 9 miles per hour and the slower skier averages 6 miles per hour. The faster skier completes the trail $\frac{1}{4}$ hour before the slower skier. How long is the trail?   4.5 miles

43. A snowstorm causes a bus driver to decrease the usual average rate along a 60-mile route by 15 miles per hour. As a result, the bus takes two hours longer than usual to complete the route. At what average rate does the bus usually cover the 60-mile route?   30 miles per hour

44. One pipe can fill a swimming pool in 2 hours, a second can fill the pool in 3 hours, and a third pipe can fill the pool in 4 hours. How many minutes, to the nearest minute, would it take to fill the pool with all three pipes operating?   approximately 55 minutes

45. Ben can prepare a company report in 3 hours. Shane can prepare the report in 4.2 hours. How long will it take them, working together, to prepare *four* company reports?   7 hours

46. An experienced carpenter can panel a room 3 times faster than an apprentice can. Working together, they can panel the room in 6 hours. How long would it take each person working alone to do the job?   It will take the experienced carpenter 8 hours and the apprentice 24 hours.

47. It normally takes 2 hours to fill a swimming pool. The pool has developed a slow leak. If the pool were full, it would take 10 hours for all the water to leak out. If the pool is empty, how long will it take to fill it?   2.5 hours

48. Two investments have interest rates that differ by 1%. An investment for 1 year at the lower rate earns $175. The same principal invested for a year at the higher rate earns $200. What are the two interest rates?   7% and 8%

Review Exercises

49. Factor: $25x^2 - 81$. (Section 6.4, Example 1)
$(5x + 9)(5x - 9)$

50. Solve: $x^2 - 12x + 36 = 0$. (Section 6.6, Example 4) $\{6\}$

51. Graph: $y = -\dfrac{2}{3}x + 4$. (Section 3.4, Example 3)   See graphing answer section.

SECTION
7.8

Objectives

1 Solve direct variation problems.

2 Solve inverse variation problems.

3 Solve combined variation problems.

4 Solve problems involving joint variation.

MODELING USING VARIATION

Have you ever wondered how telecommunication companies estimate the number of phone calls expected per day between two cities? The formula

$$C = \frac{0.02P_1P_2}{d^2}$$

shows that the daily number of phone calls, C, increases as the populations of the cities, P_1 and P_2, in thousands, increase and decreases as the distance, d, between the cities increases.

Certain formulas occur so frequently in applied situations that they are given special names. Variation formulas show how one quantity changes in relation to other quantities. Quantities can vary *directly*, *inversely*, or *jointly*. In this section, we look at situations that can be modeled by each of these kinds of variation. And think of this. The next time you get one of those "all-circuits-are-busy" messages, you will be able to use a variation formula to estimate how many other callers you're competing with for those precious 5-cent minutes.

1 Solve direct variation problems.

Direct Variation When you swim underwater, the pressure in your ears depends on the depth at which you are swimming. The formula

$$p = 0.43d$$

describes the water pressure, p, in pounds per square inch, at a depth of d feet. We can use this linear function to determine the pressure in your ears at various depths:

If $d = 20$, $p = 0.43(20) = 8.6$.   At a depth of 20 feet, water pressure is 8.6 pounds per square inch.

Doubling the depth doubles the pressure.

If $d = 40$, $p = 0.43(40) = 17.2$.   At a depth of 40 feet, water pressure is 17.2 pounds per square inch.

Doubling the depth doubles the pressure.

If $d = 80$, $p = 0.43(80) = 34.4$.   At a depth of 80 feet, water pressure is 34.4 pounds per square inch.

The formula $p = 0.43d$ illustrates that water pressure is a constant multiple of your underwater depth. If your depth is doubled, the pressure is doubled; if your depth is tripled, the pressure is triped; and so on. Because of this, the pressure in your ears is said to **vary directly** as your underwater depth. The **equation of variation** is

$$p = 0.43d.$$

Generalizing, we obtain the following statement:

DIRECT VARIATION   If a situation is described by an equation in the form

$$y = kx$$

where k is a constant, we say that y **varies directly as x**. The number k is called the **constant of variation**.

Can you see that the direct variation equation, $y = kx$, is a special case of the linear equation $y = mx + b$? When $m = k$ and $b = 0$, $y = mx + b$ becomes $y = kx$. Thus, the slope of a direct variation equation is k, the constant of variation. Because b, the y-intercept, is 0, the graph of a direct variation equation is a line through the origin. This is illustrated in Figure 7.7, which shows the graph of $p = 0.43d$: Water pressure varies directly as depth.

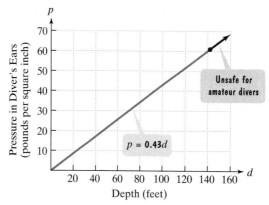

FIGURE 7.7   Water pressure at various depths

Problems involving direct variation can be solved using the following procedure. This procedure applies to direct variation problems, as well as to the other kinds of variation problems that we will discuss.

SOLVING VARIATION PROBLEMS

1. Write an equation that describes the given English statement.

2. Substitute the given pair of values into the equation in step 1 and find the value of k.

3. Substitute the value of k into the equation in step 1.

4. Use the equation from step 3 to answer the problem's question.

EXAMPLE 1   Solving a Direct Variation Problem

Many areas of Northern California depend on the snowpack of the Sierra Nevada mountain range for their water supply. The volume of water produced from melting snow varies directly as the volume of snow. Meteorologists have determined that 250 cubic centimeters of snow will melt to 28 cubic centimeters of water. How much water does 1200 cubic centimeters of melting snow produce?

SOLUTION

Step 1. **Write an equation.** We know that y varies directly as x is expressed as

$$y = kx.$$

By changing letters, we can write an equation that describes the following English statement: Volume of water, W, varies directly as volume of snow, S.

$$W = kS$$

Step 2. **Use the given values to find k.** We are told that 250 cubic centimeters of snow will melt to 28 cubic centimeters of water. Substitute 28 for W and 250 for S in the direct variation equation. Then solve for k.

$W = kS$	Volume of water varies directly as volume of melting snow.
$28 = k(250)$	250 cubic centimeters of snow melt to 28 cubic centimeters of water.
$\dfrac{28}{250} = \dfrac{k(250)}{250}$	Divide both sides by 250.
$0.112 = k$	Simplify.

Step 3. **Substitute the value of k into the equation.**

$W = kS$	This is the equation from step 1.
$W = 0.112S$	Replace k, the constant of variation, with 0.112.

Step 4. **Answer the problem's question.** How much water does 1200 cubic centimeters of melting snow produce? Substitute 1200 for S in $W = 0.112S$ and solve for W.

$W = 0.112S$	Use the equation from step 3.
$W = 0.112(1200)$	Substitute 1200 for S.
$W = 134.4$	Multiply.

A snowpack measuring 1200 cubic centimeters will produce 134.4 cubic centimeters of water.

✔ CHECK POINT 1 The number of gallons of water, W, used when taking a shower varies directly as the time, t, in minutes, in the shower. A shower lasting 5 minutes uses 30 gallons of water. How much water is used in a shower lasting 11 minutes?   66 gallons.

The direct variation equation $y = kx$ is a linear equation. If $k > 0$, then the slope of the line is positive. Consequently, as x increases, y also increases.

② Solve inverse variation problems.

Inverse Variation   The distance from San Francisco to Los Angeles is 420 miles. The time that it takes to drive from San Francisco to Los Angeles depends on the rate at which one drives and is given by

$$\text{Time} = \frac{420}{\text{Rate}}.$$

For example, if you average 30 miles per hour, the time for the drive is

$$\text{Time} = \frac{420}{30} = 14,$$

or 14 hours. If you average 50 miles per hour, the time for the drive is

$$\text{Time} = \frac{420}{50} = 8.4,$$

FIGURE 7.8

or 8.4 hours. As your rate (or speed) increases, the time for the trip decreases and vice versa. This is illustrated by the graph in Figure 7.8.

We can express the time for the San Francisco–Los Angeles trip using t for time and r for rate:

$$t = \frac{420}{r}.$$

This equation is an example of an **inverse variation** equation. Time, t, **varies inversely** as rate, r. When two quantities vary inversely, one quantity increases as the other decreases, and vice versa.

Generalizing, we obtain the following statement:

INVERSE VARIATION   If a situation is described by an equation in the form

$$y = \frac{k}{x}$$

where k is a constant, we say that y **varies inversely as x**. The number k is called the **constant of variation**.

Notice that the inverse variation equation

$$y = \frac{k}{x}$$

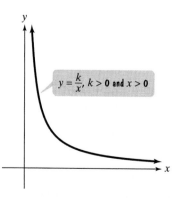

FIGURE 7.9   The graph of the inverse variation equation

involves a rational expression, $\dfrac{k}{x}$. For $k > 0$ and $x > 0$, the graph of the equation takes on the shape shown in Figure 7.9. Under these conditions, as x increases, y decreases.

We use the same procedure to solve inverse variation problems as we did to solve direct variation problems. Example 2 illustrates this procedure.

P $2P$

$2V$

V

Doubling the pressure halves the volume.

EXAMPLE 2 Solving an Inverse Variation Problem

When you use a spray can and press the valve at the top, you decrease the pressure of the gas in the can. This decrease of pressure causes the volume of the gas in the can to increase. Because the gas needs more room than is provided in the can, it expands in spray form through the small hole near the valve. In general, if the temperature is constant, the pressure, P, of a gas in a container varies inversely as the volume, V, of the container. The pressure of a gas sample in a container whose volume is 8 cubic inches is 12 pounds per square inch. If the sample expands to a volume of 22 cubic inches, what is the new pressure of the gas?

SOLUTION

Step 1. **Write an equation.** We know that y varies inversely as x is expressed as

$$y = \frac{k}{x}.$$

By changing letters, we can write an equation that describes the following English statement: The pressure, P, of a gas in a container varies inversely as the volume, V.

$$P = \frac{k}{V}.$$

Step 2. **Use the given values to find k.** The pressure of a gas sample in a container whose volume is 8 cubic inches is 12 pounds per square inch. Substitute 12 for P and 8 for V in the inverse variation equation. Then solve for k.

$$P = \frac{k}{V} \qquad \text{Pressure varies inversely as volume.}$$

$$12 = \frac{k}{8} \qquad \text{The pressure in an 8-cubic-inch container is 12 pounds per square inch.}$$

$$12 \cdot 8 = \frac{k}{8} \cdot 8 \qquad \text{Multiply both sides by 8.}$$

$$96 = k \qquad \text{Simplify.}$$

Step 3. **Substitute the value of k into the equation.**

$$P = \frac{k}{V} \qquad \text{Use the equation from step 1.}$$

$$P = \frac{96}{V} \qquad \text{Replace } k \text{, the constant of variation, with 96.}$$

Step 4. **Answer the problem's question.** We need to find the pressure when the volume expands to 22 cubic inches. Substitute 22 for V and solve for P.

$$P = \frac{96}{V} = \frac{96}{22} = 4\frac{4}{11}$$

When the volume is 22 cubic inches, the pressure of the gas is $4\frac{4}{11}$ pounds per square inch.

✔ **CHECK POINT 2** The length of a violin string varies inversely as the frequency of its vibrations. A violin string 8 inches long vibrates at a frequency of 640 cycles per second. What is the frequency of a 10-inch string?   512 cycles per second

---

3 Solve combined variation problems.

Combined Variation   In **combined variation**, direct and inverse variation occur at the same time. For example, as the advertising budget, A, of a company increases, its monthly sales, S, also increase. Monthly sales vary directly as the advertising budget:

$$S = kA.$$

By contrast, as the price of the company's product, P, increases, its monthly sales, S, decrease. Monthly sales vary inversely as the price of the product:

$$S = \frac{k}{P}.$$

We can combine these two variation equations into one combined equation:

$$S = \frac{kA}{P}.$$

Monthly sales , S, vary directly as the advertising budget, A, and inversely as the price of the product, P.

The following example illustrates an application of combined variation:

EXAMPLE 3   Solving a Combined Variation Problem

The owners of Rollerblades Now determine that the monthly sales, S, of its skates vary directly as its advertising budget, A, and inversely as the price of the skates, P. When $60,000 is spent on advertising and the price of the skates is $40, the monthly sales are 12,000 pairs of rollerblades.

   a. Write an equation of variation that describes this situation.
   b. Determine monthly sales if the amount of the advertising budget is increased to $70,000.

SOLUTION

   a. Write an equation.

$$S = \frac{kA}{P}.$$

Translate "sales vary directly as the advertising budget and inversely as the skates' price."

Use the given values to find k.

$$12,000 = \frac{k(60,000)}{40}$$

When $60,000 is spent on advertising ($A = 60,000$) and the price is $40 ($P = 40$), monthly sales are 12,000 units ($S = 12,000$).

$$12,000 = k \cdot 1500$$

Divide 60,000 by 40.

$$\frac{12,000}{1500} = \frac{k \cdot 1500}{1500}$$

Divide both sides of the equation by 1500.

$$8 = k$$

Simplify.

Therefore, the equation of variation that describes monthly sales is

$$S = \frac{8A}{P}.$$ Substitute 8 for k in $S = \frac{kA}{P}$.

b. The advertising budget is increased to $70,000, so $A = 70,000$. The skates' price is still $40, so $P = 40$.

$$S = \frac{8A}{P}$$ This is the equation from part (a).

$$S = \frac{8(70{,}000)}{40}$$ Substitute 70,000 for A and 40 for P.

$$S = 14{,}000$$ Simplify.

With a $70,000 advertising budget and $40 price, the company can expect to sell 14,000 pairs of rollerblades in a month (up from 12,000). ■

 CHECK POINT 3 The number of minutes needed to solve an exercise set of variation problems varies directly as the number of problems and inversely as the number of people working to solve the problems. It takes 4 people 32 minutes to solve 16 problems. How many minutes will it take 8 people to solve 24 problems?
It will take 24 minutes for 8 people to solve 24 problems.

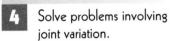

4 Solve problems involving joint variation.

Joint Variation   Joint variation is a variation in which a variable varies directly as the product of two or more other variables. Thus, the equation $y = kxz$ is read "y varies jointly as x and z."

Joint variation plays a critical role in Isaac Newton's formula for gravitation:

$$F = G\frac{m_1 m_2}{d^2}.$$

The formula states that the force of gravitation, F, between two bodies varies jointly as the product of their masses, m_1 and m_2, and inversely as the square of the distance between them, d. (G is the gravitational constant.) The formula indicates that gravitational force exists between any two objects in the universe, increasing as the distance between the bodies decreases. One practical result is that the pull of the moon on the oceans is greater on the side of Earth closer to the moon. This gravitational imbalance is what produces tides.

EXAMPLE 4   Modeling Centrifugal Force

The centrifugal force, C, of a body moving in a circle varies jointly with the radius of the circular path, r, and the body's mass, m, and inversely with the square of the time, t, it takes to move about one full circle. A 6-gram body moving in a circle with radius 100 centimeters at a rate of 1 revolution in 2 seconds has a centrifugal force of 6000 dynes. Find the centrifugal force of an 18-gram body moving in a circle with radius 100 centimeters at a rate of 1 revolution in 3 seconds.

SOLUTION

$$C = \frac{krm}{t^2}$$

Translate "Centrifugal force, C, varies jointly with radius, r, and mass, m, and inversely with the square of time, t."

$$6000 = \frac{k(100)(6)}{2^2}$$

If $r = 100$, $m = 6$, and $t = 2$, then $C = 6000$.

$$6000 = 150k$$

Simplify.

$$40 = k$$

Divide both sides by 150 and solve for k.

$$C = \frac{40rm}{t^2}$$

Substitute 40 for k in the model for centrifugal force.

$$C = \frac{40(100)(18)}{3^2}$$

Find C when $r = 100$, $m = 18$, and $t = 3$.

$$= 8000$$

Simplify.

The centrifugal force is 8000 dynes. ∎

 CHECK POINT 4 The volume of a cone, V, varies jointly as its height, h, and the square of its radius r. A cone with a radius measuring 6 feet and a height measuring 10 feet has a volume of 120π cubic feet. Find the volume of a cone having a radius of 12 feet and a height of 2 feet. The volume of a cone having a radius of 12 feet and a height of 2 feet is 96π cubic feet.

7.8 EXERCISE SET

Student Solutions Manual   CD/Video   PH Math/Tutor Center   MathXL Tutorials on CD   MathXL®   MyMathLab   Interactmath.com

Practice Exercises

Use the four-step procedure for solving variation problems given on page 509 to solve Exercises 1–10.

1. y varies directly as x. $y = 65$ when $x = 5$. Find y when $x = 12$. $y = 156$

2. y varies directly as x. $y = 45$ when $x = 5$. Find y when $x = 13$. $y = 117$

3. y varies inversely as x. $y = 12$ when $x = 5$. Find y when $x = 2$. $y = 30$

4. y varies inversely as x. $y = 6$ when $x = 3$. Find y when $x = 9$. $y = 2$

5. y varies directly as x and inversely as the square of z. $y = 20$ when $x = 50$ and $z = 5$. Find y when $x = 3$ and $z = 6$. $y = \frac{5}{6}$

6. a varies directly as b and inversely as the square of c. $a = 7$ when $b = 9$ and $c = 6$. Find a when $b = 4$ and $c = 8$. $a = \frac{7}{4}$

7. y varies jointly as x and z. $y = 25$ when $x = 2$ and $z = 5$. Find y when $x = 8$ and $z = 12$. $y = 240$

8. C varies jointly as A and T. $C = 175$ when $A = 2100$ and $T = 4$. Find C when $A = 2400$ and $T = 6$. $C = 300$

9. y varies jointly as a and b and inversely as the square root of c. $y = 12$ when $a = 3, b = 2$, and $c = 25$. Find y when $a = 5, b = 3$, and $c = 9$. $y = 50$

10. y varies jointly as m and the square of n and inversely as p. $y = 15$ when $m = 2, n = 1$, and $p = 6$. Find y when $m = 3, n = 4$, and $p = 10$. $y = 216$

Practice Plus

In Exercises 11–20, write an equation that expresses each relationship. Then solve the equation for y.

11. x varies jointly as y and z. $x = kyz$; $y = \frac{x}{kz}$

12. x varies jointly as y and the square of z. $x = kyz^2$; $y = \frac{x}{kz^2}$

13. x varies directly as the cube of z and inversely as y. $x = \frac{kz^3}{y}$; $y = \frac{kz^3}{x}$

14. x varies directly as the cube root of z and inversely as y. $x = \frac{k\sqrt[3]{z}}{y}$; $y = \frac{k\sqrt[3]{z}}{x}$

15. x varies jointly as y and z and inversely as the square root of w. $x = \dfrac{kyz}{\sqrt{w}}$; $y = \dfrac{x\sqrt{w}}{kz}$

16. x varies jointly as y and z and inversely as the square of w.
$x = \dfrac{kyz}{w^2}$; $y = \dfrac{xw^2}{kz}$

17. x varies jointly as z and the sum of y and w.
$x = kz(y + w)$; $y = \dfrac{x - kzw}{kz}$

18. x varies jointly as z and the difference between y and w.
$x = kz(y - w)$; $y = \dfrac{x + kzw}{kz}$

19. x varies directly as z and inversely as the difference between y and w. $x = \dfrac{kz}{y - w}$; $y = \dfrac{xw + kz}{x}$

20. x varies directly as z and inversely as the sum of y and w.
$x = \dfrac{kz}{y + w}$; $y = \dfrac{kz - xw}{x}$

Application Exercises

Use the four-step procedure for solving variation problems given on page 509 to solve Exercises 21–36.

21. An alligator's tail length, T, varies directly as its body length, B. An alligator with a body length of 4 feet has a tail length of 3.6 feet. What is the tail length of an alligator whose body length is 6 feet? The tail length is 5.4 feet.

|← —— Body length, B —— →|← —— Tail length, T —— →|

22. An object's weight on the moon, M, varies directly as its weight on Earth, E. Neil Armstrong, the first person to step on the moon on July 20, 1969, weighed 360 pounds on Earth (with all of his equipment on) and 60 pounds on the moon. What is the moon weight of a person who weighs 186 pounds on Earth? A person who weighs 186 pounds on Earth weighs 31 pounds on the moon.

23. The height that a ball bounces varies directly as the height from which it was dropped. A tennis ball dropped from 12 inches bounces 8.4 inches. From what height was the tennis ball dropped if it bounces 56 inches? It was dropped from 80 inches.

24. The distance that a spring will stretch varies directly as the force applied to the spring. A force of 12 pounds is needed to stretch a spring 9 inches. What force is required to stretch the spring 15 inches? A force of 20 pounds is needed.

25. If all men had identical body types, their weight would vary directly as the cube of their height. Shown at the top of the next column is Robert Wadlow, who reached a record height of 8 feet 11 inches (107 inches) before his death at age 22. If a man who is 5 feet 10 inches tall (70 inches) with

the same body type as Mr. Wadlow weighs 170 pounds, what was Robert Wadlow's weight shortly before his death? Robert Wadlow's weight was about 607 pounds.

26. On a dry asphalt road, a car's stopping distance varies directly as the square of its speed. A car traveling at 45 miles per hour can stop in 67.5 feet. What is the stopping distance for a car traveling at 60 miles per hour? The stopping distance is 120 feet.

27. The figure shows that a bicyclist tips the cycle when making a turn. The angle B, formed by the vertical direction and the bicycle, is called the banking angle. The banking angle varies inversely as the cycle's turning radius. When the turning radius is 4 feet, the banking angle is 28°. What is the banking angle when the turning radius is 3.5 feet? The banking angle is 32° when the turning radius is 3.5 feet.

28. The water temperature of the Pacific Ocean varies inversely as the water's depth. At a depth of 1000 meters, the water temperature is 4.4° Celsius. What is the water temperature at a depth of 5000 meters? The water temperature is 0.88° Celsius at a depth of 5000 meters.

29. Radiation machines, used to treat tumors, produce an intensity of radiation that varies inversely as the square of the distance from the machine. At 3 meters, the radiation intensity is 62.5 milliroentgens per hour. What is the intensity at a distance of 2.5 meters?   The intensity is 90 milliroentgens per hour.

30. The illumination provided by a car's headlight varies inversely as the square of the distance from the headlight. A car's head-light produces an illumination of 3.75 footcandles at a distance of 40 feet. What is the illumination when the distance is 50 feet? The illumination is 2.4 footcandles at a distance of 50 feet.

31. Body-mass index, or BMI, takes both weight and height into account when assessing whether an individual is underweight or overweight. BMI varies directly as one's weight, in pounds, and inversely as the square of one's height, in inches. In adults, normal values for the BMI are between 20 and 25, inclusive. Values below 20 indicate that an individual is underweight and values above 30 indicate that an individual is obese. A person who weighs 180 pounds and is 5 feet, or 60 inches, tall has a BMI of 35.15. What is the BMI, to the nearest tenth, for a 170 pound person who is 5 feet 10 inches tall. Is this person over-weight?   This person has a BMI of 24.4 and is not overweight.

32. One's intelligence quotient, or IQ, varies directly as a person's mental age and inversely as that person's chronological age. A person with a mental age of 25 and a chronological age of 20 has an IQ of 125. What is the chronological age of a person with a mental age of 40 and an IQ of 80?   The chronological age is 50.

33. The heat loss of a glass window varies jointly as the window's area and the difference between the outside and inside temper-atures. A window 3 feet wide by 6 feet long loses 1200 Btu per hour when the temperature outside is 20° colder than the tem-perature inside. Find the heat loss through a glass window that is 6 feet wide by 9 feet long when the temperature outside is 10° colder than the temperature inside.   The heat loss is 1800 Btu.

34. Kinetic energy varies jointly as the mass and the square of the velocity. A mass of 8 grams and velocity of 3 centimeters per second has a kinetic energy of 36 ergs. Find the kinetic energy for a mass of 4 grams and velocity of 6 centimeters per second.   A mass of 4 grams and velocity of 6 centimeters per second has a kinetic energy of 72 ergs.

35. Sound intensity varies inversely as the square of the distance from the sound source. If you are in a movie theater and you change your seat to one that is twice as far from the speakers, how does the new sound intensity compare to that of your orig-inal seat?   The sound intensity is $\frac{1}{4}$ of what it was originally.

36. Many people claim that as they get older, time seems to pass more quickly. Suppose that the perceived length of a period of time is inversely proportional to your age. How long will a year seem to be when you are three times as old as you are now?   A year will seem to be $\frac{1}{3}$ of a year.

37. The average number of daily phone calls, C, between two cities varies jointly as the product of their populations, P_1 and P_2, and inversely as the square of the distance, d, between them.

    a. Write an equation that expresses this relationship.
    $C = \dfrac{kP_1P_2}{d^2}$

    b. The distance between San Francisco (population: 777,000) and Los Angeles (population: 3,695,000) is 420 miles. If the average number of daily phone calls between the cities is 326,000, find the value of k to two decimal places and write the equation of variation. $k \approx 0.02$; $C = \dfrac{0.02\,P_1P_2}{d^2}$

    c. Memphis (population: 650,000) is 400 miles from New Orleans (population: 490,000). Find the average number of daily phone calls, to the nearest whole number, between these cities.   approximately 39,813 daily phone calls

38. The force of wind blowing on a window positioned at a right angle to the direction of the wind varies jointly as the area of the window and the square of the wind's speed. It is known that a wind of 30 miles per hour blowing on a window mea-suring 4 feet by 5 feet exerts a force of 150 pounds. During a storm with winds of 60 miles per hour, should hurricane shut-ters be placed on a window that measures 3 feet by 4 feet and is capable of withstanding 300 pounds of force?   Yes, the wind will exert a force of 360 pounds on the window.

39. The table shows the values for the current, I, in an electric circuit and the resistance, R, of the circuit.

I (amperes)	0.5	1.0	1.5	2.0	2.5	3.0	4.0	5.0
R (ohms)	12	6.0	4.0	3.0	2.4	2.0	1.5	1.2

    a. Graph the ordered pairs in the table of values, with values of I along the x-axis and values of R along the y-axis. Con-nect the eight points with a smooth curve.   See graphing answer section.

    b. Does current vary directly or inversely as resistance? Use your graph and explain how you arrived at your answer. Current varies inversely as resistance. Answers will vary.

    c. Write an equation of variation for I and R, using one of the ordered pairs in the table to find the constant of vari-ation. Then use your variation equation to verify the other seven ordered pairs in the table. $R = \dfrac{6}{I}$

Writing in Mathematics

40. What does it mean if two quantities vary directly?   40–44. Answers will vary.

41. In your own words, explain how to solve a variation problem.

42. What does it mean if two quantities vary inversely?

43. Explain what is meant by combined variation. Give an example with your explanation.

44. Explain what is meant by joint variation. Give an example with your explanation.

In Exercises 45–46, describe in words the variation shown by the given equation.

45. $z = \dfrac{k\sqrt{x}}{y^2}$ z varies directly as the square root of x and inversely as the square of y.

46. $z = kx^2\sqrt{y}$ z varies jointly as the square of x and the square root of y.

47. We have seen that the daily number of phone calls between two cities varies jointly as their populations and inversely as the square of the distance between them. This model, used by telecommunication companies to estimate the line capacities needed among various cities, is called the *gravity model*. Compare the model to Newton's formula for gravitation on page 513 and describe why the name *gravity model* is appropriate.   Answers will vary.

Technology Exercise

48. Use a graphing utility to graph any three of the variation equations in Exercises 21–30. Then TRACE along each curve and identify the point that corresponds to the problem's solution.   Answers will vary.

Critical Thinking Exercises

49. In a hurricane, the wind pressure varies directly as the square of the wind velocity. If wind pressure is a measure of a hurricane's destructive capacity, what happens to this destructive power when the wind speed doubles?   If wind speed doubles, the wind pressure is 4 times more destructive.

50. The heat generated by a stove element varies directly as the square of the voltage and inversely as the resistance. If the voltage remains constant, what needs to be done to triple the amount of heat generated?   To triple the amount of heat, the resistance should be divided by 3.

51. Galileo's telescope brought about revolutionary changes in astronomy. A comparable leap in our ability to observe the universe took place as a result of the Hubble Space Telescope. The space telescope can see stars and galaxies whose brightness is $\frac{1}{50}$ of the faintest objects now observable using ground-based telescopes. Use the fact that the brightness of a point source, such as a star, varies inversely as the square of its distance from an observer to show that the space telescope can see about seven times farther than a ground-based telescope.   Distance is increased by $\sqrt{50}$, or about 7.07, for the space telescope.

Review Exercises

52. Solve:

$$8(2 - x) = -5x.$$

(Section 2.3, Example 2) $\left\{ \frac{16}{3} \right\}$

53. Divide:

$$\frac{27x^3 - 8}{3x + 2}.$$

(Section 5.6, Example 3) $9x^2 - 6x + 4 - \dfrac{16}{3x + 2}$

54. Factor:

$$6x^3 - 6x^2 - 120x.$$

(Section 6.5, Example 2) $6x(x - 5)(x + 4)$

GROUP PROJECT

CHAPTER

7

Additional group projects can be found in the *Instructor's Resource Manual.*

A cost-benefit analysis compares the estimated costs of a project with the benefits that will be achieved. Costs and benefits are given monetary values and compared using a benefit-cost ratio. As shown in the figure, a favorable ratio for a project means that the benefits outweigh the costs and the project is cost-effective. As a group, select an environmental project that was implemented in your area of the country. Research the cost and benefit graphs that resulted in the implementation of this project. How were the benefits converted into monetary terms? Is there an equation for either the cost model or the benefit model? Group members may need to interview members of environmental groups and businesses that were part of this project. You may wish to consult an environmental science textbook to find out more about cost-benefit analyses. After doing your research, the group should write or present a report explaining why the cost-benefit analysis resulted in the project's implementation.

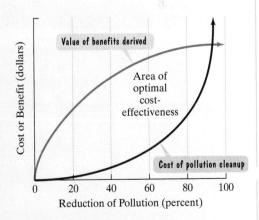

CHAPTER 7 SUMMARY

Definitions and Concepts	Examples

Section 7.1 Rational Expressions and their Simplification

A rational expression is the quotient of two polynomials. To find values for which a rational expression is undefined, set the denominator equal to 0 and solve.

Find all numbers for which

$$\frac{7x}{x^2 - 3x - 4}$$

is undefined.

$$x^2 - 3x - 4 = 0$$

$$(x - 4)(x + 1) = 0$$

$$x - 4 = 0 \quad \text{or} \quad x + 1 = 0$$

$$x = 4 \qquad\qquad x = -1$$

Undefined at 4 and −1

To simplify a rational expression:

1. Factor the numerator and the denominator completely.

2. Divide the numerator and the denominator by any common factors.

If factors in the numerator and denominator are opposites, their quotient is −1.

Simplify: $\dfrac{3x + 18}{x^2 - 36}$.

$$\frac{3x + 18}{x^2 - 36} = \frac{3\overset{1}{\cancel{(x + 6)}}}{\underset{1}{\cancel{(x + 6)}}(x - 6)} = \frac{3}{x - 6}$$

Section 7.2 Multiplying and Dividing Rational Expressions

Multiplying Rational Expressions

1. Factor completely.

2. Divide numerators and denominators by common factors.

3. Multiply remaining factors in the numerators and multiply the remaining factors in the denominators.

$$\frac{x^2 + 3x - 10}{x^2 - 2x} \cdot \frac{x^2}{x^2 - 25}$$

$$= \frac{\overset{1}{\cancel{(x + 5)}}\,\overset{1}{\cancel{(x - 2)}}}{\underset{1}{\cancel{x}}\,\underset{1}{\cancel{(x - 2)}}} \cdot \frac{\overset{1}{\cancel{x}} \cdot x}{\underset{1}{\cancel{(x + 5)}}(x - 5)}$$

$$= \frac{x}{x - 5}$$

Dividing Rational Expressions

Invert the divisor and multiply.

$$\frac{3y + 3}{(y + 2)^2} \div \frac{y^2 - 1}{y + 2}$$

$$= \frac{3y + 3}{(y + 2)^2} \cdot \frac{y + 2}{y^2 - 1}$$

$$= \frac{3\overset{1}{\cancel{(y + 1)}}}{(y + 2)\underset{1}{\cancel{(y + 2)}}} \cdot \frac{\overset{1}{\cancel{(y + 2)}}}{\underset{1}{\cancel{(y + 1)}}(y - 1)}$$

$$= \frac{3}{(y + 2)(y - 1)}$$

Definitions and Concepts	Examples

Section 7.3 Adding and Subtracting Rational Expressions with the Same Denominator

To add or subtract rational expressions with the same denominator, add or subtract the numerators and place the result over the common denominator. If possible, simplify the resulting expression.	$$\frac{y^2 - 3y + 4}{y^2 + 8y + 15} - \frac{y^2 - 5y - 2}{y^2 + 8y + 15}$$ $$= \frac{y^2 - 3y + 4 - (y^2 - 5y - 2)}{y^2 + 8y + 15}$$ $$= \frac{y^2 - 3y + 4 - y^2 + 5y + 2}{y^2 + 8y + 15}$$ $$= \frac{2y + 6}{(y + 5)(y + 3)}$$ $$= \frac{2\overset{1}{\cancel{(y + 3)}}}{(y + 5)\underset{1}{\cancel{(y + 3)}}} = \frac{2}{y + 5}$$
To add or subtract rational expressions with opposite denominators, multiply either rational expression by $\frac{-1}{-1}$ to obtain a common denominator.	$$\frac{7}{x - 6} + \frac{x + 4}{6 - x}$$ $$= \frac{7}{x - 6} + \frac{(-1)}{(-1)} \cdot \frac{x + 4}{6 - x}$$ $$= \frac{7}{x + 6} + \frac{-x - 4}{x - 6}$$ $$= \frac{7 - x - 4}{x - 6} = \frac{3 - x}{x - 6}$$

Section 7.4 Adding and Subtracting Rational Expressions with Different Denominators

Finding the Least Common Denominator (LCD) 1. Factor denominators completely. 2. List factors of the first denominator. 3. Add to the list factors of the second denominator that are not already in the list. 4. The LCD is the product of factors in step 3.	Find the LCD of $$\frac{x + 1}{2x - 2} \quad \text{and} \quad \frac{2x}{x^2 + 2x - 3}.$$ $$2x - 2 = 2(x - 1)$$ $$x^2 + 2x - 3 = (x - 1)(x + 3)$$ Factors of first denominator: $2, x - 1$ Factors of second denominator not in the list: $x + 3$ LCD: $2(x - 1)(x + 3)$
Adding and Subtracting Rational Expressions with Different Denominators 1. Find the LCD. 2. Rewrite each rational expression as an equivalent expression with the LCD. 3. Add or subtract numerators, placing the resulting expression over the LCD. 4. If possible, simplify.	$$\frac{x + 1}{2x - 2} - \frac{2x}{x^2 + 2x - 3}$$ $$= \frac{x + 1}{2(x - 1)} - \frac{2x}{(x - 1)(x + 3)}$$ LCD is $2(x - 1)(x + 3)$. $$= \frac{(x + 1)(x + 3)}{2(x - 1)(x + 3)} - \frac{2x \cdot 2}{2(x - 1)(x + 3)}$$ $$= \frac{x^2 + 4x + 3 - 4x}{2(x - 1)(x + 3)}$$ $$= \frac{x^2 + 3}{2(x - 1)(x + 3)}$$

Definitions and Concepts	Examples

Section 7.5 Complex Rational Expressions

Complex rational expressions have numerators or denominators containing one or more rational expressions. Complex rational expressions can be simplified by obtaining single expressions in the numerator and denominator and then dividing. They can also be simplified by multiplying the numerator and denominator by the LCD of all rational expressions within the complex rational expression.

Simplify by dividing: $\dfrac{\dfrac{1}{x} + 5}{\dfrac{1}{x} - \dfrac{1}{3}}$.

$$= \frac{\dfrac{1}{x} + \dfrac{5x}{x}}{\dfrac{3}{3x} - \dfrac{x}{3x}} = \frac{\dfrac{1 + 5x}{x}}{\dfrac{3 - x}{3x}} = \frac{1 + 5x}{\overset{1}{\cancel{x}}} \cdot \frac{\overset{1}{\cancel{3x}}}{3 - x}$$

$$= \frac{3(1 + 5x)}{3 - x} \quad \text{or} \quad \frac{3 + 15x}{3 - x}$$

Simplify by the LCD method: $\dfrac{\dfrac{1}{x} + 5}{\dfrac{1}{x} - \dfrac{1}{3}}$.

LCD is $3x$.

$$\frac{3x}{3x} \cdot \frac{\left(\dfrac{1}{x} + 5\right)}{\left(\dfrac{1}{x} - \dfrac{1}{3}\right)} = \frac{3x \cdot \dfrac{1}{x} + 3x \cdot 5}{3x \cdot \dfrac{1}{x} - 3x \cdot \dfrac{1}{3}}$$

$$= \frac{3 + 15x}{3 - x}$$

Section 7.6 Solving Rational Equations

A rational equation is an equation containing one or more rational expressions.

Solving Rational Equations

1. List restrictions on the variable.

2. Clear fractions by multiplying both sides by the LCD.

3. Solve the resulting equation.

4. Reject any proposed solution in the list of restrictions. Check other proposed solutions in the original equation.

Solve: $\dfrac{7x}{x^2 - 4} + \dfrac{5}{x - 2} = \dfrac{2x}{x^2 - 4}$

$$\frac{7x}{(x + 2)(x - 2)} + \frac{5}{x - 2} = \frac{2x}{(x + 2)(x - 2)}$$

Denominators would equal 0 if $x = -2$ or $x = 2$.
Restrictions: $x \neq -2$ and $x \neq 2$.

LCD is $(x + 2)(x - 2)$.

$$(x + 2)(x - 2)\left[\frac{7x}{(x + 2)(x - 2)} + \frac{5}{x - 2}\right]$$

$$= (x + 2)(x - 2) \cdot \frac{2x}{(x + 2)(x - 2)}$$

$$7x + 5(x + 2) = 2x$$
$$7x + 5x + 10 = 2x$$
$$12x + 10 = 2x$$
$$10 = -10x$$
$$-1 = x$$

The proposed solution, -1, is not part of the restriction $x \neq -2$ and $x \neq 2$. It checks. The solution is -1 and the solution set is $\{-1\}$.

Definitions and Concepts	Examples

Section 7.7 Applications Using Rational Equations and Proportions

Motion problems involving time are solved using

$$t = \frac{d}{r}.$$

$$\text{Time traveled} = \frac{\text{Distance traveled}}{\text{Rate of travel}}$$

It takes a cyclist who averages 16 miles per hour in still air the same time to travel 48 miles with the wind as 16 miles against the wind. What is the wind's rate?

$$x = \text{wind's rate}$$

$$16 + x = \text{cyclist's rate with wind}$$

$$16 - x = \text{cyclist's rate against wind}$$

	Distance	Rate	Time $= \dfrac{\text{Distance}}{\text{Rate}}$
With wind	48	$16 + x$	$\dfrac{48}{16 + x}$
Against wind	16	$16 - x$	$\dfrac{9}{16 - x}$

Two times are equal

$$\frac{48}{16 + x} = \frac{16}{16 - x}$$

$$(16 + x)(16 - x) \cdot \frac{48}{16 + x} = \frac{16}{16 - x} \cdot (16 + x)(16 - x)$$

$$48(16 - x) = 16(16 + x)$$

Solving this equation, $x = 8$.
The wind's rate is 8 miles per hour.

Work problems are solved using the following condition:

Fraction of job done by the first $+$ fraction of job done by the second $=$ 1.

One pipe fills a pool in 20 hours and a second pipe in 15 hours. How long will it take to fill the pool using both pipes?

$$x = \text{time using both pipes}$$

Fraction of pool filled by pipe 1 in x hours $+$ fraction of pool filled by pipe 2 in x hours $=$ 1.

$$\frac{x}{20} \qquad + \qquad \frac{x}{15} \qquad = \qquad 1$$

$$60\left(\frac{x}{20} + \frac{x}{15}\right) = 60 \cdot 1$$

$$3x + 4x = 60$$

$$7x = 60$$

$$x = \frac{60}{7} = 8\frac{4}{7} \text{ hours}$$

It will take $8\frac{4}{7}$ hours for both pipes to fill the pool.

Definitions and Concepts	Examples

Section 7.7 Applications Using Rational Equations and Proportions (continued)

A proportion is a statement in the form $\frac{a}{b} = \frac{c}{d}$. The cross-products principle states that if $\frac{a}{b} = \frac{c}{d}$, then $ad = bc$ ($b \neq 0$ and $d \neq 0$).

Solving Applied Problems Using Proportions

1. Read the problem and represent the unknown quantity by x (or any letter).

2. Set up a proportion by listing the given ratio on one side and the ratio with the unknown quantity on the other side.

3. Drop units and apply the cross-products principle.

4. Solve for x and answer the question.

30 elk are tagged and released. Sometime later, a sample of 80 elk are observed and 10 are tagged. How many elk are there?

$$x = \text{number of elk}$$

$$\text{Tagged} \quad \frac{30}{x} = \frac{10}{80} \quad \text{Total}$$

$$10x = 30 \cdot 80$$

$$10x = 2400$$

$$x = 240$$

There are 240 elk.

Similar triangles have the same shape, but not necessarily the same size. Corresponding angles have the same measure, and corresponding sides are proportional. If the measures of two angles of one triangle are equal to those of two angles of a second triangle, then the two triangles are similar.

Find x for these similar triangles.

Corresponding sides are proportional:

$$\frac{7}{x} = \frac{10}{5}. \quad \left(\text{or } \frac{7}{x} = \frac{14}{7}\right)$$

$$5x \cdot \frac{7}{x} = \frac{10}{5} \cdot 5x$$

$$35 = 10x$$

$$x = \frac{35}{10} = 3.5$$

Definitions and Concepts	Examples

Section 7.8 Modeling Using Variation

English Statement	Equation
y varies directly as x.	$y = kx$
y varies directly as x^n.	$y = kx^n$
y varies inversely as x.	$y = \dfrac{k}{x}$
y varies inversely as x^n.	$y = \dfrac{k}{x^n}$
y varies directly as x and inversely as z.	$y = \dfrac{kx}{z}$
y varies jointly as x and z.	$y = kxz$

The time that it takes you to drive a certain distance varies inversely as your driving rate. Averaging 40 miles per hour, it takes you 10 hours to drive the distance. How long would the trip take averaging 50 miles per hour?

1. $t = \dfrac{k}{r}$ **Time, t, varies inversely as rate, r.**

2. It takes 10 hours at 40 miles per hour.

$$10 = \dfrac{k}{40}$$

$$k = 10(40) = 400$$

3. $t = \dfrac{400}{r}$

4. How long at 50 miles per hour? Substitute 50 for r.

$$t = \dfrac{400}{50} = 8$$

It takes 8 hours at 50 miles per hour.

CHAPTER 7   REVIEW EXERCISES

7.1 *In Exercises 1–4, find all numbers for which each rational expression is undefined. If the rational expression is defined for all real numbers, so state.*

1. $\dfrac{5x}{6x - 24}$

$x = 4$

2. $\dfrac{x + 3}{(x - 2)(x + 5)}$

$x = 2$ and $x = -5$

3. $\dfrac{x^2 + 3}{x^2 - 3x + 2}$

$x = 1$ and $x = 2$

4. $\dfrac{7}{x^2 + 81}$

defined for all real numbers

In Exercises 5–12, simplify each rational expression. If the rational expression cannot be simplified, so state.

5. $\dfrac{16x^2}{12x}$ $\dfrac{4x}{3}$

6. $\dfrac{x^2 - 4}{x - 2}$ $x + 2$

7. $\dfrac{x^3 + 2x^2}{x + 2}$ x^2

8. $\dfrac{x^2 + 3x - 18}{x^2 - 36}$ $\dfrac{x - 3}{x - 6}$

9. $\dfrac{x^2 - 4x - 5}{x^2 + 8x + 7}$ $\dfrac{x - 5}{x + 7}$

10. $\dfrac{y^2 + 2y}{y^2 + 4y + 4}$ $\dfrac{y}{y + 2}$

11. $\dfrac{x^2}{x^2 + 4}$ cannot be simplified

12. $\dfrac{2x^2 - 18y^2}{3y - x}$ $-2(x + 3y)$

7.2 *In Exercises 13–17, multiply as indicated.*

13. $\dfrac{x^2 - 4}{12x} \cdot \dfrac{3x}{x + 2}$ $\dfrac{x - 2}{4}$

14. $\dfrac{5x + 5}{6} \cdot \dfrac{3x}{x^2 + x}$ $\dfrac{5}{2}$

15. $\dfrac{x^2 + 6x + 9}{x^2 - 4} \cdot \dfrac{x - 2}{x + 3}$ $\dfrac{x + 3}{x + 2}$

16. $\dfrac{y^2 - 2y + 1}{y^2 - 1} \cdot \dfrac{2y^2 + y - 1}{5y - 5}$ $\dfrac{2y - 1}{5}$

17. $\dfrac{2y^2 + y - 3}{4y^2 - 9} \cdot \dfrac{3y + 3}{5y - 5y^2}$ $\dfrac{-3(y + 1)}{5y(2y - 3)}$

In Exercises 18–22, divide as indicated.

18. $\dfrac{x^2 + x - 2}{10} \div \dfrac{2x + 4}{5}$ $\dfrac{x - 1}{4}$

19. $\dfrac{6x + 2}{x^2 - 1} \div \dfrac{3x^2 + x}{x - 1}$ $\dfrac{2}{x(x + 1)}$

20. $\dfrac{1}{y^2 + 8y + 15} \div \dfrac{7}{y + 5}$ $\dfrac{1}{7(y + 3)}$

21. $\dfrac{y^2 + y - 42}{y - 3} \div \dfrac{y + 7}{(y - 3)^2}$ $(y - 3)(y - 6)$

22. $\dfrac{8x + 8y}{x^2} \div \dfrac{x^2 - y^2}{x^2}$ $\dfrac{8}{x - y}$

7.3 In Exercises 23–28, add or subtract as indicated. Simplify the result, if possible.

23. $\dfrac{4x}{x+5} + \dfrac{20}{x+5}$ 4

24. $\dfrac{8x-5}{3x-1} + \dfrac{4x+1}{3x-1}$ 4

25. $\dfrac{3x^2+2x}{x-1} - \dfrac{10x-5}{x-1}$ $3x-5$

26. $\dfrac{6y^2-4y}{2y-3} - \dfrac{12-3y}{2y-3}$ $3y+4$

27. $\dfrac{x}{x-2} + \dfrac{x-4}{2-x}$ $\dfrac{4}{x-2}$

28. $\dfrac{x+5}{x-3} - \dfrac{x}{3-x}$ $\dfrac{2x+5}{x-3}$

7.4 In Exercises 29–31, find the least common denominator of the rational expressions.

29. $\dfrac{7}{9x^3}$ and $\dfrac{5}{12x}$ $36x^3$

30. $\dfrac{3}{x^2(x-1)}$ and $\dfrac{11}{x(x-1)^2}$ $x^2(x-1)^2$

31. $\dfrac{x}{x^2+4x+3}$ and $\dfrac{17}{x^2+10x+21}$ $(x+3)(x+1)(x+7)$

In Exercises 32–42, add or subtract as indicated. Simplify the result, if possible.

32. $\dfrac{7}{3x} + \dfrac{5}{2x^2}$ $\dfrac{14x+15}{6x^2}$

33. $\dfrac{5}{x+1} + \dfrac{2}{x}$ $\dfrac{7x+2}{x(x+1)}$

34. $\dfrac{7}{x+3} + \dfrac{4}{(x+3)^2}$ $\dfrac{7x+25}{(x+3)^2}$

35. $\dfrac{6y}{y^2-4} - \dfrac{3}{y+2}$ $\dfrac{3}{y-2}$

36. $\dfrac{y-1}{y^2-2y+1} - \dfrac{y+1}{y-1}$ $-\dfrac{y}{y-1}$

37. $\dfrac{x+y}{y} - \dfrac{x-y}{x}$ $\dfrac{x^2+y^2}{xy}$

38. $\dfrac{2x}{x^2+2x+1} + \dfrac{x}{x^2-1}$ $\dfrac{3x^2-x}{(x+1)^2(x-1)}$

39. $\dfrac{5x}{x+1} - \dfrac{2x}{1-x^2}$ $\dfrac{5x^2-3x}{(x+1)(x-1)}$

40. $\dfrac{4}{x^2-x-6} - \dfrac{4}{x^2-4}$ $\dfrac{4}{(x+2)(x-2)(x-3)}$

41. $\dfrac{7}{x+3} + 2$ $\dfrac{2x+13}{x+3}$

42. $\dfrac{2y-5}{6y+9} - \dfrac{4}{2y^2+3y}$ $\dfrac{y-4}{3y}$

7.5 In Exercises 43–47, simplify each complex rational expression.

43. $\dfrac{\frac{1}{2} + \frac{3}{8}}{\frac{3}{4} - \frac{1}{2}}$ $\dfrac{7}{2}$

44. $\dfrac{\frac{1}{x}}{1 - \frac{1}{x}}$ $\dfrac{1}{x-1}$

45. $\dfrac{\frac{1}{x} + \frac{1}{y}}{\frac{1}{xy}}$ $x+y$

46. $\dfrac{\frac{1}{x} - \frac{1}{2}}{\frac{1}{3} - \frac{x}{6}}$ $\dfrac{3}{x}$

47. $\dfrac{3 + \frac{12}{x}}{1 - \frac{16}{x^2}}$ $\dfrac{3x}{x-4}$

7.6 In Exercises 48–55, solve each rational equation. If an equation has no solution, so state.

48. $\dfrac{3}{x} - \dfrac{1}{6} = \dfrac{1}{x}$ $\{12\}$

49. $\dfrac{3}{4x} = \dfrac{1}{x} + \dfrac{1}{4}$ $\{-1\}$

50. $x + 5 = \dfrac{6}{x}$ $\{-6, 1\}$

51. $4 - \dfrac{x}{x+5} = \dfrac{5}{x+5}$ no solution or $\varnothing$

52. $\dfrac{2}{x-3} = \dfrac{4}{x+3} + \dfrac{8}{x^2-9}$ $\{5\}$

53. $\dfrac{2}{x} = \dfrac{2}{3} + \dfrac{x}{6}$ $\{-6, 2\}$

54. $\dfrac{13}{y-1} - 3 = \dfrac{1}{y-1}$ $\{5\}$

55. $\dfrac{1}{x+3} - \dfrac{1}{x-1} = \dfrac{x+1}{x^2+2x-3}$ $\{-5\}$

56. Park rangers introduce 50 elk into a wildlife preserve. The formula

$$P = \dfrac{250(3t+5)}{t+25}$$

models the elk population, P, after t years. How many years will it take for the population to increase to 125 elk?   3 years

57. The formula

$$S = \dfrac{C}{1-r}$$

describes the selling price, S, of a product in terms of its cost to the retailer, C, and its markup, r, usually expressed as a percent. A small television cost a retailer $140 and was sold for $200. Find the markup. Express the answer as a percent.   30%

In Exercises 58–62, solve each formula for the specified variable.

58. $P = \dfrac{R - C}{n}$ for C $C = R - nP$

59. $\dfrac{P_1 V_1}{T_1} = \dfrac{P_2 V_2}{T_2}$ for T_1 $T_1 = \dfrac{P_1 V_1 T_2}{P_2 V_2}$

60. $T = \dfrac{A - P}{Pr}$ for P $P = \dfrac{A}{rT + 1}$

61. $\dfrac{1}{R} = \dfrac{1}{R_1} + \dfrac{1}{R_2}$ for R $R = \dfrac{R_1 R_2}{R_2 + R_1}$

62. $I = \dfrac{nE}{R + nr}$ for n $n = \dfrac{IR}{E - Ir}$

7.7

63. In still water, a paddle boat averages 20 miles per hour. It takes the boat the same amount of time to travel 72 miles downstream, with the current, as 48 miles upstream, against the current. What is the rate of the water's current?   4 miles per hour

64. A car travels 60 miles in the same time that a car traveling 10 miles per hour faster travels 90 miles. What is the rate of each car?   Slower car's rate is 20 miles per hour; faster car's rate is 30 miles per hour.

65. A painter can paint a fence around a house in 6 hours. Working alone, the painter's apprentice can paint the same fence in 12 hours. How many hours would it take them to do the job if they worked together?   4 hours

66. If a school board determines that there should be 3 teachers for every 50 students, how many teachers are needed for an enrollment of 5400 students?   324 teachers

67. To determine the number of trout in a lake, a conservationist catches 112 trout, tags them, and returns them to the lake. Later, 82 trout are caught, and 32 of them are found to be tagged. How many trout are in the lake?   287 trout

68. The triangles shown in the figure are similar. Find the length of the side marked with an x.   5 feet

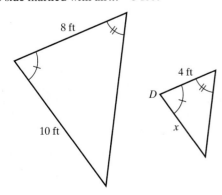

69. Find the height of the lamppost in the figure. $12\frac{1}{2}$ feet

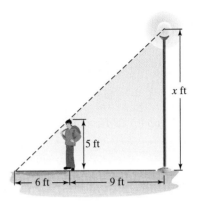

7.8 *Solve the variation problems in Exercises 70–75.*

70. A company's profit varies directly as the number of products it sells. The company makes a profit of $1175 on the sale of 25 products. What is the company's profit when it sells 105 products?   The company's profit is $4935.

71. The distance that a body falls from rest varies directly as the square of the time of the fall. If skydivers fall 144 feet in 3 seconds, how far will they fall in 10 seconds?   Skydivers will fall 1600 feet in 10 seconds.

72. The pitch of a musical tone varies inversely as its wavelength. A tone has a pitch of 660 vibrations per second and a wavelength of 1.6 feet. What is the pitch of a tone that has a wavelength of 2.4 feet?   The tone's pitch is 440 vibrations per second.

73. The loudness of a stereo speaker, measured in decibels, varies inversely as the square of your distance from the speaker. When you are 8 feet from the speaker, the loudness is 28 decibels. What is the loudness when you are 4 feet from the speaker?   The loudness at 4 feet from the speaker is 112 decibels.

74. The time required to assemble computers varies directly as the number of computers assembled and inversely as the number of workers. If 30 computers can be assembled by 6 workers in 10 hours, how long would it take 5 workers to assemble 40 computers?   It will take 16 hours for 5 workers to assemble 40 computers.

75. The volume of a pyramid varies jointly as its height and the area of its base. A pyramid with a height of 15 feet and a base with an area of 35 square feet has a volume of 175 cubic feet. Find the volume of a pyramid with a height of 20 feet and a base with an area of 120 square feet.   The volume of a pyramid with a height of 20 feet and a base with an area of 120 square feet is 800 cubic feet.

CHAPTER 7 TEST

Remember to use your Chapter Test Prep Video CD to see the worked-out solutions to the test questions you want to review.

1. Find all numbers for which

$$\frac{x + 7}{x^2 + 5x - 36}$$

is undefined. $x = -9$ and $x = 4$

In Exercises 2–3, simplify each rational expression.

2. $\dfrac{x^2 + 2x - 3}{x^2 - 3x + 2}$ $\dfrac{x + 3}{x - 2}$

3. $\dfrac{4x^2 - 20x}{x^2 - 4x - 5}$ $\dfrac{4x}{x + 1}$

In Exercises 4–16, perform the indicated operations. Simplify the result, if possible.

4. $\dfrac{x^2 - 16}{10} \cdot \dfrac{5}{x + 4}$ $\dfrac{x - 4}{2}$

5. $\dfrac{x^2 - 7x + 12}{x^2 - 4x} \cdot \dfrac{x^2}{x^2 - 9}$ $\dfrac{x}{x + 3}$

6. $\dfrac{2x + 8}{x - 3} \div \dfrac{x^2 + 5x + 4}{x^2 - 9}$ $\dfrac{2x + 6}{x + 1}$

7. $\dfrac{5y + 5}{(y - 3)^2} \div \dfrac{y^2 - 1}{y - 3}$ $\dfrac{5}{(y - 1)(y - 3)}$

8. $\dfrac{2y^2 + 5}{y + 3} + \dfrac{6y - 5}{y + 3}$ $2y$

9. $\dfrac{y^2 - 2y + 3}{y^2 + 7y + 12} - \dfrac{y^2 - 4y - 5}{y^2 + 7y + 12}$ $\dfrac{2}{y + 3}$

10. $\dfrac{x}{x + 3} + \dfrac{5}{x - 3}$ $\dfrac{x^2 + 2x + 15}{(x + 3)(x - 3)}$

11. $\dfrac{2}{x^2 - 4x + 3} + \dfrac{6}{x^2 + x - 2}$ $\dfrac{8x - 14}{(x + 2)(x - 1)(x - 3)}$

12. $\dfrac{4}{x - 3} + \dfrac{x + 5}{3 - x}$ $\dfrac{-x - 1}{x - 3}$

13. $1 + \dfrac{3}{x - 1}$ $\dfrac{x + 2}{x - 1}$

14. $\dfrac{2x + 3}{x^2 - 7x + 12} - \dfrac{2}{x - 3}$ $\dfrac{11}{(x - 3)(x - 4)}$

15. $\dfrac{8y}{y^2 - 16} - \dfrac{4}{y - 4}$ $\dfrac{4}{y + 4}$

16. $\dfrac{(x - y)^2}{x + y} \div \dfrac{x^2 - xy}{3x + 3y}$ $\dfrac{3x - 3y}{x}$

In Exercises 17–18, simplify each complex rational expression.

17. $\dfrac{5 + \dfrac{5}{x}}{2 + \dfrac{1}{x}}$ $\dfrac{5x + 5}{2x + 1}$

18. $\dfrac{\dfrac{1}{x} - \dfrac{1}{y}}{\dfrac{1}{x}}$ $\dfrac{y - x}{y}$

In Exercises 19–21, solve each rational equation.

19. $\dfrac{5}{x} + \dfrac{2}{3} = 2 - \dfrac{2}{x} - \dfrac{1}{6}$ 6 or $\{6\}$

20. $\dfrac{3}{y + 5} - 1 = \dfrac{4 - y}{2y + 10}$ -8 or $\{-8\}$

21. $\dfrac{2}{x - 1} = \dfrac{3}{x^2 - 1} + 1$ 0 and 2, or $\{0, 2\}$

22. Solve: $R = \dfrac{as}{a + s}$ for a. $a = \dfrac{Rs}{s - R}$ or $-\dfrac{Rs}{R - s}$

23. In still water, a boat averages 30 miles per hour. It takes the boat the same amount of time to travel 16 miles downstream, with the current, as 14 miles upstream, against the current. What is the rate of the water's current?   2 miles per hour

24. One pipe can fill a hot tub in 20 minutes and a second pipe can fill it in 30 minutes. If the hot tub is empty, how long will it take both pipes to fill it?   12 minutes

25. Park rangers catch, tag, and release 200 tule elk back into a wildlife refuge. Two weeks later they observe a sample of 150 elk, of which 5 are tagged. Assuming that the ratio of tagged elk in the sample holds for all elk in the refuge, how many elk are there in the park?   6000 tule elk

26. The triangles in the figure are similar. Find the length of the side marked with an x.   3.2 inches

27. The amount of current flowing in an electrical circuit varies inversely as the resistance in the circuit. When the resistance in a particular circuit is 5 ohms, the current is 42 amperes. What is the current when the resistance is 4 ohms?   52.5 amperes

Note to Instructor: The Chapter 7 Test file in the TestGen program provides algorithms specifically matched to these problems so you can easily replicate this chapter test for additional practice or assessment purposes.

CUMULATIVE REVIEW EXERCISES (CHAPTERS 1–7)

In Exercises 1–6, solve each equation, inequality, or system of equations.

1. $2(x - 3) + 5x = 8(x - 1)$ $\{2\}$

2. $-3(2x - 4) > 2(6x - 12)$ $\{x \mid x < 2\}$

3. $x^2 + 3x = 18$ $\{-6, 3\}$

4. $\dfrac{2x}{x^2 - 4} + \dfrac{1}{x - 2} = \dfrac{2}{x + 2}$ $\{-6\}$

5. $\begin{aligned} y &= 2x - 3 \\ x + 2y &= 9 \end{aligned}$ $\{(3, 3)\}$

6. $\begin{aligned} 3x + 2y &= -2 \\ -4x + 5y &= 18 \end{aligned}$ $\{(-2, 2)\}$

In Exercises 7–9, graph each equation in a rectangular coordinate system. **7–9.** See graphing answer section.

7. $3x - 2y = 6$ **8.** $y = -2x + 3$ **9.** $y = -3$

In Exercises 10–12, simplify each expression.

10. $-21 - 16 - 3(2 - 8)$ -19

11. $\left(\dfrac{4x^5}{2x^2}\right)^3$ $8x^9$

12. $\dfrac{\frac{1}{x} - 2}{4 - \frac{1}{x}}$ $\dfrac{1 - 2x}{4x - 1}$

In Exercises 13–15, factor completely.

13. $4x^2 - 13x + 3$ $(4x - 1)(x - 3)$

14. $4x^2 - 20x + 25$ $(2x - 5)^2$

15. $3x^2 - 75$ $3(x + 5)(x - 5)$

In Exercises 16–18, perform the indicated operations.

16. $(4x^2 - 3x + 2) - (5x^2 - 7x - 6)$ $-x^2 + 4x + 8$

17. $\dfrac{-8x^6 + 12x^4 - 4x^2}{4x^2}$ $-2x^4 + 3x^2 - 1$

18. $\dfrac{x + 6}{x - 2} + \dfrac{2x + 1}{x + 3}$ $\dfrac{3x^2 + 6x + 16}{(x - 2)(x + 3)}$

19. You invested $4000, part at 5% and the remainder at 9% annual interest. At the end of the year, the total interest from these investments was $311. How much was invested at each rate? $1225 at 5% and $2775 at 9%

20. A 68-inch board is to be cut into two pieces. If one piece must be three times as long as the other, find the length of each piece. 17 inches and 51 inches

Algebra is cumulative. This means that **your performance in intermediate algebra depends heavily on the skills you acquired in introductory algebra**. Do you need a quick review of introductory algebra topics before starting the intermediate algebra portion of this book? This mid-textbook Check Point provides a fast way to review and practice the prerequisite skills needed in intermediate algebra.

STUDY TIP

You can quickly review the major topics of introductory algebra by studying the review grids for each of the first seven chapters in this book. Each chart summarizes the definitions and concepts in every section of the chapter. Examples that illustrate the key concepts are also included in the chart. The review charts, with worked-out examples, for each of the book's first seven chapters begin on pages 92, 174, 236, 298, 375, 435, and 518.

A Diagnostic Test for Your Introductory Algebra Skills. You can use the 36 exercises in this mid-textbook Check Point to test your understanding of introductory algebra topics. These exercises cover the fundamental algebra skills upon which the intermediate algebra portion of this book is based. Here are some suggestions for using these exercises as a diagnostic test:

1. Work through all 36 items at your own pace.
2. Use the answer section in the back of the book to check your work.
3. If your answer differs from that in the answer section or if you are not certain how to proceed with a particular item, turn to the section and the worked-out example given in parentheses at the end of each exercise. Study the step-by-step solution of the example that parallels the exercise and then try working the exercise again. If you feel that you need more assistance, study the entire section in which the example appears and work on a selected group of exercises in the exercise set for that section.

In Exercises 1–7, solve each equation, inequality, or system of equations.

1. $2 - 4(x + 2) = 5 - 3(2x + 1)$ (Section 2.3, Example 3)   4 or $\{4\}$

2. $\dfrac{x}{2} - 3 = \dfrac{x}{5}$ (Section 2.3, Example 4)   10 or $\{10\}$

3. $3x + 9 \geq 5(x - 1)$ (Section 2.7, Example 7) $\{x | x \leq 7\}$

4. $2x + 3y = 6$
   $x + 2y = 5$ (Section 4.3, Example 2) $(-3, 4)$ or $\{(-3, 4)\}$

5. $3x - 2y = 1$
   $y = 10 - 2x$ (Section 4.2, Example 1) $(3, 4)$ or $\{(3, 4)\}$

6. $\dfrac{3}{x + 5} - 1 = \dfrac{4 - x}{2x + 10}$ (Section 7.6, Example 4) -8 or $\{-8\}$

7. $x + \dfrac{6}{x} = -5$ (Section 7.6, Example 3) -3 and -2, or $\{-3, -2\}$

In Exercises 8–16, perform the indicated operations. If possible, simplify the answer.

8. $\dfrac{12x^3}{3x^{12}}$ (Section 5.7, Example 4) $\dfrac{4}{x^9}$

9. $4 \cdot 6 \div 2 \cdot 3 + (-5)$ (Section 1.8, Example 4)   31

10. $(6x^2 - 8x + 3) - (-4x^2 + x - 1)$ (Section 5.1, Example 3) $10x^2 - 9x + 4$

11. $(7x + 4)(3x - 5)$ (Section 5.3, Example 2) $21x^2 - 23x - 20$

12. $(5x - 2)^2$ (Section 5.3, Example 6) $25x^2 - 20x + 4$

13. $(x + y)(x^2 - xy + y^2)$ (Section 5.4, Example 8) $x^3 + y^3$

14. $\dfrac{x^2 + 6x + 8}{x^2} \div (3x^2 + 6x)$ (Section 7.2, Example 7) $\dfrac{x + 4}{3x^3}$

15. $\dfrac{x}{x^2 + 2x - 3} - \dfrac{x}{x^2 - 5x + 4}$ (Section 7.4, Example 7) $\dfrac{-7x}{(x + 3)(x - 1)(x - 4)}$

16. $\dfrac{x - \dfrac{1}{5}}{5 - \dfrac{1}{x}}$ (Section 7.5, Examples 2 and 5) $\dfrac{x}{5}$

In Exercises 17–22, factor completely.

17. $4x^2 - 49$ (Section 6.4, Example 1) $(2x + 7)(2x - 7)$

18. $x^3 + 3x^2 - x - 3$ (Section 6.5, Example 4) $(x + 3)(x + 1)(x - 1)$

19. $2x^2 + 8x - 42$ (Section 6.5, Example 2) $2(x - 3)(x + 7)$

20. $x^5 - 16x$ (Section 6.4, Example 4) $x(x^2 + 4)(x + 2)(x - 2)$

21. $x^3 - 10x^2 + 25x$ (Section 6.4, Example 5) $x(x-5)^2$

22. $x^3 - 8$ (Section 6.4, Example 8) $(x-2)(x^2+2x+4)$

In Exercises 23–25, graph each equation in a rectangular coordinate system.

23. $y = \dfrac{1}{3}x - 1$ (Section 3.4, Example 3)

   23–25. See graphing answer section.

24. $3x + 2y = -6$ (Section 3.2, Example 4)

25. $y = -2$ (Section 3.2, Example 7)

26. Find the slope of the line passing through the points $(-1, 3)$ and $(2, -3)$. (Section 3.3, Example 1) -2

27. Write the point-slope form of the equation of the line passing through the points $(1, 2)$ and $(3, 6)$. Then use the point-slope equation to write the slope-intercept form of the line's equation. (Section 3.5, Example 2) $y - 2 = 2(x - 1)$ or $y - 6 = 2(x - 3)$; $y = 2x$

In Exercises 28–36, use an equation or a system of equations to solve each problem.

28. Seven subtracted from five times a number is 208. Find the number. (Section 2.5, Example 2)   43

29. After a 20% reduction, a digital camera sold for $256. What was the price before the reduction? (Section 2.5, Example 7) $320

30. A rectangular field is three times as long as it is wide. If the perimeter of the field is 400 yards, what are the field's dimensions? (Section 2.5, Example 6)   length is 150 yards; width is 50 yards.

31. You invested $20,000 in two accounts paying 7% and 9% annual interest, respectively. If the total interest earned for the year was $1550, how much was invested at each rate? (Section 4.4, Example 4) $12,500 at 7% and $7500 at 9%

32. A chemist needs to mix a 40% acid solution with a 70% acid solution to obtain 12 liters of a 50% acid solution. How many liters of each solution should be used? (Section 4.4, Example 5)   8 liters of 40% and 4 liters of 70%

33. A sailboat has a triangular sail with an area of 120 square feet and a base that is 15 feet long. Find the height of the sail. (Section 2.6, Example 1)   16 feet

34. In a triangle, the measure of the first angle is 10° more than the measure of the second angle. The measure of the third angle is 20° more than four times that of the second angle. What is the measure of each angle? (Section 2.6, Example 6)   1ˢᵗ is 35°; 2ⁿᵈ is 25°; 3ʳᵈ is 120°.

35. A salesperson works in the TV and stereo department of an electronics store. One day she sold 3 TVs and 4 stereos for $2530. The next day, she sold 4 of the same TVs and 3 of the same stereos for $2510. Find the price of a TV and a stereo. (Section 4.4, Example 1)   TV price is $350 and stereo price is $370.

36. The length of a rectangle is 6 meters more than the width. The area is 55 square meters. Find the rectangle's dimensions. (Section 6.6, Example 8)   length is 11 meters; width is 5 meters.

8

'Tis the season and you've waited until the last minute to mail your holiday gifts. Your only option is overnight express mail. You realize that the cost of mailing a gift depends on its weight, but the mailing costs seem somewhat odd. Your packages that weigh 1.1 pounds, 1.5 pounds, and 2 pounds cost $15.75 each to send overnight. Packages that weigh 2.01 pounds and 3 pounds cost you $18.50 each. Finally, your heaviest gift is barely over 3 pounds and its mailing cost is $21.25. What sort of system is this in which costs increase by $2.75, stepping from $15.75 to $18.50 and from $18.50 to $21.25?

Graphs that ascend in steps are explored in Exercises 77–80 in Exercise Set 8.1.

Basics of Functions

8.1 Introduction to Functions

8.2 The Algebra of Functions

8.3 Composite and Inverse Functions

The cost of mailing a package depends on its weight. The probability that you and another person in a room share the same birthday depends on the number of people in the room. In both of these situations, the relationship between variables can be illustrated with the notion of a *function*. Understanding this concept will give you a new perspective on many ordinary situations. Much of our work in the intermediate algebra portion of this book will be devoted to the important topic of functions and how they model your world.

SECTION 8.1

Objectives

1 Find the domain and range of a relation.

2 Determine whether a relation is a function.

3 Evaluate a function.

4 Use the vertical line test to identify functions.

5 Obtain information about a function from its graph.

6 Identify the domain and range of a function from its graph.

INTRODUCTION TO FUNCTIONS

Jerry Orbach	$34,000
Charles Shaughnessy	$31,800
Andy Richter	$29,400
Norman Schwarzkopf	$28,000
Jon Stewart	$28,000

The answer: See the above list. The question: Who are *Celebrity Jeopardy's* five all-time highest earners? The list indicates a correspondence between the five all-time highest earners and their winnings. We can write this correspondence using a set of ordered pairs:

{(Orbach, $34,000), (Shaughnessy, $31,800), (Richter, $29,400), (Schwarzkopf, $28,000), (Stewart, $28,000)}.

The mathematical term for a set of ordered pairs is a *relation*.

> **DEFINITION OF A RELATION** A **relation** is any set of ordered pairs. The set of all first components of the ordered pairs is called the **domain** of the relation and the set of all second components is called the **range** of the relation.

---

1 Find the domain and range of a relation.

EXAMPLE 1 Finding the Domain and Range of a Relation

Find the domain and range of the relation:

{(Orbach, $34,000), (Shaughnessy, $31,800), (Richter, $29,400), (Schwarzkopf, $28,000), (Stewart, $28,000)}.

SOLUTION The domain is the set of all first components. Thus, the domain is

{Orbach, Shaughnessy, Richter, Schwarzkopf, Stewart}.

The range is the set of all second components. Thus, the range is

{$34,000, $31,800, $29,400, $28,000}.

> Although Schwarzkopf and Stewart both won $28,000, it is not necessary to list $28,000 twice.

✔ **CHECK POINT 1** Find the domain and the range of the relation:
{(5, 12.8), (10, 16.2), (15, 18.9), (20, 20.7), (25, 21.8)}.
domain: {5, 10, 15, 20, 25}; range: {12.8, 16.2, 18.9, 20.7, 21.8}

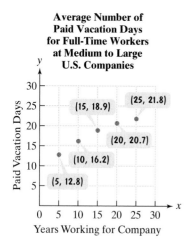

FIGURE 8.1 The graph of a relation showing a correspondence between years with a company and paid vacation days

Source: Bureau of Labor Statistics

As you worked Check Point 1, did you wonder if there was a rule that assigned the "inputs" in the domain to the "outputs" in the range? For example, for the ordered pair (15, 18.9), how does the output 18.9 depend on the input 15? Think paid vacation days! The first number in each ordered pair is the number of years that a full-time employee has been employed by a medium to large U.S. company. The second number is the average number of paid vacation days each year. Consider, for example, the ordered pair (15, 18.9).

The relation in the vacation-days example can be pictured as follows:

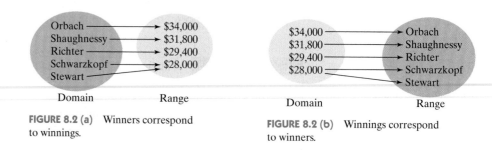

The five points in Figure 8.1 are another way to visually represent the relation.

Functions   Shown, again, in the margin are *Celebrity Jeopardy's* five all-time highest winners and their winnings. We've used this information to define two relations. Figure 8.2(a) shows a correspondence between winners and their winnings. Figure 8.2(b) shows a correspondence between winnings and winners.

2	Determine whether a relation is a function.

Jerry Orbach	$34,000
Charles Shaughnessy	$31,800
Andy Richter	$29,400
Norman Schwarzkopf	$28,000
Jon Stewart	$28,000

FIGURE 8.2 (a)   Winners correspond to winnings.

FIGURE 8.2 (b)   Winnings correspond to winners.

A relation in which each member of the domain corresponds to exactly one member of the range is a **function**. Can you see that the relation in Figure 8.2(a) is a function? Each winner in the domain corresponds to exactly one winning amount in the range. If we know the winner, we can be sure of the amount won. Notice that more than one element in the domain can correspond to the same element in the range. (Schwarzkopf and Stewart both won $28,000.)

Is the relation in Figure 8.2(b) a function? Does each member of the domain correspond to precisely one member of the range? This relation is not a function because there is a member of the domain that corresponds to two different members of the range:

($28,000, Schwarzkopf)   ($28,000, Stewart).

The member of the domain, $28,000, corresponds to both Schwarzkopf and Stewart in the range. If we know the amount won, $28,000, we cannot be sure of the winner. Because **a function is a relation in which no two ordered pairs have the same first**

component and different second components, the ordered pairs($28,000, Schwarzkopf) and ($28,000, Stewart) are not ordered pairs of a function.

> Same first component

($28,000, Schwarzkopf)   ($28,000, Stewart)

> Different second components

> **DEFINITION OF A FUNCTION**   A **function** is a correspondence from a first set, called the **domain**, to a second set, called the **range**, such that each element in the domain corresponds to *exactly one* element in the range.

Example 2 illustrates that not every correspondence between sets is a function.

> ### EXAMPLE 2   Determining Whether a Relation Is a Function

Determine whether each relation is a function:

a. $\{(1,5), (2,5), (3,7), (4,8)\}$ **b.** $\{(5,1), (5,2), (7,3), (8,4)\}$.

SOLUTION   We begin by making a figure for each relation that shows the domain and the range (Figure 8.3).

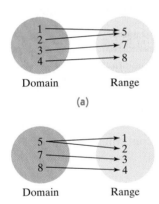

a. Figure 8.3(a) shows that every element in the domain corresponds to exactly one element in the range. The element 1 in the domain corresponds to the element 5 in the range. Furthermore, 2 corresponds to 5, 3 corresponds to 7, and 4 corresponds to 8. No two ordered pairs in the given relation have the same first component and different second components. Thus, the relation is a function.

b. Figure 8.3(b) shows that 5 corresponds to both 1 and 2. If any element in the domain corresponds to more than one element in the range, the relation is not a function. This relation is not a function because two ordered pairs have the same first component and different second components.

> Same first component

(5, 1)   (5, 2)

> Different second components

FIGURE 8.3

STUDY TIP

If a relation is a function, reversing the components in each of its ordered pairs may result in a relation that is not a function.

Look at Figure 8.3(a) again. The fact that 1 and 2 in the domain correspond to the same number, 5, in the range does not violate the definition of a function. **A function can have two different first components with the same second component.** By contrast, a relation is not a function when two different ordered pairs have the same first component and different second components. Thus, the relation in Example 2(b) is not a function.

 CHECK POINT 2 Determine whether each relation is a function:

a. $\{(1,2), (3,4), (5,6), (5,7)\}$ not a function

b. $\{(1,2), (3,4), (6,5), (7,5)\}$.   function

3 Evaluate a function.

Input *x*

f

Output
f(x)

FIGURE 8.4 A "function machine" with inputs and outputs

Functions as Equations and Function Notation Functions are usually given in terms of equations rather than as sets of ordered pairs. For example, here is an equation that models paid vacation days each year as a function of years working for a company:

$$y = -0.016x^2 + 0.93x + 8.5.$$

The variable *x* represents years working for a company. The variable *y* represents the average number of vacation days each year. The variable *y* is a function of the variable *x*. For each value of *x*, there is one and only one value of *y*. The variable *x* is called the **independent variable** because it can be assigned any value from the domain. Thus, *x* can be assigned any positive integer representing years working for a company. The variable *y* is called the **dependent variable** because its value depends on *x*. Paid vacation days depend on years working for a company. The value of the dependent variable, *y*, is calculated after selecting a value for the independent variable, *x*.

If an equation in *x* and *y* gives one and only one value of *y* for each value of *x*, then the variable *y* is a function of the variable *x*. When an equation represents a function, the function is often named by a letter such as f, g, h, F, G, or H. Any letter can be used to name a function. Suppose that *f* names a function. Think of the domain as the set of the function's inputs and the range as the set of the function's outputs. As shown in Figure 8.4, the input is represented by *x* and the output by $f(x)$. The special notation $f(x)$, read "*f* of *x*" or "*f* at *x*," represents the **value of the function at the number *x*.**

Let's make this clearer by considering a specific example. We know that the equation

$$y = -0.016x^2 + 0.93x + 8.5$$

defines *y* as a function of *x*. We'll name the function *f*. Now, we can apply our new function notation.

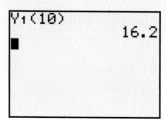
	We read this equation as "*f* of *x* equals $-0.016x^2 + 0.93x + 8.5$."
Input	Output	Equation
x	$f(x)$	$f(x) = -0.016x^2 + 0.93x + 8.5$

STUDY TIP

The notation $f(x)$ does *not* mean "*f* times *x*." The notation describes the value of the function at *x*.

Suppose we are interested in finding $f(10)$, the function's output when the input is 10. To find the value of the function at 10, we substitute 10 for *x*. We are **evaluating the function** at 10.

$f(x) = -0.016x^2 + 0.93x + 8.5$	This is the given function.
$f(10) = -0.016(10)^2 + 0.93(10) + 8.5$	Replace each occurrence of *x* with 10.
$= -0.016(100) + 0.93(10) + 8.5$	Evaluate the exponential expression: $10^2 = 100$.
$= -1.6 + 9.3 + 8.5$	Perform the multiplications.
$= 16.2$	Add from left to right.

The statement $f(10) = 16.2$, read "*f* of 10 equals 16.2," tells us that the value of the function at 10 is 16.2. When the function's input is 10, its output is 16.2. (After 10 years, workers average 16.2 vacation days each year.) To find other function values, such as $f(15), f(20)$, or $f(23)$, substitute the specified input values for *x* into the function's equation.

If a function is named *f* and *x* represents the independent variable, the notation $f(x)$ corresponds to the *y*-value for a given *x*. Thus,

$$f(x) = -0.016x^2 + 0.93x + 8.5 \quad \text{and} \quad y = -0.016x^2 + 0.93x + 8.5$$

define the same function. This function may be written as

$$y = f(x) = -0.016x^2 + 0.93x + 8.5.$$

| EXAMPLE 3 | Using Function Notation |

Find the indicated function value:

a. $f(4)$ for $f(x) = 2x + 3$ **b.** $g(-2)$ for $g(x) = 2x^2 - 1$

c. $h(-5)$ for $h(r) = r^3 - 2r^2 + 5$ **d.** $F(a + h)$ for $F(x) = 5x + 7$.

SOLUTION

a. $f(x) = 2x + 3$ This is the given function.

         $f(4) = 2 \cdot 4 + 3$ To find f of 4, replace x with 4.

            $= 8 + 3$ Multiply: $2 \cdot 4 = 8$.

            $= 11$ *f of 4 is 11.*   Add.

b. $g(x) = 2x^2 - 1$ This is the given function.

      $g(-2) = 2(-2)^2 - 1$ To find g of −2, replace x with −2.

            $= 2(4) - 1$ Evaluate the exponential expression: $(-2)^2 = 4$.

            $= 8 - 1$ Multiply: $2(4) = 8$.

            $= 7$ *g of −2 is 7.*   Subtract.

c. $h(r) = r^3 - 2r^2 + 5$ The function's name is h and r represents the independent variable.

   $h(-5) = (-5)^3 - 2(-5)^2 + 5$ To find h of −5, replace each occurrence of r with −5.

            $= -125 - 2(25) + 5$ Evaluate exponential expressions.

            $= -125 - 50 + 5$ Multiply.

            $= -170$ *h of −5 is −170.* $-125 - 50 = -175$ and $-175 + 5 = -170$.

d. $F(x) = 5x + 7$ This is the given function.

    $F(a + h) = 5(a + h) + 7$ Replace x with $a + h$.

            $= 5a + 5h + 7$ Apply the distributive property.   ■

     F of a + h is 5a + 5h + 7.

✔ **CHECK POINT 3** Find the indicated function value:

a. $f(6)$ for $f(x) = 4x + 5$ 29

b. $g(-5)$ for $g(x) = 3x^2 - 10$ 65

c. $h(-4)$ for $h(r) = r^2 - 7r + 2$ 46

d. $F(a + h)$ for $F(x) = 6x + 9$. $6a + 6h + 9$

STUDY TIP

Equations of the form $y = mx + b$, now expressed as $f(x) = mx + b$, were presented in Section 3.4. You can find a brief review of the slope-intercept form of the equation of a line in the Section 3.4 summary on page 238.

     The functions in Check Point 3(a) and 3(d), $f(x) = 4x + 5$ and $F(x) = 6x + 9$, are examples of *linear functions*. **Linear functions** have equations of the form $f(x) = mx + b$. By contrast, the functions in Check Point 3(b) and 3(c), $g(x) = 3x^2 - 10$ and $h(r) = r^2 - 7r + 2$, are examples of *quadratic functions*. **Quadratic functions** have equations of the form $f(x) = ax^2 + bx + c, a \neq 0$.

4 Use the vertical line test to identify functions.

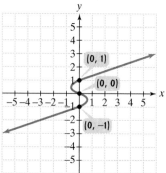

FIGURE 8.5 *y* is not a function of *x* because 0 is paired with three values of *y*, namely, 1, 0, and −1.

Graphs of Functions and the Vertical Line Test The **graph of a function** is the graph of its ordered pairs. For example, the graph of $f(x) = 3x + 1$ is the set of points (x, y) in the rectangular coordinate system satisfying the equation $y = 3x + 1$. Thus, the graph of f is a line with slope 3 and *y*-intercept 1. Similarly, the graph of $f(x) = x^2 + 3x + 5$ is the set of points (x, y) in the rectangular coordinate system satisfying the equation $y = x^2 + 3x + 5$.

Not every graph in the rectangular coordinate system is the graph of a function. The definition of a function specifies that no value of *x* can be paired with two or more different values of *y*. Consequently, if a graph contains two or more different points with the same first coordinate, the graph cannot represent a function. This is illustrated in Figure 8.5. Observe that points sharing a common first coordinate are vertically above or below each other.

This observation is the basis of a useful test for determining whether a graph defines *y* as a function of *x*. The test is called the **vertical line test**.

> **THE VERTICAL LINE TEST FOR FUNCTIONS** If any vertical line intersects a graph in more than one point, the graph does not define *y* as a function of *x*.

EXAMPLE 4 Using the Vertical Line Test

Use the vertical line test to identify graphs in which *y* is a function of *x*.

a.

b.

c.

d.
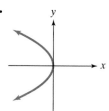

SOLUTION *y* is a function of *x* for the graphs in (b) and (c).

a.

y **is not a function** of *x*. Two values of *y* correspond to one *x*-value.

b.

y **is a function** of *x*.

c.

y **is a function** of *x*.

d.
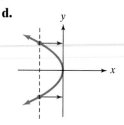

y **is not a function** of *x*. Two values of *y* correspond to one *x*-value. ■

> ✓ **CHECK POINT 4** Use the vertical line test to identify graphs in which *y* is a function of *x*.
>
> **a.** function   **b.** function   **c.** not a function
>
>

5 Obtain information about a function from its graph.

Obtaining Information from Graphs You can obtain information about a function from its graph. At the right or left of a graph, you will often find closed dots, open dots, or arrows.

- A closed dot indicates that the graph does not extend beyond this point and the point belongs to the graph.
- An open dot indicates that the graph does not extend beyond this point and the point does not belong to the graph.
- An arrow indicates that the graph extends indefinitely in the direction in which the arrow points.

EXAMPLE 5 Analyzing the Graph of a Function

The function
$$f(x) = -0.016x^2 + 0.93x + 8.5$$
models the average number of paid vacation days each year, $f(x)$, for full-time workers at medium to large U.S. companies after x years. The graph of f is shown in Figure 8.6.

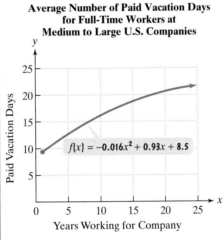

Average Number of Paid Vacation Days for Full-Time Workers at Medium to Large U.S. Companies

$f(x) = -0.016x^2 + 0.93x + 8.5$

Paid Vacation Days / Years Working for Company

FIGURE 8.6

Source: Bureau of Labor Statistics.

a. Explain why f represents the graph of a function.

b. Use the graph to find a reasonable estimate of $f(5)$.

c. For what value of x is $f(x) = 20$?

d. Describe the general trend shown by the graph.

SOLUTION

a. No vertical line intersects the graph of f more than once. By the vertical line test, f represents the graph of a function.

b. To find $f(5)$, or f of 5, we locate 5 on the x-axis. The figure shows the point on the graph of f for which 5 is the first coordinate. From this point, we look to the y-axis to find the corresponding y-coordinate. A reasonable estimate of the y-coordinate is 13. Thus, $f(5) \approx 13$. After 5 years, a worker can expect approximately 13 paid vacation days.

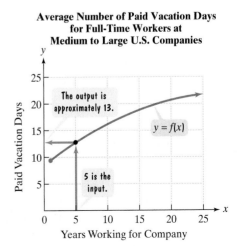

Average Number of Paid Vacation Days for Full-Time Workers at Medium to Large U.S. Companies

The output is approximately 13.

$y = f(x)$

5 is the input.

Paid Vacation Days / Years Working for Company

c. To find the value of x for which $f(x) = 20$, we locate 20 on the y-axis. The figure shows that there is one point on the graph of f for which 20 is the second coordinate. From this point, we look to the x-axis to find the corresponding x-coordinate. A reasonable estimate of the x-coordinate is 18. Thus, $f(x) = 20$ for $x \approx 18$. A worker with 20 paid vacation days has been with the company approximately 18 years.

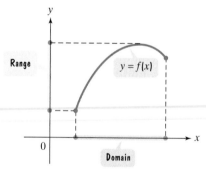

Average Number of Paid Vacation Days for Full-Time Workers at Medium to Large U.S. Companies

d. The graph of f is rising from left to right. This shows that paid vacation days increase as time with the company increases. However, the rate of increase is slowing down as the graph moves to the right. This means that the increase in paid vacation days takes place more slowly the longer an employee is with the company. ■

 CHECK POINT 5

a. Use the graph of f in Figure 8.6 on page 537 to find a reasonable estimate of $f(10)$. $f(10) \approx 16$

b. For what value of x is $f(x) = 15$? Round to the nearest whole number.
$x \approx 8$

6 Identify the domain and range of a function from its graph.

Identifying Domain and Range from a Function's Graph   Figure 8.7 illustrates how the graph of a function is used to determine the function's domain and its range.

Domain: set of inputs

> Found on the x-axis

Range: set of outputs

> Found on the y-axis

FIGURE 8.7   Domain and range of f

Let's apply these ideas to the graph of the function shown in Figure 8.8. To find the domain, look for all the inputs on the x-axis that correspond to points on the graph. Can you see that they extend from -4 to 2, inclusive? Using set-builder notation, the function's domain can be represented as follows:

$$\{\, x \mid -4 \leq x \leq 2 \,\}.$$

> The set of all x such that x is greater than or equal to -4 and less than or equal to 2.

To find the range, look for all the outputs on the y-axis that correspond to points on the graph. They extend from 1 to 4, inclusive. Using set-builder notation, the function's range can be represented as follows:

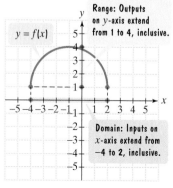

FIGURE 8.8   Domain and range of f

$$\{\, y \mid 1 \le y \le 4 \,\}$$

| The set of all y | such that | y is greater than or equal to 1 and less than or equal to 4. |

EXAMPLE 6 Identifying the Domain and Range of a Function from Its Graph

Use the graph of each function to identify its domain and its range.

a.

b.

c.

d.

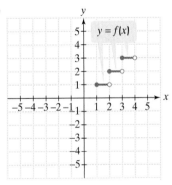

SOLUTION For the graph of each function, the domain is highlighted in blue on the x-axis and the range is highlighted in green on the y-axis.

a.

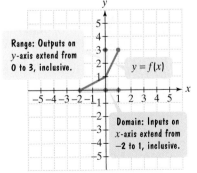

Range: Outputs on y-axis extend from 0 to 3, inclusive.

Domain: Inputs on x-axis extend from −2 to 1, inclusive.

Domain = $\{x \mid -2 \le x \le 1\}$
Range = $\{y \mid 0 \le y \le 3\}$

b.

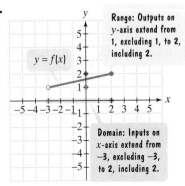

Range: Outputs on y-axis extend from 1, excluding 1, to 2, including 2.

Domain: Inputs on x-axis extend from −3, excluding −3, to 2, including 2.

Domain = $\{x \mid -3 < x \le 2\}$
Range = $\{y \mid 1 < y \le 2\}$

c.

Range: Outputs on y-axis include real numbers greater than or equal to 0.

$y = f(x)$

Domain: Inputs on x-axis include real numbers less than or equal to 4.

Domain $= \{x | x \leq 4\}$
Range $= \{y | y \geq 0\}$

d.

Range: Outputs on y-axis "step" from 1 to 2 to 3.

$y = f(x)$

Domain: Inputs on x-axis extend from 1, including 1, to 4, excluding 4.

Domain $= \{x | 1 \leq x < 4\}$
Range $= \{y | y = 1, 2, 3\}$

 CHECK POINT 6 Use the graph of each function to identify its domain and its range.

a.

$y = f(x)$

Domain $= \{x | -2 \leq x \leq 1\}$
Range $= \{y | 0 \leq y \leq 3\}$

b.

$y = f(x)$

Domain $= \{x | -2 < x \leq 1\}$
Range $= \{y | -1 \leq y < 2\}$

c.

$y = f(x)$

Domain $= \{x | -3 \leq x < 0\}$
Range $= \{y | y = -3, -2, -1\}$

8.1 EXERCISE SET

Student Solutions Manual   CD/Video   PH Math/Tutor Center   MathXL Tutorials on CD   MathXL®   MyMathLab   Interactmath.com

Practice Exercises

In Exercises 1–8, determine whether each relation is a function. Give the domain and range for each relation.

1. $\{(1, 2), (3, 4), (5, 5)\}$ function; domain: $\{1, 3, 5\}$; range: $\{2, 4, 5\}$

2. $\{(4, 5), (6, 7), (8, 8)\}$ function; domain: $\{4, 6, 8\}$; range: $\{5, 7, 8\}$

3. $\{(3, 4), (3, 5), (4, 4), (4, 5)\}$ not a function; domain: $\{3, 4\}$; range: $\{4, 5\}$

4. $\{(5, 6), (5, 7), (6, 6), (6, 7)\}$ not a function; domain: $\{5, 6\}$; range: $\{6, 7\}$

5. $\{(-3, -3), (-2, -2), (-1, -1), (0, 0)\}$ function; domain: $\{-3, -2, -1, 0\}$; range: $\{-3, -2, -1, 0\}$

6. $\{(-7, -7), (-5, -5), (-3, -3), (0, 0)\}$ function; domain: $\{-7, -5, -3, 0\}$; range: $\{-7, -5, -3, 0\}$

7. $\{(1, 4), (1, 5), (1, 6)\}$ not a function; domain: $\{1\}$; range: $\{4, 5, 6\}$

8. $\{(4, 1), (5, 1), (6, 1)\}$ function; domain: $\{4, 5, 6\}$; range: $\{1\}$

In Exercises 9–18, find the indicated function values.

9. $f(x) = x + 1$

   a. $f(0)$ 1   **b.** $f(5)$ 6   **c.** $f(-8)$ -7

   d. $f(2a)$ $2a + 1$ **e.** $f(a + 2)$ $a + 3$

10. $f(x) = x + 3$

   a. $f(0)$ 3   **b.** $f(5)$ 8   **c.** $f(-8)$ -5

   d. $f(2a)$ $2a + 3$ **e.** $f(a + 2)$ $a + 5$

11. $g(x) = 3x - 2$

   a. $g(0)$ -2 **b.** $g(-5)$ -17 **c.** $g\left(\dfrac{2}{3}\right)$ 0

   d. $g(4b)$ $12b - 2$ **e.** $g(b + 4)$ $3b + 10$

12. $g(x) = 4x - 3$

   a. $g(0)$ -3 **b.** $g(-5)$ -23 **c.** $g\left(\dfrac{3}{4}\right)$ 0

   d. $g(5b)$ $20b - 3$ **e.** $g(b + 5)$ $4b + 17$

13. $h(x) = 3x^2 + 5$

   a. $h(0)$ 5   **b.** $h(-1)$ 8   **c.** $h(4)$ 53

   d. $h(-3)$ 32   **e.** $h(4b)$ $48b^2 + 5$

14. $h(x) = 2x^2 - 4$

   a. $h(0)$ -4 **b.** $h(-1)$ -2 **c.** $h(5)$ 46

   d. $h(-3)$ 14   **e.** $h(5b)$ $50b^2 - 4$

15. $f(x) = 2x^2 + 3x - 1$

   a. $f(0)$ -1 **b.** $f(3)$ 26   **c.** $f(-4)$ 19

   d. $f(b)$ $2b^2 + 3b - 1$

   e. $f(5a)$ $50a^2 + 15a - 1$

16. $f(x) = 3x^2 + 4x - 2$

   a. $f(0)$ -2 **b.** $f(3)$ 37   **c.** $f(-5)$ 53

   d. $f(b)$ $3b^2 + 4b - 2$

   e. $f(5a)$ $75a^2 + 20a - 2$

17. $f(x) = \dfrac{2x - 3}{x - 4}$

   a. $f(0)$ $\dfrac{3}{4}$ **b.** $f(3)$ -3 **c.** $f(-4)$ $\dfrac{11}{8}$

   d. $f(-5)$ $\dfrac{13}{9}$ **e.** $f(a + h)$ $\dfrac{2a + 2h - 3}{a + h - 4}$

   f. Why must 4 be excluded from the domain of f?
     Denominator would be zero.

18. $f(x) = \dfrac{3x - 1}{x - 5}$

   a. $f(0)$ $\dfrac{1}{5}$ **b.** $f(3)$ -4 **c.** $f(-3)$ $\dfrac{5}{4}$

   d. $f(10)$ $\dfrac{29}{5}$ **e.** $f(a + h)$ $\dfrac{3a + 3h - 1}{a + h - 5}$

   f. Why must 5 be excluded from the domain of f?
     Denominator would be zero.

In Exercises 19–26, use the vertical line test to identify graphs in which y is a function of x.

19.

function

20.

function

21.

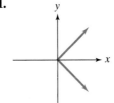

not a function

22.

function graph (arrows)

not a function

23.

function

24.

function graph (vertical segment)

not a function

25.

function graph (rectangle)

not a function

26.

function

In Exercises 27–32, use the graph of f to find each indicated function value.

27. $f(-2)$ -4

28. $f(2)$ -4

29. $f(4)$ 4

30. $f(-4)$ 4

31. $f(-3)$ 0

32. $f(-1)$ 0

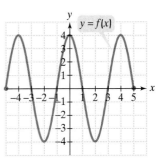

Use the graph of g to solve Exercises 33–38.

33. Find $g(-4)$.   2

34. Find $g(2)$. -2

35. Find $g(-10)$.   2

36. Find $g(10)$. -2

37. For what value of x
is $g(x) = 1$? -2

38. For what value of x
is $g(x) = -1$?   1

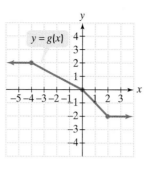

In Exercises 39–48, use the graph of each function to identify its domain and its range.

39.

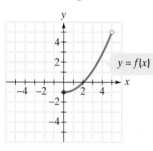

domain = $\{x|0 \le x < 5\}$;
range = $\{y|-1 \le y < 5\}$

40.

domain = $\{x|-5 < x \le 0\}$;
range = $\{y|-3 \le y < 3\}$

41.

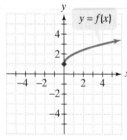

domain = $\{x|x \ge 0\}$;
range = $\{y|y \ge 1\}$

42.

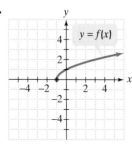

domain = $\{x|x \ge -1\}$;
range = $\{y|y \ge 0\}$

43.

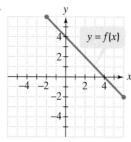

domain = $\{x|-2 \le x \le 6\}$;
range = $\{y|-2 \le y \le 6\}$

44.

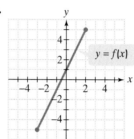

domain = $\{x|-3 \le x \le 2\}$;
range = $\{y|-5 \le y \le 5\}$

45.

domain = $\{x|x \text{ is a real number}\}$;
range = $\{y|y \le -2\}$

46.

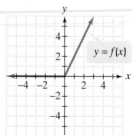

domain = $\{x|x \text{ is a real number}\}$;
range = $\{y|y \ge 0\}$

47.

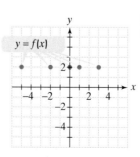

domain = $\{x|x = -5, -2, 0, 1, 3\}$;
range = $\{y|y = 2\}$

48.

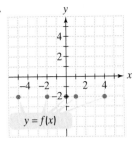

$y = f(x)$

domain $= \{x | x = -5, -2, 0, 1, 4\}$;
range $= \{y | y = -2\}$

Practice Plus

In Exercises 49–50, let $f(x) = x^2 - x + 4$ and $g(x) = 3x - 5$.

49. Find $g(1)$ and $f(g(1))$. $-2; 10$

50. Find $g(-1)$ and $f(g(-1))$. $-8; 76$

In Exercises 51–52, let f and g be defined by the following table:

x	$f(x)$	$g(x)$
-2	6	0
-1	3	4
0	-1	1
1	-4	-3
2	0	-6

51. Find $\sqrt{f(-1) - f(0)} - [g(2)]^2 + f(-2) \div g(2) \cdot g(-1)$ -38

52. Find $|f(1) - f(0)| - [g(1)]^2 + g(1) \div f(-1) \cdot g(2)$ 0

In Exercises 53–54, find $f(-x) - f(x)$ for the given function f. Then simplify the expression.

53. $f(x) = x^3 + x - 5$ $-2x^3 - 2x$

54. $f(x) = x^2 - 3x + 7$ $6x$

In Exercises 55–56, each function is defined by two equations. The equation in the first row gives the output for negative numbers in the domain. The equation in the second row gives the output for non-negative numbers in the domain. Find the indicated function values.

55. $f(x) = \begin{cases} 3x + 5 & \text{if } x < 0 \\ 4x + 7 & \text{if } x \geq 0 \end{cases}$

   a. $f(-2)$ -1 **b.** $f(0)$ 7

   c. $f(3)$ 19 **d.** $f(-100) + f(100)$ 112

56. $f(x) = \begin{cases} 6x - 1 & \text{if } x < 0 \\ 7x + 3 & \text{if } x \geq 0 \end{cases}$

   a. $f(-3)$ -19 **b.** $f(0)$ 3

   c. $f(4)$ 31 **d.** $f(-100) + f(100)$ 102

In Exercises 57–58, use the graph of each function to identify its domain and its range.

57.

Graph approaches but never touches the dashed vertical line.

$y = f(x)$

On both sides, graph never touches the x-axis.

domain $= \{x | x \neq 1\}$; range $= \{y | y \neq 0\}$

58.

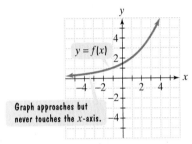

$y = f(x)$

Graph approaches but never touches the x-axis.

domain $= \{x | x \text{ is a real number}\}$; range $= \{y | y > 0\}$

Application Exercises

59. The bar graph shows the breakdown of political ideologies in the United States.

Political Ideologies in the U.S.

Percentage of Voters

- Extremely Liberal: 1%
- Liberal: 7%
- Slightly Liberal: 11%
- Moderate: 52%
- Slightly Conservative: 13%
- Conservative: 13%
- Extremely Conservative: 3%

Ideology

Source: Center for Political Studies, University of Michigan

   a. Write a set of seven ordered pairs in which political ideologies correspond to percentages. Each ordered pair should be in the form

         (ideology, percent).

   Use EL, L, SL, M, SC, C, and EC to represent the respective ideologies from left to right. $\{(\text{EL}, 1\%), (\text{L}, 7\%), (\text{SL}, 11\%), (\text{M}, 52\%), (\text{SC}, 13\%), (\text{C}, 13\%), (\text{EC}, 3\%)\}$

b. Is the relation in part (a) a function? Explain your answer.   Yes; each ideology corresponds to exactly one percentage.

c. Write a set of seven ordered pairs in which percentages correspond to political ideologies. Each ordered pair should be in the form

(percent, ideology).

{(1%, EL), (7%, L), (11%, SL), (52%, M),
(13%, SC), (13%, C), (3%, EC)}

d. Is the relation in part (c) a function? Explain your answer.   No; 13% in the domain corresponds to two ideologies in the range, SC and C.

60. The bar graph shows the percentage of teens in the United States who say they really need various technologies.

Percentage of U.S. Teens Saying They Really Need Various Technologies

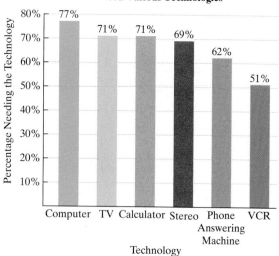

Source: Gallup Organization

a. Write a set of six ordered pairs in which technologies correspond to percentages. Each ordered pair should be in the form

(technology, percent).

{(Computer, 77%), (TV, 71%), (Calculator, 71%),
(Stereo, 69%), (Ph. Ans. M.,62%), (VCR, 51%)}

b. Is the relation in part (a) a function? Explain your answer.   Yes; each technology corresponds to exactly one percentage.

c. Write a set of six ordered pairs in which percentages correspond to technologies. Each ordered pair should be in the form

(percent, technology).

{(77%, Computer), (71%, TV), (71%, Calculator),
(69%, Stereo), (62%, Ph. Ans. M.), (51%, VCR)}

d. Is the relation in part (c) a function? Explain your answer.
No; 71% in the domain corresponds to two technologies in the range, TV and calculator.

The male minority? The graphs show enrollment in U.S. colleges, with projections through 2009. The trend indicated by the graphs is among the hottest topics of debate among college-admissions officers. Some private liberal arts colleges have quietly begun special efforts to recruit men—including admissions preferences for them.

Enrollment in U.S. Colleges

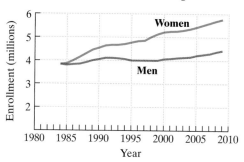

Source: Department of Education

The function

$$W(x) = 0.07x + 4.1$$

models the number of women, $W(x)$, in millions, enrolled in U.S. colleges x years after 1984. The function

$$M(x) = 0.01x + 3.9$$

models the number of men, $M(x)$, in millions, enrolled in U.S. colleges x years after 1984. Use these functions to solve Exercises 61–64.

61. Find and interpret $W(16)$. Identify this information as a point on the graph for women.   5.22; In 2000, there were 5.22 million women enrolled in U.S. colleges; (2000, 5.22).

62. Find and interpret $M(16)$. Identify this information as a point on the graph for men.   4.06; In 2000, there were 4.06 million men enrolled in U.S. colleges; (2000, 4.06).

63. Find and interpret $W(20) - M(20)$.   1.4; In 2004, there were 1.4 million more women than men enrolled in U.S. colleges.

64. Find and interpret $W(25) - M(25)$.   1.7; In 2009, there will be 1.7 million more women than men enrolled in U.S. colleges.

The function

$$f(x) = 0.4x^2 - 36x + 1000$$

models the number of accidents, $f(x)$, per 50 million miles driven as a function of a driver's age, x, in years, where x includes drivers from ages 16 through 74, inclusive. The graph of f is shown. Use the equation for f to solve Exercises 65–68.

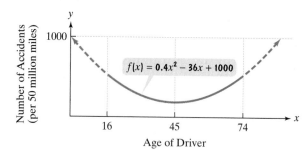

Age of Driver

65. Find and interpret $f(20)$. Identify this information as a point on the graph of f. 440; For 20-year-old drivers, there are 440 accidents per 50 million miles driven; $(20, 440)$

66. Find and interpret $f(50)$. Identify this information as a point on the graph of f. 200; For 50-year-old drivers, there are 200 accidents per 50 million miles driven; $(50, 200)$

67. For what value of x does the graph reach its lowest point? Use the equation for f to find the minimum value of y. Describe the practical significance of this minimum value. $x = 45$; $y = 190$; The minimum number of accidents is 190 per 50 million miles driven and is attributed to 45-year-old drivers.

68. Use the graph to identify two different ages for which drivers have the same number of accidents. Use the equation for f to find the number of accidents for drivers at each of these ages. Answers will vary; an example is 16 and 74 years old. For these ages, the number of accidents is 526.4 per 50 million miles.

The figure shows the percentage of Jewish Americans in the U.S. population, $f(x)$, x years after 1900. Use the graph to solve Exercises 69–76. **69–74.** Answers will vary. Example are given.

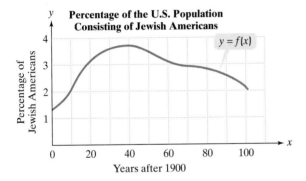

Source: American Jewish Yearbook

69. Use the graph to find a reasonable estimate of $f(60)$. What does this mean in terms of the variables in this situation? $f(60) \approx 3.1$; In 1960, Jewish Americans made up about 3.1% of the U.S. population.

70. Use the graph to find a reasonable estimate of $f(100)$. What does this mean in terms of the variables in this situation? $f(100) \approx 2.2$; In 2000, Jewish Americans made up about 2.2% of the U.S. population.

71. For what value or values of x is $f(x) = 3$? Round to the nearest year. What does this mean in terms of the variables in this situation? $x \approx 19$ and $x \approx 64$; In 1919 and in 1964, Jewish Americans made up about 3% of the U.S. population.

72. For what value or values of x is $f(x) = 2.5$? Round to the nearest year. What does this mean in terms of the variables in this situation? $x \approx 14$ and $x \approx 90$; In 1914 and in 1990, Jewish Americans made up 2.5% of the U.S. population.

73. In which year did the percentage of Jewish Americans in the U.S. population reach a maximum? What is a reasonable estimate of the percentage for that year? 1940; 3.7%

74. In which year was the percentage of Jewish Americans in the U.S. population at a minimum? What is a reasonable estimate of the percentage for that year? 1900; 1.4%

75. Explain why f represents the graph of a function. Each year corresponds to only one percentage.

76. Describe the general trend shown by the graph. As a percentage of the U.S. population, the Jewish-American population was increasing until 1940 and has been decreasing since then.

The figure shows the cost of mailing a first-class letter, $f(x)$, as a function of its weight, x, in ounces. Use the graph to solve Exercises 77–80.

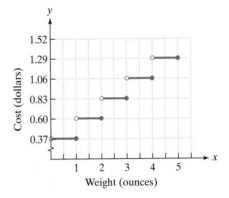

77. Find $f(3)$. What does this mean in terms of the variables in this situation? 0.83; It costs $0.83 to mail a 3-ounce first-class letter.

78. Find $f(4)$. What does this mean in terms of the variables in this situation? 1.06; It costs $1.06 to mail a 4-ounce first-class letter.

79. What is the cost of mailing a letter that weighs 1.5 ounces? $0.60

80. What is the cost of mailing a letter that weighs 1.8 ounces? $0.60

Writing in Mathematics

81. What is a relation? Describe what is meant by its domain and its range. **81–88.** Answers will vary.

82. Explain how to determine whether a relation is a function. What is a function?

83. Does $f(x)$ mean f times x when referring to function f? If not, what does $f(x)$ mean? Provide an example with your explanation.

84. What is the graph of a function?

85. Explain how the vertical line test is used to determine whether a graph represents a function.

86. Explain how to identify the domain and range of a function from its graph.

87. For people filing a single return, federal income tax is a function of adjusted gross income because for each value of adjusted gross income there is a specific tax to be paid. By contrast, the price of a house is not a function of the lot size on which the house sits because houses on same-sized lots can sell for many different prices.

   a. Describe an everyday situation between variables that is a function.

   b. Describe an everyday situation between variables that is not a function.

88. Do you believe that the trend shown by the graphs for Exercises 61–64 should be reversed by providing admissions preferences for men? Explain your position on this issue.

Technology Exercise

89. The function

$$f(x) = -0.00002x^3 + 0.008x^2 - 0.3x + 6.95$$

models the number of annual physician visits, $f(x)$, by a person of age x. Graph the function in a $[0, 100, 5]$ by $[0, 40, 2]$ viewing rectangle. What does the shape of the graph indicate about the relationship between one's age and the number of annual physician visits? Use the TRACE or minimum function capability to find the coordinates of the minimum point on the graph of the function. What does this mean? See graphing answer section; The number of annual visits decreases until about age 20 and then increases; (20.3, 4.0); The minimum is at about age 20 when there are about 4 annual physician visits.

Critical Thinking Exercises

90. Which one of the following is true? **d**

   a. All relations are functions.

   b. No two ordered pairs of a function can have the same second component and different first components.

   c. The graph of every line is a function.

   d. A horizontal line can intersect the graph of a function in more than one point.

91. If $f(x) = 3x + 7$, find $\dfrac{f(a + h) - f(a)}{h}$. 3

92. Give an example of a relation with the following characteristics: The relation is a function containing two ordered pairs. Reversing the components in each ordered pair results in a relation that is not a function. Answers will vary; an example is $\{(1, 1), (2, 1)\}$.

93. If $f(x + y) = f(x) + f(y)$ and $f(1) = 3$, find $f(2)$, $f(3)$, and $f(4)$. Is $f(x + y) = f(x) + f(y)$ for all functions? $f(2) = 6; f(3) = 9; f(4) = 12;$ no

Review Exercises

94. Simplify: $24 \div 4[2 - (5 - 2)]^2 - 6$. (Section 1.8, Example 8) 0

95. Simplify: $\left(\dfrac{3x^2 y^{-2}}{y^3}\right)^{-2}$. (Section 5.7, Example 6) $\dfrac{y^{10}}{9x^4}$

96. Solve: $\dfrac{x}{3} = \dfrac{3x}{5} + 4$. (Section 2.3, Example 4) $\{-15\}$

SECTION 8.2

THE ALGEBRA OF FUNCTIONS

Objectives

1 Find the domain of a function.

2 Find the sum of two functions.

3 Use the algebra of functions to combine functions and determine domains.

2006 Pontiac Solstice, the first two-seater from Pontiac in more than 20 years. Starting price: $30,000

America's big three automakers, GM, Ford, and Chrysler, know that consumers have always been willing to pay more for a cool design and a hot car. In 2004, Detroit gave consumers that opportunity, unleashing 40 new cars or updated models. The Big Three's discovery of the automobile in 2004 might seem odd, but from 1990 through 2003, Detroit focused much of its energy and money on SUVs and trucks, which commanded high prices and high profits, and saw less competition from imports. The line graphs in Figure 8.9 show the number, in millions, of cars and SUVs sold by the Big Three from 1990 through 2003. In this section, we will look at these data from the perspective of functions. By considering total sales of cars and SUVs, you will see that functions can be combined using procedures that will remind you of combining algebraic expressions.

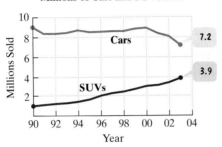

FIGURE 8.9 Sales by America's Big Three automakers

1 Find the domain of a function.

The Domain of a Function We begin with two functions that model the data in Figure 8.9.

$$C(x) = -0.14x + 9 \qquad S(x) = 0.22x + 1$$

Car sales, $C(x)$, in millions, x years after 1990

SUV sales, $S(x)$, in millions, x years after 1990

How far beyond 1990 should we extend these models? The trend in Figure 8.9 shows decreasing car sales and increasing SUV sales. Because the three automakers focused on SUVs from 1990 through 2003 and launched a fleet of new cars in 2004, the trend shown by the data changed in 2004. Thus, we should not extend the models beyond 2003. Because x represents the number of years after 1990,

$$\text{Domain of } C = \{x \mid x = 0, 1, 2, 3, \ldots, 13\}$$

and

$$\text{Domain of } S = \{x \mid x = 0, 1, 2, 3, \ldots, 13\}.$$

Functions that model data often have their domains explicitly given with the function's equation. However, for most functions, only an equation is given, and the domain is not specified. In cases like this, the domain of a function f is the largest set of real numbers for which the value of $f(x)$ is a real number. For example, consider the function

$$f(x) = \frac{1}{x - 3}.$$

Because division by 0 is undefined (and not a real number), the denominator, $x - 3$, cannot be 0. Thus, x cannot equal 3. The domain of the function consists of all real numbers other than 3, represented by

$$\text{Domain of } f = \{x \mid x \text{ is a real number and } x \neq 3\}.$$

In Chapter 10, we will be studying square root functions such as

$$g(x) = \sqrt{x}.$$

The equation tells us to take the square root of *x*. Because only nonnegative numbers have square roots that are real numbers, the expression under the square root sign, *x*, must be nonnegative. Thus,

Domain of $g = \{x \mid x \text{ is a nonnegative real number}\}$.

Equivalently,

Domain of $g = \{x \mid x \geq 0\}$.

FINDING A FUNCTION'S DOMAIN   If a function f does not model data or verbal conditions, its domain is the largest set of real numbers for which the value of $f(x)$ is a real number. Exclude from a function's domain real numbers that cause division by zero and real numbers that result in a square root of a negative number.

EXAMPLE 1   **Finding the Domain of a Function**

Find the domain of each function:

a. $f(x) = 3x + 2$ **b.** $g(x) = \dfrac{3x + 2}{x + 1}$.

SOLUTION

a. The function $f(x) = 3x + 2$ contains neither division nor a square root. For every real number, *x*, the algebraic expression $3x + 2$ is a real number. Thus, the domain of f is the set of all real numbers.

Domain of $f = \{x \mid x \text{ is a real number}\}$

b. The function $g(x) = \dfrac{3x + 2}{x + 1}$ contains division. Because division by 0 is undefined, we must exclude from the domain the value of *x* that causes $x + 1$ to be 0. Thus, *x* cannot equal −1.

Domain of $g = \{x \mid x \text{ is a real number and } x \neq -1\}$ ∎

✔ **CHECK POINT 1** Find the domain of each function:

a. $f(x) = \dfrac{1}{2}x + 3$ Domain of $f = \{x \mid x \text{ is a real number}\}$

b. $g(x) = \dfrac{7x + 4}{x + 5}$.   Domain of $g = \{x \mid x \text{ is a real number and } x \neq -5\}$

The Algebra of Functions   We return to the functions that model millions of car and SUV sales from 1990 through 2003:

$$C(x) = -0.14x + 9 \qquad S(x) = 0.22x + 1.$$

Car sales, *C*(*x*), in millions, *x* years after 1990   SUV sales, *S*(*x*), in millions, *x* years after 1990

How can we use these functions to find total sales of cars and SUVs in 2003? Because 2003 is 13 years after 1990 and 13 is in the domain of each function, we need to find the sum of two function values:

$$C(13) + S(13).$$

Here is how it's done:

$$C(13) = -0.14(13) + 9 = 7.18 \qquad S(13) = 0.22(13) + 1 = 3.86$$

| Substitute 13 for x in $C(x) = -0.14x + 9$. | 7.18 million cars were sold in 2003. | | Substitute 13 for x in $S(x) = 0.22x + 1$. | 3.86 million SUVs were sold in 2003. |

$$C(13) + S(13) = 7.18 + 3.86 = 11.04.$$

Thus, a total of 11.04 million cars and SUVs were sold in 2003.

There is a second way we can obtain this number. We can first add the functions C and S to obtain a new function, $C + S$. To do so, we add the terms to the right of the equal sign for $C(x)$ to the terms to the right of the equal sign for $S(x)$:

$$\begin{aligned}(C + S)(x) &= C(x) + S(x) \\ &= (-0.14x + 9) + (0.22x + 1) \qquad \text{Add terms for } C(x) \text{ and } S(x). \\ &= 0.08x + 10. \qquad\qquad\qquad \text{Combine like terms.}\end{aligned}$$

Thus,

$$(C + S)(x) = 0.08x + 10$$

| Total car and SUV sales, in millions, x years after 1990 |

Do you see how we can use this new function to find total car and SUV sales in 2003? Substitute 13 for x in the equation for $C + S$:

$$(C + S)(13) = 0.08(13) + 10 = 11.04$$

| Substitute 13 for x in $(C + S)(x) = 0.08x + 10$. | As we found above, a total of 11.04 million cars and SUVs were sold in 2003. |

The domain of the new function, $C + S$, consists of the numbers x that are in the domain of C and in the domain of S. Because both functions model data from 1990 through 2003,

$$\text{Domain of } C + S = \{x \mid x = 0, 1, 2, 3, \ldots, 13\}.$$

Here is a general definition for function addition:

THE SUM OF FUNCTIONS Let f and g be two functions. The **sum $f + g$** is the function defined by

$$(f + g)(x) = f(x) + g(x).$$

The domain of $f + g$ is the set of all real numbers that are common to the domain of f and the domain of g.

2 Find the sum of two functions.

EXAMPLE 2 Finding the Sum of Two Functions

Let $f(x) = x^2 - 3$ and $g(x) = 4x + 5$. Find each at the following:

  a. $(f + g)(x)$ **b.** $(f + g)(3)$.

SOLUTION

  a. $(f + g)(x) = f(x) + g(x) = (x^2 - 3) + (4x + 5) = x^2 + 4x + 2$

    Thus,

$$(f + g)(x) = x^2 + 4x + 2.$$

b. We find $(f + g)(3)$ by substituting 3 for x in the equation for $f + g$.

$$(f + g)(x) = x^2 + 4x + 2 \qquad \text{This is the equation for } f + g.$$

Substitute 3 for x.

$$(f + g)(3) = 3^2 + 4 \cdot 3 + 2 = 9 + 12 + 2 = 23 \qquad ■$$

 CHECK POINT 2 Let $f(x) = 3x^2 + 4x - 1$ and $g(x) = 2x + 7$. Find each of the following:

a. $(f + g)(x)$ $\quad 3x^2 + 6x + 6$ **b.** $(f + g)(4)$. $\quad 78$

EXAMPLE 3   Adding Functions and Determining the Domain

Let $f(x) = \dfrac{4}{x}$ and $g(x) = \dfrac{3}{x + 2}$. Find each of the following:

a. $(f + g)(x)$ **b.** the domain of $f + g$.

SOLUTION

a. $(f + g)(x) = f(x) + g(x) = \dfrac{4}{x} + \dfrac{3}{x + 2}$

(In Chapter 7, we discussed how to add these rational expressions. In our study of the algebra of functions, we will leave these fractions in the form shown.)

b. The domain of $f + g$ is the set of all real numbers that are common to the domain of f and the domain of g. Thus, we must find the domains of f and g. We will do so for f first.

Note that $f(x) = \dfrac{4}{x}$ is a function involving division. Because division by 0 is undefined, x cannot equal 0.

Domain of $f = \{x \,|\, x \text{ is a real number and } x \neq 0\}$

The function $g(x) = \dfrac{3}{x + 2}$ is also a function involving division. Because division by 0 is undefined, x cannot equal -2.

Domain of $g = \{x \,|\, x \text{ is a real number and } x \neq -2\}$

To find $f(x) + g(x)$, x must be in both domains listed. Thus,

Domain of $f + g = \{x \,|\, x \text{ is a real number and } x \neq 0 \text{ and } x \neq -2\}$.   ■

 CHECK POINT 3 Let $f(x) = \dfrac{5}{x}$ and $g(x) = \dfrac{7}{x - 8}$. Find each of the following:

a. $(f + g)(x)$ $\quad \dfrac{5}{x} + \dfrac{7}{x - 8}$ **b.** the domain of $f + g$.

Domain of $f + g = \{x \,|\, x$ is a real number and $x \neq 0$ and $x \neq 8\}$

3 Use the algebra of functions to combine functions and determine domains.

We can also combine functions using subtraction, multiplication, and division by performing operations with the algebraic expressions that appear on the right side of the equations. For example, the functions $f(x) = 2x$ and $g(x) = x - 1$ can be combined to form the difference, product, and quotient of f and g. Here's how it's done:

Difference: $f - g$

$$(f - g)(x) = f(x) - g(x)$$
$$= 2x - (x - 1) = 2x - x + 1 = x + 1$$

For each function,
$f(x) = 2x$ and
$g(x) = x - 1$.

Product: fg

$$(fg)(x) = f(x) \cdot g(x)$$
$$= 2x(x - 1) = 2x^2 - 2x$$

Quotient: $\dfrac{f}{g}$

$$\left(\frac{f}{g}\right)(x) = \frac{f(x)}{g(x)} = \frac{2x}{x - 1}, x \neq 1.$$

Just like the domain for $f + g$, the domain for each of these functions consists of all real numbers that are common to the domains of f and g. In the case of the quotient function $\dfrac{f(x)}{g(x)}$, we must remember not to divide by 0, so we add the further restriction that $g(x) \neq 0$.

The following definitions summarize our discussion:

THE ALGEBRA OF FUNCTIONS: SUM, DIFFERENCE, PRODUCT, AND QUOTIENT OF FUNCTIONS   Let f and g be two functions. The **sum** $f + g$, the **difference** $f - g$, the **product** fg, and the **quotient** $\dfrac{f}{g}$ are functions whose domains are the set of all real numbers common to the domains of f and g, defined as follows:

1. Sum: $(f + g)(x) = f(x) + g(x)$
2. Difference: $(f - g)(x) = f(x) - g(x)$
3. Product: $(fg)(x) = f(x) \cdot g(x)$
4. Quotient: $\left(\dfrac{f}{g}\right)(x) = \dfrac{f(x)}{g(x)}$, provided $g(x) \neq 0$.

EXAMPLE 4   Using the Algebra of Functions

Let $f(x) = x^2 + x$ and $g(x) = x - 5$. Find each of the following:

a. $(f + g)(4)$

b. $(f - g)(x)$ and $(f - g)(-3)$

c. $(fg)(x)$ and $(fg)(-2)$

d. $\left(\dfrac{f}{g}\right)(x)$ and $\left(\dfrac{f}{g}\right)(7)$

SOLUTION

a. We can find $(f + g)(4)$ using $f(4)$ and $g(4)$.

$$f(x) = x^2 + x \qquad g(x) = x - 5$$
$$f(4) = 4^2 + 4 = 20 \quad g(4) = 4 - 5 = -1$$

Thus,

$$(f + g)(4) = f(4) + g(4) = 20 + (-1) = 19.$$

We can also find $(f + g)(4)$ by first finding $(f + g)(x)$ and then substituting 4 for x:

$$(f + g)(x) = f(x) + g(x) \qquad \text{This is the definition of the sum } f + g.$$
$$= (x^2 + x) + (x - 5) \qquad \text{Substitute the given functions.}$$
$$= x^2 + 2x - 5. \qquad \text{Simplify.}$$

Using $(f + g)(x) = x^2 + 2x - 5$. we have

$$(f + g)(4) = 4^2 + 2 \cdot 4 - 5 = 16 + 8 - 5 = 19.$$

$f(x) = x^2 + x$
$g(x) = x - 5$

The given functions, f and g (repeated)

b. $(f - g)(x) = f(x) - g(x)$ *This is the definition of the difference $f - g$.*

$\quad\quad\quad\quad = (x^2 + x) - (x - 5)$ *Substitute the given functions.*

$\quad\quad\quad\quad = x^2 + x - x + 5$ *Remove parentheses and change the sign of each term in the second set of parentheses.*

$\quad\quad\quad\quad = x^2 + 5$ *Simplify.*

Using $(f - g)(x) = x^2 + 5$, we have

$$(f - g)(-3) = (-3)^2 + 5 = 9 + 5 = 14.$$

c. $(fg)(x) = f(x) \cdot g(x)$ *This is the definition of the product fg.*

$\quad\quad\quad = (x^2 + x)(x - 5)$ *Substitute the given functions.*

$\quad\quad\quad = x^3 - 5x^2 + x^2 - 5x$ *Multiply using the FOIL method.*

$\quad\quad\quad = x^3 - 4x^2 - 5x$ *Combine like terms: $-5x^2 + x^2 = -4x^2$.*

Using $(fg)(x) = x^3 - 4x^2 - 5x$, we have

$(fg)(-2) = (-2)^3 - 4(-2)^2 - 5(-2)$

$\quad\quad\quad = -8 - 4(4) - 5(-2)$ *Evaluate exponential expressions.*

$\quad\quad\quad = -8 - 16 + 10 = -14.$

We can also find $(fg)(-2)$ using the fact that

$$(fg)(-2) = f(-2) \cdot g(-2).$$

$f(x) = x^2 + x \quad\quad\quad\quad\quad\quad\quad\quad g(x) = x - 5$

$f(-2) = (-2)^2 + (-2) = 4 - 2 = 2 \quad\quad g(-2) = -2 - 5 = -7$

Thus,

$$(fg)(-2) = f(-2) \cdot g(-2) = 2(-7) = -14.$$

d. $\left(\dfrac{f}{g}\right)(x) = \dfrac{f(x)}{g(x)}$ *This is the definition of the quotient $\dfrac{f}{g}$.*

$\quad\quad\quad\quad = \dfrac{x^2 + x}{x - 5}$ *Substitute the given functions.*

Using $\left(\dfrac{f}{g}\right)(x) = \dfrac{x^2 + x}{x - 5}$, we have

$$\left(\frac{f}{g}\right)(7) = \frac{7^2 + 7}{7 - 5} = \frac{56}{2} = 28.$$

$\blacksquare$

✔ **CHECK POINT 4** Let $f(x) = x^2 - 2x$ and $g(x) = x + 3$. Find each of the following:

a. $(f + g)(5)$ 23

b. $(f - g)(x)$ and $(f - g)(-1)$ $x^2 - 3x - 3; 1$

c. $(fg)(x)$ and $(fg)(-4)$ $x^3 + x^2 - 6x; -24$

d. $\left(\dfrac{f}{g}\right)(x)$ and $\left(\dfrac{f}{g}\right)(7)$ $\dfrac{x^2 - 2x}{x + 3}; \dfrac{7}{2}$

8.2 EXERCISE SET

Student Solutions Manual CD/Video PH Math/Tutor Center MathXL Tutorials on CD MathXL® MyMathLab Interactmath.com

Practice Exercises

In Exercises 1–10, find the domain of each function.

1. $f(x) = 3x + 5$ Domain of $f = \{x | x \text{ is a real number}\}$

2. $f(x) = 4x + 7$ Domain of $f = \{x | x \text{ is a real number}\}$

3. $g(x) = \dfrac{1}{x + 4}$

Domain of $g = \{x | x \text{ is a real number and } x \neq -4\}$

4. $g(x) = \dfrac{1}{x + 5}$

Domain of $g = \{x | x \text{ is a real number and } x \neq -5\}$

5. $f(x) = \dfrac{2x}{x - 3}$

Domain of $f = \{x | x \text{ is a real number and } x \neq 3\}$

6. $f(x) = \dfrac{4x}{x - 2}$

Domain of $f = \{x | x \text{ is a real number and } x \neq 2\}$

7. $g(x) = x + \dfrac{3}{5 - x}$

Domain of $g = \{x | x \text{ is a real number and } x \neq 5\}$

8. $g(x) = x + \dfrac{7}{6 - x}$

Domain of $g = \{x | x \text{ is a real number and } x \neq 6\}$

9. $f(x) = \dfrac{1}{x + 7} + \dfrac{3}{x - 9}$ Domain of
$f = \{x | x \text{ is a real number and } x \neq -7 \text{ and } x \neq 9\}$

10. $f(x) = \dfrac{1}{x + 8} + \dfrac{3}{x - 10}$ Domain of
$f = \{x | x \text{ is a real number and } x \neq -8 \text{ and } x \neq 10\}$

In Exercises 11–16, find **a.** $(f + g)(x)$ **b.** $(f + g)(5)$.

11. $f(x) = 3x + 1, g(x) = 2x - 6$
$(f + g)(x) = 5x - 5; (f + g)(5) = 20$

12. $f(x) = 4x + 2, g(x) = 2x - 9$
$(f + g)(x) = 6x - 7; (f + g)(5) = 23$

13. $f(x) = x - 5, g(x) = 3x^2$
$(f + g)(x) = 3x^2 + x - 5; (f + g)(5) = 75$

14. $f(x) = x - 6, g(x) = 2x^2$
$(f + g)(x) = 2x^2 + x - 6; (f + g)(5) = 49$

15. $f(x) = 2x^2 - x - 3, g(x) = x + 1$
$(f + g)(x) = 2x^2 - 2; (f + g)(5) = 48$

16. $f(x) = 4x^2 - x - 3, g(x) = x + 1$
$(f + g)(x) = 4x^2 - 2; (f + g)(5) = 98$

In Exercises 17–28, for each pair of functions, f and g, determine the domain of f + g.

17. $f(x) = 3x + 7, g(x) = 9x + 10$
Domain of $f + g = \{x | x \text{ is a real number}\}$

18. $f(x) = 7x + 4, g(x) = 5x - 2$
Domain of $f + g = \{x | x \text{ is a real number}\}$

19. $f(x) = 3x + 7, g(x) = \dfrac{2}{x - 5}$
Domain of $f + g = \{x | x \text{ is a real number and } x \neq 5\}$

20. $f(x) = 7x + 4, g(x) = \dfrac{2}{x - 6}$
Domain of $f + g = \{x | x \text{ is a real number and } x \neq 6\}$

21. $f(x) = \dfrac{1}{x}, g(x) = \dfrac{2}{x - 5}$ Domain of
$f + g = \{x | x \text{ is a real number and } x \neq 0 \text{ and } x \neq 5\}$

22. $f(x) = \dfrac{1}{x}, g(x) = \dfrac{2}{x - 6}$ Domain of
$f + g = \{x | x \text{ is a real number and } x \neq 0 \text{ and } x \neq 6\}$

23. $f(x) = \dfrac{8x}{x - 2}, g(x) = \dfrac{6}{x + 3}$ Domain of
$f + g = \{x | x \text{ is a real number and } x \neq 2 \text{ and } x \neq -3\}$

24. $f(x) = \dfrac{9x}{x - 4}, g(x) = \dfrac{7}{x + 8}$ Domain of
$f + g = \{x | x \text{ is a real number and } x \neq 4 \text{ and } x \neq -8\}$

25. $f(x) = \dfrac{8x}{x - 2}, g(x) = \dfrac{6}{2 - x}$
Domain of $f + g = \{x | x \text{ is a real number and } x \neq 2\}$

26. $f(x) = \dfrac{9x}{x - 4}, g(x) = \dfrac{7}{4 - x}$
Domain of $f + g = \{x | x \text{ is a real number and } x \neq 4\}$

27. $f(x) = x^2, g(x) = x^3$
Domain of $f + g = \{x | x \text{ is a real number}\}$

28. $f(x) = x^2 + 1, g(x) = x^3 - 1$
Domain of $f + g = \{x | x \text{ is a real number}\}$

In Exercises 29–48, let
$$f(x) = x^2 + 4x \quad \text{and} \quad g(x) = 2 - x.$$

Find each of the following.

29. $(f + g)(x)$ and $(f + g)(3)$
$(f + g)(x) = x^2 + 3x + 2; (f + g)(3) = 20$

30. $(f + g)(x)$ and $(f + g)(4)$
$(f + g)(x) = x^2 + 3x + 2; (f + g)(4) = 30$

31. $f(-2) + g(-2)$ 0

32. $f(-3) + g(-3)$ 2

33. $(f - g)(x)$ and $(f - g)(5)$
$(f - g)(x) = x^2 + 5x - 2; (f - g)(5) = 48$

34. $(f - g)(x)$ and $(f - g)(6)$
$(f - g)(x) = x^2 + 5x - 2; (f - g)(6) = 64$

35. $f(-2) - g(-2)$ -8 **36.** $f(-3) - g(-3)$ -8

37. $(fg)(x)$ and $(fg)(2)$
$(fg)(x) = -x^3 - 2x^2 + 8x; (fg)(2) = 0$

38. $(fg)(x)$ and $(fg)(3)$
$(fg)(x) = -x^3 - 2x^2 + 8x$; $(fg)(3) = -21$

39. $(fg)(5)$ -135 **40.** $(fg)(6)$ -240

41. $\left(\dfrac{f}{g}\right)(x)$ and $\left(\dfrac{f}{g}\right)(1)$ $\left(\dfrac{f}{g}\right)(x) = \dfrac{x^2 + 4x}{2 - x}$; $\left(\dfrac{f}{g}\right)(1) = 5$

42. $\left(\dfrac{f}{g}\right)(x)$ and $\left(\dfrac{f}{g}\right)(3)$ $\left(\dfrac{f}{g}\right)(x) = \dfrac{x^2 + 4x}{2 - x}$; $\left(\dfrac{f}{g}\right)(3) = -21$

43. $\left(\dfrac{f}{g}\right)(-1)$ -1 **44.** $\left(\dfrac{f}{g}\right)(0)$ 0

45. The domain of $f + g$
Domain of $f + g = \{x | x \text{ is a real number}\}$

46. The domain of $f - g$
Domain of $f - g = \{x | x \text{ is a real number}\}$

47. The domain of $\dfrac{f}{g}$

Domain of $\dfrac{f}{g} = \{x | x \text{ is a real number and } x \neq 2\}$

48. The domain of fg Domain of $fg = \{x | x \text{ is a real number}\}$

Practice Plus

Use the graphs of f and g to solve Exercises 49–56.

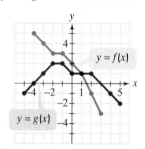

49. Find $(f + g)(-3)$.   5   **50.** Find $(g - f)(-2)$. -1

51. Find $(fg)(2)$. -1 **52.** Find $\left(\dfrac{g}{f}\right)(3)$.   0

53. Find the domain of $f + g$. $\{x | -4 \leq x \leq 3\}$

54. Find the domain of $\dfrac{f}{g}$. $\{x | -4 < x < 3\}$

55. Graph $f + g$.   **55–56.** See graphing answer section.

56. Graph $f - g$.

Use the table defining f and g to solve Exercises 57–60.

x	$f(x)$	$g(x)$
-2	5	0
-1	3	-2
0	-2	4
1	-6	-3
2	0	1

57. Find $(f + g)(1) - (g - f)(-1)$. -4

58. Find $(f + g)(-1) - (g - f)(0)$. -5

59. Find $(fg)(-2) - \left[\left(\dfrac{f}{g}\right)(1)\right]^2$. -4

60. Find $(fg)(2) - \left[\left(\dfrac{g}{f}\right)(0)\right]^2$. -4

Application Exercises

The table shows the total number of births and the total number of deaths in the United States from 1995 through 2002.

Births and Deaths in the U.S.

Year	Births	Deaths
1995	3,899,589	2,312,132
1996	3,891,494	2,314,690
1997	3,880,894	2,314,245
1998	3,941,553	2,337,256
1999	3,959,417	2,391,399
2000	4,058,814	2,403,351
2001	4,025,933	2,416,425
2002	4,022,000	2,436,000

Source: Department of Health and Human Services

The data can be modeled by the following functions:

Number of births $B(x) = 24{,}770x + 3{,}873{,}266$

Number of deaths $D(x) = 20{,}205x + 2{,}294{,}970.$

In each function, x represents the number of years after 1995. Assume that the functions apply only to the years shown in the table. Use these functions to solve Exercises 61–64.

61. Find the domain of B. $\{x | x = 0, 1, 2, \ldots, 7\}$

62. Find the domain of D. $\{x | x = 0, 1, 2, \ldots, 7\}$

63. a. Find $(B - D)(x)$. What does this function represent?
$(B - D)(x) = 4565x + 1{,}578{,}296$; change in U.S. population

  b. Use the function in part (a) to find $(B - D)(6)$. What does this mean in terms of the U.S. population and to which year does this apply? $(B - D)(6) = 1{,}605{,}686$; In 2001, U.S. population increased by 1,605,686.

  c. Use the data shown in the table to find $(B - D)(6)$. How well does the difference of functions used in part (b) model this number?   1,609,508; fairly well

64. a. Find $(B - D)(x)$. What does this function represent?
$(B - D)(x) = 4565x + 1{,}578{,}296$; change in U.S. population

  b. Use the function in part (a) to find $(B - D)(2)$. What does this mean in terms of the U.S. population and to which year does this apply? $(B - D)(2) = 1{,}587{,}426$; In 1997, U.S. population increased by 1,587,426.

c. Use the data shown in the table to find $(B - D)(2)$. How well does the difference of functions used in part (b) model this number?   1,566,649; fairly well, although answers may vary

Consider the following functions:

$f(x) =$ population of the world's more-developed regions in year x

$g(x) =$ the population of the world's less-developed regions in year x

$h(x) =$ total world population in year x.

Use these functions and the graphs shown to answer Exercises 65–68.

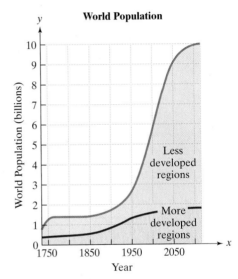

World Population

Source: Population Reference Bureau

65. What does the function $f + g$ represent? $h(x)$ or total world population in year x

66. What does the function $h - g$ represent? $f(x)$ or population of the world's more-developed regions in year x

67. Use the graph to estimate $(f + g)(2000)$.
$(f + g)(2000) \approx 5.9$ billion

68. Use the graph to estimate $(h - g)(2000)$.
$(h - g)(2000) \approx 1.6$ billion

69. A company that sells radios has yearly fixed costs of $600,000. It costs the company $45 to produce each radio. Each radio will sell for $65. The company's costs and revenue are modeled by the following functions:

$C(x) = 600,000 + 45x$ This function models the company's costs.

$R(x) = 65x$ This function models the company's revenue.

Find and interpret $(R - C)(20,000)$, $(R - C)(30,000)$, and $(R - C)(40,000)$. $(R-C)(20,000) = -200,000$: If the company produces and sells 20,000 radios, it will lose $200,000; $(R - C)(30,000) = 0$: If the company produces and sells 30,000 radios, it will break even with its costs equal to its revenue; $(R - C)(40,000) = 200,000$: If the company produces and sells 40,000 radios, it will make a profit of $200,000.

70. The function $f(t) = -0.14t^2 + 0.51t + 31.6$ models the U.S. population, $f(t)$, in millions, ages 65 and older, t years after 1990. The function $g(t) = 0.54t^2 + 12.64t + 107.1$ models the total yearly cost of Medicare, $g(t)$, in billions of dollars, t years after 1990.

a. What does the function $\dfrac{g}{f}$ represent?

the yearly cost of Medicare per person aged 65 and older t years after 1990 (in thousands of dollars)

b. Find and interpret $\dfrac{g}{f}(10)$.

In 2000, the yearly cost of Medicare per person aged 65 and older was about $12.7 thousand, or $12,700.

Writing in Mathematics

71. If a function is defined by an equation, explain how to find its domain.   **71–74.** Answers will vary.

72. If equations for functions f and g are given, explain how to find $f + g$.

73. If the equations of two functions are given, explain how to obtain the quotient function and its domain.

74. If equations for functions f and g are given, describe two ways to find $(f - g)(3)$.

Technology Exercises

75. Graph the function $\dfrac{g}{f}$ from Exercise 70 in a $[0, 15, 1]$ by $[0, 60, 1]$ viewing rectangle. What does the shape of the graph indicate about the per capita costs of Medicare for the U.S. population ages 65 and over with increasing time?   Per capita costs of Medicare are increasing more and more rapidly over time.

In Exercises 76–79, graph each of the three functions in the same $[-10, 10, 1]$ *by* $[-10, 10, 1]$ *viewing rectangle.*   **76–79.** See graphing answer section.

76. $y_1 = 2x + 3$
$y_2 = 2 - 2x$
$y_3 = y_1 + y_2$

77. $y_1 = x - 4$
$y_2 = 2x$
$y_3 = y_1 - y_2$

78. $y_1 = x$
$y_2 = x - 4$
$y_3 = y_1 \cdot y_2$

79. $y_1 = x^2 - 2x$
$y_2 = x$
$y_3 = \dfrac{y_1}{y_2}$

80. In Exercise 79, use the ⟨TRACE⟩ feature to trace along y_3. What happens at $x = 0$? Explain why this occurs.   No y-value is displayed; y_3 is undefined at $x = 0$.

Critical Thinking Exercises

81. Which one of the following is false?   **d**

    a. If $(f + g)(a) = 0$, then $f(a)$ and $g(a)$ must be opposites, or additive inverses.

    b. If $(f - g)(a) = 0$, then $f(a)$ and $g(a)$ must be equal.

    c. If $\left(\dfrac{f}{g}\right)(a) = 0$, then $f(a)$ must be 0.

    d. If $(fg)(a) = 0$, then $f(a)$ must be 0.

82. Use the graphs given in Exercises 65–68 to create a graph that shows the population, in billions, of less-developed regions from 1950 through 2050.   See graphing answer section.

Review Exercises

83. Solve the system:

$$11x + 4y = -3$$
$$-13x + y = 15.$$

(Section 4.3, Example 2) $\{(-1, 2)\}$

84. Solve: $3(6 - x) = 3 - 2(x - 4)$. (Section 2.3, Example 3)
$\{7\}$

85. If $f(x) = 6x - 4$, find $f(b + 2)$. (Section 8.1, Example 3)
$6b + 8$

✔ MID-CHAPTER CHECK POINT

CHAPTER 8

What You Know: We learned that a function is a relation in which no two ordered pairs have the same first component and different second components. We represented functions as equations and used function notation. We applied the vertical line test to identify graphs of functions. We determined the domain and range of a function from its graph, using inputs on the x-axis for the domain and outputs on the y-axis for the range. Finally, we developed an algebra of functions to combine functions and determine their domains.

In Exercises 1–6, determine whether each relation is a function. Give the domain and range for each relation.

1. $\{(2, 6), (1, 4), (2, -6)\}$

    not a function; domain: $\{1, 2\}$; range: $\{-6, 4, 6\}$

2. $\{(0, 1), (2, 1), (3, 4)\}$

    function; domain: $\{0, 2, 3\}$; range: $\{1, 4\}$

3.

function; domain:
$\{x \mid -2 \le x < 2\}$;
range: $\{y \mid 0 \le y \le 3\}$

4.

not a function; domain:
$\{x \mid -3 < x \le 4\}$; range:
$\{y \mid -1 \le y \le 2\}$

5.

not a function; domain:
$\{-2, -1, 0, 1, 2\}$;
range: $\{-2, -1, 1, 3\}$

6.

function; domain:
$\{x \mid x \le 1\}$;
range: $\{y \mid y \ge -1\}$

Use the graph of f to solve Exercises 7–12.

7. Explain why f represents the graph of a function.

    No vertical line intersects the graph of f more than once.

8. Use the graph to find $f(-4)$.   3

9. For what value or values of x is $f(x) = 4$? -2

10. For what value or values of x is $f(x) = 0$? -6 and 2

11. Find the domain of f. $\{x \mid x \text{ is a real number}\}$

12. Find the range of f. $\{y \mid y \le 4\}$

In Exercises 13–14, find the domain of each function.

13. $f(x) = (x + 2)(x - 2)$ {$x \mid x$ is a real number}

14. $g(x) = \dfrac{1}{(x + 2)(x - 2)}$

{$x \mid x$ is a real number and $x \neq -2$ and $x \neq 2$}

In Exercises 15–22, let

$$f(x) = x^2 - 3x + 8 \text{ and } g(x) = -2x - 5.$$

Find each of the following:

15. $f(0) + g(-10)$ 23   **16.** $f(-1) - g(3)$ 23

17. $f(a) + g(a + 3)$ $a^2 - 5a - 3$

18. $(f + g)(x)$ and $(f + g)(-2)$ $x^2 - 5x + 3; 17$

19. $(f - g)(x)$ and $(f - g)(5)$ $x^2 - x + 13; 33$

20. $(fg)(x)$ and $(fg)(-1)$ $-2x^3 + x^2 - x - 40; -36$

21. $\left(\dfrac{f}{g}\right)(x)$ and $\left(\dfrac{f}{g}\right)(-4)$ $\dfrac{x^2 - 3x + 8}{-2x - 5}; 12$

22. The domain of $\dfrac{f}{g}$ $\left\{ x \mid x \text{ is a real number and } x \neq -\dfrac{5}{2} \right\}$

8.3

Objectives

1 Form composite functions.

2 Verify inverse functions.

3 Find the inverse of a function.

4 Use the horizontal line test to determine if a function has an inverse function.

5 Use the graph of a one-to-one function to graph its inverse function.

COMPOSITE AND INVERSE FUNCTIONS

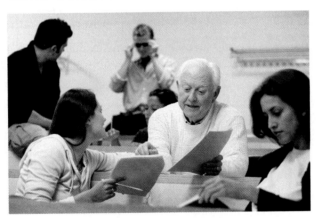

You are an over-40 student returning to college. Through consistent hard work, you've earned a near-perfect grade point average, feeling more empowered by knowledge with each academic success. Unfortunately, when it comes to technology, you are at a disadvantage compared to younger students. There were no computers in the class-room during your high-school years, and you glaze over hearing students chatting about browsers, gigabytes, RAM, and DVDs. It's time to turn things around and become computer literate.

Composite Functions Luckily, your local computer store is having a sale. The models that are on sale cost either $300 less than the regular price or 85% of the regular price. If x represents the computer's regular price, both discounts can be described with the following functions:

$$f(x) = x - 300 \qquad g(x) = 0.85x.$$

The computer is on sale for $300 less than it's regular price.

The computer is on sale for 85% of its regular price.

At the store, you bargain with the salesperson. Eventually, she makes an offer you can't refuse. The sale price will be 85% of the regular price followed by a $300 reduction:

$$0.85x - 300.$$

| 85% of the regular price | followed by a $300 reduction |

In terms of functions f and g, this offer can be obtained by taking the output of $g(x) = 0.85x$, namely $0.85x$, and using it as the input of f:

$$f(x) = x - 300$$

Replace x with $0.85x$, the output of $g(x) = 0.85x$.

$$f(0.85x) = 0.85x - 300.$$

Because $0.85x$ is $g(x)$, we can write this last equation as

$$f(g(x)) = 0.85x - 300.$$

We read this equation as "f of g of x is equal to $0.85x - 300$." We call $f(g(x))$ the **composition of the function f with g**, or a **composite function**. This composite function is written $f \circ g$. Thus,

$$(f \circ g)(x) = f(g(x)) = 0.85x - 300.$$

Like all functions, we can evaluate $f \circ g$ for a specified value of x in the function's domain. For example, here's how to find the value of this function at 1400:

$$(f \circ g)(x) = 0.85x - 300$$

This composite function describes the offer you cannot refuse.

Replace x with 1400.

$$(f \circ g)(1400) = 0.85(1400) - 300 = 1190 - 300 = 890.$$

This means that a computer that regularly sells for $1400 is on sale for $890 subject to both discounts. We can use a partial table of coordinates for each of the discount functions, g and f, to numerically verify this result.

Computer's regular price x	85% of the regular price $g(x) = 0.85x$	85% of the regular price x	$300 reduction $f(x) = x - 300$
1200	1020	1020	720
1300	1105	1105	805
1400	1190	1190	890

Using these tables, we can find $(f \circ g)(1400)$:

$$(f \circ g)(1400) = f(g(1400)) = f(1190) = 890.$$

The table for g shows that $g(1400) = 1190$. The table for f shows that $f(1190) = 890$.

This verifies that a computer that regularly sells for $1400 is on sale for $890 subject to both discounts.

Before you run out to buy a computer, let's generalize our discussion of the computer's double discount and define the composition of any two functions.

> **THE COMPOSITION OF FUNCTIONS** The **composition of the function f with g** is denoted by $f \circ g$ and is defined by the equation
>
> $$(f \circ g)(x) = f(g(x)).$$
>
> The domain of the **composite function $f \circ g$** is the set of all x such that
>
> **1.** x is in the domain of g and
>
> **2.** $g(x)$ is in the domain of f.

The composition of f with g, $f \circ g$, is pictured as a machine with inputs and outputs in Figure 8.10. The diagram indicates that the output of g, or $g(x)$, becomes the input for "machine" f. If $g(x)$ is not in the domain of f, it cannot be input into machine f, and so $g(x)$ must be discarded.

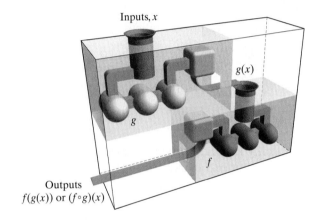

Inputs, x

$g(x)$

g

f

Outputs
$f(g(x))$ or $(f \circ g)(x)$

FIGURE 8.10 Inputting one function into a second function

EXAMPLE 1 Forming Composite Functions

Given $f(x) = 3x - 4$ and $g(x) = x^2 + 6$, find each of the following composite functions:

   a. $(f \circ g)(x)$ **b.** $(g \circ f)(x)$.

SOLUTION

  a. We begin with $(f \circ g)(x)$, the composition of f with g. Because $(f \circ g)(x)$ means $f(g(x))$, we must replace each occurrence of x in the equation for f with $g(x)$.

$$f(x) = 3x - 4 \qquad \text{This is the given equation for f.}$$

Replace x with $g(x)$.

$$(f \circ g)(x) = f(g(x)) = 3g(x) - 4$$

$$= 3(x^2 + 6) - 4 \qquad \text{Because } g(x) = x^2 + 6, \text{ replace } g(x) \text{ with } x^2 + 6.$$

$$= 3x^2 + 18 - 4 \qquad \text{Use the distributive property.}$$

$$= 3x^2 + 14 \qquad \text{Simplify.}$$

Thus, $(f \circ g)(x) = 3x^2 + 14$.

b. Next, we find $(g \circ f)(x)$, the composition of g with f. Because $(g \circ f)(x)$ means $g(f(x))$, we must replace each occurrence of x in the equation for g with $f(x)$.

$$g(x) = x^2 + 6 \qquad \text{This is the given equation for } g.$$

Replace x with $f(x)$.

$$(g \circ f)(x) = g(f(x)) = (f(x))^2 + 6$$

$$= (3x - 4)^2 + 6 \qquad \begin{array}{l}\text{Because } f(x) = 3x - 4, \text{ replace} \\ f(x) \text{ with } 3x - 4.\end{array}$$

$$= 9x^2 - 24x + 16 + 6 \qquad \begin{array}{l}\text{Use } (A - B)^2 = A^2 - 2AB + B^2 \\ \text{to square } 3x - 4.\end{array}$$

$$= 9x^2 - 24x + 22 \qquad \text{Simplify.}$$

Thus, $(g \circ f)(x) = 9x^2 - 24x + 22$. **Notice that $f \circ g$ is not the same function as $g \circ f$.** ■

 CHECK POINT **1** Given $f(x) = 5x + 6$ and $g(x) = x^2 - 1$, find each of the following composite functions:

a. $(f \circ g)(x)$ $5x^2 + 1$ **b.** $(g \circ f)(x)$. $25x^2 + 60x + 35$

Inverse Functions Here are two functions that describe situations related to the price, x, of a computer:

$$f(x) = x - 300 \qquad g(x) = x + 300.$$

Function f subtracts \$300 from the computer's price and function g adds \$300 to the computer's price. Let's see what $f(g(x))$ does. Put $g(x)$ into f:

$$f(x) = x - 300 \qquad \text{This is the given equation for } f.$$

Replace x with $g(x)$.

$$f(g(x)) = g(x) - 300$$

$$= x + 300 - 300 \qquad \begin{array}{l}\text{Because } g(x) = x + 300, \\ \text{replace } g(x) \text{ with } x + 300.\end{array}$$

$$= x.$$

This is the computer's original price.

By putting $g(x)$ into f and finding $f(g(x))$, we see that the computer's price, x, went through two changes: the first, an increase; the second, a decrease:

$$x + 300 - 300.$$

The final price of the computer, x, is identical to its starting price, x.

In general, if the changes made to x by function g are undone by the changes made by function f, then

$$f(g(x)) = x.$$

Assume, also, that this "undoing" takes place in the other direction:

$$g(f(x)) = x.$$

Under these conditions, we say that each function is the **inverse function** of the other. The fact that g is the inverse of f is expressed by renaming g as f^{-1}, read "f-inverse." For example, the inverse functions

$$f(x) = x - 300 \qquad g(x) = x + 300$$

are usually named as follows:

$$f(x) = x - 300 \qquad f^{-1}(x) = x + 300.$$

We can use partial tables of coordinates for f and f^{-1} to gain numerical insight into the relationship between a function and its inverse function.

Computer's regular price	$300 reduction		Price with $300 reduction	$300 price increase
x	$f(x) = x - 300$		x	$f^{-1}(x) = x + 300$
1200	900		900	1200
1300	1000		1000	1300
1400	1100		1100	1400

Ordered pairs for f:
(1200, 900), (1300, 1000), (1400, 1100)

Ordered pairs for f^{-1}:
(900, 1200), (1000, 1300), (1100, 1400)

The tables illustrate that if a function f is the set of ordered pairs (x, y), then the inverse, f^{-1}, is the set of ordered pairs (y, x). Using these tables, we can see how one function's changes to x are undone by the other function:

$$(f^{-1} \circ f)(1300) = f^{-1}(f(1300)) = f^{-1}(1000) = 1300.$$

The table for f shows that $f(1300) = 1000$.

The table for f^{-1} shows that $f^{-1}(1000) = 1300$.

The final price of the computer, $1300, is identical to its starting price, $1300.

With these ideas in mind, we present the formal definition of the inverse of a function:

DEFINITION OF THE INVERSE OF A FUNCTION Let f and g be two functions such that

$$f(g(x)) = x \qquad \text{for every } x \text{ in the domain of } g$$

and

$$g(f(x)) = x \qquad \text{for every } x \text{ in the domain of } f.$$

The function g is the **inverse of the function** f, and is denoted by f^{-1} (read "f-inverse"). Thus, $f(f^{-1}(x)) = x$ and $f^{-1}(f(x)) = x$. The domain of f is equal to the range of f^{-1}, and vice versa.

2 Verify inverse functions.

EXAMPLE 2 Verifying Inverse Functions

Show that each function is the inverse of the other:

$$f(x) = 5x \qquad \text{and} \qquad g(x) = \frac{x}{5}.$$

SOLUTION To show that f and g are inverses of each other, we must show that $f(g(x)) = x$ and $g(f(x)) = x$. We begin with $f(g(x))$.

$$f(x) = 5x \qquad \text{This is the given equation for } f.$$

Replace x with $g(x)$.

$$f(g(x)) = 5g(x) = 5\left(\frac{x}{5}\right) = x$$

Because $g(x) = \dfrac{x}{5}$, replace $g(x)$ with $\dfrac{x}{5}$.

Then simplify.

Next, we find $g(f(x))$.

$$g(x) = \frac{x}{5}$$

This is the given equation for g.

Replace x with $f(x)$.

$$g(f(x)) = \frac{f(x)}{5} = \frac{5x}{5} = x$$

Because f(x) = 5x, replace f(x) with 5x. Then simplify.

Because g is the inverse of f (and vice versa), we can use inverse notation and write

$$f(x) = 5x \qquad \text{and} \qquad f^{-1}(x) = \frac{x}{5}.$$

Notice how f^{-1} undoes the change produced by f: f changes x by multiplying by 5 and f^{-1} undoes this change by dividing by 5. ■

✔ **CHECK POINT 2** Show that each function is the inverse of the other:

$$f(g(x)) = 7\left(\frac{x}{7}\right) = x \quad f(x) = 7x \qquad \text{and} \qquad g(x) = \frac{x}{7}. \quad g(f(x)) = \frac{7x}{7} = x$$

EXAMPLE 3   Verifying Inverse Functions

Show that each function is the inverse of the other:

$$f(x) = 3x + 2 \qquad \text{and} \qquad g(x) = \frac{x - 2}{3}.$$

SOLUTION   To show that f and g are inverses of each other, we must show that $f(g(x)) = x$ and $g(f(x)) = x$. We begin with $f(g(x))$.

$$f(x) = 3x + 2$$

This is the equation for f.

Replace x with $g(x)$.

$$f(g(x)) = 3g(x) + 2 = 3\left(\frac{x - 2}{3}\right) + 2 = x - 2 + 2 = x$$

$$g(x) = \frac{x-2}{3}$$

Next, we find $g(f(x))$.

$$g(x) = \frac{x - 2}{3}$$

This is the equation for g.

Replace x with $f(x)$.

$$g(f(x)) = \frac{f(x) - 2}{3} = \frac{(3x + 2) - 2}{3} = \frac{3x}{3} = x$$

$$f(x) = 3x + 2$$

Because g is the inverse of f (and vice versa), we can use inverse notation and write

$$f(x) = 3x + 2 \qquad \text{and} \qquad f^{-1}(x) = \frac{x - 2}{3}.$$

Notice how f^{-1} undoes the changes produced by f: f changes x by *multiplying* by 3 and *adding* 2, and f^{-1} undoes this by *subtracting* 2 and *dividing* by 3. This "undoing" process is illustrated in Figure 8.11.

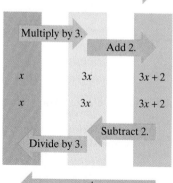

FIGURE 8.11 f^{-1} undoes the changes produced by f.

✔ **CHECK POINT 3** Show that each function is the inverse of the other:

$$f(x) = 4x - 7 \quad \text{and} \quad g(x) = \frac{x + 7}{4}.$$

$$f(g(x)) = 4\left(\frac{x + 7}{4}\right) - 7 = x + 7 - 7 = x; \quad g(f(x)) = \frac{(4x - 7) + 7}{4} = \frac{4x}{4} = x$$

3 Find the inverse of a function.

Finding the Inverse of a Function The definition of the inverse of a function tells us that the domain of f is equal to the range of f^{-1}, and vice versa. This means that if the function f is the set of ordered pairs (x, y), then the inverse of f is the set of ordered pairs (y, x). If a function is defined by an equation, we can obtain the equation for f^{-1}, the inverse of f, by interchanging the roles of x and y in the equation for the function f.

FINDING THE INVERSE OF A FUNCTION The equation for the inverse of a function f can be found as follows:

1. Replace $f(x)$ with y in the equation for $f(x)$.
2. Interchange x and y.
3. Solve for y. If this equation does not define y as a function of x, the function f does not have an inverse function and the procedure ends. If this equation does define y as a function of x, the function f has an inverse function.
4. If f has an inverse function, replace y in step 3 with $f^{-1}(x)$. We can verify our result by showing that $f(f^{-1}(x)) = x$ and $f^{-1}(f(x)) = x$.

STUDY TIP

The procedure for finding a function's inverse uses a *switch-and-solve* strategy. Switch x and y, then solve for y.

EXAMPLE 4 Finding the Inverse of a Function

Find the inverse of $f(x) = 7x - 5$.

SOLUTION

Step 1. **Replace $f(x)$ with y:**

$$y = 7x - 5.$$

Step 2. **Interchange x and y:**

$$x = 7y - 5. \quad \text{This is the inverse function.}$$

Step 3. **Solve for y:**

$$x + 5 = 7y \qquad \text{Add 5 to both sides.}$$

$$\frac{x + 5}{7} = y \qquad \text{Divide both sides by 7.}$$

Step 4. **Replace y with $f^{-1}(x)$:**

$$f^{-1}(x) = \frac{x + 5}{7}. \qquad \text{The equation is written with } f^{-1} \text{ on the left.}$$

Thus, the inverse of $f(x) = 7x - 5$ is $f^{-1}(x) = \frac{x + 5}{7}$.

The inverse function, f^{-1}, undoes the changes produced by f. f changes x by multiplying by 7 and subtracting 5. f^{-1} undoes this by adding 5 and dividing by 7. ■

DISCOVER FOR YOURSELF

In Example 4, we found that if $f(x) = 7x - 5$, then

$$f^{-1}(x) = \frac{x + 5}{7}.$$

Verify this result by showing that

$$f(f^{-1}(x)) = x$$

and

$$f^{-1}(f(x)) = x.$$

✔ **CHECK POINT 4** Find the inverse of $f(x) = 2x + 7$. $\quad f^{-1}(x) = \dfrac{x - 7}{2}$

4 Use the horizontal line test to determine if a function has an inverse function.

The Horizontal Line Test and One-To-One Functions Some functions do not have inverses that are functions. For example, consider the function $f(x) = x^2$, or $y = x^2$. We can use a few of the solutions of $y = x^2$ to illustrate numerically that this function does not have an inverse:

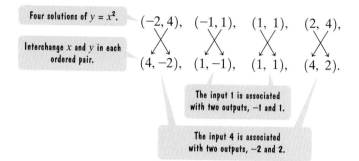

Four solutions of $y = x^2$.

$(-2, 4), \quad (-1, 1), \quad (1, 1), \quad (2, 4),$

Interchange x and y in each ordered pair.

$(4, -2), \quad (1, -1), \quad (1, 1), \quad (4, 2).$

The input 1 is associated with two outputs, −1 and 1.

The input 4 is associated with two outputs, −2 and 2.

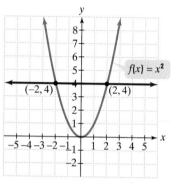

FIGURE 8.12 The horizontal line intersects the graph twice.

A function provides exactly one output for each input. Thus, the ordered pairs in the bottom row do not define a function.

Can we look at the graph of a function and tell if it represents a function with an inverse? Yes. The graph of the function $f(x) = x^2$ is shown in Figure 8.12. Four units above the x-axis, a horizontal line is drawn. This line intersects the graph at two of its points, $(-2, 4)$ and $(2, 4)$. Inverse functions have ordered pairs with the coordinates reversed. We just saw what happened when we interchanged x and y. We obtained $(4, -2)$ and $(4, 2)$, and these ordered pairs do not define a function.

If any horizontal line, such as the one in Figure 8.12, intersects a graph at two or more points, the set of these points will not define a function when their coordinates are reversed. This suggests the **horizontal line test** for inverse functions.

DISCOVER FOR YOURSELF

How might you restrict the domain of $f(x) = x^2$, graphed in Figure 8.12, so that the remaining portion of the graph passes the horizontal line test?

THE HORIZONTAL LINE TEST FOR INVERSE FUNCTIONS A function f has an inverse that is a function, f^{-1}, if there is no horizontal line that intersects the graph of the function f at more than one point.

EXAMPLE 5 Applying the Horizontal Line Test

Which of the following graphs represent functions that have inverse functions?

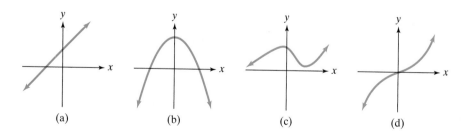

SOLUTION Notice that horizontal lines can be drawn in graphs (b) and (c) that intersect the graphs more than once. These graphs do not pass the horizontal line test. These are not the graphs of functions with inverse functions. By contrast, no horizontal line can be drawn in graphs (a) and (d) that intersects the graphs more than once.

These graphs pass the horizontal line test. Thus, the graphs in parts (a) and (d) represent functions that have inverse functions.

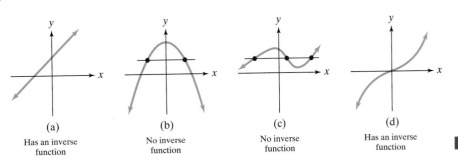

| (a) | (b) | (c) | (d) |
| Has an inverse function | No inverse function | No inverse function | Has an inverse function |

CHECK POINT **5** Which of the following graphs represent functions that have inverse functions?   Graphs (b) and (c) have inverse functions.

(a)   (b)   (c)

A function passes the horizontal line test when no two different ordered pairs have the same second component. This means that if $x_1 \neq x_2$, then $f(x_1) \neq f(x_2)$. Such a function is called a **one-to-one function**. Thus, a one-to-one function is a function in which no two different ordered pairs have the same second component. Only one-to-one functions have inverse functions. Any function that passes the horizontal line test is a one-to-one function. Any one-to-one function has a graph that passes the horizontal line test.

5 Use the graph of a one-to-one function to graph its inverse function.

Graphs of f and f^{-1}   There is a relationship between the graph of a one-to-one function, f, and its inverse, f^{-1}. Because inverse functions have ordered pairs with the coordinates reversed, if the point (a, b) is on the graph of f, then the point (b, a) is on the graph of f^{-1}. The points (a, b) and (b, a) are symmetric with respect to the line $y = x$. Thus, **the graph of f^{-1} is a reflection of the graph of f about the line $y = x$.** This is illustrated in Figure 8.13.

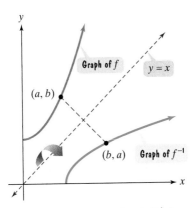

FIGURE 8.13 The graph of f^{-1} is a reflection of the graph of f about $y = x$.

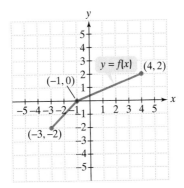

FIGURE 8.14

EXAMPLE 6   Graphing the Inverse Function

Use the graph of f in Figure 8.14 to draw the graph of its inverse function.

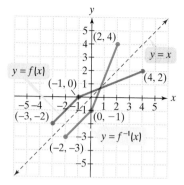

FIGURE 8.15 The graphs of f and f^{-1}

SOLUTION We begin by noting that no horizontal line intersects the graph of f, shown again in blue in Figure 8.15, at more than one point, so f does have an inverse function. Because the points $(-3, -2)$, $(-1, 0)$, and $(4, 2)$ are on the graph of f, the graph of the inverse function, f^{-1}, has points with these ordered pairs reversed. Thus, $(-2, -3)$, $(0, -1)$, and $(2, 4)$ are on the graph of f^{-1}. We can use these points to graph f^{-1}. The graph of f^{-1} is shown in green in Figure 8.15. Note that the green graph of f^{-1} is the reflection of the blue graph of f about the line $y = x$. ∎

 CHECK POINT 6 Use the graph of f in the figure below to draw the graph of its inverse function.   See graphing answer section.

8.3 EXERCISE SET

Student Solutions Manual   CD/Video   PH Math/Tutor Center   MathXL Tutorials on CD   MathXL   MathXL®   MyMathLab   MyMathLab   Interactmath.com

Practice Exercises

In Exercises 1–14, find

    a. $(f \circ g)(x)$;
    b. $(g \circ f)(x)$;
    c. $(f \circ g)(2)$.

1. $f(x) = 2x$, $g(x) = x + 7$
  a. $2x + 14$ **b.** $2x + 7$ **c.** 18

2. $f(x) = 3x$, $g(x) = x - 5$
  a. $3x - 15$ **b.** $3x - 5$ **c.** -9

3. $f(x) = x + 4$, $g(x) = 2x + 1$
  a. $2x + 5$ **b.** $2x + 9$ **c.** 9

4. $f(x) = 5x + 2$, $g(x) = 3x - 4$
  a. $15x - 18$ **b.** $15x + 2$ **c.** 12

5. $f(x) = 4x - 3$, $g(x) = 5x^2 - 2$
  a. $20x^2 - 11$ **b.** $80x^2 - 120x + 43$ **c.** 69

6. $f(x) = 7x + 1$, $g(x) = 2x^2 - 9$
  a. $14x^2 - 62$ **b.** $98x^2 + 28x - 7$ **c.** -6

7. $f(x) = x^2 + 2$, $g(x) = x^2 - 2$
  a. $x^4 - 4x^2 + 6$ **b.** $x^4 + 4x^2 + 2$ **c.** 6

8. $f(x) = x^2 + 1$, $g(x) = x^2 - 3$
  a. $x^4 - 6x^2 + 10$ **b.** $x^4 + 2x^2 - 2$ **c.** 2

9. $f(x) = \sqrt{x}$, $g(x) = x - 1$
  a. $\sqrt{x - 1}$ **b.** $\sqrt{x} - 1$ **c.** 1

10. $f(x) = \sqrt{x}$, $g(x) = x + 2$
  a. $\sqrt{x + 2}$ **b.** $\sqrt{x} + 2$ **c.** 2

11. $f(x) = 2x - 3$, $g(x) = \dfrac{x + 3}{2}$ **a.** x **b.** x **c.** 2

12. $f(x) = 6x - 3$, $g(x) = \dfrac{x + 3}{6}$ **a.** x **b.** x **c.** 2

13. $f(x) = \dfrac{1}{x}$, $g(x) = \dfrac{1}{x}$ **a.** x **b.** x **c.** 2

14. $f(x) = \dfrac{1}{x}$, $g(x) = \dfrac{2}{x}$ **a.** $\dfrac{x}{2}$ **b.** $2x$ **c.** 1

In Exercises 15–24, find $f(g(x))$ and $g(f(x))$ and determine whether each pair of functions f and g are inverses of each other.

15. $f(x) = 4x$ and $g(x) = \dfrac{x}{4}$
  $f(g(x)) = x$; $g(f(x)) = x$;   inverses

16. $f(x) = 6x$ and $g(x) = \dfrac{x}{6}$
  $f(g(x)) = x$; $g(f(x)) = x$; inverses

17. $f(x) = 3x + 8$ and $g(x) = \dfrac{x - 8}{3}$
  $f(g(x)) = x$; $g(f(x)) = x$;   inverses

18. $f(x) = 4x + 9$ and $g(x) = \dfrac{x - 9}{4}$
  $f(g(x)) = x$; $g(f(x)) = x$;   inverses

19. $f(x) = 5x - 9$ and $g(x) = \dfrac{x + 5}{9}$
  $f(g(x)) = \dfrac{5x - 56}{9}$; $g(f(x)) = \dfrac{5x - 4}{9}$;   not inverses

20. $f(x) = 3x - 7$ and $g(x) = \dfrac{x + 3}{7}$

$f(g(x)) = \dfrac{3x - 40}{7}$; $g(f(x)) = \dfrac{3x - 4}{7}$; not inverses

21. $f(x) = \dfrac{3}{x - 4}$ and $g(x) = \dfrac{3}{x} + 4$

$f(g(x)) = x$; $g(f(x)) = x$; inverses

22. $f(x) = \dfrac{2}{x - 5}$ and $g(x) = \dfrac{2}{x} + 5$

$f(g(x)) = x$; $g(f(x)) = x$; inverses

23. $f(x) = -x$ and $g(x) = -x$

$f(g(x)) = x$; $g(f(x)) = x$; inverses

24. $f(x) = -x$ and $g(x) = x$

$f(g(x)) = -x$; $g(f(x)) = -x$; not inverses

The functions in Exercises 25–34 are all one-to-one. For each function,

a. *Find an equation for $f^{-1}(x)$, the inverse function.*

b. *Verify that your equation is correct by showing that $f(f^{-1}(x)) = x$ and $f^{-1}(f(x)) = x$.*

25. $f(x) = x + 3$ **a.** $f^{-1}(x) = x - 3$

b. $f(f^{-1}(x)) = (x - 3) + 3 = x$ and
$f^{-1}(f(x)) = (x + 3) - 3 = x$

26. $f(x) = x + 5$ **a.** $f^{-1}(x) = x - 5$

b. $f(f^{-1}(x)) = (x - 5) + 5 = x$ and
$f^{-1}(f(x)) = (x + 5) - 5 = x$

27. $f(x) = 2x$ **a.** $f^{-1}(x) = \dfrac{x}{2}$

b. $f(f^{-1}(x)) = 2\left(\dfrac{x}{2}\right) = x$ and $f^{-1}(f(x)) = \dfrac{2x}{2} = x$

28. $f(x) = 4x$ **a.** $f^{-1}(x) = \dfrac{x}{4}$

b. $f(f^{-1}(x)) = 4\left(\dfrac{x}{4}\right) = x$ and $f^{-1}(f(x)) = \dfrac{4x}{4} = x$

29. $f(x) = 2x + 3$ **a.** $f^{-1}(x) = \dfrac{x - 3}{2}$

b. $f(f^{-1}(x)) = 2\left(\dfrac{x - 3}{2}\right) + 3 = (x - 3) + 3 = x$ and

$f^{-1}(f(x)) = \dfrac{(2x + 3) - 3}{2} = \dfrac{2x}{2} = x$

30. $f(x) = 3x - 1$ **a.** $f^{-1}(x) = \dfrac{x + 1}{3}$

b. $f(f^{-1}(x)) = 3\left(\dfrac{x + 1}{3}\right) - 1 = (x + 1) - 1 = x$ and

$f^{-1}(f(x)) = \dfrac{(3x - 1) + 1}{3} = \dfrac{3x}{3} = x$

31. $f(x) = \dfrac{1}{x}$ **a.** $f^{-1}(x) = \dfrac{1}{x}$

b. $f(f^{-1}(x)) = \dfrac{1}{\frac{1}{x}} = x$ and $f^{-1}(f(x)) = \dfrac{1}{\frac{1}{x}} = x$

32. $f(x) = \dfrac{2}{x}$ **a.** $f^{-1}(x) = \dfrac{2}{x}$

b. $f(f^{-1}(x)) = \dfrac{2}{\frac{2}{x}} = 2 \cdot \dfrac{x}{2} = x$ and $f^{-1}(f(x)) = \dfrac{2}{\frac{2}{x}} = 2 \cdot \dfrac{x}{2} = x$

33. $f(x) = \dfrac{2x + 1}{x - 3}$ **a.** $f^{-1}(x) = \dfrac{3x + 1}{x - 2}$

b. $f(f^{-1}(x)) = \dfrac{2\left(\dfrac{3x + 1}{x - 2}\right) + 1}{\left(\dfrac{3x + 1}{x - 2}\right) - 3} = \dfrac{2(3x + 1) + 1(x - 2)}{(3x + 1) - 3(x - 2)}$

$= \dfrac{7x}{7} = x$ and $f^{-1}(f(x)) = \dfrac{3\left(\dfrac{2x + 1}{x - 3}\right) + 1}{\left(\dfrac{2x + 1}{x - 3}\right) - 2}$

$= \dfrac{3(2x + 1) + 1(x - 3)}{(2x + 1) - 2(x - 3)} = \dfrac{7x}{7} = x$

34. $f(x) = \dfrac{2x - 3}{x + 1}$ **a.** $f^{-1}(x) = -\dfrac{x + 3}{x - 2}$

b. $f(f^{-1}(x)) = \dfrac{2\left(-\dfrac{x + 3}{x - 2}\right) - 3}{\left(-\dfrac{x + 3}{x - 2}\right) + 1} = \dfrac{-2(x + 3) - 3(x - 2)}{-(x + 3) + 1(x - 2)}$

$= \dfrac{-5x}{-5} = x$ and $f^{-1}(f(x)) = -\dfrac{\left(\dfrac{2x - 3}{x + 1}\right) + 3}{\left(\dfrac{2x - 3}{x + 1}\right) - 2}$

$= -\dfrac{(2x - 3) + 3(x + 1)}{(2x - 3) - 2(x + 1)} = \dfrac{-5x}{-5} = x$

Which graphs in Exercises 35–40 represent functions that have inverse functions?

35.

no inverse function

36.

inverse function

37.

no inverse function

38.

no inverse function

39.

inverse function

40.

inverse function

In Exercises 41–44, use the graph of f to draw the graph of its inverse function. **41–44.** See graphing answer section.

41.

42.

43.

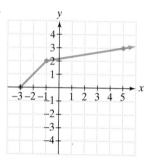 The graph approaches, but never touches the negative portion of the x-axis.

44.

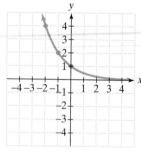 The graph approaches, but never touches the positive portion of the x-axis.

Practice Plus

In Exercises 45–50, f and g are defined by the following tables. Use the tables to evaluate each composite function.

x	$f(x)$
−1	1
0	4
1	5
2	−1

x	$g(x)$
−1	0
1	1
4	2
10	−1

45. $f(g(1))$ 5
46. $f(g(4))$ −1
47. $(g \circ f)(-1)$ 1
48. $(g \circ f)(0)$ 2
49. $f^{-1}(g(10))$ 2
50. $f^{-1}(g(1))$ −1

In Exercises 51–54, use the graphs of f and g to evaluate each composite function.

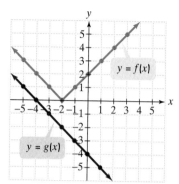

51. $(f \circ g)(-1)$ 1
52. $(f \circ g)(1)$ 3
53. $(g \circ f)(0)$ −6
54. $(g \circ f)(-1)$ −5

In Exercises 55–60, let

$$f(x) = 2x - 5$$
$$g(x) = 4x - 1$$
$$h(x) = x^2 + x + 2.$$

Evaluate the indicated function without finding an equation for the function.

55. $(f \circ g)(0)$ −7
56. $(g \circ f)(0)$ −21
57. $f^{-1}(1)$ 3
58. $g^{-1}(7)$ 2
59. $g(f[h(1)])$ 11
60. $f(g[h(1)])$ 25

Application Exercises

61. The regular price of a computer is x dollars. Let $f(x) = x - 400$ and $g(x) = 0.75x$.

a. Describe what the functions f and g model in terms of the price of the computer.   *f* represents the price after a $400 discount and *g* represents the price after a 25% discount (75% of the regular price).

b. Find $(f \circ g)(x)$ and describe what this models in terms of the price of the computer. $0.75x - 400$; $f \circ g$ represents an additional $400 discount on a price that has already been reduced by 25%.

c. Repeat part (b) for $(g \circ f)(x)$. $0.75(x - 400) = 0.75x - 300$; $g \circ f$ represents an additional 25% discount on a price that has already been reduced by $400.

d. Which composite function models the greater discount on the computer, $f \circ g$ or $g \circ f$? Explain. $f \circ g$; $0.75x - 400 < 0.75x - 300$, so $f \circ g$ represents the lower price after the two discounts.

e. Find f^{-1} and describe what this models in terms of the price of the computer. $f^{-1}(x) = x + 400$; f^{-1} represents the regular price, since the value of x here is the price after a $400 discount.

62. The regular price of a pair of jeans is x dollars. Let $f(x) = x - 5$ and $g(x) = 0.6x$.

a. Describe what functions f and g model in terms of the price of the jeans. *f represents the price after a $5 discount, and g represents the price after a 40% discount (60% of the regular price).*

b. Find $(f \circ g)(x)$ and describe what this models in terms of the price of the jeans. *0.6x − 5; f ∘ g represents an additional $5 discount on a price that has already been reduced by 40%.*

c. Repeat part (b) for $(g \circ f)(x)$. *0.6(x − 5) = 0.6x − 3; g ∘ f represents an additional 60% discount on a price that has already been reduced by $5.*

d. Which composite function models the greater discount on the jeans, $f \circ g$ or $g \circ f$? Explain. *f ∘ g; 0.6x − 5 < 0.6x − 3, so f ∘ g represents the lower price after the two discounts.*

e. Find f^{-1} and describe what this models in terms of the price of the jeans. *f⁻¹(x) = x + 5; f⁻¹ represents the regular price, since the value of x here is the price after a $5 discount.*

In most societies, women say they prefer to marry men who are older than themselves, whereas men say they prefer women who are younger. Evolutionary psychologists attribute these preferences to female concern with a partner's material resources and male concern with a partner's fertility (Source: Davie M. Buss, Psychological Inquiry, 6, 1–30). The graph shows the preferred age in a mate in five selected countries. Use the information in the graph to solve Exercises 63–64.

63. a. Consider a function, f, whose domain is the set of the five countries shown in the graph. Let the range be the set of the average number of years women in each of the respective countries prefer men who are older than themselves. Write function f as a set of ordered pairs. *f: {(Zambia, 4.2), (Colombia, 4.5), (Poland, 3.3), (Italy, 3.3), (U.S., 2.5)}*

b. Write the relation that is the inverse of f as a set of ordered pairs. Is this relation a function? Explain your answer. *{(4.2, Zambia), (4.5, Colombia), (3.3, Poland), (3.3, Italy), (2.5, U.S.)}; No; The input 3.3 is associated with two outputs, Poland and Italy.*

64. a. Consider a function, f, whose domain is the set of the five countries shown in the graph. Let the range be the set of the average number of years men in each of the respective countries prefer women who are younger than themselves. Write f as a set of ordered pairs. *f: {(Zambia, −7.3), (Colombia, −4.5), (Poland, −2.8), (Italy, −2.8), (U.S., −1.9)}*

b. Write the relation that is the inverse of f as a set of ordered pairs. Is this relation a function? Explain your answer. *{(−7.3, Zambia), (−4.5, Colombia), (−2.8, Poland), (−2.8, Italy), (−1.9, U.S.)}; No; The input −2.8 is associated with two outputs, Poland and Italy.*

65. The graph represents the probability that two people in the same room share a birthday as a function of the number of people in the room. Call the function f.

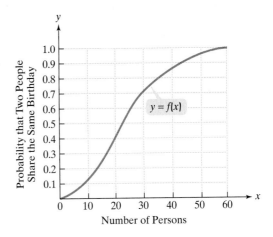

a. Explain why f has an inverse that is a function. *No horizontal line intersects the graph of f in more than one point.*

b. Describe in practical terms the meanings of $f^{-1}(0.25)$, $f^{-1}(0.5)$, and $f^{-1}(0.7)$. *f⁻¹(0.25), or approximately 15, represents the number of people who would have to be in the room so that the probability of two sharing a birthday would be 0.25; f⁻¹(0.5), or approximately 23, represents the number of people so that the probability would be 0.5; f⁻¹(0.7), or approximately 30, represents the number of people so that the probability would be 0.7.*

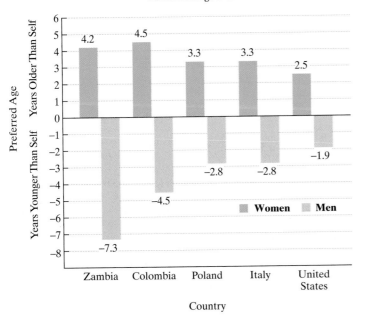

Preferred Age in a Mate

Source: Carole Wade and Carol Tavris, Psychology, 6th Edition, Prentice Hall, 2000

66. The formula

$$y = f(x) = \frac{9}{5}x + 32$$

is used to convert from x degrees Celsius to y degrees Fahrenheit. The formula

$$y = g(x) = \frac{5}{9}(x - 32)$$

is used to convert from x degrees Fahrenheit to y degrees Celsius. Show that f and g are inverse functions.

$$f(g(x)) = \frac{9}{5}\left[\frac{5}{9}(x - 32)\right] + 32 = x \quad \text{and}$$

$$g(f(x)) = \frac{5}{9}\left[\left(\frac{9}{5}x + 32\right) - 32\right] = x$$

Writing in Mathematics

67. Describe a procedure for finding $(f \circ g)(x)$.
67–72. Answers will vary.

68. Explain how to determine if two functions are inverses of each other.

69. Describe how to find the inverse of a one-to-one function.

70. What is the horizontal line test and what does it indicate?

71. Describe how to use the graph of a one-to-one function to draw the graph of its inverse function.

72. How can a graphing utility be used to visually determine if two functions are inverses of each other?

Technology Exercises

In Exercises 73–80, use a graphing utility to graph each function. Use the graph to determine whether the function has an inverse that is a function (that is, whether the function is one-to-one).
73–80. See graphing answer section.

73. $f(x) = x^2 - 1$ no inverse function

74. $f(x) = \sqrt[3]{2 - x}$ inverse function

75. $f(x) = \frac{x^3}{2}$ inverse function

76. $f(x) = \frac{x^4}{4}$ no inverse function

77. $f(x) = |x - 2|$ no inverse function

78. $f(x) = (x - 1)^3$ inverse function

79. $f(x) = -\sqrt{16 - x^2}$ no inverse function

80. $f(x) = x^3 + x + 1$ inverse function

In Exercises 81–83, use a graphing utility to graph f and g in the same viewing rectangle. In addition, graph the line $y = x$ and visually determine if f and g are inverses. 81–83. See graphing answer section.

81. $f(x) = 4x + 4, \quad g(x) = 0.25x - 1$ inverses

82. $f(x) = \frac{1}{x} + 2, \quad g(x) = \frac{1}{x - 2}$ inverses

83. $f(x) = \sqrt[3]{x} - 2, \quad g(x) = (x + 2)^3$ inverses

Critical Thinking Exercises

84. Which one of the following is true?   d
   a. The inverse of $\{(1, 4), (2, 7)\}$ is $\{(2, 7), (1, 4)\}$.
   b. The function $f(x) = 5$ is one-to-one.
   c. If $f(x) = 3x$, then $f^{-1}(x) = \frac{1}{3x}$.
   d. The domain of f is the same as the range of f^{-1}.

85. If $h(x) = \sqrt{3x^2 + 5}$, find functions f and g so that $h(x) = (f \circ g)(x)$.   Answers will vary; An example is $f(x) = \sqrt{x + 5}$ and $g(x) = 3x^2$.

86. If $f(x) = 3x$ and $g(x) = x + 5$, find $(f \circ g)^{-1}(x)$ and $(g^{-1} \circ f^{-1})(x)$.

$(f \circ g)^{-1}(x) = \dfrac{x - 15}{3}; (g^{-1} \circ f^{-1})(x) = \dfrac{x}{3} - 5 = \dfrac{x - 15}{3}$

87. Show that

$$f(x) = \frac{3x - 2}{5x - 3}$$

is its own inverse.

$$f(f(x)) = \frac{3\left(\dfrac{3x - 2}{5x - 3}\right) - 2}{5\left(\dfrac{3x - 2}{5x - 3}\right) - 3} = \frac{3(3x - 2) - 2(5x - 3)}{5(3x - 2) - 3(5x - 3)}$$

$$= \frac{9x - 6 - 10x + 6}{15x - 10 - 15x + 9} = \frac{-x}{-1} = x$$

88. Consider the two functions defined by $f(x) = m_1x + b_1$ and $g(x) = m_2x + b_2$. Prove that the slope of the composite function of f with g is equal to the product of the slopes of the two functions.

$(f \circ g)(x) = m_1(m_2x + b_2) + b_1 = m_1m_2x + m_1b_2 + b_1$;
The slope of $f \circ g$ is m_1m_2, which is the product of the slopes of f and g.

Review Exercises

89. Divide and write the quotient in scientific notation:

$$\frac{4.3 \times 10^5}{8.6 \times 10^{-4}}.$$

(Section 5.7, Example 9) 5×10^8

90. Divide: $\dfrac{x^3 + 7x^2 - 2x + 3}{x - 2}$. (Section 5.6, Example 2)

$x^2 + 9x + 16 + \dfrac{35}{x - 2}$

91. Solve:

$$3x + 2y = 6$$

$$8x - 3y = 1.$$

(Section 4.3, Example 4) $\left\{\left(\dfrac{4}{5}, \dfrac{9}{5}\right)\right\}$

GROUP PROJECT

The bar graphs in Exercises 63 and 64 in Exercise Set 8.3 (see page 569) illustrate that if a relation is a function, reversing the components in each of its ordered pairs may result in a relation that is no longer a function. Group members should find examples of bar graphs, like the ones in Exercises 63 and 64, that illustrate this idea. Consult almanacs, newspapers, magazines, or the Internet. The group should select the graph with the most intriguing data. For the graph selected, write and solve a problem with four parts similar to Exercise 63 or 64.

Additional group projects can be found in the *Instructor's Resource Manual.*

CHAPTER 8 SUMMARY

Definitions and Concepts	Examples

Section 8.1 Introduction to Functions

A relation is any set of ordered pairs. The set of first components of the ordered pairs is the domain and the set of second components is the range. A function is a relation in which each member of the domain corresponds to exactly one member of the range. No two ordered pairs of a function can have the same first component and different second components.

The domain of the relation $\{(1, 2), (3, 4), (3, 7)\}$ is $\{1, 3\}$. The range is $\{2, 4, 7\}$. The relation is not a function: 3, in the domain, corresponds to both 4 and 7 in the range.

If a function is defined by an equation, the notation $f(x)$, read "f of x" or "f at x," describes the value of the function at the number, or input, x.

If $f(x) = 7x - 5$, then
$$f(a + 2) = 7(a + 2) - 5$$
$$= 7a + 14 - 5$$
$$= 7a + 9.$$

The graph of a function is the graph of its ordered pairs.

The Vertical Line Test for Functions
If any vertical line intersects a graph in more than one point, the graph does not define y as a function of x.

At the left or right of a function's graph, you will often find closed dots, open dots, or arrows. A closed dot shows that the graph ends and the point belongs to the graph. An open dot shows that the graph ends and the point does not belong to the graph. An arrow indicates that the graph extends indefinitely.

The graph of a function can be used to determine the function's domain and its range. To find the domain, look for all the inputs on the x-axis that correspond to points on the graph. To find the range, look for all the outputs on the y-axis that correspond to points on the graph.

Not the graph of a function   The graph of a function

(0, 2) belongs to the graph of f; (0, 0) does not. $y = f(x)$

To find $f(2)$, locate 2 on the x-axis. The graph shows $f(2) = 4$.
Domain of $f = \{x \mid x \text{ is a real number}\}$
Range of $f = \{y \mid y > 0\}$

Definitions and Concepts	Examples

Section 8.2 The Algebra of Functions

A Function's Domain

If a function f does not model data or verbal conditions, its domain is the largest set of real numbers for which the value of $f(x)$ is a real number. Exclude from a function's domain real numbers that cause division by zero and real numbers that result in a square root of a negative number.

$$f(x) = 7x + 13$$

Domain of $f = \{x | x \text{ is a real number}\}$

$$g(x) = \frac{7x}{12 - x}$$

Domain of $g = \{x | x \text{ is a real number and } x \neq 12\}$

The Algebra of Functions

Let f and g be two functions. The sum $f + g$, the difference $f - g$, the product fg, and the quotient $\dfrac{f}{g}$ are functions whose domains are the set of all real numbers common to the domains of f and g, defined as follows:

1. Sum: $(f + g)(x) = f(x) + g(x)$
2. Difference: $(f - g)(x) = f(x) - g(x)$
3. Product: $(fg)(x) = f(x) \cdot g(x)$
4. Quotient: $\left(\dfrac{f}{g}\right)(x) = \dfrac{f(x)}{g(x)}, g(x) \neq 0.$

Let $f(x) = x^2 + 2x$ and $g(x) = 4 - x$.

- $(f + g)(x) = (x^2 + 2x) + (4 - x) = x^2 + x + 4$

  $(f + g)(-2) = (-2)^2 + (-2) + 4 = 4 - 2 + 4 = 6$

- $(f - g)(x) = (x^2 + 2x) - (4 - x) = x^2 + 2x - 4 + x$
  $$= x^2 + 3x - 4$$

  $(f - g)(5) = 5^2 + 3 \cdot 5 - 4 = 25 + 15 - 4 = 36$

- $(fg)(x) = (x^2 + 2x)(4 - x) = 4x^2 - x^3 + 8x - 2x^2$
  $$= -x^3 + 2x^2 + 8x$$

  $(fg)(1) = -1^3 + 2 \cdot 1^2 + 8 \cdot 1 = -1 + 2 + 8 = 9$

- $\left(\dfrac{f}{g}\right)(x) = \dfrac{x^2 + 2x}{4 - x}, x \neq 4$

  $\left(\dfrac{f}{g}\right)(3) = \dfrac{3^2 + 2 \cdot 3}{4 - 3} = \dfrac{9 + 6}{1} = 15$

Section 8.3 Composite and Inverse Functions

Composite Functions

The composite function $f \circ g$ is defined by

$$(f \circ g)(x) = f(g(x)).$$

The composite function $g \circ f$ is defined by

$$(g \circ f)(x) = g(f(x)).$$

Let $f(x) = x^2 + x$ and $g(x) = 2x + 1$.

- $(f \circ g)(x) = f(g(x)) = (g(x))^2 + g(x)$

  Replace x with $g(x)$.

  $= (2x + 1)^2 + (2x + 1) = 4x^2 + 4x + 1 + 2x + 1$

  $= 4x^2 + 6x + 2$

- $(g \circ f)(x) = g(f(x)) = 2f(x) + 1$

  Replace x with $f(x)$.

  $= 2(x^2 + x) + 1 = 2x^2 + 2x + 1$

Definitions and Concepts	Examples

Section 8.3 Composite and Inverse Functions (continued)

Inverse Functions

If $f(g(x)) = x$ and $g(f(x)) = x$, function g is the inverse of function f, denoted f^{-1} and read "f inverse." The procedure for finding a function's inverse uses a switch-and-solve strategy. Switch x and y, then solve for y.

If $f(x) = 2x - 5$, find $f^{-1}(x)$.

$$y = 2x - 5 \qquad \text{Replace } f(x) \text{ with } y.$$

$$x = 2y - 5 \qquad \text{Exchange } x \text{ and } y.$$

$$x + 5 = 2y \qquad \text{Solve for } y.$$

$$\frac{x + 5}{2} = y$$

$$f^{-1}(x) = \frac{x + 5}{2} \qquad \text{Replace } y \text{ with } f^{-1}(x).$$

The Horizontal Line Test for Inverse Functions

A function, f, has an inverse that is a function, f^{-1}, if there is no horizontal line that intersects the graph of f at more than one point. A one-to-one function is one in which no two different ordered pairs have the same second component. Only one-to-one functions have inverse functions. If the point (a, b) is on the graph of f, then the point (b, a) is on the graph of f^{-1}. The graph of f^{-1} is a reflection of the graph of f about the line $y = x$.

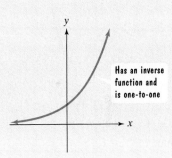

Does not have an inverse function and is not one-to-one

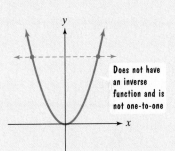

Has an inverse function and is one-to-one

CHAPTER 8 REVIEW EXERCISES

8.1 *In Exercises 1–3, determine whether each relation is a function. Give the domain and range for each relation.*

1. $\{(3, 10), (4, 10), (5, 10)\}$
   function; domain: $\{3, 4, 5\}$; range: $\{10\}$

2. $\{(1, 12), (2, 100), (3, \pi), (4, -6)\}$
   function; domain: $\{1, 2, 3, 4\}$; range: $\{12, 100, \pi, -6\}$

3. $\{(13, 14), (15, 16), (13, 17)\}$
   not a function; domain: $\{13, 15\}$; range: $\{14, 16, 17\}$

In Exercises 4–5, find the indicated function values.

4. $f(x) = 7x - 5$
   a. $f(0)$ -5 **b.** $f(3)$ 16 **c.** $f(-10)$ -75
   d. $f(2a)$ $14a - 5$ **e.** $f(a + 2)$ $7a + 9$

5. $g(x) = 3x^2 - 5x + 2$
   a. $g(0)$ 2 **b.** $g(5)$ 52 **c.** $g(-4)$ 70
   d. $g(b)$ $3b^2 - 5b + 2$ **e.** $g(4a)$ $48a^2 - 20a + 2$

In Exercises 6–11, use the vertical line test to identify graphs in which y is a function of x.

6.

not a function

7.

function

8.

function

9.

not a function

10.

not a function

11.

function

Use the graph of f to solve Exercises 12–16.

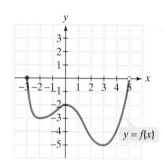

$y = f(x)$

12. Find $f(-2)$. -3

13. Find $f(0)$. -2

14. For what value of x is $f(x) = -5$? 3

15. Find the domain of f. $\{x|-3 \le x < 5\}$

16. Find the range of f. $\{y|-5 \le y \le 0\}$

17. The graph shows the height, in meters, of a vulture in terms of its time, in seconds, in flight.

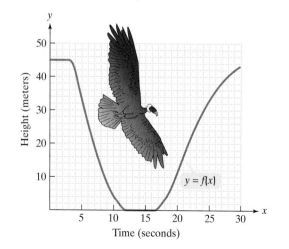

$y = f(x)$

Time (seconds)

a. Use the graph to explain why the vulture's height is a function of its time in flight.   For each time, there is only one height.

b. Find $f(15)$. Describe what this means in practical terms.   0; The vulture was on the ground after 15 seconds.

c. What is a reasonable estimate of the vulture's maximum height?   45 meters

d. For what values of x is $f(x) = 20$? Describe what this means in practical terms.   7 and 22; After 7 and 22 seconds, the vulture's height is 20 meters.

e. Use the graph of the function to write a description of the vulture's flight.   Answers will vary.

8.2 *In Exercises 18–20, find the domain of each function.*

18. $f(x) = 7x - 3$ Domain of $f = \{x|x$ is a real number$\}$

19. $g(x) = \dfrac{1}{x + 8}$

Domain of $g = \{x|x$ is a real number and $x \ne -8\}$

20. $f(x) = x + \dfrac{3x}{x - 5}$

Domain of $f = \{x|x$ is a real number and $x \ne 5\}$

*In Exercises 21–22, find **a.** $(f + g)(x)$ and **b.** $(f + g)(3)$.*

21. $f(x) = 4x - 5$, $g(x) = 2x + 1$ **a.** $6x - 4$ **b.** 14

22. $f(x) = 5x^2 - x + 4$, $g(x) = x - 3$ **a.** $5x^2 + 1$ **b.** 46

In Exercises 23–24, for each pair of functions, f and g, determine the domain of f + g.

23. $f(x) = 3x + 4$, $g(x) = \dfrac{5}{4 - x}$

Domain of $f + g = \{x|x$ is a real number and $x \ne 4\}$

24. $f(x) = \dfrac{7x}{x + 6}$, $g(x) = \dfrac{4}{x + 1}$ Domain of

$f + g = \{x|x$ is a real number and $x \ne -6$ and $x \ne -1\}$

In Exercises 25–32, let

$$f(x) = x^2 - 2x \quad \text{and} \quad g(x) = x - 5.$$

Find each of the following.

25. $(f + g)(x)$ and $(f + g)(-2)$ $x^2 - x - 5; 1$

26. $f(3) + g(3)$ 1

27. $(f - g)(x)$ and $(f - g)(1)$ $x^2 - 3x + 5; 3$

28. $f(4) - g(4)$ 9

29. $(fg)(x)$ and $(fg)(-3)$ $x^3 - 7x^2 + 10x; -120$

30. $\left(\dfrac{f}{g}\right)(x)$ and $\left(\dfrac{f}{g}\right)(4)$ $\dfrac{x^2 - 2x}{x - 5}; -8$

31. The domain of $f - g$

Domain of $f - g = \{x|x$ is a real number$\}$

32. The domain of $\dfrac{f}{g}$

Domain of $\dfrac{f}{g} = \{x|x$ is a real number and $x \ne 5\}$

8.3 *In Exercises 33–34, find* **a.** $(f \circ g)(x)$; **b.** $(g \circ f)(x)$; **c.** $(f \circ g)(3)$.

33. $f(x) = x^2 + 3$, $g(x) = 4x - 1$ **a.** $16x^2 - 8x + 4$
  b. $4x^2 + 11$ **c.** 124

34. $f(x) = \sqrt{x}$, $g(x) = x + 1$ **a.** $\sqrt{x + 1}$ **b.** $\sqrt{x} + 1$
  c. 2

In Exercises 35–36, find $f(g(x))$ and $g(f(x))$ and determine whether each pair of functions f and g are inverses of each other.

35. $f(x) = \dfrac{3}{5}x + \dfrac{1}{2}$ and $g(x) = \dfrac{5}{3}x - 2$

  $f(g(x)) = x - \frac{7}{10}$; $g(f(x)) = x - \frac{7}{6}$; not inverses

36. $f(x) = 2 - 5x$ and $g(x) = \dfrac{2 - x}{5}$

  $f(g(x)) = x$; $g(f(x)) = x$; f and g are inverses of each other.

The functions in Exercises 37–38 are all one-to-one. For each function,

  a. Find an equation of $f^{-1}(x)$, the inverse function.
  b. Verify that your equation is correct by showing that $f(f^{-1}(x)) = x$ and $f^{-1}(f(x)) = x$.

37. $f(x) = 4x - 3$ **a.** $f^{-1}(x) = \dfrac{x + 3}{4}$

  b. $f(f^{-1}(x)) = 4\left(\dfrac{x + 3}{4}\right) - 3 = (x + 3) - 3 = x$

  and $f^{-1}(f(x)) = \dfrac{(4x - 3) + 3}{4} = \dfrac{4x}{4} = x$

38. $f(x) = -\dfrac{1}{x}$ **a.** $f^{-1}(x) = -\dfrac{1}{x}$

  b. $f(f^{-1}(x)) = -\dfrac{1}{\left(-\dfrac{1}{x}\right)} = x$

  and $f^{-1}(f(x)) = -\dfrac{1}{\left(-\dfrac{1}{x}\right)} = x$

Which graphs in Exercises 39–42 represent functions that have inverse functions?

39.

inverse function

40.

no inverse function

41.

inverse function

42.

no inverse function

43. Use the graph of f in the figure shown to draw the graph of its inverse function. See graphing answer section.

CHAPTER 8 TEST

Remember to use your Chapter Test Prep Video CD to see the worked-out solutions to the test questions you want to review.

In Exercises 1–2, determine whether each relation is a function. Give the domain and range for each relation.

1. $\{(1, 2), (3, 4), (5, 6), (6, 6)\}$
function; domain: $\{1, 3, 5, 6\}$; range: $\{2, 4, 6\}$

2. $\{(2, 1), (4, 3), (6, 5), (6, 6)\}$
not a function; domain: $\{2, 4, 6\}$; range: $\{1, 3, 5, 6\}$

3. If $f(x) = 3x - 2$, find $f(a + 4)$. $3a + 10$

4. If $f(x) = 4x^2 - 3x + 6$, find $f(-2)$. 28

In Exercises 5–6, identify the graph or graphs in which y is a function of x.

5.

function

6.

not a function

Use the graph of f to solve Exercises 7–10.

7. Find $f(6)$. -3

8. List two values of x for which $f(x) = 0$. -2 and 3

9. Find the domain of f. $\{x \mid x \text{ is a real number}\}$

10. Find the range of f. $\{y \mid y \le 3\}$

11. Find the domain of $f(x) = \dfrac{6}{10 - x}$.
Domain of $f = \{x \mid x \text{ is a real number and } x \neq 10\}$

In Exercises 12–16, let
$$f(x) = x^2 + 4x \quad \text{and} \quad g(x) = x + 2.$$
Find each of the following.

12. $(f + g)(x)$ and $(f + g)(3)$ $x^2 + 5x + 2; 26$

13. $(f - g)(x)$ and $(f - g)(-1)$ $x^2 + 3x - 2; -4$

14. $(fg)(x)$ and $(fg)(-5)$ $x^3 + 6x^2 + 8x; -15$

15. $\left(\dfrac{f}{g}\right)(x)$ and $\left(\dfrac{f}{g}\right)(2)$ $\dfrac{x^2 + 4x}{x + 2}; 3$

16. The domain of $\dfrac{f}{g}$
Domain of $\dfrac{f}{g} = \{x \mid x \text{ is a real number and } x \neq -2\}$

17. If $f(x) = x^2 + x$ and $g(x) = 3x - 1$, find $(f \circ g)(x)$ and $(g \circ f)(x)$. $9x^2 - 3x; 3x^2 + 3x - 1$

18. If $f(x) = 5x - 7$, find $f^{-1}(x)$. $f^{-1}(x) = \dfrac{x + 7}{5}$

19. A function f models the amount given to charity as a function of income. The graph of f is shown in the figure.

a. Explain why f has an inverse that is a function.
No horizontal line intersects the graph of f in more than one point.

b. Find $f(80)$. 2000

c. Describe in practical terms the meaning of $f^{-1}(2000)$.
$f^{-1}(2000)$ represents the income, $80 thousand, of a family that gives $2000 to charity.

Note to Instructor: The Chapter 8 Test file in the TestGen program provides algorithms specifically matched to these problems so you can easily replicate this chapter test for additional practice or assessment purposes.

CUMULATIVE REVIEW EXERCISES (CHAPTERS 1–8)

In Exercises 1–6, solve each equation or system of equations.

1. $2x + 3x - 5 + 7 = 10x + 3 - 6x - 4$ $\{-3\}$

2. $2x^2 + 5x = 12$ $\{-4, \frac{3}{2}\}$

3. $8x - 5y = -4$
$2x + 15y = -66$ $\{(-3, -4)\}$

4. $\dfrac{15}{x} - 4 = \dfrac{6}{x} + 3$ $\left\{\dfrac{9}{7}\right\}$

5. $-3x - 7 = 8$ $\{-5\}$

6. If $f(x) = 2x^2 - 5x + 2$ and $g(x) = x^2 - 2x + 3$, find $(f - g)(x)$ and $(f - g)(3)$.
$(f - g)(x) = x^2 - 3x - 1; -1$

In Exercises 7–11, simplify each expression.

7. $\dfrac{8x^3}{-4x^7} \quad -\dfrac{2}{x^4}$

8. $-8 - (-3) \cdot 4 \quad 4$

9. $\dfrac{\dfrac{1}{x} - \dfrac{1}{2}}{\dfrac{1}{3} - \dfrac{x}{6}} \quad \dfrac{3}{x}$

10. $\dfrac{4 - x^2}{3x^2 - 5x - 2} \quad -\dfrac{2 + x}{3x + 1}$

11. $-5 - (-8) - (4 - 6) \quad 5$

In Exercises 12–13, factor completely.

12. $x^2 - 18x + 77 \quad (x - 7)(x - 11)$

13. $x^3 - 25x \quad x(x + 5)(x - 5)$

In Exercises 14–17, perform the indicated operations. If possible, simplify the answer.

14. $\dfrac{6x^3 - 19x^2 + 16x - 4}{x - 2} \quad 6x^2 - 7x + 2$

15. $(2x - 3)(4x^2 + 6x + 9) \quad 8x^3 - 27$

16. $\dfrac{3x}{x^2 + x - 2} - \dfrac{2}{x + 2} \quad \dfrac{1}{x - 1}$

17. $\dfrac{5x^2 - 6x + 1}{x^2 - 1} \div \dfrac{16x^2 - 9}{4x^2 + 7x + 3} \quad \dfrac{5x - 1}{4x - 3}$

18. Solve the system:

$$\begin{aligned} x + 3y - z &= 5 \\ -x + 2y + 3z &= 13 \\ 2x - 5y - z &= -8. \end{aligned} \qquad \{(3, 2, 4)\}$$

In Exercises 19–20, graph each equation in a rectangular coordinate system.

19. $2x - y = 4$ **19–20.** See graphing answer section.

20. $y = -\dfrac{2}{3}x$

21. Is $\{(1, 5), (2, 5), (3, 5), (4, 5), (6, 5)\}$ a function? Give the relation's domain and range. function; domain: $\{1, 2, 3, 4, 6\}$; range: $\{5\}$

22. Find the slope of the line through $(-1, 5)$ and $(2, -3)$ $\quad -\dfrac{8}{3}$

23. Write the point-slope form of the equation of the line with slope 5, passing through $(-2, -3)$. Then use the point-slope equation to write the slope-intercept form of the line's equation. $\quad y + 3 = 5(x + 2); y = 5x + 7$

24. Multiply and write the answer in scientific notation:

$$(7 \times 10^{-8})(3 \times 10^2). \quad 2.1 \times 10^{-5}$$

25. Find the domain of $f(x) = \dfrac{1}{15 - x}$. Domain of $f = \{x \mid x \text{ is a real number and } x \neq 15\}$

Had a good workout lately? If so, could you tell if you were overdoing it or not pushing yourself hard enough?

Problems involving target heart rate ranges for various exercise goals appear as Exercises 59–62 in Exercise Set 9.4.

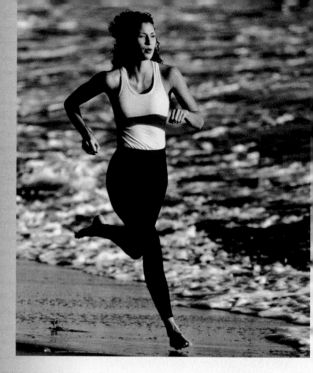

Inequalities and Problem Solving

Experts suggest a method for determining the heart rate you should maintain during aerobic activities, such as running, biking, or swimming. Subtract your age from 220; find 70% of that figure for the low end of the range and 80% for the high end. Your heart rate, in beats per minute, should stay between these two figures if you want to get the most cardiovascular benefit from your training.

The phrase "should stay between" indicates an algebraic inequality, a symbolic statement containing an inequality symbol such as $<$ (is less than). In this chapter, you will learn methods for modeling your world with inequalities. You will even see how inequalities are used to describe some of the most magnificent places in our nation's landscape.

SECTION 9.1

INTERVAL NOTATION AND BUSINESS APPLICATIONS USING LINEAR INEQUALITIES

Objectives

1 Use interval notation.

2 Review how to solve linear inequalities.

3 Use linear inequalities to solve problems involving revenue, cost, and profit.

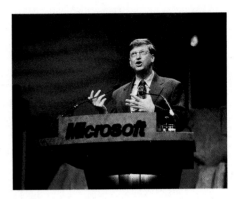

Driving through your neighborhood, you see kids selling lemonade. Would it surprise you to know that this activity can be analyzed using linear inequalities? By doing so, you will view profit and loss in the business world in a new way. In this section, we use linear inequalities to solve problems and model business ventures.

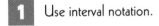

Use interval notation.

Interval Notation Recall from Chapter 2 that any inequality in the form $ax + b \leq c$ is called a **linear inequality in one variable**. The symbol between $ax + b$ and c can be $\leq$ (is less than or equal to), $<$ (is less than), $\geq$ (is greater than or equal to), or $>$ (is greater than). The greatest exponent on the variable in such an inequality is 1.

Solving an inequality is the process of finding the set of numbers that make the inequality a true statement. These numbers are called the **solutions** of the inequality and we say that they **satisfy** the inequality. The set of all solutions is called the **solution set** of the inequality.

Graphs of solutions to linear inequalities are shown on a number line by shading all points representing numbers that are solutions. For example, the solutions of $x > 4$ are all real numbers that are greater than 4. They are graphed on a number line by shading all points to the right of 4. The open dot at 4 indicates that 4 is not a solution:

Throughout the remainder of this book, a convenient notation, called *interval notation*, will also be used to represent solution sets of inequalities. To help understand this notation, a different graphing notation will be used. Parentheses, rather than open dots, indicate endpoints that are not solutions. Thus, we graph the solutions of $x > 4$ as

Square brackets, rather than closed dots, indicate endpoints that are solutions. Thus, we graph the solutions of $x \geq 4$ as

EXAMPLE 1 Graphing Inequalities

Graph the solutions of each inequality:

a. $x < 2$ **b.** $x \geq -3$ **c.** $-2 < x \leq 4$.

SOLUTION

a. The solutions of $x < 2$ are all real numbers that are less than 2. They are graphed on a number line by shading all points to the left of 2. The parenthesis at 2 indicates that 2 is not a solution, but numbers such as 1.9999 and 1.6 are. The arrow shows that the graph extends indefinitely to the left.

b. The solutions of $x \geq -3$ are all real numbers that are greater than or equal to -3. We shade all points to the right of -3 and the point for -3 itself. The bracket at -3 shows that -3 is a solution of the given inequality. The arrow shows that the graph extends indefinitely to the right.

c. The inequality $-2 < x \leq 4$ is read "-2 is less than x *and* x is less than or equal to 4," or "x is greater than -2 *and* less than or equal to 4." The solutions of $-2 < x \leq 4$ are all real numbers between -2 and 4, not including -2 but including 4. The parenthesis at -2 indicates that -2 is not a solution. By contrast, the bracket at 4 shows that 4 is a solution. Shading indicates the other solutions.

■

✔ **CHECK POINT 1** Graph the solutions of each inequality:

a. $x \leq 3$ **b.** $x > -5$ **c.** $-4 \leq x < 1$.   See graphing answer section.

In Chapter 2, we used set-builder notation to represent solution sets of inequalities. Using this method, the solution set of $x > -4$ can be expressed as

$$\{x \mid x > -4\}.$$

The set of all x such that x is greater than -4.

We read this as "the set of all real numbers x such that x is greater than -4."

As previously mentioned, another method used to represent solution sets of inequalities is **interval notation**. Using this notation, the solution set of $x > -4$ is expressed as $(-4, \infty)$. The parenthesis at -4 indicates that -4 is not included in the interval. The infinity symbol, ∞, does not represent a real number. It indicates that the interval extends indefinitely to the right.

Table 9.1 lists nine possible types of intervals used to describe sets of real numbers.

Table 9.1	Intervals on the Real Number Line	

Let a and b be real numbers such that $a < b$.

Interval Notation	Set-Builder Notation	Graph
(a, b)	$\{x \mid a < x < b\}$	
$[a, b]$	$\{x \mid a \leq x \leq b\}$	
$[a, b)$	$\{x \mid a \leq x < b\}$	
$(a, b]$	$\{x \mid a < x \leq b\}$	
(a, ∞)	$\{x \mid x > a\}$	
$[a, \infty)$	$\{x \mid x \geq a\}$	
$(-\infty, b)$	$\{x \mid x < b\}$	
$(-\infty, b]$	$\{x \mid x \leq b\}$	
$(-\infty, \infty)$	$\{x \mid x$ is a real number$\}$ or $\mathbb{R}$ (set of all real numbers)	

EXAMPLE 2 Intervals and Inequalities

Express each interval in set-builder notation and graph:

    a. $(-1, 4]$ **b.** $[2.5, 4]$ **c.** $(-3, \infty)$.

SOLUTION

    a. $(-1, 4] = \{x \mid -1 < x \leq 4\}$

    b. $[2.5, 4] = \{x \mid 2.5 \leq x \leq 4\}$

    c. $(-3, \infty) = \{x \mid x > -3\}$

See graphing answer section.

 CHECK POINT 2 Express each interval in set-builder notation and graph:

    a. $[-2, 5)$ **b.** $[1, 3.5]$ **c.** $(-\infty, -1)$.

    $\{x \mid -2 \leq x < 5\}$ $\{x \mid 1 \leq x \leq 3.5\}$ $\{x \mid x < -1\}$

2 Review how to solve linear inequalities.

Solving Linear Inequalities Recall from Chapter 2 the following procedure for solving a linear inequality. The only difference is that we will now use two notations, set-builder notation and interval notation, to express solution sets.

SOLVING A LINEAR INEQUALITY

1. Simplify the algebraic expression on each side.
2. Use the addition property of inequality to collect all the variable terms on one side and all the constant terms on the other side.
3. Use the multiplication property of inequality to isolate the variable and solve. Reverse the sense of the inequality when multiplying or dividing both sides by a negative number.
4. Express the solution set in set-builder or interval notation and graph the solution set on a number line.

STUDY TIP

If you need a more detailed presentation on solving linear inequalities, see Section 2.7, Examples 2 through 7, on pages 164 through 168.

EXAMPLE 3 Solving a Linear Inequality

Solve and graph the solution set on a number line:

$$2 - 12x \leq 7(1 - x).$$

SOLUTION

Step 1. **Simplify each side.** We use the distributive property to remove parentheses on the right side.

$$2 - 12x \leq 7(1 - x)$$ This is the given inequality.
$$2 - 12x \leq 7 - 7x$$ Use the distributive property.

STUDY TIP

You can solve

$$2 - 12x \leq 7 - 7x$$

by isolating x on the right side. Add $12x$ to both sides:

$$2 - 12x + 12x$$
$$\leq 7 - 7x + 12x$$
$$2 \leq 7 + 5x.$$

Now subtract 7 from both sides:

$$2 - 7 \leq 7 + 5x - 7$$
$$-5 \leq 5x.$$

Finally, divide both sides by 5:

$$\frac{-5}{5} \leq \frac{5x}{5}$$
$$-1 \leq x.$$

This last inequality means the same thing as

$$x \geq -1.$$

Step 2. **Collect variable terms on one side and constant terms on the other side.** We will collect variable terms on the left and constant terms on the right.

$$2 - 12x + 7x \leq 7 - 7x + 7x$$ Add $7x$ to both sides.
$$2 - 5x \leq 7$$ Simplify.
$$2 - 5x - 2 \leq 7 - 2$$ Subtract 2 from both sides.
$$-5x \leq 5$$ Simplify.

Step 3. **Isolate the variable and solve.** We isolate the variable, x, by dividing both sides by 5. Because we are dividing by a negative number, we must reverse the inequality symbol.

$$\frac{5x}{-5} \geq \frac{5}{-5}$$ Divide both sides by -5. and reverse the sense of the inequality.
$$x \geq -1$$ Simplify.

Step 4. **Express the solution set in set-builder or interval notation and graph the set on a number line.** The solution set consists of all real numbers that are greater than or equal to -1, expressed in set-builder notation as $\{x | x \geq -1\}$. The interval notation for this solution set is $[-1, \infty)$. The graph of the solution set is shown as follows:

✔ CHECK POINT **3** Solve and graph the solution set on a number line:

$$-12 - 8x \le 4(6 - x). \qquad \{x \,|\, x \ge -9\} \text{ or } [-9, \infty);$$
$$\text{See graphing answer section.}$$

If an inequality contains fractions, begin by multiplying both sides by the least common denominator. This will clear the inequality of fractions.

EXAMPLE 4 Solving a Linear Inequality

Solve and graph the solution set on a number line:

$$\frac{x + 3}{4} > \frac{x - 2}{3} + \frac{1}{4}.$$

SOLUTION The denominators are 4, 3, and 4. The least common denominator is 12. We begin by multiplying both sides of the inequality by 12.

$$\frac{x + 3}{4} > \frac{x - 2}{3} + \frac{1}{4} \qquad \text{This is the given inequality.}$$

$$12\left(\frac{x + 3}{4}\right) > 12\left(\frac{x - 2}{3} + \frac{1}{4}\right) \qquad \begin{array}{l}\text{Multiply both sides by 12. Multiply-} \\ \text{ing by a positive number preserves} \\ \text{the sense of the inequality.}\end{array}$$

$$\frac{12}{1} \cdot \frac{x + 3}{4} > \frac{12}{1} \cdot \frac{x - 2}{3} + \frac{12}{1} \cdot \frac{1}{4} \qquad \begin{array}{l}\text{Use the distributive property and} \\ \text{multiply each term by 12.}\end{array}$$

$$\frac{\overset{3}{\cancel{12}}}{1} \cdot \frac{x + 3}{\underset{1}{\cancel{4}}} > \frac{\overset{4}{\cancel{12}}}{1} \cdot \frac{x - 2}{\underset{1}{\cancel{3}}} + \frac{\overset{3}{\cancel{12}}}{1} \cdot \frac{1}{\underset{1}{\cancel{4}}} \qquad \begin{array}{l}\text{Divide out common factors in each} \\ \text{multiplication.}\end{array}$$

$$3(x + 3) > 4(x - 2) + 3 \qquad \text{The fractions are now cleared.}$$

Now that fractions are cleared, we follow the four steps that we used in the previous example.

Step 1. Simplify each side.

$$3(x + 3) > 4(x - 2) + 3 \qquad \begin{array}{l}\text{This is the inequality with the} \\ \text{fractions cleared.}\end{array}$$

$$3x + 9 > 4x - 8 + 3 \qquad \text{Use the distributive property.}$$

$$3x + 9 > 4x - 5 \qquad \text{Simplify.}$$

Step 2. Collect variable terms on one side and constant terms on the other side. We will collect variable terms on the left and constant terms on the right.

$$3x + 9 - 4x > 4x - 5 - 4x \qquad \text{Subtract 4x from both sides.}$$

$$-x + 9 > -5 \qquad \text{Simplify.}$$

$$-x + 9 - 9 > -5 - 9 \qquad \text{Subtract 9 from both sides.}$$

$$-x > -14 \qquad \text{Simplify.}$$

Step 3.   **Isolate the variable and solve.** To isolate x in $-x > -14$ we must eliminate the negative sign in front of the x. Because $-x$ means $-1x$, we can do this by multiplying (or dividing) both sides of the inequality by -1. We are multiplying by a negative number. Thus, we must reverse the inequality symbol.

$$(-1)(-x) > (-1)(-14)$$ Multiply both sides by -1 and reverse the sense of the inequality.

$$x > 14$$ Simplify.

Step 4.   **Express the solution set in set-builder or interval notation and graph the set on a number line.** The solution set consists of all real numbers that are less than 14, expressed in set-builder notation as $\{x \mid x < 14\}$. The interval notation for this solution set is $(-\infty, 14)$. The graph of the solution set is shown as follows:

5   6   7   8   9   10   11   12   13   14   15

 CHECK POINT 4 Solve and graph the solution set on a number line:

$$\frac{x-4}{2} > \frac{x-2}{3} + \frac{5}{6}.$$ $\{x \mid x > 13\}$ or $(13, \infty)$;
See graphing answer section.

---

3 Use linear inequalities to solve problems involving revenue, cost, and profit.

Functions of Business and Linear Inequalities   As a young entrepreneur, did you ever try selling lemonade in your front yard? Suppose that you charged 55 cents for each cup and you sold 45 cups. Your **revenue** is your income from selling these 45 units, or $\$0.55(45) = \$24.75$. Your *revenue function* from selling x cups is

$$R(x) = 0.55x.$$

This is the unit price: 55¢ for each cup.   This is the number of units sold.

For any business, the **revenue function**, $R(x)$, is the money generated by selling x units of the product:

$$R(x) = px.$$

Price per unit x units sold

Back to selling lemonade and energizing the neighborhood with white sugar: Is your revenue for the afternoon also your profit? No. We need to consider the cost of the business. You estimate that the lemons, white sugar, and bottled water cost 5 cents per cup. Furthermore, mommy dearest is charging you a $10 rental fee for use of your (her?) front yard. Your *cost function* for selling x cups of lemonade is

$$C(x) = 10 + 0.05x.$$

This is your $10 fixed cost.   This is your variable cost: 5¢ for each cup produced.

For any business, the **cost function**, $C(x)$, is the cost of producing x units of the product:

$$C(x) = (\text{fixed cost}) + cx.$$

Cost per unit x units produced

The term on the right, cx, represents **variable cost**, because it varies based on the number of units produced. Thus, the cost function is the sum of the fixed cost and the variable cost.

REVENUE AND COST FUNCTIONS   A company produces and sells x units of a product.

Revenue Function

$$R(x) = (\text{price per unit sold})x$$

Cost Function

$$C(x) = \text{fixed cost} + (\text{cost per unit produced})x$$

What does every entrepreneur, from a kid selling lemonade to Donald Trump, want to do? Generate profit, of course. The *profit* made is the money taken in, or the revenue, minus the money spent, or the cost. This relationship between revenue and cost allows us to define the *profit function*, $P(x)$.

THE PROFIT FUNCTION   The profit, $P(x)$, generated after producing and selling x units of a product is given by the **profit function**

$$P(x) = R(x) - C(x)$$

where R and C are the revenue and cost functions, respectively.

Figure 9.1 shows the graphs of the revenue and cost functions for the lemonade business. Similar graphs and models apply no matter how small or large a business venture may be.

$$R(x) = 0.55x \qquad C(x) = 10 + 0.05x$$

Revenue is 55¢ times the number of cups sold.

Cost is $10 plus 5¢ times the number of cups produced.

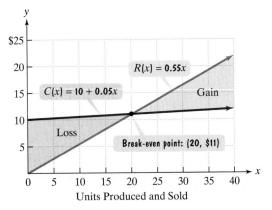

FIGURE 9.1

The lines intersect at the point $(20, 11)$. This means that when 20 cups are produced and sold, both cost and revenue are $11. In business, this point of intersection is called the **break-even point**. At the break-even point, the money coming in is equal to

FIGURE 9.1 (repeated)

the money going out. Can you see what happens for *x*-values less than 20? The red cost graph is above the blue revenue graph: $C(x) > R(x)$. The cost is greater than the revenue and the business is losing money. Thus, if you sell fewer than 20 cups of lemonade, the result is a *loss*. By contrast, look at what happens for *x*-values greater than 20. The blue revenue graph is above the red cost graph: $R(x) > C(x)$. The revenue is greater than the cost and the business is making money. Thus, if you sell more than 20 cups of lemonade, the result is a *profit*.

EXAMPLE 5 Writing a Profit Function to Determine a Profit

a. Use the revenue and cost functions shown in Figure 9.1,

$$R(x) = 0.55x \quad \text{and} \quad C(x) = 10 + 0.05x,$$

to write the profit function for producing and selling *x* cups of lemonade.

b. Use and solve a linear inequality to answer this question: More than how many cups of lemonade must be sold to have a profit?

SOLUTION

a. The profit function is the difference between the revenue function and the cost function.

$$
\begin{aligned}
P(x) &= R(x) - C(x) && \text{This is the definition of the profit function.} \\
&= 0.55x - (10 + 0.05x) && \text{Substitute the given functions.} \\
&= 0.55x - 10 - 0.05x && \text{Distribute } -1 \text{ to each term in parentheses.} \\
&= 0.50x - 10 && \text{Simplify.}
\end{aligned}
$$

The profit function is $P(x) = 0.50x - 10$.

b. A business makes money, or experiences a profit, when $P(x) > 0$. A business loses money, or experiences a loss, when $P(x) < 0$. Because we are interested in a profit, this occurs when $P(x) > 0$.

$$
\begin{aligned}
0.50x - 10 &> 0 && \text{This inequality models a profit.} \\
0.50x &> 10 && \text{Add 10 to both sides.} \\
\frac{0.50x}{0.50} &> \frac{10}{0.50} && \text{Divide both sides by 0.50. Division by a positive number does not reverse the sense of the inequality.} \\
x &> 20 && \text{Simplify.}
\end{aligned}
$$

More than 20 cups of lemonade must be sold to have a profit. ∎

DISCOVER FOR YOURSELF

A profit occurs when revenue exceeds cost:

$$R(x) > C(x).$$

Use $R(x) = 0.55x$ and $C(x) = 10 + 0.05x$ to verify that more than 20 cups of lemonade must be sold to have a profit.

The graph of the profit function, $P(x) = 0.50x - 10$, is shown in Figure 9.2. The red portion of the graph lies below the *x*-axis and shows a loss when fewer than 20 units are sold. Thus, if $x < 20$ then $P(x) < 0$. The lemonade business is "in the red." The black portion of the graph lies above the *x*-axis and shows a gain when more than 20 units are sold. Thus, if $x > 20$ then $P(x) > 0$. The lemonade business is "in the black."

FIGURE 9.2 In the red and in the black

CHECK POINT 5

a. Use the revenue and cost functions

$$R(x) = 200x$$

$$C(x) = 160,000 + 75x$$

to write the profit function for producing and selling x units.

$P(x) = 125x - 160,000$

b. More than how many units must be produced and sold to have a profit?   more than 1280 units

---

Summary of Given Information

Fixed cost: $500,000
Cost to produce each
wheelchair: $400
Selling price per
wheelchair: $600

| EXAMPLE 6 | Writing a Profit Function to Determine a Profit |

Technology is now promising to bring light, fast, and beautiful wheelchairs to millions of disabled people. A company is planning to manufacture these radically different wheelchairs. Fixed cost will be $500,000 and it will cost $400 to produce each wheelchair. Each wheelchair will be sold for $600.

a. Write the cost function, C, of producing x wheelchairs.

b. Write the revenue function, R, from the sale of x wheelchairs.

c. Write the profit function, P, from producing and selling x wheelchairs.

d. More than how many wheelchairs must be produced and sold to have a profit?

SOLUTION

a. The cost function is the sum of the fixed cost and variable cost.

Fixed cost of $500,000   plus   Variable cost: $400 for each chair produced

$$C(x) = 500,000 + 400x$$

b. The revenue function is the money generated from the sale of x wheelchairs.

Revenue per chair, $600, times   the number of chairs sold

$$R(x) = 600x$$

c. The profit function is the difference between the revenue function and the cost function.

$$P(x) = R(x) - C(x) \qquad \text{This is the definition of the profit function.}$$

$$= 600x - (500,000 + 400x) \qquad \text{Substitute the given functions.}$$

$$= 600x - 500,000 - 400x \qquad \text{Distribute } -1 \text{ to each term in parentheses.}$$

$$= 200x - 500,000 \qquad \text{Simplify.}$$

The profit function is $P(x) = 200x - 500,000$.

The graphs of the wheelchair company's cost and revenue functions

$$C(x) = 500,000 + 400x$$

and $R(x) = 600x$

are shown in a

$$[0, 5000, 1000] \text{ by}$$
$$[0, 3,000,000, 1,000,000]$$

viewing rectangle. To the right of the intersection point, the graph of the revenue function lies above that of the cost function. This confirms that producing and selling more than 2500 wheelchairs results in a profit.

$y_2 = 600x$

$y_1 = 500,000 + 400x$

$(2500, 1,500,000)$

d. A profit occurs when $P(x) > 0$. We use the profit function $P(x) = 200x - 500,000$ from part (c).

$200x - 500,000 > 0$	Set the profit function greater than 0.
$200x > 500,000$	Add 500,000 to both sides.
$\dfrac{200x}{200} > \dfrac{500,000}{200}$	Divide both sides by 200.
$x > 2500$	Simplify.

More than 2500 wheelchairs must be produced and sold to have a profit. ■

 CHECK POINT 6 A company that manufactures running shoes has a fixed cost of $300,000. Additionally, it costs $30 to produce each pair of shoes. They are sold at $80 per pair.

a. Write the cost function, C, of producing x pairs of running shoes.
$C(x) = 300,000 + 30x$

b. Write the revenue function, R, from the sale of x pairs of running shoes.
$R(x) = 80x$

c. Write the profit function, P, from producing and selling x pairs of running shoes. $P(x) = 50x - 300,000$

d. More than how many pairs of running shoes must be produced and sold to have a profit? _more than 6000 pairs_

9.1 EXERCISE SET

Student Solutions Manual   CD/Video   PH Math/Tutor Center   MathXL Tutorials on CD   MathXL®   MyMathLab   Interactmath.com

Practice Exercises

In Exercises 1–14, express each interval in set-builder notation and graph the interval on a number line. **1–14.** See graphing answer section.

1. $(1, 6]$ $\{x | 1 < x \le 6\}$
2. $(-2, 4]$ $\{x | -2 < x \le 4\}$
3. $[-5, 2)$ $\{x | -5 \le x < 2\}$
4. $[-4, 3)$ $\{x | -4 \le x < 3\}$
5. $[-3, 1]$ $\{x | -3 \le x \le 1\}$
6. $[-2, 5]$ $\{x | -2 \le x \le 5\}$
7. $(2, \infty)$ $\{x | x > 2\}$
8. $(3, \infty)$ $\{x | x > 3\}$
9. $[-3, \infty)$ $\{x | x \ge -3\}$
10. $[-5, \infty)$ $\{x | x \ge -5\}$
11. $(-\infty, 3)$ $\{x | x < 3\}$
12. $(-\infty, 2)$ $\{x | x < 2\}$
13. $(-\infty, 5.5)$ $\{x | x < 5.5\}$
14. $(-\infty, 3.5]$ $\{x | x \le 3.5\}$

In Exercises 15–40, solve each linear inequality and graph the solution set on a number line. **15–40.** See graphing answer section.

15. $5x + 11 < 26$
$\{x | x < 3\}$ or $(-\infty, 3)$
16. $2x + 5 < 17$
$\{x | x < 6\}$ or $(-\infty, 6)$
17. $3x - 8 \ge 13$
$\{x | x \ge 7\}$ or $[7, \infty)$
18. $8x - 2 \ge 14$
$\{x | x \ge 2\}$ or $[2, \infty)$

19. $-9x \ge 36$
$\{x | x \le -4\}$ or $(-\infty, -4]$
20. $-5x \le 30$
$\{x | x \ge -6\}$ or $[-6, \infty)$
21. $8x - 11 \le 3x - 13$
$\{x | x \le -\frac{2}{5}\}$ or $(-\infty, -\frac{2}{5}]$
22. $18x + 45 \le 12x - 8$
$\{x | x \le -\frac{53}{6}\}$ or $(-\infty, -\frac{53}{6}]$
23. $4(x + 1) + 2 \ge 3x + 6$ $\{x | x \ge 0\}$ or $[0, \infty)$
24. $8x + 3 > 3(2x + 1) + x + 5$ $\{x | x > 5\}$ or $(5, \infty)$
25. $2x - 11 < -3(x + 2)$ $\{x | x < 1\}$ or $(-\infty, 1)$
26. $-4(x + 2) > 3x + 20$ $\{x | x < -4\}$ or $(-\infty, -4)$
27. $1 - (x + 3) \ge 4 - 2x$ $\{x | x \ge 6\}$ or $[6, \infty)$
28. $5(3 - x) \le 3x - 1$ $\{x | x \ge 2\}$ or $[2, \infty)$
29. $\dfrac{x}{4} - \dfrac{1}{2} \le \dfrac{x}{2} + 1$ $\{x | x \ge -6\}$ or $[-6, \infty)$
30. $\dfrac{3x}{10} + 1 \ge \dfrac{1}{5} - \dfrac{x}{10}$ $\{x | x \ge -2\}$ or $[-2, \infty)$
31. $1 - \dfrac{x}{2} > 4$ $\{x | x < -6\}$ or $(-\infty, -6)$

32. $7 - \dfrac{4}{5}x < \dfrac{3}{5}$ $\{x \mid x > 8\}$ or $(8, \infty)$

33. $\dfrac{x-4}{6} \geq \dfrac{x-2}{9} + \dfrac{5}{18}$ $\{x \mid x \geq 13\}$ or $[13, \infty)$

34. $\dfrac{4x-3}{6} + 2 \geq \dfrac{2x-1}{12}$ $\{x \mid x \geq -\frac{19}{6}\}$ or $\left[-\frac{19}{6}, \infty\right)$

35. $7(y+4) - 13 < 12 + 13(3+y)$ $\{y \mid y > -6\}$ or $(-6, \infty)$

36. $-3[7y - (2y-3)] > -2(y+1)$ $\{y \mid y < -\frac{7}{13}\}$ or $\left(-\infty, -\frac{7}{13}\right)$

37. $6 - \dfrac{2}{3}(3x-12) \leq \dfrac{2}{5}(10x+50)$ $\{x \mid x \geq -1\}$ or $[-1, \infty)$

38. $\dfrac{2}{7}(7-21x) - 4 > 10 - \dfrac{3}{11}(11x-11)$
$\{x \mid x < -5\}$ or $(-\infty, -5)$

39. $3[3(y+5) + 8y + 7] + 5[3(y-6)$
$- 2(3y-5)] < 2(4y+3)$ $\{y \mid y < -2\}$ or $(-\infty, -2)$

40. $5[3(2-3y) - 2(5-y)] - 6[5(y-2)$
$- 2(4y-3)] < 3y + 19$ $\{y \mid y > -\frac{3}{4}\}$ or $\left(-\frac{3}{4}, \infty\right)$

41. Let $f(x) = 3x + 2$ and $g(x) = 5x - 8$. Find all values of x for which $f(x) > g(x)$. $\{x \mid x < 5\}$ or $(-\infty, 5)$

42. Let $f(x) = 2x - 9$ and $g(x) = 5x + 4$. Find all values of x for which $f(x) > g(x)$. $\{x \mid x < -\frac{13}{3}\}$ or $\left(-\infty, -\frac{13}{3}\right)$

43. Let $f(x) = \dfrac{2}{5}(10x+15)$ and $g(x) = \dfrac{1}{4}(8-12x)$. Find all values of x for which $g(x) \leq f(x)$. $\{x \mid x \geq -\frac{4}{7}\}$ or $\left[-\frac{4}{7}, \infty\right)$

44. Let $f(x) = \dfrac{3}{5}(10x-15) + 9$ and $g(x) = \dfrac{3}{8}(16-8x) - 7$. Find all values of x for which $g(x) \leq f(x)$. $\{x \mid x \geq -\frac{1}{9}\}$ or $\left[-\frac{1}{9}, \infty\right)$

In Exercises 45–48, cost and revenue functions for producing and selling x units of a product are given. Cost and revenue are expressed in dollars.

a. Write the profit function from producing and selling x units of the product.

b. More than how many units must be produced and sold to have a profit?

45. $C(x) = 25{,}500 + 15x$
$R(x) = 32x$
a. $P(x) = 17x - 25{,}500$
b. more than 1500 units

46. $C(x) = 15{,}000 + 12x$
$R(x) = 32x$
a. $P(x) = 20x - 15{,}000$
b. more than 750 units

47. $C(x) = 105x + 70{,}000$
$R(x) = 245x$
a. $P(x) = 140x - 70{,}000$
b. more than 500 units

48. $C(x) = 1.2x + 1500$
$R(x) = 1.7x$
a. $P(x) = 0.5x - 1500$
b. more than 3000 units

Practice Plus

In Exercises 49–50, solve each linear inequality and graph the solution set on a number line.   49–50. See graphing answer section.

49. $2(x+3) > 6 - \{4[x - (3x-4) - x] + 4\}$
$\{x \mid x < 2\}$ or $(-\infty, 2)$

50. $3(4x-6) < 4 - \{5x - [6x - (4x - (3x+2))]\}$
$\{x \mid x < 2\}$ or $(-\infty, 2)$

In Exercises 51–52, write an inequality with x isolated on the left side that is equivalent to the given inequality.

51. $ax + b > c$; Assume $a < 0$. $x < \dfrac{c-b}{a}$

52. $\dfrac{ax+b}{c} > b$; Assume $a > 0$ and $c < 0$. $x < \dfrac{bc - b}{a}$

In Exercises 53–54, use the graphs of y_1 and y_2 to solve each inequality.

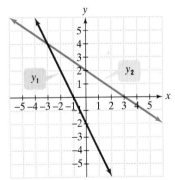

53. $y_1 \geq y_2$ $\{x \mid x \leq -3\}$ or $(-\infty, -3]$
54. $y_1 \leq y_2$ $\{x \mid x \geq -3\}$ or $[-3, \infty)$

In Exercises 55–56, use the table of values for the linear functions y_1 and y_2 to solve each inequality.

X	Y1	Y2
-1.8	9.4	7.4
-1.7	9.1	7.6
-1.6	8.8	7.8
-1.5	8.5	8
-1.4	8.2	8.2
-1.3	7.9	8.4
-1.2	7.6	8.6

X=-1.8

55. $y_1 < y_2$ $\{x \mid x > -1.4\}$ or $(-1.4, \infty)$
56. $y_1 > y_2$ $\{x \mid x < -1.4\}$ or $(-\infty, -1.4)$

Application Exercises

The graphs show that the three components of love, namely passion, intimacy, and commitment, progress differently over time. Passion peaks early in a relationship and then declines. By contrast, intimacy and commitment build gradually. Use the graphs to solve Exercises 57–64.

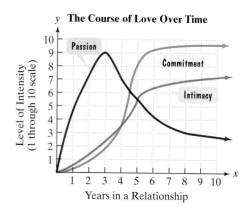

The Course of Love Over Time

Source: R. J. Sternberg, A Triangular Theory of Love, *Psychological Review*, 93, 119–135.

57. Use interval notation to write an inequality that expresses for which years in a relationship intimacy is greater that commitment. $(0, 4)$

58. Use interval notation to write an inequality that expresses for which years in a relationship passion is greater than or equal to intimacy. $[0, 5]$

59. What is the relationship between passion and intimacy on the interval $[5, 7)$? intimacy $\geq$ passion or passion $\leq$ intimacy

60. What is the relationship between intimacy and commitment on the interval $[4, 7)$? commitment $\geq$ intimacy or intimacy $\leq$ commitment

61. What is the relationship between passion and commitment for $\{x | 6 < x < 8\}$? commitment > passion or passion < commitment

62. What is the relationship between passion and commitment for $\{x | 7 < x < 9\}$? commitment > passion or passion < commitment

63. What is the maximum level of intensity for passion? After how many years in a relationship does this occur? 9; after 3 years

64. After approximately how many years do levels of intensity for commitment exceed the maximum level of intensity for passion? after approximately $5\frac{1}{2}$ years

65. The percentage, P, of U.S. voters who used electronic voting systems, such as optical scans, in national elections can be modeled by the formula

$$P = 3.1x + 25.8,$$

where x is the number of years after 1994. In which years will more than 63% of U.S. voters use electronic systems? voting years after 2006

66. The percentage, P, of U.S. voters who used punch cards or lever machines in national elections can be modeled by the formula

$$P = -2.5x + 63.1,$$

where x is the number of years after 1994. In which years will fewer than 38.1% of U.S. voters use punch cards or lever machines? voting years after 2004

The Olympic 500-meter speed skating times have generally been decreasing over time. The formulas

$$W = -0.19t + 57 \quad and \quad M = -0.15t + 50$$

model the winning times, in seconds, for women, W, and men, M, t years after 1900. Use these models to solve Exercises 67–68.

67. Find values of t such that $W < M$. Describe what this means in terms of winning times. $\{t | t > 175\}$ or $(175, \infty)$; The women's speed skating times will be less than the men's after the year 2075.

68. Find values of t such that $W > M$. Describe what this means in terms of winning times. $\{t | t < 175\}$ or $(-\infty, 175)$; The women's speed skating times will be greater than the men's until the year 2075.

Exercises 69–72 describe a number of business ventures. For each exercise,

a. *Write the cost function, C.*

b. *Write the revenue function, R.*

c. *Write the profit function, P.*

d. *More than how many units must be produced and sold to have a profit?*

69. A company that manufactures small canoes has a fixed cost of $18,000. It costs $20 to produce each canoe. The selling price is $80 per canoe. (In solving this exercise, let x represent the number of canoes produced and sold.)

    a. $C(x) = 18,000 + 20x$ **b.** $R(x) = 80x$

    c. $P(x) = 60x - 18,000$ **d.** more than 300 canoes

70. A company that manufactures bicycles has a fixed cost of $100,000. It costs $100 to produce each bicycle. The selling price is $300 per bike. (In solving this exercise, let x represent the number of bicycles produced and sold.)

    a. $C(x) = 100,000 + 100x$ **b.** $R(x) = 300x$

    c. $P(x) = 200x - 100,000$ **d.** more than 500 bicycles

71. You invest in a new play. The cost includes an overhead of $30,000, plus production costs of $2500 per performance. A sold-out performance brings in $3125. (In solving this exercise, let x represent the number of sold-out performances.)

    a. $C(x) = 30,000 + 2500x$ **b.** $R(x) = 3125x$

    c. $P(x) = 625x - 30,000$ **d.** more than 48 sold-out performances

72. You invested $30,000 and started a business writing greeting cards. Supplies cost 2¢ per card and you are selling each card for 50¢. (In solving this exercise, let x represent the number of cards produced and sold.)
 a. $C(x) = 30,000 + 0.02x$ **b.** $R(x) = 0.5x$
 c. $P(x) = 0.48x - 30,000$ **d.** more than 62,500 cards

73. A company manufactures and sells blank audiocassette tapes. The weekly fixed cost is $10,000 and it costs $0.40 to produce each tape. The selling price is $2.00 per tape. How many tapes must be produced and sold each week for the company to have a profit?   More than 6250 tapes must be produced and sold each week for a profit.

74. A company manufactures and sells personalized stationery. The weekly fixed cost is $3000 and it costs $3.00 to produce each package of stationery. The selling price is $5.50 per package. How many packages of stationery must be produced and sold each week for the company to have a profit?   More than 1200 packages of stationery must be produced and sold each week for a profit.

75. You are choosing between two long-distance telephone plans. Plan A has a monthly fee of $15 with a charge of $0.08 per minute for all long-distance calls. Plan B has a monthly fee of $3 with a charge of $0.12 per minute for all long-distance calls. How many minutes of long-distance calls in a month make plan A the better deal?   More than 300 minutes make plan A the better deal.

76. A city commission has proposed two tax bills. The first bill requires that a homeowner pay $1800 plus 3% of the assessed home value in taxes. The second bill requires taxes of $200 plus 8% of the assessed home value. What price range of home assessment would make the first bill a better deal?   A home assessment of greater than $32,000 would make the first bill a better deal.

Writing in Mathematics

77. When graphing the solutions of an inequality, what does a parenthesis signify? What does a bracket signify?
 77–81. Answers will vary.

78. Describe a revenue function for a business venture.

79. Describe a cost function for a business venture. What are the two kinds of costs that are modeled by this function?

80. What is the profit function for a business venture and how is it determined?

81. If the profit function for a business venture is known, explain how to find how many units must be produced and sold to have a profit.

Technology Exercises

In Exercises 82–83, solve each inequality using a graphing utility. Graph each side separately. Then determine the values of x for which the graph on the left side lies above the graph on the right side.

 82. $-3(x - 6) > 2x - 2$ $\{x | x < 4\}$ or $(-\infty, 4)$
 82–83. See graphing answer section.

83. $-2(x + 4) > 6x + 16$ $\{x | x < -3\}$ or $(-\infty, -3)$

84. Use a graphing utility's ⬚TABLE⬚ feature to verify your work in Exercises 82–83.   See graphing answer section.

Use the same technique employed in Exercises 82–83 to solve each inequality in Exercises 85–86. In each case, what conclusion can you draw? What happens if you try solving the inequalities algebraically?

 85. $12x - 10 > 2(x - 4) + 10x$ $\varnothing$

 86. $2x + 3 > 3(2x - 4) - 4x$
  $\{x | x \text{ is a real number}\}$ or $\mathbb{R}$ or $(-\infty, \infty)$

87. A bank offers two checking account plans. Plan A has a base service charge of $4.00 per month plus 10¢ per check. Plan B charges a base service charge of $2.00 per month plus 15¢ per check.

 a. Write models for the total monthly costs for each plan if x checks are written.   plan A: $4 + 0.10x$; plan B: $2 + 0.15x$

 b. Use a graphing utility to graph the models in the same $[0, 50, 1]$ by $[0, 10, 1]$ viewing rectangle.   See graphing answer section.

 c. Use the graphs (and the intersection feature) to determine for what number of checks per month plan A will be better than plan B.   **c–d.** more than 40 checks per month

 d. Verify the result of part (c) algebraically by solving an inequality.

Critical Thinking Exercises

88. Which one of the following is true?   **d**

 a. The inequality $3x > 6$ is equivalent to $2 > x$.

 b. The smallest real number in the solution set of $2x > 6$ is 4.

 c. If x is at least 7, then $x > 7$.

 d. The inequality $-3x > 6$ is equivalent to $-2 > x$.

89. What's wrong with this argument? Suppose x and y represent two real numbers, where $x > y$.

$2 > 1$	This is a true statement.
$2(y - x) > 1(y - x)$	Multiply both sides by $y - x$.
$2y - 2x > y - x$	Use the distributive property.
$y - 2x > -x$	Subtract y from both sides.
$y > x$	Add $2x$ to both sides.

The final inequality, $y > x$, is impossible because we were initially given $x > y$. Since $x > y$, $y - x < 0$. Thus, when both sides were multiplied by $y - x$, the sense of the inequality should have been changed.

90. The figure shows a square with four identical equilateral triangles attached. If the perimeter of the figure is at least 24 centimeters, what are the possible values for k?

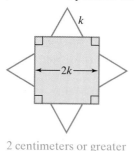

2 centimeters or greater

Review Exercises

91. If $f(x) = x^2 - 2x + 5$, find $f(-4)$.
(Section 8.1, Example 3)   29

92. Solve the system:

$$2x - y - z = -3$$
$$3x - 2y - 2z = -5$$
$$-x + y + 2z = 4.$$

(Section 4.5, Example 2) $\{(-1, -1, 2)\}$

93. Factor: $25x^2 - 81$. (Section 6.4, Example 1)
$(5x + 9)(5x - 9)$

SECTION

9.2

Objectives

1 Find the intersection of two sets.

2 Solve compound inequalities involving *and*.

3 Find the union of two sets.

4 Solve compound inequalities involving *or*.

COMPOUND INEQUALITIES

Boys and girls differ in their toy preferences. These differences are functions of our gender stereotypes—that is, our widely shared beliefs about males' and females' abilities, personality traits, and social behavior. Generally, boys have less leeway to play with "feminine" toys than girls do with "masculine" toys.

Which toys are requested by more than 40% of boys *and* more than 10% of girls? Which toys are requested by more than 40% of boys *or* more than 10% of girls? These questions are not the same. One involves inequalities joined by the word "and"; the other involves inequalities joined by the word "or." A **compound inequality** is formed by joining two inequalities with the word *and* or the word *or*.

Examples of Compound Inequalities

- $x - 3 < 5$ and $2x + 4 < 14$
- $3x - 5 \leq 13$ or $5x + 2 > -3$

In this section, you will learn to solve compound inequalities. With this skill, you will be able to analyze childrens' toy preferences in the exercise set.

1 Find the intersection of two sets.

Compound Inequalities Involving *And*   You need to determine whether there is sufficient support on campus to have a blood drive. You take a survey to obtain information, asking students two questions:

Would you be willing to donate blood?

Would you be willing to help serve a free breakfast to blood donors?

Part of your report involves the number of students willing to both donate blood *and* serve breakfast. You must focus on the set containing all students who are common to both the set of donors and the set of breakfast servers. This set is called the *intersection* of the two sets.

> **DEFINITION OF THE INTERSECTION OF SETS** The **intersection** of sets A and B, written $A \cap B$, is the set of elements common to both set A **and** set B. This definition can be expressed in set-builder notation as follows:
>
> $$A \cap B = \{x \mid x \in A \text{ AND } x \in B\}.$$

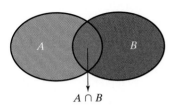

FIGURE 9.3 Picturing the intersection of two sets

Figure 9.3 shows a useful way of picturing the intersection of sets A and B. The figure indicates that $A \cap B$ contains those elements that belong to both A and B at the same time.

EXAMPLE 1 Finding the Intersection of Two Sets

Find the intersection: $\{7, 8, 9, 10, 11\} \cap \{6, 8, 10, 12\}$.

SOLUTION   The elements common to $\{7, 8, 9, 10, 11\}$ and $\{6, 8, 10, 12\}$ are 8 and 10. Thus,

$$\{7, 8, 9, 10, 11\} \cap \{6, 8, 10, 12\} = \{8, 10\}.$$ ∎

✔ **CHECK POINT 1** Find the intersection: $\{3, 4, 5, 6, 7\} \cap \{3, 7, 8, 9\}$. $\{3, 7\}$

---

2 Solve compound inequalities involving *and*.

A number is a **solution of a compound inequality formed by the word *and*** if it is a solution of both inequalities. For example, the solution set of the compound inequality

$$x \le 6 \quad \text{and} \quad x \ge 2$$

is the set of values of x that satisfy $x \le 6$ and $x \ge 2$. Thus, the solution set is the intersection of the solution sets of the two inequalities.

What are the numbers that satisfy both $x \le 6$ and $x \ge 2$? These numbers are easier to see if we graph the solution set to each inequality on a number line. These graphs are shown in Figure 9.4. The intersection is shown in the third graph.

FIGURE 9.4 Numbers satisfying both $x \le 6$ and $x \ge 2$

The numbers common to both sets are those that are less than or equal to 6 and greater than or equal to 2. This set is $\{x \mid 2 \le x \le 6\}$, or, in interval notation, $[2, 6]$.

Here is a procedure for finding the solution set of a compound inequality containing the word *and*.

SOLVING COMPOUND INEQUALITIES INVOLVING *AND*

1. Solve each inequality separately.

2. Graph the solution set to each inequality on a number line and take the intersection of these solution sets.

> Keep in mind that this intersection appears as the portion of the number line that the two graphs have in common.

EXAMPLE 2   Solving a Compound Inequality with *And*

Solve: $x - 3 < 5$ and $2x + 4 < 14$.

SOLUTION

Step 1.   Solve each inequality separately.

$$x - 3 < 5 \quad \text{and} \quad 2x + 4 < 14$$
$$x < 8 \qquad\qquad 2x < 10$$
$$x < 5$$

Step 2.   Take the intersection of the solution sets of the two inequalities. We graph the solution sets of $x < 8$ and $x < 5$. The intersection is shown in the third graph.

The numbers common to both sets are those that are less than 5. The solution set is $\{x \mid x < 5\}$, or, in interval notation, $(-\infty, 5)$. Take a moment to check that any number in $(-\infty, 5)$ satisfies both of the original inequalities. ■

✔ **CHECK POINT 2** Solve: $x + 2 < 5$ and $2x - 4 < -2$. $\{x \mid x < 1\}$ or $(-\infty, 1)$

EXAMPLE 3   Solving a Compound Inequality with *And*

Solve: $2x - 7 > 3$ and $5x - 4 < 6$.

SOLUTION

Step 1.   Solve each inequality separately.

$$2x - 7 > 3 \quad \text{and} \quad 5x - 4 < 6$$
$$2x > 10 \qquad\qquad 5x < 10$$
$$x > 5 \qquad\qquad\quad x < 2$$

Step 2. **Take the intersection of the solution sets of the two inequalities.** We graph the solution sets of $x > 5$ and $x < 2$. We use these graphs to find their intersection.

$\{x \mid x > 5\}$ $(5, \infty)$

$\{x \mid x < 2\}$ $(-\infty, 2)$

$\{x \mid x > 5\} \cap \{x \mid x < 2\}$ $= \varnothing$ $\varnothing$

There is no number that is both greater than 5 and at the same time less than 2. Thus, the solution set is the empty set, $\varnothing$. ∎

✓ **CHECK POINT 3** Solve: $4x - 5 > 7$ and $5x - 2 < 3$. $\varnothing$

If $a < b$, the compound inequality

$$a < x \text{ and } x < b$$

can be written in the shorter form

$$a < x < b.$$

For example, the compound inequality

$$-3 < 2x + 1 \text{ and } 2x + 1 < 3$$

can be abbreviated

$$-3 < 2x + 1 < 3.$$

The word "and" does not appear when the inequality is written in the shorter form, although it is implied. The shorter form enables us to solve both inequalities at once. By performing the same operations on all three parts of the inequality, our goal is to **isolate x in the middle**.

EXAMPLE 4 Solving a Compound Inequality

Solve and graph the solution set:

$$-3 < 2x + 1 \leq 3.$$

SOLUTION We would like to isolate x in the middle. We can do this by first subtracting 1 from all three parts of the compound inequality. Then we isolate x from $2x$ by dividing all three parts of the inequality by 2.

$$-3 < 2x + 1 \leq 3 \qquad \text{This is the given inequality.}$$
$$-3 - 1 < 2x + 1 - 1 \leq 3 - 1 \qquad \text{Subtract 1 from all three parts.}$$
$$-4 < 2x \leq 2 \qquad \text{Simplify.}$$
$$\frac{-4}{2} < \frac{2x}{2} \leq \frac{2}{2} \qquad \text{Divide each part by 2.}$$
$$-2 < x \leq 1 \qquad \text{Simplify.}$$

The solution set consists of all real numbers greater than -2 and less than or equal to 1, represented by $\{x \mid -2 < x \leq 1\}$ in set-builder notation and $(-2, 1]$ in interval notation. The graph is shown as follows:

✔ CHECK POINT 4 Solve and graph the solution set: $1 \le 2x + 3 < 11$.

$\{x | -1 \le x < 4\}$ or $[-1, 4)$; See graphing answer section.

---

3 Find the union of two sets.

$A \cup B$

FIGURE 9.5 Picturing the union of two sets

Compound Inequalities Involving *Or* Earlier, we considered a survey asking students if they were willing to donate blood or serve breakfast to donors. You continue to sort the results. Part of your report involves the number of students willing to donate blood *or* serve breakfast *or* do both. You must focus on the set containing donors or breakfast servers or both. This set is called the *union* of two sets.

> **DEFINITION OF THE UNION OF SETS** The **union** of sets A and B, written $A \cup B$, is the set of elements that are members of set A **or** of set B or of both sets. This definition can be expressed in set-builder notation as follows:
>
> $$A \cup B = \{x | x \in A \text{ OR } x \in B\}.$$

Figure 9.5 shows a useful way of picturing the union of sets A and B. The figure indicates that $A \cup B$ is formed by joining the sets together.

We can find the union of set A and set B by listing the elements of set A. Then, we include any elements of set B that have not already been listed. Enclose all elements that are listed with braces. This shows that the union of two sets is also a set.

STUDY TIP

When finding the union of two sets, do not list twice any elements that appear in both sets.

EXAMPLE 5 Finding the Union of Two Sets

Find the union: $\{7, 8, 9, 10, 11\} \cup \{6, 8, 10, 12\}$.

SOLUTION   To find $\{7, 8, 9, 10, 11\} \cup \{6, 8, 10, 12\}$, start by listing all the elements from the first set, namely 7, 8, 9, 10, and 11. Now list all the elements from the second set that are not in the first set, namely 6 and 12. The union is the set consisting of all these elements. Thus,

$$\{7, 8, 9, 10, 11\} \cup \{6, 8, 10, 12\} = \{6, 7, 8, 9, 10, 11, 12\}. \qquad \blacksquare$$

✔ CHECK POINT 5 Find the union: $\{3, 4, 5, 6, 7\} \cup \{3, 7, 8, 9\}$. $\{3, 4, 5, 6, 7, 8, 9\}$

---

4 Solve compound inequalities involving *or*.

A number is a **solution of a compound inequality formed by the word *or*** if it is a solution of either inequality. Thus, the solution set of a compound inequality formed by the word *or* is the union of the solution sets of the two inequalities.

> **SOLVING COMPOUND INEQUALITIES INVOLVING *OR***
>
> **1.** Solve each inequality separately.
> **2.** Graph the solution set to each inequality on a number line and take the union of these solution sets.
>
> > Keep in mind that this union appears as the portion of the number line representing the total collection of numbers in the two graphs.

EXAMPLE 6 Solving a Compound Inequality with *Or*

Solve: $2x - 3 < 7$ or $35 - 4x \le 3$.

SOLUTION

Step 1.   **Solve each inequality separately.**

$$2x - 3 < 7 \quad \text{or} \quad 35 - 4x \le 3$$
$$2x < 10 \qquad\qquad -4x \le -32$$
$$x < 5 \qquad\qquad\quad x \ge 8$$

Step 2.   **Take the union of the solution sets of the two inequalities.** We graph the solution sets of $x < 5$ and $x \ge 8$. We use these graphs to find their union.

The solution set consists of all numbers that are less than 5 or greater than or equal to 8. The solution set is $\{x \mid x < 5 \text{ or } x \ge 8\}$, or, in interval notation, $(-\infty, 5) \cup [8, \infty)$. There is no shortcut way to express this union when interval notation is used. ∎

✔ **CHECK POINT 6** Solve: $3x - 5 \le -2$ or $10 - 2x < 4$. $\{x \mid x \le 1 \text{ or } x > 3\}$ or $(-\infty, 1] \cup (3, \infty)$

EXAMPLE 7 Solving a Compound Inequality with *Or*

Solve: $3x - 5 \le 13$ or $5x + 2 > -3$.

SOLUTION

Step 1.   **Solve each inequality separately.**

$$3x - 5 \le 13 \quad \text{or} \quad 5x + 2 > -3$$
$$3x \le 18 \qquad\qquad 5x > -5$$
$$x \le 6 \qquad\qquad\quad x > -1$$

Step 2.   **Take the union of the solution sets of the two inequalities.** We graph the solution sets of $x \le 6$ and $x > -1$. We use these graphs to find their union.

Because all real numbers are less than or equal to 6 or greater than -1, the union of the two sets fills the entire number line. Thus, the solution set is $\{x \mid x \text{ is a real number}\}$, or $\mathbb{R}$. The solution set in interval notation is $(-\infty, \infty)$. Any real number that you select will satisfy either or both of the original inequalities. ∎

CHECK POINT 7 Solve: $2x + 5 \geq 3$ or $2x + 3 < 3$. $\{x \mid x \text{ is a real number}\}$ or $\mathbb{R}$ or $(-\infty, \infty)$

9.2 EXERCISE SET

Student Solutions Manual   CD/Video   PH Math/Tutor Center   MathXL Tutorials on CD   MathXL®   MyMathLab   Interactmath.com

Practice Exercises

In Exercises 1–6, find the intersection of the sets.

1. $\{1, 2, 3, 4\} \cap \{2, 4, 5\}$ $\{2, 4\}$

2. $\{1, 3, 7\} \cap \{2, 3, 8\}$ $\{3\}$

3. $\{1, 3, 5, 7\} \cap \{2, 4, 6, 8, 10\}$ $\varnothing$

4. $\{0, 1, 3, 5\} \cap \{-5, -3, -1\}$ $\varnothing$

5. $\{a, b, c, d\} \cap \varnothing$ $\varnothing$

6. $\{w, y, z\} \cap \varnothing$ $\varnothing$

In Exercises 7–24, solve each compound inequality. Use graphs to show the solution set to each of the two given inequalities, as well as a third graph that shows the solution set of the compound inequality. Except for the empty set, express the solution set in both set-builder and interval notations. **7–24.** See graphing answer section.

7. $x > 3$ and $x > 6$ $\{x \mid x > 6\}$; $(6, \infty)$

8. $x > 2$ and $x > 4$ $\{x \mid x > 4\}$; $(4, \infty)$

9. $x \leq 5$ and $x \leq 1$ $\{x \mid x \leq 1\}$; $(-\infty, 1]$

10. $x \leq 6$ and $x \leq 2$ $\{x \mid x \leq 2\}$; $(-\infty, 2]$

11. $x < 2$ and $x \geq -1$ $\{x \mid -1 \leq x < 2\}$; $[-1, 2)$

12. $x < 3$ and $x \geq -1$ $\{x \mid -1 \leq x < 3\}$; $[-1, 3)$

13. $x > 2$ and $x < -1$ $\varnothing$

14. $x > 3$ and $x < -1$ $\varnothing$

15. $5x < -20$ and $3x > -18$ $\{x \mid -6 < x < -4\}$; $(-6, -4)$

16. $3x \leq 15$ and $2x > -6$ $\{x \mid -3 < x \leq 5\}$; $(-3, 5]$

17. $x - 4 \leq 2$ and $3x + 1 > -8$ $\{x \mid -3 < x \leq 6\}$; $(-3, 6]$

18. $3x + 2 > -4$ and $2x - 1 < 5$ $\{x \mid -2 < x < 3\}$; $(-2, 3)$

19. $2x > 5x - 15$ and $7x > 2x + 10$ $\{x \mid 2 < x < 5\}$; $(2, 5)$

20. $6 - 5x > 1 - 3x$ and $4x - 3 > x - 9$
$\{x \mid -2 < x < \frac{5}{2}\}$; $(-2, \frac{5}{2})$

21. $4(1 - x) < -6$ and $\dfrac{x - 7}{5} \leq -2$ $\varnothing$

22. $5(x - 2) > 15$ and $\dfrac{x - 6}{4} \leq -2$ $\varnothing$

23. $x - 1 \leq 7x - 1$ and $4x - 7 < 3 - x$
$\{x \mid 0 \leq x < 2\}$; $[0, 2)$

24. $2x + 1 > 4x - 3$ and $x - 1 \geq 3x + 5$
$\{x \mid x \leq -3\}$; $(-\infty, -3]$

In Exercises 25–32, solve each inequality and graph the solution set on a number line. Express the solution set in both set-builder and interval notations. **25–32.** See graphing answer section.

25. $6 < x + 3 < 8$ $\{x \mid 3 < x < 5\}$; $(3, 5)$

26. $7 < x + 5 < 11$ $\{x \mid 2 < x < 6\}$; $(2, 6)$

27. $-3 \leq x - 2 < 1$ $\{x \mid -1 \leq x < 3\}$; $[-1, 3)$

28. $-6 < x - 4 \leq 1$ $\{x \mid -2 < x \leq 5\}$; $(-2, 5]$

29. $-11 < 2x - 1 \leq -5$ $\{x \mid -5 < x \leq -2\}$; $(-5, -2]$

30. $3 \leq 4x - 3 < 19$ $\{x \mid \frac{3}{2} \leq x < \frac{11}{2}\}$; $[\frac{3}{2}, \frac{11}{2})$

31. $-3 \leq \dfrac{2x}{3} - 5 < -1$ $\{x \mid 3 \leq x < 6\}$; $[3, 6)$

32. $-6 \leq \dfrac{x}{2} - 4 < -3$ $\{x \mid -4 \leq x < 2\}$; $[-4, 2)$

In Exercises 33–38, find the union of the sets.

33. $\{1, 2, 3, 4\} \cup \{2, 4, 5\}$ $\{1, 2, 3, 4, 5\}$

34. $\{1, 3, 7, 8\} \cup \{2, 3, 8\}$ $\{1, 2, 3, 7, 8\}$

35. $\{1, 3, 5, 7\} \cup \{2, 4, 6, 8, 10\}$ $\{1, 2, 3, 4, 5, 6, 7, 8, 10\}$

36. $\{0, 1, 3, 5\} \cup \{2, 4, 6\}$ $\{0, 1, 2, 3, 4, 5, 6\}$

37. $\{a, e, i, o, u\} \cup \varnothing$ $\{a, e, i, o, u\}$

38. $\{e, m, p, t, y\} \cup \varnothing$ $\{e, m, p, t, y\}$

In Exercises 39–54, solve each compound inequality. Use graphs to show the solution set to each of the two given inequalities, as well as a third graph that shows the solution set of the compound inequality. Express the solution set in both set-builder and interval notations. **39–54.** See graphing answer section.

39. $x > 3$ or $x > 6$ $\{x \mid x > 3\}$; $(3, \infty)$

40. $x > 2$ or $x > 4$ $\{x \mid x > 2\}$; $(2, \infty)$

41. $x \leq 5$ or $x \leq 1$ $\{x \mid x \leq 5\}$; $(-\infty, 5]$

42. $x \le 6$ or $x \le 2$ $\{x|x \le 6\}; (-\infty, 6]$

43. $x < 2$ or $x \ge -1$ $\{x|x \text{ is a real number}\}; (-\infty, \infty)$

44. $x < 3$ or $x \ge -1$ $\{x|x \text{ is a real number}\}; (-\infty, \infty)$

45. $x \ge 2$ or $x < -1$ $\{x|x < -1 \text{ or } x \ge 2\}; (-\infty, -1) \cup [2, \infty)$

46. $x \ge 3$ or $x < -1$ $\{x|x < -1 \text{ or } x \ge 3\}; (-\infty, -1) \cup [3, \infty)$

47. $3x > 12$ or $2x < -6$ $\{x|x < -3 \text{ or } x > 4\}; (-\infty, -3) \cup (4, \infty)$

48. $3x < 3$ or $2x > 10$ $\{x|x < 1 \text{ or } x > 5\}; (-\infty, 1) \cup (5, \infty)$

49. $3x + 2 \le 5$ or $5x - 7 \ge 8$
$\{x|x \le 1 \text{ or } x \ge 3\}; (-\infty, 1] \cup [3, \infty)$

50. $2x - 5 \le -11$ or $5x + 1 \ge 6$
$\{x|x \le -3 \text{ or } x \ge 1\}; (-\infty, -3] \cup [1, \infty)$

51. $4x + 3 < -1$ or $2x - 3 \ge -11$
$\{x|x \text{ is a real number}\}; (-\infty, \infty)$

52. $2x + 1 < 15$ or $3x - 4 \ge -1$
$\{x|x \text{ is a real number}\}; (-\infty, \infty)$

53. $-2x + 5 > 7$ or $-3x + 10 > 2x$ $\{x|x < 2\}; (-\infty, 2)$

54. $16 - 3x \ge -8$ or $13 - x > 4x + 3$ $\{x|x \le 8\}; (-\infty, 8]$

55. Let $f(x) = 2x + 3$ and $g(x) = 3x - 1$. Find all values of x for which $f(x) \ge 5$ and $g(x) > 11$. $\{x|x > 4\}$ or $(4, \infty)$

56. Let $f(x) = 4x + 5$ and $g(x) = 3x - 4$. Find all values of x for which $f(x) \ge 5$ and $g(x) \le 2$. $\{x|0 \le x \le 2\}$ or $[0, 2]$

57. Let $f(x) = 3x - 1$ and $g(x) = 4 - x$. Find all values of x for which $f(x) < -1$ or $g(x) < -2$.
$\{x|x < 0 \text{ or } x > 6\}$ or $(-\infty, 0) \cup (6, \infty)$

58. Let $f(x) = 2x - 5$ and $g(x) = 3 - x$. Find all values of x for which $f(x) \ge 3$ or $g(x) < 0$. $\{x|x > 3\}$ or $(3, \infty)$

Practice Plus

In Exercises 59–60, write an inequality with x isolated in the middle that is equivalent to the given inequality. Assume $a > 0, b > 0,$ and $c > 0$.

59. $-c < ax - b < c$ $\dfrac{b - c}{a} < x < \dfrac{b + c}{a}$

60. $-2 < \dfrac{ax - b}{c} < 2$ $\dfrac{b - 2c}{a} < x < \dfrac{b + 2c}{a}$

In Exercises 61–62, use the graphs of y_1, y_2, and y_3, to solve each compound inequality.

61. $-3 \le 2x - 1 \le 5$

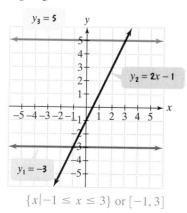

$\{x|-1 \le x \le 3\}$ or $[-1, 3]$

62. $x - 2 < 2x - 1 < x + 2$

$[-5, 5, 1]$ by $[-5, 8, 1]$

$\{x|-1 < x < 3\}$ or $(-1, 3)$

63. Solve $x - 2 < 2x - 1 < x + 2$, the inequality in Exercise 62, using algebraic methods. (*Hint*: Rewrite the inequality as $2x - 1 > x - 2$ and $2x - 1 < x + 2$.) $\{x|-1 < x < 3\}$ or $(-1, 3)$

64. Use the hint given in Exercise 63 to solve $x \le 3x - 10 \le 2x$. $\{x|5 \le x \le 10\}$ or $[5, 10]$

In Exercises 65–66, use the table to solve each inequality.

65. $-2 \le 5x + 3 < 13$

$\{x|-1 \le x < 2\}$ or $[-1, 2)$

66. $-3 < 2x - 5 \le 3$

$\{x|1 < x \le 4\}$ or $(-1, 4]$

In Exercises 67–68, use the roster method to find the set of negative integers that are solutions of each inequality.

67. $5 - 4x \ge 1$ and $3 - 7x < 31$ $\{-3, -2, -1\}$

68. $-5 < 3x + 4 \le 16$ $\{-2, -1\}$

Application Exercises

As a result of cultural expectations about what is appropriate behavior for each gender, boys and girls differ substantially in their toy preferences. The graph on the next page shows the percentage of boys and girls asking for various types of toys in letters to Santa Claus. Use the information in the graph to solve Exercises 69–74.

Toys Requested by Children

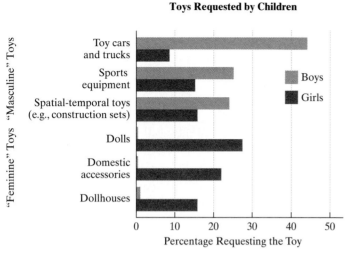

Source: Richardson, J. G., & Simpson, C. H. (1982). Children, gender and social structure: An analysis of the contents of letters to Santa Claus. *Child Development, 53,* 429–436.

69. Which toys were requested by more than 10% of the boys *and* less than 20% of the girls?   spatial-temporal toys, sports equipment, and toy cars and trucks

70. Which toys were requested by fewer than 5% of the boys *and* fewer than 20% of the girls?   dollhouses

71. Which toys were requested by more than 10% of the boys *or* less than 20% of the girls?   dollhouses, spatial-temporal toys, sports equipment, and toy cars and trucks

72. Which toys were requested by fewer than 5% of the boys *or* fewer than 20% of the girls?   dollhouses, domestic accessories, dolls, spatial-temporal toys, sports equipment, and toy cars and trucks

73. Which toys were requested by more than 40% of the boys *and* more than 10% of the girls?   none of the toys

74. Which toys were requested by more than 40% of the boys *or* more than 10% of the girls?   dollhouses, domestic accessories, dolls, spatial-temporal toys, sports equipment, and toy cars and trucks

75. A basic cellular phone plan costs $20 per month for 60 calling minutes. Additional time costs $0.40 per minute. The formula

$$C = 20 + 0.40(x - 60)$$

gives the monthly cost for this plan, *C*, for *x* calling minutes, where *x* > 60. How many calling minutes are possible for a monthly cost of at least $28 and at most $40?   between 80 and 110 minutes, inclusive

76. The formula for converting Fahrenheit temperature, *F*, to Celsius temperature, *C*, is

$$C = \frac{5}{9}(F - 32).$$

If Celsius temperature ranges from 15° to 35°, inclusive, what is the range for the Fahrenheit temperature? Use interval notation to express this range.   [59°F, 95°F]

77. On the first of four exams, your grades are 70, 75, 87, and 92. There is still one more exam, and you are hoping to earn a B in the course. This will occur if the average of your five exam grades is greater than or equal to 80 and less than 90. What range of grades on the fifth exam will result in earning a B? Use interval notation to express this range.   [76, 126); If the highest grade is 100, then [76, 100].

78. On the first of four exams, your grades are 82, 75, 80, and 90. There is still a final exam, and it counts as two grades. You are hoping to earn a B in the course. This will occur if the average of your six exam grades is greater than or equal to 80 and less than 90. What range of grades on the final exam will result in earning a B? Use interval notation to express this range.   [76.5, 106.5); If the highest grade is 100, then [76.5, 100].

79. The toll to a bridge is $3.00. A three-month pass costs $7.50 and reduces the toll to $0.50. A six-month pass costs $30 and permits crossing the bridge for no additional fee. How many crossings per three-month period does it take for the three-month pass to be the best deal?   more than 3 and less than 15 crossings per three-month period

80. Parts for an automobile repair cost $175. The mechanic charges $34 per hour. If you receive an estimate for at least $226 and at most $294 for fixing the car, what is the time interval that the mechanic will be working on the job?   The mechanic will be working on the job from 1.5 to 3.5 hours, inclusive.

Writing in Mathematics

81. Describe what is meant by the intersection of two sets. Give an example.   81–87. Answers will vary.

82. Explain how to solve a compound inequality involving "and."

83. Why is $1 < 2x + 3 < 9$ a compound inequality? Where are the two inequalities and what is the word that joins them?

84. Explain how to solve $1 < 2x + 3 < 9$.

85. Describe what is meant by the union of two sets. Give an example.

86. Explain how to solve a compound inequality involving "or."

87. How many Christmas trees are sold each holiday season? The function $T(x) = 0.9x + 32$ models the number of trees sold, $T(x)$, in millions, *x* years after 1990. Use the function to write a problem that can be solved using a compound inequality. Then solve the problem.

Technology Exercises

In Exercises 88–91, solve each inequality using a graphing utility. Graph each of the three parts of the inequality separately in the same viewing rectangle. The solution set consists of all values of x for which the graph of the linear function in the middle lies between the graphs of the constant functions on the left and the right.
88–91. See graphing answer section.

88. $1 < x + 3 < 9$ $\{x | -2 < x < 6\}$ or $(-2, 6)$

89. $-1 < \dfrac{x + 4}{2} < 3$ $\{x | -6 < x < 2\}$ or $(-6, 2)$

90. $1 \le 4x - 7 \le 3$ $\left\{x | 2 \le x \le \frac{5}{2}\right\}$ or $\left[2, \frac{5}{2}\right]$

91. $2 \le 4 - x \le 7$ $\{x | -3 \le x \le 2\}$ or $[-3, 2]$

Critical Thinking Exercises

92. Which one of the following is true?   a

   a. $(-\infty, -1] \cap [-4, \infty) = [-4, -1]$

   b. $(-\infty, 3) \cup (-\infty, -2) = (-\infty, -2)$

   c. The union of two sets can never give the same result as the intersection of those sets.

   d. The solution set of $x < a$ and $x > a$ is the set of all real numbers excluding a.

93. Solve and express the solution set in interval notation:

$$-7 \le 8 - 3x \le 20 \text{ and } -7 < 6x - 1 < 41.   (-1, 5]$$

The graphs of $f(x) = \sqrt{4 - x}$ and $g(x) = \sqrt{x + 1}$ are shown in a $[-3, 10, 1]$ by $[-2, 5, 1]$ viewing rectangle.

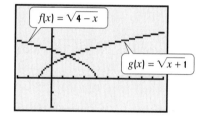

In Exercises 94–97, use the graphs and interval notation to express the domain of the given function.

94. The domain of f $(-\infty, 4]$

95. The domain of g $[-1, \infty)$

96. The domain of $f + g$ $[-1, -4]$

97. The domain of $\dfrac{f}{g}$ $(-1, 4]$

98. At the end of the day, the change machine at a laundrette contained at least \$3.20 and at most \$5.45 in nickels, dimes, and quarters. There were 3 fewer dimes than twice the number of nickels and 2 more quarters than twice the number of nickels. What was the least possible number and the greatest possible number of nickels? The least possible number of nickels was 4 and the greatest possible number of nickels was 7.

Review Exercises

99. If $f(x) = x^2 - 3x + 4$ and $g(x) = 2x - 5,$ find $(g - f)(x)$ and $(g - f)(-1)$. (Section 8.2, Example 4) $(g - f)(x) = -x^2 + 5x - 9; -15$

100. Use function notation to write the equation of the line passing through $(4, 2)$ and perpendicular to the line whose equation is $4x - 2y = 8$. (Section 3.5, Examples 3 and 4) $f(x) = -\frac{1}{2}x + 4$

101. Simplify: $4 - [2(x - 4) - 5]$. (Section 1.8, Example 11) $17 - 2x$

SECTION

9.3

Objectives

1 Solve equations involving absolute value.

2 Solve inequalities involving absolute value.

3 Recognize absolute value inequalities with no solution or all real numbers as solutions.

4 Solve problems using absolute value inequalities.

EQUATIONS AND INEQUALITIES INVOLVING ABSOLUTE VALUE

$M*A*S*H$ was set in the early 1950s during the Korean War. By the final episode, the show had lasted four times as long as the Korean War.

At the end of the twentieth century, there were 94 million households in the United States with television sets. The television program viewed by the greatest percentage of such households in that century was the final episode of $M*A*S*H$. Over 50 million American households watched this program.

Numerical information, such as the number of households watching a television program, is often given with a margin of error. Inequalities involving absolute value are used to describe errors in polling, as well as errors of measurement in manufacturing, engineering, science, and other fields. In this section, you will learn to solve equations and inequalities containing absolute value. With these skills, you will be able to analyze the percentage of households that watched the final episode of $M*A*S*H$.

Equations Involving Absolute Value We have seen that the absolute value of a, denoted $|a|$, is the distance from 0 to a on a number line. Now consider **absolute value equations**, such as

$$|x| = 2.$$

This means that we must determine real numbers whose distance from the origin on a number line is 2. Figure 9.6 shows that there are two numbers such that $|x| = 2$, namely, 2 and -2. We write $x = 2$ or $x = -2$. This observation can be generalized as follows:

1 Solve equations involving absolute value.

FIGURE 9.6 If $|x| = 2$, then $x = 2$ or $x = -2$.

> **REWRITING AN ABSOLUTE VALUE EQUATION WITHOUT ABSOLUTE VALUE BARS** If c is a positive real number and X represents any algebraic expression, then $|X| = c$ is equivalent to $X = c$ or $X = -c$.

EXAMPLE 1 Solving an Equation Involving Absolute Value

Solve: $|2x - 3| = 11$.

SOLUTION

$|2x - 3| = 11$	This is the given equation.
$2x - 3 = 11$ or $2x - 3 = -11$	Rewrite the equation without absolute value bars.
$2x = 14 \qquad\qquad 2x = -8$	Add 3 to both sides of each equation.
$x = 7 \qquad\qquad\quad x = -4$	Divide both sides of each equation by 2.

You can use a graphing utility to verify the solution set of an absolute value equation. Consider, for example,

$$|2x - 3| = 11.$$

Graph $y_1 = |2x - 3|$ and $y_2 = 11$. The graphs are shown in a $[-10, 10, 1]$ by $[-1, 15, 1]$ viewing rectangle. The x-coordinates of the intersection points are -4 and 7, verifying that $\{-4, 7\}$ is the solution set.

$[-10, 10, 1]$ by $[-1, 15, 1]$

Check 7:	**Check -4:**	
$|2x - 3| = 11$	$|2x - 3| = 11$	This is the original equation.
$|2(7) - 3| \stackrel{?}{=} 11$	$|2(-4) - 3| \stackrel{?}{=} 11$	Substitute the proposed solutions.
$|14 - 3| \stackrel{?}{=} 11$	$|-8 - 3| \stackrel{?}{=} 11$	Perform operations inside the absolute value bars.
$|11| \stackrel{?}{=} 11$	$|-11| \stackrel{?}{=} 11$	
$11 = 11$, true	$11 = 11$, true	These true statements indicate that 7 and -4 are solutions.

The solutions are -4 and 7. We can also say that the solution set is $\{-4, 7\}$. ∎

✔ **CHECK POINT 1** Solve: $|2x - 1| = 5$. -2 and 3, or $\{-2, 3\}$

The absolute value of a number is never negative. Thus, if X is an algebraic expression and c is a negative number, then $|X| = c$ has no solution. For example, the equation $|3x - 6| = -2$ has no solution because $|3x - 6|$ cannot be negative. The solution set is $\varnothing$, the empty set.

The absolute value of 0 is 0. Thus, if X is an algebraic expression and $|X| = 0$, the solution is found by solving $X = 0$. For example, the solution of $|x - 2| = 0$ is obtained by solving $x - 2 = 0$. The solution is 2 and the solution set is $\{2\}$.

To solve some absolute value equations, it is necessary to first isolate the expression containing the absolute value symbols. For example, consider the equation

$$3|2x - 3| - 8 = 25.$$

> We need to isolate $|2x-3|$.

How can we isolate $|2x - 3|$? Add 8 to both sides of the equation and then divide both sides by 3.

$3|2x - 3| - 8 = 25$	This is the given equation.
$3|2x - 3| = 33$	Add 8 to both sides.
$|2x - 3| = 11$	Divide both sides by 3.

This results in the equation we solved in Example 1.

Some equations have two absolute value expressions, such as

$$|3x - 1| = |x + 5|.$$

These absolute value expressions are equal when the expressions inside the absolute value bars are equal to or opposites of each other.

> **REWRITING AN ABSOLUTE VALUE EQUATION WITH TWO ABSOLUTE VALUES WITHOUT ABSOLUTE VALUE BARS** If $|X_1| = |X_2|$, then $X_1 = X_2$ or $X_1 = -X_2$.

EXAMPLE 2 Solving an Absolute Value Equation with Two Absolute Values

Solve: $|3x - 1| = |x + 5|.$

SOLUTION We rewrite the equation without absolute value bars.

$$|X_1| = |X_2| \quad \text{means} \quad X_1 = X_2 \quad \text{or} \quad X_1 = -X_2$$

$$|3x - 1| = |x + 5| \quad \text{means} \quad 3x - 1 = x + 5 \quad \text{or} \quad 3x - 1 = -(x + 5).$$

We now solve the two equations that do not contain absolute value bars.

$$\begin{array}{ll}
3x - 1 = x + 5 & \text{or} \quad 3x - 1 = -(x + 5) \\
2x - 1 = 5 & \qquad 3x - 1 = -x - 5 \\
2x = 6 & \qquad 4x - 1 = -5 \\
x = 3 & \qquad 4x = -4 \\
& \qquad x = -1
\end{array}$$

Take a moment to complete the solution process by checking the two proposed solutions in the original equation. The solutions are -1 and 3, and the solution set is $\{-1, 3\}$. ■

✔ **CHECK POINT 2** Solve: $|2x - 7| = |x + 3|.$ $\frac{4}{3}$ and 10, or $\left\{\frac{4}{3}, 10\right\}$

Inequalities Involving Absolute Value Absolute value can also arise in inequalities. Consider, for example,

$$|x| < 2.$$

This means that the distance of x from 0 is *less than* 2, as shown in Figure 9.7. The interval shows values of x that lie less than 2 units from 0. Thus, x can lie between -2 and 2. That is, x is greater than -2 and less than 2: $-2 < x < 2$.

2 Solve inequalities involving absolute value.

−4 −3 −2 −1 0 1 2 3 4

FIGURE 9.7 $|x| < 2$, so $-2 < x < 2$.

FIGURE 9.8 $|x| > 2$, so $x < -2$ or $x > 2$.

Some absolute value inequalities use the "greater than" symbol. For example, $|x| > 2$ means that the distance of x from 0 is *greater than* 2, as shown in Figure 9.8. Thus, x can be less than -2 *or* greater than 2: $x < -2$ or $x > 2$.

These observations suggest the following principles for solving inequalities with absolute value.

SOLVING AN ABSOLUTE VALUE INEQUALITY If X is an algebraic expression and c is a positive number,

  1. The solutions of $|X| < c$ are the numbers that satisfy $-c < X < c$.

  2. The solutions of $|X| > c$ are the numbers that satisfy $X < -c$ or $X > c$.

These rules are valid if $<$ is replaced by $\leq$ and $>$ is replaced by $\geq$.

EXAMPLE 3   Solving an Absolute Value Inequality with $<$

Solve and graph the solution set on a number line: $|x - 4| < 3$.

SOLUTION   We rewrite the inequality without absolute value bars.

$$|X| \; < \; c \;\; \text{means} \;\; -c \; < \; X \; < \; c.$$

$$|x - 4| < 3 \quad \text{means} \quad -3 < x - 4 < 3.$$

We solve the compound inequality by adding 4 to all three parts.

$$-3 < x - 4 < 3$$
$$-3 + 4 < x - 4 + 4 < 3 + 4$$
$$1 < x < 7$$

The solution set is all real numbers greater than 1 and less than 7, denoted by $\{x | 1 < x < 7\}$ or $(1, 7)$. The graph of the solution set is shown as follows:

FIGURE 9.9 The solution set of $|x - 4| < 3$ is $(1, 7)$.

We can use the rectangular coordinate system to visualize the solution set of

$$|x - 4| < 3.$$

Figure 9.9 shows the graphs of $f(x) = |x - 4|$ and $g(x) = 3$. The solution set of $|x - 4| < 3$ consists of all values of x for which the blue graph of f lies below the red graph of g. These x-values make up the interval $(1, 7)$, which is the solution set.

✓ CHECK POINT 3 Solve and graph the solution set on a number line: $|x - 2| < 5$.
$\{x | -3 < x < 7\}$ or $(-3, 7)$; See graphing answer section.

EXAMPLE 4   Solving an Absolute Value Inequality with $\geq$

Solve and graph the solution set on a number line: $|2x + 3| \geq 5$.

SOLUTION   We rewrite the inequality without absolute value bars.

$$|X| \; \geq \; c \quad \text{means} \quad X \leq -c \quad \text{or} \quad X \geq c.$$

$$|2x + 3| \geq 5 \quad \text{means} \quad 2x + 3 \leq -5 \quad \text{or} \quad 2x + 3 \geq 5.$$

We solve this compound inequality by solving each of these inequalities separately. Then we take the union of their solution sets.

$$2x + 3 \leq -5 \quad \text{or} \quad 2x + 3 \geq 5 \qquad \text{These are the inequalities without absolute value bars.}$$

$$2x \leq -8 \qquad\qquad 2x \geq 2 \qquad \text{Subtract 3 from both sides.}$$

$$x \leq -4 \qquad\qquad x \geq 1 \qquad \text{Divide both sides by 2.}$$

The solution set consists of all numbers that are less than or equal to -4 or greater than or equal to 1. The solution set is $\{x | x \leq -4 \text{ or } x \geq 1\}$, or, in interval notation, $(-\infty, -4] \cup [1, \infty)$. The graph of the solution set is shown as follows:

$$\xleftarrow{\quad} \begin{array}{ccccccccccc} & & &] & & & & & [& & \\ -7 & -6 & -5 & -4 & -3 & -2 & -1 & 0 & 1 & 2 & 3 \end{array} \xrightarrow{\quad} \qquad ■$$

✔ **CHECK POINT 4** Solve and graph the solution set on a number line: $|2x - 5| \geq 3$. $\{x | x \leq 1 \text{ or } x \geq 4\}$ or $(-\infty, 1] \cup [4, \infty)$; See graphing answer section.

3 Recognize absolute value inequalities with no solution or all real numbers as solutions.

Absolute Value Inequalities with Unusual Solution Sets If c is a positive number, the solutions of $|X| < c$ comprise a single interval. By contrast, the solutions of $|X| > c$ make up two intervals.

The solutions of $|X| < c$ comprise the single interval $-c < X < c$.

The solutions of $|X| > c$ comprise two intervals: $X < -c$ or $X > c$.

Now let's see what happens to these inequalities if c is a negative number. Consider, for example, $|x| < -2$. Because $|x|$ always has a value that is greater than or equal to 0, there is no number whose absolute value is less than -2. The inequality $|x| < -2$ has no solution. The solution set is $\varnothing$.

Now consider the inequality $|x| > -2$. Because $|x|$ is never negative, all numbers have an absolute value that is greater than -2. All real numbers satisfy the inequality $|x| > -2$. The solution set is $(-\infty, \infty)$.

> **ABSOLUTE VALUE INEQUALITIES WITH UNUSUAL SOLUTION SETS**   If X is an algebraic expression and c is a negative number,
>
> **1.** the inequality $|X| < c$ has no solution.
>
> **2.** the inequality $|X| > c$ is true for all real numbers for which X is defined.

Table 9.2

What Is Bad about Being a Kid?

Kids Say

Getting bossed around	17%
School, homework	15%
Can't do everything I want	11%
Chores	9%
Being grounded	9%

Source: Penn, Schoen, and Berland using 1172 interviews with children ages 6 to 14 from May 14 to June 1, 1999, Margin of error: ±2.9%

Note the margin of error.

Applications   When you were between the ages of 6 and 14, how would you have responded to this question:

What is bad about being a kid?

In a random sample of 1172 children ages 6 to 14, 17% of the children responded, "Getting bossed around." The problem is that this is a single random sample. Do 17% of kids in the entire population of children ages 6 to 14 think that getting bossed around is a bad thing?

If you look at the results of a poll like the one in Table 9.2, you will observe that a **margin of error** is reported. The margin of error is ±2.9%. This means that the actual percentage of children who feel getting bossed around is a bad thing is at most 2.9% greater than or less than 17%. If x represents the percentage of children in the population who think that getting bossed around is a bad thing, then the poll's margin of error can be expressed as an absolute value inequality:

$$|x - 17| \leq 2.9.$$

4 Solve problems using absolute value inequalities.

EXAMPLE 5 Analyzing a Poll's Margin of Error

The inequality

$$|x - 9| \le 2.9$$

describes the percentage of children in the population who think that being grounded is a bad thing about being a kid. (See Table 9.2.) Solve the inequality and interpret the solution.

SOLUTION We rewrite the inequality without absolute value bars.

$$|X| \le c \quad \text{means} \quad -c \le X \le c.$$

$$|x - 9| \le 2.9 \quad \text{means} \quad -2.9 \le x - 9 \le 2.9.$$

We solve the compound inequality by adding 9 to all three parts.

$$-2.9 \le x - 9 \le 2.9$$
$$-2.9 + 9 \le x - 9 + 9 \le 2.9 + 9$$
$$6.1 \le x \le 11.9$$

The percentage of children in the population who think that being grounded is a bad thing is somewhere between a low of 6.1% and a high of 11.9%. Notice that these percents are always within 2.9% above and below the given 9%, and that 2.9% is the poll's margin of error. ■

 CHECK POINT 5 Solve the inequality:

$$|x - 11| \le 2.9.$$

Interpret the solution in terms of the information in Table 9.2 on page 605.

$\{x | 8.1 \le x \le 13.9\}$ or $[8.1, 13.9]$; The percentage of children in the population who think that not being able to do everything they want is a bad thing is between a low of 8.1% and a high of 13.9%.

9.3 EXERCISE SET

 Student Solutions Manual CD/Video PH Math/Tutor Center MathXL Tutorials on CD   MathXL®   MyMathLab   Interactmath.com

Practice Exercises

In Exercises 1–38, find the solution set for each equation.

1. $|x| = 8$ $\{-8, 8\}$
2. $|x| = 6$ $\{-6, 6\}$

3. $|x - 2| = 7$ $\{-5, 9\}$
4. $|x + 1| = 5$ $\{-6, 4\}$

5. $|2x - 1| = 7$ $\{-3, 4\}$
6. $|2x - 3| = 11$ $\{-4, 7\}$

7. $\left|\dfrac{4x - 2}{3}\right| = 2$ $\{-1, 2\}$
8. $\left|\dfrac{3x - 1}{5}\right| = 1$ $\left\{-\dfrac{4}{3}, 2\right\}$

9. $|x| = -8$ $\varnothing$
10. $|x| = -6$ $\varnothing$

11. $|x + 3| = 0$ $\{-3\}$
12. $|x + 2| = 0$ $\{-2\}$

13. $2|y + 6| = 10$ $\{-11, -1\}$
14. $3|y + 5| = 12$ $\{-9, -1\}$

15. $3|2x - 1| = 21$ $\{-3, 4\}$
16. $2|3x - 2| = 14$ $\left\{-\dfrac{5}{3}, 3\right\}$

17. $|6y - 2| + 4 = 32$ $\left\{-\dfrac{13}{3}, 5\right\}$

18. $|3y - 1| + 10 = 25$ $\left\{-\dfrac{14}{3}, \dfrac{16}{3}\right\}$

19. $7|5x| + 2 = 16$ $\left\{-\dfrac{2}{5}, \dfrac{2}{5}\right\}$
20. $7|3x| + 2 = 16$ $\left\{-\dfrac{2}{3}, \dfrac{2}{3}\right\}$

21. $|x + 1| + 5 = 3$ $\varnothing$
22. $|x + 1| + 6 = 2$ $\varnothing$

23. $|4y + 1| + 10 = 4$ $\varnothing$
24. $|3y - 2| + 8 = 1$ $\varnothing$

25. $|2x - 1| + 3 = 3$ $\left\{\dfrac{1}{2}\right\}$
26. $|3x - 2| + 4 = 4$ $\left\{\dfrac{2}{3}\right\}$

27. $|5x - 8| = |3x + 2|$ $\left\{\dfrac{3}{4}, 5\right\}$

28. $|4x - 9| = |2x + 1|$ $\left\{\dfrac{4}{3}, 5\right\}$

29. $|2x - 4| = |x - 1|$ $\left\{\dfrac{5}{3}, 3\right\}$
30. $|6x| = |3x - 9|$ $\{-3, 1\}$

31. $|2x - 5| = |2x + 5|$ $\{0\}$ **32.** $|3x - 5| = |3x + 5|$ $\{0\}$

33. $|x - 3| = |5 - x|$ $\{4\}$ **34.** $|x - 3| = |6 - x|$ $\left\{\frac{9}{2}\right\}$

35. $|2y - 6| = |10 - 2y|$ $\{4\}$ **36.** $|4y + 3| = |4y + 5|$ $\{-1\}$

37. $\left|\dfrac{2x}{3} - 2\right| = \left|\dfrac{x}{3} + 3\right|$ $\{-1, 15\}$

38. $\left|\dfrac{x}{2} - 2\right| = \left|x - \dfrac{1}{2}\right|$ $\left\{-3, \frac{5}{3}\right\}$

In Exercises 39–70, solve and graph the solution set on a number line.
39–70. See graphing answer section.

39. $|x| < 3$ $\{x | -3 < x < 3\}$ or $(-3, 3)$

40. $|x| < 5$ $\{x | -5 < x < 5\}$ or $(-5, 5)$

41. $|x - 2| < 1$ $\{x | 1 < x < 3\}$ or $(1, 3)$

42. $|x - 1| < 5$ $\{x | -4 < x < 6\}$ or $(-4, 6)$

43. $|x + 2| \le 1$ $\{x | -3 \le x \le -1\}$ or $[-3, -1]$

44. $|x + 1| \le 5$ $\{x | -6 \le x \le 4\}$ or $[-6, 4]$

45. $|2x - 6| < 8$ $\{x | -1 < x < 7\}$ or $(-1, 7)$

46. $|3x + 5| < 17$ $\{x | -\frac{22}{3} < x < 4\}$ or $\left(-\frac{22}{3}, 4\right)$

47. $|x| > 3$ $\{x | x < -3 \text{ or } x > 3\}$ or $(-\infty, -3) \cup (3, \infty)$

48. $|x| > 5$ $\{x | x < -5 \text{ or } x > 5\}$ or $(-\infty, -5) \cup (5, \infty)$

49. $|x + 3| > 1$ $\{x | x < -4 \text{ or } x > -2\}$ or $(-\infty, -4) \cup (-2, \infty)$

50. $|x - 2| > 5$ $\{x | x < -3 \text{ or } x > 7\}$ or $(-\infty, -3) \cup (7, \infty)$

51. $|x - 4| \ge 2$ $\{x | x \le 2 \text{ or } x \ge 6\}$ or $(-\infty, 2] \cup [6, \infty)$

52. $|x - 3| \ge 4$ $\{x | x \le -1 \text{ or } x \ge 7\}$ or $(-\infty, -1] \cup [7, \infty)$

53. $|3x - 8| > 7$ $\left\{x | x < \frac{1}{3} \text{ or } x > 5\right\}$ or $\left(-\infty, \frac{1}{3}\right) \cup (5, \infty)$

54. $|5x - 2| > 13$ $\left\{x | x < -\frac{11}{5} \text{ or } x > 3\right\}$ or $\left(-\infty, -\frac{11}{5}\right) \cup (3, \infty)$

55. $|2(x - 1) + 4| \le 8$ $\{x | -5 \le x \le 3\}$ or $[-5, 3]$

56. $|3(x - 1) + 2| \le 20$ $\left\{x | -\frac{19}{3} \le x \le 7\right\}$ or $\left[-\frac{19}{3}, 7\right]$

57. $\left|\dfrac{2x + 6}{3}\right| < 2$ $\{x | -6 < x < 0\}$ or $(-6, 0)$

58. $\left|\dfrac{3x - 3}{4}\right| < 6$ $\{x | -7 < x < 9\}$ or $(-7, 9)$

59. $\left|\dfrac{2x + 2}{4}\right| \ge 2$ $\{x | x \le -5 \text{ or } x \ge 3\}$ or $(-\infty, -5] \cup [3, \infty)$

60. $\left|\dfrac{3x - 3}{9}\right| \ge 1$ $\{x | x \le -2 \text{ or } x \ge 4\}$ or $(-\infty, -2] \cup [4, \infty)$

61. $\left|3 - \dfrac{2x}{3}\right| > 5$ $\{x | x < -3 \text{ or } x > 12\}$ or $(-\infty, -3) \cup (12, \infty)$

62. $\left|3 - \dfrac{3x}{4}\right| > 9$ $\{x | x < -8 \text{ or } x > 16\}$ or $(-\infty, -8) \cup (16, \infty)$

63. $|x - 2| < -1$ $\varnothing$ **64.** $|x - 3| < -2$ $\varnothing$

65. $|x + 6| > -10$ $\{x | x \text{ is a real number}\}$ or $(-\infty, \infty)$

66. $|x + 4| > -12$ $\{x | x \text{ is a real number}\}$ or $(-\infty, \infty)$

67. $|x + 2| + 9 \le 16$ $\{x | -9 \le x \le 5\}$ or $[-9, 5]$

68. $|x - 2| + 4 \le 5$ $\{x | 1 \le x \le 3\}$ or $[1, 3]$

69. $2|2x - 3| + 10 > 12$
$\{x | x < 1 \text{ or } x > 2\}$ or $(-\infty, 1) \cup (2, \infty)$

70. $3|2x - 1| + 2 > 8$
$\left\{x | x < -\frac{1}{2} \text{ or } x > \frac{3}{2}\right\}$ or $\left(-\infty, -\frac{1}{2}\right) \cup \left(\frac{3}{2}, \infty\right)$

71. Let $f(x) = |5 - 4x|$. Find all values of x for which $f(x) = 11$. $-\frac{3}{2}$ and 4

72. Let $f(x) = |2 - 3x|$. Find all values of x for which $f(x) = 13$. $-\frac{11}{3}$ and 5

73. Let $f(x) = |3 - x|$ and $g(x) = |3x + 11|$. Find all values of x for which $f(x) = g(x)$. -7 and -2

74. Let $f(x) = |3x + 1|$ and $g(x) = |6x - 2|$. Find all values of x for which $f(x) = g(x)$. $\frac{1}{9}$ and 1

75. Let $g(x) = |-1 + 3(x + 1)|$. Find all values of x for which $g(x) \le 5$. $\left\{x | -\frac{7}{3} \le x \le 1\right\}$ or $\left[-\frac{7}{3}, 1\right]$

76. Let $g(x) = |-3 + 4(x + 1)|$. Find all values of x for which $g(x) \le 3$. $\left\{x | -1 \le x \le \frac{1}{2}\right\}$ or $\left[-1, \frac{1}{2}\right]$

77. Let $h(x) = |2x - 3| + 1$. Find all values of x for which $h(x) > 6$. $\{x | x < -1 \text{ or } x > 4\}$ or $(-\infty, -1) \cup (4, \infty)$

78. Let $h(x) = |2x - 4| - 6$. Find all values of x for which $h(x) > 18$. $\{x | x < -10 \text{ or } x > 14\}$ or $(-\infty, -10) \cup (14, \infty)$

Practice Plus

79. When 3 times a number is subtracted from 4, the absolute value of the difference is at least 5. Use interval notation to express the set of all numbers that satisfy this condition. $\left(-\infty, -\frac{1}{3}\right] \cup [3, \infty)$

80. When 4 times a number is subtracted from 5, the absolute value of the difference is at most 13. Use interval notation to express the set of all numbers that satisfy this condition. $\left[-2, \frac{9}{2}\right]$

In Exercises 81–82, solve each inequality. Assume that $a > 0$ and $c > 0$. Use set-builder notation to express each solution set.

81. $|ax + b| < c$ $\left\{x | \dfrac{-c - b}{a} < x < \dfrac{c - b}{a}\right\}$

82. $|ax + b| \ge c$ $\left\{x | x \le \dfrac{-c - b}{a} \text{ or } x \ge \dfrac{c - b}{a}\right\}$

In Exercises 83–84, use the graph of $f(x) = |4 - x|$ to solve each equation or inequality.

83. $|4 - x| = 1$ 3 and 5

84. $|4 - x| < 5$ $\{x | -1 < x < 9\}$ or $(-1, 9)$

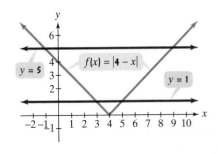

In Exercises 85–86, use the table to solve each inequality.

85. $|2x + 1| \leq 3$ $\{x|-2 \leq x \leq 1\}$ or $[-2, 1]$

86. $|2x + 1| \geq 3$ $\{x|x \leq -2$ or $x \geq 1\}$ or $(-\infty, -2] \cup [1, \infty)$

Application Exercises

The three television programs viewed by the greatest percentage of U.S. households in the twentieth century are shown in the table. The data are from a random survey of 4000 TV households by Nielsen Media Research. In Exercises 87–88, let x represent the actual viewing percentage in the U.S. population.

Program	Viewing Percentage in Survey
TV Programs with the Greatest U.S. Audience Viewing Percentage of the Twentieth Century	
1. "M*A*S*H" Feb. 28, 1983	60.2%
2. "Dallas" Nov. 21, 1980	53.3%
3. "Roots" Part 8 Jan. 30, 1977	51.1%

Source: Nielsen Media Research

87. Solve the inequality: $|x - 60.2| \leq 1.6$. Interpret the solution in terms of the information in the table. What is the margin of error? $\{x|58.6 \leq x \leq 61.8\}$ or $[58.6, 61.8]$; The percentage of the U.S. population that watched M*A*S*H is between a low of 58.6% and a high of 61.8%. The margin of error is 1.6%.

88. Solve the inequality: $|x - 51.1| \leq 1.6$. Interpret the solution in terms of the information in the table. What is the margin of error? $\{x|49.5 \leq x \leq 52.7\}$ or $[49.5, 52.7]$; The percentage of the U.S. population that watched "Roots" is between a low of 49.5% and a high of 52.7%. The margin of error is 1.6%.

89. The inequality $|T - 57| \leq 7$ describes the range of monthly average temperature, T, in degrees Fahrenheit, for San Francisco, California. Solve the inequality and interpret the solution. $\{T|50 \leq T \leq 64\}$ or $[50, 64]$; The monthly average temperature for San Francisco, California is between a low of 50°F and a high of 64°F.

90. The inequality $|T - 50| \leq 22$ describes the range of monthly average temperature, T, in degrees Fahrenheit, for Albany, New York. Solve the inequality and interpret the solution. $\{T|28 \leq T \leq 72\}$ or $[28, 72]$; The monthly average temperature for Albany, New York is between a low of 28°F and a high of 72°F.

The specifications for machine parts are given with tolerance limits that describe a range of measurements for which the part is acceptable. In Exercises 91–92, x represents the length of a machine part, in centimeters. The tolerance limit is 0.01 centimeter.

91. Solve: $|x - 8.6| \leq 0.01$. If the length of the machine part is supposed to be 8.6 centimeters, interpret the solution. $\{x|8.59 \leq x \leq 8.61\}$ or $[8.59, 8.61]$; A machine part that is supposed to be 8.6 centimeters is acceptable between a low of 8.59 and a high of 8.61 centimeters.

92. Solve: $|x - 9.4| \leq 0.01$. If the length of the machine part is supposed to be 9.4 centimeters, interpret the solution. $\{x|9.39 \leq x \leq 9.41\}$ or $[9.39, 9.41]$; A machine part that is supposed to be 9.4 centimeters is acceptable between a low of 9.39 and a high of 9.41 centimeters.

93. If a coin is tossed 100 times, we would expect approximately 50 of the outcomes to be heads. It can be demonstrated that a coin is unfair if h, the number of outcomes that result in heads, satisfies $\left|\dfrac{h - 50}{5}\right| \geq 1.645$. Describe the number of outcomes that determine an unfair coin that is tossed 100 times.   If the number of outcomes that result in heads is 41 or less or 59 or more, then the coin is unfair.

Writing in Mathematics

94. Explain how to solve an equation containing one absolute value expression.   **94–101.** Answers will vary.

95. Explain why the procedure that you described in Exercise 94 does not apply to the equation $|x - 5| = -3$. What is the solution set of this equation?

96. Describe how to solve an absolute value equation with two absolute values.

97. Describe how to solve an absolute value inequality involving the symbol $<$. Give an example.

98. Explain why the procedure that you described in Exercise 97 does not apply to the inequality $|x - 5| < -3$. What is the solution set of this inequality?

99. Describe how to solve an absolute value inequality involving the symbol $>$. Give an example.

100. Explain why the procedure that you described in Exercise 99 does not apply to the inequality $|x - 5| > -3$. What is the solution set of this inequality?

101. The final episode of $M*A*S*H$ was viewed by more than 58% of U.S. television households. Is it likely that a popular television series in the twenty-first century will achieve a 58% market share? Explain your answer.

Technology Exercises

In Exercises 102–104, solve each equation using a graphing utility. Graph each side separately in the same viewing rectangle. The solutions are the x-coordinates of the intersection points.

102. $|x + 1| = 5$ -6 and 4 **102–110.** See graphing answer section.

103. $|3(x + 4)| = 12$ -8 and 0

104. $|2x - 3| = |9 - 4x|$ 2 and 3

In Exercises 105–107, solve each inequality using a graphing utility. Graph each side separately in the same viewing rectangle. The solution set consists of all values of x for which the the graph of the left side lies below the graph of the right side.

105. $|2x + 3| < 5$ $\{x | -4 < x < 1\}$ or $(-4, 1)$

106. $\left|\dfrac{2x - 1}{3}\right| < \dfrac{5}{3}$ $\{x | -2 < x < 3\}$ or $(-2, 3)$

107. $|x + 4| < -1$ $\varnothing$

In Exercises 108–110, solve each inequality using a graphing utility. Graph each side separately in the same viewing rectangle. The solution set consists of all values of x for which the graph of the left side lies above the graph of the right side.

108. $|2x - 1| > 7$ $\{x | x < -3 \text{ or } x > 4\}$ or $(-\infty, -3) \cup (4, \infty)$

109. $|0.1x - 0.4| + 0.4 > 0.6$ $\{x | x < 2 \text{ or } x > 6\}$ or $(-\infty, 2) \cup (6, \infty)$

110. $|x + 4| > -1$ $\{x | x \text{ is a real number}\}$ or $(-\infty, \infty)$

111. Use a graphing utility to verify the solution sets for any five equations or inequalities that you solved by hand in Exercises 1–70.   Answers will vary.

Critical Thinking Exercises

112. Which one of the following is true?   **d**

    a. All absolute value equations have two solutions.

    b. The equation $|x| = -6$ is equivalent to $x = 6$ or $x = -6$.

    c. We can rewrite the inequality $x > 5$ or $x < -5$ more compactly as $-5 < x < 5$.

    d. Absolute value inequalities in the form $|ax + b| < c$ translate into *and* compound statements, which may be written as three-part inequalities.

113. Write an absolute value inequality for which the interval shown is the solution.

Solutions lie within 3 units of 4.

    a. $|x - 4| < 3$ **b.** $|x - 4| \geq 3$

114. The percentage, p, of defective products manufactured by a company is given as $|p - 0.3\%| \leq 0.2\%$. If 100,000 products are manufactured and the company offers a \$5 refund for each defective product, describe the company's cost for refunds.   The cost for refunds will be between a low of \$500 and a high of \$2500.

115. Solve: $|2x + 5| = 3x + 4$. 1

Review Exercises

116–118. See graphing answer section.

In Exercises 116–118, graph each linear function.

116. $3x - 5y = 15$ (Section 3.2, Example 5)

117. $f(x) = -\dfrac{2}{3}x$ or $y = -\dfrac{2}{3}x$ (Section 3.4, Example 4)

118. $f(x) = -2$ or $y = -2$ (Section 3.2, Example 7)

✔ **MID-CHAPTER CHECK POINT**

CHAPTER **9**

What You Know: We reviewed how to solve linear inequalities, expressing solution sets in set-builder and interval notations. We know that it is necessary to reverse the sense of an inequality when multiplying or dividing both sides by a negative number. We solved compound inequalities with *and* by finding the intersection of solution sets and with *or* by finding the union of solution sets. Finally, we solved equations and inequalities involving absolute value by carefully rewriting the given equation or inequality without absolute value bars. For positive values of c, we wrote $|X| = c$ as $X = c$ or $X = -c$, we wrote $|X| < c$ as $-c < X < c$, and we wrote $|X| > c$ as $X < -c$ or $X > c$.

In Exercises 1–17, solve each inequality or equation.

1. $4 - 3x \geq 12 - x$ $\{x | x \leq -4\}; (-\infty, -4]$

2. $5 \leq 2x - 1 < 9$ $\{x | 3 \leq x < 5\}; [3, 5)$

3. $|4x - 7| = 5$ $\left\{\frac{1}{2}, 3\right\}$

4. $-10 - 3(2x + 1) > 8x + 1$ $\{x | x < -1\}; (-\infty, -1)$

5. $2x + 7 < -11$ or $-3x - 2 < 13$
$\{x | x < -9 \text{ or } x > -5\}; (-\infty, -9) \cup (-5, \infty)$

6. $|3x - 2| \le 4$ $\{x | -\frac{2}{3} \le x \le 2\}; \left[-\frac{2}{3}, 2\right]$

7. $|x + 5| = |5x - 8|$ $\left\{\frac{1}{2}, \frac{13}{4}\right\}$

8. $5 - 2x \ge 9$ and $5x + 3 > -17$
$\{x | -4 < x \le -2\}$ or $(-4, -2]$

9. $3x - 2 > -8$ or $2x + 1 < 9$
$\{x | x \text{ is a real number}\}; (-\infty, \infty)$

10. $\dfrac{x}{2} + 3 \le \dfrac{x}{3} + \dfrac{5}{2}$ $\{x | x \le -3\}$ or $(-\infty, -3]$

11. $\dfrac{2}{3}(6x - 9) + 4 > 5x + 1$ $\{x | x < -3\}$ or $(-\infty, -3)$

12. $|5x + 3| > 2$ $\left\{x \middle| x < -1 \text{ or } x > -\frac{1}{5}\right\}; (-\infty, -1) \cup \left(-\frac{1}{5}, \infty\right)$

13. $7 - \left|\dfrac{x}{2} + 2\right| \le 4$ $\{x | x \le -10 \text{ or } x \ge 2\}; (-\infty, -10] \cup [2, \infty)$

14. $\dfrac{x + 3}{4} < \dfrac{1}{3}$ $\left\{x \middle| x < -\frac{5}{3}\right\}; \left(-\infty, -\frac{5}{3}\right)$

15. $5x + 1 \ge 4x - 2$ and $2x - 3 > 5$
$\{x | x > 4\}; (4, \infty)$

16. $3 - |2x - 5| = -6$ $\{-2, 7\}$

17. $3 + |2x - 5| = -6$ $\varnothing$

18. A company that manufactures compact discs has fixed monthly overhead costs of $60,000. Each disc costs $0.18 to produce and sells for $0.30.
  a. Write the cost function, C, of producing x discs per month. $C(x) = 60,000 + 0.18x$
  b. Write the revenue function, R, from the sale of x discs per month. $R(x) = 0.30x$
  c. Write the profit function, P, from producing and selling x discs per month. $P(x) = 0.12x - 60,000$
  d. How many discs should be produced and sold each month for the company to have a profit of at least $30,000? at least 750,000 discs each month

19. A car rental agency rents a certain car for $40 per day with unlimited mileage or $24 per day plus $0.20 per mile. How far can a customer drive this car per day for the $24 option to cost no more than the unlimited mileage option? no more than 80 miles per day

20. To receive a B in a course, you must have an average of at least 80% but less than 90% on five exams. Your grades on the first four exams were 95%, 79%, 91%, and 86%. What range of grades on the fifth exam will result in a B for the course? [49%, 99%)

21. A retiree requires an annual income of at least $9000 from an investment paying 7.5% annual interest. How much should the retiree invest to achieve the desired return? at least $120,000

LINEAR INEQUALITIES IN TWO VARIABLES

Objectives

1 Graph a linear inequality in two variables.

2 Graph a system of linear inequalities.

3 Solve applied problems involving systems of inequalities.

This book was written in Point Reyes National Seashore, 40 miles north of San Francisco. The park consists of 75,000 acres with miles of pristine surf-washed beaches, forested ridges, and bays bordered by white cliffs.

Like your author, many people are kept inspired and energized surrounded by nature's unspoiled beauty. In this section, you will see how systems of inequalities model whether a region's natural beauty manifests itself in forests, grasslands, or deserts.

Linear Inequalities in Two Variables and Their Solutions   We have seen that the graphs of equations in the form $Ax + By = C$ are straight lines. If we change the $=$ sign to $>$, $<$, $\geq$, or $\leq$, we obtain a **linear inequality in two variables**. Some examples of linear inequalities in two variables are $x + y > 2$, $3x - 5y \leq 15$, and $2x - y < 4$.

A **solution of an inequality in two variables**, x and y, is an ordered pair of real numbers with the following property: When the x-coordinate is substituted for x and the y-coordinate is substituted for y in the inequality, we obtain a true statement. For example, $(3, 2)$ is a solution of the inequality $x + y > 1$. When 3 is substituted for x and 2 is substituted for y, we obtain the true statement $3 + 2 > 1$, or $5 > 1$. Because there are infinitely many pairs of numbers that have a sum greater than 1, the inequality $x + y > 1$ has infinitely many solutions. Each ordered-pair solution is said to **satisfy** the inequality. Thus, $(3, 2)$ satisfies the inequality $x + y > 1$.

The Graph of a Linear Inequality in Two Variables   We know that the graph of an equation in two variables is the set of all points whose coordinates satisfy the equation. Similarly, the **graph of an inequality in two variables** is the set of all points whose coordinates satisfy the inequality.

Let's use Figure 9.10 to get an idea of what the graph of a linear inequality in two variables looks like. Part of the figure shows the graph of the linear equation $x + y = 2$. The line divides the points in the rectangular coordinate system into three sets. First, there is the set of points along the line, satisfying $x + y = 2$. Next, there is the set of points in the green region above the line. Points in the green region satisfy the linear inequality $x + y > 2$. Finally, there is the set of points in the purple region below the line. Points in the purple region satisfy the linear inequality $x + y < 2$.

A **half-plane** is the set of all the points on one side of a line. In Figure 9.10, the green region is a half-plane. The purple region is also a half-plane. A half-plane is the graph of a linear inequality that involves $>$ or $<$. The graph of an inequality that involves $\geq$ or $\leq$ is a half-plane and a line. A solid line is used to show that a line is part of a graph. A dashed line is used to show that a line is not part of a graph.

1 Graph a linear inequality in two variables.

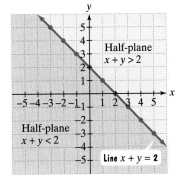

FIGURE 9.10

GRAPHING A LINEAR INEQUALITY IN TWO VARIABLES

1. Replace the inequality symbol with an equal sign and graph the corresponding linear equation. Draw a solid line if the original inequality contains a $\leq$ or $\geq$ symbol. Draw a dashed line if the original inequality contains a $<$ or $>$ symbol.

2. Choose a test point in one of the half-planes that is not on the line. Substitute the coordinates of the test point into the inequality.

3. If a true statement results, shade the half-plane containing this test point. If a false statement results, shade the half-plane not containing this test point.

EXAMPLE 1 Graphing a Linear Inequality in Two Variables

Graph: $2x - 3y \geq 6$.

SOLUTION

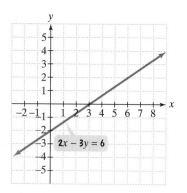

FIGURE 9.11 Preparing to graph $2x - 3y \geq 6$

Step 1. **Replace the inequality symbol by = and graph the linear equation.** We need to graph $2x - 3y = 6$. We can use intercepts to graph this line.

We set $y = 0$ to find the x-intercept:	We set $x = 0$ to find the y-intercept:
$2x - 3y = 6$	$2x - 3y = 6$
$2x - 3 \cdot 0 = 6$	$2 \cdot 0 - 3y = 6$
$2x = 6$	$-3y = 6$
$x = 3.$	$y = -2.$

The x-intercept is 3, so the line passes through $(3, 0)$. The y-intercept is -2, so the line passes through $(0, -2)$. Using the intercepts, the line is shown in Figure 9.11 as a solid line. This is because the inequality $2x - 3y \geq 6$ contains a $\geq$ symbol, in which equality is included.

Step 2. **Choose a test point in one of the half-planes that is not on the line. Substitute its coordinates into the inequality.** The line $2x - 3y = 6$ divides the plane into three parts—the line itself and two half-planes. The points in one half-plane satisfy $2x - 3y > 6$. The points in the other half-plane satisfy $2x - 3y < 6$. We need to find which half-plane belongs to the solution of $2x - 3y \geq 6$. To do so, we test a point from either half-plane. The origin, $(0, 0)$, is the easiest point to test.

$$2x - 3y \geq 6 \qquad \text{This is the given inequality.}$$
$$2 \cdot 0 - 3 \cdot 0 \overset{?}{\geq} 6 \qquad \text{Test } (0, 0) \text{ by substituting 0 for } x \text{ and 0 for } y.$$
$$0 - 0 \overset{?}{\geq} 6 \qquad \text{Multiply.}$$
$$0 \geq 6 \qquad \text{This statement is false.}$$

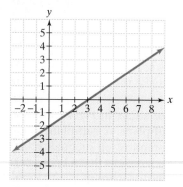

FIGURE 9.12 The graph of $2x - 3y \geq 6$

Step 3. **If a false statement results, shade the half-plane not containing the test point.** Because 0 is not greater than or equal to 6, the test point, $(0, 0)$, is not part of the solution set. Thus, the half-plane below the solid line $2x - 3y = 6$ is part of the solution set. The solution set is the line and the half-plane that does not contain the point $(0, 0)$, indicated by shading this half-plane. The graph is shown using green shading and a blue line in Figure 9.12. ∎

✔ **CHECK POINT 1** Graph: $4x - 2y \geq 8$. See graphing answer section.

When graphing a linear inequality, test a point that lies in one of the half-planes and *not on the line dividing the half-planes*. The test point $(0, 0)$ is convenient because it is easy to calculate when 0 is substituted for each variable. However, if $(0, 0)$ lies on the dividing line and not in a half-plane, a different test point must be selected.

EXAMPLE 2 Graphing a Linear Inequality in Two Variables

Graph: $y > -\dfrac{2}{3}x$.

SOLUTION

Step 1. **Replace the inequality symbol by = and graph the linear equation.** Because we are interested in graphing $y > -\frac{2}{3}x$, we begin by graphing $y = -\frac{2}{3}x$. We can use the slope and the y-intercept to graph this linear function.

The y-intercept is 0, so the line passes through $(0, 0)$. Using the y-intercept and the slope, the line is shown in Figure 9.13 as a dashed line. This is because the inequality $y > -\frac{2}{3}x$ contains a > symbol, in which equality is not included.

Step 2. **Choose a test point in one of the half-planes that is not on the line. Substitute its coordinates into the inequality.** We cannot use $(0, 0)$ as a test point because it lies on the line and not in a half-plane. Let's use $(1, 1)$, which lies in the half-plane above the line.

$$y > -\frac{2}{3}x \qquad \text{This is the given inequality.}$$

$$1 \overset{?}{>} -\frac{2}{3} \cdot 1 \qquad \text{Test } (1, 1) \text{ by substituting 1 for } x \text{ and 1 for } y.$$

$$1 > -\frac{2}{3} \qquad \text{This statement is true.}$$

Step 3. **If a true statement results, shade the half-plane containing the test point.** Because 1 is greater than $-\frac{2}{3}$, the test point $(1, 1)$ is part of the solution set. All the points on the same side of the line $y = -\frac{2}{3}x$ as the point $(1, 1)$ are members of the solution set. The solution set is the half-plane that contains the point $(1, 1)$, indicated by shading this half-plane. The graph is shown using green shading and a dashed blue line in Figure 9.13. ■

FIGURE 9.13 The graph of $y > -\frac{2}{3}x$

USING TECHNOLOGY

Most graphing utilities can graph inequalities in two variables with the $\boxed{\text{SHADE}}$ feature. The procedure varies by model, so consult your manual. For most graphing utilities, you must first solve for y if it is not already isolated. The figure shows the graph of $y > -\frac{2}{3}x$. Most displays do not distinguish between dashed and solid boundary lines.

 CHECK POINT 2 Graph: $y > -\frac{3}{4}x$.   *See graphing answer section.*

Graphing Linear Inequalities without Using Test Points You can graph inequalities in the form $y > mx + b$ or $y < mx + b$ without using test points. The inequality symbol indicates which half-plane to shade.

- If $y > mx + b$, shade the half-plane above the line $y = mx + b$.
- If $y < mx + b$, shade the half-plane below the line $y = mx + b$.

It is also not necessary to use test points when graphing inequalities involving half-planes on one side of a vertical or a horizontal line.

For the Vertical Line $x = a$:

- If $x > a$, shade the half-plane to the right of $x = a$.
- If $x < a$, shade the half-plane to the left of $x = a$.

For the Horizontal Line $y = b$:

- If $y > b$, shade the half-plane above $y = b$.
- If $y < b$, shade the half-plane below $y = b$.

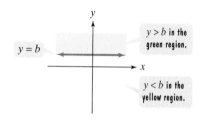

EXAMPLE 3 Graphing Inequalities without Using Test Points

Graph each inequality in a rectangular coordinate system:

a. $y \leq -3$ **b.** $x > 2$.

SOLUTION

a. $y \leq -3$

Graph $y = -3$, a horizontal line with y-intercept -3. The line is solid because equality is included in $y \leq -3$. Because of the less than part of $\leq$, shade the half-plane below the horizontal line.

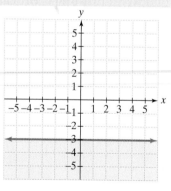

b. $x > 2$

Graph $x = 2$, a vertical line with x-intercept 2. The line is dashed because equality is not included in $x > 2$. Because of $>$, the greater than symbol, shade the half-plane to the right of the vertical line.

 CHECK POINT 3 Graph each inequality in a rectangular coordinate system:

a. $y > 1$ **b.** $x \leq -2$.   See graphing answer section.

2 Graph a system of linear inequalities.

Systems of Linear Inequalities   Just as two linear equations make up a system of linear equations, two or more linear inequalities make up a **system of linear inequalities**. Here is an example of a system of linear inequalities:

$$x - y < 1$$
$$2x + 3y \geq 12.$$

A **solution of a system of linear inequalities** in two variables is an ordered pair that satisfies each inequality in the system. The set of all such ordered pairs is the **solution set** of the system. Thus, to graph a system of inequalities in two variables, begin by graphing each individual inequality in the same rectangular coordinate system. Then find the region, if there is one, that is common to every graph in the system. This region of intersection gives a picture of the system's solution set.

EXAMPLE 4 Graphing a System of Linear Inequalities

Graph the solution set of the system:

$$x - y < 1$$

$$2x + 3y \geq 12.$$

SOLUTION Replacing each inequality symbol with an equal sign indicates that we need to graph $x - y = 1$ and $2x + 3y = 12$. We can use intercepts to graph these lines.

$x - y = 1$		$2x + 3y = 12$

Set $y = 0$ in each equation.

x-intercept: $x - 0 = 1$ x-intercept: $2x + 3 \cdot 0 = 12$
            $x = 1$ $2x = 12$
The line passes through $(1, 0)$. $x = 6$
                              The line passes through $(6, 0)$.

Set $x = 0$ in each equation.

y-intercept: $0 - y = 1$ y-intercept: $2 \cdot 0 + 3y = 12$
            $-y = 1$ $3y = 12$
             $y = -1$ $y = 4$
The line passes through $(0, -1)$ The line passes through $(0, 4)$.

Now we are ready to graph the solution set of the system of linear inequalities.

Graph $x - y < 1$. The blue line, $x - y = 1$, is dashed: Equality is not included in $x - y < 1$. Because $(0, 0)$ makes the inequality true $(0 - 0 < 1$, or $0 < 1$, is true), shade the half-plane containing $(0, 0)$ in yellow.

Add the graph of $2x + 3y \geq 12$. The red line, $2x + 3y = 12$, is solid: Equality is included in $2x + 3y \geq 12$. Because $(0, 0)$ makes the inequality false $(2 \cdot 0 + 3 \cdot 0 \geq 12$, or $0 \geq 12$, is false), shade the half-plane not containing $(0, 0)$ using green vertical shading.

The solution set of the system is graphed as the intersection (the overlap) of the two half-planes. This is the region in which the yellow shading and the green vertical shading overlap.

$x - y = 1$: passes through $(1, 0)$ and $(0, -1)$

The graph of $x - y < 1$

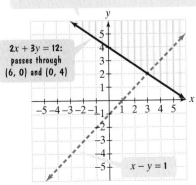

$2x + 3y = 12$: passes through $(6, 0)$ and $(0, 4)$

$x - y = 1$

Adding the graph of $2x + 3y \geq 12$

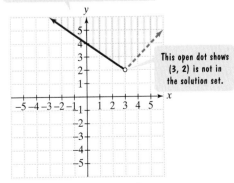

This open dot shows $(3, 2)$ is not in the solution set.

The graph of $x - y < 1$ and $2x + 3y \geq 12$

CHECK POINT 4 Graph the solution set of the system: See graphing answer section.

$$x - 3y < 6$$
$$2x + 3y \geq -6.$$

A system of inequalities has no solution if there are no points in the rectangular coordinate system that simultaneously satisfy each inequality in the system. For example, the system

$$2x + 3y \geq 6$$
$$2x + 3y \leq 0$$

whose separate graphs are shown in Figure 9.14 has no overlapping region. Thus, the system has no solution. The solution set is $\varnothing$, the empty set.

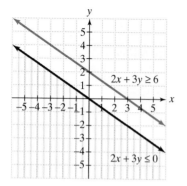

FIGURE 9.14 A system of inequalities with no solution

EXAMPLE 5 Graphing a System of Inequalities

Graph the solution set:

$$x - y < 2$$
$$-2 \leq x < 4$$
$$y < 3.$$

SOLUTION We begin by graphing $x - y < 2$, the first given inequality. The line $x - y = 2$ has an x-intercept of 2 and a y-intercept of -2. The test point $(0, 0)$ makes the inequality $x - y < 2$ true, and its graph is shown in Figure 9.15.

Now, let's consider the second given inequality, $-2 \leq x < 4$. Replacing the inequality symbols by $=$, we obtain $x = -2$ and $x = 4$, graphed as red vertical lines in Figure 9.16. The line of $x = 4$ is not included. Using $(0, 0)$ as a test point and substituting the x-coordinate, 0, into $-2 \leq x < 4$, we obtain the true statement $-2 \leq 0 < 4$. We therefore shade the region between the vertical lines. We must intersect this region with the yellow region in Figure 9.15. The resulting region is shown in yellow and green vertical shading in Figure 9.16.

Finally, let's consider the third given inequality, $y < 3$. Replacing the inequality symbol by $=$, we obtain $y = 3$, which graphs as a horizontal line. Because of the less than symbol in $y < 3$, the graph consists of the half-plane below the line $y = 3$. We must intersect this half-plane with the region in Figure 9.16. The resulting region is shown in yellow and green vertical shading in Figure 9.17. This region represents the graph of the solution set of the given system.

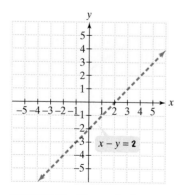

FIGURE 9.15 The graph of $x - y < 2$

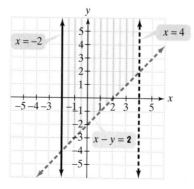

FIGURE 9.16 The graph of $x - y < 2$ and $-2 \leq x < 4$

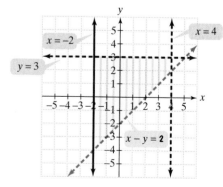

FIGURE 9.17 The graph of $x - y < 2$ and $-2 \leq x < 4$ and $y < 3$

✓ CHECK POINT **5** Graph the solution set:

$$x + y < 2$$
$$-2 \leq x < 1$$
$$y > -3. \qquad \text{See graphing answer section.}$$

3 Solve applied problems involving systems of inequalities.

Applications Temperature and precipitation affect whether or not trees and forests can grow. At certain levels of precipitation and temperature, only grasslands and deserts will exist. Figure 9.18 shows three kinds of regions —deserts, grasslands, and forests—that result from various ranges of temperature, T, and precipitation, P.

Systems of inequalities can be used to describe where forests, grasslands, and deserts occur. Because these regions occur when the average annual temperature, T, is 35°F or greater, each system contains the inequality $T \geq 35$.

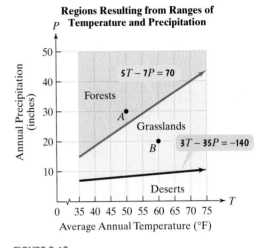

Regions Resulting from Ranges of Temperature and Precipitation

FIGURE 9.18

Source: A. Miller and J. Thompson, *Elements of Meteorology*

Forests occur if	**Grasslands occur if**	**Deserts occur if**
$T \geq 35$	$T \geq 35$	$T \geq 35$
$5T - 7P < 70.$	$5T - 7P \geq 70$	$3T - 35P > -140.$
	$3T - 35P \leq -140.$	

EXAMPLE 6 Forests and Systems of Inequalities

Show that point A in Figure 9.18 is a solution of the system of inequalities that describes where forests occur.

SOLUTION Point A has coordinates $(50, 30)$. This means that if a region has an average annual temperature of 50°F and an average annual precipitation of 30 inches, a forest occurs. We can show that $(50, 30)$ satisfies the system of inequalities for forests by substituting 50 for T and 30 for P in each inequality in the system.

$$
\begin{array}{ll}
T \geq 35 & 5T - 7P < 70 \\
50 \geq 35, \ \text{true} & 5 \cdot 50 - 7 \cdot 30 \stackrel{?}{<} 70 \\
& 250 - 210 \stackrel{?}{<} 70 \\
& 40 < 70, \ \text{true}
\end{array}
$$

The coordinates $(50, 30)$ make each inequality true. Thus, $(50, 30)$ satisfies the system for forests. ∎

 CHECK POINT **6** Show that point B in Figure 9.18 is a solution of the system of inequalities that describes where grasslands occur. Point $B = (60, 20)$; Using $T = 60$ and $P = 20$, each of the three inequalities for grasslands is true: $60 \geq 35$, true; $5(60) - 7(20) \geq 70$, true; $3(60) - 35(20) \leq -140$, true.

9.4 EXERCISE SET

Student Solutions Manual   CD/Video   PH Math/Tutor Center   MathXL Tutorials on CD   MathXL®   MyMathLab   Interactmath.com

Practice Exercises

In Exercises 1–22, graph each inequality.
1–46. See graphing answer section.

1. $x + y \geq 3$

2. $x + y \geq 2$

3. $x - y < 5$

4. $x - y < 6$

5. $x + 2y > 4$

6. $2x + y > 6$

7. $3x - y \leq 6$

8. $x - 3y \leq 6$

9. $\dfrac{x}{2} + \dfrac{y}{3} < 1$

10. $\dfrac{x}{4} + \dfrac{y}{2} < 1$

11. $y > \dfrac{1}{3}x$

12. $y > \dfrac{1}{4}x$

13. $y \leq 3x + 2$

14. $y \leq 2x - 1$

15. $y < -\dfrac{1}{4}x$

16. $y < -\dfrac{1}{3}x$

17. $x \leq 2$

18. $x \leq -4$

19. $y > -4$

20. $y > -2$

21. $y \geq 0$

22. $x \leq 0$

In Exercises 23–46 graph the solution set of each system of inequalities or indicate that the system has no solution.

23. $3x + 6y \leq 6$
$2x + y \leq 8$

24. $x - y \geq 4$
$x + y \leq 6$

25. $2x - 5y \leq 10$
$3x - 2y > 6$

26. $2x - y \leq 4$
$3x + 2y > -6$

27. $y > 2x - 3$
$y < -x + 6$

28. $y < -2x + 4$
$y < x - 4$

29. $x + 2y \leq 4$
$y \geq x - 3$

30. $x + y \leq 4$
$y \geq 2x - 4$

31. $x \leq 2$
$y \geq -1$

32. $x \leq 3$
$y \leq -1$

33. $-2 \leq x < 5$

34. $-2 < y \leq 5$

35. $x - y \leq 1$
$x \geq 2$

36. $4x - 5y \geq -20$
$x \geq -3$

37. $x + y > 4$
$x + y < -1$ $\varnothing$

38. $x + y > 3$
$x + y < -2$ $\varnothing$

39. $x + y > 4$
$x + y > -1$

40. $x + y > 3$
$x + y > -2$

41. $x - y \leq 2$
$x \geq -2$
$y \leq 3$

42. $3x + y \leq 6$
$x \geq -2$
$y \leq 4$

43. $x \geq 0$
$y \geq 0$
$2x + 5y \leq 10$
$3x + 4y \leq 12$

44. $x \geq 0$
$y \geq 0$
$2x + y \leq 4$
$2x - 3y \leq 6$

45. $3x + y \leq 6$
$2x - y \leq -1$
$x \geq -2$
$y \leq 4$

46. $2x + y \leq 6$
$x + y \geq 2$
$1 \leq x \leq 2$
$y \leq 3$

Practice Plus

In Exercises 47–48, write each sentence as a linear inequality in two variables. Then graph the inequality.
47–50. See graphing answer section.

47. The y-variable is at least 4 more than the product of -2 and the x-variable. $y \geq -2x + 4$

48. The y-variable is at least 2 more than the product of -3 and the x-variable. $y \geq -3x + 2$

In Exercises 49–50, write the given sentences as a system of linear inequalities in two variables. Then graph the system.

49. The sum of the x-variable and the y-variable is at most 4. The y-variable added to the product of 3 and the x-variable does not exceed 6. $x + y \leq 4$ and $3x + y \leq 6$

50. The sum of the x-variable and the y-variable is at most 3. The y-variable added to the product of 4 and the x-variable does not exceed 6. $x + y \leq 3$ and $4x + y \leq 6$

In Exercises 51–52, rewrite each inequality in the system without absolute value bars. Then graph the rewritten system in rectangular coordinates.   51–52. See graphing answer section.

51. $|x| \leq 2$ $-2 \leq x \leq 2$
$|y| \leq 3$ $-3 \leq y \leq 3$

52. $|x| \leq 1$ $-1 \leq x \leq 1$
$|y| \leq 2$ $-2 \leq y \leq 2$

*The graphs of solution sets of systems of inequalities involve finding the intersection of the solution sets of two or more inequalities. By contrast, in Exercises 53–54 you will be graphing the **union** of the solution sets of two inequalities.*   53–54. See graphing answer section.

53. Graph the union of $y > \dfrac{3}{2}x - 2$ and $y < 4$.

54. Graph the union of $x - y \geq -1$ and $5x - 2y \leq 10$.

Without graphing, in Exercises 55–58, determine if each system has no solution or infinitely many solutions.

55. $3x + y < 9$
$3x + y > 9$
no solution

56. $6x - y \leq 24$
$6x - y > 24$
no solution

57. $3x + y \leq 9$
$3x + y \geq 9$
infinitely many solutions

58. $6x - y \leq 24$
$6x - y \geq 24$
infinitely many solutions

Application Exercises

Maximum heart rate, H, in beats per minute is a function of age, a, modeled by the formula

$$H = 220 - a,$$

where $10 \le a \le 70$. The bar graph shows the target heart rate ranges for four types of exercise goals.

Target Heart Rate Ranges for Exercise Goals

Exercise Goal

Boost performance as a competitive athlete

Improve cardiovascular conditioning

Lose weight

Improve overall health and reduce risk of heart attack

40% 50% 60% 70% 80% 90% 100%
Percentage of Maximum Heart Rate

Source: Vitality

In Exercises 59–62, systems of inequalities will be used to model three of the target heart rate ranges shown in the bar graph. We begin with the target heart rate range for cardiovascular conditioning, modeled by the following system of inequalities:

$10 \le a \le 70$ — Heart rate ranges apply to ages 10 through 70, inclusive.

$H \ge 0.7(220 - a)$ — Target heart rate range is greater than or equal to 70% of maximum heart rate

$H \le 0.8(220 - a)$ — and less than or equal to 80% of maximum heart rate.

The graph of this system is shown in the figure. Use the graph to solve Exercises 59–60.

Target Heart Rate Range for Cardiovascular Conditioning

$a = 10$

$H = 0.8(220 - a)$

$a = 70$

$H = 0.7(220 - a)$

Heart Rate (beats per minute)

Age

59. a. What are the coordinates of point *A* and what does this mean in terms of age and heart rate? $A = (20, 150)$; A 20-year-old with a heart rate of 150 beats per minute is within the target range.

  b. Show that point *A* is a solution of the system of inequalities. $10 \le 20 \le 70$, true; $150 \ge 0.7(220 - 20)$, true; $150 \le 0.8(220 - 20)$, true

60. a. What are the coordinates of point *B* and what does this mean in terms of age and heart rate? $B = (40, 130)$; A 40-year-old with a heart rate of 130 beats per minute is within the target range.

  b. Show that point B is a solution of the system of inequalities. $10 \le 40 \le 70$, true; $130 \ge 0.7(220 - 40)$, true; $130 \le 0.8(220 - 40)$, true

61. Write a system of inequalities that models the target heart rate range for the goal of losing weight. $10 \le a \le 70$, $H \ge 0.6(220 - a)$, $H \le 0.7(220 - a)$

62. Write a system of inequalities that models the target heart rate range for improving overall health. $10 \le a \le 70$, $H \ge 0.5(220 - a)$, $H \le 0.6(220 - a)$

63. On your next vacation, you will divide lodging between large resorts and small inns. Let *x* represent the number of nights spent in large resorts. Let *y* represent the number of nights spent in small inns.

  a. Write a system of inequalities that models the following conditions:

    You want to stay at least 5 nights. At least one night should be spent at a large resort. Large resorts average $200 per night and small inns average $100 per night. Your budget permits no more than $700 for lodging. $y \ge 0, x + y \ge 5, x \ge 1, 200x + 100y \le 700$

  b. Graph the solution set of the system of inequalities in part (a).   See graphing answer section.

  c. Based on your graph in part (b), how many nights could you spend at a large resort and still stay within your budget?   2 nights

64. a. An elevator can hold no more than 2000 pounds. If children average 80 pounds and adults average 160 pounds, write a system of inequalities that models when the elevator holding *x* children and *y* adults is overloaded. $x \ge 0, y \ge 0, 80x + 160y > 2000$

  b. Graph the solution set of the system of inequalities in part (a).   See graphing answer section.

Writing in Mathematics

65. What is a linear inequality in two variables? Provide an example with your description.   **65–74.** Answers will vary.

66. How do you determine if an ordered pair is a solution of an inequality in two variables, *x* and *y*?

67. What is a half-plane?

68. What does a solid line mean in the graph of an inequality?

69. What does a dashed line mean in the graph of an inequality?

70. Explain how to graph $x - 2y < 4$.

71. What is a system of linear inequalities?

72. What is a solution of a system of linear inequalities?

73. Explain how to graph the solution set of a system of inequalities.

74. What does it mean if a system of linear inequalities has no solution?

Technology Exercises

Graphing utilities can be used to shade regions in the rectangular coordinate system, thereby graphing an inequality in two variables. Read the section of the user's manual for your graphing utility that describes how to shade a region. Then use your graphing utility to graph the inequalities in Exercises 75–78. **75–78.** See graphing answer section.

75. $y \leq 4x + 4$

76. $y \geq \dfrac{2}{3}x - 2$

77. $2x + y \leq 6$

78. $3x - 2y \geq 6$

79. Does your graphing utility have any limitations in terms of graphing inequalities? If so, what are they? **79–81.** Answers will vary.

80. Use a graphing utility with a $\boxed{\text{SHADE}}$ feature to verify any five of the graphs that you drew by hand in Exercises 1–22.

81. Use a graphing utility with a $\boxed{\text{SHADE}}$ feature to verify any five of the graphs that you drew by hand for the systems in Exercises 23–46.

Critical Thinking Exercises

82. Write a linear inequality in two variables whose graph is shown. $x - y < 2$ or $y > x - 2$

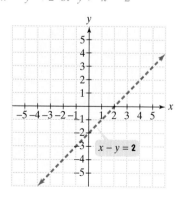

In Exercises 83–84, write a system of inequalities for each graph.

83. $x \geq -2, y > -1$

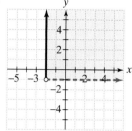

84. $y > x - 3, y \leq x$

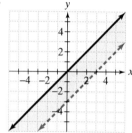

85. Write a system of inequalities whose solution set includes every point in the rectangular coordinate system.   Answers will vary.

86. Sketch the graph of the solution set for the following system of inequalities: See graphing answer section.

$$y \geq nx + b \ (n < 0, b > 0)$$

$$y \leq mx + b \ (m > 0, b > 0).$$

Review Exercises

87. Solve the system:

$$3x - y = 8$$
$$x - 5y = -2.$$

(Section 4.3, Example 2) $\{(3, 1)\}$

88. Solve by graphing:

$$y = 3x - 2$$
$$y = -2x + 8.$$

(Section 4.1, Example 3) $\{(2, 4)\}$

89. Factor completely: $2x^6 + 20x^5y + 50x^4y^2$.
(Section 6.5, Example 8) $2x^4(x + 5y)^2$

SECTION

9.5

Objectives

1 Write an objective function describing a quantity that must be maximized or minimized.

2 Use inequalities to describe limitations in a situation.

3 Use linear programming to solve problems.

LINEAR PROGRAMMING

West Berlin children at Tempelhof Airport watch fleets of U.S. airplanes bringing in supplies to circumvent the Soviet blockade. The airlift began June 28, 1948 and continued for 15 months.

The Berlin Airlift (1948–1949) was an operation by the United States and Great Britain in response to military action by the former Soviet Union: Soviet troops closed all roads and rail lines between West Germany and Berlin, cutting off supply routes to the city. The Allies used a mathematical technique developed during World War II to maximize the amount of supplies transported. During the 15-month airlift, 278,228 flights provided basic necessities to blockaded Berlin, saving one of the world's great cities.

In this section, we will look at an important application of systems of linear inequalities. Such systems arise in **linear programming**, a method for solving problems in which a particular quantity that must be maximized or minimized is limited by other factors. Linear programming is one of the most widely used tools in management science. It helps businesses allocate resources to manufacture products in a way that will maximize profit. Linear programming accounts for more than 50% and perhaps as much as 90% of all computing time used for management decisions in business. The Allies used linear programming to save Berlin.

Objective Functions in Linear Programming Many problems involve quantities that must be maximized or minimized. Businesses are interested in maximizing profit. An operation in which bottled water and medical kits are shipped to earthquake victims needs to maximize the number of victims helped by this shipment. An **objective function** is an algebraic expression in two or more variables describing a quantity that must be maximized or minimized.

1 Write an objective function describing a quantity that must be maximized or minimized.

EXAMPLE 1   Writing an Objective Function

Bottled water and medical supplies are to be shipped to victims of an earthquake by plane. Each container of bottled water will serve 10 people and each medical kit will aid 6 people. If x represents the number of bottles of water to be shipped and y represents the number of medical kits, write the objective function that describes the number of people that can be helped.

SOLUTION   Because each bottle of water serves 10 people and each medical kit aids 6 people, we have

The number of people helped	is	10 times the number of bottles of water	plus	6 times the number of medical kits.
$=$		$10x$	$+$	$6y.$

Using z to represent the number of people helped, the objective function is

$$z = 10x + 6y.$$

Unlike the functions that we have seen so far, the objective function is an equation in three variables. For a value of x and a value of y, there is one and only one value of z. Thus, z is a function of x and y. ■

 CHECK POINT 1 A company manufactures bookshelves and desks for computers. Let x represent the number of bookshelves manufactured daily and y the number of desks manufactured daily. The company's profits are $25 per bookshelf and $55 per desk. Write the objective function that describes the company's total daily profit, z, from x bookshelves and y desks. (Check Points 2 through 4 are related to this situation, so keep track of your answers.) $z = 25x + 55y$

2 Use inequalities to describe limitations in a situation.

Constraints in Linear Programming Ideally, the number of earthquake victims helped in Example 1 should increase without restriction so that every victim receives water and medical kits. However, the planes that ship these supplies are subject to weight and volume restrictions. In linear programming problems, such restrictions are called **constraints**. Each constraint is expressed as a linear inequality. The list of constraints forms a system of linear inequalities.

EXAMPLE 2 Writing a Constraint

Each plane can carry no more than 80,000 pounds. The bottled water weighs 20 pounds per container and each medical kit weighs 10 pounds. Let x represent the number of bottles of water to be shipped and y the number of medical kits. Write an inequality that describes this constraint.

SOLUTION Because each plane can carry no more than 80,000 pounds, we have

| The total weight of the water bottles | plus | the total weight of the medical kits | must be less than or equal to | 80,000 pounds. |

$$20x \quad + \quad 10y \quad \leq \quad 80{,}000.$$

    ⌐Each bottle   ⌐Each kit
    weighs   weighs
    20 pounds.   10 pounds.

The plane's weight constraint is described by the inequality

$$20x + 10y \leq 80{,}000.$$ ■

 CHECK POINT 2 To maintain high quality, the company in Check Point 1 should not manufacture more than a total of 80 bookshelves and desks per day. Write an inequality that describes this constraint. $x + y \leq 80$

In addition to a weight constraint on its cargo, each plane has a limited amount of space in which to carry supplies. Example 3 demonstrates how to express this constraint.

EXAMPLE 3 Writing a Constraint

Planes can carry a total volume for supplies that does not exceed 6000 cubic feet. Each water bottle is 1 cubic foot and each medical kit also has a volume of 1 cubic foot. With x still representing the number of water bottles and y the number of medical kits, write an inequality that describes this second constraint.

SOLUTION Because each plane can carry a volume of supplies that does not exceed 6000 cubic feet, we have

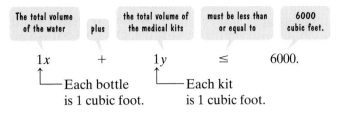

$$1x \quad + \quad 1y \quad \leq \quad 6000.$$

└─ Each bottle   └─ Each kit
is 1 cubic foot.   is 1 cubic foot.

The plane's volume constraint is described by the inequality $x + y \leq 6000$. ∎

In summary, here's what we have described so far in this aid-to-earthquake-victims situation:

$$z = 10x + 6y$$ *This is the objective function describing the number of people helped with x bottles of water and y medical kits.*

$$20x + 10y \leq 80,000$$
$$x + y \leq 6000.$$
These are the constraints based on each plane's weight and volume limitations.

> ✔ **CHECK POINT 3** To meet customer demand, the company in Check Point 1 must manufacture between 30 and 80 bookshelves per day, inclusive. Furthermore, the company must manufacture at least 10 and no more than 30 desks per day. Write an inequality that describes each of these sentences. Then summarize what you have described about this company by writing the objective function for its profits, and the three constraints.

$30 \leq x \leq 80$
$10 \leq y \leq 30;$
$z = 25x + 55y$
$x + y \leq 80$
$30 \leq x \leq 80$
$10 \leq y \leq 30$

3 Use linear programming to solve problems.

Solving Problems with Linear Programming The problem in the earthquake situation described previously is to maximize the number of victims who can be helped, subject to the planes' weight and volume constraints. The process of solving this problem is called *linear programming*, based on a theorem that was proven during World War II.

> **SOLVING A LINEAR PROGRAMMING PROBLEM** Let $z = ax + by$ be an objective function that depends on x and y. Furthermore, z is subject to a number of constraints on x and y. If a maximum or minimum value of z exists, it can be determined as follows:
>
> **1.** Graph the system of inequalities representing the constraints.
> **2.** Find the value of the objective function at each corner, or **vertex**, of the graphed region. The maximum and minimum of the objective function occur at one or more of the corner points.

EXAMPLE 4 Solving a Linear Programming Problem

Determine how many bottles of water and how many medical kits should be sent on each plane to maximize the number of earthquake victims who can be helped.

SOLUTION We must maximize $z = 10x + 6y$ subject to the following constraints:

$$20x + 10y \leq 80,000$$
$$x + y \leq 6000.$$

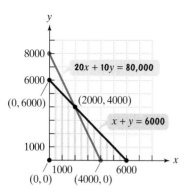

FIGURE 9.19 The region in quadrant I representing the constraints $20x + 10y \leq 80,000$ and $x + y \leq 6000$

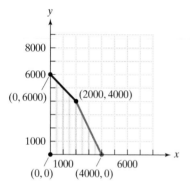

FIGURE 9.20

Step 1. **Graph the system of inequalities representing the constraints.** Because x (the number of bottles of water per plane) and y (the number of medical kits per plane) must be nonnegative, we need to graph the system of inequalities in quadrant I and its boundary only. To graph the inequality $20x + 10y \leq 80,000$, we graph the equation $20x + 10y = 80,000$ as a solid blue line (Figure 9.19). Setting $y = 0$, the x-intercept is 4000 and setting $x = 0$, the y-intercept is 8000. Using $(0, 0)$ as a test point, the inequality is satisfied, so we shade below the blue line, as shown in yellow in Figure 9.19. Now we graph $x + y \leq 6000$ by first graphing $x + y = 6000$ as a solid red line. Setting $y = 0$, the x-intercept is 6000. Setting $x = 0$, the y-intercept is 6000. Using $(0, 0)$ as a test point, the inequality is satisfied, so we shade below the red line, as shown using green vertical shading in Figure 9.19.

We use the addition method to find where the lines $20x + 10y = 80,000$ and $x + y = 6000$ intersect.

$$
\begin{array}{ll}
20x + 10y = 80,000 & \xrightarrow{\text{No change}} \quad 20x + 10y = 80,000 \\
x + y = 6000 & \xrightarrow{\text{Multiply by } -10.} \quad -10x - 10y = -60,000 \\
& \qquad\qquad\qquad \text{Add:} \quad 10x = 20,000 \\
& \qquad\qquad\qquad\qquad\quad\; x = 2000
\end{array}
$$

Back-substituting 2000 for x in $x + y = 6000$, we find $y = 4000$, so the intersection point is $(2000, 4000)$.

The system of inequalities representing the constraints is shown by the region in which the yellow shading and the green vertical shading overlap in Figure 9.19. The graph of the system of inequalities is shown again in Figure 9.20. The red and blue line segments are included in the graph.

Step 2. **Find the value of the objective function at each corner of the graphed region. The maximum and minimum of the objective function occur at one or more of the corner points.** We must evaluate the objective function, $z = 10x + 6y$, at the four corners, or vertices, of the region in Figure 9.20.

Corner (x, y)	Objective Function $z = 10x + 6y$
$(0, 0)$	$z = 10(0) + 6(0) = 0$
$(4000, 0)$	$z = 10(4000) + 6(0) = 40,000$
$(2000, 4000)$	$z = 10(2000) + 6(4000) = 44,000$ ← maximum
$(0, 6000)$	$z = 10(0) + 6(6000) = 36,000$

Thus, the maximum value of z is 44,000 and this occurs when $x = 2000$ and $y = 4000$. In practical terms, this means that the maximum number of earthquake victims who can be helped with each plane shipment is 44,000. This can be accomplished by sending 2000 water bottles and 4000 medical kits per plane. ∎

 CHECK POINT 4 For the company in Check Points 1–3, how many bookshelves and how many desks should be manufactured per day to obtain maximum profit? What is the maximum daily profit? To maximize profit, 50 bookshelves and 30 desks should be manufactured per day. The maximum daily profit is $2900.

EXAMPLE 5 Solving a Linear Programming Problem

Find the maximum value of the objective function

$$z = 2x + y$$

subject to the following constraints:

$$x \geq 0, y \geq 0$$

$$x + 2y \leq 5$$

$$x - y \leq 2.$$

SOLUTION We begin by graphing the region in quadrant I ($x \geq 0$, $y \geq 0$) formed by the constraints. The graph is shown in Figure 9.21.

Now we evaluate the objective function at the four vertices of this region.

Objective function: $z = 2x + y$

At $(0, 0)$: $z = 2 \cdot 0 + 0 = 0$

At $(2, 0)$: $z = 2 \cdot 2 + 0 = 4$

At $(3, 1)$: $z = 2 \cdot 3 + 1 = 7$ ⟵ **Maximum value of z**

At $(0, 2.5)$: $z = 2 \cdot 0 + 2.5 = 2.5$

Thus, the maximum value of z is 7, and this occurs when $x = 3$ and $y = 1$. ∎

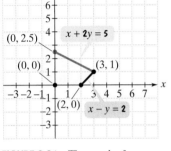

FIGURE 9.21 The graph of $x + 2y \leq 5$ and $x - y \leq 2$ in quadrant I

We can see why the objective function in Example 5 has a maximum value that occurs at a vertex by solving the equation for y.

$$z = 2x + y \qquad \text{This is the objective function of Example 5.}$$

$$y = -2x + z \qquad \text{Solve for } y. \text{ Recall that the slope-intercept form of a line is } y = mx + b.$$

Slope $= -2$ y-intercept $= z$

In this form, z represents the y-intercept of the objective function. The equation describes infinitely many parallel lines, each with slope -2. The process in linear programming involves finding the maximum z-value for all lines that intersect the region determined by the constraints. Of all the lines whose slope is -2, we're looking for the one with the greatest y-intercept that intersects the given region. As we see in Figure 9.22, such a line will pass through one (or possibly more) of the vertices of the region.

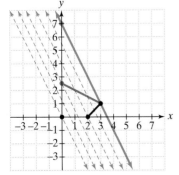

FIGURE 9.22 The line with slope -2 with the greatest y-intercept that intersects the shaded region passes through one of its vertices.

 CHECK POINT 5 Find the maximum value of the objective function $z = 3x + 5y$ subject to the constraints $x \geq 0$, $y \geq 0$, $x + y \geq 1$, $x + y \leq 6$.   30

9.5 EXERCISE SET

Student Solutions Manual   CD/Video   PH Math/Tutor Center   MathXL Tutorials on CD   MathXL®   MyMathLab   Interactmath.com

Practice Exercises

In Exercises 1–4, find the value of the objective function at each corner of the graphed region. What is the maximum value of the objective function? What is the minimum value of the objective function?

1. Objective Function
$z = 5x + 6y$

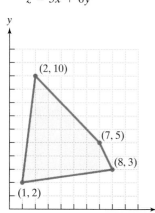

(1, 2): 17; (2, 10): 70;
(7, 5): 65; (8, 3): 58;
maximum: 70;
minimum: 17

2. Objective Function
$z = 3x + 2y$

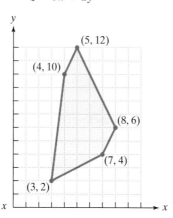

(3, 2): 13; (4, 10): 32;
(5, 12): 39; (8, 6): 36; (7, 4): 29;
maximum: 39;
minimum: 13

3. Objective Function
$z = 40x + 50y$

(0, 0): 0; (0, 8): 400;
(4, 9): 610; (8, 0): 320;
maximum: 610;
minimum: 0

4. Objective Function
$z = 30x + 45y$

(0, 0): 0; (0, 9): 405;
(4, 4): 300; (3, 0): 90;
maximum: 405;
minimum: 0

In Exercises 5–14, an objective function and a system of linear inequalities representing constraints are given.

a. *Graph the system of inequalities representing the constraints.*
  5–14a. See graphing answer section.

b. *Find the value of the objective function at each corner of the graphed region.*

c. *Use the values in part (b) to determine the maximum value of the objective function and the values of x and y for which the maximum occurs.*

5. Objective Function $z = 3x + 2y$
Constraints $x \ge 0, y \ge 0$
    $2x + y \le 8$
    $x + y \ge 4$

b. (0, 4): 8;
(0, 8): 16;
(4, 0): 12

c. maximum: 16; at (0, 8)

6. Objective Function $z = 2x + 3y$
Constraints $x \ge 0, y \ge 0$
    $2x + y \le 8$
    $2x + 3y \le 12$

b. (0, 0): 0;
(0, 4): 12;
(3, 2): 12;
(4, 0): 8;

c. maximum: 12; at (0, 4) and (3, 2)

7. Objective Function $z = 4x + y$
Constraints $x \ge 0, y \ge 0$
    $2x + 3y \le 12$
    $x + y \ge 3$

b. (0, 3): 3;
(0, 4): 4;
(6, 0): 24;
(3, 0): 12

c. maximum: 24; at (6, 0)

8. Objective Function $z = x + 6y$
Constraints $x \ge 0, y \ge 0$
    $2x + y \le 10$
    $x - 2y \ge -10$

b. (0, 0): 0;
(0, 5): 30;
(2, 6): 38;
(5, 0): 5

c. maximum: 38; at (2, 6)

9. Objective Function $z = 3x - 2y$
Constraints $1 \le x \le 5$
    $y \ge 2$
    $x - y \ge -3$

b. (1, 2): −1;
(1, 4): −5;
(5, 8): −1;
(5, 2): 11;

c. maximum: 11; at (5, 2)

10. Objective Function $z = 5x - 2y$
Constraints $0 \le x \le 5$
    $0 \le y \le 3$
    $x + y \ge 2$

b. (0, 2): −4;
(0, 3): −6;
(5, 3): 19;
(5, 0): 25;
(2, 0): 10

c. maximum: 25; at (5, 0)

11. Objective Function $z = 4x + 2y$
Constraints $x \ge 0, y \ge 0$
    $2x + 3y \le 12$
    $3x + 2y \le 12$
    $x + y \ge 2$

b. (0, 2): 4;
(0, 4): 8;
$\left(\frac{12}{5}, \frac{12}{5}\right): \frac{72}{5} = 14.4$;
(4, 0): 16;
(2, 0): 8

c. maximum: 16; at (4, 0)

12. Objective Function $z = 2x + 4y$
Constraints $x \ge 0, y \ge 0$
    $x + 3y \ge 6$
    $x + y \ge 3$
    $x + y \le 9$

b. $\left(\frac{3}{2}, \frac{3}{2}\right): 9$;
(0, 3): 12;
(0, 9): 36;
(9, 0): 18;
(6, 0): 12

c. maximum: 36; at (0, 9)

13. Objective Function $\quad z = 10x + 12y$
Constraints $\quad\quad\quad x \geq 0, y \geq 0$
$\quad\quad\quad\quad\quad\quad\quad\quad x + y \leq 7$
$\quad\quad\quad\quad\quad\quad\quad\quad 2x + y \leq 10$
$\quad\quad\quad\quad\quad\quad\quad\quad 2x + 3y \leq 18$

b. $(0, 0): 0;$
$(0, 6): 72;$
$(3, 4): 78;$
$(5, 0): 50$

c. maximum: 78; at $(3, 4)$

14. Objective Function $\quad z = 5x + 6y$
Constraints $\quad\quad\quad x \geq 0, y \geq 0$
$\quad\quad\quad\quad\quad\quad\quad\quad 2x + y \geq 10$
$\quad\quad\quad\quad\quad\quad\quad\quad x + 2y \geq 10$
$\quad\quad\quad\quad\quad\quad\quad\quad x + y \leq 10$

b. $\left(\frac{10}{3}, \frac{10}{3}\right): \frac{110}{3} = 36\frac{2}{3};$
$(0, 10): 60;$
$(10, 0): 50$

c. maximum: 60; at $(0, 10)$

Application Exercises

15. A television manufacturer makes console and wide-screen televisions. The profit per unit is $125 for the console televisions and $200 for the wide-screen televisions.

a. Let $x =$ the number of consoles manufactured in a month and $y =$ the number of wide-screens manufactured in a month. Write the objective function that describes the total monthly profit. $\quad z = 125x + 200y$

b. The manufacturer is bound by the following constraints:

- Equipment in the factory allows for making at most 450 console televisions in one month.

- Equipment in the factory allows for making at most 200 wide-screen televisions in one month.

- The cost to the manufacturer per unit is $600 for the console televisions and $900 for the wide-screen televisions. Total monthly costs cannot exceed $360,000.

Write a system of three inequalities that describes these constraints. $\quad x \leq 450; y \leq 200; 600x + 900y \leq 360,000$

c. Graph the system of inequalities in part (b). Use only the first quadrant and its boundary, because x and y must both be nonnegative. See graphing answer section.

d. Evaluate the objective function for total monthly profit at each of the five vertices of the graphed region. [The vertices should occur at $(0, 0), (0, 200), (300, 200),$ $(450, 100),$ and $(450, 0)$.] $\quad 0; 40,000; 77,500; 76,250; 56,250$

e. Complete the missing portions of this statement: The television manufacturer will make the greatest profit by manufacturing 300 console televisions each month and 200 wide-screen televisions each month. The maximum monthly profit is $77,500.

16. a. A student earns $10 per hour for tutoring and $7 per hour as a teacher's aid. Let $x =$ the number of hours each week spent tutoring and $y =$ the number of hours each week spent as a teacher's aid. Write the objective function that describes total weekly earnings. $\quad z = 10x + 7y$

b. The student is bound by the following constraints:

- To have enough time for studies, the student can work no more than 20 hours a week.

- The tutoring center requires that each tutor spend at least three hours a week tutoring.

- The tutoring center requires that each tutor spend no more than eight hours a week tutoring.

Write a system of three inequalities that describes these constraints. $\quad x + y \leq 20; x \geq 3; x \leq 8$

c. Graph the system of inequalities in part (b). Use only the first quadrant and its boundary, because x and y are nonnegative. See graphing answer section.

d. Evaluate the objective function for total weekly earnings at each of the four vertices of the graphed region. [The vertices should occur at $(3, 0), (8, 0), (3, 17),$ and $(8, 12)$.] $30; 80; 149; 164$

e. Complete the missing portions of this statement: The student can earn the maximum amount per week by tutoring for 8 hours per week and working as a teacher's aid for 12 hours per week. The maximum amount that the student can earn each week is $164.

Use the two steps for solving a linear programming problem, given in the box on page 623, to solve the problems in Exercises 17–23.

17. A manufacturer produces two models of mountain bicycles. The times (in hours) required for assembling and painting each model are given in the following table:

	Model A	Model B
Assembling	5	4
Painting	2	3

The maximum total weekly hours available in the assembly department and the paint department are 200 hours and 108 hours, respectively. The profits per unit are $25 for model A and $15 for model B. How many of each type should be produced to maximize profit? 40 of model A and 0 of model B

18. A large institution is preparing lunch menus containing foods A and B. The specifications for the two foods are given in the following table:

Food	Units of Fat per Ounce	Units of Carbohydrates per Ounce	Units of Protein per Ounce
A	1	2	1
B	1	1	1

Each lunch must provide at least 6 units of fat per serving, no more than 7 units of protein, and at least 10 units of carbohydrates. The institution can purchase food A for $0.12 per ounce and food B for $0.08 per ounce. How many ounces of each food should a serving contain to meet the dietary requirements at the least cost? 4 ounces of A and 2 ounces of B

19. Food and clothing are shipped to victims of a hurricane. Each carton of food will feed 12 people, while each carton of clothing will help 5 people. Each 20-cubic-foot box of food weighs 50 pounds and each 10-cubic-foot box of clothing weighs 20 pounds. The commercial carriers transporting food and clothing are bound by the following constraints:

• The total weight per carrier cannot exceed 19,000 pounds.

• The total volume must be less than 8000 cubic feet.

How many cartons of food and how many cartons of clothing should be sent with each plane shipment to maximize the number of people who can be helped?   300 boxes of food and 200 boxes of clothing

20. On June 24, 1948, the former Soviet Union blocked all land and water routes through East Germany to Berlin. A gigantic airlift was organized using American and British planes to bring food, clothing, and other supplies to the more than 2 million people in West Berlin. The cargo capacity was 30,000 cubic feet for an American plane and 20,000 cubic feet for a British plane. To break the Soviet blockade, the Western Allies had to maximize cargo capacity, but were subject to the following restrictions:

• No more than 44 planes could be used.

• The larger American planes required 16 personnel per flight, double that of the requirement for the British planes. The total number of personnel available could not exceed 512.

• The cost of an American flight was $9000 and the cost of a British flight was $5000. Total weekly costs could not exceed $300,000.

Find the number of American planes and the number of British planes that were used to maximize cargo capacity.   20 American planes and 24 British planes

21. A theater is presenting a program on drinking and driving for students and their parents. The proceeds will be donated to a local alcohol information center. Admission is $2.00 for parents and $1.00 for students. However, the situation has two constraints: The theater can hold no more than 150 people and every two parents must bring at least one student. How many parents and students should attend to raise the maximum amount of money?   100 parents and 50 students

22. You are about to take a test that contains computation problems worth 6 points each and word problems worth 10 points each. You can do a computation problem in 2 minutes and a word problem in 4 minutes. You have 40 minutes to take the test and may answer no more than 12 problems. Assuming you answer all the problems attempted correctly, how many of each type of problem must you do to maximize your score? What is the maximum score?   4 computation and 8 word problems; a score of 104

23. In 1978, a ruling by the Civil Aeronautics Board allowed Federal Express to purchase larger aircraft. Federal Express's options included 20 Boeing 727s that United Airlines was retiring and/or the French-built Dassault Fanjet Falcon 20. To aid in their decision, executives at Federal Express analyzed the data at the top of the next column.

	Boeing 727	Falcon 20
Direct Operating Cost	$1400 per hour	$500 per hour
Payload	42,000 pounds	6000 pounds

Federal Express was faced with the following constraints:

• Hourly operating cost was limited to $35,000.

• Total payload had to be at least 672,000 pounds.

• Only twenty 727s were available.

Given the constraints, how many of each kind of aircraft should Federal Express have purchased to maximize the number of aircraft?   10 Boeing 727s and 42 Falcon 20s

Writing in Mathematics

24. What kinds of problems are solved using the linear programming method?   24–28. Answers will vary.

25. What is an objective function in a linear programming problem?

26. What is a constraint in a linear programming problem? How is a constraint represented?

27. In your own words, describe how to solve a linear programming problem.

28. Describe a situation in your life in which you would like to maximize something, but are limited by at least two constraints. Can linear programming be used in this situation? Explain your answer.

Critical Thinking Exercises

29. Suppose that you inherit $10,000. The will states how you must invest the money. Some (or all) of the money must be invested in stocks and bonds. The requirements are that at least $3000 be invested in bonds, with expected returns of $0.08 per dollar, and at least $2000 be invested in stocks, with expected returns of $0.12 per dollar. Because the stocks are medium risk, the final stipulation requires that the investment in bonds should never be less than the investment in stocks. How should the money be invested so as to maximize your expected returns? $5000 in stocks and $5000 in bonds

30. Consider the objective function $z = Ax + By$ ($A > 0$ and $B > 0$) subject to the following constraints: $2x + 3y \leq 9$, $x - y \leq 2, x \geq 0$, and $y \geq 0$. Prove that the objective function will have the same maximum value at the vertices $(3, 1)$ and $(0, 3)$ if $A = \frac{2}{3}B$.   If $A = \frac{2}{3}B$, then $z = \frac{2}{3}Bx + By$. Thus, $z = \frac{2}{3}B(3) + B(1) = 3B$ at the vertex $(3, 1)$, and $z = \frac{2}{3}B(0) + B(3) = 3B$ at the vertex $(0, 3)$. At $(0, 0)$, $z = 0$, and at $(2, 0)$, $z = \frac{4}{3}B$.

Review Exercises

31. Solve: $x^2 - 12x + 36 = 0$
(Section 6.6, Example 4) $\{6\}$

32. Divide: $\dfrac{1}{x^2 - 17x + 30} \div \dfrac{1}{x^2 + 7x - 18}$.

(Section 7.2, Example 6) $\dfrac{x + 9}{x - 15}$

33. If $f(x) = x^3 + 2x^2 - 5x + 4$, find $f(-1)$.
(Section 8.1, Example 3)   10

GROUP PROJECT

CHAPTER **9**

Additional group projects can be found in the *Instructor's Resource Manual.*

Each group member should research one situation that provides two different pricing options. These can involve areas such as public transportation options (with or without coupon books) or long-distance telephone plans or anything of interest. Be sure to bring in all the details for each option. At the group meeting, select the two pricing situations that are most interesting and relevant. Using each situation, write a word problem about selecting the better of the two options. The word problem should be one that can be solved using a linear inequality. The group should turn in the two problems and their solutions.

CHAPTER 9   SUMMARY

Definitions and Concepts	Examples

Section 9.1 Interval Notation and Business Applications Using Linear Inequalities

A linear inequality in one variable can be written in the form $ax + b < c$, $ax + b \leq c$, $ax + b > c$, or $ax + b \geq c$. The set of all numbers that make the inequality a true statement is its solution set. Graphs of solution sets are shown on a number line by shading all points representing numbers that are solutions. Parentheses indicate endpoints that are not solutions. Square brackets indicate endpoints that are solutions.

• $(-2, 1] = \{x \mid -2 < x \leq 1\}$

$$\underset{-4\ -3\ -2\ -1\ \ 0\ \ 1\ \ 2\ \ 3\ \ 4}{\longleftrightarrow}$$

• $[-2, \infty) = \{x \mid x \geq -2\}$

$$\underset{-4\ -3\ -2\ -1\ \ 0\ \ 1\ \ 2\ \ 3\ \ 4}{\longleftrightarrow}$$

Solving a Linear Inequality

1. Simplify each side.
2. Collect variable terms on one side and constant terms on the other side.
3. Isolate the variable and solve.

If an inequality is multiplied or divided by a negative number, the inequality symbol must be reversed.

Solve: $2(x + 3) - 5x \leq 15.$

$$2x + 6 - 5x \leq 15$$

$$-3x + 6 \leq 15$$

$$-3x \leq 9$$

$$\frac{-3x}{-3} \geq \frac{9}{-3}$$

$$x \geq -3$$

Solution set: $\{x \mid x \geq -3\}$ or $[-3 \,\infty)$

Definitions and Concepts	Examples

Section 9.1 Interval Notation and Business Applications Using Linear Inequalities (continued)

Functions of Business

A company produces and sells x units of a product.

Revenue Function

$$R(x) = (\text{price per unit sold})x$$

Cost Function

$$C(x) = \text{fixed cost} + (\text{cost per unit produced})x$$

Profit Function

$$P(x) = R(x) - C(x)$$

A business makes money, or has a profit, when $P(x) > 0$.
A business loses money, or has a loss, when $P(x) < 0$.

A company that manufactures lamps has a fixed cost of $80,000 and it costs $20 to produce each lamp. Lamps are sold for $70.

a. Write the cost function.

$$C(x) = 80,000 + 20x$$

Fixed cost   Variable cost: $20 per lamp

b. Write the revenue function.

$$R(x) = 70x$$

Revenue per lamp, $70, times number of lamps sold

c. Write the profit function.

$$P(x) = R(x) - C(x)$$

$$= 70x - (80,000 + 20x)$$

$$= 50x - 80,000$$

d. More than how many lamps must be produced and sold to have a profit?

Solve $\qquad P(x) > 0.$

$$50x - 80,000 > 0$$

$$50x > 80,000$$

$$x > 1600$$

More than 1600 lamps must be produced and sold to have a profit.

Section 9.2 Compound Inequalities

Intersection ($\cap$) and Union ($\cup$)

$A \cap B$ is the set of elements common to both set A and set B.
$A \cup B$ is the set of elements that are members of set A or set B or of both sets.

$\{1, 3, 5, 7\} \cap \{5, 7, 9, 11\} = \{5, 7\}$

$\{1, 3, 5, 7\} \cup \{5, 7, 9, 11\} = \{1, 3, 5, 7, 9, 11\}$

A compound inequality is formed by joining two inequalities with the word *and* or *or*. When the connecting word is *and*, graph each inequality separately and take the intersection of their solution sets.

Solve: $x + 1 > 3$ and $x + 4 \leq 8.$

$\qquad x > 2$ and $\qquad x \leq 4$

$$-4\,-3\,-2\,-1\ \ 0\ \ 1\ \ 2\ \ 3\ \ 4$$

Solution set: $\{x \mid 2 < x \leq 4\}$ or $(2, 4]$

Definitions and Concepts	Examples

Section 9.2 Compound Inequalities (continued)

The compound inequality $a < x < b$ means $a < x$ and $x < b$. Solve by isolating the variable in the middle.	Solve: $-1 < \dfrac{2x+1}{3} \le 2$. $-3 < 2x + 1 \le 6$ Multiply by 3. $-4 < 2x \le 5$ Subtract 1. $-2 < x \le \dfrac{5}{2}$ Divide by 2. Solution set: $\left\{ x \mid -2 < x \le \dfrac{5}{2} \right\}$ or $\left(-2, \dfrac{5}{2} \right]$ $\xleftarrow{\qquad} {\underset{-4\,-3\,-2\,-1\ \ 0\ \ 1\ \ 2\ \ 3\ \ 4}{\longrightarrow}}$
When the connecting word in a compound inequality is *or*, graph each inequality separately and take the union of their solution sets.	Solve: $x - 2 > -3$ or $2x \le -6$. $x > -1$ or $x \le -3$ $\xleftarrow{\qquad} {\underset{-6\,-5\,-4\,-3\,-2\,-1\ \ 0\ \ 1\ \ 2}{\longrightarrow}}$ Solution set: $\{ x \mid x \le -3 \text{ or } x > -1 \}$ or $(\infty, -3] \cup (-1, \infty)$

Section 9.3 Equations and Inequalities Involving Absolute Value

Absolute Value Equations **1.** If $c > 0$, then $\lvert X \rvert = c$ means $X = c$ or $X = -c$. **2.** If $c < 0$, then $\lvert X \rvert = c$ has no solution. **3.** If $c = 0$, then $\lvert X \rvert = 0$ means $X = 0$.	Solve: $\lvert 2x - 7 \rvert = 3$. $2x - 7 = 3$ or $2x - 7 = -3$ $2x = 10$ $2x = 4$ $x = 5$ $x = 2$ The solution set is $\{2, 5\}$.
Absolute Value Equations with Two Absolute Value Bars If $\lvert X_1 \rvert = \lvert X_2 \rvert$, then $X_1 = X_2$ or $X_1 = -X_2$.	Solve: $\lvert x - 6 \rvert = \lvert 2x + 1 \rvert$. $x - 6 = 2x + 1$ or $x - 6 = -(2x + 1)$ $-x - 6 = 1$ $x - 6 = -2x - 1$ $-x = 7$ $3x - 6 = -1$ $x = -7$ $3x = 5$ $x = \dfrac{5}{3}$ The solutions are -7 and $\frac{5}{3}$, and the solution set is $\left\{ -7, \frac{5}{3} \right\}$.
If c is a positive number, then to solve $\lvert X \rvert < c$, solve the compound inequality $-c < X < c$. If c is negative, $\lvert X \rvert < c$ has no solution.	Solve: $\lvert x - 4 \rvert < 3$. $-3 < x - 4 < 3$ $1 < x < 7$ The solution set is $\{ x \mid 1 < x < 7 \}$ or $(1, 7)$. $\xleftarrow{\qquad} {\underset{0\ \ 1\ \ 2\ \ 3\ \ 4\ \ 5\ \ 6\ \ 7\ \ 8}{\longrightarrow}}$

Definitions and Concepts	**Examples**

Section 9.3 Equations and Inequalities Involving Absolute Value (continued)

If c is a positive number, then to solve $\|X\| > c$, solve the compound inequality $X < -c$ or $X > c$. If c is negative, $\|X\| > c$ is true for all real numbers for which X is defined.	Solve: $\left\|\dfrac{x}{3} - 1\right\| \geq 2$.

$$\frac{x}{3} - 1 \leq -2 \quad \text{or} \quad \frac{x}{3} - 1 \geq 2.$$

$$x - 3 \leq -6 \quad \text{or} \quad x - 3 \geq 6 \qquad \text{Multiply by 3.}$$

$$x \leq -3 \quad \text{or} \qquad x \geq 9 \qquad \text{Add 3.}$$

The solution set is $\{x \mid x \leq -3 \text{ or } x \geq 9\}$ or $(-\infty, -3] \cup [9, \infty)$.

$$\overset{\longleftarrow \,+\,\rule[0.4ex]{0.8em}{0.12ex}\,+\,+\,+\,+\,+\,\rule[0.4ex]{0.8em}{0.12ex}\,+\longrightarrow}{{-6}\,{-4}\,{-2}\ \ 0\ \ 2\ \ 4\ \ 6\ \ 8\ \ 10}$$

Section 9.4 Linear Inequalities in Two Variables

If the equal sign in $Ax + By = C$ is replaced with an inequality symbol, the result is a linear inequality in two variables. Its graph is the set of all points whose coordinates satisfy the inequality. To obtain the graph,	Graph: $x - 2y \leq 4$.

1. Replace the inequality symbol with an equal sign and graph the boundary line. Use a solid line for $\leq$ or $\geq$ and a dashed line for $<$ or $>$.

2. Choose a test point not on the line and substitute its coordinates into the inequality.

3. If a true statement results, shade the half-plane containing the test point. If a false statement results, shade the half-plane not containing the test point.

1. Graph $x - 2y = 4$. Use a solid line because the inequality symbol is $\leq$.

2. Test $(0, 0)$.

$$x - 2y \leq 4$$
$$0 - 2 \cdot 0 \overset{?}{\leq} 4$$
$$0 \leq 4, \qquad \text{true}$$

3. The inequality is true. Shade the half-plane containing $(0, 0)$.

Two or more linear inequalities make up a system of linear inequalities. A solution is an ordered pair satisfying all inequalities in the system. To graph a system of inequalities, graph each inequality in the system. The overlapping region represents the solutions of the system.	Graph the solutions of the system: $$y \leq -2x$$ $$x - y \geq 3.$$

Definitions and Concepts	Examples

Section 9.5 Linear Programming

Linear programming is a method for solving problems in which a particular quantity that must be maximized or minimized is limited. An objective function is an algebraic expression in three variables describing a quantity that must be maximized or minimized. Constraints are restrictions, expressed as linear inequalities.

Solving a Linear Programming Problem

1. Graph the system of inequalities representing the constraints.

2. Find the value of the objective function at each corner, or vertex, of the graphed region. The maximum and minimum of the objective function occur at one or more vertices.

Find the maximum value of the objective function $z = 3x + 2y$ subject to the following constraints: $x \geq 0, y \geq 0, 2x + 3y \leq 18, 2x + y \leq 10$.

1. Graph the system of inequalities representing the constraints.

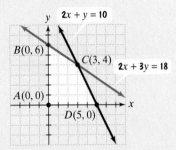

2. Evaluate the objective function at each vertex.

Vertex	$z = 3x + 2y$
$A(0, 0)$	$z = 3(0) + 2(0) = 0$
$B(0, 6)$	$z = 3(0) + 2(6) = 12$
$C(3, 4)$	$z = 3(3) + 2(4) = 17$
$D(5, 0)$	$z = 3(5) + 2(0) = 15$

The maximum value of the objective function is 17.

CHAPTER 9 REVIEW EXERCISES

9.1

In Exercises 1–3, express each interval in set-builder notation and graph the interval on a number line.   **1–3.** See graphing answer section.

1. $(-2, 3]$ $\{x | -2 < x \leq 3\}$
2. $[-1.5, 2]$ $\{x | -1.5 \leq x \leq 2\}$
3. $[-1, \infty)$ $\{x | x > -1\}$

In Exercises 4–8, solve each linear inequality and graph the solution set on a number line. Express the solution set in both set-builder and interval notations.   **4–8.** See graphing answer section.

4. $-6x + 3 \leq 15$ $\{x | x \geq -2\}; [-2, \infty)$
5. $6x - 9 \geq -4x - 3$ $\{x | x \geq \frac{3}{5}\}; [\frac{3}{5}, \infty)$
6. $\frac{x}{3} - \frac{3}{4} - 1 > \frac{x}{2}$ $\{x | x < -\frac{21}{2}\}; (-\infty, -\frac{21}{2})$
7. $6x + 5 > -2(x - 3) - 25$ $\{x | x > -3\}; (-3, \infty)$
8. $3(2x - 1) - 2(x - 4) \geq 7 + 2(3 + 4x)$ $\{x | x \leq -2\}; (-\infty, -2]$
9. The cost and revenue functions for producing and selling x units of a toaster oven are

$$C(x) = 40x + 357{,}000 \quad \text{and} \quad R(x) = 125x.$$

a. Write the profit function, P, from producing and selling x toaster ovens. $P(x) = 85x - 357{,}000$

b. More than how many toaster ovens must be produced and sold to have profit?   more than 4200

Use this information to solve Exercises 10–13: A company is planning to produce and sell a new line of computers. The fixed cost will be \$360,000 and it will cost \$850 to produce each computer. Each computer will be sold for \$1150.

10. Write the cost function, C, of producing x computers. $C(x) = 360{,}000 + 850x$

11. Write the revenue function, R, from the sale of x computers. $R(x) = 1150x$

12. Write the profit function, P, from producing and selling x computers. $P(x) = 300x - 360{,}000$

13. More than how many computers must be produced and sold to have a profit?   more than 1200

14. A person can choose between two charges on a checking account. The first method involves a fixed cost of \$11 per month plus 6¢ for each check written. The second method involves a fixed cost of \$4 per month plus 20¢ for each check written. How many checks should be written to make the first method a better deal?   more than 50 checks

15. A salesperson earns \$500 per month plus a commission of 20% of sales. Describe the sales needed to receive a total income that exceeds \$3200 per month.   more than \$13,500 in sales

9.2 *In Exercises 16–19, let* $A = \{a, b, c\}, B = \{a, c, d, e\}$ *and* $C = \{a, d, f, g\}$. *Find the indicated set.*

16. $A \cap B$ $\{a, c\}$

17. $A \cap C$ $\{a\}$

18. $A \cup B$ $\{a, b, c, d, e\}$

19. $A \cup C$ $\{a, b, c, d, f, g\}$

In Exercises 20–30, solve each compound inequality. Except for the empty set, express the solution set in both set-builder and interval notations. Graph the solution set on a number line.
20–30. See graphing answer section.

20. $x \le 3$ and $x < 6$ $\{x | x \le 3\}; (-\infty, 3]$

21. $x \le 3$ or $x < 6$ $\{x | x < 6\}; (-\infty, 6)$

22. $-2x < -12$ and $x - 3 < 5$ $\{x | 6 < x < 8\}; (6, 8)$

23. $5x + 3 \le 18$ and $2x - 7 \le -5$ $\{x | x \le 1\}; (-\infty, 1]$

24. $2x - 5 > -1$ and $3x < 3$ $\varnothing$

25. $2x - 5 > -1$ or $3x < 3$
$\{x | x < 1 \text{ or } x > 2\}; (-\infty, 1) \cup (2, \infty)$

26. $x + 1 \le -3$ or $-4x + 3 < -5$
$\{x | x \le -4 \text{ or } x > 2\}; (-\infty, -4] \cup (2, \infty)$

27. $5x - 2 \le -22$ or $-3x - 2 > 4$ $\{x | x < -2\}; (-\infty, -2)$

28. $5x + 4 \ge -11$ or $1 - 4x \ge 9$
$\{x | x \text{ is a real number}\}; (-\infty, \infty)$

29. $-3 < x + 2 \le 4$ $\{x | -5 < x \le 2\}; (-5, 2]$

30. $-1 \le 4x + 2 \le 6$ $\{x | -\frac{3}{4} \le x \le 1\}; [-\frac{3}{4}, 1]$

31. On the first of four exams, your grades are 72, 73, 94, and 80. There is still one more exam, and you are hoping to earn a B in the course. This will occur if the average of your five exam grades is greater than or equal to 80 and less than 90. What range of grades on the fifth exam will result in receiving a B? Use interval notation to express this range.   [81, 131); If the highest grade is 100, then [81, 100].

9.3 *In Exercises 32–35, find the solution set for each equation.*

32. $|2x + 1| = 7$ $\{-4, 3\}$

33. $|3x + 2| = -5$ $\varnothing$

34. $2|x - 3| - 7 = 10$ $\{-\frac{11}{2}, \frac{23}{2}\}$

35. $|4x - 3| = |7x + 9|$ $\{-4, -\frac{6}{11}\}$

In Exercises 36–39, solve and graph the solution set on a number line. Except for the empty set, express the solution set in both set-builder and interval notations.   36–39. See graphing answer section.

36. $|2x + 3| \le 15$ $\{x | -9 \le x \le 6\}; [-9, 6]$

37. $\left| \dfrac{2x + 6}{3} \right| > 2$ $\{x | x < -6 \text{ or } x > 0\}; (-\infty, -6) \cup (0, \infty)$

38. $|2x + 5| - 7 < -6$ $\{x | -3 < x < -2\}; (-3, -2)$

39. $|2x - 3| + 4 \le -10$ $\varnothing$

40. Approximately 90% of the population sleeps h hours daily, where h is modeled by the inequality $|h - 6.5| \le 1$. Write a sentence describing the range for the number of hours that most people sleep. Do *not* use the phrase "absolute value" in your description.   Approximately 90% of the population sleeps between 5.5 hours and 7.5 hours daily, inclusive.

9.4 *In Exercises 41–46, graph each inequality in a rectangular coordinate system.*   41–46. See graphing answer section.

41. $3x - 4y > 12$ **42.** $x - 3y \le 6$

43. $y \le -\dfrac{1}{2}x + 2$ **44.** $y > \dfrac{3}{5}x$

45. $x \le 2$ **46.** $y > -3$

In Exercises 47–55, graph the solution set of each system of inequalities or indicate that the system has no solution.
47–55. See graphing answer section.

47. $2x - y \le 4$
  $x + y \ge 5$

48. $y < -x + 4$
  $y > x - 4$

49. $-3 \le x < 5$

50. $-2 < y \le 6$

51. $x \ge 3$
  $y \le 0$

52. $2x - y > -4$
  $x \ge 0$

53. $x + y \le 6$
  $y \ge 2x - 3$

54. $3x + 2y \ge 4$
  $x - y \le 3$
  $x \ge 0, y \ge 0$

55. $2x - y > 2$
  $2x - y < -2$ $\varnothing$

9.5

56. Find the value of the objective function $z = 2x + 3y$ at each corner of the graphed region shown. What is the maximum value of the objective function? What is the minimum value of the objective function? $(\frac{1}{2}, \frac{1}{2}): \frac{5}{2}; (2, 2): 10; (4, 0): 8;$ $(1, 0): 2;$ maximum: 10; minimum: 2

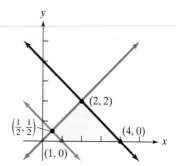

In Exercises 57–59, graph the region determined by the constraints. Then find the maximum value of the given objective function, subject to the constraints.   57–59. See graphing answer section.

57. Objective Function
  Constraints

$z = 2x + 3y$
$x \ge 0, y \ge 0$
$x + y \le 8$
$3x + 2y \ge 6$

maximum: 24

58. Objective Function $\quad z = x + 4y$

Constraints $\qquad 0 \le x \le 5, 0 \le y \le 7$

maximum: 33 $\qquad\qquad x + y \ge 3$

59. Objective Function $\quad z = 5x + 6y$

Constraints $\qquad x \ge 0, y \ge 0$

$\qquad\qquad\qquad y \le x$

$\qquad\qquad\qquad 2x + y \le 12$

maximum: 44 $\qquad\qquad 2x + 3y \ge 6$

60. A paper manufacturing company converts wood pulp to writing paper and newsprint. The profit on a unit of writing paper is \$500 and the profit on a unit of newsprint is \$350.

a. Let x represent the number of units of writing paper produced daily. Let y represent the number of units of newsprint produced daily. Write the objective function that models total daily profit. $\quad z = 500x + 350y$

b. The manufacturer is bound by the following constraints:

- Equipment in the factory allows for making at most 200 units of paper (writing paper and newsprint) in a day.
- Regular customers require at least 10 units of writing paper and at least 80 units of newsprint daily.

Write a system of inequalities that models these constraints. $\quad x + y \le 200; x \ge 10; y \ge 80$

c. Graph the inequalities in part (b). Use only the first quadrant, because x and y must both be positive. (*Suggestion*: Let each unit along the x- and y-axes represent 20.) See graphing answer section.

d. Evaluate the objective profit function at each of the three vertices of the graphed region. $(10, 80)$: 33,000; $(10, 190)$: 71,500; $(120, 80)$: 88,000

e. Complete the missing portions of this statement: The company will make the greatest profit by producing <u>120</u> units of writing paper and <u>80</u> units of newsprint each day. The maximum daily profit is \$<u>88,000</u>.

61. A manufacturer of lightweight tents makes two models whose specifications are given in the following table.

	Cutting Time per Tent	Assembly Time per Tent
Model A	0.9 hour	0.8 hour
Model B	1.8 hours	1.2 hours

Each month, the manufacturer has no more than 864 hours of labor available in the cutting department and at most 672 hours in the assembly division. The profits come to \$25 per tent for model A and \$40 per tent for model B. How many of each should be manufactured monthly to maximize the profit? 480 of model A and 240 of model B

CHAPTER 9 TEST Remember to use your Chapter Test Prep Video CD to see the worked-out solutions to the test questions you want to review.

In Exercises 1–2, express each interval in set-builder notation and graph the interval on a number line. 1–2. See graphing answer section.

1. $[-3, 2)$ $\quad \{x \mid -3 \le x < 2\}$ **2.** $(-\infty, -1]$ $\quad \{x \mid x \le -1\}$

In Exercises 3–4, solve and graph the solution set on a number line. Express the solution set in both set-builder and interval notations. 3–4. See graphing answer section.

3. $3(x + 4) \ge 5x - 12$ $\quad \{x \mid x \le 12\}; (-\infty, 12]$

4. $\dfrac{x}{6} + \dfrac{1}{8} \le \dfrac{x}{2} - \dfrac{3}{4}$ $\quad \left\{ x \mid x \ge \dfrac{21}{8} \right\}; \left[\dfrac{21}{8}, \infty \right)$

5. A company is planning to manufacture computer desks. The fixed cost will be \$60,000 and it will cost \$200 to produce each desk. Each desk will be sold for \$450.

a. Write the cost function, C, of producing x desks. $C(x) = 60,000 + 200x$

b. Write the revenue function, R, from the sale of x desks. $R(x) = 450x$

c. Write the profit function, P, from producing and selling x desks. $P(x) = 250x - 60,000$

d. More than how many desks must be produced and sold to have a profit? more than 240

6. Find the intersection: $\{2, 4, 6, 8, 10\} \cap \{4, 6, 12, 14\}$. $\{4, 6\}$

7. Find the union: $\{2, 4, 6, 8, 10\} \cup \{4, 6, 12, 14\}$. $\{2, 4, 6, 8, 10, 12, 14\}$

In Exercises 8–12, solve each compound inequality. Except for the empty set, express the solution set in both set-builder and interval notations. Graph the solution set on a number line. 8–12. See graphing answer section.

8. $2x + 4 < 2$ and $x - 3 > -5$

$\{x \mid -2 < x < -1\}; (-2, -1)$

9. $x + 6 \ge 4$ and $2x + 3 \ge -2$ $\quad \{x \mid x \ge -2\}; [-2, \infty)$

10. $2x - 3 < 5$ or $3x - 6 \le 4$ $\quad \{x \mid x < 4\}; (-\infty, 4)$

11. $x + 3 \le -1$ or $-4x + 3 < -5$

$\{x \mid x \le -4 \text{ or } x > 2\}; (-\infty, -4] \cup (2, \infty)$

12. $-3 \le \dfrac{2x + 5}{3} < 6$ $\quad \left\{ x \mid -7 \le x < \dfrac{13}{2} \right\}; \left[-7, \dfrac{13}{2} \right)$

In Exercises 13–14, find the solution set for each equation.

13. $|5x + 3| = 7$ $\quad \left\{ -2, \dfrac{4}{5} \right\}$

14. $|6x + 1| = |4x + 15|$ $\quad \left\{ -\dfrac{8}{5}, 7 \right\}$

In Exercises 15–16, solve and graph the solution set on a number line. Express the solution set in both set-builder and interval notations. 15–16. See graphing answer section.

15. $|2x - 1| < 7$ $\quad \{x \mid -3 < x < 4\}; (-3, 4)$

16. $|2x - 3| \ge 5$ $\quad \{x \mid x \le -1 \text{ or } x \ge 4\}; (-\infty, -1] \cup [4, \infty)$

17. The inequality $|b - 98.6| > 8$ describes a person's body temperature, b, in degrees Fahrenheit, when hyperthermia (extremely high body temperature) or hypothermia (extremely low body temperature) occurs. Solve the inequality and interpret the solution. $\{b|b < 90.6 \text{ or } b > 106.6\}$ or $(-\infty, 90.6) \cup (106.6, \infty)$; Hypothermia: Body temperature below 90.6°F; Hyperthermia: Body temperature above 106.6°F

In Exercises 18–20, graph each inequality in a rectangular coordinate system.   18–20. See graphing answer section.

18. $3x - 2y < 6$ **19.** $y \ge \dfrac{1}{2}x - 1$ **20.** $y \le -1$

In Exercises 21–23, graph the solution set of each system of inequalities.   21–23. See graphing answer section.

21. $x + y \ge 2$ **22.** $3x + y \le 9$ **23.** $-2 < x \le 4$
  $x - y \ge 4$ $2x + 3y \ge 6$
                           $x \ge 0, y \ge 0$

24. Find the maximum value of the objective function $z = 3x + 5y$ subject to the following constraints: $x \ge 0, y \ge 0, x + y \le 6, x \ge 2$.   maximum: 26

25. A manufacturer makes two types of jet skis, regular and deluxe. The profit on a regular jet ski is $200 and the profit on the deluxe model is $250. To meet customer demand, the company must manufacture at least 50 regular jet skis per week and at least 75 deluxe models. To maintain high quality, the total number of both models of jet skis manufactured by the company should not exceed 150 per week. How many jet skis of each type should be manufactured per week to obtain maximum profit? What is the maximum weekly profit?   50 regular and 100 deluxe; $35,000

Note to Instructor: The Chapter 9 Test file in the TestGen program provides algorithms specifically matched to these problems so you can easily replicate this chapter test for additional practice or assessment purposes.

CUMULATIVE REVIEW EXERCISES (CHAPTERS 1–9)

In Exercises 1–2, solve each equation.

1. $5(x + 1) + 2 = x - 3(2x + 1)$ $\{-1\}$

2. $\dfrac{2(x + 6)}{3} = 1 + \dfrac{4x - 7}{3}$ $\{8\}$

3. Simplify: $\dfrac{-10x^2 y^4}{15x^7 y^{-3}} \cdot -\dfrac{2y^7}{3x^5}$

4. If $f(x) = x^2 - 3x + 4$, find $f(-3)$ and $f(2a)$.
  22; $4a^2 - 6a + 4$

5. If $f(x) = 3x^2 - 4x + 1$ and $g(x) = x^2 - 5x - 1$, find $(f - g)(x)$ and $(f - g)(2)$. $2x^2 + x + 2$; 12

6. Use function notation to write the equation of the line passing through $(2, 3)$ and perpendicular to the line whose equation is $y = 2x - 3$. $f(x) = -\tfrac{1}{2}x + 4$

In Exercises 7–10, graph each equation or inequality in a rectangular coordinate system.   7–10. See graphing answer section.

7. $f(x) = 2x + 1$ **8.** $y > 2x$

9. $2x - y \ge 6$ **10.** $f(x) = -1$

11. Solve the system:
$$3x - y + z = -15$$
$$x + 2y - z = 1.$$
$$2x + 3y - 2z = 0 \quad (-4, 2, -1)$$

12. If $f(x) = \dfrac{x}{3} - 4$, find $f^{-1}(x)$. $f^{-1}(x) = 3x + 12$

13. If $f(x) = 3x^2 - 1$ and $g(x) = x + 2$, find $f(g(x))$ and $g(f(x))$. $f(g(x)) = 3x^2 + 12x + 11; g(f(x)) = 3x^2 + 1$

14. A motel with 60 rooms charges $90 per night for rooms with kitchen facilities and $80 per night for rooms without kitchen facilities. When all rooms are occupied, the nightly revenue is $5260. How many rooms of each kind are there?   46 rooms with kitchen facilities and 14 without kitchen facilities

15. Which of the following are functions?

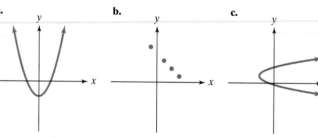

a. and b. are functions.

In Exercises 16–20, solve and graph the solution set on a number line. Express the solution set in both set-builder and interval notations.   16–20. See graphing answer section.

16. $\dfrac{x}{4} - \dfrac{3}{4} - 1 \le \dfrac{x}{2}$ $\{x|x \ge -7\}; [-7, \infty)$

17. $2x + 5 \le 11$ and $-3x > 18$ $\{x|x < -6\}; (-\infty, -6)$

18. $x - 4 \ge 1$ or $-3x + 1 \ge -5 - x$
  $\{x|x \le 3 \text{ or } x \ge 5\}; (-\infty, 3] \cup [5, \infty)$

19. $|2x + 3| \le 17$ $\{x|-10 \le x \le 7\}; [-10, 7]$

20. $|3x - 8| > 7$ $\{x|x < \tfrac{1}{3} \text{ or } x > 5\}; \left(\infty, \tfrac{1}{3}\right) \cup (5, \infty)$

You enjoy the single life. You have a terrific career, a support group of close friends, and your own home. Still, you feel the pressure to marry. Your parents belong to a generation that believes that you are not complete until you have found your "other half." They see your staying unattached as a "failure to marry." Perhaps you will someday enter into a permanent relationship, but at the moment you are satisfied with your life and career, and plan to stay single into the future.

This discussion is developed algebraically in the Section 10.6 opener and in Example 6 in Section 10.6.

Radicals, Radical Functions, and Rational Exponents

An increase in the median age at which Americans first marry and the increased rate of divorce have contributed to the growth of our single population. Data indicate that the number of Americans living alone increased quite rapidly toward the end of the twentieth century. However, now this growth rate is beginning to slow down. In this chapter, you will see why radical functions with square roots are used to describe phenomena that are continuing to grow, but whose growth is leveling off. By learning about radicals and radical functions, you will have new algebraic tools for describing your world.

SECTION

10.1

Objectives

1 Evaluate square roots.

2 Evaluate square root functions.

3 Find the domain of square root functions.

4 Use models that are square root functions.

5 Simplify expressions of the form $\sqrt{a^2}$.

6 Evaluate cube root functions.

7 Simplify expressions of the form $\sqrt[3]{a^3}$.

8 Find even and odd roots.

9 Simplify expressions of the form $\sqrt[n]{a^n}$.

RADICAL EXPRESSIONS AND FUNCTIONS

S = Sail area

L = Length

D = Displacement

The America's Cup is the supreme event in ocean sailing. Competition is fierce and the costs are huge. Competitors look to mathematics to provide the critical innovation that can make the difference between winning and losing. The basic dimensions of competitors' yachts must satisfy an inequality containing square roots and cube roots:

$$L + 1.25\sqrt{S} - 9.8\sqrt[3]{D} \leq 16.296.$$

In the inequality, L is the yacht's length, in meters, S is its sail area, in square meters, and D is its displacement, in cubic meters.

In this section, we introduce a new category of expressions and functions that contain roots. You will see why square root functions are used to describe phenomena that are continuing to grow but whose growth is leveling off.

1   Evaluate square roots.

Square Roots From our earlier work with exponents, we are aware that the square of both 5 and −5 is 25:

$$5^2 = 25 \quad \text{and} \quad (-5)^2 = 25.$$

The reverse operation of squaring a number is finding the *square root* of the number. For example,

- One square root of 25 is 5 because $5^2 = 25$.

- Another square root of 25 is −5 because $(-5)^2 = 25$.

In general, **if $b^2 = a$, then b is a square root of a.**

The symbol $\sqrt{}$ is used to denote the *positive* or *principal square root* of a number. For example,

- $\sqrt{25} = 5$ because $5^2 = 25$ and 5 is positive.

- $\sqrt{100} = 10$ because $10^2 = 100$ and 10 is positive.

The symbol $\sqrt{}$ that we use to denote the principal square root is called a **radical sign**. The number under the radical sign is called the **radicand**. Together we refer to the radical sign and its radicand as a **radical expression**.

Radical sign $\sqrt{a}$ Radicand

Radical expression

DEFINITION OF THE PRINCIPAL SQUARE ROOT If a is a nonnegative real number, the nonnegative number b such that $b^2 = a$, denoted by $b = \sqrt{a}$, is the **principal square root** of a.

The symbol $-\sqrt{}$ is used to denote the negative square root of a number. For example,

- $-\sqrt{25} = -5$ because $(-5)^2 = 25$ and -5 is negative.
- $-\sqrt{100} = -10$ because $(-10)^2 = 100$ and -10 is negative.

EXAMPLE 1 Evaluating Square Roots

Evaluate:

a. $\sqrt{81}$ **b.** $-\sqrt{9}$ **c.** $\sqrt{\dfrac{4}{49}}$

d. $\sqrt{0.0064}$ **e.** $\sqrt{36 + 64}$ **f.** $\sqrt{36} + \sqrt{64}$.

SOLUTION

a. $\sqrt{81} = 9$ — The principal square root of 81 is 9 because $9^2 = 81$.

b. $-\sqrt{9} = -3$ — The negative square root of 9 is -3 because $(-3)^2 = 9$.

c. $\sqrt{\dfrac{4}{49}} = \dfrac{2}{7}$ — The principal square root of $\dfrac{4}{49}$ is $\dfrac{2}{7}$ because $\left(\dfrac{2}{7}\right)^2 = \dfrac{4}{49}$.

d. $\sqrt{0.0064} = 0.08$ — The principal square root of 0.0064 is 0.08 because $(0.08)^2 = (0.08)(0.08) = 0.0064$.

e. $\sqrt{36 + 64} = \sqrt{100}$ — Simplify the radicand.
$\quad = 10$ — Take the principal square root of 100, which is 10.

f. $\sqrt{36} + \sqrt{64} = 6 + 8$ — $\sqrt{36} = 6$ because $6^2 = 36$. $\sqrt{64} = 8$
$\quad = 14$ — because $8^2 = 64$. ∎

STUDY TIP

In Example 1, parts (e) and (f), observe that $\sqrt{36 + 64}$ is not equal to $\sqrt{36} + \sqrt{64}$. In general,

$$\sqrt{a + b} \neq \sqrt{a} + \sqrt{b}$$

and

$$\sqrt{a - b} \neq \sqrt{a} - \sqrt{b}.$$

CHECK POINT 1 Evaluate:

a. $\sqrt{64}$ 8 **b.** $-\sqrt{49}$ -7 **c.** $\sqrt{\dfrac{16}{25}}$ $\dfrac{4}{5}$

c. $\sqrt{0.0081}$ 0.09 **e.** $\sqrt{9 + 16}$ 5 **f.** $\sqrt{9} + \sqrt{16}$. 7

Let's see what happens to the radical expression $\sqrt{x}$ if x is a negative number. Is the square root of a negative number a real number? For example, consider $\sqrt{-25}$. Is there a real number whose square is -25? No. Thus, $\sqrt{-25}$ is not a real number. In general, **a square root of a negative number is not a real number**.

2 Evaluate square root functions.

Square Root Functions

Because each nonnegative real number, x, has precisely one principal square root, $\sqrt{x}$, there is a **square root function** defined by

$$f(x) = \sqrt{x}.$$

The domain of this function is $[0, \infty)$. We can graph $f(x) = \sqrt{x}$ by selecting nonnegative real numbers for x. It is easiest to choose perfect squares, numbers that have rational square roots. Table 10.1 shows five such choices for x and the calculations for the corresponding outputs. We plot these ordered pairs as points in the rectangular coordinate system and connect the points with a smooth curve. The graph of $f(x) = \sqrt{x}$ is shown in Figure 10.1.

Table 10.1

x	$f(x) = \sqrt{x}$	(x, y) or $(x, f(x))$
0	$f(0) = \sqrt{0} = 0$	$(0, 0)$
1	$f(1) = \sqrt{1} = 1$	$(1, 1)$
4	$f(4) = \sqrt{4} = 2$	$(4, 2)$
9	$f(9) = \sqrt{9} = 3$	$(9, 3)$
16	$f(16) = \sqrt{16} = 4$	$(16, 4)$

FIGURE 10.1 The graph of the square root function $f(x) = \sqrt{x}$

Is it possible to choose values of x for Table 10.1 that are not squares of integers, or perfect squares? Yes. For example, we can let $x = 3$. Thus, $f(3) = \sqrt{3}$. Because 3 is not a perfect square, $\sqrt{3}$ is an irrational number, one that cannot be expressed as a quotient of integers. We can use a calculator to find a decimal approximation of $\sqrt{3}$.

Many Scientific Calculators	**Many Graphing Calculators**
3 $\boxed{\sqrt{}}$	$\boxed{\sqrt{}}$ 3 $\boxed{\text{ENTER}}$

Rounding the displayed number to two decimal places, $\sqrt{3} \approx 1.73$. This information is shown visually as a point, approximately $(3, 1.73)$, on the graph of $f(x) = \sqrt{x}$ in Figure 10.2.

FIGURE 10.2 Visualizing $\sqrt{3}$ as a point on the graph of $f(x) = \sqrt{x}$

Approximately $(3, 1.73)$

$f(x) = \sqrt{x}$

$\sqrt{3} \approx 1.73$

To evaluate a square root function, we use substitution, just as we did to evaluate other functions.

EXAMPLE 2 Evaluating Square Root Functions

For each function, find the indicated function value:

a. $f(x) = \sqrt{5x - 6}; f(2)$ **b.** $g(x) = -\sqrt{64 - 8x}; g(-3)$.

SOLUTION

a. $f(2) = \sqrt{5 \cdot 2 - 6}$ Substitute 2 for x in $f(x) = \sqrt{5x - 6}$.

$= \sqrt{4} = 2$ Simplify the radicand and take the square root.

b. $g(-3) = -\sqrt{64 - 8(-3)}$ Substitute -3 for x in $g(x) = -\sqrt{64 - 8x}$.

$= -\sqrt{88} \approx -9.38$ Simplify the radicand:

$64 - 8(-3) = 64 - (-24) = 64 + 24 = 88$.

Then use a calculator to approximate $\sqrt{88}$.

CHECK POINT 2 For each function, find the indicated function value:

a. $f(x) = \sqrt{12x - 20}; f(3)$ $f(3) = 4$

b. $g(x) = -\sqrt{9 - 3x}; g(-5).$ $g(-5) = -\sqrt{24} \approx -4.90$

3 Find the domain of square root functions.

We have seen that the domain of a function f is the largest set of real numbers for which the value of $f(x)$ is a real number. Because only nonnegative numbers have real square roots, the domain of a square root function is the set of real numbers for which the radicand is nonnegative.

EXAMPLE 3   Finding the Domain of a Square Root Function

Find the domain of

$$f(x) = \sqrt{3x + 12}.$$

SOLUTION   The domain is the set of real numbers, x, for which the radicand, $3x + 12$, is nonnegative. We set the radicand greater than or equal to 0 and solve the resulting inequality.

$$3x + 12 \geq 0$$
$$3x \geq -12$$
$$x \geq -4$$

The domain of f is $\{x | x \geq -4\}$ or $[-4, \infty)$. ∎

$f(x) = \sqrt{3x + 12}$

FIGURE 10.3

Figure 10.3 shows the graph of $f(x) = \sqrt{3x + 12}$ in a $[-10, 10, 1]$ by $[-10, 10, 1]$ viewing rectangle. The graph appears only for $x \geq -4$, verifying $[-4, \infty)$ as the domain. Can you see how the graph also illustrates this square root function's range? The graph only appears for nonnegative values of y. Thus, the range is $\{y | y \geq 0\}$ or $[0, \infty)$.

 CHECK POINT 3 Find the domain of

$$f(x) = \sqrt{9x - 27}.$$ Domain of f is $\{x | x \geq 3\}$ or $[3, \infty)$.

4 Use models that are square root functions.

The graph of the square root function $f(x) = \sqrt{x}$ is increasing from left to right. However, the rate of increase is slowing down as the graph moves to the right. This is why square root functions are often used to model growing phenomena with growth that is leveling off.

EXAMPLE 4   Modeling with a Square Root Function

The graph in Figure 10.4 shows the percentage of the U.S. population who used the Internet from 2000 through 2003. Although this percent grew from year to year, the growth was leveling off. The data can be modeled by the function

$$P(x) = 6.7\sqrt{x} + 44.4,$$

where $P(x)$ is the percentage of the U.S. population who used the Internet x years after 2000. According to the model, what percentage used the Internet in 2003? Round to the nearest tenth of a percent. How well does the model describe the actual data?

Percentage of U.S. Population Using the Internet

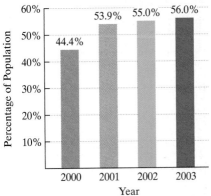

FIGURE 10.4

Source: U.S. Department of Commerce

SOLUTION Because 2003 is 3 years after 2000, we substitute 3 for x and evaluate the function at 3.

$$P(x) = 6.7\sqrt{x} + 44.4 \quad \text{Use the given function.}$$
$$P(3) = 6.7\sqrt{3} + 44.4 \quad \text{Substitute 3 for x.}$$
$$\approx 56.0 \quad \text{Use a calculator.}$$

The model indicates that approximately 56% of the U.S. population used the Internet in 2003. Figure 10.4 on the previous page shows 56%, so our model provides an excellent description of the actual data value for 2003. ∎

 CHECK POINT 4 Use the square root function in Example 4 to find the percentage of Internet users in 2002. Round to the nearest tenth of a percent. How well does the model describe the actual data? Approximately 53.9% used the Internet in 2002. This is the percent given for 2001, not 2002, so the model does not provide a totally accurate description of the data.

---

5 Simplify expressions of the form $\sqrt{a^2}$.

Simplifying Expressions of the Form $\sqrt{a^2}$ You may think that $\sqrt{a^2} = a$. However, this is not necessarily true. Consider the following examples:

$$\sqrt{4^2} = \sqrt{16} = 4$$
$$\sqrt{(-4)^2} = \sqrt{16} = 4.$$

The result is not −4, but rather the absolute value of −4, or 4.

Here is a rule for simplifying expressions of the form $\sqrt{a^2}$:

SIMPLIFYING $\sqrt{a^2}$ For any real number a,
$$\sqrt{a^2} = |a|.$$

In words, the principal square root of a^2 is the absolute value of a.

EXAMPLE 5 Simplifying Radical Expressions

Simplify each expression:

a. $\sqrt{(-6)^2}$ **b.** $\sqrt{(x + 5)^2}$ **c.** $\sqrt{25x^6}$ **d.** $\sqrt{x^2 - 4x + 4}$.

SOLUTION The principal square root of an expression squared is the absolute value of that expression. In parts (a) and (b), we are given squared radicands. In parts (c) and (d), it will first be necessary to express the radicand as an expression that is squared.

a. $\sqrt{(-6)^2} = |-6| = 6$

b. $\sqrt{(x + 5)^2} = |x + 5|$

c. To simplify $\sqrt{25x^6}$, first write $25x^6$ as an expression that is squared: $25x^6 = (5x^3)^2$. Then simplify.

$$\sqrt{25x^6} = \sqrt{(5x^3)^2} = |5x^3| \text{ or } 5|x^3|$$

d. To simplify $\sqrt{x^2 - 4x + 4}$, first write $x^2 - 4x + 4$ as an expression that is squared by factoring the perfect square trinomial: $x^2 - 4x + 4 = (x - 2)^2$. Then simplify.

$$\sqrt{x^2 - 4x + 4} = \sqrt{(x - 2)^2} = |x - 2| \qquad \blacksquare$$

 CHECK POINT 5 Simplify each expression:

a. $\sqrt{(-7)^2}$ 7

b. $\sqrt{(x + 8)^2}$ $|x + 8|$

c. $\sqrt{49x^{10}}$ $|7x^5|$ or $7|x^5|$

d. $\sqrt{x^2 - 6x + 9}$. $|x - 3|$

In some situations, we are told that no radicands involve negative quantities raised to even powers. When the expression being squared is nonnegative, it is not necessary to use absolute value when simplifying $\sqrt{a^2}$. For example, assuming that no radicands contain negative quantities that are squared,

$$\sqrt{x^8} = \sqrt{(x^4)^2} = x^4$$

$$\sqrt{25x^2 + 10x + 1} = \sqrt{(5x + 1)^2} = 5x + 1.$$

Cube Roots and Cube Root Functions   Finding the square root of a number reverses the process of squaring a number. Similarly, finding the cube root of a number reverses the process of cubing a number. For example, $2^3 = 8$, and so the cube root of 8 is 2. The notation that we use is $\sqrt[3]{8} = 2$.

6 Evaluate cube root functions.

DEFINITION OF THE CUBE ROOT OF A NUMBER   The **cube root** of a real number a is written $\sqrt[3]{a}$.

$$\sqrt[3]{a} = b \quad \text{means that} \quad b^3 = a.$$

For example,

$$\sqrt[3]{64} = 4 \qquad \text{because} \qquad 4^3 = 64.$$

$$\sqrt[3]{-27} = -3 \quad \text{because} \quad (-3)^3 = -27.$$

In contrast to square roots, the cube root of a negative number is a real number. All real numbers have cube roots. The cube root of a positive number is positive. The cube root of a negative number is negative.

Because every real number, x, has precisely one cube root, $\sqrt[3]{x}$, there is a **cube root function** defined by

$$f(x) = \sqrt[3]{x}.$$

The domain of this function is the set of all real numbers. We can graph $f(x) = \sqrt[3]{x}$ by selecting perfect cubes, numbers that have rational cube roots, for x.

STUDY TIP

Some cube roots occur so frequently that you might want to memorize them.

$$\sqrt[3]{1} = 1$$
$$\sqrt[3]{8} = 2$$
$$\sqrt[3]{27} = 3$$
$$\sqrt[3]{64} = 4$$
$$\sqrt[3]{125} = 5$$
$$\sqrt[3]{216} = 6$$
$$\sqrt[3]{1000} = 10$$

Table 10.2 shows five such choices for x and the calculations for the corresponding outputs. We plot these ordered pairs as points in the rectangular coordinate system and connect the points with a smooth curve. The graph of $f(x) = \sqrt[3]{x}$ is shown in Figure 10.5.

Table 10.2

x	$f(x) = \sqrt[3]{x}$	(x, y) or $(x, f(x))$
-8	$f(-8) = \sqrt[3]{-8} = -2$	$(-8, -2)$
-1	$f(-1) = \sqrt[3]{-1} = -1$	$(-1, -1)$
0	$f(0) = \sqrt[3]{0} = 0$	$(0, 0)$
1	$f(1) = \sqrt[3]{1} = 1$	$(1, 1)$
8	$f(8) = \sqrt[3]{8} = 2$	$(8, 2)$

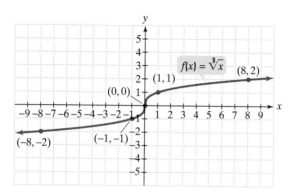

FIGURE 10.5 The graph of the cube root function $f(x) = \sqrt[3]{x}$

Notice that both the domain and the range of $f(x) = \sqrt[3]{x}$ are the set of all real numbers.

EXAMPLE 6 Evaluating Cube Root Functions

For each function, find the indicated function value:

a. $f(x) = \sqrt[3]{x - 2};\ \ f(127)$ **b.** $g(x) = \sqrt[3]{8x - 8};\ \ g(-7)$.

SOLUTION

a. $f(127) = \sqrt[3]{127 - 2}$ Substitute 127 for x in $f(x) = \sqrt[3]{x - 2}$.

$= \sqrt[3]{125}$ Simplify the radicand.

$= 5$ $\sqrt[3]{125} = 5$ because $5^3 = 125$.

b. $g(-7) = \sqrt[3]{8(-7) - 8}$ Substitute -7 for x in $g(x) = \sqrt[3]{8x - 8}$.

$= \sqrt[3]{-64}$ Simplify the radicand: $8(-7) - 8 = -56 - 8 = -64$.

$= -4$ $\sqrt[3]{-64} = -4$ because $(-4)^3 = -64$. ∎

CHECK POINT 6 For each function, find the indicated function value:

a. $f(x) = \sqrt[3]{x - 6};\ \ f(33)$ $f(33) = 3$

b. $g(x) = \sqrt[3]{2x + 2};\ \ g(-5)$. $g(-5) = -2$

7 Simplify expressions of the form $\sqrt[3]{a^3}$.

Because the cube root of a positive number is positive and the cube root of a negative number is negative, absolute value is not needed to simplify expressions of the form $\sqrt[3]{a^3}$.

SIMPLIFYING $\sqrt[3]{a^3}$ For any real number a,

$$\sqrt[3]{a^3} = a.$$

In words, the cube root of any expression cubed is that expression.

EXAMPLE 7   Simplifying a Cube Root

Simplify: $\sqrt[3]{-64x^3}$.

SOLUTION   Begin by expressing the radicand as an expression that is cubed: $-64x^3 = (-4x)^3$. Then simplify.

$$\sqrt[3]{-64x^3} = \sqrt[3]{(-4x)^3} = -4x$$

We can check our answer by cubing $-4x$:

$$(-4x)^3 = (-4)^3x^3 = -64x^3.$$

By obtaining the original radicand, we know that our simplification is correct. ∎

 CHECK POINT 7 Simplify: $\sqrt[3]{-27x^3}$. $-3x$

8 Find even and odd roots.

Even and Odd nth Roots   Up to this point, we have focused on square roots and cube roots. Other radical expressions have different roots. For example, the fifth root of a, written $\sqrt[5]{a}$, is the number b for which $b^5 = a$. Thus,

$$\sqrt[5]{32} = 2 \quad \text{because} \quad 2^5 = 2 \cdot 2 \cdot 2 \cdot 2 \cdot 2 = 32.$$

The radical expression $\sqrt[n]{a}$ represents the **nth root** of a. The number n is called the **index**. An index of 2 represents a square root and is not written. An index of 3 represents a cube root.

If the index n in $\sqrt[n]{a}$ is an odd number, a root is said to be an **odd root**. A cube root is an odd root. Other odd roots have the same characteristics as cube roots. Every real number has exactly one real root when n is odd. The (odd) nth root of a, $\sqrt[n]{a}$, is the number b for which $b^n = a$. An odd root of a positive number is positive and an odd root of a negative number is negative. For example,

$$\sqrt[5]{243} = 3 \quad \text{because} \quad 3^5 = 3 \cdot 3 \cdot 3 \cdot 3 \cdot 3 = 243$$

and $\sqrt[5]{-243} = -3$ because $(-3)^5 = (-3)(-3)(-3)(-3)(-3) = -243.$

If the index n in $\sqrt[n]{a}$ is an even number, a root is said to be an **even root**. A square root is an even root. Other even roots have the same characteristics as square roots. Every positive real number has two real roots when n is even. One root is positive and one is negative. The positive root, called the **principal nth root** and represented by $\sqrt[n]{a}$, is the nonnegative number b for which $b^n = a$. The symbol $-\sqrt[n]{a}$ is used to denote the negative nth root. **An even root of a negative number is not a real number.**

STUDY TIP

Some higher even and odd roots occur so frequently that you might want to memorize them.

Fourth Roots	Fifth Roots
$\sqrt[4]{1} = 1$	$\sqrt[5]{1} = 1$
$\sqrt[4]{16} = 2$	$\sqrt[5]{32} = 2$
$\sqrt[4]{81} = 3$	$\sqrt[5]{243} = 3$
$\sqrt[4]{256} = 4$	
$\sqrt[4]{625} = 5$	

EXAMPLE 8   Finding Even and Odd Roots

Find the indicated root, or state that the expression is not a real number:

    a. $\sqrt[4]{81}$ **b.** $-\sqrt[4]{81}$ **c.** $\sqrt[4]{-81}$ **d.** $\sqrt[5]{-32}.$

SOLUTION

    a. $\sqrt[4]{81} = 3$ The principal fourth root of 81 is 3 because
                       $3^4 = 3 \cdot 3 \cdot 3 \cdot 3 = 81.$

    b. $-\sqrt[4]{81} = -3$ The negative fourth root of 81 is -3 because
                       $(-3)^4 = (-3)(-3)(-3)(-3) = 81.$

c. $\sqrt[4]{-81}$ is not a real number because the index, 4, is even and the radicand, -81, is negative. No real number can be raised to the fourth power to give a negative result such as -81. Real numbers to even powers can only result in nonnegative numbers.

d. $\sqrt[5]{-32} = -2$ because $(-2)^5 = (-2)(-2)(-2)(-2)(-2) = -32$. An odd root of a negative real number is always negative. ■

CHECK POINT 8 Find the indicated root, or state that the expression is not a real number:

a. $\sqrt[4]{16}$ 2   **b.** $-\sqrt[4]{16}$ -2

c. $\sqrt[4]{-16}$ not a real number   **d.** $\sqrt[5]{-1}$. -1

---

9 Simplify expressions of the form $\sqrt[n]{a^n}$.

Simplifying Expressions of the Form $\sqrt[n]{a^n}$ We have seen that

$$\sqrt{a^2} = |a| \quad \text{and} \quad \sqrt[3]{a^3} = a.$$

Expressions of the form $\sqrt[n]{a^n}$ can be simplified in the same manner. Unless a is known to be nonnegative, absolute value notation is needed when n is even. When the index is odd, absolute value bars are not necessary.

> **SIMPLIFYING $\sqrt[n]{a^n}$**   For any real number a,
>
> **1.** If n is even, $\sqrt[n]{a^n} = |a|$.
>
> **2.** If n is odd, $\sqrt[n]{a^n} = a$.

EXAMPLE 9   **Simplifying Radical Expressions**

Simplify:

     a. $\sqrt[4]{(x-3)^4}$ **b.** $\sqrt[5]{(2x+7)^5}$ **c.** $\sqrt[6]{(-5)^6}$.

SOLUTION   Each expression involves the nth root of a radicand raised to the nth power. Thus, each radical expression can be simplified. Absolute value bars are necessary in parts (a) and (c) because the index, n, is even.

     a. $\sqrt[4]{(x-3)^4} = |x-3|$ $\sqrt[n]{a^n} = |a|$ if n is even.

     b. $\sqrt[5]{(2x+7)^5} = 2x+7$ $\sqrt[n]{a^n} = a$ if n is odd.

     c. $\sqrt[6]{(-5)^6} = |-5| = 5$ $\sqrt[n]{a^n} = |a|$ if n is even. ■

CHECK POINT 9 Simplify:

     a. $\sqrt[4]{(x+6)^4}$ $|x+6|$

     b. $\sqrt[5]{(3x-2)^5}$ $3x-2$

     c. $\sqrt[6]{(-8)^6}$.   8

10.1 EXERCISE SET

Student Solutions Manual   CD/Video   PH Math/Tutor Center   MathXL Tutorials on CD   MathXL®   MyMathLab   Interactmath.com

Practice Exercises

In Exercises 1–20, evaluate each expression, or state that the expression is not a real number.

1. $\sqrt{36}$ 6

2. $\sqrt{16}$ 4

3. $-\sqrt{36}$ −6

4. $-\sqrt{16}$ −4

5. $\sqrt{-36}$ not a real number   **6.** $\sqrt{-16}$ not a real number

7. $\sqrt{\dfrac{1}{25}}$ $\dfrac{1}{5}$

8. $\sqrt{\dfrac{1}{49}}$ $\dfrac{1}{7}$

9. $-\sqrt{\dfrac{9}{16}}$ $-\dfrac{3}{4}$

10. $-\sqrt{\dfrac{4}{25}}$ $-\dfrac{2}{5}$

11. $\sqrt{0.81}$ 0.9

12. $\sqrt{0.49}$ 0.7

13. $-\sqrt{0.04}$ −0.2

14. $-\sqrt{0.64}$ −0.8

15. $\sqrt{25 - 16}$ 3

16. $\sqrt{144 + 25}$ 13

17. $\sqrt{25} - \sqrt{16}$ 1

18. $\sqrt{144} + \sqrt{25}$ 17

19. $\sqrt{16 - 25}$ not a real number

20. $\sqrt{25 - 144}$ not a real number

In Exercises 21–26, find the indicated function values for each function. If necessary, round to two decimal places. If the function value is not a real number and does not exist, so state.

21. $f(x) = \sqrt{x - 2}$; $f(18), f(3), f(2), f(-2)$
4; 1; 0; not a real number

22. $f(x) = \sqrt{x - 3}$; $f(28), f(4), f(3), f(-1)$
5; 1; 0; not a real number

23. $g(x) = -\sqrt{2x + 3}$; $g(11), g(1), g(-1), g(-2)$
$-5; -\sqrt{5} \approx -2.24; -1$; not a real number

24. $g(x) = -\sqrt{2x + 1}$; $g(4), g(1), g\left(-\dfrac{1}{2}\right), g(-1)$
$-3; -\sqrt{3} \approx -1.73; 0$; not a real number

25. $h(x) = \sqrt{(x - 1)^2}$; $h(5), h(3), h(0), h(-5)$ 4; 2; 1; 6

26. $h(x) = \sqrt{(x - 2)^2}$; $h(5), h(3), h(0), h(-5)$ 3; 1; 2; 7

In Exercises 27–32, find the domain of each square root function. Then use the domain to match the radical function with its graph. [The graphs are labeled (a) through (f) and are shown in $[-10, 10, 1]$ by $[-10, 10, 1]$ viewing rectangles.]

27. $f(x) = \sqrt{x - 3}$ Domain of f is $\{x|x \ge 3\} = [3, \infty)$; c

28. $f(x) = \sqrt{x + 2}$ Domain of f is $\{x|x \ge -2\} = [-2, \infty)$; f

29. $f(x) = \sqrt{3x + 15}$ Domain of f is $\{x|x \ge -5\} = [-5, \infty)$; d

30. $f(x) = \sqrt{3x - 15}$ Domain of f is $\{x|x \ge 5\} = [5, \infty)$; a

31. $f(x) = \sqrt{6 - 2x}$ Domain of f is $\{x|x \le 3\} = (-\infty, 3]$; e

32. $f(x) = \sqrt{8 - 2x}$ Domain of f is $\{x|x \le 4\} = (-\infty, 4]$; b

a.

b.

c.

d.

e.

f.

In Exercises 33–46, simplify each expression.

33. $\sqrt{5^2}$ 5

34. $\sqrt{7^2}$ 7

35. $\sqrt{(-4)^2}$ 4

36. $\sqrt{(-10)^2}$ 10

37. $\sqrt{(x-1)^2}$ $|x-1|$

38. $\sqrt{(x-2)^2}$ $|x-2|$

39. $\sqrt{36x^4}$ $|6x^2|$ or $6x^2$

40. $\sqrt{81x^4}$ $|9x^2|$ or $9x^2$

41. $-\sqrt{100x^6}$ $-|10x^3|$ or $-10|x^3|$

42. $-\sqrt{49x^6}$ $-|7x^3|$ or $-7|x^3|$

43. $\sqrt{x^2+12x+36}$ $|x+6|$

44. $\sqrt{x^2+14x+49}$ $|x+7|$

45. $-\sqrt{x^2-8x+16}$ $-|x-4|$

46. $-\sqrt{x^2-10x+25}$ $-|x-5|$

In Exercises 47–52, find each cube root.

47. $\sqrt[3]{27}$ 3

48. $\sqrt[3]{64}$ 4

49. $\sqrt[3]{-27}$ -3

50. $\sqrt[3]{-64}$ -4

51. $\sqrt[3]{\dfrac{1}{125}}$ $\dfrac{1}{5}$

52. $\sqrt[3]{\dfrac{1}{1000}}$ $\dfrac{1}{10}$

In Exercises 53–56, find the indicated function values for each function.

53. $f(x) = \sqrt[3]{x-1}$; $f(28), f(9), f(0), f(-63)$ 3; 2; -1; -4

54. $f(x) = \sqrt[3]{x-3}$; $f(30), f(11), f(2), f(-122)$ 3; 2; -1; -5

55. $g(x) = -\sqrt[3]{8x-8}$; $g(2), g(1), g(0)$ -2; 0; 2

56. $g(x) = -\sqrt[3]{2x+1}$; $g(13), g(0), g(-63)$ -3; -1; 5

In Exercises 57–74, find the indicated root, or state that the expression is not a real number.

57. $\sqrt[4]{1}$ 1

58. $\sqrt[5]{1}$ 1

59. $\sqrt[4]{16}$ 2

60. $\sqrt[4]{81}$ 3

61. $-\sqrt[4]{16}$ -2

62. $-\sqrt[4]{81}$ -3

63. $\sqrt[4]{-16}$ not a real number

64. $\sqrt[4]{-81}$ not a real number

65. $\sqrt[5]{-1}$ -1

66. $\sqrt[7]{-1}$ -1

67. $\sqrt[6]{-1}$ not a real number

68. $\sqrt[8]{-1}$ not a real number

69. $-\sqrt[4]{256}$ -4

70. $-\sqrt[4]{10,000}$ -10

71. $\sqrt[6]{64}$ 2

72. $\sqrt[5]{32}$ 2

73. $-\sqrt[5]{32}$ -2

74. $-\sqrt[6]{64}$ -2

In Exercises 75–88, simplify each expression. Include absolute value bars where necessary.

75. $\sqrt[3]{x^3}$ x

76. $\sqrt[5]{x^5}$ x

77. $\sqrt[4]{y^4}$ $|y|$

78. $\sqrt[6]{y^6}$ $|y|$

79. $\sqrt[3]{-8x^3}$ $-2x$

80. $\sqrt[3]{-125x^3}$ $-5x$

81. $\sqrt[3]{(-5)^3}$ -5

82. $\sqrt[3]{(-6)^3}$ -6

83. $\sqrt[4]{(-5)^4}$ 5

84. $\sqrt[6]{(-6)^6}$ 6

85. $\sqrt[4]{(x+3)^4}$ $|x+3|$

86. $\sqrt[4]{(x+5)^4}$ $|x+5|$

87. $\sqrt[5]{-32(x-1)^5}$ $-2(x-1)$

88. $\sqrt[5]{-32(x-2)^5}$ $-2(x-2)$

Practice Plus

In Exercises 89–92, complete each table and graph the given function. Identify the function's domain and range. 89–92. See graphing answer section.

89. $f(x) = \sqrt{x} + 3$

x	$f(x) = \sqrt{x} + 3$
0	3
1	4
4	5
9	6

Domain: $\{x \mid x \geq 0\} = [0, \infty)$; Range: $\{y \mid y \geq 3\} = [3, \infty)$

90. $f(x) = \sqrt{x} - 2$

x	$f(x) = \sqrt{x} - 2$
0	-2
1	-1
4	0
9	1

Domain: $\{x \mid x \geq 0\} = [0, \infty)$; Range: $\{y \mid y \geq -2\} = [-2, \infty)$

91. $f(x) = \sqrt{x-3}$

x	$f(x) = \sqrt{x-3}$
3	0
4	1
7	2
12	3

Domain: $\{x \mid x \geq 3\} = [3, \infty)$; Range: $\{y \mid y \geq 0\} = [0, \infty)$

92. $f(x) = \sqrt{4-x}$

x	$f(x) = \sqrt{4-x}$
-5	3
0	2
3	1
4	0

Domain: $\{x \mid x \leq 4\} = (-\infty, 4]$; Range: $\{y \mid y \geq 0\} = [0, \infty)$

In Exercises 93–96, find the domain of each function.

93. $f(x) = \dfrac{\sqrt[3]{x}}{\sqrt{30-2x}}$ $\{x \mid x < 15\} = (-\infty, 15)$

94. $f(x) = \dfrac{\sqrt[3]{x}}{\sqrt{80-5x}}$ $\{x \mid x < 16\} = (-\infty, 16)$

95. $f(x) = \dfrac{\sqrt{x-1}}{\sqrt{3-x}}$ $\{x \mid 1 \le x < 3\} = [1, 3)$

96. $f(x) = \dfrac{\sqrt{x-2}}{\sqrt{7-x}}$ $\{x \mid 2 \le x < 7\} = [2, 7)$

In Exercises 97–98, evaluate each expression.

97. $\sqrt[3]{\sqrt[4]{16} + \sqrt{625}}$ 3

98. $\sqrt[3]{\sqrt{\sqrt{169} + \sqrt[3]{9}} + \sqrt{\sqrt[3]{1000} + \sqrt[3]{216}}}$ 2

Application Exercises

The table shows the median, or average, heights for boys of various ages in the United States, from birth through 60 months, or five years. The data can be modeled by the radical function

$$f(x) = 2.9\sqrt{x} + 20.1,$$

where $f(x)$ is the median height, in inches, of boys who are x months of age. Use the function to solve Exercises 99–100.

Boys' Median Heights	
Age (months)	Height (inches)
0	20.5
6	27.0
12	30.8
18	32.9
24	35.0
36	37.5
48	40.8
60	43.4

Source: The Portable Pediatrician for Parents, by Laura Walther Nathanson, M.D., FAAP

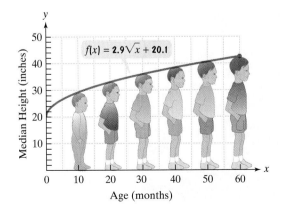

99. According to the model, what is the median height of boys who are 48 months, or four years, old? Use a calculator and round to the nearest tenth of an inch. How well does the model describe the actual median height shown in the table? According to the model, the median height of boys who are 48 months old is approximately 40.2 inches. The model describes actual data well.

100. According to the model, what is the median height of boys who are 60 months, or five years, old? Use a calculator and round to the nearest tenth of an inch. How well does the model describe the actual median height shown in the table? According to the model, the median height of boys who are 60 months old is approximately 42.6 inches. The model describes actual data well.

Police use the function $f(x) = \sqrt{20x}$ to estimate the speed of a car, $f(x)$, in miles per hour, based on the length, x, in feet, of its skid marks upon sudden braking on a dry asphalt road. Use the function to solve Exercises 101–102.

101. A motorist is involved in an accident. A police officer measures the car's skid marks to be 245 feet long. Estimate the speed at which the motorist was traveling before braking. If the posted speed limit is 50 miles per hour and the motorist tells the officer he was not speeding, should the officer believe him? Explain. The motorist was traveling at 70 miles per hour before braking. The officer should not believe the motorist; Answers will vary.

102. A motorist is involved in an accident. A police officer measures the car's skid marks to be 45 feet long. Estimate the speed at which the motorist was traveling before braking. If the posted speed limit is 35 miles per hour and the motorist tells the officer she was not speeding, should the officer believe her? Explain. The motorist was traveling at 30 miles per hour before braking. The officer should believe the motorist; Answers will vary.

Writing in Mathematics

103. What are the square roots of 36? Explain why each of these numbers is a square root.   103–113. Answers will vary.

104. What does the symbol $\sqrt{}$ denote? Which of your answers in Exercise 103 is given by this symbol? Write the symbol needed to obtain the other answer.

105. Explain why $\sqrt{-1}$ is not a real number.

106. Explain how to find the domain of a square root function.

107. Explain how to simplify $\sqrt{a^2}$. Give an example with your explanation.

108. Explain why $\sqrt[3]{8}$ is 2. Then describe what is meant by the cube root of a real number.

109. Describe two differences between odd and even roots.

110. Explain how to simplify $\sqrt[n]{a^n}$ if n is even and if n is odd. Give examples with your explanations.

111. Explain the meaning of the words "radical," "radicand," and "index." Give an example with your explanation.

112. Describe the trend in a boy's growth from birth through five years, shown in the table for Exercises 99–100. Why is a square root function a useful model for the data?

113. The function $f(x) = 6.75\sqrt{x} + 12$ models the amount of new student loans, $f(x)$, in billions of dollars, x years after 1993. Write a word problem using the function. Then solve the problem.

Technology Exercises

114. Use a graphing utility to graph $y_1 = \sqrt{x}$, $y_2 = \sqrt{x+4}$, and $y_3 = \sqrt{x-3}$ in the same $[-5, 10, 1]$ by $[0, 6, 1]$ viewing rectangle. Describe one similarity and one difference that you observe among the graphs. Use the word "shift" in your response.   114–117. See graphing answer section.

115. Use a graphing utility to graph $y = \sqrt{x}$, $y = \sqrt{x}+4$, and $y = \sqrt{x}-3$ in the same $[-1, 10, 1]$ by $[-10, 10, 1]$ viewing rectangle. Describe one similarity and one difference that you observe among the graphs.

116. Use a graphing utility to graph $f(x) = \sqrt{x}$, $g(x) = -\sqrt{x}$, $h(x) = \sqrt{-x}$, and $k(x) = -\sqrt{-x}$ in the same $[-10, 10, 1]$ by $[-4, 4, 1]$ viewing rectangle. Use the graphs to describe the domains and the ranges of functions $f, g, h,$ and k. Domain of f is $\{x \mid x \ge 0\} = [0, \infty)$; Range of f is $\{y \mid y \ge 0\} = [0, \infty)$. Domain of g is $\{x \mid x \ge 0\} = [0, \infty)$; Range of g is $\{y \mid y \le 0\} = (-\infty, 0]$. Domain of h is $\{x \mid x \le 0\} = (-\infty, 0]$; Range of h is $\{y \mid y \ge 0\} = [0, \infty)$. Domain of k is $\{x \mid x \le 0\} = (-\infty, 0]$; Range of k is $\{y \mid y \le 0\} = (-\infty, 0]$.

117. Use a graphing utility to graph $y_1 = \sqrt{x^2}$ and $y_2 = -x$ in the same viewing rectangle.
  a. For what values of x is $\sqrt{x^2} = -x$? $\{x \mid x \le 0\}$
  b. For what values of x is $\sqrt{x^2} \ne -x$? $\{x \mid x > 0\}$

Critical Thinking Exercises

118. Which one of the following is true?   d
  a. The domain of $f(x) = \sqrt[3]{x} - 4$ is $[4, \infty)$.
  b. If n is odd and b is negative, then $\sqrt[n]{b}$ is not a real number.
  c. The expression $\sqrt[n]{4}$ represents increasingly larger numbers for $n = 2, 3, 4, 5, 6,$ and so on.
  d. None of the above is true.

119. Write a function whose domain is $(-\infty, 5]$.   Answers will vary; an example is $f(x) = \sqrt{15 - 3x}$.

120. Let $f(x) = \sqrt{x-3}$ and $g(x) = \sqrt{x+1}$. Find the domain of $f + g$ and $\dfrac{f}{g}$.   Domain of $f + g$ is $\{x \mid x \ge 3\} = [3, \infty)$ and domain of $\dfrac{f}{g}$ is $\{x \mid x \ge 3\} = [3, \infty)$.

121. Simplify: $\sqrt{(2x+3)^{10}}$. $|(2x+3)^5|$

In Exercises 122–123, graph each function by hand. Then describe the relationship between the function that you graphed and the graph of $f(x) = \sqrt{x}$.   122–123. See graphing answer section.

122. $g(x) = \sqrt{x} + 2$ The graph of g is the graph of f shifted up 2 units.

123. $h(x) = \sqrt{x+3}$ The graph of h is the graph of f shifted left 3 units.

Review Exercises

124. Simplify: $3x - 2[x - 3(x + 5)]$. (Section 1.8, Example 11) $7x + 30$

125. Simplify: $(-3x^{-4}y^3)^{-2}$. (Section 5.7, Example 6) $\dfrac{x^8}{9y^6}$

126. Solve: $|3x - 4| > 11$. (Section 9.3, Example 4) $\left\{x \mid x < -\tfrac{7}{3} \text{ or } x > 5\right\}$ or $\left(-\infty, -\tfrac{7}{3}\right) \cup (5, \infty)$

10.2

RATIONAL EXPONENTS

Objectives

1 Use the definition of $a^{\frac{1}{n}}$.

2 Use the definition of $a^{\frac{m}{n}}$.

3 Use the definition of $a^{-\frac{m}{n}}$.

4 Use models that contain rational exponents.

5 Simplify expressions with rational exponents.

6 Simplify radical expressions using rational exponents.

Marine iguanas of the Galápagos Islands

The Galápagos Islands are a chain of volcanic islands lying 600 miles west of Ecuador. They are famed for over 5000 species of plants and animals, including a rare flightless cormorant, marine iguanas, and giant tortoises weighing more than 600 pounds. Early in 2001, the plants and wildlife that live in the Galápagos were at risk from a massive oil spill that flooded 150,000 gallons of toxic fuel into one of the world's most fragile ecosystems. The long-term danger of the accident is that fuel sinking to the ocean floor will destroy algae that is vital to the food chain. Any imbalance in the food chain could threaten the rare Galápagos plant and animal species that have evolved for thousands of years in isolation with little human intervention.

At risk on these ecologically vulnerable islands are unique flora and fauna that helped to inspire Charles Darwin's theory of evolution. Darwin made an enormous collection of the islands' plant species. The function

$$f(x) = 29x^{\frac{1}{3}}$$

models the number of plant species, $f(x)$, on the various islands of the Galápagos in terms of the area, x, in square miles, of a particular island. But x to the *what* power? How can we interpret the information given by this function? In this section, we turn our attention to rational exponents such as $\frac{1}{3}$ and their relationship to roots of real numbers.

Defining Rational Exponents We define rational exponents so that their properties are the same as the properties for integer exponents. For example, suppose that $x = 7^{\frac{1}{3}}$. We know that exponents are multiplied when an exponential expression is raised to a power. For this to be true,

$$x^3 = \left(7^{\frac{1}{3}}\right)^3 = 7^{\frac{1}{3} \cdot 3} = 7^1 = 7.$$

We see that $x^3 = 7$. This means that x is the number whose cube is 7. Thus, $x = \sqrt[3]{7}$. Remember that we began with $x = 7^{\frac{1}{3}}$. This means that

$$7^{\frac{1}{3}} = \sqrt[3]{7}.$$

We can generalize this idea with the following definition:

1 Use the definition of $a^{\frac{1}{n}}$.

THE DEFINITION OF $a^{\frac{1}{n}}$ If $\sqrt[n]{a}$ represents a real number and $n \geq 2$ is an integer, then

$$a^{\frac{1}{n}} = \sqrt[n]{a}.$$

> The denominator of the rational exponent is the radical's index.

If a is negative, n must be odd. If a is nonnegative, n can be any index.

EXAMPLE 1 Using the Definition of $a^{\frac{1}{n}}$

Use radical notation to rewrite each expression. Simplify, if possible:

a. $64^{\frac{1}{2}}$ **b.** $(-125)^{\frac{1}{3}}$ **c.** $(6x^2y)^{\frac{1}{5}}$.

SOLUTION

a. $64^{\frac{1}{2}} = \sqrt{64} = 8$

> The denominator is the index.

b. $(-125)^{\frac{1}{3}} = \sqrt[3]{-125} = -5$ **c.** $(6x^2y)^{\frac{1}{5}} = \sqrt[5]{6x^2y}$

CHECK POINT 1 Use radical notation to rewrite each expression. Simplify, if possible:

a. $25^{\frac{1}{2}}$ $\sqrt{25} = 5$ **b.** $(-8)^{\frac{1}{3}}$ $\sqrt[3]{-8} = -2$ **c.** $(5xy^2)^{\frac{1}{4}}$. $\sqrt[4]{5xy^2}$

In our next example, we begin with radical notation and rewrite the expression with rational exponents.

The radical's index becomes the exponent's denominator.

$$\sqrt[n]{a} = a^{\frac{1}{n}}$$

The radicand becomes the base.

EXAMPLE 2   Using the Definition of $a^{\frac{1}{n}}$

Rewrite with rational exponents:

a. $\sqrt[5]{13ab}$ **b.** $\sqrt[7]{\dfrac{xy^2}{17}}$.

SOLUTION   Parentheses are needed to show that the entire radicand becomes the base.

a. $\sqrt[5]{13ab} = (13ab)^{\frac{1}{5}}$

The index is the exponent's denominator.

b. $\sqrt[7]{\dfrac{xy^2}{17}} = \left(\dfrac{xy^2}{17}\right)^{\frac{1}{7}}$

■

 CHECK POINT 2 Rewrite with rational exponents:

a. $\sqrt[4]{5xy}$ $(5xy)^{\frac{1}{4}}$ **b.** $\sqrt[5]{\dfrac{a^3b}{2}}$. $\left(\dfrac{a^3b}{2}\right)^{\frac{1}{5}}$

2 Use the definition of $a^{\frac{m}{n}}$.

Can rational exponents have numerators other than 1? The answer is yes. If the numerator is some other integer, we still want to multiply exponents when raising a power to a power. For this reason,

$$a^{\frac{2}{3}} = (a^{\frac{1}{3}})^2 \text{ and } a^{\frac{2}{3}} = (a^2)^{\frac{1}{3}}.$$

This means $(\sqrt[3]{a})^2$.   This means $\sqrt[3]{a^2}$.

Thus,

$$a^{\frac{2}{3}} = (\sqrt[3]{a})^2 = \sqrt[3]{a^2}.$$

Do you see that the denominator, 3, of the rational exponent is the same as the index of the radical? The numerator, 2, of the rational exponent serves as an exponent in each of the two radical forms. We generalize these ideas with the following definition:

THE DEFINITION OF $a^{\frac{m}{n}}$ If $\sqrt[n]{a}$ represents a real number, $\dfrac{m}{n}$ is a positive rational number reduced to lowest terms, and $n \geq 2$ is an integer, then

$$a^{\frac{m}{n}} = \left(\sqrt[n]{a}\right)^m$$

> First take the nth root of a.

and

$$a^{\frac{m}{n}} = \sqrt[n]{a^m}.$$

> First raise a to the m power.

The first form of the definition shown in the box involves taking the root first. This form is often preferable because smaller numbers are involved. Notice that the rational exponent consists of two parts, indicated by the following voice balloons:

> The numerator is the exponent.

$$a^{\frac{m}{n}} = \left(\sqrt[n]{a}\right)^m.$$

> The denominator is the radical's index.

EXAMPLE 3   Using the Definition of $a^{\frac{m}{n}}$

Use radical notation to rewrite each expression and simplify:

a. $1000^{\frac{2}{3}}$ **b.** $16^{\frac{3}{2}}$ **c.** $-32^{\frac{3}{5}}$.

SOLUTION

a. $(1000)^{\frac{2}{3}} = \left(\sqrt[3]{1000}\right)^2 = 10^2 = 100$

> The denominator of $\frac{2}{3}$ is the root and the numerator is the exponent.

b. $16^{\frac{3}{2}} = \left(\sqrt{16}\right)^3 = 4^3 = 64$

c. $-32^{\frac{3}{5}} = -\left(\sqrt[5]{32}\right)^3 = -2^3 = -8$

> The base is 32 and the negative sign is not affected by the exponent.

CHECK POINT 3 Use radical notation to rewrite each expression and simplify:

a. $8^{\frac{4}{3}}$ $\left(\sqrt[3]{8}\right)^4 = 16$ **b.** $25^{\frac{3}{2}}$ $\left(\sqrt{25}\right)^3 = 125$ **c.** $-81^{\frac{3}{4}}$. $-\left(\sqrt[4]{81}\right)^3 = -27$

In our next example, we begin with radical notation and rewrite the expression with rational exponents. When changing from radical form to exponential form, the index becomes the denominator of the rational exponent.

EXAMPLE 4   Using the Definition of $a^{\frac{m}{n}}$

Rewrite with rational exponents:

a. $\sqrt[3]{7^5}$ **b.** $\left(\sqrt[4]{13xy}\right)^9$.

TECHNOLOGY

Here are the calculator keystroke sequences for $1000^{\frac{2}{3}}$:

Many Scientific Calculators

$1000 \;\boxed{y^x}\; \boxed{(}\; 2 \;\boxed{\div}\; 3 \;\boxed{)}\; \boxed{=}$

Many Graphing Calculators

$1000 \;\boxed{\wedge}\; \boxed{(}\; 2 \;\boxed{\div}\; 3 \;\boxed{)}\; \boxed{\text{ENTER}}$

SOLUTION

a. $\sqrt[3]{7^5} = 7^{\frac{5}{3}}$

The index is the exponent's denominator.

b. $\left(\sqrt[4]{13xy}\right)^9 = (13xy)^{\frac{9}{4}}$

■

✔ **CHECK POINT 4** Rewrite with rational exponents:

a. $\sqrt[3]{6^4}$ $6^{\frac{4}{3}}$ **b.** $\left(\sqrt[5]{2xy}\right)^7$. $(2xy)^{\frac{7}{5}}$

3 Use the definition of $a^{-\frac{m}{n}}$.

Can a rational exponent be negative? Yes. The way that negative rational exponents are defined is similar to the way that negative integer exponents are defined.

> **THE DEFINITION OF** $a^{-\frac{m}{n}}$ If $a^{\frac{m}{n}}$ is a nonzero real number, then
>
> $$a^{-\frac{m}{n}} = \frac{1}{a^{\frac{m}{n}}}.$$

EXAMPLE 5   Using the Definition of $a^{-\frac{m}{n}}$

Rewrite each expression with a positive exponent. Simplify, if possible:

a. $36^{-\frac{1}{2}}$ **b.** $125^{-\frac{1}{3}}$ **c.** $16^{-\frac{3}{4}}$ **d.** $(7xy)^{-\frac{4}{7}}$.

SOLUTION

a. $36^{-\frac{1}{2}} = \frac{1}{36^{\frac{1}{2}}} = \frac{1}{\sqrt{36}} = \frac{1}{6}$

b. $125^{-\frac{1}{3}} = \frac{1}{125^{\frac{1}{3}}} = \frac{1}{\sqrt[3]{125}} = \frac{1}{5}$

c. $16^{-\frac{3}{4}} = \frac{1}{16^{\frac{3}{4}}} = \frac{1}{\left(\sqrt[4]{16}\right)^3} = \frac{1}{2^3} = \frac{1}{8}$

d. $(7xy)^{-\frac{4}{7}} = \frac{1}{(7xy)^{\frac{4}{7}}}$

■

✔ **CHECK POINT 5** Rewrite each expression with a positive exponent. Simplify, if possible:

a. $100^{-\frac{1}{2}}$ $\frac{1}{100^{\frac{1}{2}}} = \frac{1}{10}$ **b.** $8^{-\frac{1}{3}}$ $\frac{1}{8^{\frac{1}{3}}} = \frac{1}{2}$

c. $32^{-\frac{3}{5}}$ $\frac{1}{32^{\frac{3}{5}}} = \frac{1}{8}$ **d.** $(3xy)^{-\frac{5}{9}}$. $\frac{1}{(3xy)^{\frac{5}{9}}}$

4 Use models that contain rational exponents.

Applications   Now that you know the meaning of rational exponents, you can work with mathematical models that contain these exponents.

EXAMPLE 6 Calculating Windchill

The way that we perceive the temperature on a cold day depends on both air temperature and wind speed. The windchill is what the air temperature would have to be with no wind to achieve the same chilling effect on the skin. In 2002, the National Weather Service issued new windchill temperatures, shown in Table 10.3. (One reason for this new windchill index is that the wind speed is now calculated at 5 feet, the average height of the human body's face, rather than 33 feet, the height of the standard anemometer, an instrument that calculates wind speed.)

Table 10.3 New Windchill Temperature Index

Air Temperature (°F)

Wind Speed (miles per hour)	30	25	20	15	10	5	0	−5	−10	−15	−20	−25
5	25	19	13	7	1	−5	−11	−16	−22	−28	−34	−40
10	21	15	9	3	−4	−10	−16	−22	−28	−35	−41	−47
15	19	13	6	0	−7	−13	−19	−26	−32	−39	−45	−51
20	17	11	4	−2	−9	−15	−22	−29	−35	−42	−48	−55
25	16	9	3	−4	−11	−17	−24	−31	−37	−44	−51	−58
30	15	8	1	−5	−12	−19	−26	−33	−39	−46	−53	−60
35	14	7	0	−7	−14	−21	−27	−34	−41	−48	−55	−62
40	13	6	−1	−8	−15	−22	−29	−36	−43	−50	−57	−64
45	12	5	−2	−9	−16	−23	−30	−37	−44	−51	−58	−65
50	12	4	−3	−10	−17	−24	−31	−38	−45	−52	−60	−67
55	11	4	−3	−11	−18	−25	−32	−39	−46	−54	−61	−68
60	10	3	−4	−11	−19	−26	−33	−40	−48	−55	−62	−69

▒ Frostbite occurs in 15 minutes or less.

Source: National Weather Service

The windchill temperatures shown in Table 10.3 can be calculated using

$$C = 35.74 + 0.6215t - 35.74v^{\frac{4}{25}} + 0.4275tv^{\frac{4}{25}}$$

in which C is the windchill, in degrees Fahrenheit, t is the air temperature, in degrees Fahrenheit, and v is the wind speed, in miles per hour. Use the formula to find the windchill temperature, to the nearest degree, when the air temperature is 25°F and the wind speed is 30 miles per hour.

SOLUTION Because we are interested in the windchill when the air temperature is 25°F and the wind speed is 30 miles per hour, substitute 25 for t and 30 for v. Then calculate C.

$C = 35.74 + 0.6215t - 35.74v^{\frac{4}{25}} + 0.4275tv^{\frac{4}{25}}$ *This is the given formula.*

$C = 35.74 + 0.6215(25) - 35.74(30)^{\frac{4}{25}} + 0.4275(25)(30)^{\frac{4}{25}}$ *Substitute 25 for t and 30 for v.*

≈ 8 *Use a calculator.*

When the air temperature is 25°F and the wind speed is 30 miles per hour, the windchill temperature is 8°. In other words, a 30-mile-per-hour wind makes 25° feel like 8°. ■

 CHECK POINT 6 Use the windchill formula in Example 6 to find the windchill temperature, to the nearest degree, when the air temperature is 35°F and the wind speed is 15 miles per hour. The windchill temperature is 25°.

5 Simplify expressions with rational exponents.

Properties of Rational Exponents The same properties apply to rational exponents as to integer exponents. The following is a summary of these properties:

> **PROPERTIES OF RATIONAL EXPONENTS** If m and n are rational exponents, and a and b are real numbers for which the following expressions are defined, then
>
> **1.** $b^m \cdot b^n = b^{m+n}$ — When multiplying exponential expressions with the same base, add the exponents. Use this sum as the exponent of the common base.
>
> **2.** $\dfrac{b^m}{b^n} = b^{m-n}$ — When dividing exponential expressions with the same base, subtract the exponents. Use this difference as the exponent of the common base.
>
> **3.** $(b^m)^n = b^{mn}$ — When an exponential expression is raised to a power, multiply the exponents. Place the product of the exponents on the base and remove the parentheses.
>
> **4.** $(ab)^n = a^n b^n$ — When a product is raised to a power, raise each factor to that power and multiply.
>
> **5.** $\left(\dfrac{a}{b}\right)^n = \dfrac{a^n}{b^n}$ — When a quotient is raised to a power, raise the numerator to that power and divide by the denominator to that power.

We can use these properties to simplify exponential expressions with rational exponents. As with integer exponents, an expression with rational exponents is **simplified** when:

- No parentheses appear.
- No powers are raised to powers.
- Each base occurs only once.
- No negative or zero exponents appear.

EXAMPLE 7 Simplifying Expressions with Rational Exponents

Simplify:

a. $6^{\frac{1}{7}} \cdot 6^{\frac{4}{7}}$ **b.** $\dfrac{32x^{\frac{1}{2}}}{16x^{\frac{3}{4}}}$ **c.** $\left(8.3^{\frac{3}{4}}\right)^{\frac{2}{3}}$ **d.** $\left(x^{-\frac{2}{5}}y^{\frac{1}{3}}\right)^{\frac{1}{2}}$.

SOLUTION

a. $6^{\frac{1}{7}} \cdot 6^{\frac{4}{7}} = 6^{\frac{1}{7}+\frac{4}{7}}$ *To multiply with the same base, add exponents.*

$\qquad\qquad = 6^{\frac{5}{7}}$ *Simplify:* $\dfrac{1}{7} + \dfrac{4}{7} = \dfrac{5}{7}.$

b. $\dfrac{32x^{\frac{1}{2}}}{16x^{\frac{3}{4}}} = \dfrac{32}{16}x^{\frac{1}{2}-\frac{3}{4}}$ *Divide coefficients. To divide with the same base, subtract exponents.*

$\qquad\qquad = 2x^{\frac{2}{4}-\frac{3}{4}}$ *Write exponents in terms of the LCD, 4.*

$\qquad\qquad = 2x^{-\frac{1}{4}}$ *Subtract:* $\dfrac{2}{4} - \dfrac{3}{4} = -\dfrac{1}{4}.$

$\qquad\qquad = \dfrac{2}{x^{\frac{1}{4}}}$ *Rewrite with a positive exponent:* $a^{-\frac{m}{n}} = \dfrac{1}{a^{\frac{m}{n}}}.$

c. $\left(8.3^{\frac{3}{4}}\right)^{\frac{2}{3}} = 8.3^{\left(\frac{3}{4}\right)\left(\frac{2}{3}\right)}$ To raise a power to a power, multiply exponents.

$= 8.3^{\frac{1}{2}}$ Multiply: $\dfrac{3}{4} \cdot \dfrac{2}{3} = \dfrac{6}{12} = \dfrac{1}{2}$.

d. $\left(x^{-\frac{2}{5}}y^{\frac{1}{3}}\right)^{\frac{1}{2}} = \left(x^{-\frac{2}{5}}\right)^{\frac{1}{2}}\left(y^{\frac{1}{3}}\right)^{\frac{1}{2}}$ To raise a product to a power, raise each factor to the power.

$= x^{-\frac{1}{5}}y^{\frac{1}{6}}$ Multiply: $-\dfrac{2}{5} \cdot \dfrac{1}{2} = -\dfrac{1}{5}$ and $\dfrac{1}{3} \cdot \dfrac{1}{2} = \dfrac{1}{6}$.

$= \dfrac{y^{\frac{1}{6}}}{x^{\frac{1}{5}}}$ Rewrite with positive exponents.

 CHECK POINT 7 Simplify:

a. $7^{\frac{1}{2}} \cdot 7^{\frac{1}{3}}$ $7^{\frac{5}{6}}$ **b.** $\dfrac{50x^{\frac{1}{3}}}{10x^{\frac{4}{3}}}$ $\dfrac{5}{x}$ **c.** $\left(9.1^{\frac{2}{5}}\right)^{\frac{3}{4}}$ $9.1^{\frac{3}{10}}$ **d.** $\left(x^{-\frac{3}{5}}y^{\frac{1}{4}}\right)^{\frac{1}{3}}.$ $\dfrac{y^{\frac{1}{12}}}{x^{\frac{1}{5}}}$

6 Simplify radical expressions using rational exponents.

Using Rational Exponents to Simplify Radical Expressions Some radical expressions can be simplified using rational exponents. We will use the following procedure:

SIMPLIFYING RADICAL EXPRESSIONS USING RATIONAL EXPONENTS

1. Rewrite each radical expression as an exponential expression with a rational exponent.
2. Simplify using properties of rational exponents.
3. Rewrite in radical notation if rational exponents still appear.

EXAMPLE 8 Simplifying Radical Expressions Using Rational Exponents

Use rational exponents to simplify:

a. $\sqrt[10]{x^5}$ **b.** $\sqrt[3]{27a^{15}}$ **c.** $\sqrt[4]{x^6y^2}$ **d.** $\sqrt{x} \cdot \sqrt[3]{x}$ **e.** $\sqrt[3]{\sqrt{x}}.$

SOLUTION

a. $\sqrt[10]{x^5} = x^{\frac{5}{10}}$ Rewrite as an exponential expression.

$= x^{\frac{1}{2}}$ Simplify the exponent.

$= \sqrt{x}$ Rewrite in radical notation.

b. $\sqrt[3]{27a^{15}} = (27a^{15})^{\frac{1}{3}}$ Rewrite as an exponential expression.

$= 27^{\frac{1}{3}}(a^{15})^{\frac{1}{3}}$ Raise each factor in parentheses to the $\frac{1}{3}$ power.

$= \sqrt[3]{27} \cdot a^{15\left(\frac{1}{3}\right)}$ To raise a power to a power, multiply exponents.

$= 3a^5$ $\sqrt[3]{27} = 3$. Multiply exponents: $15 \cdot \frac{1}{3} = 5$.

c. $\sqrt[4]{x^6 y^2} = (x^6 y^2)^{\frac{1}{4}}$ Rewrite as an exponential expression.

$= (x^6)^{\frac{1}{4}}(y^2)^{\frac{1}{4}}$ Raise each factor in parentheses to the $\frac{1}{4}$ power.

$= x^{\frac{6}{4}} y^{\frac{2}{4}}$ To raise powers to powers, multiply.

$= x^{\frac{3}{2}} y^{\frac{1}{2}}$ Simplify.

$= (x^3 y)^{\frac{1}{2}}$ $a^n b^n = (ab)^n$

$= \sqrt{x^3 y}$ Rewrite in radical notation.

d. $\sqrt{x} \cdot \sqrt[3]{x} = x^{\frac{1}{2}} \cdot x^{\frac{1}{3}}$ Rewrite as exponential expressions.

$= x^{\frac{1}{2} + \frac{1}{3}}$ To multiply with the same base, add exponents.

$= x^{\frac{3}{6} + \frac{2}{6}}$ Write exponents in terms of the LCD, 6.

$= x^{\frac{5}{6}}$ Add: $\frac{3}{6} + \frac{2}{6} = \frac{5}{6}$.

$= \sqrt[6]{x^5}$ Rewrite in radical notation.

e. $\sqrt[3]{\sqrt{x}} = \sqrt[3]{x^{\frac{1}{2}}}$ Write the radicand as an exponential expression.

$= \left(x^{\frac{1}{2}}\right)^{\frac{1}{3}}$ Write the entire expression in exponential form.

$= x^{\frac{1}{6}}$ To raise powers to powers, multiply: $\frac{1}{2} \cdot \frac{1}{3} = \frac{1}{6}$.

$= \sqrt[6]{x}$ Rewrite in radical notation.

✔ **CHECK POINT 8** Use rational exponents to simplify:

a. $\sqrt[6]{x^3}$ **b.** $\sqrt[3]{8a^{12}}$ $2a^4$ **c.** $\sqrt[8]{x^4 y^2}$ **d.** $\dfrac{\sqrt{x}}{\sqrt[3]{x}}$ **e.** $\sqrt{\sqrt[3]{x}}$.

10.2 EXERCISE SET

Student Solutions Manual   CD/Video   PH Math/Tutor Center   MathXL Tutorials on CD   MathXL®   MyMathLab   Interactmath.com

Practice Exercises

In Exercises 1–20, use radical notation to rewrite each expression. Simplify, if possible.

1. $49^{\frac{1}{2}}$ $\sqrt{49} = 7$

2. $100^{\frac{1}{2}}$ $\sqrt{100} = 10$

3. $(-27)^{\frac{1}{3}}$ $\sqrt[3]{-27} = -3$

4. $(-64)^{\frac{1}{3}}$ $\sqrt[3]{-64} = -4$

5. $-16^{\frac{1}{4}}$ $-\sqrt[4]{16} = -2$

6. $-81^{\frac{1}{4}}$ $-\sqrt[4]{81} = -3$

7. $(xy)^{\frac{1}{3}}$ $\sqrt[3]{xy}$

8. $(xy)^{\frac{1}{4}}$ $\sqrt[4]{xy}$

9. $(2xy^3)^{\frac{1}{5}}$ $\sqrt[5]{2xy^3}$

10. $(3xy^4)^{\frac{1}{5}}$ $\sqrt[5]{3xy^4}$

11. $81^{\frac{3}{2}}$ $\left(\sqrt{81}\right)^3 = 729$

12. $25^{\frac{3}{2}}$ $\left(\sqrt{25}\right)^3 = 125$

13. $125^{\frac{2}{3}}$ $\left(\sqrt[3]{125}\right)^2 = 25$

14. $1000^{\frac{2}{3}}$ $\left(\sqrt[3]{1000}\right)^2 = 100$

15. $(-32)^{\frac{3}{5}}$ $\left(\sqrt[5]{-32}\right)^3 = -8$

16. $(-27)^{\frac{2}{3}}$ $\left(\sqrt[3]{-27}\right)^2 = 9$

17. $27^{\frac{2}{3}} + 16^{\frac{3}{4}}$
$\left(\sqrt[3]{27}\right)^2 + \left(\sqrt[4]{16}\right)^3 = 17$

18. $4^{\frac{5}{2}} - 8^{\frac{2}{3}}$
$\left(\sqrt{4}\right)^5 - \left(\sqrt[3]{8}\right)^2 = 28$

19. $(xy)^{\frac{4}{7}}$ $\sqrt[7]{(xy)^4}$

20. $(xy)^{\frac{4}{9}}$ $\sqrt[9]{(xy)^4}$

In Exercises 21–38, rewrite each expression with rational exponents.

21. $\sqrt{7}$ $7^{\frac{1}{2}}$

22. $\sqrt{13}$ $13^{\frac{1}{2}}$

23. $\sqrt[3]{5}$ $5^{\frac{1}{3}}$

24. $\sqrt[3]{6}$ $6^{\frac{1}{3}}$

25. $\sqrt[5]{11x}$ $(11x)^{\frac{1}{5}}$

26. $\sqrt[5]{13x}$ $(13x)^{\frac{1}{5}}$

27. $\sqrt{x^3}$ $x^{\frac{3}{2}}$

28. $\sqrt{x^5}$ $x^{\frac{5}{2}}$

29. $\sqrt[5]{x^3}$ $x^{\frac{3}{5}}$

30. $\sqrt[7]{x^4}$ $x^{\frac{4}{7}}$

31. $\sqrt[5]{x^2 y}$ $(x^2 y)^{\frac{1}{5}}$

32. $\sqrt[7]{xy^3}$ $(xy^3)^{\frac{1}{7}}$

33. $\left(\sqrt{19xy}\right)^3$ $(19xy)^{\frac{3}{2}}$ **34.** $\left(\sqrt{11xy}\right)^3$ $(11xy)^{\frac{3}{2}}$

35. $\left(\sqrt[6]{7xy^2}\right)^5$ $(7xy^2)^{\frac{5}{6}}$ **36.** $\left(\sqrt[6]{9x^2y}\right)^5$ $(9x^2y)^{\frac{5}{6}}$

37. $2x\sqrt[3]{y^2}$ $2xy^{\frac{2}{3}}$ **38.** $4x\sqrt[5]{y^2}$ $4xy^{\frac{2}{5}}$

In Exercises 39–54, rewrite each expression with a positive rational exponent. Simplify, if possible.

39. $49^{-\frac{1}{2}}$ $\dfrac{1}{49^{\frac{1}{2}}} = \dfrac{1}{7}$ **40.** $9^{-\frac{1}{2}}$ $\dfrac{1}{9^{\frac{1}{2}}} = \dfrac{1}{3}$

41. $27^{-\frac{1}{3}}$ $\dfrac{1}{27^{\frac{1}{3}}} = \dfrac{1}{3}$ **42.** $125^{-\frac{1}{3}}$ $\dfrac{1}{125^{\frac{1}{3}}} = \dfrac{1}{5}$

43. $16^{-\frac{3}{4}}$ $\dfrac{1}{16^{\frac{3}{4}}} = \dfrac{1}{8}$ **44.** $81^{-\frac{5}{4}}$ $\dfrac{1}{81^{\frac{5}{4}}} = \dfrac{1}{243}$

45. $8^{-\frac{2}{3}}$ $\dfrac{1}{8^{\frac{2}{3}}} = \dfrac{1}{4}$ **46.** $32^{-\frac{4}{5}}$ $\dfrac{1}{32^{\frac{4}{5}}} = \dfrac{1}{16}$

47. $\left(\dfrac{8}{27}\right)^{-\frac{1}{3}}$ $\left(\dfrac{27}{8}\right)^{\frac{1}{3}} = \dfrac{3}{2}$ **48.** $\left(\dfrac{8}{125}\right)^{-\frac{1}{3}}$ $\left(\dfrac{125}{8}\right)^{\frac{1}{3}} = \dfrac{5}{2}$

49. $(-64)^{-\frac{2}{3}}$ $\dfrac{1}{(-64)^{\frac{2}{3}}} = \dfrac{1}{16}$ **50.** $(-8)^{-\frac{2}{3}}$ $\dfrac{1}{(-8)^{\frac{2}{3}}} = \dfrac{1}{4}$

51. $(2xy)^{-\frac{7}{10}}$ $\dfrac{1}{(2xy)^{\frac{7}{10}}}$ **52.** $(4xy)^{-\frac{4}{7}}$ $\dfrac{1}{(4xy)^{\frac{4}{7}}}$

53. $5xz^{-\frac{1}{3}}$ $\dfrac{5x}{z^{\frac{1}{3}}}$ **54.** $7xz^{-\frac{1}{4}}$ $\dfrac{7x}{z^{\frac{1}{4}}}$

In Exercises 55–78, use properties of rational exponents to simplify each expression. Assume that all variables represent positive numbers.

55. $3^{\frac{3}{4}} \cdot 3^{\frac{1}{4}}$ 3 **56.** $5^{\frac{2}{3}} \cdot 5^{\frac{1}{3}}$ 5

57. $\dfrac{16^{\frac{3}{4}}}{16^{\frac{1}{4}}}$ 4 **58.** $\dfrac{100^{\frac{3}{4}}}{100^{\frac{1}{4}}}$ 10

59. $x^{\frac{1}{2}} \cdot x^{\frac{1}{3}}$ $x^{\frac{5}{6}}$ **60.** $x^{\frac{1}{2}} \cdot x^{\frac{2}{3}}$ $x^{\frac{7}{6}}$

61. $\dfrac{x^{\frac{4}{5}}}{x^{\frac{1}{5}}}$ $x^{\frac{3}{5}}$ **62.** $\dfrac{x^{\frac{3}{7}}}{x^{\frac{1}{7}}}$ $x^{\frac{2}{7}}$

63. $\dfrac{x^{\frac{1}{3}}}{x^{\frac{3}{4}}}$ $\dfrac{1}{x^{\frac{5}{12}}}$ **64.** $\dfrac{x^{\frac{1}{4}}}{x^{\frac{3}{5}}}$ $\dfrac{1}{x^{\frac{7}{20}}}$

65. $\left(5^{\frac{2}{3}}\right)^3$ 25 **66.** $\left(3^{\frac{4}{5}}\right)^5$ 81

67. $\left(y^{-\frac{2}{3}}\right)^{\frac{1}{4}}$ $\dfrac{1}{y^{\frac{1}{6}}}$ **68.** $\left(y^{-\frac{3}{4}}\right)^{\frac{1}{6}}$ $\dfrac{1}{y^{\frac{1}{8}}}$

69. $\left(2x^{\frac{1}{5}}\right)^5$ $32x$ **70.** $\left(2x^{\frac{1}{4}}\right)^4$ $16x$

71. $(25x^4y^6)^{\frac{1}{2}}$ $5x^2y^3$ **72.** $(125x^9y^6)^{\frac{1}{3}}$ $5x^3y^2$

73. $\left(x^{\frac{1}{2}}y^{-\frac{3}{5}}\right)^{\frac{1}{2}}$ $\dfrac{x^{\frac{1}{4}}}{y^{\frac{3}{10}}}$ **74.** $\left(x^{\frac{1}{4}}y^{-\frac{2}{5}}\right)^{\frac{1}{3}}$ $\dfrac{x^{\frac{1}{12}}}{y^{\frac{2}{15}}}$

75. $\dfrac{3^{\frac{1}{2}} \cdot 3^{\frac{3}{4}}}{3^{\frac{1}{4}}}$ 3 **76.** $\dfrac{5^{\frac{3}{4}} \cdot 5^{\frac{1}{2}}}{5^{\frac{1}{4}}}$ 5

77. $\dfrac{\left(3y^{\frac{1}{4}}\right)^3}{y^{\frac{1}{12}}}$ $27y^{\frac{2}{3}}$ **78.** $\dfrac{\left(2y^{\frac{1}{5}}\right)^4}{y^{\frac{3}{10}}}$ $16y^{\frac{1}{2}}$

In Exercises 79–112, use rational exponents to simplify each expression. If rational exponents appear after simplifying, write the answer in radical notation. Assume that all variables represent positive numbers.

79. $\sqrt[8]{x^2}$ $\sqrt[4]{x}$ **80.** $\sqrt[10]{x^2}$ $\sqrt[5]{x}$

81. $\sqrt[3]{8a^6}$ $2a^2$ **82.** $\sqrt[3]{27a^{12}}$ $3a^4$

83. $\sqrt[5]{x^{10}y^{15}}$ x^2y^3 **84.** $\sqrt[5]{x^{15}y^{20}}$ x^3y^4

85. $(\sqrt[3]{xy})^{18}$ x^6y^6 **86.** $(\sqrt[3]{xy})^{21}$ x^7y^7

87. $\sqrt[10]{(3y)^2}$ $\sqrt[5]{3y}$ **88.** $\sqrt[12]{(3y)^2}$ $\sqrt[6]{3y}$

89. $\left(\sqrt[6]{2a}\right)^4$ $\sqrt[3]{4a^2}$ **90.** $\left(\sqrt[8]{2a}\right)^6$ $\sqrt[4]{8a^3}$

91. $\sqrt[9]{x^6y^3}$ $\sqrt[3]{x^2y}$ **92.** $\sqrt[4]{x^2y^6}$ $\sqrt{xy^3}$

93. $\sqrt{2} \cdot \sqrt[3]{2}$ $\sqrt[6]{2^5}$ or $\sqrt[6]{32}$ **94.** $\sqrt{3} \cdot \sqrt[3]{3}$ $\sqrt[6]{3^5}$ or $\sqrt[6]{243}$

95. $\sqrt[5]{x^2} \cdot \sqrt{x}$ $\sqrt[10]{x^9}$ **96.** $\sqrt[7]{x^2} \cdot \sqrt{x}$ $\sqrt[14]{x^{11}}$

97. $\sqrt[4]{a^2b} \cdot \sqrt[3]{ab}$ $\sqrt[12]{a^{10}b^7}$ **98.** $\sqrt[6]{ab^2} \cdot \sqrt[3]{a^2b}$ $\sqrt[6]{a^5b^4}$

99. $\dfrac{\sqrt[4]{x}}{\sqrt[5]{x}}$ $\sqrt[20]{x}$ **100.** $\dfrac{\sqrt[3]{x}}{\sqrt[4]{x}}$ $\sqrt[12]{x}$

101. $\dfrac{\sqrt[3]{y^2}}{\sqrt[6]{y}}$ $\sqrt{y}$ **102.** $\dfrac{\sqrt[5]{y^2}}{\sqrt[10]{y^3}}$ $\sqrt[10]{y}$

103. $\sqrt[4]{\sqrt{x}}$ $\sqrt[8]{x}$ **104.** $\sqrt[5]{\sqrt{x}}$ $\sqrt[10]{x}$

105. $\sqrt{\sqrt{x^2y}}$ $\sqrt[4]{x^2y}$ **106.** $\sqrt{\sqrt{xy^2}}$ $\sqrt[4]{xy^2}$

107. $\sqrt[4]{\sqrt[3]{2x}}$ $\sqrt[12]{2x}$ **108.** $\sqrt[5]{\sqrt[3]{2x}}$ $\sqrt[15]{2x}$

109. $\left(\sqrt[4]{x^3y^5}\right)^{12}$ x^9y^{15} **110.** $\left(\sqrt[5]{x^4y^2}\right)^{20}$ $x^{16}y^8$

111. $\dfrac{\sqrt[4]{a^5b^5}}{\sqrt{ab}}$ $\sqrt[4]{a^3b^3}$ **112.** $\dfrac{\sqrt[4]{a^3b^3}}{\sqrt{ab}}$ $\sqrt[4]{ab}$

Practice Plus

In Exercises 113–116, use the distributive property or the FOIL method to perform each multiplication.

113. $x^{\frac{1}{3}}\left(x^{\frac{1}{3}} - x^{\frac{2}{3}}\right)$ $x^{\frac{2}{3}} - x$

114. $x^{-\frac{1}{4}}\left(x^{\frac{9}{4}} - x^{\frac{1}{4}}\right)$ $x^2 - 1$

115. $\left(x^{\frac{1}{2}} - 3\right)\left(x^{\frac{1}{2}} + 5\right)$ $x + 2x^{\frac{1}{2}} - 15$

116. $\left(x^{\frac{1}{3}} - 2\right)\left(x^{\frac{1}{3}} + 6\right)$ $x^{\frac{2}{3}} + 4x^{\frac{1}{3}} - 12$

In Exercises 117–120, factor out the greatest common factor from each expression.

117. $6x^{\frac{1}{2}} + 2x^{\frac{3}{2}}$ $2x^{\frac{1}{2}}(3 + x)$

118. $8x^{\frac{1}{4}} + 4x^{\frac{5}{4}}$ $4x^{\frac{1}{4}}(2 + x)$

119. $15x^{\frac{1}{3}} - 60x$ $15x^{\frac{1}{3}}\left(1 - 4x^{\frac{2}{3}}\right)$

120. $7x^{\frac{1}{3}} - 70x$ $7x^{\frac{1}{3}}\left(1 - 10x^{\frac{2}{3}}\right)$

In Exercises 121–124, simplify each expression. Assume that all variables represent positive numbers.

121. $(49x^{-2}y^4)^{-\frac{1}{2}}\left(xy^{\frac{1}{2}}\right)$ $\dfrac{x^2}{7y^2}$

122. $(8x^{-6}y^3)^{\frac{1}{3}}\left(x^{\frac{5}{6}}y^{-\frac{1}{3}}\right)^6$ $\dfrac{2x^3}{y}$

123. $\left(\dfrac{x^{-\frac{5}{4}}y^{\frac{1}{3}}}{x^{-\frac{3}{4}}}\right)^{-6}$ $\dfrac{x^3}{y^2}$

124. $\left(\dfrac{x^{\frac{1}{2}}y^{-\frac{7}{4}}}{y^{-\frac{5}{4}}}\right)^{-4}$ $\dfrac{y^2}{x^2}$

Application Exercises

The Galápagos Islands, lying 600 miles west of Ecuador, are famed for their extraordinary wildlife. The function

$$f(x) = 29x^{\frac{1}{3}}$$

models the number of plant species, $f(x)$, on the various islands of the Galápagos chain in terms of the area, x, in square miles, of a particular island. Use the function to solve Exercises 125–126.

125. How many species of plants are on a Galápagos island that has an area of 8 square miles?   58 species of plants

126. How many species of plants are on a Galápagos island that has an area of 27 square miles?   87 species of plants

The function

$$f(x) = 70x^{\frac{3}{4}}$$

models the number of calories per day, $f(x)$, a person needs to maintain life in terms of that person's weight, x, in kilograms. (1 kilogram is approximately 2.2 pounds.) Use this model and a calculator to solve Exercises 127–128. Round answers to the nearest calorie.

127. How many calories per day does a person who weighs 80 kilograms (approximately 176 pounds) need to maintain life?   about 1872 calories per day

128. How many calories per day does a person who weighs 70 kilograms (approximately 154 pounds) need to maintain life?   about 1694 calories per day

We have seen that when the air temperature is t degrees Fahrenheit and the wind speed is v miles per hour, the windchill temperature, C, in degrees Fahrenheit, is given by the formula

$$C = 35.74 + 0.6215t - 35.74v^{\frac{4}{25}} + 0.4275tv^{\frac{4}{25}}.$$

Use this formula to solve Exercises 129–130.

129. a. Substitute 0 for t and write a function $C(v)$ that gives the windchill temperature as a function of wind speed for an air temperature of 0 °F.

$C(v) = 35.74 - 35.74v^{\frac{4}{25}}$

   b. Find and interpret $C(25)$. Use a calculator and round to the nearest degree. $C(25) \approx -24$; When the air temperature is 0°F and the wind speed is 25 miles per hour, the windchill temperature is −24°.

   c. Identify your solution to part (b) on the graph shown.   the point $(25, -24)$

Windchill temperature as a function of air speed for an air temperature of 0°F

$[0, 60, 5]$ by $[-35, 0, 5]$

130. a. Substitute 30 for t and write a function $C(v)$ that gives the windchill temperature as a function of wind speed for an air temperature of 30°F. Simplify the function's formula so that it contains exactly two terms.

$C(v) = 54.385 - 22.915v^{\frac{4}{25}}$

   b. Find and interpret $C(40)$. Use a calculator and round to the nearest degree. $C(40) \approx 13$; When the air temperature is 30°F and the wind speed is 40 miles per hour, the windchill temperature is 13°.

   c. Identify your solution to part (b) on the graph shown.   the point $(40, 13)$

Windchill temperature as a function of air speed for an air temperature of 30°F

Air Speed (miles per hour)
$[0, 60, 5]$ by $[0, 25, 5]$

Your job is to determine whether or not yachts are eligible for the America's Cup, the supreme event in ocean sailing. The basic dimensions of competitors' yachts must satisfy

$$L + 1.25\sqrt{S} - 9.8\sqrt[3]{D} \le 16.296,$$

where L is the yacht's length, in meters, S is its sail area, in square meters, and D is its displacement, in cubic meters. Use this information to solve Exercises 131–132.

131. a. Rewrite the inequality using rational exponents.

$$L + 1.25S^{\frac{1}{2}} - 9.8D^{\frac{1}{3}} \le 16.296$$

 b. Use your calculator to determine if a yacht with length 20.85 meters, sail area 276.4 square meters, and displacement 18.55 cubic meters is eligible for the America's Cup.   eligible

132. a. Rewrite the inequality using rational exponents.

$$L + 1.25S^{\frac{1}{2}} - 9.8D^{\frac{1}{3}} \le 16.296$$

 b. Use your calculator to determine if a yacht with length 22.85 meters, sail area 312.5 square meters, and displacement 22.34 cubic meters is eligible for the America's Cup.   not eligible

Writing in Mathematics

133. What is the meaning of $a^{\frac{1}{n}}$? Give an example to support your explanation.   **133–140.** Answers will vary.

134. What is the meaning of $a^{\frac{m}{n}}$? Give an example.

135. What is the meaning of $a^{-\frac{m}{n}}$? Give an example.

136. Explain why $a^{\frac{1}{n}}$ is negative when n is odd and a is negative. What happens if n is even and a is negative? Why?

137. In simplifying $36^{\frac{3}{2}}$, is it better to use $a^{\frac{m}{n}} = \sqrt[n]{a^m}$ or $a^{\frac{m}{n}} = (\sqrt[n]{a})^m$? Explain.

138. How can you tell if an expression with rational exponents is simplified?

139. Explain how to simplify $\sqrt[3]{x} \cdot \sqrt{x}$.

140. Explain how to simplify $\sqrt[3]{\sqrt{x}}$.

Technology Exercises

141. Use a scientific or graphing calculator to verify your results in Exercises 15–18.

142. Use a scientific or graphing calculator to verify your results in Exercises 45–50.

Exercises 143–145 show a number of simplifications, not all of which are correct. Enter the left side of each equation as y_1 and the right side as y_2. Then use your graphing utility's TABLE *feature to determine if the simplification is correct. If it is not, correct the right side and use the* TABLE *feature to verify your simplification.*   **143–145.** See graphing answer section.

143. $x^{\frac{3}{5}} \cdot x^{-\frac{1}{10}} = x^{\frac{1}{2}}$ simplified correctly

144. $\left(x^{-\frac{1}{2}} \cdot x^{\frac{3}{4}}\right)^{-2} = x^{\frac{1}{2}}$ Right side should be $\dfrac{1}{x^{\frac{1}{2}}}$.

145. $\dfrac{x^{\frac{1}{4}}}{x^{\frac{1}{2}} \cdot x^{-\frac{3}{4}}} = \dfrac{1}{x^{\frac{1}{2}}}$ Right side should be $x^{\frac{1}{2}}$.

Critical Thinking Exercises

146. Which one of the following is true?   **a**

 a. If n is odd, then $(-b)^{\frac{1}{n}} = -b^{\frac{1}{n}}$.

 b. $(a + b)^{\frac{1}{n}} = a^{\frac{1}{n}} + b^{\frac{1}{n}}$

 c. $8^{-\frac{2}{3}} = -4$

 d. None of the above is true.

147. A mathematics professor recently purchased a birthday cake for her son with the inscription

$$\text{Happy}\ \left(2^{\frac{5}{2}} \cdot 2^{\frac{3}{4}} \div 2^{\frac{1}{4}}\right)\text{th Birthday.}$$

How old is the son?   8 years old

148. The birthday boy in Exercise 147, excited by the inscription on the cake, tried to wolf down the whole thing. Professor Mom, concerned about the possible metamorphosis of her son into a blimp, exclaimed, "Hold on! It is your birthday, so why not take $\dfrac{8^{-\frac{4}{3}} + 2^{-2}}{16^{-\frac{3}{4}} + 2^{-1}}$ of the cake? I'll eat half of what's left over." How much of the cake did the professor eat? $\frac{1}{4}$ of the cake

149. Simplify: $\left[3 + \left(27^{\frac{2}{3}} + 32^{\frac{2}{5}}\right)^{\frac{3}{2}}\right] - 9^{\frac{1}{2}}$.   61

150. Find the domain of $f(x) = (x - 3)^{\frac{1}{2}}(x + 4)^{-\frac{1}{2}}$.   Domain of f is $\{x | x \ge 3\}$ or $[3, \infty)$.

Review Exercises

151. Write the equation of the linear function whose graph passes through $(5, 1)$ and $(4, 3)$. (Section 3.5, Example 2) $y = -2x + 11$ or $f(x) = -2x + 11$

152. Graph $y \le -\dfrac{3}{2}x + 3$. (Section 9.4, Example 2)

 See graphing answer section.

153. If $f(x) = 3x^2 - 5x + 4$, find $f(a + h)$. (Section 8.1, Example 3) $3a^2 + 6ah + 3h^2 - 5a - 5h + 4$

SECTION

10.3

MULTIPLYING AND SIMPLIFYING RADICAL EXPRESSIONS

Objectives

1 Use the product rule to multiply radicals.

2 Use factoring and the product rule to simplify radicals.

3 Multiply radicals and then simplify.

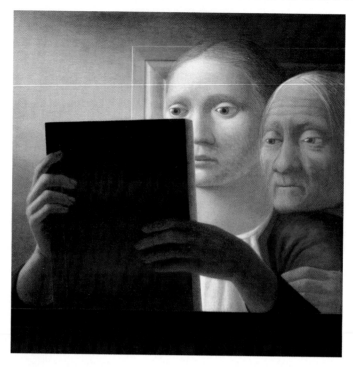

George Tooker (b. 1920) "Mirror II" 1963, egg tempera on gesso panel, 1968.4.

A difficult challenge of middle adulthood is confronting the aging process. Middle-aged adults are forced to acknowledge their mortality as they witness the deaths of parents, colleagues, and friends. In addition, there are a number of physical transformations, including changes in vision that require glasses for reading, the onset of wrinkles and sagging skin, and a decrease in heart response. A change in heart response occurs fairly early; after 20, our hearts become less adept at accelerating in response to exercise. In this section, you will see how a radical function models changes in heart function throughout the aging process, as we turn to multiplying and simplifying radical expressions.

1 Use the product rule to multiply radicals.

The Product Rule for Radicals A rule for multiplying radicals can be generalized by comparing $\sqrt{25} \cdot \sqrt{4}$ and $\sqrt{25 \cdot 4}$. Notice that

$$\sqrt{25} \cdot \sqrt{4} = 5 \cdot 2 = 10 \quad \text{and} \quad \sqrt{25 \cdot 4} = \sqrt{100} = 10.$$

Because we obtain 10 in both situations, the original radical expressions must be equal. That is,

$$\sqrt{25} \cdot \sqrt{4} = \sqrt{25 \cdot 4}.$$

This result is a special case of the **product rule for radicals** that can be generalized as follows:

> **THE PRODUCT RULE FOR RADICALS** If $\sqrt[n]{a}$ and $\sqrt[n]{b}$ are real numbers, then
> $$\sqrt[n]{a} \cdot \sqrt[n]{b} = \sqrt[n]{ab}.$$

The product of two nth roots is the nth root of the product.

STUDY TIP

The product rule can be used only when the radicals have the same index. If indices differ, rational exponents can be used, as in $\sqrt{x} \cdot \sqrt[3]{x}$, which was Example 8(d) in the previous section.

EXAMPLE 1   Using the Product Rule for Radicals

Multiply:

a. $\sqrt{3} \cdot \sqrt{7}$ **b.** $\sqrt{x+7} \cdot \sqrt{x-7}$ **c.** $\sqrt[3]{7} \cdot \sqrt[3]{9}$ **d.** $\sqrt[8]{10x} \cdot \sqrt[8]{8x^4}$.

SOLUTION   In each problem, the indices are the same. Thus, we multiply by multiplying the radicands.

a. $\sqrt{3} \cdot \sqrt{7} = \sqrt{3 \cdot 7} = \sqrt{21}$

b. $\sqrt{x+7} \cdot \sqrt{x-7} = \sqrt{(x+7)(x-7)} = \sqrt{x^2 - 49}$

> This is not equal to $\sqrt{x^2} - \sqrt{49}$.

c. $\sqrt[3]{7} \cdot \sqrt[3]{9} = \sqrt[3]{7 \cdot 9} = \sqrt[3]{63}$

d. $\sqrt[8]{10x} \cdot \sqrt[8]{8x^4} = \sqrt[8]{10x \cdot 8x^4} = \sqrt[8]{80x^5}$ ∎

✔ **CHECK POINT 1** Multiply:

a. $\sqrt{5} \cdot \sqrt{11}$ $\sqrt{55}$ **b.** $\sqrt{x+4} \cdot \sqrt{x-4}$ $\sqrt{x^2 - 16}$

c. $\sqrt[3]{6} \cdot \sqrt[3]{10}$ $\sqrt[3]{60}$ **d.** $\sqrt[7]{2x} \cdot \sqrt[7]{6x^3}$. $\sqrt[7]{12x^4}$

2   Use factoring and the product rule to simplify radicals.

USING TECHNOLOGY

You can use a calculator to provide numerical support that $\sqrt{300} = 10\sqrt{3}$. First find an approximation for $\sqrt{300}$:

300 $\boxed{\sqrt{}}$ ≈ 17.32

or

$\boxed{\sqrt{}}$ 300 $\boxed{\text{ENTER}}$ ≈ 17.32.

Now find an approximation for $10\sqrt{3}$:

10 $\boxed{\times}$ 3 $\boxed{\sqrt{}}$ ≈ 17.32

or

10 $\boxed{\sqrt{}}$ 3 $\boxed{\text{ENTER}}$ ≈ 17.32.

Correct to two decimal places, $\sqrt{300} \approx 17.32$ and $10\sqrt{3} \approx 17.32$.

This verifies that
$$\sqrt{300} = 10\sqrt{3}.$$

Use this technique to support the numerical results for the answers in this section. *Caution:* A simplified radical does not mean a decimal approximation.

Using Factoring and the Product Rule to Simplify Radicals   A number that is the square of an integer is a **perfect square**. For example, 100 is a perfect square because $100 = 10^2$. A number is a **perfect cube** if it is the cube of an integer. Thus, 125 is a perfect cube because $125 = 5^3$. In general, a number is a **perfect nth power** if it is the nth power of an integer. Thus, p is a perfect nth power if there is an integer q such that $p = q^n$.

A radical of index n is **simplified** when its radicand has no factors other than 1 that are perfect nth powers. For example, $\sqrt{300}$ is not simplified because it can be expressed as $\sqrt{100 \cdot 3}$ and 100 is a perfect square. We can use the product rule in the form

$$\sqrt[n]{ab} = \sqrt[n]{a} \cdot \sqrt[n]{b}$$

to simplify $\sqrt[n]{ab}$ when $\sqrt[n]{a}$ or $\sqrt[n]{b}$ is a perfect nth power. Consider $\sqrt{300}$. To simplify, we factor 300 so that one of its factors is the greatest perfect square possible.

$$\sqrt{300} = \sqrt{100 \cdot 3}$$
Factor 300. 100 is the greatest perfect square factor.

$$= \sqrt{100} \cdot \sqrt{3}$$
Use the product rule: $\sqrt[n]{ab} = \sqrt[n]{a} \cdot \sqrt[n]{b}$.

$$= 10\sqrt{3}$$
Write $\sqrt{100}$ as 10. We read $10\sqrt{3}$ as "ten times the square root of three."

SIMPLIFYING RADICAL EXPRESSIONS BY FACTORING   A radical expression whose index is n is **simplified** when its radicand has no factors that are perfect nth powers. To simplify, use the following procedure:

1. Write the radicand as the product of two factors, one of which is the greatest perfect nth power.

2. Use the product rule to take the nth root of each factor.

3. Find the nth root of the perfect nth power.

EXAMPLE 2 Simplifying Radicals by Factoring

Simplify by factoring:

a. $\sqrt{75}$ **b.** $\sqrt[3]{54}$ **c.** $\sqrt[5]{64}$ **d.** $\sqrt{500xy^2}$.

SOLUTION

a. $\sqrt{75} = \sqrt{25 \cdot 3}$ 25 is the greatest perfect square that is a factor of 75.

$= \sqrt{25} \cdot \sqrt{3}$ Take the square root of each factor: $\sqrt[n]{ab} = \sqrt[n]{a} \cdot \sqrt[n]{b}$.

$= 5\sqrt{3}$ Write $\sqrt{25}$ as 5.

b. $\sqrt[3]{54} = \sqrt[3]{27 \cdot 2}$ 27 is the greatest perfect cube that is a factor of 54: $27 = 3^3$.

$= \sqrt[3]{27} \cdot \sqrt[3]{2}$ Take the cube root of each factor: $\sqrt[n]{ab} = \sqrt[n]{a} \cdot \sqrt[n]{b}$.

$= 3\sqrt[3]{2}$ Write $\sqrt[3]{27}$ as 3.

c. $\sqrt[5]{64} = \sqrt[5]{32 \cdot 2}$ 32 is the greatest perfect fifth power that is a factor of 64: $32 = 2^5$.

$= \sqrt[5]{32} \cdot \sqrt[5]{2}$ Take the fifth root of each factor: $\sqrt[n]{ab} = \sqrt[n]{a} \cdot \sqrt[n]{b}$.

$= 2\sqrt[5]{2}$ Write $\sqrt[5]{32}$ as 2.

d. $\sqrt{500xy^2} = \sqrt{100y^2 \cdot 5x}$ $100y^2$ is the greatest perfect square that is a factor of $500xy^2$: $100y^2 = (10y)^2$.

$= \sqrt{100y^2} \cdot \sqrt{5x}$ Factor into two radicals.

$= 10|y|\sqrt{5x}$ Take the square root of $100y^2$. ■

 CHECK POINT 2 Simplify by factoring:

a. $\sqrt{80}$ $4\sqrt{5}$ **b.** $\sqrt[3]{40}$ $2\sqrt[3]{5}$

c. $\sqrt[4]{32}$ $2\sqrt[4]{2}$ **d.** $\sqrt{200x^2y}$. $10|x|\sqrt{2y}$

EXAMPLE 3 Simplifying a Radical Function

If

$$f(x) = \sqrt{2x^2 + 4x + 2},$$

express the function, f, in simplified form.

SOLUTION   Begin by factoring the radicand. The GCF is 2. Simplification is possible if we obtain a factor that is a perfect square.

$f(x) = \sqrt{2x^2 + 4x + 2}$ This is the given function.

$= \sqrt{2(x^2 + 2x + 1)}$ Factor out the GCF.

$= \sqrt{2(x + 1)^2}$ Factor the perfect square trinomial: $A^2 + 2AB + B^2 = (A + B)^2$.

$= \sqrt{2} \cdot \sqrt{(x + 1)^2}$ Take the square root of each factor. The factor $(x + 1)^2$ is a perfect square.

$= \sqrt{2}\,|x + 1|$ Take the square root of $(x + 1)^2$.

In simplified form,

$$f(x) = \sqrt{2}\,|x + 1|.$$ ■

The graphs of

$$f(x) = \sqrt{2x^2 + 4x + 2}, \quad g(x) = \sqrt{2}|x + 1|, \quad \text{and} \quad h(x) = \sqrt{2}(x + 1)$$

are shown in three separate $[-5, 5, 1]$ by $[-5, 5, 1]$ viewing rectangles. The graphs in parts (a) and (b) are identical. This verifies that our simplification in Example 3 is correct: $\sqrt{2x^2 + 4x + 2} = \sqrt{2}|x + 1|$. Now compare the graphs in parts (a) and (c). Can you see that they are not the same? This illustrates the importance of not leaving out absolute value bars: $\sqrt{2x^2 + 4x + 2} \neq \sqrt{2}(x + 1)$.

$y = \sqrt{2x^2 + 4x + 2}$

(a)

$y = \sqrt{2}|x + 1|$

(b)

$y = \sqrt{2}(x + 1)$

(c)

✔ **CHECK POINT 3** If $f(x) = \sqrt{3x^2 - 12x + 12}$, express the function, f, in simplified form. $f(x) = \sqrt{3}|x - 2|$

For the remainder of this chapter, in situations that do not involve functions, we will **assume that no radicands involve negative quantities raised to even powers. Based upon this assumption, absolute value bars are not necessary when taking even roots.**

> **SIMPLIFYING WHEN VARIABLES TO EVEN POWERS IN A RADICAND ARE NONNEGATIVE QUANTITIES**   For any nonnegative real number a,
> $$\sqrt[n]{a^n} = a.$$

In simplifying an nth root, how do we find variable factors in the radicand that are perfect nth powers? The **perfect nth powers have exponents that are divisible by n**. Simplification is possible by observation or by using rational exponents. Here are some examples:

- $\sqrt{x^6} = \sqrt{(x^3)^2} = x^3$ or $\sqrt{x^6} = (x^6)^{\frac{1}{2}} = x^3$

  6 is divisible by the index, 2. Thus, x^6 is a perfect square.

- $\sqrt[3]{y^{21}} = \sqrt[3]{(y^7)^3} = y^7$ or $\sqrt[3]{y^{21}} = (y^{21})^{\frac{1}{3}} = y^7$

  21 is divisible by the index, 3. Thus, y^{21} is a perfect cube.

- $\sqrt[6]{z^{24}} = \sqrt[6]{(z^4)^6} = z^4$ or $\sqrt[6]{z^{24}} = (z^{24})^{\frac{1}{6}} = z^4$.

  24 is divisible by the index, 6. Thus, z^{24} is a perfect 6th power.

EXAMPLE 4   Simplifying a Radical by Factoring

Simplify: $\sqrt{x^5 y^{13} z^7}$.

SOLUTION   We write the radicand as the product of the greatest perfect square factor and another factor. Because the index is 2, variables that have exponents that are divisible by 2 are part of the perfect square factor. We use the greatest exponents that are divisible by 2.

$$\sqrt{x^5 y^{13} z^7} = \sqrt{x^4 \cdot x \cdot y^{12} \cdot y \cdot z^6 \cdot z} \quad \text{Use the greatest even power of each variable.}$$

$$= \sqrt{(x^4 y^{12} z^6)(xyz)} \quad \text{Group the perfect square factors.}$$

$$= \sqrt{x^4 y^{12} z^6} \cdot \sqrt{xyz} \quad \text{Factor into two radicals.}$$

$$= x^2 y^6 z^3 \sqrt{xyz} \quad \sqrt{x^4 y^{12} z^6} = (x^4 y^{12} z^6)^{\frac{1}{2}} = x^2 y^6 z^3 \quad \blacksquare$$

✔ **CHECK POINT 4** Simplify: $\sqrt{x^9 y^{11} z^3}$. $\quad x^4 y^5 z \sqrt{xyz}$

EXAMPLE 5 Simplifying a Radical by Factoring

Simplify: $\sqrt[3]{32 x^8 y^{16}}$.

SOLUTION We write the radicand as the product of the greatest perfect cube factor and another factor. Because the index is 3, variables that have exponents that are divisible by 3 are part of the perfect cube factor. We use the greatest exponents that are divisible by 3.

$$\sqrt[3]{32 x^8 y^{16}} = \sqrt[3]{8 \cdot 4 \cdot x^6 \cdot x^2 \cdot y^{15} \cdot y} \quad \text{Identify perfect cube factors.}$$

$$= \sqrt[3]{(8 x^6 y^{15})(4 x^2 y)} \quad \text{Group the perfect cube factors.}$$

$$= \sqrt[3]{8 x^6 y^{15}} \cdot \sqrt[3]{4 x^2 y} \quad \text{Factor into two radicals.}$$

$$= 2 x^2 y^5 \sqrt[3]{4 x^2 y} \quad \sqrt[3]{8} = 2 \text{ and } \sqrt[3]{x^6 y^{15}} = (x^6 y^{15})^{\frac{1}{3}} = x^2 y^5. \quad \blacksquare$$

✔ **CHECK POINT 5** Simplify: $\sqrt[3]{40 x^{10} y^{14}}$. $\quad 2 x^3 y^4 \sqrt[3]{5 x y^2}$

EXAMPLE 6 Simplifying a Radical by Factoring

Simplify: $\sqrt[5]{64 x^3 y^7 z^{29}}$.

SOLUTION We write the radicand as the product of the greatest perfect 5th power and another factor. Because the index is 5, variables that have exponents that are divisible by 5 are part of the perfect fifth factor. We use the greatest exponents that are divisible by 5.

$$\sqrt[5]{64 x^3 y^7 z^{29}} = \sqrt[5]{32 \cdot 2 \cdot x^3 \cdot y^5 \cdot y^2 \cdot z^{25} \cdot z^4} \quad \text{Identify perfect fifth factors.}$$

$$= \sqrt[5]{(32 y^5 z^{25})(2 x^3 y^2 z^4)} \quad \text{Group the perfect fifth factors.}$$

$$= \sqrt[5]{32 y^5 z^{25}} \cdot \sqrt[5]{2 x^3 y^2 z^4} \quad \text{Factor into two radicals.}$$

$$= 2 y z^5 \sqrt[5]{2 x^3 y^2 z^4} \quad \sqrt[5]{32} = 2 \text{ and } \sqrt[5]{y^5 z^{25}} = (y^5 z^{25})^{\frac{1}{5}} = y z^5. \quad \blacksquare$$

✔ **CHECK POINT 6** Simplify: $\sqrt[5]{32 x^{12} y^2 z^8}$. $\quad 2 x^2 z \sqrt[5]{x^2 y^2 z^3}$

---

3 Multiply radicals and then simplify.

Multiplying and Simplifying Radicals We have seen how to use the product rule when multiplying radicals with the same index. Sometimes after multiplying, we can simplify the resulting radical.

EXAMPLE 7 Multiplying Radicals and Then Simplifying

Multiply and simplify:

a. $\sqrt{15} \cdot \sqrt{3}$ **b.** $7\sqrt[3]{4} \cdot 5\sqrt[3]{6}$ **c.** $\sqrt[4]{8x^3y^2} \cdot \sqrt[4]{8x^5y^3}$.

SOLUTION

a.
$$\sqrt{15} \cdot \sqrt{3} = \sqrt{15 \cdot 3}$$
$$= \sqrt{45} = \sqrt{9 \cdot 5}$$

Use the product rule.

9 is the greatest perfect square factor of 45.

$$= \sqrt{9} \cdot \sqrt{5} = 3\sqrt{5}$$

b.
$$7\sqrt[3]{4} \cdot 5\sqrt[3]{6} = 35\sqrt[3]{4 \cdot 6}$$
$$= 35\sqrt[3]{24} = 35\sqrt[3]{8 \cdot 3}$$

Use the product rule.

8 is the greatest perfect cube factor of 24.

$$= 35\sqrt[3]{8} \cdot \sqrt[3]{3} = 35 \cdot 2 \cdot \sqrt[3]{3}$$
$$= 70\sqrt[3]{3}$$

c. $\sqrt[4]{8x^3y^2} \cdot \sqrt[4]{8x^5y^3} = \sqrt[4]{8x^3y^2 \cdot 8x^5y^3}$

Use the product rule.

$$= \sqrt[4]{64x^8y^5}$$

Multiply.

$$= \sqrt[4]{16 \cdot 4 \cdot x^8 \cdot y^4 \cdot y}$$

Identify perfect fourth factors.

$$= \sqrt[4]{(16x^8y^4)(4y)}$$

Group the perfect fourth factors.

$$= \sqrt[4]{16x^8y^4} \cdot \sqrt[4]{4y}$$

Factor into two radicals.

$$= 2x^2y\sqrt[4]{4y}$$

$\sqrt[4]{16} = 2$ and $\sqrt[4]{x^8y^4} = (x^8y^4)^{\frac{1}{4}} = x^2y$. ∎

✔ CHECK POINT 7 Multiply and simplify:

a. $\sqrt{6} \cdot \sqrt{2}$ $\quad 2\sqrt{3}$

b. $10\sqrt[3]{16} \cdot 5\sqrt[3]{2}$ $\quad 100\sqrt[3]{4}$

c. $\sqrt[4]{4x^2y} \cdot \sqrt[4]{8x^6y^3}$. $\quad 2x^2y\sqrt[4]{2}$

10.3 EXERCISE SET

Student Solutions Manual CD/Video PH Math/Tutor Center MathXL Tutorials on CD MathXL® MyMathLab Interactmath.com

Practice Exercises

In Exercises 1–20, use the product rule to multiply.

1. $\sqrt{3} \cdot \sqrt{5}$ $\quad \sqrt{15}$

2. $\sqrt{7} \cdot \sqrt{5}$ $\quad \sqrt{35}$

3. $\sqrt[3]{2} \cdot \sqrt[3]{9}$ $\quad \sqrt[3]{18}$

4. $\sqrt[3]{5} \cdot \sqrt[3]{4}$ $\quad \sqrt[3]{20}$

5. $\sqrt[4]{11} \cdot \sqrt[4]{3}$ $\quad \sqrt[4]{33}$

6. $\sqrt[5]{9} \cdot \sqrt[5]{3}$ $\quad \sqrt[5]{27}$

7. $\sqrt{3x} \cdot \sqrt{11y}$ $\quad \sqrt{33xy}$

8. $\sqrt{5x} \cdot \sqrt{11y}$ $\quad \sqrt{55xy}$

9. $\sqrt[5]{6x^3} \cdot \sqrt[5]{4x}$ $\quad \sqrt[5]{24x^4}$

10. $\sqrt[4]{6x^2} \cdot \sqrt[4]{3x}$ $\quad \sqrt[4]{18x^3}$

11. $\sqrt{x+3} \cdot \sqrt{x-3}$ $\quad \sqrt{x^2-9}$

12. $\sqrt{x+6} \cdot \sqrt{x-6}$ $\quad \sqrt{x^2-36}$

13. $\sqrt[6]{x-4} \cdot \sqrt[6]{(x-4)^4}$ $\quad \sqrt[6]{(x-4)^5}$

14. $\sqrt[6]{x-5} \cdot \sqrt[6]{(x-5)^4}$ $\quad \sqrt[6]{(x-5)^5}$

15. $\sqrt{\dfrac{2x}{3}} \cdot \sqrt{\dfrac{3}{2}}$ $\quad \sqrt{x}$

16. $\sqrt{\dfrac{2x}{5}} \cdot \sqrt{\dfrac{5}{2}}$ $\quad \sqrt{x}$

17. $\sqrt[4]{\dfrac{x}{7}} \cdot \sqrt[4]{\dfrac{3}{y}}$ $\quad \sqrt[4]{\dfrac{3x}{7y}}$

18. $\sqrt[4]{\dfrac{x}{3}} \cdot \sqrt[4]{\dfrac{7}{y}}$ $\quad \sqrt[4]{\dfrac{7x}{3y}}$

19. $\sqrt[7]{7x^2y} \cdot \sqrt[7]{11x^3y^2}$ $\quad \sqrt[7]{77x^5y^3}$

20. $\sqrt[9]{12x^2y^3} \cdot \sqrt[9]{3x^3y^4}$ $\quad \sqrt[9]{36x^5y^7}$

In Exercises 21–32, simplify by factoring.

21. $\sqrt{50}$ $\quad 5\sqrt{2}$

22. $\sqrt{27}$ $\quad 3\sqrt{3}$

23. $\sqrt{45}$ $\quad 3\sqrt{5}$

24. $\sqrt{28}$ $\quad 2\sqrt{7}$

25. $\sqrt{75x}$ $\quad 5\sqrt{3x}$

26. $\sqrt{40x}$ $\quad 2\sqrt{10x}$

27. $\sqrt[3]{16}$ $\quad 2\sqrt[3]{2}$

28. $\sqrt[3]{54}$ $\quad 3\sqrt[3]{2}$

29. $\sqrt[3]{27x^3}$ $\quad 3x$

30. $\sqrt[3]{250x^3}$ $\quad 5x\sqrt[3]{2}$

31. $\sqrt[3]{-16x^2y^3}$ $\quad -2y\sqrt[3]{2x^2}$ **32.** $\sqrt[3]{-32x^2y^3}$ $\quad -2y\sqrt[3]{4x^2}$

In Exercises 33–38, express the function, f, in simplified form. Assume that x can be any real number.

33. $f(x) = \sqrt{36(x+2)^2}$ $\quad 6|x+2|$

34. $f(x) = \sqrt{81(x-2)^2}$ $\quad 9|x-2|$

35. $f(x) = \sqrt[3]{32(x+2)^3}$ $\quad 2(x+2)\sqrt[3]{4}$

36. $f(x) = \sqrt[3]{48(x-2)^3}$ $\quad 2(x-2)\sqrt[3]{6}$

37. $f(x) = \sqrt{3x^2 - 6x + 3}$ $\quad |x-1|\sqrt{3}$

38. $f(x) = \sqrt{5x^2 - 10x + 5}$ $\quad |x-1|\sqrt{5}$

In Exercises 39–60, simplify by factoring. Assume that all variables in a radicand represent positive real numbers and no radicands involve negative quantities raised to even powers.

39. $\sqrt{x^7}$ $\quad x^3\sqrt{x}$

40. $\sqrt{x^5}$ $\quad x^2\sqrt{x}$

41. $\sqrt{x^8y^9}$ $\quad x^4y^4\sqrt{y}$

42. $\sqrt{x^6y^7}$ $\quad x^3y^3\sqrt{y}$

43. $\sqrt{48x^3}$ $\quad 4x\sqrt{3x}$

44. $\sqrt{40x^3}$ $\quad 2x\sqrt{10x}$

45. $\sqrt[3]{y^8}$ $\quad y^2\sqrt[3]{y^2}$

46. $\sqrt[3]{y^{11}}$ $\quad y^3\sqrt[3]{y^2}$

47. $\sqrt[3]{x^{14}y^3z}$ $\quad x^4y\sqrt[3]{x^2z}$

48. $\sqrt[3]{x^3y^{17}z^2}$ $\quad xy^5\sqrt[3]{y^2z^2}$

49. $\sqrt[3]{81x^8y^6}$ $\quad 3x^2y^2\sqrt[3]{3x^2}$ **50.** $\sqrt[3]{32x^9y^{17}}$ $\quad 2x^3y^5\sqrt[3]{4y^2}$

51. $\sqrt[3]{(x+y)^5}$ $\quad (x+y)\sqrt[3]{(x+y)^2}$

52. $\sqrt[3]{(x+y)^4}$ $\quad (x+y)\sqrt[3]{x+y}$

53. $\sqrt[5]{y^{17}}$ $\quad y^3\sqrt[5]{y^2}$

54. $\sqrt[5]{y^{18}}$ $\quad y^3\sqrt[5]{y^3}$

55. $\sqrt[5]{64x^6y^{17}}$ $\quad 2xy^3\sqrt[5]{2xy^2}$ **56.** $\sqrt[5]{64x^7y^{16}}$ $\quad 2xy^3\sqrt[5]{2x^2y}$

57. $\sqrt[4]{80x^{10}}$ $\quad 2x^2\sqrt[4]{5x^2}$

58. $\sqrt[4]{96x^{11}}$ $\quad 2x^2\sqrt[4]{6x^3}$

59. $\sqrt[4]{(x-3)^{10}}$ $\quad (x-3)^2\sqrt[4]{(x-3)^2}$ or $(x-3)^2\sqrt{x-3}$

60. $\sqrt[4]{(x-2)^{14}}$ $\quad (x-2)^3\sqrt[4]{(x-2)^2}$ or $(x-2)^3\sqrt{x-2}$

In Exercises 61–82, multiply and simplify. Assume that all variables in a radicand represent positive real numbers and no radicands involve negative quantities raised to even powers.

61. $\sqrt{12} \cdot \sqrt{2}$ $\quad 2\sqrt{6}$

62. $\sqrt{3} \cdot \sqrt{6}$ $\quad 3\sqrt{2}$

63. $\sqrt{5x} \cdot \sqrt{10y}$ $\quad 5\sqrt{2xy}$

64. $\sqrt{8x} \cdot \sqrt{10y}$ $\quad 4\sqrt{5xy}$

65. $\sqrt{12x} \cdot \sqrt{3x}$ $\quad 6x$

66. $\sqrt{20x} \cdot \sqrt{5x}$ $\quad 10x$

67. $\sqrt{50xy} \cdot \sqrt{4xy^2}$ $\quad 10xy\sqrt{2y}$

68. $\sqrt{5xy} \cdot \sqrt{10xy^2}$ $\quad 5xy\sqrt{2y}$

69. $2\sqrt{5} \cdot 3\sqrt{40}$ $\quad 60\sqrt{2}$ **70.** $3\sqrt{15} \cdot 5\sqrt{6}$ $\quad 45\sqrt{10}$

71. $\sqrt[3]{12} \cdot \sqrt[3]{4}$ $\quad 2\sqrt[3]{6}$ **72.** $\sqrt[4]{4} \cdot \sqrt[4]{8}$ $\quad 2\sqrt[4]{2}$

73. $\sqrt{5x^3} \cdot \sqrt{8x^2}$ $\quad 2x^2\sqrt{10x}$

74. $\sqrt{2x^7} \cdot \sqrt{12x^4}$ $\quad 2x^5\sqrt{6x}$

75. $\sqrt[3]{25x^4y^2} \cdot \sqrt[3]{5xy^{12}}$ $\quad 5xy^4\sqrt[3]{x^2y^2}$

76. $\sqrt[3]{6x^7y} \cdot \sqrt[3]{9x^4y^{12}}$ $\quad 3x^3y^4\sqrt[3]{2x^2y}$

77. $\sqrt[4]{8x^2y^3z^6} \cdot \sqrt[4]{2x^4yz}$ $\quad 2xyz\sqrt[4]{x^2z^3}$

78. $\sqrt[4]{4x^2y^3z^3} \cdot \sqrt[4]{8x^4yz^6}$ $\quad 2xyz^2\sqrt[4]{2x^2z}$

79. $\sqrt[5]{8x^4y^6z^2} \cdot \sqrt[5]{8xy^7z^4}$ $\quad 2xy^2z\sqrt[5]{2y^3z}$

80. $\sqrt[5]{8x^4y^3z^3} \cdot \sqrt[5]{8xy^9z^8}$ $\quad 2xy^2z^2\sqrt[5]{2y^2z}$

81. $\sqrt[3]{x-y} \cdot \sqrt[3]{(x-y)^7}$ $\quad (x-y)^2\sqrt[3]{(x-y)^2}$

82. $\sqrt[3]{x-6} \cdot \sqrt[3]{(x-6)^7}$ $\quad (x-6)^2\sqrt[3]{(x-6)^2}$

Practice Plus

In Exercises 83–92, simplify each expression. Assume that all variables in a radicand represent positive real numbers and no radicands involve negative quantities raised to even powers.

83. $-2x^2y\left(\sqrt[3]{54x^3y^7z^2}\right)$ $\quad -6x^3y^3\sqrt[3]{2yz^2}$

84. $\dfrac{-x^2y^7}{2}\left(\sqrt[3]{-32x^4y^9z^7}\right)$ $\quad x^3y^{10}z^2\sqrt[3]{4xz}$

85. $-3y\left(\sqrt[5]{64x^3y^6}\right)$ $\quad -6y^2\sqrt[5]{2x^3y}$

86. $-4x^2y^7\left(\sqrt[5]{-32x^{11}y^{17}}\right)$ $\quad 8x^4y^{10}\sqrt[5]{xy^2}$

87. $\left(-2xy^2\sqrt{3x}\right)\left(xy\sqrt{6x}\right)$ $\quad -6x^3y^3\sqrt{2}$

88. $\left(-5x^2y^3z\sqrt{2xyz}\right)\left(-x^4z\sqrt{10xz}\right)$ $\quad 10x^7y^3z^3\sqrt{5y}$

89. $\left(2x^2y\sqrt[4]{8xy}\right)\left(-3xy^2\sqrt[4]{2x^2y^3}\right)$ $\quad -12x^3y^4\sqrt[4]{x^3}$

90. $\left(5a^2b\sqrt[4]{8a^2b}\right)\left(4ab\sqrt[4]{4a^3b^2}\right)$ $\quad 40a^4b^2\sqrt[4]{2ab^3}$

91. $\sqrt[5]{8x^4y^6}\left(\sqrt[5]{2xy^7} + \sqrt[5]{4x^6y^9}\right)$ $\quad xy^2\sqrt[5]{16y^3} + 2x^2y^3$

92. $\sqrt[4]{2x^3y^2}\left(\sqrt[4]{6x^5y^6} + \sqrt[4]{8x^5y^7}\right)$ $\quad x^2y^2\sqrt[4]{12} + 2x^2y^2\sqrt[4]{y}$

Application Exercises

The function

$$d(x) = \sqrt{\dfrac{3x}{2}}$$

models the distance, $d(x)$, in miles, that a person h feet high can see to the horizon. Use this function to solve Exercises 93–94.

93. The pool deck on a cruise ship is 72 feet above the water. How far can passengers on the pool deck see? Write the answer in simplified radical form. Then use the simplified radical form and a calculator to express the answer to the nearest tenth of a mile. $\quad 6\sqrt{3}$ miles; 10.4 miles

94. The captain of a cruise ship is on the star deck, which is 120 feet above the water. How far can the captain see? Write the answer in simplified radical form. Then use the simplified radical form and a calculator to express the answer to the nearest tenth of a mile. $6\sqrt{5}$ miles; 13.4 miles

Paleontologists use the function

$$W(x) = 4\sqrt{2x}$$

to estimate the walking speed of a dinosaur, $W(x)$, in feet per second, where x is the length, in feet, of the dinosaur's leg. The graph of W is shown in the figure. Use this information to solve Exercises 95–96.

Dinosaur Walking Speeds

$W(x) = 4\sqrt{2x}$

Walking Speed (feet per second)

Leg Length (feet)
$[0, 12, 1]$ by $[0, 20, 2]$

95. What is the walking speed of a dinosaur whose leg length is 6 feet? Use the function's equation and express the answer in simplified radical form. Then use the function's graph to estimate the answer to the nearest foot per second. $8\sqrt{3}$ feet per second; 14 feet per second

96. What is the walking speed of a dinosaur whose leg length is 10 feet? Use the function's equation and express the answer in simplified radical form. Then use the function's graph to estimate the answer to the nearest foot per second. $8\sqrt{5}$ feet feet per second; 18 feet per second

*Your **cardiac index** is your heart's output, in liters of blood per minute, divided by your body's surface area, in square meters. The cardiac index, $C(x)$, can be modeled by*

$$C(x) = \frac{7.644}{\sqrt[4]{x}}, \qquad 10 \le x \le 80,$$

where x is an individual's age, in years. The graph of the function is shown. Use the function to solve Exercises 97–98.

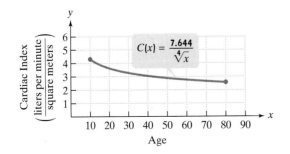

$C(x) = \dfrac{7.644}{\sqrt[4]{x}}$

Cardiac Index (liters per minute / square meters)

Age

97. a. Find the cardiac index of a 32-year-old. Express the denominator in simplified radical form and reduce the fraction. $\dfrac{7.644}{2\sqrt[4]{2}} = \dfrac{3.822}{\sqrt[4]{2}}$

b. Use the form of the answer in part (a) and a calculator to express the cardiac index to the nearest hundredth. Identify your solution as a point on the graph.   3.21 liters of blood per minute per square meter; (32, 3.21)

98. a. Find the cardiac index of an 80-year-old. Express the denominator in simplified radical form and reduce the fraction. $\dfrac{7.644}{2\sqrt[4]{5}} = \dfrac{3.822}{\sqrt[4]{5}}$

b. Use the form of the answer in part (a) and a calculator to express the cardiac index to the nearest hundredth. Identify your solution as a point on the graph.   2.56 liters of blood per minute per square meter; (80, 2.56)

Writing in Mathematics

99. What is the product rule for radicals? Give an example to show how it is used.   99–103. Answers will vary.

100. Explain why $\sqrt{50}$ is not simplified. What do we mean when we say a radical expression is simplified?

101. In simplifying an nth root, explain how to find variable factors in the radicand that are perfect nth powers.

102. Without showing all the details, explain how to simplify $\sqrt[3]{16x^{14}}$.

103. As you get older, what would you expect to happen to your heart's output? Explain how this is shown in the graph for Exercises 97–98. Is this trend taking place progressively more rapidly or more slowly over the entire interval? What does this mean about this aspect of aging?

Technology Exercises

104. Use a calculator to provide numerical support for your simplifications in Exercises 21–24 and 27–28. In each case, find a decimal approximation for the given expression. Then find a decimal approximation for your simplified expression. The approximations should be the same.

In Exercises 105–108, determine if each simplification is correct by graphing the function on each side of the equation with your graphing utility. Use the given viewing rectangle. The graphs should be the same. If they are not, correct the right side of the equation and then use your graphing utility to verify the simplification.   See graphing answer section.

105. $\sqrt{x^4} = x^2$ Graphs are the same;
$[0, 5, 1]$ by $[0, 20, 1]$ simplification is correct.

106. $\sqrt{8x^2} = 4x\sqrt{2}$ Graphs are not the same;
$[-5, 5, 1]$ by $[-5, 20, 1]$ $\sqrt{8x^2} = 2|x|\sqrt{2}$.

107. $\sqrt{3x^2 - 6x + 3} = (x - 1)\sqrt{3}$ Graphs are not the same;
$[-5, 5, 1]$ by $[-5, 5, 1]$ $\sqrt{3x^2 - 6x + 3} = |x - 1|\sqrt{3}$.

108. $\sqrt[3]{2x} \cdot \sqrt[3]{4x^2} = 4x$ Graphs are not the same;
$[-10, 10, 1]$ by $[-10, 10, 1]$ $\sqrt[3]{2x} \cdot \sqrt[3]{4x^2} = 2x$.

Critical Thinking Exercises

109. Which one of the following is true?   **d**

    a. $2\sqrt{5} \cdot 6\sqrt{5} = 12\sqrt{5}$

    b. $\sqrt[3]{4} \cdot \sqrt[3]{4} = 4$

    c. $\sqrt{12} = 2\sqrt{6}$

    d. $\sqrt[3]{3^{15}} = 243$

110. If a number is tripled, what happens to its square root?   Its square root is multiplied by $\sqrt{3}$.

111. What must be done to a number so that its cube root is tripled?   The number must be multiplied by 27.

112. If $f(x) = \sqrt[3]{2x}$ and $(fg)(x) = 2x$, find $g(x)$. $g(x) = \sqrt[3]{4x^2}$

113. Graph $f(x) = \sqrt{(x-1)^2}$ by hand.   See graphing answer section.

Review Exercises

114. Solve: $2x - 1 \le 21$ and $2x + 2 \ge 12$. (Section 9.2, Example 2) $\{x \mid 5 \le x \le 11\}$ or $[5, 11]$

115. Solve:

$$5x + 2y = \;\; 2$$
$$4x + 3y = -4.$$

    (Section 4.3, Example 4) $\{(2, -4)\}$

116. Factor: $64x^3 - 27$. (Section 6.4, Example 8) $(4x - 3)(16x^2 + 12x + 9)$

SECTION

10.4

ADDING, SUBTRACTING, AND DIVIDING RADICAL EXPRESSIONS

Objectives

1 Add and subtract radical expressions.

2 Use the quotient rule to simplify radical expressions.

3 Use the quotient rule to divide radical expressions.

The future is now: You have the opportunity to explore the cosmos in a starship traveling near the speed of light. The experience will enable you to understand the mysteries of the universe first hand, transporting you to unimagined levels of knowing and being. The down side: According to Einstein's theory of relativity, close to the speed of light, your aging rate relative to friends on Earth is nearly zero. You will return from your two-year journey to a futuristic world in which friends and loved ones are long dead. Do you explore space or stay here on Earth? In this section, in addition to learning to perform various operations with radicals, you will see how they model your return to a world where the people you knew have long since departed.

1 Add and subtract radical expressions

Adding and Subtracting Radical Expressions   Two or more radical expressions that have the same indices and radicands are called **like radicals**. Like radicals are combined in exactly the same way that we combine like terms. For example,

$$7\sqrt{11} + 6\sqrt{11} = (7 + 6)\sqrt{11} = 13\sqrt{11}.$$

> 7 square roots of 11 plus 6 square roots of 11 result in 13 square roots of 11.

EXAMPLE 1 Adding and Subtracting Like Radicals

Simplify (add or subtract) by combining like radical terms:

a. $7\sqrt{2} + 8\sqrt{2}$ **b.** $\sqrt[3]{5} - 4x\sqrt[3]{5} + 8\sqrt[3]{5}$ **c.** $8\sqrt[6]{5x} - 5\sqrt[6]{5x} + 4\sqrt[3]{5x}$.

SOLUTION

a. $7\sqrt{2} + 8\sqrt{2}$

$= (7 + 8)\sqrt{2}$ Apply the distributive property.

$= 15\sqrt{2}$ Simplify.

b. $\sqrt[3]{5} - 4x\sqrt[3]{5} + 8\sqrt[3]{5}$

$= (1 - 4x + 8)\sqrt[3]{5}$ Apply the distributive property.

$= (9 - 4x)\sqrt[3]{5}$ Simplify.

c. $8\sqrt[6]{5x} - 5\sqrt[6]{5x} + 4\sqrt[3]{5x}$

$= (8 - 5)\sqrt[6]{5x} + 4\sqrt[3]{5x}$ Apply the distributive property to the two terms with like radicals.

$= 3\sqrt[6]{5x} + 4\sqrt[3]{5x}$ The indices, 6 and 3, differ. These are not like radicals and cannot be combined. ∎

 CHECK POINT 1 Simplify by combining like radical terms:

a. $8\sqrt{13} + 2\sqrt{13}$ $10\sqrt{13}$

b. $9\sqrt[3]{7} - 6x\sqrt[3]{7} + 12\sqrt[3]{7}$ $(21 - 6x)\sqrt[3]{7}$

c. $7\sqrt[4]{3x} - 2\sqrt[4]{3x} + 2\sqrt[3]{3x}$. $5\sqrt[4]{3x} + 2\sqrt[3]{3x}$

In some cases, radical expressions can be combined once they have been simplified. For example, to add $\sqrt{2}$ and $\sqrt{8}$, we can write $\sqrt{8}$ as $\sqrt{4 \cdot 2}$ because 4 is a perfect square factor of 8.

$$\sqrt{2} + \sqrt{8} = \sqrt{2} + \sqrt{4 \cdot 2} = 1\sqrt{2} + 2\sqrt{2} = (1 + 2)\sqrt{2} = 3\sqrt{2}$$

EXAMPLE 2 Combining Radicals That First Require Simplification

Simplify by combining like radical terms, if possible:

a. $7\sqrt{18} + 5\sqrt{8}$ **b.** $4\sqrt{27x} - 8\sqrt{12x}$ **c.** $7\sqrt{3} - 2\sqrt{5}$.

SOLUTION

a. $7\sqrt{18} + 5\sqrt{8}$

$= 7\sqrt{9 \cdot 2} + 5\sqrt{4 \cdot 2}$ Factor the radicands using the greatest perfect square factors.

$= 7\sqrt{9} \cdot \sqrt{2} + 5\sqrt{4} \cdot \sqrt{2}$ Take the square root of each factor.

$= 7 \cdot 3 \cdot \sqrt{2} + 5 \cdot 2 \cdot \sqrt{2}$ $\sqrt{9} = 3$ and $\sqrt{4} = 2$.

$= 21\sqrt{2} + 10\sqrt{2}$ Multiply.

$= (21 + 10)\sqrt{2}$ Apply the distributive property.

$= 31\sqrt{2}$ Simplify.

b. $4\sqrt{27x} - 8\sqrt{12x}$

$$= 4\sqrt{9 \cdot 3x} - 8\sqrt{4 \cdot 3x}$$ Factor the radicands using the greatest perfect square factors.

$$= 4\sqrt{9} \cdot \sqrt{3x} - 8\sqrt{4} \cdot \sqrt{3x}$$ Take the square root of each factor.

$$= 4 \cdot 3 \cdot \sqrt{3x} - 8 \cdot 2 \cdot \sqrt{3x}$$ $\sqrt{9} = 3$ and $\sqrt{4} = 2$.

$$= 12\sqrt{3x} - 16\sqrt{3x}$$ Multiply.

$$= (12 - 16)\sqrt{3x}$$ Apply the distributive property.

$$= -4\sqrt{3x}$$ Simplify.

c. $7\sqrt{3} - 2\sqrt{5}$ cannot be simplified. The radical expressions have different radicands, namely 3 and 5, and are not like terms. ■

 CHECK POINT 2 Simplify by combining like radical terms, if possible:

a. $3\sqrt{20} + 5\sqrt{45}$ $21\sqrt{5}$

b. $3\sqrt{12x} - 6\sqrt{27x}$ $-12\sqrt{3x}$

c. $8\sqrt{5} - 6\sqrt{2}$.   cannot be simplified

EXAMPLE 3   Adding and Subtracting with Higher Indices

Simplify by combining like radical terms, if possible:

a. $2\sqrt[3]{16} - 4\sqrt[3]{54}$ **b.** $5\sqrt[3]{xy^2} + \sqrt[3]{8x^4y^5}$.

SOLUTION

a. $2\sqrt[3]{16} - 4\sqrt[3]{54}$

$$= 2\sqrt[3]{8 \cdot 2} - 4\sqrt[3]{27 \cdot 2}$$ Factor the radicands using the greatest perfect cube factors.

$$= 2\sqrt[3]{8} \cdot \sqrt[3]{2} - 4\sqrt[3]{27} \cdot \sqrt[3]{2}$$ Take the cube root of each factor.

$$= 2 \cdot 2 \cdot \sqrt[3]{2} - 4 \cdot 3 \cdot \sqrt[3]{2}$$ $\sqrt[3]{8} = 2$ and $\sqrt[3]{27} = 3$.

$$= 4\sqrt[3]{2} - 12\sqrt[3]{2}$$ Multiply.

$$= (4 - 12)\sqrt[3]{2}$$ Apply the distributive property.

$$= -8\sqrt[3]{2}$$ Simplify.

b. $5\sqrt[3]{xy^2} + \sqrt[3]{8x^4y^5}$

$$= 5\sqrt[3]{xy^2} + \sqrt[3]{(8x^3y^3)xy^2}$$ Factor the second radicand using the greatest perfect cube factor.

$$= 5\sqrt[3]{xy^2} + \sqrt[3]{8x^3y^3} \cdot \sqrt[3]{xy^2}$$ Take the cube root of each factor.

$$= 5\sqrt[3]{xy^2} + 2xy\sqrt[3]{xy^2}$$ $\sqrt[3]{8} = 2$ and $\sqrt[3]{x^3y^3} = (x^3y^3)^{\frac{1}{3}} = xy$.

$$= (5 + 2xy)\sqrt[3]{xy^2}$$ Apply the distributive property.   ■

 CHECK POINT 3 Simplify by combining like radical terms, if possible.

a. $3\sqrt[3]{24} - 5\sqrt[3]{81}$ $-9\sqrt[3]{3}$

b. $5\sqrt[3]{x^2y} + \sqrt[3]{27x^5y^4}$. $(5 + 3xy)\sqrt[3]{x^2y}$

Dividing Radical Expressions We have seen that the root of a product is the product of the roots. The root of a quotient can also be expressed as the quotient of roots. Here is an example:

$$\sqrt{\frac{64}{4}} = \sqrt{16} = 4 \quad \text{and} \quad \frac{\sqrt{64}}{\sqrt{4}} = \frac{8}{2} = 4.$$

| This expression is the square root of a quotient. | This expression is the quotient of two square roots. |

The two procedures produce the same result, 4. This is a special case of the **quotient rule for radicals**.

2 Use the quotient rule to simplify radical expressions.

THE QUOTIENT RULE FOR RADICALS If $\sqrt[n]{a}$ and $\sqrt[n]{b}$ are real numbers and $b \neq 0$, then

$$\sqrt[n]{\frac{a}{b}} = \frac{\sqrt[n]{a}}{\sqrt[n]{b}}.$$

The *n*th root of a quotient is the quotient of the *n*th roots.

We know that a radical is simplified when its radicand has no factors other than 1 that are perfect *n*th powers. The quotient rule can be used to simplify some radicals. Keep in mind that all variables in radicands represent positive real numbers.

EXAMPLE 4 Using the Quotient Rule to Simplify Radicals

Simplify using the quotient rule:

a. $\sqrt[3]{\frac{16}{27}}$ **b.** $\sqrt{\frac{x^2}{25y^6}}$ **c.** $\sqrt[4]{\frac{7y^5}{x^{12}}}.$

SOLUTION We simplify each expression by taking the roots of the numerator and the denominator. Then we use factoring to simplify the resulting radicals, if possible.

a. $\sqrt[3]{\frac{16}{27}} = \frac{\sqrt[3]{16}}{\sqrt[3]{27}} = \frac{\sqrt[3]{8 \cdot 2}}{3} = \frac{\sqrt[3]{8} \cdot \sqrt[3]{2}}{3} = \frac{2\sqrt[3]{2}}{3}$

b. $\sqrt{\frac{x^2}{25y^6}} = \frac{\sqrt{x^2}}{\sqrt{25y^6}} = \frac{x}{5(y^6)^{\frac{1}{2}}} = \frac{x}{5y^3}$

Try to do this step mentally.

c. $\sqrt[4]{\frac{7y^5}{x^{12}}} = \frac{\sqrt[4]{7y^5}}{\sqrt[4]{x^{12}}} = \frac{\sqrt[4]{y^4 \cdot 7y}}{\sqrt[4]{x^{12}}} = \frac{y\sqrt[4]{7y}}{x^3}$

✔ **CHECK POINT 4** Simplify using the quotient rule:

a. $\sqrt[3]{\frac{24}{125}}$ $\frac{2\sqrt[3]{3}}{5}$ **b.** $\sqrt{\frac{9x^3}{y^{10}}}$ $\frac{3x\sqrt{x}}{y^5}$ **c.** $\sqrt[3]{\frac{8y^7}{x^{12}}}.$ $\frac{2y^2\sqrt[3]{y}}{x^4}$

By reversing the two sides of the quotient rule, we obtain a procedure for dividing radical expressions.

3 Use the quotient rule to divide radical expressions.

DIVIDING RADICAL EXPRESSIONS If $\sqrt[n]{a}$ and $\sqrt[n]{b}$ are real numbers and $b \neq 0$, then

$$\frac{\sqrt[n]{a}}{\sqrt[n]{b}} = \sqrt[n]{\frac{a}{b}}.$$

To divide two radical expressions with the same index, divide the radicands and retain the common index.

EXAMPLE 5 Dividing Radical Expressions

Divide and, if possible, simplify:

a. $\dfrac{\sqrt{48x^3}}{\sqrt{6x}}$　　**b.** $\dfrac{\sqrt{45xy}}{2\sqrt{5}}$　　**c.** $\dfrac{\sqrt[3]{16x^5y^2}}{\sqrt[3]{2xy^{-1}}}.$

SOLUTION In each part of this problem, the indices in the numerator and the denominator are the same. Perform each division by dividing the radicands and retaining the common index.

a. $\dfrac{\sqrt{48x^3}}{\sqrt{6x}} = \sqrt{\dfrac{48x^3}{6x}} = \sqrt{8x^2} = \sqrt{4x^2 \cdot 2} = \sqrt{4x^2} \cdot \sqrt{2} = 2x\sqrt{2}$

b. $\dfrac{\sqrt{45xy}}{2\sqrt{5}} = \dfrac{1}{2} \cdot \sqrt{\dfrac{45xy}{5}} = \dfrac{1}{2} \cdot \sqrt{9xy} = \dfrac{1}{2} \cdot 3\sqrt{xy}$ or $\dfrac{3\sqrt{xy}}{2}$

c. $\dfrac{\sqrt[3]{16x^5y^2}}{\sqrt[3]{2xy^{-1}}} = \sqrt[3]{\dfrac{16x^5y^2}{2xy^{-1}}}$　　Divide the radicands and retain the common index.

$= \sqrt[3]{8x^{5-1}y^{2-(-1)}}$　　Divide factors in the radicand. Subtract exponents on common bases.

$= \sqrt[3]{8x^4y^3}$　　Simplify.

$= \sqrt[3]{(8x^3y^3)x}$　　Factor using the greatest perfect cube factor.

$= \sqrt[3]{8x^3y^3} \cdot \sqrt[3]{x}$　　Factor into two radicals.

$= 2xy\sqrt[3]{x}$　　Simplify.

 CHECK POINT 5 Divide and, if possible, simplify:

a. $\dfrac{\sqrt{40x^5}}{\sqrt{2x}}$　$2x^2\sqrt{5}$　　**b.** $\dfrac{\sqrt{50xy}}{2\sqrt{2}}$　$\dfrac{5\sqrt{xy}}{2}$

c. $\dfrac{\sqrt[3]{48x^7y}}{\sqrt[3]{6xy^{-2}}}.$　$2x^2y$

10.4 EXERCISE SET

 Student Solutions Manual   CD/Video   PH Math/Tutor Center   MathXL Tutorials on CD   Math XL MathXL®   MyMathLab MyMathLab   Interactmath.com

Practice Exercises

In this exercise set, assume that all variables represent positive real numbers.

In Exercises 1–10, add or subtract as indicated.

1. $8\sqrt{5} + 3\sqrt{5}$ $11\sqrt{5}$ **2.** $7\sqrt{3} + 2\sqrt{3}$ $9\sqrt{3}$

3. $9\sqrt[3]{6} - 2\sqrt[3]{6}$ $7\sqrt[3]{6}$ **4.** $9\sqrt[3]{7} - 4\sqrt[3]{7}$ $5\sqrt[3]{7}$

5. $4\sqrt[5]{2} + 3\sqrt[5]{2} - 5\sqrt[5]{2}$ $2\sqrt[5]{2}$ **6.** $6\sqrt[5]{3} + 2\sqrt[5]{3} - 3\sqrt[5]{3}$ $5\sqrt[5]{3}$

7. $3\sqrt{13} - 2\sqrt{5} - 2\sqrt{13} + 4\sqrt{5}$ $\sqrt{13} + 2\sqrt{5}$

8. $8\sqrt{17} - 5\sqrt{19} - 6\sqrt{17} + 4\sqrt{19}$ $2\sqrt{17} - \sqrt{19}$

9. $3\sqrt{5} - \sqrt[3]{x} + 4\sqrt{5} + 3\sqrt[3]{x}$ $7\sqrt{5} + 2\sqrt[3]{x}$

10. $6\sqrt{7} - \sqrt[3]{x} + 2\sqrt{7} + 5\sqrt[3]{x}$ $8\sqrt{7} + 4\sqrt[3]{x}$

In Exercises 11–28, add or subtract as indicated. You will need to simplify terms to identify the like radicals.

11. $\sqrt{3} + \sqrt{27}$ $4\sqrt{3}$ **12.** $\sqrt{5} + \sqrt{20}$ $3\sqrt{5}$

13. $7\sqrt{12} + \sqrt{75}$ $19\sqrt{3}$ **14.** $5\sqrt{12} + \sqrt{75}$ $15\sqrt{3}$

15. $3\sqrt{32x} - 2\sqrt{18x}$ $6\sqrt{2x}$ **16.** $5\sqrt{45x} - 2\sqrt{20x}$ $11\sqrt{5x}$

17. $5\sqrt[3]{16} + \sqrt[3]{54}$ $13\sqrt[3]{2}$ **18.** $3\sqrt[3]{24} + \sqrt[3]{81}$ $9\sqrt[3]{3}$

19. $3\sqrt{45x^3} + \sqrt{5x}$ $(9x + 1)\sqrt{5x}$

20. $8\sqrt{45x^3} + \sqrt{5x}$ $(24x + 1)\sqrt{5x}$

21. $\sqrt[3]{54xy^3} + y\sqrt[3]{128x}$ $7y\sqrt[3]{2x}$

22. $\sqrt[3]{24xy^3} + y\sqrt[3]{81x}$ $5y\sqrt[3]{3x}$

23. $\sqrt[3]{54x^4} - \sqrt[3]{16x}$ $(3x - 2)\sqrt[3]{2x}$

24. $\sqrt[3]{81x^4} - \sqrt[3]{24x}$ $(3x - 2)\sqrt[3]{3x}$

25. $\sqrt{9x - 18} + \sqrt{x - 2}$ $4\sqrt{x - 2}$

26. $\sqrt{4x - 12} + \sqrt{x - 3}$ $3\sqrt{x - 3}$

27. $2\sqrt[3]{x^4y^2} + 3x\sqrt[3]{xy^2}$ $5x\sqrt[3]{xy^2}$

28. $4\sqrt[3]{x^4y^2} + 5x\sqrt[3]{xy^2}$ $9x\sqrt[3]{xy^2}$

In Exercises 29–44, simplify using the quotient rule.

29. $\sqrt{\dfrac{11}{4}}$ $\dfrac{\sqrt{11}}{2}$ **30.** $\sqrt{\dfrac{19}{25}}$ $\dfrac{\sqrt{19}}{5}$

31. $\sqrt[3]{\dfrac{19}{27}}$ $\dfrac{\sqrt[3]{19}}{3}$ **32.** $\sqrt[3]{\dfrac{11}{64}}$ $\dfrac{\sqrt[3]{11}}{4}$

33. $\sqrt{\dfrac{x^2}{36y^8}}$ $\dfrac{x}{6y^4}$ **34.** $\sqrt{\dfrac{x^2}{144y^{12}}}$ $\dfrac{x}{12y^6}$

35. $\sqrt{\dfrac{8x^3}{25y^6}}$ $\dfrac{2x\sqrt{2x}}{5y^3}$ **36.** $\sqrt{\dfrac{50x^3}{81y^8}}$ $\dfrac{5x\sqrt{2x}}{9y^4}$

37. $\sqrt[3]{\dfrac{x^4}{8y^3}}$ $\dfrac{x\sqrt[3]{x}}{2y}$ **38.** $\sqrt[3]{\dfrac{x^5}{125y^3}}$ $\dfrac{x\sqrt[3]{x^2}}{5y}$

39. $\sqrt[3]{\dfrac{50x^8}{27y^{12}}}$ $\dfrac{x^2\sqrt[3]{50x^2}}{3y^4}$ **40.** $\sqrt[3]{\dfrac{81x^8}{8y^{15}}}$ $\dfrac{3x^2\sqrt[3]{3x^2}}{2y^5}$

41. $\sqrt[4]{\dfrac{9y^6}{x^8}}$ $\dfrac{y\sqrt[4]{9y^2}}{x^2}$ **42.** $\sqrt[4]{\dfrac{13y^7}{x^{12}}}$ $\dfrac{y\sqrt[4]{13y^3}}{x^3}$

43. $\sqrt[5]{\dfrac{64x^{13}}{y^{20}}}$ $\dfrac{2x^2\sqrt[5]{2x^3}}{y^4}$ **44.** $\sqrt[5]{\dfrac{64x^{14}}{y^{15}}}$ $\dfrac{2x^2\sqrt[5]{2x^4}}{y^3}$

In Exercises 45–66, divide and, if possible, simplify.

45. $\dfrac{\sqrt{40}}{\sqrt{5}}$ $2\sqrt{2}$ **46.** $\dfrac{\sqrt{200}}{\sqrt{10}}$ $2\sqrt{5}$

47. $\dfrac{\sqrt[3]{48}}{\sqrt[3]{6}}$ 2 **48.** $\dfrac{\sqrt[3]{54}}{\sqrt[3]{2}}$ 3

49. $\dfrac{\sqrt{54x^3}}{\sqrt{6x}}$ $3x$ **50.** $\dfrac{\sqrt{72x^3}}{\sqrt{2x}}$ $6x$

51. $\dfrac{\sqrt{x^5y^3}}{\sqrt{xy}}$ x^2y **52.** $\dfrac{\sqrt{x^7y^6}}{\sqrt{x^3y^2}}$ x^2y^2

53. $\dfrac{\sqrt{200x^3}}{\sqrt{10x^{-1}}}$ $2x^2\sqrt{5}$ **54.** $\dfrac{\sqrt{500x^3}}{\sqrt{10x^{-1}}}$ $5x^2\sqrt{2}$

55. $\dfrac{\sqrt{48a^8b^7}}{\sqrt{3a^{-2}b^{-3}}}$ $4a^5b^5$ **56.** $\dfrac{\sqrt{54a^7b^{11}}}{\sqrt{3a^{-4}b^{-2}}}$ $3a^5b^6\sqrt{2ab}$

57. $\dfrac{\sqrt{72xy}}{2\sqrt{2}}$ $3\sqrt{xy}$ **58.** $\dfrac{\sqrt{50xy}}{2\sqrt{2}}$ $\dfrac{5\sqrt{xy}}{2}$

59. $\dfrac{\sqrt[3]{24x^3y^5}}{\sqrt[3]{3y^2}}$ $2xy$ **60.** $\dfrac{\sqrt[3]{250x^5y^3}}{\sqrt[3]{2x^3}}$ $5y\sqrt[3]{x^2}$

61. $\dfrac{\sqrt[4]{32x^{10}y^8}}{\sqrt[4]{2x^2y^{-2}}}$ $2x^2y^2\sqrt[4]{y^2}$ or $2x^2y^2\sqrt{y}$

62. $\dfrac{\sqrt[5]{96x^{12}y^{11}}}{\sqrt[5]{3x^2y^{-2}}}$ $2x^2y^2\sqrt[5]{y^3}$ **63.** $\dfrac{\sqrt[3]{x^2 + 5x + 6}}{\sqrt[3]{x + 2}}$ $\sqrt[3]{x + 3}$

64. $\dfrac{\sqrt[3]{x^2 + 7x + 12}}{\sqrt[3]{x + 3}}$ $\sqrt[3]{x + 4}$ **65.** $\dfrac{\sqrt[3]{a^3 + b^3}}{\sqrt[3]{a + b}}$ $\sqrt[3]{a^2 - ab + b^2}$

66. $\dfrac{\sqrt[3]{a^3 - b^3}}{\sqrt[3]{a - b}}$ $\sqrt[3]{a^2 + ab + b^2}$

Practice Plus

In Exercises 67–76, perform the indicated operations.

67. $\dfrac{\sqrt{32}}{5} + \dfrac{\sqrt{18}}{7}$ $\dfrac{43\sqrt{2}}{35}$ **68.** $\dfrac{\sqrt{27}}{2} + \dfrac{\sqrt{75}}{7}$ $\dfrac{31\sqrt{3}}{14}$

69. $3x\sqrt{8xy^2} - 5y\sqrt{32x^3} + \sqrt{18x^3y^2}$ $-11xy\sqrt{2x}$

70. $6x\sqrt{3xy^2} - 4x^2\sqrt{27xy} - 5\sqrt{75x^5y}$ $6xy\sqrt{3x} - 37x^2\sqrt{3xy}$

71. $5\sqrt{2x^3} + \dfrac{30x^3\sqrt{24x^2}}{3x^2\sqrt{3x}}$ $25x\sqrt{2x}$

72. $7\sqrt{2x^3} + \dfrac{40x^3\sqrt{150x^2}}{5x^2\sqrt{3x}}$ $47x\sqrt{2x}$

73. $2x\sqrt{75xy} - \dfrac{\sqrt{81xy^2}}{\sqrt{3x^{-2}y}}$ $7x\sqrt{3xy}$

74. $5\sqrt{8x^2y^3} - \dfrac{9x^2\sqrt{64y}}{3x\sqrt{2y^{-2}}}$ $-2xy\sqrt{2y}$

75. $\dfrac{15x^4\sqrt[3]{80x^3y^2}}{5x^3\sqrt[3]{2x^2y}} - \dfrac{75\sqrt[3]{5x^3y}}{25\sqrt[3]{x^{-1}}}$ $3x\sqrt[3]{5xy}$

76. $\dfrac{16x^4\sqrt[3]{48x^3y^2}}{8x^3\sqrt[3]{3x^2y}} - \dfrac{20\sqrt[3]{2x^3y}}{4\sqrt[3]{x^{-1}}}$ $-x\sqrt[3]{2xy}$

In Exercises 77–80, find $\left(\frac{f}{g}\right)(x)$ and the domain of $\left(\frac{f}{g}\right)$. Express each quotient function in simplified form.

77. $f(x) = \sqrt{48x^5}, \quad g(x) = \sqrt{3x^2}$
$\left(\dfrac{f}{g}\right)(x) = 4x\sqrt{x}$; Domain of $\dfrac{f}{g} = \{x \mid x > 0\}$ or $(0, \infty)$

78. $f(x) = \sqrt{x^2 - 25}, \quad g(x) = \sqrt{x + 5}$
$\left(\dfrac{f}{g}\right)(x) = \sqrt{x - 5}$; Domain of $\dfrac{f}{g} = \{x \mid x \geq 5\}$ or $[5, \infty)$

79. $f(x) = \sqrt[3]{32x^6}, \quad g(x) = \sqrt[3]{2x^2}$ $\left(\dfrac{f}{g}\right)(x) = 2x\sqrt[3]{2x}$;
Domain of $\dfrac{f}{g} = \{x \mid x \text{ is a real number and } x \neq 0\}$ or $(-\infty, 0) \cup (0, \infty)$

80. $f(x) = \sqrt[3]{2x^6}, \quad g(x) = \sqrt[3]{16x}$ $\left(\dfrac{f}{g}\right)(x) = \dfrac{x\sqrt[3]{x^2}}{2}$;
Domain of $\dfrac{f}{g} = \{x \mid x \text{ is a real number and } x \neq 0\}$ or $(-\infty, 0) \cup (0, \infty)$

Application Exercises

What does travel in space have to do with radicals? Imagine that in the future we will be able to travel in starships at velocities approaching the speed of light (approximately 186,000 miles per second). According to Einstein's theory of relativity, time would pass more quickly on Earth than it would in the moving starship. The radical expression

$$R_f\frac{\sqrt{c^2 - v^2}}{\sqrt{c^2}}$$

gives the aging rate of an astronaut relative to the aging rate of a friend, R_f, on Earth. In the expression, v is the astronaut's velocity and c is the speed of light. Use the expression to solve Exercises 81–82. Imagine that you are the astronaut on the starship.

81. a. Use the quotient rule and simplify the expression that shows your aging rate relative to a friend on Earth. Working in a step-by-step manner, express your aging rate as

$$R_f\sqrt{1 - \left(\frac{v}{c}\right)^2}.$$

  b. You are moving at velocities approaching the speed of light. Substitute c, the speed of light, for v in the simplified expression from part (a). Simplify completely. Close to the speed of light, what is your aging rate relative to a friend on Earth? What does this mean?

82. a. Use the quotient rule and simplify the expression that shows your aging rate relative to a friend on Earth. Working step-by-step, express your aging rate as

$$R_f\sqrt{1 - \left(\frac{v}{c}\right)^2}.$$

  b. You are moving at 90% of the speed of light. Substitute $0.9c$ for v, your velocity, in the simplified expression from part (a). What is your aging rate, correct to two decimal places, relative to a friend on Earth? If you are gone for 44 weeks, approximately how many weeks have passed for your friend? $0.44R_f$; 100 weeks

81–82.a. $R_f\dfrac{\sqrt{c^2 - v^2}}{\sqrt{c^2}} = R_f\sqrt{\dfrac{c^2 - v^2}{c^2}} = R_f\sqrt{\dfrac{c^2}{c^2} - \dfrac{v^2}{c^2}} =$
$= R_f\sqrt{1 - \dfrac{v^2}{c^2}} = R_f\sqrt{1 - \left(\dfrac{v}{c}\right)^2}$

81.b. $R_f\sqrt{1 - \left(\dfrac{c}{c}\right)^2} = 0$; Close to the speed of light, the aging rate is 0 relative to a friend on Earth. A person moving close to the speed of light does not age relative to a friend on Earth.

In Exercises 83–84, find the perimeter and area of each rectangle. Express answers in simplified radical form.

83.

$2\sqrt{20}$ feet $P = 18\sqrt{5}$ feet
$A = 100$ square feet
$\sqrt{125}$ feet

84. $4\sqrt{20}$ feet $P = 24\sqrt{5}$ feet
$A = 160$ square feet
$\sqrt{80}$ feet

Writing in Mathematics

85. What are like radicals? Give an example with your explanation.   **85–90.** Answers will vary.

86. Explain how to add like radicals. Give an example with your explanation.

87. If only like radicals can be combined, why is it possible to add $\sqrt{2}$ and $\sqrt{8}$?

88. Explain how to simplify a radical expression using the quotient rule. Provide an example.

89. Explain how to divide radical expressions with the same index.

90. Answer the question posed in the section opener on page 670. What will you do: explore space or stay here on Earth? What are the reasons for your choice?

Technology Exercises

91. Use a calculator to provide numerical support to any four exercises that you worked from Exercises 1–66 that do not contain variables. Begin by finding a decimal approximation for the given expression. Then find a decimal approximation for your answer. The two decimal approximations should be the same. *Answers will vary.*

In Exercises 92–94, determine if each operation is performed correctly by graphing the function on each side of the equation with your graphing utility. Use the given viewing rectangle. The graphs should be the same. If they are not, correct the right side of the equation and then use your graphing utility to verify the correction. **92–94.** *See graphing answer section.*

92. $\sqrt{4x} + \sqrt{9x} = 5\sqrt{x}$

$[0, 5, 1]$ by $[0, 10, 1]$
Graphs are the same; Simplification is correct.

93. $\sqrt{16x} - \sqrt{9x} = \sqrt{7x}$

$[0, 5, 1]$ by $[0, 5, 1]$
Graphs are not the same; $\sqrt{16x} - \sqrt{9x} = \sqrt{x}.$

94. $x\sqrt{8} + x\sqrt{2} = x\sqrt{10}$

$[-5, 5, 1]$ by $[-15, 15, 1]$
Graphs are not the same; $x\sqrt{8} + x\sqrt{2} = 3x\sqrt{2}.$

Critical Thinking Exercises

95. Which one of the following is true? d
   a. $\sqrt{5} + \sqrt{5} = \sqrt{10}$
   b. $4\sqrt{3} + 5\sqrt{3} = 9\sqrt{6}$
   c. If any two radical expressions are completely simplified, they can then be combined through addition or subtraction.
   d. None of the above is true.

96. If an irrational number is decreased by $2\sqrt{18} - \sqrt{50}$, the result is $\sqrt{2}$. What is the number? $2\sqrt{2}$

97. Simplify: $\dfrac{\sqrt{20}}{3} + \dfrac{\sqrt{45}}{4} - \sqrt{80}.$ $-\dfrac{31\sqrt{5}}{12}$

98. Simplify: $\dfrac{6\sqrt{49xy}\,\sqrt{ab^2}}{7\sqrt{36x^{-3}y^{-5}}\,\sqrt{a^{-9}b^{-1}}}.$ $x^2y^3a^5b\sqrt{b}$

Review Exercises

99. Solve: $2(3x - 1) - 4 = 2x - (6 - x).$
(Section 2.3, Example 3) $\{0\}$

100. Factor: $x^2 - 8xy + 12y^2.$
(Section 6.2, Example 6) $(x - 2y)(x - 6y)$

101. Add: $\dfrac{2}{x^2 + 5x + 6} + \dfrac{3x}{x^2 + 6x + 9}.$
(Section 7.4, Example 7) $\dfrac{3x^2 + 8x + 6}{(x + 3)^2(x + 2)}$

✔ MID-CHAPTER CHECK POINT

CHAPTER 10

What You Know: We learned to find roots of numbers. We saw that the domain of a square root function is the set of real numbers for which the radicand is nonnegative. We learned to simplify radical expressions, using $\sqrt[n]{a^n} = |a|$ if n is even and $\sqrt[n]{a^n} = a$ if n is odd. The definition $a^{\frac{m}{n}} = (\sqrt[n]{a})^m = \sqrt[n]{a^m}$ connected rational exponents and radicals. Finally, we performed various operations with radicals, including multiplication, addition, subtraction, and division.

In Exercises 1–23, simplify the given expression or perform the indicated operation(s) and, if possible, simplify. Assume that all variables represent positive real numbers.

1. $\sqrt{100} - \sqrt[3]{-27}$ 13

2. $\sqrt{8x^5y^7}$ $2x^2y^3\sqrt{2xy}$

3. $3\sqrt[3]{4x^2} + 2\sqrt[3]{4x^2}$ $5\sqrt[3]{4x^2}$

4. $\left(3\sqrt[3]{4x^2}\right)\left(2\sqrt[3]{4x^2}\right)$ $12x\sqrt[3]{2x}$

5. $27^{\frac{2}{3}} + (-32)^{\frac{3}{5}}$ 1

6. $\left(64x^3y^{\frac{1}{4}}\right)^{\frac{1}{3}}$ $4xy^{\frac{1}{12}}$

7. $5\sqrt{27} - 4\sqrt{48}$ $-\sqrt{3}$

8. $\sqrt{\dfrac{500x^3}{4y^4}}$ $\dfrac{5x\sqrt{5x}}{y^2}$

9. $\dfrac{x}{\sqrt[4]{x}}$ $\sqrt[4]{x^3}$

10. $\sqrt[3]{54x^5}$ $3x\sqrt[3]{2x^2}$

11. $\dfrac{\sqrt[3]{160}}{\sqrt[3]{2}}$ $2\sqrt[3]{10}$

12. $\sqrt[5]{\dfrac{x^{10}}{y^{20}}}$ $\dfrac{x^2}{y^4}$

13. $\dfrac{\left(x^{\frac{2}{3}}\right)^2}{\left(x^{\frac{1}{4}}\right)^3}$ $x^{\frac{7}{12}}$

14. $\sqrt[6]{x^6y^4}$ $x\sqrt[3]{y^2}$

15. $\sqrt[7]{(x-2)^3} \cdot \sqrt[7]{(x-2)^6}$ $(x-2)\sqrt[7]{(x-2)^2}$

16. $\sqrt[4]{32x^{11}y^{17}}$ $2x^2y^4\sqrt[4]{2x^3y}$ **17.** $4\sqrt[3]{16} + 2\sqrt[3]{54}$ $14\sqrt[3]{2}$

18. $\dfrac{\sqrt[7]{x^4y^9}}{\sqrt[7]{x^{-5}y^7}}$ $x\sqrt[7]{x^2y^2}$ **19.** $(-125)^{-\frac{2}{3}}$ $\dfrac{1}{25}$

20. $\sqrt{2} \cdot \sqrt[3]{2}$ $\sqrt[6]{32}$

21. $\sqrt[3]{\dfrac{32x}{y^4}} \cdot \sqrt[3]{\dfrac{2x^2}{y^2}}$ $\dfrac{4x}{y^2}$

22. $\sqrt{32xy^2} \cdot \sqrt{2x^3y^5}$ $8x^2y^3\sqrt{y}$

23. $4x\sqrt{6x^4y^3} - 7y\sqrt{24x^6y}$ $-10x^3y\sqrt{6y}$

In Exercises 24–25, find the domain of each function.

24. $f(x) = \sqrt{30 - 5x}$ $\{x | x \le 6\} = (-\infty, 6]$

25. $g(x) = \sqrt[3]{3x - 15}$ $\{x | x \text{ is a real number}\} = (-\infty, \infty)$

SECTION

10.5

Objectives

1 Multiply radicals with more than one term.

2 Use polynomial special products to multiply radicals.

3 Rationalize denominators containing one term.

4 Rationalize denominators containing two terms.

5 Rationalize numerators.

MULTIPLYING WITH MORE THAN ONE TERM AND RATIONALIZING DENOMINATORS

PEANUTS reprinted by permission of United Feature Syndicate, Inc.

The late Charles Schulz, creator of the "Peanuts" comic strip, transfixed 350 million readers worldwide with the joys and angst of his hapless Charlie Brown and Snoopy, a romantic self-deluded beagle. In 18,250 comic strips that spanned nearly 50 years, mathematics was often featured. Is the discussion of radicals shown above the real thing, or is it just an algebraic scam? By the time you complete this section on multiplying and dividing radicals, you will be able to decide.

1 Multiply radicals with more than one term.

Multiplying Radicals with More Than One Term Radical expressions with more than one term are multiplied in much the same way that polynomials with more than one term are multiplied. Example 1 uses the distributive property and the FOIL method to perform multiplications.

EXAMPLE 1 Multiplying Radicals

Multiply:

a. $\sqrt{7}\left(x + \sqrt{2}\right)$ **b.** $\sqrt[3]{x}\left(\sqrt[3]{6} - \sqrt[3]{x^2}\right)$ **c.** $\left(5\sqrt{2} + 2\sqrt{3}\right)\left(4\sqrt{2} - 3\sqrt{3}\right)$.

SOLUTION

a. $\sqrt{7}\left(x + \sqrt{2}\right)$

$= \sqrt{7} \cdot x + \sqrt{7} \cdot \sqrt{2}$ Use the distributive property.

$= x\sqrt{7} + \sqrt{14}$ Multiply the radicals.

b. $\sqrt[3]{x}\left(\sqrt[3]{6} - \sqrt[3]{x^2}\right)$

$= \sqrt[3]{x} \cdot \sqrt[3]{6} - \sqrt[3]{x} \cdot \sqrt[3]{x^2}$ Use the distributive property.

$= \sqrt[3]{6x} - \sqrt[3]{x^3}$ Multiply the radicals: $\sqrt[n]{a} \cdot \sqrt[n]{b} = \sqrt[n]{ab}$.

$= \sqrt[3]{6x} - x$ Simplify: $\sqrt[3]{x^3} = x$.

c. $(5\sqrt{2} + 2\sqrt{3})(4\sqrt{2} - 3\sqrt{3})$ Use the FOIL method.

first ← last
inside
outside

$$
\begin{array}{cccc}
F & O & I & L
\end{array}
$$

$$= (5\sqrt{2})(4\sqrt{2}) + (5\sqrt{2})(-3\sqrt{3}) + (2\sqrt{3})(4\sqrt{2}) + (2\sqrt{3})(-3\sqrt{3})$$

$$= 20 \cdot 2 - 15\sqrt{6} + 8\sqrt{6} - 6 \cdot 3 \quad \text{Multiply. Note that } \sqrt{2} \cdot \sqrt{2} = \sqrt{4} = 2 \text{ and}$$
$$\sqrt{3} \cdot \sqrt{3} = \sqrt{9} = 3.$$

$$= 40 - 15\sqrt{6} + 8\sqrt{6} - 18 \quad \text{Complete the multiplications.}$$

$$= (40 - 18) + \left(-15\sqrt{6} + 8\sqrt{6}\right) \quad \text{Group like terms. Try to do this step mentally.}$$

$$= 22 - 7\sqrt{6} \quad \text{Combine numerical terms and like radicals.} \blacksquare$$

 CHECK POINT 1 Multiply:

a. $\sqrt{6}(x + \sqrt{10})$ $x\sqrt{6} + 2\sqrt{15}$

b. $\sqrt[3]{y}\left(\sqrt[3]{y^2} - \sqrt[3]{7}\right)$ $y - \sqrt[3]{7y}$

c. $\left(6\sqrt{5} + 3\sqrt{2}\right)\left(2\sqrt{5} - 4\sqrt{2}\right).$ $36 - 18\sqrt{10}$

Some radicals can be multiplied using the special products for multiplying polynomials.

2 Use polynomial special products to multiply radicals.

EXAMPLE 2   Using Special Products to Multiply Radicals

Multiply:

a. $\left(\sqrt{3} + \sqrt{7}\right)^2$ **b.** $\left(\sqrt{7} + \sqrt{3}\right)\left(\sqrt{7} - \sqrt{3}\right)$ **c.** $\left(\sqrt{a} - \sqrt{b}\right)\left(\sqrt{a} + \sqrt{b}\right).$

SOLUTION   Use the special-product formulas shown.

$$(A + B)^2 = A^2 + 2 \cdot A \cdot B + B^2$$

a. $\left(\sqrt{3} + \sqrt{7}\right)^2 = \left(\sqrt{3}\right)^2 + 2 \cdot \sqrt{3} \cdot \sqrt{7} + \left(\sqrt{7}\right)^2$ Use the special product for $(A + B)^2$.

$$= 3 + 2\sqrt{21} + 7 \quad \text{Multiply the radicals.}$$

$$= 10 + 2\sqrt{21} \quad \text{Simplify.}$$

$$(A + B) \cdot (A - B) = A^2 - B^2$$

b. $\left(\sqrt{7} + \sqrt{3}\right)\left(\sqrt{7} - \sqrt{3}\right) = \left(\sqrt{7}\right)^2 - \left(\sqrt{3}\right)^2$ Use the special product for $(A + B)(A - B)$.

$$= 7 - 3 \quad \text{Simplify: } \left(\sqrt{a}\right)^2 = a.$$

$$= 4$$

$$(A - B) \cdot (A + B) = A^2 - B^2$$

c. $\left(\sqrt{a} - \sqrt{b}\right)\left(\sqrt{a} + \sqrt{b}\right) = \left(\sqrt{a}\right)^2 - \left(\sqrt{b}\right)^2 = a - b$

STUDY TIP

Just to recap, special products for multiplying polynomials are summarized as follows:

Square of a Binomial:

$(A + B)^2 = A^2 + 2AB + B^2$

$(A - B)^2 = A^2 - 2AB + B^2$

Product of the Sum and Difference of Two Terms:

$(A + B)(A - B) = A^2 - B^2.$

Radical expressions that involve the sum and difference of the same two terms are called **conjugates**. For example,

$$\sqrt{7} + \sqrt{3} \quad \text{and} \quad \sqrt{7} - \sqrt{3}$$

are conjugates of each other. Parts (b) and (c) of Example 2 illustrate that the product of two radical expressions need not be a radical expression:

$$(\sqrt{7} + \sqrt{3})(\sqrt{7} - \sqrt{3}) = 4$$
$$(\sqrt{a} - \sqrt{b})(\sqrt{a} + \sqrt{b}) = a - b.$$

The product of conjugates does not contain a radical.

Later in this section, we will use conjugates to simplify quotients.

✔ **CHECK POINT 2** Multiply:

a. $\left(\sqrt{5} + \sqrt{6}\right)^2$ $11 + 2\sqrt{30}$ **b.** $\left(\sqrt{6} + \sqrt{5}\right)\left(\sqrt{6} - \sqrt{5}\right)$ 1

c. $\left(\sqrt{a} - \sqrt{7}\right)\left(\sqrt{a} + \sqrt{7}\right).$ $a - 7$

3 Rationalize denominators containing one term.

Rationalizing Denominators Containing One Term   You can use a calculator to compare the approximate values for $\dfrac{1}{\sqrt{3}}$ and $\dfrac{\sqrt{3}}{3}$. The two approximations are the same. This is not a coincidence:

$$\frac{1}{\sqrt{3}} = \frac{1}{\sqrt{3}} \cdot \boxed{\frac{\sqrt{3}}{\sqrt{3}}} = \frac{\sqrt{3}}{\sqrt{9}} = \frac{\sqrt{3}}{3}.$$

Any number divided by itself is 1. Multiplication by 1 does not change the value of $\dfrac{1}{\sqrt{3}}$.

This process involves rewriting a radical expression as an equivalent expression in which the denominator no longer contains any radicals. The process is called **rationalizing the denominator**. When the denominator contains a single radical with an nth root, **multiply the numerator and the denominator by a radical of index n that produces a perfect nth power in the denominator's radicand.**

EXAMPLE 3   Rationalizing Denominators

Rationalize each denominator:

a. $\dfrac{\sqrt{5}}{\sqrt{6}}$ **b.** $\sqrt[3]{\dfrac{7}{16}}.$

SOLUTION

a. If we multiply the numerator and the denominator of $\dfrac{\sqrt{5}}{\sqrt{6}}$ by $\sqrt{6}$, the denominator becomes $\sqrt{6} \cdot \sqrt{6} = \sqrt{36} = 6$. The denominator's radicand, 36, is a perfect square.

The denominator no longer contains a radical. Therefore, we multiply by 1, choosing $\dfrac{\sqrt{6}}{\sqrt{6}}$ for 1.

$$\dfrac{\sqrt{5}}{\sqrt{6}} = \dfrac{\sqrt{5}}{\sqrt{6}} \cdot \dfrac{\sqrt{6}}{\sqrt{6}} \qquad \text{Multiply the numerator and denominator by } \sqrt{6} \text{ to remove the radical in the denominator.}$$

$$= \dfrac{\sqrt{30}}{\sqrt{36}} \qquad \text{Multiply numerators and multiply denominators. The denominator's radicand, 36, is a perfect square.}$$

$$= \dfrac{\sqrt{30}}{6} \qquad \text{Simplify: } \sqrt{36} = 6.$$

b. Using the quotient rule, we can express $\sqrt[3]{\dfrac{7}{16}}$ as $\dfrac{\sqrt[3]{7}}{\sqrt[3]{16}}$. We have cube roots, so we want the denominator's radicand to be a perfect cube. Right now, the denominator's radicand is 16 or 4^2. We know that $\sqrt[3]{4^3} = 4$. If we multiply the numerator and the denominator of $\dfrac{\sqrt[3]{7}}{\sqrt[3]{16}}$ by $\sqrt[3]{4}$, the denominator becomes

$$\sqrt[3]{16} \cdot \sqrt[3]{4} = \sqrt[3]{4^2} \cdot \sqrt[3]{4} = \sqrt[3]{4^3} = 4.$$

The denominator's radicand, 4^3, is a perfect cube. The denominator no longer contains a radical. Therefore, we multiply by 1, choosing $\dfrac{\sqrt[3]{4}}{\sqrt[3]{4}}$ for 1.

$$\sqrt[3]{\dfrac{7}{16}} = \dfrac{\sqrt[3]{7}}{\sqrt[3]{16}} \qquad \text{Use the quotient rule and rewrite as the quotient of radicals.}$$

$$= \dfrac{\sqrt[3]{7}}{\sqrt[3]{4^2}} \qquad \text{Write the denominator's radicand as an exponential expression.}$$

$$= \dfrac{\sqrt[3]{7}}{\sqrt[3]{4^2}} \cdot \dfrac{\sqrt[3]{4}}{\sqrt[3]{4}} \qquad \text{Multiply the numerator and denominator by } \sqrt[3]{4} \text{ to remove the radical in the denominator.}$$

$$= \dfrac{\sqrt[3]{28}}{\sqrt[3]{4^3}} \qquad \text{Multiply numerators and denominators. The denominator's radicand, } 4^3, \text{ is a perfect cube.}$$

$$= \dfrac{\sqrt[3]{28}}{4} \qquad \text{Simplify: } \sqrt[3]{4^3} = 4.$$

✔ **CHECK POINT 3** Rationalize each denominator:

a. $\dfrac{\sqrt{3}}{\sqrt{7}} \quad \dfrac{\sqrt{21}}{7}$ **b.** $\sqrt[3]{\dfrac{2}{9}} \cdot \quad \dfrac{\sqrt[3]{6}}{3}$

Example 3 showed that it is helpful to express the denominator's radicand using exponents. In this way, we can find the extra factor or factors needed to produce a perfect nth power. For example, suppose that $\sqrt[5]{8}$ appears in the denominator. We want a perfect fifth power. By expressing $\sqrt[5]{8}$ as $\sqrt[5]{2^3}$, we would multiply the numerator and denominator by $\sqrt[5]{2^2}$ because

$$\sqrt[5]{2^3} \cdot \sqrt[5]{2^2} = \sqrt[5]{2^5} = 2.$$

EXAMPLE 4   Rationalizing Denominators

Rationalize each denominator:

a. $\sqrt{\dfrac{3x}{5y}}$ **b.** $\dfrac{\sqrt[3]{x}}{\sqrt[3]{36y}}$ **c.** $\dfrac{10y}{\sqrt[5]{4x^3y}}$.

SOLUTION   By examining each denominator, you can determine how to multiply by 1. Let's first look at the denominators. For the square root, we must produce exponents of 2 in the radicand. For the cube root, we need exponents of 3, and for the fifth root, we want exponents of 5.

- $\sqrt{5y}$
- $\sqrt[3]{36y}$ or $\sqrt[3]{6^2y}$
- $\sqrt[5]{4x^3y}$ or $\sqrt[5]{2^2x^3y}$

Multiply by $\sqrt{5y}$:
$\sqrt{5y} \cdot \sqrt{5y} = \sqrt{25y^2} = 5y.$

Multiply by $\sqrt[3]{6y^2}$:
$\sqrt[3]{6^2y} \cdot \sqrt[3]{6y^2} = \sqrt[3]{6^3y^3} = 6y.$

Multiply by $\sqrt[5]{2^3x^2y^4}$:
$\sqrt[5]{2^2x^3y} \cdot \sqrt[5]{2^3x^2y^4} = \sqrt[5]{2^5x^5y^5} = 2xy.$

a. $\sqrt{\dfrac{3x}{5y}} = \dfrac{\sqrt{3x}}{\sqrt{5y}} = \dfrac{\sqrt{3x}}{\sqrt{5y}} \cdot \dfrac{\sqrt{5y}}{\sqrt{5y}} = \dfrac{\sqrt{15xy}}{\sqrt{25y^2}} = \dfrac{\sqrt{15xy}}{5y}$

Multiply by 1. $25y^2$ is a perfect square.

b. $\dfrac{\sqrt[3]{x}}{\sqrt[3]{36y}} = \dfrac{\sqrt[3]{x}}{\sqrt[3]{6^2y}} = \dfrac{\sqrt[3]{x}}{\sqrt[3]{6^2y}} \cdot \dfrac{\sqrt[3]{6y^2}}{\sqrt[3]{6y^2}} = \dfrac{\sqrt[3]{6xy^2}}{\sqrt[3]{6^3y^3}} = \dfrac{\sqrt[3]{6xy^2}}{6y}$

Multiply by 1. 6^3y^3 is a perfect cube.

c. $\dfrac{10y}{\sqrt[5]{4x^3y}} = \dfrac{10y}{\sqrt[5]{2^2x^3y}} = \dfrac{10y}{\sqrt[5]{2^2x^3y}} \cdot \dfrac{\sqrt[5]{2^3x^2y^4}}{\sqrt[5]{2^3x^2y^4}}$

Multiply by 1.

$= \dfrac{10y\sqrt[5]{2^3x^2y^4}}{\sqrt[5]{2^5x^5y^5}} = \dfrac{10y\sqrt[5]{8x^2y^4}}{2xy} = \dfrac{5\sqrt[5]{8x^2y^4}}{x}$

$2^5x^5y^5$ is a perfect 5th power.

Simplify: Divide numerator and denominator by $2y$.

CHECK POINT 4 Rationalize each denominator:

a. $\sqrt{\dfrac{2x}{7y}}$ $\dfrac{\sqrt{14xy}}{7y}$

b. $\dfrac{\sqrt[3]{x}}{\sqrt[3]{9y}}$ $\dfrac{\sqrt[3]{3xy^2}}{3y}$

c. $\dfrac{6x}{\sqrt[5]{8x^2y^4}}$. $\dfrac{3\sqrt[5]{4x^3y}}{y}$

4 Rationalize denominators containing two terms.

Rationalizing Denominators Containing Two Terms   How can we rationalize a denominator if the denominator contains two terms with one or more square roots? **Multiply the numerator and the denominator by the conjugate of the denominator.** Here are three examples of such expressions:

$$\cdot \; \frac{8}{3\sqrt{2}+4} \qquad \cdot \; \frac{2+\sqrt{5}}{\sqrt{6}-\sqrt{3}} \qquad \cdot \; \frac{h}{\sqrt{x+h}-\sqrt{x}}$$

The conjugate of the denominator is $3\sqrt{2}-4$.

The conjugate of the denominator is $\sqrt{6}+\sqrt{3}$.

The conjugate of the denominator is $\sqrt{x+h}+\sqrt{x}$.

The product of the denominator and its conjugate is found using the formula

$$(A+B)(A-B) = A^2 - B^2.$$

The simplified product will not contain a radical.

EXAMPLE 5   Rationalizing a Denominator Containing Two Terms

Rationalize the denominator: $\dfrac{8}{3\sqrt{2}+4}$.

SOLUTION   The conjugate of the denominator is $3\sqrt{2}-4$. If we multiply the numerator and the denominator by $3\sqrt{2}-4$, the simplified denominator will not contain a radical. Therefore, we multiply by 1, choosing $\dfrac{3\sqrt{2}-4}{3\sqrt{2}-4}$ for 1.

$$\frac{8}{3\sqrt{2}+4} = \frac{8}{3\sqrt{2}+4} \cdot \frac{3\sqrt{2}-4}{3\sqrt{2}-4} \qquad \text{Multiply by 1.}$$

$$= \frac{8(3\sqrt{2}-4)}{(3\sqrt{2})^2 - 4^2} \qquad (A+B)(A-B) = A^2 - B^2$$

> Leave the numerator in factored form. This helps simplify, if possible.

$$= \frac{8(3\sqrt{2}-4)}{18-16} \qquad (3\sqrt{2})^2 = 9 \cdot 2 = 18$$

$$= \frac{8(3\sqrt{2}-4)}{2} \qquad \text{This expression can still be simplified.}$$

$$= \frac{\overset{4}{8}(3\sqrt{2}-4)}{\underset{1}{2}} \qquad \text{Divide the numerator and denominator by 2.}$$

$$= 4(3\sqrt{2}-4) \quad \text{or} \quad 12\sqrt{2}-16$$

✔ **CHECK POINT 5** Rationalize the denominator: $\dfrac{18}{2\sqrt{3}+3}$. $12\sqrt{3}-18$

EXAMPLE 6 Rationalizing a Denominator Containing Two Terms

Rationalize the denominator: $\dfrac{2 + \sqrt{5}}{\sqrt{6} - \sqrt{3}}$.

SOLUTION Multiplication of both the numerator and denominator by $\sqrt{6} + \sqrt{3}$ will rationalize the denominator.

$$\frac{2 + \sqrt{5}}{\sqrt{6} - \sqrt{3}} = \frac{2 + \sqrt{5}}{\sqrt{6} - \sqrt{3}} \cdot \frac{\sqrt{6} + \sqrt{3}}{\sqrt{6} + \sqrt{3}}$$

Multiply by 1.

$$= \frac{2\sqrt{6} + 2\sqrt{3} + \sqrt{5} \cdot \sqrt{6} + \sqrt{5} \cdot \sqrt{3}}{\left(\sqrt{6}\right)^2 - \left(\sqrt{3}\right)^2}$$

Use FOIL in the numerator and $(A - B)(A + B) = A^2 - B^2$ in the denominator.

$$= \frac{2\sqrt{6} + 2\sqrt{3} + \sqrt{30} + \sqrt{15}}{6 - 3}$$

$$= \frac{2\sqrt{6} + 2\sqrt{3} + \sqrt{30} + \sqrt{15}}{3}$$

Further simplification is not possible.

✔ **CHECK POINT 6** Rationalize the denominator: $\dfrac{3 + \sqrt{7}}{\sqrt{5} - \sqrt{2}}$.

$\dfrac{3\sqrt{5} + 3\sqrt{2} + \sqrt{35} + \sqrt{14}}{3}$

---

5 Rationalize numerators.

Rationalizing Numerators We have seen that square root functions are often used to model growing phenomena with growth that is leveling off. Figure 10.6 shows a male's height as a function of his age. The pattern of his growth suggests modeling with a square root function.

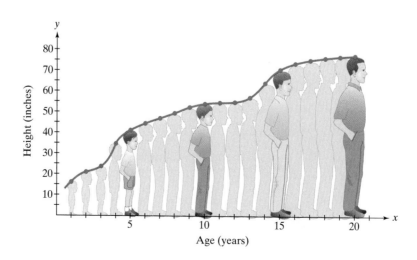

FIGURE 10.6

If we use $f(x) = \sqrt{x}$ to model height, $f(x)$, at age x, the expression

$$\frac{f(a+h) - f(a)}{h} = \frac{\sqrt{a+h} - \sqrt{a}}{h}$$

describes the man's average growth rate from age a to age $a + h$. Can you see that this expression is not defined if $h = 0$? However, to explore the man's average growth rates for successively shorter periods of time, we need to know what happens to the expression as h takes on values that get closer and closer to 0.

What happens to growth near the instant in time that the man is age a? The question is answered by first **rationalizing the numerator**. The procedure is similar to rationalizing the denominator. **To rationalize a numerator, multiply by 1 to eliminate the radical in the *numerator*.**

EXAMPLE 7   Rationalizing a Numerator

Rationalize the numerator:

$$\frac{\sqrt{a+h} - \sqrt{a}}{h}.$$

SOLUTION   The conjugate of the numerator is $\sqrt{a+h} + \sqrt{a}$. If we multiply the numerator and the denominator by $\sqrt{a+h} + \sqrt{a}$, the simplified numerator will not contain radicals. Therefore, we multiply by 1, choosing $\dfrac{\sqrt{a+h} + \sqrt{a}}{\sqrt{a+h} + \sqrt{a}}$ for 1.

$$\frac{\sqrt{a+h} - \sqrt{a}}{h} = \frac{\sqrt{a+h} - \sqrt{a}}{h} \cdot \frac{\sqrt{a+h} + \sqrt{a}}{\sqrt{a+h} + \sqrt{a}} \qquad \text{Multiply by 1.}$$

$$= \frac{\left(\sqrt{a+h}\right)^2 - \left(\sqrt{a}\right)^2}{h\left(\sqrt{a+h} + \sqrt{a}\right)} \qquad \begin{array}{l}(A - B)(A + B) = A^2 - B^2 \\ \text{Leave the denominator in} \\ \text{factored form.}\end{array}$$

$$= \frac{a + h - a}{h\left(\sqrt{a+h} + \sqrt{a}\right)} \qquad \begin{array}{l}\left(\sqrt{a+h}\right)^2 = a + h \\ \text{and } (\sqrt{a})^2 = a.\end{array}$$

$$= \frac{h}{h\left(\sqrt{a+h} + \sqrt{a}\right)} \qquad \text{Simplify the numerator.}$$

$$= \frac{1}{\sqrt{a+h} + \sqrt{a}} \qquad \begin{array}{l}\text{Simplify by dividing the numerator} \\ \text{and the denominator by } h.\end{array}$$

What happens to the rationalized expression in Example 7 as h is close to 0? The expression is close to

$$\frac{1}{\sqrt{a+0} + \sqrt{a}} = \frac{1}{\sqrt{a} + \sqrt{a}} = \frac{1}{2\sqrt{a}}.$$

The man's growth rate at one instant in time, age a, can be determined with this expression. This expression reveals how the square root function $f(x) = \sqrt{x}$ is changing, instant by instant.

 CHECK POINT 7   Rationalize the numerator: $\dfrac{\sqrt{x+3} - \sqrt{x}}{3}$. $\dfrac{1}{\sqrt{x+3} + \sqrt{x}}$

10.5 EXERCISE SET

Student Solutions Manual   CD/Video   PH Math/Tutor Center   MathXL Tutorials on CD   MathXL®   MyMathLab   Interactmath.com

Practice Exercises

In this exercise set, assume that all variables represent positive real numbers.

In Exercises 1–38, multiply as indicated. If possible, simplify any radical expressions that appear in the product.

1. $\sqrt{2}(x + \sqrt{7})$ $x\sqrt{2} + \sqrt{14}$

2. $\sqrt{5}(x + \sqrt{3})$ $x\sqrt{5} + \sqrt{15}$

3. $\sqrt{6}(7 - \sqrt{6})$ $7\sqrt{6} - 6$

4. $\sqrt{3}(5 - \sqrt{3})$ $5\sqrt{3} - 3$

5. $\sqrt{3}(4\sqrt{6} - 2\sqrt{3})$ $12\sqrt{2} - 6$

6. $\sqrt{6}(4\sqrt{6} - 3\sqrt{2})$ $24 - 6\sqrt{3}$

7. $\sqrt[3]{2}(\sqrt[3]{6} + 4\sqrt[3]{5})$ $\sqrt[3]{12} + 4\sqrt[3]{10}$

8. $\sqrt[3]{3}(\sqrt[3]{6} + 7\sqrt[3]{4})$ $\sqrt[3]{18} + 7\sqrt[3]{12}$

9. $\sqrt[3]{x}(\sqrt[3]{16x^2} - \sqrt[3]{x})$ $2x\sqrt[3]{2} - \sqrt[3]{x^2}$

10. $\sqrt[3]{x}(\sqrt[3]{24x^2} - \sqrt[3]{x})$ $2x\sqrt[3]{3} - \sqrt[3]{x^2}$

11. $(5 + \sqrt{2})(6 + \sqrt{2})$ $32 + 11\sqrt{2}$

12. $(7 + \sqrt{2})(8 + \sqrt{2})$ $58 + 15\sqrt{2}$

13. $(6 + \sqrt{5})(9 - 4\sqrt{5})$ $34 - 15\sqrt{5}$

14. $(4 + \sqrt{5})(10 - 3\sqrt{5})$ $25 - 2\sqrt{5}$

15. $(6 - 3\sqrt{7})(2 - 5\sqrt{7})$ $117 - 36\sqrt{7}$

16. $(7 - 2\sqrt{7})(5 - 3\sqrt{7})$ $77 - 31\sqrt{7}$

17. $(\sqrt{2} + \sqrt{7})(\sqrt{3} + \sqrt{5})$ $\sqrt{6} + \sqrt{10} + \sqrt{21} + \sqrt{35}$

18. $(\sqrt{3} + \sqrt{2})(\sqrt{10} + \sqrt{11})$ $\sqrt{30} + \sqrt{33} + 2\sqrt{5} + \sqrt{22}$

19. $(\sqrt{2} - \sqrt{7})(\sqrt{3} - \sqrt{5})$ $\sqrt{6} - \sqrt{10} - \sqrt{21} + \sqrt{35}$

20. $(\sqrt{3} - \sqrt{2})(\sqrt{10} - \sqrt{11})$ $\sqrt{30} - \sqrt{33} - 2\sqrt{5} + \sqrt{22}$

21. $(3\sqrt{2} - 4\sqrt{3})(2\sqrt{2} + 5\sqrt{3})$ $-48 + 7\sqrt{6}$

22. $(3\sqrt{5} - 2\sqrt{3})(4\sqrt{5} + 5\sqrt{3})$ $30 + 7\sqrt{15}$

23. $(\sqrt{3} + \sqrt{5})^2$ $8 + 2\sqrt{15}$

24. $(\sqrt{2} + \sqrt{7})^2$ $9 + 2\sqrt{14}$

25. $(\sqrt{3x} - \sqrt{y})^2$ $3x - 2\sqrt{3xy} + y$

26. $(\sqrt{2x} - \sqrt{y})^2$ $2x - 2\sqrt{2xy} + y$

27. $(\sqrt{5} + 7)(\sqrt{5} - 7)$ -44

28. $(\sqrt{6} + 2)(\sqrt{6} - 2)$ 2

29. $(2 - 5\sqrt{3})(2 + 5\sqrt{3})$ -71

30. $(3 - 5\sqrt{2})(3 + 5\sqrt{2})$ -41

31. $(3\sqrt{2} + 2\sqrt{3})(3\sqrt{2} - 2\sqrt{3})$ 6

32. $(4\sqrt{3} + 3\sqrt{2})(4\sqrt{3} - 3\sqrt{2})$ 30

33. $(3 - \sqrt{x})(2 - \sqrt{x})$ $6 - 5\sqrt{x} + x$

34. $(4 - \sqrt{x})(3 - \sqrt{x})$ $12 - 7\sqrt{x} + x$

35. $(\sqrt[3]{x} - 4)(\sqrt[3]{x} + 5)$ $\sqrt[3]{x^2} + \sqrt[3]{x} - 20$

36. $(\sqrt[3]{x} - 3)(\sqrt[3]{x} + 7)$ $\sqrt[3]{x^2} + 4\sqrt[3]{x} - 21$

37. $(x + \sqrt[3]{y^2})(2x - \sqrt[3]{y^2})$ $2x^2 + x\sqrt[3]{y^2} - y\sqrt[3]{y}$

38. $(x - \sqrt[5]{y^3})(2x + \sqrt[5]{y^3})$ $2x^2 - x\sqrt[5]{y^3} - y\sqrt[5]{y}$

In Exercises 39–64, rationalize each denominator.

39. $\dfrac{\sqrt{2}}{\sqrt{5}}$ $\dfrac{\sqrt{10}}{5}$

40. $\dfrac{\sqrt{7}}{\sqrt{3}}$ $\dfrac{\sqrt{21}}{3}$

41. $\sqrt{\dfrac{11}{x}}$ $\dfrac{\sqrt{11x}}{x}$

42. $\sqrt{\dfrac{6}{x}}$ $\dfrac{\sqrt{6x}}{x}$

43. $\dfrac{9}{\sqrt{3y}}$ $\dfrac{3\sqrt{3y}}{y}$

44. $\dfrac{12}{\sqrt{3y}}$ $\dfrac{4\sqrt{3y}}{y}$

45. $\dfrac{1}{\sqrt[3]{2}}$ $\dfrac{\sqrt[3]{4}}{2}$

46. $\dfrac{1}{\sqrt[3]{3}}$ $\dfrac{\sqrt[3]{9}}{3}$

47. $\dfrac{6}{\sqrt[3]{4}}$ $3\sqrt[3]{2}$

48. $\dfrac{10}{\sqrt[3]{5}}$ $2\sqrt[3]{25}$

49. $\sqrt[3]{\dfrac{2}{3}}$ $\dfrac{\sqrt[3]{18}}{3}$

50. $\sqrt[3]{\dfrac{3}{4}}$ $\dfrac{\sqrt[3]{6}}{2}$

51. $\dfrac{4}{\sqrt[3]{x}}$ $\dfrac{4\sqrt[3]{x^2}}{x}$

52. $\dfrac{7}{\sqrt[3]{x}}$ $\dfrac{7\sqrt[3]{x^2}}{x}$

53. $\sqrt[3]{\dfrac{2}{y^2}}$ $\dfrac{\sqrt[3]{2y}}{y}$

54. $\sqrt[3]{\dfrac{5}{y^2}}$ $\dfrac{\sqrt[3]{5y}}{y}$

55. $\dfrac{7}{\sqrt[3]{2x^2}}$ $\dfrac{7\sqrt[3]{4x}}{2x}$

56. $\dfrac{10}{\sqrt[3]{4x^2}}$ $\dfrac{5\sqrt[3]{2x}}{x}$

57. $\sqrt[3]{\dfrac{2}{xy^2}}$ $\dfrac{\sqrt[3]{2x^2y}}{xy}$

58. $\sqrt[3]{\dfrac{3}{xy^2}}$ $\dfrac{\sqrt[3]{3x^2y}}{xy}$

59. $\dfrac{3}{\sqrt[4]{x}}$ $\dfrac{3\sqrt[4]{x^3}}{x}$

60. $\dfrac{5}{\sqrt[4]{x}}$ $\dfrac{5\sqrt[4]{x^3}}{x}$

61. $\dfrac{6}{\sqrt[5]{8x^3}}$ $\dfrac{3\sqrt[5]{4x^2}}{x}$

62. $\dfrac{10}{\sqrt[5]{16x^2}}$ $\dfrac{5\sqrt[5]{2x^3}}{x}$

63. $\dfrac{2x^2y}{\sqrt[5]{4x^2y^4}}$ $x\sqrt[5]{8x^3y}$ **64.** $\dfrac{3xy^2}{\sqrt[5]{8xy^3}}$ $\dfrac{3y\sqrt[5]{4x^4y^2}}{2}$

In Exercises 65–74, simplify each radical expression and then rationalize the denominator.

65. $\dfrac{9}{\sqrt{3x^2y}}$ $\dfrac{3\sqrt{3y}}{xy}$ **66.** $\dfrac{25}{\sqrt{5x^2y}}$ $\dfrac{5\sqrt{5y}}{xy}$

67. $-\sqrt{\dfrac{75a^5}{b^3}}$ $-\dfrac{5a^2\sqrt{3ab}}{b^2}$ **68.** $-\sqrt{\dfrac{150a^3}{b^5}}$ $-\dfrac{5a\sqrt{6ab}}{b^3}$

69. $\sqrt{\dfrac{7m^2n^3}{14m^3n^2}}$ $\dfrac{\sqrt{2mn}}{2m}$ **70.** $\sqrt{\dfrac{5m^4n^6}{15m^3n^4}}$ $\dfrac{n\sqrt{3m}}{3}$

71. $\dfrac{3}{\sqrt[4]{x^5y^3}}$ $\dfrac{3\sqrt[4]{x^3y}}{x^2y}$ **72.** $\dfrac{5}{\sqrt[4]{x^2y^7}}$ $\dfrac{5\sqrt[4]{x^2y}}{xy^2}$

73. $\dfrac{12}{\sqrt[3]{-8x^5y^8}}$ $-\dfrac{6\sqrt[3]{xy}}{x^2y^3}$ **74.** $\dfrac{15}{\sqrt[3]{-27x^4y^{11}}}$ $-\dfrac{5\sqrt[3]{x^2y}}{x^2y^4}$

In Exercises 75–92, rationalize each denominator. Simplify, if possible.

75. $\dfrac{8}{\sqrt{5}+2}$ $8\sqrt{5}-16$ **76.** $\dfrac{15}{\sqrt{6}+1}$ $3\sqrt{6}-3$

77. $\dfrac{13}{\sqrt{11}-3}$ $\dfrac{13\sqrt{11}+39}{2}$ **78.** $\dfrac{17}{\sqrt{10}-2}$ $\dfrac{17\sqrt{10}+34}{6}$

79. $\dfrac{6}{\sqrt{5}+\sqrt{3}}$ $3\sqrt{5}-3\sqrt{3}$ **80.** $\dfrac{12}{\sqrt{7}+\sqrt{3}}$ $3\sqrt{7}-3\sqrt{3}$

81. $\dfrac{\sqrt{a}}{\sqrt{a}-\sqrt{b}}$ $\dfrac{a+\sqrt{ab}}{a-b}$ **82.** $\dfrac{\sqrt{b}}{\sqrt{a}-\sqrt{b}}$ $\dfrac{\sqrt{ab}+b}{a-b}$

83. $\dfrac{25}{5\sqrt{2}-3\sqrt{5}}$ $25\sqrt{2}+15\sqrt{5}$

84. $\dfrac{35}{5\sqrt{2}-3\sqrt{5}}$ $35\sqrt{2}+21\sqrt{5}$

85. $\dfrac{\sqrt{5}+\sqrt{3}}{\sqrt{5}-\sqrt{3}}$ $4+\sqrt{15}$ **86.** $\dfrac{\sqrt{11}-\sqrt{5}}{\sqrt{11}+\sqrt{5}}$ $\dfrac{8-\sqrt{55}}{3}$

87. $\dfrac{\sqrt{x}+1}{\sqrt{x}+3}$ $\dfrac{x-2\sqrt{x}-3}{x-9}$ **88.** $\dfrac{\sqrt{x}-2}{\sqrt{x}-5}$ $\dfrac{x+3\sqrt{x}-10}{x-25}$

89. $\dfrac{5\sqrt{3}-3\sqrt{2}}{3\sqrt{2}-2\sqrt{3}}$ $\dfrac{3\sqrt{6}+4}{2}$ **90.** $\dfrac{2\sqrt{6}+\sqrt{5}}{3\sqrt{6}-\sqrt{5}}$ $\dfrac{41+5\sqrt{30}}{49}$

91. $\dfrac{2\sqrt{x}+\sqrt{y}}{\sqrt{y}-2\sqrt{x}}$ $\dfrac{4\sqrt{xy}+4x+y}{y-4x}$

92. $\dfrac{3\sqrt{x}+\sqrt{y}}{\sqrt{y}-3\sqrt{x}}$ $\dfrac{6\sqrt{xy}+9x+y}{y-9x}$

In Exercises 93–104, rationalize each numerator. Simplify, if possible.

93. $\sqrt{\dfrac{3}{2}}$ $\dfrac{3}{\sqrt{6}}$ **94.** $\sqrt{\dfrac{5}{3}}$ $\dfrac{5}{\sqrt{15}}$

95. $\dfrac{\sqrt[3]{4x}}{\sqrt[3]{y}}$ $\dfrac{2x}{\sqrt[3]{2x^2y}}$ **96.** $\dfrac{\sqrt[3]{2x}}{\sqrt[3]{y}}$ $\dfrac{2x}{\sqrt[3]{4x^2y}}$

97. $\dfrac{\sqrt{x}+3}{\sqrt{x}}$ $\dfrac{x-9}{x-3\sqrt{x}}$ **98.** $\dfrac{\sqrt{x}+4}{\sqrt{x}}$ $\dfrac{x-16}{x-4\sqrt{x}}$

99. $\dfrac{\sqrt{a}+\sqrt{b}}{\sqrt{a}-\sqrt{b}}$ $\dfrac{a-b}{a-2\sqrt{ab}+b}$

100. $\dfrac{\sqrt{a}-\sqrt{b}}{\sqrt{a}+\sqrt{b}}$ $\dfrac{a-b}{a+2\sqrt{ab}+b}$

101. $\dfrac{\sqrt{x+5}-\sqrt{x}}{5}$ $\dfrac{1}{\sqrt{x+5}+\sqrt{x}}$

102. $\dfrac{\sqrt{x+7}-\sqrt{x}}{7}$ $\dfrac{1}{\sqrt{x+7}+\sqrt{x}}$

103. $\dfrac{\sqrt{x}+\sqrt{y}}{x^2-y^2}$ $\dfrac{1}{(x+y)(\sqrt{x}-\sqrt{y})}$

104. $\dfrac{\sqrt{x}-\sqrt{y}}{x^2-y^2}$ $\dfrac{1}{(x+y)(\sqrt{x}+\sqrt{y})}$

Practice Plus

In Exercises 105–112, add or subtract as indicated. Begin by rationalizing denominators for all terms in which denominators contain radicals.

105. $\sqrt{2}+\dfrac{1}{\sqrt{2}}$ $\dfrac{3\sqrt{2}}{2}$ **106.** $\sqrt{5}+\dfrac{1}{\sqrt{5}}$ $\dfrac{6\sqrt{5}}{5}$

107. $\sqrt[3]{25}-\dfrac{15}{\sqrt[3]{5}}$ $-2\sqrt[3]{25}$ **108.** $\sqrt[4]{8}-\dfrac{20}{\sqrt[4]{2}}$ $-9\sqrt[4]{8}$

109. $\sqrt{6}-\sqrt{\dfrac{1}{6}}+\sqrt{\dfrac{2}{3}}$ $\dfrac{7\sqrt{6}}{6}$

110. $\sqrt{15}-\sqrt{\dfrac{5}{3}}+\sqrt{\dfrac{3}{5}}$ $\dfrac{13\sqrt{15}}{15}$

111. $\dfrac{2}{\sqrt{2}+\sqrt{3}}+\sqrt{75}-\sqrt{50}$ $7\sqrt{3}-7\sqrt{2}$

112. $\dfrac{5}{\sqrt{2}+\sqrt{7}}-2\sqrt{32}+\sqrt{28}$ $-9\sqrt{2}+3\sqrt{7}$

113. Let $f(x)=x^2-6x-4$. Find $f\big(3-\sqrt{13}\big)$.   0

114. Let $f(x)=x^2+4x-2$. Find $f\big(-2+\sqrt{6}\big)$.   0

115. Let $f(x)=\sqrt{9+x}$. Find $f\big(3\sqrt{5}\big)\cdot f\big(-3\sqrt{5}\big)$.   6

116. Let $f(x)=x^2$. Find $f\big(\sqrt{a+1}-\sqrt{a-1}\big)$. $2a-2\sqrt{a^2-1}$

Application Exercises

The bar graph shows the percentage of U.S. households online from 1997 through 2003. The function

$$P(t) = 15.92\sqrt{t} + 19$$

models the percentage, $P(t)$, of U.S. households online t years after 1997. Use the function to solve Exercises 117–118.

Percentage of U.S. Households Online

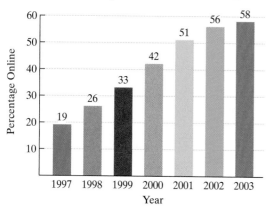

Source: U.S. Department of Commerce

117. Find $P(4)$ and describe what this means. Round to the nearest percent. How well does your answer model the actual data shown in the graph?   51; In 2001, approximately 51% of U.S. households were online; Answer models data well.

118. Find $P(6)$ and describe what this means. Use a calculator and round to the nearest percent. How well does your answer model the actual data shown in the graph?   58; In 2003, approximately 58% of U.S. households were online; Answer models data well.

In Exercises 119–120, use the data shown in the bar graph for the percentage of U.S. households online.

119. The average rate of change in the percentage of households online from 1997 through 2003 is

> Change in percent

$$\frac{\text{percentage online in 2003} - \text{percentage online in 1997}}{2003 - 1997}$$

> Change in time

Find the average yearly increase in the percentage of online households during this period.   6.5%

120. The average rate of change in the percentage of households online from 2001 through 2003 is

> Change in percent

$$\frac{\text{percentage online in 2003} - \text{percentage online in 2001}}{2003 - 2001}$$

> Change in time

Find the average yearly increase in the percentage of online households during this period.   3.5%

The algebraic expression

$$15.92\left(\frac{\sqrt{t+h} - \sqrt{t}}{h}\right)$$

models the average rate of change in the percentage of households online from t years after 1997 to $t + h$ years after 1997. Use the expression to solve Exercises 121–124.

121. Let $t = 0$ and $h = 6$. Evaluate the expression to find the average rate of change in the percentage of households online from 1997 through 2003—that is, between 0 years and 6 years after 1997. How well does your answer model the actual yearly increase in the percentage that you found in Exercise 119?   The average rate of change in percentage during this period is approximately 6.5%. This answer models the actual yearly percentage increase well.

122. Let $t = 4$ and $h = 2$. Evaluate the expression to find the average rate of change in the percentage of households online from 2001 through 2003—that is, between 4 years and $4 + 2$, or 6, years after 1997. How well does your answer model the actual yearly increase in the percentage that you found in Exercise 120?   The average rate of change in percentage during this period is approximately 3.6%. This answer models the actual yearly percentage increase well.

123. a. Rewrite the average rate of change as an equivalent expression by rationalizing the numerator of the rational expression in parentheses. $\dfrac{15.92}{\sqrt{t+h} + \sqrt{t}}$

b. Substitute 0 for h in your expression from part (a) and simplify. The resulting expression gives the rate of change in the percentage of households online precisely t years after 1997. $\dfrac{15.92}{2\sqrt{t}}$ or $\dfrac{7.96}{\sqrt{t}}$

c. Use the expression from part (b) to find the rate of change in the percentage of households online in 2003. Round to the nearest tenth of a percent.   The rate of change in percentage in 2003 was 3.2%.

124. Solve parts (a) and (b) of Exercise 123. Then find the rate of change in the percentage of households online in 2001. Round to the nearest tenth of a percent.

a. $\dfrac{15.92}{\sqrt{t+h} + \sqrt{t}}$ **b.** $\dfrac{7.96}{\sqrt{t}}$
The rate of change in percentage in 2001 was 4.0%.

In Exercises 125–126, write expressions for the perimeter and area of each figure. Then simplify these expressions. Assume that all measures are given in inches.

125. $\sqrt{8} + 1$ $\sqrt{8} - 1$ $P = 8\sqrt{2}$ inches $A = 7$ square inches

126. $2\sqrt{3} + \sqrt{2}$ $2\sqrt{3} + \sqrt{2}$ $P = 8\sqrt{3} + 4\sqrt{2}$ inches $A = 14 + 4\sqrt{6}$ square inches

127. In the "Peanuts" cartoon shown in the section opener on page 678, Woodstock appears to be working steps mentally. Fill in the missing steps that show how to go from $\dfrac{7\sqrt{2\cdot2\cdot3}}{6}$ to $\dfrac{7}{3}\sqrt{3}$. $\dfrac{7\sqrt{2\cdot2\cdot3}}{6}=\dfrac{7\cdot2\sqrt{3}}{6}=\dfrac{14\sqrt{3}}{6}=\dfrac{7\sqrt{3}}{3}=\dfrac{7}{3}\sqrt{3}$

128. Rationalize the denominator of the golden ratio

$$\frac{2}{\sqrt{5}-1},$$

discussed in the enrichment essay on page 683. Then use a calculator and find the ratio of width to height, correct to the nearest hundredth, in golden rectangles. $\dfrac{\sqrt{5}+1}{2}$; 1.62 to 1

Writing in Mathematics

129. Explain how to perform this multiplication: $\sqrt{2}(\sqrt{7}+\sqrt{10})$.
129–137. Answers will vary.

130. Explain how to perform this multiplication: $(2+\sqrt{3})(4+\sqrt{3})$.

131. Explain how to perform this multiplication: $(2+\sqrt{3})^2$.

132. What are conjugates? Give an example with your explanation.

133. Describe how to multiply conjugates.

134. Describe what it means to rationalize a denominator. Use both $\dfrac{1}{\sqrt{5}}$ and $\dfrac{1}{5+\sqrt{5}}$ in your explanation.

135. When a radical expression has its denominator rationalized, we change the denominator so that it no longer contains any radicals. Doesn't this change the value of the radical expression? Explain.

136. Square the real number $\dfrac{2}{\sqrt{3}}$. Observe that the radical is eliminated from the denominator. Explain whether this process is equivalent to rationalizing the denominator.

137. Describe the trend shown in the bar graph in Exercises 117–118 for the percentage of U.S. households online from 1997 through 2003. What explanation can you offer for this trend?

Technology Exercises

In Exercises 138–141, determine if each operation is performed correctly by graphing the function on each side of the equation with your graphing utility. Use the given viewing rectangle. The graphs should be the same. If they are not, correct the right side of the equation and then use your graphing utility to verify the correction. 138–141. See graphing answer section.

138. $(\sqrt{x}-1)(\sqrt{x}-1)=x+1$

$[0, 5, 1]$ by $[-1, 2, 1]$ Graphs are not the same; $(\sqrt{x}-1)(\sqrt{x}-1)=x-2\sqrt{x}+1$.

139. $(\sqrt{x}+2)(\sqrt{x}-2)=x^2-4$ for $x\geq0$

$[0, 10, 1]$ by $[-10, 10, 1]$ Graphs are not the same; $(\sqrt{x}+2)(\sqrt{x}-2)=x-4$ for $x\geq0$.

140. $(\sqrt{x}+1)^2=x+1$

$[0, 8, 1]$ by $[0, 15, 1]$
Graphs are not the same; $(\sqrt{x}+1)^2=x+2\sqrt{x}+1$.

141. $\dfrac{3}{\sqrt{x+3}-\sqrt{x}}=\sqrt{x+3}+\sqrt{x}$

$[0, 8, 1]$ by $[0, 6, 1]$
Graphs are the same; Simplification is correct.

Critical Thinking Exercises

142. Which one of the following is true?   c

a. $\dfrac{\sqrt{3}+7}{\sqrt{3}-2}=-\dfrac{7}{2}$

b. $\dfrac{4}{\sqrt{x}+y}=\dfrac{4\sqrt{x}-y}{x-y}$

c. $\dfrac{4\sqrt{x}}{\sqrt{x}-y}=\dfrac{4x+4y\sqrt{x}}{x-y^2}$

d. $(\sqrt{x}-7)^2=x-49$

143. Solve:

$7[(2x-5)-(x+1)]=(\sqrt{7}+2)(\sqrt{7}-2)$. $\dfrac{45}{7}$ or $\left\{\dfrac{45}{7}\right\}$

144. Simplify: $(\sqrt{2+\sqrt{3}}+\sqrt{2-\sqrt{3}})^2$. 6

145. Rationalize the denominator: $\dfrac{1}{\sqrt{2}+\sqrt{3}+\sqrt{4}}$.

$\dfrac{5\sqrt{2}+3\sqrt{3}-4\sqrt{6}+2}{23}$

Review Exercises

146. Add: $\dfrac{2}{x-2}+\dfrac{3}{x^2-4}$.

(Section 7.4, Example 5) $\dfrac{2x+7}{x^2-4}$

147. Solve: $3x-4\leq2$ and $4x+5\geq5$
(Section 9.2, Example 2) $\{x|0\leq x\leq2\}$ or $[0, 2]$

148. Is $\{(-1, 1), (1, 1), (-2, 4), (2, 4)\}$ a function? Is the inverse of this relation a function? (Section 8.1, Example 2, and Section 8.3, first paragraph on page 564) yes; no

RADICAL EQUATIONS

Objectives

1 Solve radical equations.

2 Use models that are radical functions to solve problems.

A best guess at the look of our nation in the next decades indicates that the number of men and women living alone will increase each year. By 2010, approximately 28 million men and women are projected to be living alone. The function

$$f(x) = 2.6\sqrt{x} + 11$$

models the number of Americans living alone, $f(x)$, in millions, x years after 1970. How can we predict the year when, say, 29 million of us will live alone? Substitute 29 for $f(x)$ and solve for x:

$$29 = 2.6\sqrt{x} + 11.$$

The resulting equation contains a variable in the radicand and is called a *radical equation*. A **radical equation** is an equation in which the variable occurs in a square root, cube root, or any higher root. Some examples of radical equations are

$$\sqrt{2x + 3} = 5, \quad \sqrt{3x + 1} - \sqrt{x + 4} = 1, \quad \text{and} \quad \sqrt[3]{3x - 1} + 4 = 0.$$

Variables occur in radicands.

In this section, you will learn how to solve radical equations. Solving such equations will enable you to solve new kinds of problems using radical functions.

1 Solve radical equations.

Solving Radical Equations Consider the following radical equation:

$$\sqrt{x} = 9.$$

We solve the equation by squaring both sides:

Squaring both sides eliminates the square root.

$$\left(\sqrt{x}\right)^2 = 9^2$$
$$x = 81.$$

The proposed solution, 81, can be checked in the original equation, $\sqrt{x} = 9$. Because $\sqrt{81} = 9$, the solution is 81 and the solution set is $\{81\}$.

In general, we solve radical equations with square roots by squaring both sides of the equation. We solve radical equations with nth roots by raising both sides of the equation to the nth power. Unfortunately, if n is even, all the solutions of the equation raised to the even power may not be solutions of the original equation. Consider, for example, the equation

$$x = 4.$$

If we square both sides, we obtain

$$x^2 = 16.$$

This new equation has two solutions, -4 and 4. By contrast, only 4 is a solution of the original equation, $x = 4$. For this reason, **when raising both sides of an equation to an even power, always check proposed solutions in the original equation.**

Here is a general method for solving radical equations with nth roots:

SOLVING RADICAL EQUATIONS CONTAINING nTH ROOTS

1. If necessary, arrange terms so that one radical is isolated on one side of the equation.
2. Raise both sides of the equation to the nth power to eliminate the nth root.
3. Solve the resulting equation. If this equation still contains radicals, repeat steps 1 and 2.
4. Check all proposed solutions in the original equation.

EXAMPLE 1 Solving a Radical Equation

Solve: $\sqrt{2x + 3} = 5$.

SOLUTION

Step 1. **Isolate a radical on one side.** The radical, $\sqrt{2x + 3}$, is already isolated on the left side of the equation, so we can skip this step.

Step 2. **Raise both sides to the nth power.** Because n, the index, is 2, we square both sides.

$$\sqrt{2x + 3} = 5 \qquad \text{This is the given equation.}$$
$$\left(\sqrt{2x + 3}\right)^2 = 5^2 \qquad \text{Square both sides to eliminate the radical.}$$
$$2x + 3 = 25 \qquad \text{Simplify.}$$

Step 3. **Solve the resulting equation.**

$$2x + 3 = 25 \qquad \text{The resulting equation is a linear equation.}$$
$$2x = 22 \qquad \text{Subtract 3 from both sides.}$$
$$x = 11 \qquad \text{Divide both sides by 2.}$$

Step 4. **Check the proposed, solution in the original equation.** Because both sides were raised to an even power, this check is essential.

Check 11:
$$\sqrt{2x + 3} = 5$$
$$\sqrt{2 \cdot 11 + 3} \stackrel{?}{=} 5$$
$$\sqrt{25} \stackrel{?}{=} 5$$
$$5 = 5, \qquad \text{true}$$

The solution is 11 and the solution set is $\{11\}$.

 CHECK POINT 1 Solve: $\sqrt{3x + 4} = 8$. 20 or $\{20\}$

EXAMPLE 2 Solving a Radical Equation

Solve: $\sqrt{x - 3} + 6 = 5$.

SOLUTION

Step 1. **Isolate a radical on one side.** The radical, $\sqrt{x - 3}$, can be isolated by subtracting 6 from both sides. We obtain

$$\sqrt{x - 3} = -1.$$

A principal square root cannot be negative. This equation has no solution. Let's continue the solution procedure to see what happens.

Step 2. **Raise both sides to the *n*th power.** Because n, the index, is 2, we square both sides.

$$\left(\sqrt{x - 3}\right)^2 = (-1)^2$$

$$x - 3 = 1 \qquad \text{Simplify.}$$

Step 3. **Solve the resulting equation.**

$$x - 3 = 1 \qquad \text{The resulting equation is a linear equation.}$$

$$x = 4 \qquad \text{Add 3 to both sides.}$$

Step 4. **Check the proposed solution in the original equation.**

Check 4:

$$\sqrt{x - 3} + 6 = 5$$

$$\sqrt{4 - 3} + 6 \stackrel{?}{=} 5$$

$$\sqrt{1} + 6 \stackrel{?}{=} 5$$

$$1 + 6 \stackrel{?}{=} 5$$

$$7 = 5, \qquad \text{false}$$

This false statement indicates that 4 is not a solution. Thus, the equation has no solution. The solution set is Ø, the empty set.

Example 2 illustrates that extra solutions may be introduced when you raise both sides of a radical equation to an even power. Such solutions, which are not solutions of the given equation, are called **extraneous solutions**. Thus, 4 is an extraneous solution of $\sqrt{x - 3} + 6 = 5$.

CHECK POINT **2** Solve: $\sqrt{x - 1} + 7 = 2$.   no solution or Ø

EXAMPLE 3 Solving a Radical Equation

Solve: $x + \sqrt{26 - 11x} = 4$.

SOLUTION

Step 1. **Isolate a radical on one side.** We isolate the radical, $\sqrt{26 - 11x}$, by subtracting x from both sides.

$$x + \sqrt{26 - 11x} = 4 \qquad \text{This is the given equation.}$$

$$\sqrt{26 - 11x} = 4 - x. \qquad \text{Subtract x from both sides.}$$

Step 2. **Square both sides.**

$$\left(\sqrt{26-11x}\right)^2 = (4-x)^2$$

$$26 - 11x = 16 - 8x + x^2 \quad \text{Simplify. Use the special-product formula}$$
$$(A-B)^2 = A^2 - 2AB + B^2 \text{ to square } 4 - x.$$

Step 3. **Solve the resulting equation.** Because of the x^2-term, the resulting equation is a quadratic equation. We need to write this quadratic equation in standard form. We can obtain zero on the left side by subtracting 26 and adding $11x$ on both sides.

$$26 - 26 - 11x + 11x = 16 - 26 - 8x + 11x + x^2$$

$$0 = x^2 + 3x - 10 \quad \text{Simplify.}$$

$$0 = (x+5)(x-2) \quad \text{Factor.}$$

$$x + 5 = 0 \quad \text{or} \quad x - 2 = 0 \quad \text{Set each factor equal to zero.}$$

$$x = -5 \qquad\qquad x = 2 \quad \text{Solve for x.}$$

Step 4. **Check the proposed solutions in the original equation.**

Check −5:

$$x + \sqrt{26 - 11x} = 4$$
$$-5 + \sqrt{26 - 11(-5)} \stackrel{?}{=} 4$$
$$-5 + \sqrt{81} \stackrel{?}{=} 4$$
$$-5 + 9 \stackrel{?}{=} 4$$
$$4 = 4, \quad \text{true}$$

Check 2:

$$x + \sqrt{26 - 11x} = 4$$
$$2 + \sqrt{26 - 11\cdot2} \stackrel{?}{=} 4$$
$$2 + \sqrt{4} \stackrel{?}{=} 4$$
$$2 + 2 \stackrel{?}{=} 4$$
$$4 = 4, \quad \text{true}$$

The solutions are −5 and 2, and the solution set is $\{-5, 2\}$.

USING TECHNOLOGY

You can use a graphing utility to provide a graphic check that $\{-5, 2\}$ is the solution set of $x + \sqrt{26 - 11x} = 4$.

Use the given equation

$$x + \sqrt{26 - 11x} = 4.$$

Enter $y_1 = x + \sqrt{26 - 11x}$ in the $\boxed{y=}$ screen. Enter $y_2 = 4$ in the $\boxed{y=}$ screen.

Display graphs for y_1 and y_2. The solutions are the x-coordinates of the intersection points. These x-coordinates are −5 and 2. This verifies $\{-5, 2\}$ as the solution set of $x + \sqrt{26 - 11x} = 4$.

Use the equivalent equation

$$x + \sqrt{26 - 11x} - 4 = 0.$$

Enter $y_1 = x + \sqrt{26 - 11x} - 4$ in the $\boxed{y=}$ screen.

Display the graph for y_1. The solutions are the x-intercepts. The x-intercepts are −5 and 2. This verifies $\{-5, 2\}$ as the solution set of $x + \sqrt{26 - 11x} - 4 = 0$.

[−8, 5, 1] by [0, 7, 1]

[−10, 3, 1] by [−4, 3, 1]

✔ CHECK POINT **3** Solve: $\sqrt{6x + 7} - x = 2.$ -1 and 3, or $\{-1, 3\}$

The solution of radical equations with two or more square root expressions involves isolating a radical, squaring both sides, and then repeating this process. Let's consider an equation containing two square root expressions.

EXAMPLE 4 Solving an Equation That Has Two Radicals

Solve: $\sqrt{3x + 1} - \sqrt{x + 4} = 1.$

SOLUTION

Step 1. **Isolate a radical on one side.** We can isolate the radical $\sqrt{3x + 1}$ by adding $\sqrt{x + 4}$ to both sides. We obtain

$$\sqrt{3x + 1} = \sqrt{x + 4} + 1.$$

Step 2. **Square both sides.**

$$\left(\sqrt{3x + 1}\right)^2 = \left(\sqrt{x + 4} + 1\right)^2$$

Squaring the expression on the right side of the equation can be a bit tricky. We have to use the formula

$$(A + B)^2 = A^2 + 2AB + B^2.$$

Focusing on just the right side, here is how the squaring is done:

$$(A + B)^2 \quad = \quad A^2 \; + \; 2 \; \cdot \; A \; \cdot \; B \; + \; B^2$$

$$\left(\sqrt{x + 4} + 1\right)^2 = \left(\sqrt{x + 4}\right)^2 + 2 \cdot \sqrt{x + 4} \cdot 1 + 1^2 = x + 4 + 2\sqrt{x + 4} + 1.$$

Now let's return to squaring both sides.

$\left(\sqrt{3x + 1}\right)^2 = \left(\sqrt{x + 4} + 1\right)^2$	Square both sides of the equation with an isolated radical.
$3x + 1 = x + 4 + 2\sqrt{x + 4} + 1$	$\left(\sqrt{3x + 1}\right)^2 = 3x + 1$; square the right side using the formula for $(A + B)^2$.
$3x + 1 = x + 5 + 2\sqrt{x + 4}$	Combine numerical terms on the right side: $4 + 1 = 5$.

Can you see that the resulting equation still contains a radical, namely $\sqrt{x + 4}$? Thus, we need to repeat the first two steps.

Repeat Step 1. **Isolate a radical on one side.** We isolate $2\sqrt{x + 4}$, the radical term, by subtracting $x + 5$ from both sides. We obtain

$3x + 1 = x + 5 + 2\sqrt{x + 4}$	This is the equation from our last step.
$2x - 4 = 2\sqrt{x + 4}.$	Subtract x and subtract 5 from both sides.

Although we can simplify the equation by dividing both sides by 2, this sort of simplification is not always helpful. Thus, we will work with the equation in this form.

Repeat Step 2. **Square both sides.**

> Be careful in squaring both sides.
> Use $(A - B)^2 = A^2 - 2AB + B^2$
> to square the left side. Use
> $(AB)^2 = A^2B^2$ to square the right
> side.

$$(2x - 4)^2 = (2\sqrt{x + 4})^2 \quad \text{Square both sides.}$$

$$4x^2 - 16x + 16 = 4(x + 4) \quad \text{Square both 2 and } \sqrt{x + 4} \text{ on the right side.}$$

Step 3. **Solve the resulting equation.** We solve this quadratic equation by writing it in standard form.

$$4x^2 - 16x + 16 = 4x + 16 \quad \text{Use the distributive property.}$$
$$4x^2 - 20x = 0 \quad \text{Subtract } 4x + 16 \text{ from both sides.}$$
$$4x(x - 5) = 0 \quad \text{Factor.}$$
$$4x = 0 \quad \text{or} \quad x - 5 = 0 \quad \text{Set each factor equal to zero.}$$
$$x = 0 \qquad\qquad x = 5 \quad \text{Solve for } x.$$

Step 4. **Check the proposed solutions in the original equation.**

Check 0:	**Check 5:**

$$\sqrt{3x + 1} - \sqrt{x + 4} = 1 \qquad\qquad \sqrt{3x + 1} - \sqrt{x + 4} = 1$$
$$\sqrt{3 \cdot 0 + 1} - \sqrt{0 + 4} \overset{?}{=} 1 \qquad \sqrt{3 \cdot 5 + 1} - \sqrt{5 + 4} \overset{?}{=} 1$$
$$\sqrt{1} - \sqrt{4} \overset{?}{=} 1 \qquad\qquad \sqrt{16} - \sqrt{9} \overset{?}{=} 1$$
$$1 - 2 \overset{?}{=} 1 \qquad\qquad\qquad 4 - 3 \overset{?}{=} 1$$
$$-1 = 1, \;\; \text{false} \qquad\qquad\qquad 1 = 1, \;\; \text{true}$$

The check indicates that 0 is not a solution. It is an extraneous solution brought about by squaring each side of the equation. The only solution is 5, and the solution set is $\{5\}$. ■

 CHECK POINT 4 Solve: $\sqrt{x + 5} - \sqrt{x - 3} = 2$. $\quad$ 4 or $\{4\}$

EXAMPLE 5 $\;$ Solving a Radical Equation

Solve: $(3x - 1)^{\frac{1}{3}} + 4 = 0$.

SOLUTION $\;$ Although we can rewrite the equation in radical form

$$\sqrt[3]{3x - 1} + 4 = 0,$$

it is not necessary to do so. Because the equation involves a cube root, we isolate the radical term—that is, the term with the rational exponent—and cube both sides.

$$(3x - 1)^{\frac{1}{3}} + 4 = 0 \qquad \text{This is the given equation.}$$

$$(3x - 1)^{\frac{1}{3}} = -4 \qquad \text{Subtract 4 from both sides and isolate the term with the rational exponent.}$$

$$\left[(3x - 1)^{\frac{1}{3}}\right]^3 = (-4)^3 \qquad \text{Cube both sides.}$$

$$3x - 1 = -64 \qquad \text{Multiply exponents on the left side and simplify.}$$

$$3x = -63 \qquad \text{Add 1 to both sides.}$$

$$x = -21 \qquad \text{Divide both sides by 3.}$$

Because both sides were raised to an odd power, it is not essential to check the proposed solution, -21. However, checking is always a good idea. Do so now and verify that -21 is the solution and the solution set is $\{-21\}$. ■

Example 5 illustrates that a radical equation with rational exponents can be solved by

1. isolating the expression with the rational exponent, and

2. raising both sides of the equation to a power that is the reciprocal of the rational exponent.

Keep in mind that it is essential to check when both sides have been raised to even powers. Thus, equations with rational exponents such as $\frac{1}{2}$ and $\frac{1}{4}$ must be checked.

CHECK POINT 5 Solve: $(2x - 3)^{\frac{1}{3}} + 3 = 0$. -12 or $\{-12\}$

---

2 Use models that are radical functions to solve problems.

Applications of Radical Equations   Radical equations can be solved to answer questions about variables contained in radical functions.

EXAMPLE 6   Living Alone

The bar graph in Figure 10.7 shows the number of people in the United States living alone for selected years from 1970 through 2000. The function

$$f(x) = 2.6\sqrt{x} + 11$$

models the number of Americans living alone, $f(x)$, in millions, x years after 1970. According to the model, when will 29 million Americans live alone?

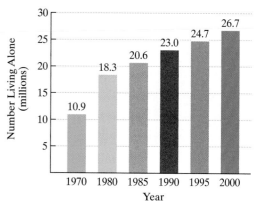

Number of Americans 15 Years and Older Living Alone

FIGURE 10.7

Source: U.S. Census Bureau

SOLUTION   To find when 29 million Americans will live alone, substitute 29 for $f(x)$ in the given function. Then solve for x, the number of years after 1970.

$$f(x) = 2.6\sqrt{x} + 11 \qquad \text{This is the given function.}$$

$$29 = 2.6\sqrt{x} + 11 \qquad \text{Substitute 29 for } f(x).$$

$$18 = 2.6\sqrt{x} \qquad \text{Subtract 11 from both sides.}$$

$$\frac{18}{2.6} = \sqrt{x} \qquad \text{Divide both sides by 2.6.}$$

$$\left(\frac{18}{2.6}\right)^2 = (\sqrt{x})^2 \qquad \text{Square both sides.}$$

$$48 \approx x \qquad \text{Use a calculator.}$$

The model indicates that 29 million Americans will live alone approximately 48 years after 1970. Because $1970 + 48 = 2018$, this is predicted to occur in 2018.

✔ CHECK POINT 6 The function $f(x) = 2.9\sqrt{x} + 11$ serves as another model for the number of Americans living alone, $f(x)$, in millions, x years after 1970. According to this model, when will 29 million Americans live alone? Use a calculator and round to the nearest year. Does this model predict faster or slower growth in the number of people living alone than the function used in Example 6?   2009; faster

10.6 EXERCISE SET

Student Solutions Manual   CD/Video   PH Math/Tutor Center   MathXL Tutorials on CD   MathXL®   MyMathLab   Interactmath.com

Practice Exercises
In Exercises 1–38, solve each radical equation.

1. $\sqrt{3x - 2} = 4$ $\{6\}$

2. $\sqrt{5x - 1} = 8$ $\{13\}$

3. $\sqrt{5x - 4} - 9 = 0$ $\{17\}$

4. $\sqrt{3x - 2} - 5 = 0$ $\{9\}$

5. $\sqrt{3x + 7} + 10 = 4$ $\varnothing$

6. $\sqrt{2x + 5} + 11 = 6$ $\varnothing$

7. $x = \sqrt{7x + 8}$ $\{8\}$

8. $x = \sqrt{6x + 7}$ $\{7\}$

9. $\sqrt{5x + 1} = x + 1$ $\{0, 3\}$

10. $\sqrt{2x + 1} = x - 7$ $\{12\}$

11. $x = \sqrt{2x - 2} + 1$ $\{1, 3\}$

12. $x = \sqrt{3x + 7} - 3$ $\{-2, -1\}$

13. $x - 2\sqrt{x - 3} = 3$ $\{3, 7\}$

14. $3x - \sqrt{3x + 7} = -5$ $\{-1\}$

15. $\sqrt{2x - 5} = \sqrt{x + 4}$ $\{9\}$

16. $\sqrt{6x + 2} = \sqrt{5x + 3}$ $\{1\}$

17. $\sqrt[3]{2x + 11} = 3$ $\{8\}$

18. $\sqrt[3]{6x - 3} = 3$ $\{5\}$

19. $\sqrt[3]{2x - 6} - 4 = 0$ $\{35\}$

20. $\sqrt[3]{4x - 3} - 5 = 0$ $\{32\}$

21. $\sqrt{x - 7} = 7 - \sqrt{x}$ $\{16\}$

22. $\sqrt{x - 8} = \sqrt{x} - 2$ $\{9\}$

23. $\sqrt{x + 2} + \sqrt{x - 1} = 3$ $\{2\}$

24. $\sqrt{x - 4} + \sqrt{x + 4} = 4$ $\{5\}$

25. $2\sqrt{4x + 1} - 9 = x - 5$ $\{2, 6\}$

26. $2\sqrt{x - 3} + 4 = x + 1$ $\{3, 7\}$

27. $(2x + 3)^{\frac{1}{3}} + 4 = 6$ $\{\frac{5}{2}\}$

28. $(3x - 6)^{\frac{1}{3}} + 5 = 8$ $\{11\}$

29. $(3x + 1)^{\frac{1}{4}} + 7 = 9$ $\{5\}$

30. $(2x + 3)^{\frac{1}{4}} + 7 = 10$ $\{39\}$

31. $(x + 2)^{\frac{1}{2}} + 8 = 4$ $\varnothing$

32. $(x - 3)^{\frac{1}{2}} + 8 = 6$ $\varnothing$

33. $\sqrt{2x - 3} - \sqrt{x - 2} = 1$ $\{2, 6\}$

34. $\sqrt{x + 2} + \sqrt{3x + 7} = 1$ $\{-2\}$

35. $3x^{\frac{1}{3}} = (x^2 + 17x)^{\frac{1}{3}}$ $\{0, 10\}$

36. $2(x - 1)^{\frac{1}{3}} = (x^2 + 2x)^{\frac{1}{3}}$ $\{2, 4\}$

37. $(x + 8)^{\frac{1}{4}} = (2x)^{\frac{1}{4}}$ $\{8\}$

38. $(x - 2)^{\frac{1}{4}} = (3x - 8)^{\frac{1}{4}}$ $\{3\}$

Practice Plus

39. If $f(x) = x + \sqrt{x + 5}$, find all values of x for which $f(x) = 7$.   4

40. If $f(x) = x - \sqrt{x - 2}$, find all values of x for which $f(x) = 4$.   6

41. If $f(x) = (5x + 16)^{\frac{1}{3}}$ and $g(x) = (x - 12)^{\frac{1}{3}}$, find all values of x for which $f(x) = g(x)$. -7

42. If $f(x) = (9x + 2)^{\frac{1}{4}}$ and $g(x) = (5x + 18)^{\frac{1}{4}}$, find all values of x for which $f(x) = g(x)$.   4

In Exercises 43–46, solve each formula for the specified variable.

43. $r = \sqrt{\dfrac{3V}{\pi h}}$ for V $V = \dfrac{\pi r^2 h}{3}$ or $\dfrac{1}{3}\pi r^2 h$

44. $r = \sqrt{\dfrac{A}{4\pi}}$ for A $A = 4\pi r^2$

45. $t = 2\pi\sqrt{\dfrac{l}{32}}$ for l $l = \dfrac{8t^2}{\pi^2}$

46. $v = \sqrt{\dfrac{FR}{m}}$ for m $\quad m = \dfrac{FR}{v^2}$

47. If 5 times a number is decreased by 4, the principal square root of this difference is 2 less than the number. Find the number(s). 8

48. If a number is decreased by 3, the principal square root of this difference is 5 less than the number. Find the number(s). 7

In Exercises 49–50, find the x-intercept(s) of the graph of each function without graphing the function.

49. $f(x) = \sqrt{x + 16} - \sqrt{x} - 2$ 9

50. $f(x) = \sqrt{2x - 3} - \sqrt{2x} + 1$ 2

Application Exercises

In 2002, the average surface temperature on Earth was 57.9°F, approximately 1.4° higher than it was one hundred years ago. Worldwide temperatures have risen only 9°F since the end of the last ice age 12,000 years ago. Most climatologists are convinced that over the next one hundred years, global temperatures will continue to increase, possibly setting off a chain of devastating events beginning with a rise in sea levels worldwide and ending with the destruction of water supplies, forests, and agriculture in many parts of the world. The graph shows global annual average temperatures from 1880 through 2002, with projections from 2002 through 2100.

Global Annual Average Temperatures and Projections through 2100

Source: National Oceanic and Atmospheric Administration

The temperature projections shown in the graph can be modeled by two functions:

$f(x) = 0.083x + 57.9$ — Models temperatures at the high end of the range

$g(x) = 0.36\sqrt{x} + 57.9.$ — Models temperatures at the low end of the range

In these functions, $f(x)$ and $g(x)$ describe projected global annual average temperatures, in degrees Fahrenheit, x years after 2002, where $0 \le x \le 98$. Use the functions to solve Exercises 51–54.

51. Use f and g to determine the temperatures at the high and low end of the range of projected global average temperatures for 2100. Round to the nearest tenth of a degree. High: 66.0°; Low: 61.5°

52. Use f and g to determine the temperatures at the high and low end of the range of projected global average temperatures for 2080. Round to the nearest tenth of a degree. High: 64.4°; Low: 61.1°

53. Use f and g to determine in which years the projected global average temperature will exceed the 2002 average of 57.9° by one degree. Round to the nearest year. Using f: 2014; Using g: 2010

54. Use f and g to determine in which years the projected global average temperature will exceed the 2002 average of 57.9° by two degrees. Using f: 2026; Using g: 2033

Out of a group of 50,000 births, the number of people, $f(x)$, surviving to age x is modeled by the function

$$f(x) = 5000\sqrt{100 - x}.$$

The graph of the function is shown. Use the function to solve Exercises 55–56.

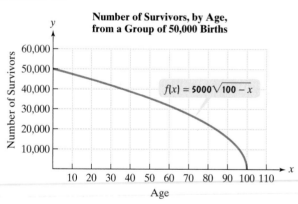

Number of Survivors, by Age, from a Group of 50,000 Births

$f(x) = 5000\sqrt{100 - x}$

55. To what age will 40,000 people in the group survive? Identify the solution as a point on the graph of the function. 40,000 people in the group will survive to age 36; (36, 40,000).

56. To what age will 35,000 people in the group survive? Identify the solution as a point on the graph of the function. 35,000 people in the group will survive to age 51; (51, 35,000).

The function

$$f(x) = 29x^{\frac{1}{3}}$$

models the number of plant species, $f(x)$, on the islands of the Galápagos in terms of the area, x, in square miles, of a particular island. Use the function to solve Exercises 57–58.

57. What is the area of a Galápagos island that has 87 species of plants? 27 square miles.

58. What is the area of a Galápagos island that has 58 species of plants? 8 square miles.

For each planet in our solar system, its year is the time it takes the planet to revolve once around the sun. The function

$$f(x) = 0.2x^{\frac{3}{2}}$$

models the number of Earth days in a planet's year, $f(x)$, where x is the average distance of the planet from the sun, in millions of kilometers. Use the function to solve Exercises 59–60.

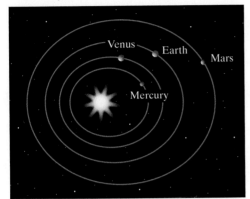

Venus   Earth   Mars   Mercury

59. We, of course, have 365 Earth days in our year. What is the average distance of Earth from the sun? Use a calculator and round to the nearest million kilometers. The average distance of Earth from the sun is approximately 149 million kilometers.

60. There are approximately 88 Earth days in the year of the planet Mercury. What is the average distance of Mercury from the sun? Use a calculator and round to the nearest million kilometers. The average distance of Mercury from the sun is approximately 58 million kilometers.

Writing in Mathematics

61. What is a radical equation?   61–67. Answers will vary.
62. In solving $\sqrt{2x - 1} + 2 = x$, why is it a good idea to isolate the radical term? What if we don't do this and simply square each side? Describe what happens.
63. What is an extraneous solution to a radical equation?
64. Explain why $\sqrt{x} = -1$ has no solution.
65. Explain how to solve a radical equation with rational exponents.
66. The radical function in Example 6 on page 696 shows a gradual year-by-year increase in the number of Americans living alone. What explanations can you offer for this trend? Describe an event that might occur in the future that could cause this trend to change.
67. Describe the trend shown by the graph of f in Exercises 55–56. When is the rate of decrease most rapid? What does this mean about survival rate by age?

Technology Exercises

In Exercises 68–72, use a graphing utility to solve each radical equation. Graph each side of the equation in the given viewing rectangle. The equation's solution is given by the x-coordinate of the point(s) of intersection. Check by substitution. 68–72. See graphing answer section.

68. $\sqrt{2x + 2} = \sqrt{3x - 5}$
$[-1, 10, 1]$ by $[-1, 5, 1]$ 7 or $\{7\}$

69. $\sqrt{x} + 3 = 5$
$[-1, 6, 1]$ by $[-1, 6, 1]$ 4 or $\{4\}$

70. $\sqrt{x^2 + 3} = x + 1$
$[-1, 6, 1]$ by $[-1, 6, 1]$ 1 or $\{1\}$

71. $4\sqrt{x} = x + 3$
$[-1, 10, 1]$ by $[-1, 14, 1]$ 1 and 9, or $\{1, 9\}$

72. $\sqrt{x} + 4 = 2$
$[-2, 18, 1]$ by $[0, 10, 1]$ no solution or $\varnothing$

Critical Thinking Exercises

73. Which one of the following is true?   c
   a. The first step in solving $\sqrt{x + 6} = x + 2$ is to square both sides, obtaining $x + 6 = x^2 + 4$.
   b. The equations $\sqrt{x + 4} = -5$ and $x + 4 = 25$ have the same solution set.
   c. If $T = 2\pi\sqrt{\dfrac{L}{32}}$, then $L = \dfrac{8T^2}{\pi^2}$.
   d. The equation $\sqrt{x^2 + 9x + 3} = -x$ has no solution because a principal square root is always nonnegative.

74. Find the length of the three sides of the right triangle shown in the figure.

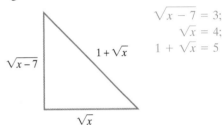

$\sqrt{x - 7} = 3$;
$\sqrt{x} = 4$;
$1 + \sqrt{x} = 5$

In Exercises 75–77, solve each equation.

75. $\sqrt[3]{x\sqrt{x}} = 9$ 81 or $\{81\}$
76. $\sqrt{\sqrt{x} + \sqrt{x + 9}} = 3$ 16 or $\{16\}$
77. $(x - 4)^{\frac{2}{3}} = 25$ 129 or $\{129\}$

Review Exercises

78. Divide using synthetic division:
$(4x^4 - 3x^3 + 2x^2 - x - 1) \div (x + 3)$.

(Section 5.6, Example 5) $4x^3 - 15x^2 + 47x - 142 + \dfrac{425}{x + 3}$

79. Divide:

$$\frac{3x^2 - 12}{x^2 + 2x - 8} \div \frac{6x + 18}{x + 4}.$$

(Section 7.2, Example 6) $\dfrac{x + 2}{2(x + 3)}$

80. Factor: $y^2 - 6y + 9 - 25x^2$.

(Section 6.5, Example 3) $(y - 3 + 5x)(y - 3 - 5x)$

COMPLEX NUMBERS

Objectives

1 Express square roots of negative numbers in terms of *i*.

2 Add and subtract complex numbers.

3 Multiply complex numbers.

4 Divide complex numbers.

5 Simplify powers of *i*.

Who is this kid warning us about our eyeballs turning black if we attempt to find the square root of −9? Don't believe what you hear on the street. Although square roots of negative numbers are not real numbers, they do play a significant role in algebra. In this section, we move beyond the real numbers and discuss square roots with negative radicands.

1 Express square roots of negative numbers in terms of *i*.

The Imaginary Unit *i* In Chapter 11 we will study equations whose solutions involve the square roots of negative numbers. Because the square of a real number is never negative, there is no real number x such that $x^2 = -1$. To provide a setting in which such equations have solutions, mathematicians invented an expanded system of numbers, the complex numbers. The *imaginary number i*, defined to be a solution of the equation $x^2 = -1$, is the basis of this new set.

> **THE IMAGINARY UNIT** *i* The **imaginary unit** *i* is defined as
> $$i = \sqrt{-1}, \quad \text{where} \quad i^2 = -1.$$

Using the imaginary unit *i*, we can express the square root of any negative number as a real multiple of *i*. For example,

$$\sqrt{-25} = \sqrt{25(-1)} = \sqrt{25}\sqrt{-1} = 5i.$$

We can check that $\sqrt{-25} = 5i$ by squaring $5i$ and obtaining -25.

$$(5i)^2 = 5^2 i^2 = 25(-1) = -25$$

THE SQUARE ROOT OF A NEGATIVE NUMBER If b is a positive real number, then
$$\sqrt{-b} = \sqrt{b(-1)} = \sqrt{b}\sqrt{-1} = \sqrt{b}i \quad \text{or} \quad i\sqrt{b}.$$

EXAMPLE 1 Expressing Square Roots of Negative Numbers as Multiples of i

Write as a multiple of i:

   a. $\sqrt{-9}$ **b.** $\sqrt{-3}$ **c.** $\sqrt{-80}$.

SOLUTION

   a. $\sqrt{-9} = \sqrt{9(-1)} = \sqrt{9}\sqrt{-1} = 3i$

> Be sure not to write i under the radical.

   b. $\sqrt{-3} = \sqrt{3(-1)} = \sqrt{3}\sqrt{-1} = \sqrt{3}\,i$

   c. $\sqrt{-80} = \sqrt{80(-1)} = \sqrt{80}\sqrt{-1} = \sqrt{16 \cdot 5}\sqrt{-1} = 4\sqrt{5}\,i$ ∎

✔ **CHECK POINT 1** Write as a multiple of i:

   a. $\sqrt{-64}$ $8i$ **b.** $\sqrt{-11}$ $\sqrt{11}i$ **c.** $\sqrt{-48}$. $4\sqrt{3}i$

A new system of numbers, called *complex numbers*, is based on adding multiples of i, such as $5i$, to the real numbers.

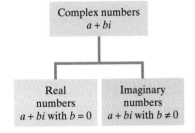

Complex numbers
$a + bi$

Real numbers	Imaginary numbers
$a + bi$ with $b = 0$	$a + bi$ with $b \neq 0$

FIGURE 10.8 The complex number system

COMPLEX NUMBERS AND IMAGINARY NUMBERS The set of all numbers in the form
$$a + bi$$
with real numbers a and b, and i, the imaginary unit, is called the set of **complex numbers**. The real number a is called the **real part**, and the real number b is called the **imaginary part** of the complex number $a + bi$. If $b \neq 0$, then the complex number is called an **imaginary number** (Figure 10.8).

Here are some examples of complex numbers. Each number can be written in the form $a + bi$.

$$-4 + 6i \qquad\qquad 2i = 0 + 2i \qquad\qquad 3 = 3 + 0i$$

a, the real part, is -4.	b, the imaginary part, is 6.	a, the real part, is 0.	b, the imaginary part, is 2.	a, the real part, is 3.	b, the imaginary part, is 0.

Can you see that b, the imaginary part, is not zero in the first two complex numbers? Because $b \neq 0$, these complex numbers are imaginary numbers. By contrast, the imaginary part of the complex number on the right is zero. This complex number is not an imaginary number. The number 3, or $3 + 0i$, is a real number.

Adding and Subtracting Complex Numbers The form of a complex number $a + bi$ is like the binomial $a + bx$. Consequently, we can add, subtract, and multiply complex numbers using the same methods we used for binomials, remembering that $i^2 = -1$.

2   Add and subtract complex numbers.

ADDING AND SUBTRACTING COMPLEX NUMBERS

1. $(a + bi) + (c + di) = (a + c) + (b + d)i$
   In words, this says that you add complex numbers by adding their real parts, adding their imaginary parts, and expressing the sum as a complex number.
2. $(a + bi) - (c + di) = (a - c) + (b - d)i$
   In words, this says that you subtract complex numbers by subtracting their real parts, subtracting their imaginary parts, and expressing the difference as a complex number.

EXAMPLE 2 Adding and Subtracting Complex Numbers

Perform the indicated operations, writing the result in the form $a + bi$:

a. $(5 - 11i) + (7 + 4i)$ **b.** $(-5 + 7i) - (-11 - 6i)$.

SOLUTION

a. $(5 - 11i) + (7 + 4i)$
  $= 5 - 11i + 7 + 4i$ Remove the parentheses.
  $= 5 + 7 - 11i + 4i$ Group real and imaginary terms.
  $= (5 + 7) + (-11 + 4)i$ Add real parts and add imaginary parts.
  $= 12 - 7i$ Simplify.

b. $(-5 + 7i) - (-11 - 6i)$
  $= -5 + 7i + 11 + 6i$ Remove the parentheses. Change signs of real and imaginary parts in the complex number being subtracted.
  $= -5 + 11 + 7i + 6i$ Group real and imaginary terms.
  $= (-5 + 11) + (7 + 6)i$ Add real parts and add imaginary parts.
  $= 6 + 13i$ Simplify. ∎

CHECK POINT 2 Add or subtract as indicated:

a. $(5 - 2i) + (3 + 3i)$ $8 + i$
b. $(2 + 6i) - (12 - 4i)$. $-10 + 10i$

STUDY TIP

The following examples, using the same integers as in Example 2, show how operations with complex numbers are just like operations with polynomials.

a. $(5 - 11x) + (7 + 4x)$
  $= 12 - 7x$
b. $(-5 + 7x) - (-11 - 6x)$
  $= -5 + 7x + 11 + 6x$
  $= 6 + 13x$

---

3 Multiply complex numbers.

Multiplying Complex Numbers Multiplication of complex numbers is performed the same way as multiplication of polynomials, using the distributive property and the FOIL method. After completing the multiplication, we replace i^2 with -1. This idea is illustrated in the next example.

EXAMPLE 3 Multiplying Complex Numbers

Find the products:

a. $4i(3 - 5i)$ **b.** $(7 - 3i)(-2 - 5i)$.

SOLUTION

a. $4i(3 - 5i)$
  $= 4i \cdot 3 - 4i \cdot 5i$ Distribute $4i$ throughout the parentheses.
  $= 12i - 20i^2$ Multiply.
  $= 12i - 20(-1)$ Replace i^2 with -1.
  $= 20 + 12i$ Simplify to $12i + 20$ and write in $a + bi$ form.

b. $(7 - 3i)(-2 - 5i)$

F   O   I   L

$$= -14 - 35i + 6i + 15i^2. \qquad \text{Use the FOIL method.}$$
$$= -14 - 35i + 6i + 15(-1) \qquad i^2 = -1$$
$$= -14 - 15 - 35i + 6i \qquad \text{Group real and imaginary terms.}$$
$$= -29 - 29i \qquad \text{Combine real and imaginary terms.}$$

■

✔ CHECK POINT 3 Find the products:

a. $7i(2 - 9i)$ $63 + 14i$ **b.** $(5 + 4i)(6 - 7i)$. $58 - 11i$

Consider the multiplication problem

$$5i \cdot 2i = 10i^2 = 10(-1) = -10.$$

This problem can also be given in terms of square roots of negative numbers:

$$\sqrt{-25} \cdot \sqrt{-4}.$$

Because the product rule for radicals only applies to real numbers, multiplying radicands is incorrect. **When performing operations with square roots of negative numbers, begin by expressing all square roots in terms of i.** Then perform the indicated operation.

CORRECT:

$$\sqrt{-25} \cdot \sqrt{-4} = \sqrt{25}\sqrt{-1} \cdot \sqrt{4}\sqrt{-1}$$
$$= 5i \cdot 2i$$
$$= 10i^2 = 10(-1) = -10$$

INCORRECT!

$$\sqrt{-25} \cdot \sqrt{-4} = \sqrt{(-25)(-4)}$$
$$= \sqrt{100}$$
$$= 10$$

EXAMPLE 4 Multiplying Square Roots of Negative Numbers

Multiply: $\sqrt{-3} \cdot \sqrt{-5}$.

SOLUTION

$$\sqrt{-3} \cdot \sqrt{-5} = \sqrt{3}\sqrt{-1} \cdot \sqrt{5}\sqrt{-1}$$
$$= \sqrt{3}i \cdot \sqrt{5}i \qquad \text{Express square roots in terms of } i.$$
$$= \sqrt{15}i^2 \qquad \sqrt{3} \cdot \sqrt{5} = \sqrt{15} \text{ and } i \cdot i = i^2.$$
$$= \sqrt{15}(-1) \qquad i^2 = -1$$
$$= -\sqrt{15}$$

■

✔ CHECK POINT 4 Multiply: $\sqrt{-5} \cdot \sqrt{-7}$. $-\sqrt{35}$

4 Divide complex numbers.

Conjugates and Division   It is possible to multiply imaginary numbers and obtain a real number. Here is an example:

F   O   I   L

$$(4 + 7i)(4 - 7i) = 16 - 28i + 28i - 49i^2$$
$$= 16 - 49i^2 = 16 - 49(-1) = 65.$$

Replace i^2 with -1.

You can also perform $(4 + 7i)(4 - 7i)$ using the formula

$$(A + B)(A - B) = A^2 - B^2.$$

A real number is obtained even faster:

$$(4 + 7i)(4 - 7i) = 4^2 - (7i)^2 = 16 - 49i^2 = 16 - 49(-1) = 65.$$

The **conjugate** of the complex number $a + bi$ is $a - bi$. The **conjugate** of the complex number $a - bi$ is $a + bi$. The multiplication problem that we just performed involved conjugates. The multiplication of conjugates always results in a real number:

$$(a + bi)(a - bi) = a^2 - (bi)^2 = a^2 - b^2i^2 = a^2 - b^2(-1) = a^2 + b^2.$$

> The product eliminates i.

Conjugates are used to divide complex numbers. By multiplying the numerator and the denominator of the division by the conjugate of the denominator, you will obtain a real number in the denominator. Here are two examples of such divisions:

- $\dfrac{7 + 4i}{2 - 5i}$

  > The conjugate of the denominator is $2 + 5i$.

- $\dfrac{5i - 4}{3i}$ or $\dfrac{5i - 4}{0 + 3i}$.

  > The conjugate of the denominator is $0 - 3i$, or $-3i$.

The procedure for dividing complex numbers, illustrated in Examples 5 and 6, should remind you of rationalizing denominators.

EXAMPLE 5   Using Conjugates to Divide Complex Numbers

Divide and simplify to the form $a + bi$:

$$\frac{7 + 4i}{2 - 5i}.$$

SOLUTION   The conjugate of the denominator is $2 + 5i$. Multiplication of both the numerator and the denominator by $2 + 5i$ will eliminate i from the denominator.

$$\frac{7 + 4i}{2 - 5i} = \frac{7 + 4i}{2 - 5i} \cdot \frac{2 + 5i}{2 + 5i}$$
Multiply by 1.

F O I L

$$= \frac{14 + 35i + 8i + 20i^2}{2^2 - (5i)^2}$$
Use FOIL in the numerator and $(A - B)(A + B) = A^2 - B^2$ in the denominator.

$$= \frac{14 + 43i + 20i^2}{4 - 25i^2}$$
Simplify.

$$= \frac{14 + 43i + 20(-1)}{4 - 25(-1)}$$
$i^2 = -1$

$$= \frac{14 + 43i - 20}{4 + 25}$$
Perform the multiplications involving -1.

$$= \frac{-6 + 43i}{29}$$
Combine real terms in the numerator and denominator.

$$= -\frac{6}{29} + \frac{43}{29}i$$
Express the answer in the form $a + bi$.

CHECK POINT **5** Divide and simplify to the form $a + bi$:

$$\frac{6 + 2i}{4 - 3i}. \qquad \frac{18}{25} + \frac{26}{25}i$$

| EXAMPLE 6 | Using Conjugates to Divide Complex Numbers |

Divide and simplify to the form $a + bi$:

$$\frac{5i - 4}{3i}.$$

SOLUTION   The conjugate of the denominator, $0 + 3i$, is $0 - 3i$. Multiplication of both the numerator and the denominator by $-3i$ will eliminate i from the denominator.

$$\frac{5i - 4}{3i} = \frac{5i - 4}{3i} \cdot \frac{-3i}{-3i} \qquad \text{Multiply by 1.}$$

$$= \frac{-15i^2 + 12i}{-9i^2} \qquad \begin{array}{l}\text{Multiply. Use the distributive property}\\ \text{in the numerator.}\end{array}$$

$$= \frac{-15(-1) + 12i}{-9(-1)} \qquad i^2 = -1$$

$$= \frac{15 + 12i}{9} \qquad \text{Perform the multiplications involving } -1.$$

$$= \frac{15}{9} + \frac{12}{9}i \qquad \text{Express the division in the form } a + bi.$$

$$= \frac{5}{3} + \frac{4}{3}i \qquad \text{Simplify real and imaginary parts.} \qquad ■$$

 CHECK POINT **6** Divide and simplify to the form $a + bi$:

$$\frac{3 - 2i}{4i}. \qquad -\frac{1}{2} - \frac{3}{4}i$$

---

5 Simplify powers of i.

Powers of i   Cycles govern many aspects of life—heartbeats, sleep patterns, seasons, and tides all follow regular, predictable cycles. Surprisingly, so do powers of i. To see how this occurs, use the fact that $i^2 = -1$ and express each power of i in terms of i^2:

$$i$$

$$i^2 = -1$$

$$i^3 = i^2 \cdot i = (-1)i = -i$$

$$i^4 = (i^2)^2 = (-1)^2 = 1$$

$$i^5 = i^4 \cdot i = (i^2)^2 \cdot i = (-1)^2 \cdot i = i$$

$$i^6 = (i^2)^3 = (-1)^3 = -1.$$

Multiples of i

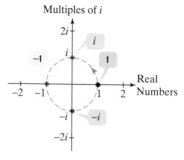

FIGURE 10.9

Can you see that the powers of i are cycling through the values i, -1, $-i$, and 1? The cycle is illustrated in Figure 10.9. In the figure, tick marks on the horizontal axis represent real numbers and tick marks along the vertical axis represent multiples of i. Using this representation, powers of i are equally spaced at 90° intervals on a circle having radius 1.

Here is a procedure for simplifying powers of i:

SIMPLIFYING POWERS OF i

1. Express the given power of i in terms of i^2.
2. Replace i^2 with -1 and simplify. Use the fact that -1 to an even power is 1 and -1 to an odd power is -1.

EXAMPLE 7 Simplifying Powers of i

Simplify:

a. i^{12} **b.** i^{39} **c.** i^{50}.

SOLUTION

a. $i^{12} = (i^2)^6 = (-1)^6 = 1$
b. $i^{39} = i^{38}i = (i^2)^{19}i = (-1)^{19}i = (-1)i = -i$
c. $i^{50} = (i^2)^{25} = (-1)^{25} = -1$

 CHECK POINT 7 Simplify:

a. i^{16} 1 **b.** i^{25} i **c.** i^{35}. $-i$

The Patterns of Chaos

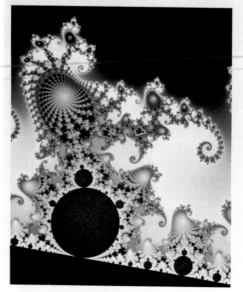

One of the new frontiers of mathematics suggests that there is an underlying order in things that appear to be random, such as the hiss and crackle of background noises as you tune a radio. Irregularities in the heartbeat, some of them severe enough to cause a heart attack, or irregularities in our sleeping patterns, such as insomnia, are examples of chaotic behavior. Chaos in the mathematical sense does not mean a complete lack of form or arrangement. In mathematics, chaos is used to describe something that appears to be random but is not actually random. The patterns of chaos appear in images like the one on the left, called the Mandelbrot set. Magnified portions of this image yield repetitions of the original structure, as well as new and unexpected patterns. The Mandelbrot set transforms the hidden structure of chaotic events into a source of wonder and inspiration.

The Mandelbrot set is made possible by opening up graphing to include complex numbers in the form $a + bi$. Although the details are beyond the scope of this text, the coordinate system that is used is shown in Figure 10.9. Plot certain complex numbers in this system, add color to the magnified boundary of the graph, and the patterns of chaos begin to appear.

R. F. Voss "29-Fold M-set Seahorse" computer-generated image

10.7 EXERCISE SET

Student Solutions Manual   CD/Video   PH Math/Tutor Center   MathXL Tutorials on CD   MathXL®   MyMathLab   Interactmath.com

Practice Exercises

In Exercises 1–16, express each number in terms of i and simplify, if possible.

1. $\sqrt{-100}$ $10i$
2. $\sqrt{-49}$ $7i$
3. $\sqrt{-23}$ $\sqrt{23}i$
4. $\sqrt{-21}$ $\sqrt{21}i$
5. $\sqrt{-18}$ $3\sqrt{2}i$
6. $\sqrt{-125}$ $5\sqrt{5}i$
7. $\sqrt{-63}$ $3\sqrt{7}i$
8. $\sqrt{-28}$ $2\sqrt{7}i$
9. $-\sqrt{-108}$ $-6\sqrt{3}i$
10. $-\sqrt{-300}$ $-10\sqrt{3}i$
11. $5 + \sqrt{-36}$ $5 + 6i$
12. $7 + \sqrt{-4}$ $7 + 2i$
13. $15 + \sqrt{-3}$ $15 + \sqrt{3}i$
14. $20 + \sqrt{-5}$ $20 + \sqrt{5}i$
15. $-2 - \sqrt{-18}$ $-2 - 3\sqrt{2}i$
16. $-3 - \sqrt{-27}$ $-3 - 3\sqrt{3}i$

In Exercises 17–32, add or subtract as indicated. Write the result in the form a + bi.

17. $(3 + 2i) + (5 + i)$ $8 + 3i$
18. $(6 + 5i) + (4 + 3i)$ $10 + 8i$
19. $(7 + 2i) + (1 - 4i)$ $8 - 2i$
20. $(-2 + 6i) + (4 - i)$ $2 + 5i$
21. $(10 + 7i) - (5 + 4i)$ $5 + 3i$
22. $(11 + 8i) - (2 + 5i)$ $9 + 3i$
23. $(9 - 4i) - (10 + 3i)$ $-1 - 7i$
24. $(8 - 5i) - (6 + 2i)$ $2 - 7i$
25. $(3 + 2i) - (5 - 7i)$ $-2 + 9i$
26. $(-7 + 5i) - (9 - 11i)$ $-16 + 16i$
27. $(-5 + 4i) - (-13 - 11i)$ $8 + 15i$
28. $(-9 + 2i) - (-17 - 6i)$ $8 + 8i$
29. $8i - (14 - 9i)$ $-14 + 17i$
30. $15i - (12 - 11i)$ $-12 + 26i$
31. $\left(2 + \sqrt{3}i\right) + \left(7 + 4\sqrt{3}i\right)$ $9 + 5\sqrt{3}i$
32. $\left(4 + \sqrt{5}i\right) + \left(8 + 6\sqrt{5}i\right)$ $12 + 7\sqrt{5}i$

In Exercises 33–62, find each product. Write imaginary results in the form a + bi.

33. $2i(5 + 3i)$ $-6 + 10i$
34. $5i(4 + 7i)$ $-35 + 20i$
35. $3i(7i - 5)$ $-21 - 15i$
36. $8i(4i - 3)$ $-32 - 24i$
37. $-7i(2 - 5i)$ $-35 - 14i$
38. $-6i(3 - 5i)$ $-30 - 18i$
39. $(3 + i)(4 + 5i)$ $7 + 19i$
40. $(4 + i)(5 + 6i)$ $14 + 29i$
41. $(7 - 5i)(2 - 3i)$ $-1 - 31i$
42. $(8 - 4i)(3 - 2i)$ $16 - 28i$
43. $(6 - 3i)(-2 + 5i)$ $3 + 36i$
44. $(7 - 2i)(-3 + 6i)$ $-9 + 48i$
45. $(3 + 5i)(3 - 5i)$ 34
46. $(2 + 7i)(2 - 7i)$ 53
47. $(-5 + 3i)(-5 - 3i)$ 34
48. $(-4 + 2i)(-4 - 2i)$ 20
49. $\left(3 - \sqrt{2}i\right)\left(3 + \sqrt{2}i\right)$ 11
50. $\left(5 - \sqrt{3}i\right)\left(5 + \sqrt{3}i\right)$ 28
51. $(2 + 3i)^2$ $-5 + 12i$
52. $(3 + 2i)^2$ $5 + 12i$
53. $(5 - 2i)^2$ $21 - 20i$
54. $(5 - 3i)^2$ $16 - 30i$
55. $\sqrt{-7} \cdot \sqrt{-2}$ $-\sqrt{14}$
56. $\sqrt{-7} \cdot \sqrt{-3}$ $-\sqrt{21}$
57. $\sqrt{-9} \cdot \sqrt{-4}$ -6
58. $\sqrt{-16} \cdot \sqrt{-4}$ -8
59. $\sqrt{-7} \cdot \sqrt{-25}$ $-5\sqrt{7}$
60. $\sqrt{-3} \cdot \sqrt{-36}$ $-6\sqrt{3}$
61. $\sqrt{-8} \cdot \sqrt{-3}$ $-2\sqrt{6}$
62. $\sqrt{-9} \cdot \sqrt{-5}$ $-3\sqrt{5}$

In Exercises 63–84, divide and simplify to the form a + bi.

63. $\dfrac{2}{3 + i}$ $\dfrac{3}{5} - \dfrac{1}{5}i$
64. $\dfrac{3}{4 + i}$ $\dfrac{12}{17} - \dfrac{3}{17}i$
65. $\dfrac{2i}{1 + i}$ $1 + i$
66. $\dfrac{5i}{2 + i}$ $1 + 2i$
67. $\dfrac{7}{4 - 3i}$ $\dfrac{28}{25} + \dfrac{21}{25}i$
68. $\dfrac{9}{1 - 2i}$ $\dfrac{9}{5} + \dfrac{18}{5}i$
69. $\dfrac{6i}{3 - 2i}$ $-\dfrac{12}{13} + \dfrac{18}{13}i$
70. $\dfrac{5i}{2 - 3i}$ $-\dfrac{15}{13} + \dfrac{10}{13}i$
71. $\dfrac{1 + i}{1 - i}$ $0 + i$ or i
72. $\dfrac{1 - i}{1 + i}$ $0 - i$ or $-i$
73. $\dfrac{2 - 3i}{3 + i}$ $\dfrac{3}{10} - \dfrac{11}{10}i$
74. $\dfrac{2 + 3i}{3 - i}$ $\dfrac{3}{10} + \dfrac{11}{10}i$
75. $\dfrac{5 - 2i}{3 + 2i}$ $\dfrac{11}{13} - \dfrac{16}{13}i$
76. $\dfrac{6 - 3i}{4 + 2i}$ $\dfrac{9}{10} - \dfrac{6}{5}i$
77. $\dfrac{4 + 5i}{3 - 7i}$ $-\dfrac{23}{58} + \dfrac{43}{58}i$
78. $\dfrac{5 - i}{3 - 2i}$ $\dfrac{17}{13} + \dfrac{7}{13}i$
79. $\dfrac{7}{3i}$ $0 - \dfrac{7}{3}i$ or $-\dfrac{7}{3}i$
80. $\dfrac{5}{7i}$ $0 - \dfrac{5}{7}i$ or $-\dfrac{5}{7}i$
81. $\dfrac{8 - 5i}{2i}$ $-\dfrac{5}{2} - 4i$
82. $\dfrac{3 + 4i}{5i}$ $\dfrac{4}{5} - \dfrac{3}{5}i$
83. $\dfrac{4 + 7i}{-3i}$ $-\dfrac{7}{3} + \dfrac{4}{3}i$
84. $\dfrac{5 + i}{-4i}$ $-\dfrac{1}{4} + \dfrac{5}{4}i$

In Exercises 85–100, simplify each expression.

85. i^{10} -1
86. i^{14} -1
87. i^{11} $-i$
88. i^{15} $-i$
89. i^{22} -1
90. i^{46} -1
91. i^{200} 1
92. i^{400} 1
93. i^{17} i
94. i^{21} i
95. $(-i)^4$ 1
96. $(-i)^6$ -1
97. $(-i)^9$ $-i$
98. $(-i)^{13}$ $-i$
99. $i^{24} + i^2$ 0
100. $i^{28} + i^{30}$ 0

Practice Plus

In Exercises 101–108, perform the indicated operation(s) and write the result in the form a + bi.

101. $(2 - 3i)(1 - i) - (3 - i)(3 + i)$ $-11 - 5i$
102. $(8 + 9i)(2 - i) - (1 - i)(1 + i)$ $23 + 10i$
103. $(2 + i)^2 - (3 - i)^2$ $-5 + 10i$
104. $(4 - i)^2 - (1 + 2i)^2$ $18 - 12i$
105. $5\sqrt{-16} + 3\sqrt{-81}$ $0 + 47i$ or $47i$
106. $5\sqrt{-8} + 3\sqrt{-18}$ $0 + 19\sqrt{2}i$ or $19\sqrt{2}i$
107. $\dfrac{i^4 + i^{12}}{i^8 - i^7}$ $1 - i$
108. $\dfrac{i^8 + i^{40}}{i^4 + i^3}$ $1 + i$
109. Let $f(x) = x^2 - 2x + 2$. Find $f(1 + i)$. 0
110. Let $f(x) = x^2 - 2x + 5$. Find $f(1 - 2i)$. 0

In Exercises 111–114, simplify each evaluation to the form a + bi.

111. Let $f(x) = x - 3i$ and $g(x) = 4x + 2i$. Find $(fg)(-1)$.
   $10 + 10i$

112. Let $f(x) = 12x - i$ and $g(x) = 6x + 3i$. Find $(fg)\left(-\dfrac{1}{3}\right)$.
   $11 - 10i$

113. Let $f(x) = \dfrac{x^2 + 19}{2 - x}$. Find $f(3i)$. $\dfrac{20}{13} + \dfrac{30}{13}i$

114. Let $f(x) = \dfrac{x^2 + 11}{3 - x}$. Find $f(4i)$. $-\dfrac{3}{5} - \dfrac{4}{5}i$

Application Exercises

Complex numbers are used in electronics to describe the current in an electric circuit. Ohm's law relates the current in a circuit, I, in amperes, the voltage of the circuit, E, in volts, and the resistance of the circuit, R, in ohms, by the formula $E = IR$. Use this formula to solve Exercises 115–116.

115. Find E, the voltage of a circuit, if $I = (4 - 5i)$ amperes and $R = (3 + 7i)$ ohms. $(47 + 13i)$ volts

116. Find E, the voltage of a circuit, if $I = (2 - 3i)$ amperes and $R = (3 + 5i)$ ohms. $(21 + i)$ volts

117. The mathematician Girolamo Cardano is credited with the first use (in 1545) of negative square roots in solving the now-famous problem, "Find two numbers whose sum is 10 and whose product is 40." Show that the complex numbers $5 + \sqrt{15}i$ and $5 - \sqrt{15}i$ satisfy the conditions of the problem. (Cardano did not use the symbolism $\sqrt{15}i$ or even $\sqrt{-15}$. He wrote R.m 15 for $\sqrt{-15}$, meaning "radix minus 15." He regarded the numbers $5 +$ R.m 15 and $5 -$ R.m 15 as "fictitious" or "ghost numbers," and considered the problem "manifestly impossible." But in a mathematically adventurous spirit, he exclaimed, "Nevertheless, we will operate.") $\left(5 + \sqrt{15}i\right) + \left(5 - \sqrt{15}i\right) = 10;$
   $\left(5 + \sqrt{15}i\right)\left(5 - \sqrt{15}i\right) = 25 - 15i^2 = 25 + 15 = 40$

Writing in Mathematics

118. What is the imaginary unit i?   118–130. Answers will vary.
119. Explain how to write $\sqrt{-64}$ as a multiple of i.
120. What is a complex number? Explain when a complex number is a real number and when it is an imaginary number. Provide examples with your explanation.
121. Explain how to add complex numbers. Give an example.
122. Explain how to subtract complex numbers. Give an example.
123. Explain how to find the product of $2i$ and $5 + 3i$.
124. Explain how to find the product of $2i + 3$ and $5 + 3i$.
125. Explain how to find the product of $2i + 3$ and $2i - 3$.
126. Explain how to find the product of $\sqrt{-1}$ and $\sqrt{-4}$. Describe a common error in the multiplication that needs to be avoided.
127. What is the conjugate of $2 + 3i$? What happens when you multiply this complex number by its conjugate?
128. Explain how to divide complex numbers. Provide an example with your explanation.

129. Explain each of the three jokes in the cartoon on page 700.
130. A stand-up comedian uses algebra in some jokes, including one about a telephone recording that announces "You have just reached an imaginary number. Please multiply by i and dial again." Explain the joke.

Explain the errors in Exercises 131–132.

131. $\sqrt{-9} + \sqrt{-16} = \sqrt{-25} = \sqrt{25}i = 5i$
   $\sqrt{-9} + \sqrt{-16} = 3i + 4i = 7i$

132. $\left(\sqrt{-9}\right)^2 = \sqrt{-9} \cdot \sqrt{-9} = \sqrt{81} = 9$
   $\left(\sqrt{-9}\right)^2 = \sqrt{-9} \cdot \sqrt{-9} = 3i \cdot 3i = 9i^2 = -9$

Critical Thinking Exercises

133. Which one of the following is true?   d
   a. Some irrational numbers are not complex numbers.
   b. $(3 + 7i)(3 - 7i)$ is an imaginary number.
   c. $\dfrac{7 + 3i}{5 + 3i} = \dfrac{7}{5}$
   d. In the complex number system, $x^2 + y^2$ (the sum of two squares) can be factored as $(x + yi)(x - yi)$.

In Exercises 134–136, perform the indicated operations and write the result in the form a + bi.

134. $\dfrac{4}{(2 + i)(3 - i)}$ $\dfrac{14}{25} - \dfrac{2}{25}i$

135. $\dfrac{1 + i}{1 + 2i} + \dfrac{1 - i}{1 - 2i}$ $\dfrac{6}{5} + 0i$ or $\dfrac{6}{5}$

136. $\dfrac{8}{1 + \dfrac{2}{i}}$ $\dfrac{8}{5} + \dfrac{16}{5}i$

Review Exercises

137. Simplify:
$$\frac{\dfrac{x}{y^2} + \dfrac{1}{y}}{\dfrac{y}{x^2} + \dfrac{1}{x}}.$$
(Section 7.5, Example 3 or Example 6) $\dfrac{x^2}{y^2}$

138. Solve for x: $\dfrac{1}{x} + \dfrac{1}{y} = \dfrac{1}{z}$.

(Section 7.6, Example 7) $x = \dfrac{yz}{y - z}$

139. If $f(x) = 2x^2 - x$ and $g(x) = x - 6$, find $(g - f)(3)$.

(Section 8.2, Example 4) -18

GROUP PROJECT

Additional group projects can be found in the *Instructor's Resource Manual*.

Group members should consult an almanac, newspaper, magazine, or the Internet and return to the group with as much data as possible that show phenomena that are continuing to grow over time, but whose growth is leveling off. Select the five data sets that you find most intriguing. Let x represent the number of years after the first year in each data set. Model the data by hand using

$$f(x) = a\sqrt{x} + b.$$

Use the first and last data points to find values for a and b. The first data point corresponds to $x = 0$. Its second coordinate gives the value of b. To find a, substitute the second data point into $f(x) = a\sqrt{x} + b$, with the value that you obtained for b. Now solve the equation and find a. Substitute a and b into $f(x) = a\sqrt{x} + b$ to obtain a square root function that models each data set. Then use the function to make predictions about what might occur in the future. Are there circumstances that might affect the accuracy of the predictions? List some of these circumstances.

CHAPTER 10 SUMMARY

Definitions and Concepts	Examples

Section 10.1 Radical Expressions and Functions

If $b^2 = a$, then b is a square root of a. The principal square root of a, designated $\sqrt{a}$, is the nonnegative number satisfying $b^2 = a$. The negative square root of a is written $-\sqrt{a}$. A square root of a negative number is not a real number. A radical function in x is a function defined by an expression containing a root of x. The domain of a square root function is the set of real numbers for which the radicand is nonnegative.	Let $f(x) = \sqrt{6 - 2x}$. $f(-15) = \sqrt{6 - 2(-15)} = \sqrt{6 + 30} = \sqrt{36} = 6$ $f(5) = \sqrt{6 - 2 \cdot 5} = \sqrt{6 - 10} = \sqrt{-4}$, not a real number Domain of f: Set the radicand greater than or equal to zero. $$6 - 2x \geq 0$$ $$-2x \geq -6$$ $$x \leq 3$$ Domain of $f = \{x \mid x \leq 3\}$ or $(-\infty, 3]$.
The cube root of a real number a is written $\sqrt[3]{a}$. $\qquad \sqrt[3]{a} = b \quad$ means that $\quad b^3 = a$. The nth root of a real number a is written $\sqrt[n]{a}$. The number n is the index. Every real number has one root when n is odd. The odd nth root of a, $\sqrt[n]{a}$, is the number b for which $b^n = a$. Every positive real number has two real roots when n is even. An even root of a negative number is not a real number. If n is even, then $\sqrt[n]{a^n} = |a|$. If n is odd, then $\sqrt[n]{a^n} = a$.	• $\sqrt[3]{-8} = -2$ because $(-2)^3 = -8$. • $\sqrt[4]{-16}$ is not a real number. • $\sqrt{x^2 - 14x + 49} = \sqrt{(x - 7)^2} = \|x - 7\|$ • $\sqrt[3]{125(x + 6)^3} = 5(x + 6)$

Definitions and Concepts	Examples

Section 10.2 Rational Exponents

- $a^{\frac{1}{n}} = \sqrt[n]{a}$

- $a^{\frac{m}{n}} = \left(\sqrt[n]{a}\right)^m$ or $\sqrt[n]{a^m}$

- $a^{-\frac{m}{n}} = \dfrac{1}{a^{\frac{m}{n}}}$

- $121^{\frac{1}{2}} = \sqrt{121} = 11$

- $64^{\frac{1}{3}} = \sqrt[3]{64} = 4$

- $27^{\frac{5}{3}} = \left(\sqrt[3]{27}\right)^5 = 3^5 = 3 \cdot 3 \cdot 3 \cdot 3 \cdot 3 = 243$

- $16^{-\frac{3}{4}} = \dfrac{1}{16^{\frac{3}{4}}} = \dfrac{1}{\left(\sqrt[4]{16}\right)^3} = \dfrac{1}{2^3} = \dfrac{1}{8}$

- $\left(\sqrt[3]{7xy}\right)^4 = (7xy)^{\frac{4}{3}}$

Properties of integer exponents are true for rational exponents. An expression with rational exponents is simplified when no parentheses appear, no powers are raised to powers, each base occurs once, and no negative or zero exponents appear.

Simplify: $\left(8x^{\frac{1}{3}}y^{-\frac{1}{2}}\right)^{\frac{1}{3}}$.

$$= 8^{\frac{1}{3}}\left(x^{\frac{1}{3}}\right)^{\frac{1}{3}}\left(y^{-\frac{1}{2}}\right)^{\frac{1}{3}}$$

$$= 2x^{\frac{1}{9}}y^{-\frac{1}{6}} = \frac{2x^{\frac{1}{9}}}{y^{\frac{1}{6}}}$$

Some radical expressions can be simplifed using rational exponents. Rewrite the expression using rational exponents, simplify, and rewrite in radical notation if rational exponents still appear.

- $\sqrt[9]{x^3} = x^{\frac{3}{9}} = x^{\frac{1}{3}} = \sqrt[3]{x}$

- $\sqrt[5]{x^2} \cdot \sqrt[4]{x} = x^{\frac{2}{5}} \cdot x^{\frac{1}{4}} = x^{\frac{2}{5}+\frac{1}{4}}$

$$= x^{\frac{8}{20}+\frac{5}{20}} = x^{\frac{13}{20}} = \sqrt[20]{x^{13}}$$

Section 10.3 Multiplying and Simplifying Radical Expressions

The product rule for radicals can be used to multiply radicals with the same indices:

$$\sqrt[n]{a} \cdot \sqrt[n]{b} = \sqrt[n]{ab}.$$

$\sqrt[3]{7x} \cdot \sqrt[3]{10y^2} = \sqrt[3]{7x \cdot 10y^2} = \sqrt[3]{70xy^2}$

The product rule for radicals can be used to simplify radicals:

$$\sqrt[n]{ab} = \sqrt[n]{a} \cdot \sqrt[n]{b}.$$

A radical expression with index n is simplified when its radicand has no factors that are perfect nth powers. To simplify, write the radicand as the product of two factors, one of which is the greatest perfect nth power. Then use the product rule to take the nth root of each factor. If all variables in a radicand are positive, then

$$\sqrt[n]{a^n} = a.$$

Some radicals can be simplified after the multiplication is performed.

- Simplify: $\sqrt[3]{54x^7y^{11}}$.

$$= \sqrt[3]{27 \cdot 2 \cdot x^6 \cdot x \cdot y^9 \cdot y^2}$$

$$= \sqrt[3]{(27x^6y^9)(2xy^2)}$$

$$= \sqrt[3]{27x^6y^9} \cdot \sqrt[3]{2xy^2} = 3x^2y^3\sqrt[3]{2xy^2}$$

- Assuming positive variables, multiply and simplify: $\sqrt[4]{4x^2y} \cdot \sqrt[4]{4xy^3}$.

$$= \sqrt[4]{4x^2y \cdot 4xy^3} = \sqrt[4]{16x^3y^4}$$

$$= \sqrt[4]{16y^4} \cdot \sqrt[4]{x^3} = 2y\sqrt[4]{x^3}$$

Definitions and Concepts	Examples

Section 10.4 Adding, Subtracting, and Dividing Radical Expressions

Like radicals have the same indices and radicands. Like radicals can be added or subtracted using the distributive property. In some cases, radicals can be combined once they have been simplified.	$4\sqrt{18} - 6\sqrt{50}$ $= 4\sqrt{9 \cdot 2} - 6\sqrt{25 \cdot 2} = 4 \cdot 3\sqrt{2} - 6 \cdot 5\sqrt{2}$ $= 12\sqrt{2} - 30\sqrt{2} = -18\sqrt{2}$
The quotient rule for radicals can be used to simplify radicals: $\sqrt[n]{\dfrac{a}{b}} = \dfrac{\sqrt[n]{a}}{\sqrt[n]{b}}.$	$\sqrt[3]{-\dfrac{8}{x^{12}}} = \dfrac{\sqrt[3]{-8}}{\sqrt[3]{x^{12}}} = -\dfrac{2}{x^4}$ $\boxed{\sqrt[3]{x^{12}} = (x^{12})^{\frac{1}{3}} = x^4}$
The quotient rule for radicals can be used to divide radicals with the same indices: $\dfrac{\sqrt[n]{a}}{\sqrt[n]{b}} = \sqrt[n]{\dfrac{a}{b}}.$ Some radicals can be simplified after the division is performed.	Assuming a positive variable, divide and simplify: $\dfrac{\sqrt[4]{64x^5}}{\sqrt[4]{2x^{-2}}} = \sqrt[4]{32x^{5-(-2)}} = \sqrt[4]{32x^7}$ $= \sqrt[4]{16 \cdot 2 \cdot x^4 \cdot x^3} = \sqrt[4]{16x^4} \cdot \sqrt[4]{2x^3}$ $= 2x\sqrt[4]{2x^3}.$

Section 10.5 Multiplying with More Than One Term and Rationalizing Denominators

Radical expressions with more than one term are multiplied in much the same way that polynomials with more than one term are multiplied.	• $\sqrt{5}(2\sqrt{6} - \sqrt{3}) = 2\sqrt{30} - \sqrt{15}$ • $(4\sqrt{3} - 2\sqrt{2})(\sqrt{3} + \sqrt{2})$ **F** **O** **I** **L** $= 4\sqrt{3} \cdot \sqrt{3} + 4\sqrt{3} \cdot \sqrt{2} - 2\sqrt{2} \cdot \sqrt{3} - 2\sqrt{2} \cdot \sqrt{2}$ $= 4 \cdot 3 + 4\sqrt{6} - 2\sqrt{6} - 2 \cdot 2$ $= 12 + 4\sqrt{6} - 2\sqrt{6} - 4 = 8 + 2\sqrt{6}$
Radical expressions that involve the sum and difference of the same two terms are called conjugates. Use $(A + B)(A - B) = A^2 - B^2$ to multiply conjugates.	$(8 + 2\sqrt{5})(8 - 2\sqrt{5})$ $= 8^2 - (2\sqrt{5})^2 = 64 - 4 \cdot 5$ $= 64 - 20 = 44$
The process of rewriting a radical expression as an equivalent expression without any radicals in the denominator is called rationalizing the denominator. When the denominator contains a single radical with an *n*th root, multiply the numerator and the denominator by a radical of index *n* that produces a perfect *n*th power in the denominator's radicand.	Rationalize the denominator: $\dfrac{7}{\sqrt[3]{2x}}.$ $= \dfrac{7}{\sqrt[3]{2x}} \cdot \dfrac{\sqrt[3]{4x^2}}{\sqrt[3]{4x^2}} = \dfrac{7\sqrt[3]{4x^2}}{\sqrt[3]{8x^3}} = \dfrac{7\sqrt[3]{4x^2}}{2x}$

Definitions and Concepts	Examples

Section 10.5 Multiplying with More Than One Term and Rationalizing Denominators (continued)

If the denominator contains two terms, rationalize the denominator by multiplying the numerator and the denominator by the conjugate of the denominator.

$$\frac{13}{5 - \sqrt{3}} = \frac{13}{5 - \sqrt{3}} \cdot \frac{5 + \sqrt{3}}{5 + \sqrt{3}}$$

$$= \frac{13(5 + \sqrt{3})}{5^2 - (\sqrt{3})^2}$$

$$= \frac{13(5 + \sqrt{3})}{25 - 3} = \frac{13(5 + \sqrt{3})}{22}$$

Section 10.6 Radical Equations

A radical equation is an equation in which the variable occurs in a radicand.

Solving Radical Equations Containing *n*th Roots

1. Isolate one radical on one side of the equation.
2. Raise both sides to the *n*th power.
3. Solve the resulting equation.
4. Check proposed solutions in the original equation. Solutions of an equation to an even power that is radical-free, but not the original equation, are called extraneous solutions.

Solve: $\sqrt{6x + 13} - 2x = 1$.

$\sqrt{6x + 13} = 2x + 1$ Isolate the radical.

$(\sqrt{6x + 13})^2 = (2x + 1)^2$ Square both sides.

$6x + 13 = 4x^2 + 4x + 1$

$0 = 4x^2 - 2x - 12$ Subtract $6x + 13$ from both sides.

$0 = 2(2x^2 - x - 6)$ Factor out the GCF.

$0 = 2(2x + 3)(x - 2)$ Factor completely.

$2x + 3 = 0$ or $x - 2 = 0$ Set variable factors equal to zero.

$2x = -3$ $x = 2$ Solve for x.

$x = -\dfrac{3}{2}$

Check both proposed solutions. 2 checks, but $-\dfrac{3}{2}$ is extraneous.

The solution is 2 and the solution set is $\{2\}$.

Section 10.7 Complex Numbers

The imaginary unit *i* is defined as

$$i = \sqrt{-1}, \quad \text{where} \quad i^2 = -1.$$

The set of numbers in the form $a + bi$ is called the set of complex numbers; *a* is the real part and *b* is the imaginary part. If $b = 0$, the complex number is a real number. If $b \neq 0$, the complex number is an imaginary number.

• $\sqrt{-81} = \sqrt{81(-1)} = \sqrt{81}\sqrt{-1} = 9i$

• $\sqrt{-75} = \sqrt{75(-1)} = \sqrt{25 \cdot 3}\sqrt{-1} = 5\sqrt{3}i$

To add or subtract complex numbers, add or subtract their real parts and add or subtract their imaginary parts.

$(2 - 4i) - (7 - 10i)$

$= 2 - 4i - 7 + 10i$

$= (2 - 7) + (-4 + 10)i = -5 + 6i$

To multiply complex numbers, multiply as if they are polynomials. After completing the multiplication, replace i^2 with -1. When performing operations with square roots of negative numbers, begin by expressing all square roots in terms of *i*. Then multiply.

F   O   I   L

• $(2 - 3i)(4 + 5i) = 8 + 10i - 12i - 15i^2$

$= 8 + 10i - 12i - 15(-1)$

$= 23 - 2i$

• $\sqrt{-36} \cdot \sqrt{-100} = \sqrt{36(-1)} \cdot \sqrt{100(-1)}$

$= 6i \cdot 10i = 60i^2 = 60(-1) = -60$

Definitions and Concepts

Examples

Section 10.7 Complex Numbers (continued)

Definitions and Concepts	Examples
The complex numbers $a + bi$ and $a - bi$ are conjugates. Conjugates can be multiplied using the formula $$(A + B)(A - B) = A^2 - B^2.$$ The multiplication of conjugates results in a real number.	$(3 + 5i)(3 - 5i)$ $= 3^2 - (5i)^2$ $= 9 - 25i^2$ $= 9 - 25(-1) = 34$
To divide complex numbers, multiply the numerator and the denominator by the conjugate of the denominator.	$\dfrac{5 + 2i}{4 - i} = \dfrac{5 + 2i}{4 - i} \cdot \dfrac{4 + i}{4 + i} = \dfrac{20 + 5i + 8i + 2i^2}{16 - i^2}$ $= \dfrac{20 + 13i + 2(-1)}{16 - (-1)}$ $= \dfrac{20 + 13i - 2}{16 + 1}$ $= \dfrac{18 + 13i}{17} = \dfrac{18}{17} + \dfrac{13}{17}i$
To simplify powers of i, rewrite the expression in terms of i^2. Then replace i^2 with -1 and simplify.	Simplify: i^{27}. $i^{27} = i^{26} \cdot i = (i^2)^{13}i$ $= (-1)^{13}i = (-1)i = -i$

CHAPTER 10 REVIEW EXERCISES

10.1 *In Exercises 1–5, find the indicated root, or state that the expression is not a real number.*

1. $\sqrt{81}$ 9

2. $-\sqrt{\dfrac{1}{100}}$ $-\dfrac{1}{10}$

3. $\sqrt[3]{-27}$ -3

4. $\sqrt[4]{-16}$ not a real number

5. $\sqrt[5]{-32}$ -2

In Exercises 6–7, find the indicated function values for each function. If necessary, round to two decimal places. If the function value is not a real number and does not exist, so state.

6. $f(x) = \sqrt{2x - 5}$; $f(15), f(4), f\left(\dfrac{5}{2}\right), f(1)$

   5; 1.73; 0; not a real number

7. $g(x) = \sqrt[3]{4x - 8}$; $g(4), g(0), g(-14)$ 2; -2; -4

In Exercises 8–9, find the domain of each square root function.

8. $f(x) = \sqrt{x - 2}$ $\{x \mid x \geq 2\}$ or $[2, \infty)$

9. $g(x) = \sqrt{100 - 4x}$ $\{x \mid x \leq 25\}$ or $(-\infty, 25]$

In Exercises 10–15, simplify each expression. Assume that each variable can represent any real number, so include absolute value bars where necessary.

10. $\sqrt{25x^2}$ $5|x|$

11. $\sqrt{(x + 14)^2}$ $|x + 14|$

12. $\sqrt{x^2 - 8x + 16}$ $|x - 4|$ **13.** $\sqrt[3]{64x^3}$ $4x$

14. $\sqrt[4]{16x^4}$ $2|x|$ **15.** $\sqrt[5]{-32(x + 7)^5}$ $-2(x + 7)$

10.2 *In Exercises 16–18, use radical notation to rewrite each expression. Simplify, if possible.*

16. $(5xy)^{\frac{1}{3}}$ $\sqrt[3]{5xy}$ **17.** $16^{\frac{3}{2}}$ $\left(\sqrt{16}\right)^3 = 64$

18. $32^{\frac{4}{5}}$ $\left(\sqrt[5]{32}\right)^4 = 16$

In Exercises 19–20, rewrite each expression with rational exponents.

19. $\sqrt{7x}$ $(7x)^{\frac{1}{2}}$ **20.** $\left(\sqrt[3]{19xy}\right)^5$ $(19xy)^{\frac{5}{3}}$

In Exercises 21–22, rewrite each expression with a positive rational exponent. Simplify, if possible.

21. $8^{-\frac{2}{3}}$ $\dfrac{1}{8^{\frac{2}{3}}} = \dfrac{1}{4}$ **22.** $3x(ab)^{-\frac{4}{5}}$ $\dfrac{3x}{(ab)^{\frac{4}{5}}} = \dfrac{3x}{\sqrt[5]{(ab)^4}}$

In Exercises 23–26, use properties of rational exponents to simplify each expression.

23. $x^{\frac{1}{3}} \cdot x^{\frac{1}{4}}$ $x^{\frac{7}{12}}$ **24.** $\dfrac{5^{\frac{1}{2}}}{5^{\frac{1}{3}}}$ $5^{\frac{1}{6}}$

25. $(8x^6y^3)^{\frac{1}{3}}$ $2x^2y$

26. $\left(x^{-\frac{2}{3}}y^{\frac{1}{4}}\right)^{\frac{1}{2}}$ $\dfrac{y^{\frac{1}{8}}}{x^{\frac{1}{3}}}$

In Exercises 27–31, use rational exponents to simplify each expression. If rational exponents appear after simplifying, write the answer in radical notation.

27. $\sqrt[3]{x^9y^{12}}$ x^3y^4

28. $\sqrt[9]{x^3y^9}$ $y\sqrt[3]{x}$

29. $\sqrt{x}\cdot\sqrt[3]{x}$ $\sqrt[6]{x^5}$

30. $\dfrac{\sqrt[3]{x^2}}{\sqrt[4]{x^2}}$ $\sqrt[6]{x}$

31. $\sqrt[5]{\sqrt[3]{x}}$ $\sqrt[15]{x}$

32. The function $f(x) = 350x^{\frac{2}{3}}$ models the expenditures, $f(x)$, in millions of dollars, for the U.S. National Park Service x years after 1985. According to this model, what will expenditures be in 2012?   Expenditures will be $3150 million in 2012.

10.3 *In Exercises 33–35, use the product rule to multiply.*

33. $\sqrt{3x}\cdot\sqrt{7y}$ $\sqrt{21xy}$

34. $\sqrt[5]{7x^2}\cdot\sqrt[5]{11x}$ $\sqrt[5]{77x^3}$

35. $\sqrt[6]{x-5}\cdot\sqrt[6]{(x-5)^4}$ $\sqrt[6]{(x-5)^5}$

36. If $f(x) = \sqrt{7x^2 - 14x + 7}$, express the function, f, in simplified form. Assume that x can be any real number. $f(x) = \sqrt{7}|x-1|$

In Exercises 37–39, simplify by factoring. Assume that all variables in a radicand represent positive real numbers.

37. $\sqrt{20x^3}$ $2x\sqrt{5x}$

38. $\sqrt[3]{54x^8y^6}$ $3x^2y^2\sqrt[3]{2x^2}$

39. $\sqrt[4]{32x^3y^{11}z^5}$ $2y^2z\sqrt[4]{2x^3y^3z}$

In Exercises 40–43, multiply and simplify, if possible. Assume that all variables in a radicand represent positive real numbers.

40. $\sqrt{6x^3}\cdot\sqrt{4x^2}$ $2x^2\sqrt{6x}$

41. $\sqrt[3]{4x^2y}\cdot\sqrt[3]{4xy^4}$ $2xy\sqrt[3]{2y^2}$

42. $\sqrt[5]{2x^4y^3z^4}\cdot\sqrt[5]{8xy^6z^7}$ $xyz^2\sqrt[5]{16y^4z}$

43. $\sqrt{x+1}\cdot\sqrt{x-1}$ $\sqrt{x^2-1}$

10.4 *Assume that all variables represent positive real numbers.*

In Exercises 44–47, add or subtract as indicated.

44. $6\sqrt[3]{3} + 2\sqrt[3]{3}$ $8\sqrt[3]{3}$

45. $5\sqrt{18} - 3\sqrt{8}$ $9\sqrt{2}$

46. $\sqrt[3]{27x^4} + \sqrt[3]{xy^6}$ $(3x + y^2)\sqrt[3]{x}$

47. $2\sqrt[3]{6} - 5\sqrt[3]{48}$ $-8\sqrt[3]{6}$

In Exercises 48–50, simplify using the quotient rule.

48. $\sqrt[3]{\dfrac{16}{125}}$ $\dfrac{2}{5}\sqrt[3]{2}$

49. $\sqrt{\dfrac{x^3}{100y^4}}$ $\dfrac{x\sqrt{x}}{10y^2}$

50. $\sqrt[4]{\dfrac{3y^5}{16x^{20}}}$ $\dfrac{y\sqrt[4]{3y}}{2x^5}$

In Exercises 51–54, divide and, if possible, simplify.

51. $\dfrac{\sqrt{48}}{\sqrt{2}}$ $2\sqrt{6}$

52. $\dfrac{\sqrt[3]{32}}{\sqrt[3]{2}}$ $2\sqrt[3]{2}$

53. $\dfrac{\sqrt[4]{64x^7}}{\sqrt[4]{2x^2}}$ $2x\sqrt[4]{2x}$

54. $\dfrac{\sqrt{200x^3y^2}}{\sqrt{2x^{-2}y}}$ $10x^2\sqrt{xy}$

10.5 *Assume that all variables represent positive real numbers.*

In Exercises 55–62, multiply as indicated. If possible, simplify any radical expressions that appear in the product.

55. $\sqrt{3}\left(2\sqrt{6} + 4\sqrt{15}\right)$ $6\sqrt{2} + 12\sqrt{5}$

56. $\sqrt[3]{5}\left(\sqrt[3]{50} - \sqrt[3]{2}\right)$ $5\sqrt[3]{2} - \sqrt[3]{10}$

57. $\left(\sqrt{7} - 3\sqrt{5}\right)\left(\sqrt{7} + 6\sqrt{5}\right)$ $-83 + 3\sqrt{35}$

58. $\left(\sqrt{x} - \sqrt{11}\right)\left(\sqrt{y} - \sqrt{11}\right)$ $\sqrt{xy} - \sqrt{11x} - \sqrt{11y} + 11$

59. $\left(\sqrt{5} + \sqrt{8}\right)^2$ $13 + 4\sqrt{10}$

60. $\left(2\sqrt{3} - \sqrt{10}\right)^2$ $22 - 4\sqrt{30}$

61. $\left(\sqrt{7} + \sqrt{13}\right)\left(\sqrt{7} - \sqrt{13}\right)$ -6

62. $\left(7 - 3\sqrt{5}\right)\left(7 + 3\sqrt{5}\right)$ 4

In Exercises 63–75, rationalize each denominator. Simplify, if possible.

63. $\dfrac{4}{\sqrt{6}}$ $\dfrac{2\sqrt{6}}{3}$

64. $\sqrt{\dfrac{2}{7}}$ $\dfrac{\sqrt{14}}{7}$

65. $\dfrac{12}{\sqrt[3]{9}}$ $4\sqrt[3]{3}$

66. $\sqrt{\dfrac{2x}{5y}}$ $\dfrac{\sqrt{10xy}}{5y}$

67. $\dfrac{14}{\sqrt[3]{2x^2}}$ $\dfrac{7\sqrt[3]{4x}}{x}$

68. $\sqrt[4]{\dfrac{7}{3x}}$ $\dfrac{\sqrt[4]{189x^3}}{3x}$

69. $\dfrac{5}{\sqrt[5]{32x^4y}}$ $\dfrac{5\sqrt[5]{xy^4}}{2xy}$

70. $\dfrac{6}{\sqrt{3} - 1}$ $3\sqrt{3} + 3$

71. $\dfrac{\sqrt{7}}{\sqrt{5} + \sqrt{3}}$ $\dfrac{\sqrt{35} - \sqrt{21}}{2}$

72. $\dfrac{10}{2\sqrt{5} - 3\sqrt{2}}$ $10\sqrt{5} + 15\sqrt{2}$

73. $\dfrac{\sqrt{x} + 5}{\sqrt{x} - 3}$ $\dfrac{x + 8\sqrt{x} + 15}{x - 9}$

74. $\dfrac{\sqrt{7} + \sqrt{3}}{\sqrt{7} - \sqrt{3}}$ $\dfrac{5 + \sqrt{21}}{2}$

75. $\dfrac{2\sqrt{3} + \sqrt{6}}{2\sqrt{6} + \sqrt{3}}$ $\dfrac{3\sqrt{2} + 2}{7}$

In Exercises 76–79, rationalize each numerator. Simplify, if possible.

76. $\sqrt{\dfrac{2}{7}}$ $\dfrac{2}{\sqrt{14}}$

77. $\dfrac{\sqrt[3]{3x}}{\sqrt[3]{y}}$ $\dfrac{3x}{\sqrt[3]{9x^2y}}$

78. $\dfrac{\sqrt{7}}{\sqrt{5} + \sqrt{3}}$ $\dfrac{7}{\sqrt{35} + \sqrt{21}}$

79. $\dfrac{\sqrt{7} + \sqrt{3}}{\sqrt{7} - \sqrt{3}}$ $\dfrac{2}{5 - \sqrt{21}}$

10.6 *In Exercises 80–84, solve each radical equation.*

80. $\sqrt{2x + 4} = 6$ 16 or {16}

81. $\sqrt{x - 5} + 9 = 4$ no solution or Ø

82. $\sqrt{2x - 3} + x = 3$ 2 or {2}

83. $\sqrt{x - 4} + \sqrt{x + 1} = 5$ 8 or {8}

84. $(x^2 + 6x)^{\frac{1}{3}} + 2 = 0$ −4 and −2, or {−4, −2}

85. The time, $f(x)$, in seconds, for a free-falling object to fall x feet is modeled by the function

$$f(x) = \sqrt{\frac{x}{16}}.$$

If a worker accidently drops a hammer from a building and it hits the ground after 4 seconds, from what height was the hammer dropped?   The hammer was dropped from 256 feet.

86. Out of a group of 50,000 births, the number of people, $f(x)$, surviving to age x is modeled by the function

$$f(x) = 5000\sqrt{100 - x}.$$

To what age will 20,000 people in the group survive?   20,000 people in the group will survive to 84 years old.

10.7 *In Exercises 87–89, express each number in terms of i and simplify, if possible.*

87. $\sqrt{-81}$ 9i

88. $\sqrt{-63}$ $3\sqrt{7}i$

89. $-\sqrt{-8}$ $-2\sqrt{2}i$

In Exercises 90–99, perform the indicated operation. Write the result in the form a + bi.

90. $(7 + 12i) + (5 - 10i)$ 12 + 2i

91. $(8 - 3i) - (17 - 7i)$ −9 + 4i

92. $4i(3i - 2)$ −12 − 8i

93. $(7 - 5i)(2 + 3i)$ 29 + 11i

94. $(3 - 4i)^2$ −7 − 24i

95. $(7 + 8i)(7 - 8i)$ 113 + 0i or 113

96. $\sqrt{-8} \cdot \sqrt{-3}$ $-2\sqrt{6} + 0i$ or $-2\sqrt{6}$

97. $\dfrac{6}{5 + i}$ $\dfrac{15}{13} - \dfrac{3}{13}i$

98. $\dfrac{3 + 4i}{4 - 2i}$ $\dfrac{1}{5} + \dfrac{11}{10}i$

99. $\dfrac{5 + i}{3i}$ $\dfrac{1}{3} - \dfrac{5}{3}i$

In Exercises 100–101, simplify each expression.

100. i^{16} 1

101. i^{23} −i

CHAPTER 10 TEST   Remember to use your Chapter Test Prep Video CD to see the worked-out solutions to the test questions you want to review.

1. Let $f(x) = \sqrt{8 - 2x}$.
  a. Find $f(-14)$.   6
  b. Find the domain of f. $\{x \mid x \le 4\}$ or $(-\infty, 4]$

2. Evaluate: $27^{-\frac{4}{3}}$. $\dfrac{1}{81}$

3. Simplify: $(25x^{-\frac{1}{2}}y^{\frac{1}{4}})^{\frac{1}{2}}$. $\dfrac{5\sqrt[8]{y}}{\sqrt[4]{x}}$

In Exercises 4–5, use rational exponents to simplify each expression. If rational exponents appear after simplifying, write the answer in radical notation.

4. $\sqrt[8]{x^4}$ $\sqrt{x}$

5. $\sqrt[4]{x} \cdot \sqrt[5]{x}$ $\sqrt[20]{x^9}$

In Exercises 6–9, simplify each expression. Assume that each variable can represent any real number.

6. $\sqrt{75x^2}$ $5|x|\sqrt{3}$

7. $\sqrt{x^2 - 10x + 25}$ $|x - 5|$

8. $\sqrt[3]{16x^4y^8}$ $2xy^2\sqrt[3]{2xy^2}$

9. $\sqrt[5]{-\dfrac{32}{x^{10}}}$ $-\dfrac{2}{x^2}$

In Exercises 10–17, perform the indicated operation and, if possible, simplify. Assume that all variables represent positive real numbers.

10. $\sqrt[3]{5x^2} \cdot \sqrt[3]{10y}$ $\sqrt[3]{50x^2y}$

11. $\sqrt[4]{8x^3y} \cdot \sqrt[4]{4xy^2}$ $2x\sqrt[4]{2y^3}$

12. $3\sqrt{18} - 4\sqrt{32}$ $-7\sqrt{2}$

13. $\sqrt[3]{8x^4} + \sqrt[3]{xy^6}$ $(2x + y^2)\sqrt[3]{x}$

14. $\dfrac{\sqrt[3]{16x^8}}{\sqrt[3]{2x^4}}$ $2x\sqrt[3]{x}$

15. $\sqrt{3}(4\sqrt{6} - \sqrt{5})$ $12\sqrt{2} - \sqrt{15}$

16. $(5\sqrt{6} - 2\sqrt{2})(\sqrt{6} + \sqrt{2})$ $26 + 6\sqrt{3}$

17. $(7 - \sqrt{3})^2$ $52 - 14\sqrt{3}$

In Exercises 18–20, rationalize each denominator. Simplify, if possible. Assume all variables represent positive real numbers.

18. $\sqrt{\dfrac{5}{x}}$ $\dfrac{\sqrt{5x}}{x}$

19. $\dfrac{5}{\sqrt[3]{5x^2}}$ $\dfrac{\sqrt[3]{25x}}{x}$

20. $\dfrac{\sqrt{2} - \sqrt{3}}{\sqrt{2} + \sqrt{3}}$ $-5 + 2\sqrt{6}$

In Exercises 21–23, solve each radical equation.

21. $3 + \sqrt{2x - 3} = x$ 6 or {6}

22. $\sqrt{x + 9} - \sqrt{x - 7} = 2$ 16 or {16}

23. $(11x + 6)^{\frac{1}{3}} + 3 = 0$ −3 or {−3}

24. The function

$$f(x) = 2.9\sqrt{x} + 20.1$$

models the average height, $f(x)$, in inches, of boys who are x months of age, $0 \le x \le 60$. Find the age at which the average height of boys is 40.4 inches.   49 months

25. Express in terms of i and simplify: $\sqrt{-75}$. $5\sqrt{3}i$

In Exercises 26–29, perform the indicated operation. Write the result in the form $a + bi$.

26. $(5 - 3i) - (6 - 9i)$ −1 + 6i

27. $(3 - 4i)(2 + 5i)$ 26 + 7i

28. $\sqrt{-9} \cdot \sqrt{-4}$ −6 + 0i or −6

29. $\dfrac{3 + i}{1 - 2i}$ $\dfrac{1}{5} + \dfrac{7}{5}i$

30. Simplify: i^{35}.   −i

Note to Instructor: The Chapter 10 Test file in the TestGen program provides algorithms specifically matched to these problems so you can easily replicate this chapter test for additional practice or assessment purposes.

CUMULATIVE REVIEW EXERCISES (CHAPTERS 1–10)

In Exercises 1–5, solve each equation, inequality, or system.

1. $2x - y + z = -5$

$x - 2y - 3z = 6$

$x + y - 2z = 1$ {(−2,−1,−2)}

2. $3x^2 - 11x = 4$ $\left\{-\frac{1}{3}, 4\right\}$

3. $2(x + 4) < 5x + 3(x + 2)$ $\left\{x \mid x > \frac{1}{3}\right\}$ or $\left(\frac{1}{3}, \infty\right)$

4. $\dfrac{1}{x + 2} + \dfrac{15}{x^2 - 4} = \dfrac{5}{x - 2}$ $\left\{\dfrac{3}{4}\right\}$

5. $\sqrt{x + 2} - \sqrt{x + 1} = 1$ {−1}

6. Graph the solution set of the system:

$$x + 2y < 2$$
$$2y - x > 4.$$

See graphing answer section.

In Exercises 7–15, perform the indicated operations.

7. $\dfrac{8x^2}{3x^2 - 12} \div \dfrac{40}{x - 2}$ $\dfrac{x^2}{15(x + 2)}$

8. $\dfrac{x + \dfrac{1}{y}}{y + \dfrac{1}{x}}$ $\dfrac{x}{y}$

9. $(2x - 3)(4x^2 - 5x - 2)$ $8x^3 - 22x^2 + 11x + 6$

10. $\dfrac{7x}{x^2 - 2x - 15} - \dfrac{2}{x - 5}$ $\dfrac{5x - 6}{(x - 5)(x + 3)}$

11. $7(8 - 10)^3 - 7 + 3 \div (-3)$ −64

12. $\sqrt{80x} - 5\sqrt{20x} + 2\sqrt{45x}$ 0

13. $\dfrac{\sqrt{3} - 2}{2\sqrt{3} + 5}$ $\dfrac{16 - 9\sqrt{3}}{13}$

14. $(2x^3 - 3x^2 + 3x - 4) \div (x - 2)$ $2x^2 + x + 5 + \dfrac{6}{x - 2}$

15. $\left(2\sqrt{3} + 5\sqrt{2}\right)\left(\sqrt{3} - 4\sqrt{2}\right)$ $-34 - 3\sqrt{6}$

In Exercises 16–17, factor completely.

16. $24x^2 + 10x - 4$ $2(3x + 2)(4x - 1)$

17. $16x^4 - 1$ $(4x^2 + 1)(2x + 1)(2x - 1)$

18. The amount of light provided by a light bulb varies inversely as the square of the distance from the bulb. The illumination provided is 120 lumens at a distance of 10 feet. How many lumens are provided at a distance of 15 feet?   About 53 lumens are provided at a distance of 15 feet.

19. You invested $6000 in two accounts paying 7% and 9% annual interest, respectively. At the end of the year, the total interest from these investments was $510. How much was invested at each rate? $1500 was invested at 7% and $4500 at 9%.

20. Although there are 2332 students enrolled in the college, this is 12% fewer students than there were enrolled last year. How many students were enrolled last year?   2650 students were enrolled last year.

CHAPTER

Quadratic Equations and Functions

11.1 The Square Root Property and Completing the Square; Distance and Midpoint Formulas

11.2 The Quadratic Formula

11.3 Quadratic Functions and Their Graphs

11.4 Equations Quadratic in Form

11.5 Polynomial and Rational Inequalities

Because we tend to take our food abundance for granted, it might surprise you to know that the number of U.S. farms has declined since the 1920s, as individually owned family farms have been swallowed up by huge agribusinesses owned by corporations. Currently, there are about 1.8 acres of cropland to grow food for each American. If the current trends in population growth and loss of farmland continue, there will be only 0.6 acre per American in the year 2050. This is the rate that currently exists worldwide.

Like the number of farms and the amount of cropland in the United States, many phenomena follow trends that involve growth and decline, or vice versa. In this chapter, you will learn to solve equations and graph functions that provide new ways of looking at and understanding these phenomena.

THE SQUARE ROOT PROPERTY AND COMPLETING THE SQUARE; DISTANCE AND MIDPOINT FORMULAS

Objectives

1 Solve quadratic equations using the square root property.

2 Complete the square of a binomial.

3 Solve quadratic equations by completing the square.

4 Solve problems using the square root property.

5 Find the distance between two points.

6 Find the midpoint of a line segment.

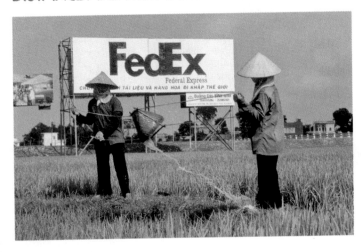

"In the future there will be two kinds of corporations; those that go global, and those that go bankrupt."

C. Michael Armstrong, CEO, AT&T

For better or worse, ours is the era of the multinational corporation. New technology that the multinational corporations control is expanding their power. And their numbers are growing fast. There were approximately 300 multinationals in 1900. By 1970, there were close to 7000, and by 1990 the number had swelled to 30,000. In 2001, more than 65,000 multinational corporations enveloped the world.

In this section, you will learn two new methods for solving quadratic equations. These methods are called the *square root method* and *completing the square*. Using these techniques, you will explore the growth of multinational corporations and make predictions about the number of global corporations in the future.

STUDY TIP

Here is a summary of what we already know about quadratic equations and quadratic functions.

1. A **quadratic equation** in x can be written in the standard form

$$ax^2 + bx + c = 0, \quad a \neq 0.$$

2. Some quadratic equations can be solved by factoring.

Solve:
$$2x^2 + 7x - 4 = 0.$$
$$(2x - 1)(x + 4) = 0$$
$$2x - 1 = 0 \quad \text{or} \quad x + 4 = 0$$
$$2x = 1 \qquad\qquad x = -4$$
$$x = \tfrac{1}{2}$$

The solutions are -4 and $\tfrac{1}{2}$, and the solution set is $\left\{-4, \tfrac{1}{2}\right\}$.

| x-intercept is −4. |
| x-intercept is $\frac{1}{2}$. |
| $f(x) = 2x^2 + 7x - 4$ |

FIGURE 11.1

3. A polynomial function of the form

$$f(x) = ax^2 + bx + c, \quad a \neq 0$$

is a **quadratic function**. Graphs of quadratic functions have cuplike shapes with the same behavior at each end.

4. The real solutions of $ax^2 + bx + c = 0$ correspond to the x-intercepts for the graph of the quadratic function $f(x) = ax^2 + bx + c$. For example, the solutions of the equation $2x^2 + 7x - 4 = 0$ are -4 and $\tfrac{1}{2}$. Figure 11.1 shows that the solutions appear as x-intercepts on the graph of the quadratic function $f(x) = 2x^2 + 7x - 4$.

Now that we've summarized what we know, let's look at where we go. How do we solve a quadratic equation, $ax^2 + bx + c = 0$, if the trinomial $ax^2 + bx + c$ cannot be factored? Methods other than factoring are needed. In this section, we look at other ways of solving quadratic equations.

1 Solve quadratic equations using the square root property.

The Square Root Property Let's begin with a relatively simple quadratic equation:

$$x^2 = 9.$$

The value of x must be a number whose square is 9. There are two numbers whose square is 9:

$$x = \sqrt{9} = 3 \quad \text{or} \quad x = -\sqrt{9} = -3.$$

Thus, the solutions of $x^2 = 9$ are 3 and -3. This is an example of the **square root property**.

DISCOVER FOR YOURSELF

Solve $x^2 = 9$, or

$$x^2 - 9 = 0,$$

by factoring. What is the advantage of using the square root property?

> **THE SQUARE ROOT PROPERTY** If u is an algebraic expression and d is a nonzero real number, then $u^2 = d$ has exactly two solutions:
>
> $$\text{If } u^2 = d, \quad \text{then } u = \sqrt{d} \text{ or } u = -\sqrt{d}.$$
>
> Equivalently,
>
> $$\text{If } u^2 = d, \quad \text{then } u = \pm\sqrt{d}.$$

Notice that $u = \pm\sqrt{d}$ is a shorthand notation to indicate that $u = \sqrt{d}$ or $u = -\sqrt{d}$. Although we usually read $u = \pm\sqrt{d}$ as "u equals plus or minus the square root of d," we actually mean that u is the positive square root of d or the negative square root of d.

EXAMPLE 1 Solving a Quadratic Equation by the Square Root Property

Solve: $3x^2 = 18$.

SOLUTION To apply the square root property, we need a squared expression by itself on one side of the equation.

$$3x^2 = 18$$

We want x^2 by itself.

We can get x^2 by itself if we divide both sides by 3.

$$3x^2 = 18 \qquad \text{This is the original equation.}$$

$$\frac{3x^2}{3} = \frac{18}{3} \qquad \text{Divide both sides by 3.}$$

$$x^2 = 6 \qquad \text{Simplify.}$$

$$x = \sqrt{6} \quad \text{or} \quad x = -\sqrt{6} \qquad \text{Apply the square root property.}$$

Now let's check these proposed solutions in the original equation.

$$\textbf{Check } \sqrt{6}: \qquad\qquad \textbf{Check } -\sqrt{6}:$$

$$3x^2 = 18 \qquad\qquad 3x^2 = 18$$

$$3\left(\sqrt{6}\right)^2 \overset{?}{=} 18 \qquad\qquad 3\left(-\sqrt{6}\right)^2 \overset{?}{=} 18$$

$$3\cdot 6 \overset{?}{=} 18 \qquad\qquad 3\cdot 6 \overset{?}{=} 18$$

$$18 = 18, \quad\text{true} \qquad\qquad 18 = 18, \quad\text{true}$$

The solutions are $-\sqrt{6}$ and $\sqrt{6}$. The solution set is $\left\{-\sqrt{6}, \sqrt{6}\right\}$ or $\left\{\pm\sqrt{6}\right\}$. ■

✔ **CHECK POINT 1** Solve: $4x^2 = 28$. $-\sqrt{7}$ and $\sqrt{7}$, or $\left\{\pm\sqrt{7}\right\}$

In this section, we will express irrational solutions in simplified radical form, rationalizing denominators when possible.

EXAMPLE 2 Solving a Quadratic Equation by the Square Root Property

Solve: $2x^2 - 7 = 0$.

SOLUTION To solve by the square root property, we isolate the squared expression on one side of the equation.

$$2x^2 - 7 = 0$$

We want x^2 by itself.

$$2x^2 - 7 = 0 \qquad\text{This is the original equation.}$$

$$2x^2 = 7 \qquad\text{Add 7 to both sides.}$$

$$x^2 = \frac{7}{2} \qquad\text{Divide both sides by 2.}$$

$$x = \sqrt{\frac{7}{2}} \quad\text{or}\quad x = -\sqrt{\frac{7}{2}} \qquad\text{Apply the square root property.}$$

Because the proposed solutions are opposites, we can rationalize both denominators at once:

$$\pm\sqrt{\frac{7}{2}} = \pm\frac{\sqrt{7}}{\sqrt{2}}\cdot\frac{\sqrt{2}}{\sqrt{2}} = \pm\frac{\sqrt{14}}{2}.$$

Substitute these values into the original equation and verify that the solutions are $-\dfrac{\sqrt{14}}{2}$ and $\dfrac{\sqrt{14}}{2}$. The solution set is $\left\{-\dfrac{\sqrt{14}}{2}, \dfrac{\sqrt{14}}{2}\right\}$ or $\left\{\pm\dfrac{\sqrt{14}}{2}\right\}$. ■

✔ **CHECK POINT 2** Solve: $3x^2 - 11 = 0$. $\pm\frac{\sqrt{33}}{3}$ or $\left\{\pm\frac{\sqrt{33}}{3}\right\}$

Some quadratic equations have solutions that are imaginary numbers.

EXAMPLE 3 Solving a Quadratic Equation by the Square Root Property

Solve: $9x^2 + 25 = 0$.

SOLUTION

$$9x^2 + 25 = 0$$

| We need to isolate x^2. |

This is the original equation.

$$9x^2 = -25$$

Subtract 25 from both sides.

$$x^2 = -\frac{25}{9}$$

Divide both sides by 9.

$$x = \sqrt{-\frac{25}{9}} \quad \text{or} \quad x = -\sqrt{-\frac{25}{9}}$$

Apply the square root property.

$$x = \sqrt{\frac{25}{9}}\sqrt{-1} \qquad x = -\sqrt{\frac{25}{9}}\sqrt{-1}$$

$$x = \frac{5}{3}i \qquad\qquad x = -\frac{5}{3}i \qquad\qquad \sqrt{-1} = i$$

Because the equation has an x^2-term and no x-term, we can check both proposed solutions, $\pm\frac{5}{3}i$, at once.

USING TECHNOLOGY

The graph of

$$f(x) = 9x^2 + 25$$

has no x-intercepts. This shows that

$$9x^2 + 25 = 0$$

has no real solutions. Example 3 algebraically establishes that the solutions are imaginary numbers.

[−3, 3, 1] by [−5, 100, 5]

Check $\frac{5}{3}i$ and $-\frac{5}{3}i$:

$$9x^2 + 25 = 0$$

$$9\left(\pm\frac{5}{3}i\right)^2 + 25 \stackrel{?}{=} 0$$

$$9\left(\frac{25}{9}i^2\right) + 25 \stackrel{?}{=} 0$$

$$25i^2 + 25 \stackrel{?}{=} 0$$

| $i^2 = -1$ |

$$25(-1) + 25 \stackrel{?}{=} 0$$

$$0 = 0, \quad \text{true}$$

The solutions are $-\frac{5}{3}i$ and $\frac{5}{3}i$. The solution set is $\left\{-\frac{5}{3}i, \frac{5}{3}i\right\}$ or $\left\{\pm\frac{5}{3}i\right\}$. ■

 CHECK POINT 3 Solve: $4x^2 + 9 = 0$. $\pm\frac{3}{2}i$ or $\left\{\pm\frac{3}{2}i\right\}$

Can we solve an equation such as $(x - 1)^2 = 5$ using the square root property? Yes. The equation is in the form $u^2 = d$, where u^2, the squared expression, is by itself on the left side.

$$(x - 1)^2 \quad = \quad 5$$

| This is u^2 in $u^2 = d$ with $u = x - 1$. | | This is d in $u^2 = d$ with $d = 5$. |

Try solving

$$(x - 1)^2 = 5$$

by writing the equation in standard form and factoring. What problem do you encounter?

EXAMPLE 4 Solving a Quadratic Equation by the Square Root Property

Solve by the square root property: $(x - 1)^2 = 5$.

SOLUTION

$$(x - 1)^2 = 5 \qquad \text{This is the original equation.}$$

$$x - 1 = \sqrt{5} \quad \text{or} \quad x - 1 = -\sqrt{5} \qquad \text{Apply the square root property.}$$

$$x = 1 + \sqrt{5} \qquad\qquad x = 1 - \sqrt{5} \qquad \text{Add 1 to both sides in each equation.}$$

Check $1 + \sqrt{5}$:

$$(x - 1)^2 = 5$$

$$\left(1 + \sqrt{5} - 1\right)^2 \overset{?}{=} 5$$

$$\left(\sqrt{5}\right)^2 \overset{?}{=} 5$$

$$5 = 5, \ \text{true}$$

Check $1 - \sqrt{5}$:

$$(x - 1)^2 = 5$$

$$\left(1 - \sqrt{5} - 1\right)^2 \overset{?}{=} 5$$

$$\left(-\sqrt{5}\right)^2 \overset{?}{=} 5$$

$$5 = 5, \ \text{true}$$

The solutions are $1 \pm \sqrt{5}$, and the solution set is $\left\{1 + \sqrt{5}, 1 - \sqrt{5}\right\}$ or $\left\{1 \pm \sqrt{5}\right\}$. ■

✔ **CHECK POINT 4** Solve: $(x - 3)^2 = 10$. $\quad 3 \pm \sqrt{10} \quad$ or $\quad \left\{3 \pm \sqrt{10}\right\}$

Completing the Square We return to the question that opened this section: How do we solve a quadratic equation, $ax^2 + bx + c = 0$, if the trinomial $ax^2 + bx + c$ cannot be factored? We can convert the equation into an equivalent equation that can be solved using the square root property. This is accomplished by **completing the square**.

2 Complete the square of a binomial.

COMPLETING THE SQUARE If $x^2 + bx$ is a binomial, then by adding $\left(\dfrac{b}{2}\right)^2$, which is the square of half the coefficient of x, a perfect square trinomial will result. That is,

$$x^2 + bx + \left(\frac{b}{2}\right)^2 = \left(x + \frac{b}{2}\right)^2.$$

EXAMPLE 5 Completing the Square

What term should be added to each binomial so that it becomes a perfect square trinomial? Write and factor the trinomial.

a. $x^2 + 8x$ **b.** $x^2 - 7x$ **c.** $x^2 + \dfrac{3}{5}x$

SOLUTION To complete the square, we must add a term to each binomial. The term that should be added is the square of half the coefficient of x.

$$x^2 + 8x \qquad\qquad x^2 - 7x \qquad\qquad x^2 + \frac{3}{5}x$$

Add $\left(\frac{8}{2}\right)^2 = 4^2$.
Add 16 to complete the square.

Add $\left(\frac{-7}{2}\right)^2$, or $\frac{49}{4}$, to complete the square.

Add $\left(\frac{1}{2} \cdot \frac{3}{5}\right)^2 = \left(\frac{3}{10}\right)^2$.
Add $\frac{9}{100}$ to complete the square.

Visualizing Completing the Square

This figure, with area $x^2 + 8x$, is not a complete square. The bottom-right corner is missing.

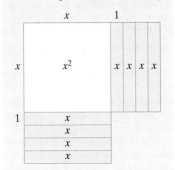

Area: $x^2 + 8x$

Add 16 square units to the missing portion and you, literally, complete the square.

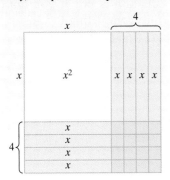

Area: $x^2 + 8x + 16 = (x + 4)^2$

a. The coefficient of the x-term in $x^2 + 8x$ is 8. Half of 8 is 4, and $4^2 = 16$. Add 16. The result is a perfect square trinomial.

$$x^2 + 8x + 16 = (x + 4)^2$$

b. The coefficient of the x-term in $x^2 - 7x$ is -7. Half of -7 is $-\dfrac{7}{2}$, and $\left(-\dfrac{7}{2}\right)^2 = \dfrac{49}{4}$. Add $\dfrac{49}{4}$. The result is a perfect square trinomial.

$$x^2 - 7x + \frac{49}{4} = \left(x - \frac{7}{2}\right)^2$$

c. The coefficient of the x-term in $x^2 + \dfrac{3}{5}x$ is $\dfrac{3}{5}$. Half of $\dfrac{3}{5}$ is $\dfrac{1}{2} \cdot \dfrac{3}{5}$, or $\dfrac{3}{10}$, and $\left(\dfrac{3}{10}\right)^2 = \dfrac{9}{100}$. Add $\dfrac{9}{100}$. The result is a perfect square trinomial.

$$x^2 + \frac{3}{5}x + \frac{9}{100} = \left(x + \frac{3}{10}\right)^2$$

STUDY TIP

You may not be accustomed to factoring perfect square trinomials in which fractions are involved. The constant in the factorization is always half the coefficient of x.

$$x^2 - 7x + \frac{49}{4} = \left(x - \frac{7}{2}\right)^2 \qquad x^2 + \frac{3}{5}x + \frac{9}{100} = \left(x + \frac{3}{10}\right)^2$$

Half the coefficient of x, -7, is $-\frac{7}{2}$.   Half the coefficient of x, $\frac{3}{5}$, is $\frac{3}{10}$.

 CHECK POINT 5 What term should be added to each binomial so that it becomes a perfect square trinomial? Write and factor the trinomial.

a. $x^2 + 10x$ 25; $x^2 + 10x + 25 = (x + 5)^2$

b. $x^2 - 3x$ $\dfrac{9}{4}$; $x^2 - 3x + \dfrac{9}{4} = \left(x - \dfrac{3}{2}\right)^2$

c. $x^2 + \dfrac{3}{4}x$. $\dfrac{9}{64}$; $x^2 + \dfrac{3}{4}x + \dfrac{9}{64} = \left(x + \dfrac{3}{8}\right)^2$

3   Solve quadratic equations by completing the square.

Solving Quadratic Equations by Completing the Square   We can solve any quadratic equation by completing the square. If the coefficient of the x^2-term is one, we add the square of half the coefficient of x to both sides of the equation. **When you add a constant term to one side of the equation to complete the square, be certain to add the same constant to the other side of the equation.** These ideas are illustrated in Example 6.

EXAMPLE 6   Solving a Quadratic Equation by Completing the Square

Solve by completing the square: $x^2 - 6x + 4 = 0$.

SOLUTION   We begin by subtracting 4 from both sides. This is done to isolate the binomial $x^2 - 6x$ so that we can complete the square.

$$x^2 - 6x + 4 = 0 \qquad \text{This is the original equation.}$$
$$x^2 - 6x = -4 \qquad \text{Subtract 4 from both sides.}$$

Next, we work with $x^2 - 6x = -4$ and complete the square. Find half the coefficient of the x-term and square it. The coefficient of the x-term is -6. Half of -6 is -3 and $(-3)^2 = 9$. Thus, we add 9 to both sides of the equation.

$$x^2 - 6x + 9 = -4 + 9 \qquad \text{Add 9 to both sides to complete the square.}$$

$$(x - 3)^2 = 5 \qquad \text{Factor and simplify.}$$

$$x - 3 = \sqrt{5} \quad \text{or} \quad x - 3 = -\sqrt{5} \qquad \text{Apply the square root property.}$$

$$x = 3 + \sqrt{5} \qquad x = 3 - \sqrt{5} \qquad \text{Add 3 to both sides in each equation.}$$

The solutions are $3 \pm \sqrt{5}$, and the solution set is $\{3 + \sqrt{5}, 3 - \sqrt{5}\}$ or $\{3 \pm \sqrt{5}\}$. ∎

If you solve a quadratic equation by completing the square and the solutions are rational numbers, the equation can also be solved by factoring. By contrast, quadratic equations with irrational solutions cannot be solved by factoring. However, all quadratic equations can be solved by completing the square.

✔ **CHECK POINT 6** Solve by completing the square: $x^2 + 4x - 1 = 0$.
$-2 \pm \sqrt{5}$ or $\{-2 \pm \sqrt{5}\}$

If the coefficient of the x^2-term in a quadratic equation is not 1, you must divide each side of the equation by this coefficient before completing the square. For example, to solve $9x^2 - 6x - 4 = 0$ by completing the square, first divide every term by 9:

$$\frac{9x^2}{9} - \frac{6x}{9} - \frac{4}{9} = \frac{0}{9}$$

$$x^2 - \frac{6}{9}x - \frac{4}{9} = 0$$

$$x^2 - \frac{2}{3}x - \frac{4}{9} = 0.$$

Now that the coefficient of the x^2-term is 1, we can solve by completing the square.

 EXAMPLE 7 Solving a Quadratic Equation by Completing the Square

Solve by completing the square: $9x^2 - 6x - 4 = 0$.

SOLUTION

$$9x^2 - 6x - 4 = 0 \qquad \text{This is the original equation.}$$

$$x^2 - \frac{2}{3}x - \frac{4}{9} = 0 \qquad \text{Divide both sides by 9.}$$

$$x^2 - \frac{2}{3}x = \frac{4}{9} \qquad \text{Add } \tfrac{4}{9} \text{ to both sides to isolate the binomial.}$$

$$x^2 - \frac{2}{3}x + \frac{1}{9} = \frac{4}{9} + \frac{1}{9} \qquad \text{Complete the square: Half of } -\tfrac{2}{3} \text{ is } -\tfrac{2}{6}, \text{ or } -\tfrac{1}{3}, \text{ and } \left(-\tfrac{1}{3}\right)^2 = \tfrac{1}{9}.$$

$$\left(x - \frac{1}{3}\right)^2 = \frac{5}{9} \qquad \text{Factor and simplify.}$$

Obtain a decimal approximation for each solution of $9x^2 - 6x - 4 = 0$, the equation in Example 7.

$$\frac{1 + \sqrt{5}}{3} \approx 1.1$$

$$\frac{1 - \sqrt{5}}{3} \approx -0.4$$

$[-2, 2, 1]$ by $[-10, 10, 1]$

The x-intercepts of $f(x) = 9x^2 - 6x - 4$ verify the solutions.

---

4 Solve problems using the square root property.

$$x - \frac{1}{3} = \sqrt{\frac{5}{9}} \quad \text{or} \quad x - \frac{1}{3} = -\sqrt{\frac{5}{9}} \qquad \text{Apply the square root property.}$$

$$x - \frac{1}{3} = \frac{\sqrt{5}}{3} \qquad x - \frac{1}{3} = -\frac{\sqrt{5}}{3} \qquad \sqrt{\frac{5}{9}} = \frac{\sqrt{5}}{\sqrt{9}} = \frac{\sqrt{5}}{3}$$

$$x = \frac{1}{3} + \frac{\sqrt{5}}{3} \qquad x = \frac{1}{3} - \frac{\sqrt{5}}{3} \qquad \text{Add } \tfrac{1}{3} \text{ to both sides and solve for x.}$$

$$x = \frac{1 + \sqrt{5}}{3} \qquad x = \frac{1 - \sqrt{5}}{3} \qquad \text{Express solutions with a common denominator.}$$

The solutions are $\dfrac{1 \pm \sqrt{5}}{3}$, and the solution set is $\left\{ \dfrac{1 \pm \sqrt{5}}{3} \right\}$. ■

✔ **CHECK POINT 7** Solve by completing the square: $2x^2 + 3x - 4 = 0$.
$$\frac{-3 \pm \sqrt{41}}{4} \quad \text{or} \quad \left\{ \frac{-3 \pm \sqrt{41}}{4} \right\}$$

Applications We all want a wonderful life with fulfilling work, good health, and loving relationships. And let's be honest: Financial security wouldn't hurt! Achieving this goal depends on understanding how money in a savings account grows in remarkable ways as a result of *compound interest*. **Compound interest** is interest computed on your original investment as well as on any accumulated interest. For example, suppose you deposit $1000, the principal, in a savings account at a rate of 5%. Table 11.1 shows how the investment grows if the interest earned is automatically added on to the principal.

Table 11.1 **Compound Interest on $1000**

Year	Starting Balance	Interest Earned: $I = Pr$	New Balance
1	$1000	$1000 × 0.05 = $50	$1050
2	$1050	$1050 × 0.05 = $52.50	$1102.50
3	$1102.50	$1102.50 × 0.05 ≈ $55.13	$1157.63

A faster way to determine the amount, A, in an account subject to compound interest is to use the following formula:

A FORMULA FOR COMPOUND INTEREST Suppose that an amount of money, P, is invested at interest rate r, compounded annually. In t years, the amount, A, or balance, in the account is given by the formula

$$A = P(1 + r)^t.$$

Some compound interest problems can be solved using quadratic equations.

EXAMPLE 8 Solving a Compound Interest Problem

You invested $1000 in an account whose interest is compounded annually. After 2 years, the amount, or balance, in the account is $1210. Find the annual interest rate.

SOLUTION   We are given that

$$P \quad \text{(the amount invested)} = \$1000$$

$$t \quad \text{(the time of the investment)} = 2 \text{ years}$$

$$A \quad \text{(the amount, or balance, in the account)} = \$1210.$$

We are asked to find the annual interest rate, r. We substitute the three given values into the compound interest formula and solve for r.

$A = P(1 + r)^t$	Use the compound interest formula.
$1210 = 1000(1 + r)^2$	Substitute the given values.
$\dfrac{1210}{1000} = (1 + r)^2$	Divide both sides by 1000.
$\dfrac{121}{100} = (1 + r)^2$	Simplify the fraction.
$1 + r = \sqrt{\dfrac{121}{100}} \quad \text{or} \quad 1 + r = -\sqrt{\dfrac{121}{100}}$	Apply the square root property.

$$1 + r = \frac{11}{10} \qquad\qquad 1 + r = -\frac{11}{10}$$

$$\sqrt{\frac{121}{100}} = \frac{\sqrt{121}}{\sqrt{100}} = \frac{11}{10}$$

$$r = \frac{11}{10} - 1 \qquad\qquad r = -\frac{11}{10} - 1 \qquad \text{Subtract 1 from both sides and solve for } r.$$

$$r = \frac{1}{10} \qquad\qquad\qquad r = -\frac{21}{10}$$

$$\frac{11}{10} - 1 = \frac{11}{10} - \frac{10}{10} = \frac{1}{10} \text{ and}$$

$$-\frac{11}{10} - 1 = -\frac{11}{10} - \frac{10}{10} = -\frac{21}{10}.$$

Because the interest rate cannot be negative, we reject $-\dfrac{21}{10}$. Thus, the annual interest rate is $\dfrac{1}{10} = 0.10 = 10\%$.

We can check this answer using the formula $A = P(1 + r)^t$. If \$1000 is invested for 2 years at 10% interest, compounded annually, the balance in the account is

$$A = \$1000(1 + 0.10)^2 = \$1000(1.10)^2 = \$1210.$$

Because this is precisely the amount given by the problem's conditions, the annual interest rate is, indeed, 10% compounded annually.   ■

  CHECK POINT 8 You invested \$3000 in an account whose interest is compounded annually. After 2 years, the amount, or balance, in the account is \$4320. Find the annual interest rate.   The annual interest rate is 20%.

The Pythagorean Theorem and the Square Root Property   The ancient Greek philosopher and mathematician Pythagoras (approximately 582–500 B.C.) founded a school whose motto was "All is number." Pythagoras is best remembered for his work with the **right triangle**, a triangle with one angle measuring 90°. The side opposite the 90° angle is called the **hypotenuse**. The other sides are called **legs**. Pythagoras found that if he constructed squares on each of the legs, as well as a larger square on the hypotenuse, the sum of the areas of the smaller squares is equal to the area of the larger square. This is illustrated in Figure 11.2.

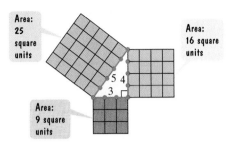

Area:
25
square
units

Area:
16 square
units

Area:
9 square
units

5 4
3

FIGURE 11.2 The area of the large square equals the sum of the areas of the smaller squares.

This relationship is usually stated in terms of the lengths of the three sides of a right triangle and is called the **Pythagorean Theorem**.

THE PYTHAGOREAN THEOREM The sum of the squares of the lengths of the legs of a right triangle equals the square of the length of the hypotenuse.

If the legs have lengths a and b, and the hypotenuse has length c, then

$$a^2 + b^2 = c^2.$$

| EXAMPLE 9 | Using the Pythagorean Theorem and the Square Root Property |

A supporting wire is to be attached to the top of an 80-foot antenna, as shown in Figure 11.3. Because of surrounding trees, the wire must be anchored 80 feet from the base of the antenna. What length of wire is required?

SOLUTION Let w = the wire's length. We can find w in the right triangle using the Pythagorean Theorem.

(Leg)²	plus	(Leg)²	equals	(Hypotenuse)².
80^2	$+$	80^2	$=$	$w^2.$

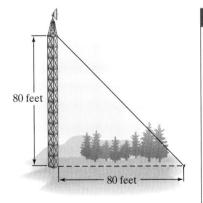

80 feet

80 feet

FIGURE 11.3

We solve this equation using the square root property.

$$6400 + 6400 = w^2 \qquad \text{Square 80.}$$

$$12{,}800 = w^2 \qquad \text{Add.}$$

$$w = \sqrt{12{,}800} \quad \text{or} \quad w = -\sqrt{12{,}800} \qquad \text{Apply the square root property.}$$

$$w = \sqrt{6400}\sqrt{2} \quad \text{or} \quad w = -\sqrt{6400}\sqrt{2} \qquad \text{6400 is the greatest perfect square factor of 12,800.}$$

$$w = 80\sqrt{2} \quad \text{or} \quad w = -80\sqrt{2}$$

Because w represents the wire's length, we reject the negative value. Thus, a wire of length $80\sqrt{2}$ feet is needed to support the antenna. Using a calculator, this is approximately 113.1 feet of wire.

✔ **CHECK POINT 9** A 50-foot supporting wire is to be attached to an antenna. The wire is anchored 20 feet from the base of the antenna. How high up the antenna is the wire attached? Express the answer in simplified radical form. Then find a decimal approximation to the nearest tenth of a foot. *The wire is attached $10\sqrt{21}$ feet, or about 45.8 feet, up the antenna.*

50 feet

x feet

20 feet

5 Find the distance between two points.

The Distance Formula Using the Pythagorean theorem, we can find the distance between the two points $P_1(x_1, y_1)$ and $P_2(x_2, y_2)$ in the rectangular coordinate system. The two points are illustrated in Figure 11.4.

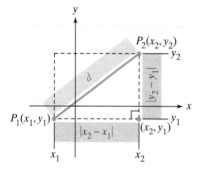

FIGURE 11.4

The distance that we want to find is represented by d and shown in blue. Notice that the distance between two points on the dashed horizontal line is the absolute value of the difference between the x-coordinates of the two points. This distance, $|x_2 - x_1|$, is shown in pink. Similarly, the distance between two points on the dashed vertical line is the absolute value of the difference between the y-coordinates of the two points. This distance, $|y_2 - y_1|$, is also shown in pink.

Because the dashed lines in Figure 11.4 are horizontal and vertical, a right triangle is formed. Thus, we can use the Pythagorean theorem to find the distance d. Squaring the lengths of the triangle's sides results in positive numbers, so absolute value notation is not necessary.

$$d^2 = (x_2 - x_1)^2 + (y_2 - y_1)^2$$ Apply the Pythagorean theorem to the right triangle in Figure 11.4.

$$d = \pm\sqrt{(x_2 - x_1)^2 + (y_2 - y_1)^2}$$ Apply the square root property.

$$d = \sqrt{(x_2 - x_1)^2 + (y_2 - y_1)^2}$$ Because distance is nonnegative, write only the principal square root.

This result is called the **distance formula**.

THE DISTANCE FORMULA The distance, d, between the points (x_1, y_1) and (x_2, y_2) in the rectangular coordinate system is

$$d = \sqrt{(x_2 - x_1)^2 + (y_2 - y_1)^2}.$$

When using the distance formula, it does not matter which point you call (x_1, y_1) and which you call (x_2, y_2).

EXAMPLE 10   Using the Distance Formula

Find the distance between $(-1, -3)$ and $(2, 3)$.

SOLUTION   Letting $(x_1, y_1) = (-1, -3)$ and $(x_2, y_2) = (2, 3)$, we obtain

$$d = \sqrt{(x_2 - x_1)^2 + (y_2 - y_1)^2} \qquad \text{Use the distance formula.}$$

$$= \sqrt{[2 - (-1)]^2 + [3 - (-3)]^2} \qquad \text{Substitute the given values.}$$

$$= \sqrt{(2 + 1)^2 + (3 + 3)^2} \qquad \text{Perform subtractions within the grouping symbols.}$$

$$= \sqrt{3^2 + 6^2} \qquad \text{Perform the resulting additions.}$$

$$= \sqrt{9 + 36} \qquad \text{Square 3 and 6.}$$

$$= \sqrt{45} \qquad \text{Add.}$$

$$= 3\sqrt{5} \approx 6.71. \qquad \sqrt{45} = \sqrt{9 \cdot 5} = \sqrt{9}\,\sqrt{5} = 3\sqrt{5}$$

The distance between the given points is $3\sqrt{5}$ units, or approximately 6.71 units. The situation is illustrated in Figure 11.5. ∎

CHECK POINT 10 Find the distance between $(2, -2)$ and $(5, 2)$.   *5 units*

FIGURE 11.5 Finding the distance between two points

6 Find the midpoint of a line segment.

The Midpoint Formula   The distance formula can be used to derive a formula for finding the midpoint of a line segment between two given points. The formula is given as follows:

THE MIDPOINT FORMULA   Consider a line segment whose endpoints are (x_1, y_1) and (x_2, y_2). The coordinates of the segment's midpoint are

$$\left(\frac{x_1 + x_2}{2}, \frac{y_1 + y_2}{2} \right).$$

To find the midpoint, take the average of the two x-coordinates and the average of the two y-coordinates.

STUDY TIP

The midpoint formula requires finding the *sum* of coordinates. By contrast, the distance formula requires finding the *difference* of coordinates:

Midpoint: Sum of coordinates

Distance: Difference of coordinates

$$\left(\frac{x_1 + x_2}{2}, \frac{y_1 + y_2}{2} \right) \qquad \sqrt{(x_2 - x_1)^2 + (y_2 - y_1)^2}$$

It's easy to confuse the two formulas. Be sure to use addition, not subtraction, when applying the midpoint formula.

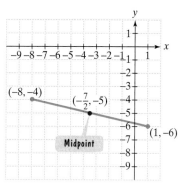

FIGURE 11.6 Finding a line segment's midpoint

EXAMPLE 11 Using the Midpoint Formula

Find the midpoint of the line segment with endpoints $(1, -6)$ and $(-8, -4)$.

SOLUTION To find the coordinates of the midpoint, we average the coordinates of the endpoints.

$$\text{Midpoint} = \left(\frac{1 + (-8)}{2}, \frac{-6 + (-4)}{2}\right) = \left(\frac{-7}{2}, \frac{-10}{2}\right) = \left(-\frac{7}{2}, -5\right)$$

Average the x-coordinates. Average the y-coordinates.

Figure 11.6 illustrates that the point $\left(-\frac{7}{2}, -5\right)$ is midway between the points $(1, -6)$ and $(-8, -4)$. ■

 CHECK POINT 11 Find the midpoint of the line segment with endpoints $(1, 2)$ and $(7, -3)$. $\left(4, -\frac{1}{2}\right)$

11.1 EXERCISE SET

Practice Exercises

In Exercises 1–22, solve each equation by the square root property. If possible, simplify radicals or rationalize denominators. Express imaginary solutions in the form a + bi.

1. $3x^2 = 75$ $\{\pm 5\}$

2. $5x^2 = 20$ $\{\pm 2\}$

3. $7x^2 = 42$ $\{\pm\sqrt{6}\}$

4. $8x^2 = 40$ $\{\pm\sqrt{5}\}$

5. $16x^2 = 25$ $\{\pm\frac{5}{4}\}$

6. $4x^2 = 49$ $\{\pm\frac{7}{2}\}$

7. $3x^2 - 2 = 0$ $\left\{\pm\frac{\sqrt{6}}{3}\right\}$

8. $3x^2 - 5 = 0$ $\left\{\pm\frac{\sqrt{15}}{3}\right\}$

9. $25x^2 + 16 = 0$ $\{\pm\frac{4}{5}i\}$

10. $4x^2 + 49 = 0$ $\{\pm\frac{7}{2}i\}$

11. $(x + 7)^2 = 9$ $\{-10, -4\}$

12. $(x + 3)^2 = 64$ $\{-11, 5\}$

13. $(x - 3)^2 = 5$ $\{3 \pm \sqrt{5}\}$

14. $(x - 4)^2 = 3$ $\{4 \pm \sqrt{3}\}$

15. $2(x + 2)^2 = 16$ $\{-2 \pm 2\sqrt{2}\}$

16. $3(x + 2)^2 = 36$ $\{-2 \pm 2\sqrt{3}\}$

17. $(x - 5)^2 = -9$ $\{5 \pm 3i\}$

18. $(x - 5)^2 = -4$ $\{5 \pm 2i\}$

19. $\left(x + \frac{3}{4}\right)^2 = \frac{11}{16}$ $\left\{\frac{-3 \pm \sqrt{11}}{4}\right\}$

20. $\left(x + \frac{2}{5}\right)^2 = \frac{7}{25}$ $\left\{\frac{-2 \pm \sqrt{7}}{5}\right\}$

21. $x^2 - 6x + 9 = 36$ $\{-3, 9\}$

22. $x^2 - 6x + 9 = 49$ $\{-4, 10\}$

In Exercises 23–34, determine the constant that should be added to the binomial so that it becomes a perfect square trinomial. Then write and factor the trinomial.

23. $x^2 + 2x$ $1; x^2 + 2x + 1 = (x + 1)^2$

24. $x^2 + 4x$ $4; x^2 + 4x + 4 = (x + 2)^2$

25. $x^2 - 14x$ $49; x^2 - 14x + 49 = (x - 7)^2$

26. $x^2 - 10x$ $25; x^2 - 10x + 25 = (x - 5)^2$

27. $x^2 + 7x$ $\frac{49}{4}; x^2 + 7x + \frac{49}{4} = \left(x + \frac{7}{2}\right)^2$

28. $x^2 + 9x$ $\frac{81}{4}; x^2 + 9x + \frac{81}{4} = \left(x + \frac{9}{2}\right)^2$

29. $x^2 - \frac{1}{2}x$ $\frac{1}{16}; x^2 - \frac{1}{2}x + \frac{1}{16} = \left(x - \frac{1}{4}\right)^2$

30. $x^2 - \frac{1}{3}x$ $\frac{1}{36}; x^2 - \frac{1}{3}x + \frac{1}{36} = \left(x - \frac{1}{6}\right)^2$

31. $x^2 + \dfrac{4}{3}x \quad \dfrac{4}{9}; x^2 + \dfrac{4}{3}x + \dfrac{4}{9} = \left(x + \dfrac{2}{3}\right)^2$

32. $x^2 + \dfrac{4}{5}x \quad \dfrac{4}{25}; x^2 + \dfrac{4}{5}x + \dfrac{4}{25} = \left(x + \dfrac{2}{5}\right)^2$

33. $x^2 - \dfrac{9}{4}x \quad \dfrac{81}{64}; x^2 - \dfrac{9}{4}x + \dfrac{81}{64} = \left(x - \dfrac{9}{8}\right)^2$

34. $x^2 - \dfrac{9}{5}x \quad \dfrac{81}{100}; x^2 - \dfrac{9}{5}x + \dfrac{81}{100} = \left(x - \dfrac{9}{10}\right)^2$

In Exercises 35–56, solve each quadratic equation by completing the square.

35. $x^2 + 4x = 32$ $\{-8, 4\}$ **36.** $x^2 + 6x = 7$ $\{-7, 1\}$

37. $x^2 + 6x = -2$ $\{-3 \pm \sqrt{7}\}$

38. $x^2 + 2x = 5$ $\{-1 \pm \sqrt{6}\}$

39. $x^2 - 8x + 1 = 0$ $\{4 \pm \sqrt{15}\}$

40. $x^2 + 8x - 5 = 0$ $\{-4 \pm \sqrt{21}\}$

41. $x^2 + 2x + 2 = 0$ $\{-1 \pm i\}$

42. $x^2 - 4x + 8 = 0$ $\{2 \pm 2i\}$

43. $x^2 + 3x - 1 = 0$ $\left\{\dfrac{-3 \pm \sqrt{13}}{2}\right\}$

44. $x^2 - 3x - 5 = 0$ $\left\{\dfrac{3 \pm \sqrt{29}}{2}\right\}$

45. $x^2 + \dfrac{4}{7}x + \dfrac{3}{49} = 0$ $\left\{-\dfrac{3}{7}, -\dfrac{1}{7}\right\}$

46. $x^2 + \dfrac{6}{5}x + \dfrac{8}{25} = 0$ $\left\{-\dfrac{4}{5}, -\dfrac{2}{5}\right\}$

47. $x^2 + x - 1 = 0$ $\left\{\dfrac{-1 \pm \sqrt{5}}{2}\right\}$

48. $x^2 + 3x - 1 = 0$ $\left\{\dfrac{-3 \pm \sqrt{13}}{2}\right\}$

49. $2x^2 + 3x - 5 = 0$ $\left\{-\dfrac{5}{2}, 1\right\}$

50. $2x^2 + 5x - 3 = 0$ $\left\{-3, \dfrac{1}{2}\right\}$

51. $3x^2 + 6x + 1 = 0$ $\left\{\dfrac{-3 \pm \sqrt{6}}{3}\right\}$

52. $3x^2 - 6x + 2 = 0$ $\left\{\dfrac{3 \pm \sqrt{3}}{3}\right\}$

53. $3x^2 - 8x + 1 = 0$ $\left\{\dfrac{4 \pm \sqrt{13}}{3}\right\}$

54. $2x^2 + 3x - 4 = 0$ $\left\{\dfrac{-3 \pm \sqrt{41}}{4}\right\}$

55. $8x^2 - 4x + 1 = 0$ $\left\{\dfrac{1}{4} \pm \dfrac{1}{4}i\right\}$

56. $9x^2 - 6x + 5 = 0$ $\left\{\dfrac{1}{3} \pm \dfrac{2}{3}i\right\}$

57. If $f(x) = (x - 1)^2$, find all values of x for which $f(x) = 36$. $-5, 7$

58. If $f(x) = (x + 2)^2$, find all values of x for which $f(x) = 25$. $-7, 3$

59. If $g(x) = \left(x - \dfrac{2}{5}\right)^2$, find all values of x for which $g(x) = \dfrac{9}{25}$. $-\dfrac{1}{5}, 1$

60. If $g(x) = \left(x + \dfrac{1}{3}\right)^2$, find all values of x for which $g(x) = \dfrac{4}{9}$. $-1, \dfrac{1}{3}$

61. If $h(x) = 5(x + 2)^2$, find all values of x for which $h(x) = -125$. $-2 \pm 5i$

62. If $h(x) = 3(x - 4)^2$, find all values of x for which $h(x) = -12$. $4 \pm 2i$

In Exercises 63–80, find the distance between each pair of points. If necessary, round answers to two decimal places.

63. $(2, 3)$ and $(14, 8)$ 13 units

64. $(5, 1)$ and $(8, 5)$ 5 units

65. $(4, 1)$ and $(6, 3)$ $2\sqrt{2}$ or 2.83 units

66. $(2, 3)$ and $(3, 5)$ $\sqrt{5}$ or 2.24 units

67. $(0, 0)$ and $(-3, 4)$ 5 units

68. $(0, 0)$ and $(3, -4)$ 5 units

69. $(-2, -6)$ and $(3, -4)$ $\sqrt{29}$ or 5.39 units

70. $(-4, -1)$ and $(2, -3)$ $2\sqrt{10}$ or 6.32 units

71. $(0, -3)$ and $(4, 1)$ $4\sqrt{2}$ or 5.66 units

72. $(0, -2)$ and $(4, 3)$ $\sqrt{41}$ or 6.40 units

73. $(3.5, 8.2)$ and $(-0.5, 6.2)$ $2\sqrt{5}$ or 4.47 units

74. $(2.6, 1.3)$ and $(1.6, -5.7)$ $5\sqrt{2}$ or 7.07 units

75. $\left(0, -\sqrt{3}\right)$ and $\left(\sqrt{5}, 0\right)$ $2\sqrt{2}$ or 2.83 units

76. $\left(0, -\sqrt{2}\right)$ and $\left(\sqrt{7}, 0\right)$ 3 units

77. $\left(3\sqrt{3}, \sqrt{5}\right)$ and $\left(-\sqrt{3}, 4\sqrt{5}\right)$ $\sqrt{93}$ or 9.64 units

78. $\left(2\sqrt{3}, \sqrt{6}\right)$ and $\left(-\sqrt{3}, 5\sqrt{6}\right)$ $\sqrt{123}$ or 11.09 units

79. $\left(\dfrac{7}{3}, \dfrac{1}{5}\right)$ and $\left(\dfrac{1}{3}, \dfrac{6}{5}\right)$ $\sqrt{5}$ or 2.24 units

80. $\left(-\dfrac{1}{4}, -\dfrac{1}{7}\right)$ and $\left(\dfrac{3}{4}, \dfrac{6}{7}\right)$ $\sqrt{2}$ or 1.41 units

In Exercises 81–92, find the midpoint of the line segment with the given endpoints.

81. $(6, 8)$ and $(2, 4)$ $(4, 6)$ **82.** $(10, 4)$ and $(2, 6)$ $(6, 5)$

83. $(-2, -8)$ and $(-6, -2)$ $(-4, -5)$

84. $(-4, -7)$ and $(-1, -3)$ $\left(-\dfrac{5}{2}, -5\right)$

85. $(-3, -4)$ and $(6, -8)$ $\left(\dfrac{3}{2}, -6\right)$

86. $(-2, -1)$ and $(-8, 6)$ $\left(-5, \frac{5}{2}\right)$

87. $\left(-\frac{7}{2}, \frac{3}{2}\right)$ and $\left(-\frac{5}{2}, -\frac{11}{2}\right)$ $(-3, -2)$

88. $\left(-\frac{2}{5}, \frac{7}{15}\right)$ and $\left(-\frac{2}{5}, -\frac{4}{15}\right)$ $\left(-\frac{2}{5}, \frac{1}{10}\right)$

89. $\left(8, 3\sqrt{5}\right)$ and $\left(-6, 7\sqrt{5}\right)$ $\left(1, 5\sqrt{5}\right)$

90. $\left(7\sqrt{3}, -6\right)$ and $\left(3\sqrt{3}, -2\right)$ $\left(5\sqrt{3}, -4\right)$

91. $\left(\sqrt{18}, -4\right)$ and $\left(\sqrt{2}, 4\right)$ $\left(2\sqrt{2}, 0\right)$

92. $\left(\sqrt{50}, -6\right)$ and $\left(\sqrt{2}, 6\right)$ $\left(3\sqrt{2}, 0\right)$

Practice Plus

93. Three times the square of the difference between a number and 2 is -12. Find the number(s). $2 + 2i$ and $2 - 2i$

94. Three times the square of the difference between a number and 9 is -27. Find the number(s). $9 + 3i$ and $9 - 3i$

In Exercises 95–98, solve the formula for the specified variable. Because each variable is nonnegative, list only the principal square root. If possible, simplify radicals or rationalize denominators.

95. $h = \dfrac{v^2}{2g}$ for v $v = \sqrt{2gh}$

96. $s = \dfrac{kwd^2}{l}$ for d $d = \dfrac{\sqrt{slkw}}{kw}$

97. $A = P(1 + r)^2$ for r $r = \dfrac{\sqrt{AP}}{P} - 1$

98. $C = \dfrac{kP_1P_2}{d^2}$ for d $d = \dfrac{\sqrt{kP_1P_2C}}{C}$

In Exercises 99–102, solve each quadratic equation by completing the square.

99. $\dfrac{x^2}{3} + \dfrac{x}{9} - \dfrac{1}{6} = 0$ $\left\{\dfrac{-1 \pm \sqrt{19}}{6}\right\}$

100. $\dfrac{x^2}{2} - \dfrac{x}{6} - \dfrac{3}{4} = 0$ $\left\{\dfrac{1 \pm \sqrt{55}}{6}\right\}$

101. $x^2 - bx = 2b^2$ $\{-b, 2b\}$

102. $x^2 - bx = 6b^2$ $\{-2b, 3b\}$

Application Exercises

In Exercises 103–106, use the compound interest formula

$$A = P(1 + r)^t$$

to find the annual interest rate, r.

103. In 2 years, an investment of $2000 grows to $2880.   20%

104. In 2 years, an investment of $2000 grows to $2420.   10%

105. In 2 years, an investment of $1280 grows to $1445.   6.25%

106. In 2 years, an investment of $80,000 grows to $101,250.   12.5%

Of the one hundred largest economies in the world, 53 are multinational corporations. In 1970, there were approximately 7000 multinational corporations. By 2001, more than 65,000 corporations enveloped the world. The graph shows this rapid growth from 1970 through 2001, including the starting dates of some notable corporations.

Number of Multinational Corporations in the World

Source: Medard Gabel, *Global Inc.*, The New Press, 2003

The data shown can be modeled by the function

$$f(x) = 62.2x^2 + 7000,$$

where $f(x)$ represents the number of multinational corporations in the world x years after 1970. Use this function and the square root property to solve Exercises 107–108.

107. In which year will there be 92,000 multinational corporations?   In approximately 2007, there will be 92,000 corporations.

108. In which year were there 46,000 multinational corporations? How well does the function model the actual number of corporations for that year shown in the graph?   In approximately 1995, there were 46,000 corporations. The function models the actual number quite well.

The function $s(t) = 16t^2$ models the distance, $s(t)$, in feet, that an object falls in t seconds. Use this function and the square root property to solve Exercises 109–110. Express answers in simplified radical form. Then use your calculator and find a decimal approximation to the nearest tenth of a second.

109. A sky diver jumps from an airplane and falls for 4800 feet before opening a parachute. For how many seconds was the diver in a free fall?   The diver was in a free fall $10\sqrt{3}$ seconds, or about 17.3 seconds.

110. A sky diver jumps from an airplane and falls for 3200 feet before opening a parachute. For how many seconds was the diver in a free fall?   The diver was in a free fall $10\sqrt{2}$ seconds, or about 14.1 seconds.

Use the Pythagorean theorem and the square root property to solve Exercises 111–116. Express answers in simplified radical form. Then find a decimal approximation to the nearest tenth.

111. A rectangular park is 6 miles long and 3 miles wide. How long is a pedestrian route that runs diagonally across the park?   The route is $3\sqrt{5}$ miles, or about 6.7 miles long.

112. A rectangular park is 4 miles long and 2 miles wide. How long is a pedestrian route that runs diagonally across the park?   The route is $2\sqrt{5}$ miles, or about 4.5 miles long.

113. The base of a 30-foot ladder is 10 feet from the building. If the ladder reaches the flat roof, how tall is the building?

30 ft
x
10 ft

The building is $20\sqrt{2}$ feet tall, or about 28.3 feet.

114. The doorway into a room is 4 feet wide and 8 feet high. What is the length of the longest rectangular panel that can be taken through this doorway diagonally?

4 ft
8 ft

The length is $4\sqrt{5}$ feet, or about 8.9 feet.

115. A supporting wire is to be attached to the top of a 50-foot antenna. If the wire must be anchored 50 feet from the base of the antenna, what length of wire is required?   The wire needs to be $50\sqrt{2}$ feet, or about 70.7 feet.

116. A supporting wire is to be attached to the top of a 70-foot antenna. If the wire must be anchored 70 feet from the base of the antenna, what length of wire is required?   The wire needs to be $70\sqrt{2}$ feet, or about 99.0 feet.

117. A square flower bed is to be enlarged by adding 2 meters on each side. If the larger square has an area of 196 square meters, what is the length of a side of the original square?   10 meters

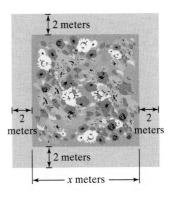

2 meters
2 meters
2 meters
2 meters
x meters

118. A square flower bed is to be enlarged by adding 4 feet on each side. If the larger square has an area of 225 square feet, what is the length of a side of the original square?   7 feet

119. A rectangular coordinate system with coordinates in miles is placed on the map in the figure shown. Bangkok has coordinates $(-115, 170)$ and Phnom Penh has coordinates $(65, 70)$. How long will it take a plane averaging 400 miles per hour to fly directly from one city to the other? Round to the nearest tenth of an hour. Approximately how many minutes is the flight?   0.5 hour; 30 minutes

y
Hanoi
LAOS
VIETNAM
THAILAND
$A(-115, 170)$
Bangkok
Cambodia
Phnom Penh
x
$B(65, 70)$

Writing in Mathematics

120. What is the square root property?   120–128. Answers will vary.

121. Explain how to solve $(x - 1)^2 = 16$ using the square root property.

122. Explain how to complete the square for a binomial. Use $x^2 + 6x$ to illustrate your explanation.

123. Explain how to solve $x^2 + 6x + 8 = 0$ by completing the square.

124. What is compound interest?

125. State the Pythagorean theorem.

126. Describe how to find the distance between two points in the rectangular coordinate system.

127. Describe how to find the midpoint of a line segment if its endpoints are known.

128. A basketball player's hang time is the time spent in the air when shooting a basket. The function $f(x) = 4x^2$ models the player's vertical leap, $f(x)$, in feet, in terms of the hang time, x, measured in seconds. Use this function to write a problem that can be solved by the square root property. Then solve the problem. Be realistic: It is unlikely that hang time can exceed 1 second.

Technology Exercises

129. Use a graphing utility to solve $4 - (x + 1)^2 = 0$. Graph $y = 4 - (x + 1)^2$ in a $[-5, 5, 1]$ by $[-5, 5, 1]$ viewing rectangle. The equation's solutions are the graph's x-intercepts. Check by substitution in the given equation. -3 and 1, or $\{-3, 1\}$ **129–131.** See graphing answer section.

130. Use a graphing utility to solve $(x - 1)^2 - 9 = 0$. Graph $y = (x - 1)^2 - 9$ in a $[-5, 5, 1]$ by $[-9, 3, 1]$ viewing rectangle. The equation's solutions are the graph's x-intercepts. Check by substitution in the given equation. -2 and 4, or $\{-2, 4\}$

131. Use a graphing utility and x-intercepts to verify any of the real solutions that you obtained for five of the quadratic equations in Exercises 35–56.   Answers will vary.

Critical Thinking Exercises

132. Which one of the following is true?   d

a. The equation $(x - 5)^2 = 12$ is equivalent to $x - 5 = 2\sqrt{3}$.

b. In completing the square for $2x^2 - 6x = 5$, we should add 9 to both sides.

c. Although not every quadratic equation can be solved by completing the square, they can all be solved by factoring.

d. The graph of $y = (x - 2)^2 + 3$ cannot have x-intercepts.

133. Solve for y: $\dfrac{x^2}{a^2} + \dfrac{y^2}{b^2} = 1$. $y = \pm\dfrac{b\sqrt{a^2 - x^2}}{a}$

134. Solve by completing the square:

$$x^2 + x + c = 0. \quad \left\{\dfrac{-1 \pm \sqrt{1 - 4c}}{2}\right\}$$

135. Solve by completing the square:

$$x^2 + bx + c = 0. \quad \left\{\dfrac{-b \pm \sqrt{b^2 - 4c}}{2}\right\}$$

136. Solve: $x^4 - 8x^2 + 15 = 0$. $\left\{\pm\sqrt{5}, \pm\sqrt{3}\right\}$

137. Show that the points $A(1, 1 + d)$, $B(3, 3 + d)$, and $C(6, 6 + d)$ are collinear (lie along a straight line) by showing that the distance from A to B plus the distance from B to C equals the distance from A to C.   The distance from A to B is $2\sqrt{2}$ and the distance from B to C is $3\sqrt{2}$. The distance from A to C is $5\sqrt{2}$, which is equal to $2\sqrt{2} + 3\sqrt{2}$.

138. Prove the midpoint formula by using the following procedure.

a. Show that the distance between (x_1, y_1) and

$\left(\dfrac{x_1 + x_2}{2}, \dfrac{y_1 + y_2}{2}\right)$ is equal to the distance between

(x_2, y_2) and $\left(\dfrac{x_1 + x_2}{2}, \dfrac{y_1 + y_2}{2}\right)$.

$$\sqrt{\left(x_1 - \dfrac{x_1 + x_2}{2}\right)^2 + \left(y_1 - \dfrac{y_1 + y_2}{2}\right)^2} = \dfrac{1}{2}\sqrt{(x_1 - x_2)^2 + (y_1 - y_2)^2}$$

$$\sqrt{\left(x_2 - \dfrac{x_1 + x_2}{2}\right)^2 + \left(y_2 - \dfrac{y_1 + y_2}{2}\right)^2} = \dfrac{1}{2}\sqrt{(x_1 - x_2)^2 + (y_1 - y_2)^2}$$

b. Use the procedure from Exercise 137 and the distances from part (a) to show that the points (x_1, y_1), $\left(\dfrac{x_1 + x_2}{2}, \dfrac{y_1 + y_2}{2}\right)$, and (x_2, y_2) are collinear.   The distance from (x_1, y_1) to (x_2, y_2) is $\sqrt{(x_1 - x_2)^2 + (y_1 - y_2)^2}$, which equals the sum of the distances from part (a).

139. Find all points with y-coordinate 2 so that the distance between $(x, 2)$ and $(2, -1)$ is 5. $(-2, 2)$ and $(6, 2)$

Review Exercises

140. Simplify: $4x - 2 - 3[4 - 2(3 - x)]$. (Section 1.8, Example 11) $4 - 2x$

141. Factor: $1 - 8x^3$. (Section 6.4, Example 8) $(1 - 2x)(1 + 2x + 4x^2)$

142. Divide: $(x^4 - 5x^3 + 2x^2 - 6) \div (x - 3)$. (Section 5.6, Example 2 or Example 5) $x^3 - 2x^2 - 4x - 12 - \dfrac{42}{x - 3}$

SECTION

11.2

Objectives

1 Solve quadratic equations using the quadratic formula.

2 Use the discriminant to determine number and type of solutions.

3 Determine the most efficient method to use when solving a quadratic equation.

4 Write quadratic equations from solutions.

5 Use the quadratic formula to solve problems.

THE QUADRATIC FORMULA

America is a nation of immigrants. Since 1820, more than 40 million people have immigrated to the United States from all over the world. They chose to come for various reasons, such as to live in freedom, to practice religion freely, to escape poverty or oppression, and to make better lives for themselves and their children. As a result, a substantial percentage of the U.S. population is foreign-born. The graph in Figure 11.7 shows the percentage of foreign-born Americans from 1930 through 2002.

**Percentage of U.S. Population
That Was Foreign-Born, 1930-2002**

FIGURE 11.7

Source: U.S. Census Bureau

The data shown in Figure 11.7 can be modeled by the quadratic function

$$f(x) = 0.0049x^2 - 0.361x + 11.79,$$

where $f(x)$ is the percentage of the U.S. population that was foreign-born x years after 1930. If the trend shown by the data continues, in which year will 15% of the U.S. population be foreign-born? To answer the question, it is necessary to substitute 15 for $f(x)$ in the given function and solve for x, the number of years after 1930:

$$15 = 0.0049x^2 - 0.361x + 11.79.$$

In this section, we will derive a formula that will enable you to solve quadratic equations more quickly than the method of completing the square. Using this formula, we will work with the function that models the percentage of foreign-born Americans and determine the year in which 15% of our population will be foreign-born.

1 Solve quadratic equations using the quadratic formula.

Solving Quadratic Equations Using the Quadratic Formula We can use the method of completing the square to derive a formula that can be used to solve all quadratic equations. The derivation given here also shows a particular quadratic equation, $3x^2 - 2x - 4 = 0$, to specifically illustrate each of the steps.

Deriving the Quadratic Formula

Standard Form of a Quadratic Equation	Comment	A Specific Example
$ax^2 + bx + c = 0, a > 0$	This is the given equation.	$3x^2 - 2x - 4 = 0$
$x^2 + \dfrac{b}{a}x + \dfrac{c}{a} = 0$	Divide both sides by the coefficient of x^2.	$x^2 - \dfrac{2}{3}x - \dfrac{4}{3} = 0$
$x^2 + \dfrac{b}{a}x = -\dfrac{c}{a}$	Isolate the binomial by adding $-\dfrac{c}{a}$ on both sides.	$x^2 - \dfrac{2}{3}x = \dfrac{4}{3}$
$x^2 + \dfrac{b}{a}x + \left(\dfrac{b}{2a}\right)^2 = -\dfrac{c}{a} + \left(\dfrac{b}{2a}\right)^2$ $\underbrace{}_{(\text{half})^2}$	Complete the square. Add the square of half the coefficient of x to both sides.	$x^2 - \dfrac{2}{3}x + \left(-\dfrac{1}{3}\right)^2 = \dfrac{4}{3} + \left(-\dfrac{1}{3}\right)^2$ $\underbrace{}_{(\text{half})^2}$
$x^2 + \dfrac{b}{a}x + \dfrac{b^2}{4a^2} = -\dfrac{c}{a} + \dfrac{b^2}{4a^2}$		$x^2 - \dfrac{2}{3}x + \dfrac{1}{9} = \dfrac{4}{3} + \dfrac{1}{9}$
$\left(x + \dfrac{b}{2a}\right)^2 = -\dfrac{c}{a}\cdot\dfrac{4a}{4a} + \dfrac{b^2}{4a^2}$	Factor on the left side and obtain a common denominator on the right side.	$\left(x - \dfrac{1}{3}\right)^2 = \dfrac{4}{3}\cdot\dfrac{3}{3} + \dfrac{1}{9}$
$\left(x + \dfrac{b}{2a}\right)^2 = \dfrac{-4ac + b^2}{4a^2}$	Add fractions on the right side.	$\left(x - \dfrac{1}{3}\right)^2 = \dfrac{12 + 1}{9}$
$\left(x + \dfrac{b}{2a}\right)^2 = \dfrac{b^2 - 4ac}{4a^2}$		$\left(x - \dfrac{1}{3}\right)^2 = \dfrac{13}{9}$
$x + \dfrac{b}{2a} = \pm\sqrt{\dfrac{b^2 - 4ac}{4a^2}}$	Apply the square root property.	$x - \dfrac{1}{3} = \pm\sqrt{\dfrac{13}{9}}$
$x + \dfrac{b}{2a} = \pm\dfrac{\sqrt{b^2 - 4ac}}{2a}$	Take the square root of the quotient, simplifying the denominator.	$x - \dfrac{1}{3} = \pm\dfrac{\sqrt{13}}{3}$
$x = \dfrac{-b}{2a} \pm \dfrac{\sqrt{b^2 - 4ac}}{2a}$	Solve for x by subtracting $\dfrac{b}{2a}$ from both sides.	$x = \dfrac{1}{3} \pm \dfrac{\sqrt{13}}{3}$
$x = \dfrac{-b \pm \sqrt{b^2 - 4ac}}{2a}$	Combine fractions on the right side.	$x = \dfrac{1 \pm \sqrt{13}}{3}$

The formula shown at the bottom of the left column is called the *quadratic formula*. A similar proof shows that the same formula can be used to solve quadratic equations if a, the coefficient of the x^2-term, is negative.

> **THE QUADRATIC FORMULA** The solutions of a quadratic equation in standard form $ax^2 + bx + c = 0$, with $a \neq 0$, are given by the **quadratic formula**
>
> $$x = \frac{-b \pm \sqrt{b^2 - 4ac}}{2a}.$$
>
> *x equals negative b plus or minus the square root of $b^2 - 4ac$, all divided by 2a.*

To use the quadratic formula, write the quadratic equation in standard form if necessary. Then determine the numerical values for a (the coefficient of the x^2-term), b (the coefficient of the x-term), and c (the constant term). Substitute the values of a, b, and c into the quadratic formula and evaluate the expression. The $\pm$ sign indicates that there are two solutions of the equation.

> ### EXAMPLE 1 Solving a Quadratic Equation Using the Quadratic Formula

Solve using the quadratic formula: $8x^2 + 2x - 1 = 0$.

SOLUTION The given equation is in standard form. Begin by identifying the values for a, b, and c.

$$8x^2 + 2x - 1 = 0$$

$a = 8$ $\quad b = 2$ $\quad c = -1$

Substituting these values into the quadratic formula and simplifying gives the equation's solutions.

$$x = \frac{-b \pm \sqrt{b^2 - 4ac}}{2a}$$ Use the quadratic formula.

$$x = \frac{-2 \pm \sqrt{2^2 - 4(8)(-1)}}{2(8)}$$ Substitute the values for a, b, and c: $a = 8, b = 2$, and $c = -1$.

$$= \frac{-2 \pm \sqrt{4 - (-32)}}{16}$$ $2^2 - 4(8)(-1) = 4 - (-32)$

$$= \frac{-2 \pm \sqrt{36}}{16}$$ $4 - (-32) = 4 + 32 = 36$

$$= \frac{-2 \pm 6}{16}$$ $\sqrt{36} = 6$

Now we will evaluate this expression in two different ways to obtain the two solutions. On the left, we will *add* 6 to -2. On the right, we will *subtract* 6 from -2.

$$x = \frac{-2 + 6}{16} \quad \text{or} \quad x = \frac{-2 - 6}{16}$$

$$= \frac{4}{16} = \frac{1}{4} \qquad\qquad = \frac{-8}{16} = -\frac{1}{2}$$

The solutions are $-\frac{1}{2}$ and $\frac{1}{4}$, and the solution set is $\left\{-\frac{1}{2}, \frac{1}{4}\right\}$. ∎

In Example 1, the solutions of $8x^2 + 2x - 1 = 0$ are rational numbers. This means that the equation can also be solved by factoring. The reason that the solutions are rational numbers is that $b^2 - 4ac$, the radicand in the quadratic formula, is 36, which is a perfect square. If a, b, and c are rational numbers, all quadratic equations for which $b^2 - 4ac$ is a perfect square have rational solutions.

> ### ✔ CHECK POINT 1 Solve using the quadratic formula: $2x^2 + 9x - 5 = 0$.
> -5 and $\frac{1}{2}$, or $\left\{-5, \frac{1}{2}\right\}$

USING TECHNOLOGY

The graph of the quadratic function

$$y = 8x^2 + 2x - 1$$

has x-intercepts at $-\frac{1}{2}$ and $\frac{1}{4}$. This verifies that $\left\{-\frac{1}{2}, \frac{1}{4}\right\}$ is the solution set of the quadratic equation

$$8x^2 + 2x - 1 = 0.$$

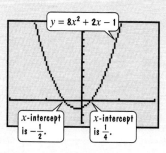

$[-2, 2, 1]$ by $[-3, 10, 1]$

| EXAMPLE 2 | Solving a Quadratic Equation Using the Quadratic Formula |

Solve using the quadratic formula:

$$2x^2 = 4x + 1.$$

SOLUTION The quadratic equation must be in standard form to identify the values for a, b, and c. To move all terms to one side and obtain zero on the right, we subtract $4x + 1$ from both sides. Then we can identify the values for a, b, and c.

$$2x^2 = 4x + 1 \qquad \text{This is the given equation.}$$

$$2x^2 - 4x - 1 = 0 \qquad \text{Subtract } 4x + 1 \text{ from both sides.}$$

$$a = 2 \quad b = -4 \quad c = -1$$

Substituting these values into the quadratic formula and simplifying gives the equation's solutions.

$$x = \frac{-b \pm \sqrt{b^2 - 4ac}}{2a} \qquad \text{Use the quadratic formula.}$$

$$x = \frac{-(-4) \pm \sqrt{(-4)^2 - 4(2)(-1)}}{2(2)} \qquad \begin{array}{l} \text{Substitute the values for } a, b, \text{ and } c: \\ a = 2, b = -4, \text{ and } c = -1. \end{array}$$

$$= \frac{4 \pm \sqrt{16 - (-8)}}{4} \qquad (-4)^2 - 4(2)(-1) = 16 - (-8)$$

$$= \frac{4 \pm \sqrt{24}}{4} \qquad 16 - (-8) = 16 + 8 = 24$$

$$= \frac{4 \pm 2\sqrt{6}}{4} \qquad \sqrt{24} = \sqrt{4 \cdot 6} = \sqrt{4} \cdot \sqrt{6} = 2\sqrt{6}$$

$$= \frac{2(2 \pm \sqrt{6})}{4} \qquad \text{Factor out 2 from the numerator.}$$

$$= \frac{2 \pm \sqrt{6}}{2} \qquad \text{Divide the numerator and denominator by 2.}$$

The solutions are $\dfrac{2 \pm \sqrt{6}}{2}$, and the solution set is $\left\{ \dfrac{2 + \sqrt{6}}{2}, \dfrac{2 - \sqrt{6}}{2} \right\}$ or $\left\{ \dfrac{2 \pm \sqrt{6}}{2} \right\}$.

USING TECHNOLOGY

You can use a graphing utility to verify that the solutions of $2x^2 - 4x - 1 = 0$ are $\dfrac{2 \pm \sqrt{6}}{2}$. Begin by entering $y_1 = 2x^2 - 4x - 1$ in the $\boxed{Y=}$ screen. Then evaluate this function at each of the proposed solutions.

```
Y₁((2+√(6))/2)
              0
Y₁((2-√(6))/2)
              0
■
```

In each case, the function value is 0, verifying that the solutions satisfy $2x^2 - 4x - 1 = 0$.

In Example 2, the solutions of $2x^2 = 4x + 1$ are irrational numbers. This means that the equation cannot be solved by factoring. The reason that the solutions are irrational numbers is that $b^2 - 4ac$, the radicand in the quadratic formula, is 24, which is not a perfect square. Notice, too, that the solutions, $\dfrac{2 + \sqrt{6}}{2}$ and $\dfrac{2 - \sqrt{6}}{2}$, are conjugates.

STUDY TIP

Many students use the quadratic formula correctly until the last step, where they make an error in simplifying the solutions. Be sure to factor the numerator before dividing the numerator and the denominator by the greatest common factor.

$$\frac{4 \pm 2\sqrt{6}}{4} = \frac{2(2 \pm \sqrt{6})}{4} = \frac{\overset{1}{2}(2 \pm \sqrt{6})}{\underset{2}{4}} = \frac{2 \pm \sqrt{6}}{2}$$

You cannot divide just one term in the numerator and the denominator by their greatest common factor.

Incorrect!

$$\frac{\overset{1}{4} \pm 2\sqrt{6}}{\underset{1}{4}} = 1 \pm 2\sqrt{6} \qquad \frac{4 \pm \overset{1}{2}\sqrt{6}}{\underset{2}{4}} = \frac{4 \pm \sqrt{6}}{2}$$

Can all irrational solutions of quadratic equations be simplified? No. The following solutions cannot be simplified:

$$\frac{5 \pm 2\sqrt{7}}{2} \qquad \qquad \frac{-4 \pm 3\sqrt{7}}{2}.$$

Other than 1, terms in each numerator have no common factor.

CHECK POINT 2 Solve using the quadratic formula: $2x^2 = 6x - 1$.

 or $\left\{ \dfrac{3 \pm \sqrt{7}}{2} \right\}$

EXAMPLE 3 Solving a Quadratic Equation Using the Quadratic Formula

Solve using the quadratic formula:

$$3x^2 + 2 = -4x.$$

SOLUTION Begin by writing the quadratic equation in standard form.

$$3x^2 + 2 = -4x \qquad \text{This is the given equation.}$$

$$3x^2 + 4x + 2 = 0 \qquad \text{Add 4x to both sides.}$$

$$a = 3 \qquad b = 4 \qquad c = 2$$

Substituting these values into the quadratic formula and simplifying gives the equation's solutions.

$$x = \frac{-b \pm \sqrt{b^2 - 4ac}}{2a}$$ Use the quadratic formula.

$$x = \frac{-4 \pm \sqrt{4^2 - 4 \cdot 3 \cdot 2}}{2 \cdot 3}$$ Substitute the values for a, b, and c: $a = 3$, $b = 4$, and $c = 2$.

$$= \frac{-4 \pm \sqrt{16 - 24}}{6}$$ Multiply under the radical.

$$= \frac{-4 \pm \sqrt{-8}}{6}$$ Subtract under the radical.

$$= \frac{-4 \pm 2\sqrt{2}i}{6}$$ $\sqrt{-8} = \sqrt{8(-1)} = \sqrt{8}\sqrt{-1} = \sqrt{8}i$
$= \sqrt{4 \cdot 2}i = 2\sqrt{2}i$

$$= \frac{2(-2 \pm \sqrt{2}i)}{6}$$ Factor out 2 from the numerator.

$$= \frac{-2 \pm \sqrt{2}i}{3}$$ Divide the numerator and denominator by 2.

$$= -\frac{2}{3} \pm \frac{\sqrt{2}}{3}i$$ Write in the form $a + bi$.

The solutions are $-\frac{2}{3} \pm \frac{\sqrt{2}}{3}i$, and the solution set is $\left\{ -\frac{2}{3} + \frac{\sqrt{2}}{3}i, -\frac{2}{3} - \frac{\sqrt{2}}{3}i \right\}$ or

$\left\{ -\frac{2}{3} \pm \frac{\sqrt{2}}{3}i \right\}$. ■

USING TECHNOLOGY

The graph of the quadratic function

$$y = 3x^2 + 4x + 2$$

has no x-intercepts. This verifies that the equation in Example 3

$$3x^2 + 2 = -4x, \quad \text{or}$$
$$3x^2 + 4x + 2 = 0$$

has imaginary solutions.

$y = 3x^2 + 4x + 2$

$[-3, 3, 1]$ by $[-2, 10, 1]$

In Example 3, the solutions of $3x^2 + 2 = -4x$ are imaginary numbers. This means that the equation cannot be solved using factoring. The reason that the solutions are imaginary numbers is that $b^2 - 4ac$, the radicand in the quadratic formula, is -8, which is negative. Notice, too, that the solutions are complex conjugates.

 CHECK POINT 3 Solve using the quadratic formula: $3x^2 + 5 = -6x$.

$-1 \pm \frac{\sqrt{6}}{3}i$ or $\left\{ -1 \pm \frac{\sqrt{6}}{3}i \right\}$

Some rational equations can be solved using the quadratic formula. For example, consider the equation

$$3 + \frac{4}{x} = -\frac{2}{x^2}.$$

The denominators are x and x^2. The least common denominator is x^2. We clear fractions by multiplying both sides of the equation by x^2. Notice that x cannot equal zero.

$$x^2\left(3 + \frac{4}{x}\right) = x^2\left(-\frac{2}{x^2}\right), \quad x \neq 0$$

$$3x^2 + \frac{4}{x} \cdot x^2 = x^2\left(-\frac{2}{x^2}\right)$$ Use the distributive property.

$$3x^2 + 4x = -2$$ Simplify.

By adding 2 to both sides of $3x^2 + 4x = -2$, we obtain the standard form of the quadratic equation:

$$3x^2 + 4x + 2 = 0.$$

This is the equation that we solved in Example 3. The two imaginary solutions are not part of the restriction that $x \neq 0$.

The Discriminant The quantity $b^2 - 4ac$, which appears under the radical sign in the quadratic formula, is called the **discriminant**. The discriminant of the quadratic equation $ax^2 + bx + c = 0$ determines the number and type of solutions. In Example 1, the discriminant was 36, a positive number that is a perfect square. The equation had two solutions that were rational numbers. In Example 2, the discriminant was 24, a positive number that is not a perfect square. The equation had two solutions that were irrational numbers. Finally, in Example 3, the discriminant was -8, a negative number. The equation had solutions that were imaginary numbers involving i. In this case, the graph of the corresponding quadratic function had no x-intercepts.

These observations are generalized in Table 11.2.

2 Use the discriminant to determine the number and type of solutions.

Table 11.2 The Discriminant and the Kinds of Solutions to $ax^2 + bx + c = 0$

Discriminant $b^2 - 4ac$	Kinds of Solutions to $ax^2 + bx + c = 0$	Graph of $y = ax^2 + bx + c$
$b^2 - 4ac > 0$	**Two unequal real solutions;** If a, b, and c are rational numbers and the discriminant is a perfect square, the solutions are rational. If the discriminant is not a perfect square, the solutions are irrational conjugates.	 Two x-intercepts
$b^2 - 4ac = 0$	**One real solution (a repeated solution);** If a, b, and c are rational numbers, the repeated solution is also a rational number.	 One x-intercept
$b^2 - 4ac < 0$	**No real solution; two imaginary solutions;** The solutions are complex conjugates.	 No x-intercepts

EXAMPLE 4 Using the Discriminant

For each equation, compute the discriminant. Then determine the number and type of solutions:

a. $3x^2 + 4x - 5 = 0$ **b.** $9x^2 - 6x + 1 = 0$ **c.** $3x^2 - 8x + 7 = 0.$

SOLUTION Begin by identifying the values for a, b, and c in each equation. Then compute $b^2 - 4ac$, the discriminant.

a. $3x^2 + 4x - 5 = 0$

$a = 3$ $b = 4$ $c = -5$

Substitute and compute the discriminant:

$$b^2 - 4ac = 4^2 - 4 \cdot 3(-5) = 16 - (-60) = 16 + 60 = 76.$$

The discriminant, 76, is a positive number that is not a perfect square. Thus, there are two irrational solutions. (These solutions are conjugates of each other.)

b. $9x^2 - 6x + 1 = 0$

$a = 9$ $b = -6$ $c = 1$

Substitute and compute the discriminant:

$$b^2 - 4ac = (-6)^2 - 4 \cdot 9 \cdot 1 = 36 - 36 = 0.$$

The discriminant, 0, shows that there is only one real solution. This real solution is a rational number.

c. $3x^2 - 8x + 7 = 0$

$a = 3$ $b = -8$ $c = 7$

$$b^2 - 4ac = (-8)^2 - 4 \cdot 3 \cdot 7 = 64 - 84 = -20$$

The negative discriminant, -20, shows that there are two imaginary solutions. (These solutions are complex conjugates of each other.) ∎

✓ **CHECK POINT 4** For each equation, compute the discriminant. Then determine the number and type of solutions:

a. $x^2 + 6x + 9 = 0$ 0; one real solution

b. $2x^2 - 7x - 4 = 0$ 81; two rational solutions

c. $3x^2 - 2x + 4 = 0$. -44; two imaginary solutions that are complex conjugates

3 Determine the most efficient method to use when solving a quadratic equation.

Determining Which Method to Use All quadratic equations can be solved by the quadratic formula. However, if an equation is in the form $u^2 = d$, such as $x^2 = 5$ or $(2x + 3)^2 = 8$, it is faster to use the square root property, taking the square root of both sides. If the equation is not in the form $u^2 = d$, write the quadratic equation in standard form $(ax^2 + bx + c = 0)$. Try to solve the equation by factoring. If $ax^2 + bx + c$ cannot be factored, then solve the quadratic equation by the quadratic formula.

Because we used the method of completing the square to derive the quadratic formula, we no longer need it for solving quadratic equations. However, we will use completing the square in Chapter 13 to help graph certain kinds of equations.

Table 11.3 summarizes our observations about which technique to use when solving a quadratic equation.

Table 11.3	Determining the Most Efficient Technique to Use when Solving a Quadratic Equation	
Description and Form of the Quadratic Equation	**Most Efficient Solution Method**	**Example**
$ax^2 + bx + c = 0$ and $ax^2 + bx + c$ can be factored easily.	Factor and use the zero-product principle.	$3x^2 + 5x - 2 = 0$ $(3x - 1)(x + 2) = 0$ $3x - 1 = 0$ or $x + 2 = 0$ $x = \dfrac{1}{3}$ $\qquad x = -2$
$ax^2 + c = 0$ The quadratic equation has no x-term. ($b = 0$)	Solve for x^2 and apply the square root property.	$4x^2 - 7 = 0$ $4x^2 = 7$ $x^2 = \dfrac{7}{4}$ $x = \pm\dfrac{\sqrt{7}}{2}$
$u^2 = d$; u is a first-degree polynomial.	Use the square root property.	$(x + 4)^2 = 5$ $x + 4 = \pm\sqrt{5}$ $x = -4 \pm \sqrt{5}$
$ax^2 + bx + c = 0$ and $ax^2 + bx + c$ cannot be factored or the factoring is too difficult.	Use the quadratic formula: $x = \dfrac{-b \pm \sqrt{b^2 - 4ac}}{2a}.$	$x^2 - 2x - 6 = 0$ $a = 1 \quad b = -2 \quad c = -6$ $x = \dfrac{-(-2) \pm \sqrt{(-2)^2 - 4(1)(-6)}}{2(1)}$ $= \dfrac{2 \pm \sqrt{4 - 4(1)(-6)}}{2(1)}$ $= \dfrac{2 \pm \sqrt{28}}{2} = \dfrac{2 \pm \sqrt{4}\sqrt{7}}{2}$ $= \dfrac{2 \pm 2\sqrt{7}}{2} = \dfrac{2(1 \pm \sqrt{7})}{2}$ $= 1 \pm \sqrt{7}$

4 Write quadratic equations from solutions.

Writing Quadratic Equations from Solutions Using the zero-product principle, the equation $(x - 3)(x + 5) = 0$ has two solutions, 3 and -5. By applying the zero-product principle in reverse, we can find a quadratic equation that has two given numbers as its solutions.

THE ZERO-PRODUCT PRINCIPLE IN REVERSE If $A = 0$ or $B = 0$, then $AB = 0$.

Jasper Johns, "Zero". © Jasper Johns/Licensed by VAGA, New York, NY.

The special properties of zero make it possible to write a quadratic equation from its solutions.

EXAMPLE 5 Writing Equations from Solutions

Write a quadratic equation with the given solution set:

a. $\left\{-\dfrac{5}{3}, \dfrac{1}{2}\right\}$ **b.** $\{-5i, 5i\}$.

SOLUTION

a. Because the solution set is $\left\{-\dfrac{5}{3}, \dfrac{1}{2}\right\}$, then

$$x = -\frac{5}{3} \quad \text{or} \quad x = \frac{1}{2}.$$

$$x + \frac{5}{3} = 0 \quad \text{or} \quad x - \frac{1}{2} = 0 \qquad \text{Obtain zero on one side of each equation.}$$

$$3x + 5 = 0 \quad \text{or} \quad 2x - 1 = 0 \qquad \text{Clear fractions, multiplying by 3 and 2, respectively.}$$

$$(3x + 5)(2x - 1) = 0 \qquad \text{Use the zero-product principle in reverse: If A = 0 or B = 0, then AB = 0.}$$

$$6x^2 - 3x + 10x - 5 = 0 \qquad \text{Use the FOIL method to multiply.}$$

$$6x^2 + 7x - 5 = 0. \qquad \text{Combine like terms.}$$

Thus, one equation is $6x^2 + 7x - 5 = 0$. Many other quadratic equations have $\left\{-\dfrac{5}{3}, \dfrac{1}{2}\right\}$ for their solution sets. These equations can be obtained by multiplying both sides of $6x^2 + 7x - 5 = 0$ by any nonzero real number.

b. Because the solution set is $\{-5i, 5i\}$, then

$$x = -5i \quad \text{or} \quad x = 5i$$

$$x + 5i = 0 \quad \text{or} \quad x - 5i = 0 \qquad \text{Obtain zero on one side of each equation.}$$

$$(x + 5i)(x - 5i) = 0 \qquad \text{Use the zero-product principle in reverse: If A = 0 or B = 0, then AB = 0.}$$

$$x^2 - (5i)^2 = 0 \qquad \text{Multiply conjugates using } (A + B)(A - B) = A^2 - B^2.$$

$$x^2 - 25i^2 = 0 \qquad (5i)^2 = 5^2 i^2 = 25 i^2$$

$$x^2 - 25(-1) = 0 \qquad i^2 = -1$$

$$x^2 + 25 = 0. \qquad \text{This is the required equation.}$$ ∎

 CHECK POINT 5 Write a quadratic equation with the given solution set:

a. $\left\{-\dfrac{3}{5}, \dfrac{1}{4}\right\}$ $20x^2 + 7x - 3 = 0$ **b.** $\{-7i, 7i\}$. $x^2 + 49 = 0$

5 Use the quadratic formula to solve problems.

Applications Quadratic equations can be solved to answer questions about variables contained in quadratic functions.

EXAMPLE 6 Making Predictions about the U.S. Foreign-Born Population

The quadratic function

$$f(x) = 0.0049x^2 - 0.361x + 11.79$$

models the percentage of the U.S. population, $f(x)$, that was foreign-born x years after 1930. According to this model, in which year will 15% of the U.S. population be foreign-born?

SOLUTION Because we are interested in when the foreign-born percentage will reach 15%, we substitute 15 for $f(x)$ in the given function. Then we solve for x, the number of years after 1930.

$$f(x) = 0.0049x^2 - 0.361x + 11.79 \qquad \text{This is the given function.}$$

$$15 = 0.0049x^2 - 0.361x + 11.79 \qquad \text{Substitute 15 for } f(x).$$

$$0 = 0.0049x^2 - 0.361x - 3.21 \qquad \text{Subtract 15 from both sides and write the quadratic equation in standard form.}$$

$$\boxed{a = 0.0049} \qquad \boxed{b = -0.361} \qquad \boxed{c = -3.21}$$

Because the trinomial on the right side of the equation is not easily factored, if it can be factored at all, we solve using the quadratic formula.

$$x = \frac{-b \pm \sqrt{b^2 - 4ac}}{2a} \qquad \text{Use the quadratic formula.}$$

$$= \frac{-(-0.361) \pm \sqrt{(-0.361)^2 - 4(0.0049)(-3.21)}}{2(0.0049)} \qquad \begin{array}{l}\text{Substitute the values for } a, \\ b, \text{ and } c: a = 0.0049, \\ b = -0.361, c = -3.21.\end{array}$$

$$= \frac{0.361 \pm \sqrt{0.193237}}{0.0098} \qquad \begin{array}{l}\text{Perform the indicated} \\ \text{computations. Use a calculator} \\ \text{to simplify the radicand.}\end{array}$$

$$x = \frac{0.361 + \sqrt{0.193237}}{0.0098} \quad \text{or} \quad x = \frac{0.361 - \sqrt{0.193237}}{0.0098}$$

$$x \approx 82 \qquad\qquad\qquad x \approx -8$$

> Reject this solution. The model applies to years **after**, not **before**, 1930.

According to the model, approximately 82 years after 1930, in 2012, 15% of the U.S. population will be foreign-born. ∎

USING TECHNOLOGY

A graphing utility's ┃TABLE┃ feature can be used to verify the solution to Example 6.

Enter the foreign-born model:
$y_1 = 0.0049x^2 - 0.361x + 11.79$

X	Y1
79	13.852
80	14.27
81	14.698
82	15.136
83	15.583
84	16.04
85	16.508

X=79

The foreign-born population is approximately 15% of the U.S. population 82 years after 1930.

 CHECK POINT 6 According to the model in Example 6, in which year(s) was 5.4% of the U.S. population foreign-born? How well does the function describe the situation for that (these) year(s) shown in Figure 11.7 on page 735?   1960 and 1974; very well

11.2 EXERCISE SET

    Math $\cdot_{XL}$ **MyMathLab**

Student Solutions Manual   CD/Video   PH Math/Tutor Center   MathXL Tutorials on CD   MathXL®   MyMathLab   Interactmath.com

Practice Exercises

In Exercises 1–18, solve each equation using the quadratic formula. Simplify solutions, if possible.

1. $x^2 + 8x + 12 = 0$ $\{-6, -2\}$

2. $x^2 + 8x + 15 = 0$ $\{-5, -3\}$

3. $2x^2 - 7x = -5$ $\{1, \frac{5}{2}\}$

4. $5x^2 + 8x = -3$ $\{-1, -\frac{3}{5}\}$

5. $x^2 + 3x - 20 = 0$ $\left\{\dfrac{-3 \pm \sqrt{89}}{2}\right\}$

6. $x^2 + 5x - 10 = 0$ $\left\{\dfrac{-5 \pm \sqrt{65}}{2}\right\}$

7. $3x^2 - 7x = 3$ $\left\{\dfrac{7 \pm \sqrt{85}}{6}\right\}$

8. $4x^2 + 3x = 2$ $\left\{\dfrac{-3 \pm \sqrt{41}}{8}\right\}$

9. $6x^2 = 2x + 1$ $\left\{\dfrac{1 \pm \sqrt{7}}{6}\right\}$

10. $2x^2 = -4x + 5$ $\left\{\dfrac{-2 \pm \sqrt{14}}{2}\right\}$

11. $4x^2 - 3x = -6$ $\left\{\dfrac{3}{8} \pm \dfrac{\sqrt{87}}{8}i\right\}$

12. $9x^2 + x = -2$ $\left\{-\dfrac{1}{18} \pm \dfrac{\sqrt{71}}{18}i\right\}$

13. $x^2 - 4x + 8 = 0$ $\{2 \pm 2i\}$

14. $x^2 + 6x + 13 = 0$ $\{-3 \pm 2i\}$

15. $3x^2 = 8x - 7$ $\left\{\dfrac{4}{3} \pm \dfrac{\sqrt{5}}{3}i\right\}$

16. $3x^2 = 4x - 6$ $\left\{\dfrac{2}{3} \pm \dfrac{\sqrt{14}}{3}i\right\}$

17. $2x(x - 2) = x + 12$ $\{-\frac{3}{2}, 4\}$

18. $2x(x + 4) = 3x - 3$ $\{-\frac{3}{2}, -1\}$

In Exercises 19–30, compute the discriminant. Then determine the number and type of solutions for the given equation.

19. $x^2 + 8x + 3 = 0$ 52; two irrational solutions

20. $x^2 + 7x + 4 = 0$ 33; two irrational solutions

21. $x^2 + 6x + 8 = 0$ 4; two rational solutions

22. $x^2 + 2x - 3 = 0$ 16; two rational solutions

23. $2x^2 + x + 3 = 0$ −23; two imaginary solutions

24. $2x^2 - 4x + 3 = 0$ −8; two imaginary solutions

25. $2x^2 + 6x = 0$ 36; two rational solutions

26. $3x^2 - 5x = 0$ 25; two rational solutions

27. $5x^2 + 3 = 0$ −60; two imaginary solutions

28. $5x^2 + 4 = 0$ −80; two imaginary solutions

29. $9x^2 = 12x - 4$ 0; one (repeated) rational solution

30. $4x^2 = 20x - 25$ 0; one (repeated) rational solution

In Exercises 31–46, solve each equation by the method of your choice. Simplify solutions, if possible.

31. $3x^2 - 4x = 4$ $\{-\frac{2}{3}, 2\}$

32. $2x^2 - x = 1$ $\{-\frac{1}{2}, 1\}$

33. $x^2 - 2x = 1$ $\{1 \pm \sqrt{2}\}$

34. $2x^2 + 3x = 1$ $\left\{\dfrac{-3 \pm \sqrt{17}}{4}\right\}$

35. $(2x - 5)(x + 1) = 2$ $\left\{\dfrac{3 \pm \sqrt{65}}{4}\right\}$

36. $(2x + 3)(x + 4) = 1$ $\left\{\dfrac{-11 \pm \sqrt{33}}{4}\right\}$

37. $(3x - 4)^2 = 16$ $\{0, \frac{8}{3}\}$

38. $(2x + 7)^2 = 25$ $\{-6, -1\}$

39. $\dfrac{x^2}{2} + 2x + \dfrac{2}{3} = 0$ $\left\{\dfrac{-6 \pm 2\sqrt{6}}{3}\right\}$

40. $\dfrac{x^2}{3} - x - \dfrac{1}{6} = 0$ $\left\{\dfrac{3 \pm \sqrt{11}}{2}\right\}$

41. $(3x - 2)^2 = 10$ $\left\{\dfrac{2 \pm \sqrt{10}}{3}\right\}$

42. $(4x - 1)^2 = 15$ $\left\{\dfrac{1 \pm \sqrt{15}}{4}\right\}$

43. $\dfrac{1}{x} + \dfrac{1}{x + 2} = \dfrac{1}{3}$ $\{2 \pm \sqrt{10}\}$

44. $\dfrac{1}{x} + \dfrac{1}{x + 3} = \dfrac{1}{4}$ $\left\{\dfrac{5 \pm \sqrt{73}}{2}\right\}$

45. $(2x - 6)(x + 2) = 5(x - 1) - 12$ $\{1, \frac{5}{2}\}$

46. $7x(x - 2) = 3 - 2(x + 4)$ $\{\frac{5}{7}, 1\}$

In Exercises 47–60, write a quadratic equation in standard form with the given solution set.

47. $\{-3, 5\}$ $x^2 - 2x - 15 = 0$

48. $\{-2, 6\}$ $x^2 - 4x - 12 = 0$

49. $\left\{-\dfrac{2}{3}, \dfrac{1}{4}\right\}$ $12x^2 + 5x - 2 = 0$

50. $\left\{-\dfrac{5}{6}, \dfrac{1}{3}\right\}$ $18x^2 + 9x - 5 = 0$

51. $\{-6i, 6i\}$ $x^2 + 36 = 0$

52. $\{-8i, 8i\}$ $x^2 + 64 = 0$

53. $\left\{-\sqrt{2}, \sqrt{2}\right\}$ $\quad x^2 - 2 = 0$

54. $\left\{-\sqrt{3}, \sqrt{3}\right\}$ $\quad x^2 - 3 = 0$

55. $\left\{-2\sqrt{5}, 2\sqrt{5}\right\}$ $\quad x^2 - 20 = 0$

56. $\left\{-3\sqrt{5}, 3\sqrt{5}\right\}$ $\quad x^2 - 45 = 0$

57. $\{1 + i, 1 - i\}$ $\quad x^2 - 2x + 2 = 0$

58. $\{2 + i, 2 - i\}$ $\quad x^2 - 4x + 5 = 0$

59. $\left\{1 + \sqrt{2}, 1 - \sqrt{2}\right\}$ $\quad x^2 - 2x - 1 = 0$

60. $\left\{1 + \sqrt{3}, 1 - \sqrt{3}\right\}$ $\quad x^2 - 2x - 2 = 0$

Practice Plus

Exercises 61–64 describe quadratic equations. Match each description with the graph of the corresponding qudratic function. Each graph is shown in a $[-10, 10, 1]$ *by* $[-10, 10, 1]$ *viewing rectangle.*

61. A quadratic equation whose solution set contains imaginary numbers   b

62. A quadratic equation whose discriminant is 0   c

63. A quadratic equation whose solution set is $\{3 \pm \sqrt{2}\}$ a

64. A quadratic equation whose solution set contains integers   d

a.

b.

c.

d.

65. When the sum of 6 and twice a positive number is subtracted from the square of the number, 0 results. Find the number. $1 + \sqrt{7}$

66. When the sum of 1 and twice a negative number is subtracted from twice the square of the number, 0 results. Find the number. $\dfrac{1 - \sqrt{3}}{2}$

In Exercises 67–72, solve each equation by the method of your choice.

67. $\dfrac{1}{x^2 - 3x + 2} = \dfrac{1}{x + 2} + \dfrac{5}{x^2 - 4}$ $\left\{\dfrac{-1 \pm \sqrt{21}}{2}\right\}$

68. $\dfrac{x - 1}{x - 2} + \dfrac{x}{x - 3} = \dfrac{1}{x^2 - 5x + 6}$ $\left\{\dfrac{3 \pm \sqrt{5}}{2}\right\}$

69. $\sqrt{2}x^2 + 3x - 2\sqrt{2} = 0$ $\left\{-2\sqrt{2}, \dfrac{\sqrt{2}}{2}\right\}$

70. $\sqrt{3}x^2 + 6x + 7\sqrt{3} = 0$ $\left\{-\sqrt{3} \pm 2i\right\}$

71. $\left|x^2 + 2x\right| = 3$ $\left\{-3, 1, -1 \pm \sqrt{2}i\right\}$

72. $\left|x^2 + 3x\right| = 2$ $\left\{-2, -1, \dfrac{-3 \pm \sqrt{17}}{2}\right\}$

Application Exercises

A driver's age has something to do with his or her chance of getting into a fatal car crash. The bar graph shows the number of fatal vehicle crashes per 100 million miles driven for drivers of various age groups. For example, 25-year-old drivers are involved in 4.1 fatal crashes per 100 million miles driven. Thus, when a group of 25-year-old Americans have driven a total of 100 million miles, approximately 4 have been in accidents in which someone died.

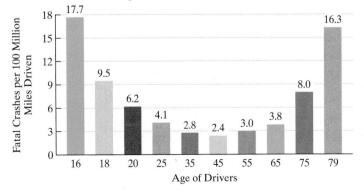

Age of U.S. Drivers and Fatal Crashes

Source: Insurance Institute for Highway Safety

The number of fatal vehicle crashes per 100 million miles, $f(x)$, for drivers of age x can be modeled by the quadratic function

$$f(x) = 0.013x^2 - 1.19x + 28.24.$$

Use the function to solve Exercises 73–74.

73. What age groups are expected to be involved in 3 fatal crashes per 100 million miles driven? How well does the function model the trend in the actual data shown in the bar graph on the previous page?   33-year-olds and 58-year-olds; The function models the actual data well.

74. What age groups are expected to be involved in 10 fatal crashes per 100 million miles driven? How well does the function model the trend in the actual data shown in the bar graph on the previous page?   19-year-olds and 72-year-olds; Answers will vary.

Throwing events in track and field include the shot put, the discus throw, the hammer throw, and the javelin throw. The distance that an athlete can achieve depends on the initial velocity of the object thrown and the angle above the horizontal at which the object leaves the hand.

In Exercises 75–76, an athlete whose event is the shot put releases the shot with the same initial velocity, but at different angles.

75. When the shot is released at an angle of 35°, its path can be modeled by the function

$$f(x) = -0.01x^2 + 0.7x + 6.1$$

in which x is the shot's horizontal distance, in feet, and $f(x)$ is its height, in feet. This function is shown by one of the graphs, (a) or (b), in the figure. Use the function to determine the shot's maximum distance. Use a calculator and round to the nearest tenth of a foot. Which graph, (a) or (b), shows the shot's path?   77.8 feet; (b)

Horizontal Distance
[0, 80, 10] by [0, 40, 10]

76. When the shot is released at an angle of 65°, its path can be modeled by the function

$$f(x) = -0.04x^2 + 2.1x + 6.1$$

in which x is the shot's horizontal distance, in feet, and $f(x)$ is its height, in feet. This function is shown by one of the graphs, (a) or (b), in the figure at the bottom of the previous column. Use the function to determine the shot's maximum distance. Use a calculator and round to the nearest tenth of a foot. Which graph, (a) or (b), shows the shot's path?   55.3 feet; (a)

77. The length of a rectangle is 4 meters longer than the width. If the area is 8 square meters, find the rectangle's dimensions. Round to the nearest tenth of a meter.   1.5 meters by 5.5 meters

78. The length of a rectangle exceeds twice its width by 3 inches. If the area is 10 square inches, find the rectangle's dimensions. Round to the nearest tenth of an inch.   1.6 inches by 6.2 inches

79. One leg of a right triangle exceeds the shorter leg by 1 inch, and the hypotenuse exceeds the longer leg by 7 inches. Find the lengths of the legs. Round to the nearest tenth of a inch.   17.6 inches and 18.6 inches

80. The hypotenuse of a right triangle is 6 feet long. One leg is 2 feet shorter than the other. Find the lengths of the legs. Round to the nearest tenth of a foot.   5.1 feet and 3.1 feet

81. A rain gutter is made from sheets of aluminum that are 20 inches wide. As shown in the figure, the edges are turned up to form right angles. Determine the depth of the gutter that will allow a cross-sectional area of 13 square inches. Show that there are two different solutions to the problem. Round to the nearest tenth of an inch.

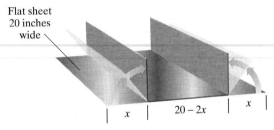

9.3 inches and 0.7 inch

82. A piece of wire is 8 inches long. The wire is cut into two pieces and then each piece is bent into a square. Find the length of each piece if the sum of the areas of these squares is to be 2 square inches.

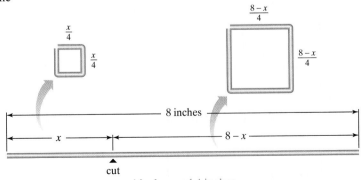

4 inches and 4 inches

83. Working together, two people can mow a large lawn in 4 hours. One person can do the job alone 1 hour faster than the other person. How long does it take each person working alone to mow the lawn? Round to the nearest tenth of an hour.   7.5 hours and 8.5 hours

84. A pool has an inlet pipe to fill it and an outlet pipe to empty it. It takes 2 hours longer to empty the pool than it does to fill it. The inlet pipe is turned on to fill the pool, but the outlet pipe is accidentally open. Despite this, the pool fills in 8 hours. How long does it take the outlet pipe to empty the pool? Round to the nearest tenth of an hour.   5.1 hours

Writing in Mathematics

85. What is the quadratic formula and why is it useful?
85–92. Answers will vary.

86. Without going into specific details for every step, describe how the quadratic formula is derived.

87. Explain how to solve $x^2 + 6x + 8 = 0$ using the quadratic formula.

88. If a quadratic equation has imaginary solutions, how is this shown on the graph of the corresponding quadratic function?

89. What is the discriminant and what information does it provide about a quadratic equation?

90. If you are given a quadratic equation, how do you determine which method to use to solve it?

91. Explain how to write a quadratic equation from its solution set. Give an example with your explanation.

92. Describe the trend shown by the data for the percentage of the U.S. population that was foreign-born from 1930 through 2002. (See Figure 11.7 on page 735.) Do you believe that this trend is likely to continue or might something occur that would make it impossible to extend the model in Example 6 on page 744 into the future? Explain your answer.

Technology Exercises

93. Use a graphing utility to graph the quadratic function related to any five of the quadratic equations in Exercises 19–30. How does each graph illustrate what you determined algebraically using the discriminant?   Answers will vary.

94. Reread Exercise 81. The cross-sectional area of the gutter is given by the quadratic function

$$f(x) = x(20 - 2x).$$

Graph the function in a $[0, 10, 1]$ by $[0, 60, 5]$ viewing rectangle. Then ⎡TRACE⎤ along the curve or use the maximum function feature to determine the depth of the gutter that will maximize its cross-sectional area and allow the greatest amount of water to flow. What is the maximum area? Does the situation described in Exercise 81 take full advantage of the sheets of aluminum?   See graphing answer section. Depth: 5 inches; maximum area: 50 square inches; Answers will vary.

Critical Thinking Exercises

95. Which one of the following is true?   **d**

   a. The quadratic formula is developed by applying factoring and the zero-product principle to the quadratic equation $ax^2 + bx + c = 0$.

   b. In using the quadratic formula to solve the quadratic equation $5x^2 = 2x - 7$, we have $a = 5, b = 2$, and $c = -7$.

   c. The quadratic formula cannot be used to solve the equation $x^2 - 9 = 0$.

   d. Any quadratic equation that can be solved by completing the square can be solved by the quadratic formula.

96. Solve for t: $s = -16t^2 + v_0 t$. $t = \dfrac{v_0 \pm \sqrt{v_0^2 - 64s}}{32}$

97. A rectangular swimming pool is 12 meters long and 8 meters wide. A tile border of uniform width is to be built around the pool using 120 square meters of tile. The tile is from a discontinued stock (so no additional materials are available) and all 120 square meters are to be used. How wide should the border be? Round to the nearest tenth of a meter. If zoning laws require at least a 2-meter-wide border around the pool, can this be done with the available tile?   The border should be about 2.4 meters wide. Yes, the available tile meets the zoning laws.

98. The area of the shaded region outside the rectangle and inside the triangle is 10 square yards. Find the triangle's height, represented by $2x$. Round to the nearest tenth of a yard.

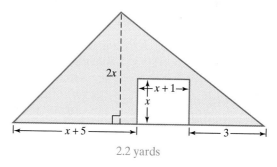

2.2 yards

Review Exercises

99. Solve: $|5x + 2| = |4 - 3x|$. (Section 9.3, Example 2)
   $\left\{-3, \frac{1}{4}\right\}$

100. Solve: $\sqrt{2x - 5} - \sqrt{x - 3} = 1$. (Section 10.6, Example 4)
   $\{3, 7\}$

101. Rationalize the denominator: $\dfrac{5}{\sqrt{3} + x}$.

   (Section 10.5, Example 5) $\dfrac{5\sqrt{3} - 5x}{3 - x^2}$

11.3

Objectives

1 Recognize characteristics of parabolas.

2 Graph parabolas in the form $f(x) = a(x - h)^2 + k$.

3 Graph parabolas in the form $f(x) = ax^2 + bx + c$.

4 Solve problems involving minimizing or maximizing quadratic functions.

QUADRATIC FUNCTIONS AND THEIR GRAPHS

We have a long history of throwing things. Before 400 B.C., the Greeks competed in games that included discus throwing. In the seventeenth century, English soldiers organized cannonball-throwing competitions. In 1827, a Yale University student, disappointed over failing an exam, took out his frustrations at the passing of a collection plate in chapel. Seizing the monetary tray, he flung it in the direction of a large open space on campus. Yale students see this act of frustration as the origin of the Frisbee.

In this section, we study quadratic functions and their graphs. By graphing functions that model the paths of the things we throw, you will be able to determine both the maximum height and the distance of these objects.

Graphs of Quadratic Functions   The graph of the quadratic function

$$f(x) = ax^2 + bx + c, \quad a \neq 0,$$

1 Recognize characteristics of parabolas.

is called a **parabola**. Parabolas are shaped like cups, as shown in Figure 11.8. If the coefficient of x^2 (the value of a in $ax^2 + bx + c$) is positive, the parabola opens upward. If the coefficient of x^2 is negative, the parabola opens downward. The **vertex** (or turning point) of the parabola is the lowest point on the graph when it opens upward and the highest point on the graph when it opens downward.

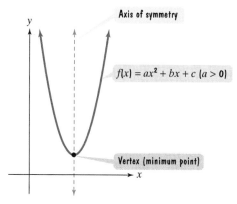

$a > 0$: Parabola opens upward.

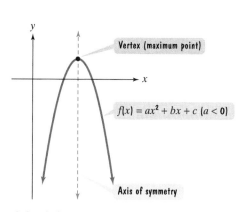

$a < 0$: Parabola opens downward.

FIGURE 11.8   Characteristics of graphs of quadratic functions

Look at the unusual image of the word "mirror" shown below. The artist, Scott Kim, has created the image so that the two halves of the whole are mirror images of each other. A parabola shares this kind of symmetry, in which a line through the vertex divides the figure in half. Parabolas are symmetric with respect to this line, called the **axis of symmetry**. If a parabola is folded along its axis of symmetry, the two halves match exactly.

Graphing Quadratic Functions in the Form $f(x) = a(x - h)^2 + k$ One way to obtain the graph of a quadratic function is to use point plotting. Let's begin by graphing the functions $f(x) = x^2, g(x) = 2x^2,$ and $h(x) = \frac{1}{2}x^2$ in the same rectangular coordinate system. Select integers for x, starting with -3 and ending with 3. A partial table of coordinates for each function is shown below. The three parabolas are shown in Figure 11.9.

2 Graph parabolas in the form $f(x) = a(x - h)^2 + k$.

x	$f(x) = x^2$	(x, y) or $(x, f(x))$
-3	$f(-3) = (-3)^2 = 9$	$(-3, 9)$
-2	$f(-2) = (-2)^2 = 4$	$(-2, 4)$
-1	$f(-1) = (-1)^2 = 1$	$(-1, 1)$
0	$f(0) = 0^2 = 0$	$(0, 0)$
1	$f(1) = 1^2 = 1$	$(1, 1)$
2	$f(2) = 2^2 = 4$	$(2, 4)$
3	$f(3) = 3^2 = 9$	$(3, 9)$

x	$g(x) = 2x^2$	(x, y) or $(x, g(x))$
-3	$g(-3) = 2(-3)^2 = 18$	$(-3, 18)$
-2	$g(-2) = 2(-2)^2 = 8$	$(-2, 8)$
-1	$g(-1) = 2(-1)^2 = 2$	$(-1, 2)$
0	$g(0) = 2 \cdot 0^2 = 0$	$(0, 0)$
1	$g(1) = 2 \cdot 1^2 = 2$	$(1, 2)$
2	$g(2) = 2 \cdot 2^2 = 8$	$(2, 8)$
3	$g(3) = 2 \cdot 3^2 = 18$	$(3, 18)$

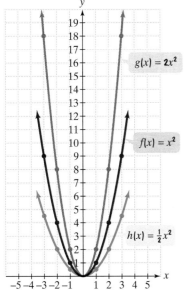

FIGURE 11.9

x	$h(x) = \frac{1}{2}x^2$	(x, y) or $(x, h(x))$
-3	$h(-3) = \frac{1}{2}(-3)^2 = \frac{9}{2}$	$\left(-3, \frac{9}{2}\right)$
-2	$h(-2) = \frac{1}{2}(-2)^2 = 2$	$(-2, 2)$
-1	$h(-1) = \frac{1}{2}(-1)^2 = \frac{1}{2}$	$\left(-1, \frac{1}{2}\right)$
0	$h(0) = \frac{1}{2} \cdot 0^2 = 0$	$(0, 0)$
1	$h(1) = \frac{1}{2} \cdot 1^2 = \frac{1}{2}$	$\left(1, \frac{1}{2}\right)$
2	$h(2) = \frac{1}{2} \cdot 2^2 = 2$	$(2, 2)$
3	$h(3) = \frac{1}{2} \cdot 3^2 = \frac{9}{2}$	$\left(3, \frac{9}{2}\right)$

Can you see that the graphs of f, g, and h all have the same vertex, $(0, 0)$? They also have the same axis of symmetry, the y-axis, or $x = 0$. This is true for all graphs of the form $f(x) = ax^2$. However, the blue graph of $g(x) = 2x^2$ is a narrower parabola than the red graph of $f(x) = x^2$. By contrast, the green graph of $h(x) = \frac{1}{2}x^2$ is a flatter parabola than the red graph of $f(x) = x^2$.

Is there a more efficient method than point plotting to obtain the graph of a quadratic function? The answer is yes. The method is based on comparing graphs of the form $g(x) = a(x - h)^2 + k$ to those of the form $f(x) = ax^2$.

In Figure 11.10(a), the graph of $f(x) = ax^2$ for $a > 0$ is shown in black. The parabola's vertex is $(0, 0)$ and it opens upward. In Figure 11.10(b), the graph of $f(x) = ax^2$ for $a < 0$ is shown in black. The parabola's vertex is $(0, 0)$ and it opens downward.

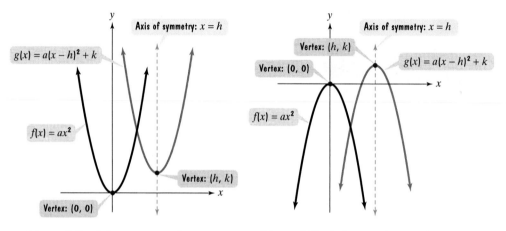

(a) $a > 0$: Parabola opens upward. **(b)** $a < 0$: Parabola opens downward.

FIGURE 11.10 Moving, or shifting, the graph of $f(x) = ax^2$

Figure 11.10 also shows the graph of $g(x) = a(x - h)^2 + k$ in blue. Compare these graphs to those of $f(x) = ax^2$. Observe that h causes a horizontal move, or shift, and k causes a vertical shift of the graph of $f(x) = ax^2$. Consequently, the vertex $(0, 0)$ on the graph of $f(x) = ax^2$ moves to the point (h, k) on the graph of $g(x) = a(x - h)^2 + k$. The axis of symmetry is the vertical line whose equation is $x = h$.

The form of the expression for g is convenient because it immediately identifies the vertex of the parabola as (h, k). The sign of a in $g(x) = a(x - h)^2 + k$ determines whether the parabola opens upward or downward. Furthermore, if $|a|$ is small, the parabola opens more flatly than if $|a|$ is large.

> **GRAPHING QUADRATIC FUNCTIONS WITH EQUATIONS IN THE FORM $f(x) = a(x - h)^2 + k$** To graph $f(x) = a(x - h)^2 + k$,
>
> 1. Determine whether the parabola opens upward or downward. If $a > 0$, it opens upward. If $a < 0$, it opens downward.
> 2. Determine the vertex of the parabola. The vertex is (h, k).
> 3. Find any x-intercepts by replacing $f(x)$ with 0. Solve the resulting quadratic equation for x.
> 4. Find the y-intercept by replacing x with 0.
> 5. Plot the intercepts and vertex and additional points as necessary. Connect these points with a smooth curve that is shaped like a cup.

EXAMPLE 1 Graphing a Quadratic Function in the Form $f(x) = a(x - h)^2 + k$

Graph the quadratic function $f(x) = -2(x - 3)^2 + 8$.

SOLUTION We can graph this function by following the steps in the preceding box. We begin by identifying values for $a, h,$ and k.

$$f(x) = a(x - h)^2 + k$$

$$a = -2 \quad h = 3 \quad k = 8$$

$$f(x) = -2(x - 3)^2 + 8$$

Step 1. **Determine how the parabola opens.** Note that a, the coefficient of x^2, is -2. Thus, $a < 0$; this negative value tells us that the parabola opens downward.

Step 2. **Find the vertex.** The vertex of the parabola is at (h, k). Because $h = 3$ and $k = 8$, the parabola has its vertex at $(3, 8)$.

Step 3. **Find the x-intercepts.** Replace $f(x)$ with 0 in $f(x) = -2(x - 3)^2 + 8$.

$0 = -2(x - 3)^2 + 8$	Find x-intercepts, setting f(x) equal to 0.
$2(x - 3)^2 = 8$	Solve for x. Add 2(x − 3)² to both sides of the equation.
$(x - 3)^2 = 4$	Divide both sides by 2.
$x - 3 = \sqrt{4} \quad$ or $\quad x - 3 = -\sqrt{4}$	Apply the square root property.
$x - 3 = 2 \qquad\qquad x - 3 = -2$	$\sqrt{4} = 2$
$x = 5 \qquad\qquad\quad x = 1$	Add 3 to both sides in each equation.

The x-intercepts are 5 and 1. The parabola passes through $(5, 0)$ and $(1, 0)$.

Step 4. **Find the y-intercept.** Replace x with 0 in $f(x) = -2(x - 3)^2 + 8$.

$$f(0) = -2(0 - 3)^2 + 8 = -2(-3)^2 + 8 = -2(9) + 8 = -10$$

The y-intercept is -10. The parabola passes through $(0, -10)$.

Step 5. **Graph the parabola.** With a vertex at $(3, 8)$, x-intercepts at 5 and 1, and a y-intercept at -10, the graph of f is shown in Figure 11.11. The axis of symmetry is the vertical line whose equation is $x = 3$. ∎

FIGURE 11.11 The graph of $f(x) = -2(x - 3)^2 + 8$

✔ **CHECK POINT 1** Graph the quadratic function $f(x) = -(x - 1)^2 + 4$.
See graphing answer section.

EXAMPLE 2 Graphing a Quadratic Function in the Form $f(x) = a(x - h)^2 + k$

Graph the quadratic function $f(x) = (x + 3)^2 + 1$.

SOLUTION We begin by finding values for $a, h,$ and k.

$$f(x) = a(x - h)^2 + k \quad \text{Form of quadratic function}$$

$$f(x) = (x + 3)^2 + 1 \quad \text{Given function}$$

or

$$f(x) = 1(x - (-3))^2 + 1$$

$$a = 1 \quad h = -3 \quad k = 1$$

Axis of symmetry: $x = -3$

y-intercept: 10

Vertex: (−3, 1)

FIGURE 11.12 The graph of $f(x) = (x + 3)^2 + 1$

Step 1. **Determine how the parabola opens.** Note that a, the coefficient of x^2, is 1. Thus, $a > 0$; this positive value tells us that the parabola opens upward.

Step 2. **Find the vertex.** The vertex of the parabola is at $(h, k.)$ Because $h = -3$ and $k = 1$, the parabola has its vertex at $(-3, 1)$.

Step 3. **Find the x-intercepts.** Replace $f(x)$ with 0 in $f(x) = (x + 3)^2 + 1$. Because the vertex is at $(-3, 1)$, which lies above the x-axis, and the parabola opens upward, it appears that this parabola has no x-intercepts. We can verify this observation algebraically.

$$0 = (x + 3)^2 + 1$$

Find possible x-intercepts, setting f(x) equal to 0.

$$-1 = (x + 3)^2$$

Solve for x. Subtract 1 from both sides.

$$x + 3 = \sqrt{-1} \quad \text{or} \quad x + 3 = -\sqrt{-1}$$

Apply the square root property.

$$x + 3 = i \qquad\qquad x + 3 = -i$$

$\sqrt{-1} = i$

$$x = -3 + i \qquad\qquad x = -3 - i$$

The solutions are $-3 \pm i$.

Because this equation has no real solutions, the parabola has no x-intercepts.

Step 4. **Find the y-intercept.** Replace x with 0 in $f(x) = (x + 3)^2 + 1$.

$$f(0) = (0 + 3)^2 + 1 = 3^2 + 1 = 9 + 1 = 10$$

The y-intercept is 10. The parabola passes through $(0, 10)$.

Step 5. **Graph the parabola.** With a vertex at $(-3, 1)$, no x-intercepts, and a y-intercept at 10, the graph of f is shown in Figure 11.12. The axis of symmetry is the vertical line whose equation is $x = -3$. ∎

 CHECK POINT 2 Graph the quadratic function $f(x) = (x - 2)^2 + 1$.
See graphing answer section.

---

3 Graph parabolas in the form $f(x) = ax^2 + bx + c$.

Graphing Quadratic Functions in the Form $f(x) = ax^2 + bx + c$ Quadratic functions are frequently expressed in the form $f(x) = ax^2 + bx + c$. How can we identify the vertex of a parabola whose equation is in this form? Completing the square provides the answer to this question.

$$f(x) = ax^2 + bx + c$$

$$= a\left(x^2 + \frac{b}{a}x\right) + c$$

Factor out a from $ax^2 + bx$.

$$= a\left(x^2 + \frac{b}{a}x + \frac{b^2}{4a^2}\right) + c - a\left(\frac{b^2}{4a^2}\right)$$

Complete the square by adding the square of half the coefficient of x.

By completing the square, we added $a \cdot \frac{b^2}{4a^2}$. To avoid changing the function's equation, we must subtract this term.

$$= a\left(x + \frac{b}{2a}\right)^2 + c - \frac{b^2}{4a}$$

Write the trinomial as the square of a binomial and simplify the constant term.

Now let's compare the form of the equation on the bottom of the previous page with a quadratic function in the form $f(x) = a(x - h)^2 + k$.

The form we know how to graph $\qquad f(x) = a(x - h)^2 + k$

$$h = -\frac{b}{2a} \qquad k = c - \frac{b^2}{4a}$$

Equation under discussion $\qquad f(x) = a\left(x - \left(-\frac{b}{2a}\right)\right)^2 + c - \frac{b^2}{4a}$

The important part of this observation is that h, the x-coordinate of the vertex, is $-\dfrac{b}{2a}$.

The y-coordinate can be found by evaluating the function at $-\dfrac{b}{2a}$.

THE VERTEX OF A PARABOLA WHOSE EQUATION IS $f(x) = ax^2 + bx + c$

Consider the parabola defined by the quadratic function $f(x) = ax^2 + bx + c$.

The parabola's vertex is $\left(-\dfrac{b}{2a}, f\left(-\dfrac{b}{2a}\right)\right)$.

We can apply our five-step procedure and graph parabolas in the form $f(x) = ax^2 + bx + c$. The only step that is different is how we determine the vertex.

EXAMPLE 3 Graphing a Quadratic Function in the Form $f(x) = ax^2 + bx + c$

Graph the quadratic function $f(x) = -x^2 - 2x + 1$. Use the graph to identify the function's domain and its range.

SOLUTION

Step 1. **Determine how the parabola opens.** Note that a, the coefficient of x^2, is -1. Thus, $a < 0$; this negative value tells us that the parabola opens downward.

Step 2. **Find the vertex.** We know that the x-coordinate of the vertex is $x = -\dfrac{b}{2a}$. We identify $a, b,$ and c in $f(x) = ax^2 + bx + c$.

$$f(x) = -x^2 - 2x + 1$$

$$a = -1 \qquad b = -2 \qquad c = 1$$

Substitute the values of a and b into the equation for the x-coordinate:

$$x = -\frac{b}{2a} = -\frac{-2}{2(-1)} = -\left(\frac{-2}{-2}\right) = -1.$$

The x-coordinate of the vertex is -1. We substitute -1 for x in the equation of the function to find the y-coordinate:

$$f(-1) = -(-1)^2 - 2(-1) + 1 = -1 + 2 + 1 = 2.$$

The vertex is $(-1, 2)$.

Step 3. **Find the x-intercepts.** Replace $f(x)$ with 0 in $f(x) = -x^2 - 2x + 1$. We obtain $0 = -x^2 - 2x + 1$. This equation cannot be solved by factoring. We will use the quadratic formula to solve it.

$$-x^2 - 2x + 1 = 0$$

$$\boxed{a = -1} \quad \boxed{b = -2} \quad \boxed{c = 1}$$

$$x = \frac{-b + \sqrt{b^2 - 4ac}}{2a} = \frac{-(-2) \pm \sqrt{(-2)^2 - 4(-1)(1)}}{2(-1)} = \frac{2 \pm \sqrt{4 - (-4)}}{-2}$$

> To locate the x-intercepts, we need decimal approximations. Thus, there is no need to simplify the radical form of the solutions.

$$x = \frac{2 + \sqrt{8}}{-2} \approx -2.4 \quad \text{or} \quad x = \frac{2 - \sqrt{8}}{-2} \approx 0.4$$

The x-intercepts are approximately -2.4 and 0.4. The parabola passes through $(-2.4, 0)$ and $(0.4, 0)$.

Step 4. **Find the y-intercept.** Replace x with 0 in $f(x) = -x^2 - 2x + 1$.

$$f(0) = -0^2 - 2 \cdot 0 + 1 = 1$$

The y-intercept is 1. The parabola passes through $(0, 1)$.

Step 5. **Graph the parabola.** With a vertex at $(-1, 2)$, x-intercepts at -2.4 and 0.4, and a y-intercept at 1, the graph of f is shown in Figure 11.13(a). The axis of symmetry is the vertical line whose equation is $x = -1$.

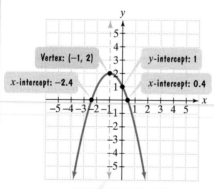

Vertex: $(-1, 2)$ — y-intercept: 1 — x-intercept: -2.4 — x-intercept: 0.4

Axis of symmetry: $x = -1$

FIGURE 11.13(a) The graph of $f(x) = -x^2 - 2x + 1$

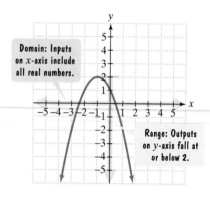

Domain: Inputs on x-axis include all real numbers.

Range: Outputs on y-axis fall at or below 2.

FIGURE 11.13(b) Determining the domain and range of $f(x) = -x^2 - 2x + 1$

STUDY TIP

The domain of any quadratic function includes all real numbers. If the vertex is the graph's highest point, the range includes all real numbers at or below the y-coordinate of the vertex. If the vertex is the graph's lowest point, the range includes all real numbers at or above the y-coordinate of the vertex.

Now we are ready to determine the domain and range of $f(x) = -x^2 - 2x + 1$. We can use the parabola, shown again in Figure 11.13(b), to do so. To find the domain, look for all the inputs on the x-axis that correspond to points on the graph. As the graph widens and continues to fall at both ends, can you see that these inputs include all real numbers?

Domain of f is $\{x \mid x$ is a real number$\}$ or $(-\infty, \infty)$.

To find the range, look for all the outputs on the y-axis that correspond to points on the graph. Figure 11.13(b) shows that the parabola's vertex, $(-1, 2)$, is the highest point on the graph. Because the y-coordinate of the vertex is 2, outputs on the y-axis fall at or below 2.

Range of f is $\{y \mid y \leq 2\}$ or $(-\infty, 2]$.

✔ CHECK POINT **3** Graph the quadratic function $f(x) = -x^2 + 4x + 1$. Use the graph to identify the function's domain and its range. See graphing answer section.
Domain = $\{x | x$ is a real number$\}$ or $(-\infty, \infty)$, range = $\{y | y \leq 5\}$ or $(-\infty, 5]$.

4 Solve problems involving minimizing or maximizing quadratic functions.

Applications of Quadratic Functions When do the things we throw reach their maximum height and what is that height? How many hours of sleep per night minimize the risk of death? How many units of a product should a business manufacture to maximize its profits? The answers to these questions involve finding the maximum or minimum value of quadratic functions.

Consider the quadratic function $f(x) = ax^2 + bx + c$. If $a > 0$, the parabola opens upward and the vertex is its lowest point. If $a < 0$, the parabola opens downward and the vertex is its highest point. The x-coordinate of the vertex is $-\dfrac{b}{2a}$. Thus, we can find the minimum or maximum value of f by evaluating the quadratic function at $x = -\dfrac{b}{2a}$.

MINIMUM AND MAXIMUM: QUADRATIC FUNCTIONS Consider $f(x) = ax^2 + bx + c$.

1. If $a > 0$, then f has a minimum that occurs at $x = -\dfrac{b}{2a}$.
   This minimum value is $f\left(-\dfrac{b}{2a}\right)$.

2. If $a < 0$, then f has a maximum that occurs at $x = -\dfrac{b}{2a}$.
   This maximum value is $f\left(-\dfrac{b}{2a}\right)$.

EXAMPLE 4 Parabolic Paths of a Shot Put

An athlete whose event is the shot put releases the shot with the same initial velocity, but at different angles. Figure 11.14 shows the parabolic paths for shots released at angles of 35° and 65°.

Maximum height

$g(x) = -0.04x^2 + 2.1x + 6.1$
Shot released at 65°

Maximum height

Distance of throw or maximum horizontal distance

$f(x) = -0.01x^2 + 0.7x + 6.1$
Shot released at 35°

Path's Vertical Distance (feet)

Path's Horizontal Distance (feet)

FIGURE 11.14 Two paths of a shot put

Throwing to Achieve Maximum Distance

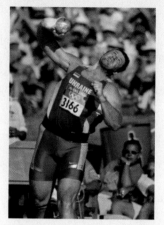

Although Example 4 focuses on maximum height of a propelled object, the goal of the shot put throw is to achieve the greatest possible distance. If an object is propelled from ground level, an angle of 45° produces the maximum distance for a given initial velocity. An angle slightly less than 45° is better when an object is thrown from above the ground. For the shot in Example 4, released from 6.1 feet above the ground, an angle of approximately 43° produces the maximum distance.

When the shot is released at an angle of 35°, its path can be modeled by the function

$$f(x) = -0.01x^2 + 0.7x + 6.1$$

in which x is the shot's horizontal distance, in feet, and $f(x)$ is its height, in feet. What is the maximum height of this shot's path?

SOLUTION The quadratic function is in the form $f(x) = ax^2 + bx + c$, with $a = -0.01$ and $b = 0.7$. Because $a < 0$, the function has a maximum that occurs at $x = -\dfrac{b}{2a}$.

$$x = -\frac{b}{2a} = -\frac{0.7}{2(-0.01)} = -(-35) = 35$$

This means that the shot's maximum height occurs when its horizontal distance is 35 feet. Can you see how this is shown by the blue graph of f in Figure 11.14 on the previous page? The maximum height of this path is

$$f(35) = -0.01(35)^2 + 0.7(35) + 6.1 = 18.35$$

or 18.35 feet. ∎

 CHECK POINT 4 Use function g, whose equation and graph are shown in Figure 11.14, to find the maximum height, to the nearest tenth of a foot, when the shot is released at an angle of 65°.   33.7 feet

Quadratic functions can also be formed from verbal conditions. Once we have obtained a quadratic function, we can then use the x-coordinate of the vertex to determine its maximum or minimum value. Here is a step-by-step strategy for solving these kinds of problems:

STRATEGY FOR SOLVING PROBLEMS INVOLVING MAXIMIZING OR MINIMIZING QUADRATIC FUNCTIONS

1. Read the problem carefully and decide which quantity is to be maximized or minimized.
2. Use the conditions of the problem to express the quantity as a function in one variable.
3. Rewrite the function in the form $f(x) = ax^2 + bx + c$.
4. Calculate $-\dfrac{b}{2a}$. If $a > 0$, f has a minimum at $x = -\dfrac{b}{2a}$. This minimum value is $f\left(-\dfrac{b}{2a}\right)$. If $a < 0$, f has a maximum at $x = -\dfrac{b}{2a}$. This maximum value is $f\left(-\dfrac{b}{2a}\right)$.
5. Answer the question posed in the problem.

EXAMPLE 5 Minimizing a Product

Among all pairs of numbers whose difference is 10, find a pair whose product is as small as possible. What is the minimum product?

SOLUTION

Step 1. **Decide what must be maximized or minimized.** We must minimize the product of two numbers. Calling the numbers x and y, and calling the product P, we must minimize

$$P = xy.$$

Step 2. **Express this quantity as a function in one variable.** In the formula $P = xy$, P is expressed in terms of two variables, x and y. However, because the difference of the numbers is 10, we can write

$$x - y = 10.$$

We can solve this equation for y in terms of x (or vice versa), substitute the result into $P = xy$, and obtain P as a function of one variable.

$$-y = -x + 10 \qquad \text{Subtract } x \text{ from both sides of } x - y = 10.$$

$$y = x - 10 \qquad \text{Multiply both sides of the equation by } -1 \text{ and solve for } y$$

Now we substitute $x - 10$ for y in $P = xy$.

$$P = xy = x(x - 10).$$

Because P is now a function of x, we can write

$$P(x) = x(x - 10).$$

Step 3. **Write the function in the form $f(x) = ax^2 + bx + c$.** We apply the distributive property to obtain

$$P(x) = x(x - 10) = x^2 - 10x$$

$$\boxed{a = 1} \quad \boxed{b = -10}$$

Step 4. **Calculate $-\dfrac{b}{2a}$. If $a > 0$, the function has a minimum at this value.** The voice balloons show that $a = 1$ and $b = -10$.

$$x = -\frac{b}{2a} = -\frac{-10}{2(1)} = -(-5) = 5$$

This means that the product, P, of two numbers whose difference is 10 is a minimum when one of the numbers, x, is 5.

Step 5. **Answer the question posed by the problem.** The problem asks for the two numbers and the minimum product. We found that one of the numbers, x, is 5. Now we must find the second number, y.

$$y = x - 10 = 5 - 10 = -5.$$

The number pair whose difference is 10 and whose product is as small as possible is 5, -5. The minimum product is $5(-5)$, or -25. ∎

CHECK POINT 5 Among all pairs of numbers whose difference is 8, find a pair whose product is as small as possible. What is the minimum product? $4, -4; -16$

USING TECHNOLOGY

The TABLE feature of a graphing utility can be used to verify our work in Example 5.

Enter $y_1 = x^2 - 10x$, the
function for the product,
when one of the numbers
is x.

X	Y1
2	-16
3	-21
4	-24
5	-25
6	-24
7	-21
8	-16

X=2

The product is a
minimum, −25, when
one of the numbers
is 5.

EXAMPLE 6 Maximizing Area

You have 100 yards of fencing to enclose a rectangular region. Find the dimensions of the rectangle that maximize the enclosed area. What is the maximum area?

SOLUTION

Step 1. **Decide what must be maximized or minimized.** We must maximize area. What we do not know are the rectangle's dimensions, x and y.

Step 2. **Express this quantity as a function in one variable.** Because we must maximize area, we have $A = xy$. We need to transform this into a function in which A is represented by one variable. Because you have 100 yards of fencing, the perimeter of the rectangle is 100 yards. This means that

$$2x + 2y = 100.$$

We can solve this equation for y in terms of x, substitute the result into $A = xy$, and obtain A as a function in one variable. We begin by solving for y.

$$2y = 100 - 2x \qquad \text{Subtract 2x from both sides.}$$

$$y = \frac{100 - 2x}{2} \qquad \text{Divide both sides by 2.}$$

$$y = 50 - x \qquad \text{Divide each term in the numerator by 2.}$$

Now we substitute $50 - x$ for y in $A = xy$.

$$A = xy = x(50 - x)$$

FIGURE 11.15 What value of x will maximize the rectangle's area?

The rectangle and its dimensions are illustrated in Figure 11.15. Because A is now a function of x, we can write

$$A(x) = x(50 - x).$$

This function models the area, $A(x)$, of any rectangle whose perimeter is 100 yards in terms of one of its dimensions, x.

Step 3. **Write the function in the form $f(x) = ax^2 + bx + c$.** We apply the distributive property to obtain

$$A(x) = x(50 - x) = 50x - x^2 = -x^2 + 50x.$$

$a = -1$ $b = 50$

The graph of the area function

$$A(x) = x(50 - x)$$

was obtained with a graphing utility using a $[0, 50, 2]$ by $[0, 700, 25]$ viewing rectangle. The maximum function feature verifies that a maximum area of 625 square yards occurs when one of the dimensions is 25 yards.

Step 4. **Calculate** $-\dfrac{b}{2a}$**. If $a < 0$, the function has a maximum at this value.** The voice balloons on the previous page show that $a = -1$ and $b = 50$.

$$x = -\frac{b}{2a} = -\frac{50}{2(-1)} = 25$$

This means that the area, $A(x)$, of a rectangle with perimeter 100 yards is a maximum when one of the rectangle's dimensions, x, is 25 yards.

FIGURE 11.15 (repeated)

Step 5. **Answer the question posed by the problem.** We found that $x = 25$. Figure 11.15 shows that the rectangle's other dimension is $50 - x = 50 - 25 = 25$. The dimensions of the rectangle that maximize the enclosed area are 25 yards by 25 yards. The rectangle that gives the maximum area is actually a square with an area of 25 yards $\cdot$ 25 yards, or 625 square yards. ■

CHECK POINT 6 You have 120 feet of fencing to enclose a rectangular region. Find the dimensions of the rectangle that maximize the enclosed area. What is the maximum area? *The dimensions are 30 feet by 30 feet with an area of 900 square feet.*

We've come a long way from the small nation of "embattled farmers" who launched the American Revolution. In the early days of our Republic, 95% of the population was involved in farming. The graph in Figure 11.16 shows the number of farms in the United States from 1850 through 2010 (projected). Because the graph is shaped like a cup, with an increasing number of farms from 1850 to 1910 and a decreasing number of farms from 1910 to 2010, a quadratic function is an appropriate model for the data. You can use the statistical menu of a graphing utility to enter the data in Figure 11.16. We entered the data using (number of decades after 1850, millions of U.S. farms). The data are shown to the right of Figure 11.16.

Number of U. S. Farms, 1850–2010

FIGURE 11.16

Source: U. S. Bureau of the Census

Data:

$(0, 2.3), (2, 3.3), (4, 5.1),$
$(6, 6.7), (8, 6.4), (10, 5.8),$
$(12, 3.6), (14, 2.9), (16, 2.3)$

```
QuadReg
 y=ax²+bx+c
 a=-.0643668831
 b=.9873701299
 c=2.203636364
```

Upon entering the QUADratic REGression program, we obtain the results shown in the screen. Thus, the quadratic function of best fit is approximately

$$f(x) = -0.064x^2 + 0.99x + 2.2,$$

where x represents the number of decades after 1850 and $f(x)$ represents the number of U.S. farms, in millions.

11.3 EXERCISE SET

Practice Exercises

In Exercises 1–4, the graph of a quadratic function is given. Write the function's equation, selecting from the following options:

$$f(x) = (x + 1)^2 - 1, g(x) = (x + 1)^2 + 1,$$
$$h(x) = (x - 1)^2 + 1, j(x) = (x - 1)^2 - 1.$$

1. $h(x) = (x - 1)^2 + 1$

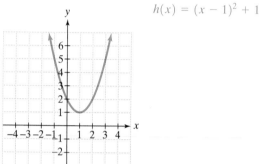

2. $g(x) = (x + 1)^2 + 1$

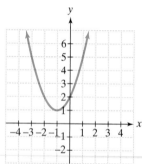

3. $j(x) = (x - 1)^2 - 1$

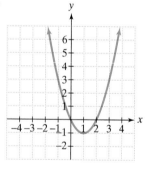

4. $f(x) = (x + 1)^2 - 1$

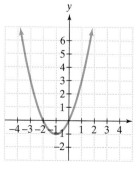

In Exercises 5–8, the graph of a quadratic function is given. Write the function's equation, selecting from the following options:

$$f(x) = x^2 + 2x + 1, g(x) = x^2 - 2x + 1,$$
$$h(x) = x^2 - 1, j(x) = -x^2 - 1.$$

5. $h(x) = x^2 - 1$

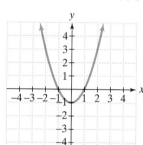

6. $f(x) = x^2 + 2x + 1$

7. $g(x) = x^2 - 2x + 1$

8. $j(x) = -x^2 - 1$

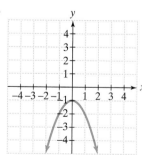

In Exercises 9–16, find the coordinates of the vertex for the parabola defined by the given quadratic function.

9. $f(x) = 2(x - 3)^2 + 1$ $(3, 1)$

10. $f(x) = -3(x - 2)^2 + 12$ $(2, 12)$

11. $f(x) = -2(x + 1)^2 + 5$ $(-1, 5)$

12. $f(x) = -2(x + 4)^2 - 8$ $(-4, -8)$

13. $f(x) = 2x^2 - 8x + 3$ $(2, -5)$

14. $f(x) = 3x^2 - 12x + 1$ $(2, -11)$

15. $f(x) = -x^2 - 2x + 8$ $(-1, 9)$

16. $f(x) = -2x^2 + 8x - 1$ $(2, 7)$

In Exercises 17–34, use the vertex and intercepts to sketch the graph of each quadratic function. Give the equation for the parabola's axis of symmetry. Use the parabola to identify the function's range.   **17–34.** See graphing answer section.

17. $f(x) = (x - 4)^2 - 1$ $x = 4; \{y | y \geq -1\}$ or $[-1, \infty)$

18. $f(x) = (x - 1)^2 - 2$ $x = 1; \{y | y \geq -2\}$ or $[-2, \infty)$

19. $f(x) = (x - 1)^2 + 2$ $x = 1; \{y | y \geq 2\}$ or $[2, \infty)$

20. $f(x) = (x - 3)^2 + 2$ $x = 3; \{y | y \geq 2\}$ or $[2, \infty)$

21. $y - 1 = (x - 3)^2$ $x = 3; \{y | y \geq 1\}$ or $[1, \infty)$

22. $y - 3 = (x - 1)^2$ $x = 1; \{y | y \geq 3\}$ or $[3, \infty)$

23. $f(x) = 2(x + 2)^2 - 1$ $x = -2; \{y | y \geq -1\}$ or $[-1, \infty)$

24. $f(x) = \dfrac{5}{4} - \left(x - \dfrac{1}{2}\right)^2$ $x = \dfrac{1}{2}; \{y | y \leq \frac{5}{4}\}$ or $\left(-\infty, \frac{5}{4}\right]$

25. $f(x) = 4 - (x - 1)^2$ $x = 1; \{y | y \leq 4\}$ or $(-\infty, 4]$

26. $f(x) = 1 - (x - 3)^2$ $x = 3; \{y | y \leq 1\}$ or $(-\infty, 1]$

27. $f(x) = x^2 - 2x - 3$ $x = 1; \{y | y \geq -4\}$ or $[-4, \infty)$

28. $f(x) = x^2 - 2x - 15$ $x = 1; \{y | y \geq -16\}$ or $[-16, \infty)$

29. $f(x) = x^2 + 3x - 10$ $x = -\frac{3}{2}; \{y | y \geq -\frac{49}{4}\}$ or $\left[-\frac{49}{4}, \infty\right)$

30. $f(x) = 2x^2 - 7x - 4$ $x = \frac{7}{4}; \{y | y \geq -\frac{81}{8}\}$ or $\left[-\frac{81}{8}, \infty\right)$

31. $f(x) = 2x - x^2 + 3$ $x = 1; \{y | y \leq 4\}$ or $(-\infty, 4]$

32. $f(x) = 5 - 4x - x^2$ $x = -2; \{y | y \leq 9\}$ or $(-\infty, 9]$

33. $f(x) = 2x - x^2 - 2$ $x = 1; \{y | y \leq -1\}$ or $(-\infty, -1]$

34. $f(x) = 6 - 4x + x^2$ $x = 2; \{y | y \geq 2\}$ or $[2, \infty)$

In Exercises 35–40, determine, without graphing, whether the given quadratic function has a minimum value or a maximum value. Then find the coordinates of the minimum or the maximum point.

35. $f(x) = 3x^2 - 12x - 1$ minimum; $(2, -13)$

36. $f(x) = 2x^2 - 8x - 3$ minimum; $(2, -11)$

37. $f(x) = -4x^2 + 8x - 3$ maximum; $(1, 1)$

38. $f(x) = -2x^2 - 12x + 3$ maximum; $(-3, 21)$

39. $f(x) = 5x^2 - 5x$ minimum; $\left(\frac{1}{2}, -\frac{5}{4}\right)$

40. $f(x) = 6x^2 - 6x$ minimum; $\left(\frac{1}{2}, -\frac{3}{2}\right)$

Practice Plus

In Exercises 41–44, give the domain and the range of each quadratic function whose graph is described.

41. The vertex is $(-1, -2)$ and the parabola opens up.
domain $= \{x | x$ is a real number$\}$ or
$(-\infty, \infty)$; range $= \{y | y \geq -2\}$ or $[-2, \infty)$

42. The vertex is $(-3, -4)$ and the parabola opens down.
domain $= \{x | x$ is a real number$\}$ or
$(-\infty, \infty)$; range $= \{y | y \leq -4\}$ or $(-\infty, -4]$

43. The maximum point on the parabola is $(10, 3)$.
domain $= \{x | x$ is a real number$\}$ or
$(-\infty, \infty)$; range $= \{y | y \leq 3\}$ or $(-\infty, 3]$

44. The minimum point on the parabola is $(5, 12)$.
domain $= \{x | x$ is a real number$\}$ or
$(-\infty, \infty)$; range $= \{y | y \geq 12\}$ or $[12, \infty)$

In Exercises 45–48, write an equation of the parabola that has the same shape as the graph of $f(x) = 2x^2$, but with the given point as the vertex.

45. $(5, 3)$ $f(x) = 2(x - 5)^2 + 3$

46. $(7, 4)$ $f(x) = 2(x - 7)^2 + 4$

47. $(-10, -5)$ $f(x) = 2(x + 10)^2 - 5$

48. $(-8, -6)$ $f(x) = 2(x + 8)^2 - 6$

In Exercises 49–52, write an equation of the parabola that has the same shape as the graph of $f(x) = 3x^2$ or $g(x) = -3x^2$, but with the given point as a maximum or minimum point.

49. maximum: $(-2, 4)$ $f(x) = -3(x + 2)^2 + 4$

50. maximum: $(5, -7)$ $f(x) = -3(x - 5)^2 - 7$

51. minimum: $(11, 0)$ $f(x) = 3(x - 11)^2$

52. minimum: $(9, 0)$ $f(x) = 3(x - 9)^2$

Application Exercises

In the United States, HCV, or hepatitis C virus, is four times as widespread as HIV. Few of the nation's three to four million carriers have any idea they are infected. The graph shows the projected mortality, in thousands, from the virus. Use the information in the graph to solve Exercises 53–54.

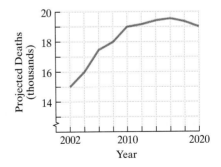

Projected Hepatitis C Deaths in the U. S.

Source: American Journal of Public Health

53. Suppose that a quadratic function is used to model the data shown with ordered pairs representing (number of years after 2002, thousands of hepatitis C deaths). Determine, without obtaining an actual quadratic function that models the data, the approximate coordinates of the vertex for the function's graph. Describe what this means in practical terms. Use the word "maximum" in your description. $(14, 19.5)$; In 2016, the projected mortality will reach a maximum of 19,500.

54. Why is a quadratic function an appropriate model for the data shown in the graph on the previous page? The graph has the shape of a parabola.

55. A person standing close to the edge on the top of a 160-foot building throws a baseball vertically upward. The quadratic function

$$s(t) = -16t^2 + 64t + 160$$

models the ball's height above the ground, $s(t)$, in feet, t seconds after it was thrown.

a. After how many seconds does the ball reach its maximum height? What is the maximum height? 2 seconds; 224 feet

b. How many seconds does it take until the ball finally hits the ground? Round to the nearest tenth of a second. 5.7 seconds

c. Find $s(0)$ and describe what this means. 160; 160 feet is the height of the building.

d. Use your results from parts (a) through (c) to graph the quadratic function. Begin the graph with $t = 0$ and end with the value of t for which the ball hits the ground. See graphing answer section.

56. A person standing close to the edge on the top of a 200-foot building throws a baseball vertically upward. The quadratic function

$$s(t) = -16t^2 + 64t + 200$$

models the ball's height above the ground, $s(t)$, in feet, t seconds after it was thrown.

a. After how many seconds does the ball reach its maximum height? What is the maximum height? 2 seconds; 264 feet

b. How many seconds does it take until the ball finally hits the ground? Round to the nearest tenth of a second. 6.1 seconds

c. Find $s(0)$ and describe what this means. 200; 200 feet is the height of the building.

d. Use your results from parts (a) through (c) to graph the quadratic function. Begin the graph with $t = 0$ and end with the value of t for which the ball hits the ground. See graphing answer section.

57. The function

$$f(x) = 104.5x^2 - 1501.5x + 6016$$

models the death rate per year per 100,000 males, $f(x)$, for U.S. men who average x hours of sleep each night. How many hours of sleep, to the nearest tenth of an hour, corresponds to the minimum death rate? What is this minimum death rate, to the nearest whole number? 7.2 hours; 622 per 100,000 males

58. The function

$$f(x) = -0.064x^2 + 0.99x + 2.2$$

models the number of U.S. farms, $f(x)$, in millions, x decades after 1850. How many decades after 1850, to the nearest decade, did the number of U.S. farms reach a maximum? How many farms, to the nearest tenth of a million, were there in that year? How well does this model the actual data shown in Figure 11.16 on page 761? 8 decades; 6.0 million farms; Answers will vary.

59. Among all pairs of numbers whose sum is 16, find a pair whose product is as large as possible. What is the maximum product? The maximum product is 64 when the numbers are 8 and 8.

60. Among all pairs of numbers whose sum is 20, find a pair whose product is as large as possible. What is the maximum product? The maximum product is 100 when the numbers are 10 and 10.

61. Among all pairs of numbers whose difference is 16, find a pair whose product is as small as possible. What is the minimum product? The minimum product is -64 when the numbers are 8 and -8.

62. Among all pairs of numbers whose difference is 24, find a pair whose product is as small as possible. What is the minimum product? The minimum product is -144 when the numbers are 12 and -12.

63. You have 600 feet of fencing to enclose a rectangular plot that borders on a river. If you do not fence the side along the river, find the length and width of the plot that will maximize the area. What is the largest area that can be enclosed?

The maximum area is 45,000 square feet when the length is 300 feet and the width is 150 feet.

64. You have 200 feet of fencing to enclose a rectangular plot that borders on a river. If you do not fence the side along the river, find the length and width of the plot that will maximize the area. What is the largest area that can be enclosed?

The maximum area is 5000 square feet when the length is 100 feet and the width is 50 feet.

65. You have 50 yards of fencing to enclose a rectangular region. Find the dimensions of the rectangle that maximize the enclosed area. What is the maximum area?   The maximum area is 156.25 square yards when the dimensions are 12.5 yards by 12.5 yards.

66. You have 80 yards of fencing to enclose a rectangular region. Find the dimensions of the rectangle that maximize the enclosed area. What is the maximum area?   The maximum area is 400 square yards when the dimensions are 20 yards by 20 yards.

67. A rain gutter is made from sheets of aluminum that are 20 inches wide by turning up the edges to form right angles. Determine the depth of the gutter that will maximize its cross-sectional area and allow the greatest amount of water to flow. What is the maximum cross-sectional area?

Flat sheet 20 inches wide

x $20 - 2x$ x

The maximum cross-sectional area is 50 square inches when the depth is 5 inches.

68. A rain gutter is made from sheets of aluminum that are 12 inches wide by turning up the edges to form right angles. Determine the depth of the gutter that will maximize its cross-sectional area and allow the greatest amount of water to flow. What is the maximum cross-sectional area?   The maximum cross-sectional area is 18 square inches when the depth is 3 inches.

In Chapter 9, we saw that the profit, P(x), generated after producing and selling x units of a product is given by the function

$$P(x) = R(x) - C(x),$$

where R and C are the revenue and cost functions, respectively. Use these functions to solve Exercises 69–70.

69. Hunky Beef, a local sandwich store, has a fixed weekly cost of $525.00, and variable costs for making a roast beef sandwich are $0.55.

a. Let x represent the number of roast beef sandwiches made and sold each week. Write the weekly cost function, C, for Hunky Beef. $C(x) = 525 + 0.55x$

b. The function $R(x) = -0.001x^2 + 3x$ describes the money that Hunky Beef takes in each week from the sale of x roast beef sandwiches. Use this revenue function and the cost function from part (a) to write the store's weekly profit function, P. $P(x) = -0.001x^2 + 3x - (525 + 0.55x)$ $= -0.001x^2 + 2.45x - 525$

c. Use the store's profit function to determine the number of roast beef sandwiches it should make and sell each week to maximize profit. What is the maximum weekly profit?   The maximum weekly profit is $975.63 when 1225 sandwiches are made and sold each week.

70. Virtual Fido is a company that makes electronic virtual pets. The fixed weekly cost is $3000, and variable costs for each pet are $20.

a. Let x represent the number of virtual pets made and sold each week. Write the weekly cost function, C, for Virtual Fido. $C(x) = 3000 + 20x$

b. The function $R(x) = -x^2 + 1000x$ describes the money that Virtual Fido takes in each week from the sale of x virtual pets. Use this revenue function and the cost function from part (a) to write the weekly profit function, P. $P(x) = -x^2 + 1000x - (3000 + 20x) = -x^2 + 980x - 3000$

c. Use the profit function to determine the number of virtual pets that should be made and sold each week to maximize profit. What is the maximum weekly profit?   The maximum weekly profit is $237,100 when 490 virtual pets are made and sold each week.

Writing in Mathematics

71. What is a parabola? Describe its shape.   **71–75.** Answers will vary.

72. Explain how to decide whether a parabola opens upward or downward.

73. Describe how to find a parabola's vertex if its equation is in the form $f(x) = a(x - h)^2 + k$. Give an example.

74. Describe how to find a parabola's vertex if its equation is in the form $f(x) = ax^2 + bx + c$. Use $f(x) = x^2 - 6x + 8$ as an example.

75. A parabola that opens upward has its vertex at $(1, 2)$. Describe as much as you can about the parabola based on this information. Include in your discussion the number of x-intercepts (if any) for the parabola.

Technology Exercises

76. Use a graphing utility to verify any five of your hand-drawn graphs in Exercises 17–34.   Answers will vary.

77. a. Use a graphing utility to graph $y = 2x^2 - 82x + 720$ in a standard viewing rectangle. What do you observe?

b. Find the coordinates of the vertex for the given quadratic function. $(20.5, -120.5)$

c. The answer to part (b) is $(20.5, -120.5)$. Because the leading coefficient, 2, of the given function is positive, the vertex is a minimum point on the graph. Use this fact to help find a viewing rectangle that will give a relatively complete picture of the parabola. With an axis of symmetry at $x = 20.5$, the setting for x should extend past this, so try Xmin = 0 and Xmax = 30. The setting for y should include (and probably go below) the y-coordinate of the graph's minimum point, so try Ymin = −130. Experiment with Ymax until your utility shows the parabola's major features.   See graphing answer section.

d. In general, explain how knowing the coordinates of a parabola's vertex can help determine a reasonable viewing rectangle on a graphing utility for obtaining a complete picture of the parabola.   Answer will vary.

In Exercises 78–81, find the vertex for each parabola. Then determine a reasonable viewing rectangle on your graphing utility and use it to graph the parabola.   See graphing answer section

78. $y = -0.25x^2 + 40x$ Vertex = $(80, 1600)$

79. $y = -4x^2 + 20x + 160$ Vertex = $(2.5, 185)$

80. $y = 5x^2 + 40x + 600$ Vertex = $(-4, 520)$

81. $y = 0.01x^2 + 0.6x + 100$ Vertex = $(-30, 91)$

82. The following data show fuel efficiency, in miles per gallon, for all U.S. automobiles in the indicated year.

x (Years after 1940)	y (Average Number of Miles per Gallon for U.S. Automobiles)
1940: 0	14.8
1950: 10	13.9
1960: 20	13.4
1970: 30	13.5
1980: 40	15.9
1990: 50	20.2
2000: 60	22.0

Source: U.S. Department of Transportation

a. Use a graphing utility to draw a scatter plot of the data. Explain why a quadratic function is appropriate for modeling these data.   See graphing answer section. The values of y decrease, then increase.

b. Use the quadratic regression feature to find the quadratic function that best fits the data.
$y = 0.005x^2 - 0.170x + 14.817$

c. Use the equation in part (b) to determine the worst year for automobile fuel efficiency. What was the average number of miles per gallon for that year?   The worst year for fuel efficiency was 1957 with about 13.372 miles per gallon.

d. Use a graphing utility to draw a scatter plot of the data and graph the quadratic function of best fit on the scatter plot.   See graphing answer section.

Critical Thinking Exercises

83. Which one of the following is true?   a

a. No quadratic functions have a range of $(-\infty, \infty)$.

b. The vertex of the parabola described by $f(x) = 2(x - 5)^2 - 1$ is at $(5, 1)$.

c. The graph of $f(x) = -2(x + 4)^2 - 8$ has one y-intercept and two x-intercepts.

d. The maximum value of y for the quadratic function $f(x) = -x^2 + x + 1$ is 1.

In Exercises 84–85, find the axis of symmetry for each parabola whose equation is given. Use the axis of symmetry to find a second point on the parabola whose y-coordinate is the same as the given point.

84. $f(x) = 3(x + 2)^2 - 5$; $(-1, -2)$ $x = -2$; $(-3, -2)$

85. $f(x) = (x - 3)^2 + 2$; $(6, 11)$ $x = 3$; $(0, 11)$

In Exercises 86–87, write the equation of each parabola in $f(x) = a(x - h)^2 + k$ form.

86. Vertex: $(-3, -4)$; The graph passes through the point $(1, 4)$. $f(x) = \frac{1}{2}(x + 3)^2 - 4$

87. Vertex: $(-3, -1)$; The graph passes through the point $(-2, -3)$. $f(x) = -2(x + 3)^2 - 1$

88. A rancher has 1000 feet of fencing to construct six corrals, as shown in the figure. Find the dimensions that maximize the enclosed area. What is the maximum area?

125 feet by about 166.7 feet: about 20,833 square feet

89. The annual yield per lemon tree is fairly constant at 320 pounds when the number of trees per acre is 50 or fewer. For each additional tree over 50, the annual yield per tree for all trees on the acre decreases by 4 pounds due to overcrowding. Find the number of trees that should be planted on an acre to produce the maximum yield. How many pounds is the maximum yield?   65 trees; 16,900 pounds

Review Exercises

90. Solve: $\dfrac{2}{x + 5} + \dfrac{1}{x - 5} = \dfrac{16}{x^2 - 25}$. (Section 7.6, Example 5)
$\{7\}$

91. Simplify: $\dfrac{1 + \dfrac{2}{x}}{1 - \dfrac{4}{x^2}}$. (Section 7.5, Example 2 or Example 5)
$\dfrac{x}{x - 2}$

92. Solve the system:

$$2x + 3y = 6$$
$$x - 4y = 14.$$

(Section 4.3, Example 2) $\{(6, -2)\}$

MID-CHAPTER CHECK POINT

CHAPTER 11

What You Know: We saw that not all quadratic equations can be solved by factoring. We learned three new methods for solving these equations: the square root property, completing the square, and the quadratic formula. We saw that the discriminant of $ax^2 + bx + c$, namely $b^2 - 4ac$, determines the number and type of the equation's solutions. We graphed quadratic functions using vertices, intercepts, and additional points, as necessary. We learned that the vertex of $f(x) = a(x - h)^2 + k$ is (h, k) and the vertex of $f(x) = ax^2 + bx + c$ is $\left(-\dfrac{b}{2a}, f\left(-\dfrac{b}{2a}\right)\right)$. We used the vertex to solve problems that involved minimizing or maximizing quadratic functions.

In Exercises 1–13, solve each equation by the method of your choice. Simplify solutions, if possible.

1. $(3x - 5)^2 = 36$ $\left\{-\frac{1}{3}, \frac{11}{3}\right\}$

2. $5x^2 - 2x = 7$ $\left\{-1, \frac{7}{5}\right\}$

3. $3x^2 - 6x - 2 = 0$ $\left\{\dfrac{3 \pm \sqrt{15}}{3}\right\}$

4. $x^2 + 6x = -2$ $\left\{-3 \pm \sqrt{7}\right\}$

5. $5x^2 + 1 = 37$ $\left\{\pm\dfrac{6\sqrt{5}}{5}\right\}$

6. $x^2 - 5x + 8 = 0$ $\left\{\dfrac{5}{2} \pm \dfrac{\sqrt{7}}{2}i\right\}$

7. $2x^2 + 26 = 0$ $\left\{\pm\sqrt{13}i\right\}$

8. $(2x + 3)(x + 2) = 10$ $\left\{-4, \frac{1}{2}\right\}$

9. $(x + 3)^2 = 24$ $\left\{-3 \pm 2\sqrt{6}\right\}$

10. $\dfrac{1}{x^2} - \dfrac{4}{x} + 1 = 0$ $\left\{2 \pm \sqrt{3}\right\}$

11. $x(2x - 3) = -4$ $\left\{\dfrac{3}{4} \pm \dfrac{\sqrt{23}}{4}i\right\}$

12. $\dfrac{x^2}{3} + \dfrac{x}{2} = \dfrac{2}{3}$ $\left\{\dfrac{-3 \pm \sqrt{41}}{4}\right\}$

13. $\dfrac{2x}{x^2 + 6x + 8} = \dfrac{x}{x + 4} - \dfrac{2}{x + 2}$ $\{-2, 4\}$

14. Solve by completing the square: $x^2 + 10x - 3 = 0$. $\left\{-5 \pm 2\sqrt{7}\right\}$

In Exercises 15–16, find the length (in simplified radical form and rounded to two decimal places) and the midpoint of the line segment with the given endpoints.

15. $(2, -2)$ and $(-2, 2)$ $4\sqrt{2} \approx 5.66$ units; $(0, 0)$

16. $(-5, 8)$ and $(-10, 14)$ $\sqrt{61} \approx 7.81$ units; $\left(-\dfrac{15}{2}, 11\right)$

In Exercises 17–20, graph the given quadratic function. Give each function's domain and range.

17–20. See graphing answer section.

17. $f(x) = (x - 3)^2 - 4$ domain: $\{x | x \text{ is a real number}\}$ or $(-\infty, \infty)$; range: $\{y | y \geq -4\}$ or $[-4, \infty)$

18. $g(x) = 5 - (x + 2)^2$ domain: $\{x | x \text{ is a real number}\}$ or $(-\infty, \infty)$; range: $\{y | y \leq 5\}$ or $(-\infty, 5]$

19. $h(x) = -x^2 - 4x + 5$ domain: $\{x | x \text{ is a real number}\}$ or $(-\infty, \infty)$; range: $\{y | y \leq 9\}$ or $(-\infty, 9]$

20. $f(x) = 3x^2 - 6x + 1$ domain: $\{x | x \text{ is a real number}\}$ or $(-\infty, \infty)$; range: $\{y | y \geq -2\}$ or $[-2, \infty)$

In Exercises 21–22, without solving the equation, determine the number and type of solutions.

21. $2x^2 + 5x + 4 = 0$ two imaginary solutions

22. $10x(x + 4) = 15x - 15$ two rational solutions

In Exercises 23–24, write a quadratic equation in standard form with the given solution set.

23. $\left\{-\dfrac{1}{2}, \dfrac{3}{4}\right\}$ $8x^2 - 2x - 3 = 0$

24. $\left\{-2\sqrt{3}, 2\sqrt{3}\right\}$ $x^2 - 12 = 0$

25. A company manufactures and sells bath cabinets. The function
$$P(x) = -x^2 + 150x - 4425$$
models the company's daily profit, $P(x)$, when x cabinets are manufactured and sold per day. How many cabinets should be manufactured and sold per day to maximize the company's profit? What is the maximum daily profit? 75 cabinets per day; $1200

26. Among all pairs of numbers whose sum is -18, find a pair whose product is as large as possible. What is the maximum product? -9 and -9; 81

27. The base of a triangle measures 40 inches minus twice the measure of its height. For what measure of the height does the triangle have a maximum area? What is the maximum area? 10 inches; 100 square inches

SECTION 11.4

Objective

1 Solve equations that are quadratic in form.

EQUATIONS QUADRATIC IN FORM

"My husband asked me if we have any cheese puffs. Like he can't go and lift the couch cushion up himself."—Roseanne

Photo by Michael Lutzky. © 2000, *The Washington Post*. Reprinted with permission.

How important is it for you to have a clean house? The percentage of people who find this to be quite important varies by age. In the exercise set, you will work with a function that models this phenomenon. Your work will be based on equations that are not quadratic, but that can be written as quadratic equations using an appropriate substitution. Here are some examples:

Given Equation	Substitution	New Equation
$x^4 - 10x^2 + 9 = 0$ or $(x^2)^2 - 10x^2 + 9 = 0$	$t = x^2$	$t^2 - 10t + 9 = 0$
$5x^{\frac{2}{3}} + 11x^{\frac{1}{3}} + 2 = 0$ or $5\left(x^{\frac{1}{3}}\right)^2 + 11x^{\frac{1}{3}} + 2 = 0$	$t = x^{\frac{1}{3}}$	$5t^2 + 11t + 2 = 0$

An equation that is **quadratic in form** is one that can be expressed as a quadratic equation using an appropriate substitution. Both of the preceding given equations are quadratic in form.

1 Solve equations that are quadratic in form.

Equations that are quadratic in form contain an expression to a power, the same expression to that power squared, and a constant term. By letting t equal the expression to the power, a quadratic equation in t will result. Now it's easy. Solve this quadratic equation for t. Finally, use your substitution to find the values for the variable in the given equation. Example 1 shows how this is done.

EXAMPLE 1 Solving an Equation Quadratic in Form

Solve: $x^4 - 10x^2 + 9 = 0$.

SOLUTION Notice that the equation contains an expression to a power, x^2, the same expression to that power squared, x^4 or $(x^2)^2$, and a constant term, 9. We let t equal the expression to the power. Thus,

$$\text{let } t = x^2.$$

The graph of

$$y = x^4 - 10x^2 + 9$$

has x-intercepts at $-3, -1, 1,$ and 3. This verifies that the solution set of

$$x^4 - 10x^2 + 9 = 0$$

is $\{-3, -1, 1, 3\}$.

$[-5, 5, 1]$ by $[-20, 10, 5]$

Now we write the given equation as a quadratic equation in t and solve for t.

$x^4 - 10x^2 + 9 = 0$	This is the given equation.
$(x^2)^2 - 10x^2 + 9 = 0$	The given equation contains x^2 and x^2 squared.
$t^2 - 10t + 9 = 0$	Let $t = x^2$. Replace x^2 with t.
$(t - 9)(t - 1) = 0$	Factor.
$t - 9 = 0$ or $t - 1 = 0$	Set each factor equal to zero.
$t = 9$ $t = 1$	Solve for t.

We're not done! Why not? We were asked to solve for x and we have values for t. We use the original substitution, $t = x^2$, to solve for x. Replace t with x^2 in each equation shown.

$x^2 = 9$	or	$x^2 = 1$	
$x = \pm\sqrt{9}$		$x = \pm\sqrt{1}$	Apply the square root property.
$x = \pm 3$		$x = \pm 1$	

Substitute these values into the given equation and verify that the solutions are $-3, 3, -1,$ and 1. The solution set is $\{-3, 3, -1, 1\}$. You may prefer to express the solution set in numerical order as $\{-3, -1, 1, 3\}$. ∎

✔ **CHECK POINT 1** Solve: $x^4 - 17x^2 + 16 = 0$. $-4, 4, -1,$ and 1, or $\{-4, 4, -1, 1\}$

If checking proposed solutions is not overly cumbersome, you should do so either algebraically or with a graphing utility. The Using Technology box shows a check of the four solutions in Example 1. Are there situations when solving equations quadratic in form where a check is essential? Yes. **If at any point in the solution process both sides of an equation are raised to an even power, a check is required.** Extraneous solutions that are not solutions of the given equation may have been introduced.

EXAMPLE 2   Solving an Equation Quadratic in Form

Solve: $2x - \sqrt{x} - 10 = 0$.

SOLUTION   To identify exponents on the terms, let's rewrite $\sqrt{x}$ as $x^{\frac{1}{2}}$. The equation can be expressed as

$$2x^1 - x^{\frac{1}{2}} - 10 = 0.$$

Notice that the equation contains an expression to a power, $x^{\frac{1}{2}}$, the same expression to that power squared, x^1 or $\left(x^{\frac{1}{2}}\right)^2$, and a constant term, -10. We let t equal the expression to the power. Thus,

$$\text{let } t = x^{\frac{1}{2}}.$$

Now we write the given equation as a quadratic equation in t and solve for t.

$2x^1 - x^{\frac{1}{2}} - 10 = 0$	This is the given equation in exponential form.
$2\left(x^{\frac{1}{2}}\right)^2 - x^{\frac{1}{2}} - 10 = 0$	The equation contains $x^{\frac{1}{2}}$ and $x^{\frac{1}{2}}$ squared.
$2t^2 - t - 10 = 0$	Let $t = x^{\frac{1}{2}}$. Replace $x^{\frac{1}{2}}$ with t.
$(2t - 5)(t + 2) = 0$	Factor.
$2t - 5 = 0$ or $t + 2 = 0$	Set each factor equal to zero.
$t = \dfrac{5}{2}$ $t = -2$	Solve for t.

Use the original substitution, $t = x^{\frac{1}{2}}$, to solve for x. Replace t with $x^{\frac{1}{2}}$ in each of the preceding equations, $t = \frac{5}{2}$ and $t = -2$.

$$x^{\frac{1}{2}} = \frac{5}{2} \quad \text{or} \quad x^{\frac{1}{2}} = -2 \qquad \text{Replace } t \text{ with } x^{\frac{1}{2}}.$$

$$\left(x^{\frac{1}{2}}\right)^2 = \left(\frac{5}{2}\right)^2 \quad \left(x^{\frac{1}{2}}\right)^2 = (-2)^2 \qquad \text{Solve for } x \text{ by squaring both sides of each equation.}$$

> Both sides are raised to even powers. We must check.

$$x = \frac{25}{4} \qquad\qquad x = 4 \qquad \text{Square } \frac{5}{2} \text{ and } -2.$$

It is essential to check both proposed solutions in the original equation.

Check $\frac{25}{4}$:

$$2x - \sqrt{x} - 10 = 0$$

$$2 \cdot \frac{25}{4} - \sqrt{\frac{25}{4}} - 10 \overset{?}{=} 0$$

$$\frac{25}{2} - \frac{5}{2} - 10 \overset{?}{=} 0$$

$$\frac{20}{2} - 10 \overset{?}{=} 0$$

$$0 = 0, \quad \text{true}$$

Check 4:

$$2x - \sqrt{x} - 10 = 0$$

$$2 \cdot 4 - \sqrt{4} - 10 \overset{?}{=} 0$$

$$8 - 2 - 10 \overset{?}{=} 0$$

$$6 - 10 \overset{?}{=} 0$$

$$-4 = 0, \quad \text{false}$$

The check indicates that 4 is not a solution. It is an extraneous solution brought about by squaring each side of the equation. The only solution is $\frac{25}{4}$, and the solution set is $\left\{\frac{25}{4}\right\}$.

✔ **CHECK POINT 2** Solve: $x - 2\sqrt{x} - 8 = 0$.   16 or {16}

The equations in Examples 1 and 2 can be solved by methods other than using substitutions.

$$x^4 - 10x^2 + 9 = 0 \qquad\qquad 2x - \sqrt{x} - 10 = 0$$

> This equation can be solved directly by factoring:
> $(x^2 - 9)(x^2 - 1) = 0$.

> This equation can be solved by isolating the radical term:
> $2x - 10 = \sqrt{x}$.
> Then square both sides.

In the examples that follow, solving the equations by methods other than first introducing a substitution becomes increasingly difficult.

EXAMPLE 3 Solving an Equation Quadratic in Form

Solve: $(x^2 - 5)^2 + 3(x^2 - 5) - 10 = 0$.

SOLUTION This equation contains $x^2 - 5$ and $x^2 - 5$ squared. We

$$\text{let} \quad t = x^2 - 5.$$

$(x^2 - 5)^2 + 3(x^2 - 5) - 10 = 0$	This is the given equation.
$t^2 + 3t - 10 = 0$	Let $t = x^2 - 5$.
$(t + 5)(t - 2) = 0$	Factor.
$t + 5 = 0 \quad \text{or} \quad t - 2 = 0$	Set each factor equal to zero.
$t = -5 \qquad\qquad t = 2$	Solve for t.

Use the original substitution, $t = x^2 - 5$, to solve for x. Replace t with $x^2 - 5$ in each of the preceding equations.

$x^2 - 5 = -5 \quad \text{or} \quad x^2 - 5 = 2$	Replace t with $x^2 - 5$.
$x^2 = 0 \qquad\qquad x^2 = 7$	Solve for x by isolating x^2.
$x = 0 \qquad\qquad x = \pm\sqrt{7}$	Apply the square root property.

Although we did not raise both sides of an equation to an even power, checking the three proposed solutions in the original equation is a good idea. Do this now and verify that the solutions are $-\sqrt{7}$, 0, and $\sqrt{7}$, and the solution set is $\left\{-\sqrt{7}, 0, \sqrt{7}\right\}$. ∎

 CHECK POINT 3 Solve: $(x^2 - 4)^2 + (x^2 - 4) - 6 = 0$.
$-\sqrt{6}, -1, 1, \sqrt{6}$ or $\left\{-\sqrt{6}, -1, 1, \sqrt{6}\right\}$

EXAMPLE 4 Solving an Equation Quadratic in Form

Solve: $10x^{-2} + 7x^{-1} + 1 = 0$.

SOLUTION Notice that the equation contains an expression to a power, x^{-1}, the same expression to that power squared, x^{-2} or $(x^{-1})^2$, and a constant term, 1. We let t equal the expression to the power. Thus,

$$\text{let } t = x^{-1}.$$

Now we write the given equation as a quadratic equation in t and solve for t.

$10x^{-2} + 7x^{-1} + 1 = 0$	This is the given equation.
$10(x^{-1})^2 + 7x^{-1} + 1 = 0$	The equation contains x^{-1} and x^{-1} squared.
$10t^2 + 7t + 1 = 0$	Let $t = x^{-1}$.
$(5t + 1)(2t + 1) = 0$	Factor.
$5t + 1 = 0 \quad \text{or} \quad 2t + 1 = 0$	Set each factor equal to zero.
$5t = -1 \qquad\qquad 2t = -1$	Solve each equation for t.
$t = -\dfrac{1}{5} \qquad\qquad t = -\dfrac{1}{2}$	

Use the original substitution, $t = x^{-1}$, to solve for x. Replace t with x^{-1} in each of the preceding equations, $t = -\frac{1}{5}$ and $t = -\frac{1}{2}$.

$$x^{-1} = -\frac{1}{5} \quad \text{or} \quad x^{-1} = -\frac{1}{2} \qquad \text{Replace } t \text{ with } x^{-1}.$$

$$(x^{-1})^{-1} = \left(-\frac{1}{5}\right)^{-1} \qquad (x^{-1})^{-1} = \left(-\frac{1}{2}\right)^{-1} \qquad \text{Solve for } x \text{ by raising both sides of} \\ \text{each equation to the } -1 \text{ power.}$$

$$x = -5 \qquad\qquad\qquad x = -2$$

$$\left(-\tfrac{1}{5}\right)^{-1} = \tfrac{1}{-\frac{1}{5}} = -5 \qquad\qquad \left(-\tfrac{1}{2}\right)^{-1} = \tfrac{1}{-\frac{1}{2}} = -2$$

We did not raise both sides of an equation to an even power. A check will show that both -5 and -2 are solutions of the original equation. The solution set is $\{-5, -2\}$. ∎

 CHECK POINT 4 Solve: $2x^{-2} + x^{-1} - 1 = 0$. -1 and 2, or $\{-1, 2\}$

EXAMPLE 5 Solving an Equation Quadratic in Form

Solve: $5x^{\frac{2}{3}} + 11x^{\frac{1}{3}} + 2 = 0$.

SOLUTION   Notice that the equation contains an expression to a power, $x^{\frac{1}{3}}$, the same expression to that power squared, $x^{\frac{2}{3}}$ or $\left(x^{\frac{1}{3}}\right)^2$, and a constant term, 2. We let t equal the expression to the power. Thus,

$$\text{let } t = x^{\frac{1}{3}}.$$

Now we write the given equation as a quadratic equation in t and solve for t.

$$5x^{\frac{2}{3}} + 11x^{\frac{1}{3}} + 2 = 0 \qquad \text{This is the given equation.}$$

$$5\left(x^{\frac{1}{3}}\right)^2 + 11\left(x^{\frac{1}{3}}\right) + 2 = 0 \qquad \text{The given equation contains } x^{\frac{1}{3}} \text{ and } x^{\frac{1}{3}} \text{ squared.}$$

$$5t^2 + 11t + 2 = 0 \qquad \text{Let } t = x^{\frac{1}{3}}.$$

$$(5t + 1)(t + 2) = 0 \qquad \text{Factor.}$$

$$5t + 1 = 0 \quad \text{or} \quad t + 2 = 0 \qquad \text{Set each factor equal to 0.}$$

$$t = -\tfrac{1}{5} \qquad\qquad t = -2 \qquad \text{Solve for } t.$$

Use the original substitution, $t = x^{\frac{1}{3}}$, to solve for x. Replace t with $x^{\frac{1}{3}}$ in each of the preceding equations.

$$x^{\frac{1}{3}} = -\frac{1}{5} \quad \text{or} \quad x^{\frac{1}{3}} = -2 \qquad \text{Replace } t \text{ with } x^{\frac{1}{3}}.$$

$$\left(x^{\frac{1}{3}}\right)^3 = \left(-\frac{1}{5}\right)^3 \qquad \left(x^{\frac{1}{3}}\right)^3 = (-2)^3 \qquad \text{Solve for } x \text{ by cubing both sides of each} \\ \text{equation.}$$

$$x = -\frac{1}{125} \qquad\qquad x = -8$$

We did not raise both sides of an equation to an even power. A check will show that both -8 and $-\frac{1}{125}$ are solutions of the original equation. The solution set is $\left\{-8, -\frac{1}{125}\right\}$. ∎

 CHECK POINT 5 Solve: $3x^{\frac{2}{3}} - 11x^{\frac{1}{3}} - 4 = 0$. $-\frac{1}{27}$ and 64, or $\left\{-\frac{1}{27}, 64\right\}$

11.4 EXERCISE SET

Student Solutions Manual   CD/Video   PH Math/Tutor Center   MathXL Tutorials on CD   MathXL®   MyMathLab   Interactmath.com

Practice Exercises

In Exercises 1–32, solve each equation by making an appropriate substitution. If at any point in the solution process both sides of an equation are raised to an even power, a check is required.

1. $x^4 - 5x^2 + 4 = 0$ $\{-2, 2, -1, 1\}$

2. $x^4 - 13x^2 + 36 = 0$ $\{-3, 3, -2, 2\}$

3. $x^4 - 11x^2 + 18 = 0$ $\{-3, 3, -\sqrt{2}, \sqrt{2}\}$

4. $x^4 - 9x^2 + 20 = 0$ $\{-\sqrt{5}, \sqrt{5}, -2, 2\}$

5. $x^4 + 2x^2 = 8$ $\{-2i, 2i, -\sqrt{2}, \sqrt{2}\}$

6. $x^4 + 4x^2 = 5$ $\{-\sqrt{5}i, \sqrt{5}i, -1, 1\}$

7. $x + \sqrt{x} - 2 = 0$ $\{1\}$

8. $x + \sqrt{x} - 6 = 0$ $\{4\}$

9. $x - 4x^{\frac{1}{2}} - 21 = 0$ $\{49\}$

10. $x - 6x^{\frac{1}{2}} + 8 = 0$ $\{4, 16\}$

11. $x - 13\sqrt{x} + 40 = 0$ $\{25, 64\}$

12. $2x - 7\sqrt{x} - 30 = 0$ $\{36\}$

13. $(x - 5)^2 - 4(x - 5) - 21 = 0$ $\{2, 12\}$

14. $(x + 3)^2 + 7(x + 3) - 18 = 0$ $\{-12, -1\}$

15. $(x^2 - 1)^2 - (x^2 - 1) = 2$ $\{-\sqrt{3}, 0, \sqrt{3}\}$

16. $(x^2 - 2)^2 - (x^2 - 2) = 6$ $\{-\sqrt{5}, 0, \sqrt{5}\}$

17. $(x^2 + 3x)^2 - 8(x^2 + 3x) - 20 = 0$ $\{-5, -2, -1, 2\}$

18. $(x^2 - 2x)^2 - 11(x^2 - 2x) + 24 = 0$ $\{-2, -1, 3, 4\}$

19. $x^{-2} - x^{-1} - 20 = 0$ $\{-\frac{1}{4}, \frac{1}{5}\}$

20. $x^{-2} - x^{-1} - 6 = 0$ $\{-\frac{1}{2}, \frac{1}{3}\}$

21. $2x^{-2} - 7x^{-1} + 3 = 0$ $\{\frac{1}{3}, 2\}$

22. $20x^{-2} + 9x^{-1} + 1 = 0$ $\{-5, -4\}$

23. $x^{-2} - 4x^{-1} = 3$ $\left\{\dfrac{-2 \pm \sqrt{7}}{3}\right\}$

24. $x^{-2} - 6x^{-1} = -4$ $\left\{\dfrac{3 \pm \sqrt{5}}{4}\right\}$

25. $x^{\frac{2}{3}} - x^{\frac{1}{3}} - 6 = 0$ $\{-8, 27\}$

26. $x^{\frac{2}{3}} + 2x^{\frac{1}{3}} - 3 = 0$ $\{-27, 1\}$

27. $x^{\frac{2}{5}} + x^{\frac{1}{5}} - 6 = 0$ $\{-243, 32\}$

28. $x^{\frac{2}{5}} + x^{\frac{1}{5}} - 2 = 0$ $\{-32, 1\}$

29. $2x^{\frac{1}{2}} - x^{\frac{1}{4}} = 1$ $\{1\}$

30. $2x^{\frac{1}{2}} - 5x^{\frac{1}{4}} = 3$ $\{81\}$

31. $\left(x - \dfrac{8}{x}\right)^2 + 5\left(x - \dfrac{8}{x}\right) - 14 = 0$ $\{-8, -2, 1, 4\}$

32. $\left(x - \dfrac{10}{x}\right)^2 + 6\left(x - \dfrac{10}{x}\right) - 27 = 0$ $\{-10, -2, 1, 5\}$

In Exercises 33–38, find the x-intercepts of the given function, f. Then use the x-intercepts to match each function with its graph. [The graphs are labeled (a) through (f).]

33. $f(x) = x^4 - 5x^2 + 4$ $-2, 2, -1,$ and 1; **c**

34. $f(x) = x^4 - 13x^2 + 36$ $-3, 3, -2,$ and 2; **a**

35. $f(x) = x^{\frac{1}{3}} + 2x^{\frac{1}{6}} - 3$ 1; **e**

36. $f(x) = x^{-2} - x^{-1} - 6$ $-\frac{1}{2}$ and $\frac{1}{3}$; **b**

37. $f(x) = (x + 2)^2 - 9(x + 2) + 20$ 2 and 3; **f**

38. $f(x) = 2(x + 2)^2 + 5(x + 2) - 3$ -5 and $-\frac{3}{2}$; **d**

a.

$[-5, 5, 1]$ by $[-10, 40, 5]$

b.

$[-3, 3, 1]$ by $[-10, 10, 1]$

c.

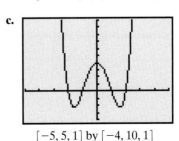

$[-5, 5, 1]$ by $[-4, 10, 1]$

d.

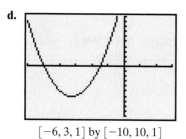

$[-6, 3, 1]$ by $[-10, 10, 1]$

e.

$[-1, 10, 1]$ by $[-3, 3, 1]$

f.

$[-1, 6, 1]$ by $[-1, 10, 1]$

Practice Plus

39. Let $f(x) = (x^2 + 3x - 2)^2 - 10(x^2 + 3x - 2)$. Find all x such that $f(x) = -16$. $-5, -4, 1,$ and 2

40. Let $f(x) = (x^2 + 2x - 2)^2 - 7(x^2 + 2x - 2)$. Find all x such that $f(x) = -6$. $-4, -3, 1,$ and 2

41. Let $f(x) = 3\left(\dfrac{1}{x} + 1\right)^2 + 5\left(\dfrac{1}{x} + 1\right)$. Find all x such that $f(x) = 2$. $-\frac{3}{2}$ and $-\frac{1}{3}$

42. Let $f(x) = 2x^{\frac{2}{3}} + 3x^{\frac{1}{3}}$. Find all x such that $f(x) = 2$. -8 and $\frac{1}{8}$

43. Let $f(x) = \dfrac{x}{x - 4}$ and $g(x) = 13\sqrt{\dfrac{x}{x - 4}} - 36$. Find all x such that $f(x) = g(x)$. $\frac{81}{20}$ and $\frac{64}{15}$

44. Let $f(x) = \dfrac{x}{x - 2} + 10$ and $g(x) = -11\sqrt{\dfrac{x}{x - 2}}$. Find all x such that $f(x) = g(x)$. No values of x satisfy $f(x) = g(x)$.

45. Let $f(x) = 3(x - 4)^{-2}$ and $g(x) = 16(x - 4)^{-1}$. Find all x such that $f(x)$ exceeds $g(x)$ by 12. $\frac{5}{2}$ and $\frac{25}{6}$

46. Let $f(x) = 6\left(\dfrac{2x}{x - 3}\right)^2$ and $g(x) = 5\left(\dfrac{2x}{x - 3}\right)$. Find all x such that $f(x)$ exceeds $g(x)$ by 6. -9 and $\frac{3}{4}$

Application Exercises

How important is it for you to have a clean house? The bar graph indicates that the percentage of people who find this to be quite important varies by age. The percentage, $P(x)$, who find having a clean house very important can be modeled by the function

$$P(x) = 0.04(x + 40)^2 - 3(x + 40) + 104,$$

where x is the number of years a person's age is above or below 40. Thus, x is positive for people over 40 and negative for people under 40. Use the function to solve Exercises 47–48.

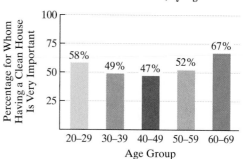

The Importance of Having a Clean House, by Age

Source: Soap and Detergent Association

47. According to the model, at which ages do 60% of us feel that having a clean house is very important? Substitute 60 for $P(x)$ and solve the quadratic-in-form equation. How well does the function model the data shown in the bar graph? ages 20 and 55; The function models the data well.

48. According to the model, at which ages do 50% of us feel that having a clean house is very important? Substitute 50 for $P(x)$ and solve the quadratic-in-form equation. How well does the function model the data shown in the bar graph? ages 30 and 45; The function models the data well.

Writing in Mathematics

49. Explain how to recognize an equation that is quadratic in form. Provide two original examples with your explanation. 49–51. Answers will vary.

50. Describe two methods for solving this equation:

$$x - 5\sqrt{x} + 4 = 0.$$

51. In the twenty-first century, collecting and presenting data are big business. Because statisticians both record and influence our behavior, you should always ask yourself if the person or group presenting the data has any special case to make for or against the displayed information. Using this criterion, describe your impressions of the data in the bar graph for Exercises 47–48.

Technology Exercises

52. Use a graphing utility to verify the solutions of any five equations in Exercises 1–32 that you solved algebraically. The real solutions should appear as x-intercepts on the graph of the function related to the given equation.   Answers will vary.

Use a graphing utility to solve the equations in Exercises 53–60. Check by direct substitution.

53. $x^6 - 7x^3 - 8 = 0$ $\{-1, 2\}$

54. $3(x - 2)^{-2} - 4(x - 2)^{-1} + 1 = 0$ $\{3, 5\}$

55. $x^4 - 10x^2 + 9 = 0$ $\{-1, 1, -3, 3\}$

56. $2x + 6\sqrt{x} = 8$ $\{1\}$

57. $2(x + 1)^2 = 5(x + 1) + 3$ $\{-\frac{3}{2}, 2\}$

58. $(x^2 - 3x)^2 + 2(x^2 - 3x) - 24 = 0$ $\{-1, 4\}$

59. $x^{\frac{1}{2}} + 4x^{\frac{1}{4}} = 5$ $\{1\}$

60. $x^{\frac{2}{3}} - 3x^{\frac{1}{3}} + 2 = 0$ $\{1, 8\}$

Critical Thinking Exercises

61. Which one of the following is true?   b

    a. If an equation is quadratic in form, there is only one method that can be used to obtain its solution.

    b. An equation that is quadratic in form must have a variable factor in one term that is the square of the variable factor in another term.

    c. Because x^6 is the square of x^3, the equation $x^6 - 5x^3 + 6x = 0$ is quadratic in form.

    d. To solve $x - 9\sqrt{x} + 14 = 0$, we let $\sqrt{t} = x$.

In Exercises 62–64, use a substitution to solve each equation.

62. $x^4 - 5x^2 - 2 = 0$ $\left\{ \pm \dfrac{\sqrt{10 + 2\sqrt{33}}}{2}, \pm \dfrac{\sqrt{10 - 2\sqrt{33}}}{2} \right\}$

63. $5x^6 + x^3 = 18$ $\left\{ \sqrt[3]{-2}, \dfrac{\sqrt[3]{225}}{5} \right\}$

64. $\sqrt{\dfrac{x + 4}{x - 1}} + \sqrt{\dfrac{x - 1}{x + 4}} = \dfrac{5}{2} \left(\text{Let } t = \sqrt{\dfrac{x + 4}{x - 1}}. \right)$ $\left\{ -\dfrac{17}{3}, \dfrac{8}{3} \right\}$

Review Exercises

65. Simplify:

$$\frac{2x^2}{10x^3 - 2x^2}.$$

(Section 7.1, Example 3) $\dfrac{1}{5x - 1}$

66. Divide: $\dfrac{2 + i}{1 - i}$. (Section 10.7, Example 5) $\dfrac{1}{2} + \dfrac{3}{2}i$

67. If $f(x) = \sqrt{x + 1}$, find $f(3) - f(24)$.
(Section 8.1, Example 3) -3

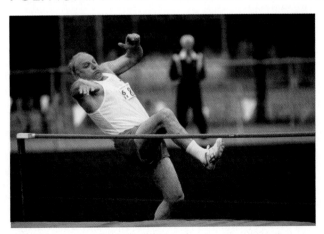

SECTION 11.5

POLYNOMIAL AND RATIONAL INEQUALITIES

Objectives

1 Solve polynomial inequalities.

2 Solve rational inequalities.

3 Solve problems modeled by polynomial or rational inequalities.

People are going to live longer in the twenty-first century. This will put added pressure on the Social Security and Medicare systems. The bar graph in Figure 11.17 shows the cost of Medicare, in billions of dollars, through 2005.

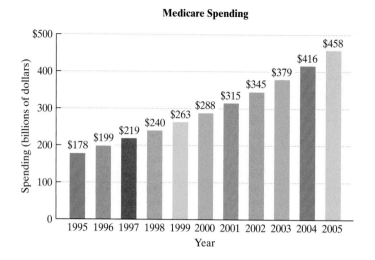

FIGURE 11.17

Source: Congressional Budget Office

Medicare spending, $f(x)$, in billions of dollars, x years after 1995 can be modeled by the quadratic function

$$f(x) = 1.2x^2 + 15.2x + 181.4.$$

To determine in which years Medicare spending will exceed $500 billion, we must solve the inequality

$$1.2x^2 + 15.2x + 181.4 > 500.$$

Medicare spending   exceeds $500 billion.

We begin by subtracting 500 from both sides. This will give us zero on the right:

$$1.2x^2 + 15.2x + 181.4 - 500 > 500 - 500$$
$$1.2x^2 + 15.2x - 318.6 > 0.$$

The form of this inequality is $ax^2 + bx + c > 0$. Such an inequality is called a *polynomial inequality*.

> **DEFINITION OF A POLYNOMIAL INEQUALITY** A polynomial inequality is any inequality that can be put in one of the forms
>
> $$f(x) < 0, \quad f(x) > 0, \quad f(x) \le 0, \quad \text{or} \quad f(x) \ge 0,$$
>
> where f is a polynomial function.

In this section, we establish the basic techniques for solving polynomial inequalities. We will use these techniques to solve inequalities involving rational functions.

Solving Polynomial Inequalities Graphs can help us visualize the solutions of polynomial inequalities. For example, the graph of $f(x) = x^2 - 7x + 10$ is shown in Figure 11.18. The x-intercepts, 2 and 5, are **boundary points** between where the graph lies above the x-axis, shown in blue, and where the graph lies below the x-axis, shown in red.

Locating the x-intercepts of a polynomial function, f, is an important step in finding the solution set for polynomial inequalities in the form $f(x) < 0$ or $f(x) > 0$.

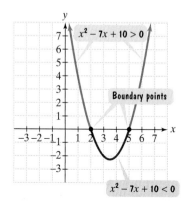

FIGURE 11.18

We use the x-intercepts of f as boundary points that divide the real number line into intervals. On each interval, the graph of f is either above the x-axis $[f(x) > 0]$ or below the x-axis $[f(x) < 0]$. For this reason, x-intercepts play a fundamental role in solving polynomial inequalities. The x-intercepts are found by solving the equation $f(x) = 0$.

1 Solve polynomial inequalities.

PROCEDURE FOR SOLVING POLYNOMIAL INEQUALITIES

1. Express the inequality in the form

$$f(x) < 0 \quad \text{or} \quad f(x) > 0,$$

where f is a polynomial function

2. Solve the equation $f(x) = 0$. The real solutions are the **boundary points**.

3. Locate these boundary points on a number line, thereby dividing the number line into intervals.

4. Choose one representative number within each interval and evaluate f at that number.

   a. If the value of f is positive, then $f(x) > 0$ for all numbers, x, in the interval.

   b. If the value of f is negative, then $f(x) < 0$ for all numbers, x, in the interval.

5. Write the solution set, selecting the interval(s) that satisfy the given inequality.

This procedure is valid if $<$ is replaced by $\leq$ or $>$ is replaced by $\geq$. However, if the inequality involves $\leq$ or $\geq$, include the boundary points [the solutions of $f(x) = 0$] in the solution set.

EXAMPLE 1   Solving a Polynomial Inequality

Solve and graph the solution set on a real number line: $2x^2 + x > 15$.

SOLUTION

Step 1. **Express the inequality in the form $f(x) < 0$ or $f(x) > 0$.** We begin by rewriting the inequality so that 0 is on the right side.

$$
\begin{array}{ll}
2x^2 + x > 15 & \text{This is the given inequality.} \\
2x^2 + x - 15 > 15 - 15 & \text{Subtract 15 from both sides.} \\
2x^2 + x - 15 > 0 & \text{Simplify.}
\end{array}
$$

This inequality is equivalent to the one we wish to solve. It is in the form $f(x) > 0$, where $f(x) = 2x^2 + x - 15$.

Step 2. **Solve the equation $f(x) = 0$.** We find the x-intercepts of $f(x) = 2x^2 + x - 15$ by solving the equation $2x^2 + x - 15 = 0$.

$$
\begin{array}{ll}
2x^2 + x - 15 = 0 & \text{This polynomial equation is a} \\
& \text{quadratic equation.} \\
(2x - 5)(x + 3) = 0 & \text{Factor.} \\
2x - 5 = 0 \quad \text{or} \quad x + 3 = 0 & \text{Set each factor equal to 0.} \\
x = \frac{5}{2} \qquad\qquad x = -3 & \text{Solve for x.}
\end{array}
$$

The x-intercepts of f are -3 and $\frac{5}{2}$. We will use these x-intercepts as boundary points on a number line.

Step 3. **Locate the boundary points on a number line and separate the line into intervals.** The number line with the boundary points is shown as follows:

The boundary points divide the number line into three intervals:

$$(-\infty, -3) \quad \left(-3, \tfrac{5}{2}\right) \quad \left(\tfrac{5}{2}, \infty\right).$$

Step 4. **Choose one representative number within each interval and evaluate f at that number.**

Interval	Representative Number	Substitute into $f(x) = 2x^2 + x - 15$	Conclusion
$(-\infty, -3)$	-4	$f(-4) = 2(-4)^2 + (-4) - 15$ $= 13$, positive	$f(x) > 0$ for all x in $(-\infty, -3)$.
$\left(-3, \tfrac{5}{2}\right)$	0	$f(0) = 2 \cdot 0^2 + 0 - 15$ $= -15$, negative	$f(x) < 0$ for all x in $\left(-3, \tfrac{5}{2}\right)$.
$\left(\tfrac{5}{2}, \infty\right)$	3	$f(3) = 2 \cdot 3^2 + 3 - 15$ $= 6$, positive	$f(x) > 0$ for all x in $\left(\tfrac{5}{2}, \infty\right)$.

USING TECHNOLOGY

The solution set for

$$2x^2 + x > 15$$

or, equivalently,

$$2x^2 + x - 15 > 0$$

can be verified with a graphing utility. The graph of $f(x) = 2x^2 + x - 15$ was obtained using a $[-10, 10, 1]$ by $[-16, 6, 1]$ viewing rectangle. The graph lies above the x-axis, representing $>$, for all x in $(-\infty, -3)$ or $\left(\tfrac{5}{2}, \infty\right)$.

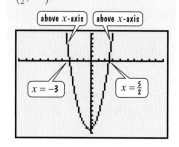

Step 5. **Write the solution set, selecting the interval(s) that satisfy the given inequality.** Based on our work in step 4, we see that $f(x) > 0$ for all x in $(-\infty, -3)$ or $\left(\tfrac{5}{2}, \infty\right)$. Thus, the solution set of the given inequality, $2x^2 + x > 15$, or, equivalently, $2x^2 + x - 15 > 0$, is

$$(-\infty, -3) \cup \left(\tfrac{5}{2}, \infty\right) \text{ or } \left\{x \mid x < -3 \text{ or } x > \tfrac{5}{2}\right\}.$$

The graph of the solution set on a number line is shown as follows:

✔ **CHECK POINT 1** Solve and graph the solution set: $x^2 - x > 20$.
See graphing answer section. $(-\infty, -4) \cup (5, \infty)$ or $\{x \mid x < -4 \text{ or } x > 5\}$

EXAMPLE 2 Solving a Polynomial Inequality

Solve and graph the solution set on a real number line: $x^3 + x^2 \le 4x + 4$.

SOLUTION

Step 1. **Express the inequality in the form $f(x) \le 0$ or $f(x) \ge 0$.** We begin by rewriting the inequality so that 0 is on the right side.

$$x^3 + x^2 \le 4x + 4 \qquad \text{This is the given inequality.}$$
$$x^3 + x^2 - 4x - 4 \le 4x + 4 - 4x - 4 \qquad \text{Subtract } 4x + 4 \text{ from both sides.}$$
$$x^3 + x^2 - 4x - 4 \le 0 \qquad \text{Simplify.}$$

This inequality is equivalent to the one we wish to solve. It is in the form $f(x) \le 0$, where $f(x) = x^3 + x^2 - 4x - 4$.

The solution set for
$$x^3 + x^2 \leq 4x + 4$$
or, equivalently,
$$x^3 + x^2 - 4x - 4 \leq 0$$
can be verified with a graphing utility. The graph of $f(x) = x^3 + x^2 - 4x - 4$ lies on or below the x-axis, representing $\leq$, for all x in $(-\infty, -2]$ or $[-1, 2]$.

$[-4, 4, 1]$ by $[-7, 3, 1]$

Step 2. **Solve the equation $f(x) = 0$.** We find the x-intercepts of $f(x) = x^3 + x^2 - 4x - 4$ by solving the equation $x^3 + x^2 - 4x - 4 = 0$.

$$x^3 + x^2 - 4x - 4 = 0 \qquad \text{This polynomial equation is of degree 3.}$$
$$x^2(x + 1) - 4(x + 1) = 0 \qquad \text{Factor } x^2 \text{ from the first two terms and } -4$$
$$\text{from the last two terms.}$$
$$(x + 1)(x^2 - 4) = 0 \qquad \text{A common factor of } x + 1 \text{ is factored from}$$
$$\text{the expression.}$$
$$x + 1 = 0 \quad \text{or} \quad x^2 - 4 = 0 \qquad \text{Set each factor equal to 0.}$$
$$x = -1 \qquad\qquad x^2 = 4 \qquad \text{Solve for x.}$$
$$x = \pm 2 \qquad \text{Use the square root property.}$$

The x-intercepts of f are -2, -1, and 2. We will use these x-intercepts as boundary points on a number line.

Step 3. **Locate the boundary points on a number line and separate the line into intervals.** The number line with the boundary points is shown as follows:

The boundary points divide the number line into four intervals:
$$(-\infty, -2) \quad (-2, -1) \quad (-1, 2) \quad (2, \infty).$$

Step 4. **Choose one representative number within each interval and evaluate f at that number.**

Interval	Representative Number	Substitute into $f(x) = x^3 + x^2 - 4x - 4$	Conclusion
$(-\infty, -2)$	-3	$f(-3) = (-3)^3 + (-3)^2 - 4(-3) - 4$ $= -10$, negative	$f(x) < 0$ for all x in $(-\infty, -2)$.
$(-2, -1)$	-1.5	$f(-1.5) = (-1.5)^3 + (-1.5)^2 - 4(-1.5) - 4$ $= 0.875$, positive	$f(x) > 0$ for all x in $(-2, -1)$.
$(-1, 2)$	0	$f(0) = 0^3 + 0^2 - 4 \cdot 0 - 4$ $= -4$, negative	$f(x) < 0$ for all x in $(-1, 2)$.
$(2, \infty)$	3	$f(3) = 3^3 + 3^2 - 4 \cdot 3 - 4$ $= 20$, positive	$f(x) > 0$ for all x in $(2, \infty)$.

Step 5. **Write the solution set, selecting the interval(s) that satisfy the given inequality.** Based on our work in step 4, we see that $f(x) < 0$ for all x in $(-\infty, -2)$ or $(-1, 2)$. However, because the inequality involves $\leq$ (less than or *equal to*), we must also include the solutions of $x^3 + x^2 - 4x - 4 = 0$, namely -2, -1, and 2, in the solution set. Thus, the solution set of the given inequality, $x^3 + x^2 \leq 4x + 4$, or, equivalently, $x^3 + x^2 - 4x - 4 \leq 0$, is

$$(-\infty, -2] \cup [-1, 2]$$

or $\quad \{x | x \leq -2 \text{ or } -1 \leq x \leq 2\}.$

The graph of the solution set on a number line is shown as follows:

✔ **CHECK POINT 2** Solve and graph the solution set on a real number line: $x^3 + 3x^2 \le x + 3$.   See graphing answer section.

$(-\infty, -3] \cup [-1, 1]$ or $\{x \mid x \le -3 \text{ or } -1 \le x \le 1\}$

2 Solve rational inequalities.

Solving Rational Inequalities   A **rational inequality** is any inequality that can be put in one of the forms

$$f(x) < 0, \quad f(x) > 0, \quad f(x) \le 0, \quad \text{or} \quad f(x) \ge 0,$$

where f is a rational function. An example of a rational inequality is

$$\frac{3x + 3}{2x + 4} > 0.$$

This inequality is in the form $f(x) > 0$, where f is the rational function given by

$$f(x) = \frac{3x + 3}{2x + 4}.$$

The graph of f is shown in Figure 11.19.

We can find the x-intercept of f by setting the numerator equal to 0:

$$3x + 3 = 0$$
$$3x = -3$$
$$x = -1.$$

> f has an x-intercept at −1 and passes through (−1, 0).

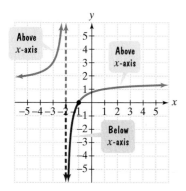

FIGURE 11.19   The graph of $f(x) = \dfrac{3x + 3}{2x + 4}$

We can determine where f is undefined by setting the denominator equal to 0:

$$2x + 4 = 0$$
$$2x = -4$$
$$x = -2.$$

> f is undefined at −2. Figure 11.19 shows that the function's vertical asymptote is $x = -2$.

By setting the numerator and the denominator of f equal to 0, we obtained the solutions −2 and −1. These numbers separate the x-axis into three intervals: $(-\infty, -2), (-2, -1),$ and $(-1, \infty)$. On each interval, the graph of f is either above the x-axis $[f(x) > 0]$ or below the x-axis $[f(x) < 0]$.

Examine the graph in Figure 11.19 carefully. Can you see that it is above the x-axis for all x in $(-\infty, -2)$ or $(-1, \infty)$, shown in blue? Thus, the solution set of $\dfrac{3x + 3}{2x + 4} > 0$ is $(-\infty, -2) \cup (-1, \infty)$. By contrast, the graph of f lies below the x-axis for all x in $(-2, -1)$, shown in red. Thus, the solution set of $\dfrac{3x + 3}{2x + 4} < 0$ is $(-2, -1)$.

The first step in solving a rational inequality is to bring all terms to one side, obtaining zero on the other side. Then express the rational function on the nonzero side as a single quotient. The second step is to set the numerator and the denominator of f equal to zero. The solutions of these equations serve as boundary points that separate the real number line into intervals. At this point, the procedure is the same as the one we used for solving polynomial inequalities.

EXAMPLE 3   Solving a Rational Inequality

Solve and graph the solution set: $\dfrac{x + 1}{x + 3} \ge 2.$

SOLUTION

Step 1. **Express the inequality so that one side is zero and the other side is a single quotient.** We subtract 2 from both sides to obtain zero on the right.

$$\frac{x+1}{x+3} \geq 2 \qquad \text{This is the given inequality.}$$

$$\frac{x+1}{x+3} - 2 \geq 0 \qquad \text{Subtract 2 from both sides, obtaining 0 on the right.}$$

$$\frac{x+1}{x+3} - \frac{2(x+3)}{x+3} \geq 0 \qquad \text{The least common denominator is } x+3. \text{ Express 2 in terms of this denominator.}$$

$$\frac{x+1-2(x+3)}{x+3} \geq 0 \qquad \text{Subtract rational expressions.}$$

$$\frac{x+1-2x-6}{x+3} \geq 0 \qquad \text{Apply the distributive property.}$$

$$\frac{-x-5}{x+3} \geq 0 \qquad \text{Simplify.}$$

This inequality is equivalent to the one we wish to solve. It is in the form $f(x) \geq 0$, where $f(x) = \dfrac{-x-5}{x+3}$.

Step 2. **Set the numerator and the denominator of f equal to zero.** The real solutions are the boundary points.

$$-x - 5 = 0 \qquad x + 3 = 0 \qquad \text{Set the numerator and denominator equal to 0. These are the values that make the previous quotient zero or undefined.}$$

$$x = -5 \qquad x = -3 \qquad \text{Solve for } x.$$

We will use these solutions as boundary points on a number line.

Step 3. **Locate the boundary points on a number line and separate the line into intervals.** The number line with the boundary points is shown as follows:

The boundary points divide the number line into three intervals:

$$(-\infty, -5) \quad (-5, -3) \quad (-3, \infty).$$

Step 4. **Choose one representative number within each interval and evaluate f at that number.**

Interval	Representative Number	Substitute into $f(x) = \dfrac{-x-5}{x+3}$	Conclusion
$(-\infty, -5)$	-6	$f(-6) = \dfrac{-(-6)-5}{-6+3}$ $= -\frac{1}{3}$, negative	$f(x) < 0$ for all x in $(-\infty, -5)$.
$(-5, -3)$	-4	$f(-4) = \dfrac{-(-4)-5}{-4+3}$ $= 1$, positive	$f(x) > 0$ for all x in $(-5, -3)$.
$(-3, \infty)$	0	$f(0) = \dfrac{-0-5}{0+3}$ $= -\frac{5}{3}$, negative	$f(x) < 0$ for all x in $(-3, \infty)$.

Step 5. **Write the solution set, selecting the interval(s) that satisfy the given inequality.** Based on our work in step 4, we see that $f(x) > 0$ for all x in $(-5, -3)$. However, because the inequality involves $\geq$ (greater than or *equal to*), we must also include the solution of $f(x) = 0$, namely the value that we obtained when we set the numerator of f equal to zero. Thus, we must include -5 in the solution set. The solution set of the given inequality is

$$[-5, -3) \text{ or } \{x | -5 \leq x < -3\}.$$

The graph of the solution set on a number line is shown as follows:

The solution set for

$$\frac{x + 1}{x + 3} \geq 2$$

or, equivalently,

$$\frac{-x - 5}{x + 3} \geq 0$$

can be verified with a graphing utility. The graph of $f(x) = \dfrac{-x - 5}{x + 3}$ lies on or above the x-axis, representing $\geq$, for all x in $[-5, -3)$.

Graph lies on or above the x-axis.

$x = -5$ $x = -3$

$[-8, 8, 1]$ by $[-3, 3, 1]$

DISCOVER FOR YOURSELF

Because $(x + 3)^2$ is positive, it is possible so solve

$$\frac{x + 1}{x + 3} \geq 2$$

by first multiplying both sides by $(x + 3)^2$ (where $x \neq -3$). This will not reverse the sense of the inequality and will clear the fraction. Try using this solution method and compare it to the one shown on pages 780–782.

 CHECK POINT 3 Solve and graph the solution set: $\dfrac{2x}{x + 1} \geq 1$.

See graphing answer section. $(-\infty, -1) \cup [1, \infty)$ or $\{x | x < -1 \text{ or } x \geq 1\}$

3 Solve problems modeled by polynomial or rational inequalities.

Applications Polynomial inequalities can be solved to answer questions about variables contained in polynomial functions.

> **EXAMPLE 4** Modeling the Position of a Free-Falling Object

A ball is thrown vertically upward from the top of the Leaning Tower of Pisa (190 feet high) with an initial velocity of 96 feet per second (Figure 11.20). The function

$$s(t) = -16t^2 + 96t + 190$$

models the ball's height above the ground, $s(t)$, in feet, t seconds after it was thrown. During which time period will the ball's height exceed that of the tower?

SOLUTION   Using the problem's question and the given model for the ball's height, $s(t) = -16t^2 + 96t + 190$, we obtain a polynomial inequality.

$$-16t^2 + 96t + 190 > 190$$

exceed that   of the tower?

When will the ball's height

190 feet

FIGURE 11.20 Throwing a ball from the top of the Leaning Tower of Pisa

$-16t^2 + 96t + 190 > 190$	This is the inequality that models the problem's question. We must find t.
$-16t^2 + 96t > 0$	Subtract 190 from both sides. This inequality is in the form $f(t) > 0$, where $f(t) = -16t^2 + 96t$.
$-16t^2 + 96t = 0$	Solve the equation $f(t) = 0$.
$-16t(t - 6) = 0$	Factor.
$-16t = 0$ or $t - 6 = 0$	Set each factor equal to 0.
$t = 0$ $t = 6$	Solve for t. The boundary points are 0 and 6.

$$\overset{\ }{\underset{-2\ -1\ \ \ 0\ \ \ 1\ \ \ 2\ \ \ 3\ \ \ 4\ \ \ 5\ \ \ 6\ \ \ 7\ \ \ 8}{\longleftrightarrow}}$$

Locate these values on a number line, with $t \geq 0$.

The intervals are $(-\infty, 0)$, $(0, 6)$ and $(6, \infty)$. For our purposes, the mathematical model is useful only from $t = 0$ until the ball hits the ground. (By setting $-16t^2 + 96t + 190$ equal to zero, we find $t \approx 7.57$; the ball hits the ground after approximately 7.57 seconds.) Thus, we use $(0, 6)$ and $(6, 7.57)$ for our intervals.

Interval	Representative Number	Substitute into $f(t) = -16t^2 + 96t$	Conclusion
$(0, 6)$	1	$f(1) = -16 \cdot 1^2 + 96 \cdot 1$ $= 80$, positive	$f(t) > 0$ for all t in $(0, 6)$.
$(6, 7.57)$	7	$f(7) = -16 \cdot 7^2 + 96 \cdot 7$ $= -112$, negative	$f(t) < 0$ for all t in $(6, 7.57)$.

We see that $f(t) > 0$ for all t in $(0, 6)$. This means that the ball's height exceeds that of the tower between 0 and 6 seconds. ∎

USING TECHNOLOGY

The graphs of

$$y_1 = -16x^2 + 96x + 190$$

and

$$y_2 = 190$$

are shown in a

$$[0, 8, 1] \text{ by } [0, 360, 36]$$

viewing rectangle. The graphs show that the ball's height exceeds that of the tower between 0 and 6 seconds.

Height of Ball:
$y_1 = -16x^2 + 96x + 190$

360

Height (feet)

Height of Tower:
$y_2 = 190$

Ball hits ground after 7.57 seconds.

0   6   8

Seconds in Motion
$[0, 8, 1]$ by $[0, 360, 36]$

 CHECK POINT 4 An object is propelled straight up from ground level with an initial velocity of 80 feet per second. Its height at time t is modeled by

$$s(t) = -16t^2 + 80t,$$

where the height, $s(t)$, is measured in feet and the time, t, is measured in seconds. In which time interval will the object be more than 64 feet above the ground?
The object will be more than 64 feet above ground between 1 and 4 seconds, excluding $t = 1$ and $t = 4$.

11.5 EXERCISE SET

Student Solutions Manual   CD/Video   PH Math/Tutor Center   MathXL Tutorials on CD   MathXL®   MyMathLab   Interactmath.com

Practice Exercises

Solve each polynomial inequality in Exercises 1–36, and graph the solution set on a real number line.
1–36. See graphing answer section.

1. $(x - 4)(x + 2) > 0$
$\{x|x < -2 \text{ or } x > 4\}$ or $(-\infty, -2) \cup (4, \infty)$

2. $(x + 3)(x - 5) > 0$
$\{x|x < -3 \text{ or } x > 5\}$ or $(-\infty, -3) \cup (5, \infty)$

3. $(x - 7)(x + 3) \leq 0$ $\{x|-3 \leq x \leq 7\}$ or $[-3, 7]$

4. $(x + 1)(x - 7) \leq 0$ $\{x|-1 \leq x \leq 7\}$ or $[-1, 7]$

5. $x^2 - 5x + 4 > 0$
$\{x|x < 1 \text{ or } x > 4\}$ or $(-\infty, -1) \cup (4, \infty)$

6. $x^2 - 4x + 3 < 0$ $\{x|1 < x < 3\}$ or $(1, 3)$

7. $x^2 + 5x + 4 > 0$
$\{x|x < -4 \text{ or } x > -1\}$ or $(-\infty, -4) \cup (-1, \infty)$

8. $x^2 + x - 6 > 0$
$\{x|x < -3 \text{ or } x > 2\}$ or $(-\infty, -3) \cup (2, \infty)$

9. $x^2 - 6x + 8 \leq 0$ $\{x|2 \leq x \leq 4\}$ or $[2, 4]$

10. $x^2 - 2x - 3 \geq 0$
  $\{x | x \leq -1 \text{ or } x \geq 3\}$ or $(-\infty, -1] \cup [3, \infty)$

11. $3x^2 + 10x - 8 \leq 0$ $\{x | -4 \leq x \leq \frac{2}{3}\}$ or $\left[-4, \frac{2}{3}\right]$

12. $9x^2 + 3x - 2 \geq 0$
  $\{x | x \leq -\frac{2}{3} \text{ or } x \geq \frac{1}{3}\}$ or $\left(-\infty, -\frac{2}{3}\right] \cup \left[\frac{1}{3}, \infty\right)$

13. $2x^2 + x < 15$ $\{x | -3 < x < \frac{5}{2}\}$ or $\left(-3, \frac{5}{2}\right)$

14. $6x^2 + x > 1$
  $\{x | x < -\frac{1}{2} \text{ or } x > \frac{1}{3}\}$ or $\left(-\infty, -\frac{1}{2}\right) \cup \left(\frac{1}{3}, \infty\right)$

15. $4x^2 + 7x < -3$ $\{x | -1 < x < -\frac{3}{4}\}$ or $\left(-1, -\frac{3}{4}\right)$

16. $3x^2 + 16x < -5$ $\{x | -5 < x < -\frac{1}{3}\}$ or $\left(-5, -\frac{1}{3}\right)$

17. $x^2 - 4x \geq 0$ $\{x | x \leq 0 \text{ or } x \geq 4\}$ or $(-\infty, 0] \cup [4, \infty)$

18. $x^2 + 2x < 0$ $\{x | -2 < x < 0\}$ or $(-2, 0)$

19. $2x^2 + 3x > 0$
  $\{x | x < -\frac{3}{2} \text{ or } x > 0\}$ or $\left(-\infty, -\frac{3}{2}\right) \cup (0, \infty)$

20. $3x^2 - 5x \leq 0$ $\{x | 0 \leq x \leq \frac{5}{3}\}$ or $\left[0, \frac{5}{3}\right]$

21. $-x^2 + x \geq 0$ $\{x | 0 \leq x \leq 1\}$ or $[0, 1]$

22. $-x^2 + 2x \geq 0$ $\{x | 0 \leq x \leq 2\}$ or $[0, 2]$

23. $x^2 \leq 4x - 2$
  $\{x | 2 - \sqrt{2} \leq x \leq 2 + \sqrt{2}\}$ or $\left[2 - \sqrt{2}, 2 + \sqrt{2}\right]$

24. $x^2 \leq 2x + 2$
  $\{x | 1 - \sqrt{3} \leq x \leq 1 + \sqrt{3}\}$ or $\left[1 - \sqrt{3}, 1 + \sqrt{3}\right]$

25. $x^2 - 6x + 9 < 0$ no solution or $\emptyset$

26. $4x^2 - 4x + 1 \geq 0$
  $\{x | x \text{ is a real number}\}$ or $\mathbb{R}$ or $(-\infty, \infty)$

27. $(x - 1)(x - 2)(x - 3) \geq 0$
  $\{x | 1 \leq x \leq 2 \text{ or } x \geq 3\}$ or $[1, 2] \cup [3, \infty)$

28. $(x + 1)(x + 2)(x + 3) \geq 0$
  $\{x | -3 \leq x \leq -2 \text{ or } x \geq -1\}$ or $[-3, -2] \cup [-1, \infty)$

29. $x^3 + 2x^2 - x - 2 \geq 0$
  $\{x | -2 \leq x \leq -1 \text{ or } x \geq 1\}$ or $[-2, -1] \cup [1, \infty)$

30. $x^3 + 2x^2 - 4x - 8 \geq 0$
  $\{x | x = -2 \text{ or } x \geq 2\}$ or $\{-2\} \cup [2, \infty)$

31. $x^3 - 3x^2 - 9x + 27 < 0$ $\{x | x < -3\}$ or $(-\infty, -3)$

32. $x^3 + 7x^2 - x - 7 < 0$
  $\{x | x < -7 \text{ or } -1 < x < 1\}$ or $(-\infty, -7) \cup (-1, 1)$

33. $x^3 + x^2 + 4x + 4 > 0$ $\{x | x > -1\}$ or $(-1, \infty)$

34. $x^3 - x^2 + 9x - 9 > 0$ $\{x | x > 1\}$ or $(1, \infty)$

35. $x^3 \geq 9x^2$ $\{x | x = 0 \text{ or } x \geq 9\}$ or $\{0\} \cup [9, \infty)$

36. $x^3 \leq 4x^2$ $\{x | x \leq 4\}$ or $(-\infty, 4]$

Solve each rational inequality in Exercises 37–52 and graph the solution set on a real number line.   **37–52.** See graphing answer section.

37. $\dfrac{x - 4}{x + 3} > 0$ $\{x | x < -3 \text{ or } x > 4\}$ or $(-\infty, -3) \cup (4, \infty)$

38. $\dfrac{x + 5}{x - 2} > 0$ $\{x | x < -5 \text{ or } x > 2\}$ or $(-\infty, -5) \cup (2, \infty)$

39. $\dfrac{x + 3}{x + 4} < 0$ $\{x | -4 < x < -3\}$ or $(-4, -3)$

40. $\dfrac{x + 5}{x + 2} < 0$ $\{x | -5 < x < -2\}$ or $(-5, -2)$

41. $\dfrac{-x + 2}{x - 4} \geq 0$ $\{x | 2 \leq x < 4\}$ or $[2, 4)$

42. $\dfrac{-x - 3}{x + 2} \leq 0$ $\{x | x \leq -3 \text{ or } x > -2\}$ or $(-\infty, -3] \cup (-2, \infty)$

43. $\dfrac{4 - 2x}{3x + 4} \leq 0$ $\{x | x < -\frac{4}{3} \text{ or } x \geq 2\}$ or $\left(-\infty, -\frac{4}{3}\right) \cup [2, \infty)$

44. $\dfrac{3x + 5}{6 - 2x} \geq 0$ $\{x | -\frac{5}{3} \leq x < 3\}$ or $\left[-\frac{5}{3}, 3\right)$

45. $\dfrac{x}{x - 3} > 0$ $\{x | x < 0 \text{ or } x > 3\}$ or $(-\infty, 0) \cup (3, \infty)$

46. $\dfrac{x + 4}{x} > 0$ $\{x | x < -4 \text{ or } x > 0\}$ or $(-\infty, -4) \cup (0, \infty)$

47. $\dfrac{x + 1}{x + 3} < 2$ $\{x | x < -5 \text{ or } x > -3\}$ or $(-\infty, -5) \cup (-3, \infty)$

48. $\dfrac{x}{x - 1} > 2$ $\{x | 1 < x < 2\}$ or $(1, 2)$

49. $\dfrac{x + 4}{2x - 1} \leq 3$ $\{x | x < \frac{1}{2} \text{ or } x \geq \frac{7}{5}\}$ or $\left(-\infty, \frac{1}{2}\right) \cup \left[\frac{7}{5}, \infty\right)$

50. $\dfrac{1}{x - 3} < 1$ $\{x | x < 3 \text{ or } x > 4\}$ or $(-\infty, 3) \cup (4, \infty)$

51. $\dfrac{x - 2}{x + 2} \leq 2$ $\{x | x \leq -6 \text{ or } x > -2\}$ or $(-\infty, -6] \cup (-2, \infty)$

52. $\dfrac{x}{x + 2} \geq 2$ $\{x | -4 \leq x < -2\}$ or $[-4, -2)$

In Exercises 53–56, use the given functions to find all values of x that satisfy the required inequality.

53. $f(x) = 2x^2, g(x) = 5x - 2; f(x) \geq g(x)$
  $\{x | x \leq \frac{1}{2} \text{ or } x \geq 2\}$ or $\left(-\infty, \frac{1}{2}\right] \cup [2, \infty)$

54. $f(x) = 4x^2, g(x) = 9x - 2; f(x) < g(x)$
  $\{x | \frac{1}{4} < x < 2\}$ or $\left(\frac{1}{4}, 2\right)$

55. $f(x) = \dfrac{2x}{x + 1}, g(x) = 1; f(x) < g(x)$
  $\{x | -1 < x < 1\}$ or $(-1, 1)$

56. $f(x) = \dfrac{x}{2x - 1}, g(x) = 1; f(x) \geq g(x)$
  $\{x | \frac{1}{2} < x \leq 1\}$ or $\left(\frac{1}{2}, 1\right]$

Practice Plus

Solve each inequality in Exercises 57–62 and graph the solution set on a real number line. 57–62. See graphing answer section.

57. $|x^2 + 2x - 36| > 12$ $\{x | x < -8 \text{ or } -6 < x < 4 \text{ or } x > 6\}$
or $(-\infty, -8) \cup (-6, 4) \cup (6, \infty)$

58. $|x^2 + 6x + 1| > 8$
$\{x | x < -7 \text{ or } x > 1\}$ or $(-\infty, -7) \cup (1, \infty)$

59. $\dfrac{3}{x + 3} > \dfrac{3}{x - 2}$ $\{x | -3 < x < 2\}$ or $(-3, 2)$

60. $\dfrac{1}{x + 1} > \dfrac{2}{x - 1}$
$\{x | x < -3 \text{ or } -1 < x < 1\}$ or $(-\infty, -3) \cup (-1, 1)$

61. $\dfrac{x^2 - x - 2}{x^2 - 4x + 3} > 0$ $\{x | x < -1 \text{ or } 1 < x < 2 \text{ or } x > 3\}$
or $(-\infty, -1) \cup (1, 2) \cup (3, \infty)$

62. $\dfrac{x^2 - 3x + 2}{x^2 - 2x - 3} > 0$ $\{x | x < -1 \text{ or } 1 < x < 2 \text{ or } x > 3\}$
or $(-\infty, -1) \cup (1, 2) \cup (3, \infty)$

In Exercises 63–64, use the graph of the polynomial function to solve each inequality.

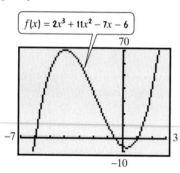

$f(x) = 2x^3 + 11x^2 - 7x - 6$

63. $2x^3 + 11x^2 \geq 7x + 6$
$\{x | -6 \leq x \leq -\frac{1}{2} \text{ or } x \geq 1\}$ or $\left[-6, -\frac{1}{2}\right] \cup [1, \infty)$

64. $2x^3 + 11x^2 < 7x + 6$
$\{x | x < -6 \text{ or } -\frac{1}{2} < x < 1\}$ or $(-\infty, -6) \cup \left(\frac{1}{2}, 1\right)$

In Exercises 65–66, use the graph of the rational function to solve each inequality.

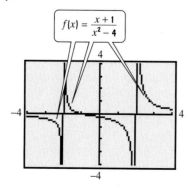

$f(x) = \dfrac{x + 1}{x^2 - 4}$

65. $\dfrac{1}{4(x + 2)} \leq -\dfrac{3}{4(x - 2)}$
$\{x | x < -2 \text{ or } -1 \leq x < 2\}$ or $(-\infty, -2) \cup [-1, 2)$

66. $\dfrac{1}{4(x + 2)} > -\dfrac{3}{4(x - 2)}$
$\{x | -2 < x < -1 \text{ or } x > 2\}$ or $(-2, -1) \cup (2, \infty)$

Application Exercises

67. You throw a ball straight up from a rooftop 160 feet high with an initial speed of 48 feet per second. The function

$$s(t) = -16t^2 + 48t + 160$$

models the ball's height above the ground, $s(t)$, in feet, t seconds after it was thrown. During which time period will the ball's height exceed that of the rooftop? 0 and 3 seconds, excluding $t = 0$ and $t = 3$

68. Divers in Acapulco, Mexico, dive headfirst from the top of a cliff 87 feet above the Pacific Ocean. The function

$$s(t) = -16t^2 + 8t + 87$$

models a diver's height above the ocean, $s(t)$, in feet, t seconds after leaping. During which time period will the diver's height exceed that of the cliff? 0 and $\frac{1}{2}$ second, excluding $t = 0$ and $t = \frac{1}{2}$

The bar graph in Figure 11.17 on page 776 shows the cost of Medicare, in billions of dollars, through 2005. Using the regression feature of a graphing utility, these data can be modeled by

a linear function, $f(x) = 27x + 163$;
a quadratic function, $g(x) = 1.2x^2 + 15.2x + 181.4$.

In each function, x represents the number of years after 1995. Use these functions to solve Exercises 69–72.

69. The graph indicates that Medicare spending reached $379 billion in 2003. Find the amount predicted by each of the functions, f and g, for that year. How well do the functions model the value in the graph? f: $379 billion; g: $379.8 billion; The functions model the value in the graph well.

70. The graph indicates that Medicare spending reached $458 billion in 2005. Find the amount predicted by each of the functions, f and g, for that year. How well do the functions model the value in the graph? Which function serves as a better model for that year? f: $433 billion; g: $453.4 billion; g serves as a better model for 2005.

71. For which years does the quadratic model indicate that Medicare spending will exceed $536.6 billion? years after 2007

72. For which years does the quadratic model indicate that Medicare spending will exceed $629.4 billion? years after 2009

A company manufactures wheelchairs. The average cost function, $\overline{C}$, of producing x wheelchairs per month is given by

$$\overline{C}(x) = \frac{500,000 + 400x}{x}.$$

The graph of the rational function is shown. Use the function to solve Exercises 73–74.

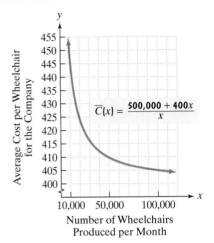

Average Cost per Wheelchair for the Company

Number of Wheelchairs Produced per Month

$\overline{C}(x) = \dfrac{500,000 + 400x}{x}$

73. Describe the company's production level so that the average cost of producing each wheelchair does not exceed $425. Use a rational inequality to solve the problem. Then explain how your solution is shown on the graph.   The company's production level must be at least 20,000 wheelchairs per month. For values of x greater than or equal to 20,000, the graph lies on or below the line $y = 425$.

74. Describe the company's production level so that the average cost of producing each wheelchair does not exceed $410. Use a rational inequality to solve the problem. Then explain how your solution is shown on the graph.   The company's production level must be at least 50,000 wheelchairs per month. For values of x greater than or equal to 50,000, the graph lies on or below the line $y = 410$.

75. The perimeter of a rectangle is 50 feet. Describe the possible length of a side if the area of the rectangle is not to exceed 114 square feet.   The length of the shorter side cannot exceed 6 feet.

76. The perimeter of a rectangle is 180 feet. Describe the possible lengths of a side if the area of the rectangle is not to exceed 800 square feet.   The length of the shorter side cannot exceed 10 feet.

Writing in Mathematics

77. What is a polynomial inequality?   77–79. Answers will vary.

78. What is a rational inequality?

79. Describe similarities and differences between the solutions of

$$(x - 2)(x + 5) \geq 0 \quad \text{and} \quad \frac{x - 2}{x + 5} \geq 0.$$

Technology Exercises

Solve each inequality in Exercises 80–85 using a graphing utility.

80. $x^2 + 3x - 10 > 0$ $\{x | x < -5 \text{ or } x > 2\}$ or $(-\infty, -5) \cup (2, \infty)$

81. $2x^2 + 5x - 3 \leq 0$ $\{x | -3 \leq x \leq \frac{1}{2}\}$ or $[-3, \frac{1}{2}]$

82. $\dfrac{x - 4}{x - 1} \leq 0$ $\{x | 1 < x \leq 4\}$ or $(1, 4]$

83. $\dfrac{x + 2}{x - 3} \leq 2$ $\{x | x < 3 \text{ or } x \geq 8\}$ or $(-\infty, 3) \cup [8, \infty)$

84. $\dfrac{1}{x + 1} \leq \dfrac{2}{x + 4}$
$\{x | -4 < x < -1 \text{ or } x \geq 2\}$ or $(-4, -1) \cup [2, \infty)$

85. $x^3 + 2x^2 - 5x - 6 > 0$
$\{x | -3 < x < -1 \text{ or } x > 2\}$ or $(-3, -1) \cup (2, \infty)$

Critical Thinking Exercises

86. Which one of the following is true?   d

a. The solution set of $x^2 > 25$ is $(5, \infty)$.

b. The inequality $\dfrac{x - 2}{x + 3} < 2$ can be solved by multiplying both sides by $x + 3$, resulting in the equivalent inequality $x - 2 < 2(x + 3)$.

c. $(x + 3)(x - 1) \geq 0$ and $\dfrac{x + 3}{x - 1} \geq 0$ have the same solution set.

d. None of the above statements is true.

87. Write a quadratic inequality whose solution set is $[-3, 5]$. Answers will vary; An example is $x^2 - 2x - 15 \leq 0$.

88. Write a rational inequality whose solution set is $(-\infty, -4) \cup [3, \infty)$.   Answers will vary; An example is $\dfrac{x - 3}{x + 4} \geq 0$.

In Exercises 89–92, use inspection to describe each inequality's solution set. Do not solve any of the inequalities.

89. $(x - 2)^2 > 0$ $\{x | x < 2 \text{ or } x > 2\}$ or $(-\infty, 2) \cup (2, \infty)$

90. $(x - 2)^2 \leq 0$ $\{2\}$

91. $(x - 2)^2 < -1$ $\varnothing$

92. $\dfrac{1}{(x - 2)^2} > 0$ $\{x | x < 2 \text{ or } x > 2\}$ or $(-\infty, 2) \cup (2, \infty)$

93. The graphing calculator screen shows the graph of $y = 4x^2 - 8x + 7$.

$y = 4x^2 - 8x + 7$

$[-2, 6, 1]$ by $[-2, 8, 1]$

a. Use the graph to describe the solution set for $4x^2 - 8x + 7 > 0$.   all real numbers

b. Use the graph to describe the solution set for $4x^2 - 8x + 7 < 0$.   empty set

c. Use an algebraic approach to verify each of your descriptions in parts (a) and (b).

94. The graphing calculator screen shows the graph of $y = \sqrt{27 - 3x^2}$. Write and solve a quadratic inequality that explains why the graph only appears for $-3 \le x \le 3$.

$y = \sqrt{27 - 3x^2}$

$[-5, 5, 1]$ by $[0, 6, 1]$

$27 - 3x^2 \ge 0$; $\{x | -3 \le x \le 3\}$ or $[-3, 3]$

Review Exercises

95. Solve: $\left| \dfrac{x - 5}{3} \right| < 8$. (Section 9.3, Example 3)

$\{x | -19 < x < 29\}$ or $(-19, 29)$

96. Divide:

$$\frac{2x + 6}{x^2 + 8x + 16} \div \frac{x^2 - 9}{x^2 + 3x - 4}.$$

(Section 7.2, Example 6) $\dfrac{2(x - 1)}{(x + 4)(x - 3)}$

97. Factor completely: $x^4 - 16y^4$. (Section 6.5, Example 7)

$(x^2 + 4y^2)(x + 2y)(x - 2y)$

GROUP PROJECT

CHAPTER 11

Additional group projects can be found in the *Instructor's Resource Manual*.

Throughout the chapter, we have considered functions that model the position of free-falling objects. Any object that is falling, or vertically projected into the air, has its height above the ground, $s(t)$, in feet, after t seconds in motion, modeled by the quadratic function

$$s(t) = -16t^2 + v_o t + s_o,$$

where v_o is the original velocity (initial velocity) of the object, in feet per second, and s_o is the original height (initial height) of the object, in feet, above the ground. In this exercise, group members will be working with this position function. The exercise is appropriate for groups of three to five people.

a. Drop a ball from a height of 3 feet, 6 feet, and 12 feet. Record the number of seconds it takes for the ball to hit the ground.

b. For each of the three initial positions, use the position function to determine the time required for the ball to hit the ground.

c. What factors might result in differences between the times that you recorded and the times indicated by the function?

d. What appears to be happening to the time required for a free-falling object to hit the ground as its initial height is doubled? Verify this observation algebraically and with a graphing utility.

e. Repeat part (a) using a sheet of paper rather than a ball. What differences do you observe? What factor seems to be ignored in the position function?

f. What is meant by the acceleration of gravity and how does this number appear in the position function for a free-falling object?

CHAPTER 11 SUMMARY

Definitions and Concepts	Examples

Section 11.1 The Square Root Property and Completing the Square; Distance and Midpoint Formulas

The Square Root Property

If u is an algebraic expression and d is a real number, then

If $u^2 = d$, then $u = \sqrt{d}$ or $u = -\sqrt{d}$.

Equivalently,

If $u^2 = d$, then $u = \pm\sqrt{d}$.

Solve: $(x - 6)^2 = 50$.

$$x - 6 = \pm\sqrt{50}$$
$$x - 6 = \pm\sqrt{25 \cdot 2}$$
$$x - 6 = \pm 5\sqrt{2}$$
$$x = 6 \pm 5\sqrt{2}$$

The solutions are $6 \pm 5\sqrt{2}$ and the solution set is $\left\{6 \pm 5\sqrt{2}\right\}$.

Completing the Square

If $x^2 + bx$ is a binomial, then by adding $\left(\dfrac{b}{2}\right)^2$, the square of half the coefficient of x, a perfect square trinomial will result. That is,

$$x^2 + bx + \left(\frac{b}{2}\right)^2 = \left(x + \frac{b}{2}\right)^2.$$

Complete the square:

$$x^2 + \frac{2}{7}x.$$

Half of $\frac{2}{7}$ is $\frac{1}{2} \cdot \frac{2}{7} = \frac{1}{7}$ and $\left(\frac{1}{7}\right)^2 = \frac{1}{49}$.

$$x^2 + \frac{2}{7}x + \frac{1}{49} = \left(x + \frac{1}{7}\right)^2$$

Solving Quadratic Equations by Completing the Square

1. If the coefficient of x^2 is not 1, divide both sides by this coefficient.
2. Isolate variable terms on one side.
3. Complete the square by adding the square of half the coefficient of x to both sides.
4. Factor the perfect square trinomial.
5. Solve by applying the square root property.

Solve by completing the square.

$$2x^2 + 16x - 6 = 0.$$

$$\frac{2x^2}{2} + \frac{16x}{2} - \frac{6}{2} = \frac{0}{2} \qquad \text{Divide by 2.}$$

$$x^2 + 8x - 3 = 0 \qquad \text{Simplify.}$$

$$x^2 + 8x = 3 \qquad \text{Add 3.}$$

The coefficient of x is 8. Half of 8 is 4 and $4^2 = 16$. Add 16 to both sides.

$$x^2 + 8x + 16 = 3 + 16$$
$$(x + 4)^2 = 19$$
$$x + 4 = \pm\sqrt{19}$$
$$x = -4 \pm \sqrt{19}$$

The Distance Formula

The distance, d, between the points (x_1, y_1) and (x_2, y_2) is given by

$$d = \sqrt{(x_2 - x_1)^2 + (y_2 - y_1)^2}.$$

Find the distance between $(-3, -5)$ and $(6, -2)$.

$$d = \sqrt{[6 - (-3)]^2 + [-2 - (-5)]^2}$$

$$= \sqrt{9^2 + 3^2} = \sqrt{81 + 9} = \sqrt{90} = 3\sqrt{10} \approx 9.49$$

The Midpoint Formula

The midpoint of the line segment whose endpoints are (x_1, y_1) and (x_2, y_2) is the point with coordinates

$$\left(\frac{x_1 + x_2}{2}, \frac{y_1 + y_2}{2}\right).$$

Find the midpoint of the line segment whose endpoints are $(-3, 6)$ and $(4, 1)$.

$$\text{midpoint} = \left(\frac{-3 + 4}{2}, \frac{6 + 1}{2}\right) = \left(\frac{1}{2}, \frac{7}{2}\right)$$

Definitions and Concepts	Examples

Section 11.2 The Quadratic Formula

The solutions of a quadratic equation in standard form

$$ax^2 + bx + c = 0, \quad a \neq 0,$$

are given by the quadratic formula

$$x = \frac{-b \pm \sqrt{b^2 - 4ac}}{2a}.$$

Solve using the quadratic formula:

$$2x^2 = 6x - 3.$$

First write the equation in standard form by subtracting $6x$ and adding 3 on both sides.

$$2x^2 - 6x + 3 = 0$$

$$a = 2 \quad b = -6 \quad c = 3$$

$$x = \frac{-(-6) \pm \sqrt{(-6)^2 - 4 \cdot 2 \cdot 3}}{2 \cdot 2} = \frac{6 \pm \sqrt{36 - 24}}{4}$$

$$= \frac{6 \pm \sqrt{12}}{4} = \frac{6 \pm \sqrt{4 \cdot 3}}{4} = \frac{6 \pm 2\sqrt{3}}{4}$$

$$= \frac{\cancel{2}(3 \pm \sqrt{3})}{\cancel{2} \cdot 2} = \frac{3 \pm \sqrt{3}}{2}$$

The Discriminant
The discriminant, $b^2 - 4ac$, of the quadratic equation $ax^2 + bx + c = 0$ determines the number and type of solutions.

Discriminant	Solutions
Positive perfect square, with a, b, and c rational numbers	2 rational solutions
Positive and not a perfect square	2 irrational solutions
Zero, with a, b, and c rational numbers	1 rational solution
Negative	2 imaginary solutions

• $2x^2 - 7x - 4 = 0$

$$a = 2 \quad b = -7 \quad c = -4$$

$$b^2 - 4ac = (-7)^2 - 4(2)(-4)$$
$$= 49 - (-32) = 49 + 32 = 81$$

Positive perfect square

The equation has 2 rational solutions.

Writing Quadratic Equations from Solutions
The zero-product principle in reverse makes it possible to write a quadratic equation from solutions:

If $A = 0$ or $B = 0$, then $AB = 0$.

Write a quadratic equation with the solution set $\{-2\sqrt{3}, 2\sqrt{3}\}$.

$$x = -2\sqrt{3} \quad \text{or} \quad x = 2\sqrt{3}$$

$$x + 2\sqrt{3} = 0 \quad \text{or} \quad x - 2\sqrt{3} = 0$$

$$(x + 2\sqrt{3})(x - 2\sqrt{3}) = 0$$

$$x^2 - (2\sqrt{3})^2 = 0$$

$$x^2 - 12 = 0$$

Definitions and Concepts

Examples

Section 11.3 Quadratic Functions and Their Graphs

The graph of the quadratic function

$$f(x) = a(x - h)^2 + k, \quad a \neq 0,$$

is called a parabola. The vertex, or turning point, is (h, k). The graph opens upward if a is positive and downward if a is negative. The axis of symmetry is a vertical line passing through the vertex. The graph can be obtained using the vertex, x-intercepts, if any, (set $f(x)$ equal to zero), and the y-intercept (set $x = 0$).

Graph: $f(x) = -(x + 3)^2 + 1$.

$$f(x) = -1(x - (-3))^2 + 1$$

$\boxed{a = -1}$ $\boxed{h = -3}$ $\boxed{k = 1}$

- Vertex: $(h, k) = (-3, 1)$.
- Opens downward because $a < 0$
- x-intercepts: Set $f(x) = 0$.

$$0 = -(x + 3)^2 + 1$$
$$(x + 3)^2 = 1$$
$$x + 3 = \pm\sqrt{1}$$
$$x + 3 = 1 \quad \text{or} \quad x + 3 = -1$$
$$x = -2 \qquad\qquad x = -4$$

- y-intercept: Set $x = 0$.

$$f(0) = -(0 + 3)^2 + 1 = -9 + 1 = -8$$

Vertex: $(-3, 1)$
$(-4, 0)$ $(-2, 0)$
$(0, -8)$
Axis of symmetry: $x = -3$

A parabola whose equation is in the form

$$f(x) = ax^2 + bx + c, \quad a \neq 0,$$

has its vertex at

$$\left(-\frac{b}{2a}, f\left(-\frac{b}{2a}\right)\right).$$

The parabola is graphed as described in the left column on the previous page. The only difference is how we determine the vertex. If $a > 0$, then f has a minimum that occurs at $x = -\frac{b}{2a}$. This minimum value is $f\left(-\frac{b}{2a}\right)$. If $a < 0$, then f has a maximum that occurs at $x = -\frac{b}{2a}$. This maximum value is $f\left(-\frac{b}{2a}\right)$.

Graph: $f(x) = x^2 - 6x + 5$.

$\boxed{a = 1}$ $\boxed{b = -6}$ $\boxed{c = 5}$

- Vertex: $x = -\frac{b}{2a} = -\frac{-6}{2 \cdot 1} = 3$

$$f(3) = 3^2 - 6 \cdot 3 + 5 = -4$$
Vertex is at $(3, -4)$.

- Opens upward because $a > 0$.
- x-intercepts: Set $f(x) = 0$.

$$x^2 - 6x + 5 = 0$$
$$(x - 1)(x - 5) = 0$$
$$x = 1 \quad \text{or} \quad x = 5$$

- y-intercept: Set $x = 0$.

$$f(0) = 0^2 - 6 \cdot 0 + 5 = 5$$

Axis of symmetry: $x = 3$
$(0, 5)$
$(1, 0)$ $(5, 0)$
Vertex: $(3, -4)$

Definitions and Concepts	Examples

Section 11.4 Equations Quadratic in Form

An equation that is quadratic in form is one that can be expressed as a quadratic equation using an appropriate substitution. These equations contain an expression to a power, the same expression to that power squared, and a constant term: Let t = the expression to the power. If at any point in the solution process both sides of an equation are raised to an even power, a check is required.

Solve:
$$x^{\frac{2}{3}} - 3x^{\frac{1}{3}} + 2 = 0.$$

$$\left(x^{\frac{1}{3}}\right)^2 - 3x^{\frac{1}{3}} + 2 = 0$$

Let $t = x^{\frac{1}{3}}$.

$$t^2 - 3t + 2 = 0$$

$$(t - 1)(t - 2) = 0$$

$$t - 1 = 0 \quad \text{or} \quad t - 2 = 0$$

$$t = 1 \qquad\qquad t = 2$$

$$x^{\frac{1}{3}} = 1 \qquad\qquad x^{\frac{1}{3}} = 2$$

$$\left(x^{\frac{1}{3}}\right)^3 = 1^3 \qquad \left(x^{\frac{1}{3}}\right)^3 = 2^3$$

$$x = 1 \qquad\qquad x = 8$$

The solutions are 1 and 8, and the solution set is $\{1, 8\}$.

Section 11.5 Polynomial and Rational Inequalities

Solving Polynomial Inequalities

1. Express the inequality in the form
$$f(x) < 0 \quad \text{or} \quad f(x) > 0,$$
where f is a polynomial function.

2. Solve the equation $f(x) = 0$. The real solutions are the boundary points.

3. Locate these boundary points on a number line, thereby dividing the number line into intervals.

4. Choose one representative number within each interval and evaluate f at that number.

   a. If the value of f is positive, then $f(x) > 0$ for all x in the interval.

   b. If the value of f is negative, then $f(x) < 0$ for all x in the interval.

5. Write the solution set, selecting the interval(s) that satisfy the given inequality.

This procedure is valid if $<$ is replaced by $\leq$ and $>$ is replaced by $\geq$. In these cases, include the boundary points in the solution set.

Solve: $2x^2 + x - 6 > 0$.
The form of the inequality is $f(x) > 0$ with $f(x) = 2x^2 + x - 6$. Solve $f(x) = 0$.

$$2x^2 + x - 6 = 0$$

$$(2x - 3)(x + 2) = 0$$

$$2x - 3 = 0 \quad \text{or} \quad x + 2 = 0$$

$$x = \frac{3}{2} \qquad\qquad x = -2$$

$$f(-3) = 2(-3)^2 + (-3) - 6 = 9, \text{ positive}$$

$$f(x) > 0 \text{ for all } x \text{ in } (-\infty, -2).$$

$$f(0) = 2 \cdot 0^2 + 0 - 6 = -6, \text{ negative}$$

$$f(x) < 0 \text{ for all } x \text{ in } \left(-2, \frac{3}{2}\right).$$

$$f(2) = 2 \cdot 2^2 + 2 - 6 = 4, \text{ positive}$$

$$f(x) > 0 \text{ for all } x \text{ in } \left(\frac{3}{2}, \infty\right).$$

The solution set is $\left\{x \mid x < -2 \text{ or } x > \frac{3}{2}\right\}$ or $(-\infty, -2) \cup \left(\frac{3}{2}, \infty\right)$.

Definitions and Concepts	Examples

Section 11.5 Polynomial and Rational Inequalities (continued)

Solving Rational Inequalities

1. Express the inequality in the form
$$f(x) < 0 \quad \text{or} \quad f(x) > 0,$$
where f is a rational function.

2. Set the numerator and the denominator of f equal to zero. The real zeros are the boundary points.

3. Locate these boundary points on a number line, thereby dividing the number line into intervals.

4. Choose one representative number within each interval and evaluate f at that number.

   a. If the value of f is positive, then $f(x) > 0$ for all x in the interval.

   b. If the value of f is negative, then $f(x) < 0$ for all x in the interval.

5. Write the solution set, selecting the interval(s) that satisfy the given inequality.

This procedure is valid if $<$ is replaced by $\leq$ and $>$ is replaced by $\geq$. In these cases, include any values that make the numerator of f zero. Always exclude any values that make the denominator zero.

Solve: $\dfrac{x}{x+4} \geq 2$.

$$\frac{x}{x+4} - \frac{2(x+4)}{x+4} \geq 0$$

$$\frac{-x-8}{x+4} \geq 0$$

The form of the inequality is $f(x) \geq 0$ with $f(x) = \dfrac{-x-8}{x+4}$.

Set the numerator and the denominator equal to zero.

$$-x - 8 = 0 \qquad x + 4 = 0$$
$$-8 = x \qquad x = -4$$

$$f(-9) = \frac{-(-9)-8}{-9+4} = \frac{1}{-5}, \text{ negative}$$
$$f(x) < 0 \text{ for all } x \text{ in } (-\infty, -8).$$

$$f(-7) = \frac{-(-7)-8}{-7+4} = \frac{-1}{-3} = \frac{1}{3}, \text{ positive}$$
$$f(x) > 0 \text{ for all } x \text{ in } (-8, -4)$$

$$f(-3) = \frac{-(-3)-8}{-3+4} = \frac{-5}{1} = -5, \text{ negative}$$
$$f(x) < 0 \text{ for all } x \text{ in } (-4, \infty).$$

Because of $\geq$, include -8, the value that makes the numerator zero, in the solution set.
The solution set is $\{x \mid -8 \leq x < -4\}$ or $[-8, -4)$.

CHAPTER 11 REVIEW EXERCISES

11.1 *In Exercises 1–5, solve each equation by the square root property. If possible, simplify radicals or rationalize denominators. Express imaginary solutions in the form $a + bi$.*

1. $2x^2 - 3 = 125$ $\{\pm 8\}$

2. $3x^2 - 150 = 0$ $\{\pm 5\sqrt{2}\}$

3. $3x^2 - 2 = 0$ $\left\{\pm \dfrac{\sqrt{6}}{3}\right\}$

4. $(x - 4)^2 = 18$ $\{4 \pm 3\sqrt{2}\}$

5. $(x + 7)^2 = -36$ $\{-7 \pm 6i\}$

In Exercises 6–7, determine the constant that should be added to the binomial so that it becomes a perfect square trinomial. Then write and factor the trinomial.

6. $x^2 + 20x$ $100; x^2 + 20x + 100 = (x + 10)^2$

7. $x^2 - 3x$ $\frac{9}{4}; x^2 - 3x + \frac{9}{4} = \left(x - \frac{3}{2}\right)^2$

In Exercises 8–10, solve each quadratic equation by completing the square.

8. $x^2 - 12x + 27 = 0$ $\{3, 9\}$

9. $x^2 - 7x - 1 = 0$ $\left\{\dfrac{7 \pm \sqrt{53}}{2}\right\}$

10. $2x^2 + 3x - 4 = 0$ $\left\{\dfrac{-3 \pm \sqrt{41}}{4}\right\}$

11. In 2 years, an investment of \$2500 grows to \$2916. Use the compound interest formula
$$A = P(1 + r)^t$$
to find the annual interest rate, r. The annual interest rate is 8%.

12. The function $W(t) = 3t^2$ models the weight of a human fetus, $W(t)$, in grams, after t weeks, where $0 \leq t \leq 39$. After how many weeks does the fetus weigh 588 grams? After 14 weeks, the fetus weighs 588 grams.

13. A building casts a shadow that is double the length of the building's height. If the distance from the end of the shadow to the top of the building is 300 meters, how high is the building? Express the answer in simplified radical form. Then find a decimal approximation to the nearest tenth of a meter. The building is $60\sqrt{5}$ meters, or about 134.2 meters high.

In Exercises 14–15, find the distance between each pair of points. If necessary, round answers to two decimal places.

14. $(-2, -3)$ and $(3, 9)$ 13 units

15. $(-4, 3)$ and $(-2, 5)$ $2\sqrt{2}$ or 2.83 units

In Exercises 16–17, find the midpoint of the line segment with the given endpoints.

16. $(2, 6)$ and $(-12, 4)$ $(-5, 5)$

17. $(4, -6)$ and $(-15, 2)$ $\left(-\frac{11}{2}, -2\right)$

11.2 *In Exercises 18–20, solve each equation using the quadratic formula. Simplify solutions, if possible.*

18. $x^2 = 2x + 4$ $\{1 \pm \sqrt{5}\}$

19. $x^2 - 2x + 19 = 0$ $\{1 \pm 3\sqrt{2}i\}$

20. $2x^2 = 3 - 4x$ $\left\{\dfrac{-2 \pm \sqrt{10}}{2}\right\}$

In Exercises 21–23, without solving the given quadratic equation, determine the number and type of solutions.

21. $x^2 - 4x + 13 = 0$ two imaginary solutions

22. $9x^2 = 2 - 3x$ two rational solutions

23. $2x^2 + 4x = 3$ two irrational solutions

In Exercises 24–30, solve each equation by the method of your choice. Simplify solutions, if possible.

24. $3x^2 - 10x - 8 = 0$ $\left\{-\frac{2}{3}, 4\right\}$

25. $(2x - 3)(x + 2) = x^2 - 2x + 4$ $\{-5, 2\}$

26. $5x^2 - x - 1 = 0$ $\left\{\dfrac{1 \pm \sqrt{21}}{10}\right\}$

27. $x^2 - 16 = 0$ $\{-4, 4\}$

28. $(x - 3)^2 - 8 = 0$ $\{3 \pm 2\sqrt{2}\}$

29. $3x^2 - x + 2 = 0$ $\left\{\dfrac{1}{6} \pm \dfrac{\sqrt{23}}{6}i\right\}$

30. $\dfrac{5}{x + 1} + \dfrac{x - 1}{4} = 2$ $\{4 \pm \sqrt{5}\}$

In Exercises 31–33, write a quadratic equation in standard form with the given solution set.

31. $\left\{-\dfrac{1}{3}, \dfrac{3}{5}\right\}$ $15x^2 - 4x - 3 = 0$

32. $\{-9i, 9i\}$ $x^2 + 81 = 0$

33. $\{-4\sqrt{3}, 4\sqrt{3}\}$ $x^2 - 48 = 0$

34. The bar graph shows the number of convictions of police officers throughout the United States for seven years. The number of police officers convicted of felonies, $f(x)$, x years after 1990 can be modeled by the quadratic function

$$f(x) = 23x^2 - 259x + 816.$$

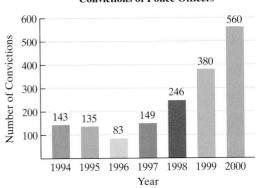

Convictions of Police Officers

Source: F.B.I.

In which year after 1990 were 1020 police officers convicted of felonies?   2002

35. A baseball is hit by a batter. The function

$$s(t) = -16t^2 + 140t + 3$$

models the ball's height above the ground, $s(t)$, in feet, t seconds after it is hit. How long will it take for the ball to strike the ground? Round to the nearest tenth of a second.   It will take about 8.8 seconds for the ball to strike the ground.

11.3 *In Exercises 36–39, use the vertex and intercepts to sketch the graph of each quadratic function. Give the equation for the parabola's axis of symmetry.*
36–39. See graphing answer section.

36. $f(x) = -(x + 1)^2 + 4$ $x = -1$

37. $f(x) = (x + 4)^2 - 2$ $x = -4$

38. $f(x) = -x^2 + 2x + 3$ $x = 1$

39. $f(x) = 2x^2 - 4x - 6$ $x = 1$

40. The function

$$f(x) = -0.02x^2 + x + 1$$

models the yearly growth of a young redwood tree, $f(x)$, in inches, with x inches of rainfall per year. How many inches of rainfall per year result in maximum tree growth? What is the maximum yearly growth?   The maximum yearly growth is 13.5 inches with 25 inches of rainfall per year.

41. A model rocket is launched upward from a platform 40 feet above the ground. The quadratic function

$$s(t) = -16t^2 + 400t + 40$$

models the rocket's height above the ground, $s(t)$, in feet, t seconds after it was launched. After how many seconds does the rocket reach its maximum height? What is the maximum height?   The maximum height is 2540 feet after 12.5 seconds.

42. Suppose that a quadratic function is used to model the data shown in the graph using

(number of years after 1960, divorce rate per 1000 population).

U. S. Divorce Rate

Source: National Center for Health Statistics

Determine, without obtaining an actual quadratic function that models the data, the approximate coordinates of the vertex for the function's graph. Describe what this means in practical terms. $(25, 5)$; The maximum divorce rate of 5 per 1000 population was in 1985.

43. A field bordering a straight stream is to be enclosed. The side bordering the stream is not to be fenced. If 1000 yards of fencing material is to be used, what are the dimensions of the largest rectangular field that can be fenced? What is the maximum area?

The maximum area is 125,000 square yards when the dimensions are 250 by 500 yards.

44. Among all pairs of numbers whose difference is 14, find a pair whose product is as small as possible. What is the minimum product?   The minimum product is -49 when the numbers are -7 and 7.

11.4 *In Exercises 45–50, solve each equation by making an appropriate substitution. When necessary, check proposed solutions.*

45. $x^4 - 6x^2 + 8 = 0$ $\{-\sqrt{2}, \sqrt{2}, -2, 2\}$

46. $x + 7\sqrt{x} - 8 = 0$ $\{1\}$

47. $(x^2 + 2x)^2 - 14(x^2 + 2x) = 15$ $\{-5, -1, 3\}$

48. $x^{-2} + x^{-1} - 56 = 0$ $\left\{-\frac{1}{8}, \frac{1}{7}\right\}$

49. $x^{\frac{2}{3}} - x^{\frac{1}{3}} - 12 = 0$ $\{-27, 64\}$

50. $x^{\frac{1}{2}} + 3x^{\frac{1}{4}} - 10 = 0$ $\{16\}$

11.5 *In Exercises 51–55, solve each inequality and graph the solution set on a real number line.*   **51–55.** See graphing answer section.

51. $2x^2 + 5x - 3 < 0$ $\left\{x \mid -3 < x < \frac{1}{2}\right\}$ or $\left(-3, \frac{1}{2}\right)$

52. $2x^2 + 9x + 4 \geq 0$ $\left\{x \mid x \leq -4 \text{ or } x \geq -\frac{1}{2}\right\}$ or $(-\infty, -4] \cup \left[-\frac{1}{2}, \infty\right)$

53. $x^3 + 2x^2 > 3x$ $\{x \mid -3 < x < 0 \text{ or } x > 1\}$ or $(-3, 0) \cup (1, \infty)$

54. $\dfrac{x - 6}{x + 2} > 0$ $\{x \mid x < -2 \text{ or } x > 6\}$ or $(-\infty, -2) \cup (6, \infty)$

55. $\dfrac{x + 3}{x - 4} \leq 5$ $\left\{x \mid x < 4 \text{ or } x \geq \frac{23}{4}\right\}$ or $\left(-\infty, 4\right) \cup \left[\frac{23}{4}, \infty\right)$

56. A model rocket is launched from ground level. The function

$$s(t) = -16t^2 + 48t$$

models the rocket's height above the ground, $s(t)$, in feet, t seconds after it was launched. During which time period will the rocket's height exceed 32 feet?   The rocket's height will exceed 32 feet between 1 and 2 seconds, excluding $t = 1$ and $t = 2$.

57. The function

$$H(x) = \frac{15}{8}x^2 - 30x + 200$$

models heart rate, $H(x)$, in beats per minute, x minutes after a strenuous workout.

a. What is the heart rate immediately following the workout?   The heart rate is 200 beats per minute immediately following the workout.

b. According to the model, during which intervals of time after the strenuous workout does the heart rate exceed 110 beats per minute? For which of these intervals has model breakdown occurred? Which interval provides a more realistic answer? How did you determine this?   The heart rate exceeds 110 beats per minute between 0 and 4 minutes after the workout and more than 12 minutes after the workout. Between 0 and 4 minutes provides a more realistic answer; Answers will vary.

CHAPTER 11 TEST

Remember to use your Chapter Test Prep Video CD to see the worked-out solutions to the test questions you want to review.

Express solutions to all equations in simplified form. Rationalize denominators, if possible.

In Exercises 1–2, solve each equation by the square root property.

1. $2x^2 - 5 = 0$ $\pm\dfrac{\sqrt{10}}{2}$ or $\left\{\pm\dfrac{\sqrt{10}}{2}\right\}$

2. $(x - 3)^2 = 20$ $3 \pm 2\sqrt{5}$ or $\left\{3 \pm 2\sqrt{5}\right\}$

In Exercises 3–4, determine the constant that should be added to the binomial so that it becomes a perfect square trinomial. Then write and factor the trinomial.

3. $x^2 - 16x$ 64; $x^2 - 16x + 64 = (x - 8)^2$

4. $x^2 + \dfrac{2}{5}x$ $\dfrac{1}{25}$; $x^2 + \dfrac{2}{5}x + \dfrac{1}{25} = \left(x + \dfrac{1}{5}\right)^2$

5. Solve by completing the square: $x^2 - 6x + 7 = 0$. $3 \pm \sqrt{2}$ or $\left\{3 \pm \sqrt{2}\right\}$

6. Use the measurements determined by the surveyor to find the width of the pond. Express the answer in simplified radical form.

50 feet   50 feet

The width of the pond is $50\sqrt{2}$ feet.

7. Find the distance between $(-1, 5)$ and $(2, -3)$. If necessary, round the answer to two decimal places. $\sqrt{73}$ or 8.54 units

8. Find the midpoint of the line segment whose endpoints are $(-5, -2)$ and $(12, -6)$. $\left(\dfrac{7}{2}, -4\right)$

In Exercises 9–10, without solving the given quadratic equation, determine the number and type of solutions.

9. $3x^2 + 4x - 2 = 0$ two irrational solutions

10. $x^2 = 4x - 8$ two imaginary solutions

In Exercises 11–14, solve each equation by the method of your choice.

11. $2x^2 + 9x = 5$ -5 and $\dfrac{1}{2}$, or $\left\{-5, \dfrac{1}{2}\right\}$

12. $x^2 + 8x + 5 = 0$ $-4 \pm \sqrt{11}$ or $\left\{-4 \pm \sqrt{11}\right\}$

13. $(x + 2)^2 + 25 = 0$ $-2 \pm 5i$ or $\{-2 \pm 5i\}$

14. $2x^2 - 6x + 5 = 0$ $\dfrac{3}{2} \pm \dfrac{1}{2}i$ or $\left\{\dfrac{3}{2} \pm \dfrac{1}{2}i\right\}$

In Exercises 15–16, write a quadratic equation in standard form with the given solution set.

15. $\{-3, 7\}$ $x^2 - 4x - 21 = 0$

16. $\{-10i, 10i\}$ $x^2 + 100 = 0$

17. The function
$$f(x) = -0.5x^2 + 4x + 19$$
models the number of people in the United States, $f(x)$, in millions, receiving food stamps x years after 1990. In which year(s) were 20 million people receiving food stamps? Use a calculator and round to the nearest year(s). 20 million people were receiving food stamps in 1990 and 1998.

In Exercises 18–19, use the vertex and intercepts to sketch the graph of each quadratic function. Give the equation for the parabola's axis of symmetry. **18–19.** See graphing answer section.

18. $f(x) = (x + 1)^2 + 4$ $x = -1$

19. $f(x) = x^2 - 2x - 3$ $x = 1$

A baseball player hits a pop fly into the air. The function
$$s(t) = -16t^2 + 64t + 5$$
models the ball's height above the ground, $s(t)$, in feet, t seconds after it is hit. Use the function to solve Exercises 20–21.

20. When does the baseball reach its maximum height? What is that height? The baseball reaches its maximum height of 69 feet after 2 seconds.

21. After how many seconds does the baseball hit the ground? Round to the nearest tenth of a second. The baseball hits the ground after about 4.1 seconds.

22. The function $f(x) = -x^2 + 46x - 360$ models the daily profit, $f(x)$, in hundreds of dollars, for a company that manufactures x computers daily. How many computers should be manufactured each day to maximize profit? What is the maximum daily profit? The maximum daily profit is $169 hundreds, or $16,900, when 23 computers are manufactured each day.

In Exercises 23–25, solve each equation by making an appropriate substitution. When necessary, check proposed solutions.

23. $(2x - 5)^2 + 4(2x - 5) + 3 = 0$ 1 and 2, or $\{1, 2\}$

24. $x^4 - 13x^2 + 36 = 0$ $-3, 3, -2$, and 2, or $\{-3, 3, -2, 2\}$

25. $x^{\frac{2}{3}} - 9x^{\frac{1}{3}} + 8 = 0$ 1 and 512, or $\{1, 512\}$

In Exercises 26–27, solve each inequality and graph the solution set on a real number line.   **26–27.** See graphing answer section.

26. $x^2 - x - 12 < 0$ $\{x | -3 < x < 4\}$ or $(-3, 4)$

27. $\dfrac{2x + 1}{x - 3} \le 3$ $\{x | x < 3 \text{ or } x \ge 10\}$ or $(-\infty, 3) \cup [10, \infty)$

Note to Instructor: The Chapter 11 Test file in the TestGen program provides algorithms specifically matched to these problems so you can easily replicate this chapter test for additional practice or assessment purposes.

CUMULATIVE REVIEW EXERCISES (CHAPTERS 1–11)

In Exercises 1–7, solve each equation, inequality, or system.

1. $8 - (4x - 5) = x - 7$ 4 or $\{4\}$

2. $5x + 4y = 22$
$3x - 8y = -18$ $\{(2, 3)\}$

3. $-3x + 2y + 4z = 6$
$7x - y + 3z = 23$
$2x + 3y + z = 7$ $\{(2, 0, 3)\}$

4. $|x - 1| > 3$ $\{x | x < -2 \text{ or } x > 4\}$ or $(-\infty, -2) \cup (4, \infty)$

5. $\sqrt{x + 4} - \sqrt{x - 4} = 2$ $\{5\}$

6. $x - 4 \ge 0$ and $-3x \le -6$ $\{x | x \ge 4\}$ or $[4, \infty)$

7. $2x^2 = 3x - 2$ $\left\{\dfrac{3}{4} \pm \dfrac{\sqrt{7}}{4}i\right\}$

In Exercises 8–12, graph each function, equation, or inequality in a rectangular coordinate system.   **8–12.** See graphing answer section.

8. $3x = 15 + 5y$

9. $2x - 3y > 6$

10. $f(x) = -\dfrac{1}{2}x + 1$

11. $f(x) = x^2 + 6x + 8$

12. $f(x) = (x - 3)^2 - 4$

13. If $f(x) = \sqrt[3]{x + 1}$, find $f^{-1}(x)$. $f^{-1}(x) = x^3 - 1$

14. Solve for c: $A = \dfrac{cd}{c + d}$. $c = \dfrac{Ad}{d - A}$

15. Write the equation of the linear function whose graph contains the point $(-2, 4)$ and is perpendicular to the line whose equation is $2x + y = 10$. $y = \frac{1}{2}x + 5$ or $f(x) = \frac{1}{2}x + 5$

In Exercises 16–20, perform the indicated operations and simplify, if possible.

16. $\dfrac{-5x^3y^7}{15x^4y^{-2}}$ $-\dfrac{y^9}{3x}$

17. $(4x^2 - 5y)^2$ $16x^4 - 40x^2y + 25y^2$

18. $(5x^3 - 24x^2 + 9) \div (5x + 1)$ $x^2 - 5x + 1 + \dfrac{8}{5x + 1}$

19. $\dfrac{\sqrt[3]{32xy^{10}}}{\sqrt[3]{2xy^2}}$ $2y^2\sqrt[3]{2y^2}$

20. $\dfrac{x + 2}{x^2 - 6x + 8} + \dfrac{3x - 8}{x^2 - 5x + 6}$ $\dfrac{4x - 13}{(x - 3)(x - 4)}$

In Exercises 21–22, factor completely.

21. $x^4 - 4x^3 + 8x - 32$ $(x - 4)(x + 2)(x^2 - 2x + 4)$

22. $2x^2 + 12xy + 18y^2$ $2(x + 3y)^2$

23. The length of a rectangular carpet is 4 feet greater than twice its width. If the area is 48 square feet, find the carpet's length and width.   The carpet's length is 12 feet and its width is 4 feet.

24. Working alone, you can mow the lawn in 2 hours and your sister can do it in 3 hours. How long will it take you to do the job if you work together?   It will take $\frac{6}{5}$ hours, or 1 hour and 12 minutes, to do the job if you work together.

25. Your motorboat can travel 15 miles per hour in still water. Traveling with the river's current, the boat can cover 20 miles in the same time it takes to go 10 miles against the current. Find the rate of the current.   The rate of the current is 5 miles per hour.

MATRIX SOLUTIONS TO LINEAR SYSTEMS

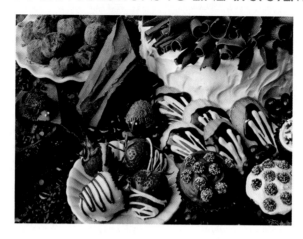

Yes, we overindulged, but it was delicious. Anyway, a few hours of moderate activity and we'll just burn off those extra calories. The chart below should help. It shows that the number of calories burned per hour depends on our weight. Four hours of tennis and we'll be as good as new!

How Fast You Burn Off Calories

	Weight (pounds)					
	110	132	154	176	187	209
Activity	Calories Burned per Hour					
Housework	175	210	245	285	300	320
Cycling	190	215	245	270	280	295
Tennis	335	380	425	470	495	520
Watching TV	60	70	80	85	90	95

The 24 numbers inside the red brackets are arranged in four rows and six columns. This rectangular array of 24 numbers, arranged in rows and columns and placed in brackets, is an example of a **matrix** (plural: **matrices**). The numbers inside the brackets are called **elements** of the matrix. Matrices are used to display information and to solve systems of linear equations.

1 Write the augmented matrix for a linear system.

Solving Linear Systems by Using Matrices A matrix gives us a shortened way of writing a system of equations. The first step in solving a system of linear equations using matrices is to write the *augmented matrix*. An **augmented matrix** has a vertical bar separating the columns of the matrix into two groups. The coefficients of each variable are placed to the left of the vertical line and the constants are placed to the right. If any variable is missing, its coefficient is 0. Here are two examples:

System of Linear Equations	**Augmented Matrix**
$\begin{aligned} x + 3y &= 5 \\ 2x - y &= -4 \end{aligned}$	$\begin{bmatrix} 1 & 3 & \bigm| & 5 \\ 2 & -1 & \bigm| & -4 \end{bmatrix}$
$\begin{aligned} x + 2y - 5z &= -19 \\ y + 3z &= 9 \\ z &= 4 \end{aligned}$	$\begin{bmatrix} 1 & 2 & -5 & \bigm| & -19 \\ 0 & 1 & 3 & \bigm| & 9 \\ 0 & 0 & 1 & \bigm| & 4 \end{bmatrix}$

Notice how the second augmented matrix contains 1s down the diagonal from upper left to lower right and 0s below the 1s. This arrangement makes it easy to find the solution of the system of equations, as Example 1 shows.

EXAMPLE 1 Solving a System Using a Matrix

Write the solution for a system of equations represented by the matrix

$$\begin{bmatrix} 1 & 2 & -5 & | & -19 \\ 0 & 1 & 3 & | & 9 \\ 0 & 0 & 1 & | & 4 \end{bmatrix}.$$

SOLUTION   The system represented by the given matrix is

$$\begin{bmatrix} 1 & 2 & -5 & | & -19 \\ 0 & 1 & 3 & | & 9 \\ 0 & 0 & 1 & | & 4 \end{bmatrix} \rightarrow \begin{matrix} 1x + 2y - 5z = -19 \\ 0x + 1y + 3z = 9 \\ 0x + 0y + 1z = 4 \end{matrix}.$$

This system can be simplified as follows.

$$\begin{aligned} x + 2y - 5z &= -19 \quad &\text{Equation 1}\\ y + 3z &= 9 \quad &\text{Equation 2}\\ z &= 4 \quad &\text{Equation 3} \end{aligned}$$

The value of z is known. We can find y by back-substitution.

$$\begin{aligned} y + 3z &= 9 \quad &\text{Equation 2}\\ y + 3(4) &= 9 \quad &\text{Substitute 4 for z.}\\ y + 12 &= 9 \quad &\text{Multiply.}\\ y &= -3 \quad &\text{Subtract 12 from both sides.} \end{aligned}$$

With values for y and z, we can now use back-substitution to find x.

$$\begin{aligned} x + 2y - 5z &= -19 \quad &\text{Equation 1}\\ x + 2(-3) - 5(4) &= -19 \quad &\text{Substitute } -3 \text{ for y and 4 for z.}\\ x - 6 - 20 &= -19 \quad &\text{Multiply.}\\ x - 26 &= -19 \quad &\text{Add.}\\ x &= 7 \quad &\text{Add 26 to both sides.} \end{aligned}$$

We see that $x = 7$, $y = -3$, and $z = 4$. The solution is $(7, -3, 4)$ and the solution set for the system is $\{(7, -3, 4)\}$. ∎

 CHECK POINT 1 Write the solution for a system of equations represented by the matrix

$$\begin{bmatrix} 1 & -1 & 1 & | & 8 \\ 0 & 1 & -12 & | & -15 \\ 0 & 0 & 1 & | & 1 \end{bmatrix}. \quad (4, -3, 1) \text{ or } \{(4, -3, 1)\}$$

Our goal in solving a linear system using matrices is to produce a matrix with 1s down the diagonal from upper left to lower right, called the **main diagonal**, and 0s below the 1s. In general, the matrix will be one of the following forms.

This is the desired form for systems with two equations.
$$\begin{bmatrix} 1 & a & | & b \\ 0 & 1 & | & c \end{bmatrix}$$

$$\begin{bmatrix} 1 & a & b & | & c \\ 0 & 1 & d & | & e \\ 0 & 0 & 1 & | & f \end{bmatrix}$$
This is the desired form for systems with three equations.

The last row of these matrices gives us the value of one variable. The other variables can then be found by back-substitution.

2 Perform matrix row operations.

A matrix with 1s down the main diagonal and 0s below the 1s is said to be in **row-echelon form**. How do we produce a matrix in this form? We use **row operations** on the augmented matrix. These row operations are just like what you did when solving a linear system by the addition method. The difference is that we no longer write the variables, usually represented by x, y, and z.

> **MATRIX ROW OPERATIONS** The following row operations produce matrices that represent systems with the same solution:
>
> **1.** Two rows of a matrix may be interchanged. This is the same as interchanging two equations in the linear system.
> **2.** The elements in any row may be multiplied by a nonzero number. This is the same as multiplying both sides of an equation by a nonzero number.
> **3.** The elements in any row may be multiplied by a nonzero number, and these products may be added to the corresponding elements in any other row. This is the same as multiplying an equation by a nonzero number and then adding equations to eliminate a variable.
>
> Two matrices are **row equivalent** if one can be obtained from the other by a sequence of row operations.

Each matrix row operation in the preceding box can be expressed symbolically as follows:

1. Interchange the elements in the ith and jth rows: $R_i \leftrightarrow R_j$.
2. Multiply each element in the ith row by k: kR_i.
3. Add k times the elements in row i to the corresponding elements in row j: $kR_i + R_j$.

STUDY TIP

When performing the row operation

$$kR_i + R_j$$

you use row i to find the products. However, **elements in row i do not change. It is the elements in row j that change:** Add k times the elements in row i to the corresponding elements in row j. Replace elements in row j by these sums.

STUDY TIP

As you read the solution, keep looking back at the given matrix

$$\begin{bmatrix} 3 & 18 & -12 & | & 21 \\ 1 & 2 & -3 & | & 5 \\ -2 & -3 & 4 & | & -6 \end{bmatrix}.$$

EXAMPLE 2 Performing Matrix Row Operations

Use the matrix

$$\begin{bmatrix} 3 & 18 & -12 & | & 21 \\ 1 & 2 & -3 & | & 5 \\ -2 & -3 & 4 & | & -6 \end{bmatrix}$$

and perform each indicated row operation:

a. $R_1 \leftrightarrow R_2$ **b.** $\dfrac{1}{3}R_1$ **c.** $2R_2 + R_3$.

SOLUTION

a. The notation $R_1 \leftrightarrow R_2$ means to interchange the elements in row 1 and row 2. This results in the row-equivalent matrix

$$\begin{bmatrix} 1 & 2 & -3 & | & 5 \\ 3 & 18 & -12 & | & 21 \\ -2 & -3 & 4 & | & -6 \end{bmatrix}.$$

This was row 2; now it's row 1.

This was row 1; now it's row 2.

b. The notation $\frac{1}{3}R_1$ means to multiply each element in row 1 by $\frac{1}{3}$. This results in the row-equivalent matrix

$$\begin{bmatrix} \frac{1}{3}(3) & \frac{1}{3}(18) & \frac{1}{3}(-12) & | & \frac{1}{3}(21) \\ 1 & 2 & -3 & | & 5 \\ -2 & -3 & 4 & | & -6 \end{bmatrix} = \begin{bmatrix} 1 & 6 & -4 & | & 7 \\ 1 & 2 & -3 & | & 5 \\ -2 & -3 & 4 & | & -6 \end{bmatrix}.$$

c. The notation $2R_2 + R_3$ means to add 2 times the elements in row 2 to the corresponding elements in row 3. Replace the elements in row 3 by these sums. First, we find 2 times the elements in row 2, namely, $1, 2, -3$ and 5:

$$2(1) \text{ or } 2, \qquad 2(2) \text{ or } 4, \qquad 2(-3) \text{ or } -6, \qquad 2(5) \text{ or } 10.$$

Now we add these products to the corresponding elements in row 3. Although we use row 2 to find the products, row 2 does not change. It is the elements in row 3 that change, resulting in the row-equivalent matrix

$$\begin{bmatrix} 3 & 18 & -12 & 21 \\ 1 & 2 & -3 & 5 \\ -2+2=0 & -3+4=1 & 4+(-6)=-2 & -6+10=4 \end{bmatrix} = \begin{bmatrix} 3 & 18 & -12 & 21 \\ 1 & 2 & -3 & 5 \\ 0 & 1 & -2 & 4 \end{bmatrix}. \qquad \blacksquare$$

✔ **CHECK POINT 2** Use the matrix

$$\begin{bmatrix} 4 & 12 & -20 & 8 \\ 1 & 6 & -3 & 7 \\ -3 & -2 & 1 & -9 \end{bmatrix}$$

and perform each indicated row operation:

a. $R_1 \leftrightarrow R_2$ $\quad \begin{bmatrix} 1 & 6 & -3 & 7 \\ 4 & 12 & -20 & 8 \\ -3 & -2 & 1 & -9 \end{bmatrix}$ **b.** $\dfrac{1}{4}R_1$ $\quad \begin{bmatrix} 1 & 3 & -5 & 2 \\ 1 & 6 & -3 & 7 \\ -3 & -2 & 1 & -9 \end{bmatrix}$

c. $3R_2 + R_3$. $\quad \begin{bmatrix} 4 & 12 & -20 & 8 \\ 1 & 6 & -3 & 7 \\ 0 & 16 & -8 & 12 \end{bmatrix}$

3 Use matrices to solve systems.

The process that we use to solve linear systems using matrix row operations is often called **Gaussian elimination**, after the German mathematician Carl Friedrich Gauss (1777–1855). Here are the steps used in solving systems with matrices:

SOLVING LINEAR SYSTEMS USING MATRICES

1. Write the augmented matrix for the system.

2. Use matrix row operations to simplify the matrix to one with 1s down the main diagonal from upper left to lower right, and 0s below the 1s.

3. Write the system of linear equations corresponding to the matrix in step 2 and use back-substitution to find the system's solution.

EXAMPLE 3 Using Matrices to Solve a Linear System

Use matrices to solve the system:

$$\begin{aligned} 4x - 3y &= -15 \\ x + 2y &= -1. \end{aligned}$$

SOLUTION

Step 1. **Write the augmented matrix for the system.**

Linear System	**Augmented Matrix**
$\begin{aligned} 4x - 3y &= -15 \\ x + 2y &= -1 \end{aligned}$	$\begin{bmatrix} 4 & -3 & -15 \\ 1 & 2 & -1 \end{bmatrix}$

Step 2. **Use matrix row operations to simplify the matrix to one with 1s down the diagonal from upper left to lower right, and 0s below the 1s.** Our goal is to obtain a matrix of the form

$$\begin{bmatrix} 1 & a & | & b \\ 0 & 1 & | & c \end{bmatrix}.$$

Our first step in achieving this goal is to get 1 in the top position of the first column.

We want 1 in this position. $\begin{bmatrix} 4 & -3 & | & -15 \\ 1 & 2 & | & -1 \end{bmatrix}$

To get 1 in this position, we interchange rows 1 and 2: $R_1 \leftrightarrow R_2$.

$\begin{bmatrix} 1 & 2 & | & -1 \\ 4 & -3 & | & -15 \end{bmatrix}$ This was row 2; now it's row 1.

This was row 1; now it's row 2.

Now we want 0 below the 1 in the first column.

We want 0 in this position. $\begin{bmatrix} 1 & 2 & | & -1 \\ 4 & -3 & | & -15 \end{bmatrix}$

Let's get a 0 where there is now a 4. If we multiply the top row of numbers by -4 and add these products to the second row of numbers, we will get 0 in this position: $-4R_1 + R_2$. *We change only row 2.*

$-4R_1 + R_2$ $\begin{bmatrix} 1 & 2 & | & -1 \\ -4(1) + 4 & -4(2) + (-3) & | & -4(-1) + (-15) \end{bmatrix} = \begin{bmatrix} 1 & 2 & | & -1 \\ 0 & -11 & | & -11 \end{bmatrix}$

row 1 element   row 2 element   row 1 element   row 2 element   row 1 element   row 2 element

We move on to the second column. We want 1 in the second row, second column.

We want 1 in this position. $\begin{bmatrix} 1 & 2 & | & -1 \\ 0 & -11 & | & -11 \end{bmatrix}$

To get 1 in the desired position, we multiply -11 by its multiplicative inverse, or reciprocal, $-\frac{1}{11}$. Therefore, we multiply all the numbers in the second row by $-\frac{1}{11}$: $-\frac{1}{11}R_2$.

$-\frac{1}{11}R_2$ $\begin{bmatrix} 1 & 2 & | & -1 \\ -\frac{1}{11}(0) & -\frac{1}{11}(-11) & | & -\frac{1}{11}(-11) \end{bmatrix} = \begin{bmatrix} 1 & 2 & | & -1 \\ 0 & 1 & | & 1 \end{bmatrix}$

We now have the desired matrix with 1s down the main diagonal and 0 below the 1.

Step 3. **Write the system of linear equations corresponding to the matrix in step 2 and use back-substitution to find the system's solution.** The system represented by the matrix in step 2 is

$\begin{bmatrix} 1 & 2 & | & -1 \\ 0 & 1 & | & 1 \end{bmatrix} \rightarrow \begin{matrix} 1x + 2y = -1 \\ 0x + 1y = 1 \end{matrix}$ or $\begin{matrix} x + 2y = -1 \\ y = 1. \end{matrix}$

We immediately see that the value for y is 1. To find x, we back-substitute 1 for y in the first equation.

$$x + 2y = -1 \qquad \text{Equation 1}$$
$$x + 2 \cdot 1 = -1 \qquad \text{Substitute 1 for } y.$$
$$x = -3 \qquad \text{Solve for } x.$$

With $x = -3$ and $y = 1$, the proposed solution is $(-3, 1)$. Take a moment to show that $(-3, 1)$ satisfies both equations. The solution is $(-3, 1)$ and the solution set is $\{(-3, 1)\}$. ■

 CHECK POINT 3 Use matrices to solve the system:

$$2x - y = -4$$
$$x + 3y = 5. \quad (-1, 2) \text{ or } \{(-1, 2)\}$$

When solving a system of linear equations in three variables, use matrix row operations in the following order to obtain a matrix with 1s down the main diagonal and 0s below the 1s:

$$\begin{bmatrix} 1 & * & * & | & * \\ * & * & * & | & * \\ * & * & * & | & * \end{bmatrix} \rightarrow \begin{bmatrix} 1 & * & * & | & * \\ 0 & * & * & | & * \\ 0 & * & * & | & * \end{bmatrix} \rightarrow \begin{bmatrix} 1 & * & * & | & * \\ 0 & 1 & * & | & * \\ 0 & * & * & | & * \end{bmatrix} \rightarrow \begin{bmatrix} 1 & * & * & | & * \\ 0 & 1 & * & | & * \\ 0 & 0 & * & | & * \end{bmatrix} \rightarrow \begin{bmatrix} 1 & * & * & | & * \\ 0 & 1 & * & | & * \\ 0 & 0 & 1 & | & * \end{bmatrix}$$

Get 1 in the upper left-hand corner. — Use the 1 in the first column to get 0s below it. — Get 1 in the second row, second column position. — Use the 1 in the second column to get 0 below it. — Get 1 in the third row, third column position.

EXAMPLE 4 Using Matrices to Solve a Linear System

Use matrices to solve the system:

$$3x + y + 2z = 31$$
$$x + y + 2z = 19$$
$$x + 3y + 2z = 25.$$

SOLUTION

Step 1. Write the augmented matrix for the system.

Linear System	Augmented Matrix

$$\begin{array}{rcl} 3x + y + 2z &=& 31 \\ x + y + 2z &=& 19 \\ x + 3y + 2z &=& 25 \end{array} \qquad \begin{bmatrix} 3 & 1 & 2 & | & 31 \\ 1 & 1 & 2 & | & 19 \\ 1 & 3 & 2 & | & 25 \end{bmatrix}$$

Step 2. Use matrix row operations to simplify the matrix to one with 1s down the diagonal from upper left to lower right, and 0s below the 1s. Our first step in achieving this goal is to get 1 in the top position of the first column.

We want 1 in this position.
$$\begin{bmatrix} 3 & 1 & 2 & | & 31 \\ 1 & 1 & 2 & | & 19 \\ 1 & 3 & 2 & | & 25 \end{bmatrix}$$

To get 1 in this position, we interchange rows 1 and 2: $R_1 \leftrightarrow R_2$. (We could also interchange rows 1 and 3 to attain our goal.)

$$\begin{bmatrix} 1 & 1 & 2 & | & 19 \\ 3 & 1 & 2 & | & 31 \\ 1 & 3 & 2 & | & 25 \end{bmatrix}$$

This was row 2; now it's row 1.

This was row 1; now it's row 2.

Now we want to get 0s below the 1 in the first column.

We want 0 in these positions.
$$\begin{bmatrix} 1 & 1 & 2 & | & 19 \\ 3 & 1 & 2 & | & 31 \\ 1 & 3 & 2 & | & 25 \end{bmatrix}$$

To get a 0 where there is now a 3, multiply the top row of numbers by -3 and add these products to the second row of numbers: $-3R_1 + R_2$. To get a 0 where there is now a 1, multiply the top row of numbers by -1 and add these products to the third row of numbers: $-1R_1 + R_3$. Although we are using row 1 to find the products, the numbers in row 1 do not change.

$$
\begin{array}{c}
\\
-3R_1 + R_2 \\
\\
-1R_1 + R_3
\end{array}
\left[
\begin{array}{ccc|c}
1 & 1 & 2 & 19 \\
-3(1) + 3 & -3(1) + 1 & -3(2) + 2 & -3(19) + 31 \\
-1(1) + 1 & -1(1) + 3 & -1(2) + 2 & -1(19) + 25
\end{array}
\right]
$$

$$
=
\begin{bmatrix}
1 & 1 & 2 & 19 \\
0 & -2 & -4 & -26 \\
0 & 2 & 0 & 6
\end{bmatrix}
$$

We move on to the second column. We want 1 in the second row, second column.

We want 1 in this position.
$$
\begin{bmatrix}
1 & 1 & 2 & 19 \\
0 & -2 & -4 & -26 \\
0 & 2 & 0 & 6
\end{bmatrix}
$$

To get 1 in the desired position, we multiply -2 by its reciprocal, $-\frac{1}{2}$. Therefore, we multiply all the numbers in the second row by $-\frac{1}{2}$: $-\frac{1}{2}R_2$.

$$
-\tfrac{1}{2}R_2
\left[
\begin{array}{ccc|c}
1 & 1 & 2 & 19 \\
-\frac{1}{2}(0) & -\frac{1}{2}(-2) & -\frac{1}{2}(-4) & -\frac{1}{2}(-26) \\
0 & 2 & 0 & 6
\end{array}
\right]
=
\begin{bmatrix}
1 & 1 & 2 & 19 \\
0 & 1 & 2 & 13 \\
0 & 2 & 0 & 6
\end{bmatrix}.
$$

We want 0 in this position.

We are not yet done with the second column. The voice balloon shows that we want to get a 0 where there is now a 2. If we multiply the second row of numbers by -2 and add these products to the third row of numbers, we will get 0 in this position: $-2R_2 + R_3$. Although we are using the numbers in row 2 to find the products, the numbers in row 2 do not change.

$$
-2R_2 + R_3
\left[
\begin{array}{ccc|c}
1 & 1 & 2 & 19 \\
0 & 1 & 2 & 13 \\
-2(0) + 0 & -2(1) + 2 & -2(2) + 0 & -2(13) + 6
\end{array}
\right]
=
\begin{bmatrix}
1 & 1 & 2 & 19 \\
0 & 1 & 2 & 13 \\
0 & 0 & -4 & -20
\end{bmatrix}
$$

We move on to the third column. We want 1 in the third row, third column.

We want 1 in this position.
$$
\begin{vmatrix}
1 & 1 & 2 & 19 \\
0 & 1 & 2 & 13 \\
0 & 0 & -4 & -20
\end{vmatrix}
$$

To get 1 in the desired position, we multiply -4 by its reciprocal, $-\frac{1}{4}$. Therefore, we multiply all the numbers in the third row by $-\frac{1}{4}$: $-\frac{1}{4}R_3$.

$$
-\tfrac{1}{4}R_3
\left[
\begin{array}{ccc|c}
1 & 1 & 2 & 19 \\
0 & 1 & 2 & 13 \\
-\frac{1}{4}(0) & -\frac{1}{4}(0) & -\frac{1}{4}(-4) & -\frac{1}{4}(-20)
\end{array}
\right]
=
\begin{bmatrix}
1 & 1 & 2 & 19 \\
0 & 1 & 2 & 13 \\
0 & 0 & 1 & 5
\end{bmatrix}.
$$

We now have the desired matrix with 1s down the main diagonal and 0s below the 1s.

USING TECHNOLOGY

Most graphing utilities can convert an augmented matrix to one with 1s down the diagonal from upper left to lower right, and 0s below the 1s. Recall that this is called row-echelon form. However, row-echelon form is not unique. Your graphing utility might give a row-echelon form different from the one you obtained by hand. However, all row-echelon forms for a given system's augmented matrix produce the same solution to the system. Enter the augmented matrix and name it A. Then use the $\boxed{\text{REF}}$ (row-echelon form) command on matrix A.

This is the augmented matrix in Example 4

This matrix is equivalent to the row-echelon form we obtained by hand. It produces the same solution: $z = 5$, $y = 3$, and $x = 6$.

---

4 Use matrices to identify inconsistent and dependent systems.

Step 3. **Write the system of linear equations corresponding to the matrix in step 2 and use back-substitution to find the system's solution.** The system represented by the matrix in step 2 is

$$\begin{bmatrix} 1 & 1 & 2 & | & 19 \\ 0 & 1 & 2 & | & 13 \\ 0 & 0 & 1 & | & 5 \end{bmatrix} \rightarrow \begin{matrix} 1x + 1y + 2z = 19 \\ 0x + 1y + 2z = 13 \\ 0x + 0y + 1z = 5 \end{matrix} \quad \text{or} \quad \begin{matrix} x + y + 2z = 19 & (1) \\ y + 2z = 13. & (2) \\ z = 5 & (3) \end{matrix}$$

We immediately see from equation (3) that the value for z is 5. To find y, we back-substitute 5 for z in the second equation.

$$y + 2z = 13 \quad \text{Equation (2)}$$
$$y + 2(5) = 13 \quad \text{Substitute 5 for } z.$$
$$y = 3 \quad \text{Solve for } y.$$

Finally, back-substitute 3 for y and 5 for z in the first equation.

$$x + y + 2z = 19 \quad \text{Equation (1)}$$
$$x + 3 + 2(5) = 19 \quad \text{Substitute 3 for } y \text{ and 5 for } z.$$
$$x + 13 = 19 \quad \text{Multiply and add.}$$
$$x = 6 \quad \text{Subtract 13 both sides.}$$

The solution of the original system is $(6, 3, 5)$ and the solution set is $\{(6, 3, 5)\}$. Check to see that the solution satisfies all three equations in the given system. ■

 CHECK POINT **4** Use matrices to solve the system:

$$\begin{matrix} 2x + y + 2z = 18 \\ x - y + 2z = 9 \\ x + 2y - z = 6. \end{matrix} \quad (5, 2, 3) \text{ or } \{(5, 2, 3)\}$$

Modern supercomputers are capable of solving systems with more than 600,000 variables. The augmented matrices for such systems are huge, but the solution using matrices is exactly like what we did in Example 4. Work with the augmented matrix, one column at a time. First, get 1 in the desired position. Then get 0s below the 1.

Inconsistent Systems and Systems with Dependent Equations When solving a system using matrices, you might obtain a matrix with a row in which the numbers to the left of the vertical bar are all zeros, but a nonzero number number appears on the right. In such a case, the system is inconsistent and has no solution. For example, a system of equations that yields the following matrix is an inconsistent system:

$$\begin{bmatrix} 1 & -2 & | & 3 \\ 0 & 0 & | & -4 \end{bmatrix}.$$

The second row of the matrix represents the equation $0x + 0y = -4$, which is false for all values of x and y.

If you obtain a matrix in which a 0 appears across an entire row, the system contains dependent equations and has infinitely many solutions. This row of zeros represents $0x + 0y = 0$ or $0x + 0y + 0z = 0$. These equations are satisfied by infinitely many ordered pairs or triples.

A EXERCISE SET

Student Solutions Manual   CD/Video   PH Math/Tutor Center   MathXL Tutorials on CD   MathXL®   MyMathLab   Interactmath.com

Practice Exercises

In Exercises 1–8, write the system of linear equations represented by the augmented matrix. Use x, y, and, if necessary, z, for the variables. Once the system is written, use back-substitution to find its solution.

1. $\begin{bmatrix} 1 & -3 & | & 11 \\ 0 & 1 & | & -3 \end{bmatrix}$
$x - 3y = 11$
$y = -3$
$(2, -3)$

2. $\begin{bmatrix} 1 & 3 & | & 5 \\ 0 & 1 & | & 2 \end{bmatrix}$
$x + 3y = 5$
$y = 2$
$(-1, 2)$

3. $\begin{bmatrix} 1 & -3 & | & 1 \\ 0 & 1 & | & -1 \end{bmatrix}$
$x - 3y = 1$
$y = -1$
$(-2, -1)$

4. $\begin{bmatrix} 1 & 2 & | & 13 \\ 0 & 1 & | & 4 \end{bmatrix}$
$x + 2y = 13$
$y = 4$
$(5, 4)$

5. $\begin{bmatrix} 1 & 0 & -4 & | & 5 \\ 0 & 1 & -12 & | & 13 \\ 0 & 0 & 1 & | & -\frac{1}{2} \end{bmatrix}$
$x - 4z = 5$
$y - 12z = 13$
$z = -\frac{1}{2}$
$\left(3, 7, -\frac{1}{2}\right)$

6. $\begin{bmatrix} 1 & 2 & 1 & | & 0 \\ 0 & 1 & 0 & | & -2 \\ 0 & 0 & 1 & | & 3 \end{bmatrix}$
$x + 2y + z = 0$
$y = -2$
$z = 3$
$(1, -2, 3)$

7. $\begin{bmatrix} 1 & \frac{1}{2} & 1 & | & \frac{11}{2} \\ 0 & 1 & \frac{3}{2} & | & 7 \\ 0 & 0 & 1 & | & 4 \end{bmatrix}$
$x + \frac{1}{2}y + z = \frac{11}{2}$
$y + \frac{3}{2}z = 7$
$z = 4$
$(1, 1, 4)$

8. $\begin{bmatrix} 1 & 1 & 0 & | & 3 \\ 0 & 1 & \frac{3}{2} & | & -2 \\ 0 & 0 & 1 & | & 0 \end{bmatrix}$
$x + y = 3$
$y + \frac{3}{2}z = -2$
$z = 0$
$(5, -2, 0)$

In Exercises 9–22, perform each matrix row operation and write the new matrix.

9. $\begin{bmatrix} 2 & 2 & | & 5 \\ 1 & -\frac{3}{2} & | & 5 \end{bmatrix} R_1 \leftrightarrow R_2$ $\begin{bmatrix} 1 & -\frac{3}{2} & | & 5 \\ 2 & 2 & | & 5 \end{bmatrix}$

10. $\begin{bmatrix} -6 & 9 & | & 4 \\ 1 & -\frac{3}{2} & | & 4 \end{bmatrix} R_1 \leftrightarrow R_2$ $\begin{bmatrix} 1 & -\frac{3}{2} & | & 4 \\ -6 & 9 & | & 4 \end{bmatrix}$

11. $\begin{bmatrix} -6 & 8 & | & -12 \\ 3 & 5 & | & -2 \end{bmatrix} -\frac{1}{6}R_1$ $\begin{bmatrix} 1 & -\frac{4}{3} & | & 2 \\ 3 & 5 & | & -2 \end{bmatrix}$

12. $\begin{bmatrix} -2 & 3 & | & -10 \\ 4 & 2 & | & 5 \end{bmatrix} -\frac{1}{2}R_1$ $\begin{bmatrix} 1 & -\frac{3}{2} & | & 5 \\ 4 & 2 & | & 5 \end{bmatrix}$

13. $\begin{bmatrix} 1 & -3 & | & 5 \\ 2 & 6 & | & 4 \end{bmatrix} -2R_1 + R_2$ $\begin{bmatrix} 1 & -3 & | & 5 \\ 0 & 12 & | & -6 \end{bmatrix}$

14. $\begin{bmatrix} 1 & -3 & | & 1 \\ 2 & 1 & | & -5 \end{bmatrix} -2R_1 + R_2$ $\begin{bmatrix} 1 & -3 & | & 1 \\ 0 & 7 & | & -7 \end{bmatrix}$

15. $\begin{bmatrix} 1 & -\frac{3}{2} & | & \frac{7}{2} \\ 3 & 4 & | & 2 \end{bmatrix} -3R_1 + R_2$ $\begin{bmatrix} 1 & -\frac{3}{2} & | & \frac{7}{2} \\ 0 & \frac{17}{2} & | & -\frac{17}{2} \end{bmatrix}$

16. $\begin{bmatrix} 1 & -\frac{2}{5} & | & \frac{3}{4} \\ 4 & 2 & | & -1 \end{bmatrix} -4R_1 + R_2$ $\begin{bmatrix} 1 & -\frac{2}{5} & | & \frac{3}{4} \\ 0 & \frac{18}{5} & | & -4 \end{bmatrix}$

17. $\begin{bmatrix} 2 & -6 & 4 & | & 10 \\ 1 & 5 & -5 & | & 0 \\ 3 & 0 & 4 & | & 7 \end{bmatrix} \frac{1}{2}R_1$ $\begin{bmatrix} 1 & -3 & 2 & | & 5 \\ 1 & 5 & -5 & | & 0 \\ 3 & 0 & 4 & | & 7 \end{bmatrix}$

18. $\begin{bmatrix} 3 & -12 & 6 & | & 9 \\ 1 & -4 & 4 & | & 0 \\ 2 & 0 & 7 & | & 4 \end{bmatrix} \frac{1}{3}R_1$ $\begin{bmatrix} 1 & -4 & 2 & | & 3 \\ 1 & -4 & 4 & | & 0 \\ 2 & 0 & 7 & | & 4 \end{bmatrix}$

19. $\begin{bmatrix} 1 & -3 & 2 & | & 0 \\ 3 & 1 & -1 & | & 7 \\ 2 & -2 & 1 & | & 3 \end{bmatrix} -3R_1 + R_2$ $\begin{bmatrix} 1 & -3 & 2 & | & 0 \\ 0 & 10 & -7 & | & 7 \\ 2 & -2 & 1 & | & 3 \end{bmatrix}$

20. $\begin{bmatrix} 1 & -1 & 5 & | & -6 \\ 3 & 3 & -1 & | & 10 \\ 1 & 3 & 2 & | & 5 \end{bmatrix} -3R_1 + R_2$ $\begin{bmatrix} 1 & -1 & 5 & | & -6 \\ 0 & 6 & -16 & | & 28 \\ 1 & 3 & 2 & | & 5 \end{bmatrix}$

21. $\begin{bmatrix} 1 & 1 & -1 & | & 6 \\ 2 & -1 & 1 & | & -3 \\ 3 & -1 & -1 & | & 4 \end{bmatrix} \begin{matrix} -2R_1 + R_2 \\ \text{and} \\ -3R_1 + R_3 \end{matrix}$ $\begin{bmatrix} 1 & 1 & -1 & | & 6 \\ 0 & -3 & 3 & | & -15 \\ 0 & -4 & 2 & | & -14 \end{bmatrix}$

22. $\begin{bmatrix} 1 & 2 & 1 & | & 2 \\ -2 & -1 & 2 & | & 5 \\ 1 & 3 & -2 & | & -8 \end{bmatrix} \begin{matrix} 2R_1 + R_2 \\ \text{and} \\ -1R_1 + R_3 \end{matrix}$ $\begin{bmatrix} 1 & 2 & 1 & | & 2 \\ 0 & 3 & 4 & | & 9 \\ 0 & 1 & -3 & | & -10 \end{bmatrix}$

In Exercises 23–46, solve each system using matrices. If there is no solution or if there are infinitely many solutions and a system's equations are dependent, so state.

23. $x + y = 6$
$x - y = 2$ $\{(4, 2)\}$

24. $x + 2y = 11$
$x - y = -1$ $\{(3, 4)\}$

25. $2x + y = 3$
$x - 3y = 12$ $\{(3, -3)\}$

26. $3x - 5y = 7$
$x - y = 1$ $\{(-1, -2)\}$

27. $5x + 7y = -25$
$11x + 6y = -8$ $\{(2, -5)\}$

28. $3x - 5y = 22$
$4x - 2y = 20$ $\{(4, -2)\}$

29. $4x - 2y = 5$
$-2x + y = 6$ no solution or $\emptyset$

30. $-3x + 4y = 12$
$6x - 8y = 16$ no solution or $\emptyset$

31. $\begin{aligned} x - 2y &= 1 \\ -2x + 4y &= -2 \end{aligned}$ infinitely many solutions; dependent equations

32. $3x - 6y = 1$

$2x - 4y = \dfrac{2}{3}$ infinitely many solutions; dependent equations

33. $\begin{aligned} x + y - z &= -2 \\ 2x - y + z &= 5 \\ -x + 2y + 2z &= 1 \end{aligned}$ $\{(1, -1, 2)\}$

34. $\begin{aligned} x - 2y - z &= 2 \\ 2x - y + z &= 4 \\ -x + y - 2z &= -4 \end{aligned}$ $\{(1, -1, 1)\}$

35. $\begin{aligned} x + 3y &= 0 \\ x + y + z &= 1 \\ 3x - y - z &= 11 \end{aligned}$ $\{(3, -1, -1)\}$

36. $\begin{aligned} 3y - z &= -1 \\ x + 5y - z &= -4 \\ -3x + 6y + 2z &= 11 \end{aligned}$ $\{(-3, 0, 1)\}$

37. $\begin{aligned} 2x + 2y + 7z &= -1 \\ 2x + y + 2z &= 2 \\ 4x + 6y + z &= 15 \end{aligned}$ $\{(1, 2, -1)\}$

38. $\begin{aligned} 3x + 2y + 3z &= 3 \\ 4x - 5y + 7z &= 1 \\ 2x + 3y - 2z &= 6 \end{aligned}$ $\{(2, 0, -1)\}$

39. $\begin{aligned} x + y + z &= 6 \\ x - z &= -2 \\ y + 3z &= 11 \end{aligned}$ $\{(1, 2, 3)\}$

40. $\begin{aligned} x + y + z &= 3 \\ -y + 2z &= 1 \\ -x + z &= 0 \end{aligned}$ $\{(1, 1, 1)\}$

41. $\begin{aligned} x - y + 3z &= 4 \\ 2x - 2y + 6z &= 7 \\ 3x - y + 5z &= 14 \end{aligned}$ no solution or $\varnothing$

42. $\begin{aligned} 3x - y + 2z &= 4 \\ -6x + 2y - 4z &= 1 \\ 5x - 3y + 8z &= 0 \end{aligned}$ no solution or $\varnothing$

43. $\begin{aligned} x - 2y + z &= 4 \\ 5x - 10y + 5z &= 20 \\ -2x + 4y - 2z &= -8 \end{aligned}$ infinitely many solutions; dependent equations

44. $\begin{aligned} x - 3y + z &= 2 \\ 4x - 12y + 4z &= 8 \\ -2x + 6y - 2z &= -4 \end{aligned}$ infinitely many solutions; dependent equations

45. $\begin{aligned} x + y &= 1 \\ y + 2z &= -2 \\ 2x - z &= 0 \end{aligned}$ $\{(-1, 2, -2)\}$

46. $\begin{aligned} x + 3y &= 3 \\ y + 2z &= -8 \\ x - z &= 7 \end{aligned}$ $\{(3, 0, -4)\}$

Practice Plus

In Exercises 47–48, write the system of linear equations represented by the augmented matrix. Use w, x, y, and z, for the variables. Once the system is written, use back-substitution to find its solution, (w, x, y, z).

47. $\begin{bmatrix} 1 & -1 & 1 & 1 & | & 3 \\ 0 & 1 & -2 & -1 & | & 0 \\ 0 & 0 & 1 & 6 & | & 17 \\ 0 & 0 & 0 & 1 & | & 3 \end{bmatrix}$

$\begin{aligned} x - 2y - z &= 0 \\ y + 6z &= 17 \\ z &= 3 \end{aligned}$

$w - x + y + z = 3; (2, 1, -1, 3)$

48. $\begin{bmatrix} 1 & 2 & -1 & 0 & | & 2 \\ 0 & 1 & 1 & -2 & | & -3 \\ 0 & 0 & 1 & -1 & | & -2 \\ 0 & 0 & 0 & 1 & | & 3 \end{bmatrix}$

$\begin{aligned} w + 2x - y &= 2; (-1, 2, 1, 3) \\ x + y - 2z &= -3 \\ y - z &= -2 \\ z &= 3 \end{aligned}$

In Exercises 49–50, perform each matrix row operation and write the new matrix.

49. $\begin{bmatrix} 1 & -1 & 1 & 1 & | & 3 \\ 0 & 1 & -2 & -1 & | & 0 \\ 2 & 0 & 3 & 4 & | & 11 \\ 5 & 1 & 2 & 4 & | & 6 \end{bmatrix} \begin{array}{l} \\ \\ -2R_1 + R_3 \\ -5R_1 + R_4 \end{array}$ $\begin{bmatrix} 1 & -1 & 1 & 1 & | & 3 \\ 0 & 1 & -2 & -1 & | & 0 \\ 0 & 2 & 1 & 2 & | & 5 \\ 0 & 6 & -3 & -1 & | & -9 \end{bmatrix}$

50. $\begin{bmatrix} 1 & -5 & 2 & -2 & | & 4 \\ 0 & 1 & -3 & -1 & | & 0 \\ 3 & 0 & 2 & -1 & | & 6 \\ -4 & 1 & 4 & 2 & | & -3 \end{bmatrix} \begin{array}{l} \\ \\ -3R_1 + R_3 \\ 4R_1 + R_4 \end{array}$ $\begin{bmatrix} 1 & -5 & 2 & -2 & | & 4 \\ 0 & 1 & -3 & -1 & | & 0 \\ 0 & 15 & -4 & 5 & | & -6 \\ 0 & -19 & 12 & -6 & | & 13 \end{bmatrix}$

In Exercises 51–52, solve each system using matrices. You will need to use matrix row operations to obtain matrices like those in Exercises 47 and 48, with 1s down the main diagonal and 0s below the 1s. Express the solution as (w, x, y, z).

51. $\begin{aligned} w + x + y + z &= 4 \\ 2w + x - 2y - z &= 0 \\ w - 2x - y - 2z &= -2 \\ 3w + 2x + y + 3z &= 4 \end{aligned}$ $\{(1, 2, 3, -2)\}$

52. $\begin{aligned} w + x + y + z &= 5 \\ w + 2x - y - 2z &= -1 \\ w - 3x - 3y - z &= -1 \\ 2w - x + 2y - z &= -2 \end{aligned}$ $\{(2, 1, -1, 3)\}$

APPENDIX

B

Objectives

1 Evaluate a second-order determinant.

2 Solve a system of linear equations in two variables using Cramer's rule.

3 Evaluate a third-order determinant.

4 Solve a system of linear equations in three variables using Cramer's rule.

5 Use determinants to identify inconsistent and dependent systems.

DETERMINANTS AND CRAMER'S RULE

A portion of Charles Babbage's unrealized difference engine

As cyberspace absorbs more and more of our work, play, shopping, and socializing, where will it all end? Which activities will still be offline in 2025?

Our technologically transformed lives can be traced back to the English inventor Charles Babbage (1791–1871). Babbage knew of a method for solving linear systems called *Cramer's rule*, in honor of the Swiss geometer Gabriel Cramer (1704–1752). Cramer's rule was simple, but involved numerous multiplications for large systems. Babbage designed a machine, called the "difference engine," that consisted of toothed wheels on shafts for performing these multiplications. Despite the fact that only one-seventh of the functions ever worked, Babbage's invention demonstrated how complex calculations could be handled mechanically. In 1944, scientists at IBM used the lessons of the difference engine to create the world's first computer.

Those who invented computers hoped to relegate the drudgery of repeated computation to a machine. In this section, we look at a method for solving linear systems that played a critical role in this process. The method uses arrays of numbers, called *determinants*. As with matrix methods, solutions are obtained by writing down the coefficients and constants of a linear system and performing operations with them.

1 Evaluate a second-order determinant.

The Determinant of a 2 × 2 Matrix A matrix of **order m × n** has m rows and n columns. If $m = n$, a matrix has the same number of rows as columns and is called a **square matrix**. Associated with every square matrix is a real number, called its **determinant**. The determinant for a 2×2 square matrix is defined as follows:

STUDY TIP

To evaluate a second-order determinant, find the difference of the product of the two diagonals.

$$\begin{vmatrix} a_1 & b_1 \\ a_2 & b_2 \end{vmatrix} = a_1b_2 - a_2b_1$$

DEFINITION OF THE DETERMINANT OF A 2 × 2 MATRIX The determinant of the matrix $\begin{bmatrix} a_1 & b_1 \\ a_2 & b_2 \end{bmatrix}$ is denoted by $\begin{vmatrix} a_1 & b_1 \\ a_2 & b_2 \end{vmatrix}$ and is defined by

$$\begin{vmatrix} a_1 & b_1 \\ a_2 & b_2 \end{vmatrix} = a_1b_2 - a_2b_1.$$

We also say that the **value** of the **second-order determinant** $\begin{vmatrix} a_1 & b_1 \\ a_2 & b_2 \end{vmatrix}$ is $a_1b_2 - a_2b_1$.

Example 1 illustrates that the determinant of a matrix may be positive or negative. The determinant can also have 0 as its value.

EXAMPLE 1 Evaluating the Determinant of a 2 × 2 Matrix

Evaluate the determinant of each of the following matrices:

a. $\begin{bmatrix} 5 & 6 \\ 7 & 3 \end{bmatrix}$ **b.** $\begin{bmatrix} 2 & 4 \\ -3 & -5 \end{bmatrix}.$

SOLUTION   We multiply and subtract as indicated.

a. $\begin{vmatrix} 5 & 6 \\ 7 & 3 \end{vmatrix} = 5 \cdot 3 - 7 \cdot 6 = 15 - 42 = -27$ *The value of the second-order determinant is −27.*

b. $\begin{vmatrix} 2 & 4 \\ -3 & -5 \end{vmatrix} = 2(-5) - (-3)(4) = -10 + 12 = 2$ *The value of the second-order determinant is 2.* ∎

> ✔ **CHECK POINT 1** Evaluate the determinant of each of the following matrices:
>
> **a.** $\begin{bmatrix} 10 & 9 \\ 6 & 5 \end{bmatrix}$ −4   **b.** $\begin{bmatrix} 4 & 3 \\ -5 & -8 \end{bmatrix}.$ −17

2 Solve a system of linear equations in two variables using Cramer's rule.

Solving Systems of Linear Equations in Two Variables Using Determinants

Determinants can be used to solve a linear system in two variables. In general, such a system appears as

$$a_1 x + b_1 y = c_1$$
$$a_2 x + b_2 y = c_2.$$

Let's first solve this system for x using the addition method. We can solve for x by eliminating y from the equations. Multiply the first equation by b_2 and the second equation by $-b_1$. Then add the two equations:

$$
\begin{array}{rl}
a_1 x + b_1 y = c_1 & \xrightarrow{\text{Multiply by } b_2.} \quad a_1 b_2 x + b_1 b_2 y = c_1 b_2 \\
a_2 x + b_2 y = c_2 & \xrightarrow{\text{Multiply by } -b_1.} \underline{-a_2 b_1 x - b_1 b_2 y = -c_2 b_1} \\
& \text{Add: } (a_1 b_2 - a_2 b_1)x = c_1 b_2 - c_2 b_1 \\
& x = \dfrac{c_1 b_2 - c_2 b_1}{a_1 b_2 - a_2 b_1}
\end{array}
$$

Because

$$\begin{vmatrix} c_1 & b_1 \\ c_2 & b_2 \end{vmatrix} = c_1 b_2 - c_2 b_1 \quad \text{and} \quad \begin{vmatrix} a_1 & b_1 \\ a_2 & b_2 \end{vmatrix} = a_1 b_2 - a_2 b_1$$

we can express our answer for x as the quotient of two determinants:

$$x = \dfrac{\begin{vmatrix} c_1 & b_1 \\ c_2 & b_2 \end{vmatrix}}{\begin{vmatrix} a_1 & b_1 \\ a_2 & b_2 \end{vmatrix}}.$$

Similarly, we could use the addition method to solve our system for y, again expressing y as the quotient of two determinants. This method of using determinants to solve the linear system, called **Cramer's rule**, is summarized in the box.

SOLVING A LINEAR SYSTEM IN TWO VARIABLES USING DETERMINANTS

Cramer's Rule

If

$$a_1x + b_1y = c_1$$
$$a_2x + b_2y = c_2$$

then

$$x = \frac{\begin{vmatrix} c_1 & b_1 \\ c_2 & b_2 \end{vmatrix}}{\begin{vmatrix} a_1 & b_1 \\ a_2 & b_2 \end{vmatrix}} \quad \text{and} \quad y = \frac{\begin{vmatrix} a_1 & c_1 \\ a_2 & c_2 \end{vmatrix}}{\begin{vmatrix} a_1 & b_1 \\ a_2 & b_2 \end{vmatrix}},$$

where

$$\begin{vmatrix} a_1 & b_1 \\ a_2 & b_2 \end{vmatrix} \neq 0.$$

Here are some helpful tips when solving

$$a_1x + b_1y = c_1$$
$$a_2x + b_2y = c_2$$

using determinants:

1. Three different determinants are used to find x and y. The determinants in the denominators for x and y are identical. The determinants in the numerators for x and y differ. In abbreviated notation, we write

$$x = \frac{D_x}{D} \quad \text{and} \quad y = \frac{D_y}{D}, \text{where } D \neq 0.$$

2. The elements of D, the determinant in the denominator, are the coefficients of the variables in the system.

$$D = \begin{vmatrix} a_1 & b_1 \\ a_2 & b_2 \end{vmatrix}$$

3. D_x, the determinant in the numerator of x, is obtained by replacing the x-coefficients in D, a_1 and a_2, with the constants on the right sides of the equations, c_1 and c_2.

$$D = \begin{vmatrix} a_1 & b_1 \\ a_2 & b_2 \end{vmatrix} \quad \text{and} \quad D_x = \begin{vmatrix} c_1 & b_1 \\ c_2 & b_2 \end{vmatrix}$$ Replace the column with a_1 and a_2 with the constants c_1 and c_2 to get D_x.

4. D_y, the determinant in the numerator for y, is obtained by replacing the y-coefficients in D, b_1 and b_2, with the constants on the right sides of the equations, c_1 and c_2.

$$D = \begin{vmatrix} a_1 & b_1 \\ a_2 & b_2 \end{vmatrix} \quad \text{and} \quad D_y = \begin{vmatrix} a_1 & c_1 \\ a_2 & c_2 \end{vmatrix}$$ Replace the column with b_1 and b_2 with the constants c_1 and c_2 to get D_y.

EXAMPLE 2 Using Cramer's Rule to Solve a Linear System

Use Cramer's rule to solve the system:

$$5x - 4y = 2$$

$$6x - 5y = 1.$$

SOLUTION Because

$$x = \frac{D_x}{D} \quad \text{and} \quad y = \frac{D_y}{D},$$

we will set up and evaluate the three determinants D, D_x, and D_y.

1. D, the determinant in both denominators, consists of the x- and y-coefficients.

$$D = \begin{vmatrix} 5 & -4 \\ 6 & -5 \end{vmatrix} = (5)(-5) - (6)(-4) = -25 + 24 = -1$$

Because this determinant is not zero, we continue to use Cramer's rule to solve the system.

2. D_x, the determinant in the numerator for x, is obtained by replacing the x-coefficients in D, 5 and 6, by the constants on the right sides of the equations, 2 and 1.

$$D_x = \begin{vmatrix} 2 & -4 \\ 1 & -5 \end{vmatrix} = (2)(-5) - (1)(-4) = -10 + 4 = -6$$

3. D_y, the determinant in the numerator for y, is obtained by replacing the y-coefficients in D, -4 and -5, by the constants on the right sides of the equations, 2 and 1.

$$D_y = \begin{vmatrix} 5 & 2 \\ 6 & 1 \end{vmatrix} = (5)(1) - (6)(2) = 5 - 12 = -7$$

4. Thus,

$$x = \frac{D_x}{D} = \frac{-6}{-1} = 6 \quad \text{and} \quad y = \frac{D_y}{D} = \frac{-7}{-1} = 7.$$

As always, the ordered pair $(6, 7)$ should be checked by substituting these values into the original equations. The solution is $(6, 7)$ and the solution set is $\{(6, 7)\}$. ∎

CHECK POINT **2** Use Cramer's rule to solve the system:

$$5x + 4y = 12$$

$$3x - 6y = 24. \quad (4, -2) \text{ or } \{(4, -2)\}$$

3 Evaluate a third-order determinant.

The Determinant of a 3 × 3 Matrix   The determinant for a 3 × 3 matrix is defined in terms of second-order determinants:

DEFINITION OF THE DETERMINANT OF A 3 × 3 MATRIX   A third-order determinant is defined by

$$\begin{vmatrix} a_1 & b_1 & c_1 \\ a_2 & b_2 & c_2 \\ a_3 & b_3 & c_3 \end{vmatrix} = a_1 \begin{vmatrix} b_2 & c_2 \\ b_3 & c_3 \end{vmatrix} \overset{\text{Subtract.}}{-} a_2 \begin{vmatrix} b_1 & c_1 \\ b_3 & c_3 \end{vmatrix} \overset{\text{Add.}}{+} a_3 \begin{vmatrix} b_1 & c_1 \\ b_2 & c_2 \end{vmatrix}$$

Each a on the right comes from the first column.

Here are some tips that should be helpful when evaluating the determinant of a 3 × 3 matrix:

EVALUATING THE DETERMINANT OF A 3 × 3 MATRIX

1. Each of the three terms in the definition contains two factors—a numerical factor and a second-order determinant.

2. The numerical factor in each term is an element from the first column of the third-order determinant.

3. The minus sign precedes the second term.

4. The second-order determinant that appears in each term is obtained by crossing out the row and the column containing the numerical factor.

$$a_1 \begin{vmatrix} b_2 & c_2 \\ b_3 & c_3 \end{vmatrix} - a_2 \begin{vmatrix} b_1 & c_1 \\ b_3 & c_3 \end{vmatrix} + a_3 \begin{vmatrix} b_1 & c_1 \\ b_2 & c_2 \end{vmatrix}$$

$$\begin{vmatrix} a_1 & b_1 & c_1 \\ a_2 & b_2 & c_2 \\ a_3 & b_3 & c_3 \end{vmatrix} \quad \begin{vmatrix} a_1 & b_1 & c_1 \\ a_2 & b_2 & c_2 \\ a_3 & b_3 & c_3 \end{vmatrix} \quad \begin{vmatrix} a_1 & b_1 & c_1 \\ a_2 & b_2 & c_2 \\ a_3 & b_3 & c_3 \end{vmatrix}$$

The **minor** of an element is the determinant that remains after deleting the row and column of that element. For this reason, we call this method **expansion by minors**.

EXAMPLE 3   Evaluating the Determinant of a 3 × 3 Matrix

Evaluate the determinant of the following matrix:

$$\begin{bmatrix} 4 & 1 & 0 \\ -9 & 3 & 4 \\ -3 & 8 & 1 \end{bmatrix}.$$

SOLUTION   We know that each of the three terms in the determinant contains a numerical factor and a second-order determinant. The numerical factors are from the first column of the given matrix. They are shown in red in the following matrix:

$$\begin{bmatrix} 4 & 1 & 0 \\ -9 & 3 & 4 \\ -3 & 8 & 1 \end{bmatrix}.$$

We find the minor for each numerical factor by deleting the row and column of that element:

$$
\begin{bmatrix} 4 & 1 & 0 \\ -9 & 3 & 4 \\ -3 & 8 & 1 \end{bmatrix}
\begin{bmatrix} 4 & 1 & 0 \\ -9 & 3 & 4 \\ -3 & 8 & 1 \end{bmatrix}
\begin{bmatrix} 4 & 1 & 0 \\ -9 & 3 & 4 \\ -3 & 8 & 1 \end{bmatrix}
$$

The minor for
4 is $\begin{vmatrix} 3 & 4 \\ 8 & 1 \end{vmatrix}$.

The minor for
-9 is $\begin{vmatrix} 1 & 0 \\ 8 & 1 \end{vmatrix}$.

The minor for
-3 is $\begin{vmatrix} 1 & 0 \\ 3 & 4 \end{vmatrix}$.

Now we have three numerical factors, 4, −9, and −3, and three second-order determinants. We multiply each numerical factor by its second-order determinant to find the three terms of the third-order determinant:

$$
4\begin{vmatrix} 3 & 4 \\ 8 & 1 \end{vmatrix}, \quad -9\begin{vmatrix} 1 & 0 \\ 8 & 1 \end{vmatrix}, \quad -3\begin{vmatrix} 1 & 0 \\ 3 & 4 \end{vmatrix}.
$$

Based on the preceding definition, we subtract the second term from the first term and add the third term:

Don't forget to
supply the minus sign.

$$
\begin{vmatrix} 4 & 1 & 0 \\ -9 & 3 & 4 \\ -3 & 8 & 1 \end{vmatrix} = 4\begin{vmatrix} 3 & 4 \\ 8 & 1 \end{vmatrix} - (-9)\begin{vmatrix} 1 & 0 \\ 8 & 1 \end{vmatrix} - 3\begin{vmatrix} 1 & 0 \\ 3 & 4 \end{vmatrix}
$$

$= 4(3 \cdot 1 - 8 \cdot 4) + 9(1 \cdot 1 - 8 \cdot 0) - 3(1 \cdot 4 - 3 \cdot 0)$ Begin evaluating the three second-order determinants.

$= 4(3 - 32) + 9(1 - 0) - 3(4 - 0)$ Multiply within parentheses.

$= 4(-29) + 9(1) - 3(4)$ Subtract within parentheses.

$= -116 + 9 - 12$ Multiply.

$= -119$ Add and subtract as indicated. ■

USING TECHNOLOGY

A graphing utility can be used to evaluate the determinant of a matrix. Enter the matrix and call it *A*. Then use the determinant command. The screen below verifies our result in Example 3.

```
[A]
       [[4  1  0]
        [-9  3  4]
        [-3  8  1]]
det([A])
              -119
■
```

 CHECK POINT 3 Evaluate the determinant of the following matrix:

$$
\begin{bmatrix} 2 & 1 & 7 \\ -5 & 6 & 0 \\ -4 & 3 & 1 \end{bmatrix}. \quad 80
$$

4 Solve a system of linear equations in three variables using Cramer's rule.

Solving Systems of Linear Equations in Three Variables Using Determinants

Cramer's rule can be applied to solving systems of linear equations in three variables. The determinants in the numerator and denominator of all variables are third-order determinants.

SOLVING THREE EQUATIONS IN THREE VARIABLES USING DETERMINANTS

Cramer's Rule

If

$$a_1x + b_1y + c_1z = d_1$$
$$a_2x + b_2y + c_2z = d_2$$
$$a_3x + b_3y + c_3z = d_3$$

then

$$x = \frac{D_x}{D}, \quad y = \frac{D_y}{D}, \quad \text{and} \quad z = \frac{D_z}{D}.$$

These four third-order determinants are given by

$$D = \begin{vmatrix} a_1 & b_1 & c_1 \\ a_2 & b_2 & c_2 \\ a_3 & b_3 & c_3 \end{vmatrix}$$ These are the coefficients of the variables x, y, and z, $D \neq 0$.

$$D_x = \begin{vmatrix} d_1 & b_1 & c_1 \\ d_2 & b_2 & c_2 \\ d_3 & b_3 & c_3 \end{vmatrix}$$ Replace x-coefficients in D with the constants on the right of the three equations.

$$D_y = \begin{vmatrix} a_1 & d_1 & c_1 \\ a_2 & d_2 & c_2 \\ a_3 & d_3 & c_3 \end{vmatrix}$$ Replace y-coefficients in D with the constants on the right of the three equations.

$$D_z = \begin{vmatrix} a_1 & b_1 & d_1 \\ a_2 & b_2 & d_2 \\ a_3 & b_3 & d_3 \end{vmatrix}.$$ Replace z-coefficients in D with the constants on the right of the three equations.

EXAMPLE 4   Using Cramer's Rule to Solve a Linear System in Three Variables

Use Cramer's rule to solve:

$$x + 2y - z = -4$$
$$x + 4y - 2z = -6$$
$$2x + 3y + z = 3.$$

SOLUTION   Because

$$x = \frac{D_x}{D}, \quad y = \frac{D_y}{D}, \quad \text{and} \quad z = \frac{D_z}{D},$$

we need to set up and evaluate four determinants.

Step 1.   Set up the determinants.

1. D, the determinant in all three denominators, consists of the x-, y-, and z-coefficients.

$$D = \begin{vmatrix} 1 & 2 & -1 \\ 1 & 4 & -2 \\ 2 & 3 & 1 \end{vmatrix}$$

$$x + 2y - z = -4$$
$$x + 4y - 2z = -6$$
$$2x + 3y + z = 3$$

The system we are solving (repeated)

2. D_x, the determinant in the numerator for x, is obtained by replacing the x-coefficients in D, 1, 1, and 2, with the constants on the right sides of the equations, -4, -6, and 3.

$$D_x = \begin{vmatrix} -4 & 2 & -1 \\ -6 & 4 & -2 \\ 3 & 3 & 1 \end{vmatrix}$$

3. D_y, the determinant in the numerator for y, is obtained by replacing the y-coefficients in D, 2, 4, and 3, with the constants on the right sides of the equations, -4, -6, and 3.

$$D_y = \begin{vmatrix} 1 & -4 & -1 \\ 1 & -6 & -2 \\ 2 & 3 & 1 \end{vmatrix}$$

4. D_z, the determinant in the numerator for z, is obtained by replacing the z-coefficients in D, -1, -2, and 1, with the constants on the right sides of the equations, -4, -6, and 3.

$$D_z = \begin{vmatrix} 1 & 2 & -4 \\ 1 & 4 & -6 \\ 2 & 3 & 3 \end{vmatrix}$$

Step 2. Evaluate the four determinants.

$$D = \begin{vmatrix} 1 & 2 & -1 \\ 1 & 4 & -2 \\ 2 & 3 & 1 \end{vmatrix} = 1 \begin{vmatrix} 4 & -2 \\ 3 & 1 \end{vmatrix} - 1 \begin{vmatrix} 2 & -1 \\ 3 & 1 \end{vmatrix} + 2 \begin{vmatrix} 2 & -1 \\ 4 & -2 \end{vmatrix}$$

$$= 1(4 + 6) - 1(2 + 3) + 2(-4 + 4)$$
$$= 1(10) - 1(5) + 2(0) = 5$$

Using the same technique to evaluate each determinant, we obtain

$$D_x = -10, \quad D_y = 5, \quad \text{and} \quad D_z = 20.$$

STUDY TIP

To find D_x, D_y, and D_z, you'll need to apply the evaluation process for a 3×3 determinant three times. The values of D_x, D_y, and D_z cannot be obtained from the numbers that occur in the computation of D.

Step 3. Substitute these four values and solve the system.

$$x = \frac{D_x}{D} = \frac{-10}{5} = -2$$

$$y = \frac{D_y}{D} = \frac{5}{5} = 1$$

$$z = \frac{D_z}{D} = \frac{20}{5} = 4$$

The ordered triple $(-2, 1, 4)$ can be checked by substitution into the original three equations. The solution is $(-2, 1, 4)$ and the solution set is $\{(-2, 1, 4)\}$. ■

CHECK POINT 4 Use Cramer's rule to solve the system:

$$3x - 2y + z = 16$$
$$2x + 3y - z = -9 \qquad (2, -3, 4) \text{ or } \{(2, -3, 4)\}$$
$$x + 4y + 3z = 2.$$

5 Use determinants to identify inconsistent and dependent systems.

Cramer's Rule with Inconsistent and Dependent Systems If D, the determinant in the denominator, is 0, the variables described by the quotient of determinants are not real numbers. However, when $D = 0$, this indicates that the system is inconsistent or contains dependent equations. This gives rise to the following two situations:

DETERMINANTS: INCONSISTENT AND DEPENDENT SYSTEMS

1. If $D = 0$ and at least one of the determinants in the numerator is not 0, then the system is inconsistent. The solution set is $\varnothing$.

2. If $D = 0$ and all the determinants in the numerators are 0, then the equations in the system are dependent. The system has infinitely many solutions.

Although we have focused on applying determinants to solve linear systems, they have other applications, some of which we consider in the exercise set that follows.

B EXERCISE SET

Student Solutions Manual   CD/Video   PH Math/Tutor Center   MathXL Tutorials on CD   MathXL®   MyMathLab   Interactmath.com

Practice Exercises

Evaluate each determinant in Exercises 1–10.

1. $\begin{vmatrix} 5 & 7 \\ 2 & 3 \end{vmatrix}$ 1

2. $\begin{vmatrix} 4 & 8 \\ 5 & 6 \end{vmatrix}$ -16

3. $\begin{vmatrix} -4 & 1 \\ 5 & 6 \end{vmatrix}$ -29

4. $\begin{vmatrix} 7 & 9 \\ -2 & -5 \end{vmatrix}$ -17

5. $\begin{vmatrix} -7 & 14 \\ 2 & -4 \end{vmatrix}$ 0

6. $\begin{vmatrix} 1 & -3 \\ -8 & 2 \end{vmatrix}$ -22

7. $\begin{vmatrix} -5 & -1 \\ -2 & -7 \end{vmatrix}$ 33

8. $\begin{vmatrix} \frac{1}{5} & \frac{1}{6} \\ -6 & 5 \end{vmatrix}$ 2

9. $\begin{vmatrix} \frac{1}{2} & \frac{1}{2} \\ \frac{1}{8} & -\frac{3}{4} \end{vmatrix}$ $-\frac{7}{16}$

10. $\begin{vmatrix} \frac{2}{3} & \frac{1}{3} \\ -\frac{1}{2} & \frac{3}{4} \end{vmatrix}$ $\frac{2}{3}$

For Exercises 11–26, use Cramer's rule to solve each system or to determine that the system is inconsistent or contains dependent equations.

11. $x + y = 7$
    $x - y = 3$ $\{(5, 2)\}$

12. $2x + y = 3$
    $x - y = 3$ $\{(2, -1)\}$

13. $12x + 3y = 15$
    $2x - 3y = 13$ $\{(2, -3)\}$

14. $x - 2y = 5$
    $5x - y = -2$ $\{(-1, -3)\}$

15. $4x - 5y = 17$
    $2x + 3y = 3$ $\{(3, -1)\}$

16. $3x + 2y = 2$
    $2x + 2y = 3$ $\{(-1, \frac{5}{2})\}$

17. $x - 3y = 4$
    $3x - 4y = 12$ $\{(4, 0)\}$

18. $2x - 9y = 5$
    $3x - 3y = 11$ $\{(4, \frac{1}{3})\}$

19. $3x - 4y = 4$
    $2x + 2y = 12$ $\{(4, 2)\}$

20. $3x = 7y + 1$
    $2x = 3y - 1$ $\{(-2, -1)\}$

21. $2x = 3y + 2$
    $5x = 51 - 4y$ $\{(7, 4)\}$

22. $y = -4x + 2$
    $2x = 3y + 8$ $\{(1, -2)\}$

23. $3x = 2 - 3y$
    $2y = 3 - 2x$
    inconsistent; no solution or $\varnothing$

24. $x + 2y - 3 = 0$
    $12 = 8y + 4x$
    dependent equations; infinitely many solutions

25. $4y = 16 - 3x$
    $6x = 32 - 8y$
    dependent equations; infinitely many solutions

26. $2x = 7 + 3y$
    $4x - 6y = 3$
    inconsistent; no solution or $\varnothing$

Evaluate each determinant in Exercises 27–32.

27. $\begin{vmatrix} 3 & 0 & 0 \\ 2 & 1 & -5 \\ 2 & 5 & -1 \end{vmatrix}$ 72

28. $\begin{vmatrix} 4 & 0 & 0 \\ 3 & -1 & 4 \\ 2 & -3 & 5 \end{vmatrix}$ 28

29. $\begin{vmatrix} 3 & 1 & 0 \\ -3 & 4 & 0 \\ -1 & 3 & -5 \end{vmatrix}$ -75

30. $\begin{vmatrix} 2 & -4 & 2 \\ -1 & 0 & 5 \\ 3 & 0 & 4 \end{vmatrix}$ -76

31. $\begin{vmatrix} 1 & 1 & 1 \\ 2 & 2 & 2 \\ -3 & 4 & -5 \end{vmatrix}$ 0

32. $\begin{vmatrix} 1 & 2 & 3 \\ 2 & 2 & -3 \\ 3 & 2 & 1 \end{vmatrix}$ -20

In Exercises 33–40, use Cramer's rule to solve each system.

33. $\begin{aligned} x + y + z &= 0 \\ 2x - y + z &= -1 \\ -x + 3y - z &= -8 \end{aligned}$
$\{(-5, -2, 7)\}$

34. $\begin{aligned} x - y + 2z &= 3 \\ 2x + 3y + z &= 9 \\ -x - y + 3z &= 11 \end{aligned}$
$\{(-2, 3, 4)\}$

35. $\begin{aligned} 4x - 5y - 6z &= -1 \\ x - 2y - 5z &= -12 \\ 2x - y &= 7 \end{aligned}$
$\{(2, -3, 4)\}$

36. $\begin{aligned} x - 3y + z &= -2 \\ x + 2y &= 8 \\ 2x - y &= 1 \end{aligned}$
$\{(2, 3, 5)\}$

37. $\begin{aligned} x + y + z &= 4 \\ x - 2y + z &= 7 \\ x + 3y + 2z &= 4 \end{aligned}$
$\{(3, -1, 2)\}$

38. $\begin{aligned} 2x + 2y + 3z &= 10 \\ 4x - y + z &= -5 \\ 5x - 2y + 6z &= 1 \end{aligned}$
$\{(-1, 3, 2)\}$

39. $\begin{aligned} x + 2z &= 4 \\ 2y - z &= 5 \\ 2x + 3y &= 13 \end{aligned}$
$\{(2, 3, 1)\}$

40. $\begin{aligned} 3x + 2z &= 4 \\ 5x - y &= -4 \\ 4y + 3z &= 22 \end{aligned}$
$\{(0, 4, 2)\}$

Practice Plus

In Exercises 41–42, evaluate each determinant.

41.
$\begin{vmatrix} \begin{vmatrix} 3 & 1 \\ -2 & 3 \end{vmatrix} & \begin{vmatrix} 7 & 0 \\ 1 & 5 \end{vmatrix} \\ \begin{vmatrix} 3 & 0 \\ 0 & 7 \end{vmatrix} & \begin{vmatrix} 9 & -6 \\ 3 & 5 \end{vmatrix} \end{vmatrix}$
-42

42.
$\begin{vmatrix} \begin{vmatrix} 5 & 0 \\ 4 & -3 \end{vmatrix} & \begin{vmatrix} -1 & 0 \\ 0 & -1 \end{vmatrix} \\ \begin{vmatrix} 7 & -5 \\ 4 & 6 \end{vmatrix} & \begin{vmatrix} 4 & 1 \\ -3 & 5 \end{vmatrix} \end{vmatrix}$
-407

In Exercises 43–44, write the system of linear equations for which Cramer's rule yields the given determinants.

43. $D = \begin{vmatrix} 2 & -4 \\ 3 & 5 \end{vmatrix}, \quad D_x = \begin{vmatrix} 8 & -4 \\ -10 & 5 \end{vmatrix}$
$\begin{aligned} 2x - 4y &= 8 \\ 3x + 5y &= -10 \end{aligned}$

44. $D = \begin{vmatrix} 2 & -3 \\ 5 & 6 \end{vmatrix}, \quad D_x = \begin{vmatrix} 8 & -3 \\ 11 & 6 \end{vmatrix}$
$\begin{aligned} 2x - 3y &= 8 \\ 5x + 6y &= 11 \end{aligned}$

In Exercises 45–48, solve each equation for x.

45. $\begin{vmatrix} -2 & x \\ 4 & 6 \end{vmatrix} = 32$ -11

46. $\begin{vmatrix} x + 3 & -6 \\ x - 2 & -4 \end{vmatrix} = 28$ 26

47. $\begin{vmatrix} 1 & x & -2 \\ 3 & 1 & 1 \\ 0 & -2 & 2 \end{vmatrix} = -8$ 4

48. $\begin{vmatrix} 2 & x & 1 \\ -3 & 1 & 0 \\ 2 & 1 & 4 \end{vmatrix} = 39$ 3

Application Exercises

Determinants are used to find the area of a triangle whose vertices are given by three points in a rectangular coordinate system. The area of a triangle with vertices (x_1, y_1), (x_2, y_2), and (x_3, y_3) is

$$\text{Area} = \pm\frac{1}{2}\begin{vmatrix} x_1 & y_1 & 1 \\ x_2 & y_2 & 1 \\ x_3 & y_3 & 1 \end{vmatrix}.$$

where the $\pm$ symbol indicates that the appropriate sign should be chosen to yield a positive area. Use this information to work Exercises 49–50.

49. Use determinants to find the area of the triangle whose vertices are $(3, -5)$, $(2, 6)$, and $(-3, 5)$.   28 square units

50. Use determinants to find the area of the triangle whose vertices are $(1, 1)$, $(-2, -3)$, and $(11, -3)$.   26 square units

Determinants are used to show that three points lie on the same line (are collinear). If

$$\begin{vmatrix} x_1 & y_1 & 1 \\ x_2 & y_2 & 1 \\ x_3 & y_3 & 1 \end{vmatrix} = 0,$$

then the points (x_1, y_1), (x_2, y_2), and (x_3, y_3) are collinear. If the determinant does not equal 0, then the points are not collinear. Use this information to work Exercises 51–52.

51. Are the points $(3, -1)$, $(0, -3)$, and $(12, 5)$ collinear?   Yes, the points are collinear.

52. Are the points $(-4, -6)$, $(1, 0)$, and $(11, 12)$ collinear?   Yes, the points are collinear.

Determinants are used to write an equation of a line passing through two points. An equation of the line passing through the distinct points (x_1, y_1) and (x_2, y_2) is given by

$$\begin{vmatrix} x & y & 1 \\ x_1 & y_1 & 1 \\ x_2 & y_2 & 1 \end{vmatrix} = 0.$$

Use this information to work Exercises 53–54.

53. Use the determinant to write an equation for the line passing through $(3, -5)$ and $(-2, 6)$. Then expand the determinant, expressing the line's equation in slope-intercept form.
$\begin{vmatrix} x & y & 1 \\ 3 & -5 & 1 \\ -2 & 6 & 1 \end{vmatrix} = 0; \ y = -\dfrac{11}{5}x + \dfrac{8}{5}$

54. Use the determinant to write an equation for the line passing through $(-1, 3)$ and $(2, 4)$. Then expand the determinant, expressing the line's equation in slope-intercept form.
$\begin{vmatrix} x & y & 1 \\ -1 & 3 & 1 \\ 2 & 4 & 1 \end{vmatrix} = 0; \ y = \dfrac{1}{3}x + \dfrac{10}{3}$

THE CIRCLE

Objectives

1 Write the standard form of a circle's equation.

2 Give the center and radius of a circle whose equation is in standard form.

3 Convert the general form of a circle's equation to standard form.

It's a good idea to know your way around a circle. Clocks, angles, maps, and compasses are based on circles. Circles occur everywhere in nature: in ripples on water, patterns on a butterfly's wings, and cross sections of trees. Some consider the circle to be the most pleasing of all shapes.

The rectangular coordinate system gives us a unique way of knowing a circle. It enables us to translate a circle's geometric definition into an algebraic equation. We begin with this geometric definition.

DEFINITION OF A CIRCLE A **circle** is the set of all points in a plane that are equidistant from a fixed point, called the **center**. The fixed distance from the circle's center to any point on the circle is called the **radius**.

Figure A.1 is our starting point for obtaining a circle's equation. We've placed the circle into a rectangular coordinate system. The circle's center is (h, k) and its radius is r. We let (x, y) represent the coordinates of any point on the circle.

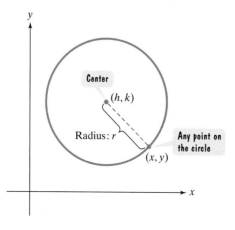

FIGURE A.1 A circle centered at (h, k) with radius r

What does the geometric definition of a circle tell us about the point (x, y) in Figure A.1? The point is on the circle if and only if its distance from the center is r. We can use the distance formula to express this idea algebraically:

The distance between (x, y) and (h, k) is always r.

$$\sqrt{(x - h)^2 + (y - k)^2} = r.$$

Squaring both sides of this equation yields the *standard form of the equation of a circle*.

1 Write the standard form of a circle's equation.

THE STANDARD FORM OF THE EQUATION OF A CIRCLE The **standard form of the equation of a circle** with center (h, k) and radius r is

$$(x - h)^2 + (y - k)^2 = r^2.$$

EXAMPLE 1 Finding the Standard Form of a Circle's Equation

Write the standard form of the equation of the circle with center $(0, 0)$ and radius 2. Graph the circle.

SOLUTION The center is $(0, 0)$. Because the center is represented as (h, k) in the standard form of the equation, $h = 0$ and $k = 0$. The radius is 2, so we will let $r = 2$ in the equation.

$$(x - h)^2 + (y - k)^2 = r^2 \qquad \text{This is the standard form of a circle's equation.}$$
$$(x - 0)^2 + (y - 0)^2 = 2^2 \qquad \text{Substitute 0 for } h, \text{0 for } k, \text{ and 2 for } r.$$
$$x^2 + y^2 = 4 \qquad \text{Simplify.}$$

The standard form of the equation of the circle is $x^2 + y^2 = 4$. Figure A.2 shows the graph. ■

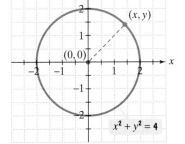

FIGURE A.2 The graph of $x^2 + y^2 = 4$

✔ **CHECK POINT 1** Write the standard form of the equation of the circle with center $(0, 0)$ and radius 4. $x^2 + y^2 = 16$

USING TECHNOLOGY

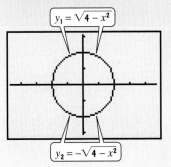

$y_1 = \sqrt{4 - x^2}$

$y_2 = -\sqrt{4 - x^2}$

To graph a circle with a graphing utility, first solve the equation for y.

$$x^2 + y^2 = 4$$
$$y^2 = 4 - x^2$$
$$y = \pm\sqrt{4 - x^2}$$

Graph the two equations

$$y_1 = \sqrt{4 - x^2} \quad \text{and} \quad y_2 = -\sqrt{4 - x^2}$$

in the same viewing rectangle. The graph of $y_1 = \sqrt{4 - x^2}$ is the top semicircle because y is always positive. The graph of $y_2 = -\sqrt{4 - x^2}$ is the bottom semicircle because y is always negative. Use a ZOOM SQUARE setting so that the circle looks like a circle. (Many graphing utilities have problems connecting the two semicircles because the segments directly to the left and to the right of the center become nearly vertical.)

Example 1 and Check Point 1 involved circles centered at the origin. The standard form of the equation of all such circles is $x^2 + y^2 = r^2$, where r is the circle's radius. Now, let's consider a circle whose center is not at the origin.

EXAMPLE 2 Finding the Standard Form of a Circle's Equation

Write the standard form of the equation of the circle with center $(-2, 3)$ and radius 4.

SOLUTION The center is $(-2, 3)$. Because the center is represented as (h, k) in the standard form of the equation, $h = -2$ and $k = 3$. The radius is 4, so we will let $r = 4$ in the equation.

$$(x - h)^2 + (y - k)^2 = r^2 \qquad \text{This is the standard form of a circle's equation.}$$

$$[x - (-2)]^2 + (y - 3)^2 = 4^2 \qquad \text{Substitute } -2 \text{ for } h, 3 \text{ for } k, \text{ and } 4 \text{ for } r.$$

$$(x + 2)^2 + (y - 3)^2 = 16 \qquad \text{Simplify.}$$

The standard form of the equation of the circle is $(x + 2)^2 + (y - 3)^2 = 16$. ∎

 CHECK POINT 2 Write the standard form of the equation of the circle with center $(5, -6)$ and radius 10. $(x - 5)^2 + (y + 6)^2 = 100$

---

2 Give the center and radius of a circle whose equation is in standard form.

EXAMPLE 3 Using the Standard Form of a Circle's Equation to Graph the Circle

Find the center and radius of the circle whose equation is

$$(x - 2)^2 + (y + 4)^2 = 9$$

and graph the equation.

SOLUTION To graph the circle, we have to know its center, (h, k), and its radius, r. We can find the values for h, k, and r by comparing the given equation to the standard form of the equation of a circle, $(x - h)^2 + (y - k)^2 = r^2$.

$$(x - 2)^2 + (y + 4)^2 = 9$$

$$(x - 2)^2 + (y - (-4))^2 = 3^2$$

| This is $(x - h)^2$, with $h = 2$. | This is $(y - k)^2$, with $k = -4$. | This is r^2, with $r = 3$. |

We see that $h = 2$, $k = -4$, and $r = 3$. Thus, the circle has center $(h, k) = (2, -4)$ and a radius of 3 units. To graph this circle, first plot the center $(2, -4)$. Because the radius is 3, you can locate at least four points on the circle by going out three units to the right, to the left, up, and down from the center.

The points three units to the right and to the left of $(2, -4)$ are $(5, -4)$ and $(-1, -4)$, respectively. The points three units up and down from $(2, -4)$ are $(2, -1)$ and $(2, -7)$, respectively.

Using these points, we obtain the graph in Figure A.3. ∎

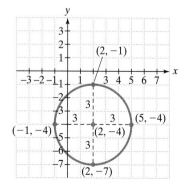

FIGURE A.3 The graph of $(x - 2)^2 + (y + 4)^2 = 9$

✔ CHECK POINT **3** Find the center and radius of the circle whose equation is

$$(x + 3)^2 + (y - 1)^2 = 4$$

and graph the equation.   The center is $(-3, 1)$ and the radius is 2 units. See graphing answer section.

If we square $x - 2$ and $y + 4$ in the standard form of the equation of Example 3, we obtain another form for the circle's equation.

$$(x - 2)^2 + (y + 4)^2 = 9 \qquad \text{This is the standard form of the equation from Example 3.}$$

$$x^2 - 4x + 4 + y^2 + 8y + 16 = 9 \qquad \text{Square } x - 2 \text{ and } y + 4.$$

$$x^2 + y^2 - 4x + 8y + 20 = 9 \qquad \text{Combine numerical terms and rearrange terms.}$$

$$x^2 + y^2 - 4x + 8y + 11 = 0 \qquad \text{Subtract 9 from both sides.}$$

This result suggests that an equation in the form $x^2 + y^2 + Dx + Ey + F = 0$ can represent a circle. This is called the *general form of the equation of a circle*.

> **THE GENERAL FORM OF THE EQUATION OF A CIRCLE**   The **general form of the equation of a circle** is
> $$x^2 + y^2 + Dx + Ey + F = 0.$$

3 Convert the general form of a circle's equation to standard form.

We can convert the general form of the equation of a circle to the standard form $(x - h)^2 + (y - k)^2 = r^2$. We do so by completing the square on x and y. Let's see how this is done.

EXAMPLE 4   Converting the General Form of a Circle's Equation to Standard Form and Graphing the Circle

Write in standard form and graph: $x^2 + y^2 + 4x - 6y - 23 = 0$.

SOLUTION   Because we plan to complete the square on both x and y, let's rearrange the terms so that x-terms are arranged in descending order, y-terms are arranged in descending order, and the constant term appears on the right.

$$x^2 + y^2 + 4x - 6y - 23 = 0 \qquad \text{This is the given equation.}$$

$$(x^2 + 4x \quad) + (y^2 - 6y \quad) = 23 \qquad \text{Rewrite in anticipation of completing the square.}$$

$$(x^2 + 4x + 4) + (y^2 - 6y + 9) = 23 + 4 + 9 \qquad \text{Complete the square on x:} \ \tfrac{1}{2} \cdot 4 = 2 \text{ and } 2^2 = 4, \text{ so add 4 to both sides. Complete the square on } y: \tfrac{1}{2}(-6) = -3 \text{ and } (-3)^2 = 9, \text{ so add 9 to both sides.}$$

Remember that numbers added on the left side must also be added on the right side.

$$(x + 2)^2 + (y - 3)^2 = 36 \qquad \text{Factor on the left and add on the right.}$$

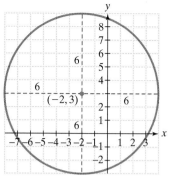

FIGURE A.4 The graph of $(x + 2)^2 + (y - 3)^2 = 36$

This last equation is in standard form. We can identify the circle's center and radius by comparing this equation to the standard form of the equation of a circle, $(x - h)^2 + (y - k)^2 = r^2$.

$$(x + 2)^2 + (y - 3)^2 = 36$$

$$(x - (-2))^2 + (y - 3)^2 = 6^2$$

This is $(x - h)^2$, with $h = -2$.　　This is $(y - k)^2$, with $k = 3$.　　This is r^2, with $r = 6$.

We use the center, $(h, k) = (-2, 3)$, and the radius, $r = 6$, to graph the circle. The graph is shown in Figure A.4.

CHECK POINT 4 Write in standard form and graph:

$$x^2 + y^2 + 4x - 4y - 1 = 0.$$

$(x + 2)^2 + (y - 2)^2 = 9$; See graphing answer section.

C EXERCISE SET

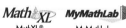

Student Solutions Manual　CD/Video　PH Math/Tutor Center　MathXL Tutorials on CD　MathXL®　MyMathLab　Interactmath.com

Practice Exercises

In Exercises 1–10, write the standard form of the equation of the circle with the given center and radius.

1. Center $(0, 0)$, $r = 7$　$x^2 + y^2 = 49$

2. Center $(0, 0)$, $r = 8$　$x^2 + y^2 = 64$

3. Center $(3, 2)$, $r = 5$　$(x - 3)^2 + (y - 2)^2 = 25$

4. Center $(2, -1)$, $r = 4$　$(x - 2)^2 + (y + 1)^2 = 16$

5. Center $(-1, 4)$, $r = 2$　$(x + 1)^2 + (y - 4)^2 = 4$

6. Center $(-3, 5)$, $r = 3$　$(x + 3)^2 + (y - 5)^2 = 9$

7. Center $(-3, -1)$, $r = \sqrt{3}$　$(x + 3)^2 + (y + 1)^2 = 3$

8. Center $(-5, -3)$, $r = \sqrt{5}$　$(x + 5)^2 + (y + 3)^2 = 5$

9. Center $(-4, 0)$, $r = 10$　$(x + 4)^2 + y^2 = 100$

10. Center $(-2, 0)$, $r = 6$　$(x + 2)^2 + y^2 = 36$

In Exercises 11–18, give the center and radius of the circle described by the equation and graph each equation. **11–18.** See graphing answer section.

11. $x^2 + y^2 = 16$　center: $(0, 0)$; $r = 4$

12. $x^2 + y^2 = 49$　center: $(0, 0)$; $r = 7$

13. $(x - 3)^2 + (y - 1)^2 = 36$　center: $(3, 1)$; $r = 6$

14. $(x - 2)^2 + (y - 3)^2 = 16$　center: $(2, 3)$; $r = 4$

15. $(x + 3)^2 + (y - 2)^2 = 4$　center: $(-3, 2)$; $r = 2$

16. $(x + 1)^2 + (y - 4)^2 = 25$　center: $(-1, 4)$; $r = 5$

17. $(x + 2)^2 + (y + 2)^2 = 4$　center: $(-2, -2)$; $r = 2$

18. $(x + 4)^2 + (y + 5)^2 = 36$　center: $(-4, -5)$; $r = 6$

In Exercises 19–26, complete the square and write the equation in standard form. Then give the center and radius of each circle and graph the equation.　**19–26.** See graphing answer section.

19. $x^2 + y^2 + 6x + 2y + 6 = 0$
$(x + 3)^2 + (y + 1)^2 = 4$; center: $(-3, -1)$; $r = 2$

20. $x^2 + y^2 + 8x + 4y + 16 = 0$
$(x + 4)^2 + (y + 2)^2 = 4$; center: $(-4, -2)$; $r = 2$

21. $x^2 + y^2 - 10x - 6y - 30 = 0$
$(x - 5)^2 + (y - 3)^2 = 64$; center: $(5, 3)$; $r = 8$

22. $x^2 + y^2 - 4x - 12y - 9 = 0$
$(x - 2)^2 + (y - 6)^2 = 49$; center: $(2, 6)$; $r = 7$

23. $x^2 + y^2 + 8x - 2y - 8 = 0$
$(x + 4)^2 + (y - 1)^2 = 25$; center: $(-4, 1)$; $r = 5$

24. $x^2 + y^2 + 12x - 6y - 4 = 0$
$(x + 6)^2 + (y - 3)^2 = 49$; center: $(-6, 3)$; $r = 7$

25. $x^2 - 2x + y^2 - 15 = 0$
$(x - 1)^2 + y^2 = 16$; center: $(1, 0)$; $r = 4$

26. $x^2 + y^2 - 6y - 7 = 0$
$x^2 + (y - 3)^2 = 16$; center: $(0, 3)$; $r = 4$

Practice Plus

In Exercises 27–30, find the solution set for each system by graphing both of the system's equations in the same rectangular coordinate system and finding points of intersection. Check all solutions in both equations.   **27–30.** See graphing answer section.

27. $x^2 + y^2 = 16$

$\quad\quad x - y = 4$ $\{(0, -4), (4, 0)\}$

28. $x^2 + y^2 = 9$

$\quad\quad x - y = 3$ $\{(0, -3), (3, 0)\}$

29. $(x - 2)^2 + (y + 3)^2 = 4$

$\quad\quad\quad\quad\quad y = x - 3$ $\{(0, -3), (2, -1)\}$

30. $(x - 3)^2 + (y + 1)^2 = 9$

$\quad\quad\quad\quad\quad y = x - 1$ $\{(0, -1), (3, 2)\}$

In Exercises 31–34, write the standard form of the equation of the circle with the given graph.

31.

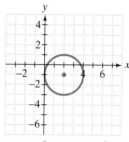

$(x - 2)^2 + (y + 1)^2 = 4$

32.

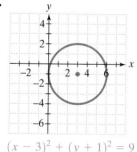

$(x - 3)^2 + (y + 1)^2 = 9$

33.

$(x + 3)^2 + (y + 2)^2 = 1$

34.

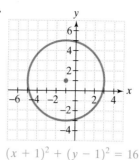

$(x + 1)^2 + (y - 1)^2 = 16$

In Exercises 35–36, a line segment through the center of each circle intersects the circle at the points shown.

a. *Find the coordinates of the circle's center.*

b. *Find the radius of the circle.*

c. *Use your answers from parts (a) and (b) to write the standard form of the circle's equation.*

35.

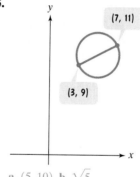

a. $(5, 10)$ **b.** $\sqrt{5}$

c. $(x - 5)^2 + (y - 10)^2 = 5$

36.

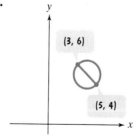

a. $(4, 5)$

b. $\sqrt{2}$

c. $(x - 4)^2 + (y - 5)^2 = 2$

SUMMATION NOTATION AND THE BINOMIAL THEOREM

Summation Notation   There is a compact notation for expressing a sum of terms.

> SUMMATION NOTATION   The sum $a_1 + a_2 + a_3 + a_4 + \cdots + a_n$ is represented by the **summation notation**
>
> $$\sum_{i=1}^{n} a_i = a_1 + a_2 + a_3 + a_4 + \cdots + a_n,$$
>
> where i is the **index of summation**, n is the **upper limit of summation**, and 1 is the **lower limit of summation.**

Any letter can be used for the index of summation. The letters i, j, and k are used commonly. Furthermore, the lower limit of summation can be an integer other than 1.

When we write out a sum that is given in summation notation, we are **expanding the summation notation**. Example 1 shows how to do this.

1   Use summation notation.

EXAMPLE 1   Using Summation Notation

Expand and evaluate the sum:

$$\textbf{a. } \sum_{i=1}^{6} (i^2 + 1) \qquad \textbf{b. } \sum_{k=4}^{7} [(-2)^k - 5] \qquad \textbf{c. } \sum_{i=1}^{5} 3.$$

SOLUTION

a. We must replace i in the expression $i^2 + 1$ with all consecutive integers from 1 to 6, inclusive. Then we add.

$$\sum_{i=1}^{6} (i^2 + 1) = (1^2 + 1) + (2^2 + 1) + (3^2 + 1) + (4^2 + 1)$$
$$+ (5^2 + 1) + (6^2 + 1)$$
$$= 2 + 5 + 10 + 17 + 26 + 37$$
$$= 97$$

b. This time the index of summation is k. First we evaluate $(-2)^k - 5$ for all consecutive integers from 4 through 7, inclusive. Then we add.

$$\sum_{k=4}^{7} [(-2)^k - 5] = [(-2)^4 - 5] + [(-2)^5 - 5]$$
$$+ [(-2)^6 - 5] + [(-2)^7 - 5]$$
$$= (16 - 5) + (-32 - 5) + (64 - 5) + (-128 - 5)$$
$$= 11 + (-37) + 59 + (-133)$$
$$= -100$$

c. To find $\sum_{i=1}^{5} 3$, we observe that every term of the sum is 3. The notation $i = 1$ through 5 indicates that we must add the first five terms in which every term is 3.

$$\sum_{i=1}^{5} 3 = 3 + 3 + 3 + 3 + 3 = 15$$

✔ CHECK POINT **1** Expand and evaluate the sum:

a. $\displaystyle\sum_{i=1}^{6} 2i^2$ $2(1)^2 + 2(2)^2 + 2(3)^2 + 2(4)^2 + 2(5)^2 + 2(6)^2 = 182$

b. $\displaystyle\sum_{k=3}^{5} (2^k - 3)$ $(2^3 - 3) + (2^4 - 3) + (2^5 - 3) = 47$

c. $\displaystyle\sum_{i=1}^{5} 4.$ $4 + 4 + 4 + 4 + 4 = 20$

---

2   Use factorial notation.

Factorial Notation   Products of consecutive positive integers occur in the Binomial Theorem. These products can be expressed in a special notation, called **factorial notation**.

> **FACTORIAL NOTATION**   If n is a positive integer, the notation $n!$ (read "n factorial") is the product of all positive integers from n down through 1.
>
> $$n! = n(n - 1)(n - 2)\cdots(3)(2)(1)$$
>
> $0!$ (zero factorial), by definition, is 1.
>
> $$0! = 1$$

The values of $n!$ for the first six positive integers are

$$1! = 1$$
$$2! = 2 \cdot 1 = 2$$
$$3! = 3 \cdot 2 \cdot 1 = 6$$
$$4! = 4 \cdot 3 \cdot 2 \cdot 1 = 24$$
$$5! = 5 \cdot 4 \cdot 3 \cdot 2 \cdot 1 = 120$$
$$6! = 6 \cdot 5 \cdot 4 \cdot 3 \cdot 2 \cdot 1 = 720.$$

EXAMPLE 2   Using Summation and Factorial Notations

Expand and evaluate the sum:

$$\sum_{j=3}^{6} 4(j - 3)!.$$

SOLUTION   We replace j in the expression $4(j - 3)!$ with all consecutive integers from 3 to 6, inclusive. Then we use factorial notation to evaluate terms. Finally, we add the terms.

$$\sum_{j=3}^{6} 4(j - 3)!$$

$$= 4(3 - 3)! + 4(4 - 3)! + 4(5 - 3)! + 4(6 - 3)!$$
$$= 4 \cdot 0! \qquad + 4 \cdot 1! \qquad + 4 \cdot 2! \qquad + 4 \cdot 3!$$
$$= 4 \cdot 1 \qquad + 4 \cdot 1 \qquad + 4 \cdot 2 \cdot 1 \qquad + 4 \cdot 3 \cdot 2 \cdot 1$$
$$= 4 \qquad\quad + 4 \qquad\quad + 8 \qquad\quad + 24$$
$$= 40$$

CHECK POINT **2** Expand and evaluate the sum:

$$\sum_{k=2}^{5} k^2(k-1)!.$$ 718

3 Evaluate a binomial coefficient.

Binomial Coefficients   Before turning to the Binomial Theorem, we introduce a special notation that uses factorials.

> **DEFINITION OF A BINOMIAL COEFFICIENT** $\binom{n}{r}$ For nonnegative integers n and r, with $n \geq r$, the expression $\binom{n}{r}$ (read "n above r") is called a **binomial coefficient** and is defined by
>
> $$\binom{n}{r} = \frac{n!}{r!(n-r)!}.$$

The symbol $_nC_r$ is often used in place of $\binom{n}{r}$ to denote binomial coefficients.

Can you see that the definition of a binomial coefficient involves a fraction with factorials in the numerator and the denominator? When evaluating such an expression, try to reduce the fraction before performing the multiplications. For example, consider $\frac{26!}{21!}$. Rather than writing out 26! as the product of all integers from 26 down to 1, we can express 26! as

$$26! = 26 \cdot 25 \cdot 24 \cdot 23 \cdot 22 \cdot 21!.$$

In this way, we can divide both the numerator and the denominator by the common factor, 21!.

$$\frac{26!}{21!} = \frac{26 \cdot 25 \cdot 24 \cdot 23 \cdot 22 \cdot \cancel{21!}}{\cancel{21!}} = 26 \cdot 25 \cdot 24 \cdot 23 \cdot 22 = 7,893,600$$

EXAMPLE 3   Evaluating Binomial Coefficients

Evaluate: **a.** $\binom{6}{2}$ **b.** $\binom{3}{0}$ **c.** $\binom{9}{3}$ **d.** $\binom{4}{4}$.

SOLUTION   In each case, we apply the definition of the binomial coefficient.

a. $\binom{6}{2} = \frac{6!}{2!(6-2)!} = \frac{6!}{2!4!} = \frac{6 \cdot 5 \cdot \cancel{4!}}{2 \cdot 1 \cdot \cancel{4!}} = 15$

b. $\binom{3}{0} = \frac{3!}{0!(3-0)!} = \frac{\cancel{3!}}{0!\cancel{3!}} = \frac{1}{1} = 1$

> Remember that 0! = 1.

c. $\binom{9}{3} = \frac{9!}{3!(9-3)!} = \frac{9!}{3!6!} = \frac{9 \cdot 8 \cdot 7 \cdot \cancel{6!}}{3 \cdot 2 \cdot 1 \cdot \cancel{6!}} = 84$

d. $\binom{4}{4} = \frac{4!}{4!(4-4)!} = \frac{\cancel{4!}}{\cancel{4!}0!} = \frac{1}{1} = 1$

✔ CHECK POINT **3** Evaluate:

a. $\dbinom{6}{3}$ 20 **b.** $\dbinom{6}{0}$ 1 **c.** $\dbinom{8}{2}$ 28 **d.** $\dbinom{3}{3}$. 1

4 Expand a binomial raised to a power.

The Binomial Theorem When we write out the *binomial expression* $(a + b)^n$, where n is a positive integer, several patterns begin to appear.

$$(a + b)^1 = a + b$$
$$(a + b)^2 = a^2 + 2ab + b^2$$
$$(a + b)^3 = a^3 + 3a^2b + 3ab^2 + b^3$$
$$(a + b)^4 = a^4 + 4a^3b + 6a^2b^2 + 4ab^3 + b^4$$
$$(a + b)^5 = a^5 + 5a^4b + 10a^3b^2 + 10a^2b^3 + 5ab^4 + b^5$$

Observe the following patterns:

1. The first term is a^n. The exponents on a decrease by 1 in each successive term.

2. The exponents on b increase by 1 in each successive term. In the first term, the exponent on b is 0. (Because $b^0 = 1$, b is not shown in the first term.) The last term is b^n.

3. The sum of the exponents on the variables in any term is equal to n, the exponent on $(a + b)^n$.

4. There is one more term in the polynomial expansion than there is in the power of the binomial, n. There are $n + 1$ terms in the expanded form of $(a + b)^n$.

Using these observations, the variable parts of the expansion of $(a + b)^6$ are

$$a^6, \quad a^5b, \quad a^4b^2, \quad a^3b^3, \quad a^2b^4, \quad ab^5, \quad b^6.$$

The first term is a^6, with the exponents on a decreasing by 1 in each successive term. The exponents on b increase from 0 to 6, with the last term being b^6. The sum of the exponents in each term is equal to 6.

We can generalize from these observations to obtain the variable parts of the expansion of $(a + b)^n$. They are

$$a^n, \quad a^{n-1}b, \quad a^{n-2}b^2, \quad a^{n-3}b^3, \dots, \quad ab^{n-1}, \quad b^n.$$

Exponents on a are decreasing by 1.
Exponents on b are increasing by 1.

Sum of exponents:
$n - 1 + 1 = n$

Sum of exponents:
$n - 3 + 3 = n$

Sum of exponents:
$1 + n - 1 = n$

If we use binomial coefficients and the pattern for the variable part of each term, a formula called the **Binomial Theorem** can be used to expand any positive integral power of a binomial.

A FORMULA FOR EXPANDING BINOMIALS: THE BINOMIAL THEOREM

For any positive integer n,

$$(a + b)^n = \binom{n}{0}a^n + \binom{n}{1}a^{n-1}b + \binom{n}{2}a^{n-2}b^2 + \binom{n}{3}a^{n-3}b^3 + \cdots + \binom{n}{n}b^n$$

$$= \sum_{r=0}^{n} \binom{n}{r}a^{n-r}b^r.$$

EXAMPLE 4 Using the Binomial Theorem

Expand: $(x + 2)^4$.

SOLUTION   We use the Binomial Theorem

$$(a + b)^n = \binom{n}{0}a^n + \binom{n}{1}a^{n-1}b + \binom{n}{2}a^{n-2}b^2 + \binom{n}{3}a^{n-3}b^3 + \cdots + \binom{n}{n}b^n$$

to expand $(x + 2)^4$. In $(x + 2)^4$, $a = x$, $b = 2$, and $n = 4$. In the expansion, powers of x are in descending order, starting with x^4. Powers of 2 are in ascending order, starting with 2^0. (Because $2^0 = 1$, a 2 is not shown in the first term.) The sum of the exponents on x and 2 in each term is equal to 4, the exponent in the expression $(x + 2)^4$.

$$(x + 2)^4 = \binom{4}{0}x^4 + \binom{4}{1}x^3 \cdot 2 + \binom{4}{2}x^2 \cdot 2^2 + \binom{4}{3}x \cdot 2^3 + \binom{4}{4}2^4$$

These binomial coefficients are evaluated using $\binom{n}{r} = \frac{n!}{r!(n-r)!}$.

$$= \frac{4!}{0!4!}x^4 + \frac{4!}{1!3!}x^3 \cdot 2 + \frac{4!}{2!2!}x^2 \cdot 4 + \frac{4!}{3!1!}x \cdot 8 + \frac{4!}{4!0!} \cdot 16$$

$\frac{4!}{2!2!} = \frac{4 \cdot 3 \cdot 2!}{2! \cdot 2 \cdot 1} = \frac{12}{2} = 6$

Take a few minutes to verify the other factorial evaluations.

$$= 1 \cdot x^4 + 4x^3 \cdot 2 + 6x^2 \cdot 4 + 4x \cdot 8 + 1 \cdot 16$$

$$= x^4 + 8x^3 + 24x^2 + 32x + 16 \qquad \blacksquare$$

✔ **CHECK POINT 4** Expand: $(x + 1)^4$. $x^4 + 4x^3 + 6x^2 + 4x + 1$

EXAMPLE 5 Using the Binomial Theorem

Expand: $(2x - y)^5$.

SOLUTION   Because the Binomial Theorem involves the addition of two terms raised to a power, we rewrite $(2x - y)^5$ as $[2x + (-y)]^5$. We use the Binomial Theorem

$$(a + b)^n = \binom{n}{0}a^n + \binom{n}{1}a^{n-1}b + \binom{n}{2}a^{n-2}b^2 + \binom{n}{3}a^{n-3}b^3 + \cdots + \binom{n}{n}b^n$$

to expand $[2x + (-y)]^5$. In $[2x + (-y)]^5$, $a = 2x$, $b = -y$, and $n = 5$. In the expansion, powers of $2x$ are in descending order, starting with $(2x)^5$. Powers of $-y$ are in ascending order, starting with $(-y)^0$. [Because $(-y)^0 = 1$, a $-y$ is not shown in the first term.] The sum of the exponents on $2x$ and $-y$ in each term is equal to 5, the exponent in the expression $(2x - y)^5$.

$$(2x - y)^5 = [2x + (-y)]^5$$

$$= \binom{5}{0}(2x)^5 + \binom{5}{1}(2x)^4(-y) + \binom{5}{2}(2x)^3(-y)^2 + \binom{5}{3}(2x)^2(-y)^3 + \binom{5}{4}(2x)(-y)^4 + \binom{5}{5}(-y)^5$$

Evaluate binomial coefficients using $\binom{n}{r} = \frac{n!}{r!(n-r)!}$.

$$= \frac{5!}{0!5!}(2x)^5 + \frac{5!}{1!4!}(2x)^4(-y) + \frac{5!}{2!3!}(2x)^3(-y)^2 + \frac{5!}{3!2!}(2x)^2(-y)^3 + \frac{5!}{4!1!}(2x)(-y)^4 + \frac{5!}{5!0!}(-y)^5$$

$\frac{5!}{2!3!} = \frac{5 \cdot 4 \cdot 3!}{2 \cdot 1 \cdot 3!} = 10$

Take a few minutes to verify the other factorial evaluations.

$$= 1(2x)^5 + 5(2x)^4(-y) + 10(2x)^3(-y)^2 + 10(2x)^2(-y)^3 + 5(2x)(-y)^4 + 1(-y)^5$$

Raise both factors in these parentheses to the indicated powers.

$$= 1(32x^5) + 5(16x^4)(-y) + 10(8x^3)(-y)^2 + 10(4x^2)(-y)^3 + 5(2x)(-y)^4 + 1(-y)^5$$

Now raise $-y$ to the indicated powers.

$$= 1(32x^5) + 5(16x^4)(-y) + 10(8x^3)y^2 + 10(4x^2)(-y^3) + 5(2x)y^4 + 1(-y^5)$$

Multiplying factors in each of the six terms gives us the desired expansion:
$$(2x - y)^5 = 32x^5 - 80x^4y + 80x^3y^2 - 40x^2y^3 + 10xy^4 - y^5.$$

 CHECK POINT 5 Expand: $(x - 2y)^5$. $\quad x^5 - 10x^4y + 40x^3y^2 - 80x^2y^3 + 80xy^4 - 32y^5$

 D EXERCISE SET

Student Solutions Manual · CD/Video · PH Math/Tutor Center · MathXL Tutorials on CD · MathXL® · MyMathLab · Interactmath.com

In Exercises 1–14, find each indicated sum.

1. $\sum_{i=1}^{6} 5i$ $\quad 105$

2. $\sum_{i=1}^{6} 7i$ $\quad 147$

3. $\sum_{i=1}^{4} 2i^2$ $\quad 60$

4. $\sum_{i=1}^{5} i^3$ $\quad 225$

5. $\sum_{k=1}^{5} k(k+4)$ $\quad 115$

6. $\sum_{k=1}^{4} (k-3)(k+2)$ $\quad -4$

7. $\sum_{i=1}^{4} \left(-\frac{1}{2}\right)^i$ $\quad -\frac{5}{16}$

8. $\sum_{i=2}^{4} \left(-\frac{1}{3}\right)^i$ $\quad \frac{7}{81}$

9. $\sum_{i=5}^{9} 11$ $\quad 55$

10. $\sum_{i=3}^{7} 12$ $\quad 60$

11. $\sum_{i=0}^{4} \frac{(-1)^i}{i!}$ $\quad \frac{3}{8}$

12. $\sum_{i=0}^{4} \frac{(-1)^{i+1}}{(i+1)!}$ $\quad -\frac{19}{30}$

13. $\sum_{i=1}^{5} \frac{i!}{(i-1)!}$ $\quad 15$

14. $\sum_{i=1}^{5} \frac{(i+2)!}{i!}$ $\quad 110$

In Exercises 15–22, evaluate the given binomial coefficient.

15. $\binom{8}{3}$ $\quad 56$

16. $\binom{7}{2}$ $\quad 21$

17. $\binom{12}{1}$ $\quad 12$

18. $\binom{11}{1}$ $\quad 11$

19. $\binom{6}{6}$ $\quad 1$

20. $\binom{15}{2}$ $\quad 105$

21. $\binom{100}{2}$ $\quad 4950$

22. $\binom{100}{98}$ $\quad 4950$

In Exercises 23–44, use the Binomial Theorem to expand each binomial and express the result in simplified form.

23. $(x + 2)^3$ $x^3 + 6x^2 + 12x + 8$

24. $(x + 4)^3$ $x^3 + 12x^2 + 48x + 64$

25. $(3x + y)^3$ $27x^3 + 27x^2y + 9xy^2 + y^3$

26. $(x + 3y)^3$ $x^3 + 9x^2y + 27xy^2 + 27y^3$

27. $(5x - 1)^3$ $125x^3 - 75x^2 + 15x - 1$

28. $(4x - 1)^3$ $64x^3 - 48x^2 + 12x - 1$

29. $(2x + 1)^4$ $16x^4 + 32x^3 + 24x^2 + 8x + 1$

30. $(3x + 1)^4$ $81x^4 + 108x^3 + 54x^2 + 12x + 1$

31. $(x^2 + 2y)^4$ $x^8 + 8x^6y + 24x^4y^2 + 32x^2y^3 + 16y^4$

32. $(x^2 + y)^4$ $x^8 + 4x^6y + 6x^4y^2 + 4x^2y^3 + y^4$

33. $(y - 3)^4$ $y^4 - 12y^3 + 54y^2 - 108y + 81$

34. $(y - 4)^4$ $y^4 - 16y^3 + 96y^2 - 256y + 256$

35. $(2x^3 - 1)^4$ $16x^{12} - 32x^9 + 24x^6 - 8x^3 + 1$

36. $(2x^5 - 1)^4$ $16x^{20} - 32x^{15} + 24x^{10} - 8x^5 + 1$

37. $(c + 2)^5$ $c^5 + 10c^4 + 40c^3 + 80c^2 + 80c + 32$

38. $(c + 3)^5$ $c^5 + 15c^4 + 90c^3 + 270c^2 + 405c + 243$

39. $(x - 1)^5$ $x^5 - 5x^4 + 10x^3 - 10x^2 + 5x - 1$

40. $(x - 2)^5$ $x^5 - 10x^4 + 40x^3 - 80x^2 + 80x - 32$

41. $(3x - y)^5$ $243x^5 - 405x^4y + 270x^3y^2 - 90x^2y^3 + 15xy^4 - y^5$

42. $(x - 3y)^5$ $x^5 - 15x^4y + 90x^3y^2 - 270x^2y^3 + 405xy^4 - 243y^5$

43. $(2a + b)^6$

   $64a^6 + 192a^5b + 240a^4b^2 + 160a^3b^3 + 60a^2b^4 + 12ab^5 + b^6$

44. $(a + 2b)^6$

   $a^6 + 12a^5b + 60a^4b^2 + 160a^3b^3 + 240a^2b^4 + 192ab^5 + 64b^6$

Answers to Selected Exercises

CHAPTER 1

Section 1.1

Check Point Exercises

1. $\frac{21}{8}$ **2.** $1\frac{2}{3}$ **3.** $2\cdot2\cdot3\cdot3$ **4. a.** $\frac{2}{3}$ **b.** $\frac{7}{4}$ **c.** $\frac{13}{15}$ **d.** $\frac{1}{5}$ **5. a.** $\frac{8}{33}$ **b.** $\frac{18}{5}$ or $3\frac{3}{5}$ **c.** $\frac{2}{7}$ **d.** $\frac{51}{10}$ or $5\frac{1}{10}$ **6. a.** $\frac{10}{3}$ or $3\frac{1}{3}$

b. $\frac{2}{9}$ **c.** $\frac{3}{2}$ or $1\frac{1}{2}$ **7. a.** $\frac{5}{11}$ **b.** $\frac{2}{3}$ **c.** $\frac{9}{4}$ or $2\frac{1}{4}$ **8.** $\frac{14}{21}$ **9. a.** $\frac{11}{10}$ or $1\frac{1}{10}$ **b.** $\frac{7}{12}$ **c.** $\frac{5}{4}$ or $1\frac{1}{4}$ **10.** $\frac{53}{60}$ **11.** $\frac{29}{96}$ **12.** $\frac{1}{4}$

Exercise Set 1.1

1. $\frac{19}{8}$ **3.** $\frac{38}{5}$ **5.** $\frac{135}{16}$ **7.** $4\frac{3}{5}$ **9.** $8\frac{4}{9}$ **11.** $35\frac{11}{20}$ **13.** $2\cdot11$ **15.** $2\cdot2\cdot5$ **17.** prime **19.** $2\cdot2\cdot3\cdot3$ **21.** $2\cdot2\cdot5\cdot7$

23. prime **25.** $3\cdot3\cdot3\cdot3$ **27.** $2\cdot2\cdot2\cdot2\cdot3\cdot5$ **29.** $\frac{5}{8}$ **31.** $\frac{5}{6}$ **33.** $\frac{7}{10}$ **35.** $\frac{2}{5}$ **37.** $\frac{22}{25}$ **39.** $\frac{60}{43}$ **41.** $\frac{2}{15}$ **43.** $\frac{21}{88}$ **45.** $\frac{36}{7}$

47. $\frac{1}{12}$ **49.** $\frac{15}{14}$ **51.** 6 **53.** $\frac{15}{16}$ **55.** $\frac{9}{5}$ **57.** $\frac{5}{9}$ **59.** 3 **61.** $\frac{7}{10}$ **63.** $\frac{1}{2}$ **65.** 6 **67.** $\frac{6}{11}$ **69.** $\frac{2}{3}$ **71.** $\frac{5}{4}$ **73.** $\frac{1}{6}$

75. 2 **77.** $\frac{7}{10}$ **79.** $\frac{9}{10}$ **81.** $\frac{19}{24}$ **83.** $\frac{7}{18}$ **85.** $\frac{7}{12}$ **87.** $\frac{41}{80}$ **89.** $1\frac{5}{12}$ or $\frac{17}{12}$ **91.** $\frac{3a}{20}$ **93.** $\frac{20}{x}$ **95.** $\frac{4}{15}$ **97. a.** $\frac{11}{20}$ **b.** 240

99. $\frac{2}{5}$ **101.** $1\frac{1}{5}$ cups **103.** $\frac{23}{20}$ mi; $\frac{7}{20}$ mi farther **105.** 38 mi **117.** d

Section 1.2

Check Point Exercises

1. a. -500 **b.** -282 **2.**   **3.**

4. a. 0.375 **b.** $0.\overline{45}$ **5. a.** $\sqrt{9}$ **b.** $0, \sqrt{9}$ **c.** $-9, 0, \sqrt{9}$ **d.** $-9, -1.3, 0, 0.\overline{3}, \sqrt{9}$ **e.** $\frac{\pi}{2}, \sqrt{10}$ **f.** $-9, -1.3, 0, 0.\overline{3}, \frac{\pi}{2}, \sqrt{9}, \sqrt{10}$

6. a. $>$ **b.** $<$ **c.** $<$ **d.** $<$ **7. a.** true **b.** true **c.** false **8. a.** 4 **b.** 6 **c.** $\sqrt{2}$

Exercise Set 1.2

1. -20 **3.** 8 **5.** -3000 **7.** -4 billion **9–19.**   **21.** 0.75 **23.** 0.35 **25.** 0.875 **27.** $0.\overline{81}$ **29.** -0.5 **31.** $-0.8\overline{3}$ **33. a.** $\sqrt{100}$ **b.** $0, \sqrt{100}$ **c.** $-9, 0, \sqrt{100}$ **d.** $-9, -\frac{4}{5}, 0, 0.25, 9.2, \sqrt{100}$ **e.** $\sqrt{3}$

f. $-9, -\frac{4}{5}, 0, 0.25, \sqrt{3}, 9.2, \sqrt{100}$ **35. a.** $\sqrt{64}$ **b.** $0, \sqrt{64}$ **c.** $-11, 0, \sqrt{64}$ **d.** $-11, -\frac{5}{6}, 0, 0.75, \sqrt{64}$ **e.** $\sqrt{5}, \pi$

f. $-11, -\frac{5}{6}, 0, 0.75, \sqrt{5}, \pi, \sqrt{64}$ **37.** 0 **39.** Answers will vary; $\frac{1}{2}$ is an example. **41.** Answers will vary; 6 is an example. **43.** Answers will vary; π is an example. **45.** $<$ **47.** $>$ **49.** $>$ **51.** $<$ **53.** $>$ **55.** $<$ **57.** $<$ **59.** $>$ **61.** $>$ **63.** true **65.** true **67.** true **69.** false **71.** 6 **73.** 7 **75.** $\frac{5}{6}$ **77.** $\sqrt{11}$ **79.** $>$ **81.** $=$ **83.** $<$ **85.** $=$ **87.** -2 **89.** 1997, 2002; budget deficit **103.** c **105.** -7 and -6 **107.** 1.732; 1 and 2 **109.** -0.414; -1 and 0

Section 1.3

Check Point Exercises

1.   **2.** $E(-4, -2), F(-2, 0), G(6, 0)$ **3.** $B(8, 200)$; After 8 sec, the watermelon is 200 ft above ground. **4.** $D(8.8, 0)$; After approximately 8.8 sec, the watermelon is 0 ft above ground. Equivalently, the watermelon splatters on the ground after approximately 8.8 sec. **5.** 30 thousand **6. a.** 22 **b.** United States, France, Canada, Sweden

Exercise Set 1.3

1. **3.** **5.** **7.**

9–23.

25. $(5, 2)$ **27.** $(-6, 5)$ **29.** $(-2, -3)$ **31.** $(5, -3)$ **33.** I and II **35.** I and III

37. Answers will vary; examples are $(-2, 2), (-1, 1), (1, 1), (2, 2)$.

39. $4\frac{1}{2}$ or $\frac{9}{2}$ **41.** $(2, 7)$; The football is 7 ft above ground when it is 2 yd from the quarterback. **43.** $(6, 9.25)$ **45.** 12 ft; 15 yd

47. $(91, 125)$; In 1991, 125 thousand acres were used for cultivation.

49. 2001; 25 thousand acres **51.** 1991 and 1992 **53.** 7.1 million barrels per day **55.** 1990; 7.2 million barrels per day

57. 1970; 9.8 million barrels per day **59.** 1.7 million barrels per day **61.** \$120 million **63.** *No Way Out, The Postman, Dragonfly*

65. 48 yr old **67.** approx 17 yr **69.** \$22 thousand **71.** \$7 thousand **73.** 270; 14 **75.** 2002 **77.** 1972, 1982, 1992 **79.** $\frac{1}{20}$

89. a.

b. A; B; Points line up vertically when x-coordinates are associated with more than one y-coordinate. **91.** $\frac{23}{20}$

92. $<$ **93.** 5.83

Section 1.4

Check Point Exercises

1. 226.5; In 1980, the population of the United States was 226.5 million. **2. a.** 3 terms **b.** 6 **c.** 11 **d.** $6x$ and $2x$
3. a. $14 + x$ **b.** $y7$ **4. a.** $17 + 5x$ **b.** $x5 + 17$ **5. a.** $20 + x$ or $x + 20$ **b.** $30x$ **6.** $12 + x$ or $x + 12$ **7.** $5x + 15$
8. $24y + 42$ **9. a.** $10x$ **b.** $5a$ **10. a.** $18x + 10$ **b.** $14x + 8y$ **11.** $25x + 21$ **12.** $38x + 23y$ **13.** 108 beats/min

Exercise Set 1.4

1. 15 **3.** 80 **5.** 46 **7.** 35 **9.** 25 **11. a.** 2 **b.** 3 **c.** 5 **d.** no **13. a.** 3 **b.** 1 **c.** 2 **d.** yes; x and $5x$
15. a. 3 **b.** 4 **c.** 1 **d.** no **17.** $4 + y$ **19.** $3x + 5$ **21.** $5y + 4x$ **23.** $5(3 + x)$ **25.** $x9$ **27.** $x + 6y$ **29.** $x7 + 23$
31. $(x + 3)5$ **33.** $(7 + 5) + x = 12 + x$ **35.** $(7 \cdot 4)x = 28x$ **37.** $3x + 15$ **39.** $16x + 24$ **41.** $4 + 2r$ **43.** $5x + 5y$
45. $3x - 6$ **47.** $8x - 10$ **49.** $\frac{5}{2}x - 6$ **51.** $8x + 28$ **53.** $6x + 18 + 12y$ **55.** $15x - 10 + 20y$ **57.** $17x$ **59.** $8a$ **61.** $14 + x$
63. $11y + 3$ **65.** $9x + 1$ **67.** $14a + 14$ **69.** $15x + 6$ **71.** $15x + 2$ **73.** $41a + 24b$ **75.** Commutative property of addition
77. Associative property of addition **79.** false **81.** true **83.** 300; You can stay in the sun for 300 min without burning with a number 15 lotion.
85. 8805; In 2002, credit card debt per U.S. household was \$8805; fairly well. **87. a.** $0.38x + 0.02$ **b.** 1.92 million; very well
101. c **103. a.** \$50.50; \$5.50; \$1.00 **b.** No; When producing 2000 clocks, the average cost is \$3. This is more than the \$1.50 maximum price at which the manufacturer can sell the clocks.
104. $0.\overline{4}$ **105.** **106.** $\frac{1}{5}$

Mid-Chapter 1 Check Point

1. $24a + 1$ **2.** $\frac{1}{6}$ **3.** $\frac{1}{2}$ **4.** $65x + 16$ **5.** $\frac{25}{66}$ **6.** $\frac{2}{3}$ **7.** $\frac{1}{5}$ **8.** $\frac{25}{27}$ **9.** $\frac{37}{18}$ or $2\frac{1}{18}$ **10.** IV

11. $(x + 3)5$ **12.** $5(3 + x)$ **13.** $5x + 15$ **14.** 7% **15.** 2002, 2003, 2004 **16.** $<$

17. $-11, -\frac{3}{7}, 0, 0.45, \sqrt{25}$ **18.** 13 **19.** 4

20. $(67; 20)$; In 1967, 20% of grades of undergraduate college students were A's. **21.** 1970; 23% **22.** 40% **23.** 19.3 **24.** $0.\overline{09}$ **25.** $56x - 80 + 24y$

Section 1.5

Check Point Exercises

1. $4 + (-7) = -3$

2. a. $-1 + (-3) = -4$

b. $-5 + 3 = -2$

3. a. -35 **b.** -1.5 **c.** $-\dfrac{5}{6}$ **4. a.** -13 **b.** 1.2 **c.** $-\dfrac{1}{2}$ **5. a.** $-17x$

b. $-7y + 6z$ **c.** $20 - 20x$ **6.** The water level is down 3 ft at the end of 5 months.

Exercise Set 1.5

1. 4 **3.** -7 **5.** -4

7. 0

9. -7 **11.** 0 **13.** -60 **15.** -18 **17.** -1.3 **19.** -1 **21.** -5 **23.** 4 **25.** -3
27. -1.5 **29.** -5.7 **31.** $\dfrac{3}{10}$ **33.** $\dfrac{1}{8}$ **35.** $-\dfrac{43}{35}$ **37.** -8 **39.** 62 **41.** 8 **43.** -21
45. 22.1 **47.** $-8x$ **49.** $13y$ **51.** $-23a$ **53.** $1 - 6x$ **55.** $-4 + 6b$ **57.** $-2x - 3y$
59. $20x - 6$ **61.** $24 + 3y$ **63.** $47 - 33a$ **65.** 12 **67.** -30 **69.** $>$ **71.** The high temperature was 44°.
73. The elevation is 600 ft below sea level. **75.** The temperature at 4:00 P.M. was 3°F. **77.** The ball was at the 25-yard line at the end of the
fourth play. **79.** $455 billion deficit **89.** d **91.** $-18y$ **94.** true **95. a.** $\sqrt{4}$ **b.** $0, \sqrt{4}$ **c.** $-6, 0, \sqrt{4}$
d. $-6, 0, 0.\overline{7}, \sqrt{4}$ **e.** $-\pi, \sqrt{3}$ **f.** $-6, -\pi, 0, 0.\overline{7}, \sqrt{3}, \sqrt{4}$ **96.** IV

Section 1.6

Check Point Exercises

1. a. -8 **b.** 9 **c.** -5 **2. a.** 9.2 **b.** $-\dfrac{14}{15}$ **c.** 7π **3.** 15 **4.** $-6, 4a, -7ab$ **5. a.** $4 - 7x$ **b.** $-9x + 4y$ **6.** 19,763 m

Exercise Set 1.6

1. a. -12 **b.** $5 + (-12)$ **3. a.** 7 **b.** $5 + 7$ **5.** 6 **7.** -6 **9.** 23 **11.** 11 **13.** -11 **15.** -38 **17.** 0 **19.** 0 **21.** 26
23. -13 **25.** 13 **27.** $-\dfrac{2}{7}$ **29.** $\dfrac{4}{5}$ **31.** -1 **33.** $-\dfrac{3}{5}$ **35.** $\dfrac{3}{4}$ **37.** $\dfrac{1}{4}$ **39.** 7.6 **41.** -2 **43.** 2.6 **45.** 0 **47.** 3π
49. 13π **51.** 19 **53.** -3 **55.** -15 **57.** 0 **59.** -52 **61.** -187 **63.** $\dfrac{7}{6}$ **65.** -4.49 **67.** $-\dfrac{3}{8}$ **69.** $-3x, -8y$
71. $12x, -5xy, -4$ **73.** $-6x$ **75.** $4 - 10y$ **77.** $5 - 7a$ **79.** $-4 - 9b$ **81.** $24 + 11x$ **83.** $3y - 8x$ **85.** 9 **87.** $\dfrac{7}{8}$
89. -2 **91.** 19,757 ft **93.** 1039 thousand jobs **95.** 268 thousand jobs **97.** 21°F **99.** 3°F
101. The maximum concentration is 0.05 and it occurs during the 3rd hour. **103.** 0.015 **105.** from 0 to 3 hr **113.** 711 yr
117.

```
        -4.5
 +--+--+--●--+--+--+--+--+--+--+--+-->
-8 -7 -6 -5 -4 -3 -2 -1  0  1  2
```

118. $10(4 + a)$ **119.** Answers will vary; -1 is an example.

Section 1.7

Check Point Exercises

1. a. -40 **b.** $-\dfrac{4}{21}$ **c.** 36 **d.** 5.5 **e.** 0 **2. a.** 24 **b.** -30 **3. a.** $\dfrac{1}{7}$ **b.** 8 **c.** $-\dfrac{1}{6}$ **d.** $-\dfrac{13}{7}$ **4. a.** -4 **b.** 8
5. a. 8 **b.** $-\dfrac{8}{15}$ **c.** -7.3 **d.** 0 **6. a.** $-20x$ **b.** $10x$ **c.** $-b$ **d.** $-21x + 28$ **e.** $-7y + 6$ **7.** $-y - 26$ **8. a.** $330
b. $60 **c.** $33

Exercise Set 1.7

1. -45 **3.** 24   **5.** -21 **7.** 19   **9.** 0   **11.** -12 **13.** 9   **15.** $\dfrac{12}{35}$ **17.** $-\dfrac{14}{27}$ **19.** -3.6 **21.** 0.12   **23.** 30   **25.** -72

27. 24   **29.** -27 **31.** 90   **33.** 0   **35.** $\dfrac{1}{4}$ **37.** 5   **39.** $-\dfrac{1}{10}$ **41.** $-\dfrac{5}{2}$ **43. a.** $-32 \cdot \left(\dfrac{1}{4}\right)$ **b.** -8 **45. a.** $-60 \cdot \left(-\dfrac{1}{5}\right)$ **b.** 12

47. -3 **49.** -7 **51.** 30   **53.** 0   **55.** undefined   **57.** -5 **59.** -12 **61.** 6   **63.** 0   **65.** undefined   **67.** -4.3 **69.** $\dfrac{5}{6}$

71. $-\dfrac{16}{9}$ **73.** -1 **75.** -15 **77.** $-10x$ **79.** $3y$ **81.** $9x$ **83.** $-4x$ **85.** $-b$ **87.** $3y$ **89.** $-8x + 12$ **91.** $6x - 12$

93. $-2y + 5$ **95.** $y - 14$ **97.** $4(-10) + 8 = -32$ **99.** $(-9)(-3) - (-2) = 29$ **101.** $\dfrac{-18}{-15 + 12} = 6$ **103.** $-6 - \left(\dfrac{12}{-4}\right) = -3$

105. 6.3 million; fairly well, but slight overestimation   **107. a.** 11 Latin words   **b.** 11 Latin words; it is the same.

109. a. \$2,000,000 (or \$200 tens of thousands)   **b.** \$8,000,000 (or \$800 tens of thousands)   **c.** Cost increases.   **119.** b   **121.** $5x$ **123.** $\dfrac{x}{12}$

127. $1.144x + 2.5$ **129.** -9 **130.** -3 **131.** 2

Section 1.8

Check Point Exercises

1. a. 36   **b.** -64 **c.** 1   **d.** -1 **2. a.** $21x^2$ **b.** $8x^3$ **c.** cannot simplify   **3.** 15   **4.** 32   **5. a.** 36   **b.** 12   **6.** $\dfrac{3}{4}$ **7.** -40

8. -31 **9.** $\dfrac{5}{7}$ **10.** -5 **11.** $7x^2 + 15$ **12.** For 40-year-old drivers, there are 200 accidents per 50 million miles driven.   **13.** $30°C$

Exercise Set 1.8

1. 81   **3.** 64   **5.** 16   **7.** -64 **9.** 625   **11.** -625 **13.** -100 **15.** $19x^2$ **17.** $15x^3$ **19.** $9x^4$ **21.** $-x^2$ **23.** x^3

25. cannot be simplified   **27.** 0   **29.** 25   **31.** 27   **33.** 12   **35.** 5   **37.** 45   **39.** -24 **41.** 300   **43.** 0   **45.** -32 **47.** 64

49. 30   **51.** $\dfrac{4}{3}$ **53.** 2   **55.** 2   **57.** 3   **59.** 88   **61.** -60 **63.** -36 **65.** 14   **67.** $-\dfrac{3}{4}$ **69.** $-\dfrac{9}{40}$ **71.** $-\dfrac{37}{36}$ or $-1\dfrac{1}{36}$

73. 24   **75.** 28   **77.** 9   **79.** -7 **81.** $15x - 27$ **83.** $15 - 3y$ **85.** $16y - 25$ **87.** $-2x^2 - 9$ **89.** $-10 - (-2)^3 = -2$

91. $[2(7 - 10)]^2 = 36$ **93.** $x - (5x + 8) = -4x - 8$ **95.** $5(x^3 - 4) = 5x^3 - 20$ **97.** 135 beats/min; (40, 135) on blue graph

99. \$287.4 billion; fairly well   **101.** 65 thousand; fairly well, but slight overestimation   **103.** $20°C$ **105.** $-30°C$ **113.** $-\dfrac{79}{4}$

115. $\left(2 \cdot 5 - \dfrac{1}{2} \cdot 10\right) \cdot 9 = 45$ **117.** Answers will vary.   **119.** 6   **120.** -24 **121.** Answers will vary; -3 is an example.

Review Exercises

1. $\dfrac{23}{7}$ **2.** $\dfrac{64}{11}$ **3.** $1\dfrac{8}{9}$ **4.** $5\dfrac{2}{5}$ **5.** $2 \cdot 2 \cdot 3 \cdot 5$ **6.** $3 \cdot 3 \cdot 7$ **7.** prime   **8.** $\dfrac{5}{11}$ **9.** $\dfrac{8}{15}$ **10.** $\dfrac{21}{50}$ **11.** $\dfrac{8}{3}$ **12.** $\dfrac{1}{4}$ **13.** $\dfrac{2}{3}$

14. $\dfrac{29}{18}$ **15.** $\dfrac{37}{60}$ **16.** $\dfrac{5}{12}$ **17.**   **18.**   **19.** 0.625   **20.** $0.\overline{27}$

21. a. $\sqrt{81}$ **b.** $0, \sqrt{81}$ **c.** $-17, 0, \sqrt{81}$ **d.** $-17, -\dfrac{9}{13}, 0, 0.75, \sqrt{81}$ **e.** $\sqrt{2}, \pi$ **f.** $-17, -\dfrac{9}{13}, 0, 0.75, \sqrt{2}, \pi, \sqrt{81}$

22. Answers will vary; -2 is an example.   **23.** Answers will vary; $\dfrac{1}{2}$ is an example.   **24.** Answers will vary; π is an example.   **25.** $<$ **26.** $>$

27. $>$ **28.** $<$ **29.** false   **30.** true   **31.** 58   **32.** 2.75

33. IV   **34.** IV   **35.** I   **36.** II

37. $A(5, 6)$; $B(-3, 0)$; $C(-5, 2)$; $D(-4, -2)$; $E(0, -5)$; $F(3, -1)$ **38.** 65   **39.** 80   **40.** 1989–1993   **41.** 1981–1985   **42.** 1977   **43.** 85

44. Finland, Germany, Austria   **45.** 73   **46.** 40   **47.** $13y + 7$ **48.** $(x + 7)9$ **49.** $(6 + 4) + y = 10 + y$ **50.** $(7 \cdot 10)x = 70x$

51. $24x - 12 + 30y$ **52.** $7a + 2$ **53.** $28x + 19$ **54.** 1800; The sale price at 25% off is \$1800 for a \$2400 computer.

55. 2   **56.** -3 **57.** $-\dfrac{11}{20}$ **58.** -7 **59.** $5y - 4x$ **60.** $40 - 2y$ **61.** 800 ft below sea level

62. The reservoir's level at the end of 5 months is 23 ft. **63.** $9 + (-13)$ **64.** 4 **65.** $-\dfrac{6}{5}$ **66.** -1.5 **67.** -7 **68.** -3

69. $-5 - 8a$ **70.** 27,150 ft **71.** 84 **72.** $-\dfrac{3}{11}$ **73.** -120 **74.** -9 **75.** undefined **76.** 2 **77.** $3x$ **78.** $-x - 1$ **79.** 36

80. -36 **81.** -32 **82.** $6x^3$ **83.** cannot be simplified **84.** -16 **85.** -16 **86.** 10 **87.** -2 **88.** 17 **89.** -88 **90.** 14

91. $-\dfrac{20}{3}$ **92.** $-\dfrac{2}{5}$ **93.** 6 **94.** 10 **95.** $28a - 20$ **96.** $6y - 12$ **97.** 13 lb; (4, 13) **98.** 16 lb; (6, 16)

99. 805 million; fairly well, but slight overestimation **100.** Sales increase from 1998 to 2000 and decrease from 2000 to 2002.

Chapter 1 Test

1. 4 **2.** -11 **3.** -51 **4.** $\dfrac{1}{5}$ **5.** $-\dfrac{35}{6}$ or $-5\dfrac{5}{6}$ **6.** -5 **7.** 1 **8.** -4 **9.** -32 **10.** 1 **11.** $4x + 4$ **12.** $13x - 19y$

13. $10 - 6x$ **14.** $-7, -\dfrac{4}{5}, 0, 0.25, \sqrt{4}, \dfrac{22}{7}$ **15.** $>$ **16.** 12.8 **17.** II;

18. $(-5, -2)$ **19.** -15 **20.** 150

21. $2(3 + x)$ **22.** $(-6 \cdot 4)x = -24x$

23. $35x - 7 + 14y$ **24.** $(30, 200)$; After 30 yr, there are 200 elk. **25.** 50

26. 725 million **27.** 16 sec **28.** 17,030 ft **29.** \$135 thousand **30.** \$134.7 thousand

CHAPTER 2

Section 2.1

Check Point Exercises

1. a. not a solution **b.** solution **2.** 17 or {17} **3.** 2.29 or {2.29} **4.** $\dfrac{1}{4}$ or $\left\{\dfrac{1}{4}\right\}$ **5.** 13 or {13} **6.** 12 or {12} **7.** 11 or {11} **8.** 2100 words

Exercise Set 2.1

1. {23} **3.** {−20} **5.** {−16} **7.** {−12} **9.** {4} **11.** {−11} **13.** {2} **15.** $\left\{-\dfrac{17}{12}\right\}$ **17.** $\left\{\dfrac{21}{4}\right\}$ **19.** $\left\{-\dfrac{11}{20}\right\}$

21. {4.3} **23.** $\left\{-\dfrac{21}{4}\right\}$ **25.** {18} **27.** $\left\{\dfrac{9}{10}\right\}$ **29.** {−310} **31.** {4.3} **33.** {0} **35.** {11} **37.** {5} **39.** {−13}

41. {6} **43.** {−12} **45.** $x = \triangle + \square$ **47.** $\triangle - \square = x$ **49.** $x - 12 = -2; 10$ **51.** $\dfrac{2}{5}x - 8 = \dfrac{7}{5}x; -8$ **53.** \$1700 **55.** 525,000 deaths

57. \$17 billion **65.** Answers will vary; example: $x - 100 = -101$. **67.** {2.7529} **68.** II **69.** -12 **70.** $6 - 9x$

Section 2.2

Check Point Exercises

1. 36 or {36} **2. a.** 21 or {21} **b.** -4 or {−4} **c.** -3.1 or {−3.1} **3. a.** 24 or {24} **b.** -16 or {−16}
4. a. -5 or {−5} **b.** 3 or {3} **5.** 6 or {6} **6.** -10 or {−10} **7.** 6 or {6} **8.** 2020

Exercise Set 2.2

1. {30} **3.** {−33} **5.** {7} **7.** {−9} **9.** $\left\{-\dfrac{7}{2}\right\}$ **11.** {6} **13.** $\left\{-\dfrac{3}{4}\right\}$ **15.** {0} **17.** {18} **19.** {−8} **21.** {−17}

23. {47} **25.** {45} **27.** {−5} **29.** {5} **31.** {6} **33.** {−1} **35.** {−2} **37.** $\left\{\dfrac{9}{4}\right\}$ **39.** {−6} **41.** {−3} **43.** {−3}

45. {4} **47.** $\left\{-\dfrac{3}{2}\right\}$ **49.** {2} **51.** {−4} **53.** {−6} **55.** $x = \square \cdot \triangle$ **57.** $-\triangle = x$ **59.** $6x = 10; \dfrac{5}{3}$ **61.** $\dfrac{x}{-9} = 5; -45$

63. 10 sec **65.** 1502.2 mph **67.** 60 yd **73.** d **75.** Answers will vary; example: $\dfrac{5}{4}x = -20$. **77.** {6.5} **78.** 100 **79.** -100 **80.** 3

Section 2.3

Check Point Exercises

1. 6 or {6} **2.** 2 or {2} **3.** 5 or {5} **4.** -2 or {−2} **5.** inconsistent, no solution; { } or ∅ **6.** all real numbers or $\mathbb{R}$ **7.** 124 lb

Exercise Set 2.3

1. $\{3\}$ **3.** $\{-1\}$ **5.** $\{4\}$ **7.** $\{4\}$ **9.** $\left\{\dfrac{7}{2}\right\}$ **11.** $\{-3\}$ **13.** $\{6\}$ **15.** $\{8\}$ **17.** $\{4\}$ **19.** $\{1\}$ **21.** $\{-4\}$ **23.** $\{5\}$

25. $\{6\}$ **27.** $\{1\}$ **29.** $\{-57\}$ **31.** $\{-10\}$ **33.** $\{18\}$ **35.** $\left\{\dfrac{7}{4}\right\}$ **37.** $\{1\}$ **39.** $\{24\}$ **41.** $\{-6\}$ **43.** $\{20\}$

45. $\{-7\}$ **47.** no solution; $\{\ \}$ or $\varnothing$ **49.** all real numbers; $\mathbb{R}$ **51.** $\left\{\dfrac{2}{3}\right\}$ **53.** all real numbers; $\mathbb{R}$ **55.** no solution; $\{\ \}$ or $\varnothing$ **57.** $\{0\}$

59. no solution; $\{\ \}$ or $\varnothing$ **61.** $\{0\}$ **63.** $\left\{\dfrac{4}{3}\right\}$ **65.** $x = \square\$ - \square\triangle$ **67.** 240 **69.** $\dfrac{x}{5} + \dfrac{x}{3} = 16; 30$ **71.** $\dfrac{3x}{4} - 3 = \dfrac{x}{2}; 12$

73. 85 mph **75.** 3.7; point (3.7, 10) on low-humor graph **77.** 409.2 ft **83.** c **85.** $\{3\}$ **87.** $\{-4.2\}$ **89.** $<$ **90.** $<$ **91.** -10

Section 2.4

Check Point Exercises

1. $l = \dfrac{A}{w}$ **2.** $l = \dfrac{P - 2w}{2}$ **3.** $m = \dfrac{T - D}{p}$ **4.** $x = 15 + 12y$ **5.** 2.3% **6. a.** 0.67 **b.** 2.5 **7.** 4.5 **8.** 15 **9.** 36%
10. 35% **11. a.** $1152 **b.** 4% decrease

Exercise Set 2.4

1. $r = \dfrac{d}{t}$ **3.** $P = \dfrac{I}{rt}$ **5.** $r = \dfrac{C}{2\pi}$ **7.** $m = \dfrac{E}{c^2}$ **9.** $m = \dfrac{y - b}{x}$ **11.** $p = \dfrac{T - D}{m}$ **13.** $b = \dfrac{2A}{h}$ **15.** $n = 5M$ **17.** $c = 4F - 160$

19. $a = 2A - b$ **21.** $r = \dfrac{S - P}{Pt}$ **23.** $b = \dfrac{2A}{h} - a$ **25.** $x = \dfrac{C - By}{A}$ **27.** 89% **29.** 0.2% **31.** 478% **33.** 10,000%

35. 0.27 **37.** 0.634 **39.** 1.7 **41.** 0.03 **43.** 0.005 **45.** 6 **47.** 7.2 **49.** 5 **51.** 170 **53.** 20% **55.** 12% **57.** 60%

59. 75% **61.** $x = \dfrac{y}{a + b}$ **63.** $x = \dfrac{y - 5}{a - b}$ **65.** $x = \dfrac{y}{c + d}$ **67.** $x = \dfrac{y + C}{A + B}$ **69. a.** $z = 3A - x - y$ **b.** 96% **71. a.** $t = \dfrac{d}{r}$
b. 2.5 hr **73.** 408 **75.** 2,369,200 **77.** 59% **79.** 12.5% **81.** $9 **83. a.** $1008 **b.** $17,808 **85. a.** $103.20 **b.** $756.80
87. 15% **89.** no; 2% loss **97.** d **99.** 1.5 sec and 60 ft **100.** $\{12\}$ **101.** $\{20\}$ **102.** 0.7x

Mid-Chapter 2 Check Point

1. $\{16\}$ **2.** $\{-3\}$ **3.** $C = \dfrac{825H}{E}$ **4.** 8.4 **5.** $\{30\}$ **6.** $\left\{-\dfrac{8}{9}\right\}$ **7.** $r = \dfrac{S}{2\pi h}$ **8.** 40 **9.** $\{20\}$ **10.** $\{-1\}$

11. $x = \dfrac{By + C}{A}$ **12.** no solution; $\{\ \}$ or $\varnothing$ **13.** 2008 **14.** $\{40\}$ **15.** 12.5% **16.** $\left\{-\dfrac{6}{5}\right\}$ **17.** 25% **18.** all real numbers; $\mathbb{R}$

Section 2.5

Check Point Exercises

1. a. $4x + 6$ **b.** $\dfrac{x - 4}{9}$ **2.** 12 **3.** MRI scan: $575; Acupuncture: $112 **4.** pages 96 and 97 **5.** 32; 4 mi **6.** 40 ft wide and 120 ft long **7.** $940

Exercise Set 2.5

1. $x + 7$ **3.** $25 - x$ **5.** $9 - 4x$ **7.** $\dfrac{83}{x}$ **9.** $2x + 40$ **11.** $9x - 93$ **13.** $8(x + 14)$ **15.** $x + 60 = 410; 350$ **17.** $x - 23 = 214; 237$

19. $7x = 126; 18$ **21.** $\dfrac{x}{19} = 5; 95$ **23.** $4 + 2x = 56; 26$ **25.** $5x - 7 = 178; 37$ **27.** $x + 5 = 2x; 5$ **29.** $2(x + 4) = 36; 14$

31. $9x = 3x + 30; 5$ **33.** $\dfrac{3x}{5} + 4 = 34; 50$ **35.** *Titanic*: $200 million; *Waterworld*: $175 million **37.** Liberals: 39.6%; Conservatives: 17.6%
39. pages 314 and 315 **41.** Rolling Stones: $121 million; Springsteen: $120 million **43.** 32 and 34 **45.** 800 mi **47.** 6 months **49.** 50 yd wide
and 200 yd long **51.** 160 ft wide and 360 ft long **53.** 12 ft long and 4 ft high **55.** $400 **57.** $67,000 **59.** $14,500 **61.** 11 hr **67.** d

69. 5 ft 7 in. **71.** The uncle is 60 years old and the woman is 20 years old. **73.** -20 **74.** 0 **75.** $w = \dfrac{3V}{lh}$

Section 2.6

Check Point Exercises

1. 12 ft **2.** 400π ft$^2 \approx 1256$ ft^2; 40π ft ≈ 126 ft **3.** large pizza **4.** 2 times **5.** No; About 32 more cubic inches are needed.
6. $120°, 40°, 20°$ **7.** $60°$

Exercise Set 2.6

1. 18 m; 18 m^2 **3.** 56 in.^{2} **5.** 91 m^2 **7.** 50 ft **9.** 8 ft **11.** 50 cm **13.** 16π cm$^2 \approx 50$ cm^2; 8π cm ≈ 25 cm
15. 36π yd$^2 \approx 113$ yd^2; 12π yd ≈ 38 yd **17.** 7 in.; 14 in. **19.** 36 in.^{3} **21.** 150π cm$^3 \approx 471$ cm^3 **23.** 972π cm$^3 \approx 3054$ cm^3

25. 48π m$^3 \approx 151$ m^3 **27.** $h = \dfrac{V}{\pi r^2}$ **29.** 9 times **31.** $50°, 50°, 80°$ **33.** $4x = 76; 3x + 4 = 61; 2x + 5 = 43; 76°, 61°, 43°$

35. $40°, 80°, 60°$ **37.** $32°$ **39.** $2°$ **41.** $48°$ **43.** $90°$ **45.** $75°$ **47.** $135°$ **49.** $50°$ **51.** 72 m^2 **53.** 70.5 cm^2 **55.** 168 cm^3
57. $698.18 **59.** large pizza **61.** $2262 **63.** approx 19.7 ft **65.** $21,000 \text{ yd}^3$ **67.** the can with diameter of 6 in. and height of 5 in.
69. Yes, the water tank is a little over one cubic foot too small. **81.** 2.25 times **83.** Volume increases 8 times. **85.** $35°$
86. $s = \dfrac{P - b}{2}$ **87.** $\{8\}$ **88.** 0

Section 2.7

Check Point Exercises

1. a. [number line: open circle at 4, shaded left] **b.** [number line: closed circle at -2, shaded right] **c.** [number line: closed circle at -4 and open circle at 1, shaded between]

2. $\{x \mid x < 3\}$ [number line: open circle at 3, shaded left] **3.** $\{x \mid x \geq -2\}$ [number line: closed circle at -2, shaded right]

4. a. $\{x \mid x < 8\}$ [number line: open circle at 8, shaded left] **b.** $\{x \mid x > -3\}$ [number line: open circle at -3, shaded right]

5. $\{y \mid y \geq 4\}$ [number line: closed circle at 4, shaded right] **6.** $\{x \mid x \geq 1\}$ [number line: closed circle at 1, shaded right]

7. $\{x \mid x \geq 1\}$ [number line: closed circle at 1, shaded right] **8.** no solution; $\emptyset$ **9.** $\{x \mid x \text{ is a real number}\}$ or $\mathbb{R}$ **10.** at least 83%

Exercise Set 2.7

1. [number line: open circle at 5, shaded left] **3.** [number line: open circle at -3, shaded left] **5.** [number line: closed circle at -4, shaded right]

7. [number line: closed circle at 4, shaded left] **9.** [number line: open circle at -2, shaded right] **11.** [number line: open circle at -1, shaded right]

13. $\{x \mid x > -2\}$ **15.** $\{x \mid x \geq 4\}$ **17.** $\{x \mid x \geq 3\}$

19. $\{x \mid x > 7\}$ [number line: open circle at 7, shaded right] **21.** $\{x \mid x \leq 6\}$ [number line: closed circle at 6, shaded left]

23. $\{y \mid y < 2\}$ [number line: open circle at 2, shaded left] **25.** $\{x \mid x \leq 3\}$ [number line: closed circle at 3, shaded left]

27. $\{x \mid x < 16\}$ [number line: open circle at 16, shaded left] **29.** $\{x \mid x > 4\}$ [number line: open circle at 4, shaded right]

31. $\left\{x \mid x > \dfrac{7}{6}\right\}$ [number line: open circle at $\frac{7}{6}$, shaded right] **33.** $\left\{y \mid y \leq -\dfrac{3}{8}\right\}$ [number line: closed circle at $-\frac{3}{8}$, shaded left]

35. $\{y \mid y > 0\}$ [number line: open circle at 0, shaded right] **37.** $\{x \mid x < 8\}$ [number line: open circle at 8, shaded left]

39. $\{x \mid x > -6\}$ [number line: open circle at -6, shaded right] **41.** $\{x \mid x < 5\}$ [number line: open circle at 5, shaded left]

43. $\{x \mid x \geq -7\}$ [number line: closed circle at -7, shaded right] **45.** $\{x \mid x > -5\}$ [number line: open circle at -5, shaded right]

47. $\{x \mid x \leq -5\}$ [number line: closed circle at -5, shaded left] **49.** $\{x \mid x < 3\}$ [number line: open circle at 3, shaded left]

51. $\left\{y \mid y \geq -\dfrac{1}{8}\right\}$ [number line: closed circle at $-\frac{1}{8}$, shaded right] **53.** $\{x \mid x > -4\}$ [number line: open circle at -4, shaded right]

55. $\{x \mid x > 5\}$ [number line: open circle at 5, shaded right] **57.** $\{x \mid x < 5\}$ [number line: open circle at 5, shaded left]

59. $\{x \mid x \geq -2\}$ [number line: closed circle at -2, shaded right] **61.** $\{x \mid x > -3\}$ [number line: open circle at -3, shaded right]

63. $\{x \mid x \geq 4\}$ [number line: closed circle at 4, shaded right] **65.** $\left\{x \mid x > \dfrac{11}{3}\right\}$ [number line: open circle at $\frac{11}{3}$, shaded right]

67. $\{y \mid y > 2\}$ [number line: open circle at 2, shaded right] **69.** $\{y \mid y < 2\}$ [number line: open circle at 2, shaded left]

71. $\{x|x < 3\}$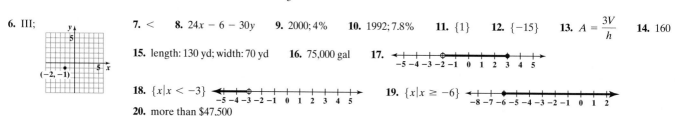

73. $\left\{x|x > \dfrac{5}{3}\right\}$

75. $\{x|x \geq 9\}$

77. $\{x|x < -6\}$

79. no solution; $\emptyset$ **81.** $\{x|x$ is a real number$\}$ or $\mathbb{R}$ **83.** no solution; $\emptyset$ **85.** $\{x|x$ is a real number$\}$ or $\mathbb{R}$ **87.** $\{x|x \leq 0\}$ **89.** $x > \dfrac{b - a}{3}$

91. $\dfrac{y - b}{m} \geq x$ **93.** x is between -2 and 2; $|x| < 2$ **95.** x is greater than 2 or less than -2; $|x| > 2$ **97.** Denmark, Netherlands, Norway

99. Japan, Mexico   **101.** Netherlands, Norway, Canada, U.S.   **103.** 20 yr; from 2009 onward   **105. a.** at least 96   **b.** if you get less than 66 on the final   **107.** up to 1280 mi   **109.** up to 29 bags of cement   **111.** An open dot indicates an endpoint that is not a solution and a closed dot indicates an endpoint that is a solution.   **117.** more than 720 mi   **119.** $\{x|x < 0.4\}$ **121.** 20   **122.** length: 11 in.; width: 6 in.   **123.** $\{4\}$

Review Exercises

1. $\{32\}$ **2.** $\{-22\}$ **3.** $\{12\}$ **4.** $\{-22\}$ **5.** $\{5\}$ **6.** $\{80\}$ **7.** $\{-56\}$ **8.** $\{11\}$ **9.** $\{4\}$ **10.** $\{-15\}$ **11.** $\{-12\}$
12. $\{-25\}$ **13.** $\{10\}$ **14.** $\{6\}$ **15.** $\{-5\}$ **16.** $\{-10\}$ **17.** $\{2\}$ **18.** $\{1\}$ **19.** 16 yr; 2006   **20.** $\{-1\}$ **21.** $\{12\}$
22. $\{-13\}$ **23.** $\{-3\}$ **24.** $\{-10\}$ **25.** $\{2\}$ **26.** $\{2\}$ **27.** no solution or $\emptyset$ **28.** all real numbers or $\mathbb{R}$ **29.** 20 yr old
30. $r = \dfrac{I}{P}$ **31.** $h = \dfrac{3V}{B}$ **32.** $w = \dfrac{P - 2l}{2}$ **33.** $B = 2A - C$ **34.** $m = \dfrac{T - D}{P}$ **35.** 72%   **36.** 0.35%   **37.** 0.65   **38.** 1.5
39. 0.03   **40.** 9.6   **41.** 200   **42.** 48%   **43.** 100%   **44.** 40%   **45.** 12.5%   **46.** no; 1%   **47. a.** $h = 7r$ **b.** 5 ft 3 in.
48. 350 gallons   **49.** 10   **50.** New York: 55 days; Los Angeles: 213 days   **51.** pages 46 and 47   **52.** Streisand: 49 albums; Madonna: 47 albums
53. 9 yr; 2012   **54.** 18 checks   **55.** length: 150 yd; width: 50 yd   **56.** $240   **57.** 32.5 ft^2 **58.** 50 cm^2 **59.** 135 yd^2
60. 20π m ≈ 63 m; 100π m$^2 \approx 314$ m^2 **61.** 6 ft   **62.** 156 ft^2 **63.** $1890   **64.** medium pizza   **65.** 60 cm^3 **66.** 128π yd$^3 \approx 402$ yd^3
67. 288π m$^3 \approx 905$ m^3 **68.** 4800 m^3 **69.** 16 fish   **70.** $x = 30, 3x = 90, 2x = 60$; $30°, 60°, 90°$ **71.** $85°, 35°, 60°$ **72.** $33°$ **73.** $105°$

74. $57.5°$ **75.** $45°$ and $135°$ **76.**   **77.**   **78.** $\{x|x > 4\}$

79. $\{x|x \leq -3\}$ **80.** $\{x|x < 4\}$ **81.** $\{x|x > -8\}$

82. $\{x|x \geq -3\}$ **83.** $\{x|x > 6\}$

84. $\{x|x \geq 4\}$ **85.** $\{x|x \leq 2\}$ **86.** $\{x|x$ is a real number$\}$ or $\mathbb{R}$

87. no solution; $\emptyset$ **88.** at least 64   **89.** no more than 99 min

Chapter 2 Test

1. $\left\{\dfrac{9}{2}\right\}$ **2.** $\{-5\}$ **3.** $\left\{-\dfrac{4}{3}\right\}$ **4.** $\{2\}$ **5.** $\{-20\}$ **6.** $\left\{-\dfrac{5}{3}\right\}$ **7.** 60 yr; 2020   **8.** $h = \dfrac{V}{\pi r^2}$ **9.** $w = \dfrac{P - 2l}{2}$ **10.** 8.4   **11.** 150   **12.** 5%
13. 63   **14.** fitness trainer: $50,950; teacher: $28,080   **15.** 600 min   **16.** length: 150 yd; width: 75 yd   **17.** $35   **18.** 517 m^2 **19.** 525 in.^{2}
20. 18 in.^{3} **21.** 175π cm$^3 \approx 550$ cm^3 **22.** $650   **23.** 14 ft   **24.** $126°, 42°, 12°$ **25.** $53°$ **26.**

27.   **28.** $\{x|x \leq -1\}$ **29.** $\{x|x < -6\}$

30. $\{x|x \leq -3\}$ **31.** $\{x|x > 7\}$ **32.** at least 92
33. widths greater than 8 inches

Cumulative Review Exercises (Chapters 1–2)

1. -4 **2.** -2 **3.** -128 **4.** $-103 - 20x$ **5.** $-4, -\dfrac{1}{3}, 0, \sqrt{4}, 1063$

6. III;   **7.** $<$ **8.** $24x - 6 - 30y$ **9.** 2000; 4%   **10.** 1992; 7.8%   **11.** $\{1\}$ **12.** $\{-15\}$ **13.** $A = \dfrac{3V}{h}$ **14.** 160

15. length: 130 yd; width: 70 yd   **16.** 75,000 gal   **17.**

18. $\{x|x < -3\}$ **19.** $\{x|x \geq -6\}$

20. more than $47,500

CHAPTER 3

Section 3.1

Check Point Exercises

1. a. solution   **b.** not a solution   **2.** $(-2, -4), (-1, -1), (0, 2), (1, 5),$ and $(2, 8)$

3.   **4.**   **5.**

6.

7. a.

$y = 2x$		$y = 10 + x$	
x	(x, y)	x	(x, y)
0	$(0, 0)$	0	$(0, 10)$
2	$(2, 4)$	2	$(2, 12)$
4	$(4, 8)$	4	$(4, 14)$
6	$(6, 12)$	6	$(6, 16)$
8	$(8, 16)$	8	$(8, 18)$
10	$(10, 20)$	10	$(10, 20)$
12	$(12, 24)$	12	$(12, 22)$

b.

c. $(10, 20)$; If the bridge is used 10 times in a month, the total monthly cost without the coupon book is the same as the monthly cost with the coupon book, namely $20.

Exercise Set 3.1

1. $(2, 3)$ and $(3, 2)$ are not solutions; $(-4, -12)$ is a solution.   **3.** $(-5, -20)$ is not a solution; $(0, 0)$ and $(9, -36)$ are solutions.

5. $(2, -2)$ is not a solution; $(0, 6)$ and $(-3, 0)$ are solutions.   **7.** $(0, 5)$ is not a solution; $(-5, 6)$ and $(10, -3)$ are solutions.

9. $\left(1, \frac{1}{3}\right)$ is not a solution; $(0, 0)$ and $\left(2, -\frac{2}{3}\right)$ are solutions.   **11.** $(3, 4)$ and $(0, -4)$ are not solutions; $(4, 7)$ is a solution.

13.

x	(x, y)
-2	$(-2, -24)$
-1	$(-1, -12)$
0	$(0, 0)$
1	$(1, 12)$
2	$(2, 24)$

15.

x	(x, y)
-2	$(-2, 20)$
-1	$(-1, 10)$
0	$(0, 0)$
1	$(1, -10)$
2	$(2, -20)$

17.

x	(x, y)
-2	$(-2, -21)$
-1	$(-1, -13)$
0	$(0, -5)$
1	$(1, 3)$
2	$(2, 11)$

19.

x	(x, y)
-2	$(-2, 13)$
-1	$(-1, 10)$
0	$(0, 7)$
1	$(1, 4)$
2	$(2, 1)$

21.

23.   **25.**   **27.**   **29.**   **31.**

33.   **35.**   **37.**   **39.**   **41.**

43.   **45.**   **47.**   **49.**   **51.**

53.

55. a. $8x + 6y = 14.50$ **b.** $1.25   **57. a.**

African Americans	
x	(x, y)
0	(0, 5.2)
10	(10, 7.6)
20	(20, 10)
30	(30, 12.4)

b.

Whites	
x	(x, y)
0	(0, 3.4)
10	(10, 5.6)
20	(20, 7.8)
30	(30, 10)

c.

Hispanics	
x	(x, y)
0	(0, 4.2)
10	(10, 5.5)
20	(20, 6.8)
30	(30, 8.1)

59. a. 40; Both rental companies have the same cost when the truck is used for 40 miles.   **b.** 55
c. 54; Both rental companies have the same cost of \$54 when the truck is used for 40 miles.

61. a.

x	(x, y)
0	(0, 30,000)
10	(10, 30,500)
20	(20, 31,000)
30	(30, 31,500)
40	(40, 32,000)

b.

67.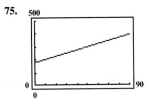

69. a. $(0, 0.6), (1, 0.3), (2, 0.2), (3, 0.3), (4, 0.6), (5, 1.1)$ **b.** between 10:00 A.M. and noon

71.

Answers will vary.

73.

Answers will vary.

75.

U.S. population increases.

76. 2   **77.** 1   **78.** $h = \dfrac{3V}{A}$

Section 3.2

Check Point Exercises

1. a. x-intercept is -3; y-intercept is 5.   **b.** y-intercept is 4; no x-intercept.   **c.** x-intercept is 0; y-intercept is 0.   **2.** 3   **3.** -4

4.   **5.**   **6.**   **7.**   **8.**

Exercise Set 3.2

1. a. 3   **b.** 4   **3. a.** -4 **b.** -2 **5. a.** 0   **b.** 0   **7. a.** no x-intercept   **b.** -2 **9.** x-intercept is 10; y-intercept is 4.

11. x-intercept is $\dfrac{15}{2}$, or $7\dfrac{1}{2}$; y-intercept is -5.   **13.** x-intercept is 8; y-intercept is $-\dfrac{8}{3}$, or $-2\dfrac{2}{3}$.   **15.** x-intercept is 0; y-intercept is 0.

17. x-intercept is $-\dfrac{11}{2}$, or $-5\dfrac{1}{2}$; y-intercept is $\dfrac{11}{3}$, or $3\dfrac{2}{3}$.

19.   **21.**   **23.**   **25.**   **27.**

29.   **31.**   **33.**   **35.**   **37.**

39.
$2x - 3y = -11$
$(-4, 1)$
$\left(-\frac{11}{2}, 0\right)$
$\left(0, \frac{11}{3}\right)$

41. $y = 3$ **43.** $x = -3$ **45.** $y = 0$ **47.**
$y = 4$
$(-2, 4)$ $(0, 4)$ $(3, 4)$

49.
$y = -2$
$(0, -2)$
$(-2, -2)$ $(3, -2)$

51. $x = 2$
$(2, 3)$
$(2, 0)$
$(2, -2)$

53. $x + 1 = 0$
$(-1, 3)$
$(-1, 0)$
$(-1, -2)$

55. $y - 3.5 = 0$
$(0, 3.5)$
$(-2, 3.5)$ $(3, 3.5)$

57. $x = 0$
$(0, 3)$
$(0, 0)$
$(0, -2)$

59. $3y = 9$
$(0, 3)$
$(-2, 3)$ $(3, 3)$

61. $12 - 3x = 0$
$(4, 3)$
$(4, 0)$
$(4, -2)$

63. Exercise 4 **65.** Exercise 7 **67.** Exercise 1 **69.**
$x + y = 11$
$(0, 11)$
$(11, 0)$

71. from 3 to 12 sec **73.** 45; The vulture was 45 m above the ground when the observation started.
75. 12, 13, 14, 15, 16; The vulture is on the ground at this time.

77. 8:00 A.M. to 11:00 A.M. **79.** 11:00 A.M. to 1:00 P.M. **81. a.** 80; In 1994, carbonated beverages had 80% of the market share. **b.** very well **c.** 140; In 2134, there will be no consumption of carbonated beverages. Model breakdown: model extended too far into the future. **93.** $y = 6$
95. -6; 3 **97.**

x-intercept: 2;
y-intercept: 4

99.

x-intercept: 15;
y-intercept: 10

101. 13.4 **102.** $4x + 5$

103.

Section 3.3

Check Point Exercises

1. a. 6 **b.** $-\frac{7}{5}$ **2. a.** 0 **b.** undefined **3.** Both slopes equal 2, so the lines are parallel. **4.** $\frac{2}{15} \approx 0.13$; The slope indicates that the number of U.S. men living alone is projected to increase by 0.13 million each year. The rate of change is 0.13 million men per year.

Exercise Set 3.3

1. $\frac{3}{4}$; rises **3.** $\frac{1}{4}$; rises **5.** 0; horizontal **7.** -5; falls **9.** undefined; vertical **11.** $\frac{1}{2}$ **13.** $-\frac{1}{3}$
15. $-\frac{1}{2}$ **17.** $-\frac{2}{3}$ **19.** 0 **21.** undefined **23.** parallel **25.** not parallel

27.
$m = 2$
$m = \frac{1}{3}$
$m = 0$
$m = -\frac{1}{2}$
$m = -1$

29. Slopes of corresponding opposite sides are equal: $-\frac{2}{5}$ and $\frac{4}{3}$. **31.** 2 **33.** collinear
35. 250; The amount spent online per U.S. online household increased by $250 each year from 1999 to 2001.
37. -0.4; The percentage of men is decreasing by 0.4% per year. **39.** 2900; Mean income is increasing $2900 per year.
41. 0.40; Cost increases by $0.40 per mile driven. **43.** pitch $= \frac{1}{3}$ **45.** 8.3% **53.** b

55. b_2, b_1, b_4, b_3 **57.** -3 **59.** $\frac{3}{4}$ **61.** 12 in. and 24 in. **62.** -42 **63.** $\{x | x \leq 4\}$;

Section 3.4

Check Point Exercises

1. a. 5; -3 **b.** $\frac{2}{3}$; 4 **c.** -7; 6 **2.** $y = 3x - 2$
$(1, 1)$
$(0, -2)$

3. $y = \frac{3}{5}x + 1$
$(0, 1)$ $(5, 4)$

4. $3x + 4y = 0$
$(0, 0)$
$(4, -3)$

Exercise Set 3.4

1. $3; 2$ **3.** $3; -5$ **5.** $-\dfrac{1}{2}; 5$ **7.** $7; 0$ **9.** $0; 10$ **11.** $-1; 4$ **13.** $y = 5x + 7; 5; 7$ **15.** $y = -x + 6; -1; 6$ **17.** $y = -6x; -6; 0$

19. $y = 2x; 2; 0$ **21.** $y = -\dfrac{2}{7}x; -\dfrac{2}{7}; 0$ **23.** $y = -\dfrac{3}{2}x + \dfrac{3}{2}; -\dfrac{3}{2}; \dfrac{3}{2}$ **25.** $y = \dfrac{3}{4}x - 3; \dfrac{3}{4}; -3$

27. $y = 2x + 4$ **29.** $y = -3x + 5$ **31.** $y = \dfrac{1}{2}x + 1$ **33.** $y = \dfrac{2}{3}x - 5$ **35.** $y = -\dfrac{3}{4}x + 2$

37. $y = -\dfrac{5}{3}x$ **39. a.** $y = -3x$ **b.** $-3; 0$ **41. a.** $y = \dfrac{4}{3}x$ **b.** $\dfrac{4}{3}; 0$ **43. a.** $y = -2x + 3$ **b.** $-2; 3$

c. $y = -3x$ **c.** $y = \dfrac{4}{3}x$ **c.** $y = -2x + 3$

45. a. $y = -\dfrac{7}{2}x + 7$ **b.** $-\dfrac{7}{2}; 7$ **c.** $y = -\dfrac{7}{2}x + 7$ **47.** **49.**

parallel; The slopes are equal.   not parallel; The slopes are not equal.

51.

53. $y = -3x + 5$ **55.** $y = -x + 2$ **57.** $y = x$ **59. a.** 10; The number of endangered species in the United States has been increasing by 10 species per year. **b.** 166; In 1980, there were 166 endangered animal species in the United States. **c.** $166; 216; 266; 316; 366; 396$; Answers will vary. **61. a.** 21; In 1997, 21 million people in sub-Saharan Africa were living with AIDS.
b. 1.6; Number of people living with AIDS in sub-Saharan Africa is increasing by 1.6 million per year.
c. $y = 1.6x + 21$ **d.** 35.4 million **67.** $F = \dfrac{9}{5}C + 32$ **69.** 8 **70.** 0 **71.** 56

parallel; The slopes are equal.

Mid-Chapter 3 Check Point

1. a. 4 **b.** 2 **c.** $-\dfrac{1}{2}$ **2. a.** -5 **b.** no y-intercept **c.** undefined slope **3. a.** 0 **b.** 0 **c.** $\dfrac{3}{5}$

4. $y = -2x$ **5.** $y = -2$ **6.** $x + y = -2$ **7.** $y = \dfrac{1}{3}x - 2$ **8.** $x = 3.5$

9. $4x - 2y = 8$ **10.** $y = 3x + 2$ **11.** $3x + y = 0$ **12.** $y = x^2 - 4$ **13.** $y = x - 4$

14. $5y = -3x$ **15.** $5y = 20$ **16.** $m = \dfrac{5}{2}$; y-intercept is -5. **17.** parallel; Both have slopes of $\dfrac{4}{5}$.

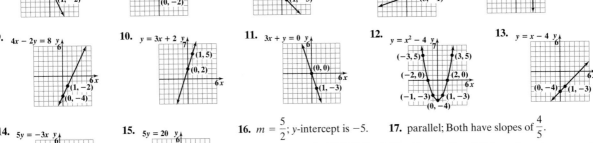

18. a. 33 **b.** In 1995, 33% of U.S. colleges offered distance learning. **c.** 7.8
d. For the years 1995 through 2002, the percentage of colleges that offered distance learning increased at a rate of 7.8% per year.

Section 3.5
Check Point Exercises

1. $y + 5 = 6(x - 2); y = 6x - 17$ **2. a.** $y + 1 = -5(x + 2),$ or $y + 6 = -5(x + 1)$ **b.** $y = -5x - 11$
3. $y - 5 = 3(x + 2); y = 3x + 11$ or $f(x) = 3x + 11$ **4.** 3   **5.** $y = 0.28x + 27.2; 41.2$

Exercise Set 3.5

1. $y - 5 = 3(x - 2); y = 3x - 1$ **3.** $y - 6 = 5(x + 2); y = 5x + 16$ **5.** $y + 2 = -8(x + 3); y = -8x - 26$

7. $y - 0 = -12(x + 8); y = -12x - 96$ **9.** $y + 2 = -1\left(x + \dfrac{1}{2}\right); y = -x - \dfrac{5}{2}$ **11.** $y - 0 = \dfrac{1}{2}(x - 0); y = \dfrac{1}{2}x$

13. $y + 2 = -\dfrac{2}{3}(x - 6); y = -\dfrac{2}{3}x + 2$ **15.** $y - 2 = 2(x - 1),$ or $y - 10 = 2(x - 5); y = 2x$

17. $y - 0 = 1(x + 3),$ or $y - 3 = 1(x - 0); y = x + 3$ **19.** $y + 1 = 1(x + 3),$ or $y - 4 = 1(x - 2); y = x + 2$

21. $y + 1 = \dfrac{5}{7}(x + 4),$ or $y - 4 = \dfrac{5}{7}(x - 3); y = \dfrac{5}{7}x + \dfrac{13}{7}$ **23.** $y + 1 = 0(x + 3),$ or $y + 1 = 0(x - 4); y = -1$

25. $y - 4 = 1(x - 2),$ or $y - 0 = 1(x + 2); y = x + 2$ **27.** $y - 0 = 8\left(x + \dfrac{1}{2}\right),$ or $y - 4 = 8(x - 0); y = 8x + 4$ **29. a.** 5   **b.** $-\dfrac{1}{5}$

31. a. -7 **b.** $\dfrac{1}{7}$ **33. a.** $\dfrac{1}{2}$ **b.** -2 **35. a.** $-\dfrac{2}{5}$ **b.** $\dfrac{5}{2}$ **37. a.** -4 **b.** $\dfrac{1}{4}$ **39. a.** $-\dfrac{1}{2}$ **b.** 2   **41. a.** $\dfrac{2}{3}$ **b.** $-\dfrac{3}{2}$

43. a. undefined   **b.** 0   **45.** $y - 2 = 2(x - 4); y = 2x - 6$ **47.** $y - 4 = -\dfrac{1}{2}(x - 2); y = -\dfrac{1}{2}x + 5$

49. $y + 10 = -4(x + 8); y = -4x - 42$ **51.** $y + 3 = -5(x - 2); y = -5x + 7$ **53.** $y - 2 = \dfrac{2}{3}(x + 2); y = \dfrac{2}{3}x + \dfrac{10}{3}$

55. $y + 7 = -2(x - 4); y = -2x + 1$ **57.** $y = 3x - 2$ **59.** $y = -5x - 20$ **61.** $y = 5$ **63.** $y = -\dfrac{1}{2}x + 1$ **65.** $y = -\dfrac{2}{3}x - 2$ **67.** $-\dfrac{A}{B}$

69. a. $y - 162 = 1(x - 2),$ or $y - 168 = 1(x - 8)$ **b.** $y = x + 160$ **c.** 180 lb

71. Answers will vary. Example: $y - 3 = -\dfrac{2}{3}(x - 12); y = -\dfrac{2}{3}x + 11; 6.3.$

73. a–d. Answers will vary.   **77.** c

81. a.

b.

c. $a = -0.6867924528; b = 11.01132075; r = -0.9214983162;$

d.
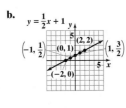

82. at most 12 sheets of paper   **83.** $1, \sqrt{4}$ **84.**

Review Exercises

1. $(-3, 3)$ is not a solution; $(0, 6)$ and $(1, 9)$ are solutions.   **2.** $(0, 4)$ and $(-1, 15)$ are not solutions; $(4, 0)$ is a solution.

3. a.

x	(x, y)
-2	(-2, -7)
-1	(-1, -5)
0	(0, -3)
1	(1, -1)
2	(2, 1)

b.

4. a.

x	(x, y)
-2	(-2, 0)
-1	$\left(-1, \dfrac{1}{2}\right)$
0	(0, 1)
1	$\left(1, \dfrac{3}{2}\right)$
2	(2, 2)

b.

5.

6. a.

x	(x, y)
10	$(10, 9)$
12	$(12, 19)$
14	$(14, 29)$
16	$(16, 39)$

b. Answers will vary.

7. a. -2 **b.** -4

8. a. no x-intercept **b.** 2

9. a. 0 **b.** 0

10.

11.

12.

13.

14.

15.

16.

17.

18. a. 5:00 P.M.; $-4°$F. **b.** 8:00 P.M.; $16°$F **c.** 4 and 6; At 4:00 P.M. and 6:00 P.M, the temperature was $0°$F. **d.** 12; At noon, the temperature was $12°$F.

e. The temperature stayed the same, $12°$F. **19.** $-\dfrac{1}{2}$; falls **20.** 3; rises **21.** 0; horizontal **22.** undefined; vertical **23.** $\dfrac{3}{5}$

24. undefined **25.** $-\dfrac{1}{3}$ **26.** 0 **27.** not parallel **28.** parallel **29. a.** $m = -44$; The number of new AIDS diagnoses decreased at a rate of 44 each year from 1999 to 2001. **b.** $m = 909$; The number of new AIDS diagnoses increased at a rate of 909 each year from 2001 to 2003.

c. $m = 432.5$; yes; Answers will vary. **30.** 5; -7 **31.** -4; 6 **32.** 0; 3 **33.** $-\dfrac{2}{3}$; 2

34.

35.

36.

37.

38.

39.

Yes, they are parallel since both have slopes of $-\dfrac{1}{2}$ and different y-intercepts.

40. a. 25; In 1990, the average age of U.S. Hispanics was 25.
b. 0.3; The average age for U.S. whites increased at a rate of about 0.3 each year from 1990 to 2000.
c. $y = 0.3x + 35$ **d.** 41 years old
41. $y - 7 = 6(x + 4)$; $y = 6x + 31$
42. $y - 4 = 3(x - 3)$, or $y - 1 = 3(x - 2)$; $y = 3x - 5$
43. $y + 7 = -3(x - 4)$; $y = -3x + 5$
44. $y - 6 = -3(x + 2)$; $y = -3x$
45. a. $y - 1.5 = 0.95(x - 1)$, or $y - 3.4 = 0.95(x - 3)$ **b.** $y = 0.95x + 0.55$ **c.** \$10.05 billion

Chapter 3 Test

1. $(-2, 1)$ is not a solution; $(0, -5)$ and $(4, 3)$ are solutions.

2.

x	(x, y)
-2	$(-2, -5)$
-1	$(-1, -2)$
0	$(0, 1)$
1	$(1, 4)$
2	$(2, 7)$

3. (graph $y = x^2 - 1$)

4. a. 2 **b.** -3 **5.** (graph $4x - 2y = -8$) **6.** (graph $y = 4$)

7. 3; rises **8.** undefined; vertical **9.** $\dfrac{3}{2}$ **10.** parallel **11.** -1; 10 **12.** -2; 6

13. (graph $y = \dfrac{2}{3}x - 1$) **14.** (graph $y = -2x + 3$) **15.** $y - 4 = -2(x + 1)$;
$y = -2x + 2$
16. $y - 1 = 3(x - 2)$, or $y + 8 = 3(x + 1)$;
$y = 3x - 5$
17. $y - 3 = 2(x + 2)$; $y = 2x + 7$

18. $y + 4 = -\dfrac{1}{2}(x - 6)$; $y = -\dfrac{1}{2}x - 1$
19. 106; Spending per pupil increased at a rate of about \$106 each year.
20. a. $y - 2.3 = 0.3(x - 1)$ or
$y - 3.2 = 0.3(x - 4)$
b. $y = 0.3x + 2$
c. 4.4 million

Cumulative Review Exercises (Chapters 1–3)

1. 2   **2.** $4 - 6x$ **3.** $\sqrt{5}$ **4.** $\{19\}$ **5.** $\left\{\dfrac{5}{4}\right\}$ **6.** $x = \dfrac{y - b}{m}$ **7.** 800   **8.** 40 mph   **9.** $\{x \mid x \le -2\}$

10. $\{x \mid x < 0\}$ **11.** 39   **12.** $<$ **13.** 6° F   **14.** 2015   **15.** width: 53 m; lenght: 120 m

16. 200 lb   **17.** 6 hr   **18.** $40°, 60°, 80°$ **19.**   **20.**

CHAPTER 4

Section 4.1

Check Point Exercises

1. a. solution   **b.** not a solution   **2.** $(1, 4)$ or $\{(1, 4)\}$ **3.** $(3, 3)$ or $\{(3, 3)\}$ **4.** no solution or $\varnothing$
5. infinitely many solutions; $\{(x, y) \mid x + y = 3\}$ or $\{(x, y) \mid 2x + 2y = 6\}$

Exercise Set 4.1

1. solution   **3.** solution   **5.** not a solution   **7.** solution   **9.** not a solution   **11.** $\{(4, 2)\}$ **13.** $\{(-1, 2)\}$ **15.** $\{(3, 0)\}$
17. $\{(1, 0)\}$ **19.** $\{(-1, 4)\}$ **21.** $\{(2, 4)\}$ **23.** $\{(2, -1)\}$ **25.** no solution; $\varnothing$ **27.** $\{(-2, 6)\}$
29. infinitely many solutions; $\{(x, y) \mid x - 2y = 4\}$ or $\{(x, y) \mid 2x - 4y = 8\}$ **31.** $\{(2, 3)\}$ **33.** no solution; $\varnothing$

35. infinitely many solutions; $\{(x, y) \mid x - y = 0\}$ or $\{(x, y) \mid y = x\}$ **37.** $\{(2, 4)\}$ **39.** no solution; $\varnothing$ **41.** no solution; $\varnothing$
43. $m = \dfrac{1}{2}; b = -3$ and -5; no solution   **45.** $m = -\dfrac{1}{2}$ and 3; $b = 4$; one solution   **47.** $m = 3; b = -6$; infinite number of solutions
49. $m = -3; b = 0$ and 1; no solution   **51. a.** Answers will vary. Example: (1996, 41); Mothers 30 years and older had 41 thousand births in 1996.
b. There are more births to mothers 30 years and older.   **61.** c   **67.** $\{(6, -1)\}$ **69.** $\{(3, 0)\}$ **71.** $\{(2, -1)\}$ **73.** $\{(-4, 4)\}$
74. -12 **75.** 6   **76.** 27

Section 4.2

Check Point Exercises

1. $(3, 2)$ or $\{(3, 2)\}$ **2.** $(1, -2)$ or $\{(1, -2)\}$ **3.** no solution or $\varnothing$
4. infinitely many solutions; $\{(x, y) \mid y = 3x - 4\}$ or $\{(x, y) \mid 9x - 3y = 12\}$ **5.** $\$30$; 400 units

Exercise Set 4.2

1. $\{(1, 3)\}$ **3.** $\{(5, 1)\}$ **5.** $\{(2, 1)\}$ **7.** $\{(-1, 3)\}$ **9.** $\{(4, 5)\}$ **11.** $\left\{\left(-\dfrac{2}{5}, -\dfrac{11}{5}\right)\right\}$ **13.** no solution or $\varnothing$

15. infinitely many solutions; $\{(x, y) \mid y = 3x - 5\}$ or $\{(x, y) \mid 21x - 35 = 7y\}$ **17.** $\{(0, 0)\}$ **19.** $\left\{\left(\dfrac{17}{7}, -\dfrac{8}{7}\right)\right\}$ **21.** no solution; $\varnothing$

23. $\left\{\left(\dfrac{43}{5}, -\dfrac{1}{5}\right)\right\}$ **25.** $\{(200, 700)\}$ **27.** $\{(7, 3)\}$ **29.** $\{(-1, -1)\}$ **31.** $\{(5, 4)\}$ **33.** $x + y = 81, y = x + 41$; 20 and 61

35. $x - y = 5, 4x = 6y$; 10 and 15   **37.** $x - y = 1, x + 2y = 7$; 2 and 3   **39.** $(2, 8)$ **41. a.** 6500 sold; 6200 supplied   **b.** $\$50$; 6250 tickets
43. 2700 gal   **45.** 2032; about 0.6 deaths per 1000 live births or less than one death per 1000 live births   **53.** $x = 1, y = -3, z = 5$

55.   **56.** 12   **57.** $-73, 0, \dfrac{3}{1}$

Section 4.3

Check Point Exercises

1. $(7, -2)$ or $\{(7, -2)\}$ **2.** $(6, 2)$ or $\{(6, 2)\}$ **3.** $(2, -1)$ or $\{(2, -1)\}$ **4.** $\left(\dfrac{60}{17}, -\dfrac{11}{17}\right)$ or $\left\{\left(\dfrac{60}{17}, -\dfrac{11}{17}\right)\right\}$ **5.** no solution or $\varnothing$
6. infinitely many solutions; $\{(x, y) \mid x - 5y = 7\}$ or $\{(x, y) \mid 3x - 15y = 21\}$

Exercise Set 4.3

1. $\{(4, -7)\}$ **3.** $\{(3, 0)\}$ **5.** $\{(-3, 5)\}$ **7.** $\{(3, 1)\}$ **9.** $\{(2, 1)\}$ **11.** $\{(-2, 2)\}$ **13.** $\{(-7, -1)\}$ **15.** $\{(2, 0)\}$

17. $\{(1, -2)\}$ **19.** $\{(-1, 1)\}$ **21.** $\{(3, 1)\}$ **23.** $\{(-5, -2)\}$ **25.** $\left\{\left(\dfrac{11}{12}, -\dfrac{7}{6}\right)\right\}$ **27.** $\left\{\left(\dfrac{23}{16}, \dfrac{3}{8}\right)\right\}$ **29.** no solution; $\emptyset$

31. infinitely many solutions; $\{(x, y) | x + 3y = 2\}$ or $\{(x, y) | 3x + 9y = 6\}$ **33.** no solution; $\emptyset$ **35.** $\left\{\left(\dfrac{1}{2}, -\dfrac{1}{2}\right)\right\}$

37. infinitely many solutions; $\{(x, y) | x = 5 - 3y\}$ or $\{(x, y) | 2x + 6y = 10\}$ **39.** $\left\{\left(\dfrac{1}{3}, 1\right)\right\}$ **41.** $\{(-10, 21)\}$ **43.** $\{(0, 1)\}$ **45.** $\{(2, -1)\}$

47. $\{(1, -3)\}$ **49.** $\{(4, 3)\}$ **51.** no solution; $\emptyset$ **53.** infinitely many solutions; $\{(x, y) | 2(x + 2y) = 6\}$ or $\{(x, y) | 3(x + 2y - 3) = 0\}$

55. $\{(3, 2)\}$ **57.** $\{(-1, 2)\}$ **59.** $\{(-1, 0)\}$ **61.** $\{(3, 1)\}$ **63.** dependent   **65.** 2002; 48% for and 48% against

73. $x = -a - b; y = -2a - b$ **77.** $\{(5, 1)\}$ **79.** 10   **80.** II   **81.** 26

Mid-Chapter 4 Check Point

1. $\{(2, 0)\}$ **2.** $\{(1, 1)\}$ **3.** no solution or $\emptyset$ **4.** $\{(5, 8)\}$ **5.** $\{(2, -1)\}$ **6.** $\{(2, 9)\}$ **7.** $\{(-2, -3)\}$

8. $\{(10, -1)\}$ **9.** no solution or $\emptyset$ **10.** $\{(-12, -1)\}$ **11.** $\{(-7, -23)\}$ **12.** $\left\{\left(\dfrac{16}{17}, -\dfrac{12}{17}\right)\right\}$

13. infinitely many solutions; $\{(x, y) | y - 2x = 7\}$ or $\{(x, y) | 4x = 2y - 14\}$ **14.** $\{(7, 11)\}$ **15.** $\{(6, 10)\}$

Section 4.4

Check Point Exercises

1. A bustard weighs 46 lb; a condor weighs 27 lb.   **2.** A Quarter Pounder has 420 cal; a Whopper with cheese has 589 cal.   **3.** 17.5 yr; $24,250   **4.** $3150 at 9%; $1850 at 11%   **5.** 12% solution: 100 oz; 20% solution: 60 oz   **6.** boat: 35 mph; current: 7 mph

Exercise Set 4.4

1. $x + y = 17, x - y = -3$; 7 and 10   **3.** $3x - y = -1, x + 2y = 23$; 3 and 10   **5.** 5.8 million lb of potato chips; 4.6 million lb of tortilla chips
7. pan pizza: 1120 calories; beef burrito: 430 calories   **9.** scrambled eggs: 366 mg; Double Beef Whopper: 175 mg   **11.** sweater: $12; shirt: $10
13. a. 300 min; $35   **b.** plan B; Answers will vary.   **15.** $600 of merchandise; $580   **17.** 26 yr; 2011; college grads: $1158; high school grads: $579
19. 2 servings of macaroni and 4 servings of broccoli   **21.** $2000 at 6% and $5000 at 8%   **23.** first fund: $8000; second fund: $6000
25. $17,000 at 12%; $3000 at a 5% loss   **27.** California: 100 gal; French: 100 gal   **29.** 18-karat gold: 96 grams; 12-karat gold: 204 grams
31. cheaper candy: 30 lb; more expensive candy: 45 lb   **33.** 8 nickels and 7 dimes   **35.** plane: 130 mph; wind: 30 mph
37. crew: 6 km/hr; current: 2 km/hr   **39.** in still water: 4.5 mph; current: 1.5 mph   **45.** 10 birds and 20 lions

47. There are 5 people downstairs and 7 people upstairs.   **50.** $-\dfrac{1}{4}$ **51.** $-\dfrac{11}{20}$ **52.**

Section 4.5

Check Point Exercises

1. $(-1) - 2(-4) + 3(5) = 22; 2(-1) - 3(-4) - 5 = 5; 3(-1) + (-4) - 5(5) = -32$ **2.** $(1, 4, -3)$ or $\{(1, 4, -3)\}$
3. $(4, 5, 3)$ or $\{(4, 5, 3)\}$ **4.** $y = 3x^2 - 12x + 13$

Exercise Set 4.5
1. not a solution   **3.** solution   **5.** $\{(2, 3, 3)\}$ **7.** $\{(2, -1, 1)\}$ **9.** $\{(1, 2, 3)\}$ **11.** $\{(3, 1, 5)\}$ **13.** $\{(1, 0, -3)\}$ **15.** $\{(1, -5, -6)\}$

17. no solution or $\emptyset$ **19.** infinitely many solutions; dependent equations   **21.** $\left\{\left(\dfrac{1}{2}, \dfrac{1}{3}, -1\right)\right\}$ **23.** $y = 2x^2 - x + 3$ **25.** $y = 2x^2 + x - 5$

27. 7, 4, and 5   **29.** $\{(4, 8, 6)\}$ **31.** $y = -\dfrac{3}{4}x^2 + 6x - 11$ **33.** $\left(\dfrac{8}{a}, -\dfrac{3}{b}, -\dfrac{5}{c}\right)$ **35. a.** $(0, 13.6), (70, 4.7), (102, 11.5)$

b. $0a + 0b + c = 13.6; 4900a + 70b + c = 4.7; 10{,}404a + 102b + c = 11.5$ **37. a.** $y = -16x^2 + 40x + 200$ **b.** 0; After 5 seconds, the ball hits the ground.
39. Carnegie: $100 billion; Vanderbilt: $96 billion; Gates: $60 billion   **41.** $1200 at 8%; $2000 at 10%; $3500 at 12%
43. 200 $8 tickets; 150 $10 tickets; 50 $12 tickets   **45.** 4 oz of food A; 0.5 oz of food B; 1 oz of food C
55. c   **57.** $60°, 55°, 65°$ **59.** $f(x) = -\dfrac{3}{4}x + 3$ **60.** $-2x + y = 6$ **61.** $f(x) = -5$

Review Exercises

1. solution **2.** not a solution **3.** no; $(-1, 3)$ does not satisfy $2x + y = -5$. **4.** $\{(4, -2)\}$ **5.** $\{(3, -2)\}$ **6.** $\{(2, 0)\}$ **7.** $\{(2, 1)\}$
8. $\{(4, -1)\}$ **9.** no solution; $\varnothing$ **10.** infinitely many solutions; $\{(x, y)|2x - 4y = 8\}$ or $\{(x, y)|x - 2y = 4\}$ **11.** no solution; $\varnothing$
12. $\{(-2, -6)\}$ **13.** $\{(2, 5)\}$ **14.** no solution; $\varnothing$ **15.** $\{(2, -1)\}$ **16.** $\{(5, 4)\}$ **17.** $\{(-2, -1)\}$ **18.** $\{(1, -4)\}$ **19.** $\{(20, -21)\}$
20. infinitely many solutions; $\{(x, y)|4x + y = 5\}$ or $\{(x, y)|12x + 3y = 15\}$ **21.** no solution; $\varnothing$ **22.** $\{(4, 18)\}$ **23.** $\left\{\left(-1, -\dfrac{1}{2}\right)\right\}$
24. \$12.50; 250 copies **25.** $\{(2, 4)\}$ **26.** $\{(-1, -1)\}$ **27.** $\{(2, -1)\}$ **28.** $\{(3, 2)\}$ **29.** $\{(2, 1)\}$ **30.** $\{(0, 0)\}$ **31.** $\left\{\left(\dfrac{17}{7}, -\dfrac{15}{7}\right)\right\}$
32. no solution; $\varnothing$ **33.** infinitely many solutions; $\{(x, y)|3x - 4y = -1\}$ or $\{(x, y)|-6x + 8y = 2\}$ **34.** $\{(4, -2)\}$ **35.** $\{(-8, -6)\}$
36. $\{(-4, 1)\}$ **37.** $\left\{\left(\dfrac{5}{2}, 3\right)\right\}$ **38.** $\{(3, 2)\}$ **39.** $\left\{\left(\dfrac{1}{2}, -2\right)\right\}$ **40.** no solution; $\varnothing$ **41.** $\{(3, 7)\}$ **42.** Japan: 73.6 yr; Switzerland: 72.8 yr
43. gorilla: 485 lb; orangutan: 165 lb **44.** shrimp: 42 mg; scallops: 15 mg **45.** 9 ft by 5 ft **46.** room: \$80; car: \$60 **47.** 200 min; \$25
48. \$2500 at 4%; \$6500 at 7% **49.** 10 ml of 34%; 90 ml of 4% **50.** plane: 630 mph; wind: 90 mph **51.** no **52.** $\{(0, 1, 2)\}$
53. $\{(2, 1, -1)\}$ **54.** infinitely many solutions; dependent equations **55.** $y = 3x^2 - 4x + 5$ **56.** per dog: \$129; per horse: \$97; per cat: \$81

Chapter 4 Test

1. solution **2.** not a solution **3.** $\{(2, 4)\}$ **4.** $\{(2, 4)\}$ **5.** $\{(1, -3)\}$ **6.** $\{(2, -3)\}$ **7.** no solution; $\varnothing$ **8.** $\{(-1, 4)\}$
9. $\{(-4, 3)\}$ **10.** infinitely many solutions; $\{(x, y)|3x - 2y = 2\}$ or $\{(x, y)|-9x + 6y = -6\}$ **11.** Mary: 2.6%; Patricia: 1.1%
12. 500 min; \$40 **13.** \$2000 at 6% and \$7000 at 7% **14.** 6% solution: 12 oz; 9% solution: 24 oz **15.** boat: 14 mph; current: 2 mph
16. $(1, 3, 2)$ or $\{(1, 3, 2)\}$

Cumulative Review Exercises (Chapters 1–4)

1. -36 **2.** $17x - 11$ **3.** $\{-6\}$ **4.** $\{20\}$ **5.** $t = \dfrac{A - P}{Pr}$ **6.** $\{x|x > 2\}$;

7. **8.** **9.**

10. $\{(0, -2)\}$ **11.** $\left\{\left(\dfrac{3}{2}, -2\right)\right\}$ **12.** 1
13. $y - 6 = -4(x + 1)$; $y = -4x + 2$ **14.** 10 ft
15. pen: \$0.80; pad: \$1.20 **16.** $-93, 0, \dfrac{7}{1}, \sqrt{100}$ **17.** 20%
18. one computer; The percentage of households with one computer is increasing faster than the percentage for multiple computers.
19. 8 yr, 2005 **20.** 2012

CHAPTER 5

Section 5.1

Check Point Exercises

1. $5x^3 + 4x^2 - 8x - 20$ **2.** $5x^3 + 4x^2 - 8x - 20$ **3.** $7x^2 + 11x + 4$ **4.** $7x^3 + 3x^2 + 12x - 8$ **5.** $3y^3 - 10y^2 - 11y - 8$
6. approximately 4 per thousand; approximately $(40, 4)$

Exercise Set 5.1

1. binomial, 1 **3.** binomial, 3 **5.** monomial, 2 **7.** monomial, 0 **9.** trinomial, 2 **11.** trinomial, 4 **13.** binomial, 3
15. monomial, 23 **17.** $-8x + 13$ **19.** $12x^2 + 15x - 9$ **21.** $10x^2 - 12x$ **23.** $5x^2 - 3x + 13$ **25.** $4y^3 + 10y^2 + y - 2$
27. $3x^3 + 2x^2 - 9x + 7$ **29.** $-2y^3 + 4y^2 + 13y + 13$ **31.** $-3y^6 + 8y^4 + y^2$ **33.** $10x^3 + 1$ **35.** $-\dfrac{2}{5}x^4 + x^3 - \dfrac{1}{8}x^2$
37. $0.01x^5 + x^4 - 0.1x^3 + 0.3x + 0.33$ **39.** $11y^3 - 3y^2$ **41.** $-2x^2 - x + 1$ **43.** $-\dfrac{1}{4}x^4 - \dfrac{7}{15}x^3 - 0.3$ **45.** $-y^3 + 8y^2 - 3y - 14$
47. $-5x^3 - 6x^2 + x - 4$ **49.** $7x^4 - 2x^3 + 4x - 2$ **51.** $8x^2 + 7x - 5$ **53.** $9x^3 - 4.9x^2 + 11.1$ **55.** $-2x - 10$ **57.** $-5x^2 - 9x - 12$
59. $-5x^2 - x$ **61.** $-4x^2 - 4x - 6$ **63.** $-2y - 6$ **65.** $6y^3 + y^2 + 7y - 20$ **67.** $n^3 + 2$ **69.** $y^6 - y^3 - y^2 + y$
71. $26x^4 + 9x^2 + 6x$ **73.** $\dfrac{5}{7}x^3 - \dfrac{9}{20}x$ **75.** $4x + 6$ **77.** $10x^2 - 7$ **79.** $-4y^2 - 7y + 5$ **81.** $9x^3 + 11x^2 - 8$
83. $-y^3 + 8y^2 + y + 14$ **85.** $7x^4 - 2x^3 + 3x^2 - x + 2$ **87.** $0.05x^3 + 0.02x^2 + 1.02x$ **89.** $x^2 + 12x$ **91.** $y^2 - 19y + 16$
93. $2x^3 + 3x^2 + 7x - 5$ **95.** $-10y^3 + 2y^2 + y + 3$ **97.** 5 billion, 7.25 billion, 17 billion, 25.25 billion, and 5 billion; 3 days; 4 days
99. 2029 cigarettes; Answers will vary. **101. a.** 73% **b.** 72.8% **c.** 73.3%; Answers will vary. **103. a.** 42 human years
b. 41.8 human years **105.** 3 dog years **117.** $-3x^2 - x - 2$ **120.** -10 **121.** 5.6 **122.** $\{-4\}$

Section 5.2

Check Point Exercises

1. a. 2^6 or 64   **b.** x^{10} **c.** y^8 **d.** y^9 **2. a.** 3^{20} **b.** x^{90} **c.** $(-5)^{21}$ **3. a.** $16x^4$ **b.** $-64y^6$ **4. a.** $70x^3$ **b.** $-20x^9$
5. a. $3x^2 + 15x$ **b.** $30x^5 - 12x^3 + 18x^2$ **6. a.** $x^2 + 9x + 20$ **b.** $10x^2 - 29x - 21$ **7.** $5x^3 - 18x^2 + 7x + 6$
8. $6x^5 - 19x^4 + 22x^3 - 8x^2$

Exercise Set 5.2

1. x^{18} **3.** y^{12} **5.** x^{11} **7.** 7^{19} **9.** 6^{90} **11.** x^{45} **13.** $(-20)^9$ **15.** $8x^3$ **17.** $25x^2$ **19.** $16x^6$ **21.** $16y^{24}$ **23.** $-32x^{35}$
25. $14x^2$ **27.** $24x^3$ **29.** $-15y^7$ **31.** $\frac{1}{8}a^5$ **33.** $-48x^7$ **35.** $4x^2 + 12x$ **37.** $x^2 - 3x$ **39.** $2x^2 - 12x$ **41.** $-12y^2 - 20y$
43. $4x^3 + 8x^2$ **45.** $2y^4 + 6y^3$ **47.** $6y^4 - 8y^3 + 14y^2$ **49.** $6x^4 + 8x^3$ **51.** $-2x^3 - 10x^2 + 6x$ **53.** $12x^4 - 3x^3 + 15x^2$
55. $x^2 + 8x + 15$ **57.** $2x^2 + 9x + 4$ **59.** $x^2 - 2x - 15$ **61.** $x^2 - 2x - 99$ **63.** $2x^2 + 3x - 20$ **65.** $\frac{3}{16}x^2 + \frac{11}{4}x - 4$
67. $x^3 + 3x^2 + 5x + 3$ **69.** $y^3 - 6y^2 + 13y - 12$ **71.** $2a^3 - 9a^2 + 19a - 15$ **73.** $x^4 + 3x^3 + 5x^2 + 7x + 4$
75. $4x^4 - 4x^3 + 6x^2 - \frac{17}{2}x + 3$ **77.** $x^4 + x^3 + x^2 + 3x + 2$ **79.** $x^3 + 3x^2 - 37x + 24$ **81.** $2x^3 - 9x^2 + 27x - 27$
83. $2x^4 + 9x^3 + 6x^2 + 11x + 12$ **85.** $12z^4 - 14z^3 + 19z^2 - 22z + 8$ **87.** $21x^5 - 43x^4 + 38x^3 - 24x^2$
89. $4y^6 - 2y^5 - 6y^4 + 5y^3 - 5y^2 + 8y - 3$ **91.** $x^4 + 6x^3 - 11x^2 - 4x + 3$ **93.** $2x - 2$ **95.** $15x^5 + 42x^3 - 8x^2$
97. $2y^3$ **99.** $16y + 32$ **101.** $2x^2 + 7x - 15$ ft^2 **103. a.** $(2x + 1)(x + 2)$ **b.** $2x^2 + 5x + 2$ **c.** $(2x + 1)(x + 2) = 2x^2 + 5x + 2$
115. $8x + 16$ **117.** $-8x^4$ **118.** $\{x | x < -1\}$ **119.** $3x - 2y = 6$ **120.** $-\frac{2}{3}$

Section 5.3

Check Point Exercises

1. $x^2 + 11x + 30$ **2.** $28x^2 - x - 15$ **3.** $6x^2 - 22x + 20$ **4. a.** $49y^2 - 64$ **b.** $16x^2 - 25$ **c.** $4a^6 - 9$ **5. a.** $x^2 + 20x + 100$
b. $25x^2 + 40x + 16$ **6. a.** $x^2 - 18x + 81$ **b.** $49x^2 - 42x + 9$

Exercise Set 5.3

1. $x^2 + 10x + 24$ **3.** $y^2 - 4y - 21$ **5.** $2x^2 + 7x - 15$ **7.** $4y^2 - y - 3$ **9.** $10x^2 - 9x - 9$ **11.** $12y^2 - 43y + 35$
13. $-15x^2 - 32x + 7$ **15.** $6y^2 - 28y + 30$ **17.** $15x^4 - 47x^2 + 28$ **19.** $-6x^2 + 17x - 10$ **21.** $x^3 + 5x^2 + 3x + 15$
23. $8x^5 + 40x^3 + 3x^2 + 15$ **25.** $x^2 - 9$ **27.** $9x^2 - 4$ **29.** $9r^2 - 16$ **31.** $9 - r^2$ **33.** $25 - 49x^2$ **35.** $4x^2 - \frac{1}{4}$ **37.** $y^4 - 1$
39. $r^6 - 4$ **41.** $1 - y^8$ **43.** $x^{20} - 25$ **45.** $x^2 + 4x + 4$ **47.** $4x^2 + 20x + 25$ **49.** $x^2 - 6x + 9$ **51.** $9y^2 - 24y + 16$
53. $16x^4 - 8x^2 + 1$ **55.** $49 - 28x + 4x^2$ **57.** $4x^2 + 2x + \frac{1}{4}$ **59.** $16y^2 - 2y + \frac{1}{16}$ **61.** $x^{16} + 6x^8 + 9$ **63.** $x^3 - 1$
65. $x^2 - 2x + 1$ **67.** $9y^2 - 49$ **69.** $12x^4 + 3x^3 + 27x^2$ **71.** $70y^2 + 2y - 12$ **73.** $x^4 + 2x^2 + 1$ **75.** $x^4 + 3x^2 + 2$
77. $x^4 - 16$ **79.** $4 - 12x^5 + 9x^{10}$ **81.** $\frac{3}{16}x^4 + 7x^2 - 96$ **83.** $x^2 + 2x + 1$ **85.** $4x^2 - 9$ **87.** $6x + 22$
89. $16x^4 - 72x^2 + 81$ **91.** $16x^4 - 1$ **93.** $x^3 + 6x^2 + 12x + 8$ **95.** $x^2 + 6x + 9 - y^2$ **97.** $(x + 1)(x + 2)$ yd^2 **99.** 56 yd^2; (6, 56)
101. $(x^2 + 4x + 4)$ in^2 **109.** $(x - 10)$ and $(x + 2)$ **111.** $x^2 + 2x$ **113.** Change $x^2 + 2x + 4$ to $x^2 + 4x + 4$.
115. Graphs coincide.   **116.** $(2, -1)$ **117.** $(1, 1)$ **118.** $\left\{\left(\frac{3}{2}, -1, \frac{1}{2}\right)\right\}$

Section 5.4

Check Point Exercises

1. -9 **2.** polynomial degree: 9;

Term	Coefficient	Degree
$8x^4y^5$	8	9
$-7x^3y^2$	-7	5
$-x^2y$	-1	3
$-5x$	-5	1
11	11	0

3. $2x^2y + 2xy - 4$ **4.** $3x^3 + 2x^2y + 5xy^2 - 10$ **5.** $60x^5y^5$ **6.** $60x^5y^7 - 12x^3y^3 + 18xy^2$
7. a. $21x^2 - 25xy + 6y^2$ **b.** $4x^2 + 16xy + 16y^2$ **8. a.** $36x^2y^4 - 25x^2$ **b.** $x^3 - y^3$

Exercise Set 5.4

1. 1 **3.** -47 **5.** -6 **7.** polynomial degree: 9;

Term	Coefficient	Degree
x^3y^2	1	5
$-5x^2y^7$	-5	9
$6y^2$	6	2
-3	-3	0

9. $7x^2y - 4xy$ **11.** $2x^2y + 13xy + 13$ **13.** $-11x^4y^2 - 11x^2y^2 + 2xy$ **15.** $-5x^3 + 8xy - 9y^2$ **17.** $x^4y^2 + 8x^3y + y - 6x$
19. $5x^3 + x^2y - xy^2 - 4y^3$ **21.** $-3x^2y^2 + xy^2 + 5y^2$ **23.** $8a^2b^4 + 3ab^2 + 8ab$ **25.** $-30x + 37y$ **27.** $40x^3y^2$ **29.** $-24x^5y^9$
31. $45x^2y + 18xy^2$ **33.** $50x^3y^2 - 15xy^3$ **35.** $28a^3b^5 + 8a^2b^3$ **37.** $-a^2b + ab^2 - b^3$ **39.** $7x^2 + 38xy + 15y^2$ **41.** $2x^2 + xy - 21y^2$
43. $15x^2y^2 + xy - 2$ **45.** $4x^2 + 12xy + 9y^2$ **47.** $x^2y^2 - 6xy + 9$ **49.** $x^4 + 2x^2y^2 + y^4$ **51.** $x^4 - 4x^2y^2 + 4y^4$ **53.** $9x^2 - y^2$
55. $a^2b^2 - 1$ **57.** $x^2 - y^4$ **59.** $9a^4b^2 - a^2$ **61.** $9x^2y^4 - 16y^2$ **63.** $a^3 - ab^2 + a^2b - b^3$ **65.** $x^3 + 4x^2y + 4xy^2 + y^3$
67. $x^3 - 4x^2y + 4xy^2 - y^3$ **69.** $x^2y^2 - a^2b^2$ **71.** $x^6y + x^4y + x^4 + 2x^2 + 1$ **73.** $x^4y^4 - 6x^2y^2 + 9$ **75.** $x^2 + 2xy + y^2 - 1$
77. $3x^2 + 8xy + 5y^2$ **79.** $2xy + y^2$ **81.** $x^{12}y^{12} - 2x^6y^6 + 1$ **83.** $x^4y^4 - 18x^2y^2 + 81$ **85.** $x^2 - y^2 - 2yz - z^2$
87. no; need 120 more board feet **89.** 192 ft **91.** 0 ft; The ball hits the ground. **93.** 2.5 to 6 sec **95.** $(2, 192)$
97. 2.5 sec; 196 ft **101.** c **103.** $5x^2 - 16y^2$ **105.** $W = \dfrac{2R - L}{3}$ **106.** 3.8 **107.** $y - 5 = 3(x + 2); y = 3x + 11$

Mid-Chapter 5 Check Point

1. $-55x^4y^6$ **2.** $6x^2y^3$ **3.** $12x^2 - x - 35$ **4.** $-x + 12$ **5.** $2x^3 - 11x^2 + 17x - 5$ **6.** $x^2 - x - 4$ **7.** $64x^2 - 48x + 9$ **8.** $70x^9$
9. $x^4 - 4$ **10.** $x^4 + 4x^2 + 4$ **11.** $18a^2 - 11ab - 10b^2$ **12.** $70x^5 - 14x^3 + 21x^2$ **13.** $5a^2b^3 + 2ab - b^2$ **14.** $18y^2 - 50$ **15.** $2x^3 - x^2 + x$
16. $10x^2 - 5xy - 3y^2$ **17.** $-4x^5 + 7x^4 - 10x + 23$ **18.** $x^3 + 27y^3$ **19.** $10x^7 - 5x^4 + 8x^3 - 4$ **20.** $y^2 - 12yz + 36z^2$ **21.** $-21x^2 + 7$

Section 5.5

Check Point Exercises

1. a. 5^8 **b.** x^7 **c.** y^{19} **2. a.** 1 **b.** 1 **c.** -1 **d.** 20 **e.** 1 **3. a.** $\dfrac{x^2}{25}$ **b.** $\dfrac{x^{12}}{8}$ **c.** $\dfrac{16a^{40}}{b^{12}}$

4. a. $-2x^8$ **b.** $\dfrac{1}{5}$ **c.** $3x^5y^3$ **5.** $-5x^7 + 2x^3 - 3x$ **6.** $5x^6 - \dfrac{7}{5}x + 2$ **7.** $3x^6y^4 - xy + 10$

Exercise Set 5.5

1. 3^{15} **3.** x^4 **5.** y^8 **7.** $5^3 \cdot 2^4$ **9.** $x^{75}y^{40}$ **11.** 1 **13.** 1 **15.** -1 **17.** 100 **19.** 1 **21.** 0 **23.** -2 **25.** $\dfrac{x^2}{9}$ **27.** $\dfrac{x^6}{64}$
29. $\dfrac{4x^6}{25}$ **31.** $-\dfrac{64}{27a^9}$ **33.** $-\dfrac{32a^{35}}{b^{20}}$ **35.** $\dfrac{x^8y^{12}}{16z^4}$ **37.** $3x^5$ **39.** $-2x^{20}$ **41.** $-\dfrac{1}{2}y^3$ **43.** $\dfrac{7}{5}y^{12}$ **45.** $6x^5y^4$ **47.** $-\dfrac{1}{2}x^{12}$ **49.** $\dfrac{9}{7}$
51. $-\dfrac{1}{10}x^8y^9z^4$ **53.** $5x^4 + x^3$ **55.** $2x^3 - x^2$ **57.** $y^6 - 9y + 1$ **59.** $-8x^2 + 5x$ **61.** $6x^3 + 2x^2 + 3x$ **63.** $3x^3 - 2x^2 + 10x$
65. $4x - 6$ **67.** $-6z^2 - 2z$ **69.** $4x^2 + 3x - 1$ **71.** $5x^4 - 3x^2 - x$ **73.** $-9x^3 + \dfrac{9}{2}x^2 - 10x + 5$ **75.** $4xy + 2x - 5y$

77. $-4x^5y^3 + 3xy + 2$ **79.** $4x^2 - x + 6$ **81.** $-xy^2$ **83.** $y + 5$ **85.** $3x^{12n} - 6x^{9n} + 2$ **87. a.** $\dfrac{6t^4 - 207t^3 + 2128t^2 - 6622t + 15{,}220}{28t^4 - 711t^3 + 5963t^2 - 1695t + 27{,}424}$

b. No; the divisor is not a monomial. **97.** $18x^8 - 27x^6 + 36x^4$ **99.** $\dfrac{3x^{14} - 6x^{12} - 9x^7}{-3x^7}$ **100.** 20.3 **101.** 0.875 **102.**

$y = \dfrac{1}{3}x + 2$

Section 5.6

Check Point Exercises

1. $x + 5$ **2.** $4x - 3 - \dfrac{3}{2x + 3}$ **3.** $x^2 + x + 1$ **4.** $2x^2 + 7x + 14 + \dfrac{21x - 10}{x^2 - 2x}$ **5.** $x^2 - 2x - 3$

Exercise Set 5.6

1. $x + 4$ **3.** $2x + 5$ **5.** $x - 2$ **7.** $2y + 1$ **9.** $x - 5 + \dfrac{14}{x + 2}$ **11.** $y + 3 + \dfrac{4}{y + 2}$ **13.** $x^2 - 5x + 2$ **15.** $6y - 1$

17. $2a + 3$ **19.** $y^2 - y + 2$ **21.** $3x + 5 - \dfrac{5}{2x - 5}$ **23.** $x^2 + 2x + 8 + \dfrac{13}{x - 2}$ **25.** $2y^2 + y + 1 + \dfrac{6}{2y + 3}$

27. $2y^2 - 3y + 2 + \dfrac{1}{3y + 2}$ **29.** $9x^2 + 3x + 1$ **31.** $y^3 - 9y^2 + 27y - 27$ **33.** $2y + 4 + \dfrac{4}{2y - 1}$ **35.** $y^3 + y^2 - y - 1 + \dfrac{4}{y - 1}$

37. $4x^2 + 3x - 8 + \dfrac{18}{x^2 + 3}$ **39.** $5x^2 + x + 3 + \dfrac{x + 7}{3x^2 - 1}$ **41.** $2x + 5$ **43.** $3x - 8 + \dfrac{20}{x + 5}$ **45.** $4x^2 + x + 4 + \dfrac{3}{x - 1}$

47. $6x^4 + 12x^3 + 22x^2 + 48x + 93 + \dfrac{187}{x - 2}$ **49.** $x^3 - 10x^2 + 51x - 260 + \dfrac{1300}{5 + x}$ **51.** $3x^2 + 3x - 3$ **53.** $x^4 + x^3 + 2x^2 + 2x + 2$

55. $x^3 + 4x^2 + 16x + 64$ **57.** $2x^4 - 7x^3 + 15x^2 - 31x + 64 - \dfrac{129}{x + 2}$ **59.** $x^3 - x^2y + xy^2 - y^3 + \dfrac{2y^4}{x + y}$ **61.** $3x^2 + 2x - 1$

63. $4x - 7 + \dfrac{4x + 8}{x^2 + x + 1}$ **65.** $x^3 + x^2 - x - 3 + \dfrac{-x + 5}{x^2 - x + 2}$ **67.** $4x^2 + 5xy - y^2$ **69.** $x^2 + 2x + 3$ **71.** $x^2 + 2x + 3$ units

73. a. $\dfrac{30{,}000x^3 - 30{,}000}{x - 1}$ **b.** $30{,}000x^2 + 30{,}000x + 30{,}000$ **c.** \$94,575 **79.** b **81.** -3 **83.** Graphs coincide.

85. $2x + 3$ should be $2x + 23 + \dfrac{130}{x - 5}$. **87.** $x^2 - 2x + 3$ should be $x^2 + 2x + 3$. **88.** $\{(5, 3)\}$ **89.** 1.2 **90.** -6

Section 5.7

Check Point Exercises

1. a. $\dfrac{1}{6^2} = \dfrac{1}{36}$ **b.** $\dfrac{1}{5^3} = \dfrac{1}{125}$ **c.** $\dfrac{1}{(-3)^4} = \dfrac{1}{81}$ **d.** $-\dfrac{1}{3^4} = -\dfrac{1}{81}$ **e.** $\dfrac{1}{8^1} = \dfrac{1}{8}$ **2. a.** $\dfrac{7^2}{2^3} = \dfrac{49}{8}$ **b.** $\dfrac{5^2}{4^2} = \dfrac{25}{16}$ **c.** $\dfrac{y^2}{7}$ **d.** $\dfrac{y^8}{x^1} = \dfrac{y^8}{x}$

3. $\dfrac{1}{x^{10}}$ **4. a.** $\dfrac{1}{x^8}$ **b.** $\dfrac{15}{x^6}$ **c.** $-\dfrac{2}{y^6}$ **5.** $\dfrac{36}{x^3}$ **6.** $\dfrac{1}{x^{20}}$ **7. a.** 7,400,000,000 **b.** 0.000003017 **8. a.** 7.41×10^9 **b.** 9.2×10^{-8}

9. a. 6×10^{10} **b.** 2.1×10^{11} **c.** 6.4×10^{-5} **10.** 5.2×10^5 mi

Exercise Set 5.7

1. $\dfrac{1}{8^2} = \dfrac{1}{64}$ **3.** $\dfrac{1}{5^3} = \dfrac{1}{125}$ **5.** $\dfrac{1}{(-6)^2} = \dfrac{1}{36}$ **7.** $-\dfrac{1}{6^2} = -\dfrac{1}{36}$ **9.** $\dfrac{1}{4^1} = \dfrac{1}{4}$ **11.** $\dfrac{1}{2^1} + \dfrac{1}{3^1} = \dfrac{1}{2} + \dfrac{1}{3} = \dfrac{5}{6}$ **13.** $3^2 = 9$ **15.** $(-3)^2 = 9$

17. $\dfrac{8^2}{2^3} = 8$ **19.** $\dfrac{4^2}{1^2} = 16$ **21.** $\dfrac{5^3}{3^3} = \dfrac{125}{27}$ **23.** $\dfrac{x^5}{6}$ **25.** $\dfrac{y^1}{x^8} = \dfrac{y}{x^8}$ **27.** $3 \cdot (-5)^3 = -375$ **29.** $\dfrac{1}{x^5}$ **31.** $\dfrac{8}{x^3}$ **33.** $\dfrac{1}{x^6}$ **35.** $\dfrac{1}{y^{99}}$

37. $\dfrac{3}{z^5}$ **39.** $-\dfrac{4}{x^4}$ **41.** $-\dfrac{1}{3a^3}$ **43.** $\dfrac{7}{5w^8}$ **45.** $\dfrac{1}{x^5}$ **47.** $\dfrac{1}{y^{11}}$ **49.** $\dfrac{16}{x^2}$ **51.** $216y^{17}$ **53.** $\dfrac{1}{x^6}$ **55.** $\dfrac{1}{16x^{12}}$ **57.** $\dfrac{x^2}{9}$ **59.** $-\dfrac{y^3}{8}$

61. $\dfrac{2x^6}{5}$ **63.** x^8 **65.** $16y^6$ **67.** $\dfrac{1}{y^2}$ **69.** $\dfrac{1}{y^{50}}$ **71.** $\dfrac{1}{a^{12}b^{15}}$ **73.** $\dfrac{a^8}{b^{24}}$ **75.** $\dfrac{4}{x^4}$ **77.** $\dfrac{y^9}{x^6}$ **79.** 870 **81.** 923,000 **83.** 3.4

85. 0.79 **87.** 0.0215 **89.** 0.000786 **91.** 3.24×10^4 **93.** 2.2×10^8 **95.** 7.13×10^2 **97.** 6.751×10^3 **99.** 2.7×10^{-3}

101. 2.02×10^{-5} **103.** 5×10^{-3} **105.** 3.14159×10^0 **107.** 6×10^5 **109.** 1.6×10^9 **111.** 3×10^4 **113.** 3×10^6 **115.** 3×10^{-6}

117. 9×10^4 **119.** 2.5×10^6 **121.** 1.25×10^8 **123.** 8.1×10^{-7} **125.** 2.5×10^{-7} **127.** 1 **129.** $\dfrac{y}{16x^8z^6}$ **131.** $\dfrac{1}{x^{12}y^{16}z^{20}}$

133. $\dfrac{x^{18}y^6}{4}$ **135.** 2.5×10^{-3} **137.** 8×10^{-5} **139.** 9.2×10^3 **141.** 2.5×10^{-16} **143.** 6×10^8 **145.** 3.99×10^{11}

147. $\$4 \times 10^{11}$ **149.** $\$3.48 \times 10^{10}$ **151.** 1.25 sec **159.** b **167.** $\{x \mid x < 2\}$ **168.** 5 **169.** $0, \sqrt{16}$

Review Exercises

1. binomial, 4 **2.** trinomial, 2 **3.** monomial, 1 **4.** $8x^3 + 10x^2 - 20x - 4$ **5.** $13y^3 - 8y^2 + 7y - 5$ **6.** $11y^2 - 4y - 4$
7. $8x^4 - 5x^3 + 6$ **8.** $-14x^4 - 13x^2 + 16x$ **9.** $7y^4 - 5y^3 + 3y^2 - y - 4$ **10.** $3x^2 - 7x + 9$ **11.** $10x^3 - 9x^2 + 2x + 11$
12. 1451; 1451 per 100,000 is the death rate for men averaging 10 hr of sleep each night. **13.** x^{23} **14.** y^{14} **15.** x^{100}
16. $100y^2$ **17.** $-64x^{30}$ **18.** $50x^4$ **19.** $-36y^{11}$ **20.** $30x^{12}$ **21.** $21x^3 + 63x$ **22.** $20x^5 - 55x^4$ **23.** $-21y^4 + 9y^3 - 18y^2$
24. $16y^8 - 20y^7 + 2y^5$ **25.** $x^3 - 2x^2 - 13x + 6$ **26.** $12y^3 + y^2 - 21y + 10$ **27.** $3y^3 - 17y^2 + 41y - 35$
28. $8x^4 + 8x^3 - 18x^2 - 20x - 3$ **29.** $x^2 + 8x + 12$ **30.** $6y^2 - 7y - 5$ **31.** $4x^4 - 14x^2 + 6$ **32.** $25x^2 - 16$
33. $49 - 4y^2$ **34.** $y^4 - 1$ **35.** $x^2 + 6x + 9$ **36.** $9y^2 + 24y + 16$ **37.** $y^2 - 2y + 1$ **38.** $25y^2 - 20y + 4$
39. $x^4 + 8x^2 + 16$ **40.** $x^4 - 16$ **41.** $x^4 - x^2 - 20$ **42.** $x^2 + 7x + 12$ **43.** $x^2 + 50x + 600$ yd^2 **44.** 28

45. polynomial degree: 5;

Term	Coefficient	Degree
$4x^2y$	4	3
$9x^3y^2$	9	5
$-17x^4$	-17	4
-12	-12	0

46. $-x^2 - 17xy + 5y^2$ **47.** $2x^3y^2 + x^2y - 6x^2 - 4$ **48.** $-35x^6y^9$ **49.** $15a^3b^5 - 20a^2b^3$ **50.** $3x^2 + 16xy - 35y^2$
51. $36x^2y^2 - 31xy + 3$ **52.** $9x^2 + 30xy + 25y^2$ **53.** $x^2y^2 - 14xy + 49$ **54.** $49x^2 - 16y^2$ **55.** $a^3 - b^3$ **56.** 6^{30} **57.** x^{15}

58. 1 **59.** -1 **60.** 400 **61.** $\dfrac{x^{12}}{8}$ **62.** $\dfrac{81}{16y^{24}}$ **63.** $-5y^6$ **64.** $8x^7y^3$ **65.** $3x^3 - 2x + 6$ **66.** $-6x^5 + 5x^4 + 8x^2$

67. $9x^2y - 3x - 6y$ **68.** $2x + 7$ **69.** $x^2 - 3x + 5$ **70.** $x^2 + 5x + 2 + \dfrac{7}{x - 7}$ **71.** $y^2 + 3y + 9$ **72.** $2x^2 + 3x - 1$

73. $4x^2 - 7x + 5 - \dfrac{4}{x+1}$ **74.** $3x^3 + 6x^2 + 10x + 10$ **75.** $x^3 - 4x^2 + 16x - 64 + \dfrac{272}{x+4}$ **76.** $\dfrac{1}{7^2} = \dfrac{1}{49}$ **77.** $\dfrac{1}{(-4)^3} = -\dfrac{1}{64}$

78. $\dfrac{1}{2^1} + \dfrac{1}{4^1} = \dfrac{1}{2} + \dfrac{1}{4} = \dfrac{3}{4}$ **79.** $5^2 = 25$ **80.** $\dfrac{5^3}{2^3} = \dfrac{125}{8}$ **81.** $\dfrac{1}{x^6}$ **82.** $\dfrac{6}{y^2}$ **83.** $\dfrac{30}{x^5}$ **84.** x^8 **85.** $81y^2$ **86.** $\dfrac{1}{y^{19}}$

87. $\dfrac{x^3}{8}$ **88.** $\dfrac{1}{x^6}$ **89.** y^{20} **90.** $23{,}000$ **91.** 0.00176 **92.** 0.9 **93.** 7.39×10^7 **94.** 6.2×10^{-4} **95.** 3.8×10^{-1} **96.** 3.8×10^0

97. 9×10^3 **98.** 5×10^4 **99.** 1.6×10^{-3} **100.** 1000 nanosec   **101.** 1.26×10^{10} people

Chapter 5 Test

1. trinomial, 2   **2.** $13x^3 + x^2 - x - 24$ **3.** $5x^3 + 2x^2 + 2x - 9$ **4.** $-35x^{11}$ **5.** $48x^5 - 30x^3 - 12x^2$ **6.** $3x^3 - 10x^2 - 17x - 6$

7. $6y^2 - 13y - 63$ **8.** $49x^2 - 25$ **9.** $x^4 + 6x^2 + 9$ **10.** $25x^2 - 30x + 9$ **11.** 30   **12.** $2x^2y^3 + 3xy + 12y^2$

13. $12a^2 - 13ab - 35b^2$ **14.** $4x^2 + 12xy + 9y^2$ **15.** $-5x^{12}$ **16.** $3x^3 - 2x^2 + 5x$ **17.** $x^2 - 2x + 3 + \dfrac{1}{2x+1}$ **18.** $3x^3 - 4x^2 + 7$

19. $\dfrac{1}{10^2} = \dfrac{1}{100}$ **20.** $4^3 = 64$ **21.** $-27x^6$ **22.** $\dfrac{4}{x^5}$ **23.** $-\dfrac{21}{x^6}$ **24.** $16y^4$ **25.** $\dfrac{x^8}{25}$ **26.** $\dfrac{1}{x^{15}}$ **27.** 0.00037

28. 7.6×10^6 **29.** 1.23×10^{-2} **30.** 2.1×10^8 **31.** $x^2 + 10x + 16$

Cumulative Review Exercises (Chapters 1–5)

1. $\left\{\dfrac{35}{9}\right\}$ **2.** $\{-128\}$ **3.** 15,050 ft   **4.** $-\dfrac{2}{3}$ **5.** $-\dfrac{5}{3}$ **6.** 4 ft by 10 ft   **7.** $\{x \mid x \geq 6\}$;

8. $\$3400$ at 12%; $\$2600$ at 14%   **9.** 15 l of 70%; 5 l of 30%;

10.   **11.**   **12.** $-\dfrac{6}{5}$; falling   **13.** $y + 1 = -2(x - 3); y = -2x + 5$ **14.** $\{(0, 5)\}$ **15.** $\{(3, -4)\}$

16. 500 min; $\$40$ **17.** 2.4×10^{-3} **18.** $3x^5 - 6x^3 + 9x + 2$ **19.** $x^2 + 2x + 3$ **20.** $\dfrac{81}{x^2}$

CHAPTER 6

Section 6.1

Check Point Exercises

1. a. $3x^2$ **b.** $4x^2$ **c.** x^2y **2.** $6(x^2 + 3)$ **3.** $5x^2(5 + 7x)$ **4.** $3x^3(5x^2 + 4x - 9)$ **5.** $2xy(4x^2y - 7x + 1)$
6. a. $(x + 1)(x^2 + 7)$ **b.** $(y + 4)(x - 7)$ **7.** $(x + 5)(x^2 + 2)$ **8.** $(y + 3)(x - 5)$

Exercise Set 6.1

1. Answers will vary; 3 examples are: $(2x)(4x^2), (4x)(2x^2),$ and $(8x)(x^2)$.   **3.** Answers will vary; 3 examples are: $(-4x^3)(3x^2), (2x^2)(-6x^3),$ and
$(-3)(4x^5)$.   **5.** Answers will vary; 3 examples are: $(6x^2)(6x^2), (-2x)(-18x^3),$ and $(4x^3)(9x)$.   **7.** 4   **9.** $4x$ **11.** $2x^3$ **13.** $3y$
15. xy **17.** $4x^4y^3$ **19.** $8(x + 1)$ **21.** $4(y - 1)$ **23.** $5(x + 6)$ **25.** $6(5x - 2)$ **27.** $x(x + 5)$ **29.** $6(3y^2 + 2)$
31. $7x^2(2x + 3)$ **33.** $y(13y - 25)$ **35.** $9y^4(1 + 3y^2)$ **37.** $4x^2(2 - x^2)$ **39.** $4(3y^2 + 4y - 2)$ **41.** $3x^2(3x^2 + 6x + 2)$
43. $50y^2(2y^3 - y + 2)$ **45.** $5x(2 - 4x + x^2)$ **47.** cannot be factored   **49.** $3xy(2x^2y + 3)$ **51.** $10xy(3xy^2 - y + 2)$
53. $8x^2y(4xy - 3x - 2)$ **55.** $(x + 5)(x + 3)$ **57.** $(x + 2)(x - 4)$ **59.** $(y + 6)(x - 7)$ **61.** $(x + y)(3x - 1)$
63. $(3x + 1)(4x + 1)$ **65.** $(5x + 4)(7x^2 + 1)$ **67.** $(x + 2)(x + 4)$ **69.** $(x - 5)(x + 3)$ **71.** $(x^2 + 5)(x - 2)$
73. $(x^2 + 2)(x - 1)$ **75.** $(y + 5)(x + 9)$ **77.** $(y - 1)(x + 5)$ **79.** $(x - 2y)(3x + 5y)$ **81.** $(3x - 2)(x^2 - 2)$ **83.** $(x - a)(x - b)$
85. $6x^2yz(4xy^2z^2 + 5y + 3z)$ **87.** $(x^3 - 4)(1 + 3y)$ **89.** $2x^2(x + 1)(2x^3 - 3x - 4)$ **91.** $(x - 1)(3x^4 + x^2 + 5)$
93. $36x^2 - 4\pi x^2; 4x^2(9 - \pi)$ **95. a.** 48 ft   **b.** $16x(4 - x)$ **c.** 48 ft; yes; no; Answers will vary.   **97.** $x^3 - 2$ **105.** d
109. $-3(x - 2)$ should be $-3(x + 2)$.   **111.** $x(x + 2) + 1$ should be $(x + 2)(x + 1)$.   **112.** $x^2 + 17x + 70$ **113.** $\{(-3, -2)\}$
114. $y - 2 = 1(x + 7),$ or $y - 5 = 1(x + 4); y = x + 9$

Section 6.2

Check Point Exercises

1. $(x + 2)(x + 3)$ **2.** $(x - 2)(x - 4)$ **3.** $(x + 5)(x - 2)$ **4.** $(y - 9)(y + 3)$ **5.** cannot factor over the integers; prime
6. $(x - 3y)(x - y)$ **7.** $2x(x - 4)(x + 7)$

Exercise Set 6.2

1. $(x + 6)(x + 1)$ **3.** $(x + 2)(x + 5)$ **5.** $(x + 1)(x + 10)$ **7.** $(x - 4)(x - 3)$ **9.** $(x - 6)(x - 6)$ **11.** $(y - 3)(y - 5)$
13. $(x + 5)(x - 2)$ **15.** $(y + 13)(y - 3)$ **17.** $(x - 5)(x + 3)$ **19.** $(x - 4)(x + 2)$ **21.** prime **23.** $(y - 4)(y - 12)$ **25.** prime
27. $(w - 32)(w + 2)$ **29.** $(y - 5)(y - 13)$ **31.** $(r + 3)(r + 9)$ **33.** prime **35.** $(x + 6y)(x + y)$ **37.** $(x - 3y)(x - 5y)$
39. $(x - 6y)(x + 3y)$ **41.** $(a - 15b)(a - 3b)$ **43.** $3(x + 2)(x + 3)$ **45.** $4(y - 2)(y + 1)$ **47.** $10(x - 10)(x + 6)$
49. $3(x - 2)(x - 9)$ **51.** $2r(r + 2)(r + 1)$ **53.** $4x(x + 6)(x - 3)$ **55.** $2r(r + 8)(r - 4)$ **57.** $y^2(y + 10)(y - 8)$
59. $x^2(x - 5)(x + 2)$ **61.** $2w^2(w - 16)(w + 3)$ **63.** $15x(y - 1)(y + 4)$ **65.** $x^3(x - y)(x + 4y)$ **67.** $2x^2(y - 15z)(y - z)$

69. $(a + b)(x + 5)(x - 4)$ **71.** $(x + 0.3)(x + 0.2)$ **73.** $\left(x - \dfrac{1}{5}\right)\left(x - \dfrac{1}{5}\right)$ **75.** $-(x + 8)(x - 5)$ **77. a.** $-16(t - 2)(t + 1)$

b. 0; yes; After 2 seconds, you hit the water. **83.** c **85.** $3, 4$ **87.** $x(x + 1)(x + 2)$; the product of three consecutive integers
89. correctly factored **91.** $(x + 1)(x - 1)$ should be $(x - 1)(x - 1)$. **93.** $2x^2 - x - 6$ **94.** $9x^2 + 15x + 4$ **95.** $\{13\}$

Section 6.3

Check Point Exercises

1. $(5x - 4)(x - 2)$ **2.** $(3x - 1)(2x + 7)$ **3.** $(3x - y)(x - 4y)$ **4.** $(3x + 5)(x - 2)$ **5.** $(2x - 1)(4x - 3)$ **6.** $y^2(5y + 3)(y + 2)$

Exercise Set 6.3

1. $(2x + 3)(x + 1)$ **3.** $(3x + 1)(x + 4)$ **5.** $(2x + 3)(x + 4)$ **7.** $(5y - 1)(y - 3)$ **9.** $(3y + 4)(y - 1)$ **11.** $(3x - 2)(x + 5)$
13. $(3x - 1)(x - 7)$ **15.** $(5y - 1)(y - 3)$ **17.** $(3x - 2)(x - 5)$ **19.** $(3w - 4)(2w - 1)$ **21.** $(8x + 1)(x + 4)$
23. $(5x - 2)(x + 7)$ **25.** $(7y - 3)(2y + 3)$ **27.** prime **29.** $(5z - 3)(5z - 3)$ **31.** $(3y + 1)(5y - 2)$ **33.** prime
35. $(5y - 1)(2y + 9)$ **37.** $(4x + 1)(2x - 1)$ **39.** $(3y - 1)(3y - 2)$ **41.** $(5x + 8)(4x - 1)$ **43.** $(2x + y)(x + y)$
45. $(3x + 2y)(x + y)$ **47.** $(2x - 3y)(x - 3y)$ **49.** $(2x - 3y)(3x + 2y)$ **51.** $(3x - 2y)(5x + 7y)$ **53.** $(2a + 5b)(a + b)$
55. $(3a - 2b)(5a + 3b)$ **57.** $(3x - 4y)(4x - 3y)$ **59.** $2(2x + 3)(x + 5)$ **61.** $3(3x + 4)(x - 2)$ **63.** $2(2y - 5)(y + 3)$
65. $3(3y - 4)(y + 5)$ **67.** $x(3x + 1)(x + 1)$ **69.** $x(2x - 5)(x + 1)$ **71.** $3y(3y - 1)(y - 4)$ **73.** $5z(6z + 1)(2z + 1)$
75. $3x^2(5x - 3)(x - 2)$ **77.** $x^3(2x - 3)(5x - 1)$ **79.** $3(2x + 3y)(x - 2y)$ **81.** $2(2x - y)(3x + 4y)$ **83.** $2y(4x - 7)(x + 6)$
85. $2b(2a - 7b)(3a - b)$ **87.** $10(y + 1)(x + 1)(3x - 2)$ **89.** $-4y^4(8x + 3)(x - 1)$ **91. a.** $(2x + 1)(x - 3)$ **b.** $(2y + 3)(y - 2)$
93. $(x - 2)(3x - 2)(x - 1)$ **95. a.** $x^2 + 3x + 2$ **b.** $(x + 2)(x + 1)$ **c.** $x^2 + 3x + 2 = (x + 2)(x + 1)$ **101.** a **103.** $5, 7, -5, -7$
105. $(2x^n + 1)(x^n - 4)$ **106.** $81x^2 - 100$ **107.** $16x^2 + 40xy + 25y^2$ **108.** $x^3 + 8$

Mid-Chapter 6 Check Point

1. $x^4(x + 1)$ **2.** $(x + 9)(x - 2)$ **3.** $x^2y(y^2 - y + 1)$ **4.** prime **5.** $(7x - 1)(x - 3)$ **6.** $(x^2 + 3)(x + 5)$ **7.** $x(2x - 1)(x - 5)$
8. $(x - 4)(y - 7)$ **9.** $(x - 15y)(x - 2y)$ **10.** $(5x + 2)(5x - 7)$ **11.** $2(8x - 3)(x - 4)$ **12.** $(3x + 7y)(x + y)$

Section 6.4

Check Point Exercises

1. a. $(x + 9)(x - 9)$ **b.** $(6x + 5)(6x - 5)$ **2. a.** $(5 + 2x^5)(5 - 2x^5)$ **b.** $(10x + 3y)(10x - 3y)$ **3. a.** $2x(3x + 1)(3x - 1)$
b. $18(2 + x)(2 - x)$ **4.** $(9x^2 + 4)(3x + 2)(3x - 2)$ **5. a.** $(x + 7)^2$ **b.** $(x - 3)^2$ **c.** $(4x - 7)^2$ **6.** $(2x + 3y)^2$
7. $(x + 3)(x^2 - 3x + 9)$ **8.** $(1 - y)(1 + y + y^2)$ **9.** $(5x + 2)(25x^2 - 10x + 4)$

Exercise Set 6.4

1. $(x + 5)(x - 5)$ **3.** $(y + 1)(y - 1)$ **5.** $(2x + 3)(2x - 3)$ **7.** $(5 + x)(5 - x)$ **9.** $(1 + 7x)(1 - 7x)$ **11.** $(3 + 5y)(3 - 5y)$
13. $(x^2 + 3)(x^2 - 3)$ **15.** $(7y^2 + 4)(7y^2 - 4)$ **17.** $(x^5 + 3)(x^5 - 3)$ **19.** $(5x + 4y)(5x - 4y)$ **21.** $(x^2 + y^5)(x^2 - y^5)$
23. $(x^2 + 4)(x + 2)(x - 2)$ **25.** $(4x^2 + 9)(2x + 3)(2x - 3)$ **27.** $2(x + 3)(x - 3)$ **29.** $2x(x + 6)(x - 6)$ **31.** prime
33. $3x(x^2 + 9)$ **35.** $2(3 + y)(3 - y)$ **37.** $3y(y + 4)(y - 4)$ **39.** $2x(3x + 1)(3x - 1)$ **41.** $(x + 1)^2$ **43.** $(x - 7)^2$ **45.** $(x - 1)^2$
47. $(x + 11)^2$ **49.** $(2x + 1)^2$ **51.** $(5y - 1)^2$ **53.** prime **55.** $(x + 7y)^2$ **57.** $(x - 6y)^2$ **59.** prime **61.** $(4x - 5y)^2$
63. $3(2x - 1)^2$ **65.** $x(3x + 1)^2$ **67.** $2(y - 1)^2$ **69.** $2y(y + 7)^2$ **71.** $(x + 1)(x^2 - x + 1)$ **73.** $(x - 3)(x^2 + 3x + 9)$
75. $(2y - 1)(4y^2 + 2y + 1)$ **77.** $(3x + 2)(9x^2 - 6x + 4)$ **79.** $(xy - 4)(x^2y^2 + 4xy + 16)$ **81.** $y(3y + 2)(9y^2 - 6y + 4)$

83. $2(3 - 2y)(9 + 6y + 4y^2)$ **85.** $(4x + 3y)(16x^2 - 12xy + 9y^2)$ **87.** $(5x - 4y)(25x^2 + 20xy + 16y^2)$ **89.** $\left(5x + \dfrac{2}{7}\right)\left(5x - \dfrac{2}{7}\right)$

91. $y\left(y - \dfrac{1}{10}\right)\left(y^2 + \dfrac{y}{10} + \dfrac{1}{100}\right)$ **93.** $x(0.5 + x)(0.5 - x)$ **95.** $(x + 6)(x - 4)$ **97.** $(x - 3)(x + 1)^2$ **99.** $x^2 - 25 = (x + 5)(x - 5)$
101. $x^2 - 16 = (x + 4)(x - 4)$ **107.** b **109.** $(x + y)(x - y + 3)$ **111.** $(2x^n + 3)^2$ **113.** $6, -6$
115. $(4x + 3)(4x - 3)$ should be $(2x + 3)(2x - 3)$. **117.** $(4x - 1)^2$ should be $(2x - 1)^2$. **119.** $80x^9y^{14}$ **120.** $-4x^2 + 3$ **121.** $2x + 5$

Section 6.5

Check Point Exercises

1. $5x^2(x + 3)(x - 3)$ **2.** $4(x - 6)(x + 2)$ **3.** $4x(x^2 + 4)(x + 2)(x - 2)$ **4.** $(x - 4)(x + 3)(x - 3)$ **5.** $3x(x - 5)^2$
6. $2x^2(x + 3)(x^2 - 3x + 9)$ **7.** $3y(x^2 + 4y^2)(x + 2y)(x - 2y)$ **8.** $3x(2x + 3y)^2$

Exercise Set 6.5

1. $5x(x + 2)(x - 2)$ **3.** $7x(x^2 + 1)$ **5.** $5(x - 3)(x + 2)$ **7.** $2(x^2 + 9)(x + 3)(x - 3)$ **9.** $(x + 2)(x + 3)(x - 3)$ **11.** $3x(x - 4)^2$
13. $2x^2(x + 1)(x^2 - x + 1)$ **15.** $2x(3x + 4)$ **17.** $2(y - 8)(y + 7)$ **19.** $7y^2(y + 1)^2$ **21.** prime **23.** $2(4y + 1)(2y - 1)$ **25.** $r(r - 25)$
27. $(2w + 5)(2w - 1)$ **29.** $x(x + 2)(x - 2)$ **31.** prime **33.** $(9y + 4)(y + 1)$ **35.** $(y + 2)(y + 2)(y - 2)$ **37.** $(4y + 3)^2$
39. $4y(y - 5)(y - 2)$ **41.** $y(y^2 + 9)(y + 3)(y - 3)$ **43.** $5a^2(2a + 3)(2a - 3)$ **45.** $(4y - 1)(3y - 2)$ **47.** $(3y + 8)(3y - 8)$
49. prime **51.** $(2y + 3)(y + 5)(y - 5)$ **53.** $2r(r + 17)(r - 2)$ **55.** $2x^3(2x + 1)(2x - 1)$ **57.** $3(x^2 + 81)$ **59.** $x(x + 2)(x^2 - 2x + 4)$
61. $2y^2(y - 1)(y^2 + y + 1)$ **63.** $2x(3x + 4y)$ **65.** $(y - 7)(x + 3)$ **67.** $(x - 4y)(x + y)$ **69.** $12a^2(6ab^2 + 1 - 2a^2b^2)$
71. $3(a + 6b)(a + 3b)$ **73.** $3x^2y(4x + 1)(4x - 1)$ **75.** $b(3a + 2)(2a - 1)$ **77.** $7xy(x^2 + y^2)(x + y)(x - y)$ **79.** $2xy(5x - 2y)(x - y)$
81. $2b(x + 11)^2$ **83.** $(5a + 7b)(3a - 2b)$ **85.** $2xy(9x - 2y)(2x - 3y)$ **87.** $(y - x)(a + b)(a - b)$ **89.** $ax(3x + 7)(3x - 2)$
91. $y(9x^2 + y^2)(3x + y)(3x - y)$ **93.** $(x + 1)(5x - 6)(2x + 1)$ **95.** $(x^2 + 6)(6x^2 - 1)$ **97.** $(x - 7 + 2a)(x - 7 - 2a)$
99. $(x + 4 + 5a)(x + 4 - 5a)$ **101.** $y(y^2 + 1)(y^4 - y^2 + 1)$ **103.** $16(4 + t)(4 - t)$ **105.** $\pi b^2 - \pi a^2; \pi(b + a)(b - a)$ **109.** d
111. $5y^2(y - 1)(y + 2)(y - 2)$ **113.** $(x - 5)^2$ **115.** $(4x - 3)^2$ should be $(2x - 3)^2$. **117.** correctly factored

119. $2(x + 5)(x^2 + 1)$ should be $2(x + 5)(x + 1)(x - 1)$. **120.** $(3x + 4)(3x - 4)$ **121.** **122.** $20°, 60°, 100°$

Section 6.6

Check Point Exercises

1. $-\frac{1}{2}$ and 4, or $\left\{-\frac{1}{2}, 4\right\}$ **2.** 5 and 1, or $\{1, 5\}$ **3.** 0 and $\frac{1}{2}$, or $\left\{0, \frac{1}{2}\right\}$ **4.** 5 or $\{5\}$ **5.** $-\frac{5}{4}$ and $\frac{5}{4}$, or $\left\{-\frac{5}{4}, \frac{5}{4}\right\}$ **6.** -2 and 9, or $\{-2, 9\}$
7. 1 sec and 2 sec; $(1, 192)$ and $(2, 192)$ **8.** length: 9 ft; width: 6 ft

Exercise Set 7.6

1. $\{-7, 0\}$ **3.** $\{-4, 6\}$ **5.** $\left\{-\frac{4}{5}, 9\right\}$ **7.** $\left\{-\frac{9}{2}, 4\right\}$ **9.** $\{-5, -3\}$ **11.** $\{-3, 5\}$ **13.** $\{-3, 7\}$ **15.** $\{-8, -1\}$

17. $\{-4, 0\}$ **19.** $\{0, 5\}$ **21.** $\{0, 4\}$ **23.** $\left\{0, \frac{5}{2}\right\}$ **25.** $\left\{-\frac{5}{3}, 0\right\}$ **27.** $\{-2\}$ **29.** $\{6\}$ **31.** $\left\{\frac{3}{2}\right\}$ **33.** $\left\{-\frac{1}{2}, 4\right\}$ **35.** $\left\{-2, \frac{9}{5}\right\}$

37. $\{-7, 7\}$ **39.** $\left\{-\frac{5}{2}, \frac{5}{2}\right\}$ **41.** $\left\{-\frac{5}{9}, \frac{5}{9}\right\}$ **43.** $\{-3, 7\}$ **45.** $\left\{-\frac{5}{2}, \frac{3}{2}\right\}$ **47.** $\{-6, 3\}$ **49.** $\left\{-2, -\frac{3}{2}\right\}$ **51.** $\{4\}$ **53.** $\left\{-\frac{5}{2}\right\}$

55. $\left\{\frac{3}{8}\right\}$ **57.** $\{-3, -2, 4\}$ **59.** $\{-6, 0, 6\}$ **61.** $\{-2, -1, 0\}$ **63.** $\{2, 8\}$ **65.** $\{4, 5\}$ **67.** 5 sec; Each tick represents one second.

69. 2 sec; $(2, 276)$ **71.** $\frac{1}{2}$ sec and 4 sec **73.** 1995; $(15, 1100)$ **75.** 10 yr **77.** $(10, 7250)$ **79.** 10 teams **81.** length: 15 yd; width: 12 yd

83. base: 5 cm; height: 6 cm **85. a.** $4x^2 + 44x$ **b.** 3 ft **89.** d **91.** $-4, 1,$ and 4 **93.** 4 and 1 **95.** a **97.** b **99.** -3 and 2

101. 1 **104.** $\{(2, 4, 4)\}$ **105.** $\frac{4}{x^6}$ **106.** -2

Review Exercises

1. $15(2x - 3)$ **2.** $4x(3x^2 + 4x - 100)$ **3.** $5x^2y(6x^2 + 3x + 1)$ **4.** $5(x + 3)$ **5.** $(7x^2 - 1)(x + y)$ **6.** $(x^2 + 2)(x + 3)$
7. $(x + 1)(y + 4)$ **8.** $(x^2 + 5)(x + 1)$ **9.** $(x - 2)(y + 4)$ **10.** $(x - 2)(x - 1)$ **11.** $(x - 5)(x + 4)$ **12.** $(x + 3)(x + 16)$
13. $(x - 4y)(x - 2y)$ **14.** prime **15.** $(x + 17y)(x - y)$ **16.** $3(x + 4)(x - 2)$ **17.** $3x(x - 11)(x - 1)$ **18.** $(x + 5)(3x + 2)$
19. $(y - 3)(5y - 2)$ **20.** $(2x + 5)(2x - 3)$ **21.** prime **22.** $2(2x + 3)(2x - 1)$ **23.** $x(2x - 9)(x + 8)$ **24.** $4y(3y + 1)(y + 2)$
25. $(2x - y)(x - 3y)$ **26.** $(5x + 4y)(x - 2y)$ **27.** $(2x + 1)(2x - 1)$ **28.** $(9 + 10y)(9 - 10y)$ **29.** $(5a + 7b)(5a - 7b)$
30. $(z^2 + 4)(z + 2)(z - 2)$ **31.** $2(x + 3)(x - 3)$ **32.** prime **33.** $x(3x + 1)(3x - 1)$ **34.** $2x(3y + 2)(3y - 2)$ **35.** $(x + 11)^2$
36. $(x - 8)^2$ **37.** $(3y + 8)^2$ **38.** $(4x - 5)^2$ **39.** prime **40.** $(6x + 5y)^2$ **41.** $(5x - 4y)^2$ **42.** $(x - 3)(x^2 + 3x + 9)$
43. $(4x + 1)(16x^2 - 4x + 1)$ **44.** $2(3x - 2y)(9x^2 + 6xy + 4y^2)$ **45.** $y(3x + 2)(9x^2 - 6x + 4)$ **46.** $(a + 3)(a - 3)$
47. $(a + 2b)(a - 2b)$ **48.** $A^2 + 2A + 1 = (A + 1)^2$ **49.** $x(x - 7)(x - 1)$ **50.** $(5y + 2)(2y + 1)$ **51.** $2(8 + y)(8 - y)$
52. $(3x + 1)^2$ **53.** $4x^3(5x^4 - 9)$ **54.** $(x - 3)^2(x + 3)$ **55.** prime **56.** $x(2x + 5)(x + 7)$ **57.** $3x(x - 5)^2$
58. $3x^2(x - 2)(x^2 + 2x + 4)$ **59.** $4y^2(y + 3)(y - 3)$ **60.** $5(x + 7)(x - 3)$ **61.** prime **62.** $2x^3(5x - 2)(x - 4)$
63. $(10y + 7)(10y - 7)$ **64.** $9x^4(x - 2)$ **65.** $(x^2 + 1)(x + 1)(x - 1)$ **66.** $2(y - 2)(y^2 + 2y + 4)$ **67.** $(x + 4)(x^2 - 4x + 16)$
68. $(3x - 2)(2x + 5)$ **69.** $3x^2(x + 2)(x - 2)$ **70.** $(x - 10)(x + 9)$ **71.** $(5x + 2y)(5x + 3y)$ **72.** $x(x + 5)(x^2 - 5x + 25)$
73. $2y(4y + 3)(4y + 1)$ **74.** $2(y - 4)^2$ **75.** $(x + 5y)(x - 7y)$ **76.** $(x + y)(x + 7)$ **77.** $(3x + 4y)^2$ **78.** $2x^2y(x + 1)(x - 1)$
79. $(10y + 7z)(10y - 7z)$ **80.** prime **81.** $3x^2y^2(x + 2y)(x - 2y)$ **82.** $\{0, 12\}$ **83.** $\left\{-\frac{9}{4}, 7\right\}$ **84.** $\{-7, 2\}$ **85.** $\{-4, 0\}$

86. $\left\{-8, \frac{1}{2}\right\}$ **87.** $\{-4, 8\}$ **88.** $\{-8, 7\}$ **89.** $\{7\}$ **90.** $\left\{-\frac{10}{3}, \frac{10}{3}\right\}$ **91.** $\{-5, -2\}$ **92.** $\left\{\frac{1}{3}, 7\right\}$ **93.** 2 sec

94. width: 5 ft; length: 8 ft **95.** 11 m by 11 m

Chapter 6 Test

1. $(x - 3)(x - 6)$ **2.** $(x - 7)^2$ **3.** $5y^2(3y - 1)(y - 2)$ **4.** $(x^2 + 3)(x + 2)$ **5.** $x(x - 9)$ **6.** $x(x + 7)(x - 1)$
7. $2(7x - 3)(x + 5)$ **8.** $(5x + 3)(5x - 3)$ **9.** $(x + 2)(x^2 - 2x + 4)$ **10.** $(x + 3)(x - 7)$ **11.** prime **12.** $3y(2y + 1)(y + 1)$
13. $4(y + 3)(y - 3)$ **14.** $4(2x + 3)^2$ **15.** $2(x^2 + 4)(x + 2)(x - 2)$ **16.** $(6x - 7)^2$ **17.** $(7x - 1)(x - 7)$ **18.** $(x^2 - 5)(x + 2)$
19. $3y(2y + 3)(2y - 5)$ **20.** $(y - 5)(y^2 + 5y + 25)$ **21.** $5(x - 3y)(x + 2y)$ **22.** $\{-6, 4\}$ **23.** $\left\{-\dfrac{1}{3}, 2\right\}$ **24.** $\{-2, 8\}$
25. $\left\{0, \dfrac{7}{2}\right\}$ **26.** $\left\{-\dfrac{9}{4}, \dfrac{9}{4}\right\}$ **27.** $\left\{-1, \dfrac{6}{5}\right\}$ **28.** $x^2 - 4 = (x + 2)(x - 2)$ **29.** 6 sec **30.** width: 5 ft; length: 11 ft

Cumulative Review Exercises (Chapters 1–6)

1. -48 **2.** $\{0\}$ **3.** $\{12\}$ **4.** $\{x | x < -2\}$; **5.** $38°, 38°, 104°$ **6.** $\$150$

7. **8.** $y + 4 = 5(x - 2)$, or $y - 1 = 5(x - 3)$; $y = 5x - 14$ **9.** $\{(2, -2, -1)\}$ **10.** $\{(3, -1)\}$ **11.** $\left\{\left(-\dfrac{2}{13}, \dfrac{23}{13}\right)\right\}$
12. $-\dfrac{13}{40}$ **13.** $2x^4 - x^3 + 3x + 9$ **14.** $6x^2 + 17xy - 45y^2$ **15.** $2x^2 + 5x - 3 - \dfrac{2}{3x - 5}$ **16.** 7.1×10^{-3}
17. $(3x + 2)(x + 3)$ **18.** $y(y^2 + 4)(y + 2)(y - 2)$ **19.** $(2x + 3)^2$ **20.** width: 4 ft; length: 6 ft

CHAPTER 7

Section 7.1

Check Point Exercises

1. a. $x = 5$ **b.** $x = -7$ and $x = 4$ **2.** $\dfrac{x + 4}{3x}$ **3.** $\dfrac{x^2}{7}$ **4.** $\dfrac{x - 1}{x + 1}$ **5.** $-\dfrac{3x + 7}{4}$

Exercise Set 7.1

1. $x = 0$ **3.** $x = 8$ **5.** $x = 4$ **7.** $x = -9$ and $x = 2$ **9.** $x = \dfrac{17}{3}$ and $x = -3$ **11.** $x = -4$ and $x = 3$
13. defined for all real numbers **15.** $y = -1$ and $y = \dfrac{3}{4}$ **17.** $y = -5$ and $y = 5$ **19.** defined for all real numbers **21.** $2x$
23. $\dfrac{x - 3}{5}$ **25.** $\dfrac{x - 4}{2x}$ **27.** $\dfrac{1}{x - 3}$ **29.** $-\dfrac{5}{x - 3}$ **31.** 3 **33.** $\dfrac{1}{x - 5}$ **35.** $\dfrac{2}{3}$ **37.** $\dfrac{1}{x - 3}$ **39.** $\dfrac{4}{x - 2}$ **41.** $\dfrac{y - 1}{y + 9}$
43. $\dfrac{y - 3}{y - 2}$ **45.** cannot be simplified **47.** $\dfrac{x + 6}{x - 6}$ **49.** $x^2 + 1$ **51.** $x^2 + 2x + 4$ **53.** $\dfrac{x - 4}{x + 4}$ **55.** cannot be simplified
57. cannot be simplified **59.** -1 **61.** -1 **63.** cannot be simplified **65.** -2 **67.** $-\dfrac{2}{x}$ **69.** $-x - 1$ **71.** $-y - 3$ **73.** $-\dfrac{1}{x}$
75. $\dfrac{x - y}{2x - y}$ **77.** $\dfrac{x - 6}{x^2 + 3x + 9}$ **79.** $\dfrac{3 + y}{3 - y}$ **81.** $\dfrac{y + 3}{x + 3}$ **83.** $\dfrac{2}{1 - 2x}$ **85. a.** It costs $\$86.67$ million to inoculate 40% of the population.
It costs $\$520$ million to inoculate 80% of the population. It costs $\$1170$ million to inoculate 90% of the population. **b.** $x = 100$
c. The cost keeps rising; No amount of money will be enough to inoculate 100% of the population. **87.** 400 mg **89. a.** $\$300$ **b.** $\$125$
c. decrease **91.** 1.5 mg per liter; $(3, 1.5)$ **93.** 20.8; no **95. a.** $\dfrac{-0.4t + 14.2}{3.7t + 257.4}$ **b.** $0.04; 4000$ **c.** fairly well; Answers will vary.
105. $\dfrac{x^2 - x - 6}{x + 2}$ **107.** $2x^2 - 1$ should be $2x + 1$. **110.** $\dfrac{3}{10}$ **111.** $\dfrac{1}{6}$ **112.** $\{(4, 2)\}$

Section 7.2

Check Point Exercises

1. $\dfrac{9x - 45}{2x + 8}$ **2.** $\dfrac{3}{8}$ **3.** $\dfrac{x + 2}{9}$ **4.** $-\dfrac{5(x + 1)}{7x(2x - 3)}$ **5.** $\dfrac{(x + 3)(x + 7)}{x - 4}$ **6.** $\dfrac{x + 3}{x - 5}$ **7.** $\dfrac{y + 1}{5y(y^2 + 1)}$

Exercise Set 7.2

1. $\dfrac{4x - 20}{9x + 27}$ **3.** $\dfrac{4x}{x + 5}$ **5.** $\dfrac{4}{5}$ **7.** $\dfrac{4}{9}$ **9.** 1 **11.** $\dfrac{x + 5}{x}$ **13.** $\dfrac{2}{y}$ **15.** $\dfrac{y + 2}{y + 4}$ **17.** $4(y + 3)$ **19.** $\dfrac{x - 1}{x + 2}$ **21.** $\dfrac{x^2 + 2x + 4}{3x}$
23. $\dfrac{x - 2}{x - 1}$ **25.** $-\dfrac{2}{x(x + 1)}$ **27.** $-\dfrac{y - 10}{y - 7}$ **29.** $(x - y)(x + y)$ **31.** $\dfrac{4(x + y)}{3(x - y)}$ **33.** $\dfrac{3x}{35}$ **35.** $\dfrac{1}{4}$ **37.** 10 **39.** $\dfrac{7}{9}$ **41.** $\dfrac{3}{4}$
43. $\dfrac{(x - 2)^2}{x}$ **45.** $\dfrac{(y - 4)(y^2 + 4)}{y - 1}$ **47.** $\dfrac{y}{3}$ **49.** $\dfrac{2(x + 3)}{3}$ **51.** $\dfrac{x - 5}{2}$ **53.** $\dfrac{y^2 + 1}{y^2}$ **55.** $\dfrac{y + 1}{y - 7}$ **57.** 6 **59.** $\dfrac{2}{3}$ **61.** $\dfrac{(x + y)^2}{32(x - y)^2}$

63. $\dfrac{y(x + y)}{(x + 1)^2}$ **65.** $\dfrac{y + 3}{y - 3}$ **67.** $\dfrac{5(x - 5)}{4}$ **69.** $\dfrac{(x + y)^2}{x - y}$ **71.** $\dfrac{2}{y^2 - by + b^2}$ **73.** $\dfrac{125x}{100 - x}$ **79.** numerator: x; denominator: $x - 4$

81. $(3x - y)(3x + y)$ **83.** correct answer **85.** $2x - 1$ should be $2x + 1$. **86.** $\{x \mid x > 18\}$ **87.** $3(x - 7)(x + 2)$ **88.** $\left\{-5, \dfrac{1}{2}\right\}$

Section 7.3

Check Point Exercises

1. $x + 2$ **2.** $\dfrac{x - 5}{x + 5}$ **3. a.** $\dfrac{3x + 5}{x + 7}$ **b.** $3x - 4$ **4.** $\dfrac{3y + 2}{y - 4}$ **5.** $x + 3$ **6.** $\dfrac{-4x^2 + 12x}{x^2 - 2x - 9}$

Exercise Set 7.3

1. $\dfrac{9x}{13}$ **3.** $\dfrac{3x}{5}$ **5.** $\dfrac{x + 3}{2}$ **7.** $\dfrac{6}{x}$ **9.** $\dfrac{7}{3x}$ **11.** $\dfrac{9}{x + 3}$ **13.** $\dfrac{5x + 5}{x - 3}$ **15.** 2 **17.** $\dfrac{2y + 3}{y - 5}$ **19.** $\dfrac{7}{5y}$ **21.** $\dfrac{2}{x + 2}$ **23.** $\dfrac{x - 2}{x - 3}$

25. $\dfrac{3x - 4}{5x - 4}$ **27.** 1 **29.** 1 **31.** $\dfrac{x}{2x - 1}$ **33.** $-\dfrac{1}{y}$ **35.** $\dfrac{y + 3}{3y + 8}$ **37.** $\dfrac{3y + 2}{y - 3}$ **39.** $\dfrac{2}{x - 3}$ **41.** $\dfrac{3x + 7}{x - 6}$ **43.** $\dfrac{3x + 1}{3x - 4}$ **45.** $x + 2$

47. 0 **49.** $\dfrac{11}{x - 1}$ **51.** $\dfrac{12}{x + 3}$ **53.** $\dfrac{y + 1}{y - 1}$ **55.** $\dfrac{x - 2}{x - 7}$ **57.** $\dfrac{2x - 4}{x^2 - 25}$ **59.** 1 **61.** $\dfrac{2}{x + y}$ **63.** $\dfrac{x - 3}{x - 1}$ **65.** $\dfrac{4b}{4b - 3}$ **67.** $\dfrac{y}{y - 5}$

69. $-\dfrac{1}{c + d}$ **71.** $\dfrac{3y^2 + 8y - 5}{(y + 1)(y - 4)}$ **73. a.** $\dfrac{100W}{L}$ **b.** round **75.** 10 m **81.** d **83.** $x - 2$ **85.** $20x - 6$ **87.** $a + 12$

89. correct answer **91.** $x + 4$ should be $x - 4$. **92.** $\dfrac{31}{45}$ **93.** $(9x^2 + 1)(3x + 1)(3x - 1)$ **94.** $3x^2 - 7x - 5$

Section 7.4

Check Point Exercises

1. $30x^2$ **2.** $(x + 3)(x - 3)$ **3.** $7x(x + 4)(x + 4)$ or $7x(x + 4)^2$ **4.** $\dfrac{9 + 14x}{30x^2}$ **5.** $\dfrac{6x + 6}{(x + 3)(x - 3)}$ **6.** $-\dfrac{5}{x + 5}$ **7.** $-\dfrac{5 + y}{5y}$ **8.** $\dfrac{x - 15}{(x + 5)(x - 5)}$

Exercise Set 7.4

1. $120x^2$ **3.** $30x^5$ **5.** $(x - 3)(x + 1)$ **7.** $7y(y + 2)$ **9.** $(x + 4)(x - 4)$ **11.** $y(y + 3)(y - 3)$ **13.** $(y + 1)(y - 1)(y - 1)$

15. $(x - 5)(x + 4)(2x - 1)$ **17.** $\dfrac{3x + 5}{x^2}$ **19.** $\dfrac{37}{18x}$ **21.** $\dfrac{8x + 7}{2x^2}$ **23.** $\dfrac{6x + 1}{x}$ **25.** $\dfrac{2 + 9x}{x}$ **27.** $\dfrac{x + 1}{2}$ **29.** $\dfrac{7x - 20}{x(x - 5)}$

31. $\dfrac{5x + 1}{(x - 1)(x + 2)}$ **33.** $\dfrac{11y + 15}{4y(y + 5)}$ **35.** $-\dfrac{7}{x + 7}$ **37.** $\dfrac{3x - 55}{(x + 5)(x - 5)}$ **39.** $\dfrac{x^2 + 6x}{(x - 4)(x + 4)}$ **41.** $\dfrac{y + 12}{(y + 3)(y - 3)}$

43. $\dfrac{7x - 10}{(x - 1)(x - 1)}$ **45.** $\dfrac{9y}{4(y - 5)}$ **47.** $\dfrac{8y + 16}{y(y + 4)}$ **49.** $\dfrac{17}{(x - 3)(x - 4)}$ **51.** $\dfrac{7x - 1}{(x + 1)(x + 1)(x - 1)}$ **53.** $\dfrac{x^2 - x}{(x + 3)(x - 2)(x + 5)}$

55. $\dfrac{y^2 + 8y + 4}{(y + 4)(y + 1)(y + 1)}$ **57.** $\dfrac{2x^2 - 4x + 34}{(x + 3)(x - 5)}$ **59.** $\dfrac{5 - 3y}{2y(y - 1)}$ **61.** $-\dfrac{x^2}{(x + 3)(x - 3)}$ **63.** $\dfrac{2}{y + 5}$ **65.** $\dfrac{4x - 11}{x - 3}$ **67.** $\dfrac{3}{y + 1}$

69. $\dfrac{-x^2 + 7x + 3}{(x - 3)(x + 2)}$ **71.** $\dfrac{x^2 + 2x - 1}{(x + 1)(x - 1)(x + 2)}$ **73.** $\dfrac{-y^2 + 8y + 9}{15y^2}$ **75.** $\dfrac{x^2 - 2x - 6}{3(x + 2)(x - 2)}$ **77.** $\dfrac{-y - 2}{(y + 1)(y - 1)}$

79. $\dfrac{x + 2xy - y}{xy}$ **81.** $\dfrac{5x + 2y}{(x + y)(x - y)}$ **83.** $\dfrac{-x + 6}{(x + 2)(x - 2)}$ **85.** $\dfrac{6x + 5}{(x - 5)(x + 5)(x - 6)}$ **87.** $\dfrac{9}{(x - 3)(x^2 + 3x + 9)}$

89. $\dfrac{3}{y - 3}$ **91.** $-\dfrac{22y}{(x + 3y)(x + y)(x - 3y)}$ **93.** $\dfrac{D}{5}$ **95.** $\dfrac{D}{24}$ **97.** no **99.** 5 yr **101.** $\dfrac{4x^2 + 14x}{(x + 3)(x + 4)}$

105. $\dfrac{3}{x + 5}$ should be $\dfrac{5 + 2x}{5x}$. **109.** $\dfrac{y^3 + y^2 - 10y - 2}{(y - 1)(y - 1)(y + 3)}$ **111.** $\dfrac{3x - 1}{x^2(x - 1)}$ **113.** $6x^2 - 11x - 35$

114. $3x - y = 3$

115. $y = x - 1$

Mid-Chapter 7 Check Point

1. $x = -2$ and $x = 4$ **2.** $\dfrac{x - 2}{2x + 1}$ **3.** $\dfrac{-3}{y - 2}$ or $\dfrac{3}{2 - y}$ **4.** $\dfrac{2}{w}$ **5.** $\dfrac{4}{x + 4}$ **6.** $\dfrac{4}{(x - 2)^2}$ **7.** $\dfrac{x + 5}{x - 2}$ **8.** $\dfrac{x + 2}{x - 4}$

9. $\dfrac{2x + 1}{(x + 2)(x + 3)(x - 1)}$ **10.** $\dfrac{9 - x}{x - 5}$ **11.** $-\dfrac{2y - 1}{3y}$ or $\dfrac{1 - 2y}{3y}$ **12.** $\dfrac{-y^2}{(y + 1)(y + 2)}$ **13.** $\dfrac{w + 1}{(w - 3)(w + 5)}$

14. $\dfrac{2z^2 - 3z + 15}{(z + 3)(z - 3)(z + 1)}$ **15.** $\dfrac{3z^2 + 5z + 3}{(3z - 1)^2}$ **16.** $\dfrac{3x - 1}{(x + 7)(x - 3)}$ **17.** x **18.** $\dfrac{2x + 1}{(x + 1)(x - 2)(x + 3)}$ **19.** $\dfrac{5x - 5y}{x}$ **20.** $\dfrac{x + 3}{x + 5}$

Section 7.5

Check Point Exercises

1. $\frac{11}{5}$ **2.** $\frac{2x-1}{2x+1}$ **3.** $y-x$ **4.** $\frac{11}{5}$ **5.** $\frac{2x-1}{2x+1}$ **6.** $y-x$

Exercise Set 7.5

1. $\frac{9}{10}$ **3.** $\frac{18}{23}$ **5.** $-\frac{4}{5}$ **7.** $\frac{3-4x}{3+4x}$ **9.** $\frac{7x-2}{5x+1}$ **11.** $\frac{2y+3}{y-7}$ **13.** $\frac{4-6y}{4+3y}$ **15.** $x-5$ **17.** $\frac{x}{x-1}$ **19.** $-\frac{1}{y}$

21. $\frac{xy+2}{x}$ **23.** $\frac{y+x}{x^2y^2}$ **25.** $\frac{x^2+y}{y(y+1)}$ **27.** $\frac{1}{y}$ **29.** $\frac{4-x}{5x-3}$ **31.** $\frac{2y}{y-3}$ **33.** $\frac{1}{x+3}$ **35.** $\frac{x^2-5x+3}{x^2-7x+2}$ **37.** $\frac{2y^2+3xy}{y^2+2x}$

39. $-\frac{6}{5}$ **41.** $\frac{1-x}{(x-3)(x+6)}$ **43.** $\frac{1}{y(y+5)}$ **45.** $\frac{1}{x-1}$ **47.** $\frac{x+1}{2x+1}$ **49.** $\frac{2r_1r_2}{r_1+r_2}$; $34\frac{2}{7}$ mph **55.** It cubes x.

57. $\frac{5y+3}{(y-3)(y+1)}$ **59.** 2 should be $1+x$. **61.** $2x(x-5)^2$ **62.** $\{-2\}$ **63.** x^3+y^3

Section 7.6

Check Point Exercises

1. 4 or $\{4\}$ **2.** 3 or $\{3\}$ **3.** -3 and -2, or $\{-3, -2\}$ **4.** 24 or $\{24\}$ **5.** no solution or $\varnothing$ **6.** 75% **7.** $x = \frac{yz}{y-z}$

Exercise Set 7.6

1. $\{12\}$ **3.** $\{0\}$ **5.** $\{-2\}$ **7.** $\{6\}$ **9.** $\{-2\}$ **11.** $\{2\}$ **13.** $\{15\}$ **15.** $\{4\}$ **17.** $\{-6, 1\}$ **19.** $\{-5, 5\}$ **21.** $\{-3, 3\}$

23. $\{-8, 1\}$ **25.** $\{-1, 16\}$ **27.** $\{3\}$ **29.** no solution or $\varnothing$ **31.** $\left\{\frac{2}{3}, 4\right\}$ **33.** no solution or $\varnothing$ **35.** $\{1\}$ **37.** $\left\{\frac{1}{5}\right\}$ **39.** $\{-3\}$

41. no solution or $\varnothing$ **43.** $\left\{-6, \frac{1}{2}\right\}$ **45.** $P_1 = \frac{P_2 V_2}{V_1}$ **47.** $f = \frac{pq}{q+p}$ **49.** $r = \frac{A-P}{P}$ **51.** $m_1 = \frac{Fd^2}{Gm_2}$ **53.** $x = \bar{x} + zs$

55. $R = \frac{E-Ir}{I}$ **57.** $f_1 = \frac{ff_2}{f_2-f}$ **59.** $\{-3\}$ **61.** $\frac{-6x-18}{(x-5)(x+4)}$ **63.** $\{1, 5\}$ **65.** $\{3\}$ **67.** 10,000 wheelchairs

69. 50% **71.** 5 yr old **73.** either 50 or approx 67 cases; (50, 350) or $\left(66\frac{2}{3}, 350\right)$ **75.** 10 more hits **83.** b **85.** no solution

87. $\{8\}$ **89.** $\{-3, -2\}$ **90.** $(x^3 - 3)(x + 2)$ **91.** $-\frac{12}{x^8}$ **92.** $-20x + 55$

Section 7.7

Check Point Exercises

1. 2 mph **2.** $2\frac{2}{3}$ hr or 2 hr 40 min **3.** $1500 **4.** 720 deer **5.** 32 in. **6.** 32 yd

Exercise Set 7.7

1. Walking rate is 6 mph; car rate is 9 mph. **3.** Downhill rate is 10 mph; uphill rate is 6 mph. **5.** Water's current is 5 mph. **7.** Walking rate is 3 mph; jogging rate is 6 mph. **9.** Still water rate is 10 mph. **11.** It will take about 8.6 min, which is enough time. **13.** It will take about 171.4 hr, which is not enough time. **15.** It will take 2.4 hr, or 2 hr 24 min. **17.** $1800 **19.** 20,489 fur seal pups **21.** $950 **23.** 154.1 in. **25.** 5 in. **27.** 6 m **29.** 16 in. **31.** 16 ft **43.** 30 mph **45.** 7 hr **47.** 2.5 hr **49.** $(5x + 9)(5x - 9)$ **50.** $\{6\}$ **51.**

Section 7.8

Check Point Exercises

1. 66 gal **2.** 512 cycles per second **3.** 24 min **4.** 96π cubic feet

Exercise Set 7.8

1. 156 **3.** 30 **5.** $\frac{5}{6}$ **7.** 240 **9.** 50 **11.** $x = kyz$; $y = \frac{x}{kz}$ **13.** $x = \frac{kz^3}{y}$; $y = \frac{kz^3}{x}$

15. $x = \frac{kyz}{\sqrt{w}}$; $y = \frac{x\sqrt{w}}{kz}$ **17.** $x = kz(y + w)$; $y = \frac{x - kzw}{kz}$ **19.** $x = \frac{kz}{y-w}$; $y = \frac{xw + kz}{x}$ **21.** 5.4 ft **23.** 80 in.

25. about 607 lb **27.** 32° **29.** 90 milliroentgens per hour **31.** This person has a BMI of 24.4 and is not overweight. **33.** 1800 Btu

35. $\frac{1}{4}$ of what it was originally **37. a.** $C = \frac{kP_1P_2}{d^2}$ **b.** $k \approx 0.02$; $C = \frac{0.02\,P_1P_2}{d^2}$ **c.** approximately 39,813 daily phone calls

39. a.

(graph with points (0.5, 12), (1, 6), (1.5, 4), (2.5, 2.4), (3, 2), (4, 1.5), (2, 3), (5, 1.2); axes R (ohms) vs I (amperes))

b. Current varies inversely as resistance. **c.** $R = \dfrac{6}{I}$

45. z varies directly as the square root of x and inversely as the square of y.

49. The wind pressure is 4 times more destructive.

51. Distance is increased by $\sqrt{50}$, or about 7.07, for the space telescope.

52. $\left\{\dfrac{16}{3}\right\}$ **53.** $9x^2 - 6x + 4 - \dfrac{16}{3x + 2}$ **54.** $6x(x - 5)(x + 4)$

Review Exercises

1. $x = 4$ **2.** $x = 2$ and $x = -5$ **3.** $x = 1$ and $x = 2$ **4.** defined for all real numbers **5.** $\dfrac{4x}{3}$ **6.** $x + 2$ **7.** x^2 **8.** $\dfrac{x - 3}{x - 6}$

9. $\dfrac{x - 5}{x + 7}$ **10.** $\dfrac{y}{y + 2}$ **11.** cannot be simplified **12.** $-2(x + 3y)$ **13.** $\dfrac{x - 2}{4}$ **14.** $\dfrac{5}{2}$ **15.** $\dfrac{x + 3}{x + 2}$ **16.** $\dfrac{2y - 1}{5}$ **17.** $\dfrac{-3(y + 1)}{5y(2y - 3)}$

18. $\dfrac{x - 1}{4}$ **19.** $\dfrac{2}{x(x + 1)}$ **20.** $\dfrac{1}{7(y + 3)}$ **21.** $(y - 3)(y - 6)$ **22.** $\dfrac{8}{x - y}$ **23.** 4 **24.** 4 **25.** $3x - 5$ **26.** $3y + 4$

27. $\dfrac{4}{x - 2}$ **28.** $\dfrac{2x + 5}{x - 3}$ **29.** $36x^3$ **30.** $x^2(x - 1)^2$ **31.** $(x + 3)(x + 1)(x + 7)$ **32.** $\dfrac{14x + 15}{6x^2}$ **33.** $\dfrac{7x + 2}{x(x + 1)}$ **34.** $\dfrac{7x + 25}{(x + 3)^2}$

35. $\dfrac{3}{y - 2}$ **36.** $-\dfrac{y}{y - 1}$ **37.** $\dfrac{x^2 + y^2}{xy}$ **38.** $\dfrac{3x^2 - x}{(x + 1)^2(x - 1)}$ **39.** $\dfrac{5x^2 - 3x}{(x + 1)(x - 1)}$ **40.** $\dfrac{4}{(x + 2)(x - 2)(x - 3)}$ **41.** $\dfrac{2x + 13}{x + 3}$

42. $\dfrac{y - 4}{3y}$ **43.** $\dfrac{7}{2}$ **44.** $\dfrac{1}{x - 1}$ **45.** $x + y$ **46.** $\dfrac{3}{x}$ **47.** $\dfrac{3x}{x - 4}$ **48.** $\{12\}$ **49.** $\{-1\}$ **50.** $\{-6, 1\}$ **51.** no solution or Ø

52. $\{5\}$ **53.** $\{-6, 2\}$ **54.** $\{5\}$ **55.** $\{-5\}$ **56.** 3 yr **57.** 30% **58.** $C = R - nP$ **59.** $T_1 = \dfrac{P_1 V_1 T_2}{P_2 V_2}$ **60.** $P = \dfrac{A}{rT + 1}$

61. $R = \dfrac{R_1 R_2}{R_2 + R_1}$ **62.** $n = \dfrac{IR}{E - Ir}$ **63.** 4 mph **64.** Slower car's rate is 20 mph; faster car's rate is 30 mph. **65.** 4 hr **66.** 324 teachers

67. 287 trout **68.** 5 ft **69.** $12\frac{1}{2}$ ft **70.** $4935 **71.** 1600 ft **72.** 440 vibrations per second **73.** 112 decibels **74.** 16 hr **75.** 800 cubic feet

Chapter 7 Test

1. $x = -9$ and $x = 4$ **2.** $\dfrac{x + 3}{x - 2}$ **3.** $\dfrac{4x}{x + 1}$ **4.** $\dfrac{x - 4}{2}$ **5.** $\dfrac{x}{x + 3}$ **6.** $\dfrac{2x + 6}{x + 1}$ **7.** $\dfrac{5}{(y - 1)(y - 3)}$ **8.** $2y$

9. $\dfrac{2}{y + 3}$ **10.** $\dfrac{x^2 + 2x + 15}{(x + 3)(x - 3)}$ **11.** $\dfrac{8x - 14}{(x + 2)(x - 1)(x - 3)}$ **12.** $\dfrac{-x - 1}{x - 3}$ **13.** $\dfrac{x + 2}{x - 1}$ **14.** $\dfrac{11}{(x - 3)(x - 4)}$ **15.** $\dfrac{4}{y + 4}$

16. $\dfrac{3x - 3y}{x}$ **17.** $\dfrac{5x + 5}{2x + 1}$ **18.** $\dfrac{y - x}{y}$ **19.** 6 or $\{6\}$ **20.** -8 or $\{-8\}$ **21.** 0 and 2, or $\{0, 2\}$ **22.** $a = -\dfrac{Rs}{s - R}$ or $a = -\dfrac{RS}{R - s}$

23. Water's current is 2 mph. **24.** 12 min **25.** 6000 tule elk **26.** 3.2 in. **27.** 52.5 amp

Cumulative Review Exercises (Chapters 1–7)

1. $\{2\}$ **2.** $\{x \mid x < 2\}$ **3.** $\{-6, 3\}$ **4.** $\{-6\}$ **5.** $\{(3, 3)\}$ **6.** $\{(-2, 2)\}$ **7.** $3x - 2y = 6$ *(graph)* **8.** $y = -2x + 3$ *(graph)*

9. $y = -3$ *(graph)* **10.** -19 **11.** $8x^9$ **12.** $\dfrac{1 - 2x}{4x - 1}$ **13.** $(4x - 1)(x - 3)$

14. $(2x - 5)^2$ **15.** $3(x + 5)(x - 5)$ **16.** $-x^2 + 4x + 8$

17. $-2x^4 + 3x^2 - 1$ **18.** $\dfrac{3x^2 + 6x + 16}{(x - 2)(x + 3)}$

19. $1225 at 5% and $2775 at 9% **20.** 17 in. and 51 in.

Mid-Textbook Check Point

1. 4 or $\{4\}$ **2.** 10 or $\{10\}$ **3.** $\{x \mid x \le 7\}$ **4.** $(-3, 4)$ or $\{(-3, 4)\}$ **5.** $(3, 4)$ or $\{(3, 4)\}$ **6.** -8 or $\{-8\}$

7. -3 and -2, or $\{-3, -2\}$ **8.** $\dfrac{4}{x^9}$ **9.** 31 **10.** $10x^2 - 9x + 4$ **11.** $21x^2 - 23x - 20$ **12.** $25x^2 - 20x + 4$ **13.** $x^3 + y^3$

14. $\dfrac{x + 4}{3x^3}$ **15.** $\dfrac{-7x}{(x + 3)(x - 1)(x - 4)}$ **16.** $\dfrac{x}{5}$ **17.** $(2x + 7)(2x - 7)$ **18.** $(x + 3)(x + 1)(x - 1)$ **19.** $2(x - 3)(x + 7)$

20. $x(x^2 + 4)(x + 2)(x - 2)$ **21.** $x(x - 5)^2$ **22.** $(x - 2)(x^2 + 2x + 4)$ **23.** $y = \frac{1}{3}x - 1$ *(graph)* **24.** $3x + 2y = -6$ *(graph)*

25. $y = -2$ *(graph)* **26.** -2 **27.** $y - 2 = 2(x - 1)$ or $y - 6 = 2(x - 3)$; $y = 2x$

28. 43 **29.** $320 **30.** length: 150 yd; width: 50 yd

31. $12,500 at 7% and $7500 at 9% **32.** 8 I of 40% and 4 I of 70% **33.** 16 ft **34.** 1st: 35°; 2nd: 25°; 3rd: 120°

35. TV: $350; stereo: $370 **36.** length: 11m; width: 5 m

CHAPTER 8

Section 8.1

Check Point Exercises

1. domain: $\{5, 10, 15, 20, 25\}$; range: $\{12.8, 16.2, 18.9, 20.7, 21.8\}$ **2. a.** not a function   **b.** function   **3. a.** 29   **b.** 65   **c.** 46
d. $6a + 6h + 9$ **4. a.** function   **b.** function   **c.** not a function   **5. a.** 16   **b.** 8
6. a. Domain $= \{x | -2 \le x \le 1\}$; Range $= \{y | 0 \le y \le 3\}$ **b.** Domain $= \{x | -2 < x \le 1\}$; Range $= \{y | -1 \le y < 2\}$
c. Domain $= \{x | -3 \le x < 0\}$; Range $= \{y | y = -3, -2, -1\}$

Exercise Set 8.1

1. function; domain: $\{1, 3, 5\}$; range: $\{2, 4, 5\}$ **3.** not a function; domain: $\{3, 4\}$; range: $\{4, 5\}$ **5.** function; domain: $\{-3, -2, -1, 0\}$;
range: $\{-3, -2, -1, 0\}$ **7.** not a function; domain: $\{1\}$; range: $\{4, 5, 6\}$ **9. a.** 1   **b.** 6   **c.** -7 **d.** $2a + 1$ **e.** $a + 3$
11. a. -2 **b.** -17 **c.** 0   **d.** $12b - 2$ **e.** $3b + 10$ **13. a.** 5   **b.** 8   **c.** 53   **d.** 32   **e.** $48b^2 + 5$

15. a. -1 **b.** 26   **c.** 19   **d.** $2b^2 + 3b - 1$ **e.** $50a^2 + 15a - 1$ **17. a.** $\dfrac{3}{4}$ **b.** -3 **c.** $\dfrac{11}{8}$ **d.** $\dfrac{13}{9}$ **e.** $\dfrac{2a + 2h - 3}{a + h - 4}$

f. Denominator would be zero.   **19.** function   **21.** not a function   **23.** function   **25.** not a function   **27.** -4 **29.** 4   **31.** 0   **33.** 2
35. 2   **37.** -2 **39.** Domain $= \{x | 0 \le x < 5\}$; Range $= \{y | -1 \le y < 5\}$ **41.** Domain $= \{x | x \ge 0\}$; Range $= \{y | y \ge 1\}$
43. Domain $= \{x | -2 \le x \le 6\}$; Range $= \{y | -2 \le y \le 6\}$ **45.** Domain $= \{x | x$ is a real number$\}$; Range $= \{y | y \le -2\}$
47. Domain $= \{x | x = -5, -2, 0, 1, 3\}$; Range $= \{y | y = 2\}$ **49.** $-2; 10$ **51.** -38 **53.** $-2x^3 - 2x$ **55. a.** -1 **b.** 7   **c.** 19   **d.** 112
57. Domain $= \{x | x \ne 1\}$; Range $= \{y | y \ne 0\}$ **59. a.** $\{(\text{EL}, 1\%), (\text{L}, 7\%), (\text{SL}, 11\%), (\text{M}, 52\%), (\text{SC}, 13\%), (\text{C}, 13\%), (\text{EC}, 3\%)\}$
b. Yes; Each ideology corresponds to exactly one percentage.   **c.** $\{(1\%, \text{EL}), (7\%, \text{L}), (11\%, \text{SL}), (52\%, \text{M}), (13\%, \text{SC}), (13\%, \text{C}), (3\%, \text{EC})\}$
d. No; 13% in the domain corresponds to two ideologies in the range, SC and C   **61.** 5.22; In 2000, there were 5.22 million women enrolled in U.S.
colleges; (2000, 5.22).   **63.** 1.4; In 2004, there were 1.4 million more women than men enrolled in U.S. colleges.   **65.** 440; For 20-year-old drivers,
there are 440 accidents per 50 million miles driven; (20, 440).   **67.** $x = 45$; $y = 190$; The minimum number of accidents is 190 per 50 million miles
driven and is attributed to 45-year-old drivers.   **69.** 3.1; In 1960, Jewish Americans made up about 3.1% of the U.S. population.   **71.** 19 and 64;
In 1919 and in 1964, Jewish Americans made up about 3% of the U.S. population.   **73.** 1940; 3.7%   **75.** Each year corresponds to only one percentage.
77. 0.83; It costs \$0.83 to mail a 3-ounce first-class letter.   **79.** \$0.60   **91.** 3   **93.** $f(2) = 6; f(3) = 9; f(4) = 12;$ no   **94.** 0
95. $\dfrac{y^{10}}{9x^4}$ **96.** -15

Section 8.2

Check Point Exercises

1. a. $\{x | x$ is a real number$\}$ **b.** $\{x | x$ is a real number and $x \ne -5\}$ **2. a.** $3x^2 + 6x + 6$ **b.** 78   **3. a.** $\dfrac{5}{x} + \dfrac{7}{x - 8}$

b. $\{x | x$ is a real number and $x \ne 0$ and $x \ne 8\}$ **4. a.** 23   **b.** $x^2 - 3x - 3; 1$ **c.** $x^3 + x^2 - 6x; -24$ **d.** $\dfrac{x^2 - 2x}{x + 3}; \dfrac{7}{2}$

Exercise Set 8.2

1. $\{x | x$ is a real number$\}$ **3.** $\{x | x$ is a real number and $x \ne -4\}$ **5.** $\{x | x$ is a real number and $x \ne 3\}$ **7.** $\{x | x$ is a real number and $x \ne 5\}$
9. $\{x | x$ is a real number and $x \ne -7$ and $x \ne 9\}$ **11. a.** $5x - 5$ **b.** 20   **13. a.** $3x^2 + x - 5$ **b.** 75   **15. a.** $2x^2 - 2$ **b.** 48
17. $\{x | x$ is a real number$\}$ **19.** $\{x | x$ is a real number and $x \ne 5\}$ **21.** $\{x | x$ is a real number and $x \ne 0$ and $x \ne 5\}$
23. $\{x | x$ is a real number and $x \ne 2$ and $x \ne -3\}$ **25.** $\{x | x$ is a real number and $x \ne 2\}$ **27.** $\{x | x$ is a real number$\}$

29. $x^2 + 3x + 2; 20$ **31.** 0   **33.** $x^2 + 5x - 2; 48$ **35.** -8 **37.** $-x^3 - 2x^2 + 8x; 0$ **39.** -135 **41.** $\dfrac{x^2 + 4x}{2 - x}; 5$ **43.** -1
45. $\{x | x$ is a real number$\}$ **47.** $\{x | x$ is a real number and $x \ne 2\}$ **49.** 5   **51.** -1 **53.** $\{x | -4 \le x \le 3\}$
55. 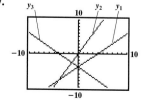   **57.** -4 **59.** -4 **61.** $\{x | x = 0, 1, 2, \ldots, 7\}$ **63. a.** $(B - D)(x) = 4565x + 1,578,296$; change in U.S. population
b. $(B - D)(6) = 1,605,686$; In 2001, U.S. population increased by 1,605,686.   **c.** 1,609,508; fairly well
65. $h(x)$ or total world population in year x **67.** 5.9 billion   **69.** $-200,000; 0; 200,000$; The company has a loss of
\$200,000, breaks even, and a profit of \$200,000 when 20,000, 30,000, and 40,000 radios are produced, respectively.
75. Per capita costs of Medicare are increasing more and more rapidly over time.
77. **79.** 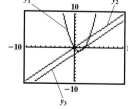   **81.** d   **83.** $\{(-1, 2)\}$ **84.** 7   **85.** $6b + 8$

Mid-Chapter 8 Check Point

1. not a function; Domain: $\{1, 2\}$; Range: $\{-6, 4, 6\}$ **2.** function; Domain: $\{0, 2, 3\}$; Range: $\{1, 4\}$ **3.** function; Domain: $\{x \mid -2 \le x < 2\}$; Range: $\{y \mid 0 \le y \le 3\}$ **4.** not a function; Domain: $\{x \mid -3 < x \le 4\}$; Range: $\{y \mid -1 \le y \le 2\}$ **5.** not a function; Domain: $\{-2, -1, 0, 1, 2\}$; Range: $\{-2, -1, 1, 3\}$ **6.** function; Domain: $\{x \mid x \le 1\}$; Range: $\{y \mid y \ge -1\}$ **7.** No vertical line intersects the graph of f more than once.
8. 3 **9.** -2 **10.** -6 and 2 **11.** $\{x \mid x \text{ is a real number}\}$ **12.** $\{y \mid y \le 4\}$ **13.** $\{x \mid x \text{ is a real number}\}$
14. $\{x \mid x \text{ is a real number and } x \ne -2 \text{ and } x \ne 2\}$ **15.** 23 **16.** 23 **17.** $a^2 - 5a - 3$ **18.** $x^2 - 5x + 3$; 17 **19.** $x^2 - x + 13$; 33
20. $-2x^3 + x^2 - x - 40$; -36 **21.** $\dfrac{x^2 - 3x + 8}{-2x - 5}$; 12 **22.** $\left\{ x \mid x \text{ is a real number and } x \ne -\dfrac{5}{2} \right\}$

Section 8.3

Check Point Exercises

1. a. $5x^2 + 1$ **b.** $25x^2 + 60x + 35$ **2.** $f(g(x)) = 7\left(\dfrac{x}{7}\right) = x; g(f(x)) = \dfrac{7x}{7} = x$

3. $f(g(x)) = 4\left(\dfrac{x + 7}{4}\right) - 7 = x + 7 - 7 = x; g(f(x)) = \dfrac{(4x - 7) + 7}{4} = \dfrac{4x}{4} = x$ **4.** $f^{-1}(x) = \dfrac{x - 7}{2}$

5. (b) and (c) **6.**

Exercise Set 8.3

1. a. $2x + 14$ **b.** $2x + 7$ **c.** 18 **3. a.** $2x + 5$ **b.** $2x + 9$ **c.** 9 **5. a.** $20x^2 - 11$ **b.** $80x^2 - 120x + 43$ **c.** 69
7. a. $x^4 - 4x^2 + 6$ **b.** $x^4 + 4x^2 + 2$ **c.** 6 **9. a.** $\sqrt{x - 1}$ **b.** $\sqrt{x} - 1$ **c.** 1 **11. a.** x **b.** x **c.** 2 **13. a.** x **b.** x **c.** 2
15. $f(g(x)) = x; g(f(x)) = x$; inverses **17.** $f(g(x)) = x; g(f(x)) = x$; inverses **19.** $f(g(x)) = \dfrac{5x - 56}{9}; g(f(x)) = \dfrac{5x - 4}{9}$; not inverses
21. $f(g(x)) = x; g(f(x)) = x$; inverses **23.** $f(g(x)) = x; g(f(x)) = x$; inverses **25. a.** $f^{-1}(x) = x - 3$ **b.** $f(f^{-1}(x)) = (x - 3) + 3 = x$
and $f^{-1}(f(x)) = (x + 3) - 3 = x$ **27. a.** $f^{-1}(x) = \dfrac{x}{2}$ **b.** $f(f^{-1}(x)) = 2\left(\dfrac{x}{2}\right) = x$ and $f^{-1}(f(x)) = \dfrac{2x}{2} = x$ **29. a.** $f^{-1}(x) = \dfrac{x - 3}{2}$
b. $f(f^{-1}(x)) = 2\left(\dfrac{x - 3}{2}\right) + 3 = x$ and $f^{-1}(f(x)) = \dfrac{(2x + 3) - 3}{2} = x$ **31. a.** $f^{-1}(x) = \dfrac{1}{x}$ **b.** $f(f^{-1}(x)) = \dfrac{1}{\frac{1}{x}} = x$ and $f^{-1}(f(x)) = \dfrac{1}{\frac{1}{x}} = x$

33. a. $f^{-1}(x) = \dfrac{3x + 1}{x - 2}$ **b.** $f(f^{-1}(x)) = \dfrac{2\left(\dfrac{3x + 1}{x - 2}\right) + 1}{\left(\dfrac{3x + 1}{x - 2}\right) - 3} = x$ and $f^{-1}(f(x)) = \dfrac{3\left(\dfrac{2x + 1}{x - 3}\right) + 1}{\left(\dfrac{2x + 1}{x - 3}\right) - 2} = x$ **35.** no inverse **37.** no inverse

39. inverse function **41.** **43.** **45.** 5 **47.** 1
49. 2 **51.** 1
53. -6 **55.** -7
57. 3 **59.** 11

61. a. f represents the price after a \$400 discount and g represents the price after a 25% discount (75% of the regular price).
 b. $0.75x - 400$; $f \circ g$ represents an additional \$400 discount on a price that has already been reduced by 25%.
 c. $0.75(x - 400) = 0.75x - 300$; $g \circ f$ represents an additional 25% discount on a price that has already been reduced \$400.
 d. $f \circ g$; $0.75x - 400 < 0.75x - 300$, so $f \circ g$ represents the lower price after the two discounts.
 e. $f^{-1}(x) = x + 400$; f^{-1} represents the regular price, since the value of x here is the price after a \$400 discount.
63. a. f: $\{(\text{Zambia}, 4.2), (\text{Colombia}, 4.5), (\text{Poland}, 3.3), (\text{Italy}, 3.3), (\text{U.S.}, 2.5)\}$
 b. $\{(4.2, \text{Zambia}), (4.5, \text{Colombia}), (3.3, \text{Poland}), (3.3, \text{Italy}), (2.5, \text{U.S.})\}$; no; The input 3.3 is associated with two outputs, Poland and Italy.
65. a. No horizontal line intersects the graph of f in more than one point.
 b. $f^{-1}(0.25)$, or approximately 15, represents the number of people who would have to be in the room so that the probability of two sharing a birthday would be 0.25; $f^{-1}(0.5)$, or approximately 23, represents the number of people so that the probability would be 0.5; $f^{-1}(0.7)$, or approximately 30, represents the number of people so that the probability would be 0.7.

73.

no inverse function

75.

inverse function

77.

no inverse function

79.

no inverse function

81.

inverses

83.

inverses

85. Answers will vary; Examples are $f(x) = \sqrt{x + 5}$ and $g(x) = 3x^2$.

87. $f(f(x)) = \dfrac{3\left(\dfrac{3x - 2}{5x - 3}\right) - 2}{5\left(\dfrac{3x - 2}{5x - 3}\right) - 3} = \dfrac{3(3x - 2) - 2(5x - 3)}{5(3x - 2) - 3(5x - 3)} = \dfrac{9x - 6 - 10x + 6}{15x - 10 - 15x + 9} = \dfrac{-x}{-1} = x$ **89.** 5×10^8 **90.** $x^2 + 9x + 16 + \dfrac{35}{x - 2}$

91. $\left\{\left(\dfrac{4}{5}, \dfrac{9}{5}\right)\right\}$

Review Exercises

1. function; domain: $\{3, 4, 5\}$; range: $\{10\}$ **2.** function; domain: $\{1, 2, 3, 4\}$; range: $\{12, 100, \pi, -6\}$ **3.** not a function; domain: $\{13, 15\}$; range: $\{14, 16, 17\}$ **4. a.** -5 **b.** 16 **c.** -75 **d.** $14a - 5$ **e.** $7a + 9$ **5. a.** 2 **b.** 52 **c.** 70 **d.** $3b^2 - 5b + 2$ **e.** $48a^2 - 20a + 2$ **6.** not a function   **7.** function   **8.** function   **9.** not a function   **10.** not a function   **11.** function   **12.** -3 **13.** -2 **14.** 3 **15.** $\{x \mid -3 \le x < 5\}$ **16.** $\{y \mid -5 \le y \le 0\}$ **17. a.** For each time, there is only one height.   **b.** 0; The vulture was on the ground after 15 seconds.   **c.** 45 m   **d.** 7 and 22; After 7 and 22 seconds, the vulture's height is 20 meters.   **e.** Answers will vary.   **18.** $\{x \mid x \text{ is a real number}\}$ **19.** $\{x \mid x \text{ is a real number and } x \ne -8\}$ **20.** $\{x \mid x \text{ is a real number and } x \ne 5\}$ **21. a.** $6x - 4$ **b.** 14 **22. a.** $5x^2 + 1$ **b.** 46 **23.** $\{x \mid x \text{ is a real number and } x \ne 4\}$ **24.** $\{x \mid x \text{ is a real number and } x \ne -6 \text{ and } x \ne -1\}$ **25.** $x^2 - x - 5; 1$ **26.** 1

27. $x^2 - 3x + 5; 3$ **28.** 9 **29.** $x^3 - 7x^2 + 10x; -120$ **30.** $\dfrac{x^2 - 2x}{x - 5}; -8$ **31.** $\{x \mid x \text{ is a real number}\}$ **32.** $\{x \mid x \text{ is a real number and } x \ne 5\}$

33. a. $16x^2 - 8x + 4$ **b.** $4x^2 + 11$ **c.** 124 **34. a.** $\sqrt{x + 1}$ **b.** $\sqrt{x} + 1$ **c.** 2 **35.** $f(g(x)) = x - \dfrac{7}{10}; g(f(x)) = x - \dfrac{7}{6}$; not inverses

36. $f(g(x)) = x; g(f(x)) = x$; inverses   **37. a.** $f^{-1}(x) = \dfrac{x + 3}{4}$ **b.** $f(f^{-1}(x)) = 4\left(\dfrac{x + 3}{4}\right) - 3 = x$ and $f^{-1}(f(x)) = \dfrac{(4x - 3) + 3}{4} = x$

38. a. $f^{-1}(x) = -\dfrac{1}{x}$ **b.** $f(f^{-1}(x)) = -\dfrac{1}{\left(-\frac{1}{x}\right)} = x$ and $f^{-1}(f(x)) = -\dfrac{1}{\left(-\frac{1}{x}\right)} = x$ **39.** inverse function   **40.** no inverse function   **41.** inverse function   **42.** no inverse function   **43.**

Chapter 8 Test

1. function; domain: $\{1, 3, 5, 6\}$; range: $\{2, 4, 6\}$ **2.** not a function; domain: $\{2, 4, 6\}$; range: $\{1, 3, 5, 6\}$ **3.** $3a + 10$ **4.** 285. function   **6.** not a function   **7.** -3 **8.** -2 and 3   **9.** $\{x \mid x \text{ is a real number}\}$ **10.** $\{y \mid y \le 3\}$ **11.** $\{x \mid x \text{ is a real number and } x \ne 10\}$ **12.** $x^2 + 5x + 2; 26$ **13.** $x^2 + 3x - 2; -4$ **14.** $x^3 + 6x^2 + 8x; -15$ **15.** $\dfrac{x^2 + 4x}{x + 2}; 3$ **16.** $\{x \mid x \text{ is a real number and } x \ne -2\}$ **17.** $(f \circ g)(x) = 9x^2 - 3x; (g \circ f)(x) = 3x^2 + 3x - 1$ **18.** $f^{-1}(x) = \dfrac{x + 7}{5}$

19. a. No horizontal line intersects the graph of f in more than one point.   **b.** 2000   **c.** $f^{-1}(2000)$ represents the income, \$80 thousand, of a family that gives \$2000 to charity.

Cumulative Review Exercises (Chapters 1–8)

1. $\{-3\}$ **2.** $\left\{-4, \dfrac{3}{2}\right\}$ **3.** $\{(-3, 4)\}$ **4.** $\left\{\dfrac{9}{7}\right\}$ **5.** $\{-5\}$ **6.** $x^2 - 3x - 1; -1$ **7.** $-\dfrac{2}{x^4}$ **8.** 4 **9.** $\dfrac{3}{x}$ **10.** $-\dfrac{2 + x}{3x + 1}$

11. 5 **12.** $(x - 7)(x - 11)$ **13.** $x(x + 5)(x - 5)$ **14.** $6x^2 - 7x + 2$ **15.** $8x^3 - 27$ **16.** $\dfrac{1}{x - 1}$ **17.** $\dfrac{5x - 1}{4x - 3}$ **18.** $\{(3, 2, 4)\}$

19. $2x - y = 4$

20. $y = -\dfrac{2}{3}x$

21. function; domain: $\{1, 2, 3, 4, 6\}$; range: $\{5\}$ **22.** $-\dfrac{8}{3}$

23. $y + 3 = 5(x + 2); y = 5x + 7$ **24.** 2.1×10^{-5}

25. $\{x \mid x \text{ is a real number and } x \ne 15\}$

CHAPTER 9

Section 9.1

Check Point Exercises

1. a. **b.** **c.**

2. a. $\{x|-2 \le x < 5\}$ **b.** $\{x|1 \le x \le 3.5\}$ **c.** $\{x|x < -1\}$

3. $\{x|x \ge -9\}$ or $[-9, \infty)$ **4.** $\{x|x > 13\}$ or $(13, \infty)$ **5. a.** $P(x) = 125x - 160,000$ **b.** more than 1280 units

6. a. $C(x) = 300,000 + 30x$ **b.** $R(x) = 80x$ **c.** $P(x) = 50x - 300,000$ **d.** more than 6000 pairs

Exercise Set 9.1

1. $\{x|1 < x \le 6\}$

3. $\{x|-5 \le x < 2\}$

5. $\{x|-3 \le x \le 1\}$

7. $\{x|x > 2\}$

9. $\{x|x \ge -3\}$

11. $\{x|x < 3\}$

13. $\{x|x < 5.5\}$

15. $\{x|x < 3\}$ or $(-\infty, 3)$

17. $\{x|x \ge 7\}$ or $[7, \infty)$

19. $\{x|x \le -4\}$ or $(-\infty, -4]$

21. $\left\{x|x \le -\dfrac{2}{5}\right\}$ or $\left(-\infty, -\dfrac{2}{5}\right]$

23. $\{x|x \ge 0\}$ or $[0, \infty)$

25. $\{x|x < 1\}$ or $(-\infty, 1)$

27. $\{x|x \ge 6\}$ or $[6, \infty)$

29. $\{x|x \ge -6\}$ or $[-6, \infty)$

31. $\{x|x < -6\}$ or $(-\infty, -6)$

33. $\{x|x \ge 13\}$ or $[13, \infty)$

35. $\{y|y > -6\}$ or $(-6, \infty)$

37. $\{x|x \ge -1\}$ or $[-1, \infty)$

39. $\{y|y < -2\}$ or $(-\infty, -2)$

41. $\{x|x < 5\}$ or $(-\infty, 5)$

43. $\left\{x|x \ge -\dfrac{4}{7}\right\}$ or $\left[-\dfrac{4}{7}, \infty\right)$

45. a. $P(x) = 17x - 25,500$ **b.** more than 1500 units **47. a.** $P(x) = 140x - 70,000$ **b.** more than 500 units

49. $\{x|x < 2\}$ or $(-\infty, 2)$

51. $x < \dfrac{c - b}{a}$ **53.** $\{x|x \le -3\}$ or $(-\infty, -3]$ **55.** $\{x|x > -1.4\}$ or $(-1.4, \infty)$ **57.** $(0, 4)$

59. intimacy $\ge$ passion or passion $\le$ intimacy **61.** commitment $>$ passion or passion $<$ commitment

63. 9; after 3 years **65.** voting years after 2006

67. $\{t|t > 175\}$ or $(175, \infty)$; The women's speed skating times will be less than the men's after the year 2075.

69. a. $C(x) = 18,000 + 20x$ **b.** $R(x) = 80x$ **c.** $P(x) = 60x - 18,000$ **d.** more than 300 canoes

71. a. $C(x) = 30,000 + 2500x$ **b.** $R(x) = 3125x$ **c.** $P(x) = 625x - 30,000$ **d.** more than 48 sold-out performances

73. more than 6250 tapes **75.** more than 300 minutes **83.** $\{x|x < -3\}$ or $(-\infty, -3)$

$y_1 = -2(x + 4)$
$y_2 = 6x + 16$

85. ∅   **87. a.** plan A: $4 + 0.10x$; plan B: $2 + 0.15x$ **b.**

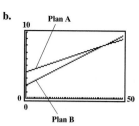

c–d. more than 40 checks per month

89. Since $x > y$, $y - x < 0$. Thus, when both sides were multiplied by $y - x$, the sense of the inequality should have been changed.

91. 29   **92.** $(-1, -1, 2)$ **93.** $(5x + 9)(5x - 9)$

Section 9.2

Check Point Exercises

1. $\{3, 7\}$ **2.** $\{x | x < 1\}$ or $(-\infty, 1)$ **3.** ∅   **4.** $\{x | -1 \le x < 4\}$ or $[-1, 4)$;

5. $\{3, 4, 5, 6, 7, 8, 9\}$ **6.** $\{x | x \le 1 \text{ or } x > 3\}$ or $(-\infty, 1] \cup (3, \infty)$ **7.** $\{x | x \text{ is a real number}\}$ or $\mathbb{R}$ or $(-\infty, \infty)$

Exercise Set 9.2

1. $\{2, 4\}$ **3.** ∅   **5.** ∅

7. $\{x | x > 6\}$; $(6, \infty)$

9. $\{x | x \le 1\}$; $(-\infty, 1]$

11. $\{x | -1 \le x < 2\}$; $[-1, 2)$

13. ∅

15. $\{x | -6 < x < -4\}$; $(-6, -4)$

17. $\{x | -3 < x \le 6\}$; $(-3, 6]$

19. $\{x | 2 < x < 5\}$; $(2, 5)$

21. ∅

23. $\{x | 0 \le x < 2\}$; $[0, 2)$

25. $\{x | 3 < x < 5\}$; $(3, 5)$

27. $\{x | -1 \le x < 3\}$; $[-1, 3)$

29. $\{x | -5 < x \le -2\}$; $(-5, -2]$

31. $\{x | 3 \le x < 6\}$; $[3, 6)$

33. $\{1, 2, 3, 4, 5\}$

35. $\{1, 2, 3, 4, 5, 6, 7, 8, 10\}$

37. $\{a, e, i, o, u\}$

39. $\{x | x > 3\}$; $(3, \infty)$

41. $\{x | x \le 5\}$; $(-\infty, 5]$

43. $\{x | x \text{ is a real number}\}$; $(-\infty, \infty)$

45. $\{x | x < -1 \text{ or } x \ge 2\}$; $(-\infty, -1) \cup [2, \infty)$

47. $\{x|x < -3 \text{ or } x > 4\}; (-\infty, -3) \cup (4, \infty)$ **49.** $\{x|x \le 1 \text{ or } x \ge 3\}; (-\infty, 1] \cup [3, \infty)$ **51.** $\{x|x \text{ is a real number}\}; (-\infty, \infty)$

53. $\{x|x < 2\}; (-\infty, 2)$

55. $\{x|x > 4\} \text{ or } (4, \infty)$ **57.** $\{x|x < 0 \text{ or } x > 6\} \text{ or } (-\infty, 0) \cup (6, \infty)$

59. $\dfrac{b-c}{a} < x < \dfrac{b+c}{a}$ **61.** $\{x|-1 \le x \le 3\} \text{ or } [-1, 3]$ **63.** $\{x|-1 < x < 3\} \text{ or } (-1, 3)$

65. $\{x|-1 \le x < 2\} \text{ or } [-1, 2)$ **67.** $\{-3, -2, -1\}$ **69.** spatial-temporal toys, sports equipment, and toy cars and trucks   **71.** dollhouses, spatial-temporal toys, sports equipment, and toy cars and trucks   **73.** none of the toys   **75.** between 80 and 110 minutes, inclusive

77. $[76, 126)$; If the highest grade is 100, then $[76, 100]$.   **79.** more than 3 and less than 15 crossings per 3-month period

89. $\{x|-6 < x < 2\} \text{ or } (-6, 2)$ **91.** $\{x|-3 \le x \le 2\} \text{ or } [-3, 2]$

93. $(-1, 5]$ **95.** $[-1, \infty)$ **97.** $(-1, 4]$ **99.** $-x^2 + 5x - 9; -15$ **100.** $f(x) = -\dfrac{1}{2}x + 4$ **101.** $17 - 2x$

Section 9.3

Check Point Exercises

1. -2 and 3, or $\{-2, 3\}$ **2.** $\dfrac{4}{3}$ and 10, or $\left\{\dfrac{4}{3}, 10\right\}$

3. $\{x|-3 < x < 7\} \text{ or } (-3, 7)$ **4.** $\{x|x \le 1 \text{ or } x \ge 4\} \text{ or } (-\infty, 1] \cup [4, \infty)$

5. $\{x|8.1 \le x \le 13.9\} \text{ or } [8.1, 13.9]$; The percentage of children in the population who think that not being able to do everything they want is a bad thing is between a low of 8.1% and a high of 13.9%.

Exercise Set 9.3

1. $\{-8, 8\}$ **3.** $\{-5, 9\}$ **5.** $\{-3, 4\}$ **7.** $\{-1, 2\}$ **9.** $\varnothing$ **11.** $\{-3\}$ **13.** $\{-11, -1\}$ **15.** $\{-3, 4\}$ **17.** $\left\{-\dfrac{13}{3}, 5\right\}$

19. $\left\{-\dfrac{2}{5}, \dfrac{2}{5}\right\}$ **21.** $\varnothing$ **23.** $\varnothing$ **25.** $\left\{\dfrac{1}{2}\right\}$ **27.** $\left\{\dfrac{3}{4}, 5\right\}$ **29.** $\left\{\dfrac{5}{3}, 3\right\}$ **31.** $\{0\}$ **33.** $\{4\}$ **35.** $\{4\}$ **37.** $\{-1, 15\}$

39. $\{x|-3 < x < 3\} \text{ or } (-3, 3)$ **41.** $\{x|1 < x < 3\} \text{ or } (1, 3)$ **43.** $\{x|-3 \le x \le -1\} \text{ or } [-3, -1]$

45. $\{x|-1 < x < 7\} \text{ or } (-1, 7)$ **47.** $\{x|x < -3 \text{ or } x > 3\} \text{ or }$ $(-\infty, -3) \cup (3, \infty)$ **49.** $\{x|x < -4 \text{ or } x > -2\} \text{ or }$ $(-\infty, -4) \cup (-2, \infty)$

51. $\{x|x \le 2 \text{ or } x \ge 6\} \text{ or }$ $(-\infty, 2] \cup [6, \infty)$ **53.** $\left\{x|x < \dfrac{1}{3} \text{ or } x > 5\right\} \text{ or } \left(-\infty, \dfrac{1}{3}\right) \cup (5, \infty)$ **55.** $\{x|-5 \le x \le 3\} \text{ or } [-5, 3]$

57. $\{x|-6 < x < 0\}$ or $(-6, 0)$

59. $\{x|x \le -5 \text{ or } x \ge 3\}$ or

$(-\infty, -5] \cup [3, \infty)$

61. $\{x|x < -3 \text{ or } x > 12\}$ or

$(-\infty, -3) \cup (12, \infty)$

63. $\varnothing$

65. $\{x|x \text{ is a real number}\}$ or $\mathbb{R}$ or $(-\infty, \infty)$

67. $\{x|-9 \le x \le 5\}$ or $[-9, 5]$

69. $\{x|x < 1 \text{ or } x > 2\}$ or $(-\infty, 1) \cup (2, \infty)$

71. $-\dfrac{3}{2}$ and 4   **73.** -7 and -2 **75.** $\left\{x \middle| -\dfrac{7}{3} \le x \le 1\right\}$ or $\left[-\dfrac{7}{3}, 1\right]$ **77.** $\{x|x < -1 \text{ or } x > 4\}$ or $(-\infty, -1) \cup (4, \infty)$

79. $\left(-\infty, -\dfrac{1}{3}\right] \cup [3, \infty)$ **81.** $\left\{x \middle| \dfrac{-c-b}{a} < x < \dfrac{c-b}{a}\right\}$ **83.** 3 and 5   **85.** $\{x|-2 \le x \le 1\}$ or $[-2, 1]$

87. $\{x|58.6 \le x \le 61.8\}$ or $[58.6, 61.8]$; The percentage of the U.S. population that watched M*A*S*H is between a low of 58.6% and a high of 61.8%.; 1.6%   **89.** $\{T|50 \le T \le 64\}$ or $[50, 64]$; The monthly average temperature for San Francisco, CA is between a low of 50°F and a high of 64°F.
91. $\{x|8.59 \le x \le 8.61\}$ or $[8.59, 8.61]$; A machine part that is supposed to be 8.6 centimeters is acceptable between a low of 8.59 and a high of 8.61 centimeters.   **93.** If the number of outcomes that result in heads is 41 or less or 59 or more, then the coin is unfair.
103. -8 and 0 **105.** $\{x|-4 < x < 1\}$ or $(-4, 1)$ **107.** $\varnothing$ **109.**

$\{x|x < 2 \text{ or } x > 6\}$ or $(-\infty, 2) \cup (6, \infty)$

113. a. $|x - 4| < 3$ **b.** $|x - 4| \ge 3$ **115.** 1   **116.**

117. $f(x) = -\dfrac{2}{3}x$

118. $f(x) = -2$

Mid-Chapter 9 Check Point

1. $\{x|x \le -4\}$; $(-\infty, -4]$ **2.** $\{x|3 \le x < 5\}$; $[3, 5)$ **3.** $\left\{\dfrac{1}{2}, 3\right\}$ **4.** $\{x|x < -1\}$; $(-\infty, -1)$

5. $\{x|x < -9 \text{ or } x > -5\}$; $(-\infty, -9) \cup (-5, \infty)$ **6.** $\left\{x \middle| -\dfrac{2}{3} \le x \le 2\right\}$; $\left[-\dfrac{2}{3}, 2\right]$ **7.** $\left\{\dfrac{1}{2}, \dfrac{13}{4}\right\}$ **8.** $\{x|-4 < x \le -2\}$; $(-4, -2]$

9. $\{x|x \text{ is real number}\}$; $(-\infty, \infty)$ **10.** $\{x|x \le -3\}$; $(-\infty, -3]$ **11.** $\{x|x < -3\}$; $(-\infty, -3)$

12. $\left\{x \middle| x < -1 \text{ or } x > -\dfrac{1}{5}\right\}$; $(-\infty, -1) \cup \left(-\dfrac{1}{5}, \infty\right)$ **13.** $\{x|x \le -10 \text{ or } x \ge 2\}$; $(-\infty, -10] \cup [2, \infty)$ **14.** $\left\{x \middle| x < -\dfrac{5}{3}\right\}$; $\left(-\infty, -\dfrac{5}{3}\right)$

15. $\{x|x > 4\}$; $(4, \infty)$ **16.** $\{-2, 7\}$ **17.** $\varnothing$ **18. a.** $C(x) = 60{,}000 + 0.18x$ **b.** $R(x) = 0.30x$ **c.** $P(x) = 0.12x - 60{,}000$
d. at least 750,000 discs each month   **19.** no more than 80 miles per day   **20.** $[49\%, 99\%)$ **21.** at least $120,000

Section 9.4

Check Point Exercises

1. $4x - 2y \ge 8$

2. $y > -\dfrac{3}{4}x$

3. a. $y > 1$

b. $x \le -2$

4. $x - 3y < 6$
$2x + 3y \ge -6$

5. $x + y < 2$
$-2 \le x < 1$
$y > -3$

6. $B = (60, 20)$; Using $T = 60$ and $P = 20$, each of the three inequalities for grasslands is true: $60 \ge 35$, true; $5(60) - 7(20) \ge 70$, true; $3(60) - 35(20) \le -140$, true.

Exercise Set 9.4

1. $x + y \geq 3$

3. $x - y < 5$

5. $x + 2y > 4$

7. $3x - y \leq 6$

9. $\frac{x}{2} + \frac{y}{3} < 1$

11. $y > \frac{1}{3}x$

13. $y \leq 3x + 2$

15. $y < -\frac{1}{4}x$

17. $x \leq 2$

19. $y > -4$

21. $y \geq 0$

23. $3x + 6y \leq 6$
$2x + y \leq 8$

25. $2x - 5y \leq 10$
$3x - 2y > 6$

27. $y > 2x - 3$
$y < -x + 6$

29. $x + 2y \leq 4$
$y \geq x - 3$

31. $x \leq 2$
$y \geq -1$

33. $-2 \leq x < 5$

35. $x - y \leq 1$
$x \geq 2$

37. $\emptyset$ **39.** $x + y > 4$
$x + y > -1$

41. $x - y \leq 2$
$x \geq -2$
$y \leq 3$

43. $x \geq 0$
$y \geq 0$
$2x + 5y \leq 10$
$3x + 4y \leq 12$

45. $3x + y \leq 6$
$2x - y \leq -1$
$x \geq -2$
$y \leq 4$

47. $y \geq -2x + 4$

49. $x + y \leq 4$ and $3x + y \leq 6$

$x + y \leq 4$
$3x + y \leq 6$

51. $-2 \leq x \leq 2; -3 \leq y \leq 3$

$-2 \leq x \leq 2$
$-3 \leq y \leq 3$

53. $y > \frac{3}{2}x - 2$ or $y < 4$

55. no solution   **57.** infinitely many solutions

59. a. $A = (20, 150)$; A 20-year-old with a heart rate of 150 beats per minute is within the target range.

b. $10 \leq 20 \leq 70$, true; $150 \geq 0.7(220 - 20)$, true; $150 \leq 0.8(220 - 20)$, true

61. $10 \leq a \leq 70; H \geq 0.6(220 - a); H \leq 0.7(220 - a)$

63. a. $y \geq 0; x + y \geq 5; x \geq 1; 200x + 100y \leq 700$

b. $y \geq 0$
$x + y \geq 5$
$x \geq 1$
$200x + 100y \leq 700$

c. 2 nights

75.

77.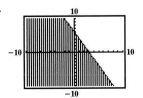

83. $x \geq -2; y > -1$ **85.** Answers will vary.   **87.** $\{(3, 1)\}$ **88.** $\{(2, 4)\}$ **89.** $2x^4(x + 5y)^2$

Section 9.5

Check Point Exercises

1. $z = 25x + 55y$ **2.** $x + y \leq 80$ **3.** $30 \leq x \leq 80, 10 \leq y \leq 30; z = 25x + 55y, x + y \leq 80, 30 \leq x \leq 80, 10 \leq y \leq 30$
4. 50 bookshelves and 30 desks; \$2900   **5.** 30

Exercise Set 9.5

1. $(1, 2)$: 17; $(2, 10)$: 70; $(7, 5)$: 65; $(8, 3)$: 58; maximum: 70; minimum: 17   **3.** $(0, 0)$: 0; $(0, 8)$: 400; $(4, 9)$: 610; $(8, 0)$: 320; maximum: 610; minimum: 0

5. a.

b. $(0, 4)$: 8; $(0, 8)$: 16; $(4, 0)$: 12

c. maximum: 16; at $(0, 8)$

7. a.

b. $(0, 3)$: 3; $(0, 4)$: 4; $(6, 0)$: 24; $(3, 0)$: 12

c. maximum: 24; at $(6, 0)$

9. a.

b. $(1, 2)$: −1; $(1, 4)$: −5; $(5, 8)$: −1; $(5, 2)$: 11

c. maximum: 11; at $(5, 2)$

11. a.

b. $(0, 2)$: 4; $(0, 4)$: 8; $\left(\dfrac{12}{5}, \dfrac{12}{5}\right)$: $\dfrac{72}{5} = 14.4$; $(4, 0)$: 16; $(2, 0)$: 8

c. maximum: 16; at $(4, 0)$

13. a.

b. $(0, 0)$: 0; $(0, 6)$: 72; $(3, 4)$: 78; $(5, 0)$: 50

c. maximum: 78; at $(3, 4)$

15. a. $z = 125x + 200y$

b. $x \leq 450; y \leq 200; 600x + 900y \leq 360,000$

c.

d. 0; 40,000; 77,500; 76,250; 56,250
e. 300; 200; 77,500

17. 40 of model A and 0 of model B   **19.** 300 boxes of food and 200 boxes of clothing   **21.** 100 parents and 50 students

23. 10 Boeing 727s and 42 Falcon 20s   **29.** \$5000 in stocks and \$5000 in bonds   **31.** {6}   **32.** $\dfrac{x + 9}{x - 15}$ **33.** 10

Review Exercises

1. $\{x | -2 < x \leq 3\}$

2. $\{x | -1.5 \leq x \leq 2\}$

3. $\{x | x > -1\}$

4. $\{x | x \geq -2\}; [-2, \infty)$

5. $\left\{x \,\middle|\, x \geq \dfrac{3}{5}\right\}; \left[\dfrac{3}{5}, \infty\right)$

6. $\left\{x \,\middle|\, x < -\dfrac{21}{2}\right\}; \left(-\infty, -\dfrac{21}{2}\right)$

7. $\{x | x > -3\}; (-3, \infty)$

8. $\{x | x \leq -2\}; (-\infty, -2]$

9. a. $P(x) = 85x - 37,000$ **b.** more than 4200 toaster ovens   **10.** $C(x) = 360,000 + 850x$ **11.** $R(x) = 1150x$
12. $P(x) = 300x - 360,000$ **13.** more than 4200 computers   **14.** more than 50 checks   **15.** more than \$13,500 in sales   **16.** $\{a, c\}$
17. $\{a\}$ **18.** $\{a, b, c, d, e\}$ **19.** $\{a, b, c, d, f, g\}$

20. $\{x | x \leq 3\}; (-\infty, 3]$

21. $\{x | x < 6\}; (-\infty, 6)$

22. $\{x | 6 < x < 8\}; (6, 8)$

23. $\{x \mid x \le 1\}; (-\infty, 1]$

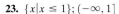

24. $\varnothing$ **25.** $\{x \mid x < 1 \text{ or } x > 2\}; (-\infty, 1) \cup (2, \infty)$ **26.** $\{x \mid x \le -4 \text{ or } x > 2\}; (-\infty, -4] \cup (2, \infty)$

27. $\{x \mid x < -2\}; (-\infty, -2)$

28. $\{x \mid x \text{ is a real number}\}; (-\infty, \infty)$

29. $\{x \mid -5 < x \le 2\}; (-5, 2]$

30. $\left\{ x \mid -\dfrac{3}{4} \le x \le 1 \right\}; \left[-\dfrac{3}{4}, 1 \right]$

31. $[81, 131);$ If the highest grade is 100, then $[81, 100].$

32. $\{-4, 3\}$ **33.** $\varnothing$ **34.** $\left\{ -\dfrac{11}{2}, \dfrac{23}{2} \right\}$ **35.** $\left\{ -4, -\dfrac{6}{11} \right\}$

36. $\{x \mid -9 \le x \le 6\}; [-9, 6]$

37. $\{x \mid x < -6 \text{ or } x > 0\}; (-\infty, -6) \cup (0, \infty)$

38. $\{x \mid -3 < x < -2\}; (-3, -2)$

39. $\varnothing$ **40.** Approximately 90% of the population sleeps between 5.5 hours and 7.5 hours daily, inclusive.

41. $3x - 4y > 12$ **42.** $x - 3y \le 6$ **43.** $y \le -\dfrac{1}{2}x + 2$ **44.** $y > \dfrac{3}{5}x$ **45.** $x \le 2$

46. $y > -3$ **47.** $2x - y \le 4$ $x + y \ge 5$ **48.** $y < -x + 4$ $y > x - 4$ **49.** $-3 \le x < 5$ **50.** $-2 < y \le 6$

51. $x \ge 3$ $y \le 0$ **52.** $2x - y > -4$ $x \ge 0$ **53.** $x + y \le 6$ $y \ge 2x - 3$ **54.** $3x + 2y \ge 4$ $x - y \le 3$ $x \ge 0, y \ge 0$

55. $\varnothing$

56. $\left(\dfrac{1}{2}, \dfrac{1}{2} \right): \dfrac{5}{2};$

$(2, 2): 10; (4, 0): 8; (1, 0): 2;$

maximum: 10; minimum: 2

57.

maximum: 24

58.

maximum: 33

59.

maximum: 44

60. a. $z = 500x + 350y$ **b.** $x + y \le 200; x \ge 10; y \ge 80$

c.

d. $(10, 80): 33,000; (10, 190): 71,500; (120, 80): 88,000$

e. 120; 80; 88,000   **61.** 480 of model A and 240 of model B

Chapter 9 Test

1. $\{x \mid -3 \le x < 2\}$

2. $\{x \mid x \le -1\}$

3. $\{x \mid x \le 12\}; (-\infty, 12]$

4. $\left\{x \mid x \ge \dfrac{21}{8}\right\}; \left[\dfrac{21}{8}, \infty\right)$

5. a. $C(x) = 600,000 + 200x$ **b.** $R(x) = 450x$
c. $P(x) = 250x - 60,000$
d. more than 240 desks

6. $\{4, 6\}$ **7.** $\{2, 4, 6, 8, 10, 12, 14\}$

8. $\{x \mid -2 < x < -1\}; (-2, -1)$

9. $\{x \mid x \ge -2\}; [-2, \infty)$

10. $\{x \mid x < 4\}; (-\infty, 4)$

11. $\{x \mid x \le -4 \text{ or } x > 2\}; (-\infty, -4] \cup (2, \infty)$

12. $\left\{x \mid -7 \le x < \dfrac{13}{2}\right\}; \left[-7, \dfrac{13}{2}\right)$

13. $\left\{-2, \dfrac{4}{5}\right\}$ **14.** $\left\{-\dfrac{8}{5}, 7\right\}$

15. $\{x \mid -3 < x < 4\}; (-3, 4)$

16. $\{x \mid x \le -1 \text{ or } x \ge 4\}; (-\infty, -1] \cup [4, \infty)$

17. $\{b \mid b < 90.6 \text{ or } b > 106.6\}$ or $(-\infty, 90.6) \cup (106.6, \infty)$; Hypothermia: Body temperature below 90.6°F; Hyperthermia: Body temperature above 106.6°F

18. $3x - 2y < 6$

19. $y \ge \dfrac{1}{2}x - 1$

20. $y \le -1$

21. $x + y \ge 2$
$x - y \ge 4$

22. $3x + y \le 9$
$2x + 3y \ge 6$
$x \ge 0$
$y \ge 0$

23. $-2 < x \le 4$

24. maximum: 26

25. 50 regular and 100 deluxe; $35,000

Cumulative Review Exercises (Chapters 1–9)

1. $\{-1\}$ **2.** $\{8\}$ **3.** $-\dfrac{2y^7}{3x^5}$ **4.** $22; 4a^2 - 6a + 4$ **5.** $2x^2 + x + 2; 12$ **6.** $f(x) = -\dfrac{1}{2}x + 4$

7. $f(x) = 2x + 1$

8. $y > 2x$

9. $2x - y \ge 6$

10. $f(x) = -1$

11. $(-4, 2, -1)$ **12.** $f^{-1}(x) = 3x + 12$
13. $f(g(x)) = 3x^2 + 12x + 11$;
$g(f(x)) = 3x^2 + 1$
14. 46 rooms with kitchen facilities and 14 without kitchen facilities
15. a. and b. are functions.

16. $\{x \mid x \ge -7\}; [-7, \infty)$

17. $\{x \mid x < -6\}; (-\infty, -6)$

18. $\{x \mid x \le 3 \text{ or } x \ge 5\}; (-\infty, 3] \cup [5, \infty)$

19. $\{x \mid -10 \le x \le 7\}; [-10, 7]$

20. $\left\{x \mid x < \dfrac{1}{3} \text{ or } x > 5\right\}; \left(-\infty, \dfrac{1}{3}\right) \cup (5, \infty)$

CHAPTER **10**

Section 10.1

Check Point Exercises

1. a. 8   **b.** -7 **c.** $\dfrac{4}{5}$ **d.** 0.09   **e.** 5   **f.** 7   **2. a.** 4   **b.** $-\sqrt{24} \approx -4.90$ **3.** $\{x|x \geq 3\}$ or $[3, \infty)$

4. 53.9%; This is the percent given for 2001, not 2002, so the model does not provide a totally accurate description of the data.

5. a. 7   **b.** $|x + 8|$ **c.** $|7x^5|$ or $7|x^5|$ **d.** $|x - 3|$ **6. a.** 3   **b.** -2 **7.** $-3x$ **8. a.** 2   **b.** -2 **c.** not a real number   **d.** -1

9. a. $|x + 6|$ **b.** $3x - 2$ **c.** 8

Exercise Set 10.1

1. 6   **3.** -6 **5.** not a real number   **7.** $\dfrac{1}{5}$ **9.** $-\dfrac{3}{4}$ **11.** 0.9   **13.** -0.2 **15.** 3   **17.** 1   **19.** not a real number

21. 4; 1; 0; not a real number   **23.** -5; $-\sqrt{5} \approx -2.24$; -1; not a real number   **25.** 4; 2; 1; 6   **27.** $\{x|x \geq 3\}$ or $[3, \infty)$; c

29. $\{x|x \geq -5\}$ or $[-5, \infty)$; d   **31.** $\{x|x \leq 3\}$ or $(-\infty, 3]$; e   **33.** 5   **35.** 4   **37.** $|x - 1|$ **39.** $|6x^2|$ or $6x^2$

41. $-|10x^3|$ or $-10|x^3|$ **43.** $|x + 6|$ **45.** $-|x - 4|$ **47.** 3   **49.** -3 **51.** $\dfrac{1}{5}$ **53.** 3; 2; -1; -4 **55.** -2; 0; 2

57. 1   **59.** 2   **61.** -2 **63.** not a real number   **65.** -1 **67.** not a real number   **69.** -4 **71.** 2   **73.** -2

75. x **77.** $|y|$ **79.** $-2x$ **81.** -5 **83.** 5   **85.** $|x + 3|$ **87.** $-2(x - 1)$

89.

Domain: $\{x|x \geq 0\} = [0, \infty)$;

Range: $\{y|y \geq 3\} = [3, \infty)$

91.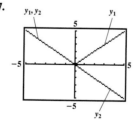

Domain: $\{x|x \geq 3\} = [3, \infty)$;

Range: $\{y|y \geq 0\} = [0, \infty)$

93. $\{x|x < 15\} = (-\infty, 15)$
95. $\{x|1 \leq x < 3\} = [1, 3)$
97. 3
99. 40.2 in.; The model describes actual data well.
101. 70 mph; The officer should not believe the motorist.; Answers will vary.

115. ; Answers will   **117.**

a. $\{x|x \leq 0\}$ **b.** $\{x|x > 0\}$
119. Answers will vary; an example is $f(x) = \sqrt{15 - 3x}$.
121. $|(2x + 3)^5|$

123. The graph of h is the graph of f shifted left by 3 units.

124. $7x + 30$ **125.** $\dfrac{x^8}{9y^6}$ **126.** $\left\{x \middle| x < -\dfrac{7}{3} \text{ or } x > 5\right\}$ or $\left(-\infty, -\dfrac{7}{3}\right) \cup (5, \infty)$

Section 10.2

Check Point Exercises

1. a. 5   **b.** -2 **c.** $\sqrt[4]{5xy^2}$ **2. a.** $(5xy)^{1/4}$ **b.** $\left(\dfrac{a^3b}{2}\right)^{1/5}$ **3. a.** $(\sqrt[3]{8})^4 = 16$ **b.** $(\sqrt{25})^3 = 125$ **c.** $-(\sqrt[4]{81})^3 = -27$ **4. a.** $6^{4/3}$ **b.** $(2xy)^{7/5}$

5. a. $\dfrac{1}{100^{1/2}} = \dfrac{1}{10}$ **b.** $\dfrac{1}{8^{1/3}} = \dfrac{1}{2}$ **c.** $\dfrac{1}{32^{3/5}} = \dfrac{1}{8}$ **d.** $\dfrac{1}{(3xy)^{5/9}}$ **6.** 25° **7. a.** $7^{5/6}$ **b.** $\dfrac{5}{x}$ **c.** $9.1^{3/10}$ **d.** $\dfrac{y^{1/12}}{x^{1/5}}$

8. a. $\sqrt{x}$ **b.** $2a^4$ **c.** $\sqrt[4]{x^2y}$ **d.** $\sqrt[6]{x}$ **e.** $\sqrt[6]{x}$

Exercise Set 10.2

1. $\sqrt{49} = 7$ **3.** $\sqrt[3]{-27} = -3$ **5.** $-\sqrt[4]{16} = -2$ **7.** $\sqrt[3]{xy}$ **9.** $\sqrt[5]{2xy^3}$ **11.** $(\sqrt{81})^3 = 729$ **13.** $(\sqrt[3]{125})^2 = 25$ **15.** $(\sqrt[5]{-32})^3 = -8$

17. $(\sqrt[3]{27})^2 + (\sqrt[4]{16})^3 = 17$ **19.** $\sqrt[7]{(xy)^4}$ **21.** $7^{1/2}$ **23.** $5^{1/3}$ **25.** $(11x)^{1/5}$ **27.** $x^{3/2}$ **29.** $x^{3/5}$ **31.** $(x^2y)^{1/5}$ **33.** $(19xy)^{3/2}$

35. $(7xy^2)^{5/6}$ **37.** $2xy^{2/3}$ **39.** $\dfrac{1}{49^{1/2}} = \dfrac{1}{7}$ **41.** $\dfrac{1}{27^{1/3}} = \dfrac{1}{3}$ **43.** $\dfrac{1}{16^{3/4}} = \dfrac{1}{8}$ **45.** $\dfrac{1}{8^{2/3}} = \dfrac{1}{4}$ **47.** $\left(\dfrac{27}{8}\right)^{1/3} = \dfrac{3}{2}$ **49.** $\dfrac{1}{(-64)^{2/3}} = \dfrac{1}{16}$

51. $\dfrac{1}{(2xy)^{7/10}}$ **53.** $\dfrac{5x}{z^{1/3}}$ **55.** 3 **57.** 4 **59.** $x^{5/6}$ **61.** $x^{3/5}$ **63.** $\dfrac{1}{x^{5/12}}$ **65.** 25 **67.** $\dfrac{1}{y^{1/6}}$ **69.** $32x$ **71.** $5x^2y^3$ **73.** $\dfrac{x^{1/4}}{y^{3/10}}$

75. 3 **77.** $27y^{2/3}$ **79.** $\sqrt[4]{x}$ **81.** $2a^2$ **83.** x^2y^3 **85.** x^6y^6 **87.** $\sqrt[3]{3y}$ **89.** $\sqrt[3]{4a^2}$ **91.** $\sqrt[3]{x^2y}$ **93.** $\sqrt[6]{2^5}$ or $\sqrt[6]{32}$ **95.** $\sqrt[10]{x^9}$

97. $\sqrt[12]{a^{10}b^7}$ **99.** $\sqrt[20]{x}$ **101.** $\sqrt{y}$ **103.** $\sqrt[8]{x}$ **105.** $\sqrt[4]{x^2y}$ **107.** $\sqrt[12]{2x}$ **109.** x^9y^{15} **111.** $\sqrt{a^3b^3}$ **113.** $x^{2/3} - x$ **115.** $x + 2x^{1/2} - 15$

117. $2x^{1/2}(3 + x)$ **119.** $15x^{1/3}(1 - 4x^{2/3})$ **121.** $\dfrac{x^2}{7y^{3/2}}$ **123.** $\dfrac{x^3}{y^2}$ **125.** 58 species of plants   **127.** about 1872 calories per day

129. a. $C(v) = 35.74 - 35.74v^{4/25}$ **b.** $C(25) \approx -24$; When the air temperature is $0°F$ and the wind speed is 25 miles per hour, the windchill temperature is $-24°$.   **c.** the point $(25, -24)$ **131. a.** $L + 1.25S^{1/2} - 9.8D^{1/3} \le 16.296$ **b.** eligible

143. simplified correctly   **145.** Right side should be $x^{1/2}$.

147. 8 years old   **149.** 61   **151.** $y = -2x + 11$ or $f(x) = -2x + 11$ **152.** $y \le -\dfrac{3}{2}x + 3$ **153.** $3a^2 + 6ah + 3h^2 - 5a - 5h + 4$

Section 10.3

Check Point Exercises

1. a. $\sqrt{55}$ **b.** $\sqrt{x^2 - 16}$ **c.** $\sqrt[3]{60}$ **d.** $\sqrt[7]{12x^4}$ **2. a.** $4\sqrt{5}$ **b.** $2\sqrt[3]{5}$ **c.** $2\sqrt[4]{2}$ **d.** $10|x|\sqrt{2y}$
3. $f(x) = \sqrt{3}|x - 2|$ **4.** $x^4y^5z\sqrt{xyz}$ **5.** $2x^3y^4\sqrt[3]{5xy^2}$ **6.** $2x^2z\sqrt[5]{x^2y^2z^3}$ **7. a.** $2\sqrt{3}$ **b.** $100\sqrt[3]{4}$ **c.** $2x^2y\sqrt[4]{2}$

Exercise Set 10.3

1. $\sqrt{15}$ **3.** $\sqrt[3]{18}$ **5.** $\sqrt[4]{33}$ **7.** $\sqrt{33xy}$ **9.** $\sqrt[5]{24x^4}$ **11.** $\sqrt{x^2 - 9}$ **13.** $\sqrt[6]{(x - 4)^5}$ **15.** $\sqrt{x}$ **17.** $\sqrt[4]{\dfrac{3x}{7y}}$ **19.** $\sqrt[7]{77x^5y^3}$

21. $5\sqrt{2}$ **23.** $3\sqrt{5}$ **25.** $5\sqrt{3x}$ **27.** $2\sqrt[3]{2}$ **29.** $3x$ **31.** $-2y\sqrt[3]{2x^2}$ **33.** $6|x + 2|$ **35.** $2(x + 2)\sqrt[3]{4}$ **37.** $|x - 1|\sqrt{3}$

39. $x^3\sqrt{x}$ **41.** $x^4y^4\sqrt{y}$ **43.** $4x\sqrt{3x}$ **45.** $y^2\sqrt[3]{y^2}$ **47.** $x^4y\sqrt[3]{x^2z}$ **49.** $3x^2y^2\sqrt[3]{3x^2}$ **51.** $(x + y)\sqrt[3]{(x + y)^2}$ **53.** $y^3\sqrt[5]{y^2}$

55. $2xy^3\sqrt[3]{2xy^2}$ **57.** $2x^2\sqrt[4]{5x^2}$ **59.** $(x - 3)^2\sqrt[4]{(x - 3)^2}$ or $(x - 3)^2\sqrt{x - 3}$ **61.** $2\sqrt{6}$ **63.** $5\sqrt{2xy}$ **65.** $6x$ **67.** $10xy\sqrt{2y}$ **69.** $60\sqrt{2}$

71. $2\sqrt[3]{6}$ **73.** $2x^2\sqrt{10x}$ **75.** $5xy^4\sqrt[3]{x^2y^2}$ **77.** $2xyz\sqrt[4]{x^2z^3}$ **79.** $2xy^2z\sqrt[5]{2y^3z}$ **81.** $(x - y)^2\sqrt[3]{(x - y)^2}$ **83.** $-6x^3y^3\sqrt[3]{2yz^2}$

85. $-6y^2\sqrt[5]{2x^3y}$ **87.** $-6x^3y^3\sqrt{2}$ **89.** $-12x^3y^4\sqrt[4]{x^3}$ **91.** $xy^2\sqrt[5]{16y^3} + 2x^2y^3$ **93.** $6\sqrt{3}$ miles; 10.4 miles   **95.** $8\sqrt{3}$ ft per sec; 14 ft per sec

97. a. $\dfrac{7.644}{2\sqrt[4]{2}} = \dfrac{3.822}{\sqrt[4]{2}}$ **b.** 3.21 liters of blood per minute per square meter; $(32, 3.21)$

105. Graphs are the same; simplification is correct.   **107.** Graphs are not the same; $\sqrt{3x^2 - 6x + 3} = |x - 1|\sqrt{3}$.   **109.** d

111. The number must be multiplied by 27.

113.
$f(x) = \sqrt{(x-1)^2}$

114. $\{x \mid 5 \le x \le 11\}$ or $[5, 11]$

115. $\{(2, -4)\}$

116. $(4x - 3)(16x^2 + 12x + 9)$

Section 10.4

Check Point Exercises

1. a. $10\sqrt{13}$ **b.** $(21 - 6x)\sqrt[3]{7}$ **c.** $5\sqrt[4]{3x} + 2\sqrt[3]{3x}$ **2. a.** $21\sqrt{5}$ **b.** $-12\sqrt{3x}$ **c.** cannot be simplified

3. a. $-9\sqrt[3]{3}$ **b.** $(5 + 3xy)\sqrt[3]{x^2 y}$ **4. a.** $\dfrac{2\sqrt[3]{3}}{5}$ **b.** $\dfrac{3x\sqrt{x}}{y^5}$ **c.** $\dfrac{2y^2\sqrt[3]{y}}{x^4}$ **5. a.** $2x^2\sqrt{5}$ **b.** $\dfrac{5\sqrt{xy}}{2}$ **c.** $2x^2 y$

Exercise Set 10.4

1. $11\sqrt{5}$ **3.** $7\sqrt[3]{6}$ **5.** $2\sqrt[5]{2}$ **7.** $\sqrt{13} + 2\sqrt{5}$ **9.** $7\sqrt{5} + 2\sqrt[3]{x}$ **11.** $4\sqrt{3}$ **13.** $19\sqrt{3}$ **15.** $6\sqrt{2x}$ **17.** $13\sqrt[3]{2}$ **19.** $(9x + 1)\sqrt{5x}$

21. $7y\sqrt[3]{2x}$ **23.** $(3x - 2)\sqrt[3]{2x}$ **25.** $4\sqrt{x - 2}$ **27.** $5x\sqrt[3]{xy^2}$ **29.** $\dfrac{\sqrt{11}}{2}$ **31.** $\dfrac{\sqrt[3]{19}}{3}$ **33.** $\dfrac{x}{6y^4}$ **35.** $\dfrac{2x\sqrt{2x}}{5y^3}$ **37.** $\dfrac{x\sqrt[3]{x}}{2y}$ **39.** $\dfrac{x^2\sqrt[3]{50x^2}}{3y^4}$

41. $\dfrac{y\sqrt[4]{9y^2}}{x^2}$ **43.** $\dfrac{2x^2\sqrt[5]{2x^3}}{y^4}$ **45.** $2\sqrt{2}$ **47.** 2 **49.** $3x$ **51.** $x^2 y$ **53.** $2x^2\sqrt{5}$ **55.** $4a^5 b^5$ **57.** $3\sqrt{xy}$ **59.** $2xy$ **61.** $2x^2 y^2\sqrt[4]{y^2}$ or $2x^2 y^2\sqrt{y}$

63. $\sqrt[3]{x + 3}$ **65.** $\sqrt[3]{a^2 - ab + b^2}$ **67.** $\dfrac{43\sqrt{2}}{35}$ **69.** $-11xy\sqrt{2x}$ **71.** $25x\sqrt{2x}$ **73.** $7x\sqrt{3xy}$ **75.** $3x\sqrt[3]{5xy}$ **77.** $\left(\dfrac{f}{g}\right)(x) = 4x\sqrt{x};$

Domain of $\dfrac{f}{g} = \{x \mid x > 0\}$ or $(0, \infty)$ **79.** $\left(\dfrac{f}{g}\right)(x) = 2x\sqrt[3]{2x};$ Domain of $\dfrac{f}{g} = \{x \mid x \text{ is a real number and } x \neq 0\}$ or $(-\infty, 0) \cup (0, \infty)$

81. a. $R_f\dfrac{\sqrt{c^2 - v^2}}{\sqrt{c^2}} = R_f\sqrt{\dfrac{c^2 - v^2}{c^2}} = R_f\sqrt{\dfrac{c^2}{c^2} - \dfrac{v^2}{c^2}} = R_f\sqrt{1 - \dfrac{v^2}{c^2}} = R_f\sqrt{1 - \left(\dfrac{v}{c}\right)^2}$ **b.** $R_f\sqrt{1 - \left(\dfrac{c}{c}\right)^2} = 0$; Aging rate is 0 relative to a friend on Earth;

A person moving at the speed of light does not age relative to a friend on Earth. **83.** $P = 18\sqrt{5}$ ft ; $A = 100$ sq ft

93. Graphs are not the same.; $\sqrt{16x} - \sqrt{9x} = \sqrt{x}$. **95.** d **97.** $-\dfrac{31\sqrt{5}}{12}$

$y = \sqrt{7x}$ $y = \sqrt{16x} - \sqrt{9x}$

99. $\{0\}$

100. $(x - 2y)(x - 6y)$

101. $\dfrac{3x^2 + 8x + 6}{(x + 3)^2(x + 2)}$

Mid-Chapter 10 Check Point

1. 13 **2.** $2x^2 y^3\sqrt{2xy}$ **3.** $5\sqrt[3]{4x^2}$ **4.** $12x\sqrt[3]{2x}$ **5.** 1 **6.** $4xy^{1/12}$ **7.** $-\sqrt{3}$ **8.** $\dfrac{5x\sqrt{5x}}{y^2}$ **9.** $\sqrt[4]{x^3}$ **10.** $3x\sqrt[3]{2x^2}$ **11.** $2\sqrt[3]{10}$

12. $\dfrac{x^2}{y^4}$ **13.** $x^{7/12}$ **14.** $x\sqrt[3]{y^2}$ **15.** $(x - 2)\sqrt[4]{(x - 2)^2}$ **16.** $2x^2 y^4\sqrt[4]{2x^3 y}$ **17.** $14\sqrt[3]{2}$ **18.** $x\sqrt[4]{x^2 y^2}$ **19.** $\dfrac{1}{25}$ **20.** $\sqrt[6]{32}$

21. $\dfrac{4x}{y^2}$ **22.** $8x^2 y^3\sqrt{y}$ **23.** $-10x^3 y\sqrt{6y}$ **24.** $\{x \mid x \le 6\} = (-\infty, 6]$ **25.** $\{x \mid x \text{ is a real number}\} = (-\infty, \infty)$

Section 10.5

Check Point Exercises

1. a. $x\sqrt{6} + 2\sqrt{15}$ **b.** $y - \sqrt[3]{7y}$ **c.** $36 - 18\sqrt{10}$ **2. a.** $11 + 2\sqrt{30}$ **b.** 1 **c.** $a - 7$ **3. a.** $\dfrac{\sqrt{21}}{7}$ **b.** $\dfrac{\sqrt[3]{6}}{3}$

4. a. $\dfrac{\sqrt{14xy}}{7y}$ **b.** $\dfrac{\sqrt[3]{3xy^2}}{3y}$ **c.** $\dfrac{3\sqrt[5]{4x^3 y}}{y}$ **5.** $12\sqrt{3} - 18$ **6.** $\dfrac{3\sqrt{5} + 3\sqrt{2} + \sqrt{35} + \sqrt{14}}{3}$ **7.** $\dfrac{1}{\sqrt{x + 3} + \sqrt{x}}$

Exercise Set 10.5

1. $x\sqrt{2} + \sqrt{14}$ **3.** $7\sqrt{6} - 6$ **5.** $12\sqrt{2} - 6$ **7.** $\sqrt[3]{12} + 4\sqrt[3]{10}$ **9.** $2x\sqrt[3]{2} - \sqrt[3]{x^2}$ **11.** $32 + 11\sqrt{2}$ **13.** $34 - 15\sqrt{5}$

15. $117 - 36\sqrt{7}$ **17.** $\sqrt{6} + \sqrt{10} + \sqrt{21} + \sqrt{35}$ **19.** $\sqrt{6} - \sqrt{10} - \sqrt{21} + \sqrt{35}$ **21.** $-48 + 7\sqrt{6}$ **23.** $8 + 2\sqrt{15}$

25. $3x - 2\sqrt{3xy} + y$ **27.** -44 **29.** -71 **31.** 6 **33.** $6 - 5\sqrt{x} + x$ **35.** $\sqrt[3]{x^2} + \sqrt[3]{x} - 20$ **37.** $2x^2 + x\sqrt[3]{y^2} - y\sqrt[3]{y}$ **39.** $\dfrac{\sqrt{10}}{5}$

41. $\dfrac{\sqrt{11x}}{x}$ **43.** $\dfrac{3\sqrt{3y}}{y}$ **45.** $\dfrac{\sqrt[3]{4}}{2}$ **47.** $3\sqrt[3]{2}$ **49.** $\dfrac{\sqrt[3]{18}}{3}$ **51.** $\dfrac{4\sqrt[3]{x^2}}{x}$ **53.** $\dfrac{\sqrt[3]{2y}}{y}$ **55.** $\dfrac{7\sqrt[3]{4x}}{2x}$ **57.** $\dfrac{\sqrt[3]{2x^2y}}{xy}$ **59.** $\dfrac{3\sqrt[4]{x^3}}{x}$

61. $\dfrac{3\sqrt[5]{4x^2}}{x}$ **63.** $x\sqrt[5]{8x^3y}$ **65.** $\dfrac{3\sqrt{3y}}{xy}$ **67.** $-\dfrac{5a^2\sqrt{3ab}}{b^2}$ **69.** $\dfrac{\sqrt{2mn}}{2m}$ **71.** $\dfrac{3\sqrt[4]{x^3y}}{x^2y}$ **73.** $-\dfrac{6\sqrt[3]{xy}}{x^2y^3}$ **75.** $8\sqrt{5} - 16$ **77.** $\dfrac{13\sqrt{11} + 39}{2}$

79. $3\sqrt{5} - 3\sqrt{3}$ **81.** $\dfrac{a + \sqrt{ab}}{a - b}$ **83.** $25\sqrt{2} + 15\sqrt{5}$ **85.** $4 + \sqrt{15}$ **87.** $\dfrac{x - 2\sqrt{x} - 3}{x - 9}$ **89.** $\dfrac{3\sqrt{6} + 4}{2}$ **91.** $\dfrac{4\sqrt{xy} + 4x + y}{y - 4x}$ **93.** $\dfrac{3}{\sqrt{6}}$

95. $\dfrac{2x}{\sqrt[3]{2x^2y}}$ **97.** $\dfrac{x - 9}{x - 3\sqrt{x}}$ **99.** $\dfrac{a - b}{a - 2\sqrt{ab} + b}$ **101.** $\dfrac{1}{\sqrt{x+5} + \sqrt{x}}$ **103.** $\dfrac{1}{(x + y)(\sqrt{x} - \sqrt{y})}$ **105.** $\dfrac{3\sqrt{2}}{2}$ **107.** $-2\sqrt[3]{25}$ **109.** $\dfrac{7\sqrt{6}}{6}$

111. $7\sqrt{3} - 7\sqrt{2}$ **113.** 0 **115.** 6 **117.** 51; In 2001, approximately 51% of U.S. households were online.; Answer models data well. **119.** 6.5%

121. approximately 6.5%; This answer models the actual yearly percentage increase well. **123. a.** $\dfrac{15.92}{\sqrt{t+h} + \sqrt{t}}$ **b.** $\dfrac{15.92}{2\sqrt{t}}$ or $\dfrac{7.96}{\sqrt{t}}$ **c.** 3.2%

125. $P = 8\sqrt{2}$ in.; $A = 7$ sq in. **127.** $\dfrac{7\sqrt{2\cdot 2\cdot 3}}{6} = \dfrac{7\cdot 2\sqrt{3}}{6} = \dfrac{14\sqrt{3}}{6} = \dfrac{7\sqrt{3}}{3} = \dfrac{7}{3}\sqrt{3}$

139.

Graphs are not the same;
$(\sqrt{x} + 2)(\sqrt{x} - 2) = x - 4$ for $x \geq 0$.

141.

Graphs are the same; Simplification is correct.

143. $\dfrac{45}{7}$ or $\left\{\dfrac{45}{7}\right\}$

145. $\dfrac{5\sqrt{2} + 3\sqrt{3} - 4\sqrt{6} + 2}{23}$

146. $\dfrac{2x + 7}{x^2 - 4}$

147. $\{x | 0 \leq x \leq 2\}$ or $[0, 2]$

148. yes; no

Section 10.6

Check Point Exercises

1. 20 or $\{20\}$ **2.** no solution or $\varnothing$ **3.** -1 and 3, or $\{-1, 3\}$ **4.** 4 or $\{4\}$ **5.** -12 or $\{-12\}$ **6.** 2009; faster

Exercise Set 10.6

1. $\{6\}$ **3.** $\{17\}$ **5.** $\varnothing$ **7.** $\{8\}$ **9.** $\{0, 3\}$ **11.** $\{1, 3\}$ **13.** $\{3, 7\}$ **15.** $\{9\}$ **17.** $\{8\}$ **19.** $\{35\}$ **21.** $\{16\}$ **23.** $\{2\}$ **25.** $\{2, 6\}$

27. $\left\{\dfrac{5}{2}\right\}$ **29.** $\{5\}$ **31.** $\varnothing$ **33.** $\{2, 6\}$ **35.** $\{0, 10\}$ **37.** $\{8\}$ **39.** 4 **41.** -7 **43.** $V = \dfrac{\pi r^2 h}{3}$ or $\dfrac{1}{3}\pi r^2 h$ **45.** $l = \dfrac{8t^2}{\pi^2}$ **47.** 8

49. 9 **51.** High: $66.0°$; Low: $61.5°$ **53.** Using f: 2014; Using g: 2010 **55.** 36 years old; $(36, 40{,}000)$ **57.** 27 sq mi **59.** 149 million km

69. 4 or $\{4\}$ **71.** 1 and 9, or $\{1, 9\}$ **73. c** **75.** 81 or $\{81\}$ **77.** 129 or $\{129\}$

78. $4x^3 - 15x^2 + 47x - 142 + \dfrac{425}{x + 3}$

79. $\dfrac{x + 2}{2(x + 3)}$

80. $(y - 3 + 5x)(y - 3 - 5x)$

Section 10.7

Check Point Exercises

1. a. $8i$ **b.** $\sqrt{11}i$ **c.** $4\sqrt{3}i$ **2. a.** $8 + i$ **b.** $-10 + 10i$ **3. a.** $63 + 14i$ **b.** $58 - 11i$

4. $-\sqrt{35}$ **5.** $\dfrac{18}{25} + \dfrac{26}{25}i$ **6.** $-\dfrac{1}{2} - \dfrac{3}{4}i$ **7. a.** 1 **b.** i **c.** $-i$

Exercise Set 10.7

1. $10i$ **3.** $\sqrt{23}i$ **5.** $3\sqrt{2}i$ **7.** $3\sqrt{7}i$ **9.** $-6\sqrt{3}i$ **11.** $5 + 6i$ **13.** $15 + \sqrt{3}i$ **15.** $-2 - 3\sqrt{2}i$ **17.** $8 + 3i$ **19.** $8 - 2i$
21. $5 + 3i$ **23.** $-1 - 7i$ **25.** $-2 + 9i$ **27.** $8 + 15i$ **29.** $-14 + 17i$ **31.** $9 + 5\sqrt{3}i$ **33.** $-6 + 10i$ **35.** $-21 - 15i$
37. $-35 - 14i$ **39.** $7 + 19i$ **41.** $-1 - 31i$ **43.** $3 + 36i$ **45.** 34 **47.** 34 **49.** 11 **51.** $-5 + 12i$ **53.** $21 - 20i$ **55.** $-\sqrt{14}$
57. -6 **59.** $-5\sqrt{7}$ **61.** $-2\sqrt{6}$ **63.** $\frac{3}{5} - \frac{1}{5}i$ **65.** $1 + i$ **67.** $\frac{28}{25} + \frac{21}{25}i$ **69.** $-\frac{12}{13} + \frac{18}{13}i$ **71.** $0 + i$ or i **73.** $\frac{3}{10} - \frac{11}{10}i$
75. $\frac{11}{13} - \frac{16}{13}i$ **77.** $-\frac{23}{58} + \frac{43}{58}i$ **79.** $0 - \frac{7}{3}i$ or $-\frac{7}{3}i$ **81.** $-\frac{5}{2} - 4i$ **83.** $-\frac{7}{3} + \frac{4}{3}i$ **85.** -1 **87.** $-i$ **89.** -1 **91.** 1
93. i **95.** 1 **97.** $-i$ **99.** 0 **101.** $-11 - 5i$ **103.** $-5 + 10i$ **105.** $0 + 47i$ or $47i$ **107.** $1 - i$ **109.** 0 **111.** $10 + 10i$
113. $\frac{20}{13} + \frac{30}{13}i$ **115.** $(47 + 13i)$ volts   **117.** $(5 + \sqrt{15}i) + (5 - \sqrt{15}i) = 10; (5 + \sqrt{15}i)(5 - \sqrt{15}i) = 25 - 15i^2 = 25 + 15 = 40$
131. $\sqrt{-9} + \sqrt{-16} = 3i + 4i = 7i$ **133.** d   **135.** $\frac{6}{5} + 0i$ or $\frac{6}{5}$ **137.** $\frac{x^2}{y^2}$ **138.** $x = \frac{yz}{y - z}$ **139.** -18

Review Exercises

1. 9 **2.** $-\frac{1}{10}$ **3.** -3 **4.** not a real number   **5.** -2 **6.** $5; 1.73; 0;$ not a real number   **7.** $2; -2; -4$ **8.** $\{x | x \geq 2\}$ or $[2, \infty)$
9. $\{x | x \leq 25\}$ or $(-\infty, 25]$ **10.** $5|x|$ **11.** $|x + 14|$ **12.** $|x - 4|$ **13.** $4x$ **14.** $2|x|$ **15.** $-2(x + 7)$ **16.** $\sqrt[3]{5xy}$
17. $(\sqrt{16})^3 = 64$ **18.** $(\sqrt[5]{32})^4 = 16$ **19.** $(7x)^{1/2}$ **20.** $(19xy)^{5/3}$ **21.** $\frac{1}{8^{2/3}} = \frac{1}{4}$ **22.** $\frac{3x}{(ab)^{4/5}} = \frac{3x}{\sqrt[5]{(ab)^4}}$ **23.** $x^{7/12}$ **24.** $5^{1/6}$
25. $2x^2y$ **26.** $\frac{y^{1/8}}{x^{1/3}}$ **27.** x^3y^4 **28.** $y\sqrt[3]{x}$ **29.** $\sqrt[6]{x^5}$ **30.** $\sqrt[6]{x}$ **31.** $\sqrt[15]{x}$ **32.** $\$3150$ million   **33.** $\sqrt{21xy}$ **34.** $\sqrt[5]{77x^3}$
35. $\sqrt[6]{(x - 5)^5}$ **36.** $f(x) = \sqrt{7}|x - 1|$ **37.** $2x\sqrt{5x}$ **38.** $3x^2y^2\sqrt[3]{2x^2}$ **39.** $2y^2z\sqrt[4]{2x^3y^3z}$ **40.** $2x^2\sqrt{6x}$ **41.** $2xy\sqrt[3]{2y^2}$
42. $xyz^2\sqrt[5]{16y^4z}$ **43.** $\sqrt{x^2 - 1}$ **44.** $8\sqrt[3]{3}$ **45.** $9\sqrt{2}$ **46.** $(3x + y^2)\sqrt[3]{x}$ **47.** $-8\sqrt[3]{6}$ **48.** $\frac{2}{5}\sqrt[3]{2}$ **49.** $\frac{x\sqrt{x}}{10y^2}$ **50.** $\frac{y\sqrt[4]{3y}}{2x^5}$
51. $2\sqrt{6}$ **52.** $2\sqrt[3]{2}$ **53.** $2x\sqrt[4]{2x}$ **54.** $10x^2\sqrt{xy}$ **55.** $6\sqrt{2} + 12\sqrt{5}$ **56.** $5\sqrt[3]{2} - \sqrt[3]{10}$ **57.** $-83 + 3\sqrt{35}$
58. $\sqrt{xy} - \sqrt{11x} - \sqrt{11y} + 11$ **59.** $13 + 4\sqrt{10}$ **60.** $22 - 4\sqrt{30}$ **61.** -6 **62.** 4 **63.** $\frac{2\sqrt{6}}{3}$ **64.** $\frac{\sqrt{14}}{7}$ **65.** $4\sqrt[3]{3}$
66. $\frac{\sqrt{10xy}}{5y}$ **67.** $\frac{7\sqrt[3]{4x}}{x}$ **68.** $\frac{\sqrt[4]{189x^3}}{3x}$ **69.** $\frac{5\sqrt[5]{xy^4}}{2xy}$ **70.** $3\sqrt{3} + 3$ **71.** $\frac{\sqrt{35} - \sqrt{21}}{2}$ **72.** $10\sqrt{5} + 15\sqrt{2}$ **73.** $\frac{x + 8\sqrt{x} + 15}{x - 9}$
74. $\frac{5 + \sqrt{21}}{2}$ **75.** $\frac{3\sqrt{2} + 2}{7}$ **76.** $\frac{2}{\sqrt{14}}$ **77.** $\frac{3x}{\sqrt[3]{9x^2y}}$ **78.** $\frac{7}{\sqrt{35} + \sqrt{21}}$ **79.** $\frac{2}{5 - \sqrt{21}}$ **80.** $\{16\}$ **81.** $\varnothing$
82. $\{2\}$ **83.** $\{8\}$ **84.** $\{-4, -2\}$ **85.** 256 feet   **86.** 84 years old   **87.** $9i$ **88.** $3\sqrt{7}i$ **89.** $-2\sqrt{2}i$
90. $12 + 2i$ **91.** $-9 + 4i$ **92.** $-12 - 8i$ **93.** $29 + 11i$ **94.** $-7 - 24i$ **95.** $113 + 0i$ or 113 **96.** $-2\sqrt{6} + 0i$ or $-2\sqrt{6}$
97. $\frac{15}{13} - \frac{3}{13}i$ **98.** $\frac{1}{5} + \frac{11}{10}i$ **99.** $\frac{1}{3} - \frac{5}{3}i$ **100.** 1 **101.** $-i$

Chapter 10 Test

1. a. 6 **b.** $\{x | x \leq 4\}$ or $(-\infty, 4]$ **2.** $\frac{1}{81}$ **3.** $\frac{5\sqrt[8]{y}}{\sqrt[4]{x}}$ **4.** $\sqrt{x}$ **5.** $\sqrt[20]{x^9}$ **6.** $5|x|\sqrt{3}$ **7.** $|x - 5|$ **8.** $2xy^2\sqrt[3]{xy^2}$ **9.** $-\frac{2}{x^2}$
10. $\sqrt[3]{50x^2y}$ **11.** $2x\sqrt[4]{2y^3}$ **12.** $-7\sqrt{2}$ **13.** $(2x + y^2)\sqrt[3]{x}$ **14.** $2x\sqrt[3]{x}$ **15.** $12\sqrt{2} - \sqrt{15}$ **16.** $26 + 6\sqrt{3}$ **17.** $52 - 14\sqrt{3}$
18. $\frac{\sqrt{5x}}{x}$ **19.** $\frac{\sqrt[3]{25x}}{x}$ **20.** $-5 + 2\sqrt{6}$ **21.** 6 or $\{6\}$ **22.** 16 or $\{16\}$ **23.** -3 or $\{-3\}$ **24.** 49 months   **25.** $5\sqrt{3}i$ **26.** $-1 + 6i$
27. $26 + 7i$ **28.** $-6 + 0i$ or -6 **29.** $\frac{1}{5} + \frac{7}{5}i$ **30.** $-i$

Cumulative Review Exercises (Chapters 1–10)

1. $\{(-2, -1, -2)\}$ **2.** $\left\{-\frac{1}{3}, 4\right\}$ **3.** $\left\{x | x > \frac{1}{3}\right\}$ or $\left(\frac{1}{3}, \infty\right)$ **4.** $\left\{\frac{3}{4}\right\}$ **5.** $\{-1\}$

6.
$x + 2y < 2$
$2y - x > 4$

  7. $\frac{x^2}{15(x + 2)}$ **8.** $\frac{x}{y}$ **9.** $8x^3 - 22x^2 + 11x + 6$ **10.** $\frac{5x - 6}{(x - 5)(x + 3)}$ **11.** -64 **12.** 0 **13.** $-\frac{16 - 9\sqrt{3}}{13}$

14. $2x^2 + x + 5 + \frac{6}{x - 2}$ **15.** $-34 - 3\sqrt{6}$ **16.** $2(3x + 2)(4x - 1)$ **17.** $(4x^2 + 1)(2x + 1)(2x - 1)$

18. about 53 lumens   **19.** $\$1500$ at 7% and $\$4500$ at 9%   **20.** 2650 students

CHAPTER 11

Section 11.1

Check Point Exercises

1. $\pm\sqrt{7}$ or $\{\pm\sqrt{7}\}$ **2.** $\pm\dfrac{\sqrt{33}}{3}$ or $\left\{\pm\dfrac{\sqrt{33}}{3}\right\}$ **3.** $\pm\dfrac{3}{2}i$ or $\left\{\pm\dfrac{3}{2}i\right\}$ **4.** $3\pm\sqrt{10}$ or $\{3\pm\sqrt{10}\}$ **5. a.** 25; $x^2+10x+25=(x+5)^2$

b. $\dfrac{9}{4}$; $x^2-3x+\dfrac{9}{4}=\left(x-\dfrac{3}{2}\right)^2$ **c.** $\dfrac{9}{64}$; $x^2+\dfrac{3}{4}x+\dfrac{9}{64}=\left(x+\dfrac{3}{8}\right)^2$ **6.** $-2\pm\sqrt{5}$ or $\{-2\pm\sqrt{5}\}$ **7.** $\dfrac{-3\pm\sqrt{41}}{4}$ or $\left\{\dfrac{-3\pm\sqrt{41}}{4}\right\}$

8. 20% **9.** $10\sqrt{21}$ ft; 45.8 ft **10.** 5 units **11.** $\left(4,-\dfrac{1}{2}\right)$

Exercise Set 11.1

1. $\{\pm5\}$ **3.** $\{\pm\sqrt{6}\}$ **5.** $\left\{\pm\dfrac{5}{4}\right\}$ **7.** $\left\{\pm\dfrac{\sqrt{6}}{3}\right\}$ **9.** $\left\{\pm\dfrac{4}{5}i\right\}$ **11.** $\{-10,-4\}$ **13.** $\{3\pm\sqrt{5}\}$ **15.** $\{-2\pm2\sqrt{2}\}$ **17.** $\{5\pm3i\}$

19. $\left\{\dfrac{-3\pm\sqrt{11}}{4}\right\}$ **21.** $\{-3,9\}$ **23.** 1; $x^2+2x+1=(x+1)^2$ **25.** 49; $x^2-14x+49=(x-7)^2$ **27.** $\dfrac{49}{4}$; $x^2+7x+\dfrac{49}{4}=\left(x+\dfrac{7}{2}\right)^2$

29. $\dfrac{1}{16}$; $x^2-\dfrac{1}{2}x+\dfrac{1}{16}=\left(x-\dfrac{1}{4}\right)^2$ **31.** $\dfrac{4}{9}$; $x^2+\dfrac{4}{3}x+\dfrac{4}{9}=\left(x+\dfrac{2}{3}\right)^2$ **33.** $\dfrac{81}{64}$; $x^2-\dfrac{9}{4}x+\dfrac{81}{64}=\left(x-\dfrac{9}{8}\right)^2$ **35.** $\{-8,4\}$

37. $\{-3\pm\sqrt{7}\}$ **39.** $\{4\pm\sqrt{15}\}$ **41.** $\{-1\pm i\}$ **43.** $\left\{\dfrac{-3\pm\sqrt{13}}{2}\right\}$ **45.** $\left\{-\dfrac{3}{7},-\dfrac{1}{7}\right\}$ **47.** $\left\{\dfrac{-1\pm\sqrt{5}}{2}\right\}$ **49.** $\left\{-\dfrac{5}{2},1\right\}$

51. $\left\{\dfrac{-3\pm\sqrt{6}}{3}\right\}$ **53.** $\left\{\dfrac{4\pm\sqrt{13}}{3}\right\}$ **55.** $\left\{\dfrac{1}{4}\pm\dfrac{1}{4}i\right\}$ **57.** $-5,7$ **59.** $-\dfrac{1}{5},1$ **61.** $-2\pm5i$

63. 13 units **65.** $2\sqrt{2}$ or 2.83 units **67.** 5 units **69.** $\sqrt{29}$ or 5.39 units **71.** $4\sqrt{2}$ or 5.66 units **73.** $2\sqrt{5}$ or 4.47 units

75. $2\sqrt{2}$ or 2.83 units **77.** $\sqrt{93}$ or 9.64 units **79.** $\sqrt{5}$ or 2.24 units **81.** $(4,6)$ **83.** $(-4,-5)$ **85.** $\left(\dfrac{3}{2},-6\right)$

87. $(-3,-2)$ **89.** $(1,5\sqrt{5})$ **91.** $(2\sqrt{2},0)$ **93.** $2\pm2i$ **95.** $v=\sqrt{2gh}$

97. $r=\dfrac{\sqrt{AP}}{P}-1$ **99.** $\dfrac{-1\pm\sqrt{19}}{6}$ or $\left\{\dfrac{-1\pm\sqrt{19}}{6}\right\}$ **101.** $-b$ and $2b$, or $\{-b,2b\}$ **103.** 20% **105.** 6.25% **107.** approximately 2007

109. $10\sqrt{3}$ sec; 17.3 sec **111.** $3\sqrt{5}$ mi; 6.7 mi **113.** $20\sqrt{2}$ ft, or 28.3 ft **115.** $50\sqrt{2}$ ft; 70.7 ft **117.** 10 m **119.** 0.5 hour or 30 minutes

129.   **133.** $y=\pm\dfrac{b\sqrt{a^2-x^2}}{a}$ **135.** $\dfrac{-b\pm\sqrt{b^2-4c}}{2}$ or $\left\{\dfrac{-b\pm\sqrt{b^2-4c}}{2}\right\}$

137. The distance from A to B is $2\sqrt{2}$ and the distance from B to C is $3\sqrt{2}$. The distance from A to C is $5\sqrt{2}$, which is equal to $2\sqrt{2}+3\sqrt{2}$. **139.** $(-2,2)$ and $(6,2)$ **140.** $4-2x$ **141.** $(1-2x)(1+2x+4x^2)$

142. $x^3-2x^2-4x-12-\dfrac{42}{x-3}$

-3 and 1, or $\{-3,1\}$

Section 11.2

Check Point Exercises

1. -5 and $\dfrac{1}{2}$, or $\left\{-5,\dfrac{1}{2}\right\}$ **2.** $\dfrac{3\pm\sqrt{7}}{2}$ or $\left\{\dfrac{3\pm\sqrt{7}}{2}\right\}$ **3.** $-1\pm\dfrac{\sqrt{6}}{3}i$ or $\left\{-1\pm\dfrac{\sqrt{6}}{3}i\right\}$

4. a. 0; one real solution **b.** 81; two rational solutions **c.** -44; two imaginary solutions that are complex conjugates

5. a. $20x^2+7x-3=0$ **b.** $x^2+49=0$ **6.** 1960 and 1974; very well

Exercise Set 11.2

1. $\{-6,-2\}$ **3.** $\left\{1,\dfrac{5}{2}\right\}$ **5.** $\left\{\dfrac{-3\pm\sqrt{89}}{2}\right\}$ **7.** $\left\{\dfrac{7\pm\sqrt{85}}{6}\right\}$ **9.** $\left\{\dfrac{1\pm\sqrt{7}}{6}\right\}$ **11.** $\left\{\dfrac{3}{8}\pm\dfrac{\sqrt{87}}{8}i\right\}$ **13.** $\{2\pm2i\}$ **15.** $\left\{\dfrac{4}{3}\pm\dfrac{\sqrt{5}}{3}i\right\}$

17. $\left\{-\dfrac{3}{2},4\right\}$ **19.** 52; two irrational solutions **21.** 4; two rational solutions **23.** -23; two imaginary solutions

25. 36; two rational solutions   **27.** −60; two imaginary solutions   **29.** 0; one (repeated) rational solution   **31.** $\left\{-\dfrac{2}{3}, 2\right\}$

33. $\{1 \pm \sqrt{2}\}$ **35.** $\left\{\dfrac{3 \pm \sqrt{65}}{4}\right\}$ **37.** $\left\{0, \dfrac{8}{3}\right\}$ **39.** $\left\{\dfrac{-6 \pm 2\sqrt{6}}{3}\right\}$

41. $\left\{\dfrac{2 \pm \sqrt{10}}{3}\right\}$ **43.** $\{2 \pm \sqrt{10}\}$ **45.** $\left\{1, \dfrac{5}{2}\right\}$ **47.** $x^2 - 2x - 15 = 0$ **49.** $12x^2 + 5x - 2 = 0$

51. $x^2 + 36 = 0$ **53.** $x^2 - 2 = 0$ **55.** $x^2 - 20 = 0$ **57.** $x^2 - 2x + 2 = 0$ **59.** $x^2 - 2x - 1 = 0$ **61.** b   **63.** a   **65.** $1 + \sqrt{7}$

67. $\left\{\dfrac{-1 \pm \sqrt{21}}{2}\right\}$ **69.** $\left\{-2\sqrt{2}, \dfrac{\sqrt{2}}{2}\right\}$ **71.** $\{-3, 1, -1 \pm \sqrt{2}i\}$

73. 33-year-olds and 58-year-olds; The function models the actual data well.   **75.** 77.8 ft; (b)   **77.** 5.5 m by 1.5 m   **79.** 17.6 in. and 18.6 in.

81. 9.3 in. and 0.7 in.   **83.** 7.5 hr and 8.5 hr   **95.** d   **97.** 2.4 m; yes   **99.** $\left\{-3, \dfrac{1}{4}\right\}$

100. $\{3, 7\}$ **101.** $\dfrac{5\sqrt{3} - 5x}{3 - x^2}$

Section 11.3

Check Point Exercises

1.

2.

3.   Domain = $\{x \mid x$ is a real number$\}$ or $(-\infty, \infty)$; Range = $\{y \mid y \le 5\}$ or $(-\infty, 5]$

4. 33.7 ft   **5.** 4, −4; −16   **6.** 30 ft by 30 ft; 900 sq ft

Exercise Set 11.3

1. $h(x) = (x - 1)^2 + 1$ **3.** $j(x) = (x - 1)^2 - 1$ **5.** $h(x) = x^2 - 1$ **7.** $g(x) = x^2 - 2x + 1$

9. $(3, 1)$ **11.** $(-1, 5)$ **13.** $(2, -5)$ **15.** $(-1, 9)$

17.
$x = 4$; $\{y \mid y \ge -1\}$ or $[-1, \infty)$

19.
$x = 1$; $\{y \mid y \ge 2\}$ or $[2, \infty)$

21.
$x = 3$; $\{y \mid y \ge 1\}$ or $[1, \infty)$

23.
$x = -2$; $\{y \mid y \ge -1\}$ or $[-1, \infty)$

25.
$x = 1$; $\{y \mid y \le 4\}$ or $(-\infty, 4]$

27.
$x = 1$; $\{y \mid y \ge -4\}$ or $[-4, \infty)$

29.
$x = -\dfrac{3}{2}$; $\left\{y \mid y \ge -\dfrac{49}{4}\right\}$ or $\left[-\dfrac{49}{4}, \infty\right)$

31.
$x = 1$; $\{y \mid y \le 4\}$ or $(-\infty, 4]$

33.
$x = 1$; $\{y \mid y \le -1\}$ or $(-\infty, -1]$

35. minimum; $(2, -13)$ **37.** maximum; $(1, 1)$ **39.** minimum; $\left(\dfrac{1}{2}, -\dfrac{5}{4}\right)$

41. Domain $= \{x | x$ is a real number$\}$ or $(-\infty, \infty)$; Range $= \{y | y \geq -2\}$ or $[-2, \infty)$

43. Domain $= \{x | x$ is a real number$\}$ or $(-\infty, \infty)$; Range $= \{y | y \leq 3\}$ or $(-\infty, 3]$ **45.** $f(x) = 2(x - 5)^2 + 3$ **47.** $f(x) = 2(x + 10)^2 - 5$

49. $f(x) = -3(x + 2)^2 + 4$ **51.** $f(x) = 3(x - 11)^2$ **53.** $(14, 19.5)$; In 2016, the projected mortality will reach a maximum of 19,500.

55. a. 2 sec; 224 ft

b. 5.7 sec

c. 160; 160 feet is the height of the building.

d.
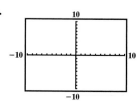

57. 7.2 hr; 622 per 100,000 males

59. 8 and 8; 64

61. 8, -8; -64

63. length: 300 ft; width: 150 ft; maximum area: 45,000 sq ft

65. 12.5 yd by 12.5 yd; 156.25 sq yd

67. 5 in.; 50 sq in.

69. a. $C(x) = 525 + 0.55x$

b. $P(x) = -0.001x^2 + 2.45x - 525$

c. 1225 sandwiches; $975.63

77. a.

b. $(20.5, -120.5)$

c.

d. Answers will vary.

79. $(2.5, 185)$

81. $(-30, 91)$

83. a   **85.** $x = 3$; $(0, 11)$

87. $f(x) = -2(x + 3)^2 - 1$

89. 65 trees; 16,900 lb

90. $\{7\}$

91. $\dfrac{x}{x - 2}$

92. $(6, -2)$ or $\{(6, -2)\}$

Mid-Chapter 11 Check Point

1. $\left\{-\dfrac{1}{3}, \dfrac{11}{3}\right\}$ **2.** $\left\{-1, \dfrac{7}{5}\right\}$ **3.** $\left\{\dfrac{3 \pm \sqrt{15}}{3}\right\}$ **4.** $\{-3 \pm \sqrt{7}\}$ **5.** $\left\{\pm\dfrac{6\sqrt{5}}{5}\right\}$ **6.** $\left\{\dfrac{5}{2} \pm \dfrac{\sqrt{7}}{2}i\right\}$ **7.** $\{\pm\sqrt{13}i\}$ **8.** $\left\{-4, \dfrac{1}{2}\right\}$

9. $\{-3 \pm 2\sqrt{6}\}$ **10.** $\{2 \pm \sqrt{3}\}$ **11.** $\left\{\dfrac{3}{4} \pm \dfrac{\sqrt{23}}{4}i\right\}$ **12.** $\left\{\dfrac{-3 \pm \sqrt{41}}{4}\right\}$ **13.** $\{-2, 4\}$ **14.** $\{-5 \pm 2\sqrt{7}\}$

15. $4\sqrt{2} \approx 5.66$ units; $(0, 0)$ **16.** $\sqrt{61} \approx 7.81$ units; $\left(-\dfrac{15}{2}, 11\right)$

17.
Domain $= \{x | x$ is a real number$\}$ or $(-\infty, \infty)$; Range $= \{y | y \geq -4\}$ or $[-4, \infty)$

18.
Domain $= \{x | x$ is a real number$\}$ or $(-\infty, \infty)$; Range $= \{y | y \leq 5\}$ or $(-\infty, 5]$

19.
Domain $= \{x | x$ is a real number$\}$ or $(-\infty, \infty)$; Range $= \{y | y \leq 9\}$ or $(-\infty, 9]$

20.
Domain $= \{x | x$ is a real number$\}$ or $(-\infty, \infty)$; Range $= \{y | y \geq -2\}$ or $[-2, \infty)$

21. two imaginary solutions   **22.** two rational solutions   **23.** $8x^2 - 2x - 3 = 0$ **24.** $x^2 - 12 = 0$ **25.** 75 cabinets per day; $1200
26. $-9, -9$; 81   **27.** 10 in.; 100 sq in.

Section 11.4

Check Point Exercises

1. $-4, 4, -1,$ and $1,$ or $\{-4, 4, -1, 1\}$ **2.** 16 or $\{16\}$ **3.** $-\sqrt{6}, -1, 1,$ and $\sqrt{6},$ or $\{-\sqrt{6}, -1, 1, \sqrt{6}\}$ **4.** -1 and $2,$ or $\{-1, 2\}$

5. $-\dfrac{1}{27}$ and $64,$ or $\left\{-\dfrac{1}{27}, 64\right\}$

Exercise Set 11.4

1. $\{-2, 2, -1, 1\}$ **3.** $\{-3, 3, -\sqrt{2}, \sqrt{2}\}$ **5.** $\{-2i, 2i, -\sqrt{2}, \sqrt{2}\}$ **7.** $\{1\}$ **9.** $\{49\}$ **11.** $\{25, 64\}$ **13.** $\{2, 12\}$

15. $\{-\sqrt{3}, 0, \sqrt{3}\}$ **17.** $\{-5, -2, -1, 2\}$ **19.** $\left\{-\dfrac{1}{4}, \dfrac{1}{5}\right\}$ **21.** $\left\{\dfrac{1}{3}, 2\right\}$ **23.** $\left\{\dfrac{-2 \pm \sqrt{7}}{3}\right\}$ **25.** $\{-8, 27\}$ **27.** $\{-243, 32\}$

29. $\{1\}$ **31.** $\{-8, -2, 1, 4\}$ **33.** $-2, 2, -1,$ and $1;$ c **35.** $1;$ e **37.** 2 and $3;$ f **39.** $-5, -4, 1,$ and 2 **41.** $-\dfrac{3}{2}$ and $-\dfrac{1}{3}$

43. $\dfrac{64}{15}$ and $\dfrac{81}{20}$ **45.** $\dfrac{5}{2}$ and $\dfrac{25}{6}$ **47.** ages 20 and 55; The function models the data well. **53.** $\{-1, 2\}$ **55.** $\{-1, 1, -3, 3\}$

57. $\left\{-\dfrac{3}{2}, 2\right\}$ **59.** $\{1\}$ **61.** b **63.** $\left\{\sqrt[3]{-2}, \dfrac{\sqrt[3]{225}}{5}\right\}$ **65.** $\dfrac{1}{5x - 1}$ **66.** $\dfrac{1}{2} + \dfrac{3}{2}i$ **67.** -5

Section 11.5

Check Point Exercises

1. $\{x | x < -4$ or $x > 5\}$ or $(-\infty, -4) \cup (5, \infty)$

2. $\{x | x \le -3$ or $-1 \le x \le 1\}$ or $(-\infty, -3] \cup [-1, 1]$

3. $\{x | x < -1$ or $x \ge 1\}$ or $(-\infty, -1) \cup [1, \infty)$

4. between 1 and 4 seconds, excluding $t = 1$ and $t = 4$

Exercise Set 11.5

1. $\{x | x < -2$ or $x > 4\}$ or $(-\infty, -2) \cup (4, \infty)$

3. $\{x | -3 \le x \le 7\}$ or $[-3, 7]$

5. $\{x | x < 1$ or $x > 4\}$ or $(-\infty, 1) \cup (4, \infty)$

7. $\{x | x < -4$ or $x > -1\}$ or $(-\infty, -4) \cup (-1, \infty)$

9. $\{x | 2 \le x \le 4\}$ or $[2, 4]$

11. $\left\{x \middle| -4 \le x \le \dfrac{2}{3}\right\}$ or $\left[-4, \dfrac{2}{3}\right]$

13. $\left\{x \middle| -3 < x < \dfrac{5}{2}\right\}$ or $\left(-3, \dfrac{5}{2}\right)$

15. $\left\{x \middle| -1 < x < -\dfrac{3}{4}\right\}$ or $\left(-1, -\dfrac{3}{4}\right)$

17. $\{x | x \le 0$ or $x \ge 4\}$ or $(-\infty, 0] \cup [4, \infty)$

19. $\left\{x \middle| x < -\dfrac{3}{2}$ or $x > 0\right\}$ or $\left(-\infty, -\dfrac{3}{2}\right) \cup (0, \infty)$

21. $\{x | 0 \le x \le 1\}$ or $[0, 1]$

23. $\{x | 2 - \sqrt{2} \le x \le 2 + \sqrt{2}\}$ or $[2 - \sqrt{2}, 2 + \sqrt{2}]$

25. no solution or ∅

27. $\{x|1 \le x \le 2 \text{ or } x \ge 3\}$ or $[1, 2] \cup [3, \infty)$

29. $\{x|-2 \le x \le -1 \text{ or } x \ge 1\}$ or $[-2, -1] \cup [1, \infty)$

31. $\{x|x < -3\}$ or $(-\infty, -3)$

33. $\{x|x > -1\}$ or $(-1, \infty)$

35. $\{x|x = 0 \text{ or } x \ge 9\}$ or $\{0\} \cup [9, \infty)$

37. $\{x|x < -3 \text{ or } x > 4\}$ or $(-\infty, -3) \cup (4, \infty)$

39. $\{x|-4 < x < -3\}$ or $(-4, -3)$

41. $\{x|2 \le x < 4\}$ or $[2, 4)$

43. $\left\{x \left| x < -\dfrac{4}{3} \text{ or } x \ge 2\right.\right\}$ or $\left(-\infty, -\dfrac{4}{3}\right) \cup [2, \infty)$

45. $\{x|x < 0 \text{ or } x > 3\}$ or $(-\infty, 0) \cup (3, \infty)$

47. $\{x|x < -5 \text{ or } x > -3\}$ or $(-\infty, -5) \cup (-3, \infty)$

49. $\left\{x \left| x < \dfrac{1}{2} \text{ or } x \ge \dfrac{7}{5}\right.\right\}$ or $\left(-\infty, \dfrac{1}{2}\right) \cup \left[\dfrac{7}{5}, \infty\right)$

51. $\{x|x \le -6 \text{ or } x > -2\}$ or $(-\infty, -6] \cup (-2, \infty)$

53. $\left\{x \left| x \le \dfrac{1}{2} \text{ or } x \ge 2\right.\right\}$ or $\left(-\infty, \dfrac{1}{2}\right] \cup [2, \infty)$

55. $\{x|-1 < x < 1\}$ or $(-1, 1)$

57. $\{x|x < -8 \text{ or } -6 < x < 4 \text{ or } x > 6\}$ or $(-\infty, -8) \cup (-6, 4) \cup (6, \infty)$

59. $\{x|-3 < x < 2\}$ or $(-3, 2)$

61. $\{x|x < -1 \text{ or } 1 < x < 2 \text{ or } x > 3\}$ or $(-\infty, -1) \cup (1, 2) \cup (3, \infty)$

63. $\left\{x \left| -6 \le x \le -\dfrac{1}{2} \text{ or } x \ge 1\right.\right\}$ or $\left[-6, -\dfrac{1}{2}\right] \cup [1, \infty)$

65. $\{x|x < 2 \text{ or } -1 \le x < 2\}$ or $(-\infty, 2) \cup [-1, 2)$ **67.** between 0 and 3 seconds, excluding $t = 0$ and $t = 3$ **69.** f: \$379 billion; g: \$379.8 billion; The functions model the value in the graph well. **71.** after 2007 **73.** The company's production level must be at least 20,000 wheelchairs per month. For values of x greater than or equal to 20,000, the graph lies on or below the line $y = 425$. **75.** The length of the shorter side cannot exceed 6 ft.

81. $\left\{x \left| -3 \le x \le \dfrac{1}{2}\right.\right\}$ or $\left[-3, \dfrac{1}{2}\right]$ **83.** $\{x|x < 3 \text{ or } x \ge 8\}$ or $(-\infty, 3) \cup [8, \infty)$ **85.** $\{x|-3 < x < -1 \text{ or } x > 2\}$ or $(-3, -1) \cup (2, \infty)$

87. Answers will vary; example: $x^2 - 2x - 15 \le 0$. **89.** $\{x|x < 2 \text{ or } x > 2\}$ or $(-\infty, 2) \cup (2, \infty)$ **91.** ∅ **93. a.** all real numbers **b.** empty set

95. $\{x|-19 < x < 29\}$ or $(-19, 29)$ **96.** $\dfrac{2(x - 1)}{(x + 4)(x - 3)}$ **97.** $(x^2 + 4y^2)(x + 2y)(x - 2y)$

Review Exercises

1. $\{\pm 8\}$ **2.** $\{\pm 5\sqrt{2}\}$ **3.** $\left\{\pm \dfrac{\sqrt{6}}{3}\right\}$ **4.** $\{4 \pm 3\sqrt{2}\}$ **5.** $\{-7 \pm 6i\}$ **6.** $100; x^2 + 20x + 100 = (x + 10)^2$

7. $\dfrac{9}{4}; x^2 - 3x + \dfrac{9}{4} = \left(x - \dfrac{3}{2}\right)^2$ **8.** $\{3, 9\}$ **9.** $\left\{\dfrac{7 + \sqrt{53}}{2}\right\}$ **10.** $\left\{\dfrac{-3 \pm \sqrt{41}}{4}\right\}$ **11.** 8% **12.** 14 weeks **13.** $60\sqrt{5}$ m; 134.2 m

14. 13 units **15.** $2\sqrt{2} \approx 2.83$ units **16.** $(-5, 5)$ **17.** $\left(-\dfrac{11}{2}, -2\right)$ **18.** $\{1 \pm \sqrt{5}\}$ **19.** $\{1 \pm 3\sqrt{2}i\}$

20. $\left\{\dfrac{-2 \pm \sqrt{10}}{2}\right\}$ **21.** two imaginary solutions **22.** two rational solutions

23. two irrational solutions   **24.** $\left\{-\dfrac{2}{3}, 4\right\}$ **25.** $\{-5, 2\}$ **26.** $\left\{\dfrac{1 \pm \sqrt{21}}{10}\right\}$

27. $\{-4, 4\}$ **28.** $\{3 \pm 2\sqrt{2}\}$ **29.** $\left\{\dfrac{1}{6} \pm \dfrac{\sqrt{23}}{6}i\right\}$ **30.** $\{4 \pm \sqrt{5}\}$

31. $15x^2 - 4x - 3 = 0$ **32.** $x^2 + 81 = 0$ **33.** $x^2 - 48 = 0$ **34.** 2002   **35.** 8.8 sec

36.   **37.**   **38.**   **39.**

40. 25 in. of rainfall per year; 13.5 in. of growth   **41.** 12.5 sec; 2540 feet   **42.** $(25, 5)$; The maximum divorce rate of about 5 per 1000 population was in 1985.   **43.** 250 yd by 500 yd; 125,000 sq yard   **44.** -7 and 7; -49 **45.** $\{-\sqrt{2}, \sqrt{2}, -2, 2\}$

46. $\{1\}$ **47.** $\{-5, -1, 3\}$ **48.** $\left\{-\dfrac{1}{8}, \dfrac{1}{7}\right\}$ **49.** $\{-27, 64\}$ **50.** $\{16\}$

51. $\left\{x \mid -3 < x < \dfrac{1}{2}\right\}$ or $\left(-3, \dfrac{1}{2}\right)$;

52. $\left\{x \mid x \le -4 \text{ or } x \ge -\dfrac{1}{2}\right\}$ or $(-\infty, -4] \cup \left[-\dfrac{1}{2}, \infty\right)$;

53. $\{x \mid -3 < x < 0 \text{ or } x > 1\}$ or $(-3, 0) \cup (1, \infty)$;

54. $\{x \mid x < -2 \text{ or } x > 6\}$ or $(-\infty, -2) \cup (6, \infty)$;

55. $\left\{x \mid x < 4 \text{ or } x \ge \dfrac{23}{4}\right\}$ or $(-\infty, 4) \cup \left[\dfrac{23}{4}, \infty\right)$;

56. between 1 and 2 seconds, excluding $t = 1$ and $t = 2$
57. a. 200 beats per minute   **b.** between 0 and 4 minutes and more than 12 minutes after the workout; between 0 and 4 minutes; Answers will vary.

Chapter 11 Test

1. $\pm\dfrac{\sqrt{10}}{2}$ or $\left\{\pm\dfrac{\sqrt{10}}{2}\right\}$ **2.** $3 \pm 2\sqrt{5}$ or $\{3 \pm 2\sqrt{5}\}$ **3.** 64; $x^2 - 16x + 64 = (x - 8)^2$ **4.** $\dfrac{1}{25}$; $x^2 + \dfrac{2}{5}x + \dfrac{1}{25} = \left(x + \dfrac{1}{5}\right)^2$ **5.** $3 \pm \sqrt{2}$ or $\{3 \pm \sqrt{2}\}$

6. $50\sqrt{2}$ ft   **7.** $\sqrt{73}$ or 8.54 units   **8.** $\left(\dfrac{7}{2}, -4\right)$ **9.** two irrational solutions   **10.** two imaginary solutions   **11.** -5 and $\dfrac{1}{2}$, or $\left\{-5, \dfrac{1}{2}\right\}$

12. $-4 \pm \sqrt{11}$ or $\{-4 \pm \sqrt{11}\}$ **13.** $-2 \pm 5i$ or $\{-2 \pm 5i\}$ **14.** $\dfrac{3}{2} \pm \dfrac{1}{2}i$ or $\left\{\dfrac{3}{2} \pm \dfrac{1}{2}i\right\}$ **15.** $x^2 - 4x - 21 = 0$ **16.** $x^2 + 100 = 0$

17. 1990 and 1998

18.   **19.**

20. after 2 sec; 69 ft   **21.** 4.1 sec   **22.** 23 computers; $169 hundreds or $16,900
23. 1 and 2, or $\{1, 2\}$ **24.** $-3, 3, -2,$ and 2, or $\{-3, 3, -2, 2\}$
25. 1 and 512, or $\{1, 512\}$
26. $\{x \mid -3 < x < 4\}$ or $(-3, 4)$

27. $\{x \mid x < 3 \text{ or } x \ge 10\}$ or $(-\infty, 3) \cup [10, \infty)$

Cumulative Review Exercises (Chapters 1–11)

1. $\{4\}$ **2.** $\{(2, 3)\}$ **3.** $\{(2, 0, 3)\}$ **4.** $\{x \mid x < -2 \text{ or } x > 4\}$ or $(-\infty, -2) \cup (4, \infty)$ **5.** $\{5\}$ **6.** $\{x \mid x \ge 4\}$ or $[4, \infty)$ **7.** $\left\{\dfrac{3}{4} \pm \dfrac{\sqrt{7}}{4}i\right\}$

8.

9.

10.

11.

12.

$f(x) = (x - 3)^2 - 4$

13. $f^{-1}(x) = x^3 - 1$ **14.** $c = \dfrac{Ad}{d - A}$ **15.** $y = \dfrac{1}{2}x + 5$ or $f(x) = \dfrac{1}{2}x + 5$ **16.** $-\dfrac{y^9}{3x}$ **17.** $16x^4 - 40x^2y + 25y^2$

18. $x^2 - 5x + 1 + \dfrac{8}{5x + 1}$ **19.** $2y^2\sqrt[3]{2y^2}$ **20.** $\dfrac{4x - 13}{(x - 3)(x - 4)}$ **21.** $(x - 4)(x + 2)(x^2 - 2x + 4)$

22. $2(x + 3y)^2$ **23.** length: 12 ft; width: 4 ft   **24.** $\dfrac{6}{5}$ hr or 1 hr and 12 min   **25.** 5 mph

Appendix A
Check Point Exercises

1. $(4, -3, 1)$ or $\{(4, -3, 1)\}$ **2. a.** $\begin{bmatrix} 1 & 6 & -3 & | & 7 \\ 4 & 12 & -20 & | & 8 \\ -3 & -2 & 1 & | & -9 \end{bmatrix}$ **b.** $\begin{bmatrix} 1 & 3 & -5 & | & 2 \\ 1 & 6 & -3 & | & 7 \\ -3 & -2 & 1 & | & -9 \end{bmatrix}$ **c.** $\begin{bmatrix} 4 & 12 & -20 & | & 8 \\ 1 & 6 & -3 & | & 7 \\ 0 & 16 & -8 & | & 12 \end{bmatrix}$

3. $(-1, 2)$ or $\{(-1, 2)\}$ **4.** $(5, 2, 3)$ or $\{(5, 2, 3)\}$

Appendix A Exercise Set

1. $x - 3y = 11$; $y = -3$; $(2, -3)$ **3.** $x - 3y = 1$; $y = -1$; $(-2, -1)$ **5.** $x - 4z = 5$; $y - 12z = 13$; $z = -\dfrac{1}{2}$; $\left(3, 7, -\dfrac{1}{2}\right)$

7. $x + \dfrac{1}{2}y + z = \dfrac{11}{2}$; $y + \dfrac{3}{2}z = 7$; $z = 4$; $(1, 1, 4)$ **9.** $\begin{bmatrix} 1 & -\dfrac{3}{2} & | & 5 \\ 2 & 2 & | & 5 \end{bmatrix}$ **11.** $\begin{bmatrix} 1 & -\dfrac{4}{3} & | & 2 \\ 3 & 5 & | & -2 \end{bmatrix}$ **13.** $\begin{bmatrix} 1 & -3 & | & 5 \\ 0 & 12 & | & -6 \end{bmatrix}$ **15.** $\begin{bmatrix} 1 & -\dfrac{3}{2} & | & \dfrac{7}{2} \\ 0 & \dfrac{17}{2} & | & -\dfrac{17}{2} \end{bmatrix}$

17. $\begin{bmatrix} 1 & -3 & 2 & | & 5 \\ 1 & 5 & -5 & | & 0 \\ 3 & 0 & 4 & | & 7 \end{bmatrix}$ **19.** $\begin{bmatrix} 1 & -3 & 2 & | & 0 \\ 0 & 10 & -7 & | & 7 \\ 2 & -2 & 1 & | & 3 \end{bmatrix}$ **21.** $\begin{bmatrix} 1 & 1 & -1 & | & 6 \\ 0 & -3 & 3 & | & -15 \\ 0 & -4 & 2 & | & -14 \end{bmatrix}$ **23.** $\{(4, 2)\}$ **25.** $\{(3, -3)\}$ **27.** $\{(2, -5)\}$ **29.** no solution or $\emptyset$ **31.** infinitely many solutions; dependent equations

33. $\{(1, -1, 2)\}$ **35.** $\{(3, -1, -1)\}$ **37.** $\{(1, 2, -1)\}$ **39.** $\{(1, 2, 3)\}$ **41.** no solution or $\emptyset$ **43.** infinitely many solutions; dependent equations   **45.** $\{(-1, 2, -2)\}$ **47.** $w - x + y + z = 3$; $x - 2y - z = 0$; $y + 6z = 17$; $z = 3$; $(2, 1, -1, 3)$

49. $\begin{bmatrix} 1 & -1 & 1 & 1 & | & 3 \\ 0 & 1 & -2 & -1 & | & 0 \\ 0 & 2 & 1 & 2 & | & 5 \\ 0 & 6 & -3 & -1 & | & -9 \end{bmatrix}$ **51.** $\{(1, 2, 3, -2)\}$

Appendix B
Check Point Exercises

1. a. -4 **b.** -17 **2.** $(4, -2)$ or $\{(4, -2)\}$ **3.** 80 **4.** $(2, -3, 4)$ or $\{(2, -3, 4)\}$

Appendix B Exercise Set

1. 1 **3.** -29 **5.** 0 **7.** 33 **9.** $-\dfrac{7}{16}$ **11.** $\{(5, 2)\}$ **13.** $\{(2, -3)\}$ **15.** $\{(3, -1)\}$ **17.** $\{(4, 0)\}$ **19.** $\{(4, 2)\}$ **21.** $\{(7, 4)\}$ **23.** inconsistent; no solution or $\emptyset$ **25.** dependent equations; infinitely many solutions   **27.** 72 **29.** -75 **31.** 0 **33.** $\{(-5, -2, 7)\}$ **35.** $\{(2, -3, 4)\}$ **37.** $\{(3, -1, 2)\}$ **39.** $\{(2, 3, 1)\}$ **41.** -42 **43.** $2x - 4y = 8$; $3x + 5y = 10$ **45.** -11 **47.** 4 **49.** 28 sq units   **51.** yes

53. $\begin{vmatrix} x & y & 1 \\ 3 & -5 & 1 \\ -2 & 6 & 1 \end{vmatrix} = 0$; $y = -\dfrac{11}{5}x + \dfrac{8}{5}$

Appendix C
Check Point Exercises

1. $x^2 + y^2 = 16$ **2.** $(x - 5)^2 + (y + 6)^2 = 100$ **3.** center: $(-3, 1)$; radius: 2 units   **4.** $(x + 2)^2 + (y - 2)^2 = 9$

$(x + 3)^2 + (y - 1)^2 = 4$

$(x + 2)^2 + (y - 2)^2 = 9$

Appendix C Exercises

1. $x^2 + y^2 = 49$ **3.** $(x - 3)^2 + (y - 2)^2 = 25$ **5.** $(x + 1)^2 + (y - 4)^2 = 4$ **7.** $(x + 3)^2 + (y + 1)^2 = 3$ **9.** $(x + 4)^2 + y^2 = 100$

11. center: $(0, 0)$; $r = 4$ **13.** center: $(3, 1)$; $r = 6$ **15.** center: $(-3, 2)$; $r = 2$ **17.** center: $(-2, -2)$; $r = 2$

$x^2 + y^2 = 16$

$(x - 3)^2 + (y - 1)^2 = 36$

$(x + 3)^2 + (y - 2)^2 = 4$

$(x + 2)^2 + (y + 2)^2 = 4$

19. $(x + 3)^2 + (y + 1)^2 = 4$; center: $(-3, -1)$; $r = 2$ **21.** $(x - 5)^2 + (y - 3)^2 = 64$; center: $(5, 3)$; $r = 8$ **23.** $(x + 4)^2 + (y - 1)^2 = 25$; center: $(-4, 1)$; $r = 5$ **25.** $(x - 1)^2 + y^2 = 16$; center: $(1, 0)$; $r = 4$

$x^2 + y^2 + 6x + 2y + 6 = 0$

$x^2 + y^2 - 10x - 6y - 30 = 0$

$x^2 + y^2 + 8x - 2y - 8 = 0$

$x^2 - 2x + y^2 - 15 = 0$

27. $\{(0, -4), (4, 0)\}$

29. $\{(0, -3), (2, -1)\}$

31. $(x - 2)^2 + (y + 1)^2 = 4$ **33.** $(x + 3)^2 + (y + 2)^2 = 1$ **35. a.** $(5, 10)$ **b.** $\sqrt{5}$ **c.** $(x - 5)^2 + (y - 10)^2 = 5$
37. 0.5 hour or 30 minutes   **39.** $(x + 2.4)^2 + (y + 2.7)^2 = 900$

Appendix D

Check Point Exercises

1. a. $2(1)^2 + 2(2)^2 + 2(3)^2 + 2(4)^2 + 2(5)^2 + 2(6)^2 = 182$ **b.** $(2^3 - 3) + (2^4 - 3) + (2^5 - 3) = 47$ **c.** $4 + 4 + 4 + 4 + 4 = 20$

2. $2^2(2 - 1)! + 3^2(3 - 1)! + 4^2(4 - 1)! + 5^2(5 - 1)! = 718$ **3. a.** 20 **b.** 1 **c.** 28 **d.** 1 **4.** $x^4 + 4x^3 + 6x^2 + 4x + 1$

5. $x^5 - 10x^4y + 40x^3y^2 - 80x^2y^3 + 80xy^4 - 32y^5$

Appendix D Exercises

1 105 **3.** 60 **5.** 115 **7.** $-\dfrac{5}{16}$ **9.** 55 **11.** $\dfrac{3}{8}$ **13.** 15 **15.** 56 **17.** 12 **19.** 1 **21.** 4950 **23.** $x^3 + 6x^2 + 12x + 8$

25. $27x^3 + 27x^2y + 9xy^2 + y^3$ **27.** $125x^3 - 75x^2 + 15x - 1$ **29.** $16x^4 + 32x^3 + 24x^2 + 8x + 1$

31. $x^8 + 8x^6y + 24x^4y^2 + 32x^2y^3 + 16y^4$ **33.** $y^4 - 12y^3 + 54y^2 - 108y + 81$ **35.** $16x^{12} - 32x^9 + 24x^6 - 8x^3 + 1$

37. $c^5 + 10c^4 + 40c^3 + 80c^2 + 80c + 32$ **39.** $x^5 - 5x^4 + 10x^3 - 10x^2 + 5x - 1$ **41.** $243x^5 - 405x^4y + 270x^3y^2 - 90x^2y^3 + 15xy^4 - y^5$

43. $64a^6 + 192a^5b + 240a^4b^2 + 160a^3b^3 + 60a^2b^4 + 12ab^5 + b^6$

Graphing Answer Section

CHAPTER 1

Exercise Set 1.1

118.

say does that   Star-span-gled   Ban-ner   yet   wave   O'er   the

Section 1.2

Check Point Exercises

2.

3.

Exercise Set 1.2

9–20.

Section 1.3

Check Point Exercises

1.

Exercise Set 1.3

1–8.

9–24.

37.

38.

89a.

90a.

Exercise Set 1.4

105.

G1

Mid-Chapter Check Point

10.

$\left(3, -5\frac{1}{2}\right)$

Section 1.5

Check Point Exercises

1.

2a.

2b.

Exercise Set 1.5

1.

2.

3.

4.

5.

6.

7.

8.

96.

$\left(\frac{7}{2}, -\frac{5}{2}\right)$

Exercise Set 1.6

117.

Review Exercises

17.

18.

33.

$(1, -5)$

34.

$(4, -3)$

35.

$\left(\frac{7}{2}, \frac{3}{2}\right)$

36.

$(-5, 2)$

55.

Chapter 1 Test

17.

$(-4, 3)$

CHAPTER **2**

Exercise Set 2.1

68.

(−3, 1)

Section 2.7

Check Point Exercises

1a.   −2 −1 0 1 2 3 4 5 6 7 8

b. −5 −4 −3 −2 −1 0 1 2 3 4 5

c. −5 −4 −3 −2 −1 0 1 2 3 4 5

2. −5 −4 −3 −2 −1 0 1 2 3 4 5

3. −5 −4 −3 −2 −1 0 1 2 3 4 5

4a. 0 1 2 3 4 5 6 7 8 9 10

b. −8 −7 −6 −5 −4 −3 −2 −1 0 1 2

5. −2 −1 0 1 2 3 4 5 6 7 8

6. −5 −4 −3 −2 −1 0 1 2 3 4 5

7. −5 −4 −3 −2 −1 0 1 2 3 4 5

Exercise Set 2.7

1. −2 −1 0 1 2 3 4 5 6 7 8

2. −5 −4 −3 −2 −1 0 1 2 3 4 5

3. −6 −5 −4 −3 −2 −1 0 1 2 3 4

4. −5 −4 −3 −2 −1 0 1 2 3 4 5

5. −8 −7 −6 −5 −4 −3 −2 −1 0 1 2

6. −8 −7 −6 −5 −4 −3 −2 −1 0 1 2

7. −2 −1 0 1 2 3 4 5 6 7 8

8. 0 1 2 3 4 5 6 7 8 9 10

9. −3 −2 −1 0 1 2 3 4 5 6 7

10. −4 −3 −2 −1 0 1 2 3 4 5 6

11. −3 −2 −1 0 1 2 3 4 5 6 7

12. −6 −5 −4 −3 −2 −1 0 1 2 3 4

19. −1 0 1 2 3 4 5 6 7 8 9

20. −2 −1 0 1 2 3 4 5 6 7 8

21. −2 −1 0 1 2 3 4 5 6 7 8

22. −2 −1 0 1 2 3 4 5 6 7 8

23. −5 −4 −3 −2 −1 0 1 2 3 4 5

24. −5 −4 −3 −2 −1 0 1 2 3 4 5

25. −2 −1 0 1 2 3 4 5 6 7 8

26. −8 −7 −6 −5 −4 −3 −2 −1 0 1 2

27. 10 11 12 13 14 15 16 17 18 19 20

28. 15 16 17 18 19 20 21 22 23 24 25

29. −2 −1 0 1 2 3 4 5 6 7 8

30. −2 −1 0 1 2 3 4 5 6 7 8

31. $\frac{7}{6}$ −5 −4 −3 −2 −1 0 1 2 3 4 5

32. $\frac{7}{6}$ −5 −4 −3 −2 −1 0 1 2 3 4 5

33. $-\frac{3}{8}$ −5 −4 −3 −2 −1 0 1 2 3 4 5

34. $\frac{5}{12}$ −5 −4 −3 −2 −1 0 1 2 3 4 5

35. −5 −4 −3 −2 −1 0 1 2 3 4 5

36. −5 −4 −3 −2 −1 0 1 2 3 4 5

37. 2 3 4 5 6 7 8 9 10 11 12

38. −2 −1 0 1 2 3 4 5 6 7 8

39. −8 −7 −6 −5 −4 −3 −2 −1 0 1 2

40. −7 −6 −5 −4 −3 −2 −1 0 1 2 3

41. −2 −1 0 1 2 3 4 5 6 7 8

42. −2 −1 0 1 2 3 4 5 6 7 8

43. −10 −9 −8 −7 −6 −5 −4 −3 −2 −1 0

44. −10 −9 −8 −7 −6 −5 −4 −3 −2 −1 0

45. −8 −7 −6 −5 −4 −3 −2 −1 0 1 2

46. −5 −4 −3 −2 −1 0 1 2 3 4 5

47. −8 −7 −6 −5 −4 −3 −2 −1 0 1 2

48. −8 −7 −6 −5 −4 −3 −2 −1 0 1 2

49. —5 —4 —3 —2 —1 0 1 2 3 4 5

50. 0 1 2 3 4 5 6 7 8 9 10

51. $-\frac{1}{8}$ —5 —4 —3 —2 —1 0 1 2 3 4 5

52. $-\frac{1}{4}$ —5 —4 —3 —2 —1 0 1 2 3 4 5

53. —8 —7 —6 —5 —4 —3 —2 —1 0 1 2

54. —5 —4 —3 —2 —1 0 1 2 3 4 5

55. —2 —1 0 1 2 3 4 5 6 7 8

56. —2 —1 0 1 2 3 4 5 6 7 8

57. —2 —1 0 1 2 3 4 5 6 7 8

58. —5 —4 —3 —2 —1 0 1 2 3 4 5

59. —5 —4 —3 —2 —1 0 1 2 3 4 5

60. —8 —7 —6 —5 —4 —3 —2 —1 0 1 2

61. —5 —4 —3 —2 —1 0 1 2 3 4 5

62. —5 —4 —3 —2 —1 0 1 2 3 4 5

63. —5 —4 —3 —2 —1 0 1 2 3 4 5

64. —2 —1 0 1 2 3 4 5 6 7 8

65. $\frac{11}{3}$ —5 —4 —3 —2 —1 0 1 2 3 4 5

66. —5 —4 —3 —2 —1 0 1 2 3 4 5

67. —5 —4 —3 —2 —1 0 1 2 3 4 5

68. —5 —4 —3 —2 —1 0 1 2 3 4 5

69. —5 —4 —3 —2 —1 0 1 2 3 4 5

70. —5 —4 —3 —2 —1 0 1 2 3 4 5

71. —5 —4 —3 —2 —1 0 1 2 3 4 5

72. —5 —4 —3 —2 —1 0 1 2 3 4 5

73. $\frac{5}{3}$ —5 —4 —3 —2 —1 0 1 2 3 4 5

74. $-\frac{5}{4}$ —5 —4 —3 —2 —1 0 1 2 3 4 5

75. 0 1 2 3 4 5 6 7 8 9 10

76. 10 11 12 13 14 15 16 17 18 19 20

77. —8 —7 —6 —5 —4 —3 —2 —1 0 1 2

78. —10 —9 —8 —7 —6 —5 —4 —3 —2 —1 0

Review Exercises

76. —5 —4 —3 —2 —1 0 1 2 3 4 5

77. —5 —4 —3 —2 —1 0 1 2 3 4 5

80. —2 —1 0 1 2 3 4 5 6 7 8

81. —10 —9 —8 —7 —6 —5 —4 —3 —2 —1 0

82. —5 —4 —3 —2 —1 0 1 2 3 4 5

83. —2 —1 0 1 2 3 4 5 6 7 8

84. —2 —1 0 1 2 3 4 5 6 7 8

85. —5 —4 —3 —2 —1 0 1 2 3 4 5

Chapter 2 Test

26. —5 —4 —3 —2 —1 0 1 2 3 4 5

27. —8 —7 —6 —5 —4 —3 —2 —1 0 1 2

29. —8 —7 —6 —5 —4 —3 —2 —1 0 1 2

30. —5 —4 —3 —2 —1 0 1 2 3 4 5

31. —2 —1 0 1 2 3 4 5 6 7 8

Cumulative Review Exercises (Chapters 1–2)

6.

17. —5 —4 —3 —2 —1 0 1 2 3 4 5

18. —5 —4 —3 —2 —1 0 1 2 3 4 5

19. —8 —7 —6 —5 —4 —3 —2 —1 0 1 2

CHAPTER **3**

Section 3.1

Check Point Exercises

3.

4.

5.

6.

7.

Exercise Set 3.1

13.

x	(x, y)
−2	(−2, −24)
−1	(−1, −12)
0	(0, 0)
1	(1, 12)
2	(2, 24)

14.

x	(x, y)
−2	(−2, −28)
−1	(−1, −14)
0	(0, 0)
1	(1, 14)
2	(2, 28)

15.

x	(x, y)
−2	(−2, 20)
−1	(−1, 10)
0	(0, 0)
1	(1, −10)
2	(2, −20)

16.

x	(x, y)
−2	(−2, 40)
−1	(−1, 20)
0	(0, 0)
1	(1, −20)
2	(2, −40)

17.

x	(x, y)
−2	(−2, −21)
−1	(−1, −13)
0	(0, −5)
1	(1, 3)
2	(2, 11)

18.

x	(x, y)
−2	(−2, −16)
−1	(−1, −10)
0	(0, −4)
1	(1, 2)
2	(2, 8)

19.

x	(x, y)
−2	(−2, 13)
−1	(−1, 10)
0	(0, 7)
1	(1, 4)
2	(2, 1)

20.

x	(x, y)
−2	(−2, 19)
−1	(−1, 14)
0	(0, 9)
1	(1, 4)
2	(2, −1)

21.

22.

23.

24.

25.

26.

27.

28.

29.

30.

31.

32.

33.
$y = -\frac{1}{4}x$

34.
$y = \frac{1}{4}x$

35.
$y = \frac{1}{3}x + 1$

36.
$y = \frac{1}{3}x - 1$

37.
$y = -\frac{3}{2}x + 1$

38.
$y = -\frac{3}{2}x + 2$

39.
$y = -\frac{5}{2}x - 1$

40.
$y = -\frac{5}{2}x + 1$

41.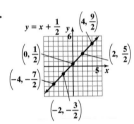
$y = x + \frac{1}{2}$

42.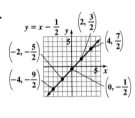
$y = x - \frac{1}{2}$

43.
$y = 4$

44.
$y = 3$

45.
$y = x^2$

46.
$y = x^2 - 2$

47.
$y = x^2 + 1$

48.
$y = x^2 + 2$

49.
$y = 4 - x^2$

50.
$y = 9 - x^2$

51a.
$y = x + 3$

52.
$y = x + 4$

53.
$y = 2x + 5$

54.
$y = 3x - 2$

57a. African Americans

x	(x, y)
0	(0, 5.2)
10	(10, 7.6)
20	(20, 10)
30	(30, 12.4)

b. Whites

x	(x, y)
0	(0, 3.4)
10	(10, 5.6)
20	(20, 7.8)
30	(30, 10)

c. Hispanics

x	(x, y)
0	(0, 4.2)
10	(10, 5.5)
20	(20, 6.8)
30	(30, 8.1)

60a.

x	(x, y)
0	(0, 1781)
5	(5, 2611)
10	(10, 3441)
15	(15, 4271)
20	(20, 5101)

b.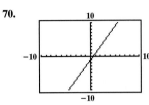

61a.

x	(x, y)
0	(0, 30,000)
10	(10, 30,500)
20	(20, 31,000)
30	(30, 31,500)
40	(40, 32,000)

b.

67.

68.

70.

71.

72.

73.

75.

Section 3.2

Check Point Exercises

4.

5.

6.

7.

8.

Exercise Set 3.2

19.

20.

21.

22.

23.

24.

25.

26.

27.

28.

29.

30.

31. $25y = 100 - 50x$
$(0, 4)$ $(1, 2)$ $(2, 0)$

32. $10y = 60 - 40x$
$(0, 6)$ $(1, 2)$ $\left(\frac{3}{2}, 0\right)$

33. $2x - 8y = 12$
$\left(0, -\frac{3}{2}\right)$ $(6, 0)$ $(2, -1)$

34. $3x - 6y = 15$
$(5, 0)$ $\left(0, -\frac{5}{2}\right)$ $(3, -1)$

35. $x + 2y = 0$
$(-2, 1)$ $(0, 0)$ $(2, -1)$

36. $2x + y = 0$
$(-1, 2)$ $(0, 0)$ $(1, -2)$

37. $y - 3x = 0$
$(1, 3)$ $(0, 0)$ $(-1, -3)$

38. $y - 4x = 0$
$(1, 4)$ $(0, 0)$ $(-1, -4)$

39. $2x - 3y = -11$
$(-4, 1)$ $\left(0, \frac{11}{3}\right)$ $\left(-\frac{11}{2}, 0\right)$

40. $3x - 2y = -7$
$(-1, 2)$ $\left(0, \frac{7}{2}\right)$ $\left(-\frac{7}{3}, 0\right)$

47. $y = 4$
$(3, 4)$ $(-2, 4)$ $(0, 4)$

48. $y = 2$
$(-2, 2)$ $(3, 2)$ $(0, 2)$

49. $y = -2$
$(0, -2)$ $(-2, -2)$ $(3, -2)$

50. $y = -3$
$(0, -3)$ $(-2, -3)$ $(3, -3)$

51. $x = 2$
$(2, 3)$ $(2, 0)$ $(2, -2)$

52. $x = 4$
$(4, 3)$ $(4, 0)$ $(4, -2)$

53. $x + 1 = 0$
$(-1, 3)$ $(-1, 0)$ $(-1, -2)$

54. $x + 5 = 0$
$(-5, 3)$ $(-5, 0)$ $(-5, -2)$

55. $y - 3.5 = 0$
$(0, 3.5)$ $(-2, 3.5)$ $(3, 3.5)$

56. $y - 2.5 = 0$
$(0, 2.5)$ $(-2, 2.5)$ $(3, 2.5)$

57. $x = 0$
$(0, 3)$ $(0, 0)$ $(0, -2)$

58. $y = 0$
$(0, 0)$ $(-2, 0)$ $(3, 0)$

59. $3y = 9$
$(0, 3)$ $(-2, 3)$ $(3, 3)$

60. $5y = 20$
$(0, 4)$ $(-2, 4)$ $(3, 4)$

61. 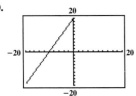 $12 - 3x = 0$
$(4, 3)$ $(4, 0)$ $(4, -2)$

62. $12 - 4x = 0$
$(3, 3)$ $(3, 0)$ $(3, -2)$

69b. $x + y = 11$
$(0, 11)$ $(11, 0)$

70b.
$(26, 9)$ $(17, 0)$

97.
x-intercept: 2; y-intercept: 4

98.
x-intercept: 3; y-intercept: −9

99.
x-intercept: 15; y-intercept: 10

100.
x-intercept: −10; y-intercept: 20

103.

Exercise Set 3.3

27.

28.

63.

Section 3.4

Check Point Exercises

2. $y = 3x - 2$

3. $y = \frac{3}{5}x + 1$

4. $3x + 4y = 0$

Exercise Set 3.4

27. $y = 2x + 4$

28. $y = 3x + 1$

29. $y = -3x + 5$

30. $y = -2x + 4$

31. $y = \frac{1}{2}x + 1$

32. $y = \frac{1}{3}x + 2$

33. $y = \frac{2}{3}x - 5$

34. $y = \frac{3}{4}x - 4$

35. $y = -\frac{3}{4}x + 2$

36. $y = -\frac{2}{3}x + 4$

37. $y = -\frac{5}{3}x$

38. $y = -\frac{4}{3}x$

39. $y = -3x$

40. $y = -2x$

41. $y = \frac{4}{3}x$

42. $y = \frac{5}{4}x$

43. $y = -2x + 3$

44. $y = -3x + 4$

45. $y = -\frac{7}{2}x + 7$

46. $y = -\frac{5}{3}x + 5$

47.

48.

49.

50.

51.

52.

Mid-Chapter 3 Check Point

4. $y = -2x$

5. $y = -2$

6. $x + y = -2$

7. $y = \frac{1}{3}x - 2$

8. $x = 3.5$

9. $4x - 2y = 8$

10. $y = 3x + 2$

11. $3x + y = 0$

12. $y = x^2 - 4$

13. $y = x - 4$

14. $5y = -3x$

15. $5y = 20$

Exercise Set 3.5

73a.

80.

81a.

81b.

81c.

81d.

84. $3x - 5y = 15$

Review Exercises

3a.

x	(x, y)
-2	$(-2, -7)$
-1	$(-1, -5)$
0	$(0, -3)$
1	$(1, -1)$
2	$(2, 1)$

b.

4a.

x	(x, y)
-2	$(-2, 0)$
-1	$\left(-1, \frac{1}{2}\right)$
0	$(0, 1)$
1	$\left(1, \frac{3}{2}\right)$
2	$(2, 2)$

b.

5.

6.

x	(x, y)
10	$(10, 9)$
12	$(12, 19)$
14	$(14, 29)$
16	$(16, 39)$

10.

11.

12.

13.

14.

15.

16.

17.

34.

35.

36.

37.

38.

39.

Chapter 3 Test

2.

x	(x, y)
-2	$(-2, -5)$
-1	$(-1, -2)$
0	$(0, 1)$
1	$(1, 4)$
2	$(2, 7)$

3.

5.

6.

13.

14.

Cumulative Review Exercises (Chapters 1–3)

9.

10.

19.

20.

CHAPTER 4

Exercise Set 4.2

55. $4x + 6y = 12$

Exercise Set 4.5

59. $f(x) = -\frac{3}{4}x + 3$

60. $-2x + y = 6$

61. $f(x) = -5$

Cumulative Review Exercises (Chapters 1–4)

6.

7. $x - 3y = 6$

8. $y = 4 - x^2$

9. $y = -\frac{3}{5}x + 2$

CHAPTER 5

Exercise Set 5.2

119. $3x - 2y = 6$

Exercise Set 5.5

102. $y = \frac{1}{3}x + 2$

Cumulative Review Exercises (Chapters 1–5)

7.

10. $y = -\frac{2}{5}x + 2$

11. $x - 2y = 4$

CHAPTER **6**

Exercise Set 6.3

96.

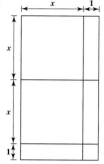

Exercise Set 6.5

121. $5x - 2y = 10$

Cumulative Review Exercises (Chapters 1–6)

4.

7. $y = -\dfrac{3}{5}x + 3$

CHAPTER **7**

Exercise Set 7.4

114. $3x - y = 3$

Exercise Set 7.7

51. $y = -\dfrac{2}{3}x + 4$

Exercise Set 7.8

39a.

Cumulative Review Exercises (Chapters 1–7)

7. $3x - 2y = 6$

8. $y = -2x + 3$

9. $y = -3$

Mid-Textbook Check Point

23. $y = \frac{1}{3}x - 1$

24. $3x + 2y = -6$

25. $y = -2$

CHAPTER 8

Exercise Set 8.2

55.

56.

76.

77.

78.

79.

82.

Population of Less Developed Regions

Section 8.3

Check Point Exercises

6.

Exercise Set 8.3

41.

42.

43.

44.

73.

74.

75.

76.

77.

78.

79.

80.

81.

82.

83.

Review Exercises

43.

Cumulative Review Exercises (Chapters 1–8)

19. $2x - y = 4$

20. $y = -\frac{2}{3}x$

CHAPTER 9

Section 9.1

Check Point Exercises

1. a.

b.

c.

3.

2. a.

b.

c.

4.

Exercise Set 9.1

1. number line, open at 1, closed at 6, segment between

2. number line, open at -2, closed at 4

3. number line, closed at -5, open at 2

4. number line, closed at -4, open at 3

5. number line, closed at -3, closed at 1

6. number line, closed at -2, closed at 5

7. number line, open at 2, ray right

8. number line, open at 3, ray right

9. number line, closed at -3, ray right

10. number line, closed at -5, ray right

11. number line, ray left, open at 2

12. number line, ray left, open at 2

13. number line, ray left, open at 6

14. number line, ray left, closed at 4

15. number line, ray left, closed at 1

16. number line, ray left, closed at 5

17. number line, closed at 7, ray right

18. number line, closed at 2, ray right

19. number line, ray left, closed at -2

20. number line, closed at -6, ray right

21. number line, ray left, closed at -1

22. number line, open at -9, ray left

23. number line, closed at -1, ray right

24. number line, open at 5, ray right

25. number line, ray left, open at 0

26. number line, ray left, closed at -5

27. number line, closed at 6, ray right

28. number line, closed at 2, ray right

29. number line, closed at -7, ray right

30. number line, closed at -2, ray right

31. number line, ray left, open at -4

32. number line, open at 7, ray right

33. number line, closed at 13, ray right

34. number line, closed at -4, ray right

35. number line, open at -6, ray right

36. number line, ray left, open at -1

37. number line, closed at -1, ray right

38. number line, ray left, open at -6

39. number line, ray left, open at -4

40. number line, open at -1, ray right

41. number line, ray left, open at 2

42. number line, ray left, open at 3

82.
$y_2 = 2x - 2$
$y_1 = -3(x - 6)$

83.
$y_1 = -2(x + 4)$
$y_2 = 6x + 16$

84. a. X=4

b. X= -3

87. b.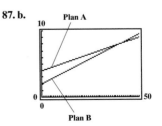
Plan A
Plan B

Section 9.2

Check Point Exercises

4.

Exercise Set 9.2

7.

8.

9.

10.

11.

12.

13.

14.

15.

16.

17.

18.

19.

20.

21.

22.

23.

24.

25.

26.

27.

28.

29.

30.

31.

32.

39.

40.

41. _[number line graphs]_

42. _[number line graphs]_

43.

44.

45.

46.

47.

48.

49.

50.

51.

52.

53.

54.

88.

89.

90.

91.

Section 9.3

Check Point Exercises

3.

4.

Exercise Set 9.3

39.

40.

41.

42.

43.

44.

45.

46.

47.

48.

49.

50.

51.

52.

53.

54.

55.

56.

57.

58.

59.

60.

61.

62.

65.

66.

67.

68.

69.

70.

102.

103.

104.

105.

106.

107.

108.

109.

110.

116.
$3x - 5y = 15$

117.
$f(x) = -\frac{2}{3}x$

118.
$f(x) = -2$

Section 9.4

Check Point Exercises

1. $4x - 2y \geq 8$

2. $y > -\frac{3}{4}x$

3. a. $y > 1$

b. $x \leq -2$

4. $x - 3y < 6$
$2x + 3y \geq -6$

5. $x + y < 2$
$-2 \leq x < 1$
$y > -3$

Exercise Set 9.4

1. $x + y \geq 3$

2. $x + y \geq 2$

3. $x - y < 5$

4. $x - y < 6$

5. $x + 2y > 4$

6. $2x + y > 6$

7. $3x - y \leq 6$

8. $x - 3y \leq 6$

9. $\frac{x}{2} + \frac{y}{3} < 1$

10. $\frac{x}{4} + \frac{y}{2} < 1$

11. $y > \frac{1}{3}x$

12. $y > \frac{1}{4}x$

13. $y \leq 3x + 2$

14. $y \leq 2x - 1$

15. $y < -\frac{1}{4}x$

16. $y < -\frac{1}{3}x$

17. $x \leq 2$

18. $x \leq -4$

19. $y > -4$

20. $y > -2$

21. $y \geq 0$

22. $x \leq 0$

23. $3x + 6y \leq 6$
$2x + y \leq 8$

24. $x - y \geq 4$
$x + y \leq 6$

25. $2x - 5y \leq 10$
$3x - 2y > 6$

26. $2x - y \leq 4$
$3x + 2y > -6$

27. $y > 2x - 3$
$y < -x + 6$

28. $y < -2x + 4$
$y < x - 4$

29. $x + 2y \leq 4$
$y \geq x - 3$

30. $x + y \leq 4$
$y \geq 2x - 4$

31. $x \leq 2$
$y \geq -1$

32. $x \leq 3$
$y \leq -1$

33. $-2 \leq x < 5$

34. $-2 < y \leq 5$

35. $x - y \leq 1$
$x \geq 2$

36. $4x - 5y \geq -20$
$x \geq -3$

39. $x + y > 4$
$x + y > -1$

40. $x + y > 3$
$x + y > -2$

41. $x - y \leq 2$
$x \geq -2$
$y \leq 3$

42. $3x + y \leq 6$
$x \geq -2$
$y \leq 4$

43. $x \geq 0$
$y \geq 0$
$2x + 5y \leq 10$
$3x + 4y \leq 12$

44. $x \geq 0$
$y \geq 0$
$2x + y \leq 4$
$2x - 3y \leq 6$

45. $3x + y \leq 6$
$2x - y \leq -1$
$x \geq -2$
$y \leq 4$

46. $2x + y \leq 6$
$x + y \geq 2$
$1 \leq x \leq 2$
$y \leq 3$
...

47. $y \geq -2x + 4$

48. $y \geq -3x + 2$

49. $x + y \leq 4$
$3x + y \leq 6$

50. $x + y \leq 3$
$4x + y \leq 6$

51. $-2 \leq x \leq 2$
$-3 \leq y \leq 3$

52. $-1 \leq x \leq 1$
$-2 \leq y \leq 2$

53. $y > \frac{3}{2}x - 2$ or $y < 4$

54. $x - y \geq -1$ or $5x - 2y \leq 10$

63. b. $y \geq 0$
$x + y \geq 5$
$x \geq 1$
$200x + 100y \leq 700$

64. b. $x \geq 0, y \geq 0$
$80x + 160y > 2000$

75.
76.
77.
78.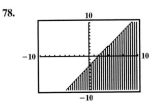

86. $y \geq nx + b$
$y \leq mx + b$

Exercise Set 9.5

5. a.
6. a.
7. a.
8. a.

9. a.
10. a.
11. a.
12. a.

13a.

14a.

15c.

16c.

Review Exercises

1.

2.

3.

4.

5.

6.

7.

8.

20.

21.

22.

23.

25.

26.

27.

28.

29

30.

36.

37.

38.

41. $3x - 4y > 12$

42. $x - 3y \le 6$

43. $y \le -\dfrac{1}{2}x + 2$

44. $y > \dfrac{3}{5}x$

45. $x \le 2$

46. $y > -3$

47. $2x - y \le 4$
$x + y \ge 5$

48. $y < -x + 4$
$y > x - 4$

49. $-3 \le x < 5$

50. $-2 < y \le 6$

51. $x \ge 3$
$y \le 0$

52. $2x - y > -4$
$x \ge 0$

53. $x + y \le 6$
$y \ge 2x - 3$

54. $3x + 2y \ge 4$
$x - y \le 3$
$x \ge 0, y \ge 0$

57.

58.

59.

60c.

Chapter 9 Test

1.

2.

3.

4.

8.

9.

10.

11.

12.

15.

16.

18. $3x - 2y < 6$

19. $y \geq \frac{1}{2}x - 1$

20. $y \leq -1$

21. $x + y \geq 2$
$x - y \geq 4$

22. $3x + y \leq 9$
$2x + 3y \geq 6$
$x \geq 0$
$y \geq 0$

23. $-2 < x \leq 4$

Cumulative Review Exercises (Chapters 1–9)

7. $f(x) = 2x + 1$

8. $y > 2x$

9. $2x - y \geq 6$

10. $f(x) = -1$

16.

17.

18.

19.

20.

CHAPTER 10

Exercise Set 10.1

89. $f(x) = \sqrt{x} + 3$

90. $f(x) = \sqrt{x} - 2$

91. $f(x) = \sqrt{x - 3}$

92. $f(x) = \sqrt{4 - x}$

114.

115.

116.

117.

122.

123.

Exercise Set 10.2

143.

144.

145.

152. $y \leq -\frac{3}{2}x + 3$

Exercise Set 10.3

105.

106.

107.

108.

113. $f(x) = \sqrt{(x-1)^2}$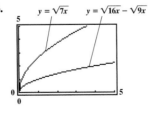

Exercise Set 10.4

92.

93.

94.

Exercise Set 10.5

138.

139.

140.

141.

$y = \sqrt{x + 3} + \sqrt{x}$

Exercise Set 10.6

68.

69.

70.

71.

72.

Cumulative Review Exercises (Chapters 1–10)

6. $x + 2y < 2$
$2y - x > 4$

CHAPTER **11**

Exercise Set 11.1

129.

130.

Exercise Set 11.2

94.

Section 11.3
Check Point Exercises

1.

$f(x) = -(x-1)^2 + 4$

2.

3.

Exercise Set 11.3

17.

$x = 4$

18.

$x = 1$

19.

$x = 1$

20.

$x = 3$

21.

$x = 3$

22.

$x = 1$

23.

24.

$x = \dfrac{1}{2}$

25.

$x = 1$

26.

$x = 3$

27.

$x = 1$

28.

$x = 1$

29.

$x = -\dfrac{3}{2}$

30.

$x = \dfrac{7}{4}$

31.

$x = 1$

32.

$x = -2$

33.

$x = 1$

34.

$x = 2$

55. d.

56. d.

77. a.   **c.**   **78.**

79.   **80.**   **81.**

82. a & d.

Mid-Chapter 11 Check Point

17. $f(x) = (x - 3)^2 - 4$

18. $g(x) = 5 - (x + 2)^2$

19. $h(x) = -x^2 - 4x + 5$

20. $f(x) = 3x^2 - 6x + 1$

Section 11.5

Check Point Exercises

1.   **2.**   **3.**

Exercise Set 11.5

1.   **2.**   **3.**

4.   **5.**   **6.**

7.
$$-5\ -4\ -3\ -2\ -1\ \ 0\ \ 1\ \ 2\ \ 3\ \ 4\ \ 5$$

8.
$$-5\ -4\ -3\ -2\ -1\ \ 0\ \ 1\ \ 2\ \ 3\ \ 4\ \ 5$$

9.
$$-5\ -4\ -3\ -2\ -1\ \ 0\ \ 1\ \ 2\ \ 3\ \ 4\ \ 5$$

10.
$$-5\ -4\ -3\ -2\ -1\ \ 0\ \ 1\ \ 2\ \ 3\ \ 4\ \ 5$$

11.
$$-5\ -4\ -3\ -2\ -1\ \ 0\ \ 1\ \ 2\ \ 3\ \ 4\ \ 5$$

12.
$$-5\ -4\ -3\ -2\ -1\ \ 0\ \ 1\ \ 2\ \ 3\ \ 4\ \ 5$$

13.
$$-5\ -4\ -3\ -2\ -1\ \ 0\ \ 1\ \ 2\ \ 3\ \ 4\ \ 5$$

14.
$$-5\ -4\ -3\ -2\ -1\ \ 0\ \ 1\ \ 2\ \ 3\ \ 4\ \ 5$$

15.
$$-1\qquad 0\qquad 1$$

16.
$$-5\ -4\ -3\ -2\ -1\ \ 0\ \ 1\ \ 2\ \ 3\ \ 4\ \ 5$$

17.
$$-5\ -4\ -3\ -2\ -1\ \ 0\ \ 1\ \ 2\ \ 3\ \ 4\ \ 5$$

18.
$$-5\ -4\ -3\ -2\ -1\ \ 0\ \ 1\ \ 2\ \ 3\ \ 4\ \ 5$$

19.
$$-5\ -4\ -3\ -2\ -1\ \ 0\ \ 1\ \ 2\ \ 3\ \ 4\ \ 5$$

20.
$$-5\ -4\ -3\ -2\ -1\ \ 0\ \ 1\ \ 2\ \ 3\ \ 4\ \ 5$$

21.
$$-5\ -4\ -3\ -2\ -1\ \ 0\ \ 1\ \ 2\ \ 3\ \ 4\ \ 5$$

22.
$$-5\ -4\ -3\ -2\ -1\ \ 0\ \ 1\ \ 2\ \ 3\ \ 4\ \ 5$$

23.
$$-5\ -4\ -3\ -2\ -1\ \ 0\ \ 1\ \ 2\ \ 3\ \ 4\ \ 5$$

24.
$$-5\ -4\ -3\ -2\ -1\ \ 0\ \ 1\ \ 2\ \ 3\ \ 4\ \ 5$$

25.
$$-5\ -4\ -3\ -2\ -1\ \ 0\ \ 1\ \ 2\ \ 3\ \ 4\ \ 5$$

26.
$$-5\ -4\ -3\ -2\ -1\ \ 0\ \ 1\ \ 2\ \ 3\ \ 4\ \ 5$$

27.
$$-5\ -4\ -3\ -2\ -1\ \ 0\ \ 1\ \ 2\ \ 3\ \ 4\ \ 5$$

28.
$$-6\ -5\ -4\ -3\ -2\ -1\ \ 0\ \ 1\ \ 2\ \ 3\ \ 4$$

29.
$$-8\ -7\ -6\ -5\ -4\ -3\ -2\ -1\ \ 0\ \ 1\ \ 2$$

30.
$$-6\ -5\ -4\ -3\ -2\ -1\ \ 0\ \ 1\ \ 2\ \ 3\ \ 4$$

31.
$$-5\ -4\ -3\ -2\ -1\ \ 0\ \ 1\ \ 2\ \ 3\ \ 4\ \ 5$$

32.
$$-8\ -7\ -6\ -5\ -4\ -3\ -2\ -1\ \ 0\ \ 1\ \ 2$$

33.
$$-9\ -8\ -7\ -6\ -5\ -4\ -3\ -2\ -1\ \ 0\ \ 1$$

34.
$$-6\ -5\ -4\ -3\ -2\ -1\ \ 0\ \ 1\ \ 2\ \ 3\ \ 4$$

35.
$$-8\ -6\ -4\ -2\ \ 0\ \ 2\ \ 4\ \ 6\ \ 8\ \ 10$$

36.
$$0\ \ 1\ \ 2\ \ 3\ \ 4\ \ 5\ \ 6\ \ 7\ \ 8\ \ 9\ \ 10$$

37.
$$-5\ -4\ -3\ -2\ -1\ \ 0\ \ 1\ \ 2\ \ 3\ \ 4\ \ 5$$

38.
$$-6\ -5\ -4\ -3\ -2\ -1\ \ 0\ \ 1\ \ 2\ \ 3\ \ 4$$

39.
$$-5\ -4\ -3\ -2\ -1\ \ 0\ \ 1\ \ 2\ \ 3\ \ 4\ \ 5$$

40.
$$-5\ -4\ -3\ -2\ -1\ \ 0\ \ 1\ \ 2\ \ 3\ \ 4\ \ 5$$

41.
$$-5\ -4\ -3\ -2\ -1\ \ 0\ \ 1\ \ 2\ \ 3\ \ 4\ \ 5$$

42.
$$-5\ -4\ -3\ -2\ -1\ \ 0\ \ 1\ \ 2\ \ 3\ \ 4\ \ 5$$

43.
$$-5\ -4\ -3\ -2\ -1\ \ 0\ \ 1\ \ 2\ \ 3\ \ 4\ \ 5$$

44.
$$-5\ -4\ -3\ -2\ -1\ \ 0\ \ 1\ \ 2\ \ 3\ \ 4\ \ 5$$

45.
$$-3\ -2\ -1\ \ 0\ \ 1\ \ 2\ \ 3\ \ 4\ \ 5\ \ 6\ \ 7$$

46.
$$-5\ -4\ -3\ -2\ -1\ \ 0\ \ 1\ \ 2\ \ 3\ \ 4\ \ 5$$

47.
$$-6\ -5\ -4\ -3\ -2\ -1\ \ 0\ \ 1\ \ 2\ \ 3\ \ 4$$

48.
$$-5\ -4\ -3\ -2\ -1\ \ 0\ \ 1\ \ 2\ \ 3\ \ 4\ \ 5$$

49.
$$-5\ -4\ -3\ -2\ -1\ \ 0\ \ 1\ \ 2\ \ 3\ \ 4\ \ 5$$

50.
$$-5\ -4\ -3\ -2\ -1\ \ 0\ \ 1\ \ 2\ \ 3\ \ 4\ \ 5$$

51.
$$-8\ -7\ -6\ -5\ -4\ -3\ -2\ -1\ \ 0\ \ 1\ \ 2$$

52.
$$-8\ -7\ -6\ -5\ -4\ -3\ -2\ -1\ \ 0\ \ 1\ \ 2$$

57.
$$-10\ -8\ -6\ -4\ -2\ \ 0\ \ 2\ \ 4\ \ 6\ \ 8\ \ 10$$

58.
$$-10\ -8\ -6\ -4\ -2\ \ 0\ \ 2\ \ 4\ \ 6\ \ 8\ \ 10$$

59.
$$-5\ -4\ -3\ -2\ -1\ \ 0\ \ 1\ \ 2\ \ 3\ \ 4\ \ 5$$

60.
$$-5\ -4\ -3\ -2\ -1\ \ 0\ \ 1\ \ 2\ \ 3\ \ 4\ \ 5$$

61.
$$-5\ -4\ -3\ -2\ -1\ \ 0\ \ 1\ \ 2\ \ 3\ \ 4\ \ 5$$

62.
$$-5\ -4\ -3\ -2\ -1\ \ 0\ \ 1\ \ 2\ \ 3\ \ 4\ \ 5$$

Review Exercises

36.

$(-1, 4)$, $f(x) = -(x + 1)^2 + 4$, $(0, 3)$, $(-3, 0)$, $(1, 0)$, $x = -1$

37.
$(0, 14)$, $f(x) = (x + 4)^2 - 2$, $(-5.4, 0)$, $(-2.6, 0)$, $(-4, -2)$, $x = -4$

38.
$(1, 4)$, $(3, 0)$, $f(x) = -x^2 + 2x + 3$, $(0, 3)$, $(-1, 0)$, $x = 1$

39.

$f(x) = 2x^2 - 4x - 6$, $(-1, 0)$, $(3, 0)$, $(0, -6)$, $(1, -8)$, $x = 1$

51.

$$-5\ -4\ -3\ -2\ -1\ \ 0\ \ 1\ \ 2\ \ 3\ \ 4\ \ 5$$

52.
$$-5\ -4\ -3\ -2\ -1\ \ 0\ \ 1\ \ 2\ \ 3\ \ 4\ \ 5$$

53.

$$-4\ -3\ -2\ -1\ \ 0\ \ 1\ \ 2\ \ 3\ \ 4\ \ 5\ \ 6$$

54.
$$-3\ -2\ -1\ \ 0\ \ 1\ \ 2\ \ 3\ \ 4\ \ 5\ \ 6\ \ 7$$

55.

$$-3\ -2\ -1\ \ 0\ \ 1\ \ 2\ \ 3\ \ 4\ \ 5\ \ 6\ \ 7$$

Chapter 11 Test

18.

$f(x) = (x + 1)^2 + 4$
$(-1, 4)$ $(0, 5)$
$x = -1$

19.

y $f(x) = x^2 - 2x - 3$
$(-1, 0)$ $(3, 0)$
$(0, -3)$ $(1, -4)$
$x = 1$

26.

27.

Cumulative Review Exercises (Chapters 1–11)

8.

$3x = 15 + 5y$

9.

$2x - 3y > 6$

10.

$f(x) = -\frac{1}{2}x + 1$

11.

$f(x) = x^2 + 6x + 8$

12.

$f(x) = (x - 3)^2 - 4$

APPENDIX C

Check Point Exercises

3.

$(x + 3)^2 + (y - 1)^2 = 4$

4.

$(x + 2)^2 + (y - 2)^2 = 9$

Appendix C Exercises

11.

$x^2 + y^2 = 16$

12.

$x^2 + y^2 = 49$

13.

$(x - 3)^2 + (y - 1)^2 = 36$

14.

$(x - 2)^2 + (y - 3)^2 = 16$

15.

$(x + 3)^2 + (y - 2)^2 = 4$

16.

$(x + 1)^2 + (y - 4)^2 = 25$

17.

$(x + 2)^2 + (y + 2)^2 = 4$

18.

$(x + 4)^2 + (y + 5)^2 = 36$

19.

$x^2 + y^2 + 6x + 2y + 6 = 0$

20.

$x^2 + y^2 + 8x + 4y + 16 = 0$

21.

$x^2 + y^2 - 10x - 6y - 30 = 0$

22.

$x^2 + y^2 - 4x - 12y - 9 = 0$

23.

$x^2 + y^2 + 8x - 2y - 8 = 0$

24.

$x^2 + y^2 + 12x - 6y - 4 = 0$

25.

$x^2 - 2x + y^2 - 15 = 0$

26.

$x^2 + y^2 - 6y - 7 = 0$

27. $x^2 + y^2 = 16$
$x - y = 4$

28. $x^2 + y^2 = 9$
$x - y = 3$

29. $(x - 2)^2 + (y + 3)^2 = 4$
$y = x - 3$

30. $(x - 3)^2 + (y + 1)^2 = 9$
$y = x - 1$

APPLICATIONS INDEX

Y

SUBJECT INDEX

A

Absolute value
   of real number, 24–25
   symbols, 81
Absolute value bars, 631, 665
   rewriting absolute value equation
      without, 602
   rewriting absolute value equation
      with two absolute values
      without, 603
   rewriting inequalities without, 604
Absolute value equations, 631
   solving, 602–3
Absolute value inequalities, 631–32
   problem solving using, 605–6
   solving, 603–5
   with unusual solution sets, 605
Addition
   associative property of, 41
   commutative property of, 40
   of complex numbers, 701–2, 712
   of functions, 549–50
   identity property of, 53, 95
   inverse property of, 53, 95
   of like radicals, 670–71
   with number line, 51–52, 53, 95
   without number line, 95
   of polynomials, 307–8, 375
   of polynomials in several variables,
      335–36
   of radical expressions, 711
   of rational expressions, 491
   of rational expressions when
      denominators are opposites, 462
   of rational expressions with
      common denominators,
      459–60, 519
   of rational expressions with
      different denominators,
      468–71, 519
   of real numbers, 51–59, 95
   simplifying series of subtractions
      and, 61
Addition method, for eliminating
      variables, 263–67, 288
Addition method, systems of linear
      equations solved by, 263–70, 299
   systems having no solution or
      infinitely many solutions, 267–68

Addition property of equality,
      103–10, 174
   to isolate variable terms, 107–8
   and multiplication property of
      equality, 114–15
   for solving equations, 104–7
Addition property of inequality, 163,
      164–68, 178
Additive inverses (or opposites), 53,
      95, 308–9, 375, 447–48
Aging process, 662
Algebraic expressions, 37–38
   applications with, 46–47, 73–75
   for complements and
      supplements, 155
   equivalent, 40
   evaluating, 38, 94
   exponents and, 80–81
   and multiplication, 71–73
   properties of real numbers and,
      40–45
   simplifying, 45–46, 55, 62, 72–73,
      80–81, 85–86, 95
   subtraction and, 61–62
   vocabulary of, 39
Algebraic translations of English
      phrases, 138–40
Algebra of functions, 548–56
   difference, 551
   product, 551
   quotient, 551
   sum, 549–50, 551
   using, 551–52
American Revolution, 761
America's Cup, 638
Analytic geometry, 28
Angles, 153
   complementary, 155, 177
   corresponding, 502, 522
   supplementary, 155, 177
   of triangle, 153–55, 177
Approximation symbol, 22
Area
   of circle, 150–52
   geometric formulas for, 149–52
   maximizing, 760–61
   of rectangle, 127–28, 149
   of triangle, 149–50
   of two-dimensional figure, 127

Armstrong, C. Michael, 718
Arrows, in graph, 537
Associative properties, 41–43, 95
   simplifying using, 42
Augmented matrix, 798–99, 800, 801,
      803, 805
Axis of symmetry, 751, 791
   of parabola, 751

B

Babbage, Charles, 808
Bar graphs, 32–33, 94
Base, 78
Berlin Airlift, 621
Berra, Yogi, 414
Binomial coefficients, 826–27
Binomial difference, square of,
      329–30, 376
Binomial expansions, Binomial
      Theorem formula for, 827
Binomials, 307, 375
   completing the square of, 722–23
   dividing polynomials by, 351–56
   FOIL method and multiplication of,
      325–27
   multiplying, 320
Binomial sum, square of, 328–29, 376
Binomial Theorem, 827–29
Body-mass index, of Miss
      America, 252
Boundary points, 776
   locating on number line, 777, 778,
      779, 781, 792, 793
Braces, 81
Brackets, 81, 83, 84, 579
Break-even point, 585–86
Business functions, 584–88, 630

C

Calculators. *See also* Graphing
      calculators/graphing utilities
   adding signed numbers on, 54
   decimal approximations for
      irrational numbers on, 21
   and decimals, 20
   to evaluate exponential
      expressions, 79

T

PHOTO CREDITS

ESSENTIALS OF INTRODUCTORY & INTERMEDIATE ALGEBRA FOR COLLEGE STUDENTS
CHAPTER TEST PREP VIDEO CDS
Robert Blitzer

0-13-186609-5
© 2006 Pearson Education, Inc.
Pearson Prentice Hall
Pearson Education, Inc.
Upper Saddle River, NJ 07458
Pearson Prentice Hall™ is a trademark of Pearson Education, Inc.

YOU SHOULD CAREFULLY READ THE TERMS AND CONDITIONS BEFORE USING THE CD-ROM PACKAGE. USING THIS CD-ROM PACKAGE INDICATES YOUR ACCEPTANCE OF THESE TERMS AND CONDITIONS.

Pearson Education, Inc. provides this program and licenses its use. You assume responsibility for the selection of the program to achieve your intended results, and for the installation, use, and results obtained from the program. This license extends only to use of the program in the United States or countries in which the program is marketed by authorized distributors.

LICENSE GRANT

You hereby accept a nonexclusive, nontransferable, permanent license to install and use the program ON A SINGLE COMPUTER at any given time. You may copy the program solely for backup or archival purposes in support of your use of the program on the single computer. You may not modify, translate, disassemble, decompile, or reverse engineer the program, in whole or in part.

TERM

The License is effective until terminated. Pearson Education, Inc. reserves the right to terminate this License automatically if any provision of the License is violated. You may terminate the License at any time. To terminate this License, you must return the program, including documentation, along with a written warranty stating that all copies in your possession have been returned or destroyed.

LIMITED WARRANTY

THE PROGRAM IS PROVIDED "AS IS" WITHOUT WARRANTY OF ANY KIND, EITHER EXPRESSED OR IMPLIED, INCLUDING, BUT NOT LIMITED TO, THE IMPLIED WARRANTIES OR MERCHANTABILITY AND FITNESS FOR A PARTICULAR PURPOSE. THE ENTIRE RISK AS TO THE QUALITY AND PERFORMANCE OF THE PROGRAM IS WITH YOU. SHOULD THE PROGRAM PROVE DEFECTIVE, YOU (AND NOT PEARSON EDUCATION, INC. OR ANY AUTHORIZED DEALER) ASSUME THE ENTIRE COST OF ALL NECESSARY SERVICING, REPAIR, OR CORRECTION. NO ORAL OR WRITTEN INFORMATION OR ADVICE GIVEN BY PEARSON EDUCATION, INC., ITS DEALERS, DISTRIBUTORS, OR AGENTS SHALL CREATE A WARRANTY OR INCREASE THE SCOPE OF THIS WARRANTY. SOME STATES DO NOT ALLOW THE EXCLUSION OF IMPLIED WARRANTIES, SO THE ABOVE EXCLUSION MAY NOT APPLY TO YOU. THIS WARRANTY GIVES YOU SPECIFIC LEGAL RIGHTS AND YOU MAY ALSO HAVE OTHER LEGAL RIGHTS THAT VARY FROM STATE TO STATE.

Pearson Education, Inc. does not warrant that the functions contained in the program will meet your requirements or that the operation of the program will be uninterrupted or error-free. However, Pearson Education, Inc. warrants the CD-ROM(s) on which the program is furnished to be free from defects in material and workmanship under normal use for a period of ninety (90) days from the date of delivery to you as evidenced by a copy of your receipt. The program should not be relied on as the sole basis to solve a problem whose incorrect solution could result in injury to person or property. If the program is employed in such a manner, it is at the user's own risk and Pearson Education, Inc. explicitly disclaims all liability for such misuse.

LIMITATION OF REMEDIES

Pearson Education, Inc.'s entire liability and your exclusive remedy shall be:
1. the replacement of any CD-ROM not meeting Pearson Education, Inc.'s "LIMITED WARRANTY" and that is returned to Pearson Education, or
2. if Pearson Education is unable to deliver a replacement CD-ROM that is free of defects in materials or workmanship, you may terminate this agreement by returning the program.

IN NO EVENT WILL PEARSON EDUCATION, INC. BE LIABLE TO YOU FOR ANY DAMAGES, INCLUDING ANY LOST PROFITS, LOST SAVINGS, OR OTHER INCIDENTAL OR CONSEQUENTIAL DAMAGES ARISING OUT OF THE USE OR INABILITY TO USE SUCH PROGRAM EVEN IF PEARSON EDUCATION, INC. OR AN AUTHORIZED DISTRIBUTOR HAS BEEN ADVISED OF THE POSSIBILITY OF SUCH DAMAGES, OR FOR ANY CLAIM BY ANY OTHER PARTY.

SOME STATES DO NOT ALLOW FOR THE LIMITATION OR EXCLUSION OF LIABILITY FOR INCIDENTAL OR CONSEQUENTIAL DAMAGES, SO THE ABOVE LIMITATION OR EXCLUSION MAY NOT APPLY TO YOU.

GENERAL

You may not sublicense, assign, or transfer the license of the program. Any attempt to sublicense, assign or transfer any of the rights, duties, or obligations hereunder is void.

This Agreement will be governed by the laws of the State of New York.

Should you have any questions concerning this Agreement, you may contact Pearson Education, Inc. by writing to:
ESM Media Development
Higher Education Division
Pearson Education, Inc.
1 Lake Street
Upper Saddle River, NJ 07458

Should you have any questions concerning technical support, you may write to:
New Media Production
Higher Education Division
Pearson Education, Inc.
1 Lake Street
Upper Saddle River, NJ 07458

YOU ACKNOWLEDGE THAT YOU HAVE READ THIS AGREEMENT, UNDERSTAND IT, AND AGREE TO BE BOUND BY ITS TERMS AND CONDITIONS. YOU FURTHER AGREE THAT IT IS THE COMPLETE AND EXCLUSIVE STATEMENT OF THE AGREEMENT BETWEEN US THAT SUPERSEDES ANY PROPOSAL OR PRIOR AGREEMENT, ORAL OR WRITTEN, AND ANY OTHER COMMUNICATIONS BETWEEN US RELATING TO THE SUBJECT MATTER OF THIS AGREEMENT.

System Requirements
- Windows:
Pentium II 300 MHz processor
Windows 98, NT, 2000, ME, or XP
64 MB RAM (128 MB RAM required for Windows XP)
4.3 available hard drive space (optional—for minimum QuickTime installation)
800 × 600 resolution
8x or faster CD-ROM drive
QuickTime 6.x
Sound Card

- Macintosh:
Power PC G3 233 MHz or better
Mac OS 9.x or 10.x
64 MB RAM
10 MB available hard drive space for Mac OS 9, 19 MB on OS X (optional—if QuickTime installation is needed)
800 × 600 resolution
8x or faster CD-ROM drive
QuickTime 6.x

Support Information
If you are having problems with this software, call (800) 677-6337 between 8:00 a.m. and 8:00 p.m. EST, Monday through Friday, and 5:00 p.m. through Midnight EST on Sundays. You can also get support by filling out the web form located at : http://247.prenhall.com/mediaform

Our technical staff will need to know certain things about your system in order to help us solve your problems more quickly and efficiently. If possible, please be at your computer when you call for support. You should have the following information ready:
• Textbook ISBN
• CD-Rom/Diskette ISBN
• corresponding product and title
• computer make and model
• Operating System (Windows or Macintosh) and Version
• RAM available
• hard disk space available
• Sound card? Yes or No
• printer make and model
• network connection
• detailed description of the problem, including the exact wording of any error messages.

NOTE: Pearson does not support and/or assist with the following:
• third-party software (i.e. Microsoft including Microsoft Office Suite, Apple, Borland, etc.)
• homework assistance
• Textbooks and CD-Rom's purchased used are not supported and are non-replaceable. To purchase a new CD-Rom contact Pearson Individual Order Copies at 1-800-282-0693

Definitions, Rules, and Formulas

The Real Numbers

Natural Numbers: $\{1, 2, 3, \ldots\}$

Whole Numbers: $\{0, 1, 2, 3, \ldots\}$

Integers: $\{\ldots, -3, -2, -1, 0, 1, 2, 3, \ldots\}$

Rational Numbers: $\left\{\frac{a}{b} \mid a \text{ and } b \text{ are integers, } b \neq 0\right\}$

Irrational Numbers: $\{x \mid x \text{ is real and not rational}\}$

Basic Rules of Algebra

Commutative: $a + b = b + a; ab = ba$

Associative: $(a + b) + c = a + (b + c);$
$\qquad\qquad (ab)c = a(bc)$

Distributive: $a(b + c) = ab + ac; a(b - c) = ab - ac$

Identity: $a + 0 = a; a \cdot 1 = a$

Inverse: $a + (-a) = 0; a \cdot \frac{1}{a} = 1 (a \neq 0)$

Multiplication Properties: $(-1)a = -a;$
$(-1)(-a) = a; a \cdot 0 = 0; (-a)(b) = (a)(-b) = -ab;$
$(-a)(-b) = ab$

Set-Builder Notation, Interval Notation, and Graphs

$(a, b) = \{x \mid a < x < b\}$

$[a, b) = \{x \mid a \leq x < b\}$

$(a, b] = \{x \mid a < x \leq b\}$

$[a, b] = \{x \mid a \leq x \leq b\}$

$(-\infty, b) = \{x \mid x < b\}$

$(-\infty, b] = \{x \mid x \leq b\}$

$(a, \infty) = \{x \mid x > a\}$

$[a, \infty) = \{x \mid x \geq a\}$

$(-\infty, \infty) = \{x \mid x \text{ is a real number}\} = \{x \mid x \in R\}$

Slope Formula

$$\text{slope } (m) = \frac{\text{change in } y}{\text{change in } x} = \frac{y_2 - y_1}{x_2 - x_1} \quad (x_1 \neq x_2)$$

Equations of Lines

1. *Slope-intercept form:* $y = mx + b$

m is the line's slope and b is its y-intercept.

2. *Standard form:* $Ax + By = C$

3. *Point-slope form:* $y - y_1 = m(x - x_1)$

m is the line's slope and (x_1, y_1) is a fixed point on the line.

4. *Horizontal line parallel to the x-axis:* $y = b$

5. *Vertical line parallel to the y-axis:* $x = a$

Systems of Equations

One solution: consistent

No solution: inconsistent

Infinitely many solutions: dependent and consistent

A system of linear equations may be solved: (a)graphically, (b) by the substitution method, (c) by the addition or elimination method, (d) by matrices, or (e) by determinants.

$$\begin{vmatrix} a_1 b_1 \\ a_2 b_2 \end{vmatrix} = a_1 b_2 - a_2 b_1$$

Cramer's Rule:

Given a system of a equations of the form

$$\begin{aligned} a_1 x + b_1 y &= c_1 \\ a_2 x + b_2 y &= c_2 \end{aligned}, \quad \text{then } x = \frac{\begin{vmatrix} c_1 b_1 \\ c_2 b_2 \end{vmatrix}}{\begin{vmatrix} a_1 b_1 \\ a_2 b_2 \end{vmatrix}} \text{ and } y = \frac{\begin{vmatrix} a_1 c_1 \\ a_2 c_2 \end{vmatrix}}{\begin{vmatrix} a_1 b_1 \\ a_2 b_2 \end{vmatrix}}.$$

Absolute Value

1. $|x| = \begin{cases} x & \text{if } x \geq 0 \\ -x & \text{if } x < 0 \end{cases}$

2. If $|x| = c$, then $x = c$ or $x = -c. (c > 0)$

3. If $|x| < c$, then $-c < x < c. (c > 0)$

4. If $|x| > c$, then $x < -c$ or $x > c. (c > 0)$

Special Factorizations

1. *Difference of two squares:*
$$A^2 - B^2 = (A + B)(A - B)$$

2. *Perfect square trinomials:*
$$A^2 + 2AB + B^2 = (A + B)^2$$
$$A^2 - 2AB + B^2 = (A - B)^2$$

3. *Sum of two cubes:*
$$A^3 + B^3 = (A + B)(A^2 - AB + B^2)$$

4. *Difference of two cubes:*
$$A^3 - B^3 = (A - B)(A^2 + AB + B^2)$$

Variation

English Statement	Equation
y varies directly as x.	$y = kx$
y varies directly as x^n.	$y = kx^n$
y varies inversely as x.	$y = \dfrac{k}{x}$
y varies inversely as x^n.	$y = \dfrac{k}{x^n}$
y varies jointly as x and z.	$y = kxz$

Exponents
Definitions of Rational Exponents

1. $a^{\frac{1}{n}} = \sqrt[n]{a}$ **2.** $a^{\frac{m}{n}} = \left(\sqrt[n]{a}\right)^m$ or $\sqrt[n]{a^m}$

3. $a^{-\frac{m}{n}} = \dfrac{1}{a^{\frac{m}{n}}}$

Properties of Rational Exponents

If m and n are rational exponents, and a and b are real numbers for which the following expressions are defined, then

1. $b^m \cdot b^n = b^{m+n}$ **2.** $\dfrac{b^m}{b^n} = b^{m-n}$

3. $\left(b^m\right)^n = b^{mn}$ **4.** $(ab)^n = a^n b^n$

5. $\left(\dfrac{a}{b}\right)^n = \dfrac{a^n}{b^n}$

Radicals

1. If n is even, then $\sqrt[n]{a^n} = |a|$.
2. If n is odd, then $\sqrt[n]{a^n} = a$.
3. The product rule: $\sqrt[n]{a} \cdot \sqrt[n]{b} = \sqrt[n]{ab}$
4. The quotient rule: $\dfrac{\sqrt[n]{a}}{\sqrt[n]{b}} = \sqrt[n]{\dfrac{a}{b}}$

Complex Numbers

1. The imaginary unit i is defined as
$$i = \sqrt{-1}, \quad \text{where} \quad i^2 = -1.$$

The set of numbers in the form $a + bi$ is called the set of complex numbers. If $b = 0$, the complex number is a real number. If $b \neq 0$, the complex number is an imaginary number.

2. The complex numbers $a + bi$ and $a - bi$ are conjugates. Conjugates can be multiplied using the formula
$$(A + B)(A - B) = A^2 - B^2.$$

The multiplication of conjugates results in a real number.

3. To simplify powers of i, rewrite the expression in terms of i^2. Then replace i^2 with -1 and simplify.

Quadratic Equations and Functions

1. The solutions of a quadratic equation in standard form
$$ax^2 + bx + c = 0, \quad a \neq 0,$$
are given by the quadratic formula
$$x = \frac{-b \pm \sqrt{b^2 - 4ac}}{2a}.$$

2. The discriminant, $b^2 - 4ac$, of the quadratic equation $ax^2 + bx + c = 0$ determines the number and type of solutions.

Discriminant	Solutions
Positive perfect square with a, b, and c rational numbers	2 rational solutions
Positive and not a perfect square	2 irrational solutions
Zero, with a, b, and c rational numbers	1 rational solution
Negative	2 imaginary solutions

3. The graph of the quadratic function
$$f(x) = a(x - h)^2 + k, \quad a \neq 0,$$
is called a parabola. The vertex, or turning point, is (h, k). The graph opens upward if a is positive and downward if a negative. The axis of symmetry is a vertical line passing through the vertex. The graph can be obtained using the vertex, x-intercepts, if any, [set $f(x)$ equal to zero], and the y-intercept (set $x = 0$).

4. A parabola whose equation is in the form
$$f(x) = ax^2 + bx + c, \quad a \neq 0,$$
has its vertex at
$$\left(-\frac{b}{2a}, f\left(-\frac{b}{2a}\right)\right).$$

If $a > 0$, then f has a minimum that occurs at $x = -\dfrac{b}{2a}$. If $a < 0$, then f has a maximum that occurs at $x = -\dfrac{b}{2a}$.

Definitions, Rules, and Formulas (continued)

Exponential and Logarithmic Functions

1. Exponential Function: $f(x) = b^x, b > 0, b \neq 1$
Graphs:

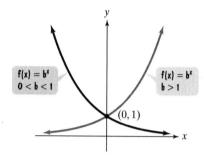

2. Logarithmic Function: $f(x) = \log_b x, b > 0, b \neq 1$
$y = \log_b x$ is equivalent to $x = b^y$.

Graphs:

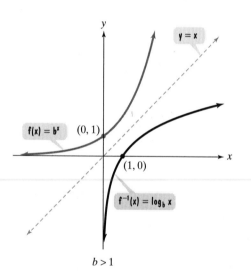

3. Properties of Logarithms

   a. $\log_b(MN) = \log_b M + \log_b N$

   b. $\log_b\left(\dfrac{M}{N}\right) = \log_b M - \log_b N$

   c. $\log_b M^p = p \log_b M$

   d. $\log_b M = \dfrac{\log_a M}{\log_a b} = \dfrac{\ln M}{\ln b} = \dfrac{\log M}{\log b}$

   e. $\log_b b^x = x; \log 10^x = x; \ln e^x = x$

   f. $b^{\log_b x} = x; 10^{\log x} = x; e^{\ln x} = x$

Distance and Midpoint Formulas

1. The distance from (x_1, y_1) to (x_2, y_2) is
$$\sqrt{(x_2 - x_1)^2 + (y_2 - y_1)^2}.$$

2. The midpoint of the line segment with endpoints (x_1, y_1) and (x_2, y_2) is
$$\left(\frac{x_1 + x_2}{2}, \frac{y_1 + y_2}{2}\right).$$

Conic Sections
Circle

Ellipse

Hyperbola

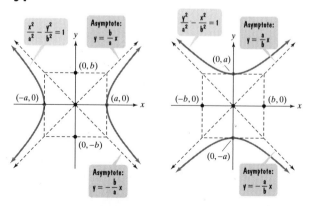